SAUNDERS MATHEMATICS BOOKS

Consulting Editor

BERNARD R. GELBAUM, University of California

MARY R. EMBRY

Associate Professor of Mathematics
University of North Carolina at Charlotte

JOSEPH F. SCHELL

Professor of Mathematics
University of North Carolina at Charlotte

J. PELHAM THOMAS

Professor of Mathematics
Western Carolina University
Cullowhee, North Carolina

CALCULUS AND LINEAR ALGEBRA
AN INTEGRATED APPROACH

W. B. SAUNDERS COMPANY

Philadelphia London Toronto

1972

W. B. Saunders Company: West Washington Square
Philadelphia, Pa. 19105

12 Dyott Street
London, WC1A 1DB

1835 Yonge Street
Toronto 7, Ontario

CALCULUS AND LINEAR ALGEBRA: An Integrated Approach   SBN 0-7216-3370-6

Print No.:   9   8   7   6   5   4   3   2   1

PREFACE

This book is the authors' attempt to provide a careful study of the calculus of one and several variables based on an early introduction and use of linear algebraic concepts. We have endeavored to motivate carefully each new concept. From the beginning we consider the general n-dimensional setting, with special attention to the first three dimensions. We do not include all the ideas usually found in a separate course in linear algebra, but restrict our scope to those which we feel are more appropriate for a first study of calculus.

The reader is expected to be quite familiar with the material summarized in Appendices I (Sets and Logic), II (The Real Number System), and III (Trigonometry). Of course, he should also possess the usual algebraic manipulative skills. A knowledge of geometry is not essential but certainly would assist in the assimilation of much of the material in Chapters 1, 2, and 7. We advise the prospective reader to begin his study with a review of the material in the appendices mentioned, giving particular attention to Appendix I. This appendix describes the logical notation employed in this text for the presentation of proofs.

There are several unique features of this book which, we hope, will contribute to mathematical pedagogy at this level.

The use of the implication scheme to present an immediate visual representation of the skeletal structure of logical argument; detailed justification is given for steps not truly obvious. (See Appendix I.)

Early introduction of n-dimensional considerations motivated by examples in two and three dimensions. (Chapters 1, 2, and 3.)

Approach to the derivative through affine approximations. (Chapter 3.)

Early treatment of differentiation and integration of $\mathcal{R} \to \mathcal{R}^m$ functions. (Chapters 3 and 12.)

Careful development of explicit, implicit, and parametric representations for point sets. (Chapter 7.)

Consistent use throughout of the notation $\int_{x=a}^{b} f(x)$—without the "dx"—for integrals, and extension of this notation to $\mathcal{R}^n$-integrals. (Chapters 8, 18, and following.)

Consistent use of $\int f$ as a symbol for a set of primitives:

$$\int f = \{F \mid F' = f\},$$

rather than the confusing (and misleading) notation $\int f = F + C$, C an "arbitrary constant." (Chapter 9 and following.)

Careful treatment of applications of the integral through the upper and lower integral approach. The key theorem is easily proved and provides an excellent motivation for a later study of the Riemann-Stieltjes integral. (Chapter 12.)

Approximation by Taylor polynomials introduced as a natural generalization of the affine approximation and then used to motivate consideration of series. (Chapter 13.)

Development of affine approximations of $\mathcal{R}^2 \to \mathcal{R}$ functions and the concept of differentiability prior to the introduction of partial derivatives. (Chapter 16.)

Careful treatment of tangent lines and planes for curves and surfaces given by explicit, implicit, and parametric representations.

Change of variables formula for multiple integrals derived in a form which does not involve the Jacobian matrix. (Jacobian form is also given.) The form used here seems more intuitive, is more readily proved, and provides motivation for the Radon-Nikodym derivative so important in functional analysis. (Chapter 19.)

Use of "big-oh" notation as the symbol for the "order dominated set" of a sequence t. (Chapter 21.)

The diagram on page viii gives the interdependency of the various chapters in the book. This diagram will be helpful in designing many possible courses.

Table I lists five sequences of courses designed to cover all the material contained herein.

The authors wish to acknowledge and to extend thanks to the many people who have contributed to the publication of this book, in particular to James H. Wahab, under whose chairmanship and direction the authors first became involved with the approach contained herein; to Mrs. Edna Stamey, who suffered through the many typescripts; to Miss Linda Biggers and Miss Helen Hargett, who assisted in proofreading; and to the faculty at the University of North Carolina at Charlotte and Western Carolina University for their many suggestions for improvement. Finally, we thank George Fleming and the editorial staff of the W. B. Saunders Company for their encouragement, assistance and patience.

MARY R. EMBRY

JOSEPH F. SCHELL

J. PELHAM THOMAS

Table I *Five Curriculum Patterns*

(1) Early rigorous treatment of limits in an *n*-dimensional setting.

   (a) Five course sequence; each three semester hours or five quarter hours.

Linear Geometry	1, 2, 14, 15
$\mathcal{R} \to \mathcal{R}^n$ Differential Calculus	3, 4, 5, 6, 7
$\mathcal{R} \to \mathcal{R}^n$ Integral Calculus	8, 9, 10, 11, 12 (selections)
Infinite Series, Improper Integrals	12, 13, 20, 21
Multivariable Calculus	14, 15, (review) 16, 17, 18, 19

   (b) Four courses; each fifteen weeks.

Linear Geometry	1, 2, 14, 15	3 hrs/week
Calculus I	3, 4, 5, 6, 8, 9	4 hrs/week
Calculus II	10, 11, 12, 13, 20, 21	4 hrs/week
Multivariable Calculus	14, 15 (review), 7,	
	16, 17, 18, 19	4 hrs/week

   (c) Four courses; each fifteen weeks.

Geometry and Calculus I	1, 2, 3, 4, 5	4 hrs/week
Geometry and Calculus II	6, 8, 9, 10, 11, 12	4 hrs/week
Geometry and Calculus III	7, 14, 15, 16, 17, 13	4 hrs/week
Geometry and Calculus IV	18, 19, 20, 21	3 hrs/week

(2) Deferred rigorous treatment of limits and continuity.

   (a) Five course sequence; three semester hours or five quarter hours each.

Calculus I	1 (Secs 1–3), 2 (Secs 1, 7–9),
	3 (Secs 1–3), 4 (Secs 1–3), 5, 6
Calculus II	8, 9, 10, 11, 12
Linear Geometry	1 (Secs 4–10), 2 (Secs 2–6),
	7, 14, 15
Calculus III	3 (Secs 4–7), 16, 17, 18, 19
Calculus IV	4 (Secs 4–5), 13, 20, 21

   (b) Four course sequence; fifteen weeks each.

$\mathcal{R} \to \mathcal{R}^n$ Calculus I	1 (Secs 1–3), 2 (Secs 1, 7–9),	
	3 (Secs 1–3), 4 (Secs 1–3),	4 hrs/week
	5, 6, 8, 9	
$\mathcal{R} \to \mathcal{R}^n$ Calculus II	10 11, 12, 4 (Secs 4–5)	
	13, 20, 21 (selections)	4 hrs/week
$\mathcal{R}^n \to \mathcal{R}^m$ Calculus I	1 (Secs 4–10), 2 (Secs 2–6)	
	3 (Secs 4–7), 7, 14, 15, 16	4 hrs/week
$\mathcal{R}^n \to \mathcal{R}^m$ Calculus II	17, 18, 19, 21	3 hrs/week

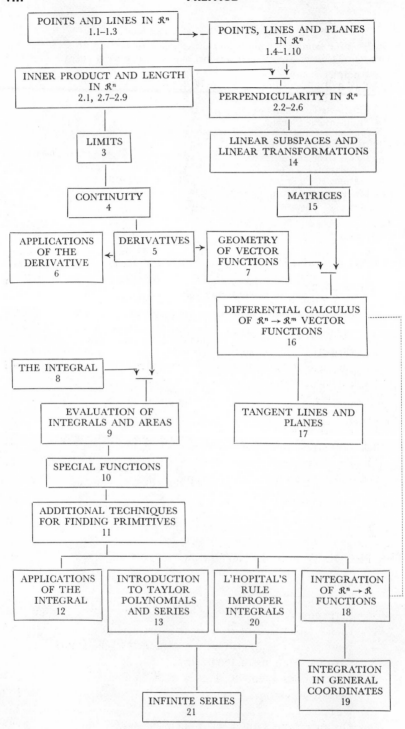

CHAPTER INTERDEPENDENCY

CONTENTS

PART II DIFFERENTIAL CALCULUS OF $\mathcal{R} \to \mathcal{R}$ FUNCTIONS

PART III INTEGRAL CALCULUS OF $\mathcal{R} \to \mathcal{R}^n$ FUNCTIONS

7

8

9

PART IV DIFFERENTIAL CALCULUS OF $\mathcal{R}^n \to \mathcal{R}^m$ FUNCTIONS

14

15

16

17

PART V INTEGRAL CALCULUS OF $\mathcal{R}^n \to \mathcal{R}$ FUNCTIONS

18

19

20

21

APPENDICES

PART I

LINEAR GEOMETRY

I

LINES AND PLANES IN $\mathcal{R}^n$

INTRODUCTION

In this chapter the real number system, denoted by $\mathcal{R}$, is employed to formulate an algebraic approach to geometry. This approach makes much of the geometry of "n-dimensions" as readily accessible as the geometry of two dimensions (the plane) and the geometry of three dimensions (space). Each concept is presented in the more general setting. However, the specialization to two and three dimensions is discussed and developed in much greater detail.

The reader need only have acquired a "geometric intuition" for the geometry of the plane and of space, such as that obtained in an elementary geometry course. No detailed knowledge of the various theorems and constructions is required.

1. REAL ORDERED PAIRS, TRIPLES, AND n-TUPLES

We begin our study of linear plane geometry with the concept of an ordered pair.

3

DEFINITION 1.1   (Real Ordered Pairs.)   Let a and b be two real numbers. The symbol

$$(a, b)$$

is called an *ordered pair of real numbers* with *first coordinate a* and *second coordinate b*.

(Equality of Ordered Pairs.)   A second ordered pair, say (c, d), shall be said to be *equal* to (a, b) if and only if corresponding coordinates are equal. That is,

$$(a, b) = (c, d) \text{ if and only if } a = c \text{ and } b = d.$$

According to this definition, the real ordered pairs $(1, 3)$ and $(3, 1)$ are different, unequal, real ordered pairs. Moreover, to say that $(x, y) = (2, 5)$ is just another way of saying in a single statement that $x = 2$ and $y = 5$.

Example 1.   Given that $(x + y, y) = (2, 1)$, we can use the definition of equality of real ordered pairs to determine x and y. Thus, we know that

$$x + y = 2$$
$$y = 1.$$

We need only solve this system for x and y. The second equation gives the value of y at once. Substituting this value into the first equation we find

$$x + 1 = 2, \quad \text{or } x = 1.$$

Problems similar to the one above will arise from time to time and some experience with them is given in the exercises. ■

Throughout this text we shall be concerned with the concept of a real ordered pair (or a generalization thereof) and sets of real ordered pairs. It is possible and helpful to use geometric language to discuss and even to picture much of what we shall do. It is the purpose of the first two chapters to make clear the way in which this language will be used.

The cornerstone on which the geometric viewpoint is laid is the geometric representation of an ordered pair itself. There are two very useful geometric representations of a real ordered pair, or rather, of the set of all real ordered pairs. Both representations, or models, have their purpose and each needs to be understood. Fortunately, both representations are quite simple and may be familiar to many of us already.

We shall assume that the reader is familiar with plane geometry and with the real number line. Our first step toward a geometrical model for the set of real ordered pairs is the introduction of a coordinate system into the plane.

DEFINITION I.2   (System of Coordinates for the Plane.)   Consider any plane $\mathcal{M}$. We say that we have introduced a coordinate system into $\mathcal{M}$ if we carry out the following procedure:

i. Select a point of the plane. Label it O and refer to it as the *origin* of the coordinate system.

ii. Through O select two distinct lines. Label one line with the Arabic numeral 1, and refer to it as the *first coordinate axis* or the 1-*axis* of the system. Label the other line with the Arabic numeral 2, and refer to it as the *second coordinate axis* or the 2-*axis* of the system.

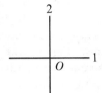

iii. Make each axis into a real number line with O as the origin on each.

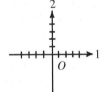

iv. For each point P in $\mathcal{M}$ determine two real numbers a and b, so that we may arrive at P by starting at O and moving
    a units in the direction of the 1-axis and
    b units in the direction of the 2-axis.

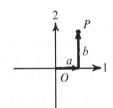

v. Refer to a as the first coordinate of P and to b as the second coordinate of P.   **D-1-2**

Certainly this is not the first time that you have seen a coordinate system for the plane. However, the definition given here is more general than it may seem on a first reading. As is usual, the origin O is quite arbitrarily selected, but so also are the two axes, except for the restriction that both must contain O. To be more specific, in step (ii) there is no requirement that the 1-axis be horizontal and that the 2-axis be vertical, or that the

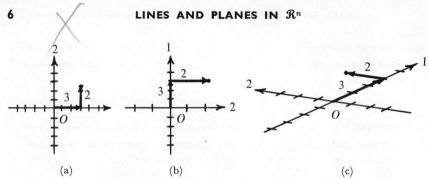

Figure 1.1 Determination of the point with first coordinate 3 and second coordinate 2 using various coordinate systems.

two axes even be perpendicular to one another. Moreover, step (iii) does not require that the same unit of length be used for the 2-axis as is used for the 1-axis, nor is the "positive" sense of each axis specified in any way.

The diagrams of Figure 1.1 are, therefore, equally acceptable coordinate systems. Figure 1.1a gives the usual coordinate system, with the 1-axis horizontal and positive to the right, the 2-axis vertical and positive to the top, and the same unit of length (scale) used on each axis. The axes of the coordinate system of Figure 1.1b are still horizontal and vertical, but the labeling is not the usual one and the two scales are not the same. Figure 1.1c illustrates an oblique coordinate system, that is, a coordinate system in which the axes form an oblique angle.

Only on rare occasions shall we use any coordinate system other than the usual rectangular one. There are times, however, when even an oblique coordinate system is useful. The labeling of the axes may be modified from time to time. At times they will be labeled x and y, at other times x and t, and in some instances still other symbols will be used. Which symbol represents the first coordinate and which denotes the second will be clear from the context.

Using any coordinate system in a plane $\mathcal{M}$, we can now specify the nature of the two geometric representations of the set of real ordered pairs. We shall refer to these representations as geometric models.

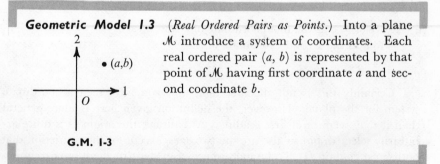

Geometric Model 1.3 (*Real Ordered Pairs as Points.*) Into a plane $\mathcal{M}$ introduce a system of coordinates. Each real ordered pair (a, b) is represented by that point of $\mathcal{M}$ having first coordinate a and second coordinate b.

G.M. 1-3

Geometric Model I.4 (*Real Ordered Pairs as Directed Line Segments.*) Into a plane $\mathcal{M}$ introduce a system of coordinates. Each real ordered pair (a, b) is represented by the directed line segment having O as initial point and the point with first coordinate a and second coordinate b as terminal point.

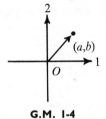

G.M. I-4

Each of the above, for a given choice of coordinate system, is a model for the entire set of real ordered pairs. A different choice of coordinate system yields a different model. In effect, instead of just two models we have two *types* of models with an unlimited number of each type, since there is an unlimited number of choices of coordinate system.

For a given choice of coordinate system we shall frequently employ both models simultaneously, and we shall use the term "real ordered pair (a, b)" interchangeably with "point (a, b)" and "directed line segment from O to (a, b)."

Consider a map of a portion of the United States. Let us place a coordinate system on this map, selecting some airport as the origin. A directed line segment (a, b) will represent the flight path and direction of flight for an airplane going from the city airport at the origin to the city which has map coordinates (a, b), and it can also be used for determining the distance and the estimated time of flight. Suppose that the estimated flying time from O to (a, b) is two hours. Further suppose that in flying along the (straight-line) flight path a forest fire is observed directly below, one hour into the flight. What map coordinates should be radioed to the ground to locate the fire? Since the fire is observed midway into the flight to (a, b), it should be evident that the fire is located at the point with map coordinates $\left(\dfrac{a}{2}, \dfrac{b}{2}\right)$ (Fig. 1.2). The directed line segment from O to the fire could be described by saying that it is just one half that from O to (a, b). Regarding (a, b) and $(a/2, b/2)$ as directed line segments, it seems reasonable to write $(a/2, b/2) = 1/2(a, b)$.

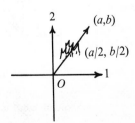

Figure I.2 Multiples of real ordered pairs.

What the preceding problem does is to suggest the need for, and the reasonableness of, defining the product of a real number, such as 1/2, and a real ordered pair (a, b). In our next definition we shall make this interpretation precise. We shall also define another operation on real ordered pairs, an operation arising from the considerations of the next paragraph.

Another problem which our mythical pilot might readily encounter arises if the city he wishes to fly to is not on the same map as the city he wishes to fly from. Suppose the flight is from city O to city B and the flight path needs to be determined. However, city B does not appear on the map with city O at the origin (Fig. 1.3a). But the map with city O at the origin, call it map O, does show city A. Fortunately, another map with A at the origin, call it map A, does show city B (Fig. 1.3b). Moreover, the coordinate axes of both maps are north-south and east-west and the scales are the same. With these two maps the pilot could fly from city O to city A and then to city B, but this would probably not be the straight-line flight from O to B. The pilot's problem is to determine the directed line segment representing a straight-line flight from O to B. The problem is readily solved by superposing map A over map O as shown in Figure 1.3c. If (a, b) are the coordinates of A on map O, and (c, d) the coordinates of B on map A, an inspection of Figure 1.3c and d should convince us that $(a + c, b + d)$ gives the directed line segment from O to B. We shall refer to this operation with two directed line segments as *addition* because of its obvious relation to real number addition.

We now formalize the two operations discussed above.

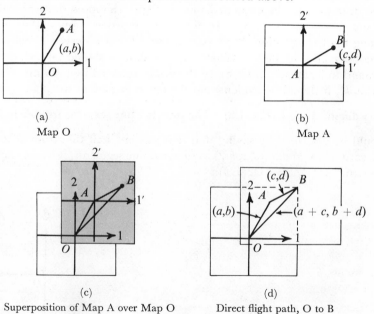

(a)
Map O

(b)
Map A

(c)
Superposition of Map A over Map O

(d)
Direct flight path, O to B

Figure 1.3   Sum of directed line segments.

DEFINITION 1.5 (Sums and Multiples of Real Ordered Pairs.) Letting (a, b) and (c, d) be two real ordered pairs and letting k be a real number, we define the following:

i. (Multiples of a Real Ordered Pair.) The multiple of (a, b) by the real number k is the real ordered pair given by

$$k(a, b) = (ka, kb).$$

ii. (Sum of Real Ordered Pairs.) The sum of (a, b) and (c, d) is the real ordered pair given by

$$(a, b) + (c, d) = (a + c, b + d).$$

These operations on real ordered pairs are referred to as (i) *multiplication by the scalar k* and (ii) *addition*, respectively.

The remainder of this section is devoted to a generalization of the concept of a real ordered pair to a *real ordered n-tuple*, together with a generalization of the operations just defined. For the special case of real ordered triples we also give geometric models.

DEFINITION 1.6 (Real Ordered Triples.) Let a, b, and c be three real numbers. The symbol

$$(a, b, c)$$

is called an *ordered triple of real numbers*, or a *real ordered triple*, with *first coordinate a*, *second coordinate b*, and *third coordinate c*.

(Equality of Ordered Triples.) A second real ordered triple, say (d, e, f), is said to be equal to (a, b, c) if and only if corresponding coordinates are equal. That is, $(a, b, c) = (d, e, f)$ if and only if $a = d$, $b = e$, and $c = f$.

(Real Ordered n-tuples.) Let $a_1, a_2, \ldots, a_n$ denote n real numbers. The symbol

$$(a_1, a_2, \ldots, a_n)$$

is called an *ordered n-tuple of real numbers*, or a *real ordered n-tuple*, with *first coordinate a_1*, *second coordinate a_2*, $\ldots$, and *n-th coordinate a_n*.

(Equality of Ordered n-tuples.) A second real ordered n-tuple, say $(b_1, b_2, \ldots, b_n)$, is said to be equal to $(a_1, a_2, \ldots, a_n)$ if and only if corresponding coordinates are equal. That is, $(a_1, a_2, \ldots, a_n) = (b_1, b_2, \ldots, b_n)$ if and only if $a_1 = b_1$, $a_2 = b_2$, $\ldots$, and $a_n = b_n$.

Sums and multiples of real ordered triples can be motivated in a way quite similar to that used previously for real ordered pairs. Such a motivation for real ordered n-tuples in general, however, requires an ability to visualize points and lines in n-space, an ability which most readers will not possess at this point if n is greater than three. Fortunately, the generalization of Definition 1.5 is straightforward and is easily accepted without further motivation.

DEFINITION 1.7   (Sums and Multiples of Real Ordered n-Tuples.)

i. (Multiples of a Real Ordered n-Tuple.)   The multiple of the real ordered n-tuple $(a_1, a_2, \ldots, a_n)$ by the real number k is the real ordered n-tuple given by

$$k(a_1, a_2, \ldots, a_n) = (ka_1, ka_2, \ldots, ka_n).$$

ii. (Sum of Ordered n-Tuples.)   The sum of $(a_1, a_2, \ldots, a_n)$ and $(b_1, b_2, \ldots, b_n)$ is the real ordered n-tuple given by

$$(a_1, a_2, \ldots, a_n) + (b_1, b_2, \ldots, b_n) = (a_1 + b_1, a_2 + b_2, \ldots, a_n + b_n).$$

These operations on real ordered n-tuples are called (i) *multiplication by the scalar k* and (ii) *addition*, respectively.

Example 2.   We present here several examples of sums and multiples of n-tuples.

(a) $2(3, 1, 2) = (6, 2, 4)$.

(b) $(2, 1) + (-1, 4) = (1, 5)$.

(c) $(x, y, 1) + (1, 2, 0) = (x + 1, y + 2, 1)$.

(d) $3(1, 1) + (-4)(6, 2) = (3, 3) + (-24, -8) = (-21, -5)$.

(e) $r(1, 1) + s(2, -1) = (r, r) + (2s, -s) = (r + 2s, r - s)$.

(f) $(1, 1, 2, 0) + 3(0, 0, 1, -2) = (1, 1, 2, 0) + (0, 0, 3, -6)$
$$= (1, 1, 5, -6). \quad \blacksquare$$

Before describing the geometrical models for real ordered triples, we define a coordinate system for space.

DEFINITION I.8 (Coordinate System for Space.) We say that we have introduced a coordinate system in space if we carry out the following procedure:

i. Select some point in space. Label it O and refer to it as the *origin* of the coordinate system.

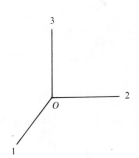

ii. Through O select three distinct lines which do not all lie in a single plane. Label one line with the Arabic numeral 1 and refer to it as the first coordinate axis or the 1-*axis* of the system. Label a second line with the Arabic numeral 2, and refer to it as the *second coordinate axis* or the 2-*axis*. Label the third line with the Arabic numeral 3 and refer to it as the *third coordinate axis* or the 3-*axis*.

iii. Make each axis into a real number line with O as the origin on each.

iv. For each point P in space determine the three real numbers a, b, and c so that we may arrive at P by a path starting at O and moving a units in the direction of the 1-axis, b units in the direction of the 2-axis, and c units in the direction of the 3-axis.

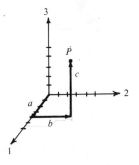

v. Refer to a, b, and c as the first, second, and third coordinate, respectively, of P.

D-I.8

You will note that in the definition just given there is no requirement that the three axes be perpendicular to one another or that the same unit of length be used on all axes. However, unless otherwise stated the usual coordinate system does employ perpendicular axes and a common scale on all axes, as indicated in the diagrams accompanying the definition.

Using a coordinate system for space, we now make precise the geometric models of real ordered triples.

Geometric Model 1.9 (*Real Ordered Triples as Points in Space.*) Introduce a coordinate system in space. Each real ordered triple (a, b, c) is represented by that point of space having a, b, and c as first, second, and third coordinate, respectively.

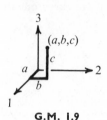

G.M. 1.9

Geometric Model 1.10 (*Real Ordered Triples as Directed Line Segments in Space.*) Introduce a coordinate system in space. Each real ordered triple (a, b, c) is represented by the directed line segment with O as the initial point and with a, b, and c as first, second, and third coordinate, respectively, of the terminal point.

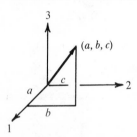

G.M. 1.10

We shall not attempt to formalize a geometric interpretation for real ordered quadruples or other n-tuples with $n > 3$. Nonetheless, we shall not hesitate to use the geometrical language when it seems to serve a purpose. We shall then on occasion refer to (a, b, c, d) as a *point in 4-space* or as a *directed line segment in 4-space*. After a time you may even feel that you can actually visualize four-space and, perhaps, n-space in general.

We close this section with a summary of the various ways in which we shall refer to symbols such as (a, b), (a, b, c), and $(a_1, a_2, \ldots, a_n)$.

Summary of Terminology 1.11 Let $a, b, c, a_1, a_2, \ldots, a_n$ be real numbers. We shall refer

    i. to (a, b) as a real ordered pair,
                    a point in a plane (or 2-space),
                    the coordinates of a point in a plane,
                    a directed line segment in a plane;

ii. to (a, b, c) as   a real ordered triple,
                 a point in 3-space,
                 the coordinates of a point in 3-space,
                 a directed line segment in 3-space;

iii. to $(a_1, a_2, \ldots, a_n)$ as   a real ordered n-tuple,
                 a point in n-space,
                 the coordinates of a point in n-space,
                 a directed line segment in n-space.

EXERCISES

In Exercises 1 to 8 determine the values of the letters such that the equality holds:

1. $(x, y) = (2, 1)$

2. $(x + 1, y - 2) = (3, 4)$

3. $(x^2, x) = (4, 1)$

4. $(y + x, y) = (2, 3)$

5. $(x, y - x) = (1, 1) + (3, 0)$

6. $r(1, 1) + s(1, -1) = (5, 0)$

7. $([x + y^2], x - y) = (1, 1)$

8. $(x, x + y, y + z) = (1, 1, 1)$

9. Trace the three coordinate systems of Figure 1.1 on a sheet of paper and on each locate the points with the following coordinates:
(a) $(2, 2)$ (b) $(-1, 2)$ (c) $(3, 0)$
(d) $(0, 0)$ (e) $(-2, -3)$ (f) $(3, -2)$

10. Carry out the indicated operations.
(a) $(1, 1) + (-2, 3)$ (b) $2(1, 7, 2)$
(c) $2(1, 3, 0, 1) + 3(2, 0, 4, 1)$ (d) $(-2)(2, 1) + 4(1, 2)$
(e) $(1, -1, 1) + (-1, 3, -1)$ (f) $3(1, 1, 2) + (-2)(0, 1, 0) +$
                                  $(0, 0, 0)$
(g) $(x, y) + 2(1, 1)$ (h) $r(-3, 1) + s(0, 2)$

11. Using graph paper set up a single coordinate system.
(a) Indicate carefully the points
      $D = (1, 2); \ -D = (-1, -2); \ 2D = (2, 4); \ 3D = (3, 6)$.
(b) Join each point with O by means of a straight line.
(c) What do you observe?

12. On the same coordinate system used in Problem 11
(a) Indicate the points $A + D$, $A + (-D)$, $A + 2D$, and $A + 3D$ if A represents the ordered pair $(-4, 1)$ and D is the ordered pair $(1, 2)$.
(b) Join each of the points just located to the point A by means of a straight line.
(c) What do you observe?

ANSWERS

1. $\begin{cases} x = 2 \\ y = 1 \end{cases}$.   3. For no value of x 5. $\begin{cases} x = 4 \\ y = 5 \end{cases}$.   7. $\begin{cases} x = 1 \\ y = 0 \end{cases}$, or $\begin{cases} x = 0 \\ y = -1 \end{cases}$.

9.

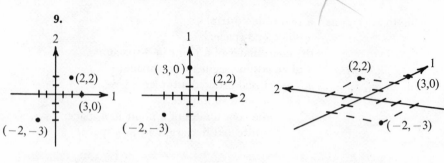

A-9

10. (a) $(-1, 4)$ (c) $(8, 6, 12, 5)$ (e) $(0, 2, 0)$ (g) $(x + 2, y + 2)$.

11. See Figure 1.6a and text on page 24.

12. See Figure 1.6b and text on page 24.

2. PROPERTIES OF n-TUPLE OPERATIONS; VECTOR SPACES $\mathcal{R}^n$

In the previous section an n-tuple was symbolized by placing n numbers in a row, separating them by commas, and enclosing them all by a pair of parentheses. There is an alternate notation that is used frequently because it provides greater clarity to computational routines. This notation merely lists the coordinates vertically in a column, rather than horizontally in a row. Thus, $(2, 1)$ and $\begin{pmatrix} 2 \\ 1 \end{pmatrix}$ are symbols for the same real ordered pair.

On many occasions we shall speak about real ordered pairs or n-tuples in general and shall not be concerned with the particular coordinates involved. At other times we shall wish to use an abbreviated notation when referring to a particular n-tuple. In such instances we shall represent an n-tuple by an upper case letter.

Notation 1.12. (*Coordinate Notation for n-Tuples.*) The n-tuple with first coordinate a_1, second coordinate $a_2, \ldots$, and n-*th* coordinate a_n will be symbolized by either the

 i. horizontal notation $(a_1, a_2, \ldots, a_n)$

or the

 ii. vertical notation $\begin{pmatrix} a_1 \\ a_2 \\ \cdot \\ \cdot \\ \cdot \\ a_n \end{pmatrix}$.

(*General, or Coordinate-free, Notation.*) When the specific values or the number of coordinates of a real ordered *n*-tuple is not of interest, or when an abbreviated notation is desired, an *n*-tuple of real numbers shall be denoted by an upper case letter such as $A, B, C, \ldots$.

Example 3. Later in this book problems of the following type will arise. The usefulness of the vertical notation will be quite evident.

(a) Express the vector $A = \begin{pmatrix} x + 2y \\ y \end{pmatrix}$ as a sum of fixed 2-tuples with variable multipliers. The coordinates of the given 2-tuple involve two variables, x and y. By using zero coefficients, the 2-tuple may be rewritten in expanded form with an x and y term present in both coordinates:

$$A = \begin{pmatrix} 1 \cdot x + 2 \cdot y \\ 0 \cdot x + 1 \cdot y \end{pmatrix}.$$

By a simple inspection, we can then write

$$A = x \begin{pmatrix} 1 \\ 0 \end{pmatrix} + y \begin{pmatrix} 2 \\ 1 \end{pmatrix}.$$

The fixed 2-tuples $\begin{pmatrix} 1 \\ 0 \end{pmatrix}$ and $\begin{pmatrix} 2 \\ 1 \end{pmatrix}$ are just the coefficients of the variables x and y in the expanded form above.

(b) Express the 2-tuple $B = \begin{pmatrix} -2x + z \\ -x + y \end{pmatrix}$ as a sum of multiples of fixed 2-tuples. The variables involved are x, y, and z. Introducing zero coefficients we may write B as

$$B = \begin{pmatrix} -2 \cdot x + 0 \cdot y + 1 \cdot z \\ -1 \cdot x + 1 \cdot y + 0 \cdot z \end{pmatrix}.$$

By inspection, we then write

$$B = x \begin{pmatrix} -2 \\ -1 \end{pmatrix} + y \begin{pmatrix} 0 \\ 1 \end{pmatrix} + z \begin{pmatrix} 1 \\ 0 \end{pmatrix}.$$

Of course, we may also express B as

$$B = -x \begin{pmatrix} 2 \\ 1 \end{pmatrix} + y \begin{pmatrix} 0 \\ 1 \end{pmatrix} + z \begin{pmatrix} 1 \\ 0 \end{pmatrix}.$$

(c) Express $C = \begin{pmatrix} 2x - y + 1 \\ 3x + 2 \end{pmatrix}$ as a sum of fixed 2-tuples with variable scalar multipliers where necessary.

As in the preceding problems we write C in expanded form

$$C = \begin{pmatrix} 2x + (-1)y + 1 \\ 3x + 0 \cdot y + 2 \end{pmatrix}.$$

A simple inspection suffices to deduce the following decomposition of C:

$$C = x \begin{pmatrix} 2 \\ 3 \end{pmatrix} + y \begin{pmatrix} -1 \\ 0 \end{pmatrix} + \begin{pmatrix} 1 \\ 2 \end{pmatrix}.$$

This problem differs from the preceding two in that one 2-tuple, namely $\begin{pmatrix} 1 \\ 2 \end{pmatrix}$, requires no variable multiplier. ■

We have already used the general (coordinate-free) notation in Exercises 11 and 12 of Section 1. At times it will be convenient to use the short notation in part of a discussion and the specific coordinates in another part. In this instance, we often write "Let $A = (a, b)$," and so forth.

We shall now examine the properties of addition of real ordered n-tuples and the properties of multiplication of real ordered n-tuples by various scalars. Since the definitions of these two operations for real ordered pairs are similar to those for general real ordered n-tuples, we can expect the properties of the respective operations to be similar. This is in fact true and the statement of the properties can be incorporated into a single general statement, valid for all n, by using the general notation.

Prior to the statement of these properties, we shall introduce a concept which is useful in all of mathematics. For each positive integer $n \geq 2$, the set of all real ordered n-tuples together with the operations of n-tuple addition and multiplication by a scalar constitute what is called a *mathematical system*, a concept of considerable generality.

DEFINITION 1.13   A *mathematical system* is a set on which one or more operations (addition and multiplication by a scalar, for example) are defined.

We are now ready to state and prove our first theorem.

 THEOREM 1.14   (Properties of n-Tuple Operations.)   Let n be a positive integer. The mathematical system of real ordered n-tuples together with the operations of addition and multiplication by scalars possesses the following properties. (A, B, and C denote n-tuples; r and s denote scalars):

   Addition Properties:

   i. (Commutative.)   For every A and B:
                                                    $A + B = B + A.$

ii. (Associative.)   For every A, B, and C:
$(A + B) + C = A + (B + C)$.

iii. (Zero.)   There is an n-tuple O_n such that for every n-tuple A:
$A + O_n = A$.

iv. (Opposites or Additive Inverses.)   For each A there is an n-tuple $-A$ such that:
$A + (-A) = O_n$.

Multiplication Properties:

v. (Unit Multiplier.)   For every A: $1 \cdot A = A$.

vi. (Mixed Associative.)   For every r, s, and A:
$r(sA) = (rs)A$.

Distributive Properties of Multiplication by a Scalar:

vii. (Over Scalar Addition.)   For every r, s, and A:
$(r + s)A = rA + sA$.

viii. (Over n-Tuple Addition.)   For every r, A, and B:
$r(A + B) = rA + rB$.

(_Note before proof._) Although the statements of this theorem employ the general notation, the proofs of these statements must employ the coordinate notation and must be proven for each n separately. This is a consequence of the fact that the definitions of the operations were made in terms of the coordinates themselves. The pattern of proof is the same whether we consider ordered pairs, ordered triples, or the general ordered n-tuple. We shall, therefore, look at the proofs of several of the properties for ordered pairs only.

Since the manner in which the details of proofs are presented in this book differs with the usual mode of presentation, the reader should be familiar with the explanation given in Appendix I on page 873.

Proof for Ordered Pairs. In the proofs which follow, we shall use both the general notation of the theorem and the coordinate notation as follows:

	General		_Coordinate_
Let	A	be the ordered pair	$\begin{pmatrix} a \\ b \end{pmatrix}$.
Let	B	be the ordered pair	$\begin{pmatrix} c \\ d \end{pmatrix}$.
Let	C	be the ordered pair	$\begin{pmatrix} e \\ f \end{pmatrix}$.

Proof of ii.
$$A + (B + C) =^1 \binom{a}{b} + \left\{ \binom{c}{d} + \binom{e}{f} \right\}$$
$$=^2 \binom{a}{b} + \binom{c+e}{d+f}$$
$$=^2 \binom{a + (c+e)}{b + (d+f)}$$
$$=^3 \binom{(a+c)+e}{(b+d)+f}$$
$$=^2 \binom{a+c}{b+d} + \binom{e}{f}$$
$$=^2 \left\{ \binom{a}{b} + \binom{c}{d} \right\} + \binom{e}{f}$$
$$=^4 (A+B) + C.$$

(1) Changing to coordinate notation.
(2) Definition 1.5 (*Sum of Ordered Pairs*), (p. 9).
(3) Applying associative property of real numbers to each coordinate and using Definition 1.1 (*Equality of Ordered Pairs*), (p. 4).
(4) Reverting to general notation.

Proof of iii. The 2-tuple $O_2 = (0, 0)$ is the zero for ordered pairs since
$$A + O_2 =^1 \binom{a}{b} + \binom{0}{0}$$
$$=^2 \binom{a+0}{b+0}$$
$$=^3 \binom{a}{b}$$
$$=^4 A.$$

(1) Changing to coordinate notation.
(2) Definition 1.5 (*Sums of Ordered Pairs*).
(3) Using the property of the real number 0 in each coordinate and equality for ordered pairs.
(4) Reverting to general notation.

Proof of iv. The ordered pair $-A = (-a, -b)$ is such that
$$A + (-A) =^1 \binom{a}{b} + \binom{-a}{-b}$$
$$=^2 \binom{a + (-a)}{b + (-b)}$$
$$=^3 \binom{0}{0}$$
$$=^4 O_2.$$

(1) Changing to coordinate notation.
(2) Definition 1.5 (*Sum of Ordered Pairs*).
(3) Using properties of additive inverses of real numbers on each coordinate and the definition of equality of ordered pairs.
(4) Definition of O_2 in Proof of (iii).

Proof of vii.
$$(r+s)A =^1 (r+s)\binom{a}{b}$$

$$=^2 \binom{(r+s)a}{(r+s)b}$$

$$=^3 \binom{ra+sa}{rb+sb}$$

$$=^4 \binom{ra}{rb}+\binom{sa}{sb}$$

$$=^2 r\binom{a}{b}+s\binom{a}{b}$$

$$=^5 rA+sA.$$

(1) Changing to coordinate notation.
(2) Definition 1.5 (*Multiples of a Real Ordered Pair*), (p. 9).
(3) Using distributive property of real numbers on each coordinate and definition of equality of ordered pairs.
(4) Definition 1.5 (*Sum of Real Ordered Pairs*).
(5) Reverting to general notation.

The proofs of the remaining properties, even for the case $n = 2$, are left as an exercise (see Exercise 17). ■

In the statement and proof of Theorem 1.14, we introduced certain special notations, namely, the symbol O_n for the zero *n*-tuple and the symbol $-A$ for the opposite or additive inverse of A. This notation is important and shall be formalized in the next definition. At the same time, we shall define what is meant by the operation of *n*-tuple subtraction.

DEFINITION 1.15 Let n be a fixed positive integer.

(Zero *n*-Tuple.) The *n*-tuple, each coordinate of which is 0, is called the *zero n-tuple* and is denoted by
$$O_n = (0, 0, \ldots , 0).$$
In particular, $\quad O_2 = (0, 0) \quad$ and $\quad O_3 = (0, 0, 0).$

(Opposite or Additive Inverse of an *n*-Tuple.) Let $A = (a_1, a_2, \ldots , a_n)$ be an *n*-tuple. The *n*-tuple whose coordinates are the additive inverses of the corresponding coordinates of A is called the *opposite* or *additive inverse* of A and is denoted by
$$-A = (-a_1, -a_2, \ldots , -a_n).$$
In particular,
$$-(a, b) = (-a, -b), \quad \text{and} \quad -(a, b, c) = (-a, -b, -c).$$

(Differences of *n*-Tuples.) The sum of an *n*-tuple A and the additive inverse of the *n*-tuple B is denoted by
$$A - B = A + (-B)$$
and is called the *difference of B from A*. The operation of determining a difference is called *subtraction*.

Exercises 18 and 19 give further useful properties of O_n and $-A$.

Throughout the remainder of this book we shall refer to the mathematical systems of real ordered n-tuples with the operations of addition and multiplication by a scalar. For ease of reference we now introduce abbreviated names and symbols for these systems.

DEFINITION 1.16 (The Vector Space $\mathfrak{R}^n$.) The mathematical system consisting of all real ordered n-tuples $(a_1, a_2, \ldots, a_n)$ together with the operations of n-tuple addition and multiplication of an n-tuple by a scalar is called the *vector space of real ordered n-tuples* and is symbolized by $\mathfrak{R}^n$. Each ordered n-tuple, when considered as part of this mathematical system, is called an *n-vector*, or simply *vector* if n is not specified.

If we set $n = 2$ or $n = 3$ in the last definition, we obtain the definition of $\mathfrak{R}^2$ and $\mathfrak{R}^3$, respectively. These are the vector spaces with which we shall primarily be concerned in this chapter and the next. In subsequent chapters we shall also be concerned with the vector space which is denoted by $\mathfrak{R}^1$, or more simply by $\mathfrak{R}$. Strictly speaking, at this point Definitions 1.6, 1.7, and 1.16 have no meaning when $n = 1$ since we have not defined a 1-tuple. This, however, can be remedied. A real ordered pair is something with two real coordinates and a real ordered n-tuple is specified by n real coordinates. Thus, a 1-tuple is reasonably considered to be specified by one real coordinate. That is, a 1-tuple of real numbers is just a single real number.

Addition of 1-tuples is just the addition of real numbers. Multiplication of a 1-tuple by a scalar is just the multiplication of one real number, the 1-tuple, by another real number, the scalar. The mathematical system of all 1-tuples and the two operations just defined are nothing more nor less than the real number system itself. Theorem 1.14 (*Properties of n-Tuple Operations*) is valid when $n = 1$ and has been stated so as to include this case.

A rereading of Definition 1.6 (*Real Ordered n-Tuples*), Definition 1.7 (*Sums and Multiples of Real Ordered n-Tuples*), and Definition 1.16 (*The Vector Space* $\mathfrak{R}^n$) will confirm the following observation.

Note 1.17. (*1-Tuple and the Vector Space* $\mathfrak{R}[\mathfrak{R}^1]$.) The content of Definitions 1.6, 1.7, and 1.16 has meaning even when $n = 1$ and thereby serves to interpret the real number system as the *vector space* $\mathfrak{R}$ and a real number as a 1-vector.

We conclude this section with a geometric description of addition and subtraction of n-vectors. Exercises 20 to 25 call for a verification of these descriptions in $\mathfrak{R}^2$.

Geometric Representation of Addition

The operation of addition of two n-vectors A and B may be carried out geometrically as follows:

i. Select a coordinate system and construct the directed line segments representing A and B (Fig. 1.4i).　　(i)

ii. Complete the parallelogram having A and B as sides (Fig. 1.4ii).　　(ii)

iii. Construct the directed line segment from O to the opposite vertex of the parallelogram. This line segment represents $A + B$ (Fig. 1.4iii).　　(iii)

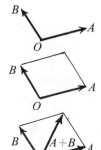

Figure 1.4

Geometric Representation of Subtraction

The operation of subtraction of the n-vector X from the n-vector Y may be carried out geometrically as follows:

i. Select a coordinate system and construct the directed line segments representing X and Y (Fig. 1.5i).　　(i)

ii. Complete the parallelogram with X as one side and Y as a diagonal (Fig. 1.5ii).　　(ii)

iii. The difference $Y - X$ is represented by the directed line segment forming the second side of the parallelogram passing through O (X forming the first side) (Fig. 1.5iii).　　(iii)

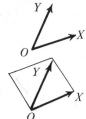

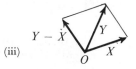

Figure 1.5

The above two representations fail if the directed line segments representing the vectors A and B or the vectors X and Y are in the same direction.

EXERCISES

In Exercises 1 to 4 determine values of the variables for which the equality holds:

1. $\begin{pmatrix} x \\ y \end{pmatrix} - 2\begin{pmatrix} x \\ x \end{pmatrix} = \begin{pmatrix} -1 \\ 3 \end{pmatrix}$.

2. $r\begin{pmatrix}1\\0\\0\end{pmatrix} + s\begin{pmatrix}1\\1\\0\end{pmatrix} + t\begin{pmatrix}1\\0\\1\end{pmatrix} = \begin{pmatrix}4\\0\\3\end{pmatrix}.$

3. $\begin{pmatrix}1\\1\end{pmatrix} + t\begin{pmatrix}2\\0\end{pmatrix} = \begin{pmatrix}1\\2\end{pmatrix} + s\begin{pmatrix}-2\\1\end{pmatrix}.$

4. $x\begin{pmatrix}x\\1\end{pmatrix} - \begin{pmatrix}1\\1\end{pmatrix} = \begin{pmatrix}1\\0\end{pmatrix}.$

In Exercises 5 to 10 express each vector as a sum of multiples of fixed vectors:

5. $\begin{pmatrix}x - y\\2x + y\end{pmatrix}.$

6. $\begin{pmatrix}x + y\\y - z\\x + z\end{pmatrix}.$

7. $\begin{pmatrix}3x + 2y - z\\4x + z\end{pmatrix}.$

8. $\begin{pmatrix}-2x + y - z\\x + y + z\\4x + 3z\\z + t\end{pmatrix}.$

9. $\begin{pmatrix}x + y - 2\\y + 3\end{pmatrix}.$

10. $\begin{pmatrix}2y - 2\\3x + 1\\-z\end{pmatrix}.$

In Exercises 11 to 16 carry out the indicated operations:

11. $3\begin{pmatrix}1\\2\\-1\end{pmatrix} - 4\begin{pmatrix}-1\\0\\-2\end{pmatrix}.$

12. $\begin{pmatrix}1\\2\end{pmatrix} + t\begin{pmatrix}0\\1\end{pmatrix}.$

13. $\begin{pmatrix}1\\2\\3\end{pmatrix} + r\begin{pmatrix}1\\0\\0\end{pmatrix} + s\begin{pmatrix}1\\0\\1\end{pmatrix}.$

14. $\begin{pmatrix}2\\0\\-4\\-10\end{pmatrix} - 6\begin{pmatrix}1\\1\\-2\\-3\end{pmatrix} + 2\begin{pmatrix}1\\2\\1\\2\end{pmatrix}.$

15. $\begin{pmatrix}3\\1\end{pmatrix} + r\begin{pmatrix}2\\0\end{pmatrix} + s\begin{pmatrix}1\\1\end{pmatrix}.$

16. $r\begin{pmatrix}1\\2\\1\end{pmatrix} + s\begin{pmatrix}4\\8\\4\end{pmatrix}.$

17. Prove properties (i), (v), (vi), and (viii) of Theorem 1.14, (p. 16), for real ordered pairs (2-vectors).

18. Let D be a nonzero n-vector and t a real number. Show that $tD = O_n$ if and only if, $t = 0$.
(This result is needed for later proofs.)

19. In view of Exercise 18, we may write $0 \cdot A = O_n$ for any n-vector A.
(a) What other symbol may be used to denote $(-1) \cdot A$?
Use the result of (a) to show that
(b) $-r \cdot A = -(r \cdot A) = r \cdot (-A)$
and (c) $sA - (rA) = (s - r)A.$

In Exercises 20 to 25, first use graph paper to determine the sum or difference geometrically and then compare with the computed result:

20. $\begin{pmatrix}2\\3\end{pmatrix} + \begin{pmatrix}4\\-1\end{pmatrix}.$

21. $\begin{pmatrix}6\\4\end{pmatrix} - \begin{pmatrix}5\\2\end{pmatrix}.$

22. $\begin{pmatrix} 2 \\ 5 \end{pmatrix} + \begin{pmatrix} 4 \\ 10 \end{pmatrix}.$ 23. $\begin{pmatrix} 3 \\ 2 \end{pmatrix} + \begin{pmatrix} 0 \\ 1 \end{pmatrix}.$

24. $\begin{pmatrix} 3 \\ 2 \end{pmatrix} + \begin{pmatrix} 6 \\ 1 \end{pmatrix}.$ 25. $\begin{pmatrix} 4 \\ 2 \end{pmatrix} - \begin{pmatrix} 2 \\ 1 \end{pmatrix}.$

ANSWERS

1. $\begin{cases} x = 1 \\ y = 5. \end{cases}$ **3.** $\begin{cases} t = 1 \\ s = -1. \end{cases}$ **5.** $x\begin{pmatrix} 1 \\ 2 \end{pmatrix} + y\begin{pmatrix} -1 \\ 1 \end{pmatrix}.$

7. $x\begin{pmatrix} 3 \\ 4 \end{pmatrix} + y\begin{pmatrix} 2 \\ 0 \end{pmatrix} + z\begin{pmatrix} -1 \\ 1 \end{pmatrix}.$ **9.** $x\begin{pmatrix} 1 \\ 0 \end{pmatrix} + y\begin{pmatrix} 1 \\ 1 \end{pmatrix} + \begin{pmatrix} -2 \\ 3 \end{pmatrix}.$ **11.** $\begin{pmatrix} 7 \\ 6 \\ 5 \end{pmatrix}.$

13. $\begin{pmatrix} 1 + r + s \\ 2 \\ 3 + s \end{pmatrix}.$ **15.** $\begin{pmatrix} 3 + 2r + s \\ 1 + s \end{pmatrix}.$

19. (a) $(-1) \cdot A = -A.$ Since $A + (-1) \cdot A = 1 \cdot A + (-1) \cdot A$
$$= (1 + (-1)) \cdot A$$
$$= O_n.$$

19. (c) $sA - (rA) = sA + [-(rA)],$ Definition 1.15 (*Subtraction*).
$$= sA + (-r)A, \quad \text{Using Exercise 19b.}$$
$$= [s + (-r)]A, \quad \text{Theorem 1.14vii.}$$
$$= (s - r)A, \quad \text{Definition of subtraction for}$$
real numbers.

22. Note that the directed line segments representing $\begin{pmatrix} 2 \\ 5 \end{pmatrix}$ and $\begin{pmatrix} 4 \\ 10 \end{pmatrix}$ are collinear. The "parallelogram" constructed with these two vectors as sides does not exist and the geometric construction fails.

3. POINTS AND LINES IN $\mathcal{R}^n$

In motivating the definition of addition of vectors and multiplication of vectors by a scalar, the ideas of geometry were used in an informal way. It is important to note that these definitions were merely *motivated* by the geometry. The definitions themselves employ only properties of the real number system.

Throughout this chapter and the next chapter, we shall proceed in reverse. We shall use the vector space concepts to formalize our geometry. We shall *define* what we mean by a point, by a line, by a plane, by parallelism, by perpendicularity, by length, and so forth in terms of the language of vectors.

The concept of a point is usually regarded as the basic concept in any geometry. In Section 1 an ordered pair was merely represented by a "point." Henceforth, when we speak of a point in $\mathcal{R}^n$ we shall not mean a dot of ink on a page, or even an imaginary "dimensionless position" in a plane or in space. By a point in $\mathcal{R}^n$ we shall mean an n-vector, an ordered n-tuple of real numbers.

> **DEFINITION 1.18**   (Points in $\mathfrak{R}^n$.)   A vector of $\mathfrak{R}^n$ is also called a *point* of $\mathfrak{R}^n$, and, conversely, by a point of $\mathfrak{R}^n$ is meant an n-vector. That is, P is a *point of* $\mathfrak{R}^n$ if, and only if, P is an n-tuple in $\mathfrak{R}^n$.

We shall, of course, continue to use diagrams to aid our discussions. The diagrams will suggest methods of proof and will also suggest ways in which other geometric terms can be introduced into discussions in $\mathfrak{R}^n$.

For example, we are now going to define what is meant by a line in $\mathfrak{R}^n$. We shall exploit our intuitive idea of a line in order that we may develop a vector space definition. It may seem that this is argument in a circle, since we

(a) assume that each of us has an idea of what a line is, and then
(b) define what a line is to be.

The evidence that we are not involved in circumlocution lies in the different ways in which the concept of a line in (a) and the concept of a line in (b) are to be regarded. In (a) the vague intuitive idea of a line which each of us has is taken as a starting point. What we seek is not some "feeling" for the concept "line," but a precise description of the concept. In (b) we agree precisely with what this description, this definition, is to be. Thereafter, we discard the informal concept of (a) and refer always to the formal definition. *It is important, then, that the formal definition of each concept be mastered as it arises and that these definitions become an integral part of our thinking.*

We now ask, "What is a line?" Rather, what shall we mean by a line in $\mathfrak{R}^n$? To answer this question let us consider our geometrical models for $\mathfrak{R}^2$. First, we shall consider lines through the origin, now denoted by O_2, of the chosen coordinate system (Fig. 1.6a). According to the usual concept of a line, a second point, say $D \neq O_2$, together with O_2 will determine a line. Consider now the points $-D$, $2D$, and $3D$. The work of Exercise 11 (p. 13), and Figure 1.6a suggest that all real multiples tD of D will be on

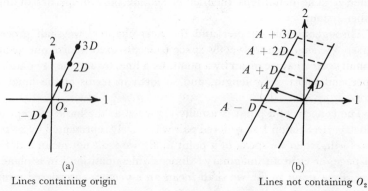

(a)   (b)
Lines containing origin   Lines not containing O_2

Figure 1.6   Lines in R^2.

this same line. The following definition of a line through O_2 is thereby suggested.

A line $\mathfrak{L}$ in $\Re^2$ through the origin O_2 is the set of all points of the form tD where t ranges over all real numbers and D is a fixed nonzero 2-vector. Symbolically

$$\mathfrak{L} = \{X \mid X = tD, \, t \in \Re\}, \text{ for some } D \neq O_2.$$

This definition, clearly, is a limited one. It only defines lines through O_2. How can we obtain other lines? Exercise 12 (p. 13) and Figure 1.6b suggest that the set of points $A + tD$, where A and D are fixed nonzero 2-vectors and where t ranges over all real numbers, will yield a line not containing O_2.

These considerations suggest the following definition of a line. Note that we have extended the above ideas to define the concept of a line in n-space, $\Re^n$, as well as in 2-space, $\Re^2$.

DEFINITION 1.19 (Line in $\Re^n$.) A set $\mathfrak{L}$ in $\Re^n$ is called a *line* if and only if there is a vector A in $\Re^n$ and a nonzero vector D in $\Re^n$ such that

$$\mathfrak{L} = \{X \mid X = A + tD, \, t \text{ a real number}\}.$$

This notation is frequently abbreviated

$$\mathfrak{L} = \{A + tD \mid t \in \Re\}.$$

(Parametric Representation for a Line.) The equation

$$X(t) = A + tD$$

is called a *parametric representation for the line $\mathfrak{L}$*. The variable t is called a *parameter*.

If we permitted $D = O_n$ in the preceding definition, the set obtained thereby would consist of a single point:

$$\{A + tD \mid t \in \Re\} =^1 \{A + tO_n \mid t \in \Re\}$$
$$=^2 \{A + O_n\}$$
$$=^3 \{A\}.$$

(1) Substituting O_n for D.
(2) Since $tO_n = O_n$ for every number t.
(3) Theorem 1.14iii.

We therefore require D to be nonzero so that our lines will contain more than one point.

Hereafter when we speak of a line, we shall mean a set $\mathfrak{L}$ of n-tuples specified by one of the above representations. The very mention of the word line should bring to your mind one of these descriptions, not a streak of chalk on a chalkboard. No picture which we use will be a line. At this point we cannot be certain that the drawing of pictures will be of any help at all. We shall remedy this situation by showing, in this and the two following sections, that lines in the sense defined above have many of the properties which we attribute to "straight streaks of graphite on a paper or of chalk on a chalkboard or of imaginary sets of 'points' in space."

Example 4.   Let $A = (1, 3, 0)$ and $D = (1, 1, 1)$ and consider the line

$$\mathfrak{L} = \{A + tD \mid t \in \mathfrak{R}\}.$$

$\mathfrak{L}$ is a line in $\mathfrak{R}^3$ since A and D are 3-vectors and $D \neq O_n$. We can determine points of this line by choosing various values for t and computing $A + tD$. If we set $t = 0$, we find that

$$A + 0 \cdot D = A + O_3, \qquad \text{Exercise 18 (p. 22)}.$$
$$= A. \qquad \text{Property of } O_3.$$

Thus A is a point of $\mathfrak{L}$, a comforting fact in view of Figure 16b.
    Setting $t = 2$, we find that

$$A + 2D = \begin{pmatrix} 1 \\ 3 \\ 0 \end{pmatrix} + 2 \begin{pmatrix} 1 \\ 1 \\ 1 \end{pmatrix} = \begin{pmatrix} 3 \\ 5 \\ 2 \end{pmatrix}$$

is a second point of $\mathfrak{L}$. Further choices of t yield other points on $\mathfrak{L}$. ∎

The preceding example suggests the following result.

THEOREM 1.20   (Points on a Line.)   Let A and $D \neq O_n$ be vectors in $\mathfrak{R}^n$. Let $\mathfrak{L}$ be the line

$$\mathfrak{L} = \{A + tD \mid t \in \mathfrak{R}\}.$$

The n-vectors X and Y are distinct points of $\mathfrak{L}$ if and only if

$$X = A + t_1 D \quad \text{and} \quad Y = A + t_2 D \quad \text{with} \quad t_1 \neq t_2.$$

Proof.   According to Definition 1.19, the two n-vectors X and Y will be points of $\mathfrak{L}$ if and only if

$$X = A + t_1 D \quad \text{and} \quad Y = A + t_2 D$$

for some choice of t_1 and t_2. We need only show that $A + t_1 D$ and $A + t_2 D$ are distinct if and only if $t_1 \neq t_2$.

We shall establish this latter fact by proving the contrapositive statement: $A + t_1D$ and $A + t_2D$ are equal if and only if t_1 and t_2 are equal. Thus,

$$A + t_1D = A + t_2D \Leftrightarrow^1 t_1D = t_2D$$

$$\Leftrightarrow^2 t_1D - t_2D = O_n$$

$$\Leftrightarrow^3 (t_1 - t_2)D = O_n$$

$$\Leftrightarrow^4 t_1 - t_2 = 0$$

$$\Leftrightarrow^5 t_1 = t_2.$$

(1) Subtracting ($\Rightarrow$) or adding ($\Leftarrow$) A from (or to) both members. See Appendix I (p.872) for significance of the symbols $\Rightarrow$, $\Leftarrow$, and $\Leftrightarrow$.
(2) Subtracting ($\Rightarrow$) or adding ($\Leftarrow$) t_2D from (or to) both members.
(3) Applying distributive property on left member (Exercise 19c of the preceding section).
(4) Using fact that $D \neq O_n$ and applying Exercise 18 (p. 22), with $t = t_1 - t_2$.
(5) Adding ($\Rightarrow$) or subtracting ($\Leftarrow$) t_2 from both members. ■

Example 4 also suggests that the point A appearing in the parametric representation $X(t) = A + tD$ for $\mathcal{L}$ is, in fact, a point of $\mathcal{L}$. The vector D also plays a special role in the description of $\mathcal{L}$. Reference to Figure 1.6b suggests that D specifies the direction of $\mathcal{L}$. On the other hand, a comparison of Figure 1.7 with Figure 1.5 suggests that the difference $Y - X$ of any two distinct points X and Y on $\mathcal{L}$ should also be regarded as giving the direction of $\mathcal{L}$.

DEFINITION 1.21   (Direction Vectors for a Line.)   Let $\mathcal{L}$ be a line in $\mathcal{R}^n$. A vector V in $\mathcal{R}^n$ is called a *direction vector for* $\mathcal{L}$ if and only if there are distinct points X and Y belonging to $\mathcal{L}$ such that

$$V = Y - X.$$

According to this definition there are infinitely many direction vectors for a given line. They are, of course, quite simply related, a fact which we now establish.

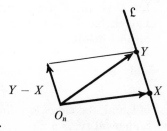

Figure 1.7   Direction vector for a line.

THEOREM 1.22   (Relation of the Vectors in a Parametric Description of a Line to the Line.) Let A and $D \neq O_n$ be in $\mathcal{R}^n$. Let $\mathcal{L}$ be the line given by

$$\mathcal{L} = \{A + tD \mid t \in \mathcal{R}\}.$$

Then

i. A is a point of $\mathcal{L}$.

ii. D is a direction vector for $\mathcal{L}$.

iii. Every direction vector for $\mathcal{L}$ is a nonzero multiple of D.

iv. Every nonzero multiple of a direction vector for $\mathcal{L}$ is again a direction vector for $\mathcal{L}$, and

v. Every direction vector for $\mathcal{L}$ is a nonzero multiple of any other direction vector for $\mathcal{L}$.

Proof $\text{Set } t = 0$

i. The proof of (i) has the same form as that given in Example 4 for a special case and is left as Exercise 24.

ii. Statement (ii) follows immediately from the fact that A and $A + D$ are points of $\mathcal{L}$ (Exercise 25).

iii. Let V be a direction vector for $\mathcal{L}$. By Definition 1.21, V is the difference $Y - X$ of two distinct points X and Y of the line $\mathcal{L}$. By Theorem 1.20 (p. 26), there are distinct real numbers t_1 and t_2 such that

$$X = A + t_1 D \quad \text{and} \quad Y = A + t_2 D \quad \text{with} \quad t_1 \neq t_2.$$

Thus,

$$V = Y - X = (A + t_2 D) - (A + t_1 D)$$
$$= t_2 D - t_1 D$$
$$= (t_2 - t_1) D.$$

The fact that $t_1 \neq t_2$ assures us that $t_1 - t_2 \neq 0$ and that V is a nonzero multiple of D.

iv. Let V be a direction vector for $\mathcal{L}$. Then there are two distinct points

$$X = A + t_1 D \quad \text{and} \quad Y = A + t_2 D \quad \text{with} \quad t_1 \neq t_2$$

such that

$$V = Y - X.$$

Let U be any nonzero multiple of V, say $U = kV$, $k \neq 0$. We leave as Exercise 26 the task of showing that U is the difference between the two distinct points

$$X' = A + kt_1 D$$

and

$$Y' = A + kt_2 D$$

of $\mathcal{L}$ and, therefore, is a direction vector for $\mathcal{L}$.

v. Statement (v) is a direct consequence of (iii) and (iv) (Exercise 27).

Example 5. Write the set description of a line through $\begin{pmatrix} 1 \\ 2 \end{pmatrix}$ with direction vector $\begin{pmatrix} 3 \\ 2 \end{pmatrix}$. The set

$$\mathfrak{L} = \left\{ \begin{pmatrix} 1 \\ 2 \end{pmatrix} + t \begin{pmatrix} 3 \\ 2 \end{pmatrix} \middle| t \in \mathfrak{R} \right\}$$

will, according to Theorem 1.22, be such a line. ∎

Example 6. Find a line in $\mathfrak{R}^3$ which contains the two points $\begin{pmatrix} 1 \\ 1 \\ 0 \end{pmatrix}$ and $\begin{pmatrix} 1 \\ 0 \\ -2 \end{pmatrix}$. A direction vector D for such a line is given by the difference

$$D = \begin{pmatrix} 1 \\ 1 \\ 0 \end{pmatrix} - \begin{pmatrix} 1 \\ 0 \\ -2 \end{pmatrix} = \begin{pmatrix} 0 \\ 1 \\ 2 \end{pmatrix}.$$

A line through $\begin{pmatrix} 1 \\ 1 \\ 0 \end{pmatrix}$ with D as a direction vector is

$$\mathfrak{L} = \left\{ \begin{pmatrix} 1 \\ 1 \\ 0 \end{pmatrix} + t \begin{pmatrix} 0 \\ 1 \\ 2 \end{pmatrix} \middle| t \in \mathfrak{R} \right\}.$$

Does this line also contain the second point $\begin{pmatrix} 1 \\ 0 \\ -2 \end{pmatrix}$? In other words, is there a value of t such that

(1.1)
$$\begin{pmatrix} 1 \\ 0 \\ -2 \end{pmatrix} = \begin{pmatrix} 1 \\ 1 \\ 0 \end{pmatrix} + t \begin{pmatrix} 0 \\ 1 \\ 2 \end{pmatrix} ?$$

Subtracting the vector $\begin{pmatrix} 1 \\ 1 \\ 0 \end{pmatrix}$ from both members of Equation 1.1, we obtain the equivalent equation

$$\begin{pmatrix} 0 \\ -1 \\ -2 \end{pmatrix} = t \begin{pmatrix} 0 \\ 1 \\ 2 \end{pmatrix}.$$

Since this equation is obviously satisfied with $t = -1$, we conclude that the line $\mathfrak{L}$ does contain each of the given points. ∎

Example 7. Which of the points $A = \begin{pmatrix} 7 \\ 4 \\ 12 \\ 2 \end{pmatrix}$ and $B = \begin{pmatrix} -2 \\ -2 \\ -3 \\ 1 \end{pmatrix}$ lie on the line

$$\mathcal{L} = \left\{ \begin{pmatrix} 1 \\ 0 \\ 2 \\ 1 \end{pmatrix} + t \begin{pmatrix} 9 \\ 6 \\ 15 \\ 3 \end{pmatrix} \,\middle|\, t \in \mathcal{R} \right\} ?$$

We note that

$$\begin{pmatrix} 7 \\ 4 \\ 12 \\ 3 \end{pmatrix} \in \mathcal{L} \Leftrightarrow^1 \begin{pmatrix} 7 \\ 4 \\ 12 \\ 3 \end{pmatrix} = \begin{pmatrix} 1 \\ 0 \\ 2 \\ 1 \end{pmatrix} + t \begin{pmatrix} 9 \\ 6 \\ 15 \\ 3 \end{pmatrix} \text{ for some } t,$$

$$\Leftrightarrow^2 \begin{pmatrix} 6 \\ 4 \\ 10 \\ 2 \end{pmatrix} = t \begin{pmatrix} 9 \\ 6 \\ 15 \\ 3 \end{pmatrix} \text{ for some } t,$$

$$\Leftrightarrow^3 \begin{cases} 6 = 9t \\ 4 = 6t \\ 10 = 15t \\ 2 = 3t \end{cases} \text{ for some } t.$$

(1) Definition of a line.
(2) Subtracting ($\Rightarrow$) or adding ($\Leftarrow$) A from (to) to both members.
(3) Definitions of multiplication by a scalar and equality of vectors.

Since this last system of equations is satisfied for $t = 2/3$, we conclude that $A \in \mathcal{L}$.

We test B in a similar manner:

$$B \in \mathcal{L} \Leftrightarrow \begin{pmatrix} -2 \\ -2 \\ -3 \\ 1 \end{pmatrix} = \begin{pmatrix} 1 \\ 0 \\ 2 \\ 1 \end{pmatrix} + t \begin{pmatrix} 9 \\ 6 \\ 15 \\ 3 \end{pmatrix} \text{ for some } t,$$

$$\Leftrightarrow \begin{pmatrix} -3 \\ -2 \\ -5 \\ 0 \end{pmatrix} = \begin{pmatrix} 9t \\ 6t \\ 15t \\ 3t \end{pmatrix} \text{ for some } t,$$

$$\Leftrightarrow \begin{cases} -3 = 9t \\ -2 = 6t \\ -5 = 15t \\ 0 = 3t \end{cases} \text{ for some } t,$$

$$\Leftrightarrow \begin{cases} t = -1/3 \\ t = -1/3 \\ t = -1/3 \\ t = 0 \end{cases} \text{ for some } t.$$

We find that $B \notin \mathcal{L}$ since the last system of equations has no solution. The value $t = -1/3$ necessary to obtain the first three coordinates of B does not give the fourth coordinate. ∎

EXERCISES

In Exercises 1 to 8 give a set description of a line through A with direction vector D, and determine a second point on the line:

1. $A = \begin{pmatrix} -2 \\ 1 \end{pmatrix}$, $D = \begin{pmatrix} 1 \\ 1 \end{pmatrix}$.

2. $A = \begin{pmatrix} 1 \\ 3 \end{pmatrix}$, $D = \begin{pmatrix} 1 \\ 3 \end{pmatrix}$.

3. $A = \begin{pmatrix} 2 \\ 0 \\ 1 \end{pmatrix}$, $D = \begin{pmatrix} 1 \\ 1 \\ 0 \end{pmatrix}$.

4. $A = \begin{pmatrix} 0 \\ 0 \end{pmatrix}$, $D = \begin{pmatrix} 1 \\ 0 \end{pmatrix}$.

5. $A = \begin{pmatrix} 2 \\ 0 \end{pmatrix}$, $D = \begin{pmatrix} 0 \\ 4 \end{pmatrix}$.

6. $A = \begin{pmatrix} 0 \\ 1 \\ 2 \end{pmatrix}$, $D = \begin{pmatrix} 0 \\ 0 \\ 1 \end{pmatrix}$.

7. $A = \begin{pmatrix} -3 \\ 1 \\ 2 \\ -2 \end{pmatrix}$, $D = \begin{pmatrix} 2 \\ 1 \\ 1 \\ 0 \end{pmatrix}$.

8. $A = \begin{pmatrix} 0 \\ 0 \\ 1 \\ 0 \end{pmatrix}$, $D = \begin{pmatrix} 0 \\ 0 \\ 1 \\ 0 \end{pmatrix}$.

In Exercises 9 to 12, determine a line containing points A and B, and show that A and B are on the line obtained:

9. $A = \begin{pmatrix} 3 \\ 1 \end{pmatrix}$, $B = \begin{pmatrix} -1 \\ 2 \end{pmatrix}$.

10. $A = \begin{pmatrix} 4 \\ 2 \end{pmatrix}$, $B = \begin{pmatrix} 0 \\ 0 \end{pmatrix}$.

11. $A = \begin{pmatrix} 1 \\ 2 \\ 3 \end{pmatrix}$, $B = \begin{pmatrix} 3 \\ 2 \\ 1 \end{pmatrix}$.

12. $A = \begin{pmatrix} 2 \\ 0 \\ 0 \\ 3 \end{pmatrix}$, $B = \begin{pmatrix} 0 \\ 1 \\ 2 \\ 0 \end{pmatrix}$.

In Exercises 13 to 16, determine which of the given points are on the given line:

13. $\mathcal{L} = \left\{ \begin{pmatrix} 2 \\ 1 \end{pmatrix} + t\begin{pmatrix} 2 \\ 1 \end{pmatrix} \middle| t \in \mathcal{R} \right\}$; $A = \begin{pmatrix} 0 \\ 0 \end{pmatrix}$, $B = \begin{pmatrix} 10 \\ 5 \end{pmatrix}$, $C = \begin{pmatrix} 7 \\ 15 \end{pmatrix}$.

14. $\mathcal{L} = \left\{ \begin{pmatrix} -1 \\ 3 \end{pmatrix} + t\begin{pmatrix} 4 \\ 5 \end{pmatrix} \middle| t \in \mathcal{R} \right\}$; $A = \begin{pmatrix} -9 \\ -7 \end{pmatrix}$, $B = \begin{pmatrix} 10 \\ 8 \end{pmatrix}$, $C = \begin{pmatrix} 19 \\ 27 \end{pmatrix}$.

15. $\mathcal{L} = \left\{ \begin{pmatrix} -1 \\ 2 \\ 1 \end{pmatrix} + t\begin{pmatrix} 2 \\ 10 \\ -2 \end{pmatrix} \middle| t \in \mathcal{R} \right\}$; $A = \begin{pmatrix} 2 \\ 17 \\ 4 \end{pmatrix}$, $B = \begin{pmatrix} -2 \\ -4 \\ -3 \end{pmatrix}$.

16. $\mathcal{L} = \left\{ \begin{pmatrix} 2 \\ 1 \\ 3 \end{pmatrix} + t\begin{pmatrix} 6 \\ -21 \\ 12 \end{pmatrix} \middle| t \in \mathcal{R} \right\}$; $A = \begin{pmatrix} 10 \\ -27 \\ 20 \end{pmatrix}$, $B = \begin{pmatrix} -6 \\ 29 \\ -12 \end{pmatrix}$.

In Exercises 17 to 18, determine whether A, B, and C are collinear (on the same line):

17. $A = \begin{pmatrix} 1 \\ 1 \end{pmatrix}$, $B = \begin{pmatrix} 2 \\ 3 \end{pmatrix}$, $C = \begin{pmatrix} 17 \\ 33 \end{pmatrix}$.

18. $A = \begin{pmatrix} 2 \\ 3 \\ -3 \end{pmatrix}$, $B = \begin{pmatrix} 1 \\ 0 \\ 2 \end{pmatrix}$, $C = \begin{pmatrix} 8 \\ 21 \\ -9 \end{pmatrix}$.

Each of the sets in Exercises 19 to 22 is a line in some $\mathcal{R}^n$. Give the value of n and write a description for the line in the form $\{A + tD \mid t \in \mathcal{R}\}$, and, if possible, give a direction vector with first coordinate 1:

19. $\mathcal{L} = \left\{ \begin{pmatrix} 2t + 1 \\ t \end{pmatrix} \middle| t \in \mathcal{R} \right\}$.

20. $\mathcal{L} = \left\{ \begin{pmatrix} 2 \\ 3 + 4t \end{pmatrix} \middle| t \in \mathcal{R} \right\}$.

21. $\mathcal{L} = \left\{ \begin{pmatrix} 0 \\ -2 - t \end{pmatrix} \middle| t \in \mathcal{R} \right\}$.

22. $\mathcal{L} = \left\{ \begin{pmatrix} 1 + 3t \\ 2 - 4t \\ 6 \end{pmatrix} \middle| t \in \mathcal{R} \right\}$.

23. $\mathcal{L} = \left\{ \begin{pmatrix} 2 \\ t - 4 \\ 2 \\ -3 + 2t \end{pmatrix} \middle| t \in \mathcal{R} \right\}$.

Exercises 24 to 27 are a completion of the proofs of the statements of Theorem 1.22 (p. 28):

24. Generalize the treatment of Example 4 (p. 26) to prove statement (i).

25. Prove statement (ii). (Hint: See comment in proof.)

26. Let $X = A + tD_1$; $Y = A + t_2D$, with $t_1 \neq t_2$; and $V = Y - X$. Show that if $U = kV$, $k \neq 0$, then U is the difference of the two distinct vectors $X' = A + kt_1D$ and $Y' = A + kt_2D$. This completes the proof of statement (iv).

27. Prove statement (v) using statements (iii) and (iv).

ANSWERS

1. $\left\{ \begin{pmatrix} -2 \\ 1 \end{pmatrix} + t\begin{pmatrix} 1 \\ 1 \end{pmatrix} \middle| t \in \mathcal{R} \right\}$, $\begin{pmatrix} -1 \\ 2 \end{pmatrix}$ is another point.

3. $\left\{ \begin{pmatrix} 2 \\ 0 \\ 1 \end{pmatrix} + t\begin{pmatrix} 1 \\ 1 \\ 0 \end{pmatrix} \middle| t \in \mathcal{R} \right\}$, $\begin{pmatrix} 1 \\ -1 \\ 1 \end{pmatrix}$ is a second point.

5. $\left\{ \begin{pmatrix} 2 \\ 0 \end{pmatrix} + t\begin{pmatrix} 0 \\ 4 \end{pmatrix} \middle| t \in \mathcal{R} \right\}$, all points are of the form $\begin{pmatrix} 2 \\ k \end{pmatrix}$, k being any real number.

7. $\left\{ \begin{pmatrix} -3 \\ 1 \\ 2 \\ -2 \end{pmatrix} + t\begin{pmatrix} 2 \\ 1 \\ 1 \\ 0 \end{pmatrix} \middle| t \in \mathcal{R} \right\}$.

9. $\mathcal{L} = \left\{ \begin{pmatrix} 3 \\ 1 \end{pmatrix} + t\begin{pmatrix} 4 \\ -1 \end{pmatrix} \middle| t \in \mathcal{R} \right\}$, $t = 0$ gives A, $t = -1$ gives B.

11. $\mathcal{L} = \left\{ \begin{pmatrix} 3 \\ 2 \\ 1 \end{pmatrix} + t\begin{pmatrix} 2 \\ 0 \\ -2 \end{pmatrix} \middle| t \in \mathcal{R} \right\}$, $t = 0$ gives B, and $t = -1$ gives A. Other answers are possible.

13. $t = -1$ gives A, $t = 4$ gives B, and C is not on line.

15. $t = \frac{3}{2}$ gives A, B not on line.   **17.** A, B, and C are collinear.

19. $\mathcal{L} = \left\{ \begin{pmatrix} 1 \\ 0 \end{pmatrix} + t \begin{pmatrix} 2 \\ 1 \end{pmatrix} \,\middle|\, t \in \mathcal{R} \right\}$, $D = \begin{pmatrix} 2 \\ 1 \end{pmatrix}$ is a direction vector, and so $\frac{1}{2} D = \begin{pmatrix} 1 \\ \frac{1}{2} \end{pmatrix}$ is a direction vector with first coordinate 1.

21. $\mathcal{L} = \left\{ \begin{pmatrix} 0 \\ -2 \end{pmatrix} + t \begin{pmatrix} 0 \\ -1 \end{pmatrix} \,\middle|\, t \in \mathcal{R} \right\}$; no direction vector has first coordinate 1.

23. $\mathcal{L} = \left\{ \begin{pmatrix} 2 \\ -4 \\ 2 \\ -3 \end{pmatrix} + t \begin{pmatrix} 0 \\ 1 \\ 0 \\ 2 \end{pmatrix} \,\middle|\, t \in \mathcal{R} \right\}$; all direction vectors have first coordinate zero.

4. PROPERTIES OF LINES: PARALLELISM

A line in $\mathcal{R}^n$, defined in the preceding section, is a set of n-vectors which can be described in a certain way, namely, as a set

$$\mathcal{L} = \{A + tD \mid t \in \mathcal{R}\}$$

for some choice of n-vector A and nonzero n-vector D. In the last theorem we proved that A is a point of $\mathcal{L}$ and that D is a direction vector for $\mathcal{L}$. The next example will pose an important question which needs to be answered.

Example 8. Consider the two lines

$$\mathcal{L}_1 = \left\{ \begin{pmatrix} 4 \\ 3 \end{pmatrix} + t \begin{pmatrix} 2 \\ 1 \end{pmatrix} \,\middle|\, t \in \mathcal{R} \right\} \quad \text{and} \quad \mathcal{L}_2 = \left\{ \begin{pmatrix} 0 \\ 1 \end{pmatrix} + s \begin{pmatrix} 2 \\ 1 \end{pmatrix} \,\middle|\, s \in \mathcal{R} \right\}.$$

According to the last theorem of the preceding section, the line $\mathcal{L}_1$ contains the point $(4, 3)$ and has $(2, 1)$ as a direction vector. But the point $(4, 3)$ is also on $\mathcal{L}_2$, a fact which can be verified by setting $s = 2$. Moreover, $\mathcal{L}_2$ has $(2, 1)$ as a direction vector. Thus, $\mathcal{L}_2$, as well as $\mathcal{L}_1$, is a line through $(4, 3)$ with $(2, 1)$ as a direction vector. Are $\mathcal{L}_1$ and $\mathcal{L}_2$ two different lines? Are $\mathcal{L}_1$ and $\mathcal{L}_2$ the same line? ■

Surely our "feeling" is this: Since $\mathcal{L}_1$ and $\mathcal{L}_2$ each contain the point $(4, 3)$ and each has the same direction vector $(2, 1)$, it must be that $\mathcal{L}_1$ and $\mathcal{L}_2$ are identical. However, we must not be content with a mere "feeling." We must show beyond any doubt that the two sets $\mathcal{L}_1$ and $\mathcal{L}_2$ are identical, though differently described. We begin by showing that each line $\mathcal{L}$ which contains a point A may be described in the form $\mathcal{L} = \{A + tD \mid t \in \mathcal{R}\}$, no matter how it was described initially.

THEOREM 1.23   (Set Description of a Line Containing Point A.) Let A be a point on the line

$$\mathfrak{L} = \{B + sD \mid s \in \mathfrak{R}\}, \qquad D \neq O_n.$$

The line $\mathfrak{L}$ may also be described as

$$\mathfrak{L} = \{A + rD \mid r \in \mathfrak{R}\}.$$

Proof.   Consider the line

(1.2) $$\mathfrak{L} = \{B + sD \mid s \in \mathfrak{R}\}, \qquad D \neq O_n.$$

Since A is a point on $\mathfrak{L}$, we have $A = B + s_0 D$ for some number s_0. The following implication chain completes the proof.

$$A = B + s_0 D \Rightarrow \quad B = A - s_0 D$$
$$\Rightarrow^1 \mathfrak{L} = \{(A - s_0 D) + sD \mid s \in \mathfrak{R}\}$$
$$\Rightarrow^2 \mathfrak{L} = \{A + (s - s_0)D \mid s \in \mathfrak{R}\}$$
$$\Rightarrow^3 \mathfrak{L} = \{A + rD \mid r \in \mathfrak{R}\}.$$

(1) Substituting for B in Equation 1.2.

(2) Using associative and distributive properties of vector (n-tuple) operations.

(3) Setting $r = s - s_0$ and noting that as s ranges over all real numbers so also will $r = s - s_0$, and, conversely, as r ranges over all real numbers so also will $s = r + s_0$. ∎

On the basis of Theorem 1.22 and Theorem 1.23 we can now show that specification of a point A and a direction vector D completely determines a unique line.

THEOREM 1.24   (Point-Direction Description of a Line).   Let A and $D \neq O_n$ be vectors in $\mathfrak{R}^n$. There is one and only one line $\mathfrak{L}$ through A with D as a direction vector. Moreover, $\mathfrak{L}$ is described by

$$\mathfrak{L} = \{A + tD \mid t \in \mathfrak{R}\}.$$

Proof.   Let A and $D \neq O_n$ be n-vectors. By Theorem 1.22 the line

(1.3) $$\mathfrak{L}_1 = \{A + tD \mid t \in \mathfrak{R}\}$$

is one line containing A and having D as a direction vector. Let $\mathfrak{L}_2$ be another such line. That is (going back to the definition of a line), let $\mathfrak{L}_2$ be a set

$$\mathfrak{L}_2 = \{B + sD_1 \mid s \in \mathfrak{R}\},$$

(where B and D_1 are n-vectors and $D_1 \neq O_n$) which contains A and has D as a direction vector. The present theorem will be proved if we show that the lines $\mathcal{L}_1$ and $\mathcal{L}_2$ are the same set of points. We do this by showing that every point of $\mathcal{L}_1$ is a point of $\mathcal{L}_2$, and, conversely, that every point of $\mathcal{L}_2$ is also a point of $\mathcal{L}_1$.

We first note that, according to Theorem 1.23, $\mathcal{L}_2$ may be described as

$$(1.4) \qquad \mathcal{L}_2 = \{A + rD_1 \mid r \in \mathcal{R}\}.$$

Moreover, by Theorem 1.22 (p. 28), D_1 is a direction vector for $\mathcal{L}_2$. Since D is also a direction vector for $\mathcal{L}_2$, we invoke Theorem 1.22(v) to conclude that

$$(1.5) \qquad D_1 = kD, \quad \text{with} \quad k \neq 0.$$

Consider now the two implication schemes:

$$X \in \mathcal{L}_1 \Rightarrow^1 X = A + tD, \text{ for some } t \in \mathcal{R},$$
$$\Rightarrow^2 X = A + \frac{t}{k}(kD)$$
$$\Rightarrow^3 X = A + rD_1, \text{ for some } r \in \mathcal{R},$$
$$\Rightarrow^4 X \in \mathcal{L}_2;$$

and

$$X \in \mathcal{L}_2 \Rightarrow^4 X = A + rD_1, \text{ for some } r \in \mathcal{R},$$
$$\Rightarrow^5 X = A + r(kD)$$
$$\Rightarrow X = A + (rk)D$$
$$\Rightarrow^6 X = A + tD, \text{ for some } t \in \mathcal{R},$$
$$\Rightarrow X \in \mathcal{L}_1.$$

(1) Using Equation 1.3 as the description for $\mathcal{L}_1$.

(2) Using the fact that $k \neq 0$ and hence $\frac{t}{k}(kD) = tD$.

(3) Using Equation 1.5 and setting $\frac{t}{k} = r$.

(4) Using Equation 1.4 as description for $\mathcal{L}_2$.
(5) Using Equation 1.5.
(6) Setting $t = rk$.

From the first implication chain we see that every point of $\mathcal{L}_1$ is also a point of $\mathcal{L}_2$. From the second chain, the converse situation clearly holds and the proof is complete. ∎

As a result of this theorem we know that the lines of Example 8 are the same. This theorem also explains why the form of your answer to Exercise 11 of the preceding section may not have agreed with the one given in the text. In Exercises 1 to 8 of the preceding section you were asked to find a line through A with direction vector D. Theorem 1.24 assures us that there is only one such line. In Exercises 9 to 12 you were asked to determine a line containing points A and B and to show that both points were on the line obtained. Because of our limited knowledge of the properties of lines it was *essential* that we show that B was a point on the line constructed. Your experience with these Exercises and Theorem 1.24 should enable you to prove the following result.

THEOREM 1.25   (Two-point Description of a Line.)   Let A and B be two distinct points of $\mathcal{R}^n$. There is one and only one line containing both A and B. It is the line

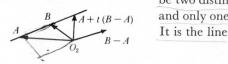

T-1.25

$$\mathcal{L} = \{A + t(B - A) \mid t \in \mathcal{R}\}.$$

The proof is left as an exercise (see Exercise 17). The content of the previous theorem may be stated in a perhaps more familiar manner: A line is determined by any two of its points.

One of the characteristic properties which we intuitively assign to the set of lines is given in the following theorem. The proof follows readily from Theorem 1.25.

THEOREM 1.26   (Intersection of Two Lines in $\mathcal{R}^n$.)   Let $\mathcal{L}_1$ and $\mathcal{L}_2$ denote two (not necessarily distinct) lines in $\mathcal{R}^n$. Exactly one of the following is true:

 i. $\mathcal{L}_1$ and $\mathcal{L}_2$ have no points in common; $\mathcal{L}_1 \cap \mathcal{L}_2 = \varnothing$.
 ii. $\mathcal{L}_1$ and $\mathcal{L}_2$ have exactly one point in common; $\mathcal{L}_1 \cap \mathcal{L}_2 = $ singleton set.
 iii. $\mathcal{L}_1$ and $\mathcal{L}_2$ coincide; $\mathcal{L}_1 \cap \mathcal{L}_2 = \mathcal{L}_1 = \mathcal{L}_2$.

Proof. It is evident that at most one of the three statements can be valid for a given choice of $\mathcal{L}_1$ and $\mathcal{L}_2$. If (i) or (ii) is satisfied, the theorem is certainly true. If neither (i) nor (ii) holds, then $\mathcal{L}_1$ and $\mathcal{L}_2$ must have at least two distinct points in common and must, by Theorem 1.25, coincide. That is if (i) and (ii) fail to hold, then (iii) must hold.   ∎

Example 9.   Find the points in $\mathcal{L}_1 \cap \mathcal{L}_2$ if

$$\mathcal{L}_1 = \left\{ \begin{pmatrix} -2 \\ 3 \end{pmatrix} + t \begin{pmatrix} 4 \\ 2 \end{pmatrix} \,\middle|\, t \in \mathcal{R} \right\} \quad \text{and} \quad \mathcal{L}_2 = \left\{ \begin{pmatrix} 2 \\ 4 \end{pmatrix} + s \begin{pmatrix} 1 \\ 1 \end{pmatrix} \,\middle|\, s \in \mathcal{R} \right\}.$$

We first note that $\mathcal{L}_1$ has direction vector $\begin{pmatrix} 4 \\ 2 \end{pmatrix}$ and $\mathcal{L}_2$ has direction vector $\begin{pmatrix} 1 \\ 1 \end{pmatrix}$. By Theorem 1.22(v), $\mathcal{L}_1$ and $\mathcal{L}_2$ cannot coincide, since the direction vector $\begin{pmatrix} 4 \\ 2 \end{pmatrix}$ for $\mathcal{L}_1$ is not a multiple of the direction vector $\begin{pmatrix} 1 \\ 1 \end{pmatrix}$ for $\mathcal{L}_2$. Thus, we have either an empty intersection or a single point C of intersection. In the latter event the point $C = \begin{pmatrix} c \\ d \end{pmatrix}$ must be on both $\mathcal{L}_1$ and $\mathcal{L}_2$, and thus we have

i. $C = \begin{pmatrix} c \\ d \end{pmatrix} = \begin{pmatrix} -2 \\ 3 \end{pmatrix} + t \begin{pmatrix} 4 \\ 2 \end{pmatrix}$, for some t, and

ii. $C = \begin{pmatrix} c \\ d \end{pmatrix} = \begin{pmatrix} 2 \\ 4 \end{pmatrix} + s \begin{pmatrix} 1 \\ 1 \end{pmatrix}$, for some s.

Consequently,

$$C \in L_1 \cap L_2 \Leftrightarrow \begin{pmatrix} -2 \\ 3 \end{pmatrix} + t \begin{pmatrix} 4 \\ 2 \end{pmatrix} = \begin{pmatrix} 2 \\ 4 \end{pmatrix} + s \begin{pmatrix} 1 \\ 1 \end{pmatrix}, \text{ for some } s \text{ and } t,$$

$$\Leftrightarrow \begin{pmatrix} -2 \\ 3 \end{pmatrix} - \begin{pmatrix} 2 \\ 4 \end{pmatrix} = -t \begin{pmatrix} 4 \\ 2 \end{pmatrix} + s \begin{pmatrix} 1 \\ 1 \end{pmatrix}$$

$$\Leftrightarrow \begin{pmatrix} -4 \\ -1 \end{pmatrix} = \begin{pmatrix} s - 4t \\ s - 2t \end{pmatrix}$$

$$\Leftrightarrow \begin{cases} s \text{ and } t \text{ is a solution of the system} \\ -4 = s - 4t \\ -1 = s - 2t. \end{cases}$$

It is readily verified that $s = 2$ and $t = \frac{3}{2}$ is a solution of the above system of equations. Using $t = \frac{3}{2}$ in (i) we obtain

$$C = \begin{pmatrix} c \\ d \end{pmatrix} = \begin{pmatrix} -2 \\ 3 \end{pmatrix} + \frac{3}{2} \begin{pmatrix} 4 \\ 2 \end{pmatrix} = \begin{pmatrix} 4 \\ 6 \end{pmatrix}.$$

As a check on our work we also use $s = 2$ in (ii) to evaluate C:

$$C = \begin{pmatrix} 2 \\ 4 \end{pmatrix} + 2 \begin{pmatrix} 1 \\ 1 \end{pmatrix} = \begin{pmatrix} 4 \\ 6 \end{pmatrix}. \quad \blacksquare$$

Example 10. Find $\mathcal{L}_1 \cap \mathcal{L}_2$ if

$$\mathcal{L}_1 = \left\{ \begin{pmatrix} 1 \\ 0 \\ 2 \end{pmatrix} + t \begin{pmatrix} 0 \\ 1 \\ 1 \end{pmatrix} \middle| t \in \mathcal{R} \right\} \quad \text{and} \quad \mathcal{L}_2 = \left\{ \begin{pmatrix} 5 \\ -5 \\ 4 \end{pmatrix} + s \begin{pmatrix} -1 \\ 2 \\ 0 \end{pmatrix} \middle| s \in \mathcal{R} \right\}.$$

Again, the direction vector $\begin{pmatrix} 0 \\ 1 \\ 1 \end{pmatrix}$ for $\mathcal{L}_1$ is not a multiple of the direction vector $\begin{pmatrix} -1 \\ 2 \\ 0 \end{pmatrix}$ for $\mathcal{L}_2$. Thus, L_1 and L_2 do not coincide, and, according to the last theorem, either $\mathcal{L}_1 \cap \mathcal{L}_2 = \varnothing$ or there is a point $C = \begin{pmatrix} c \\ d \\ e \end{pmatrix}$ such that

$\mathcal{L}_1 \cap \mathcal{L}_2 = \{C\}$. As in the preceding example we proceed thus:

There is a $C = \begin{pmatrix} c \\ d \\ e \end{pmatrix}$ in $\mathcal{L}_1 \cap \mathcal{L}_2 \Leftrightarrow \begin{cases} \begin{pmatrix} c \\ d \\ e \end{pmatrix} = \begin{pmatrix} 1 \\ 0 \\ 2 \end{pmatrix} + t \begin{pmatrix} 0 \\ 1 \\ 1 \end{pmatrix}, \text{ for some } t, \\ \text{and} \\ \begin{pmatrix} c \\ d \\ e \end{pmatrix} = \begin{pmatrix} 5 \\ -5 \\ 4 \end{pmatrix} + s \begin{pmatrix} -1 \\ 2 \\ 0 \end{pmatrix}, \text{ for some } s, \end{cases}$

$\Leftrightarrow \begin{pmatrix} 1 \\ 0 \\ 2 \end{pmatrix} + t \begin{pmatrix} 0 \\ 1 \\ 1 \end{pmatrix} = \begin{pmatrix} 5 \\ -5 \\ 4 \end{pmatrix} + s \begin{pmatrix} -1 \\ 2 \\ 0 \end{pmatrix},$

for some t and s,

$\Leftrightarrow \begin{pmatrix} -4 \\ 5 \\ -2 \end{pmatrix} = \begin{pmatrix} -s \\ 2s - t \\ -t \end{pmatrix}, \text{ for some } t \text{ and } s,$

$\Leftrightarrow \begin{cases} s = 4 \\ 2s - t = 5, \text{ for some } t \text{ and } s. \\ t = 2 \end{cases}$

The final system of equations cannot be satisfied. For if $s = 4$ and $t = 2$ we have $2s - t = 2 \cdot 4 - 2 = 6$. Thus, the values of s and t necessary to satisfy the first and last equation of the system will not satisfy the middle equation. Consequently, there is no point C in $\mathcal{L}_1 \cap \mathcal{L}_2$. Hence $\mathcal{L}_1 \cap \mathcal{L}_2 = \varnothing$. ∎

Example 11. Find $\mathcal{L}_1 \cap \mathcal{L}_2$ if

$\mathcal{L}_1 = \left\{ \begin{pmatrix} 1 \\ 0 \\ 2 \end{pmatrix} + t \begin{pmatrix} 0 \\ 1 \\ 1 \end{pmatrix} \,\middle|\, t \in \mathcal{R} \right\}$ and $\mathcal{L}_2 = \left\{ \begin{pmatrix} 1 \\ -2 \\ 0 \end{pmatrix} + s \begin{pmatrix} 0 \\ -3 \\ -3 \end{pmatrix} \,\middle|\, s \in \mathcal{R} \right\}.$

In this case $\begin{pmatrix} 0 \\ 1 \\ 1 \end{pmatrix}$ is a direction vector for both $\mathcal{L}_1$ and $\mathcal{L}_2$, since $\begin{pmatrix} 0 \\ 1 \\ 1 \end{pmatrix} = -\frac{1}{3} \begin{pmatrix} 0 \\ -3 \\ -3 \end{pmatrix}$. Thus, $\mathcal{L}_1$ and $\mathcal{L}_2$ either are identical or have an empty intersection. (See Exercise 18a of the present section.) $\mathcal{L}_1$ and $\mathcal{L}_2$ are then identical if any point of one is also a point of the other. The point $\begin{pmatrix} 1 \\ -2 \\ 0 \end{pmatrix}$ is clearly a point of $\mathcal{L}_2$. The implication chain

$\begin{pmatrix} 1 \\ -2 \\ 0 \end{pmatrix} \in \mathcal{L}_1 \Leftrightarrow \begin{pmatrix} 1 \\ -2 \\ 0 \end{pmatrix} = \begin{pmatrix} 1 \\ 0 \\ 2 \end{pmatrix} + t \begin{pmatrix} 0 \\ 1 \\ 1 \end{pmatrix}$

$$\Leftrightarrow \begin{pmatrix} 1 \\ -2 \\ 0 \end{pmatrix} - \begin{pmatrix} 1 \\ 0 \\ 2 \end{pmatrix} = t \begin{pmatrix} 0 \\ 1 \\ 1 \end{pmatrix}, \qquad \text{for some } t \text{ in } \mathcal{R},$$

$$\Leftrightarrow \begin{pmatrix} 0 \\ -2 \\ -2 \end{pmatrix} = t \begin{pmatrix} 0 \\ 1 \\ 1 \end{pmatrix}$$

shows that $\begin{pmatrix} 1 \\ -2 \\ 0 \end{pmatrix}$ is a point on $\mathcal{L}_1$ (choose $t = -2$). Therefore $\mathcal{L}_1 = \mathcal{L}_2$.

Exercise 18b gives the general principle here. ■

An important concept in the study of further properties of lines is the concept of parallelism. In a plane we say that two "lines" are parallel if they do not intersect. Figure 1.8a pictures two "lines" in space which also do not intersect. Do we wish to say that these "lines" are also parallel? Rather, do we not want "parallel lines" in space to denote lines with a "common direction," as in Figure 1.8b?

On the basis of this observation we now define the concepts of parallelism.

DEFINITION 1.27 (Parallelism.) The term *parallel* is used in the following ways:

  i. (Parallelism of Two Vectors.) Two vectors U and V in $\mathcal{R}^n$ are said to be *parallel* to one another if and only if one is a multiple of the other, that is, $U = kV$ for some real number k, or $V = mU$ for some real number m, or both.

 ii. (Parallelism of a Vector and a Line.) A vector U is said to be *parallel* to a line $\mathcal{L}$ if and only if U is parallel to a direction vector for $\mathcal{L}$.

iii. (Parallelism of Two Lines.) Two lines are said to be *parallel* to one another if and only if at least one direction vector for one line is parallel to some direction vector for the other.

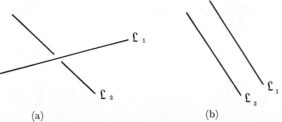

(a)	(b)
$\mathcal{L}_1$ "crosses over" $\mathcal{L}_2$	$\mathcal{L}_1$ and $\mathcal{L}_2$ have a "common direction"

Figure 1.8 Non-intersecting lines in space.

This definition contains more than is apparent on a first reading. Accordingly we make the following observations, the proofs of which are left as Exercises 19, 20, and 21.

  i. *Parallelism of Two Vectors.* Every vector U is parallel to itself. The zero vector O_n is parallel to every n-vector.

  ii. *Parallelism of a Vector and a Line.* U is parallel to the line $\mathcal{L}$ if and only if U is the zero vector O_n or U is a direction vector for $\mathcal{L}$.

  iii. *Parallelism of Two Lines.* Two lines are parallel if and only if some direction vector for one line is a direction vector for the other.

Two lines are parallel if and only if every direction vector for one line is a direction vector for the other.

If two lines are parallel, then either they do not intersect or they are identical.

Example 12. Find a line $\mathcal{L}_1$ through the point $A = \begin{pmatrix} 1 \\ 2 \\ 1 \end{pmatrix}$ and parallel to the line $\mathcal{L}_2 = \left\{ \begin{pmatrix} 3 \\ 0 \\ 2 \end{pmatrix} + t \begin{pmatrix} 1 \\ -1 \\ 1 \end{pmatrix} \middle| t \in \mathcal{R} \right\}$. Since $\mathcal{L}_1$ and $\mathcal{L}_2$ are to be parallel to one another, every direction vector for $\mathcal{L}_2$ is also a direction vector for $\mathcal{L}_1$. Therefore, since $\begin{pmatrix} 1 \\ -1 \\ 1 \end{pmatrix}$ is clearly a direction vector for $\mathcal{L}_2$, $\mathcal{L}_1$ is a line through $A = \begin{pmatrix} 1 \\ 2 \\ 1 \end{pmatrix}$ with direction vector $D = \begin{pmatrix} 1 \\ -1 \\ 1 \end{pmatrix}$. Thus,

$$\mathcal{L}_1 = \left\{ \begin{pmatrix} 1 \\ 2 \\ 1 \end{pmatrix} + t \begin{pmatrix} 1 \\ -1 \\ 1 \end{pmatrix} \middle| t \in \mathcal{R} \right\}.$$

Moreover, in view of Theorem 1.24, there is only one such line. ∎

The generalization of the problem in Example 12 to the general space $\mathcal{R}^n$ is stated as Exercise 22.

EXERCISES

In each of the Exercises 1 to 9 determine a parametric description of the line $\mathcal{L}$ having the required properties:

  1. $\mathcal{L}$ contains $\begin{pmatrix} 1 \\ 3 \end{pmatrix}$ and is parallel to the line $\left\{ \begin{pmatrix} 1 \\ 1 \end{pmatrix} + t \begin{pmatrix} 2 \\ 3 \end{pmatrix} \middle| t \in \mathcal{R} \right\}$.

  2. $\mathcal{L}$ contains $\begin{pmatrix} 1 \\ -1 \\ 2 \end{pmatrix}$ and is parallel to the line $\left\{ t \begin{pmatrix} 2 \\ 1 \\ 3 \end{pmatrix} \middle| t \in \mathcal{R} \right\}$.

  3. $\mathcal{L}$ contains $\begin{pmatrix} 2 \\ 1 \end{pmatrix}$ and $\begin{pmatrix} 3 \\ -2 \end{pmatrix}$.

4. $\mathfrak{L}$ contains $\begin{pmatrix} 2 \\ 1 \\ 3 \end{pmatrix}$ and $\begin{pmatrix} 0 \\ 0 \\ 1 \end{pmatrix}$.

5. $\mathfrak{L}$ contains O_2 and is parallel to the line through $\begin{pmatrix} 1 \\ 1 \end{pmatrix}$ and $\begin{pmatrix} 2 \\ -1 \end{pmatrix}$.

6. $\mathfrak{L}$ contains $\begin{pmatrix} 1 \\ 1 \\ 1 \end{pmatrix}$ and is parallel to the line through $\begin{pmatrix} 1 \\ 0 \\ 1 \end{pmatrix}$ and $\begin{pmatrix} 2 \\ 1 \\ 1 \end{pmatrix}$.

7. $\mathfrak{L}$ contains $\begin{pmatrix} 1 \\ 2 \\ 0 \\ 0 \end{pmatrix}$ and is parallel to $\begin{pmatrix} 1 \\ -1 \\ 0 \\ 1 \end{pmatrix}$.

8. $\mathfrak{L}$ contains $\begin{pmatrix} 2 \\ 0 \\ 0 \\ 1 \end{pmatrix}$ and $\begin{pmatrix} 0 \\ 1 \\ 1 \\ 1 \end{pmatrix}$.

9. $\mathfrak{L}$ contains $\begin{pmatrix} 2 \\ 0 \\ 0 \\ 1 \end{pmatrix}$ and is parallel to the line through $\begin{pmatrix} 1 \\ 0 \\ 0 \\ 0 \end{pmatrix}$ and $\begin{pmatrix} 0 \\ 1 \\ 0 \\ 0 \end{pmatrix}$.

In each of the Exercises 10 to 16 find all points in $\mathfrak{L}_1 \cap \mathfrak{L}_2$:

10. $\mathfrak{L}_1 = \left\{ \begin{pmatrix} 3 \\ 3 \end{pmatrix} + t \begin{pmatrix} 2 \\ 1 \end{pmatrix} \middle| t \in \mathcal{R} \right\}$, $\mathfrak{L}_2 = \left\{ \begin{pmatrix} 5 \\ -10 \end{pmatrix} + s \begin{pmatrix} -1 \\ 3 \end{pmatrix} \middle| s \in \mathcal{R} \right\}$.

11. $\mathfrak{L}_1 = \left\{ \begin{pmatrix} -1 \\ 3 \end{pmatrix} + t \begin{pmatrix} 3 \\ 5 \end{pmatrix} \middle| t \in \mathcal{R} \right\}$, $\mathfrak{L}_2 =$ the line through $\begin{pmatrix} 4 \\ -1 \end{pmatrix}$ and $\begin{pmatrix} -2 \\ -11 \end{pmatrix}$.

12. $\mathfrak{L}_1 =$ line through $\begin{pmatrix} 5 \\ 7 \end{pmatrix}$ parallel to $\begin{pmatrix} 11 \\ 3 \end{pmatrix}$, $\mathfrak{L}_2 =$ line through $\begin{pmatrix} -17 \\ 1 \end{pmatrix}$ and $\begin{pmatrix} 16 \\ 10 \end{pmatrix}$.

13. $\mathfrak{L}_1 = \left\{ \begin{pmatrix} 1 \\ 0 \\ 1 \end{pmatrix} + t \begin{pmatrix} -2 \\ 0 \\ 1 \end{pmatrix} \middle| t \in \mathcal{R} \right\}$, $\mathfrak{L}_2 = \left\{ \begin{pmatrix} 17 \\ -2 \\ -10 \end{pmatrix} + s \begin{pmatrix} 5 \\ -1 \\ -4 \end{pmatrix} \middle| s \in \mathcal{R} \right\}$.

14. $\mathfrak{L}_1 =$ line through $\begin{pmatrix} 1 \\ 12 \\ 3 \end{pmatrix}$ and $\begin{pmatrix} 0 \\ 10 \\ 1 \end{pmatrix}$, $\mathfrak{L}_2 =$ line through $\begin{pmatrix} 1 \\ -1 \\ 0 \end{pmatrix}$ and $\begin{pmatrix} 2 \\ 1 \\ 2 \end{pmatrix}$.

15. $\mathfrak{L}_1 = \left\{ \begin{pmatrix} 2 \\ -1 \\ 1 \end{pmatrix} + t \begin{pmatrix} 1 \\ 1 \\ 2 \end{pmatrix} \middle| t \in \mathcal{R} \right\}$, $\mathfrak{L}_2 =$ line through $\begin{pmatrix} -11 \\ -14 \\ -25 \end{pmatrix}$ and $\begin{pmatrix} 9 \\ 6 \\ 15 \end{pmatrix}$.

16. $\mathfrak{L}_1 =$ line through $\begin{pmatrix} 1 \\ 1 \\ 2 \\ 0 \end{pmatrix}$ and $\begin{pmatrix} 2 \\ 0 \\ 1 \\ 1 \end{pmatrix}$, $\mathfrak{L}_2 =$ line through $\begin{pmatrix} 8 \\ -4 \\ -1 \\ 5 \end{pmatrix}$ and $\begin{pmatrix} 3 \\ -4 \\ -6 \\ 5 \end{pmatrix}$.

17. Let A and B be distinct points of $\mathcal{R}^n$. Prove that there is one and only one line containing both A and B (Theorem 1.25).

18. (a) Let $\mathcal{L}_1$ and $\mathcal{L}_2$ be two lines with the same direction vector D. Show that either $\mathcal{L}_1 = \mathcal{L}_2$, or $\mathcal{L}_1 \cap \mathcal{L}_2 = \varnothing$.
    (b) Let $\mathcal{L}_1$ and $\mathcal{L}_2$ be two lines with the same direction vector D. Show that if there is one point A on $\mathcal{L}_1$ and on $\mathcal{L}_2$, then $\mathcal{L}_1 = \mathcal{L}_2$.

19. (a) Show that every vector U in $\mathcal{R}^n$ is parallel to itself.
    (b) Show that the zero vector, O_n, is parallel to every n-vector.

20. Let $\mathcal{L}$ be a line in $\mathcal{R}^n$. Show that a vector U in $\mathcal{R}^n$ is parallel to $\mathcal{L}$ if and only if $U = O_n$ or U is a direction vector for $\mathcal{L}$.

21. Let $\mathcal{L}_1$ and $\mathcal{L}_2$ be two lines in $\mathcal{R}^n$. Show that
    (a) $\mathcal{L}_1$ and $\mathcal{L}_2$ are parallel if and only if some vector D is a direction vector for both lines.
    (b) $\mathcal{L}_1$ and $\mathcal{L}_2$ are parallel if and only if every direction vector for one line is a direction vector for the other.
    (c) if $\mathcal{L}_1$ and $\mathcal{L}_2$ are parallel, then either they are identical or they have no points in common.

22. (a) Let $\mathcal{L}$ be a line in $\mathcal{R}^n$ and A a point in $\mathcal{R}^n$. Show that there is one and only one line containing A which is parallel to $\mathcal{L}$. (Hint: What does it mean to say that $\mathcal{L}$ is a line in $\mathcal{R}^n$?)
    (b) Give a parametric description of the line $\mathcal{L}_1$ in $\mathcal{R}^n$ through the point A of $\mathcal{R}^n$ and parallel to the line

$$\mathcal{L}_2 = \{B + tD \mid t \in \mathcal{R}\},$$

when B and D are in $\mathcal{R}^n$, $D \neq O_n$.

ANSWERS

1. $\mathcal{L} = \left\{ \begin{pmatrix} 1 \\ 3 \end{pmatrix} + t \begin{pmatrix} 2 \\ 3 \end{pmatrix} \,\middle|\, t \in \mathcal{R} \right\}.$ 3. $\mathcal{L} = \left\{ \begin{pmatrix} 3 \\ -2 \end{pmatrix} + t \begin{pmatrix} 1 \\ -3 \end{pmatrix} \,\middle|\, t \in \mathcal{R} \right\}.$

5. $\mathcal{L} = \left\{ t \begin{pmatrix} 1 \\ -2 \end{pmatrix} \,\middle|\, t \in \mathcal{R} \right\}.$ 7. $\mathcal{L} = \left\{ \begin{pmatrix} 1 \\ 2 \\ 0 \\ 0 \end{pmatrix} + r \begin{pmatrix} 1 \\ -1 \\ 0 \\ 1 \end{pmatrix} \,\middle|\, r \in \mathcal{R} \right\}.$

9. $\mathcal{L} = \left\{ \begin{pmatrix} 2 \\ 0 \\ 0 \\ 1 \end{pmatrix} + s \begin{pmatrix} 1 \\ -1 \\ 0 \\ 0 \end{pmatrix} \,\middle|\, s \in \mathcal{R} \right\}.$ 11. $\mathcal{L}_1 \cap \mathcal{L}_2 = \varnothing.$

13. $\mathcal{L}_1 \cap \mathcal{L}_2 = \left\{ \begin{pmatrix} 7 \\ 0 \\ -2 \end{pmatrix} \right\}.$ 15. $\mathcal{L}_1 \cap \mathcal{L}_2 = \mathcal{L}_1 = \mathcal{L}_2.$

5. LINEAR COMBINATIONS; LINEAR DEPENDENCE

We now take up a generalization of the concept of parallelism, which is applicable to sets of more than two vectors. This concept, the concept of linear dependence, is most easily described in terms of what is called a linear combination of vectors.

DEFINITION I.28 (Linear Combinations of Two or More Vectors.)
Let $X_1, X_2, X_3, \ldots, X_k$, with $k > 1$, be vectors in $\mathcal{R}^n$. Let X be any vector of $\mathcal{R}^n$.

   i. X is a *linear combination of X_1 and X_2* if and only if
      $X = r_1 X_1 + r_2 X_2$ for some real numbers r_1 and r_2.
  ii. X is a *linear combination of X_1, X_2 and X_3* if and only if
      $X = r_1 X_1 + r_2 X_2 + r_3 X_3$ for some real numbers r_1, r_2, and r_3.
 iii. X is a *linear combination of $X_1, X_2, \ldots, X_k$* if and only if
      $X = r_1 X_1 + r_2 X_2 + \cdots + r_k X_k$ for some real numbers $r_1, r_2, \ldots, r_k$.

If we set $k = 1$ in part (iii) of the above definition, we have the statement

> X is a linear combination of X_1 if and only if $X = r_1 X_1$ for some real number r_1,

which is just another way of saying that X is a multiple of X_1. Thus, we see that the statement

$$X \text{ is a linear combination of } X_1, X_2, \ldots, X_k$$

is a generalization of the statement

$$X \text{ is a multiple of } X_1.$$

Example 13. **(a)** The vector $\begin{pmatrix} 1 \\ 3 \end{pmatrix}$ is a linear combination of the vectors $\begin{pmatrix} 1 \\ 0 \end{pmatrix}$ and $\begin{pmatrix} 0 \\ 1 \end{pmatrix}$, since $\begin{pmatrix} 1 \\ 3 \end{pmatrix} = 1\begin{pmatrix} 1 \\ 0 \end{pmatrix} + 3\begin{pmatrix} 0 \\ 1 \end{pmatrix}$.

(b) On the other hand, $\begin{pmatrix} 1 \\ 3 \end{pmatrix}$ is not a linear combination of $\begin{pmatrix} 2 \\ 0 \end{pmatrix}$ and $\begin{pmatrix} 1 \\ 0 \end{pmatrix}$, since every linear combination of these latter two vectors is of the form $r\begin{pmatrix} 2 \\ 0 \end{pmatrix} + s\begin{pmatrix} 1 \\ 0 \end{pmatrix} = \begin{pmatrix} 2r + s \\ 0 \end{pmatrix}$ and has second coordinate 0.

(c) $\begin{pmatrix} 1 \\ 2 \\ 1 \end{pmatrix}$ is a linear combination of $\begin{pmatrix} 1 \\ 2 \\ 0 \end{pmatrix}$, $\begin{pmatrix} 0 \\ 0 \\ 2 \end{pmatrix}$, and $\begin{pmatrix} 1 \\ 2 \\ 2 \end{pmatrix}$. Indeed, we have $\begin{pmatrix} 1 \\ 2 \\ 1 \end{pmatrix} = 1\begin{pmatrix} 1 \\ 2 \\ 0 \end{pmatrix} + \frac{1}{2}\begin{pmatrix} 0 \\ 0 \\ 2 \end{pmatrix} + 0\begin{pmatrix} 1 \\ 2 \\ 2 \end{pmatrix}$,

as well as

$$\begin{pmatrix} 1 \\ 2 \\ 1 \end{pmatrix} = \frac{1}{2}\begin{pmatrix} 1 \\ 2 \\ 0 \end{pmatrix} + 0\begin{pmatrix} 0 \\ 0 \\ 2 \end{pmatrix} + \frac{1}{2}\begin{pmatrix} 1 \\ 2 \\ 2 \end{pmatrix}.$$

This example shows that there may be more than one choice of scalars used to express a given vector as a linear combination of other given vectors. ∎

DEFINITION 1.29 Let S be a set of two or more vectors in $\mathfrak{R}^n$.
i. (Linear Dependence of a Set.) The set S is said to be a *linearly dependent set* if and only if some vector in S is expressible as a multiple of another vector in S or as a linear combination of other vectors in S.
ii. (Linear Independence of a Set.) The set S is said to be a *linearly independent set* if and only if S is not a linearly dependent set; that is, if and only if none of the vectors in S is a multiple of another vector in S or is a linear combination of other vectors in S.

Example 14. **(a)** The set $\left\{ \begin{pmatrix} 0 \\ 1 \end{pmatrix}, \begin{pmatrix} 1 \\ 0 \end{pmatrix} \right\}$ is a linearly independent set, since neither vector is a multiple of the other.

(b) $\left\{ \begin{pmatrix} 1 \\ -1 \\ 2 \end{pmatrix}, \begin{pmatrix} -2 \\ 2 \\ -4 \end{pmatrix} \right\}$ is a linearly dependent set, since the second vector is the (-2) multiple of the first vector.

(c) $\left\{ \begin{pmatrix} 1 \\ 2 \end{pmatrix}, \begin{pmatrix} 1 \\ 0 \end{pmatrix}, \begin{pmatrix} 0 \\ 1 \end{pmatrix} \right\}$ is a linearly dependent set, since $\begin{pmatrix} 1 \\ 2 \end{pmatrix}$ is a linear combination of $\begin{pmatrix} 1 \\ 0 \end{pmatrix}$ and $\begin{pmatrix} 0 \\ 1 \end{pmatrix}$:

$$\begin{pmatrix} 1 \\ 2 \end{pmatrix} = 1 \begin{pmatrix} 1 \\ 0 \end{pmatrix} + 2 \begin{pmatrix} 0 \\ 1 \end{pmatrix}.$$

(d) $\left\{ \begin{pmatrix} 0 \\ 0 \\ 0 \end{pmatrix}, \begin{pmatrix} 1 \\ 1 \\ 0 \end{pmatrix}, \begin{pmatrix} 2 \\ -3 \\ 4 \end{pmatrix} \right\}$ is a linearly dependent set, since $\begin{pmatrix} 0 \\ 0 \\ 0 \end{pmatrix}$ is a multiple of each of the other vectors, as well as a linear combination of both:

$$\begin{pmatrix} 0 \\ 0 \\ 0 \end{pmatrix} = 0 \begin{pmatrix} 1 \\ 1 \\ 0 \end{pmatrix} + 0 \begin{pmatrix} 2 \\ -3 \\ 4 \end{pmatrix}. \quad \blacksquare$$

The reason the term "linear dependence" is used is seen in the application of the above definition to a doubleton set $\{U, V\}$. This set is said to be linearly dependent if and only if one vector is a multiple of the other. This situation occurs if and only if the "directed line segments" representing U and V are collinear, or, equivalently, if and only if U and V are parallel. Figure 1.9 illustrates the various possible cases for a doubleton set.

There is still another frequently cited reason for use of the term "linear dependence." The operations of vector addition and multiplication of a vector by a scalar are frequently referred to as *linear operations*, since they are used in the definition of a line. Since the preceding definition describes a relation which involves only these linear operations, the term "linear dependence" suggests itself.

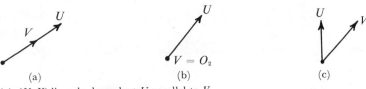

(a)   (b)   (c)

(a) $\{U, V\}$ linearly dependent U parallel to V
(b) $\{U, V\}$ linearly dependent U parallel to V
(c) $\{U, V\}$ linearly independent U not parallel to V

Figure I.9 Equivalence of linear dependence and parallelism on doubleton sets.

It is well to note that Figure 1.9b also depicts a more general situation, namely, any set of two or more n-vectors containing the zero vector O_n is a linearly dependent set. This is true since O_n is a multiple, the zero multiple, of each of the other vectors in the set.

We shall find the language of linear dependence to be quite useful in many discussions throughout this book. We shall also find it convenient to have tests for the linear dependence of a set other than that contained in the definition itself. The next theorem contains such a test.

THEOREM I.30 Let $S = \{X_1, X_2, \ldots, X_m\}$, $m \geq 2$, be a subset of $\mathcal{R}^n$. Then

i. S is a linearly dependent set if and only if there are scalars (real numbers) $r_1, r_2, \ldots, r_m$, at least one of which is not zero, such that

$$r_1 X_1 + r_2 X_2 + \cdots + r_m X_m = O_n.$$

ii. S is a linearly independent set if and only if the relation

$$r_1 X_1 + r_2 X_2 + \cdots + r_m X_m = O_n$$

holds only when each scalar $r_1, r_2, \ldots, r_m$ is zero.

Proof. We first show that statement (ii) is an immediate consequence of statement (i). Assuming that (i) can be proved, we apply Definition 1.29 (ii) to conclude that a set S is linearly independent if and only if it is *not true* that "there are scalars $r_1, r_2, \ldots, r_m$, at least one of which is not zero, such that

$$r_1 X_1 + r_2 X_2 + \cdots + r_m X_m = O_n."$$

The statement in quotation marks is false (and S is linearly independent) if and only if every choice of $r_1, r_2, \ldots, r_m$, other than the choice $r_1 = r_2 = \cdots = r_m = 0$, gives

$$r_1 X_1 + r_2 X_2 + \cdots + r_m X_m \neq O_n.$$

Statement (ii) of the theorem is a simple restatement of this result.

To complete the proof of the theorem, we must establish statement (i). We do so only for the case $m = 2$ and call for the proof when $m = 3$ in Exercise 5. The general case will be treated in Chapter 14.

Proof of i for $m = 2$. When $m = 2$, $S = \{X_1, X_2\}$. Consider the implication chain

$$\{X_1, X_2\} \text{ is linearly dependent} \Longleftrightarrow^1 \begin{cases} \text{(a) } X_1 = kX_2 \text{ for some } k, \\ \text{or} \\ \text{(b) } X_2 = lX_1 \text{ for some } l, \end{cases}$$

$$\Rightarrow \begin{cases} \text{(a) } X_1 - kX_2 = O_n \\ \text{or} \\ \text{(b) } X_2 - lX_1 = O_n \end{cases}$$

$$\Rightarrow^2 \begin{cases} r_1 X_1 + r_2 X_2 = O_n \\ \text{for some } r_1 \text{ and } r_2, \text{ at least one} \\ \text{of which is not zero.} \end{cases}$$

(1) One of the vectors is a multiple of the other, Definition 1.29i.

(2) $\begin{cases} \text{(a) Setting } r_1 = 1 \neq 0 \text{ and } r_2 = -k, \\ \text{or} \\ \text{(b) Setting } r_1 = -l \text{ and } r_2 = 1 \neq 0. \end{cases}$

This establishes the "only if" part of (i).

To establish the "if" part we consider the following implication chain:

$$\left. \begin{array}{l} r_1 X_1 + r_2 X_2 = O_n \\ \text{for some } r_1, r_2 \text{ not both zero} \end{array} \right\} \Rightarrow^3 \begin{cases} \text{(a) } r_1 \neq 0, X_1 + \dfrac{r_2}{r_1} X_2 = O_n \\ \text{or} \\ \text{(b) } r_2 \neq 0, \dfrac{r_1}{r_2} X_1 + X_2 = O_n \end{cases}$$

$$\Rightarrow \begin{cases} \text{(a) } X_1 = \dfrac{-r_2}{r_1} X_2, \\ \text{or} \\ \text{(b) } X_2 = \dfrac{-r_1}{r_2} X_1. \end{cases}$$

$$\Rightarrow^4 \{X_1, X_2\} \text{ is linearly dependent.}$$

(3) $\begin{cases} \text{(a) Multiplying by } \dfrac{1}{r_1} \text{ if } r_1 \neq 0. \\ \text{(b) Multiplying by } \dfrac{1}{r_2} \text{ if } r_2 \neq 0. \end{cases}$

(4) Since, in either case, one vector is a multiple of the other. Definition 1.29i.

Thus statement (i) is valid for $m = 2$, and, in view of previous remarks, the entire theorem holds for $m = 2$. ∎

EXERCISES

1. Show that the set $\left\{\begin{pmatrix} 1 \\ 0 \end{pmatrix}, \begin{pmatrix} 0 \\ 1 \end{pmatrix}, \begin{pmatrix} 4 \\ 3 \end{pmatrix}\right\}$ is a linearly dependent set.

2. Is the set $\left\{\begin{pmatrix} 1 \\ 0 \\ 0 \end{pmatrix}, \begin{pmatrix} 0 \\ 1 \\ 0 \end{pmatrix}, \begin{pmatrix} 0 \\ 0 \\ 1 \end{pmatrix}, \begin{pmatrix} 4 \\ -1 \\ 3 \end{pmatrix}\right\}$ linearly independent or linearly dependent? Why?

3. Show that the set $\left\{\begin{pmatrix} 2 \\ 0 \\ 0 \end{pmatrix}, \begin{pmatrix} 3 \\ 1 \\ 0 \end{pmatrix}, \begin{pmatrix} -4 \\ 21 \\ 2 \end{pmatrix}\right\}$ is a linearly independent set by using Theorem 1.30.

4. Show that the set $\left\{\begin{pmatrix} 1 \\ 0 \\ 3 \end{pmatrix}, \begin{pmatrix} 1 \\ 0 \\ 0 \end{pmatrix}, \begin{pmatrix} 1 \\ 0 \\ 2 \end{pmatrix}\right\}$ is linearly dependent by using Theorem 1.30.

5. Establish Theorem 1.30i for $m = 3$.

ANSWERS

1. $\begin{pmatrix} 4 \\ 3 \end{pmatrix} = 4\begin{pmatrix} 1 \\ 0 \end{pmatrix} + 3\begin{pmatrix} 0 \\ 1 \end{pmatrix}$; therefore, the set is linearly dependent by Definition 1.29.

2. Linearly dependent.

3. $r_1\begin{pmatrix} 2 \\ 0 \\ 0 \end{pmatrix} + r_2\begin{pmatrix} 3 \\ 1 \\ 0 \end{pmatrix} + r_3\begin{pmatrix} -4 \\ 21 \\ 2 \end{pmatrix} = \begin{pmatrix} 0 \\ 0 \\ 0 \end{pmatrix} \Leftrightarrow \begin{cases} 2r_1 + 3r_2 - 4r_3 = 0 \\ r_2 + 21r_3 = 0. \\ 2r_3 = 0 \end{cases}$ The last

equation assures us that $r_3 = 0$. Substitution into the second equation then shows that $r_2 = 0$. Finally, substituting $r_2 = 0$ and $r_3 = 0$ into the first equation shows that $r_1 = 0$. Therefore, the set must be linearly independent by Theorem 1.30.

5. Hint: $\{X_1, X_2, X_3\}$ is linearly dependent $\Leftrightarrow$ $\begin{cases} (1)\ X_1 = kX_2, \\ \text{or} \\ (2)\ X_1 = kX_2 + lX_3 \end{cases}$

for an appropriate labeling of the three vectors.

6. LINEAR DEPENDENCE IN $\mathcal{R}^2$ AND $\mathcal{R}^3$; DETERMINANTS OF ORDER TWO

Our immediate concern in this chapter is a study of the geometry of $\mathcal{R}^2$ and $\mathcal{R}^3$. We now undertake a more protracted study of linear dependence in these two spaces.

For a doubleton set $\{U, V\}$ the determination of the linear dependence or independence of the set is quite simple if the coordinates of U and V are specified. Indeed, we can determine by inspection whether one vector is a multiple of the other or not, as in Examples 14a and 14b. However, if we are not given the specific coordinates of U and V (as is the case when proving theorems involving a general U and V), other tests such as that given in Theorem 1.30 are quite useful.

We now consider a test for linear dependence of doubleton sets in $\mathcal{R}^2$. In this test the concept of a determinant is used. We first introduce the concept, then an example, and then the test itself.

DEFINITION 1.31 (Determinants of Order Two.) Let $U = \begin{pmatrix} a \\ b \end{pmatrix}$ and $V = \begin{pmatrix} c \\ d \end{pmatrix}$ be two vectors in $\mathcal{R}^2$. The determinant of order 2, denoted by $D(U, V)$ or by $\begin{vmatrix} a & c \\ b & d \end{vmatrix}$, is the real number given by

$$D(U, V) \equiv \begin{vmatrix} a & c \\ b & d \end{vmatrix} = ad - bc.$$

The notation $\begin{vmatrix} a & c \\ b & d \end{vmatrix}$ for $D(U, V)$ enables us to obtain the real number which it represents by means of the "crisscross" technique indicated below.

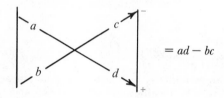

$$= ad - bc$$

The expression on the right, called the *expansion of the determinant,* is a sum of two products. The factors of the first product lie on the diagonal from top left to bottom right. The factors of the second product lie on the diagonal from bottom left to top right. Each product also carries a plus $(+)$ or a minus $(-)$ sign as indicated. The process of computing the value of $ad - bc$ for a specific choice of a, b, c, and d is called *evaluation of the determinant* $\begin{vmatrix} a & c \\ b & d \end{vmatrix}$.

Example 15. Let $U = \begin{pmatrix} 2 \\ 1 \end{pmatrix}$ and $V = \begin{pmatrix} 3 \\ -1 \end{pmatrix}$.

$$D(U, V) = \begin{vmatrix} 2 & 3 \\ 1 & -1 \end{vmatrix} = (2)(-1) - (1)(3) = -2 - 3 = -5$$

$$D(V, U) = \begin{vmatrix} 3 & 2 \\ -1 & 1 \end{vmatrix} = (3)(1) - (-1)(2) = 3 - (-2) = 5. \quad \blacksquare$$

This example illustrates the fact that $D(U, V) = -D(V, U)$ (see Exercise 15).

THEOREM 1.32 (Test for Linear Dependence of Doubleton Sets in $\mathcal{R}^2$.) Let $\{U, V\}$ be a subset of $\mathcal{R}^2$.

i. $\{U, V\}$ is a linearly dependent set if and only if

$$D(U, V) = 0.$$

ii. $\{U, V\}$ is a linearly independent set if and only if $D(U, V) \neq 0$.

Proof. Let $U = \begin{pmatrix} a \\ b \end{pmatrix}$ and $V = \begin{pmatrix} c \\ d \end{pmatrix}$.

We first prove the "only if" part of statement (i), that is, we shall assume that $\{U, V\}$ is a linearly dependent set and shall show that $D(U, V) = 0$. Since $\{U, V\}$ is a linearly dependent set, one of the vectors is a multiple of the other. We assume that $U = kV$. Thus,

$$\{U, V\} \text{ is linearly dependent} \Rightarrow U = kV, \text{ for some } k,$$

$$\Rightarrow^1 \begin{pmatrix} a \\ b \end{pmatrix} = k \begin{pmatrix} c \\ d \end{pmatrix}$$

$$\Rightarrow \begin{cases} a = kc \\ b = kd \end{cases}$$

$$\Rightarrow^2 ad - bc = kcd - kdc = 0$$

$$\Rightarrow^3 D(U, V) = 0$$

(1) Changing to coordinate notation.
(2) Substituting for a and b in $ad - bc$.

(3) Using definition of $D(U, V) = \begin{vmatrix} a & c \\ b & d \end{vmatrix} = ad - bc$.

Next we prove the "if" part of statement (i): $\{U, V\}$ is a linearly dependent set if $D(U, V) = 0$. Consequently, in this part of the proof we assume that

(1.6) $$D(U, V) = \begin{vmatrix} a & c \\ b & d \end{vmatrix} = ad - bc = 0.$$

In view of Theorem 1.30(i), $\{U, V\}$ is a linearly dependent set if we can show that there are scalars r_1 and r_2, at least one of which is not zero, such that

$$r_1 U + r_2 V = O_2$$

or in coordinate notation

(1.7) $$r_1 \begin{pmatrix} a \\ b \end{pmatrix} + r_2 \begin{pmatrix} c \\ d \end{pmatrix} = \begin{pmatrix} 0 \\ 0 \end{pmatrix}.$$

Comparison of the first coordinate $r_1 a + r_2 c = 0$ of (1.7) with (1.6) suggests that we choose $r_1 = d$ and $r_2 = -b$. Thus,

$$r_1 U + r_2 V = dU + (-b)V$$

$$= d \begin{pmatrix} a \\ b \end{pmatrix} + (-b) \begin{pmatrix} c \\ d \end{pmatrix}$$

$$= \begin{pmatrix} ad \\ bd \end{pmatrix} + \begin{pmatrix} -bc \\ -bd \end{pmatrix}$$

$$= \begin{pmatrix} ad - bc \\ bd - bd \end{pmatrix}$$

$$= \begin{pmatrix} 0 \\ 0 \end{pmatrix}, \text{ since } ad - bc = D(U, V) = 0,$$

$$= O_2.$$

Consequently, if at least one of the numbers b or d is not zero, then at least one of the numbers $r_1 = d$ and $r_2 = -b$ is not zero, and $\{U, V\}$ is a linearly dependent set by virtue of Theorem 1.30. On the other hand, if $b = d = 0$, then

$$U = \begin{pmatrix} a \\ b \end{pmatrix} = \begin{pmatrix} a \\ 0 \end{pmatrix} \quad \text{and} \quad V = \begin{pmatrix} c \\ d \end{pmatrix} = \begin{pmatrix} c \\ 0 \end{pmatrix}.$$

In this case one vector is clearly a multiple of the other. Thus, $\{U, V\}$ is a linearly dependent set in any case.

Statement (ii) is an immediate consequence of (i) and Definition 1.29ii (Linear Independence of a Set). ■

The next result is a simple but extremely useful application of determinants.

THEOREM 1.33   (Cramer's Rule in $\mathfrak{R}^2$.) Let $U = \begin{pmatrix} a \\ b \end{pmatrix}$ and $V = \begin{pmatrix} c \\ d \end{pmatrix}$ be vectors in $\mathfrak{R}^2$ such that $\{U, V\}$ is a linearly independent set. Let $K = \begin{pmatrix} k \\ l \end{pmatrix}$ be any vector of $\mathfrak{R}^2$. The system of equations in r and s

$$\begin{cases} k = ar + cs \\ l = br + ds \end{cases}$$

has a unique solution for r and s. In fact,

$$r = \frac{D(K, V)}{D(U, V)} \quad \text{and} \quad s = \frac{D(U, K)}{D(U, V)}.$$

Proof. The method of proof is straightforward. We begin with the system of equations and solve for r by eliminating s.

$$\begin{aligned} k = ar + cs \\ l = br + ds \end{aligned} \Big\} \Rightarrow^1 \begin{cases} kd = adr + cds \\ lc = bcr + dcs \end{cases}$$

$$\Rightarrow^2 \; kd - lc = (ad - bc)r$$

$$\Rightarrow^3 \; r = \frac{kd - lc}{ad - bc} = \frac{D(K, V)}{D(U, V)}.$$

(1) Multiplying both members of the first (second) equation by d (by c).
(2) Subtracting the members of the second equation from those of the first.
(3) Dividing by $ad - bc$ (note that this divisor is $D(U, V)$ and must be nonzero, since $\{U, V\}$ is a linearly independent set by Theorem 1.32ii) and using Definition 1.31 (Determinant of Order Two).

In a similar manner we solve for s and find

$$s = \frac{al - bk}{ac - bd} = \frac{D(U, K)}{D(U, V)}.$$

All that the above procedure shows is this: *If* there is a solution to the given system of equations, then r and s are uniquely determined by the expressions given in the theorem. We must still show that these values for r and s are in fact a solution. That is, we must substitute the values which we obtained above into the original system of equations and show that they are satisfied. This task is simple and is left as Exercise 16. ∎

The system of equations in the preceding theorem is equivalent to the vector equation

$$\begin{pmatrix} k \\ l \end{pmatrix} = r \begin{pmatrix} a \\ b \end{pmatrix} + s \begin{pmatrix} c \\ d \end{pmatrix}$$

or, in coordinate free notation,

$$K = rU + sV.$$

Viewed in this manner we may refer to U as the vector coefficient of r and to V as the vector coefficient of s. This terminology will assist us in remembering the solutions for r and s. You will note that the denominator for both r and s is the determinant of the coefficient vectors $D(U, V)$. Whereas the numerator for r is the determinant obtained from $D(U, V)$ by replacing the coefficient of r by K. Similarly the numerator for s is the determinant obtained from $D(U, V)$ by replacing the coefficient for s by K.

Using the above coordinate free equivalent of the system of equations of Theorem 1.33, we can reformulate that theorem in a coordinate-free fashion and also obtain an extremely important additional result.

THEOREM 1.34 (Basic Properties of Linearly Independent Sets in $\mathcal{R}^2$.) Let $\{U, V\}$ be a linearly independent set in $\mathcal{R}^2$.

i. Then any vector X in R^2 is a linear combination

$$X = rU + sV$$

of the vectors U and V. Moreover, for each vector X there is one and only one choice for the scalars r and s.

ii. Any set of three or more vectors in $\mathcal{R}^2$ is a linearly dependent set.

Proof. Statement (i) is just a restatement of Theorem 1.33 with X replacing K. We will prove statement (ii). Let S be any set of three or more vectors in $\mathcal{R}^2$. Let U and V be two vectors of S. If one of these vectors is a multiple of the other, S is linearly dependent by immediate application of Definition 1.29i. If neither of the vectors is a multiple of the other, the application of Definition 1.29ii shows that $\{U, V\}$ is a linearly independent set. Let X be a third vector in S. From statement (i) of the present theorem, X is a linear combination of U and V. Since X, U, and V are all vectors in S, a second application of Definition 1.29i again discloses S to be a linearly dependent set. ∎

Example 16. Show that $\begin{pmatrix} 7 \\ -18 \end{pmatrix}$ is a linear combination of $\begin{pmatrix} 2 \\ 6 \end{pmatrix}$ and $\begin{pmatrix} 14 \\ -49 \end{pmatrix}$. We seek scalars r and s such that

$$\begin{pmatrix} 7 \\ -18 \end{pmatrix} = r\begin{pmatrix} 2 \\ 6 \end{pmatrix} + s\begin{pmatrix} 14 \\ -49 \end{pmatrix}.$$

According to Theorems 1.33 and 1.34i, there is a unique solution for r and s since $\left\{ \begin{pmatrix} 2 \\ 6 \end{pmatrix}, \begin{pmatrix} 14 \\ -49 \end{pmatrix} \right\}$ is a linearly independent set. Morever, according to Theorem 1.33, this solution is given by

$$r = \frac{\begin{vmatrix} 7 & 14 \\ -18 & -49 \end{vmatrix}}{\begin{vmatrix} 2 & 14 \\ 6 & -49 \end{vmatrix}} = \frac{-343 + 252}{-98 - 84} = \frac{1}{2}$$

$$s = \frac{\begin{vmatrix} 2 & 7 \\ 6 & -18 \end{vmatrix}}{\begin{vmatrix} 2 & 14 \\ 6 & -49 \end{vmatrix}} = \frac{-36 - 42}{-98 - 84} = \frac{3}{7}$$

Thus,

$$\begin{pmatrix} 7 \\ -18 \end{pmatrix} = \frac{1}{2}\begin{pmatrix} 2 \\ 6 \end{pmatrix} + \frac{3}{7}\begin{pmatrix} 14 \\ -49 \end{pmatrix}. \quad ∎$$

The content of Theorem 1.34, which is concerned solely with $\mathcal{R}^2$, can be extended to a general result for $\mathcal{R}^n$. Such an extension will be covered in all its generality in Chapter 14. However, we shall presently need to use the result in $\mathcal{R}^3$ and shall state it without proof.

THEOREM 1.35

i. Let $\{U, V, W\}$ be a linearly independent set in $\mathcal{R}^3$. Then any vector X in $\mathcal{R}^3$ is a linear combination

$$X = rU + sV + tW$$

of the vectors U, V, and W. Moreover, there is one and only one choice for the scalars r, s, and t.

ii. Any set of four or more vectors in $\mathcal{R}^3$ is a linearly dependent set.

EXERCISES

Evaluate the determinants in Exercises 1 to 6:

1. $\begin{vmatrix} 1 & 3 \\ 2 & 4 \end{vmatrix}$.

2. $\begin{vmatrix} 2 & 4 \\ -1 & 6 \end{vmatrix}$.

3. $\begin{vmatrix} 0 & 1 \\ 0 & 3 \end{vmatrix}$.

4. $\begin{vmatrix} 5 & -35 \\ 3 & -21 \end{vmatrix}$.

5. $\begin{vmatrix} a & 2a \\ b & b \end{vmatrix}$.

6. $\begin{vmatrix} b & a \\ a & b \end{vmatrix}$.

In Exercises 7 to 10 determine whether the given sets are linearly independent or linearly dependent by using the results of Exercises 1 to 4:

7. $\left\{ \begin{pmatrix} 1 \\ 2 \end{pmatrix}, \begin{pmatrix} 3 \\ 4 \end{pmatrix} \right\}$.

8. $\left\{ \begin{pmatrix} 2 \\ -1 \end{pmatrix}, \begin{pmatrix} 4 \\ 6 \end{pmatrix} \right\}$.

9. $\left\{ \begin{pmatrix} 0 \\ 0 \end{pmatrix}, \begin{pmatrix} 1 \\ 3 \end{pmatrix} \right\}$.

10. $\left\{ \begin{pmatrix} 5 \\ 3 \end{pmatrix}, \begin{pmatrix} -35 \\ -21 \end{pmatrix} \right\}$.

In Exercises 11 to 14 find values of r and s such that $X = rU + sV$:

11. $X = \begin{pmatrix} 7 \\ -7 \end{pmatrix}$, $U = \begin{pmatrix} 2 \\ 1 \end{pmatrix}$, $V = \begin{pmatrix} -1 \\ 3 \end{pmatrix}$.

12. $X = \begin{pmatrix} -11 \\ 16 \end{pmatrix}$, $U = \begin{pmatrix} 3 \\ 2 \end{pmatrix}$, $V = \begin{pmatrix} -1 \\ 4 \end{pmatrix}$.

13. $X = \begin{pmatrix} -1 \\ -19 \end{pmatrix}$, $U = \begin{pmatrix} 3 \\ 2 \end{pmatrix}$, $V = \begin{pmatrix} 2 \\ 3 \end{pmatrix}$.

14. $X = \begin{pmatrix} 0 \\ -1 \end{pmatrix}$, $U = \begin{pmatrix} 5 \\ -3 \end{pmatrix}$, $V = \begin{pmatrix} -2 \\ 1 \end{pmatrix}$.

15. Let $U = \begin{pmatrix} a \\ b \end{pmatrix}$, $V = \begin{pmatrix} c \\ d \end{pmatrix}$, $W = \begin{pmatrix} e \\ f \end{pmatrix}$, and $O_2 = \begin{pmatrix} 0 \\ 0 \end{pmatrix}$ be vectors in $\mathcal{R}^2$ and k be a real number. Show that (a) $D(U, V) = -D(V, U)$; (b) $D(O_2, U) = 0$; (c) $D(kU, V) = D(U, kV) = kD(U, V)$; (d) $D(U + V, W) = D(U, W) + D(V, W)$.

16. Complete the proof of Theorem 1.33 by showing that the values obtained for r and s satisfy the given system of equations.

ANSWERS

1. -2. **3.** 0. **5.** $-ab$. **7.** Linearly independent, since $D(U, V) \neq 0$. **9.** Linearly dependent since $D(U, V) = 0$. **11.** $(r, s) = (2, -3)$ **13.** $(r, s) = (7, -11)$ **15.** (c) $D(kU, V) = D\left(k \begin{pmatrix} a \\ b \end{pmatrix}, \begin{pmatrix} c \\ d \end{pmatrix} \right) = D\left(\begin{pmatrix} ka \\ kb \end{pmatrix}, \begin{pmatrix} c \\ d \end{pmatrix} \right)$

$$= \begin{vmatrix} ka & c \\ kb & d \end{vmatrix} = kad - kbc = k(ad - bc) = kD(U, V).$$

7.   LINES IN $\mathcal{R}^2$

Theorem 1.26 (Intersection of Two Lines in $\mathcal{R}^n$) classifies the nature of the various possible intersections of two lines in $\mathcal{R}^n$. Figure 1.10 suggests that two lines in $\mathcal{R}^3$ which do not intersect are not necessarily parallel. In particular, consider the two lines

$$\mathcal{L}_1 = \left\{ t \begin{pmatrix} 1 \\ 0 \\ 0 \end{pmatrix} \middle| t \in \mathcal{R} \right\} \quad \text{and} \quad \mathcal{L}_2 = \left\{ \begin{pmatrix} 0 \\ 0 \\ 1 \end{pmatrix} + s \begin{pmatrix} 0 \\ 1 \\ 0 \end{pmatrix} \middle| s \in \mathcal{R} \right\}$$

sketched in Figure 1.10. $\mathcal{L}_1$ and $\mathcal{L}_2$ are not parallel, since the direction vector $D_1 = \begin{pmatrix} 1 \\ 0 \\ 0 \end{pmatrix}$ for $\mathcal{L}_1$ is not a multiple of the direction vector $D_2 = \begin{pmatrix} 0 \\ 1 \\ 0 \end{pmatrix}$ for $\mathcal{L}_2$. Moreover, $\mathcal{L}_1$ and $\mathcal{L}_2$ do not intersect. (The reader should verify this.) This situation is described by saying that $\mathcal{L}_1$ and $\mathcal{L}_2$ are *skew* lines.

The corresponding situation in $\mathcal{R}^2$ is different from the general case. Every pair of lines in $\mathcal{R}^2$ which do not intersect are necessarily parallel. Indeed, the following specialization of Theorem 1.26 is valid.

THEOREM 1.36   (Intersection of Two Lines in $\mathcal{R}^2$.) Let $\mathcal{L}_1$ and $\mathcal{L}_2$ denote two (not necessarily distinct) lines in $\mathcal{R}^2$. Exactly one of the following is true:

i. $\mathcal{L}_1$ is parallel to $\mathcal{L}_2$, and $\mathcal{L}_1 \cap \mathcal{L}_2 = \varnothing$.

ii. $\mathcal{L}_1$ is not parallel to $\mathcal{L}_2$, and $\mathcal{L}_1 \cap \mathcal{L}_2 = $ a singleton set.

iii. $\mathcal{L}_1$ is parallel to $\mathcal{L}_2$, and $\mathcal{L}_1$ coincides with $\mathcal{L}_2$; $\mathcal{L}_1 \cap \mathcal{L}_2 = \mathcal{L}_1 = \mathcal{L}_2$.

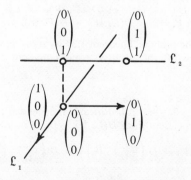

Figure 1.10   Two skew lines in $\mathcal{R}^3$.

Proof. From Theorem 1.26 we know that given two lines $\mathcal{L}_1$ and $\mathcal{L}_2$ in $\mathcal{R}^2$ exactly one of the following holds:

    i. $\mathcal{L}_1 \cap \mathcal{L}_2 = \varnothing$,

    ii. $\mathcal{L}_1 \cap \mathcal{L}_2 = $ a singleton set,

    iii. $\mathcal{L}_1 \cap \mathcal{L}_2 = \mathcal{L}_1 = \mathcal{L}_2$.

All that we need to show in addition is that whenever $\mathcal{L}_1$ and $\mathcal{L}_2$ are *not* parallel then case (ii) occurs. It follows that (i) and (iii) can occur only if $\mathcal{L}_1$ and $\mathcal{L}_2$ are parallel.

Therefore, let us assume that $\mathcal{L}_1$ and $\mathcal{L}_2$ are two nonparallel lines in $\mathcal{R}^2$. $\mathcal{L}_1$ and $\mathcal{L}_2$ will be given by

$$\mathcal{L}_1 = \{A + tU \mid t \in \mathcal{R}\} \quad \text{and} \quad \mathcal{L}_2 = \{B + sV \mid s \in \mathcal{R}\},$$

where A and B are two points in $\mathcal{R}^2$ and U and V are two nonparallel (direction) vectors in $\mathcal{R}^2$. The intersection $\mathcal{L}_1 \cap \mathcal{L}_2$ consists of all points C such that

$$C = A + tU \quad \text{and} \quad C = B + sV$$

for some s and t. Thus, there will be a point C in $\mathcal{L}_1 \cap \mathcal{L}_2$ for every choice of t and s such that

$$A + tU = B + sV,$$

or, equivalently, for every choice of t and s such that

$$A - B = (-t)U + sV.$$

According to Theorem 1.34 there is exactly one such choice of s and t, and hence exactly one point C in $\mathcal{L}_1 \cap \mathcal{L}_2$. The present theorem is thereby established. ∎

With this theorem we find that $\mathcal{R}^2$ has precisely those properties which the reader has seen ascribed to a plane in previous studies. Namely, any two distinct lines in $\mathcal{R}^2$ either intersect in a single point or are parallel and nonintersecting. We shall use this result and Theorem 1.34 to motivate the definition of a plane in $\mathcal{R}^n$. This must wait, however, until the next section.

Example 17. Find $\mathcal{L}_1 \cap \mathcal{L}_2$ if

$$\mathcal{L}_1 = \left\{ \begin{pmatrix} 1 \\ 1 \end{pmatrix} + t \begin{pmatrix} 3 \\ 4 \end{pmatrix} \,\middle|\, t \in \mathcal{R} \right\} \quad \text{and} \quad \mathcal{L}_2 = \left\{ \begin{pmatrix} 2 \\ 5 \end{pmatrix} + s \begin{pmatrix} 1 \\ 2 \end{pmatrix} \,\middle|\, s \in \mathcal{R} \right\}.$$

We first note that the lines are not parallel, since the respective direction vectors $\begin{pmatrix} 3 \\ 4 \end{pmatrix}$ and $\begin{pmatrix} 1 \\ 2 \end{pmatrix}$ are not parallel. There is then a single point C in $\mathcal{L}_1 \cap \mathcal{L}_2$, and C is given by

(1.8) $$C = \begin{pmatrix} c_1 \\ c_2 \end{pmatrix} = \begin{pmatrix} 1 \\ 1 \end{pmatrix} + t \begin{pmatrix} 3 \\ 4 \end{pmatrix} = \begin{pmatrix} 2 \\ 5 \end{pmatrix} + s \begin{pmatrix} 1 \\ 2 \end{pmatrix}$$

for some choice of t and s. From the last two members of this chain of equalities we see that t and s must satisfy the equation

$$\begin{pmatrix} 1 \\ 1 \end{pmatrix} - \begin{pmatrix} 2 \\ 5 \end{pmatrix} = (-t) \begin{pmatrix} 3 \\ 4 \end{pmatrix} + s \begin{pmatrix} 1 \\ 2 \end{pmatrix}$$

or, equivalently,

$$\begin{pmatrix} -1 \\ -4 \end{pmatrix} = t \begin{pmatrix} -3 \\ -4 \end{pmatrix} + s \begin{pmatrix} 1 \\ 2 \end{pmatrix}.$$

Applying Theorem 1.32 we find that

$$t = \frac{\begin{vmatrix} -1 & 1 \\ -4 & 2 \end{vmatrix}}{\begin{vmatrix} -3 & 1 \\ -4 & 2 \end{vmatrix}} = \frac{-2 + 4}{-6 + 4} = \frac{2}{-2} = -1$$

and

$$s = \frac{\begin{vmatrix} -3 & -1 \\ -4 & -4 \end{vmatrix}}{\begin{vmatrix} -3 & 1 \\ -4 & 2 \end{vmatrix}} = \frac{12 - 4}{-2} = \frac{8}{-2} = -4.$$

Using the value for t just computed we determine C from Equation (1.8):

$$C = \begin{pmatrix} 1 \\ 1 \end{pmatrix} + (-1) \begin{pmatrix} 3 \\ 4 \end{pmatrix} = \begin{pmatrix} -2 \\ -3 \end{pmatrix}.$$

As a check on our computations it is wise to evaluate C by using the value of $s = -4$ in Equation 1.8 as well:

$$C = \begin{pmatrix} 2 \\ 5 \end{pmatrix} + (-4) \begin{pmatrix} 1 \\ 2 \end{pmatrix} = \begin{pmatrix} -2 \\ -3 \end{pmatrix}.$$

Thus, we find that $\mathfrak{L}_1 \cap \mathfrak{L}_2 = \left\{ \begin{pmatrix} -2 \\ -3 \end{pmatrix} \right\}$. ∎

There are many ways of describing a line other than by giving its parametric representation. We shall look at other methods for n-space in general in the next chapter. Now we shall consider an alternate mode of description which is applicable only in $\mathfrak{R}^2$. We begin with the following definition.

DEFINITION 1.37   (Slope and y-Intercept of a Line in $\mathfrak{R}^2$.)   A line $\mathfrak{L}$ in $\mathfrak{R}^2$ is said

   i. to have *slope m* if and only if $\begin{pmatrix} 1 \\ m \end{pmatrix}$ is a direction vector for $\mathfrak{L}$.

   ii. to have *y-intercept b* if and only if $\begin{pmatrix} 0 \\ b \end{pmatrix}$ is a point on $\mathfrak{L}$ and is the only point on $\mathfrak{L}$ with first coordinate zero.

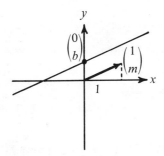

Figure 1.11 A line in $\mathfrak{R}^2$ with slope m and y-intercept b.

Figure 1.11 is a pictorial representation of the content of Definition 1.37. In this figure we have labeled the axes by the letters x and y instead of the Arabic numbers 1 and 2. The parametric equation of a line with slope m and y-intercept b is clearly

$$X(t) = \begin{pmatrix} x \\ y \end{pmatrix} = \begin{pmatrix} 0 \\ b \end{pmatrix} + t \begin{pmatrix} 1 \\ m \end{pmatrix}.$$

Thus,

$$\mathfrak{L} = \left\{ \begin{pmatrix} x \\ y \end{pmatrix} \middle| \begin{array}{l} x = t \quad \text{and} \\ y = b + tm \end{array} \right\}.$$

It is now a simple exercise (Exercise 13) to prove the following result.

THEOREM 1.38 (Explicit or Slope-Intercept Description of a Line in $\mathfrak{R}^2$.) A line in $\mathfrak{R}^2$ has slope m and y-intercept b if and only if it is described by

$$\mathfrak{L} = \left\{ \begin{pmatrix} x \\ y \end{pmatrix} \middle| y = mx + b \right\}.$$

Moreover, for each choice of m and b there is one and only one such line. Finally, a line $\mathfrak{L}$ in $\mathfrak{R}^2$ has a slope and a y-intercept if and only if $\mathfrak{L}$ is not parallel to $\begin{pmatrix} 0 \\ 1 \end{pmatrix}$.

Example 18. (a) Write a parametric description for the line $\mathfrak{L}$ given by $y = 3x + 4$. Using the preceding theorem, we note that $\mathfrak{L}$ has slope $m = 3$ and y-intercept $b = 4$. Therefore, $\begin{pmatrix} 1 \\ 3 \end{pmatrix}$ is a direction vector for $\mathfrak{L}$ and $\begin{pmatrix} 0 \\ 4 \end{pmatrix}$ is a point on $\mathfrak{L}$. Hence,

$$\mathfrak{L} = \left\{ \begin{pmatrix} 0 \\ 4 \end{pmatrix} + t \begin{pmatrix} 1 \\ 3 \end{pmatrix} \middle| t \in \mathfrak{R} \right\}$$

is a parametric representation.

(b) Write an *explicit* description for the line $\mathfrak{L} = \left\{ \begin{pmatrix} 2 \\ 0 \end{pmatrix} + t \begin{pmatrix} 2 \\ 4 \end{pmatrix} \,\middle|\, t \in \mathfrak{R} \right\}$.

We first note that $\begin{pmatrix} 2 \\ 4 \end{pmatrix}$ is a direction vector for $\mathfrak{L}$, and, according to Theorem 1.22, so is $\begin{pmatrix} 1 \\ 2 \end{pmatrix} = \dfrac{1}{2} \begin{pmatrix} 2 \\ 4 \end{pmatrix}$. Therefore, $\mathfrak{L}$ has slope 2. We now seek the y-intercept of $\mathfrak{L}$, that is, we wish that point $\begin{pmatrix} 0 \\ b \end{pmatrix}$ such that $\begin{pmatrix} 0 \\ b \end{pmatrix} \in \mathfrak{L}$. We proceed thus:

$$\begin{pmatrix} 0 \\ b \end{pmatrix} \in \mathfrak{L} \Leftrightarrow \begin{pmatrix} 0 \\ b \end{pmatrix} = \begin{pmatrix} 2 \\ 0 \end{pmatrix} + t \begin{pmatrix} 2 \\ 4 \end{pmatrix} \Leftrightarrow \begin{cases} 0 = 2 + 2t \\ b = 0 + 4t \end{cases}$$

$$\Leftrightarrow \begin{cases} t = -1 \\ b = 4t \end{cases}$$

$$\Leftrightarrow \begin{cases} t = -1 \\ b = 4(-1) = -4. \end{cases}$$

Thus, the y-intercept is $b = -4$. The explicit description of $\mathfrak{L}$ is therefore given by

$$\mathfrak{L} = \left\{ \begin{pmatrix} x \\ y \end{pmatrix} \,\middle|\, y = 2x + (-4) \right\}$$

$$= \left\{ \begin{pmatrix} x \\ y \end{pmatrix} \,\middle|\, y = 2x - 4 \right\}. \quad \blacksquare$$

EXERCISES

In Exercises 1 to 6, use the technique of Example 17 to find $\mathfrak{L}_1 \cap \mathfrak{L}_2$.

1. $\mathfrak{L}_1 = \left\{ \begin{pmatrix} 1 \\ 11 \end{pmatrix} + t \begin{pmatrix} 0 \\ 2 \end{pmatrix} \,\middle|\, t \in \mathfrak{R} \right\}$, $\mathfrak{L}_2 = \left\{ \begin{pmatrix} 4 \\ -9 \end{pmatrix} + s \begin{pmatrix} -1 \\ 4 \end{pmatrix} \,\middle|\, s \in \mathfrak{R} \right\}$.

2. $\mathfrak{L}_1 = \left\{ \begin{pmatrix} 5 \\ 33 \end{pmatrix} + t \begin{pmatrix} 1 \\ 4 \end{pmatrix} \,\middle|\, t \in \mathfrak{R} \right\}$, $\mathfrak{L}_2 = \left\{ \begin{pmatrix} 8 \\ -10 \end{pmatrix} + s \begin{pmatrix} -2 \\ 3 \end{pmatrix} \,\middle|\, s \in \mathfrak{R} \right\}$.

3. $\mathfrak{L}_1 = \left\{ \begin{pmatrix} 3 \\ 13 \end{pmatrix} + t \begin{pmatrix} 1 \\ 1 \end{pmatrix} \,\middle|\, t \in \mathfrak{R} \right\}$, $\mathfrak{L}_2 = \left\{ \begin{pmatrix} 1 \\ -5 \end{pmatrix} + s \begin{pmatrix} 3 \\ 7 \end{pmatrix} \,\middle|\, s \in \mathfrak{R} \right\}$.

4. $\mathfrak{L}_1 = \left\{ \begin{pmatrix} 9 \\ -3 \end{pmatrix} + t \begin{pmatrix} 2 \\ 7 \end{pmatrix} \,\middle|\, t \in \mathfrak{R} \right\}$, $\mathfrak{L}_2 = \left\{ \begin{pmatrix} 14 \\ 4 \end{pmatrix} + s \begin{pmatrix} 3 \\ 7 \end{pmatrix} \,\middle|\, s \in \mathfrak{R} \right\}$.

5. $\mathfrak{L}_1 = $ line through $\begin{pmatrix} 4 \\ 4 \end{pmatrix}$ and $\begin{pmatrix} -2 \\ 1 \end{pmatrix}$,

$\mathfrak{L}_2 = $ line through $\begin{pmatrix} 5 \\ 7 \end{pmatrix}$ and $\begin{pmatrix} 11 \\ 15 \end{pmatrix}$.

6. $\mathfrak{L}_1 = $ line with slope 5 and y-intercept -2,

$\mathfrak{L}_2 = $ line through $\begin{pmatrix} 3 \\ 8 \end{pmatrix}$ and $\begin{pmatrix} -5 \\ -12 \end{pmatrix}$.

In each of the Exercises 7 to 12 determine the slope and y-intercept of $\mathcal{L}$ and give the slope-intercept description of the line.

7. $\mathcal{L} = \left\{ \begin{pmatrix} 4 \\ 3 \end{pmatrix} + t \begin{pmatrix} 5 \\ 2 \end{pmatrix} \middle| t \in \mathcal{R} \right\}$.

8. $\mathcal{L} = \left\{ \begin{pmatrix} 5 \\ 7 \end{pmatrix} + t \begin{pmatrix} 3 \\ 4 \end{pmatrix} \middle| t \in \mathcal{R} \right\}$.

9. $\mathcal{L}$ = line through $\begin{pmatrix} 2 \\ 4 \end{pmatrix}$ and $\begin{pmatrix} -1 \\ 7 \end{pmatrix}$.

10. $\mathcal{L}$ = line through $\begin{pmatrix} 4 \\ 3 \end{pmatrix}$ with slope 2.

11. $\mathcal{L}$ = line through $\begin{pmatrix} 2 \\ -5 \end{pmatrix}$ parallel to $\begin{pmatrix} 7 \\ -6 \end{pmatrix}$.

12. $\mathcal{L}$ = line through $\begin{pmatrix} 5 \\ 12 \end{pmatrix}$ parallel to $\begin{pmatrix} 7 \\ 11 \end{pmatrix}$.

13. (a) Show that a line $\mathcal{L}$ in $\mathcal{R}^2$ has slope m and y-intercept b if and only if

$$\mathcal{L} = \left\{ \begin{pmatrix} x \\ y \end{pmatrix} \middle| y = mx + b \right\}.$$

    (b) Show that there is one and only one line $\mathcal{L}$ in $\mathcal{R}^2$ with a given slope m and y-intercept b.

    (c) Show that a line $\mathcal{L}$ in $\mathcal{R}^2$ has a slope and a y-intercept if and only if $\mathcal{L}$ is *not* parallel to $\begin{pmatrix} 0 \\ 1 \end{pmatrix}$.

ANSWERS

1. $\mathcal{L}_1 \cap \mathcal{L}_2 = \left\{ \begin{pmatrix} 1 \\ 3 \end{pmatrix} \right\}$ 3. $\mathcal{L}_1 \cap \mathcal{L}_2 = \left\{ \begin{pmatrix} 13 \\ 23 \end{pmatrix} \right\}$.   5. $\mathcal{L}_1 \cap \mathcal{L}_2 = \left\{ \begin{pmatrix} 2 \\ 3 \end{pmatrix} \right\}$.

7. $m = 2/5$, $b = 7/5$, $\mathcal{L} = \left\{ \begin{pmatrix} x \\ y \end{pmatrix} \middle| y = \frac{2}{5}x + \frac{7}{5}, x, y \in \mathcal{R} \right\}$.

9. $m = -1$, $b = 6$, $\mathcal{L} = \left\{ \begin{pmatrix} x \\ y \end{pmatrix} \middle| y = -1x + 6 \right\}$.

11. $m = -\frac{6}{7}$, $b = -\frac{23}{7}$, $\mathcal{L} = \left\{ \begin{pmatrix} x \\ y \end{pmatrix} \middle| y = -\frac{6}{7}x - \frac{23}{7} \right\}$

$= \left\{ \begin{pmatrix} x \\ y \end{pmatrix} \middle| 7y = -6x - 23 \right\}$.

8. PLANES IN $\mathcal{R}^n$

Theorem 1.36 (Intersection of Two Lines in $\mathcal{R}^2$) shows that the vector space $\mathcal{R}^2$ has the characteristic property that distinguishes a "plane" from "space," namely, two distinct lines either are parallel and nonintersecting or are nonparallel and intersect in exactly one point. The proof of this theorem depends solely upon the general Theorem 1.26 (Intersection of Lines in $\mathcal{R}^n$) and Theorem 1.34 (Basic Properties of Linearly Independent Sets in $\mathcal{R}^2$). This latter theorem states that any vector X in $\mathcal{R}^2$ is a unique linear combination of any two vectors U and V which form a linearly independent set $\{U, V\}$. Symbolically, if $\{U, V\}$ is a linearly independent set in $\mathcal{R}^2$, then

$$\mathcal{R}^2 = \{X \mid X = rU + sV, r \text{ and } s \in \mathcal{R}\}.$$

This fact suggests that a subset $\mathcal{M}$ in $\mathcal{R}^n$, which is defined by

$$\mathcal{M} = \{X \mid X = rU + sV, r \text{ and } s \in \mathcal{R}\},$$

where $\{U, V\}$ is a linearly independent set in $\mathcal{R}^n$, will have the properties of $\mathcal{R}^2$. Figure 1.12a lends support to this conjecture. Any such set, however, would necessarily contain the origin O_n. (Why?)

Consideration of Figure 1.12b suggests that the addition of a fixed vector A to each point of $\mathcal{M}$ would give a set $\mathcal{S}$ with the properties of a plane and which does not contain the origin. On these observations we base our formal definition of a plane of $\mathcal{R}^n$.

DEFINITION 1.39　(Planes in $\mathcal{R}^n$.)　A set $\mathcal{S}$ in $\mathcal{R}^n$ is called a *plane* if and only if there is a vector A in $\mathcal{R}^n$ and a linearly independent subset $\{U, V\}$ of $\mathcal{R}^n$ such that

$$\mathcal{S} = \{X \mid X = A + rU + sV, \text{ where } r \in \mathcal{R} \text{ and } s \in \mathcal{R}\}.$$

This notation is frequently abbreviated

$$\mathcal{S} = \{A + rU + sV \mid r, s \in \mathcal{R}\}.$$

(Parametric Representation of a Plane.)　The equation

$$X = A + rU + sV$$

is called a *parametric equation for the plane* $\mathcal{S}$. The above set description is called a *parametric representation of* $\mathcal{S}$. The variables r and s are called the *parameters* of each representation.

It is well to note the necessity of the requirement that $\{U, V\}$ be a linearly independent set. If $\{U, V\}$ were linearly dependent, one would be a multiple of the other, say $V = kU$. In this event we have

$$\mathcal{M} = \{A + rU + sV \mid r, s \in \mathcal{R}\} = \{A + rU + s(kU) \mid r, s \in \mathcal{R}\}$$
$$= \{A + (r + sk)U \mid r, s \in \mathcal{R}\}$$
$$= \{A + tU \mid t \in \mathcal{R}\}, \text{ with } t = r + sk.$$

Thus, $\mathcal{M}$ would be a line if $U \neq O_n$ or would consist of a single point if $U = O_n$.

According to the definition just given, a point X of $\mathcal{R}^n$ is on the plane $\mathcal{S}$ if and only if there are scalars r and s such that $X = A + rU + sV$. If $\mathcal{S}$ is to have all the properties of $\mathcal{R}^2$ resulting from Theorem 1.34 (Basic

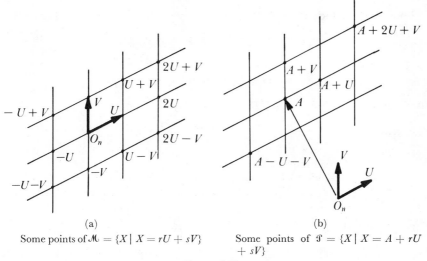

(a)

Some points of $\mathcal{M} = \{X \mid X = rU + sV\}$

(b)

Some points of $\mathcal{P} = \{X \mid X = A + rU + sV\}$

Figure I.12

Properties of Linearly Independent Sets in $\mathcal{R}^2$), the scalars r and s giving the point X of $\mathcal{P}$ must be uniquely determined by X. This happy state of affairs is guaranteed also by the linear independence of $\{U, V\}$.

THEOREM I.40 (Points of a Plane in $\mathcal{R}^n$.) Let A be a vector in $\mathcal{R}^n$ and let $\{U, V\}$ be a linearly independent subset of $\mathcal{R}^n$. Let $\mathcal{P}$ be the plane

$$\mathcal{P} = \{A + rU + sV \mid r, s \in \mathcal{R}\}.$$

The n-vectors X and Y are distinct points of $\mathcal{P}$ if and only if

$$X = A + r_1 U + s_1 V \quad \text{and} \quad Y = A + r_2 U + s_2 V$$

with the ordered pair $\begin{pmatrix} r_1 \\ s_1 \end{pmatrix}$ distinct from $\begin{pmatrix} r_2 \\ s_2 \end{pmatrix}$.

Proof. Let X and Y be points in $\mathcal{R}^n$. According to Definition 1.39 (Planes in $\mathcal{R}^n$), X and Y are points of $\mathcal{P}$ if and only if they are given by

$$X = A + r_1 U + s_1 V \quad \text{and} \quad Y = A + r_2 U + s_2 V$$

for some choice of real numbers r_1, s_1, r_2, and s_2. We need only show that X and Y are distinct points if and only if $\begin{pmatrix} r_1 \\ s_1 \end{pmatrix}$ is different from $\begin{pmatrix} r_2 \\ s_2 \end{pmatrix}$, or equivalently, that X and Y are the same point if and only if $\begin{pmatrix} r_1 \\ s_1 \end{pmatrix} = \begin{pmatrix} r_2 \\ s_2 \end{pmatrix}$.

Thus, we conclude the proof of the theorem with the implication chain:

$$X = Y \Leftrightarrow A + r_1 U + s_1 V = A + r_2 U + s_2 V$$
$$\Leftrightarrow^1 (r_1 - r_2) U + (s_1 - s_2) V = O_n$$
$$\Leftrightarrow^2 r_1 - r_2 = 0 \text{ and } s_1 - s_2 = 0$$
$$\Leftrightarrow^3 \binom{r_1}{s_1} = \binom{r_2}{s_2}. \quad \blacksquare$$

(1) Subtracting ($\Rightarrow$) or adding ($\Leftarrow$) $A + r_2 U + s_2 V$.
(2) ($\Rightarrow$) Using the linear independence of $\{U, V\}$ and Theorem 1.30ii.
    ($\Leftarrow$) Obvious implication.
(3) Using Definition 1.1 (Equality of Real Ordered Pairs).

Example 19.   Consider the plane

$$\mathfrak{I} = \left\{ \begin{pmatrix} 1 \\ 2 \\ 1 \end{pmatrix} + r \begin{pmatrix} 2 \\ 0 \\ 1 \end{pmatrix} + s \begin{pmatrix} 0 \\ 1 \\ 1 \end{pmatrix} \middle| r, s \in \mathfrak{R} \right\}.$$

Choosing $(r, s) = (0, 0)$, we find that $\begin{pmatrix} 1 \\ 2 \\ 1 \end{pmatrix}$ is a point of $\mathfrak{I}$. With $(r, s) = (1, 0)$ we obtain the point

$$X = \begin{pmatrix} 1 \\ 2 \\ 1 \end{pmatrix} + 1 \begin{pmatrix} 2 \\ 0 \\ 1 \end{pmatrix} + 0 \begin{pmatrix} 0 \\ 1 \\ 1 \end{pmatrix} = \begin{pmatrix} 3 \\ 2 \\ 2 \end{pmatrix},$$

while selecting $(r, s) = (1, 2)$ gives the point

$$Y = \begin{pmatrix} 1 \\ 2 \\ 1 \end{pmatrix} + 1 \begin{pmatrix} 2 \\ 0 \\ 1 \end{pmatrix} + 2 \begin{pmatrix} 0 \\ 1 \\ 1 \end{pmatrix} = \begin{pmatrix} 3 \\ 4 \\ 4 \end{pmatrix}. \quad \blacksquare$$

The preceding theorem and example illustrate the very close relation between $\mathfrak{R}^2$ and a plane $\mathfrak{I}$ in $\mathfrak{R}^n$. A point P in $\mathfrak{R}^2$ is given by one and only one real ordered pair; in fact, a point in $\mathfrak{R}^2$ *is* a real ordered pair $\binom{r}{s}$. Similarly, a point X in a plane $\mathfrak{I} = \{A + rU + sV \mid r, s \in \mathfrak{R}\}$ is also specified by one and only one real ordered pair $\binom{r}{s}$, even though X itself may not be a real ordered pair.

The concept of a direction vector for a plane is a straightforward generalization of the corresponding concept for lines. We shall need to augment this concept with that of a direction set in order to completely characterize a plane.

DEFINITION 1.41   (Direction Vectors for a Plane.) Let $\mathfrak{I}$ be a plane in $\mathfrak{R}^n$. A vector V in $\mathfrak{R}^n$ is called a *direction vector for* $\mathfrak{I}$ if and only if there are distinct points X and Y belonging to $\mathfrak{I}$ such that

$$V = Y - X.$$

(Direction Sets for a Plane.) A set $\mathcal{D}$ in $\mathcal{R}^n$ is called a *direction set* for $\mathcal{I}$ if and only if $\mathcal{D}$ is a linearly independent doubleton set of direction vectors for $\mathcal{I}$:

$$\mathcal{D} = \{U, V\}$$

where U, V are nonparallel direction vectors for $\mathcal{I}$.

Example 20. Again consider the plane of Example 19:

$$\mathcal{I} = \left\{ \begin{pmatrix} 1 \\ 2 \\ 1 \end{pmatrix} + r \begin{pmatrix} 2 \\ 0 \\ 1 \end{pmatrix} + s \begin{pmatrix} 0 \\ 1 \\ 1 \end{pmatrix} \middle| \; r, s \in \mathcal{R} \right\}.$$

Choosing $\begin{pmatrix} r \\ s \end{pmatrix}$ equal to $\begin{pmatrix} 0 \\ 0 \end{pmatrix}$, $\begin{pmatrix} 1 \\ 0 \end{pmatrix}$, and $\begin{pmatrix} 0 \\ 1 \end{pmatrix}$, in turn, we determine that

$$A = \begin{pmatrix} 1 \\ 2 \\ 1 \end{pmatrix}, \quad B = \begin{pmatrix} 3 \\ 2 \\ 2 \end{pmatrix}, \quad \text{and} \quad C = \begin{pmatrix} 1 \\ 3 \\ 2 \end{pmatrix}$$

are three distinct points of $\mathcal{I}$. Consequently,

$$U = B - A = \begin{pmatrix} 2 \\ 0 \\ 1 \end{pmatrix} \quad \text{and} \quad V = C - A = \begin{pmatrix} 0 \\ 1 \\ 1 \end{pmatrix}$$

are direction vectors for $\mathcal{I}$. Moreover, U and V are nonparallel direction vectors for $\mathcal{I}$. Therefore, the linearly independent set

$$\mathcal{D} = \{U, V\} = \left\{ \begin{pmatrix} 2 \\ 0 \\ 1 \end{pmatrix}, \begin{pmatrix} 0 \\ 1 \\ 1 \end{pmatrix} \right\}$$

is a direction set for $\mathcal{I}$. ∎

The last example showed that the vectors $U = \begin{pmatrix} 2 \\ 0 \\ 1 \end{pmatrix}$ and $V = \begin{pmatrix} 0 \\ 1 \\ 1 \end{pmatrix}$, which appeared in the parametric representation for the plane $\mathcal{I}$, constituted a direction set for $\mathcal{I}$ and suggests that this may be a general result. In fact, the vectors U and V of the general parametric representation $X = A + rU + sV$ bear an important relation to all direction sets for the plane so described. This relation is included in the following important theorem, a generalization of Theorem 1.22 (Relation of the Vectors of a Parametric Representation of a Line to the Line).

THEOREM 1.42   (Relation of the Vectors of a Parametric Representation of a Plane to the Plane.) Let $\mathcal{S}$ be a plane in $\mathcal{R}^n$ given by

$$\mathcal{S} = \{A + rU + sV \mid r, s \in \mathcal{R}\}$$

where A is a vector in $\mathcal{R}^n$ and $\{U, V\}$ is a linearly independent subset of $\mathcal{R}^n$, then

i. A is a point of $\mathcal{S}$.

ii. $\{U, V\}$ is a direction set for $\mathcal{S}$.

iii. Every direction vector for $\mathcal{S}$ is a linear combination of the vectors of $\{U, V\}$.

iv. Every nonzero linear combination of any two direction vectors for $\mathcal{S}$ is again a direction vector for $\mathcal{S}$.

v. Every direction vector for $\mathcal{S}$ is a linear combination of the vectors of any direction set $\mathcal{D}$ for $\mathcal{S}$.

Proof

i. The choice $\begin{pmatrix} r \\ s \end{pmatrix} = \begin{pmatrix} 0 \\ 0 \end{pmatrix}$ shows that $A = A + 0U + 0V$ is a point of $\mathcal{S}$.

ii. See Exercise 7a.

iii. See Exercise 7b.

iv. Let C and D be two direction vectors for $\mathcal{S}$, then C and D are differences of distinct points of $\mathcal{S}$. That is, there are scalars $r_1, s_1, r_2, s_2, r_3, s_3, r_4,$ and s_4 such that

$$C = (A + r_1U + s_1V) - (A + r_2U + s_2V) = (r_1 - r_2)U + (s_1 - s_2)V$$

and

$$D = (A + r_3U + s_3V) - (A + r_4U + s_4V) = (r_3 - r_4)U + (s_3 - s_4)V.$$

Let W be a nonzero linear combination of C and D:

$$W = aC + bD \neq O_n,$$

then

$$W = [A + a(r_1 - r_2)U + a(s_1 - s_2)V]$$
$$- [A + b(r_4 - r_3)U + b(s_4 - s_3)V]$$

(check this!). Hence, W is the difference of the two points

$$Y = A + a(r_1 - r_2)U + a(s_1 - s_2)V$$

and

$$X = A + b(r_4 - r_3)U + b(s_4 - s_3)V$$

which are clearly points of $\mathcal{S}$ (show this!). Moreover, X and Y are distinct since $W = Y - X$ and W is given as a nonzero vector.

v. Let $\{C, D\}$ be a direction set for $\mathcal{S}$. We are to show that every direction vector for $\mathcal{S}$ is a linear combination of C and D. We shall first show that, in particular, U and V are each linear combinations of C and D.

Using the fact that C and D are each direction vectors for $\mathcal{S}$ and applying statement (iii) of the theorem, we may write

(1.9)
$$\begin{cases} C = aU + cV \\ D = bU + dV \end{cases}$$

for some scalars a, b, c, d. Multiplying both members of the first equation by d and those of the second by c, we obtain

$$\begin{cases} dC = adU + cdV \\ cD = bcU + cdV \end{cases}$$

and, by subtraction,

$$dC - cD = (ad - bc)U.$$

We shall show below that $ad - bc \neq 0$. Assuming this result we may solve the last equation for U:

$$. \quad U = \frac{d}{ad - bc} C + \frac{-c}{ad - bc} D$$

thus showing that U is a linear combination of C and D. In a similar way we can show that

$$V = \frac{aD - bC}{ad - bc} = \frac{a}{ad - bc} D + \frac{-b}{ad - bc} C.$$

Thus, each of the vectors U and V is a linear combination of C and D:

(1.10)
$$U = kC + lD$$
$$V = mC + nD.$$

According to statement (iii), to be proved by the reader, every direction vector W for $\mathcal{S}$ is a linear combination $W = pU + qV$ of U and V. Therefore,

$$W = pU + qV =^1 p(kC + lD) + q(mC + nD)$$
$$= (pk + qm)C + (pl + qn)D.$$

(1) Substitution from Equation 1.10.
(2) From properties of the linear operations.

This shows that every direction vector W for $\mathcal{S}$ is a linear combination of C and D.

We now show that a, b, c, and d of Equation 1.9 above are such that $ad - bc \neq 0$. We do this by showing that the opposite situation contradicts the linear independence of $\{C, D\}$. Thus,

$$ad - bc = 0 \Leftrightarrow^1 \left\{ \begin{pmatrix} a \\ c \end{pmatrix}, \begin{pmatrix} b \\ d \end{pmatrix} \right\} \text{ is a linearly dependent set.}$$

$$\Leftrightarrow^2 \begin{pmatrix} a \\ c \end{pmatrix} = k \begin{pmatrix} b \\ d \end{pmatrix}, \quad \text{or} \quad \begin{pmatrix} b \\ d \end{pmatrix} = l \begin{pmatrix} a \\ c \end{pmatrix}$$

$$\Rightarrow^3 \left. \begin{array}{l} C = aU + bV \\ = kbU + kdV \\ = k(bU + dV) \\ = kD \end{array} \right\} \quad \text{or} \quad \left\{ \begin{array}{l} D = bU + dV \\ = laU + lcV \\ = l(aU + cV) \\ = lC \end{array} \right.$$

$$\Rightarrow^2 \{C, D\} \text{ is a linearly dependent set.}$$

(1) Theorem 1.32 (Determinant Test for Linear Dependence in $\mathcal{R}^2$) with

$$U = \begin{pmatrix} a \\ c \end{pmatrix} \quad \text{and} \quad V = \begin{pmatrix} b \\ d \end{pmatrix}.$$

(2) Definition 1.29i (Linear Dependence of a Set).
(3) Definition 1.1 (Equality of Ordered Pairs) and the Equations 1.9 in the proof of the present theorem. ∎

EXERCISES

In Exercises 1 to 6 give a parametric representation of a plane $\mathcal{S}$ satisfying the given conditions:

1. $\mathcal{S}$ has direction vectors $U = \begin{pmatrix} 2 \\ 1 \\ 3 \end{pmatrix}$ and $V = \begin{pmatrix} 1 \\ 2 \\ 1 \end{pmatrix}$ and contains the point $A = \begin{pmatrix} 1 \\ 0 \\ 4 \end{pmatrix}$.

2. $\mathcal{S}$ has direction vectors $U = \begin{pmatrix} 1 \\ 2 \\ 1 \\ 1 \end{pmatrix}$ and $V = \begin{pmatrix} 2 \\ 0 \\ 0 \\ 1 \end{pmatrix}$ and contains the point $A = \begin{pmatrix} 2 \\ 1 \\ 1 \\ 1 \end{pmatrix}$.

3. $\mathcal{S}$ contains the points $A = \begin{pmatrix} 1 \\ 1 \\ 2 \end{pmatrix}$, $B = \begin{pmatrix} 1 \\ 0 \\ 3 \end{pmatrix}$, and $C = \begin{pmatrix} 2 \\ -1 \\ 1 \end{pmatrix}$.

4. $\mathcal{S}$ contains the points $A = \begin{pmatrix} 1 \\ 2 \\ 1 \end{pmatrix}$ and $B = \begin{pmatrix} -1 \\ 1 \\ 1 \end{pmatrix}$ and has $U = \begin{pmatrix} 1 \\ 0 \\ 2 \end{pmatrix}$ as a direction vector.

5. $\mathcal{S}$ contains the points $A = \begin{pmatrix} 1 \\ 2 \\ 1 \\ 1 \end{pmatrix}$, $B = \begin{pmatrix} 0 \\ 0 \\ 1 \\ 1 \end{pmatrix}$, and $C = \begin{pmatrix} 2 \\ 3 \\ 1 \\ -1 \end{pmatrix}$.

6. $\mathcal{I}$ contains the points $A = \begin{pmatrix} 1 \\ 4 \\ 0 \\ 1 \end{pmatrix}$ and $B = \begin{pmatrix} 0 \\ 1 \\ 2 \\ -1 \end{pmatrix}$ and has $U = \begin{pmatrix} 1 \\ 2 \\ 1 \\ 2 \end{pmatrix}$ as a direction vector.

7. Let $\mathcal{I}$ be a plane in $\mathcal{R}^n$ given by $\mathcal{I} = \{A + rU + sV \mid r, s \in \mathcal{R}\}$, where A is a vector in $\mathcal{R}^n$ and $\{U, V\}$ is a linearly independent subset of $\mathcal{R}^n$.
   (a) Show that $\{U, V\}$ is a direction set for $\mathcal{I}$.
   (b) Show that every direction vector for $\mathcal{I}$ is a linear combination of the vectors of $\{U, V\}$.

ANSWERS

1. $\mathcal{I} = \left\{ \begin{pmatrix} 1 \\ 0 \\ 4 \end{pmatrix} + r \begin{pmatrix} 2 \\ 1 \\ 3 \end{pmatrix} + s \begin{pmatrix} 1 \\ 2 \\ 1 \end{pmatrix} \,\middle|\, r, s \in \mathcal{R} \right\}$ is one such plane.

3. $\mathcal{I} = \left\{ \begin{pmatrix} 1 \\ 1 \\ 2 \end{pmatrix} + r \begin{pmatrix} 0 \\ -1 \\ 1 \end{pmatrix} + s \begin{pmatrix} 1 \\ -1 \\ -2 \end{pmatrix} \,\middle|\, r, s \in \mathcal{R} \right\}$ is one such plane (Does it contain A, B, and C?).

5. $\mathcal{I} = \left\{ \begin{pmatrix} 0 \\ 0 \\ 1 \\ 1 \end{pmatrix} + r \begin{pmatrix} 1 \\ 2 \\ 0 \\ 0 \end{pmatrix} + s \begin{pmatrix} 1 \\ 1 \\ 0 \\ -2 \end{pmatrix} \,\middle|\, r, s \in \mathcal{R} \right\}$ is one such plane; $(r, s) =$

   $(1, 0)$, $(0, 0)$ and $(1, 1)$ give A, B and C, respectively.
7. (a) Partial answer: $Y = A + U$ and $X = A$ are points of $\mathcal{I}$. Therefore, $Y - X = (A + U) - A$ is a direction vector for $\mathcal{I}$.

9. PROPERTIES OF PLANES

There is an infinity of parametric representations for a given plane in $\mathcal{R}^n$, just as there is an infinity of such representations for a given line in $\mathcal{R}^n$. Nonetheless, there are well-defined relations among the various parametric descriptions for a plane. These relations are our first concern in this section.

THEOREM 1.43 (Parametric Description of a Plane Containing a Point A.) Let A be any point of the plane

$$\mathcal{I} = \{B + kT + lW \mid k, l \in \mathcal{R}\}$$

where $\{T, W\}$ is a direction set. The plane $\mathcal{I}$ may also be described as

$$\mathcal{I} = \{A + kT + lW \mid k, l \in \mathcal{R}\}.$$

The pattern of proof is the same as the proof of Theorem 1.23 (Parametric Description of a Line Containing a Point A). We leave the details as Exercise 13.

The point of Theorem 1.43 is that a given parametric representation for a plane $\mathcal{P}$ may be altered by substituting any point A of the plane for whatever vector B appears in the original description and that the resulting representation is merely a different description of the same plane $\mathcal{P}$. One would also expect that any direction set $\{U, V\}$ for $\mathcal{P}$ may be used in place of the direction set $\{T, W\}$. The proof of this statement is our next task.

THEOREM 1.44   (Point-Direction Set Description of a Plane.) Let $\mathcal{P}$ be a plane in $\mathcal{R}^n$. Let A be a point of $\mathcal{P}$ and $\{U, V\}$ any direction set of $\mathcal{P}$, then

$$\mathcal{P} = \{A + rU + sV \mid r, s \in \mathcal{R}\}.$$

Moreover, $\mathcal{P}$ is the only plane through A with $\{U, V\}$ as a direction set.

Proof.   Let $\mathcal{P}$ be any plane in $\mathcal{R}^n$ containing A and having $\{U, V\}$ as a direction set. According to the definition of a plane there is a vector B and a direction set $\{T, W\}$ such that

$$\mathcal{P} = \{B + kT + lW \mid k, l \in \mathcal{R}\}.$$

Consider now the plane $\mathcal{P}'$ given by

$$(1.11) \qquad \mathcal{P}' = \{A + rU + sV \mid r, s \in \mathcal{R}\}.$$

According to Theorem 1.42 i and ii, $\mathcal{P}'$ contains A and has $\{U, V\}$ as a direction set. We shall show that $\mathcal{P}'$ and $\mathcal{P}$ are the same set, thereby showing that there is only one plane $\mathcal{P}$ satisfying the requirements of the theorem. We first note that, according to Theorem 1.43, $\mathcal{P}$ may be described by

$$(1.12) \qquad \mathcal{P} = \{A + kT + lW \mid k, l \in \mathcal{R}\},$$

since A is a point of $\mathcal{P}$. By Theorem 1.42 ii, the set $\{T, W\}$ is a direction set for $\mathcal{P}$. Since U and V are each direction vectors for $\mathcal{P}$, we may apply Theorem 1.42 v to write U and V as linear combinations of T and W:

$$(1.13) \qquad \begin{cases} U = aT + bW \\ V = cT + dW \end{cases} \quad \text{for some } a, b, c, d \in \mathcal{R}.$$

In a similar way, since $\{U, V\}$ is given as a direction set for $\mathcal{P}$, and T and W are each direction vectors for $\mathcal{P}$, we express T and W as linear combinations of U and V:

$$(1.14) \qquad \begin{cases} T = pU + qV \\ W = mU + nV \end{cases} \quad \text{for some } p, q, m, n \in \mathcal{R}.$$

Consider now the two implication chains:

$$X \in \mathcal{S}' \Rightarrow^1 X = A + rU + sV \qquad \text{for some } r, s \in \mathcal{R}$$
$$\Rightarrow^2 X = A + r(aT + bW) + s(cT + dW)$$
$$\Rightarrow \ \ X = A + (ra + sc)T + (rb + sd)W$$
$$\Rightarrow^3 X = A + kT + lW, \qquad \text{for some } k, l \in \mathcal{R}$$
$$\Rightarrow^4 X \in \mathcal{S}$$

and

$$X \in \mathcal{S} \Rightarrow^4 X = A + kT + lW \qquad \text{for some } k, l \in \mathcal{R}$$
$$\Rightarrow^5 X = A + k(pU + qV) + l(mU + nV)$$
$$\Rightarrow \ \ X = A + (kp + lm)U + (kq + ln)V$$
$$\Rightarrow^6 X = A + rU + sV \qquad \text{for some } r \text{ and } s$$
$$\Rightarrow^1 X \in \mathcal{S}'.$$

(1) Using Equation 1.11 as the description of $\mathcal{S}'$.
(2) Substituting from Equation 1.13.
(3) Setting $k = ra + sc$ and $l = rb + sd$.
(4) Using Equation 1.12 as the description of $\mathcal{S}$.
(5) Substituting from Equation 1.14.
(6) Setting $r = kp + lm$ and $s = kq + ln$.

The first implication chain assures us that every point of $\mathcal{S}'$ is a point of $\mathcal{S}$. The second chain shows that every point of $\mathcal{S}$ is also a point of $\mathcal{S}'$. Thus, $\mathcal{S} = \mathcal{S}'$ and the theorem is proved. ∎

Example 21. Let $\mathcal{S}$ be the plane

$$\mathcal{S} = \left\{ \begin{pmatrix} 1 \\ 1 \\ 0 \end{pmatrix} + r \begin{pmatrix} 1 \\ 1 \\ 0 \end{pmatrix} + s \begin{pmatrix} 0 \\ 2 \\ 1 \end{pmatrix} \ \middle|\ r, s \in \mathcal{R} \right\}.$$

Choosing $\begin{pmatrix} r \\ s \end{pmatrix}$ equal to $\begin{pmatrix} 0 \\ 0 \end{pmatrix}$, $\begin{pmatrix} -1 \\ 0 \end{pmatrix}$, and $\begin{pmatrix} 0 \\ 1 \end{pmatrix}$ in turn, we find that

$$A = \begin{pmatrix} 1 \\ 1 \\ 0 \end{pmatrix}, \quad B = \begin{pmatrix} 0 \\ 0 \\ 0 \end{pmatrix}, \quad \cdot \text{ and } \quad C = \begin{pmatrix} 1 \\ 3 \\ 1 \end{pmatrix}$$

are points of $\mathcal{S}$. Consequently, $\left\{ \begin{pmatrix} 0 \\ 2 \\ 1 \end{pmatrix}, \begin{pmatrix} 1 \\ 3 \\ 1 \end{pmatrix} \right\}$ is a direction set for $\mathcal{S}$, since

$$\begin{pmatrix} 0 \\ 2 \\ 1 \end{pmatrix} = C - A \quad \text{and} \quad \begin{pmatrix} 1 \\ 3 \\ 1 \end{pmatrix} = C - B.$$

Using this direction set and point B, we may also describe $\mathcal{S}$ by

$$\mathcal{S} = \left\{ \begin{pmatrix} 0 \\ 0 \\ 0 \end{pmatrix} + k \begin{pmatrix} 0 \\ 2 \\ 1 \end{pmatrix} + l \begin{pmatrix} 1 \\ 3 \\ 1 \end{pmatrix} \,\middle|\, k, l \in \mathcal{R} \right\}$$

$$= \left\{ k \begin{pmatrix} 0 \\ 2 \\ 1 \end{pmatrix} + l \begin{pmatrix} 1 \\ 3 \\ 1 \end{pmatrix} \,\middle|\, k, l \in \mathcal{R} \right\}. \quad \blacksquare$$

The following two theorems are consequences of Theorem 1.44. Their proofs are called for as Exercise 14 and Exercise 15.

THEOREM 1.45 (Three-Point Description of a Plane.) Let A, B, and C be three points of $\mathcal{R}^n$ such that $\{B - A, C - A\}$ is a linearly independent set. There is one and only one plane $\mathcal{S}$ containing all three points. Moreover, the plane $\mathcal{S}$ has the parametric description

$$\mathcal{S} = \{A + r(B - A) + s(C - A) \mid r, s \in \mathcal{R}\}.$$

THEOREM 1.46 (A Plane Contains the Line Through Any Two of Its Points.) Let $\mathcal{S}$ be a plane in $\mathcal{R}^n$ and let A and B be points of $\mathcal{S}$. Every point of the line $\mathcal{L}$ through A and B is also a point of $\mathcal{S}$.

Example 22. To obtain the parametric description of the plane containing the points $A = \begin{pmatrix} 1 \\ 2 \\ 1 \end{pmatrix}$, $B = \begin{pmatrix} 2 \\ 0 \\ -1 \end{pmatrix}$, and $C = \begin{pmatrix} -1 \\ 1 \\ 0 \end{pmatrix}$, we first determine a direction set by computing differences of points. We may use $B - A$ and $C - A$ as given in Theorem 1.45 or we may use other differences obtained by using these three points. In the present case we may use

$$U = A - B = \begin{pmatrix} -1 \\ 2 \\ 2 \end{pmatrix} \quad \text{and} \quad V = A - C = \begin{pmatrix} 2 \\ 1 \\ 1 \end{pmatrix}.$$

Since the set $\{U, V\}$ thus obtained is a linearly independent set, there will be but one plane $\mathcal{S}$ containing A, B, and C. It is the plane described by

$$\mathcal{S} = \{A + rU + sV \mid r, s \in \mathcal{R}\} = \left\{ \begin{pmatrix} 1 \\ 2 \\ 1 \end{pmatrix} + r \begin{pmatrix} -1 \\ 2 \\ 2 \end{pmatrix} + s \begin{pmatrix} 2 \\ 1 \\ 1 \end{pmatrix} \,\middle|\, r, s \in \mathcal{R} \right\}. \quad \blacksquare$$

Example 23. Let us now determine a plane containing the points
$A = \begin{pmatrix} 1 \\ 2 \\ 1 \end{pmatrix}$, $B = \begin{pmatrix} 2 \\ 0 \\ -1 \end{pmatrix}$, and $C = \begin{pmatrix} -2 \\ 8 \\ 7 \end{pmatrix}$. We first determine direction
vectors for the plane by computing

$$B - A = \begin{pmatrix} 1 \\ -2 \\ -2 \end{pmatrix} \quad \text{and} \quad C - A = \begin{pmatrix} -3 \\ 6 \\ 6 \end{pmatrix}.$$

We now note that these two vectors are linearly dependent: $C - A = -3(B - A)$. This means that A, B, and C are collinear, that is, all three are points of the line

$$\mathcal{L} = \{A + t(B - A) \mid t \in \mathcal{R}\}$$

$$= \left\{ \begin{pmatrix} 1 \\ 2 \\ 1 \end{pmatrix} + t \begin{pmatrix} 1 \\ -2 \\ -2 \end{pmatrix} \;\middle|\; t \in \mathcal{R} \right\}.$$

Setting t equal to 0, 1, and -3, in turn, yields A, B, and C, respectively. Accordingly, there will be many planes containing all three points. In fact, any plane given by

$$\left\{ \begin{pmatrix} 1 \\ 2 \\ 1 \end{pmatrix} + t \begin{pmatrix} 1 \\ -2 \\ -2 \end{pmatrix} + s \begin{pmatrix} a \\ b \\ c \end{pmatrix} \;\middle|\; t, s \in \mathcal{R} \right\}$$

where $\begin{pmatrix} a \\ b \\ c \end{pmatrix}$ is not a multiple of $\begin{pmatrix} 1 \\ -2 \\ -2 \end{pmatrix}$, will contain all three points. In particular,

$$\mathcal{S} = \left\{ \begin{pmatrix} 1 \\ 2 \\ 1 \end{pmatrix} + t \begin{pmatrix} 1 \\ -2 \\ -2 \end{pmatrix} + s \begin{pmatrix} 1 \\ 0 \\ 0 \end{pmatrix} \;\middle|\; t, s \in \mathcal{R} \right\}$$

is one such plane. ∎

Let us now consider two planes in $\mathcal{R}^n$ and the possible ways in which they may intersect. We shall first state and prove the facts regarding the situation and then shall look at some examples, one of which may bring a surprise to many who read it.

THEOREM 1.47 (Intersection of Two Planes in $\mathcal{R}^n$.) Let $\mathcal{S}_1$ and $\mathcal{S}_2$ denote two (not necessarily distinct) planes in $\mathcal{R}^n$. Exactly one of the following is true:

  i. $\mathcal{S}_1$ and $\mathcal{S}_2$ have no points in common: $\mathcal{S}_1 \cap \mathcal{S}_2 = \varnothing$.

  ii. $\mathcal{S}_1$ and $\mathcal{S}_2$ have exactly one point in common: $\mathcal{S}_1 \cap \mathcal{S}_2 =$ singleton set.

  iii. $\mathcal{S}_1$ and $\mathcal{S}_2$ intersect in a line $\mathcal{L}$: $\mathcal{S}_1 \cap \mathcal{S}_2 = \mathcal{L}$.

  iv. $\mathcal{S}_1$ and $\mathcal{S}_2$ coincide: $\mathcal{S}_1 \cap \mathcal{S}_2 = \mathcal{S}_1 = \mathcal{S}_2$.

Proof

Clearly at most one of the four statements can be valid for a given choice of $\mathfrak{I}_1$ and $\mathfrak{I}_2$. We need to show that at least one of the statements

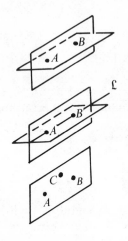

holds true. Let $\mathfrak{I}_1$ and $\mathfrak{I}_2$ be two planes in $\mathcal{R}^n$. If (i) or (ii) is satisfied, the theorem certainly is valid.

If neither (i) nor (ii) holds, then $\mathfrak{I}_1$ and $\mathfrak{I}_2$ must have two distinct points, A and B, in common. Thus, in view of Theorem 1.46 (A Plane Contains the Line Through Any Two of Its Points), $\mathfrak{I}_1$ and $\mathfrak{I}_2$ must each contain the line $\mathfrak{L}$ joining A and B. If $\mathfrak{I}_1$ and $\mathfrak{I}_2$ have no further points in common, we then have statement (iii), $\mathfrak{I}_1 \cap \mathfrak{I}_2 = \mathfrak{L}$.

If, however, $\mathfrak{I}_1$ and $\mathfrak{I}_2$ also contain a third point C not on $\mathfrak{L}$, we know that $\{B - A, C - A\}$ is a linearly independent set. (Why?) Thus, $\mathfrak{I}_1$ and $\mathfrak{I}_2$ are each planes containing points A, B, and C such that $\{B - A, C - A\}$ is a linearly independent set and must therefore be identical by Theorem 1.45 (Three Point Description of a Plane).

Thus, if statements (i) and (ii) are not true for $\mathfrak{I}_1$ and $\mathfrak{I}_2$, then either statement (iii) or statement (iv) must hold. ∎

Example 24. The planes

$$\mathfrak{I}_1 = \left\{ r \begin{pmatrix} 1 \\ 0 \\ 0 \end{pmatrix} + s \begin{pmatrix} 0 \\ 1 \\ 0 \end{pmatrix} \,\middle|\, r, s \in \mathcal{R} \right\}$$

and

$$\mathfrak{I}_2 = \left\{ \begin{pmatrix} 0 \\ 0 \\ 1 \end{pmatrix} + k \begin{pmatrix} 1 \\ 0 \\ 0 \end{pmatrix} + l \begin{pmatrix} 0 \\ 1 \\ 0 \end{pmatrix} \,\middle|\, k, l \in \mathcal{R} \right\}$$

E-24

have no points in common, since every point X of $\mathfrak{I}_1$ is given by $X = \begin{pmatrix} r \\ s \\ 0 \end{pmatrix}$ and has third coordinate zero, while every point Y of $\mathfrak{I}_2$ is given by $Y = \begin{pmatrix} k \\ l \\ 1 \end{pmatrix}$ and has third coordinate 1. ∎

Example 25. The planes

$$\mathfrak{I}_1 = \left\{ r \begin{pmatrix} 1 \\ 0 \\ 0 \end{pmatrix} + s \begin{pmatrix} 0 \\ 1 \\ 0 \end{pmatrix} \,\middle|\, r, s \in \mathcal{R} \right\}$$

and

$$\mathcal{I}_2 = \left\{ k\begin{pmatrix} 1 \\ 0 \\ 0 \end{pmatrix} + l\begin{pmatrix} 0 \\ 0 \\ 1 \end{pmatrix} \;\middle|\; k, l \in \mathcal{R} \right\}$$

have the line

$$\mathcal{L} = \left\{ t\begin{pmatrix} 1 \\ 0 \\ 0 \end{pmatrix} \;\middle|\; t \in \mathcal{R} \right\}$$

E-25

in common. If they had other points in common they would be identical.

But they are not identical since $\begin{pmatrix} 0 \\ 1 \\ 0 \end{pmatrix}$ is a point of $\mathcal{I}_1$ but not a point of $\mathcal{I}_2$. (Check this!) ■

Example 26. The planes

$$\mathcal{I}_1 = \left\{ r\begin{pmatrix} 1 \\ 0 \\ 0 \end{pmatrix} + s\begin{pmatrix} 0 \\ 1 \\ 0 \end{pmatrix} \;\middle|\; r, s \in \mathcal{R} \right\}$$

and

$$\mathcal{I}_2 = \left\{ k\begin{pmatrix} 1 \\ 1 \\ 0 \end{pmatrix} + l\begin{pmatrix} 1 \\ -1 \\ 0 \end{pmatrix} \;\middle|\; k, l \in \mathcal{R} \right\}$$

coincide. Both planes contain the

point $\begin{pmatrix} 0 \\ 0 \\ 0 \end{pmatrix}$. Moreover, $\begin{pmatrix} 1 \\ 1 \\ 0 \end{pmatrix}$ and

$\begin{pmatrix} 1 \\ -1 \\ 0 \end{pmatrix}$ are direction vectors for $\mathcal{I}_1$ since they are linear combinations of the direction vectors $\begin{pmatrix} 1 \\ 0 \\ 0 \end{pmatrix}$ and $\begin{pmatrix} 0 \\ 1 \\ 0 \end{pmatrix}$. Thus, $\mathcal{I}_1$ and $\mathcal{I}_2$ are both planes containing $\begin{pmatrix} 0 \\ 0 \\ 0 \end{pmatrix}$

and having $\left\{ \begin{pmatrix} 1 \\ 1 \\ 0 \end{pmatrix}, \begin{pmatrix} 1 \\ -1 \\ 0 \end{pmatrix} \right\}$ as a direction set. By Theorem 1.44 (Point-Direction

Set Description of a Plane), $\mathcal{I}_1$ and $\mathcal{I}_2$ must coincide. ■

Example 27. So far we have given one example for each of the cases of Theorem 1.47 except for the case in which two planes intersect in exactly one point. If you believe that such an example cannot be given, you are indeed in for a surprise. It may be of some consolation to you to know that no such examples can be given in $\mathcal{R}^3$. However, in $\mathcal{R}^4$ let us consider the two planes

$$\mathcal{I}_1 = \left\{ r\begin{pmatrix} 1 \\ 0 \\ 0 \\ 0 \end{pmatrix} + s\begin{pmatrix} 0 \\ 1 \\ 0 \\ 0 \end{pmatrix} \;\middle|\; r, s \in \mathcal{R} \right\}$$

E-26

and

$$\mathscr{T}_2 = \left\{ k \begin{pmatrix} 0 \\ 0 \\ 1 \\ 0 \end{pmatrix} + l \begin{pmatrix} 0 \\ 0 \\ 0 \\ 1 \end{pmatrix} \middle| \, k, l \in \mathscr{R} \right\}.$$

Every point X of $\mathscr{T}_1$ is given by $X = \begin{pmatrix} r \\ s \\ 0 \\ 0 \end{pmatrix}$, and every point Y of $\mathscr{T}_2$ is given

by $Y = \begin{pmatrix} 0 \\ 0 \\ k \\ l \end{pmatrix}$. If C is a point on both planes we have

$$C = \begin{pmatrix} r_0 \\ s_0 \\ 0 \\ 0 \end{pmatrix} = \begin{pmatrix} 0 \\ 0 \\ k_0 \\ l_0 \end{pmatrix} \qquad \text{for some } r_0, s_0, k_0, l_0 \in \mathscr{R}.$$

But this equality can occur only if $r_0 = s_0 = k_0 = l_0 = 0$, and thus,

$$\mathscr{T}_1 \cap \mathscr{T}_2 = \left\{ \begin{pmatrix} 0 \\ 0 \\ 0 \\ 0 \end{pmatrix} \right\}$$

is indeed a singleton set. This four-space example is oftentimes pictured in $\mathscr{R}^2$ by a sketch similar to Figure 1.13. ∎

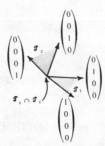

Figure 1.13 Planes in $\mathscr{R}^4$ intersecting in a single point.

Having characterized the possible ways in which two planes may intersect, let us now examine the nature of the intersection of a line and a plane.

THEOREM 1.48 (Intersection of a Line and a Plane in $\mathscr{R}^n$.) Let $\mathscr{T}$ be a plane and $\mathscr{L}$ a line in $\mathscr{R}^n$. Exactly one of the following is true:
  i. $\mathscr{T}$ and $\mathscr{L}$ have no points in common: $\mathscr{T} \cap \mathscr{L} = \varnothing$.
  ii. $\mathscr{T}$ and $\mathscr{L}$ intersect in a single point A: $\mathscr{T} \cap \mathscr{L} = \{A\}$.
  iii. $\mathscr{T}$ contains $\mathscr{L}$: $\mathscr{T} \cap \mathscr{L} = \mathscr{L}$.

Proof. Again, at most one of the statements can hold for a given plane $\mathcal{S}$ and line $\mathcal{L}$. If $\mathcal{S}$ and $\mathcal{L}$ have no points in common then (i) holds. If (i) fails to hold, then there is at least one point A in $\mathcal{S} \cap \mathcal{L}$. If there are no other points in $\mathcal{S} \cap \mathcal{L}$, then (ii) holds.

If (i) and (ii) fail to hold, then $\mathcal{S} \cap \mathcal{L}$ contains at least two points A and B. By Theorem 1.46 (A Plane Contains the Line Through Any Two of Its Points) $\mathcal{S}$ must also contain the line $\mathcal{L}$ through A and B and therefore (iii) must hold.

Thus, if neither (i) nor (ii) is true, then (iii) must be and the theorem is proved. ■

EXERCISES

In each of the Exercises 1 to 6 use the theorems of this section to show that there is but one plane $\mathcal{S}$ satisfying the conditions given in the indicated Exercise in Section 8, (p. 66):

1. Exercise 1.   2. Exercise 2.   3. Exercise 3.

4. Exercise 4.   5. Exercise 5.   6. Exercise 6.

7. Show that there is more than one plane containing the points

$$A = \begin{pmatrix} 1 \\ 2 \\ 1 \end{pmatrix}, \quad B = \begin{pmatrix} 2 \\ 0 \\ 3 \end{pmatrix}, \quad \text{and} \quad C = \begin{pmatrix} 6 \\ -8 \\ 11 \end{pmatrix}.$$

8. Show that there is more than one plane containing the points $A = \begin{pmatrix} -1 \\ 2 \\ 4 \end{pmatrix}$ and $B = \begin{pmatrix} 5 \\ -1 \\ 13 \end{pmatrix}$ and having $U = \begin{pmatrix} 2 \\ -1 \\ 3 \end{pmatrix}$ as a direction vector.

In Exercises 9 to 12 determine whether there is exactly one, or more than one, plane satisfying the given conditions.

9. Containing $A = \begin{pmatrix} 2 \\ 1 \\ 2 \\ 0 \end{pmatrix}$, $B = \begin{pmatrix} 1 \\ 2 \\ 1 \\ 0 \end{pmatrix}$, $C = \begin{pmatrix} 3 \\ 1 \\ 1 \\ 2 \end{pmatrix}$.

10. Containing $A = \begin{pmatrix} 1 \\ 2 \\ 1 \end{pmatrix}$, $B = \begin{pmatrix} 15 \\ -5 \\ 22 \end{pmatrix}$, $C = \begin{pmatrix} -17 \\ 11 \\ -26 \end{pmatrix}$.

11. Containing $A = \begin{pmatrix} 3 \\ -1 \\ 5 \end{pmatrix}$, $B = \begin{pmatrix} 42 \\ -66 \\ 96 \end{pmatrix}$, and having direction vector $U = \begin{pmatrix} 3 \\ -5 \\ 7 \end{pmatrix}$.

12. Containing $A = \begin{pmatrix} 2 \\ -1 \\ 3 \\ 5 \end{pmatrix}$, $B = \begin{pmatrix} 9 \\ 20 \\ 31 \\ 53 \end{pmatrix}$, and having direction vector $U = \begin{pmatrix} 1 \\ 3 \\ 4 \\ 7 \end{pmatrix}$.

13. Prove Theorem 1.43. (Hint: Follow the pattern of proof for Theorem 1.23, p. 34.)

14. Prove Theorem 1.45.

15. Prove Theorem 1.46.

ANSWERS

1. Theorem 1.44, since $\{U, V\} = \left\{ \begin{pmatrix} 2 \\ 1 \\ 3 \end{pmatrix}, \begin{pmatrix} 1 \\ 2 \\ 1 \end{pmatrix} \right\}$ is a linearly independent set.

3. Theorem 1.45, since $B - A = \begin{pmatrix} 0 \\ -1 \\ 1 \end{pmatrix}$ and $C - A = \begin{pmatrix} 1 \\ -2 \\ -1 \end{pmatrix}$ are not multiples of one another. **6.** Theorem 1.44, since $V = B - A = \begin{pmatrix} -1 \\ -3 \\ 2 \\ -2 \end{pmatrix}$ and $U = \begin{pmatrix} 1 \\ 2 \\ 1 \\ 2 \end{pmatrix}$ form a linearly independent set $\{U, V\}$. **7.** The line $\mathcal{L}_1$ through A and B has direction vector $B - A = \begin{pmatrix} 1 \\ -2 \\ 2 \end{pmatrix}$. The line $\mathcal{L}_2$ through A and C has direction vector $C - A = \begin{pmatrix} 5 \\ -10 \\ 10 \end{pmatrix} = 5(B - A)$.

Therefore, $\mathcal{L}_1$ and $\mathcal{L}_2$ are parallel and hence identical, since they have the point A in common. Thus, any plane containing A and B will also contain C. There are at least two such planes, namely

$$\mathcal{S}_1 = \left\{ r\begin{pmatrix} 1 \\ 2 \\ 1 \end{pmatrix} + s\begin{pmatrix} 1 \\ -2 \\ 2 \end{pmatrix} \middle| r, s \in \mathcal{R} \right\} \text{ which contains } O_3 \text{ and}$$

$$\mathcal{S}_2 = \left\{ \begin{pmatrix} 1 \\ 2 \\ 1 \end{pmatrix} + r\begin{pmatrix} 1 \\ 0 \\ 0 \end{pmatrix} + s\begin{pmatrix} 1 \\ -2 \\ 2 \end{pmatrix} \middle| r, s \in \mathcal{R} \right\} \text{ which does not contain } O_3.$$

9. Exactly one plane. **11.** More than one plane.

10. PARALLELISM AND PLANES; PLANES IN $\mathfrak{R}^3$

In this last section of Chapter 1, we extend the concept of parallelism to planes in $\mathfrak{R}^n$ and then consider the special case of planes in $\mathfrak{R}^3$. The following is an extension of the basic concept of parallelism given in Definition 1.27 (Parallelism in $\mathfrak{R}^n$).

DEFINITION 1.49 (Parallelism and Planes in $\mathcal{R}^n$.) The term _parallel_ is applied to planes in the following ways:

   i. (Parallelism of a Vector and a Plane.) A vector U is said to be parallel to a plane $\mathcal{G}$ if and only if U is parallel to a direction vector for $\mathcal{G}$.

  ii. (Parallelism of a Line and a Plane.) A line $\mathcal{L}$ is said to be parallel to a plane $\mathcal{G}$ if and only if at least one direction vector for $\mathcal{L}$ is parallel to $\mathcal{G}$.

 iii. (Parallelism of Two Planes.) A plane $\mathcal{G}_1$ is said to be parallel to a plane $\mathcal{G}_2$ if and only if there is a direction set $\{U, V\}$ for $\mathcal{G}_1$ such that U and V are both parallel to $\mathcal{G}_2$.

The reader should verify the following easily established consequences of the preceding definition.

   i. _Parallelism of a Vector and a Plane._ A vector U is parallel to a plane $\mathcal{G}$ if and only if U is the zero vector or U is a direction vector for $\mathcal{G}$.

  ii. _Parallelism of a Line and a Plane._ A line $\mathcal{L}$ is parallel to a plane $\mathcal{G}$ if and only if some direction vector for one is also a direction vector for the other.

     A line $\mathcal{L}$ is parallel to a plane $\mathcal{G}$ if and only if every direction vector for $\mathcal{L}$ is also a direction vector for $\mathcal{G}$.

     If a line $\mathcal{L}$ is parallel to $\mathcal{G}$, then either $\mathcal{G} \cap \mathcal{L} = \varnothing$ or $\mathcal{L}$ is wholly contained in $\mathcal{G}$: $\mathcal{G} \cap \mathcal{L} = \mathcal{L}$.

 iii. _Parallelism of Two Planes._ Two planes $\mathcal{G}_1$ and $\mathcal{G}_2$ are parallel if and only if at least one direction set $\{U, V\}$ for one is also a direction set for the other.

     Two planes $\mathcal{G}_1$ and $\mathcal{G}_2$ are parallel if and only if every direction set for one plane is also a direction set for the other.

     If two planes, $\mathcal{G}_1$ and $\mathcal{G}_2$, are parallel, then either they have no points in common, $\mathcal{G}_1 \cap \mathcal{G}_2 = \varnothing$, or they are identical, $\mathcal{G}_1 = \mathcal{G}_2$.

Example 28. Determine the plane $\mathcal{G}$ which contains the points $A = \begin{pmatrix} 1 \\ 2 \\ 0 \end{pmatrix}$ and $B = \begin{pmatrix} 0 \\ 1 \\ 0 \end{pmatrix}$ and is parallel to $\mathcal{L} = \left\{ \begin{pmatrix} 1 \\ 2 \\ 3 \end{pmatrix} + t \begin{pmatrix} 2 \\ 0 \\ 1 \end{pmatrix} \middle| t \in \mathcal{R} \right\}$.

Since $\mathcal{G}$ is to be parallel to $\mathcal{L}$, we know that the direction vector $D = \begin{pmatrix} 2 \\ 0 \\ 1 \end{pmatrix}$ for $\mathcal{L}$ is also a direction vector for $\mathcal{G}$. A second direction vector V for $\mathcal{G}$ is given by the difference,

$$V = A - B = \begin{pmatrix} 1 \\ 2 \\ 0 \end{pmatrix} - \begin{pmatrix} 0 \\ 1 \\ 0 \end{pmatrix} = \begin{pmatrix} 1 \\ 1 \\ 0 \end{pmatrix}.$$

The set of direction vectors $\left\{ \begin{pmatrix} 2 \\ 0 \\ 1 \end{pmatrix}, \begin{pmatrix} 1 \\ 1 \\ 0 \end{pmatrix} \right\}$ is clearly a linearly independent

set and thus forms a direction set for $\mathcal{S}$. It follows that $\mathcal{S}$ is the plane through

$A = \begin{pmatrix} 1 \\ 2 \\ 0 \end{pmatrix}$ having direction set $\left\{ \begin{pmatrix} 2 \\ 0 \\ 1 \end{pmatrix}, \begin{pmatrix} 1 \\ 1 \\ 0 \end{pmatrix} \right\}$. Thus,

$$\mathcal{S} = \left\{ \begin{pmatrix} 1 \\ 2 \\ 0 \end{pmatrix} + r \begin{pmatrix} 2 \\ 0 \\ 1 \end{pmatrix} + s \begin{pmatrix} 1 \\ 1 \\ 0 \end{pmatrix} \;\middle|\; r, s \in \mathcal{R} \right\}. \quad \blacksquare$$

Example 29. Find the plane containing the line $\mathcal{L}_1 = \left\{ \begin{pmatrix} 2 \\ -1 \\ 1 \end{pmatrix} + t \begin{pmatrix} -1 \\ 2 \\ 1 \end{pmatrix} \;\middle|\; t \in \mathcal{R} \right\}$ and parallel to $\mathcal{L}_2 = \left\{ \begin{pmatrix} 0 \\ 0 \\ 1 \end{pmatrix} + t \begin{pmatrix} 4 \\ 2 \\ 0 \end{pmatrix} \;\middle|\; t \in \mathcal{R} \right\}$. Since $\mathcal{S}$ is

to contain $\mathcal{L}_1$, we may take $A = \begin{pmatrix} 2 \\ -1 \\ 1 \end{pmatrix}$ and $U = \begin{pmatrix} -1 \\ 2 \\ 1 \end{pmatrix}$ as one direction

vector for $\mathcal{S}$. As a second direction vector for $\mathcal{S}$ we consider the direction

vector $V = \begin{pmatrix} 4 \\ 2 \\ 0 \end{pmatrix}$ of $\mathcal{L}_2$. The set $\{U, V\}$ is a linearly independent set and

therefore forms a direction set for $\mathcal{S}$. We then have

$$\mathcal{S} = \left\{ \begin{pmatrix} 2 \\ -1 \\ 1 \end{pmatrix} + r \begin{pmatrix} -1 \\ 2 \\ 1 \end{pmatrix} + s \begin{pmatrix} 4 \\ 2 \\ 0 \end{pmatrix} \;\middle|\; r, s \in \mathcal{R} \right\}. \quad \blacksquare$$

Example 30. Find a plane $\mathcal{S}$ which contains the points $A = \begin{pmatrix} 2 \\ -1 \\ 1 \end{pmatrix}$,

$B = \begin{pmatrix} 3 \\ -4 \\ 1 \end{pmatrix}$, and is parallel to the vector $U = \begin{pmatrix} 1 \\ -3 \\ 0 \end{pmatrix}$. In view of the

comments following Definition 1.49, U will be a direction vector for $\mathcal{S}$.

The difference $D = A - B = \begin{pmatrix} -1 \\ 3 \\ 0 \end{pmatrix}$ is also a direction vector for $\mathcal{S}$.

However, since $D = -U$, the set $\{U, D\}$ is not a direction set for $\mathcal{S}$. This

situation calls attention to the fact that any plane through A and B will

have $\begin{pmatrix} 1 \\ -3 \\ 0 \end{pmatrix}$ as a direction vector. There are infinitely many planes $\mathcal{S}$

satisfying the conditions of the problem. A particular plane is determined by

selecting a second vector V such that $\{U, V\}$ is a linearly independent set.

Thus,

$$\mathcal{S} = \left\{ \left. \begin{pmatrix} 2 \\ -1 \\ 1 \end{pmatrix} + r \begin{pmatrix} 1 \\ -3 \\ 0 \end{pmatrix} + s \begin{pmatrix} 1 \\ 0 \\ 0 \end{pmatrix} \right| r, s \in \mathcal{R} \right\}$$

is that plane through A and B which is parallel to $\begin{pmatrix} 1 \\ 0 \\ 0 \end{pmatrix}$. ■

Theorem 1.47 (Intersection of Two Planes in $\mathcal{R}^n$) of the preceding section lists four types of sets in which two planes may intersect in a general $\mathcal{R}^n$. If we restrict our attention to $\mathcal{R}^3$, only three of these possibilities can occur. The concept of parallelism is useful in describing the various possibilities.

THEOREM 1.50   (Intersection of Two Planes in $\mathcal{R}^3$.) Let $\mathcal{S}_1$ and $\mathcal{S}_2$ be two planes in $\mathcal{R}^3$. Exactly one of the following statements is true

i. $\mathcal{S}_1$ and $\mathcal{S}_2$ are parallel and have no points in common:

$$\mathcal{S}_1 \cap \mathcal{S}_2 = \varnothing.$$

ii. $\mathcal{S}_1$ and $\mathcal{S}_2$ intersect in a line $\mathcal{L}$: $\mathcal{S}_1 \cap \mathcal{S}_2 = \mathcal{L}.$

iii. $\mathcal{S}_1$ and $\mathcal{S}_2$ are parallel and coincide: $\mathcal{S}_1 \cap \mathcal{S}_2 = \mathcal{S}_1 = \mathcal{S}_2.$

Figure 1.14 illustrates the possible cases in $\mathcal{R}^3$. The proof of this theorem uses the property of $\mathcal{R}^3$ given in Theorem 1.35 (p. 52), namely, the property that any vector in $\mathcal{R}^3$ can be expressed as a linear combination of the vectors of a linearly independent set of three vectors $\{U, V, W\}$. The reader will find it an interesting exercise to discover a proof at this point. The material developed in Chapter 2 will enable us to construct a very short proof, and to treat specific problems more efficiently than is possible here.

We close this chapter with the specialization of Theorem 1.48 to $\mathcal{R}^3$.

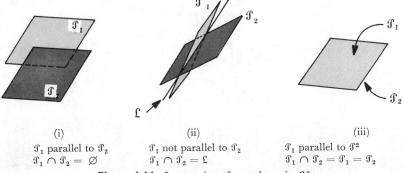

(i)   (ii)   (iii)

$\mathcal{S}_1$ parallel to $\mathcal{S}_2$ $\mathcal{S}_1$ not parallel to $\mathcal{S}_2$ $\mathcal{S}_1$ parallel to $\mathcal{S}^2$

$\mathcal{S}_1 \cap \mathcal{S}_2 = \varnothing$ $\mathcal{S}_1 \cap \mathcal{S}_2 = \mathcal{L}$ $\mathcal{S}_1 \cap \mathcal{S}_2 = \mathcal{S}_1 = \mathcal{S}_2$

Figure 1.14   Intersection of two planes in $\mathcal{R}^3$.

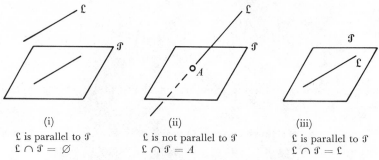

(i)   (ii)   (iii)

$\mathcal{L}$ is parallel to $\mathcal{G}$ $\mathcal{L}$ is not parallel to $\mathcal{G}$ $\mathcal{L}$ is parallel to $\mathcal{G}$

$\mathcal{L} \cap \mathcal{G} = \varnothing$ $\mathcal{L} \cap \mathcal{G} = A$ $\mathcal{L} \cap \mathcal{G} = \mathcal{L}$

Figure 1.15   Intersection of a plane and a line in $\mathcal{R}^3$.

THEOREM 1.51   (Intersection of a Line and a Plane in $\mathcal{R}^3$.)   Let $\mathcal{L}$ be a line and $\mathcal{G}$ be a plane, both in $\mathcal{R}^3$. Exactly one of the following is true:

   i. $\mathcal{L}$ and $\mathcal{G}$ are parallel and have no points in common:
$$\mathcal{L} \cap \mathcal{G} = \varnothing.$$
   ii. $\mathcal{L}$ and $\mathcal{G}$ intersect in a single point: $\mathcal{L} \cap \mathcal{G}$ = singleton set.
   iii. $\mathcal{L}$ and $\mathcal{G}$ are parallel and $\mathcal{L}$ is a subset of $\mathcal{G}$:
$$\mathcal{L} \cap \mathcal{G} = \mathcal{L}.$$

The proof of this theorem is also given in Chapter 2. Starting with Theorem 1.48 it is only necessary to show that if the line $\mathcal{L}$ is not parallel to the plane $\mathcal{G}$, then $\mathcal{L} \cap \mathcal{G}$ consists of exactly one point. It follows that $\mathcal{L}$ and $\mathcal{G}$ must be parallel in the two remaining cases. The reader may find it interesting to construct his own proof using concepts which we have developed in this chapter. Figure 1.15 illustrates the content of Theorem 1.51.

EXERCISES

In Exercises 1 to 7 determine at least one plane $\mathcal{G}$ satisfying the given conditions and find if there is more than one such plane.

1. $\mathcal{G}$ contains $A = \begin{pmatrix} 1 \\ 2 \\ -3 \end{pmatrix}$ and $B = \begin{pmatrix} 2 \\ -3 \\ -1 \end{pmatrix}$ and is parallel to $\mathcal{L} = \left\{ \begin{pmatrix} 2 \\ 1 \\ 1 \end{pmatrix} + t \begin{pmatrix} 1 \\ -1 \\ 7 \end{pmatrix} \middle| t \in \mathcal{R} \right\}.$

2. $\mathcal{G}$ contains $\mathcal{L}_1 = \left\{ t \begin{pmatrix} 1 \\ 3 \\ -4 \end{pmatrix} \middle| t \in \mathcal{R} \right\}$ and is parallel to $\mathcal{L}_2 = \left\{ \begin{pmatrix} 1 \\ 1 \\ 3 \end{pmatrix} + s \begin{pmatrix} -3 \\ 1 \\ 5 \end{pmatrix} \middle| s \in \mathcal{R} \right\}.$

3. $\mathcal{S}$ contains $A = \begin{pmatrix} 2 \\ 1 \\ 4 \\ 1 \end{pmatrix}$ and is parallel to $\mathcal{L}_1 = \left\{ \begin{pmatrix} 3 \\ 2 \\ 7 \\ 5 \end{pmatrix} + t \begin{pmatrix} 3 \\ 1 \\ 0 \\ 0 \end{pmatrix} \middle| t \in \mathcal{R} \right\}$ and

is parallel to $\mathcal{L}_2 = \left\{ s \begin{pmatrix} 4 \\ 1 \\ 2 \\ 0 \end{pmatrix} \middle| s \in \mathcal{R} \right\}$.

4. $\mathcal{S}$ contains $A = \begin{pmatrix} 1 \\ 3 \\ 4 \end{pmatrix}$ and is parallel to $\mathcal{S}_1 = \left\{ \begin{pmatrix} 1 \\ 2 \\ -1 \end{pmatrix} + r \begin{pmatrix} 2 \\ -1 \\ 3 \end{pmatrix} + s \begin{pmatrix} 7 \\ 2 \\ 5 \end{pmatrix} \middle| r, s \in \mathcal{R} \right\}$.

5. $\mathcal{S}$ contains $A = \begin{pmatrix} 9 \\ 15 \\ 3 \end{pmatrix}$ and $B = \begin{pmatrix} 2 \\ 1 \\ -4 \end{pmatrix}$ and is parallel to $\mathcal{L} = \left\{ \begin{pmatrix} 2 \\ 1 \\ 1 \end{pmatrix} + t \begin{pmatrix} 1 \\ 2 \\ 1 \end{pmatrix} \middle| t \in \mathcal{R} \right\}$.

6. $\mathcal{S}$ contains $A = \begin{pmatrix} -5 \\ 1 \\ 3 \end{pmatrix}$, is parallel to the line $\mathcal{L}_1$ through $B = \begin{pmatrix} 2 \\ 1 \\ -4 \end{pmatrix}$ and

$C = \begin{pmatrix} 3 \\ -1 \\ 4 \end{pmatrix}$, and is parallel to the line $\mathcal{L}_2$ through $D = \begin{pmatrix} 7 \\ -1 \\ 5 \end{pmatrix}$ and

$E = \begin{pmatrix} 2 \\ 1 \\ 4 \end{pmatrix}$.

7. $\mathcal{S}$ contains $\mathcal{L}_1 = \left\{ \begin{pmatrix} 1 \\ 0 \\ 2 \\ 1 \end{pmatrix} + t \begin{pmatrix} 1 \\ 2 \\ 1 \\ 3 \end{pmatrix} \middle| t \in \mathcal{R} \right\}$ and is parallel to the line through

$A = \begin{pmatrix} 2 \\ 1 \\ 4 \\ 3 \end{pmatrix}$ and $B = \begin{pmatrix} -1 \\ 5 \\ 2 \\ -7 \end{pmatrix}$.

8. Establish the seven assertions given in (i), (ii), and (iii) following Definition 1.49.

9. Show that there is exactly one plane that contains points A and B and is parallel to $\mathcal{L} = \{C + tD \mid t \in \mathcal{R}\}$ if $\{B - A, D\}$ is a linearly independent set.

10. Show that there is only one plane that contains a given point A and is parallel to a given plane $\mathcal{S} = \{B + rU + sV \mid r, s \in \mathcal{R}\}$.

11. Prove Theorem 1.50. (This is not a simple task at this point.)

12. Prove Theorem 1.51. (Not a routine exercise.)

ANSWERS

1. $\mathcal{S} = \left\{ \begin{pmatrix} 1 \\ 2 \\ -3 \end{pmatrix} + r\begin{pmatrix} 1 \\ -5 \\ 2 \end{pmatrix} + s\begin{pmatrix} 1 \\ -1 \\ 7 \end{pmatrix} \right\}$, exactly one plane.   **3.** $\mathcal{S} = \left\{ \begin{pmatrix} 2 \\ 1 \\ 4 \\ 1 \end{pmatrix} + \right.$

$\left. r\begin{pmatrix} 3 \\ 1 \\ 0 \\ 0 \end{pmatrix} + s\begin{pmatrix} 4 \\ 1 \\ 2 \\ 0 \end{pmatrix} \,\middle|\, r, \in s\,\mathcal{R} \right\}$, exactly one such plane by virtue of Theorem 1.44.

5. $\mathcal{S}_1 = \left\{ \begin{pmatrix} 2 \\ 1 \\ -4 \end{pmatrix} + r\begin{pmatrix} 1 \\ 2 \\ 1 \end{pmatrix} + s\begin{pmatrix} 1 \\ 0 \\ 0 \end{pmatrix} \,\middle|\, r, s \in \mathcal{R} \right\}$ is one plane.

$\mathcal{S}_2 = \left\{ \begin{pmatrix} 2 \\ 1 \\ -4 \end{pmatrix} + r\begin{pmatrix} 1 \\ 2 \\ 1 \end{pmatrix} + s\begin{pmatrix} 0 \\ 1 \\ 0 \end{pmatrix} \,\middle|\, r, s \in \mathcal{R} \right\}$ is another (different) plane.

7. $\mathcal{S} = \left\{ \begin{pmatrix} 5 \\ -2 \\ 5 \\ 14 \end{pmatrix} + r\begin{pmatrix} 1 \\ 2 \\ 1 \\ 3 \end{pmatrix} + s\begin{pmatrix} 3 \\ -4 \\ 2 \\ 10 \end{pmatrix} \,\middle|\, r, s \in \mathcal{R} \right\}$ is the only plane.

Your answer may be in a different form. See Theorem 1.43.

9. Use Theorem 1.44.   **11.** Hint: You will need Theorem 1.35 and a great deal of ingenuity.

2

INNER PRODUCT AND LENGTH IN $\mathfrak{R}^n$

INTRODUCTION

The concepts of ordered n-tuples and of the vector spaces $\mathfrak{R}^n$ (n any positive integer) were developed in Sections 1 and 2 of the preceding chapter. In those two sections we employed a very informal approach to give these concepts a geometrical interpretation and used these interpretations to suggest the definitions of the vector operations.

Beginning with Section 3 of Chapter 1, the point of view was reversed. We began to formalize geometry in terms of the vector space concepts. We were able to define what is meant by a point, by a line, by a plane, by parallel lines, by parallel planes, and so forth. There is a great deal of elementary geometry in Chapter 1. You will search in vain, however, for any formal discussion of the concept of perpendicularity or of the concept of distance. These two concepts are the principal concern of the present chapter. We shall find that the introduction of a single additional operation into a vector space—*the inner product*—will make it possible to define both perpendicularity and length in an acceptable manner.

1. THE INNER PRODUCT AND ITS PROPERTIES

We begin this section with the definition of an inner product of two vectors.

DEFINITION 2.1 (Inner Product of Two Vectors.)

i. (In $\mathcal{R}^2$.) Let $X = \begin{pmatrix} a \\ b \end{pmatrix}$ and $Y = \begin{pmatrix} c \\ d \end{pmatrix}$ be elements of $\mathcal{R}^2$. The *inner product* of X and Y, denoted by $X \cdot Y$, is defined by

$$X \cdot Y = \begin{pmatrix} a \\ b \end{pmatrix} \cdot \begin{pmatrix} c \\ d \end{pmatrix} = ac + bd.$$

ii. (In $\mathcal{R}^3$.) Let $X = \begin{pmatrix} a \\ b \\ c \end{pmatrix}$ and $Y = \begin{pmatrix} d \\ e \\ f \end{pmatrix}$ be elements of $\mathcal{R}^3$. The *inner product* of X and Y, denoted by $X \cdot Y$, is defined by

$$X \cdot Y = \begin{pmatrix} a \\ b \\ c \end{pmatrix} \cdot \begin{pmatrix} d \\ e \\ f \end{pmatrix} = ad + be + cf.$$

iii. (In $\mathcal{R}^n$.) Let $X = \begin{pmatrix} x_1 \\ x_2 \\ \cdot \\ \cdot \\ \cdot \\ x_n \end{pmatrix}$ and $Y = \begin{pmatrix} y_1 \\ y_2 \\ \cdot \\ \cdot \\ \cdot \\ y_n \end{pmatrix}$ be elements of $\mathcal{R}^n$. The

inner product of X and Y, denoted by $X \cdot Y$, is defined by

$$X \cdot Y = \begin{pmatrix} x_1 \\ x_2 \\ \cdot \\ \cdot \\ \cdot \\ x_n \end{pmatrix} \cdot \begin{pmatrix} y_1 \\ y_2 \\ \cdot \\ \cdot \\ \cdot \\ y_n \end{pmatrix} = x_1 y_1 + x_2 y_2 + \cdots + x_n y_n.$$

The definition of an inner product of two n-vectors is often stated verbally as: "*The inner product of two n-vectors is the sum of the products of corresponding coordinates of the two vectors.*" This verbalization and the following examples will help to fix this concept in your mind.

Example 1

(a) Let $X = \begin{pmatrix} 1 \\ 2 \end{pmatrix}$, $Y = \begin{pmatrix} 3 \\ 4 \end{pmatrix}$. The inner product $X \cdot Y$ is

$$X \cdot Y = \begin{pmatrix} 1 \\ 2 \end{pmatrix} \cdot \begin{pmatrix} 3 \\ 4 \end{pmatrix} = 1 \cdot 3 + 2 \cdot 4 = 3 + 8 = 11.$$

The inner product is obtained by determining the product of corresponding coordinates—the product $1 \cdot 3$ of the first coordinates and the product $2 \cdot 4$ of the second coordinates—and then determining the sum of these "products of corresponding coordinates."

(b) Let $X = \begin{pmatrix} -1 \\ 1 \end{pmatrix}$, $Y = \begin{pmatrix} 2 \\ 2 \end{pmatrix}$; the inner product is

$$X \cdot Y = \begin{pmatrix} -1 \\ 1 \end{pmatrix} \cdot \begin{pmatrix} 2 \\ 2 \end{pmatrix} = (-1) \cdot 2 + 1 \cdot 2 = -2 + 2 = 0.$$

(c) Let $X = \begin{pmatrix} 1 \\ 0 \\ 1 \end{pmatrix}$, $Y = \begin{pmatrix} 1 \\ 1 \\ 0 \end{pmatrix}$; the inner product is

$$X \cdot Y = \begin{pmatrix} 1 \\ 0 \\ 1 \end{pmatrix} \cdot \begin{pmatrix} 1 \\ 1 \\ 0 \end{pmatrix} = 1 \cdot 1 + 0 \cdot 1 + 1 \cdot 0 = 1 + 0 + 0 = 1.$$

(d) Let $X = \begin{pmatrix} 1 \\ -2 \\ 1 \\ 2 \end{pmatrix}$, $Y = \begin{pmatrix} 2 \\ 3 \\ 0 \\ 0 \end{pmatrix}$; the inner product is

$$X \cdot Y = \begin{pmatrix} 1 \\ -2 \\ 1 \\ 2 \end{pmatrix} \cdot \begin{pmatrix} 2 \\ 3 \\ 0 \\ 0 \end{pmatrix} = 1 \cdot 2 + (-2) \cdot 3 + 1 \cdot 0 + 2 \cdot 0$$
$$= 2 - 6 + 0 + 0 = -4. \quad \blacksquare$$

It is now appropriate to determine the properties which the inner product possesses. Note should be taken that for $n \geq 2$, the inner product of two n-vectors is a scalar and not another n-vector. It is not possible, therefore, to form the "inner product of three vectors." Thus, the inner product will not possess an associative property in $\mathcal{R}^n$ when $n \geq 2$.

THEOREM 2.2 (Properties of the Inner Product.) Let X, Y, and Z be elements of $\mathcal{R}^n$ and let r and s be real numbers.
  i. $X \cdot X \geq 0$.
  ii. $X \cdot X = 0$ if and only if $X = O_n$.
  iii. $X \cdot Y = Y \cdot X$.
  iv. $(rX) \cdot Y = X \cdot (rY) = r(X \cdot Y)$.
  v. $(X + Y) \cdot Z = (X \cdot Z) + (Y \cdot Z)$.

Proof. We give the proof only for the case $n = 2$. The proof for other values of n is essentially the same. In each of the following we set $X = \begin{pmatrix} a \\ b \end{pmatrix}$, $Y = \begin{pmatrix} c \\ d \end{pmatrix}$, and $Z = \begin{pmatrix} e \\ f \end{pmatrix}$:

Proof of i

$$X \cdot X =^1 \begin{pmatrix} a \\ b \end{pmatrix} \cdot \begin{pmatrix} a \\ b \end{pmatrix} =^2 a \cdot a + b \cdot b$$
$$=^3 a^2 + b^2$$
$$\geq^3 0.$$

Proof of ii

$$X \cdot X = 0 \Rightarrow^1 \begin{pmatrix} a \\ b \end{pmatrix} \cdot \begin{pmatrix} a \\ b \end{pmatrix} =^2 a^2 + b^2 = 0$$
$$\Rightarrow^3 a^2 = b^2 = 0$$
$$\Rightarrow^3 a = b = 0$$
$$\Rightarrow^1 X = \begin{pmatrix} a \\ b \end{pmatrix} = \begin{pmatrix} 0 \\ 0 \end{pmatrix} = O_2.$$

Conversely,

$$X = O_2 \Rightarrow X = \begin{pmatrix} 0 \\ 0 \end{pmatrix}$$
$$\Rightarrow^2 X \cdot X = \begin{pmatrix} 0 \\ 0 \end{pmatrix} \cdot \begin{pmatrix} 0 \\ 0 \end{pmatrix} = 0.$$

Proof of iii

$$X \cdot Y =^1 \begin{pmatrix} a \\ b \end{pmatrix} \cdot \begin{pmatrix} c \\ d \end{pmatrix} =^2 ac + bd$$
$$=^3 ca + db$$
$$=^2 \begin{pmatrix} c \\ d \end{pmatrix} \cdot \begin{pmatrix} a \\ b \end{pmatrix}$$
$$=^1 Y \cdot X.$$

Proof of iv

$$(rX) \cdot Y =^1 \left[r\begin{pmatrix} a \\ b \end{pmatrix} \right] \cdot \begin{pmatrix} c \\ d \end{pmatrix}$$
$$=^4 \begin{pmatrix} ra \\ rb \end{pmatrix} \cdot \begin{pmatrix} c \\ d \end{pmatrix}$$
$$=^2 (ra)c + (rb)d$$
$$=^3 a(rc) + b(rd) =^1 \begin{pmatrix} a \\ b \end{pmatrix} \cdot \begin{pmatrix} rc \\ rd \end{pmatrix} =^1 X \cdot (rY)$$
$$=^3 r(ac + bd)$$
$$=^2 r\left[\begin{pmatrix} a \\ b \end{pmatrix} \cdot \begin{pmatrix} c \\ d \end{pmatrix} \right] =^1 r(X \cdot Y).$$

Proof of v

$$(X + Y) \cdot Z =^1 \left[\binom{a}{b} + \binom{c}{d} \right] \cdot \binom{e}{f}$$

$$=^5 \binom{a+c}{b+d} \cdot \binom{e}{f}$$

$$=^2 (a+c)e + (b+d)f$$

$$=^3 ae + ce + bf + df$$

$$=^3 (ae + bf) + (ce + df)$$

$$=^2 \binom{a}{b} \cdot \binom{e}{f} + \binom{c}{d} \cdot \binom{e}{f}$$

$$=^1 X \cdot Z + Y \cdot Z.$$

(1) Substitution.
(2) Definition 2.1 (Inner Product).
(3) Properties of real numbers.
(4) Definition 1.5i (Multiples of Ordered Pairs).
(5) Definition 1.5ii (Sum of Ordered Pairs). ∎

Example 2. Compute the inner product of $X = 2\binom{1}{3} + 4\binom{0}{2}$ with $Y = \binom{-1}{1}$. We may proceed in either of two ways. Thus,

$$X \cdot Y = \left[2\binom{1}{3} + 4\binom{0}{2} \right] \cdot \binom{-1}{1}$$

$$=^1 \left[2\binom{1}{3} \right] \cdot \binom{-1}{1} + \left[4\binom{0}{2} \right] \cdot \binom{-1}{1}$$

$$=^1 2[1 \cdot (-1) + 3 \cdot 1] + 4[0 \cdot (-1) + 2 \cdot 1]$$

$$= 2(-1 + 3) + 4(0 + 2)$$

$$= 2 \cdot 2 + 4 \cdot 2$$

$$= 4 + 8$$

$$= 12.$$

or

$$X \cdot Y = \left[2\binom{1}{3} + 4\binom{0}{2} \right] \cdot \binom{-1}{1}$$

$$= \left[\binom{2}{6} + \binom{0}{8} \right] \cdot \binom{-1}{1}$$

$$= \binom{2}{14} \cdot \binom{-1}{1}$$

$$= 2 \cdot (-1) + 14 \cdot 1$$

$$= -2 + 14$$

$$= 12.$$

(1) Theorem 2.2v.
(2) Theorem 2.2iv. ∎

There is one further property of the inner product which we present here because of its relevance to the concepts of length and distance. These two concepts are developed in later sections.

THEOREM 2.3 (Cauchy-Schwartz Inequality.) Let X and Y be elements of $\mathcal{R}^n$. The inner product $X \cdot Y$ is such that

i. $(X \cdot Y)^2 \leq (X \cdot X)(Y \cdot Y)$. Moreover,

ii. $(X \cdot Y)^2 = (X \cdot X)(Y \cdot Y)$, if X and Y are parallel, and

iii. $(X \cdot Y)^2 < (X \cdot X)(Y \cdot Y)$, if X and Y are not parallel.

Proof. The proof given here is one which has no intuitive base. It is the product of the mathematician's search to develop short, easily followed proofs wherever possible. It is this aspect of mathematics which is properly referred to as the *art of mathematics*.

Proof of i. Let r be any real number and consider the vector $X - rY$. We have

$$0 \leq^1 (X - rY) \cdot (X - rY)$$
$$=^2 X \cdot (X - rY) - (rY) \cdot (X - rY)$$
$$=^3 X \cdot X - r(X \cdot Y) - r(X - rY) \cdot Y$$
$$=^2 X \cdot X - r(X \cdot Y) - r(X \cdot Y) + r^2(Y \cdot Y).$$

(1) Theorem 2.2i.
(2) Theorem 2.2v.
(3) Theorem 2.2iii, iv, and v.

Thus, for any real number r we have

(2.1) $$0 \leq X \cdot X - 2r(X \cdot Y) + r^2(Y \cdot Y).$$

Consider now the case in which $Y \neq O_n$ (or, equivalently, $Y \cdot Y \neq 0$). For such a Y the quotient $\dfrac{X \cdot Y}{Y \cdot Y}$ defines a real number and Inequality 2.1 must hold for $r = \dfrac{X \cdot Y}{Y \cdot Y}$. Substitution of this value of r into Inequality 2.1

$$0 \leq X \cdot X - \frac{2(X \cdot Y)(X \cdot Y)}{Y \cdot Y} + \frac{(X \cdot Y)^2}{(Y \cdot Y)^2}(Y \cdot Y)$$
$$= X \cdot X - 2\frac{(X \cdot Y)^2}{Y \cdot Y} + \frac{(X \cdot Y)^2}{Y \cdot Y}.$$

Therefore, for $Y \neq O_n$ we have $0 \leq X \cdot X - \dfrac{(X \cdot Y)^2}{Y \cdot Y}$, which, upon multiplication by $Y \cdot Y$, yields $0 \leq (X \cdot X)(Y \cdot Y) - (X \cdot Y)^2$ or $(X \cdot Y)^2 \leq (X \cdot X)(Y \cdot Y)$.

In Exercise 10 the reader is asked to consider the case $Y = O_n$ and to show that equality holds.

Proof of ii. The proof of statement ii is straightforward and is called for in Exercise 11.

Proof of iii. Statement iii can be established by proving the following (contrapositive) statement.

(2.2)   If $(X \cdot Y)^2 = (X \cdot X)(Y \cdot Y)$, then X is parallel to Y.

From statement 2.2 it follows at once that if X and Y are not parallel, the equality of statement i cannot hold and that statement iii must be valid.

To prove statement 2.2 we first note that if the equality holds and $Y = O_n$, then X and Y are certainly parallel. If, on the other hand, $Y \neq O_n$ we make use of portions of the proof of statement i. We showed that for every real number r

$$(X - rY) \cdot (X - rY) = X \cdot X - 2r(X \cdot Y) + r^2(Y \cdot Y).$$

Since $Y \neq O_n$, we may choose $r = \dfrac{X \cdot Y}{Y \cdot Y}$ to obtain

(2.3) $$(X - rY) \cdot (X - rY) = X \cdot X - \frac{(X \cdot Y)^2}{Y \cdot Y}.$$

Therefore, if $(X \cdot Y)^2 = (X \cdot X)(Y \cdot Y)$, substitution into the right member of Equation 2.3 shows that

$$(X - rY) \cdot (X - rY) = 0.$$

Using Theorem 2.2ii, with X replaced by $X - rY$, we conclude that $X - rY = O_n$ or, equivalently, that $X = rY$. Thus X and Y are parallel. Statement 2.2 and hence statement iii are established. ∎

EXERCISES

In Exercises 1 to 8 determine the indicated inner products:

1. $\begin{pmatrix} 3 \\ 4 \end{pmatrix} \cdot \begin{pmatrix} -1 \\ 2 \end{pmatrix}.$
    2. $\begin{pmatrix} 1 \\ 0 \\ 7 \end{pmatrix} \cdot \begin{pmatrix} -2 \\ -1 \\ 3 \end{pmatrix}.$
    3. $\begin{pmatrix} 4 \\ 3 \\ 2 \\ 1 \end{pmatrix} \cdot \begin{pmatrix} 0 \\ 7 \\ 0 \\ 11 \end{pmatrix}.$

4. $\left[\begin{pmatrix} 3 \\ 5 \end{pmatrix} + \begin{pmatrix} 2 \\ 7 \end{pmatrix} \right] \cdot \begin{pmatrix} -3 \\ -3 \end{pmatrix}.$
    5. $\left[\begin{pmatrix} 2 \\ 1 \\ 2 \end{pmatrix} + \begin{pmatrix} 6 \\ 0 \\ 1 \end{pmatrix} \right] \cdot \begin{pmatrix} 1 \\ -1 \\ 3 \end{pmatrix}.$

6. $\left[\begin{pmatrix} 2 \\ 1 \\ -1 \\ 4 \end{pmatrix} + \begin{pmatrix} 7 \\ 8 \\ -2 \\ 0 \end{pmatrix} \right] \cdot \begin{pmatrix} 7 \\ -2 \\ 1 \\ 3 \end{pmatrix}.$
    7. $\left[2 \begin{pmatrix} -3 \\ 1 \end{pmatrix} + 7 \begin{pmatrix} 2 \\ 1 \end{pmatrix} \right] \cdot \begin{pmatrix} -9 \\ 8 \end{pmatrix}.$

8. $\left[-2 \begin{pmatrix} 3 \\ -1 \\ 0 \end{pmatrix} + 3 \begin{pmatrix} 5 \\ -2 \\ -2 \end{pmatrix} \right] \cdot \begin{pmatrix} -1 \\ -5 \\ 3 \end{pmatrix}.$

9. Use Theorem 2.2 to show that $(rX + sY) \cdot Z = (rX) \cdot Z + (sY) \cdot Z = r(X \cdot Z) + s(Y \cdot Z)$.

10. Let X and Y be elements of $\mathcal{R}^n$. Let $Y = O_n$. Show that
$$(X \cdot Y)^2 \leq (X \cdot X)(Y \cdot Y)$$
by showing that in fact
$$(X \cdot Y)^2 = (X \cdot X)(Y \cdot Y).$$

11. Let X and Y be parallel vectors in $\mathcal{R}^n$. Show that
$$(X \cdot Y)^2 = (X \cdot X)(Y \cdot Y).$$

ANSWERS

1. 5.   **3.** 32.   **5.** 16.   **7.** 0.

2. PERPENDICULARITY IN $\mathcal{R}^n$

The inner product of two vectors X and Y will, of course, vary with the vectors X and Y. Thus, $\begin{pmatrix} 1 \\ 1 \end{pmatrix} \cdot \begin{pmatrix} -2 \\ 1 \end{pmatrix} = -1$ is negative, whereas $\begin{pmatrix} 2 \\ 3 \end{pmatrix} \cdot \begin{pmatrix} 0 \\ 1 \end{pmatrix} = 3$ is positive. For other choices of X and Y the inner product will be zero. For example, $\begin{pmatrix} 1 \\ 0 \end{pmatrix} \cdot \begin{pmatrix} 0 \\ 1 \end{pmatrix} = 0$ and $\begin{pmatrix} 1 \\ 2 \end{pmatrix} \cdot \begin{pmatrix} -2 \\ 1 \end{pmatrix} = 0$. The reader should take special note of the fact just illustrated, namely, the inner product of two vectors X and Y may be zero with neither X nor Y being the zero vector.

We now give a geometric meaning to a zero inner product. In the discussion which follows we use the terms "rectangular" and "perpendicular" in an intuitive and informal manner. Consider a "rectangular" coordinate system for the plane, as illustrated in Figure 2.1. A vector X in $\mathcal{R}^2$ represented by a point on the 1-axis is of the form $X = \begin{pmatrix} x \\ 0 \end{pmatrix}$. A vector Y represented by a point on the 2-axis is of the form $Y = \begin{pmatrix} 0 \\ y \end{pmatrix}$. Clearly,

$$X \cdot Y = \begin{pmatrix} x \\ 0 \end{pmatrix} \cdot \begin{pmatrix} 0 \\ y \end{pmatrix} = x \cdot 0 + 0 \cdot y = 0.$$

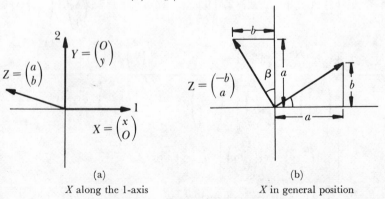

(a)   (b)

X along the 1-axis X in general position

Figure 2.1   Perpendicular vectors.

Thus, the inner product of a vector X along the 1-axis with a vector Y along the 2-axis is 0. Consider now a vector $Z = \begin{pmatrix} a \\ b \end{pmatrix}$ with $a \neq 0$. Z is represented by a point in $\mathfrak{R}^2$ which is not on the 2-axis. The inner product of X with Z is given by

$$X \cdot Z = \begin{pmatrix} x \\ 0 \end{pmatrix} \cdot \begin{pmatrix} a \\ b \end{pmatrix} = x \cdot a + 0 \cdot b = ax.$$

Since $a \neq 0$, the inner product $X \cdot Z$ is not zero unless $x = 0$, that is, unless $X = \begin{pmatrix} x \\ 0 \end{pmatrix} = \begin{pmatrix} 0 \\ 0 \end{pmatrix}$ is itself the zero vector O_2. Thus, if X denotes a nonzero vector on the 1-axis and Z denotes any other vector, then

$$X \cdot Z = 0 \text{ if } Z \text{ is a vector on the 2-axis, and}$$
$$X \cdot Z \neq 0 \text{ if } Z \text{ is not a vector on the 2-axis.}$$

In other words, the only vectors Z which have zero inner product with X are the vectors $Z = Y$ on the 2-axis, that is, those which are "perpendicular" to X. This fact suggests that two nonzero vectors X and Y will have zero inner product if and only if the directed line segments representing X and Y are "perpendicular."

Let us now consider a point $X = \begin{pmatrix} a \\ b \end{pmatrix}$ with $a \neq 0$ and $b \neq 0$. Such a choice for X is represented by a point in $\mathfrak{R}^2$ not on either coordinate axis, as shown in Figure 2.1b. Let N be the vector $N = \begin{pmatrix} -b \\ a \end{pmatrix}$. It is evident from the figure that angles α and β are "equal" and hence that N is "perpendicular" to X. Moreover, it should be intuitively obvious that all vectors Z which are "perpendicular" to X are multiples of N:

$$Z \text{ is "perpendicular" to } X \text{ if and only if } Z = kN.$$

The inner product of X with N is

$$X \cdot N = \begin{pmatrix} a \\ b \end{pmatrix} \cdot \begin{pmatrix} -b \\ a \end{pmatrix} = a \cdot (-b) + a \cdot b = -(ab) + ab = 0.$$

Therefore, for any $Z = kN$,

$$X \cdot Z = X \cdot (kN) = k(X \cdot N) = k \cdot 0 = 0.$$

Combining this result with that of the preceding paragraph, we may state: If X is any nonzero vector in $\mathfrak{R}^2$ and Z is any vector represented by a line segment "perpendicular" to that representing X, then the inner product of X with Z is zero. We could go further and show (on the informal level used above) that the only vectors Z which have a zero inner product with $X = \begin{pmatrix} a \\ b \end{pmatrix} \neq O_2$ are the vectors $Z = kN$ which are perpendicular to X.

Thus, we have

$X \cdot Z = 0$ if Z is "perpendicular" to X, and

$X \cdot Z \neq 0$ if Z is "not perpendicular to X."

The above discussion motivates the following formal definition of the concept of perpendicularity in $\Re^n$.

DEFINITION 2.4 (Perpendicular Vectors.) Let X and Y be elements of $\Re^n$. The vectors X and Y are said to be *perpendicular* to one another if and only if $X \cdot Y = 0$.

Vectors which are perpendicular to one another are also said to be *orthogonal* to one another.

It is well to note that as a consequence of this definition there is one vector in $\Re^n$ which is perpendicular to every vector. This vector is the zero vector O_n. This circumstance may seem strange at first, but the reader will feel quite at ease with it as time goes on. Of course O_n is the only such vector, as the following theorem shows.

THEOREM 2.5 Let X be an element of $\Re^n$. X is perpendicular to every element of $\Re^n$ if and only if $X = O_n$.

Proof. If $X = O_n$ and $Y = \begin{pmatrix} a \\ b \end{pmatrix}$ is any vector of $\Re^n$, it is clear from Definition 2.1 (Inner Product) that $X \cdot Y = \begin{pmatrix} 0 \\ 0 \end{pmatrix} \cdot \begin{pmatrix} a \\ b \end{pmatrix} = 0$.

Conversely, suppose that $X \cdot Y = 0$ for every vector Y in $\Re^n$. We must then have, in particular, $X \cdot X = 0$. Employing Theorem 2.2ii (Properties of Inner Product), we conclude that $X = O_n$. ∎

Example 3. Consider the vectors $X = \begin{pmatrix} x \\ 0 \\ 0 \end{pmatrix}$, $Y = \begin{pmatrix} 0 \\ y \\ 0 \end{pmatrix}$, and $Z = \begin{pmatrix} 0 \\ 0 \\ z \end{pmatrix}$ in $\Re^3$. It is clear that $X \cdot Y = 0$, $X \cdot Z = 0$, and $Y \cdot Z = 0$. Therefore, each of the vectors of the set $\{X, Y, Z\}$ is perpendicular to each of the others in the set. Figure 2.2 is a graphical illustration of this state of affairs. ∎

DEFINITION 2.6 (Orthogonal Set.) A set S of vectors in $\Re^n$ is said to be an *orthogonal set* if and only if each vector in S is orthogonal (perpendicular) to every other vector in S.

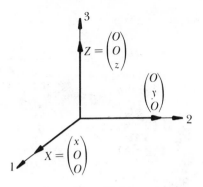

Figure 2.2

Example 4. Given one nonzero vector X in $\mathcal{R}^2$, it is quite easy to find a second nonzero vector Y in $\mathcal{R}^2$ which is perpendicular to X. If $X = \begin{pmatrix} a \\ b \end{pmatrix}$, then the vector $Y = \begin{pmatrix} -b \\ a \end{pmatrix}$ is clearly such that

$$X \cdot Y = \begin{pmatrix} a \\ b \end{pmatrix} \cdot \begin{pmatrix} -b \\ a \end{pmatrix} = 0.$$

In particular, if $X = \begin{pmatrix} 1 \\ 2 \end{pmatrix}$, then $Y = \begin{pmatrix} -2 \\ 1 \end{pmatrix}$ is perpendicular to X.

A similar technique can also be used in $\mathcal{R}^3$. Thus, if $X = \begin{pmatrix} a \\ b \\ c \end{pmatrix}$, then $Y = \begin{pmatrix} -b \\ a \\ 0 \end{pmatrix}$ is perpendicular to X, since

$$X \cdot Y = \begin{pmatrix} a \\ b \\ c \end{pmatrix} \cdot \begin{pmatrix} -b \\ a \\ 0 \end{pmatrix} = a \cdot (-b) + b \cdot a + c \cdot 0 = 0. \quad \blacksquare$$

The following theorem involving inner products will be of use later in the chapter.

THEOREM 2.7 Let N be a nonzero vector in $\mathcal{R}^n$ $(n > 1)$ and let k be an element of $\mathcal{R}$. There is a nonzero vector X in $\mathcal{R}^n$ such that

$$N \cdot X = k.$$

Proof. First assume that $k \neq 0$. Let N be a nonzero vector and apply Theorem 2.2ii to obtain $N \cdot N = v$, where $v \neq 0$. Choose X to be the vector $X = \dfrac{k}{v} N$. Since $k \neq 0$ and $N \neq O_n$ we know that $X \neq O_n$. Moreover,

$$N \cdot X = N \cdot \left(\frac{k}{v} N \right) = \frac{k}{v} (N \cdot N) = \frac{k}{v} v = k.$$

If $k = 0$, we require a nonzero vector X which is perpendicular to N, since we have

$$N \cdot X = k = 0.$$

Such a vector X can be constructed by means of the technique of Example 4 provided only that $n > 1$. ∎ *illustrate in an ex for proof*

The preceding theorem assures us that given any nonzero vector N in $\mathcal{R}^n$ ($n > 1$) there is always a vector X such that $N \cdot X$ takes on any pre-selected value k. *$N \cdot X = k = 0$*

$\begin{pmatrix} 2 \\ 3 \\ 4 \end{pmatrix} \times \begin{pmatrix} 3 \\ -2 \\ 0 \end{pmatrix}$

EXERCISES

In Exercises 1 to 5 determine whether S is an orthogonal set or not:

1. $S = \left\{ \begin{pmatrix} 1 \\ 2 \end{pmatrix}, \begin{pmatrix} -2 \\ 1 \end{pmatrix} \right\}.$

2. $S = \left\{ \begin{pmatrix} 2 \\ 1 \\ 0 \end{pmatrix}, \begin{pmatrix} -1 \\ 2 \\ 4 \end{pmatrix}, \begin{pmatrix} 4 \\ -8 \\ 5 \end{pmatrix} \right\}.$

3. $S = \left\{ \begin{pmatrix} 3 \\ 4 \end{pmatrix}, \begin{pmatrix} -8 \\ 6 \end{pmatrix}, \begin{pmatrix} -4 \\ 3 \end{pmatrix} \right\}.$

4. $S = \left\{ \begin{pmatrix} 1 \\ 2 \\ 1 \end{pmatrix}, \begin{pmatrix} 5 \\ -1 \\ -3 \end{pmatrix}, \begin{pmatrix} 2 \\ 13 \\ -1 \end{pmatrix} \right\}.$

5. $S = \left\{ \begin{pmatrix} 2 \\ -1 \\ 6 \\ 3 \end{pmatrix}, \begin{pmatrix} -6 \\ 3 \\ 2 \\ 1 \end{pmatrix}, \begin{pmatrix} 0 \\ 0 \\ 10 \\ -20 \end{pmatrix} \right\}.$

In Exercises 6 to 11 determine a nonzero vector Y which is perpendicular to the given vector X:

6. $X = \begin{pmatrix} 4 \\ 7 \end{pmatrix}.$

7. $X = \begin{pmatrix} 4 \\ -9 \\ 11 \end{pmatrix}.$

8. $X = \begin{pmatrix} 2 \\ 1 \\ 9 \\ -3 \end{pmatrix}.$

9. $X = \begin{pmatrix} 2 \\ 1 \\ 0 \end{pmatrix}.$

10. $X = \begin{pmatrix} 0 \\ 4 \\ 5 \\ 0 \end{pmatrix}.$

11. $X = \begin{pmatrix} 0 \\ -7 \\ 0 \end{pmatrix}.$

12. (a) Let $\{U, V\}$ be an orthogonal set such that $U \cdot U = V \cdot V = 1$. Let $Z = aU + bV$. Show that $Z \cdot Z = a^2 + b^2$.

    (b) Let $\{U, V, W\}$ be an orthogonal set such that $U \cdot U = V \cdot V = W \cdot W = 1$. Let $Z = aU + bV + cW$. Show that $Z \cdot Z = a^2 + b^2 + c^2$.

13. Let $D = \begin{pmatrix} a \\ b \end{pmatrix}$ be a nonzero vector in $\mathcal{R}^2$ and let N be such that $D \cdot N = 0$. Show that $N = k \begin{pmatrix} -b \\ a \end{pmatrix}$ for some real number k. $\left(\text{Hint: Let } N = \begin{pmatrix} x \\ y \end{pmatrix} \right.$ and determine all solutions $\begin{pmatrix} x \\ y \end{pmatrix}$ of the equation

$$D \cdot N = \begin{pmatrix} a \\ b \end{pmatrix} \cdot \begin{pmatrix} x \\ y \end{pmatrix} = ax + by = 0. \Big)$$

ANSWERS

1. $\mathcal{S}$ is an orthogonal set.   **3.** $\mathcal{S}$ is not an orthogonal set. $\begin{pmatrix} -8 \\ 6 \end{pmatrix} \cdot \begin{pmatrix} -4 \\ 3 \end{pmatrix} \neq 0.$

5. $\mathcal{S}$ is an orthogonal set.   **7.** $Y = \begin{pmatrix} 9 \\ 4 \\ 0 \end{pmatrix}$ is one possibility; there are many

others.   **9.** $Y = \begin{pmatrix} 0 \\ 0 \\ 1 \end{pmatrix}$ is one possibility.   **11.** $\begin{pmatrix} 1 \\ 0 \\ 0 \end{pmatrix}, \begin{pmatrix} 0 \\ 0 \\ 24 \end{pmatrix}, \begin{pmatrix} -17 \\ 0 \\ 39 \end{pmatrix}$ are

possible choices for Y.

13. $[ax + by = 0] \Rightarrow [ax = -by] \Rightarrow \left[x = \left(-\dfrac{b}{a} \right) y \text{ or } y = \left(-\dfrac{a}{b} \right) x \right]$ will get you started.

3. CHARACTERIZATIONS OF LINES IN $\mathscr{R}^2$

With the concept of perpendicularity we are able to give other descriptions (characterizations) of lines in $\mathscr{R}^2$. The following result will be helpful.

THEOREM 2.8   Let U and V be nonzero perpendicular vectors in $\mathscr{R}^2$:

  i. The set $\{U, V\}$ is a linearly independent set.
 ii. Each vector X in $\mathscr{R}^2$ is a linear combination of U and V: *by preceding theorem*

$$X = rU + sV, \text{ for some } r, s \in \mathscr{R}.$$

Proof. We shall use Theorem 1.30ii (p. 45) to show that the set $\{U, V\}$ is linearly independent. That is, we shall show that the only real numbers r and s which are such that

(2.4) $rU + sV = O_2$

are given by: $r = s = 0$.

Let r, s be any pair of real numbers for which Equation 2.4 is valid. Taking the inner product of both members of this equation with U we obtain

$$
\begin{aligned}
0 =^1 O_2 \cdot U &= (rU + sV) \cdot U \\
&=^2 r(U \cdot U) + s(V \cdot U) \\
&=^3 r(U \cdot U) + s \cdot 0 \\
&= r(U \cdot U),
\end{aligned}
$$

or

$$r(U \cdot U) = 0.$$

(1) Theorem 2.5 (p. 92).
(2) Theorem 2.2iv and v.
(3) $V \cdot U = 0$, since U and V are perpendicular.

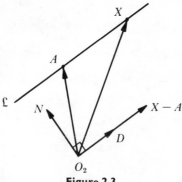

Figure 2.3

Since U is nonzero, we must have $U \cdot U \neq 0$. The last equation then requires that $r = 0$. Similarly, we can show that $s = 0$ by taking the inner product of both members of Equation 2.4 with V. This proves statement i.

Statement ii follows from statement i and Theorem 1.34.   ■

Let $\mathcal{L}$ be a line in $\mathcal{R}^2$ which contains the point A and has direction vector D. Let N be a nonzero vector which is orthogonal to D. (Is there always such a vector?) The vector N will be such that $N \cdot D = 0$. Figure 2.3 suggests that for any point X of $\mathcal{L}$ the vector $X - A$ is a multiple of D and hence that

$$(X - A) \cdot N = 0.$$

In fact, we shall show that this last equation may be used to characterize these points X in $\mathcal{R}^2$ which lie on $\mathcal{L}$. We first introduce the following terminology:

DEFINITION 2.9   (Normal to a Line in $\mathcal{R}^2$.)   Let $\mathcal{L}$ be a line in $\mathcal{R}^2$ with direction vector D. A nonzero vector N which is perpendicular to D is called a *normal* to the line $\mathcal{L}$.

There is, of course, a very close relationship between a direction vector for a line $\mathcal{L}$ in $\mathcal{R}^2$ and a normal N for $\mathcal{L}$. Indeed, the coordinates for D determine the coordinates for N, except for a scalar factor.

THEOREM 2.10   Let $\mathcal{L}$ be a line in $\mathcal{R}^2$ with direction vector $D = \begin{pmatrix} a \\ b \end{pmatrix}$. A vector N is a normal for $\mathcal{L}$ if and only if $N = k \begin{pmatrix} -b \\ a \end{pmatrix}$, where $k \neq 0$.

Proof.   Each vector N of the form $N = k\begin{pmatrix} -b \\ a \end{pmatrix} = \begin{pmatrix} -kb \\ ka \end{pmatrix}$ is such that

$$D \cdot N = \begin{pmatrix} a \\ b \end{pmatrix} \cdot \begin{pmatrix} -kb \\ ka \end{pmatrix} = a(-kb) + b(ka)$$

$$= -akb + akb$$

$$= 0.$$

Hence each such vector N is perpendicular to D. Since $D = \begin{pmatrix} a \\ b \end{pmatrix}$ is a direction vector for $\mathcal{L}$, we know that $\begin{pmatrix} a \\ b \end{pmatrix} \neq \begin{pmatrix} 0 \\ 0 \end{pmatrix}$. It follows that $N = k\begin{pmatrix} -b \\ a \end{pmatrix}$, with $k \neq 0$, is a nonzero vector perpendicular to D and is therefore a normal to $\mathcal{L}$.

Exercise 13 of the preceding section shows the converse: Every vector N which is perpendicular to D must have the form $N = k\begin{pmatrix} -b \\ a \end{pmatrix}$. Consequently, every normal to $\mathcal{L}$ must be of the form $N = k\begin{pmatrix} -b \\ a \end{pmatrix}$, with $k \neq 0$. ■

The following theorem gives a characterization for a line $\mathcal{L}$ in $\mathcal{R}^2$ in terms of a normal N.

THEOREM 2.11   (Implicit Representation of a Line in $\mathcal{R}^2$.)   Let $\mathcal{L}$ be a line in $\mathcal{R}^2$. Let A be a point of $\mathcal{L}$ and N a normal for $\mathcal{L}$. $\mathcal{L}$ can be described in either of the following *implicit* forms:

i. (Vector Implicit Form.) $\mathcal{L} = \{X \mid (X - A) \cdot N = 0 \text{ and } X \in \mathcal{R}^2\}$;

ii. (Scalar Implicit Form.) $\mathcal{L} = \left\{ \begin{pmatrix} x \\ y \end{pmatrix} \,\middle|\, ax + by + c = 0 \right\}$ where

$c = -A \cdot N$ and $N = \begin{pmatrix} a \\ b \end{pmatrix} \neq \begin{pmatrix} 0 \\ 0 \end{pmatrix}$.

Moreover, any subset $\mathcal{L}$ of $\mathcal{R}^2$ which can be described in either of these two forms is a line.

Proof.   Let $\mathcal{L}$ be a line in $\mathcal{R}^2$ with normal N. Let A be a point of $\mathcal{L}$. $\mathcal{L}$ must have a direction vector, say D, and is describable by the parametric form

(2.5) $\mathcal{L} = \{A + tD \mid t \in \mathcal{R}\}.$

We also note that $N \cdot D = 0$ and that D and N are both nonzero vectors.

Proof of i.   Let

(2.6) $\mathcal{K} = \{X \mid X \in \mathcal{R}^2 \text{ and } (X - A) \cdot N = 0\}.$

We shall prove that $\mathcal{K} = \mathcal{L}$ by showing that $\mathcal{L} \subset \mathcal{K}$ and also that $\mathcal{K} \subset \mathcal{L}$. Let X be an element of $\mathcal{L}$. Starting from Equation 2.5, we may write

$$X \in \mathcal{L} \Rightarrow^1 X = A + tD$$
$$\Rightarrow \ X - A = tD$$
$$\Rightarrow^2 (X - A) \cdot N = (tD) \cdot N, \qquad \text{for some } t \in R,$$
$$\Rightarrow^3 (X - A) \cdot N = t(D \cdot N)$$
$$\Rightarrow^4 (X - A) \cdot N = 0$$
$$\Rightarrow^5 X \in \mathcal{K}.$$

Therefore, $\mathcal{L} \subset \mathcal{K}$.

Now suppose that $X \in \mathcal{K}$. By Theorem 2.8 there are scalars r and s such that

$$(2.7) \qquad\qquad X - A = rD + sN.$$

Thus, we may write

$$X \in \mathcal{K} \Rightarrow^5 0 = \ (X - A) \cdot N$$
$$=^6 (rD + sN) \cdot N$$
$$=^7 r(D \cdot N) + s(N \cdot N),$$
$$=^4 r(0) + s(N \cdot N)$$
$$= \ s(N \cdot N)$$
$$\Rightarrow^8 s = 0$$
$$\Rightarrow^9 X - A = rD,$$
$$\Rightarrow \ X = A + rD$$
$$\Rightarrow^1 X \in \mathcal{L}.$$

Consequently, $\mathcal{K} \subset \mathcal{L}$ and the proof of Vector Implicit Form i is complete.

(1) Definition of $\mathcal{L}$ in Equation 2.5.
(2) Taking inner product with N.
(3) Theorem 2.2iv.
(4) Since $N \cdot D = 0$.
(5) Definition of $\mathcal{K}$ in Equation 2.6.
(6) Equation 2.7.
(7) Theorem 2.2iv and v.
(8) Theorem 2.2ii shows that $N \cdot N \neq 0$.
(9) Substitution into Equation 2.7.

Proof of ii. The scalar implicit form for $\mathcal{L}$ follows immediately from the vector implicit form upon setting $X = \begin{pmatrix} x \\ y \end{pmatrix}$ and $N = \begin{pmatrix} a \\ b \end{pmatrix}$ (see Exercise 17).

We must still establish the last statement of the theorem. Suppose that $\mathcal{L}$ is a subset of $\mathcal{R}^2$ described by statement i for some choice of A and nonzero vector N. By Theorem 2.7 (p. 93), there is a nonzero vector D such that

$N \cdot D = 0$. Consider now the line

$$\mathcal{M} = \{A + tD \mid t \in \mathcal{R}\}.$$

The vector N is clearly a normal to $\mathcal{M}$. By the first part of this theorem $\mathcal{M}$ may also be described by

$$\mathcal{M} = \{X \mid (X - A) \cdot N = 0 \text{ and } X \in \mathcal{R}^2\}.$$

But this is precisely the description that was assumed for $\mathcal{L}$. Hence $\mathcal{L} = \mathcal{M}$, and $\mathcal{L}$ itself must be a line in $\mathcal{R}^2$.

Finally, suppose that $\mathcal{L}$ is a subset of $\mathcal{R}^2$ described by statement ii for some choice of $\binom{a}{b} \neq O_2$ and c. From what has gone before the reader undoubtedly expects $\mathcal{L}$ to be a line in $\mathcal{R}^2$ with normal $N = \binom{a}{b}$. We shall show that this is the case by showing that $\mathcal{L}$ is also describable in the Vector Implicit Form i. We need only find a vector A such that $(-A \cdot N) = c$, or (equivalently) such that $A \cdot N = -c$. This possibility again follows from Theorem 2.7 (p. 93), with X replaced by A and k by $-c$. We summarize as follows:

$$
\begin{aligned}
\mathcal{L} &=^1 \left\{ \binom{x}{y} \,\middle|\, ax + by + c = 0 \right\} \\
&=^2 \left\{ \binom{x}{y} \,\middle|\, \binom{x}{y} \cdot \binom{a}{b} + c = 0 \right\} \\
&=^3 \{X \mid X \cdot N + c = 0, X \in \mathcal{R}^2\}, \text{ where } N = \binom{a}{b}, \\
&=^4 \{X \mid X \cdot N - A \cdot N = 0 \text{ and } X \in R^2\}, \text{ where } A \cdot N = -c, \\
&=^5 \{X \mid (X - A) \cdot N \text{ and } X \in \mathcal{R}^2\}.
\end{aligned}
$$

(1) Beginning with description ii for a subset $\mathcal{L}$.
(2) Definition 2.1 (Inner Product).
(3) Setting $X = \binom{x}{y}$ and $N = \binom{a}{b}$.
(4) Using Theorem 2.7 to find an A such that $A \cdot N = -c$.
(5) Theorem 2.2v. ■

Example 5. Find an implicit representation of the line $\mathcal{L}$ which contains the point $\binom{1}{2}$ and has direction vector $D = \binom{-1}{3}$. An examination of the statement of the last theorem reveals that two vectors need to be determined before it is possible to give an implicit representation for a line $\mathcal{L}$. We first require a point A on $\mathcal{L}$ and a normal N to $\mathcal{L}$. In this example the point A is given. We can determine N from the direction vector D using the technique of Example 4 (p. 93), or (equivalently) the content of Theorem 2.10 with $k = 1$. Thus, since $D = \binom{-1}{3}$ is a direction vector for $\mathfrak{L}$, the vector $N = \begin{pmatrix} -3 \\ -1 \end{pmatrix}$ is a normal to $\mathfrak{L}$. Hence, we may write

$$\mathfrak{L} = \{X \mid (X - A) \cdot N = 0 \text{ and } X \in \mathfrak{R}^2\}$$

$$= \left\{ X \mid \left[X - \begin{pmatrix} 1 \\ 2 \end{pmatrix} \right] \cdot \begin{pmatrix} -3 \\ -1 \end{pmatrix} = 0 \text{ and } X \in \mathfrak{R}^2 \right\}.$$

This is just one vector implicit form for $\mathfrak{L}$. There are many other such forms, since we could have made other choices for A and for N. In particular, we may choose $\begin{pmatrix} 3 \\ 1 \end{pmatrix}$ as the normal vector to obtain

$$\mathfrak{L} = \left\{ X \mid \left[X - \begin{pmatrix} 1 \\ 2 \end{pmatrix} \right] \cdot \begin{pmatrix} 3 \\ 1 \end{pmatrix} = 0 \text{ and } X \in \mathfrak{R}^2 \right\}.$$

Indeed, this latter form is more desirable to some since it contains fewer negative coordinates. From this second form we shall determine a scalar implicit representation by substituting $\begin{pmatrix} x \\ y \end{pmatrix}$ for X and computing the inner products

$$\mathfrak{L} = \left\{ \begin{pmatrix} x \\ y \end{pmatrix} \mid \left[\begin{pmatrix} x \\ y \end{pmatrix} - \begin{pmatrix} 1 \\ 2 \end{pmatrix} \right] \cdot \begin{pmatrix} 3 \\ 1 \end{pmatrix} = 0 \text{ and } \begin{pmatrix} x \\ y \end{pmatrix} \in \mathfrak{R}^2 \right\}$$

$$= \left\{ \begin{pmatrix} x \\ y \end{pmatrix} \mid \begin{pmatrix} x \\ y \end{pmatrix} \cdot \begin{pmatrix} 3 \\ 1 \end{pmatrix} - \begin{pmatrix} 1 \\ 2 \end{pmatrix} \cdot \begin{pmatrix} 3 \\ 1 \end{pmatrix} = 0; \; x, y \in \mathfrak{R} \right\}$$

$$= \left\{ \begin{pmatrix} x \\ y \end{pmatrix} \mid 3x + y - 5 = 0 \right\}. \quad \blacksquare$$

 Example 6.   Let $\mathfrak{L}$ be the line with explicit representation

$$\mathfrak{L} = \left\{ \begin{pmatrix} x \\ y \end{pmatrix} \mid y = 3x + 2 \right\}.$$

By inspection, we see that $\mathfrak{L}$ has slope 3 and y-intercept 2. Consequently, $D = \begin{pmatrix} 1 \\ 3 \end{pmatrix}$ is a direction vector for $\mathfrak{L}$ and $A = \begin{pmatrix} 0 \\ 2 \end{pmatrix}$ is a point on $\mathfrak{L}$. By Theorem 2.10 (p. 96), the vector $N = \begin{pmatrix} -3 \\ 1 \end{pmatrix}$ is a normal for the given line and $\mathfrak{L}$ can be described implicitly by

$$\mathfrak{L} = \left\{ X \mid \left[X - \begin{pmatrix} 0 \\ 2 \end{pmatrix} \right] \cdot \begin{pmatrix} -3 \\ 1 \end{pmatrix} = 0 \text{ and } X \in \mathfrak{R}^2 \right\},$$

or by

$$\mathfrak{L} = \left\{ \begin{pmatrix} x \\ y \end{pmatrix} \mid -3x + y - 2 = 0 \right\}.$$

The reader should note that the latter description is also obtainable at once from the original description. $\blacksquare$

Example 7.   Let $\mathcal{L}$ be the line given by the scalar implicit representation

$$\mathcal{L} = \left\{ \begin{pmatrix} x \\ y \end{pmatrix} \middle| 2x + 3y - 4 = 0 \right\}.$$

The explicit representation for $\mathcal{L}$ is obtained by solving the equation involved for the y-coordinate. Thus,

$$\mathcal{L} = \left\{ \begin{pmatrix} x \\ y \end{pmatrix} \middle| y = -\frac{2}{3} x + \frac{4}{3} \right\}.$$

To determine a parametric description for $\mathcal{L}$, we note, by inspection of the form just obtained, that $\mathcal{L}$ has slope $m = \dfrac{-2}{3}$ and y-intercept $b = \dfrac{4}{3}$.

Consequently, $A = \begin{pmatrix} 0 \\ 4 \\ \overline{3} \end{pmatrix}$ is a point of $\mathcal{L}$ and $\begin{pmatrix} 1 \\ -2 \\ \overline{3} \end{pmatrix}$ is a direction vector

for $\mathcal{L}$. Of course $D = 3 \begin{pmatrix} 1 \\ -2 \\ \overline{3} \end{pmatrix} = \begin{pmatrix} 3 \\ -2 \end{pmatrix}$ is also a direction vector for $\mathcal{L}$.

With the choices of A and D just obtained, we construct the parametric representation

$$\mathcal{L} = \left\{ X \middle| X = \begin{pmatrix} 0 \\ 4 \\ \overline{3} \end{pmatrix} + t\begin{pmatrix} 3 \\ -2 \end{pmatrix}, t \in \mathfrak{R} \right\}.$$

A comparison of the original description of $\mathcal{L}$ and Theorem 2.11ii shows us that $N = \begin{pmatrix} 2 \\ 3 \end{pmatrix}$ is a normal for $\mathcal{L}$. This N and the point A found above yield the following vector implicit form for $\mathcal{L}$:

$$\mathcal{L} = \left\{ X \middle| \left[X - \begin{pmatrix} 0 \\ 4 \\ \overline{3} \end{pmatrix} \right] \cdot \begin{pmatrix} 2 \\ 3 \end{pmatrix} = 0 \text{ and } X \in \mathfrak{R}^2 \right\}. \quad \blacksquare$$

We close this section with a summary of the several characterizations given in Chapters 1 and 2 for a line in $\mathfrak{R}^2$. In this summary we give the various forms which may be employed to describe a line in $\mathfrak{R}^2$. Every line in $\mathfrak{R}^2$ has a parametric description and an implicit description. Each of these descriptions may be given in a vector form (in which no coordinates appear) or in scalar form (in which all coordinates appear and all inner products have been affected). Of course only those lines which intersect the y-axis in a single point will have an explicit representation.

SUMMARY 2.12   (Representations of Lines in $\mathfrak{R}^2$.)

i. A subset $\mathcal{L}$ of $\mathfrak{R}^2$ is a line if and only if $\mathcal{L}$ is the set of all X in $\mathfrak{R}^2$ satisfying any one of the following equations:

Parametric Equation

$\begin{cases} \text{Vector Form:} \quad X = A + tD. \\ \text{Scalar Form:} \\ \quad \begin{pmatrix} x \\ y \end{pmatrix} = \begin{pmatrix} x_0 \\ y_0 \end{pmatrix} + t \begin{pmatrix} a \\ b \end{pmatrix}. \end{cases}$ $\begin{cases} \text{Point on } \mathcal{L}: \ A = \begin{pmatrix} x_0 \\ y_0 \end{pmatrix}. \\[2mm] \text{Direction Vector: } D = \begin{pmatrix} a \\ b \end{pmatrix}. \\[2mm] \text{Normal Vector: } N = \begin{pmatrix} -b \\ a \end{pmatrix}. \end{cases}$

Implicit Equation

$\begin{cases} \text{Vector Form:} \\ \quad (X - A) \cdot N = 0. \\ \text{Scalar Form:} \\ \quad a(x - x_0) + b(y - y_0) = 0. \end{cases}$ $\begin{cases} \text{Point on } \mathcal{L}: \ A = \begin{pmatrix} x_0 \\ y_0 \end{pmatrix}. \\[2mm] \text{Direction Vector: } D = \begin{pmatrix} -b \\ a \end{pmatrix}. \\[2mm] \text{Normal Vector: } N = \begin{pmatrix} a \\ b \end{pmatrix}. \end{cases}$

or

$\begin{cases} \text{Vector Form:} \\ \quad X \cdot N + c = 0. \\ \text{Scalar Form:} \\ \quad ax + by + c = 0. \end{cases}$ $\begin{cases} \text{Point on } \mathcal{L}: \ \text{any solution} \\ \quad A = \begin{pmatrix} x_0 \\ y_0 \end{pmatrix} \text{ of equation.} \\[2mm] \text{Direction Vector: } D = \begin{pmatrix} -b \\ a \end{pmatrix}. \\[2mm] \text{Normal Vector: } \ N = \begin{pmatrix} a \\ b \end{pmatrix}. \\[2mm] \quad c = -A \cdot N = -ax_0 - by_0 \end{cases}$

ii. A subset $\mathcal{L}$ of $\mathcal{R}^2$ is a line with slope m and y-intercept b if and only if $\mathcal{L}$ is the set of all X in $\mathcal{R}^2$ satisfying the following:

Explicit Equation: $y = mx + b$ $\begin{cases} \text{Point on } \mathcal{L}: \qquad A = \begin{pmatrix} 0 \\ b \end{pmatrix}. \\[2mm] \text{Direction Vector: } D = \begin{pmatrix} 1 \\ m \end{pmatrix}. \\[2mm] \text{Normal Vector: } \ N = \begin{pmatrix} -m \\ 1 \end{pmatrix}. \end{cases}$

EXERCISES

In Exercises 1 to 8 determine a vector implicit representation and a scalar implicit representation for the line $\mathcal{L}$ described:

1. $\mathcal{L} = \left\{ \begin{pmatrix} -2 \\ 5 \end{pmatrix} + t \begin{pmatrix} 3 \\ 5 \end{pmatrix} \Big| t \in \mathcal{R} \right\}.$

2. $\mathcal{L}$ is line containing $A = \begin{pmatrix} 5 \\ -5 \end{pmatrix}$ and $B = \begin{pmatrix} 0 \\ 7 \end{pmatrix}.$

3. $\mathcal{L} = \left\{ \begin{pmatrix} x \\ y \end{pmatrix} \Big| y = 7x - 11 \right\}.$ 4. $\mathcal{L} = \left\{ \begin{pmatrix} 7 \\ -3 \end{pmatrix} + t \begin{pmatrix} 4 \\ 7 \end{pmatrix} \Big| t \in \mathcal{R} \right\}.$

5. $\mathcal{L}$ is the line containing $A = \begin{pmatrix} 12 \\ 7 \end{pmatrix}$ and $B = \begin{pmatrix} 5 \\ -2 \end{pmatrix}.$

6. $\mathcal{L} = \left\{ \binom{x}{y} \,\middle|\, y = -5x + 11 \right\}.$

7. $\mathcal{L} = \left\{ \binom{x}{y} \,\middle|\, \binom{x}{y} = \binom{4}{-2} + t\binom{3}{5},\, t \in \mathcal{R} \right\}.$

8. $\mathcal{L}$ is the line containing $A = \binom{5}{-7}$ and parallel to

$$\mathcal{K} = \left\{ \left[X - \binom{2}{1} \right] \cdot \binom{4}{0} = 0 \text{ and } X \in \mathcal{R}^2 \right\}.$$

In Exercises 9 to 16 determine other representations for $\mathcal{L}$ so that $\mathcal{L}$ is described by a parametric representation, a vector implicit representation, a scalar implicit representation, and, when possible, by an explicit representation.

9. $\mathcal{L} = \left\{ \binom{x}{y} \,\middle|\, 2(x - 1) + 3(y - 2) = 0 \right\}.$

10. $\mathcal{L} = \left\{ X \,\middle|\, \left[X - \binom{1}{3} \right] \cdot \binom{1}{3} = 0,\, X \in \mathcal{R}^2 \right\}.$

11. $\mathcal{L} = \left\{ \binom{3}{7} + t\binom{7}{9} \,\middle|\, t \in \mathcal{R} \right\}.$ 12. $\mathcal{L} = \left\{ \binom{x}{y} \,\middle|\, y = 5x - 13 \right\}.$

13. $\mathcal{L} = \left\{ X \,\middle|\, X \cdot \binom{3}{2} - 4 = 0,\, X \in \mathcal{R}^2 \right\}.$

14. $\mathcal{L} = \left\{ \binom{x}{y} \,\middle|\, 2x - 3 = 0 \right\}.$ 15. $\mathcal{L} = \left\{ \binom{x}{y} \,\middle|\, y = x - 7 \right\}.$

16. $\mathcal{L} = \left\{ \binom{-5}{4} + t\binom{2}{3} \,\middle|\, t \in \mathcal{R} \right\}.$

17. Complete the proof of Theorem 2.11 by deriving the Scalar Implicit Form ii from the Vector Implicit Form i. $\left(\text{Hint: Set } X = \binom{x}{y} \text{ and } N = \binom{a}{b}. \right)$

ANSWERS

1. $\mathcal{L} = \left\{ X \,\middle|\, \left[X - \binom{-2}{5} \right] \cdot \binom{5}{-3} = 0 \text{ and } X \in \mathcal{R}^2 \right\}$

$= \left\{ \binom{x}{y} \,\middle|\, 5x - 3y + 25 = 0 \right\}.$

3. $\mathcal{L} = \left\{ \binom{x}{y} \,\middle|\, 7x - y - 11 = 0 \right\} = \left\{ X \,\middle|\, \left[X - \binom{0}{-11} \right] \cdot \binom{7}{-1} = 0 \text{ and } X \in \mathcal{R}^2 \right\}.$

5. $\mathcal{L} = \left\{ X \,\middle|\, \left[X - \binom{5}{-2} \right] \cdot \binom{-9}{7} = 0 \text{ and } X \in \mathcal{R}^2 \right\}$

$= \left\{ \binom{x}{y} \,\middle|\, -9x + 7y + 59 = 0 \right\}.$

7. $\mathcal{L} = \left\{ X \,\middle|\, \left[X - \binom{4}{-2} \right] \cdot \binom{5}{-3} = 0 \text{ and } X \in \mathcal{R}^2 \right\}$

$= \left\{ \binom{x}{y} \,\middle|\, 5x - 3y - 26 = 0 \right\}.$

9. $\mathcal{L} = \left\{ \binom{x}{y} \,\middle|\, 2x + 3y - 8 = 0 \right\} = \left\{ \binom{x}{y} \,\middle|\, y = \frac{-2}{3}x + \frac{8}{3} \right\}$

$= \left\{ X \,\middle|\, \left[X - \binom{1}{2} \right] \cdot \binom{2}{3} = 0 \text{ and } X \in \mathcal{R}^2 \right\} = \left\{ \binom{1}{2} + t\binom{-3}{2} \,\middle|\, t \in \mathcal{R} \right\}.$

11. $\mathcal{L} = \left\{ X \mid \left[X - \binom{3}{7} \right] \cdot \binom{9}{-7} = 0 \text{ and } X \in \mathcal{R}^2 \right\}$

$\quad = \left\{ \binom{x}{y} \mid 9x - 7y + 22 = 0 \right\} = \left\{ \binom{x}{y} \mid y = \dfrac{9}{7} x + \dfrac{22}{7} \right\}.$

13. $\mathcal{L} = \left\{ \binom{x}{y} \mid 3x + 2y - 4 = 0 \right\} = \left\{ \binom{x}{y} \mid y = \dfrac{-3}{2} x + 2 \right\}$

$\quad = \left\{ \binom{0}{2} + t \binom{2}{-3} \mid t \in \mathcal{R} \right\}.$

15. $\mathcal{L} = \left\{ \binom{x}{y} \mid x - y - 7 = 0 \right\}$

$\quad = \left\{ X \mid X \cdot \binom{1}{-1} - 7 = 0 \right\} = \left\{ \binom{0}{-7} + t \binom{1}{1} \mid t \in \mathcal{R} \right\}.$

4. NORMALS TO PLANES IN $\mathcal{R}^3$; CROSS-PRODUCTS OF 3-VECTORS

A plane in $\mathcal{R}^3$ can be described in a manner similar to the implicit form given for a line in $\mathcal{R}^2$. Again the basic concept required is that of a normal to the plane.

DEFINITION 2.13 Let $\mathcal{S}$ be a plane in $\mathcal{R}^3$ with direction set $\{U, V\}$. A nonzero vector N which is perpendicular to both U and V is called a *normal* to the plane $\mathcal{S}$.

Given a direction set $\{U, V\}$ for a plane in $\mathcal{R}^3$, we shall require some constructive method of obtaining a normal. To this end we introduce another vector operation in $\mathcal{R}^3$.

DEFINITION 2.14 (Cross-Product in $\mathcal{R}^3$.) Let $X = \begin{pmatrix} a \\ b \\ c \end{pmatrix}$ and $Y = \begin{pmatrix} d \\ e \\ f \end{pmatrix}$ be two vectors in $\mathcal{R}^3$. The *cross-product* of X with Y, denoted by $X \times Y$, is the vector

$$X \times Y = \begin{pmatrix} a \\ b \\ c \end{pmatrix} \times \begin{pmatrix} d \\ e \\ f \end{pmatrix} = \begin{pmatrix} \begin{vmatrix} b & e \\ c & f \end{vmatrix} \\ -\begin{vmatrix} a & d \\ c & f \end{vmatrix} \\ \begin{vmatrix} a & d \\ b & e \end{vmatrix} \end{pmatrix} = \begin{pmatrix} bf - ce \\ -(af - cd) \\ ae - bd \end{pmatrix}.$$

The cross-product of two vectors X and Y in $\mathfrak{R}^3$ is a vector in $\mathfrak{R}^3$ whose coordinates are given by second order determinants formed from the coordinates of X and Y. The coordinates of $X \times Y$ are easily obtained by writing the coordinates of X and Y in adjacent columns and proceeding as indicated in the following scheme:

$$
\begin{matrix}
X & Y \\
\downarrow & \downarrow \\
\begin{bmatrix} a & d \\ b & e \\ c & f \end{bmatrix}
\end{matrix}
\quad
\begin{bmatrix}
\text{1. Delete first coordinates of } X \text{ and } Y. \\
\text{Form determinant of remaining} \\
\text{2-vectors.}
\end{bmatrix}
\;-\;
\begin{pmatrix}
\begin{vmatrix} b & e \\ c & f \end{vmatrix} \\[2mm]
-\begin{vmatrix} a & d \\ c & f \end{vmatrix} \\[2mm]
\begin{vmatrix} a & d \\ b & e \end{vmatrix}
\end{pmatrix}
= X \times Y
$$

1. Delete first coordinates of X and Y. Form determinant of remaining 2-vectors.

2. Delete second coordinates of X and Y. Form determinant; take additive inverse.

3. Delete third coordinates of X and Y. Form determinant.

Each coordinate of $X \times Y$ is obtained by removing the corresponding coordinates from the columnar arrangement of X and Y and forming the determinant of the 2-vectors which remain, *remembering* that the second coordinate also requires the taking of an additive inverse.

Example 8

(a) Let $X = \begin{pmatrix} 1 \\ 1 \\ 2 \end{pmatrix}$ and $Y = \begin{pmatrix} -1 \\ 2 \\ 3 \end{pmatrix}$. To determine $X \times Y$ we proceed as outlined previously:

$$
\begin{matrix}
X & Y \\
\begin{bmatrix} 1 & -1 \\ 1 & 2 \\ 2 & 3 \end{bmatrix}
\end{matrix}
\quad
\begin{bmatrix}
\text{1. Delete first coordinates. Form determinant.} \\
\text{2. Delete second coordinates. Form determinants. Take additive inverse.} \\
\text{3. Delete third coordinates. Form determinant.}
\end{bmatrix}
\rightarrow
\begin{pmatrix}
\begin{vmatrix} 1 & 2 \\ 2 & 3 \end{vmatrix} \\[2mm]
-\begin{vmatrix} 1 & -1 \\ 2 & 3 \end{vmatrix} \\[2mm]
\begin{vmatrix} 1 & -1 \\ 1 & 2 \end{vmatrix}
\end{pmatrix}
= X \times Y
$$

Evaluating the determinants, we find that

$$
X \times Y = \begin{pmatrix} 3 - 4 \\ -(3 - (-2)) \\ 2 - (-1) \end{pmatrix} = \begin{pmatrix} -1 \\ -5 \\ 3 \end{pmatrix}.
$$

(b) Let us now determine $Y \times X$. We begin by writing the coordinates of Y and then the coordinates of X:

$$
\begin{matrix}
Y & X \\
\downarrow & \downarrow \\
\begin{bmatrix} -1 & 1 \\ 2 & 1 \\ 3 & 2 \end{bmatrix}
\end{matrix}
\quad\longrightarrow\quad
\begin{pmatrix}
\begin{vmatrix} 2 & 1 \\ 3 & 2 \end{vmatrix} \\[2mm]
-\begin{vmatrix} -1 & 1 \\ 3 & 2 \end{vmatrix} \\[2mm]
\begin{vmatrix} -1 & 1 \\ 2 & 1 \end{vmatrix}
\end{pmatrix}
= Y \times X.
$$

Evaluation of the determinants gives $Y \times X = \begin{pmatrix} 1 \\ 5 \\ -3 \end{pmatrix}$.

The preceding example shows that the cross-product $X \times Y$ is not the same as the cross-product $Y \times X$. Exercise 13 shows the general relationship between $X \times Y$ and $Y \times X$.

The usefulness of the cross-product of two vectors is due to the fact that $X \times Y$ is perpendicular to each factor X, Y. This and other facts concerning the cross-product are given in the following theorem:

THEOREM 2.15 Let X and Y be vectors in $\mathcal{R}^3$. The cross-product $X \times Y$ possesses the following properties:

i. $X \times Y$ is perpendicular to X and to Y: $(X \times Y) \cdot X = (X \times Y) \cdot Y = 0$.

ii. $X \times Y = O_3$ if and only if X is parallel to Y, that is, if and only if $\{X, Y\}$ is a linearly dependent set.

iii. If $\{X, Y\}$ is a linearly independent set, then the set $\{X, Y, X \times Y\}$ is also a linearly independent set.

iv. If $\{X, Y\}$ is a linearly independent set, then each vector Z in $\mathcal{R}^3$ is a linear combination of X, Y, and $X \times Y$.

The proofs of statements i and ii are straightforward and are called for in Exercises 14, 15, and 16.

Proof of iii. Let $\{X, Y\}$ be a linearly independent set in $\mathcal{R}^3$. By statement ii of the present theorem we must have

$$(2.8) \qquad X \times Y \neq O_3$$

We use this fact and Theorem 1.30ii, (p. 45), to show that $\{X, Y, X \times Y\}$ is a linearly independent set.

Let a, b, and c be three real numbers such that

$$(2.9) \qquad aX + bY + c(X \times Y) = O_3.$$

We shall show that a, b, and c must all be zero. Forming the inner-product of both members of Equation 2.9 with $(X \times Y)$ initiates the following implication chain:

$$aX \cdot (X \times Y) + bY \cdot (X \times Y) + c(X \times Y) \cdot (X \times Y) = O_3 \cdot (X \times Y)$$
$$\Rightarrow^1 a0 + b0 + c(X \times Y) \cdot (X \times Y) = 0$$
$$\Rightarrow c(X \times Y) \cdot (X \times Y) = 0$$
$$\Rightarrow^2 c = 0$$
$$\Rightarrow^3 aX + bY = 0 \text{ and } c = 0$$
$$\Rightarrow^4 a = b = c = 0.$$

(1) Using statement i of the present theorem: $X \cdot (X \times Y) = Y \cdot (X \times Y) = 0$

(2) Equation 2.8 and Theorem 2.2ii with $X \times Y$ replacing X, together with the real number property: $[c \cdot d = 0 \text{ and } d \neq 0] \Rightarrow c = 0$.

(3) Setting $c = 0$ in Equation 2.9.

(4) Using Theorem 1.30ii and the linear independence of $\{X, Y\}$.

Since the only linear combination of X, Y, and $X \times Y$ yielding the zero-vector O_3 is one with all coefficients zero, Theorem 1.30ii assures us that $\{X, Y, X \times Y\}$ is a linearly independent set.

Proof of iv. Statement iv is an immediate consequence of Statement iii and Theorem 1.35. ∎

Example 9. The first two results of the preceding theorem serve as a useful check on the evaluation of a cross-product.

(a) Let $X = \begin{pmatrix} 2 \\ 0 \\ 1 \end{pmatrix}$ and $Y = \begin{pmatrix} -1 \\ 1 \\ 4 \end{pmatrix}$. We evaluate $X \times Y$ by the technique of Example 8:

$$X \times Y = \begin{pmatrix} 2 \\ 0 \\ 1 \end{pmatrix} \times \begin{pmatrix} -1 \\ 1 \\ 4 \end{pmatrix} = \begin{pmatrix} \begin{vmatrix} 0 & 1 \\ 1 & 4 \end{vmatrix} \\ -\begin{vmatrix} 2 & -1 \\ 1 & 4 \end{vmatrix} \\ \begin{vmatrix} 2 & -1 \\ 0 & 1 \end{vmatrix} \end{pmatrix} = \begin{pmatrix} -1 \\ -9 \\ 2 \end{pmatrix}.$$

As a check we evaluate $X \cdot (X \times Y)$ and $Y \cdot (X \times Y)$. Thus,

$$X \cdot (X \times Y) = \begin{pmatrix} 2 \\ 0 \\ 1 \end{pmatrix} \cdot \begin{pmatrix} -1 \\ -9 \\ 2 \end{pmatrix} = -2 + 0 + 2 = 0.$$

$$Y \cdot (X \times Y) = \begin{pmatrix} -1 \\ 1 \\ 4 \end{pmatrix} \cdot \begin{pmatrix} -1 \\ -9 \\ 2 \end{pmatrix} = 1 - 9 + 8 = 0.$$

The verification is complete.

(b) Let $X = \begin{pmatrix} 3 \\ 12 \\ 9 \end{pmatrix}$ and $Y = \begin{pmatrix} 5 \\ 20 \\ 15 \end{pmatrix}$. We have

$$X \times Y = \begin{pmatrix} 3 \\ 12 \\ 9 \end{pmatrix} \times \begin{pmatrix} 5 \\ 20 \\ 15 \end{pmatrix} = \begin{pmatrix} \begin{vmatrix} 12 & 20 \\ 9 & 15 \end{vmatrix} \\ -\begin{vmatrix} 3 & 5 \\ 9 & 15 \end{vmatrix} \\ \begin{vmatrix} 3 & 5 \\ 12 & 20 \end{vmatrix} \end{pmatrix} = \begin{pmatrix} 0 \\ 0 \\ 0 \end{pmatrix} = O_3$$

According to statement ii of Theorem 2.15, X must be parallel to Y, that is, Y must be a multiple of X. Comparison of the first coordinates of X and Y suggests that $Y = \frac{5}{3}X$. Further computation verifies this fact. ∎

The next theorem is an important consequence of the preceding theorem. It says, simply, that any normal to a plane in $\mathfrak{R}^3$ is a nonzero multiple of the cross-product of the vectors of any direction set.

THEOREM 2.16   (Characterization of Normals to a Plane in $\mathfrak{R}^3$.) Let $\mathfrak{I}$ be a plane in $\mathfrak{R}^3$ and let $\{U, V\}$ be any direction set for $\mathfrak{I}$. A vector N in $\mathfrak{R}^3$ is a normal for $\mathfrak{I}$ if and only if

$$N = k(U \times V), \text{ for some real number } k \neq 0.$$

Proof. We first prove the "if" part. That is, we suppose that $N = k(U \times V)$ for some nonzero real number k and shall show that N is a normal to $\mathfrak{I}$. The linear independence of the direction set $\{U, V\}$ assures us that $U \times V$ is a nonzero vector (Theorem 2.15ii). Hence, the vector $N = k(U \times V)$, $k \neq 0$, is also a nonzero vector. Moreover, N is perpendicular to both U and V as shown by the following two chains of equations:

$$N \cdot U = [k(U \times V)] \cdot U =^1 k[(U \times V) \cdot U] =^2 k \cdot 0 = 0$$

and

$$N \cdot V = [k(U \times V)] \cdot V =^3 k(U \times V) \cdot V =^2 k \cdot 0 = 0.$$

(1) Theorem 2.2iv with X replaced by $U \times V$ and Y replaced by U.
(2) Since $U \times V$ is perpendicular to U and to V, Theorem 2.15i.
(3) Theorem 2.2iv with appropriate replacements.

N is therefore a normal to $\mathfrak{I}$, according to Definition 2.13.

We now prove the "only if" part. Let N be a normal to the plane $\mathfrak{I}$ with direction set $\{U, V\}$. Since $\{U, V\}$ must be a linearly independent set to be a direction set, we may use Theorem 2.15iv (with X, Y, and Z replaced by U, V, and N, respectively) to conclude that N is a linear combination of U, V, and $U \times V$:

(2.10) $N = aU + bV + k(U \times V).$

We shall show below that $a = b = 0$. Anticipating this result we then have

$$N = k(U \times V).$$

Since $N \neq O_3$, we must have $k \neq 0$ and the proof of the theorem will be complete.

To show that $a = b = 0$ in Equation 2.10, we proceed as follows: Since the vector N is a normal for $\mathfrak{I}$, it follows that N is perpendicular to U and to V (see Exercise 18). Thus,

(2.11) $N \cdot (aU + bV) = a(N \cdot U) + b(N \cdot V) = 0.$

Similarly, since $U \times V$ is perpendicular to U and to V, we have

(2.12) $(U \times V) \cdot (aU + bV) = 0.$

Now form the inner product of both members of Equation 2.10 with $(aU + bV)$ to initiate the chain:

$$0 =^1 N \cdot (aU + bV)$$
$$=^2 [(aU + bV) + k(U \times V)] \cdot (aU + bV)$$
$$=^3 (aU + bV) \cdot (aU + bV) + k(U \times V) \cdot (aU + bV)$$
$$=^4 (aU + bV) \cdot (aU + bV).$$

(1) Using Equation 2.11.
(2) Inner product of members of Equation 2.10 with $(aU + bV)$.
(3) Theorem 2.2v.
(4) Using Equation 2.12.

The above chain shows that $(aU + bV) \cdot (aU + bV) = 0$. We now employ Theorem 2.2ii with X replaced by $aU + bV$ to conclude that

$$aU + bV = O_3.$$

But $\{U, V\}$ is a linearly independent set. Theorem 1.30ii states that the only linear combination of a linearly independent set which produces the zero vector is the one in which each coefficient is zero. Thus,

$$a = b = 0$$

and our proof is complete. ∎

Example 10. Determine a normal to the plane $\mathscr{S} = \left\{ \begin{pmatrix} 1 \\ 1 \\ 3 \end{pmatrix} + r \begin{pmatrix} 2 \\ 0 \\ 1 \end{pmatrix} + s \begin{pmatrix} -1 \\ 1 \\ 4 \end{pmatrix} \middle| r, s \in \mathscr{R} \right\}$. We first note that $\left\{ \begin{pmatrix} 2 \\ 0 \\ 1 \end{pmatrix}, \begin{pmatrix} -1 \\ 1 \\ 4 \end{pmatrix} \right\}$ is a direction set for $\mathscr{S}$. Each normal to $\mathscr{S}$ is a nonzero multiple of the cross-product of these two vectors. In particular, the cross-product itself is a normal to $\mathscr{S}$. Thus,

$$N = \begin{pmatrix} 2 \\ 0 \\ 1 \end{pmatrix} \times \begin{pmatrix} -1 \\ 1 \\ 4 \end{pmatrix} = \begin{pmatrix} -1 \\ -9 \\ 2 \end{pmatrix}$$

is one normal to $\mathscr{S}$. The computation involving the cross-product should be checked. The vectors involved here are precisely those of Example 9a in which the computation was verified. We are thus certain that N is perpendicular to U and V and consequently a normal for $\mathscr{S}$. (N is obviously nonzero.) ∎

Example 11. Determine a normal to the plane $\mathscr{S}$ through the points $A = \begin{pmatrix} 2 \\ 0 \\ 3 \end{pmatrix}$ and $B = \begin{pmatrix} -1 \\ 4 \\ 2 \end{pmatrix}$ and parallel to $V = \begin{pmatrix} 3 \\ 0 \\ 1 \end{pmatrix}$.

The vectors $U = A - B = \begin{pmatrix} 3 \\ -4 \\ 1 \end{pmatrix}$ and V constitute a direction set for $\mathcal{P}$. Thus,

$$N = U \times V = \begin{pmatrix} 3 \\ -4 \\ 1 \end{pmatrix} \times \begin{pmatrix} 3 \\ 0 \\ 1 \end{pmatrix} = \begin{pmatrix} -4 \\ 0 \\ 12 \end{pmatrix}$$

is a normal to $\mathcal{P}$. This fact is verified by evaluating the inner products:

$$N \cdot U = \begin{pmatrix} -4 \\ 0 \\ 12 \end{pmatrix} \cdot \begin{pmatrix} 3 \\ -4 \\ 1 \end{pmatrix} = -12 + 0 + 12 = 0, \quad \text{and}$$

$$N \cdot V = \begin{pmatrix} -4 \\ 0 \\ 12 \end{pmatrix} \cdot \begin{pmatrix} 3 \\ 0 \\ 1 \end{pmatrix} = -12 + 0 + 12 = 0. \quad \blacksquare$$

In the next section we shall use the normal to a plane to provide the implicit mode of representation for a plane in $\mathcal{R}^3$. The principal theorem is a result of the following easily established property of a normal to a plane in $\mathcal{R}^3$.

THEOREM 2.17 Let $\mathcal{P}$ be a plane in $\mathcal{R}^3$ with direction set $\{U, V\}$. Let N be a normal to $\mathcal{P}$.
  i. The set $\{U, V, N\}$ is a linearly independent set.
  ii. Every vector Z in $\mathcal{R}^3$ is a linear combination of U, V, and N.

The proof is an immediate application of Theorem 2.15iii and iv and Theorem 2.16 (see Exercise 17).

EXERCISES

In each of the Exercises 1 to 6 compute the cross-product $X \times Y$ of the vectors X and Y. Check your computation by the method of Example 9.

1. $X = \begin{pmatrix} 1 \\ 2 \\ -1 \end{pmatrix},\ Y = \begin{pmatrix} 3 \\ 4 \\ -2 \end{pmatrix}.$ 2. $X = \begin{pmatrix} 1 \\ -1 \\ 2 \end{pmatrix},\ Y = \begin{pmatrix} 5 \\ 7 \\ 11 \end{pmatrix}.$

3. $X = \begin{pmatrix} 3 \\ 9 \\ -6 \end{pmatrix},\ Y = \begin{pmatrix} 8 \\ 24 \\ -16 \end{pmatrix}.$ 4. $X = \begin{pmatrix} 1 \\ 2 \\ 0 \end{pmatrix},\ Y = \begin{pmatrix} 0 \\ 1 \\ 1 \end{pmatrix}.$

5. $X = \begin{pmatrix} 3 \\ 5 \\ 7 \end{pmatrix},\ Y = \begin{pmatrix} 9 \\ 15 \\ 21 \end{pmatrix}.$ 6. $X = \begin{pmatrix} 7 \\ 11 \\ -9 \end{pmatrix},\ Y = \begin{pmatrix} 2 \\ -7 \\ 5 \end{pmatrix}.$

In each of the Exercises 7 to 11 determine a normal to the plane $\mathcal{F}$, and verify your computation.

7. $\mathcal{F}$ contains $A = \begin{pmatrix} 1 \\ 2 \\ -3 \end{pmatrix}$, $B = \begin{pmatrix} 2 \\ -3 \\ -1 \end{pmatrix}$ and is parallel to $\mathcal{L} = \left\{ \begin{pmatrix} 2 \\ 1 \\ 1 \end{pmatrix} + t\begin{pmatrix} 1 \\ -1 \\ 7 \end{pmatrix} \middle| t \in \mathcal{R} \right\}$.

8. $\mathcal{F}$ contains $\mathcal{L}_1 = \left\{ t\begin{pmatrix} 1 \\ 3 \\ -4 \end{pmatrix} \middle| t \in \mathcal{R} \right\}$ and is parallel to $\mathcal{L}_2 = \left\{ \begin{pmatrix} 1 \\ 1 \\ 3 \end{pmatrix} + s\begin{pmatrix} -3 \\ 1 \\ 5 \end{pmatrix} \middle| s \in \mathcal{R} \right\}$.

9. $\mathcal{F}$ contains $A = \begin{pmatrix} 1 \\ 3 \\ 4 \end{pmatrix}$ and is parallel to $\mathcal{M} = \left\{ \begin{pmatrix} 1 \\ 2 \\ 1 \end{pmatrix} + r\begin{pmatrix} 2 \\ -1 \\ 3 \end{pmatrix} + s\begin{pmatrix} 7 \\ 2 \\ 5 \end{pmatrix} \middle| r, s \in \mathcal{R} \right\}$.

10. $\mathcal{F}$ contains $A = \begin{pmatrix} -5 \\ 1 \\ 3 \end{pmatrix}$, is parallel to the line $\mathcal{L}_1$ through $B = \begin{pmatrix} 2 \\ 1 \\ -4 \end{pmatrix}$ and $C = \begin{pmatrix} 3 \\ -1 \\ 4 \end{pmatrix}$, and is parallel to the line $\mathcal{L}_2$ through $D = \begin{pmatrix} 7 \\ -1 \\ 5 \end{pmatrix}$ and $E = \begin{pmatrix} 2 \\ 1 \\ 4 \end{pmatrix}$.

11. $\mathcal{F}$ contains the points $A = \begin{pmatrix} 1 \\ 0 \\ -3 \end{pmatrix}$, $B = \begin{pmatrix} 2 \\ -3 \\ 4 \end{pmatrix}$, and $C = \begin{pmatrix} 1 \\ 1 \\ 1 \end{pmatrix}$.

12. Show that the vectors $E_1 = \begin{pmatrix} 1 \\ 0 \\ 0 \end{pmatrix}$, $E_2 = \begin{pmatrix} 0 \\ 1 \\ 0 \end{pmatrix}$, and $E_3 = \begin{pmatrix} 0 \\ 0 \\ 1 \end{pmatrix}$ are such that

$$E_1 \times E_2 = E_3, \quad E_2 \times E_3 = E_1, \quad \text{and} \quad E_3 \times E_1 = E_2.$$

In Exercises 13 to 16 X and Y denote vectors in $\mathcal{R}^3$.

13. Show that $X \times Y = -(Y \times X)$.

14. Show that $(X \times Y) \cdot X = 0$ and that $(X \times Y) \cdot Y = 0$.

15. If X is parallel to Y, show that $X \times Y = O_3$.

16. If $X \times Y = O_3$, show that X is parallel to Y.

17. Prove Theorem 2.17.

18. Let N be a normal to the plane $\mathcal{F}$. Let $\{U, V\}$ be a direction set for $\mathcal{F}$. Show that N is perpendicular to U and to V. (Hint: N is perpendicular to some direction set for $\mathcal{F}$, say $\{T, W\}$. Use Theorem 1.42iii to express U and V in terms of T and W.)

ANSWERS

1. $X \times Y = \begin{pmatrix} 0 \\ -1 \\ -2 \end{pmatrix}$. Remember to take the additive inverse for the second

coordinate. (Check: $(X \times Y) \cdot X = 0$, $(X \times Y) \cdot Y = 0$.)

3. $X \times Y = \begin{pmatrix} 0 \\ 0 \\ 0 \end{pmatrix}$. $\left(\text{Check: } Y = \dfrac{8}{3} X. \right)$ **5.** $X \times Y = 0_3$. (Check: $Y =$

$3X$.)   **7.** $N = \begin{pmatrix} 33 \\ 5 \\ -4 \end{pmatrix}$. Did you remember to take the additive inverse

for the second coordinate?   **9.** $N = \begin{pmatrix} -1 \\ 1 \\ 1 \end{pmatrix}$. You may have a different

answer, but it should be a nonzero multiple of the one given. Why?

11. $N = \begin{pmatrix} -19 \\ -4 \\ 1 \end{pmatrix}$ **13.** $\left(\text{Hint: } \text{Set } X = \begin{pmatrix} a \\ b \\ c \end{pmatrix}, \; Y = \begin{pmatrix} d \\ e \\ f \end{pmatrix} \right.$ and compute

cross-product. $\Big)$ **15.** Set $X = \begin{pmatrix} a \\ b \\ c \end{pmatrix}$. If Y is parallel to X, then $Y =$

$kX = \begin{pmatrix} ka \\ kb \\ kc \end{pmatrix}$ for some real number k or $X = 0_3$. Compute $X \times Y$.

5.   CHARACTERIZATIONS OF PLANES IN $\mathcal{R}^3$

In Section 4 we defined what is meant by a normal to a plane and found that every plane has a normal, namely, $N = U \times V$ where $\{U, V\}$ is a direction set for the plane. In this section it will be necessary to know that each nonzero vector N in $\mathcal{R}^3$ may be regarded as the normal to some plane.

THEOREM 2.18   Let $N = \begin{pmatrix} a \\ b \\ c \end{pmatrix}$ be a nonzero vector in $\mathcal{R}^3$.

i. There is always a linearly independent set $\{U, V\}$ in $\mathcal{R}^3$ such that N is perpendicular to both U and V. In fact, $\{U, V\}$ may be taken as a subset of

$$\left\{ \begin{pmatrix} -b \\ a \\ 0 \end{pmatrix}, \begin{pmatrix} -c \\ 0 \\ a \end{pmatrix}, \begin{pmatrix} 0 \\ -c \\ b \end{pmatrix} \right\}.$$

ii. N is a normal to some plane in $\mathcal{R}^3$. In fact, N is a normal to any plane with the set $\{U, V\}$ (found in part i) as a direction set.

Proof. Let $N = \begin{pmatrix} a \\ b \\ c \end{pmatrix}$ be a nonzero vector in $\mathcal{R}^3$. Each of the vectors in the set

$$\left\{ \begin{pmatrix} -b \\ a \\ 0 \end{pmatrix}, \begin{pmatrix} -c \\ 0 \\ a \end{pmatrix}, \begin{pmatrix} 0 \\ -c \\ b \end{pmatrix} \right\}$$

is easily shown to be perpendicular to N. For example,

$$N \cdot \begin{pmatrix} -b \\ a \\ 0 \end{pmatrix} = \begin{pmatrix} a \\ b \\ c \end{pmatrix} \cdot \begin{pmatrix} -b \\ a \\ 0 \end{pmatrix} = -ab + ab + 0 = 0.$$

We need only show that there are two nonparallel vectors in this set. If the vectors $\begin{pmatrix} -b \\ a \\ 0 \end{pmatrix}$ and $\begin{pmatrix} -c \\ 0 \\ a \end{pmatrix}$ are not parallel, then they may be taken as U and V, respectively. On the other hand, if these two vectors are parallel, then

$$\text{either} \quad \begin{pmatrix} -b \\ a \\ 0 \end{pmatrix} = k \begin{pmatrix} -c \\ 0 \\ a \end{pmatrix} \quad \text{or} \quad \begin{pmatrix} -c \\ 0 \\ a \end{pmatrix} = l \begin{pmatrix} -b \\ a \\ 0 \end{pmatrix}$$

for some choice of k or l. In either event we would have $a = 0$. Since $N \neq O_3$, it would then follow that either

(1) $b \neq 0$, which assures us that $\left\{ \begin{pmatrix} -b \\ a \\ 0 \end{pmatrix}, \begin{pmatrix} 0 \\ -c \\ b \end{pmatrix} \right\}$

  is a linearly independent set (why?),

or (2) $c \neq 0$, which assures us that $\left\{ \begin{pmatrix} -c \\ a \\ 0 \end{pmatrix}, \begin{pmatrix} 0 \\ -c \\ b \end{pmatrix} \right\}$

  is a linearly independent set (why?).

Statement i is thereby established. Let $\mathcal{G}$ be any plane with $\{U, V\}$ as a direction set. N is clearly a normal to $\mathcal{G}$ and statement ii follows. ∎

Example 12. Let $N = \begin{pmatrix} 1 \\ 2 \\ 3 \end{pmatrix}$. To determine a linearly independent set $\{U, V\}$ perpendicular to N, we form the tripleton set of the theorem. Note that each vector of this set is obtained by setting one coordinate equal to zero and determining the other two coordinates from the corresponding ones of N by means of an interchange and the taking of an additive inverse.

Thus,

$$\begin{pmatrix} 1 \\ 2 \\ 3 \end{pmatrix} \begin{cases} \left[\begin{array}{l} \text{Set third coordinate zero. Interchange first and second} \\ \text{coordinates of } N; \text{ take the additive inverse of one of them.} \end{array} \right] \rightarrow \begin{pmatrix} -2 \\ 1 \\ 0 \end{pmatrix} \\[2mm] \left[\begin{array}{l} \text{Set second coordinate zero. Interchange first and third} \\ \text{coordinates of } N; \text{ take the additive inverse of one of them.} \end{array} \right] \rightarrow \begin{pmatrix} -3 \\ 0 \\ 1 \end{pmatrix} \\[2mm] \left[\begin{array}{l} \text{Set first coordinate zero. Interchange second and third} \\ \text{coordinates of } N; \text{ take the additive inverse of one of them.} \end{array} \right] \rightarrow \begin{pmatrix} 0 \\ -3 \\ 2 \end{pmatrix} \end{cases}$$

Any pair of vectors from this last set will do. In particular,

$$\left\{ \begin{pmatrix} -3 \\ 0 \\ 1 \end{pmatrix}, \begin{pmatrix} 0 \\ -3 \\ 2 \end{pmatrix} \right\}$$

is a linearly independent set, each vector of which is perpendicular to N. ∎

Example 13. Let $N = \begin{pmatrix} 1 \\ 0 \\ 2 \end{pmatrix}$. To determine a linearly independent set $\{U, V\}$, each vector of which is perpendicular to N, we consider the tripleton set

$$\left\{ \begin{pmatrix} 0 \\ 1 \\ 0 \end{pmatrix}, \begin{pmatrix} -2 \\ 0 \\ 1 \end{pmatrix}, \begin{pmatrix} 0 \\ -2 \\ 0 \end{pmatrix} \right\}.$$

This time we must be careful in our selection, since the first and third vectors are parallel. Either of the other doubleton subsets are acceptable choices for $\{U, V\}$. ∎

Example 14. Let $N = \begin{pmatrix} 0 \\ 2 \\ 0 \end{pmatrix}$. The tripleton set generated by the method of Example 12 is

$$\left\{ \begin{pmatrix} -2 \\ 0 \\ 0 \end{pmatrix}, \begin{pmatrix} 0 \\ 0 \\ 0 \end{pmatrix}, \begin{pmatrix} 0 \\ 0 \\ 2 \end{pmatrix} \right\}.$$

From this set only one linearly independent doubleton subset can be chosen. Namely,

$$\left\{ \begin{pmatrix} -2 \\ 0 \\ 0 \end{pmatrix}, \begin{pmatrix} 0 \\ 0 \\ 2 \end{pmatrix} \right\}. \quad ∎$$

The reader should be aware that the technique used in Examples 12, 13, and 14 merely assures us of one linearly independent set $\{U, V\}$ whose vectors are perpendicular to the given N. There is an unlimited number of other possible choices. For instance, in Example 14 any set $\left\{ \begin{pmatrix} k \\ 0 \\ 0 \end{pmatrix}, \begin{pmatrix} 0 \\ 0 \\ l \end{pmatrix} \right\}$, with

k and l both nonzero, will suffice.

We now introduce the implicit representation for planes in $\mathcal{R}^3$.

THEOREM 2.19 (Implicit Representation of a Plane in $\mathcal{R}^3$.) Let $\mathcal{P}$ be a plane in $\mathcal{R}^3$. Let $\mathcal{P}$ contain the point A and have normal $N = \begin{pmatrix} a \\ b \\ c \end{pmatrix}$. $\mathcal{P}$ can be described in either of the following implicit forms:

i. (Vector Implicit Form.)

$$\mathcal{P} = \{X \mid (X - A) \cdot N = 0 \text{ and } X \in \mathcal{R}^3\}$$

ii. (Scalar Implicit Form.)

$$\mathcal{P} = \left\{ \begin{pmatrix} x \\ y \\ z \end{pmatrix} \middle| ax + by + cz + d = 0 \right\}, \text{ where}$$

$$d = -A \cdot N \text{ and } N = \begin{pmatrix} a \\ b \\ c \end{pmatrix} \neq O_3.$$

Moreover, any subset $\mathcal{P}$ of $\mathcal{R}^3$ that can be described in either of these two forms is a plane in $\mathcal{R}^3$ which contains A, has N as a normal, and has direction set $\{U, V\}$ where $\{U, V\}$ is any linearly independent doubleton subset of $\mathcal{R}^3$, each vector of which is perpendicular to N.

Proof. The proof is quite similar to that of Theorem 2.11 (Implicit Representation of a Line in $\mathcal{R}^2$).

Let $\mathcal{P}$ be a plane in $\mathcal{R}^3$ and let $\{U, V\}$ be a direction set for $\mathcal{P}$. In view of Theorem 2.17 the vectors U, V and the normal N constitute a linearly independent set $\{U, V, N\}$.

Proof of i. Let $\mathcal{K}$ be the subset of $\mathcal{R}^3$ described by

(2.13) $\mathcal{K} = \{X \mid (X - A) \cdot N = 0 \text{ and } X \in \mathcal{R}^3\}.$

We shall show that $\mathcal{P} \subset \mathcal{K}$ and that $\mathcal{K} \subset \mathcal{P}$, enabling us to conclude that $\mathcal{K} = \mathcal{P}$.

Let X be a point of $\mathcal{P}$. There are real numbers r and s such that

(2.14) $X = A + rU + sV.$

Starting with this equation we may write:

$$\begin{aligned}
X \in \mathcal{S} &\Rightarrow X = A + rU + sV \\
&\Rightarrow X - A = rU + sV \\
&\Rightarrow^1 (X - A) \cdot N = (rU + sV) \cdot N \\
&\Rightarrow^2 (X - A) \cdot N = r(U \cdot N) + s(V \cdot N) \\
&\Rightarrow^3 (X - A) \cdot N = 0 \\
&\Rightarrow^4 X \in \mathcal{K}
\end{aligned}$$

Therefore, $\mathcal{S} \subset \mathcal{K}$.

Now suppose that $X \in \mathcal{K}$. By Theorem 2.17ii (with $X - A$ in place of Z) there are scalars r, s, and t such that

(2.15) $X - A = rU + sV + tN.$

Now

$$\begin{aligned}
X \in \mathcal{K} \Rightarrow 0 &=^4 (X - A) \cdot N \\
&=^5 (rU + sV + tN) \cdot N \\
&=^2 r(U \cdot N) + s(V \cdot N) + t(N \cdot N) \\
&=^3 t(N \cdot N) \\
&\Rightarrow^6 t = 0 \\
&\Rightarrow^7 X - A = rU + sV \\
&\Rightarrow^8 X \in \mathcal{S}.
\end{aligned}$$

Thus, $\mathcal{K} \subset \mathcal{S}$ and the proof of Vector Implicit Form i is complete.

(1) Inner product of members of preceding equation with N.
(2) Theorem 2.2iv and v (Properties of the Inner Product).
(3) N is perpendicular to U and V: $N \cdot U = N \cdot V = 0$.
(4) Definition of $\mathcal{K}$, Equation 2.13.
(5) Substitution from Equation 2.15.
(6) Since $N \neq 0$, we have $N \cdot N \neq 0$. Hence $t(N \cdot N) = 0 \Rightarrow t = 0$.
(7) Substitution into Equation 2.15.
(8) Equation 2.14 defining $\mathcal{S}$.

Proof of ii. Statement ii is proved in a manner similar to that given for the corresponding part of Theorem 2.11 and is called for as Exercise 25.

We must still prove the last statement of the theorem. Let $\mathcal{S}$ be a subset of $\mathfrak{R}^3$ described by the Vector Implicit Form i, that is

(2.16) $\mathcal{S} = \{X \mid (X - A) \cdot N = 0 \text{ and } X \in \mathfrak{R}^3\}, \qquad N \neq O_3.$

We must show that $\mathcal{S}$ is a plane with normal N. By Theorem 2.18 we know that for a given $N \neq O_3$ there is at least one linearly independent set $\{U, V\}$ such that N is perpendicular to both U and V. Assume that we have selected any one set $\{U, V\}$. The set $\mathcal{M}$, given by

(2.17) $\mathcal{M} = \{A + rU + sV \mid r, s \in \mathfrak{R}\},$

is a plane in $\mathcal{R}^3$ and has N as a normal. Employing Statement i of the present theorem, we conclude that Equations 2.16 and 2.17 describe the same plane. That is, the set $\mathcal{S}$ described by Equation 2.16 is a plane which contains point A, has N for a normal, and has $\{U, V\}$ as a direction set.

Finally, suppose that $\mathcal{S}$ is a subset of $\mathcal{R}^3$ described by the Scalar Implicit Form ii. In this case we set $N = \begin{pmatrix} a \\ b \\ c \end{pmatrix}$ and employ the technique of the proof of Theorem 2.7 to determine a vector A such that $N \cdot A = -d$. The Scalar Implicit Form ii may then be written in Vector Implicit Form (Equation 2.16) and the proof continued as above. ∎

Example 15. Find an implicit representation of the plane $\mathcal{S}$ which contains the point $A = \begin{pmatrix} -2 \\ 3 \\ 3 \end{pmatrix}$ and has direction set $\left\{ \begin{pmatrix} 2 \\ 0 \\ 1 \end{pmatrix}, \begin{pmatrix} 1 \\ -2 \\ 1 \end{pmatrix} \right\}$. An examination of Theorem 2.19 shows that in addition to a point A on $\mathcal{S}$ we shall also need a normal N. According to Theorem 2.16, with $k = 1$, the cross-product of the vectors of a direction set for $\mathcal{S}$ will be a normal for $\mathcal{S}$. Hence, we may select

$$N = \begin{pmatrix} 2 \\ 0 \\ 1 \end{pmatrix} \times \begin{pmatrix} 1 \\ -2 \\ 1 \end{pmatrix} = \begin{pmatrix} 2 \\ -1 \\ -4 \end{pmatrix}.$$

(Check the inner products to verify that this is indeed a normal for $\mathcal{S}$.) A vector implicit representation for $\mathcal{S}$ is thus

$$\mathcal{S} = \left\{ X \,\middle|\, \left[X - \begin{pmatrix} -2 \\ 3 \\ 3 \end{pmatrix} \right] \cdot \begin{pmatrix} 2 \\ -1 \\ -4 \end{pmatrix} = 0, X \in \mathcal{R}^3 \right\}.$$

We obtain a scalar implicit representation for $\mathcal{S}$ by setting $X = \begin{pmatrix} x \\ y \\ z \end{pmatrix}$ and computing the indicated inner product:

$$\mathcal{S} = \left\{ X \,\middle|\, \left[\begin{pmatrix} x \\ y \\ z \end{pmatrix} - \begin{pmatrix} -2 \\ 3 \\ 3 \end{pmatrix} \right] \cdot \begin{pmatrix} 2 \\ -1 \\ -4 \end{pmatrix} = 0, X \in \mathcal{R}^3 \right\}$$

$$= \{ X \mid 2x - y - 4z + 19 = 0, X \in \mathcal{R}^3 \}. \quad ∎$$

Example 16. Find a parametric representation for the plane $\mathcal{S}$ with the implicit representation

(2.18) $\mathcal{S} = \{ X \mid 2x + 3y - 5z + 12 = 0, X \in \mathcal{R}^3 \}.$

In view of the last theorem, the coefficients of $x, y,$ and z are the coordinates of a normal N to $\mathcal{S}$. Thus, $N = \begin{pmatrix} 2 \\ 3 \\ -5 \end{pmatrix}$ is a normal to $\mathcal{S}$. Using the technique of Example 12, we determine a linearly independent set $\{U, V\}$ of direction vectors for $\mathcal{S}$. Hence, we consider the tripleton set

$$\left\{ \begin{pmatrix} -3 \\ 2 \\ 0 \end{pmatrix}, \begin{pmatrix} 5 \\ 0 \\ 2 \end{pmatrix}, \begin{pmatrix} 0 \\ 5 \\ 3 \end{pmatrix} \right\}.$$

From it we select the direction set $\left\{ \begin{pmatrix} -3 \\ 2 \\ 0 \end{pmatrix}, \begin{pmatrix} 5 \\ 0 \\ 2 \end{pmatrix} \right\}$ for $\mathcal{S}$. To determine a

point A, we use Equation 2.18 to determine a value x_0 such that $A = \begin{pmatrix} x_0 \\ 0 \\ 0 \end{pmatrix}$

is a point of $\mathcal{S}$. (In other problems it would perhaps be necessary to find a

value y_0 such that $\begin{pmatrix} 0 \\ y_0 \\ 0 \end{pmatrix}$ is a point of $\mathcal{S}$.) From Equation 2.18 we find that

x_0 must be such that

$$2x_0 + 3 \cdot 0 + 5 \cdot 0 + 12 = 0.$$

Thus, $x_0 = -6$ and $A = \begin{pmatrix} -6 \\ 0 \\ 0 \end{pmatrix}$. $\mathcal{S}$, therefore, has the parametric description

$$\mathcal{S} = \left\{ \begin{pmatrix} -6 \\ 0 \\ 0 \end{pmatrix} + r \begin{pmatrix} -3 \\ 2 \\ 0 \end{pmatrix} + s \begin{pmatrix} 5 \\ 0 \\ 2 \end{pmatrix} \middle| r, s \in \mathcal{R} \right\}. \quad \blacksquare$$

We close this section with a summary of various representations for planes in $\mathcal{R}^3$. Included in the summary are two implicit forms which are easily reducible to the forms discussed in this section. There is also included an *explicit* form which is applicable only to those planes which intersect the z axis in a single point.

SUMMARY 2.20   (Representations of Planes in $\mathcal{R}^3$.)

i. A subset $\mathcal{S}$ of $\mathcal{R}^3$ is a plane if and only if $\mathcal{S}$ is the set of all X in $\mathcal{R}^3$ satisfying any one of the following equations:

Parametric Equation

Vector Form:

$X = A + rU + sV.$

Scalar Form:

$$\begin{pmatrix} x \\ y \\ z \end{pmatrix} = \begin{pmatrix} x_0 \\ y_0 \\ z_0 \end{pmatrix} + r \begin{pmatrix} a \\ b \\ c \end{pmatrix} + s \begin{pmatrix} d \\ e \\ f \end{pmatrix}.$$

Point on $\mathcal{S}$: $A = \begin{pmatrix} x_0 \\ y_0 \\ z_0 \end{pmatrix}.$

Direction Set:

$$\{U, V\} = \left\{ \begin{pmatrix} a \\ b \\ c \end{pmatrix}, \begin{pmatrix} d \\ e \\ f \end{pmatrix} \right\}.$$

Normal Vector: $N = U \times V.$

Implicit Equation

$\left\{\begin{array}{l} \text{Vector Form:} \\ \quad (X - A) \cdot N = 0. \\ \\ \\ \\ \text{Scalar Form: } a(x - x_0) \\ \quad + b(y - y_0) + c(z - z_0) = 0. \end{array}\right.$ $\left|\begin{array}{l} \text{Point on } \mathscr{S}: \ A = \begin{pmatrix} x_0 \\ y_0 \\ z_0 \end{pmatrix}. \\ \\ \text{Direction Set: A subset of} \\ \quad \left\{ \begin{pmatrix} -b \\ a \\ 0 \end{pmatrix}, \begin{pmatrix} -c \\ 0 \\ a \end{pmatrix}, \begin{pmatrix} 0 \\ -c \\ b \end{pmatrix} \right\}. \\ \\ \text{Normal Vector: } N = \begin{pmatrix} a \\ b \\ c \end{pmatrix}. \end{array}\right.$

or

$\left\{\begin{array}{l} \text{Vector Form:} \\ \quad X \cdot N + d = 0. \\ \\ \\ \text{Scalar Form:} \\ \quad ax + by + cz + d = 0. \end{array}\right.$ $\left|\begin{array}{l} \text{Point on } \mathscr{S}: \ A = \text{any solution} \\ \begin{pmatrix} x_0 \\ y_0 \\ z_0 \end{pmatrix} \text{ of equation.} \\ \text{Direction Set: A subset of} \\ \quad \left\{ \begin{pmatrix} -b \\ a \\ 0 \end{pmatrix}, \begin{pmatrix} -c \\ 0 \\ a \end{pmatrix}, \begin{pmatrix} 0 \\ -c \\ b \end{pmatrix} \right\}. \\ \text{Normal Vector: } N = \begin{pmatrix} a \\ b \\ c \end{pmatrix} \text{such that} \\ d = -A \cdot N = -ax_0 - by_0 - cz_0. \end{array}\right.$

ii. A subset $\mathscr{S}$ of $\mathscr{R}^3$ is a plane with direction set $\left\{ \begin{pmatrix} 1 \\ 0 \\ m \end{pmatrix}, \begin{pmatrix} 0 \\ 1 \\ n \end{pmatrix} \right\}$ and z-intercept c if and only if $\mathscr{S}$ is the set of all X in $\mathscr{R}^3$ satisfying the

Explicit Equation

$z = mx + ny + c.$ $\left|\begin{array}{l} \text{Point on } \mathscr{S}: \ A = \begin{pmatrix} 0 \\ 0 \\ c \end{pmatrix}. \\ \\ \text{Direction Set:} \\ \quad \{U, V\} = \left\{ \begin{pmatrix} 1 \\ 0 \\ m \end{pmatrix}, \begin{pmatrix} 0 \\ 1 \\ n \end{pmatrix} \right\}. \\ \\ \text{Normal Vector: } N = \begin{pmatrix} -m \\ -n \\ 1 \end{pmatrix}. \end{array}\right.$

EXERCISES

In Exercises 1 to 6 use the technique of Examples 12, 13, and 14 to determine a direction set $\{U, V\}$ for a plane with the given normal N:

1. $N = \begin{pmatrix} 5 \\ -1 \\ 2 \end{pmatrix}$.

2. $N = \begin{pmatrix} 1 \\ 0 \\ 3 \end{pmatrix}$.

3. $N = \begin{pmatrix} 0 \\ 0 \\ 1 \end{pmatrix}$.

4. $N = \begin{pmatrix} 0 \\ 12 \\ -9 \end{pmatrix}$.

5. $N = \begin{pmatrix} 3 \\ -2 \\ -17 \end{pmatrix}$.

6. $N = \begin{pmatrix} 253 \\ 16 \\ 407 \end{pmatrix}$.

In Exercises 7 to 12 determine a vector implicit representation and a scalar implicit representation for the plane $\mathcal{S}$ described:

7. $\mathcal{S} = \left\{ \begin{pmatrix} 2 \\ 3 \\ 1 \end{pmatrix} + r \begin{pmatrix} 1 \\ 0 \\ 5 \end{pmatrix} + s \begin{pmatrix} 2 \\ -1 \\ 3 \end{pmatrix} \middle| r, s \in \mathcal{R} \right\}.$

8. $\mathcal{S} = \left\{ r \begin{pmatrix} 1 \\ 1 \\ 4 \end{pmatrix} + s \begin{pmatrix} 3 \\ -2 \\ 1 \end{pmatrix} \middle| r, s \in \mathcal{R} \right\}.$

9. $\mathcal{S} = \left\{ \begin{pmatrix} x \\ y \\ z \end{pmatrix} \middle| z = 2x - 4y + 2 \right\}.$ 10. $\mathcal{S} = \left\{ \begin{pmatrix} x \\ y \\ z \end{pmatrix} \middle| z = x - 2 \right\}.$

11. $\mathcal{S}$ is the plane which contains $\mathcal{L}_1 = \left\{ \begin{pmatrix} 1 \\ 2 \\ 1 \end{pmatrix} + t \begin{pmatrix} 2 \\ -1 \\ 1 \end{pmatrix} \middle| t \in \mathcal{R} \right\}$ and is parallel to $\mathcal{L}_2 = \left\{ r \begin{pmatrix} 4 \\ 0 \\ 2 \end{pmatrix} \middle| r \in \mathcal{R} \right\}.$

12. $\mathcal{S}$ contains the points $A = \begin{pmatrix} 1 \\ 1 \\ 1 \end{pmatrix}$, $B = \begin{pmatrix} 3 \\ 2 \\ -1 \end{pmatrix}$, and $C = \begin{pmatrix} 4 \\ -1 \\ 2 \end{pmatrix}$.

In Exercises 13 to 20 determine a parametric and, where possible, an explicit representation for the plane $\mathcal{S}$ with the given implicit description:

13. $\mathcal{S} = \left\{ X \middle| \left[X - \begin{pmatrix} 2 \\ 1 \\ 1 \end{pmatrix} \right] \cdot \begin{pmatrix} 1 \\ 2 \\ 3 \end{pmatrix} = 0, X \in \mathcal{R}^3 \right\}.$

14. $\mathcal{S} = \left\{ X \middle| X \cdot \begin{pmatrix} 3 \\ 4 \\ 1 \end{pmatrix} = 0, X \in \mathcal{R}^3 \right\}.$

15. $\mathcal{S} = \left\{ X \middle| X \cdot \begin{pmatrix} 5 \\ 4 \\ 2 \end{pmatrix} - 4 = 0, X \in \mathcal{R}^3 \right\}.$

16. $\mathcal{S} = \left\{ X \middle| X \cdot \begin{pmatrix} 2 \\ 1 \\ 3 \end{pmatrix} - 2 = 0, X \in \mathcal{R}^3 \right\}.$

17. $\mathcal{S} = \left\{ \begin{pmatrix} x \\ y \\ z \end{pmatrix} \middle| 3(x - 2) + (y - 1) = 0 \right\}.$

18. $\mathcal{S} = \left\{ \begin{pmatrix} x \\ y \\ z \end{pmatrix} \middle| 3x + 2(z - 1) = 0 \right\}.$

19. $\mathcal{S} = \left\{ \begin{pmatrix} x \\ y \\ z \end{pmatrix} \middle| 2x + 5y - z + 4 = 0 \right\}.$

20. $\mathcal{S} = \left\{ \begin{pmatrix} x \\ y \\ z \end{pmatrix} \middle| 3x + 4y = 0 \right\}.$

21. Let $N = \begin{pmatrix} a \\ b \\ c \\ d \end{pmatrix}$ be a nonzero vector in $\mathcal{R}^4$. Show that there is a linearly independent set $\{U, V, W\}$ each vector of which is perpendicular to N.

22. Let $\mathcal{S}$ be a set in $\mathcal{R}^3$ given by $\mathcal{S} = \left\{ \begin{pmatrix} x \\ y \\ z \end{pmatrix} \middle| a(x - x_0) + b(y - y_0) + c(z - z_0) = 0 \right\}.$ Show that $\mathcal{S}$ is a plane with normal $\begin{pmatrix} a \\ b \\ c \end{pmatrix}$.

23. Let $\mathcal{S}$ be a set in $\mathcal{R}^3$ given by $\mathcal{S} = \{X \mid X \cdot N + c = 0, X \in \mathcal{R}^3\}$. Show that $\mathcal{S}$ is a plane with normal N.

24. Let $\mathcal{S}$ be a set in $\mathcal{R}^3$ given by $\mathcal{S} = \left\{ \begin{pmatrix} x \\ y \\ z \end{pmatrix} \middle| z = mx + ny + c \right\}.$ Show that $\mathcal{S}$ is a plane with direction set $\left\{ \begin{pmatrix} 1 \\ 0 \\ m \end{pmatrix}, \begin{pmatrix} 0 \\ 1 \\ n \end{pmatrix} \right\}$ and with normal $N = \begin{pmatrix} -m \\ -n \\ 1 \end{pmatrix}$.

25. Complete the proof of Theorem 2.19 by showing that a plane $\mathcal{S}$ with normal $N = \begin{pmatrix} a \\ b \\ c \end{pmatrix}$ can be described by $\mathcal{S} = \left\{ \begin{pmatrix} x \\ y \\ z \end{pmatrix} \middle| ax + by + cz + d = 0 \right\},$ where $d = -A \cdot N$ for any point A of $\mathcal{S}$.

ANSWERS

1. Doubleton subset of $\left\{ \begin{pmatrix} 1 \\ 5 \\ 0 \end{pmatrix}, \begin{pmatrix} 2 \\ 0 \\ -5 \end{pmatrix}, \begin{pmatrix} 0 \\ 2 \\ 1 \end{pmatrix} \right\}.$ 3. $\left\{ \begin{pmatrix} 1 \\ 0 \\ 0 \end{pmatrix}, \begin{pmatrix} 0 \\ 1 \\ 0 \end{pmatrix} \right\}.$

5. Doubleton subset of $\left\{ \begin{pmatrix} 2 \\ 3 \\ 0 \end{pmatrix}, \begin{pmatrix} 17 \\ 0 \\ 3 \end{pmatrix}, \begin{pmatrix} 0 \\ 17 \\ -2 \end{pmatrix} \right\}.$

7. $\mathcal{S} = \left\{ X \middle| \left[X - \begin{pmatrix} 2 \\ 3 \\ 1 \end{pmatrix} \right] \cdot \begin{pmatrix} 5 \\ 7 \\ -1 \end{pmatrix} = 0, X \in \mathcal{R}^3 \right\},$ and

$\mathcal{S} = \left\{ \begin{pmatrix} x \\ y \\ z \end{pmatrix} \middle| 5x + 7y - z - 30 = 0 \right\}.$

9. $\mathcal{S} = \left\{ X \,\middle|\, \left[X - \begin{pmatrix} 0 \\ 0 \\ 2 \end{pmatrix} \right] \cdot \begin{pmatrix} 2 \\ -4 \\ -1 \end{pmatrix} = 0, X \in \mathcal{R}^3 \right\}$,

and $\mathcal{S} = \left\{ \begin{pmatrix} x \\ y \\ z \end{pmatrix} \,\middle|\, 2x - 4y - z + 2 = 0 \right\}$.

11. $\mathcal{S} = \left\{ X \,\middle|\, \left[X - \begin{pmatrix} 1 \\ 2 \\ 1 \end{pmatrix} \right] \cdot \begin{pmatrix} 2 \\ 0 \\ -4 \end{pmatrix} = 0, X \in \mathcal{R}^3 \right\}$ and $\mathcal{S} = \left\{ \begin{pmatrix} x \\ y \\ z \end{pmatrix} \,\middle|\, 2x - 4z + 2 = 0 \right\}$. **13.** $\mathcal{S} = \left\{ \begin{pmatrix} 2 \\ 1 \\ 1 \end{pmatrix} + r \begin{pmatrix} 2 \\ -1 \\ 0 \end{pmatrix} + s \begin{pmatrix} 3 \\ 0 \\ -1 \end{pmatrix} \,\middle|\, r, s \in \mathcal{R} \right\} = \left\{ \begin{pmatrix} x \\ y \\ z \end{pmatrix} \,\middle|\, z = -\frac{1}{3}x - \frac{2}{3}y + \frac{7}{3} \right\}$. **15.** $\mathcal{S} = \left\{ \begin{pmatrix} 0 \\ 1 \\ 0 \end{pmatrix} + r \begin{pmatrix} 4 \\ -5 \\ 0 \end{pmatrix} + s \begin{pmatrix} 2 \\ 0 \\ -5 \end{pmatrix} \,\middle|\, r, s \in \mathcal{R} \right\} = \left\{ \begin{pmatrix} x \\ y \\ z \end{pmatrix} \,\middle|\, z = -\frac{5}{2}x - 2y + 2 \right\}$.

17. $\mathcal{S} = \left\{ \begin{pmatrix} 2 \\ 1 \\ 0 \end{pmatrix} + r \begin{pmatrix} 1 \\ -3 \\ 0 \end{pmatrix} + s \begin{pmatrix} 0 \\ 0 \\ 1 \end{pmatrix} \,\middle|\, r, s \in \mathcal{R} \right\}$, no explicit form.

19. $\mathcal{S} = \left\{ \begin{pmatrix} 0 \\ 0 \\ 4 \end{pmatrix} + r \begin{pmatrix} 5 \\ -2 \\ 0 \end{pmatrix} + s \begin{pmatrix} 1 \\ 0 \\ 2 \end{pmatrix} \,\middle|\, r, s \in \mathcal{R} \right\} = \left\{ \begin{pmatrix} x \\ y \\ z \end{pmatrix} \,\middle|\, z = 2x + 5y + 4 \right\}$.

6. INTERSECTIONS INVOLVING PLANES IN $\mathcal{R}^3$

The implicit representation of a plane in $\mathcal{R}^3$ enables us to give rather simple proof of two theorems that were stated in Chapter 1 without proof. The theorems referred to are Theorem 1.50 (Intersection of Two Planes in $\mathcal{R}^3$), given on page 79, and Theorem 1.51 (Intersection of a Line and a Plane in $\mathcal{R}^3$), given on page 80. We shall consider the proofs of those theorems in reverse order.

Theorem 1.51 states that a line $\mathcal{L}$ and a plane $\mathcal{S}$ in $\mathcal{R}^3$ must intersect in exactly one of the following:

i. The empty set: $\mathcal{L} \cap \mathcal{S} = \varnothing$.

ii. A single point P: $\mathcal{L} \cap \mathcal{S} = \{P\}$.

iii. The entire line $\mathcal{L}$: $\mathcal{L} \cap \mathcal{S} = \mathcal{L}$.

Moreover, in the first and third cases $\mathcal{L}$ and $\mathcal{S}$ are parallel, while in case ii $\mathcal{L}$ is not parallel to $\mathcal{S}$.

Proof of Theorem 1.51. Let $\mathcal{L}$ be a line in $\mathcal{R}^3$ given by the parametric equation

(2.19) $\mathcal{L}: X = A + tD.$

Let $\mathcal{S}$ be a plane in $\mathcal{R}^3$ given by the implicit equation

(2.20) $\mathcal{S}: (X - B) \cdot N = 0.$

To determine $\mathcal{L} \cap \mathcal{S}$ we need only determine those values of t such that the point X given by Equation 2.19 is also a point of $\mathcal{S}$. Thus, from Equations 2.19 and 2.20 we have

$$X \in \mathcal{L} \cap \mathcal{S} \Leftrightarrow [(A + tD) - B] \cdot N = 0 \text{ and } X = A + tD$$
$$\Leftrightarrow [(A - B) + tD] \cdot N = 0 \text{ and } X = A + tD$$
$$\Leftrightarrow (A - B) \cdot N + t(D \cdot N) = 0 \text{ and } X = A + tD.$$

We use the last of the above two equations to determine admissible values of t. There are three cases.

i. $D \cdot N = 0$ *and* $(A - B) \cdot N = 0$. In this case the previous implication reduces to

$$X \in \mathcal{L} \cap \mathcal{S} \Leftrightarrow X = A + tD \quad \text{and} \quad 0 + t \cdot 0 = 0.$$

Since this last equation is satisfied for all values of t, it follows that every point of $\mathcal{L}$ is also a point of $\mathcal{S}$: $\mathcal{L} \cap \mathcal{S} = \mathcal{L}$. Moreover, it is easy to show that since $D \cdot N = 0$ the vector D is a direction vector for $\mathcal{S}$ (see Exercise 13). Thus, $\mathcal{L}$ is parallel to $\mathcal{S}$.

ii. $D \cdot N \neq 0$. In this case the implication chain results in

$$X \in \mathcal{L} \cap \mathcal{S} \Leftrightarrow X = A + tD \quad \text{and} \quad (A - B) \cdot N + t(D \cdot N) = 0$$
$$\Leftrightarrow X = A + tD \quad \text{and} \quad t = -\frac{(A - B) \cdot N}{D \cdot N}$$
$$\Leftrightarrow X = A - \frac{(A - B) \cdot N}{D \cdot N} D.$$

Thus, $\mathcal{L} \cap \mathcal{S}$ consists of but one point. Since $D \cdot N \neq 0$, D cannot be a direction vector for $\mathcal{S}$, and thus $\mathcal{L}$ is not parallel to $\mathcal{S}$.

iii. $D \cdot N = 0$ *and* $(A - B) \cdot N = k \neq 0$. For this case the implication chain above becomes

$$X \in \mathcal{L} \cap \mathcal{S} \Leftrightarrow X = A + tD \quad \text{and} \quad k + t \cdot 0 = 0.$$

The last equation can be satisfied for no value of t since $k + t \cdot 0 = k \neq 0$. Thus, $\mathcal{L} \cap \mathcal{S} = \varnothing$. Again, since $D \cdot N = 0$, we have $\mathcal{L}$ parallel to $\mathcal{S}$. ■

Example 17. Let $\mathcal{L} = \left\{ \begin{pmatrix} 1 \\ 2 \\ 1 \end{pmatrix} + t \begin{pmatrix} 2 \\ -3 \\ 1 \end{pmatrix} \middle| t \in \mathcal{R} \right\}$. Each point X of $\mathcal{L}$ is given by

$$(2.21) \quad \mathcal{L}: X = \begin{pmatrix} 1 \\ 2 \\ 1 \end{pmatrix} + t \begin{pmatrix} 2 \\ -3 \\ 1 \end{pmatrix},$$

where $D = \begin{pmatrix} 2 \\ -3 \\ 1 \end{pmatrix}$ is a direction vector of $\mathcal{L}$.

(a) Find $\mathfrak{L} \cap \mathfrak{S}_1$, where $\mathfrak{S}_1$ is given by the implicit equation

(2.22) $\mathfrak{S}_1: \left[X - \begin{pmatrix} -2 \\ 1 \\ 4 \end{pmatrix} \right] \cdot \begin{pmatrix} 3 \\ 5 \\ 1 \end{pmatrix} = 0,$ where $N_1 = \begin{pmatrix} 3 \\ 5 \\ 1 \end{pmatrix}$ is a normal for $\mathfrak{S}_1$.

We determine those values of t such that the point X of $\mathfrak{L}$ given by Equation 2.21 is also a point of $\mathfrak{S}$, given by Equation 2.22. Substitution from Equation 2.21 into Equation 2.22 gives

$$0 = \left[\begin{pmatrix} 1 \\ 2 \\ 1 \end{pmatrix} + t\begin{pmatrix} 2 \\ -3 \\ 1 \end{pmatrix} - \begin{pmatrix} -2 \\ 1 \\ 4 \end{pmatrix} \right] \cdot \begin{pmatrix} 3 \\ 5 \\ 1 \end{pmatrix}$$

$$= \left[\begin{pmatrix} 3 \\ 1 \\ -3 \end{pmatrix} + t\begin{pmatrix} 2 \\ -3 \\ 1 \end{pmatrix} \right] \cdot \begin{pmatrix} 3 \\ 5 \\ 1 \end{pmatrix}$$

$$= 11 + t(-8)$$

$$= 11 - 8t.$$

The coefficient of t is $D \cdot N = -8 \neq 0$. This is Case ii of Theorem 1.51. Solving for t we find that $t = 11/8$ and $\mathfrak{L} \cap \mathfrak{S}$ consists of the single point

$$X = \begin{pmatrix} 1 \\ 2 \\ 1 \end{pmatrix} + \left(\frac{11}{8}\right)\begin{pmatrix} 2 \\ -3 \\ 1 \end{pmatrix} = \begin{pmatrix} \dfrac{30}{8} \\ \dfrac{-17}{8} \\ \dfrac{19}{8} \end{pmatrix}.$$

(b) Find $\mathfrak{L} \cap \mathfrak{S}_2$, where $\mathfrak{S}_2$ is given by the implicit equation

$$\mathfrak{S}_2: \left[X - \begin{pmatrix} 2 \\ 0 \\ 1 \end{pmatrix} \right] \cdot \begin{pmatrix} 2 \\ 1 \\ -1 \end{pmatrix} = 0,$$ where $N_2 = \begin{pmatrix} 2 \\ 1 \\ -1 \end{pmatrix}$ is a normal for $\mathfrak{S}_2$.

Substitution from Equation 2.21 into the above yields

$$0 = \left[\begin{pmatrix} 1 \\ 2 \\ 1 \end{pmatrix} + t\begin{pmatrix} 2 \\ -3 \\ 1 \end{pmatrix} - \begin{pmatrix} 2 \\ 0 \\ 1 \end{pmatrix} \right] \cdot \begin{pmatrix} 2 \\ 1 \\ -1 \end{pmatrix}$$

$$= \left[\begin{pmatrix} -1 \\ 2 \\ 0 \end{pmatrix} + t\begin{pmatrix} 2 \\ -3 \\ 1 \end{pmatrix} \right] \cdot \begin{pmatrix} 2 \\ 1 \\ -1 \end{pmatrix}$$

$$= 0 + t \cdot 0$$

$$= 0.$$

This equation is satisfied for all values of t. Thus, every point of $\mathfrak{L}$ is also a point of $\mathfrak{S}_2 : \mathfrak{L} \cap \mathfrak{S}_2 = \mathfrak{L}$. In this case the coefficient of t is $D \cdot N_2 = 0$, so that $\mathfrak{L}$ is parallel to $\mathfrak{S}_2$.

(c) Find $\mathcal{L} \cap \mathcal{S}_3$, where $\mathcal{S}_3$ is given by the implicit equation

$$\mathcal{S}_3: \quad \left[X - \begin{pmatrix} 4 \\ 2 \\ 3 \end{pmatrix} \right] \cdot \begin{pmatrix} 1 \\ 1 \\ 1 \end{pmatrix} = 0, \quad \text{where} \quad N_3 = \begin{pmatrix} 1 \\ 1 \\ 1 \end{pmatrix} \text{ is a normal for } \mathcal{S}_3.$$

Substitution from Equation 2.21 gives

$$0 = \left[\begin{pmatrix} 1 \\ 2 \\ 1 \end{pmatrix} + t \begin{pmatrix} 2 \\ -3 \\ 1 \end{pmatrix} - \begin{pmatrix} 4 \\ 2 \\ 3 \end{pmatrix} \right] \cdot \begin{pmatrix} 1 \\ 1 \\ 1 \end{pmatrix}$$

$$= \left[\begin{pmatrix} -3 \\ 0 \\ -2 \end{pmatrix} + t \begin{pmatrix} 2 \\ -3 \\ 1 \end{pmatrix} \right] \cdot \begin{pmatrix} 1 \\ 1 \\ 1 \end{pmatrix}$$

$$= -5 + t \cdot 0$$

$$= -5.$$

This equation is satisfied for no value of t. Consequently, no point on $\mathcal{L}$ is also on $\mathcal{S}_3$ and $\mathcal{L} \cap \mathcal{S}_3 = \varnothing$. Again we note that the coefficient of t is $D \cdot N_3 = 0$ and $\mathcal{L}$ is parallel to $\mathcal{S}$. ∎

We now consider Theorem 1.50 which states that two planes, $\mathcal{S}_1$ and $\mathcal{S}_2$ in $\mathcal{R}^3$, must intersect in one of the following:

  i. The empty set: $\mathcal{S}_1 \cap \mathcal{S}_2 = \varnothing$. *if parallel*
  ii. A line $\mathcal{L}$: $\mathcal{S}_1 \cap \mathcal{S}_2 = \mathcal{L}$.
  iii. A plane: $\mathcal{S}_1 \cap \mathcal{S}_2 = \mathcal{S}_1 = \mathcal{S}_2$. *if parallel*

Moreover, in cases i and iii the planes must be parallel and in case ii the planes are not parallel.

Proof of Theorem 1.50. Let $\mathcal{S}_1$ and $\mathcal{S}_2$ be two planes in $\mathcal{R}^3$. Suppose that $\mathcal{S}_1$ and $\mathcal{S}_2$ are parallel. It is possible that $\mathcal{S}_1 \cap \mathcal{S}_2$ may be the empty set, as the first of the examples that follow will show. If $\mathcal{S}_1 \cap \mathcal{S}_2$ is not empty, the planes share a common point A as well as a common direction set $\{U, V\}$ and, according to Theorem 1.44 (p. 68) must be identical. In this case $\mathcal{S}_1 \cap \mathcal{S}_2 = \mathcal{S}_1 = \mathcal{S}_2$. Thus, when $\mathcal{S}_1$ and $\mathcal{S}_2$ are parallel we have either case i or case iii mentioned previously.

Suppose now that $\mathcal{S}_1$ and $\mathcal{S}_2$ are not parallel. We first show that $\mathcal{S}_1 \cap \mathcal{S}_2$ contains at least one point. Let $\{U, V\}$ be a direction set for $\mathcal{S}_1$ and let A be a point of $\mathcal{S}_1$. We may write $\mathcal{S}_1 = \{A + rU + sV \mid r, s \in \mathcal{R}\}$. At least one vector of the direction set, say the vector U, is not a direction vector for $\mathcal{S}_2$. (Otherwise $\mathcal{S}_1$ and $\mathcal{S}_2$ would be parallel.) It follows that U is not parallel to $\mathcal{S}_2$. Consequently, the line $\mathcal{K} = \{A + rU \mid r \in \mathcal{R}\}$ is not parallel to $\mathcal{S}_2$. According to Theorem 1.51 (just proved), $\mathcal{K}$ and $\mathcal{S}_2$ intersect in exactly one point, call it B. Now $\mathcal{K}$ is clearly a subset of $\mathcal{S}_1$. Thus, we conclude that B is a point of $\mathcal{S}_1 \cap \mathcal{S}_2$ (see Figure 2.4a).

Let N_1 and N_2 be normals to $\mathcal{S}_1$ and $\mathcal{S}_2$, respectively. Since $\mathcal{S}_1$ and $\mathcal{S}_2$ are not parallel, it follows that N_1 and N_2 are not parallel and that $N_1 \times N_2 \neq O_3$ (see Exercise 14). Thus, $D = N_1 \times N_2$ is a nonzero vector, which,

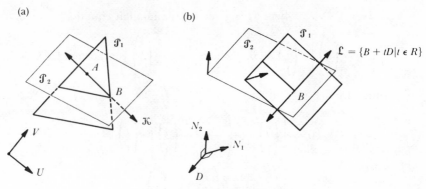

Figure 2.4

according to Theorem 2.15i, is perpendicular to N_1 and N_2. D is, therefore, a direction vector for both $\mathscr{I}_1$ and $\mathscr{I}_2$ (see Figure 2.4b and Exercise 15). The line $\mathscr{L}$ defined by

$$\mathscr{L} = \{B + tD \mid t \in \mathscr{R}\}$$

has the point B and the direction vector D in common with both $\mathscr{I}_1$ and $\mathscr{I}_2$. Thus, $\mathscr{L}$ lies in both planes and is a subset of $\mathscr{I}_1 \cap \mathscr{I}_2$. Using Theorem 1.47 (Intersection of Two Planes in $\mathscr{R}^n$) (p. 71) we note that $\mathscr{I}_1 \cap \mathscr{I}_2$ will contain points other than those of $\mathscr{L}$ only if $\mathscr{I}_1$ and $\mathscr{I}_2$ coincide. However, since $\mathscr{I}_1$ and $\mathscr{I}_2$ are not parallel, they cannot coincide. Thus, we conclude that $\mathscr{I}_1 \cap \mathscr{I}_2 = \mathscr{L}$.

We have shown that two planes $\mathscr{I}_1$ and $\mathscr{I}_2$ in $\mathscr{R}^3$ have as their intersection (i) an empty set, (ii) a line $\mathscr{L}$, or (iii) a plane. Moreover, we have shown that wherever $\mathscr{I}_1$ and $\mathscr{I}_2$ are not parallel the intersection is a line. It follows that cases i and iii can arise only if the two planes are parallel.

Example 18. Let $\mathscr{I}_1$ be the plane given by

$$(2.23) \qquad \mathscr{I}_1 = \left\{ X \;\middle|\; X \cdot \begin{pmatrix} 4 \\ -1 \\ -2 \end{pmatrix} - 5 = 0 \text{ and } X \in \mathscr{R}^3 \right\}.$$

The vector $N_1 = \begin{pmatrix} 4 \\ -1 \\ -2 \end{pmatrix}$ is clearly a normal for $\mathscr{I}_1$. We shall determine the intersection of $\mathscr{I}_1$ with each of several planes.

(a) Let $\mathscr{I}_2$ be the plane given by

$$(2.24) \qquad \mathscr{I}_2 = \left\{ \begin{pmatrix} x \\ y \\ z \end{pmatrix} \;\middle|\; x - z - 2 = 0, \text{ where } x, y, z \in \mathscr{R} \right\}.$$

The vector $N_2 = \begin{pmatrix} 1 \\ 0 \\ -1 \end{pmatrix}$, formed from the coefficients of x, y, and z, is a normal for $\mathfrak{I}_2$. Since N_1 and N_2 are not parallel, the planes $\mathfrak{I}_1$ and $\mathfrak{I}_2$ are not parallel (why?) and will intersect in a line $\mathcal{L}_1$ having $D_1 = N_1 \times N_2$ as a direction vector. We compute D_1:

$$(2.25) \qquad D_1 = N_1 \times N_2 = \begin{pmatrix} 4 \\ -1 \\ -2 \end{pmatrix} \times \begin{pmatrix} 1 \\ 0 \\ -1 \end{pmatrix} = \begin{pmatrix} 1 \\ 2 \\ 1 \end{pmatrix}$$

To complete our determination of $\mathcal{L}_1$ we must find a point on $\mathcal{L}_1$. A simple inspection of the coordinates of D_1 discloses that D_1 is not perpendicular to the vector $N = \begin{pmatrix} 0 \\ 0 \\ 1 \end{pmatrix}$. Thus, D_1 (and hence $\mathcal{L}_1$) is not parallel to the *co-ordinate plane*

$$(2.26) \qquad \mathscr{Z} = \left\{ \begin{pmatrix} x \\ y \\ z \end{pmatrix} \middle| z = 0 \right\}. \quad \textit{for convenience}$$

$$\textit{pick } z = 0$$

Consequently, Theorem 1.51 assures us that $\mathcal{L}_1$ and $\mathscr{Z}$ intersect in exactly one point. To find the point we proceed as follows:

$$X = \begin{pmatrix} x \\ y \\ z \end{pmatrix} \in \mathcal{L}_1 \cap \mathscr{Z} \Rightarrow^1 \begin{cases} X \in \mathcal{L}_1 = \mathfrak{I}_1 \cap \mathfrak{I}_2 \\ \text{and} \\ X \in \mathscr{Z} \end{cases}$$

$$\Rightarrow^1 \begin{cases} X \in \mathfrak{I}_1 \\ \text{and} \\ X \in \mathfrak{I}_2 \\ \text{and} \\ X \in \mathscr{Z} \end{cases}$$

$$\Rightarrow^2 \begin{cases} X \cdot \begin{pmatrix} 4 \\ -1 \\ -2 \end{pmatrix} - 5 = 0 \\ \text{and} \\ x - z - 2 = 0 \\ \text{and} \\ z = 0 \end{cases}$$

$$\Rightarrow^3 \begin{cases} 4x - y - 2z - 5 = 0 \\ x \qquad - z - 2 = 0 \\ \qquad\qquad z \qquad = 0 \end{cases}$$

$$\Rightarrow^4 \begin{cases} 4x - y = 5 \\ x = 2 \\ z = 0 \end{cases}$$

$$\Rightarrow^5 X = \begin{pmatrix} x \\ y \\ z \end{pmatrix} = \begin{pmatrix} 2 \\ 3 \\ 0 \end{pmatrix}$$

(1) Definition of set intersection.
(2) Substitution from Equations 2.23, 2.24, and 2.26.
(3) Carrying out the inner product.
(4) Substitution from third equation into first two.
(5) Solving first two equations for x and y.

Using the point $\begin{pmatrix} 2 \\ 3 \\ 0 \end{pmatrix}$ and the direction vector D_1 given by Equation 2.25, the line $\mathcal{L}_1$ of intersection of $\mathcal{S}_1$ and $\mathcal{S}_2$ is described by

$$\mathcal{L}_1 = \left\{ \begin{pmatrix} 2 \\ 3 \\ 0 \end{pmatrix} + t \begin{pmatrix} 1 \\ 2 \\ 1 \end{pmatrix} \,\middle|\, t \in \mathcal{R} \right\}.$$

(b) Let $\mathcal{S}_3$ be the plane given by

$$\mathcal{S}_3 = \left\{ \begin{pmatrix} 1 \\ -2 \\ 1 \end{pmatrix} + r \begin{pmatrix} 1 \\ 1 \\ 1 \end{pmatrix} + s \begin{pmatrix} 0 \\ 3 \\ -1 \end{pmatrix} \,\middle|\, r, s \in \mathcal{R} \right\}.$$

Neither of the direction vectors $\begin{pmatrix} 1 \\ 1 \\ 1 \end{pmatrix}$ and $\begin{pmatrix} 0 \\ 3 \\ -1 \end{pmatrix}$ is perpendicular to the normal N_1 for $\mathcal{S}_1$. Consequently, $\mathcal{S}_1$ and $\mathcal{S}_3$ are not parallel and must intersect in a line $\mathcal{L}_2$. To determine $\mathcal{L}_2$ we shall describe $\mathcal{S}_3$ in scalar implicit form and proceed in a manner similar to that in Part (a). The cross-product of the direction vectors introduced above, namely,

$$N_3 = \begin{pmatrix} 1 \\ 1 \\ 1 \end{pmatrix} \times \begin{pmatrix} 0 \\ 3 \\ -1 \end{pmatrix} = \begin{pmatrix} -4 \\ 1 \\ 3 \end{pmatrix}$$

is a normal for $\mathcal{S}_3$. Thus, we may represent $\mathcal{S}_3$ by

$$(2.27) \qquad \mathcal{S}_3 = \left\{ X \,\middle|\, \left[X - \begin{pmatrix} 1 \\ -2 \\ 1 \end{pmatrix} \right] \cdot \begin{pmatrix} -4 \\ 1 \\ 3 \end{pmatrix} = 0 \text{ and } X \in \mathcal{R}^3 \right\}$$

$$= \left\{ \begin{pmatrix} x \\ y \\ z \end{pmatrix} \,\middle|\, -4x + y + 3z + 3 = 0 \right\}.$$

The direction vector D_2 for $\mathcal{L}_2$ is

$$D_2 = N_3 \times N_1 = \begin{pmatrix} -4 \\ 1 \\ 3 \end{pmatrix} \times \begin{pmatrix} 4 \\ -1 \\ -2 \end{pmatrix} = \begin{pmatrix} 1 \\ 4 \\ 0 \end{pmatrix}.$$

This time we find that $D_2 \cdot \begin{pmatrix} 0 \\ 0 \\ 1 \end{pmatrix} = \begin{pmatrix} 1 \\ 4 \\ 0 \end{pmatrix} \cdot \begin{pmatrix} 0 \\ 0 \\ 1 \end{pmatrix} = 0$. Consequently, D_2 is parallel to the coordinate plane $\mathcal{Z}$ given in Equation 2.26, but it is not parallel to either of the coordinate planes

$$\mathcal{X} = \left\{ \begin{pmatrix} x \\ y \\ z \end{pmatrix} \middle| \ x = 0 \right\}$$

or

$$\mathcal{Y} = \left\{ \begin{pmatrix} x \\ y \\ z \end{pmatrix} \middle| \ y = 0 \right\}.$$

Thus, $\mathcal{L}_2$ is not parallel to either $\mathcal{X}$ or $\mathcal{Y}$. In particular, it follows that $\mathcal{L}_2 \cap \mathcal{Y}$ consists of a single point. We now determine that point X:

$$X = \begin{pmatrix} x \\ y \\ z \end{pmatrix} \in \mathcal{L}_2 \cap \mathcal{Y} \Rightarrow \begin{cases} X \in \mathcal{L}_2 \\ \text{and} \\ X \in \mathcal{Y} \end{cases}$$

$$\Rightarrow \begin{cases} X \in \mathcal{S}_1, \text{ and} \\ X \in \mathcal{S}_3, \text{ and} \\ X \in \mathcal{Y} \end{cases}$$

$$\Rightarrow \begin{cases} 4x - y - 2z - 5 = 0, \text{ and} \\ -4x + y + 3z + 3 = 0, \text{ and} \\ \qquad\qquad y \qquad\qquad = 0 \end{cases}$$

$$\Rightarrow^1 \begin{cases} 4x - 2z = 5 \\ -4x + 3z = -3 \\ \qquad y = 0 \end{cases}$$

$$\Rightarrow^2 X = \begin{pmatrix} x \\ y \\ z \end{pmatrix} = \begin{pmatrix} \frac{9}{4} \\ 0 \\ 2 \end{pmatrix}.$$

(1) Substituting from the third equation into the first two.
(2) Solving the first two equations for x and z.

Hence, we find that

$$\mathcal{L}_2 = \left\{ \begin{pmatrix} \frac{9}{4} \\ 0 \\ 2 \end{pmatrix} + t \begin{pmatrix} 1 \\ 4 \\ 0 \end{pmatrix} \middle| \ t \in \mathcal{R} \right\}.$$

If we had determined the point of $\mathcal{L}_2 \cap \mathcal{X}$, we would have obtained a second description for $\mathcal{L}_2$, namely,

$$\mathcal{L}_2 = \left\{ \begin{pmatrix} 0 \\ -9 \\ 2 \end{pmatrix} + t \begin{pmatrix} 1 \\ 4 \\ 0 \end{pmatrix} \middle| \ t \in \mathcal{R} \right\}.$$

(c) Now consider $\mathcal{S}_1 \cap \mathcal{S}_4$ where $\mathcal{S}_4$ is the plane

$$\mathcal{S}_4 = \left\{ X \middle| X \cdot \begin{pmatrix} -4 \\ 1 \\ 2 \end{pmatrix} + 7 = 0 \text{ and } X \in \mathcal{R}^3 \right\}.$$

The vector $N_4 = \begin{pmatrix} -4 \\ 1 \\ 2 \end{pmatrix}$ is a normal to $\mathcal{I}_4$ and is parallel to the normal

$N_1 = \begin{pmatrix} 4 \\ -1 \\ -2 \end{pmatrix}$ of $\mathcal{I}_1$. Consequently, $\mathcal{I}_1$ and $\mathcal{I}_4$ are parallel. Hence, they are

either identical or have an empty intersection. We shall show that $\mathcal{I}_1$ and $\mathcal{I}_4$ are not identical. Suppose X is a point of $\mathcal{I}_1$, then from Equation 2.23

$$X \cdot \begin{pmatrix} 4 \\ -1 \\ -2 \end{pmatrix} = 5.$$

Upon multiplication by -1 it follows that

$$X \cdot \begin{pmatrix} -4 \\ 1 \\ 2 \end{pmatrix} = -5 \neq -7$$

and that $X \notin \mathcal{I}_4$. Therefore, $\mathcal{I}_1 \neq \mathcal{I}_4$ and we must have $\mathcal{I}_1 \cap \mathcal{I}_4 = \varnothing$.

(d) Finally, consider the plane $\mathcal{I}_5$ given by

$$\mathcal{I}_5 = \left\{ \begin{pmatrix} -2 \\ -19 \\ 3 \end{pmatrix} + r \begin{pmatrix} 1 \\ 2 \\ 1 \end{pmatrix} + s \begin{pmatrix} 1 \\ 0 \\ 2 \end{pmatrix} \middle| r, s \in \mathcal{R} \right\}.$$

The two direction vectors $\begin{pmatrix} 1 \\ 2 \\ 1 \end{pmatrix}$ and $\begin{pmatrix} 1 \\ 0 \\ 2 \end{pmatrix}$ for $\mathcal{I}_5$ are easily shown to be per-

pendicular to the normal N_1 of $\mathcal{I}_1$. Thus, $\mathcal{I}_5$ is parallel to $\mathcal{I}_1$. Again we have the case in which either $\mathcal{I}_1 \cap \mathcal{I}_5 = \varnothing$ or $\mathcal{I}_1 = \mathcal{I}_5$. If $\mathcal{I}_1$ and $\mathcal{I}_5$ are identical,

then the point $\begin{pmatrix} -2 \\ -19 \\ 3 \end{pmatrix}$, which is on $\mathcal{I}_5$, must also be on $\mathcal{I}_1$. That $\begin{pmatrix} -2 \\ -19 \\ 3 \end{pmatrix}$

is indeed a point of $\mathcal{I}_1$ is verified by the computation

$$\begin{pmatrix} -2 \\ -19 \\ 3 \end{pmatrix} \cdot \begin{pmatrix} 4 \\ -1 \\ -2 \end{pmatrix} - 5 = (-8 + 19 - 6) - 5 = 0.$$

Thus, $\mathcal{I}_1 \cap \mathcal{I}_5 \neq \varnothing$ and we must have $\mathcal{I}_1 \cap \mathcal{I}_5 = \mathcal{I}_1 = \mathcal{I}_5$. ■

EXERCISES

In Exercises 1 to 6 determine the intersection of the given line $\mathcal{L}$ and plane $\mathcal{I}$:

1. $\mathcal{L} = \left\{ \begin{pmatrix} 3 \\ -3 \\ -7 \end{pmatrix} + t \begin{pmatrix} 1 \\ -2 \\ 3 \end{pmatrix} \middle| t \in \mathcal{R} \right\}$; $\mathcal{I} = \left\{ \begin{pmatrix} x \\ y \\ z \end{pmatrix} \middle| x + y + z - 1 = 0, \text{where} \right.$

$\left. x, y, z \in \mathcal{R} \right\}$.

2. $\mathcal{L} =$ the line through $\begin{pmatrix} 1 \\ 2 \\ -3 \end{pmatrix}$ and $\begin{pmatrix} -1 \\ -1 \\ -5 \end{pmatrix}$; $\mathscr{S} = \left\{ \begin{pmatrix} 5 \\ 6 \\ 0 \end{pmatrix} + r \begin{pmatrix} 1 \\ 0 \\ 1 \end{pmatrix} + s \begin{pmatrix} 1 \\ 1 \\ 0 \end{pmatrix} \middle| r, s \in \mathscr{R} \right\}$.

3. $\mathcal{L} = \left\{ \begin{pmatrix} 2 \\ 1 \\ -1 \end{pmatrix} + t \begin{pmatrix} 4 \\ 2 \\ -1 \end{pmatrix} \middle| t \in \mathscr{R} \right\}$; $\mathscr{S} =$ the plane containing $\begin{pmatrix} 2 \\ 1 \\ 1 \end{pmatrix}$, $\begin{pmatrix} 5 \\ 2 \\ -1 \end{pmatrix}$ and $\begin{pmatrix} 6 \\ 3 \\ 0 \end{pmatrix}$.

4. $\mathcal{L} =$ the line through $\begin{pmatrix} 3 \\ 5 \\ 3 \end{pmatrix}$ and $\begin{pmatrix} -12 \\ 2 \\ 0 \end{pmatrix}$;

$\mathscr{S} = \left\{ \begin{pmatrix} x \\ y \\ z \end{pmatrix} \middle| z = (1/2)x - (3/2)y + 18 \text{ where } x, y, z \in \mathscr{R} \right\}$.

5. $\mathcal{L} = \left\{ \begin{pmatrix} 1 \\ 1 \\ 0 \end{pmatrix} + t \begin{pmatrix} 0 \\ 1 \\ 1 \end{pmatrix} \right\}$; $\mathscr{S} =$ the plane through $\begin{pmatrix} 1 \\ -6 \\ -7 \end{pmatrix}$, $\begin{pmatrix} 11 \\ -1 \\ 3 \end{pmatrix}$ and $\begin{pmatrix} 1 \\ 14 \\ 13 \end{pmatrix}$.

6. $\mathcal{L} = \left\{ \begin{pmatrix} -5 \\ 4 \\ 2 \end{pmatrix} + t \begin{pmatrix} 2 \\ -1 \\ 1 \end{pmatrix} \middle| t \in \mathscr{R} \right\}$; $\mathscr{S} = \left\{ X \middle| X \cdot \begin{pmatrix} 1 \\ 1 \\ -2 \end{pmatrix} = 0 \text{ and } X \in \mathscr{R}^3 \right\}$.

In Exercises 7 to 12 determine the intersection of the given planes $\mathscr{S}_1$ and $\mathscr{S}_2$:

7. $\mathscr{S}_1 =$ the plane through $\begin{pmatrix} 6 \\ 3 \\ 3 \end{pmatrix}$ with normal $N_1 = \begin{pmatrix} -2 \\ 1 \\ 4 \end{pmatrix}$;

$\mathscr{S}_2 = \left\{ \begin{pmatrix} 4 \\ 3 \\ 5 \end{pmatrix} + r \begin{pmatrix} 1 \\ 2 \\ 3 \end{pmatrix} + s \begin{pmatrix} 3 \\ 2 \\ 4 \end{pmatrix} \middle| r, s \in \mathscr{R} \right\}$.

8. $\mathscr{S}_1 = \left\{ X \middle| X \cdot \begin{pmatrix} 1 \\ 3 \\ -6 \end{pmatrix} - 11 = 0 \text{ and } X \in \mathscr{R}^3 \right\}$;

$\mathscr{S}_2 = \{ X \mid x - y - 2z + 5 = 0, \text{ where } x, y, z \in \mathscr{R} \}$.

9. $\mathscr{S}_1 = \left\{ X \middle| X \cdot \begin{pmatrix} 1 \\ 1 \\ 1 \end{pmatrix} = 3 \text{ and } X \in \mathscr{R}^3 \right\}$;

$\mathscr{S}_2 = \left\{ \begin{pmatrix} 1 \\ -4 \\ 3 \end{pmatrix} + r \begin{pmatrix} 1 \\ 2 \\ -3 \end{pmatrix} + s \begin{pmatrix} 2 \\ 3 \\ -5 \end{pmatrix} \middle| r, s \in \mathscr{R} \right\}$.

10. $\mathscr{S}_1 =$ the plane through $\begin{pmatrix} 2 \\ 1 \\ 2 \end{pmatrix}$, $\begin{pmatrix} 0 \\ 1 \\ 4 \end{pmatrix}$, and $\begin{pmatrix} -1 \\ 1 \\ 2 \end{pmatrix}$;

$\mathscr{S}_2 = \left\{ X \middle| \left[X - \begin{pmatrix} 1 \\ 1 \\ 0 \end{pmatrix} \right] \cdot \begin{pmatrix} 0 \\ 3 \\ 0 \end{pmatrix} = 0 \text{ and } X \in \mathscr{R}^3 \right\}$.

11. $\mathcal{I}_1 = \{X \mid 2x - y + 2z = 1 \text{ where } x, y, z \in \mathcal{R}\};$

$$\mathcal{I}_2 = \left\{ \begin{pmatrix} 3 \\ 3 \\ -1 \end{pmatrix} + r \begin{pmatrix} 5 \\ 4 \\ -3 \end{pmatrix} + s \begin{pmatrix} 3 \\ -2 \\ -4 \end{pmatrix} \middle| r, s \in \mathcal{R} \right\}.$$

12. $\mathcal{I}_1 = \left\{ X \middle| X \cdot \begin{pmatrix} -1 \\ 0 \\ 0 \end{pmatrix} = 2 \text{ and } X \in \mathcal{R}^3 \right\};$

$$\mathcal{I}_2 = \left\{ \begin{pmatrix} x \\ y \\ z \end{pmatrix} \middle| x - y + 6 = 0 \text{ where } x, y \in \mathcal{R} \right\}.$$

13. Let N be a normal for a plane $\mathcal{I}$ in $\mathcal{R}^3$. Let D be a nonzero vector such that $D \cdot N = 0$. Show that D is a direction vector for $\mathcal{I}$.

14. Let N_1 and N_2 be normals to $\mathcal{I}_1$ and $\mathcal{I}_2$, respectively. Show that $\mathcal{I}_1$ and $\mathcal{I}_2$ are parallel if and only if N_1 and N_2 are parallel.

15. Let N_1 and N_2 be normals to $\mathcal{I}_1$ and $\mathcal{I}_2$. Assume that the vector $D = N_1 \times N_2$ is nonzero. Show that D is a direction vector for both $\mathcal{I}_1$ and $\mathcal{I}_2$.

ANSWERS

1. $\mathcal{L} \cap \mathcal{I} = \left\{ \begin{pmatrix} 7 \\ -11 \\ 5 \end{pmatrix} \right\}.$ **3.** $\mathcal{L} \cap \mathcal{I} = \varnothing.$ **5.** $\mathcal{L} \cap \mathcal{I} = \mathcal{L}.$

7. $\mathcal{I}_1 \cap \mathcal{I}_2 = \left\{ \begin{pmatrix} 1 \\ 1 \\ 1 \end{pmatrix} + t \begin{pmatrix} 2 \\ 0 \\ 1 \end{pmatrix} \middle| t \in \mathcal{R} \right\}.$ **9.** $\mathcal{I}_1 \cap \mathcal{I}_2 = \varnothing.$

11. $\mathcal{I}_1 \cap \mathcal{I}_2 = \mathcal{I}_1 = \mathcal{I}_2.$

7. LENGTH IN $\mathcal{R}^n$

We have come this far in our development of linear geometry without a formal definition of length or distance. Although we used the reader's intuition concerning length and distance to motivate the definition of multiplication of a vector by a scalar, a review of Definition 1.5 (p. 9) discloses that a concept of length is not at all required. To this point, nothing we have formulated requires the length or distance concept. We shall delay the introduction of these concepts no longer, though we could proceed still further without them.

The motivation of our definition of the length of a vector is found in the geometric representation of a vector as a directed line segment (in which an informal idea of length is involved) and in the well-known Pythagorean Theorem.

Figure 2.5a depicts a vector $V = \begin{pmatrix} a \\ b \end{pmatrix}$ in $\mathcal{R}^2$ using a "rectangular" coordinate system. The "triangle" OAV has a "right angle" at A. The geometrical representation of V, Geometrical Model 1.4 (p. 7) regards side OA as having "length" a and side AV as having "length" b. The hypotenuse, according to the Pythagorean Theorem, would then have length $l = \sqrt{a^2 + b^2}$. It seems reasonable to call this number the length of the vector $V = \begin{pmatrix} a \\ b \end{pmatrix}$.

Figure 2.5b represents a vector $V = \begin{pmatrix} a \\ b \\ c \end{pmatrix}$ in $\mathfrak{R}^3$, again referred to a "rectangular" coordinate system. In this figure there appear two "right" triangles. Triangle OAB with "right" angle at A and triangle OBV with "right" angle at B. Side OB in the first triangle is seen to have "length" $d = \sqrt{a^2 + b^2}$, where a is the "length" of OA and b is the "length" of AB. A second application of the Pythagorean Theorem to the triangle OBV gives the "length" of side OV as $l = \sqrt{d^2 + c^2} = \sqrt{a^2 + b^2 + c^2}$, where c is the "length" of side BV. Thus, in $\mathfrak{R}^3$ it seems reasonable to define the length of the vector $V = \begin{pmatrix} a \\ b \\ c \end{pmatrix}$ to be $l = \sqrt{a^2 + b^2 + c^2}$.

The following definition formalizes the concept of length suggested above for $\mathfrak{R}^2$ and $\mathfrak{R}^3$ and extends the concept to R^n. See Exercise 19 for the interpretation of this definition in $\mathfrak{R}^1$.

DEFINITION 2.21   (Length of a Vector.)

i. Let $X = \begin{pmatrix} a \\ b \end{pmatrix}$ be an element of $\mathfrak{R}^2$. The *length* of X, denoted by $\|X\|$, is defined by
$$\|X\| = \sqrt{a^2 + b^2}.$$

ii. Let $X = \begin{pmatrix} a \\ b \\ c \end{pmatrix}$ be an element of $\mathfrak{R}^3$. The *length* of X, denoted by $\|X\|$, is defined by
$$\|X\| = \sqrt{a^2 + b^2 + c^2}.$$

iii. Let $X = \begin{pmatrix} a_1 \\ a_2 \\ \cdot \\ \cdot \\ \cdot \\ a_n \end{pmatrix}$ be an element of $\mathfrak{R}^n$. The *length* of X, denoted by $\|X\|$, is defined by
$$\|X\| = \sqrt{a_1^2 + a_2^2 + \cdots + a_n^2}.$$

The more important properties of the length concept are most easily established by expressing the length of a vector in terms of an inner product and employing the properties of an inner product. The following theorem is the basis for this approach.

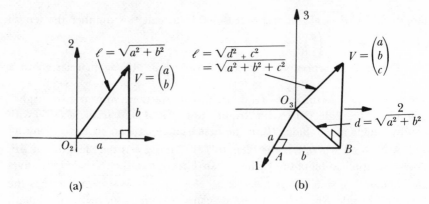

Figure 2.5

THEOREM 2.22 Let X be an element of $\mathcal{R}^n$. The length $\|X\|$ of X is such that

$$\|X\| = \sqrt{X \cdot X}$$

or, equivalently,

$$\|X\|^2 = X \cdot X.$$

Proof. Let $X = \begin{pmatrix} a_1 \\ a_2 \\ \cdot \\ \cdot \\ \cdot \\ a_n \end{pmatrix}$. The square of $\|X\|$ is given by

$$\|X\|^2 =^1 \left(\sqrt{(a_1)^2 + (a_2)^2 + \cdots + (a_n)^2} \right)^2$$
$$=^2 (a_1)^2 + (a_2)^2 + \cdots + (a_n)^2$$
$$=^2 a_1 \cdot a_1 + a_2 \cdot a_2 + \cdots + a_n \cdot a_n$$
$$=^3 \begin{pmatrix} a_1 \\ a_2 \\ \cdot \\ \cdot \\ \cdot \\ a_n \end{pmatrix} \cdot \begin{pmatrix} a_1 \\ a_2 \\ \cdot \\ \cdot \\ \cdot \\ a_n \end{pmatrix}$$
$$= X \cdot X.$$

(1) Definition 2.21 (Length).
(2) Properties of real numbers.
(3) Definition 2.1 (Inner Product).

Taking positive square roots on each side we obtain the equivalent statement

$$\|X\| = \sqrt{X \cdot X}. \quad \blacksquare$$

Example 19.

(a) Find the length of $X = \begin{pmatrix} 2 \\ 1 \end{pmatrix}$. Using Theorem 2.22 we have

$$\|X\| = \left\| \begin{pmatrix} 2 \\ 1 \end{pmatrix} \right\| = \sqrt{\begin{pmatrix} 2 \\ 1 \end{pmatrix} \cdot \begin{pmatrix} 2 \\ 1 \end{pmatrix}} = \sqrt{4 + 1} = \sqrt{5}.$$

(b) The length of $X = \begin{pmatrix} 3 \\ 0 \\ 4 \end{pmatrix}$ is given by

$$\|X\| = \sqrt{X \cdot X} = \sqrt{\begin{pmatrix} 3 \\ 0 \\ 4 \end{pmatrix} \cdot \begin{pmatrix} 3 \\ 0 \\ 4 \end{pmatrix}} = \sqrt{9 + 0 + 16} = \sqrt{25} = 5.$$

(c) The length of $X = \begin{pmatrix} 1 \\ 2 \\ 1 \\ 1 \end{pmatrix}$ is given by

$$\|X\| = \sqrt{X \cdot X} = \sqrt{\begin{pmatrix} 1 \\ 2 \\ 1 \\ 1 \end{pmatrix} \cdot \begin{pmatrix} 1 \\ 2 \\ 1 \\ 1 \end{pmatrix}} = \sqrt{1 + 4 + 1 + 1} = \sqrt{7}. \quad \blacksquare$$

In view of Theorem 2.22, the inner product of a vector X with itself is the square of the length of that vector. The content of Theorem 2.3 (Cauchy-Schwartz Inequality) can therefore be restated in a manner involving length.

THEOREM 2.23 (Cauchy-Schwartz Inequality.) Let X and Y be elements of $\mathcal{R}^n$. The inner product $X \cdot Y$ is such that

i. $|X \cdot Y| \leq \|X\| \, \|Y\|$.

Moreover,

ii. $|X \cdot Y| = \|X\| \, \|Y\|$, if and only if X and Y are parallel,

and

iii. $|X \cdot Y| < \|X\| \, \|Y\|$, if and only if X and Y are not parallel.

Proof. Let X and Y be elements of $\mathcal{R}^n$. According to Theorem 2.3 we have

$$(X \cdot Y)^2 \leq (X \cdot X)(Y \cdot Y).$$

Taking the positive square root on both sides of this inequality and using Theorem 2.22, we obtain

$$|X \cdot Y| \leq \sqrt{(X \cdot X)} \sqrt{(Y \cdot Y)}$$

$$= \|X\| \, \|Y\|.$$

Statement i is thereby established. Statements ii and iii follow from Theorem 2.3 in the same manner. ■

We now set forth the basic properties of length.

THEOREM 2.24   (Properties of Length.) Let X and Y be elements of $\mathfrak{R}^n$ and let r be a real number. Length has the following properties:

Positive Definite Property $\begin{cases} \text{i.} & \|X\| \geq 0. \\ \text{ii.} & \|X\| = 0 \text{ if and only if } X = O_n. \end{cases}$

Positive Homogeneous Property $\{$iii. $\|rX\| = |r| \, \|X\|.$

Triangle Property $\begin{cases} \text{iv.} & \|X + Y\| \leq \|X\| + \|Y\|. \\ \text{v.} & \|X + Y\| = \|X\| + \|Y\| \text{ if and} \\ & \text{only if } X = O_n \text{ or } Y = kX \text{ with} \\ & k \geq 0. \end{cases}$

Proof.   The proofs of statements i and ii follow easily from Theorem 2.2i and ii and are called for in Exercise 20.

Proof of iii

$$\|rX\| =^1 \sqrt{(rX) \cdot (rX)}$$

$$=^2 \sqrt{r^2(X \cdot X)}$$

$$=^3 \sqrt{r^2} \sqrt{X \cdot X}$$

$$=^4 |r| \, \|X\|.$$

(1) Theorem 2.22 with X replaced by rX.
(2) Theorem 2.2iv with k replaced by r and Y replaced by rX, Theorem 2.2iii, and a second application of Theorem 2.2iv.
(3) Properties of real numbers.
(4) Theorem 2.22 and the real number property $\sqrt{r^2} = |r|$.

Proof of iv

$$\|X + Y\|^2 =^1 (X + Y) \cdot (X + Y)$$

$$=^2 X \cdot X + 2(X \cdot Y) + Y \cdot Y$$

$$=^1 \|X\|^2 + 2(X \cdot Y) + \|Y\|^2$$

$$\leq^3 \|X\|^2 + 2 \|X\| \, \|Y\| + \|Y\|^2$$

$$=^4 (\|X\| + \|Y\|)^2.$$

Thus, we have $\|X + Y\|^2 \leq (\|X\| + \|Y\|)^2$. Taking the positive square root of both members we obtain statement iv.

(1) Theorem 2.22.
(2) Theorem 2.2.
(3) Theorem 2.23.
(4) Properties of real numbers.

Statement v is proved in the same manner as statement iv, with the additional observations that $X \cdot Y = \|X\| \|Y\|$ if and only if $X = O_n$ or $Y = kX$ with $k \geq 0$ (see Exercise 21). ■

A word concerning the names of the properties listed in the preceding theorem is in order. Properties i and ii together are called the *positive definite property*. They assert that the length of any vector is *positive* or zero; furthermore, the length of a nonzero vector is *definitely* positive. Property iii is called the *positive homogeneous property* since any *positive* value of r is such that r may be used as a multiplier of the vector X or as a multiplier of the length X with the same result: $\|rX\| = r \|X\|$, if $r > 0$. Properties iv and v, together, are referred to as the triangle property for reasons given in the next section.

It is often convenient to employ vectors of unit length. A short treatment is given here to facilitate later work.

DEFINITION 2.25 Let X be an element of $\mathcal{R}^n$. X is called a *unit vector* if and only if X has unit length: $\|X\| = 1$.

Example 20. Let $\mathcal{L} = \left\{ \begin{pmatrix} 2 \\ 1 \end{pmatrix} + t \begin{pmatrix} -4 \\ 3 \end{pmatrix} \,\middle|\, t \in \mathcal{R} \right\}$ be a line in $\mathcal{R}^2$. Find a unit normal to $\mathcal{L}$. Using the technique of Example 4 (p. 93), we note that since $D = \begin{pmatrix} -4 \\ 3 \end{pmatrix}$ is a direction vector for $\mathcal{L}$, $N = \begin{pmatrix} 3 \\ 4 \end{pmatrix}$ is a normal to $\mathcal{L}$. However, N has length $\|N\| = \sqrt{9 + 16} = 5$ and is therefore not a unit vector. By the positive homogeneous property of length the vector $\hat{N} = \frac{1}{5} N = \begin{pmatrix} 3/5 \\ 4/5 \end{pmatrix}$ will have length

$$\|\hat{N}\| = \left\| \frac{1}{5} N \right\| = \frac{1}{5} \|N\| = \frac{1}{5} \cdot 5 = 1.$$

Thus, $\hat{N}$ (read: "N circumflex, "N hat," or "N caret") is a unit vector. Moreover, $\hat{N}$ is a normal to $\mathcal{L}$, since it is a nonzero multiple of the normal N (see Theorem 2.10). ■

Example 21. Let $\mathcal{S} = \left\{ \begin{pmatrix} 3 \\ 1 \\ 2 \end{pmatrix} + r \begin{pmatrix} 1 \\ 2 \\ 4 \end{pmatrix} + s \begin{pmatrix} 2 \\ 0 \\ 1 \end{pmatrix} \,\middle|\, r, s \in \mathcal{R} \right\}$ be a plane in $\mathcal{R}^3$. To find a unit normal to $\mathcal{F}$ we first determine the cross-product N of the direction set obtained from the above representation.

$$N = \begin{pmatrix} 1 \\ 2 \\ 4 \end{pmatrix} \times \begin{pmatrix} 2 \\ 0 \\ 1 \end{pmatrix} = \begin{pmatrix} 2 \\ 7 \\ -4 \end{pmatrix}.$$

Now N is a normal to $\mathcal{F}$ and has length

$$\|N\| = \left\| \begin{pmatrix} 2 \\ 7 \\ -4 \end{pmatrix} \right\| = \sqrt{4 + 49 + 16} = \sqrt{69}.$$

Therefore,

$$\hat{N} = \frac{1}{\sqrt{69}} N = \frac{1}{\sqrt{69}} \begin{pmatrix} 2 \\ 7 \\ -4 \end{pmatrix}$$

is a unit normal to $\mathcal{F}$. Of course $\hat{M} = -\hat{N} = \dfrac{-1}{\sqrt{69}} \begin{pmatrix} 2 \\ 7 \\ -4 \end{pmatrix}$ is another unit normal to $\mathcal{F}$.

Notation 2.26. When it is desirable to emphasize that a vector is a unit vector we shall use the circumflex "^" notation introduced in the preceding example. Thus, $\hat{N}$ denotes a unit vector parallel to N.

EXERCISES

In each of the Exercises 1 to 10 determine the length of the given vector $\vec{V}$ and find a unit vector $\hat{V}$ which is parallel to V:

1. $V = \begin{pmatrix} 2 \\ 4 \end{pmatrix}$.　　2. $V = \begin{pmatrix} 5 \\ 12 \end{pmatrix}$.　　3. $V = \begin{pmatrix} -7 \\ 8 \end{pmatrix}$.　　4. $V = \begin{pmatrix} 10 \\ -2 \end{pmatrix}$.

5. $V = \begin{pmatrix} 1 \\ 3 \\ 2 \end{pmatrix}$.　　6. $V = \begin{pmatrix} -2 \\ 0 \\ 3 \end{pmatrix}$.　　7. $V = \begin{pmatrix} 1 \\ -2 \\ 5 \end{pmatrix}$.　　8. $V = \begin{pmatrix} 3 \\ 2 \\ -1 \\ 2 \end{pmatrix}$.

9. $V = \begin{pmatrix} 2 \\ -2 \\ -1 \\ 2 \end{pmatrix}$.　　　　　　10. $V = \begin{pmatrix} 1 \\ 1 \\ 1 \\ 1 \end{pmatrix}$.

In each of the Exercises 11 to 14 determine a unit vector N which is perpendicular to the given vector D:

11. $D = \begin{pmatrix} 2 \\ 3 \end{pmatrix}$.　　12. $D = \begin{pmatrix} -5 \\ 2 \end{pmatrix}$.　　13. $D = \begin{pmatrix} 11 \\ -3 \end{pmatrix}$.　　14. $D = \begin{pmatrix} 5 \\ 5 \end{pmatrix}$.

In each of the Exercises 15 to 18 determine a unit vector N which is perpendicular to each of the vectors of the given set $\{U, V\}$:

15. $\{U, V\} = \left\{ \begin{pmatrix} 2 \\ 1 \\ 3 \end{pmatrix}, \begin{pmatrix} 1 \\ -1 \\ 2 \end{pmatrix} \right\}.$ 16. $\{U, V\} = \left\{ \begin{pmatrix} 1 \\ 2 \\ -3 \end{pmatrix}, \begin{pmatrix} 2 \\ 0 \\ -3 \end{pmatrix} \right\}.$

17. $\{U, V\} = \left\{ \begin{pmatrix} 5 \\ 2 \\ -2 \end{pmatrix}, \begin{pmatrix} 1 \\ 1 \\ 1 \end{pmatrix} \right\}.$ 18. $\{U, V\} = \left\{ \begin{pmatrix} 7 \\ 2 \\ 5 \end{pmatrix}, \begin{pmatrix} 3 \\ -1 \\ 2 \end{pmatrix} \right\}.$

19. Let $X = (a)$ be a vector in $\mathcal{R}^1$. Show that the length of X as defined by Definition 2.21 is given by $\|X\| = \|(a)\| = |a|$.

20. Let X be an element of $\mathcal{R}^n$. Show that (a) $\|X\| \geq 0$ and that (b) $\|X\| = 0$ if and only if $X = O_n$.

21. (a) Use Theorem 2.3ii and iii to prove that $|X \cdot Y| = \|X\| \|Y\|$ if and only if $X = O_n$ or $Y = kX$.
    (b) Use the result of part a to show that $X \cdot Y = \|X\| \|Y\|$ if and only if $X = O_n$ or $Y = kX$ with $k \geq 0$.

22. Let $\{\hat{U}, \hat{V}\}$ be an orthogonal set of unit vectors. If $X = a\hat{U} + b\hat{V}$, show that $\|X\| = \sqrt{a^2 + b^2}$.

ANSWERS

1. $\hat{V} = \begin{pmatrix} \dfrac{\sqrt{5}}{5} \\ \dfrac{2\sqrt{5}}{5} \end{pmatrix}$ or $\begin{pmatrix} \dfrac{-\sqrt{5}}{5} \\ \dfrac{-2\sqrt{5}}{5} \end{pmatrix}$. 3. $\hat{V} = \begin{pmatrix} \dfrac{-7\sqrt{113}}{113} \\ \dfrac{8\sqrt{113}}{113} \end{pmatrix}$ or $\begin{pmatrix} \dfrac{7\sqrt{113}}{113} \\ \dfrac{-8\sqrt{113}}{113} \end{pmatrix}$

5. $\hat{V} = \begin{pmatrix} \dfrac{\sqrt{14}}{14} \\ \dfrac{3\sqrt{14}}{14} \\ \dfrac{\sqrt{14}}{7} \end{pmatrix}$. 7. $\hat{V} = \begin{pmatrix} \dfrac{\sqrt{30}}{30} \\ \dfrac{-\sqrt{30}}{15} \\ \dfrac{\sqrt{30}}{6} \end{pmatrix}$ or $\begin{pmatrix} \dfrac{-\sqrt{30}}{30} \\ \dfrac{\sqrt{30}}{15} \\ \dfrac{-\sqrt{30}}{6} \end{pmatrix}$.

9. $\hat{V} = \begin{pmatrix} \dfrac{2\sqrt{13}}{13} \\ \dfrac{-2\sqrt{13}}{13} \\ \dfrac{-\sqrt{13}}{13} \\ \dfrac{2\sqrt{13}}{13} \end{pmatrix}$ or $\begin{pmatrix} \dfrac{-2\sqrt{13}}{13} \\ \dfrac{2\sqrt{13}}{13} \\ \dfrac{\sqrt{13}}{13} \\ \dfrac{-2\sqrt{13}}{13} \end{pmatrix}$. 11. $\hat{N} = \dfrac{\sqrt{13}}{13}\begin{pmatrix} -3 \\ 2 \end{pmatrix}$.

13. $\hat{N} = \dfrac{\sqrt{130}}{130}\begin{pmatrix} 3 \\ 11 \end{pmatrix}$. 15. $\hat{N} = \dfrac{\sqrt{35}}{35}\begin{pmatrix} 5 \\ -1 \\ -3 \end{pmatrix}$. 17. $\dfrac{\sqrt{74}}{74}\begin{pmatrix} 4 \\ -7 \\ 3 \end{pmatrix}$.

8. DISTANCE

Given a vector X in R^n, we have defined what we mean by the length of X. Referring to Figure 2.6a, we recall that the length of a vector X was

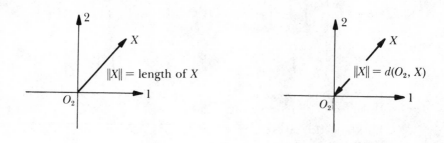

(a) $\|X\|$ as the length of X (b) $\|X\|$ as the distance $d(O_2, X)$

Figure 2.6

motivated by representing X as a *directed line segment*. $\|X\|$ then defines the *length* of this line segment. Figure 2.6b represents a different point of view arising from the representation of X as a point in the plane. From this viewpoint, $\|X\|$ defines the *distance* $d(O_2, X)$ from O_2 to X.

These two viewpoints are clearly related. The *distance* from O_2 to X is given by the *length* of the vector X. We shall extend the idea of distance to pairs of points X and Y, neither of which is the origin. Figure 2.7 gives the pictorial representation of what is involved. Since $Y = X + (Y - X)$, the vector $Y - X$ is the vector we add to X to get Y. Does this not suggest that the distance from X to Y should be the length $\|Y - X\|$? This discussion is immediately applicable in a general $\mathcal{R}^n$.

DEFINITION 2.27 (Distance.) Let X and Y be two elements of $\mathcal{R}^n$. The distance from X to Y, denoted by $d(X, Y)$, is defined to be the length of the vector $Y - X$:

$$d(X, Y) = \|Y - X\|.$$

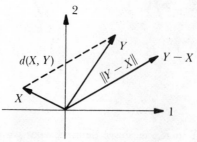

Figure 2.7 $d(X, Y) = \|Y - X\|$.

Example 22. (a) Find the distance from $X = \begin{pmatrix} 1 \\ 2 \end{pmatrix}$ to $Y = \begin{pmatrix} 3 \\ 7 \end{pmatrix}$:

$$d(X,\ Y) = \|Y - X\| = \left\| \begin{pmatrix} 3 \\ 7 \end{pmatrix} - \begin{pmatrix} 1 \\ 2 \end{pmatrix} \right\| = \left\| \begin{pmatrix} 2 \\ 5 \end{pmatrix} \right\|$$

$$= \sqrt{\begin{pmatrix} 2 \\ 5 \end{pmatrix} \cdot \begin{pmatrix} 2 \\ 5 \end{pmatrix}} = \sqrt{4 + 25} = \sqrt{29}.$$

(b) The distance from $X = \begin{pmatrix} 3 \\ -1 \\ 4 \end{pmatrix}$ to $Y = \begin{pmatrix} 4 \\ 1 \\ 6 \end{pmatrix}$ is

$$d(X,\ Y) = Y = \left\| \begin{pmatrix} 4 \\ 1 \\ 6 \end{pmatrix} - \begin{pmatrix} 3 \\ -1 \\ 4 \end{pmatrix} \right\|$$

$$= \sqrt{\begin{pmatrix} 1 \\ 2 \\ 2 \end{pmatrix} \cdot \begin{pmatrix} 1 \\ 2 \\ 2 \end{pmatrix}} = \sqrt{1 + 4 + 4} = \sqrt{9} = 3.$$

(c) The distance from $X = \begin{pmatrix} 2 \\ 3 \\ 5 \\ 7 \end{pmatrix}$ to $Y = \begin{pmatrix} 3 \\ 5 \\ 7 \\ 11 \end{pmatrix}$ is

$$d(X,\ Y) = \|Y - X\| = \left\| \begin{pmatrix} 1 \\ 2 \\ 2 \\ 4 \end{pmatrix} \right\| = \sqrt{1 + 4 + 4 + 16} = \sqrt{25} = 5. \quad \blacksquare$$

The properties of length given in Theorem 2.24 give rise to corresponding properties of distance.

THEOREM 2.28 (Properties of Distance.) Let X, Y, and Z be elements of $\mathcal{R}^n$. Distance has the following properties:

Positive Definite Property $\begin{cases} \text{i. } d(X,\ Y) \geq 0. \\ \text{ii. } d(X,\ Y) = 0 \text{ if and only if } X = Y. \end{cases}$

Symmetry Property $\{$iii. $d(X,\ Y) = d(Y,\ X).$

Triangle Property $\begin{cases} \text{iv. } d(X,\ Y) \leq d(X,\ Z) + d(Z,\ Y). \\ \text{v. } d(X,\ Y) = d(X,\ Z) + d(Z,\ Y) \text{ if and only} \\ \quad \text{if } X = Z \text{ or } Y - Z = k(Z - X), \text{ with} \\ \quad k \geq 0. \end{cases}$

Proof. The reader is asked to prove statements i, ii, and iii in Exercise 9.

Proof of iv. If X, Y, and Z are elements of $\mathcal{R}^n$, we can write

$$(2.28) \qquad Y - X = (Y - Z) + (Z - X).$$

Therefore,

$$d(X, Y) = \|Y - X\| =^1 \|(Y - Z) + (Z - X)\|$$
$$\leq^2 \|Y - Z\| + \|Z - X\|$$
$$=^3 d(Z, Y) + d(X, Z)$$
$$=^4 d(X, Z) + d(Z, Y).$$

(1) Using Equation 2.28.
(2) Theorem 2.24iv with X replaced by $Y - Z$ and Y replaced by $Z - X$.
(3) Definition 2.27 (Distance).
(4) Commutative property of real numbers.

The proof of Statement v follows a similar pattern. ■

Properties iv and v are the *triangle properties*. The reason for this name is suggested by the representations given in Figure 2.8. If X, Y, and Z are

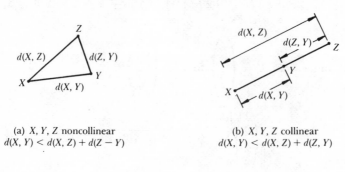

(a) X, Y, Z noncollinear
$d(X, Y) < d(X, Z) + d(Z - Y)$

(b) X, Y, Z collinear
$d(X, Y) < d(X, Z) + d(Z, Y)$

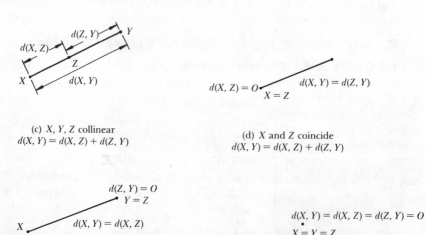

(c) X, Y, Z collinear
$d(X, Y) = d(X, Z) + d(Z, Y)$

(d) X and Z coincide
$d(X, Y) = d(X, Z) + d(Z, Y)$

(e) Y and Z coincide
$d(X, Y) = d(X, Z) + d(Z, Y)$

(f) X, Y and Z coincide
$d(X, Y) = d(X, Z) + d(Z, Y)$

Figure 2.8 The triangle inequality.

not collinear, they may be regarded as the vertices of a triangle, as in Figure 2.8a. The distances $d(X, Y)$, $d(X, Z)$, and $d(Z, Y)$ are the lengths of the sides of the triangle.

Property 2.28iv can be stated as follows: The length of one side of a triangle is less than the sum of the lengths of the other two sides. If we permit point Z to move, the triangle of Figure 2.8a can degenerate into any one of the situations pictured in Figure 2.8b, c, d, and e. Note that the equality $d(X, Y) = d(X, Z) + d(Z, Y)$ will hold if Z is a point "between" X and Y, as in Figure 2.8c, or if Z coincides with either X or Y. Note that in these three cases we have $Y - Z = k(X - Z)$, $k \geq 0$, or $X = Z$. Finally, the possibility that $X = Y = Z$ can also occur giving equality. Thus, the first two diagrams give the only instances in which the strict inequality $d(X, Y) < d(X, Z) + d(Z, Y)$ holds: that instance in which X, Y, and Z are noncollinear and that in which X, Y, and Z are distinct and collinear, with Z not lying between X and Y.

EXERCISES

In each of the Exercises 1 to 5, find the distance between the two given points:

1. $\begin{pmatrix} 1 \\ 2 \\ 3 \end{pmatrix}$ and $\begin{pmatrix} -1 \\ 2 \\ 1 \end{pmatrix}$.   2. $\begin{pmatrix} 1 \\ 3 \end{pmatrix}$ and $\begin{pmatrix} 2 \\ 7 \end{pmatrix}$.   3. $\begin{pmatrix} -1 \\ 2 \\ 3 \\ -4 \end{pmatrix}$ and $\begin{pmatrix} 1 \\ 1 \\ 2 \\ 1 \end{pmatrix}$.

4. $\begin{pmatrix} 2 \\ 0 \\ 1 \end{pmatrix}$ and $\begin{pmatrix} 0 \\ 1 \\ 4 \end{pmatrix}$.   5. $\begin{pmatrix} 2 \\ 1 \\ 3 \end{pmatrix}$ and $\begin{pmatrix} 5 \\ -7 \\ 2 \end{pmatrix}$.

6. (Pythagorean Theorem.) Letting X, Y, and Z be three distinct points such that the line $\mathcal{L}_1$ joining X and Y is perpendicular to the line $\mathcal{L}_2$ joining Y and Z, show that

$$[d(X, Z)]^2 = [d(X, Y)]^2 + [d(Y, Z)]^2.$$

7. (Converse of Pythagorean Theorem.) Let X, Y, and Z be three distinct points such that $[d(X, Z)]^2 = [d(X, Y)]^2 + [d(Y, Z)]^2$. Show that the line $\mathcal{L}_1$ joining X and Y is perpendicular to the line $\mathcal{L}_2$, joining Y and Z. (Hint: Write $X - Z = (X - Y) + (Y - Z)$ and use $[d(X, Z)]^2 = (X - Z) \cdot (X - Z)$.

8. Let $X = \begin{pmatrix} a \\ b \\ c \end{pmatrix}$ and $Y = \begin{pmatrix} d \\ e \\ f \end{pmatrix}$ be two points in $\mathcal{R}^3$. Determine the expression for $d(X, Y)$ in terms of the coordinates a, b, c, d, e, and f.

9. Let X and Y be elements of $\mathcal{R}^n$. Show that (a) $d(X, Y) \geq 0$, (b) $d(X, Y) = 0$ if and only if $X = Y$, and (c) $d(X, Y) = d(Y, X)$.

ANSWERS

1. $d = 2\sqrt{2}$. **3.** $d = \sqrt{31}$. **5.** $d = \sqrt{74}$. **7.** $[d(X, Z)]^2 = (X - Z) \cdot (X - Z) = [(X - Y) + (Y - Z)] \cdot [(X - Y) + (Y - Z)] = (X - Y) \cdot (X - Y) + 2(X - Y) \cdot (Y - Z) + (Y - Z) \cdot (Y - Z) = [d(X, Y)]^2 + 2(X - Y) \cdot (Y - Z) + [d(Y, Z)]^2;\ [d(X, Z)]^2 = [d(X, Y)]^2 + [d(Y, Z)]^2 + 2(X - Y) \cdot (Y - Z)$. Since $[d(X, Z)]^2 = [d(X, Y)]^2 + [d(Y, Z)]^2$, we must have $2(X - Y) \cdot (Y - Z) = 0$. This means that the direction vector $D_1 = X - Y$ for $\mathcal{L}_1$ is perpendicular to the direction vector $D_2 = Y - Z$ for $\mathcal{L}_2$. Therefore, $\mathcal{L}_1$ is perpendicular to $\mathcal{L}_2$.

9. DISTANCE FROM A POINT TO A LINE IN $\mathcal{R}^3$

Consider a line $\mathcal{L}$ and a point X in $\mathcal{R}^n$. We pose the following question: How far is it from X to the line $\mathcal{L}$? A look at Figure 2.9 suggests that X is at various distances from $\mathcal{L}$ depending upon whether we measure this distance from X to P_1, or from X to P_2, or from X to P_3, or from X to yet another point of $\mathcal{L}$. Figure 2.9 also suggests that there is some point P on $\mathcal{L}$ such that the distance $\|X - P\|$ is less than that from X to any other point of $\mathcal{L}$. We shall show (Theorem 2.30) that (in $\mathcal{R}^2$ at least) there is one and only one such point P and anticipate this result to make the following definition for a general $\mathcal{R}^n$.

DEFINITION 2.29 (Distance from a Point to a Line.) Let $\mathcal{L}$ be a line and X a point in $\mathcal{R}^n$ $(n > 1)$. The distance from X to $\mathcal{L}$, denoted by $d(X, \mathcal{L})$, is the distance from X to that point P of $\mathcal{L}$ such that $\|X - P\|$ is a minimum:

$$d(X, \mathcal{L}) = \text{smallest value of } \|X - P\|, \text{ where } P \in \mathcal{L}.$$

Let us now restrict our attention to $\mathcal{R}^2$ and see how $d(X, \mathcal{L})$ can be determined. Let A be a point of the line $\mathcal{L}$ and let $\hat{D}$ be a unit direction vector for $\mathcal{L}$. Any point P of $\mathcal{L}$ can then be written in the form

(2.29) $$P = A + t\hat{D}.$$

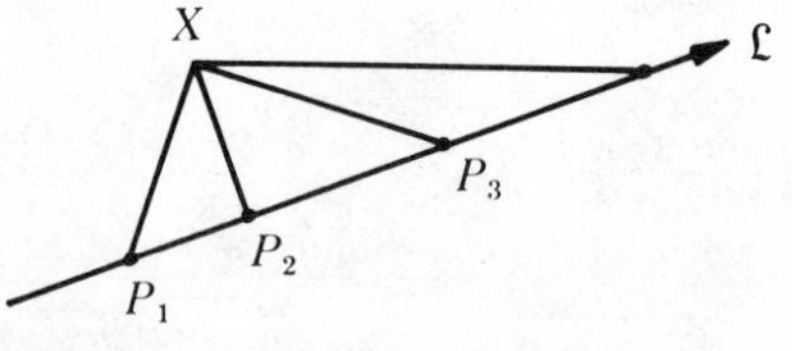

Figure 2.9 Distance from point to a line.

Thus, for a given point X in $\mathcal{R}^2$ we have

(2.30)
$$\|X - P\| = \|X - (A + t)\hat{D}\|$$
$$= \|(X - A) - t\hat{D}\|.$$

To continue with the determination $\|X - P\|$, it is convenient to express the vector $X - A$ as a linear combination of $\hat{D}$ and a unit normal $\hat{N}$ to $\mathcal{L}$. This is possible by virtue of Theorem 2.8. Thus,

(2.31)
$$X - A = a\hat{D} + n\hat{N}$$

for some choice of a and n. From Equations 2.30 and 2.31 we have

(2.32)
$$\|X - P\| = \|a\hat{D} + n\hat{N} - t\hat{D}\|$$
$$=^1 \|(a - t)\hat{D} + n\hat{N}\|$$
$$=^2 \sqrt{[(a - t)\hat{D} + n\hat{N}] \cdot [(a - t)\hat{D} + n\hat{N}]}$$
$$=^3 \sqrt{(a - t)^2(\hat{D} \cdot \hat{D}) + 2(a - t)n(\hat{D} \cdot \hat{N}) + n^2(\hat{N} \cdot \hat{N})}$$
$$=^4 \sqrt{(a - t)^2 + 2(a - t)n \cdot 0 + n^2}$$
$$= \sqrt{(a - t)^2 + n^2}.$$

(1) Associative, commutative, and distributive properties of vector algebra.
(2) Theorem 2.22.
(3) Theorem 2.2iv and v.
(4) Since $\hat{D} \cdot \hat{D} = \hat{N} \cdot \hat{N} = 1$ and $\hat{D} \cdot \hat{N} = 0$.

As t varies, the point P on $\mathcal{L}$ will vary, as will the distance $\|X - P\|$. It is clear from the last member of Equation 2.32 that the smallest value of $\|X - P\|$ will occur for precisely one value of t, namely for $t = a$. Moreover, the value of this shortest distance is given by

$$d(X, \mathcal{L}) = \sqrt{n^2} = |n|.$$

The above observations provide the major part of the proof of the following result.

THEOREM 2.30   (Distance from a Point to a Line in $\mathcal{R}^2$.)   Let $\mathcal{L}$ be a line in $\mathcal{R}^2$ which has unit direction vector $\hat{D}$, unit normal $\hat{N}$, and contains the point A. Let X be any point of $\mathcal{R}^2$ and define a and n by

$$a = (X - A) \cdot \hat{D} \quad \text{and} \quad n = (X - A) \cdot \hat{N}.$$

The distance $d(X, \mathcal{L})$ from X to $\mathcal{L}$ is given by

$$d(X, \mathcal{L}) = |n| = |(X - A) \cdot \hat{N}|.$$

Moreover, $d(X, \mathcal{L}) = \|X - P\|$ where P is the point of $\mathcal{L}$ given by

(2.33)
$$P = A + a\hat{D}$$
$$= X - n\hat{N}.$$

Proof. The expression $d(X, \mathcal{L}) = |n|$ was obtained in the discussion preceding the theorem. The real numbers a and n are such that

$$(2.31) \qquad\qquad X - A = a\hat{D} + n\hat{N}.$$

Taking the inner product of both members of this equation with $\hat{D}$ and then with $\hat{N}$ yields the expressions for a and n given in the theorem (see Exercise 9). The point P for which $d(X, \mathcal{L}) = \|X - P\|$ is found by setting $t = a$ in Equation 2.29. We find that $P = A + a\hat{D}$. From Equation 2.31 we find that $A + a\hat{D} = X - n\hat{N}$ so that we also have $P = X - n\hat{N}$. ■

The diagrams given in Figure 2.10 will assist in recalling the content of the preceding theorem. To find the shortest distance from X to $\mathcal{L}$, we determine a and n so that $X - A = a\hat{D} + n\hat{N}$ as shown on the diagram at the origin. The vector $a\hat{D}$ is referred to as the *component of $X - A$ in the direction of* $\mathcal{L}$ and the vector $n\hat{N}$ is referred to as the *component of $X - A$ normal to* $\mathcal{L}$. The coefficient n is obtained by taking the inner product of $X - A$ with $\hat{N}$. The absolute value of n gives $d(X, \mathcal{L})$. The point $P = X - n\hat{N}$ is obtained from X by subtracting the normal component of $X - A$. P is therefore the "foot of the perpendicular from X to $\mathcal{L}$" and $d(X, \mathcal{L})$ is the "perpendicular distance from X to $\mathcal{L}$."

Example 23. Find $d(X, \mathcal{L})$ if

$$X = \begin{pmatrix} 1 \\ 2 \end{pmatrix} \quad \text{and} \quad \mathcal{L} = \left\{ \begin{pmatrix} -1 \\ 3 \end{pmatrix} + t \begin{pmatrix} 1 \\ 1 \end{pmatrix} \,\middle|\, t \in \mathcal{R} \right\}$$

It is well, in this type of problem, to make a rough sketch of the geometrical representation, as in Figure 2.10. Sketch in the line $\mathcal{L}$; locate A on $\mathcal{L}$ and position X. Select P so that the line from X to P appears to be perpendicular to $\mathcal{L}$. (It is not necessary that the sketch be at all an accurate

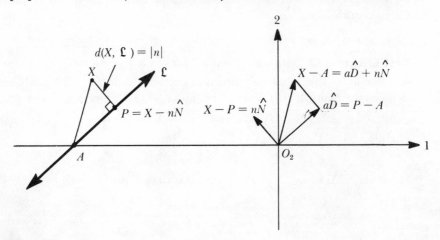

Figure 2.10

representation of the problem at hand. The sole purpose of the sketch is to recall the content of Theorem 2.30.)

The order of construction of the above sketch guides the actual algebraic process:

Select a point A on $\mathcal{L}$: $A = \begin{pmatrix} -1 \\ 3 \end{pmatrix}$.

Consider given point X: $X = \begin{pmatrix} 1 \\ 2 \end{pmatrix}$.

Determine $X - A$: $X - A = \begin{pmatrix} 1 \\ 2 \end{pmatrix} - \begin{pmatrix} -1 \\ 3 \end{pmatrix} = \begin{pmatrix} 2 \\ -1 \end{pmatrix}$.

Determine unit normal $\hat{N}$ to $\mathcal{L}$: $D = \begin{pmatrix} 1 \\ 1 \end{pmatrix}$ is a direction vector for $\mathcal{L}$

$$N = \begin{pmatrix} -1 \\ 1 \end{pmatrix} \text{ is a normal to } \mathcal{L}$$

$$\|N\| = \sqrt{1+1} = \sqrt{2}$$

$$\hat{N} = \frac{1}{\sqrt{2}} \begin{pmatrix} -1 \\ 1 \end{pmatrix}.$$

Determine component of $X - A$ perpendicular to $\mathcal{L}$:

$$n = (X - A) \cdot \hat{N}$$

$$= \begin{pmatrix} 2 \\ -1 \end{pmatrix} \cdot \frac{1}{\sqrt{2}} \begin{pmatrix} -1 \\ 1 \end{pmatrix}$$

$$= \frac{1}{\sqrt{2}} (-2 - 1) = -\frac{3}{\sqrt{2}} = -\frac{3\sqrt{2}}{2}.$$

Determine $d(X, \mathcal{L})$: $d(X, \mathcal{L}) = |n| = \left| -\frac{3\sqrt{2}}{2} \right| = \frac{3\sqrt{2}}{2}$.

Determine P: $P = X - n\hat{N}$

$$= \begin{pmatrix} 1 \\ 2 \end{pmatrix} - \left(-\frac{3\sqrt{2}}{2} \right) \cdot \frac{1}{\sqrt{2}} \begin{pmatrix} -1 \\ 1 \end{pmatrix}$$

$$= \begin{pmatrix} 1 \\ 2 \end{pmatrix} + \frac{3}{2} \begin{pmatrix} -1 \\ 1 \end{pmatrix}$$

$$= \begin{pmatrix} -1/2 \\ 7/2 \end{pmatrix}. \quad \blacksquare$$

EXERCISES

In Exercises 1 to 4 determine the distance $d(X, \mathcal{L})$ from the given point X to the given line $\mathcal{L}$:

1. $X = \begin{pmatrix} 2 \\ 3 \end{pmatrix}$; $\mathcal{L} = \left\{ \begin{pmatrix} -1 \\ 3 \end{pmatrix} + t \begin{pmatrix} 5 \\ 12 \end{pmatrix} \middle| t \in \mathcal{R} \right\}$.

2. $X = \begin{pmatrix} 0 \\ 0 \end{pmatrix}$; $\mathcal{L} = \left\{ Y \mid Y \cdot \begin{pmatrix} 7 \\ 3 \end{pmatrix} = 4; \ Y \in \mathcal{R}^2 \right\}$.

3. $X = \begin{pmatrix} 7 \\ 12 \end{pmatrix}$; $\mathcal{L} = \left\{ \begin{pmatrix} x \\ y \end{pmatrix} \mid 4x + 3y = 4; \ x, y \in \mathcal{R} \right\}$.

4. $X = \begin{pmatrix} -2 \\ 4 \end{pmatrix}$; $\mathcal{L} = \left\{ \begin{pmatrix} 1 \\ 3 \end{pmatrix} + t \begin{pmatrix} -3 \\ 1 \end{pmatrix} \mid t \in \mathcal{R} \right\}$.

In Exercises 5 to 8 determine the point P on $\mathcal{L}$ closest to the given point X:

5. $X = \begin{pmatrix} 2 \\ 3 \end{pmatrix}$; $\mathcal{L} = \left\{ \begin{pmatrix} -7 \\ 8 \end{pmatrix} + t \begin{pmatrix} 1 \\ -3 \end{pmatrix} \right\}$.

6. $X = \begin{pmatrix} 11 \\ 17 \end{pmatrix}$; $\mathcal{L} = \left\{ \begin{pmatrix} x \\ y \end{pmatrix} \mid x - 2y = 28 \right\}$.

7. $X = \begin{pmatrix} 7 \\ 13 \end{pmatrix}$; $\mathcal{L} = \left\{ Y \mid Y \cdot \begin{pmatrix} 22 \\ 32 \end{pmatrix} = 193 \right\}$.

8. $X = \begin{pmatrix} 0 \\ 0 \end{pmatrix}$; $\mathcal{L} = \left\{ \begin{pmatrix} x \\ y \end{pmatrix} \mid y = -x + 6 \right\}$.

9. Complete the proof of Theorem 2.30 by forming the inner product of both members of Equation 2.33 with $\hat{D}$ and then with $\hat{N}$ to obtain values for a and for n.

10. (a) Let $\mathcal{S}$ be a plane in $\mathcal{R}^3$ which has unit normal $\hat{N}$ and which contains the point A. Let X be any point of $\mathcal{R}^3$ and define n by

$$n = (X - A) \cdot \hat{N}.$$

Show that the point P of $\mathcal{S}$ which is closest to X is given by

$$P = X - n\hat{N},$$

and hence that the distance from X to $\mathcal{S}$ is given by

$$d(X, \mathcal{S}) = |n| = \|X - P\|.$$

(b) Use part (a) to find the distance from the point $X = \begin{pmatrix} 1 \\ 2 \\ 3 \end{pmatrix}$ to the plane $\mathcal{S} = \left\{ Y \mid Y \cdot \begin{pmatrix} 2 \\ 2 \\ 1 \end{pmatrix} = 2 \right\}$. Also find the point P of $\mathcal{S}$ which is closest to X.

ANSWERS

1. $d(X, \mathcal{L}) = \dfrac{36}{13}$.   3. $d(X, \mathcal{L}) = 12$.   5. $P = \begin{pmatrix} 5 \\ 2 \end{pmatrix}$.   7. $P = \begin{pmatrix} 3/2 \\ 5 \end{pmatrix}$.

10. (b) $P = \dfrac{1}{9} \begin{pmatrix} -5 \\ 4 \\ 20 \end{pmatrix}$.

PART II

DIFFERENTIAL CALCULUS OF $\mathcal{R} \to \mathcal{R}$ FUNCTIONS

3

LIMITS

INTRODUCTION

The two most important and basic concepts in analysis are those of function and limit. Without these concepts one cannot talk about "continuity," "differentiation," and "integration"; in effect, one cannot talk about "the calculus" without talking about functions and limits. It will be assumed that the reader has had at least an elementary introduction to functions. Consequently, only one section will be devoted to a discussion of functions as such. The remainder of the chapter will be devoted to a discussion of limits and an attempt to develop in the reader an intuitive feeling for this concept.

1. FUNCTIONS

A function is usually thought of as a correspondence between two sets, A and B. Quite frequently a simple rule will indicate the relationship between the elements of the first set and those of the second set. For example, the rule $y = x^2$ gives a simple correspondence between the set of all real numbers x and the set of all non-negative numbers y. Note that in this case there corresponds exactly one y to each x. This particular feature of a rule associating an element of A with an element of B is essential if f is to be a function from A into B. More precisely, we define a function from A into B as follows:

DEFINITION 3.1 Let A and B be sets. A set f of ordered pairs (a, b), where $a \in A$ and $b \in B$, such that each a in A is the first element of exactly one ordered pair in f, is called a *function from A into B*. A is called the *domain of f* and will be denoted by *dom f*. The set of all b in B such that $(a, b) \in f$ for at least one a in A is called the *range of f* and will be denoted by *ran f*. The notation $b = f(a)$ means that $(a, b) \in f$.

Throughout the remainder of the book we shall be concerned only with a particular class of functions, namely, those called vector functions. The reader is already familiar with several such functions: polynomials, trigonometric functions, the square root function, and so forth. Notice that each of these has both its domain and range contained in the real numbers. More generally:

DEFINITION 3.2 Let f be a function whose domain is contained in $\mathcal{R}^n$ and whose range is contained in $\mathcal{R}^m$. Then f is called a *vector function*, $\mathcal{R}^n$ is called the *domain space of f*, and $\mathcal{R}^m$ is called the *range space of f*. The notation $f : \mathcal{R}^n \to \mathcal{R}^m$ is to be read: f is a vector function with domain space $\mathcal{R}^n$ and range space $\mathcal{R}^m$. The notation "*Let f : $\mathcal{R}^n \to \mathcal{R}^m$*" is to be read: Let f be a function from $\mathcal{R}^n$ into $\mathcal{R}^m$; and the notation "Let $f : \mathcal{R}^n \to \mathcal{R}^m$ be ..." is to be read: Let f, a function from $\mathcal{R}^n$ into $\mathcal{R}^m$, be ....

The reader should be sure to make a distinction between the domain of f and the domain space of f, the range of f and the range space of f. For example, let $f : \mathcal{R} \to \mathcal{R}$, defined by $f(x) = \sqrt{1 - x^2}$. The largest possible domain of f is $\{x \mid x \in \mathcal{R}, -1 \leq x \leq 1\}$, whereas the domain space of f is the set of all real numbers. The corresponding range of f is the set

$$\{y \mid y \in \mathcal{R}, 0 \leq y \leq 1\}$$

and the range space of f is the set of all real numbers.

Let us now consider some examples of a type of function with which the reader may not be familiar: multiply-defined functions.

Example 1. Consider the function defined by

$$f(x) = \begin{cases} x^2 & \text{for } x \leq 0 \\ x & \text{for } x > 0. \end{cases}$$

Notice that the domain of f is the set of all real numbers. Figure 3.1 is a sketch of the graph of f. To obtain this sketch we divide the domain into

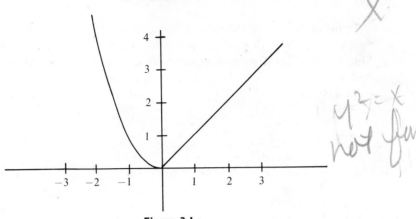

Figure 3.1

the two sets $\{x \mid x \in \mathcal{R}, x > 0\}$ and $\{x \mid x \in \mathcal{R}, x \le 0\}$. On the first set f is defined by $f(x) = x$, the graph of which is a straight line. On the second set f is defined by $f(x) = x^2$. ∎

Example 2. Consider the function defined by

$$f(x) = \begin{cases} 1 & \text{for } 0 \le x \le 1 \\ -1 & \text{for } x = 2 \\ x - 2 & \text{for } 3 \le x \le 4. \end{cases}$$

In this example the domain of f is the set

$$\{x \mid x \in \mathcal{R}, 0 \le x \le 1, \text{ or } x = 2, \text{ or } 3 \le x \le 4\}.$$

Again we divide the domain of f into parts in the obvious fashion in order to sketch the graph of f (see Figure 3.2). ∎

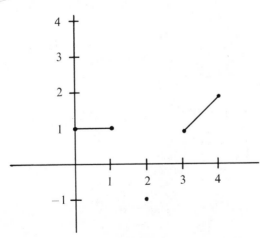

Figure 3.2

One of the most important aspects of vector functions is that frequently they can be combined in several meaningful fashions to form new functions. The reader is familiar with some of the operations, though it may never have occurred to him that he was creating a new function by combining two others.

DEFINITION 3.3 Let $f:\mathcal{R}^n \to \mathcal{R}^m$, $g:\mathcal{R}^n \to \mathcal{R}^m$, and $h:\mathcal{R}^m \to \mathcal{R}^q$.

i. (Addition.) $f + g$ is the function defined by $(f + g)X = fX + gX$ for all X in $(dom\, f) \cap (dom\, g)$.

ii. (Subtraction.) $f - g$ is the function defined by $(f - g)X = fX - gX$ for all X in $(dom\, f) \cap (dom\, g)$.

iii. (Multiplication.) fg is the function defined by $(fg)X = fX \cdot gX$ (inner product) for all X in $(dom\, f) \cap (dom\, g)$.

iv. (Scalar Multiplication.) cf is the function defined by $(cf)X = c(fX)$ for all X in $dom\, f$ and any real number c.

v. (Division.) If $m = 1$, f/g is the function defined by $(f/g)X = fX/gX$ for all X such that $X \in (dom\, f) \cap (dom\, g)$ and $gX \neq 0$.

vi. (Opposite or Additive Inverse.) $-f$ is the function defined by $(-f)X = -(fX)$ for all X in $dom\, f$.

vii. (Reciprocal.) If $m = 1$, $1/f$ is the function defined by $(1/f)X = \dfrac{1}{fX} = 1/(fX)$ for all X such that $X \in dom\, f$ and $fX \neq 0$.

viii. (Composition.) $h \circ f$ is the function defined by $(h \circ f)X = h(fX)$ for all X such that $X \in dom\, f$ and $fX \in dom\, h$.

Example 3. Let f and g be defined by $f(x,y) = \begin{pmatrix} x^2 + 1 \\ x - y \end{pmatrix}$ and $g(x,y) = \begin{pmatrix} x^2 - x \\ y^2 + 2y \end{pmatrix}$:

i. $(f + g)(x,y) = \begin{pmatrix} (x^2 + 1) \\ (x - y) \end{pmatrix} + \begin{pmatrix} (x^2 - x) \\ (y^2 + 2y) \end{pmatrix} = \begin{pmatrix} 2x^2 - x + 1 \\ y^2 + y + x \end{pmatrix}$.

ii. $(f - g)(x,y) = \begin{pmatrix} (x^2 + 1) \\ (x - y) \end{pmatrix} - \begin{pmatrix} (x^2 - x) \\ (y^2 + 2y) \end{pmatrix} = \begin{pmatrix} 1 + x \\ x - 3y - y^2 \end{pmatrix}$.

iii. $(fg)(x,y) = \begin{pmatrix} x^2 + 1 \\ x - y \end{pmatrix} \cdot \begin{pmatrix} x^2 - x \\ y^2 + 2y \end{pmatrix}$

$= (x^2 + 1)(x^2 - x) + (x - y)(y^2 + 2y)$.

iv. $(5f)(x,y) = 5\begin{pmatrix} x^2 + 1 \\ x - y \end{pmatrix} = \begin{pmatrix} 5x^2 + 5 \\ 5x - 5y \end{pmatrix}$.

vi. $(-f)(x,y) = -\begin{pmatrix} x^2 + 1 \\ x - y \end{pmatrix} = \begin{pmatrix} -x^2 - 1 \\ y - x \end{pmatrix}$. ∎

Example 4. Let f and g be defined by $f(x) = x^2 + 1$ and $g(x) = x$:

v. $(f/g)(x) = \dfrac{x^2 + 1}{x}$

$= x + \dfrac{1}{x}$ for $x \neq 0$.

vii. $(1/f)(x) = \dfrac{1}{x^2 + 1}$. ■

It will be left to the reader to verify the following properties of the addition and multiplication of vector functions: i. $f + g = g + f$; ii. $(f + g) + h = f + (g + h)$; iii. $fg = gf$; and iv. $f(g + h) = fg + fh$. What other properties can you discover about combinations of vector functions?

The operation of composition of vector functions is one of the most important. In the exercises at the end of this section emphasis will be placed on this operation.

Example 5. Let $f(x, y) = x^2 + y^2$ and $g(z) = z^3$.

Then
$$(g \circ f)(x, y) = g(f(x, y))$$
$$= g(x^2 + y^2)$$
$$= (x^2 + y^2)^3.$$

Notice that $f \circ g$ is not defined since $ran\ g \subset \mathcal{R}$ and $dom\ f \subset \mathcal{R}^2$. ■

Example 6. Let $f(x, y) = \begin{pmatrix} x - y \\ x + y \end{pmatrix}$ and $g(u, v) = \begin{pmatrix} u + v \\ 1 \end{pmatrix}$.

Then
$$(g \circ f)(x, y) = g(f(x, y))$$
$$= g\begin{pmatrix} x - y \\ x + y \end{pmatrix}$$
$$= \begin{pmatrix} (x - y) + (x + y) \\ 1 \end{pmatrix}$$
$$= \begin{pmatrix} 2x \\ 1 \end{pmatrix}.$$

Moreover, in this case $f \circ g$ is defined and
$$(f \circ g)(u, v) = f(g(u, v))$$
$$= f\begin{pmatrix} u + v \\ 1 \end{pmatrix}$$
$$= \begin{pmatrix} u + v - 1 \\ u + v + 1 \end{pmatrix}.$$

We note that $f \circ g \neq g \circ f$. ■

In the two preceding examples we handled the problem of how to form the composite of two given functions f and g. Frequently, it is of interest to be able to reverse this process, that is, given a function h, to find two functions f and g such that $g \circ f = h$.

Example 7.   Consider the function h defined by $h(u) = \sqrt{1 + u^2}$. To evaluate this function at any point u in its domain we must perform three operations: (1) square u, (2) add one to this number, and (3) take the square root of the result of (2). Hence, we see three operations at work, and we have a certain amount of freedom in choosing f and g such that $g \circ f = h$. Let us do this in two different ways:

(a) Let $f(u) = 1 + u^2$ and $g(y) = \sqrt{y}$ for $y \geq 0$. Forming the composite of f and g we have

$$(g \circ f)(u) = g(f(u))$$
$$= g(1 + u^2)$$
$$= \sqrt{1 + u^2}$$
$$= h(u).$$

(b) Let $f(u) = u^2$ and $g(y) = \sqrt{1 + y}$ for $y \geq -1$. Then

$$(g \circ f)(u) = g(f(u))$$
$$= g(u^2)$$
$$= \sqrt{1 + u^2}$$
$$= h(u). \quad \blacksquare$$

Frequently the composite of two functions f and g is a simpler function than either f or g. In the preceding example we decomposed h into two simpler functions f and g. However, notice that we could have chosen

$$f(u) = \begin{pmatrix} 1 \\ u^2 \end{pmatrix} \quad \text{and} \quad g(x, y) = \sqrt{x + y}.$$

In this case also we have

$$(g \circ f)(u) = g(f(u))$$
$$= g \begin{pmatrix} 1 \\ u^2 \end{pmatrix}$$
$$= \sqrt{1 + u^2}$$
$$= h(u).$$

EXERCISES

Calculate (a) $f + g$, (b) $f - g$, (c) fg, (d) $3f$, and (e) $-f$ in each of the following cases:

1. $\begin{cases} f(x, y) = x^2 - y \\ g(x, y) = y^3. \end{cases}$

2. $\begin{cases} f(x) = \begin{pmatrix} x \\ x^2 - 1 \end{pmatrix} \\ g(x) = \begin{pmatrix} x + 2 \\ x \end{pmatrix}. \end{cases}$

3. $\begin{cases} f(x,y) = \begin{pmatrix} x-y \\ x+y \end{pmatrix} \\ g(x,y) = \begin{pmatrix} x+y \\ y-x \end{pmatrix}. \end{cases}$

4. $\begin{cases} f(x) = \begin{pmatrix} 1 \\ x \\ x^2 \end{pmatrix} \\ g(x) = \begin{pmatrix} x^2 \\ x \\ 1 \end{pmatrix}. \end{cases}$

5. $\begin{cases} f(x,y) = \begin{pmatrix} \sin x \\ \cos y \end{pmatrix} \\ g(x,y) = \begin{pmatrix} y \\ x \end{pmatrix}. \end{cases}$

Sketch the following multiply-defined functions:

6. $f(x) = \begin{cases} 1 & \text{for } x \geq 0 \\ -1 & \text{for } x < 0. \end{cases}$

7. $f(x) = \begin{cases} x & \text{for } 0 \leq x \leq 1 \\ 2-x & \text{for } 1 < x \leq 2. \end{cases}$

8. $f(x) = \begin{cases} x^2 & \text{for } 0 \leq x \leq 1 \\ 1 & \text{for } 1 < x \leq 2 \\ x-1 & \text{for } x > 2. \end{cases}$

9. $f(x) = \begin{cases} \sqrt{1-x^2} & \text{for } -1 \leq x \leq 1 \\ 2x-2 & \text{for } x > 1 \\ -2x-2 & \text{for } x < -1. \end{cases}$

10. $f(x) = \begin{cases} 1 & \text{for } x = -1, 1 \\ x & \text{for } -1/2 \leq x \leq 1/2. \end{cases}$

Calculate $g \circ f$ in each of the following cases:

11. $\begin{cases} f(x) = x+3 \quad (x+3)^2 \\ g(y) = y^2. \end{cases}$

12. $\begin{cases} f(x,y) = x-y. \\ g(z) = z^2+1. \end{cases}$

13. $\begin{cases} f(x,y) = \begin{pmatrix} x^2+y^2-1 \\ x^2-1 \end{pmatrix} \\ g(u,v) = \begin{pmatrix} u-v \\ v^2+1 \end{pmatrix}. \end{cases}$

14. $\begin{cases} f(t) = \begin{pmatrix} \sin t \\ \cos t \end{pmatrix} \\ g(x,y) = x^2+y^2. \end{cases}$

15. $f(x,y,z) = x$

$g(t) = \begin{pmatrix} t^2 \\ t \\ t-1 \end{pmatrix}.$

16. Find a function $\mathcal{O}$ such that $\mathcal{O}:\mathcal{R}^n \to \mathcal{R}^n$ and $f + \mathcal{O} = \mathcal{O} + f = f$ for all f such that $f:\mathcal{R}^n \to \mathcal{R}^n$.

17. Let $f:\mathcal{R}^n \to \mathcal{R}^m$, $g:\mathcal{R}^m \to \mathcal{R}^q$, and $h:\mathcal{R}^q \to \mathcal{R}^p$. Prove that $h \circ (g \circ f) = (h \circ g) \circ f$ with domain $\{X \mid X \in \text{dom} f, fX \in \text{dom} g, (g \circ f)X \in \text{dom} h\}$.

18. Find an example (other than Example 6) of two functions f and g such that $f \circ g$ and $g \circ f$ are both defined, but $f \circ g \neq g \circ f$.

19. Find a function I, $I:\mathcal{R}^n \to \mathcal{R}^n$, such that $I \circ f = f \circ I = f$ for all f, $f:\mathcal{R}^n \to \mathcal{R}^n$.

For each of the following find functions f_1, g_1, f_2, and g_2 such that $f_1:\mathcal{R} \to \mathcal{R}$, $g_1:\mathcal{R} \to \mathcal{R}$, $f_2:\mathcal{R} \to \mathcal{R}^2$, and $g_2:\mathcal{R}^2 \to \mathcal{R}$ and such that $g_1 \circ f_1 = h$ and $g_2 \circ f_2 = h$.

20. $h(u) = (u-1)^3$.

21. $h(u) = \sin(u^2)$.

22. $h(u) = 1$.

23. $h(u) = \sqrt{u^4 + u^2}$.

24. $h(u) = \dfrac{1}{u^2+1}.$

ANSWERS

1a. $(f + g)(x, y) = x^2 - y + y^3$.   **1b.** $(f - g)(x, y) = x^2 - y - y^3$.
1c. $(fg)(x, y) = x^2 y^3 - y^4$.   **1d.** $(3f)(x, y) = 3x^2 - 3y$.
1e. $(-f)(x, y) = y - x^2$.   **2c.** $(fg)(x) = x^3 + x^2 + x$.
3c. $(fg)(x, y) = 0$.   **4c.** $(fg)(x, y) = 3x^2$.   **5c.** $(fg)(x, y) = y \sin x + x \cos y$.

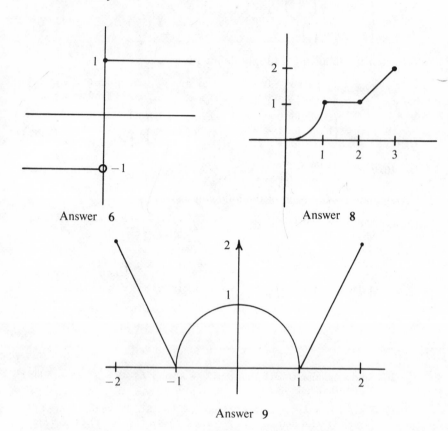

Answer **6**

Answer **8**

Answer **9**

11. $(g \circ f)(x) = (x + 3)^2$.   **12.** $(g \circ f)(x, y) = (x - y)^2 + 1$.
14. $(g \circ f)(t) = 1$.

2. BASIC DEFINITIONS

One of the major themes to be found throughout this book is that of approximation. Frequently, we shall wish to "approximate" a given point by other points or a given function by other functions. For this to be significant, we develop criteria by which we can determine whether we have a "good approximation" in a given situation.

The reader is already familiar with approximations of certain real numbers, for example, 1.4 is often used in calculations in place of $\sqrt{2}$.

However, 1.414 is a more accurate replacement for $\sqrt{2}$ in a calculation, since $|\sqrt{2} - 1.414| < |\sqrt{2} - 1.4|$. In this case we say that 1.414 is a better approximation for $\sqrt{2}$ than is 1.4.

The foundation upon which we shall start building our theory of approximation is the idea of the distance between two points in $\mathcal{R}^n$. Recall that if A and B are in $\mathcal{R}^n$, the distance between A and B is defined to be $\|A - B\|$, the length of the vector $A - B$. This section will be devoted to defining certain concepts, dependent upon the concept of distance, that will greatly facilitate our discussion of approximations.

Throughout the remainder of the book the following notation and terminology will be used:

   i. $[a, b] = \{x \mid x \in \mathcal{R}, a \leq x \leq b\}$.
  ii. $(a, b) = \{x \mid x \in \mathcal{R}, a < x < b\}$.
 iii. $[a, b) = \{x \mid x \in \mathcal{R}, a \leq x < b\}$.
  iv. $(a, b] = \{x \mid x \in \mathcal{R}, a < x \leq b\}$.
   v. $[a, \infty) = \{x \mid x \in \mathcal{R}, x \geq a\}$.
  vi. $(a, \infty) = \{x \mid x \in \mathcal{R}, x > a\}$.
 vii. $(-\infty, b] = \{x \mid x \in \mathcal{R}, x \leq b\}$.
viii. $(-\infty, b) = \{x \mid x \in \mathcal{R}, x < b\}$.
  ix. $(-\infty, \infty) = \mathcal{R}$.

The sets in i to iv are called *finite intervals with endpoints a and b*. The sets in v to ix are called *infinite intervals*. The sets $[a, b]$, $[a, \infty)$, and $(-\infty, b]$ are called *closed intervals;* the sets (a, b), (a, ∞), and $(-\infty, b)$ are called *open intervals;* the sets $(a, b]$ and $[a, b)$ are called *half-open intervals*.

DEFINITION 3.4

   i. An *open ball* in $\mathcal{R}^n$ is a set S of the form
$$S = \{X \mid X \in \mathcal{R}^n, \|X - X_0\| < r\},$$
where $X_0 \in \mathcal{R}^n$ and $r > 0$. We shall denote this set by $\mathcal{B}(X_0; r)$.
  ii. A *deleted open ball* in $\mathcal{R}^n$ is a set S of the form
$$S = \{X \mid X \in \mathcal{R}^n, X \neq X_0, \|X - X_0\| < r\},$$
where $X_0 \in \mathcal{R}^n$ and $r > 0$. We shall denote this set by $\mathcal{B}(\tilde{X}_0; r)$.
 iii. A *closed ball* in $\mathcal{R}^n$ is a set S of the form
$$S = \{X \mid X \in \mathcal{R}^n, \|X - X_0\| \leq r\},$$
where $X_0 \in \mathcal{R}^n$ and $r > 0$. We shall denote this set by $\bar{\mathcal{B}}(X_0; r)$.
  iv. A *sphere* in $\mathcal{R}^n$ is a set S of the form
$$S = \{X \mid X \in \mathcal{R}^n, \|X - X_0\| = r\},$$
where $X_0 \in \mathcal{R}^n$ and $r > 0$. We shall denote this set by $S(X_0; r)$.
   v. X_0 is called the *center* and r is called the *radius* of $\mathcal{B}(X_0; r)$, $\mathcal{B}(\tilde{X}_0; r)$, $\bar{\mathcal{B}}(X_0; r)$, or $S(X_0; r)$.

Study the following examples and figures carefully. Therein we shall explain certain conventions by which one can distinguish sketches of open, deleted open, and closed balls.

Example 8

(a) Consider first the open ball in $\mathcal{R}$ with center 1 and radius 2, $\mathcal{B}(1;2)$. By Definition 3.4i,

$$\begin{aligned}
\mathcal{B}(1;2) &= \{x \mid x \in \mathcal{R}, |x-1| < 2\} \\
&= \{x \mid x \in \mathcal{R}, -2 < x-1 < 2\} \\
&= \{x \mid x \in \mathcal{R}, -1 < x < 3\} \\
&= (-1,3).
\end{aligned}$$

Figure 3.3 is a sketch of $\mathcal{B}(1;2)$. The open circles about -1 and 3 indicate

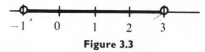

Figure 3.3

that these two points are not in the set. The solid line from -1 to 3 indicates that every point between -1 and 3 is in the set.

(b) Now consider the closed ball in $\mathcal{R}$ with center 1 and radius 2, $\bar{\mathcal{B}}(1;2)$. By Definition 3.4iii,

$$\begin{aligned}
\bar{\mathcal{B}}(1;2) &= \{x \mid x \in \mathcal{R}, |x-1| \leq 2\} \\
&= \{x \mid x \in \mathcal{R}, -2 \leq x-1 \leq 2\} \\
&= \{x \mid x \in \mathcal{R}, -1 \leq x \leq 3\} \\
&= [-1,3].
\end{aligned}$$

Figure 3.4 is a sketch of $\bar{\mathcal{B}}(1;2)$ in which the large solid dots at -1 and 3

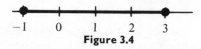

Figure 3.4

indicate that these two points are in the set and the solid line from -1 to 3 indicates that every point between -1 and 3 is also in the set.

(c) Last, let us consider the deleted open ball in $\mathcal{R}$ with center 1 and radius 2, $\mathcal{B}(\tilde{1};2)$. By Definition 3.4ii we have

$$\mathcal{B}(\tilde{1};2) = (-1,3) \setminus \{1\}.$$

(The symbol $\setminus$ denotes set difference.)
Figure 3.5 is a sketch of $\mathcal{B}(\tilde{1};2)$. As in Figure 3.3, an open circle about

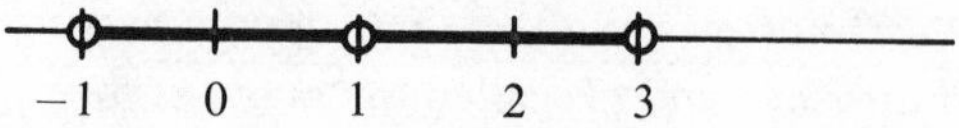

Figure 3.5

a point indicates that point is not in the set, and the solid line between two points indicates that all points between the two points are in the set. ∎

Let us now turn our attention to the problem of sketching balls in $\mathcal{R}^2$.

Example 9. Consider the open, closed, and deleted open balls in $\mathcal{R}^2$ with center $(1, 0)$ and radius 3. By Definition 3.4,

$$\mathcal{B}((1, 0); 3) = \{(x, y) \mid (x, y) \in \mathcal{R}^2, \|(x, y) - (1, 0)\| < 3\}$$
$$= \{(x, y) \mid (x, y) \in \mathcal{R}^2, (x - 1)^2 + y^2 < 9\},$$
$$\overline{\mathcal{B}}((1, 0); 3) = \{(x, y) \mid (x, y) \in \mathcal{R}^2, \|(x, y) - (1, 0)\| \leq 3\}$$
$$= \{(x, y) \mid (x, y) \in \mathcal{R}^2, (x - 1)^2 + y^2 \leq 9\},$$

and

$$\mathcal{B}((\widetilde{1, 0}); 3) = \mathcal{B}((1, 0); 3) \setminus \{(1, 0)\}.$$

Figure 3.6a is a sketch of $\mathcal{B}((1, 0); 3)$, Figure 3.6b a sketch of $\overline{\mathcal{B}}((1, 0); 3)$, and Figure 3.6c a sketch of $\mathcal{B}((\widetilde{1, 0}); 3)$. ∎

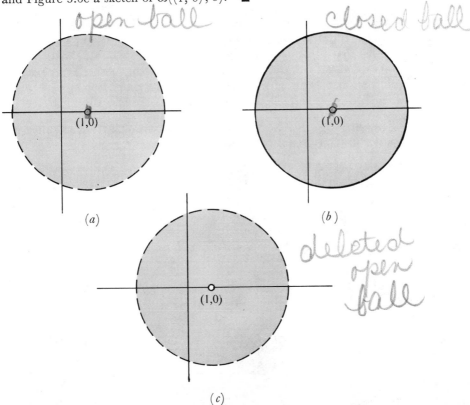

(a)

(b)

(c)

Figure 3.6

Notice that in Figure 3.6a and c the circle is broken, which indicates that the points on the circle are not in the set sketched. In Figure 3.6b the solid circle indicates that the points on the circle are in the set sketched. In Figure 3.6c the small open circle about the point $(1, 0)$ indicates that the point is not in the set. In all three sets the shading indicates that points inside the circle are in the set sketched (with the exception of the point $(1, 0)$ in Figure 3.6c).

The two concepts defined in the next definition are closely interrelated and depend heavily upon the notion of an open ball. Note carefully the distinctions between the two ideas.

DEFINITION 3.5   Let S be a subset of $\mathcal{R}^n$.
i. X_0 is an *interior point of* S if and only if there exists an $r > 0$ such that $\mathcal{B}(X_0; r) \subset S$.
ii. The *interior of* S, denoted by $\overset{\circ}{S}$, is the set of all points X_0 such that X_0 is an interior point of S.

One should note that an interior point X_0 of S is an element of S, for $X_0 \in \mathcal{B}(X_0; r)$ and $\mathcal{B}(X_0; r) \subset S$. Intuitively we regard an interior point of S as one that is completely surrounded by other points in S. If S is an arbitrary set in $\mathcal{R}^n$, two extreme situations may occur: either every point of S is an interior point of S or no point of S is an interior point of S (other possibilities exist). The following two examples illustrate these extremes:

Example 10.   Let $\mathcal{B}(X_0; r)$ be an open ball in $\mathcal{R}^n$. We shall prove that every element of $\mathcal{B}(X_0; r)$ is an interior point of $\mathcal{B}(X_0; r)$.

Proof.   Let $X_1 \in \mathcal{B}(X_0; r)$. Let $r' = r - \|X_1 - X_0\|$. We note that $r' > 0$. We shall show that $\mathcal{B}(X_1; r') \subset \mathcal{B}(X_0; r)$. Once we have done this, we know by Definition 3.5i that X_1 is an interior point of $\mathcal{B}(X_0; r)$. Figure 3.7 illustrates the situation in $\mathcal{R}^2$ and should convince the reader (intuitively)

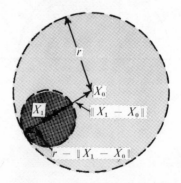

Figure 3.7

of the truth of our assertion. If Y is any element of $\mathcal{B}(X_1; r')$,

$$\|Y - X_0\| \leq \|Y - X_1\| + \|X_1 - X_0\| \quad \text{by the Triangle Inequality,}$$
$$\text{Theorem 2.28iv (p. 141),}$$
$$< r' + \|X_1 - X_0\|.$$

However, $r' + \|X_1 - X_0\| = (r - \|X_1 - X_0\|) + \|X_1 - X_0\| = r$. Consequently, $\|Y - X_0\| < r$ and $Y \in \mathcal{B}(X_0; r)$. It follows that $\mathcal{B}(X_1; r') \subset \mathcal{B}(X_0; r)$, since Y is an arbitrary element of $\mathcal{B}(X_1; r')$. Therefore, X_1 is an interior point of $\mathcal{B}(X_0; r)$; and since X_1 is an arbitrary element of $\mathcal{B}(X_0; r)$, we know that every point of $\mathcal{B}(X_0, r)$ is an interior point of $\mathcal{B}(X_0; r)$. ∎

Now let us consider an example of a set having no interior point:

Example 11. Let $S = \{X_0\}$, $X_0 \in \mathcal{R}^n$. Since S consists of a single element X_0, the only possible interior point of S is X_0. We wish to show that X_0 is not an interior point of S and hence that S has no interior points.

Let $r > 0$ and $X_0 = \begin{pmatrix} x_1 \\ x_2 \\ \cdot \\ \cdot \\ \cdot \\ x_n \end{pmatrix}$. Let $X_1 = \begin{pmatrix} x_1 + \dfrac{r}{2} \\ x_2 \\ \cdot \\ \cdot \\ \cdot \\ x_n \end{pmatrix}$. We then have

$$\|X_0 - X_1\| = \left\| \begin{pmatrix} -\dfrac{r}{2} \\ 0 \\ \cdot \\ \cdot \\ \cdot \\ 0 \end{pmatrix} \right\| = \frac{r}{2} < r. \quad \text{Hence, } X_1 \in \mathcal{B}(X_0; r), \ X_1 \notin S, \quad \text{and}$$

$\mathcal{B}(X_0; r) \not\subset S$. Thus, X_0 is not an interior point of S, and S has no interior points. ∎

A special name is given to those subsets S of $\mathcal{R}^n$ that have the property that every element of S is an interior point of S. These sets are said to be *open in* $\mathcal{R}^n$. In Example 10 we therefore proved that every open ball in $\mathcal{R}^n$ is open (by this definition) in $\mathcal{R}^n$. The everyday antonym of open is closed. Unfortunately, in mathematics the term closed is used in a more specialized sense. A subset S of $\mathcal{R}^n$ is said to be *closed in* $\mathcal{R}^n$ if and only if its complement, $\mathcal{R}^n \setminus S$, is open in $\mathcal{R}^n$. There exist many subsets of $\mathcal{R}^n$ that are neither open nor closed—hence, in their mathematical meanings these words are not antonyms.

EXERCISES

Sketch each of the following sets:

1. $\mathcal{B}(-1; 2)$.　　　　　　　　　2. $\overline{\mathcal{B}}(1/2; 3)$.

3. $\mathcal{B}(\tilde{0}; 2)$.   4. $\mathcal{S}(1; 4)$.

5. $\mathcal{B}(2; 1) \cap \mathcal{B}(3; 2)$.   6. $\mathcal{B}((1, -1); 2)$.

7. $\mathcal{B}((2, 3); 1)$.   8. $\overline{\mathcal{B}}((0, 2); 1)$.

9. $\mathcal{S}((0, 0); 3)$.   10. $\mathcal{S}((0, 1); 2) \cup \mathcal{B}((1, 2); 3)$.

Sketch each of the following sets. Determine which are the interior points of each set:

11. $[2, 4)$.   12. $\{x \mid x < 1 \text{ or } x \geq 5\}$.

13. $\{(x, y) \mid (x, y) = (1, 2) \text{ or } \|(x, y)\| > 6\}$.

14. $\{1, 2, 4\}$.

15. In $\mathcal{R}^n$ show that $\overline{\mathcal{B}}(X_0; r) \setminus \mathcal{B}(X_0; r) = \mathcal{S}(X_0; r)$ for each X_0 in $\mathcal{R}^n$ and each $r > 0$.

16. Show that for each closed ball in $\mathcal{R}^n$ every element of the complement of $\overline{\mathcal{B}}(X_0; r)$ is an interior point of the complement of $\overline{\mathcal{B}}(X_0; r)$.

DEFINITION.   Let $\mathcal{S}$ be a subset of $\mathcal{R}^n$ and $X_0 \in \mathcal{R}^n$.

i. X_0 is an *isolated point of* $\mathcal{S}$ if and only if $X_0 \in \mathcal{S}$ and there exists $r > 0$ such that

$$\mathcal{B}(X_0; r) \cap \mathcal{S} = \{X_0\}.$$

ii. X_0 is a *boundary point of* $\mathcal{S}$ if and only if for each $r > 0$

$$\mathcal{B}(X_0; r) \cap \mathcal{S} \neq \varnothing \quad \text{and} \quad \mathcal{B}(X_0; r) \cap (\mathcal{R}^n \setminus \mathcal{S}) \neq \varnothing.$$

17. Find the isolated points and the boundary points of the sets in Exercises 11 to 14.

18. Show that if X_0 is an isolated point of $\mathcal{S}$, then X_0 is a boundary point of $\mathcal{S}$.

19. Give an example of a set $\mathcal{S}$ and a point X_0 such that X_0 is a boundary point of $\mathcal{S}$, but X_0 is not an isolated point of $\mathcal{S}$.

ANSWERS

12. $\overset{\circ}{\mathcal{S}} = \{x \mid x < 1 \text{ or } x > 5\}$.   14. $\overset{\circ}{\mathcal{S}} = \varnothing$.

3. AN INTUITIVE DISCUSSION OF LIMITS

Because of the extreme importance of the concept of the limit of a function and because of the seeming sophistication of this concept, we shall devote this section to an intuitive discussion of a limit. Terms such as "close to" and "small" will be used in an imprecise manner. The context in which these words are found should explain them. In the following section of this chapter a precise mathematical definition of the term limit will be given.

We shall consider the behavior of functions $f: \mathcal{R} \to \mathcal{R}$. If f is defined and real-valued on an open interval (a, b), we know its value at each point of (a, b). However, since b may not be in the domain of f, we may not be able to talk about the value of f at b. On the other hand, we can consider the values of f at points close to b and in (a, b). For example, we could investigate the values of f at each point x of (a, b) such that $|b - x| < 1/2$, or $|b - x| < \frac{1}{10}$. For a given function f, one of two things will become evident as a result of such an investigation:

1. The values of f at points x close to b will all be close to some real number L,

or 2. There is no such real number L.

When we say that the values of f are close to L when x is close to b, we mean that $|f(x) - L|$ is small when $|b - x|$ is small. For example, let $f(x) = x^2$ for x in $(0, 2)$. Figure 3.8 is a sketch of the graph of f. Intuitively, we feel that $f(x)$ is close to 4 when x is close to 2. This can be demonstrated as follows:

$$|f(x) - 4| = |x^2 - 4|$$
$$= |x + 2|\,|x - 2|$$
$$< 4\,|x - 2|, \text{ since } x \in (0, 2).$$

It is now obvious that when $|x - 2|$ is small, $|f(x) - 4|$ is small.

If it is true that $|f(x) - L|$ is small whenever $x \neq b$ and $|x - b|$ is small, we shall say that f *approaches a limit* L *as* x *approaches* b and we shall denote this by

(3.1) $$\lim_{x \to b} f(x) = L.$$

So far we have only considered a function f defined on an interval (a, b) and its behavior at points x close to b. We could also discuss $\lim_{x \to a} f(x)$ for f defined on (a, b). However, this case is completely analogous to the preceding, and we now consider a function f that is real-valued and defined on the union of two adjacent open intervals, $(a, b) \cup (b, c)$. Again we can

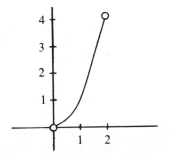

Figure 3.8

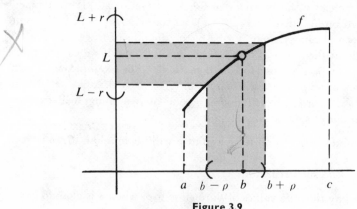

Figure 3.9

ask the question, "Does there exist a real number L such that $f(x)$ is close to L whenever $x \neq b$ and x is close to b?" If there does exist such a real number L, we again say that f approaches a limit L as x approaches b and we write $\lim\limits_{x \to b} f(x) = L$.

Figure 3.9 is a sketch of the graph of a function for which $\lim\limits_{x \to b} f(x) = L$. On the y-axis there is indicated an open interval $(L - r, L + r)$ of radius r and center L and on the x-axis an open interval $(b - \rho, b + \rho)$ of radius ρ and center b. The sketch indicates that $|f(x) - L| < r$ whenever $x \neq b$ and $|b - x| < \rho$. Although we have only exhibited one pair of intervals $(L - r, L + r)$ and $(b - \rho, b + \rho)$ having this property, it is evident from Figure 3.9 that no matter how small r is made we can always find a ρ such that the corresponding intervals also have this property.

Example 12. Let $f: \mathcal{R} \to \mathcal{R}$ be defined by

$$f(x) = \begin{cases} 1 & \text{if } x > 0 \\ -1 & \text{if } x < 0. \end{cases}$$

We shall show that f does not approach a limit as x approaches 0. To do so we assume the contrary: $\lim\limits_{x \to 0} f(x) = L$. Figure 3.10 is a sketch of the graph of f and should assist the reader in following the rest of the argument. We note that

$$|f(x) - L| = \begin{cases} |1 - L| & \text{if } x > 0 \\ |-1 - L| & \text{if } x < 0. \end{cases}$$

Since L is a real number, either $L \geq 0$ or $L \leq 0$. If $L \geq 0$ and $x < 0$, $|f(x) - L| = |-1 - L| = 1 + L \geq 1$. Similarly, if $L \leq 0$ and $x > 0$, $|f(x) - L| \geq 1$. Thus, for $x < 0$, the values of f are not close to any $L \geq 0$, and for $x > 0$, the values of f are not close to any $L \leq 0$. Consequently, we have exhausted our supply of potential values for a limit L of f, and f does not approach a limit as x approaches 0. ■

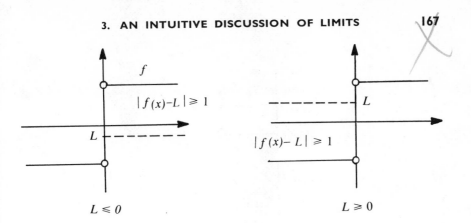

Figure 3.10

Another example of a function which does not approach a limit is as follows: let $f(x) = \dfrac{1}{x}$ for $x > 0$. We note that $f(x)$ is very large when x is very small. Thus, f does not approach a limit as x approaches 0.

Notice that when we make the statement $\lim\limits_{x \to b} f(x) = L$, we are not making *any* assertion about the value of f at b or even whether f is defined at b. We are concerned exclusively with the values $f(x)$, for x *close to but not equal to* b. Although it will frequently be the case that $\lim\limits_{x \to b} f(x) = f(b)$, the reader should not take this for granted.

We illustrate our lack of concern with the point b by the following example.

Example 13. Let $f \colon \mathcal{R} \to \mathcal{R}$ be defined by

$$f(x) = \frac{x^2 - 5x + 6}{x - 2} \quad \text{for } x \neq 2.$$

We have not defined f at $b = 2$ and further we cannot substitute $x = 2$ in the formula given for f since this would give us a zero in the denominator. The reader may feel that f is poorly behaved near $b = 2$ since the denominator becomes small when x is close to 2. However, for $x \neq 2$,

$$f(x) = \frac{(x - 2)(x - 3)}{x - 2} = x - 3.$$

Figure 3.11 is a sketch of the graph of f. Notice that the graph is a line with one point removed. It is now obvious that when x is close to b ($= 2$,) $f(x)$ is close to -1. Therefore, $\lim\limits_{x \to 2} f(x) = -1$. ∎

Consider now two functions f and g such that $\lim\limits_{x \to b} f(x) = L$ and $\lim\limits_{x \to b} g(x) = K$. Thus, if $x \neq b$ and x is close to b, $f(x)$ is close to L and $g(x)$

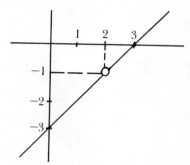

Figure 3.11   Graph $f(x) = \dfrac{x^2 - 5x + 6}{x - 2}$.

is close to K. It seems obvious that we also have $f(x) + g(x)$ close to $L + K$. This is the case as is seen by:

$$|f(x) + g(x) - (L + K)| = |(f(x) - L) + (g(x) - K)|$$
$$\leq^1 |f(x) - L| + |g(x) - K|.$$

(1) Triangle Inequality for real numbers.

It is now intuitively clear that if $|f(x) - L|$ and $|g(x) - K|$ are small, $|f(x) + g(x) - (L + K)|$ is also small.

We now state this and other properties of limits (Compare with Theorem 3.12, p. 182):

PROPERTIES OF LIMITS

Let $\lim\limits_{x \to b} f(x) = L$ and $\lim\limits_{x \to b} g(x) = K$, and let $r \in \mathfrak{R}$.

(3.2) $\lim\limits_{x \to b} (f(x) + g(x)) = L + K = \lim\limits_{x \to b} f(x) + \lim\limits_{x \to b} g(x).$

(3.3) $\lim\limits_{x \to b} (rf(x)) = rL = r \lim\limits_{x \to b} f(x).$

(3.4) $\lim\limits_{x \to b} f(x)g(x) = LK = \lim\limits_{x \to b} f(x) \lim\limits_{x \to b} g(x).$

(3.5)   If $K \neq 0$, $\lim\limits_{x \to b} \dfrac{f(x)}{g(x)} = \dfrac{L}{K} = \dfrac{\lim\limits_{x \to b} f(x)}{\lim\limits_{x \to b} g(x)}.$

(3.6) $\lim\limits_{x \to b} f(x) = L$ if and only if $\lim\limits_{h \to 0} f(b + h) = L.$

(3.7)   If $\lim\limits_{x \to b} f(x) = L$ and $f(x) \geq 0$ for all x close to b, then $L \geq 0$.

Example 14 illustrates applications of the first four theorems:

Example 14

(a) Let $f(x) = \dfrac{2x(x + 1)}{x - 2}$ for $x \neq 2$. We evaluate $\lim\limits_{x \to 3} f(x)$ as follows:

$$\lim_{x \to 3} f(x) = \lim_{x \to 3} \frac{2x(x + 1)}{x - 2}$$

$$=^1 2 \lim_{x \to 3} \frac{x(x + 1)}{x - 2}$$

$$=^2 2 \lim_{x \to 3} x \; \lim_{x \to 3} \frac{x + 1}{x - 2}$$

$$=^3 2 \lim_{x \to 3} x \; \frac{\lim\limits_{x \to 3} (x + 1)}{\lim\limits_{x \to 3} (x - 2)}$$

$$=^4 2(3) \frac{(4)}{(1)}$$

$$= 24.$$

(1) Equation 3.3.
(2) Equation 3.4.
(3) Equation 3.5.
(4) Observing that $x + 1$ is close to 4 and $x - 2$ is close to 1 when x is close to 3.

(b) Let $f(x) = x^2 + \sqrt{x}$. We shall evaluate $\lim\limits_{x \to 4} f(x)$.

$$\lim_{x \to 4} f(x) = \lim_{x \to 4} (x^2 + \sqrt{x})$$

$$=^5 \lim_{x \to 4} x^2 + \lim_{x \to 4} \sqrt{x}$$

$$=^6 \left(\lim_{x \to 4} x\right)^2 + \lim_{x \to 4} \sqrt{x}$$

$$= 4^2 + \lim_{x \to 4} \sqrt{x}$$

$$=^7 4^2 + 2$$

$$= 18.$$

(5) Equation 3.2.
(6) Equation 3.4.
(7) Intuition regarding $\lim\limits_{x \to 4} \sqrt{x}$. ∎

EXERCISES

In Exercises 1 to 15 use Equations 3.2 to 3.5 and intuition involving the sine and square root functions to find the indicated limits:

1. $\lim\limits_{x \to 4} (x + 5)$.

2. $\lim\limits_{x \to 1} (3x + 2)$.

3. $\lim_{x \to 0} (x^2 - 1)$.

4. $\lim_{x \to 4} \dfrac{x}{x + 1}$.

5. $\lim_{x \to 1} (3x^2 + 4x)$.

6. $\lim_{x \to 8} \sqrt{1 + x}$.

7. $\lim_{x \to 0} \sin x$. $= 0$

8. $\lim_{x \to 1} \dfrac{x(x^2 + 1)}{x + 2}$.

9. $\lim_{x \to 5} \dfrac{2}{\sqrt{x - 1}}$.

10. $\lim_{x \to 27} x^{2/3}$.

Study Example 13 before working Exercises 11 to 15:

11. $\lim_{x \to 1} \dfrac{x^2 - 1}{x - 1}$.

12. $\lim_{x \to 4} \dfrac{x^2 - 3x - 4}{x - 4}$.

13. $\lim_{x \to 1} \dfrac{x^2 + x - 2}{x^2 - 3x + 2}$.

14. $\lim_{x \to 0} \dfrac{x^4 - 3x^2}{2x^2}$.

15. $\lim_{x \to 1} \dfrac{x^3 - 3x^2 + 3x - 1}{x - 1}$.

In Exercises 16 to 20 indicate why the given function does not approach a limit as x approaches b.

16. $f(x) = \dfrac{1}{x^2}$ $(x \neq 0)$, $b = 0$.

17. $f(x) = \dfrac{1}{x - 1}$ $(x \neq 1)$, $b = 1$.

18. $f(x) = \begin{cases} 1 + x & \text{for } x \geq 0 \\ 2x & \text{for } x < 0 \end{cases}$, $b = 0$.

19. $f(x) = \begin{cases} x^2 & \text{for } x < 1 \\ 3 + x & \text{for } x > 1 \end{cases}$, $b = 1$.

20. $f(x) = \begin{cases} 1 \text{ for } x < -1 \\ 2 \text{ for } x > -1 \end{cases}$; $b = -1$.

ANSWERS

1. 9. **3.** -1. **5.** 7. **7.** 0. **9.** 1. **11.** 2. **13.** -3. **15.** 0.

4. DEFINITION OF LIMIT

In Section 3 we discussed in an informal manner the concept of a limit of a function $f : \mathcal{R} \to \mathcal{R}$. We are now ready to say precisely what this concept is and to extend our discussion to functions from $\mathcal{R}^n$ into $\mathcal{R}^m$.

Recall that in the preceding section we considered a function f defined on an interval (a, b) or on the union of two adjacent intervals $(a, b) \cup (b, c)$ and discussed the behavior of the values $f(x)$ for x close to but not equal to b. Since there are points x in (a, b) as close to b as we wish to specify (namely, $b - \dfrac{1}{n}$ for n sufficiently large), we can think of b as being approached by points of (a, b). Indeed, we notice that if $(b - r, b + r)$ is any open interval with center at b, there are points of (a, b) in $(b - r, b + r)$.

Let us now extend this idea of approaching a point. Since

$$(b - r, b + r) = \{x \mid x \in \mathfrak{R}, |x - b| < r\},$$

the analogue in $\mathfrak{R}^n$ of $(b - r, b + r)$ is an open ball. Consider now a non-empty set S in $\mathfrak{R}^n$ and a point X_0 of $\mathfrak{R}^n$. If it is true that each open ball $\mathcal{B}(X_0; r)$ with center at X_0 contains a point of S distinct from X_0, we can think of X_0 as being approached by points of S. Such points X_0 will be of importance in our discussion of limits. Definition 3.6 gives a special name to these points.

DEFINITION 3.6 (Cluster Point.) Let $S \subset \mathfrak{R}^n$ and $X_0 \in \mathfrak{R}^n$. X_0 *is a cluster point of* S if and only if

for each $r > 0$

there exists an X_r such that

$$X_r \in \mathcal{B}(\tilde{X}_0; r) \cap S.$$

A second way of describing a cluster point X_0 of a set S is to say that $\mathcal{B}(\tilde{X}_0; r) \cap S \neq \varnothing$ for any positive number r. Thus, a point X_0 is not a cluster point of a set S if and only if $\mathcal{B}(\tilde{X}_0; r) \cap S = \varnothing$ for at least one positive number r. Figure 3.12 illustrates a point X_0 which is not a cluster point of S.

The following examples will point out to the reader that a cluster point of a set S may or may not be an element of S.

Example 15

(a) Consider the set $S = \{1/n \mid n$ a positive integer$\}$ in $\mathfrak{R}$. We shall show that 0, which is not an element of S, is the only cluster point of S. First we note that if $r > 0$, there exists a positive integer n such that $\dfrac{1}{n} < r$; consequently, $\dfrac{1}{n} \in \mathcal{B}(\tilde{0}; r) \cap S$, and 0 is a cluster point of S. Now, let $x_0 \neq 0$. If $x_0 < 0$, $\mathcal{B}(\tilde{x}_0; |x_0|) \cap S = \varnothing$ and x_0 is not a cluster point of S. If $x_0 = \dfrac{1}{n_0}$ for some positive integer n_0, $\mathcal{B}\left(\tilde{x}_0; \dfrac{1}{n_0} - \dfrac{1}{(n_0 + 1)}\right) \cap S = \varnothing$ and x_0 is not a cluster point of S. Finally, if $x_0 > 0$ and $x_0 \notin S$,

$$\mathcal{B}\left(\tilde{x}_0; \left|x_0 - \dfrac{1}{n_0}\right|\right) \cap S = \varnothing,$$

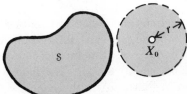

Figure 3.12

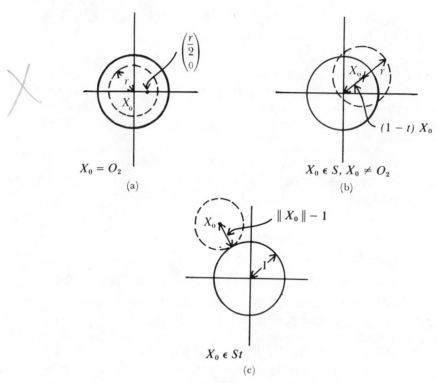

Figure 3.13 $S = \mathcal{B}(O_2; 1)$.

where $\dfrac{1}{n_0}$ is the point of S closest to x_0. Thus, if $x_0 \neq 0$, x_0 is not a cluster point of S.

(b) Consider the set $S = \bar{\mathcal{B}}(O_2; 1) = \{X \mid X \in \mathcal{R}^2, \|X\| \leq 1\}$. We shall show that every point of S is a cluster point of S and that no other point of $\mathcal{R}^2$ is a cluster point of S. Study Figure 3.13 as you follow the text.

Let $X_0 \in S$ and $r > 0$. If $X_0 = O_2$, and $r \geq 1$ then $\begin{pmatrix} 1/2 \\ 0 \end{pmatrix} \in \mathcal{B}(\tilde{X}_0; r) \cap S$.

If $X_0 = O_2$, and $r < 1$ then $\begin{pmatrix} r/2 \\ 0 \end{pmatrix} \in \mathcal{B}(\tilde{X}_0; r) \cap S$ (check this). If $X_0 \neq O_2$,

$(1 - t)X_0 \in \mathcal{B}(\tilde{X}_0; r) \cap S$ for each real t such that $0 < t < r$ and $t < 1$ (check this). Thus, each point of S is a cluster point of S. Now assume that $X_0 \notin S$ ($\|X_0\| > 1$) and let $r_0 = \|X_0\| - 1$. In this case $\mathcal{B}(\tilde{X}_0; r_0) \cap S = \varnothing$ (why?) and consequently X_0 is not a cluster point of S. ■

Consider now a function $f: \mathcal{R}^n \to \mathcal{R}^m$ and a cluster point X_0 of the domain of f. In this general situation we can ask the same question posed in the preceding section: Does there exist a point L of $\mathcal{R}^m$ such that fX is close to L when X is close (but not equal) to X_0? The following definition specifies exactly what we mean by "close to" in this context.

DEFINITION 3.7 (Limit of a Function.) Let $f: \mathcal{R}^n \to \mathcal{R}^m$; let X_0 be a cluster point of *dom f* and $L \in \mathcal{R}^m$.
Then *f approaches the limit L as X approaches X_0*
if and only if

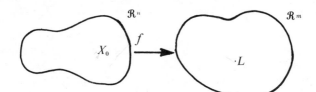

for each $r > 0$

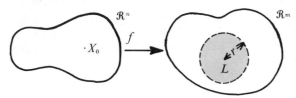

there exists a $\rho > 0$ such that

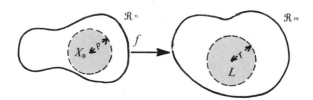

if $X \in (dom\, f) \cap \mathcal{B}(\tilde{X}_0; \rho)$, we have

$$fX \in \mathcal{B}(L; r).$$

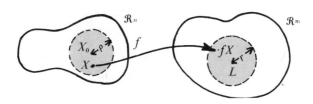

In this case we write $\lim_{X \to X_0} fX = L$.

The figures appearing in Definition 3.7 exhibit the open balls about L and X_0 of radius r and ρ, respectively. This definition makes precise the desired meaning for the expression "close to." It states, for example, that if $\lim\limits_{X \to X_0} fX = L$ and if we wish to have fX within one unit of L, we can do so for all X in some deleted open ball with center X_0. On the other hand, if $\lim\limits_{X \to X_0} fX \neq L$, then for at least one $r_0 > 0$, every open ball $\mathcal{B}(\tilde{X}_0; \rho)$ contains a point for which $\|fX - L\| \geq r_0$. In this latter case we can find points of $\mathcal{R}^n$ as close as we wish to X_0 such that fX is at least r_0 units from L (and hence not close to L).

Another way to consider the concept of a limit is as follows: If $\lim\limits_{X \to X_0} fX = L$ and g is the constant function with value L, then g is an approximation of f for X sufficiently close to X_0. This is our first example of approximating one function by a second function. Approximation of one function by another function will be a major theme in many of the following chapters.

Before considering examples of limits of functions, we pause to prove that there cannot be more than one element L of $\mathcal{R}^m$ which satisfies $\lim\limits_{X \to X_0} fX = L$.

THEOREM 3.8 Let $f: \mathcal{R}^n \to \mathcal{R}^m$ and let X_0 be a cluster point of $dom\, f$. If f approaches a limit as X approaches X_0, then this limit is unique.

Proof. Assume that $\lim\limits_{X \to X_0} fX = L$ and $K \in \mathcal{R}^m$, $K \neq L$. Let $r_0 = \frac{1}{2}\|L - K\|$ and note that $r_0 > 0$ since $K \neq L$. By Definition 3.7 there exists a $\rho_0 > 0$ such that if $X_0 \in (dom\, f) \cap \mathcal{B}(\tilde{X}_0; \rho_0)$, then $fX \in \mathcal{B}(L; r_0)$. See Figure 3.14 for an illustration of the situation.

Thus, if $X \in (dom\, f) \cap \mathcal{B}(\tilde{X}_0; \rho_0)$,

$$(3.8) \qquad\qquad \|fX - L\| < r_0.$$

Consider the following chain of inequalities in which

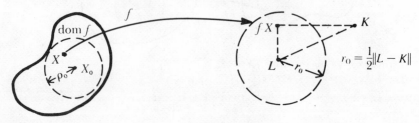

Figure 3.14

$X \in (dom\, f) \cap \mathcal{B}(\tilde{X}_0; \rho_0)$:

$$2r_0 =^1 \|L - K\|$$
$$=^2 \|(L - fX) + (fX - K)\|$$
$$\leq^3 \|L - fX\| + \|fX - K\|$$
$$<^4 r_0 + \|fX - K\|.$$

From this chain we see that $\|fX - K\| > r_0$ for all X in $(dom\, f) \cap \mathcal{B}(\tilde{X}_0; \rho_0)$. Consequently, if ρ is any positive number, there exists an X_ρ in $(dom\, f) \cap \mathcal{B}(\tilde{X}_0; \rho)$ such that $\|fX_\rho - K\| > r_0$—indeed, X_ρ may be any point of $(dom\, f) \cap \mathcal{B}(\tilde{X}_0; \rho_0) \cap \mathcal{B}(\tilde{X}_0; \rho)$. Therefore, K cannot satisfy Definition 3.7.

(1) $r_0 = \frac{1}{2} \|L - K\|$.
(2) Adding and subtracting fX.
(3) Triangle Inequality, Theorem 2.28iv (p. 141).
(4) Inequality 3.8. ∎

Example 16

(a) Let f be defined by $f(x) = x$ for x in $(1, 2)$. Intuition tells us that $\lim_{x \to 2} f(x) = 2$. To see that Definition 3.7 is satisfied we proceed as follows: Let $r > 0$. Note that for x in $(1, 2)$, $|f(x) - 2| = |x - 2|$ and consequently, $|x - 2| < r$ implies $|f(x) - 2| < r$. Thus, choosing $\rho = r$, we have $f(x) \in \mathcal{B}(2; r)$ for all x in $(dom\, f) \cap \mathcal{B}(\tilde{2}; \rho)$.

(b) Let $f(x)$ be defined by $f(x) = x^2 + 4$ for all x in $(1, 3]$. We guess that $\lim_{x \to 2} f(x) = 2^2 + 4 = 8$. To verify this guess we first let $r > 0$. We then note that if $x \in (1, 3]$,

$$|f(x) - 8| = |(x^2 + 4) - 8|$$
$$= |x^2 - 4|$$
$$= |x + 2| \, |x - 2|$$
$$\leq 5 \, |x - 2| \text{ (since } x \in (1, 3]).$$

It is obvious from this chain that if $5\,|x - 2| < r$, $|f(x) - 8| < r$ also. Equivalently, if $|x - 2| < \dfrac{r}{5}$, $|f(x) - 8| < r$. Thus, we choose $\rho = \dfrac{r}{5}$ and see that $f(x) \in \mathcal{B}(8; r)$ whenever $x \in (dom\, f) \cap \mathcal{B}(\tilde{2}; \rho)$.

(c) Let $f(x) = \sqrt{x + 1}$ for $x \geq -1$. We shall prove that $\lim_{x \to 8} f(x) = 3$. As in the last example we use algebraic manipulation:

$$|f(x) - 3| = |\sqrt{x + 1} - 3|$$
$$= \left| \frac{(\sqrt{x + 1} - 3)(\sqrt{x + 1} + 3)}{\sqrt{x + 1} + 3} \right|$$
$$= \frac{|(x + 1) - 9|}{\sqrt{x + 1} + 3}$$
$$\leq \frac{|x - 8|}{3} \text{ since } \sqrt{x + 1} \geq 0.$$

This chain implies that whenever $\dfrac{|x-8|}{3} < r$ we have $|f(x) - 3| < r$ also.

Thus, we may choose $\rho = 3r$ and Definition 3.7 is satisfied (why?). ■

The reader may wonder why we have restricted our examples to $\mathcal{R} \to \mathcal{R}$ functions when Definition 3.7 is for $\mathcal{R}^n \to \mathcal{R}^m$ functions. We do so at present because of the algebraic difficulties in proving that a function other than an $\mathcal{R} \to \mathcal{R}$ function has a certain limit. In following sections we shall develop a theory which will obviate some of these difficulties.

EXERCISES

In each of the following show that X_0 is a cluster point of $\mathcal{S}$:

1. $X_0 = 1$, $\mathcal{S} = \{1 + 1/n \mid n \text{ a positive integer}\}$.

2. $X_0 = 1$, $\mathcal{S} = \{x \mid x \in \mathcal{R}, 1 < x < 4\}$.

3. $X_0 = 3$, $\mathcal{S} = \{x \mid x \in \mathcal{R}, x > 3\}$.

4. $X_0 = \begin{pmatrix} 0 \\ 1 \end{pmatrix}$, $\mathcal{S} = \left\{ \begin{pmatrix} x \\ y \end{pmatrix} \,\middle|\, \begin{pmatrix} x \\ y \end{pmatrix} \in \mathcal{R}^2, \left\| \begin{pmatrix} x \\ y \end{pmatrix} - \begin{pmatrix} 1 \\ 1 \end{pmatrix} \right\| < 1 \right\}$.

5. $X_0 = \begin{pmatrix} 0 \\ 1 \end{pmatrix}$, $\mathcal{S} = \left\{ \begin{pmatrix} x \\ y \end{pmatrix} \,\middle|\, \begin{pmatrix} x \\ y \end{pmatrix} \in \mathcal{R}^2, \left\| \begin{pmatrix} x \\ y \end{pmatrix} - \begin{pmatrix} 1 \\ 1 \end{pmatrix} \right\| > 1 \right\}$.

For each of the following sets find the set $\mathcal{C}(\mathcal{S})$ of all cluster points of $\mathcal{S}$:

6. $\mathcal{S} = (1, 2)$.   7. $\mathcal{S} = (3, 4) \cup (4, 5)$.

8. $\mathcal{S} = \left\{ 1 - \dfrac{1}{n^2} \,\middle|\, n \text{ a positive integer} \right\}$.

9. $\mathcal{S} = \{1, 2, 3, 4\}$.

10. $\mathcal{S} = \{X \mid X \in \mathcal{R}^2, X \in \mathcal{B}(X_0; r)\}$.

11. $\mathcal{S} = \left\{ X \,\middle|\, X \in \mathcal{R}^2, X = \begin{pmatrix} 0 \\ 0 \end{pmatrix} \text{ or } X \in \mathcal{B}((1, 2); 1) \right\}$.

12. $\mathcal{S} = (-1, 1) \cup \{2\}$.   13. $\mathcal{S} = [0, 1] \cap [1, 2]$.

In Exercises 14 to 23, the interval at the right gives the domain of the function:

(a) Use intuition to find the following limits.

(b) Use techniques similar to those used in Exercise 16 to prove or disprove your guess in (a).

14. $\lim\limits_{x \to 1} (3x + 1)$, $x \in (0, 1)$.   15. $\lim\limits_{x \to -2} (1 - 2x)$, $x \in (-3, -1)$.

16. $\lim\limits_{x \to 4} x^2$, $x \in (4, 7)$.   17. $\lim\limits_{x \to -1} 2x^2$, $x \in [-5, 0)$.

18. $\lim\limits_{x \to 0} (x^3 + 1)$, $x \in (-1, 1]$.   19. $\lim\limits_{x \to 1} (4 - x^3)$, $x \in (\tfrac{1}{2}, 1)$.

20. $\lim\limits_{x \to 4} \dfrac{1}{x}$, $x \in (2, 5)$.

21. $\lim\limits_{x \to 9} 2\sqrt{x}$, $x \in [9, 20)$.

22. $\lim\limits_{x \to 1} x^{1/3}$, $x \in [0, 4]$.

23. $\lim\limits_{x \to 5} 4$, $x \in R$.

ANSWERS

7. $\mathcal{C}(\mathcal{S}) = [3, 5]$. **9.** $\mathcal{C}(\mathcal{S}) = \varnothing$. **11.** $\mathcal{C}(\mathcal{S}) = \overline{\mathcal{B}}((1, 2); 1)$. **13.** $\mathcal{C}(\mathcal{S}) = \varnothing$.
15. (a) 5. **17.** (a) 2. (b) since $x \in [-5, 0)$ we have $|2x^2 - 2| = 2|x+1|\,|x-1|$

$\leq 2(4)\,|x - 1| = 8\,|x - 1|$. Given $r > 0$, choose $\rho = \dfrac{r}{8}$. **19.** (a) 3.

(b) $|(4 - x^3) - 3| = 1 - x^3 = |1 + x + x^2|\,|1 - x| \leq 3\,|1 - x|$, since

$x \in (\tfrac{1}{2}, 1)$. Given $r > 0$, choose $\rho = \dfrac{r}{3}$. **21.** (a) 6. **23.** (a) 4.

5. THE TRANSLATION THEOREM

The algebraic manipulations encountered in the $\mathcal{R} \to \mathcal{R}$ examples and exercises of Section 4 forewarn us of even greater difficulties which may be involved in finding the limits of $\mathcal{R}^n \to \mathcal{R}^m$ functions. While continuing to rely upon intuition to tell us *what* the limits are, we shall use the following theorem as an aid in *proving* that our guesses are correct.

THEOREM 3.9 Let $f : \mathcal{R}^n \to \mathcal{R}^m$ and let X_0 be a cluster point of *dom f*. Then $\lim\limits_{X \to X_0} fX = L$ if and only if $\lim\limits_{H \to O_n} f(X_0 + H) = L$.

Proof. Suppose first that $\lim\limits_{X \to X_0} fX = L$. Let $r > 0$.
There exists a $\rho > 0$ such that
(1) $\|fX - L\| < r$ whenever $X \in (dom\,f) \cap B(\tilde{X}_0; \rho)$.
Define g by $g(H) = f(X_0 + H)$ for all H such that $X_0 + H \in dom\,f$. If $H \in (dom\,g) \cap B(\tilde{O}_n; \rho)$, then $X_0 + H \in dom\,f$; moreover,
(2) $\|(X_0 + H) - X_0\| < \rho$.
Therefore, if $H \in (dom\,g) \cap B(\tilde{O}_n; \rho)$

$$\|g(H) - L\| = \|f(X_0 + H) - L\|$$

$$< r \text{ by (1) and (2).}$$

By Definition 3.7, $\lim\limits_{H \to O_n} g(H) = L$ or $\lim\limits_{H \to O_n} f(X_0 + H) = L$. The converse can be proven in a completely analogous fashion. ∎

Now let us see how this last theorem can be useful in proving that a given function has a certain limit.

Example 17. Let f be defined by $f(x) = x^3 + 1$ for all real x. We guess that $\lim\limits_{x \to 2} f(x) = 2^3 + 1 = 9$. (The reader will discover in Section 6 of this chapter the reason why this sort of guess is mathematically valid as well as intuitively obvious.) According to Theorem 3.9, $\lim\limits_{x \to 2} f(x) = 9$ if and only if $\lim\limits_{h \to 0} f(2 + h) = 9$.

Note that

$$f(2 + h) = (2 + h)^3 + 1$$
$$= h^3 + 6h^2 + 12h + 9.$$

Thus,

$$|f(2 + h) - 9| = |h^3 + 6h^2 + 12h|$$
$$= |h|\,|h^2 + 6h + 12|$$
$$\leq |h|\,[|h|^2 + 6|h| + 12]$$
$$< 19\,|h| \text{ for } |h| < 1.$$

Let $r > 0$ and let h be any number such that $0 < |h| < r/19$ and $|h| < 1$. From the above calculation we see that

$$|f(2 + h) - 9| < 19\,|h|$$
$$< 19(r/19)$$
$$= r.$$

By Definition 3.7, $\lim\limits_{h \to 0} f(2 + h) = 9$. ∎

We shall now consider some examples of $\mathcal{R}^2 \to \mathcal{R}$ functions. The reader who can understand and imitate the technique used in the following example should be able to generalize the technique for $\mathcal{R}^n \to \mathcal{R}$ functions. Theorems in succeeding sections will enable us to handle the general $\mathcal{R}^n \to \mathcal{R}^m$ functions.

Example 18. Let $f\begin{pmatrix} x \\ y \end{pmatrix} = x^2 + xy$ for all $\begin{pmatrix} x \\ y \end{pmatrix}$ in $\mathcal{R}^2$ and let $X_0 = \begin{pmatrix} 1 \\ 2 \end{pmatrix}$. We guess that $\lim\limits_{X \to \binom{1}{2}} fX = 1^2 + 1(2) = 3$. We shall prove that this is the case by showing that $\lim\limits_{H \to O_2} f(X_0 + H) = 3$. Let $H = \begin{pmatrix} h \\ k \end{pmatrix}$ and note that

$$f(X_0 + H) = f\left(\begin{pmatrix} 1 \\ 2 \end{pmatrix} + \begin{pmatrix} h \\ k \end{pmatrix}\right)$$
$$= f\begin{pmatrix} 1 + h \\ 2 + k \end{pmatrix}$$
$$= (1 + h)^2 + (1 + h)(2 + k)$$
$$= h^2 + 4h + k + hk + 3.$$

Therefore,

$$|f(X_0 + H) - 3| = |h^2 + 4h + k + hk|$$

$$\leq^1 |h^2| + |4h| + |k| + |hk|$$

$$=^2 |h|\,|h| + 4\,|h| + |k| + |h|\,|k|$$

$$= [|h| + 4]\,|h| + [1 + |h|]\,|k|.$$

(1) Triangle Inequality for real numbers, Appendix II.
(2) Property of absolute value, Appendix II.

We now make a preliminary estimate and only consider H such that

$$(3.9) \qquad\qquad \|H\| \leq 1.$$

This results in $|h| \leq 1$ and $|k| \leq 1$ (why?). From the above calculations we see that if $\|H\| \leq 1$

$$(3.10) \qquad \begin{cases} |f(X_0 + H) - 3| \leq 5\,|h| + 2\,|k| \\ \qquad\qquad \leq^3 5[|h| + |k|] \\ \qquad\qquad \leq^4 5(2)\sqrt{h^2 + k^2} \\ \qquad\qquad =^5 10\,\|H\|. \end{cases}$$

This last number will be less than r if $\|H\| < \dfrac{r}{10}$.

(3) Since $2\,|k| \leq 5\,|k|$.
(4) $|h| + |k| \leq 2\sqrt{h^2 + k^2}$ (see Exercise 11).
(5) $\|H\| = \sqrt{h^2 + k^2}$,

Let $r > 0$ and assume that

$$(3.11) \qquad\qquad \|H\| < \frac{r}{10}.$$

From Inequality 3.10 we see that $|f(X_0 + H) - 3| < r$ for all H satisfying Inequalities 3.9 and 3.11. Thus, if ρ is the smaller of $\dfrac{r}{10}$ and 1, we have $|f(X_0 + H) - 3| < r$ for all H such that $\|H\| < \rho$, and by Definition 3.7 $\lim\limits_{H \to O_2} f(X_0 + H) = 3$. ∎

Example 19. Let $f\begin{pmatrix} x \\ y \end{pmatrix} = \dfrac{x}{y}$ for all $\begin{pmatrix} x \\ y \end{pmatrix}$ in $\mathcal{R}^2$, $y \neq 0$ and let $X_0 = \begin{pmatrix} 3 \\ -1 \end{pmatrix}$. We shall prove that $\lim\limits_{H \to O_2} f(X_0 + H) = -3$ and thus by Theorem 3.9 have $\lim\limits_{X \to X_0} fX = -3$. The reader should check each step carefully. Let

$H = \begin{pmatrix} h \\ k \end{pmatrix}$. We then have

$$|f(X_0 + H) - (-3)| = \left| f\begin{pmatrix} 3 + h \\ -1 + k \end{pmatrix} - (-3) \right|$$

$$= \left| \frac{3 + h}{-1 + k} + 3 \right|$$

$$= \left| \frac{3 + h - 3 + 3k}{k - 1} \right|$$

$$= \left| \frac{h + 3k}{k - 1} \right|$$

$$\leq \frac{|h| + 3|k|}{|k - 1|}.$$

We are now ready to make a preliminary estimate on the size of H. In order that $k - 1 \neq 0$, we require that $\|H\| \leq \frac{1}{2}$. This forces $\dfrac{1}{|k - 1|} \leq 2$ (prove this). Therefore,

$$|f(X_0 + H) - (-3)| \leq 2(|h| + 3|k|)$$

$$\leq 6(|h| + |k|)$$

$$\leq 6(2\sqrt{h^2 + k^2})$$

$$= 12\|H\|.$$

This last number will be less than r if $\|H\| < \dfrac{r}{12}$. Let $r > 0$. If we choose ρ to be the smaller of $\dfrac{1}{2}$ and $\dfrac{r}{12}$, we have $|f(X_0 + H) - (-3)| < r$ for all X in $\mathcal{B}(\tilde{O}_2; \rho) \cap dom\, f$ and consequently, $\lim\limits_{H \to O_2} f(X_0 + H) = -3$. ∎

The reader should study carefully the methods used in the two preceding examples before attempting the exercises. The following two theorems will give the reader a little practice in abstract proof involving limits, as the two proofs are left as exercises at the end of the section.

THEOREM 3.10

i. Let $f : \mathcal{R}^n \to \mathcal{R}$. If $fX \geq 0$ for all X in $dom\, f$ and $\lim\limits_{X \to X_0} fX = L$, then $L \geq 0$.

ii. Let $f : \mathcal{R} \to \mathcal{R}^n$. Then $\lim\limits_{h \to 0} \dfrac{f(h)}{|h|} = O_n$ if and only if $\lim\limits_{h \to 0} \dfrac{f(h)}{h} = O_n$.

THEOREM 3.11 Let $f: \mathcal{R}^n \to \mathcal{R}^m$ and L_1, $L_2 \in \mathcal{R}^m$, $L_1 \neq L_2$. Let X_0 be a cluster point of $dom\ f$. If for each $r > 0$ and each $\rho > 0$ there exist X_1 and X_2 in $(dom\ f) \cap \mathcal{B}(\tilde{X}_0;\ \rho)$ such that $\|fX_1 - L_1\| < r$ and $\|fX_2 - L_2\| < r$, then f does not approach a limit as X approaches X_0.

EXERCISES

In Exercises 1 to 10 use Theorem 3.9 to prove that the functions have limits:

1. $\lim\limits_{x \to 4}\ (x^2 + 5x)$.

2. $\lim\limits_{x \to -1}\ (2x^2 - 1)$.

3. $\lim\limits_{x \to 3}\ \dfrac{1}{x}$.

4. $\lim\limits_{x \to 0}\ \sqrt{x}$.

5. $\lim\limits_{x \to 1}\ (2x^3 + 5x^2 - 6)$.

6. $\lim\limits_{\binom{x}{y} \to \binom{1}{1}}\ (2x^2 + y)$.

7. $\lim\limits_{\binom{x}{y} \to \binom{2}{0}}\ (y^2 - x^2)$.

9. $\lim\limits_{\binom{x}{y} \to \binom{-1}{1}}\ \dfrac{x^2}{y}$.

10. $\lim\limits_{\binom{x}{y}{z} \to \binom{1}{2}{1}}\ xy + z$.

11. Let h and k be real numbers. Show that $|h| + |k| \leq 2\sqrt{h^2 + k^2}$ and $\sqrt{h^2 + k^2} \leq |h| + |k|$.

12. Prove Theorem 3.10.   13. Prove Theorem 3.11.

14. Use Theorem 3.11 to prove Theorem 3.8.

ANSWERS

1. 36. **3.** $\tfrac{1}{3}$.

Proof.   Let $f(x) = \dfrac{1}{x}$, $x_0 = 3$. Let $r > 0$.

$$| f(x_0 + h) - \tfrac{1}{3}| = \left| \dfrac{1}{3 + h} - \dfrac{1}{3} \right|$$

$$= \left| \dfrac{-h}{3(3 + h)} \right|$$

$$= \dfrac{|h|}{3\,|3 + h|}$$

$$\leq \dfrac{|h|}{3(2)} \text{ if } |h| \leq 1$$

$$< r \text{ if } |h| < 6r.$$

Choose ρ to be the smaller of $6r$ and 1.   **5.** 1.   **7.** -4.

Proof. Let $f\begin{pmatrix} x \\ y \end{pmatrix} = y^2 - x^2$, $H = \begin{pmatrix} h \\ k \end{pmatrix}$, $X_0 = \begin{pmatrix} 2 \\ 0 \end{pmatrix}$, and $r > 0$.

$$
\begin{aligned}
|f(X_0 + H) - (-4)| &= |k^2 - (h + 2)^2 + 4| \\
&= |k^2 - h^2 - 4h| \\
&\le |k| \, |k| + [|h| + 4] \, |h| \\
&\le |k| + 5 \, |h| \text{ if } \|H\| \le 1 \\
&\le 10\sqrt{h^2 + k^2} \\
&= 10 \, \|H\| \\
&< r \text{ if } \|H\| < \frac{r}{10}.
\end{aligned}
$$

Choose ρ to be the smaller of $\dfrac{r}{10}$ and 1. **9.** 1. **13.** *Proof of Theorem* 3.11.

Let $r_0 = \frac{1}{4} \|L_1 - L_2\| > 0$. Assume that $\lim\limits_{X \to X_0} fX = L$. There exists a $\rho_0 > 0$ such that $\|fX - L\| < r_0$ for all X in $(dom\, f) \cap \mathcal{B}(\tilde{X}_0; \rho_0)$. By hypothesis there exist X_1 and X_2 in $(dom\, f) \cap \mathcal{B}(\tilde{X}_0; \rho_0)$ such that $\|fX_1 - L_1\| < r_0$ and $\|fX_2 - L_2\| < r_0$. Find a contradiction and conclude that $\lim\limits_{X \to X_0} fX$ does not exist.

6. ALGEBRAIC PROPERTIES OF LIMITS

The theorems in this section will greatly simplify our task of determining whether certain limits exist, and indeed, of finding such limits. The reader will find it highly profitable to study the proofs carefully and to try to understand them in their entirety.

THEOREM 3.12 Let $f: \mathcal{R}^n \to \mathcal{R}^m$ and $g: \mathcal{R}^n \to \mathcal{R}^m$. Let c be a constant and let X_0 be a cluster point of $(dom\, f) \cap (dom\, g)$. Assume that

$$
\lim_{X \to X_0} fX = L \quad \text{and} \quad \lim_{X \to X_0} gX = K.
$$

Then the following are true:

i. $\lim\limits_{X \to X_0} (f + g)X = L + K$;

ii. $\lim\limits_{X \to X_0} (cf)X = cL$;

iii. $\lim\limits_{X \to X_0} (fg)X = L \cdot K$; and

iv. if $m = 1$ and $K \neq 0$, then there exists a $\rho > 0$ such that $gX \neq 0$ for each X in $(dom\, g) \cap \mathcal{B}(\tilde{X}_0; \rho)$ and $\lim\limits_{X \to X_0} (f/g)X = L/K$.

Proof of i. Recall that $dom\, (f + g) = (dom\, f) \cap (dom\, g)$ by Definition 3.3. Let $r > 0$. We wish to find a $\rho > 0$ such that $(f + g)X \in B(L + K; r)$

whenever $X \in B(\tilde{X}_0; \rho) \cap dom\,(f+g)$. If we can do this, we know that $\lim_{X \to X_0} (f+g)X = L+K$ by Definition 3.7 and Theorem 3.8.

Figure 3.15 should be of assistance to the reader in following this proof; it represents what is happening in the range space R^m.

By hypothesis $\lim_{X \to X_0} fX = L$ and $\lim_{X \to X_0} gX = K$. We know by Definition 3.7 that there exist a $\rho' > 0$ and $\rho'' > 0$ such that

(3.12) $fX \in B(L; r/2)$ whenever $X \in B(\tilde{X}_0; \rho') \cap (dom\,f)$, and

(3.13) $gX \in B(K; r/2)$ whenever $X \in B(\tilde{X}_0; \rho'') \cap (dom\,g)$.

Let ρ be the smaller of ρ' and ρ'' and note that $B(\tilde{X}_0; \rho) = B(\tilde{X}_0; \rho') \cap B(\tilde{X}_0; \rho'')$. From (3.12) and (3.13) it follows that

(3.14) $fX \in B(L; r/2)$ and $gX \in B(K; r/2)$ whenever $X \in B(\tilde{X}_0; \rho) \cap (dom\,f) \cap (dom\,g)$.

Therefore, if $X \in B(\tilde{X}_0; \rho) \cap (dom\,f) \cap (dom\,g)$, then

$$
\begin{aligned}
\|(f+g)X - (L+K)\| &= \|fX + gX - L - K\| \\
&= \|(fX - L) + (gX - K)\| \\
&\leq^1 \|fX - L\| + \|gX - K\| \\
&<^2 r/2 + r/2 \\
&= r.
\end{aligned}
$$

(1) By the Triangle Inequality, Theorem 2.28iv (p. 141).
(2) By (3.14) above.

This last series of inequalities tells us that $(f+g)X \in \mathcal{B}(L+K; r)$ whenever $X \in \mathcal{B}(\tilde{X}_0; \rho) \cap dom\,(f+g)$, and we have completed our proof.

Proof of ii. The case for which $c = 0$ is trivial. Hence let $c \neq 0$ and let $r > 0$. Since $\lim_{X \to X^0} fX = L$, we can find a $\rho > 0$ such that

(3.15) $fX \in \mathcal{B}\left(L; \dfrac{r}{|c|}\right)$ whenever $X \in \mathcal{B}(\tilde{X}_0; \rho) \cap (dom\,f)$.

Therefore, for such an X

$$
\begin{aligned}
\|(cf)X - (cL)\| &= \|c(fX) - cL\| \\
&= \|c[fX - L]\| \\
&=^1 |c|\,\|fX - L\| \\
&<^2 |c|\,\dfrac{r}{|c|} \\
&= r.
\end{aligned}
$$

(1) By Theorem 2.24iii (p. 136).
(2) By (3.15) above.

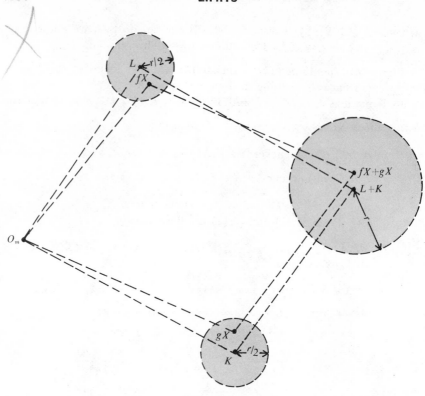

Figure 3.15

We have shown that $(cf)X \in \mathcal{B}(cL; r)$ whenever $X \in \mathcal{B}(\tilde{X}_0; \rho) \cap (dom\, f)$. Definition 3.7 and Theorem 3.8 give us the desired result:

$$\lim_{X \to X_0} (cf)\, X = cL. \quad \blacksquare$$

The reader who has followed the proofs of (i) and (ii) may well ask why we choose ρ' and ρ'' as we did. Probably the best way to answer this question is with the observation that they are chosen to be sufficiently small to do the job! The proofs of Theorem 3.12iii and iv will be left to the adventurous mind—or to the diligent student who can find the proofs elsewhere (for example, in *Methods of Real Analysis* by Richard R. Goldberg).

Theorem 3.12 will be invaluable to us in computing limits. Suppose that f is defined by $f(x) = x^2$ for all real x. Heretofore we used intuition to tell us that $\lim_{x \to 3} f(x) = 9$. However, now we have Theorem 3.12iii which tells us that $\lim_{x \to 3} f(x) = \lim_{x \to 3} x^2 = \left(\lim_{x \to 3} x\right)\left(\lim_{x \to 3} x\right) = (3) \cdot (3) = 9$.

Before looking at more applications of Theorem 3.12, let us state one further theorem. The proof of this theorem is immediate from Definition 3.7 and will be left as an exercise for the reader (see Exercise 16).

THEOREM 3.13 Let $f:\mathcal{R}^n \to \mathcal{R}^m$. If f is a constant function, and $fX = K$ for all X in $\overline{dom\,f}$, then $\lim\limits_{X \to X_0} fX = K$ whenever X_0 is a cluster point of $dom\,f$.

Example 20. Consider the following applications of Theorems 3.12 and 3.13:

(a) $\lim\limits_{t \to 4} (2t + 2) = \lim\limits_{t \to 4} 2t + \lim\limits_{t \to 4} 2$ by Theorem 3.12i

$\qquad\qquad = 2 \lim\limits_{t \to 4} t + \lim\limits_{t \to 4} 2$ by Theorem 3.12ii

$\qquad\qquad = 8 + 2$ by Theorem 3.13

$\qquad\qquad = 10.$

(b) $\lim\limits_{t \to -2} t^3 = \left(\lim\limits_{t \to -2} t \right)^3$ by Theorem 3.12iii

$\qquad\qquad = (-2)^3$

$\qquad\qquad = -8.$

(c) $\lim\limits_{x \to 1} \dfrac{x + 5}{x - 2} = \dfrac{\lim\limits_{x \to 1} (x + 5)}{\lim\limits_{x \to 1} (x - 2)}$ by Theorem 3.12iv

$\qquad\qquad = \dfrac{6}{-1}$ by Theorem 3.12i and 3.13

$\qquad\qquad = -6.$ ∎

Example 21. Consider the following applications of Theorem 3.12 to functions $f:R^n \to R$, $n > 1$:

(a) $\lim\limits_{(x,y) \to (1,2)} (x^2 + y^2) =^1 \lim\limits_{(x,y) \to (1,2)} x^2 + \lim\limits_{(x,y) \to (1,2)} y^2$

$\qquad\qquad =^2 \left(\lim\limits_{(x,y) \to (1,2)} x \right)^2 + \left(\lim\limits_{(x,y) \to (1,2)} y \right)^2$

$\qquad\qquad = (1)^2 + (2)^2$

$\qquad\qquad = 5$

(1) Theorem 3.12i.
(2) Theorem 3.12iii.

(b) $\lim\limits_{(x,y,z) \to (0,1,-1)} (xyz + z^3) =^3 \lim\limits_{(x,y,z) \to (0,1,-1)} xyz + \lim\limits_{(x,y,z) \to (0,1,-1)} z^3$

$\qquad\qquad =^4 \left(\lim\limits_{(x,y,z) \to (0,1,-1)} x \right)\left(\lim\limits_{(x,y,z) \to (0,1,-1)} y \right) \cdot$

$\qquad\qquad\quad \left(\lim\limits_{(x,y,z) \to (0,1,-1)} z \right) + \left(\lim\limits_{(x,y,z) \to (0,1,-1)} z \right)^3$

$\qquad\qquad = (0)(1)(-1) + (-1)^3$

$\qquad\qquad = -1.$

(3) Theorem 3.12i.
(4) Theorem 3.12iii. ∎

EXERCISES

Using the theorems in this section, compute the following limits. Justify each step in your computations by a theorem.

1. $\lim\limits_{t \to -1} (t^2 + 3t - 5)$.
2. $\lim\limits_{t \to 2} (t^3 - 4)$.

3. $\lim\limits_{x \to 1} \dfrac{x + 2}{x - 3}$.
4. $\lim\limits_{x \to 0} (x^2 + 1)(2x^2 - 3x + 4)$.

5. $\lim\limits_{x \to -1} \dfrac{1}{x}$.
6. $\lim\limits_{x \to 4} 7$.

(Hint: Recall the algebraic technique in Example 13 before attempting Exercises 7 to 10.)

7. $\lim\limits_{x \to 3} \dfrac{(x + 1)(x - 3)}{x - 3}$.
8. $\lim\limits_{x \to -1} \dfrac{x^2 - 1}{x + 1}$.

9. $\lim\limits_{x \to 2} \dfrac{x^3 - 6x^2 + 12x - 8}{x - 2}$.
10. $\lim\limits_{x \to 4} \dfrac{x^2 - 16}{x - 4}$.

11. $\lim\limits_{(x,y) \to (1,-2)} (x^2 + y + 1)$.
12. $\lim\limits_{(x,y) \to (0,1)} \dfrac{y^2 + 2}{x + y}$.

13. $\lim\limits_{(x,y,z) \to (1,2,3)} (x^3 + y^3 - z^2)$.
14. $\lim\limits_{(x,y,z) \to (0,0,0)} \left(\dfrac{x^2 + 3}{xyz + 1} \right)$.

15. $\lim\limits_{(x,y) \to (1,1)} \dfrac{x^2 - y^2}{x - y}$. (Hint: Use same method as in Exercises 7 to 10.)

Prove the following:

16. Theorem 3.13.

17. Let $f : \mathcal{R}^n \to \mathcal{R}$, $g : \mathcal{R}^n \to \mathcal{R}$, and let X_0 be a cluster point of $(dom\, f) \cap (dom\, g)$. Assume that $\lim\limits_{X \to X_0} fX = L \neq 0$ and $\lim\limits_{X \to X_0} gX = 0$. Then $\lim\limits_{X \to X_0} \left(\dfrac{f}{g} \right)X$ does not exist. (Hint: Assume that $\lim\limits_{X \to X_0} \left(\dfrac{f}{g} \right)X = M$ and note that $g\left(\dfrac{f}{g} \right) = f$ whenever $\dfrac{f}{g}$ is defined. Use this fact and Theorem 3.12iii to arrive at a contradiction.)

18. Let $f : \mathcal{R}^n \leftarrow \mathcal{R}^m$ and $g : \mathcal{R}^n \to \mathcal{R}^m$. Let X_0 be a cluster point of $(dom\, f) \cap (dom\, g)$. If $\lim\limits_{X \to X_0} (f + g)X = M$ and $\lim\limits_{X \to X_0} gX = K$, then $\lim\limits_{X \to X_0} fX = M - K$. (Hint: Note that $f = (f + g) + (-1)g$.)

ANSWERS

2. 4. **4.** 4. **6.** 7. **8.** −2. **10.** 8. **12.** 3. **14.** 3.

7. LIMITS OF COORDINATE FUNCTIONS

The reader may have noted that each of the Exercises 1 to 15 of the preceding section involved limits of real-valued functions. In order to handle limits of functions, for which the range space is $\mathcal{R}^m$, $m > 1$, with greater facility, we need to establish a relationship between the given function and its "coordinate functions."

DEFINITION 3.14 Let $f : \mathcal{R}^n \to \mathcal{R}^m$. For each X in *dom f* and $j = 1, \ldots, m$, define

$$f_j(X) = j\text{-th coordinate of } fX$$

f_j is called the jth *coordinate function of f*.

Example 22. Let f be defined by $f\begin{pmatrix} x \\ y \end{pmatrix} = \begin{pmatrix} x^2 + y^2 \\ x + 2 \\ xy \end{pmatrix}$. According to Definition 3.14, f has the three coordinate functions, defined by:

$$f_1(x, y) = x^2 + y^2,$$

$$f_2(x, y) = x + 2, \quad \text{and}$$

$$f_3(x, y) = xy. \quad \blacksquare$$

The reader should note that each coordinate function of f is a real-valued function with the same domain as that of f. The one theorem in this section gives us a valuable connection between limits of functions from $\mathcal{R}^n$ into $\mathcal{R}$ and limits of functions from $\mathcal{R}^n$ into $\mathcal{R}^m$. In effect it tells us that if each coordinate function approaches a limit as X approaches X_0, so does the function, and conversely.

THEOREM 3.15 Let $f : \mathcal{R}^n \to \mathcal{R}^m$ and let X_0 be a cluster point of *dom f*. Let $K = (k_1, \ldots, k_m)$. Then $\lim\limits_{X \to X_0} fX = K$ if and only if

$$\lim\limits_{X \to X_0} f_j X = k_j \quad \text{for} \quad j = 1, \ldots, m.$$

Proof

i. Assume first that $\lim\limits_{X \to X_0} fX = K$ and let $r > 0$. There exists a $\rho > 0$ such that $\| fX - K \| < r$ whenever $X \in (dom\, f) \cap B(\tilde{X}_0; \rho)$. But we also know that

$$|f_j X - K_j|$$
$$= \sqrt{(f_j X - k_j)^2}$$
$$\leq \sqrt{(f_1 X - k_1)^2 + \cdots + (f_j X - k_j)^2 + \cdots + (f_m X - k_m)^2}$$
$$=^1 \|fX - K\|.$$

(1) By definition of length and Definition 3.14.

From the preceding we see that $|f_j X - k_j| < r$ whenever $\|fX - K\| < r$. Therefore, $f_j X \in \mathcal{B}(k_j; r)$ whenever $X \in B(\tilde{X}_0; \rho)$, and $\lim\limits_{X \to X_0} f_j X = k_j$ for $j = 1, \ldots, m$.

ii. Assume now that $\lim\limits_{X \to X_0} f_j X = k_j$ for $j = 1, \ldots, m$. If $r > 0$, for each j there is a number $\rho_j > 0$ such that

$$|f_j X - k_j| < \frac{r}{\sqrt{m}} \text{ whenever } X \in dom\, f_j \cap \mathcal{B}(\tilde{X}_0; \rho_j).$$

Let ρ be the smallest of $\rho_1, \ldots, \rho_m$. We then have

$$|f_j X - k_j| < \frac{r}{\sqrt{m}} \quad \text{for} \quad j = 1, \ldots, m \quad \text{whenever} \quad X \in (dom\, f) \cap$$
$$\mathcal{B}(\tilde{X}_0; \rho).$$

Thus, for each X in $(dom\, f) \cap \mathcal{B}(\tilde{X}_0; \rho)$,

$$\|fX - K\| = \sqrt{(f_1 X - k_1)^2 + \cdots + (f_m X - k_m)^2}$$
$$< \sqrt{\left(\frac{r}{\sqrt{m}}\right)^2 + \cdots + \left(\frac{r}{\sqrt{m}}\right)^2} \text{ (m terms)}$$
$$= \sqrt{m\left(\frac{r}{\sqrt{m}}\right)^2}$$
$$= \sqrt{r^2}$$
$$= r, \text{ since } r > 0.$$

We have satisfied the condition in Definition 3.7 and have $\lim\limits_{X \to X_0} fX = K$. ∎

Example 23

(a) Let f be defined by $f(x) = \begin{pmatrix} x^2 + 1 \\ 2x - 1 \end{pmatrix}$. By Definition 3.14, f_1 and f_2 are defined by $f_1(x) = x^2 + 1$ and $f_2(x) = 2x - 1$. Using Theorems 3.12 and 3.13, we have $\lim\limits_{x \to 3} f_1(x) = 10$, $\lim\limits_{x \to 3} f_2(x) = 5$. Hence,

$$\lim\limits_{x \to 3} f(x) = \lim\limits_{x \to 3} \begin{pmatrix} x^2 + 1 \\ 2x - 1 \end{pmatrix} =^1 \begin{pmatrix} \lim\limits_{x \to 3}(x^2 + 1) \\ \lim\limits_{x \to 3}(2x - 1) \end{pmatrix} = \begin{pmatrix} 10 \\ 5 \end{pmatrix}.$$

(1) By Theorem 3.15.

(b) Let f be defined by $f\begin{pmatrix} x \\ y \end{pmatrix} = \begin{pmatrix} x^2 + 3x - y \\ 2 + y \end{pmatrix}$. Then

$$\lim_{(x,y)\to(1,-1)} f\begin{pmatrix} x \\ y \end{pmatrix} =^2 \begin{pmatrix} \lim_{(x,y)\to(1,-1)} (x^2 + 3x - y) \\ \lim_{(x,y)\to(1,-1)} (2 + y) \end{pmatrix}$$

$$=^3 \begin{pmatrix} 5 \\ 1 \end{pmatrix}.$$

(2) By Theorem 3.15.

(3) By Theorems 3.12 and 3.13. ∎

In all the examples considered up to this point, the function f and the point X_0 were such that $\lim_{X\to X_0} fX$ did exist. How does one show that a function f does *not* approach a limit as X approaches X_0, and what are some examples of such functions? If we negate the condition in Definition 3.7, we find that f does not approach a limit as X approaches X_0 if and only if

> for each L in $\mathscr{R}^m$
> there exists an $r > 0$ such that
> for each $\rho > 0$
> there exists an X in $(dom\,f) \cap \mathscr{B}(\tilde{X}_0; \rho)$ such that $fX \notin \mathscr{B}(L; r)$.

In general this is an extremely difficult condition to apply. There are, however, several techniques that are relatively simple and that in certain cases show that a given function does not approach a limit as X approaches X_0.

The first technique makes use of Exercise 17 in the preceding section. If h is a function such that $hX = \dfrac{fX}{gX}$, $\lim_{X\to X_0} fX = L \neq 0$, and $\lim_{X\to X_0} gX = 0$, then $\lim_{X\to X_0} hX$ does not exist.

Example 24. Let $h(x) = \dfrac{1}{x}$ for $x \neq 0$. Then $\lim_{x\to 0} 1 = 1$ and $\lim_{x\to 0} x = 0$. Therefore, $\lim_{x\to 0} \dfrac{1}{x}$ does not exist. ∎

We can extend this technique to certain other situations. If $h = \dfrac{f}{g}$, and $\lim_{X\to X_0} fX = 0$ and $\lim_{X\to X_0} gX = 0$, it may be the case that $\lim_{X\to X_0} hX$ exists (as in Example 13, p. 167) or it may be the case that $\lim_{X\to X_0} hX$ does not exist. The essence of the technique used above lies in the fact that if $h = \dfrac{f}{g}$, $\lim_{X\to X_0} fX = L \neq 0$ and $\lim_{X\to X_0} gX = 0$, then $\|hX\|$ gets arbitrarily large when X gets close to X_0. Consider the following example:

Example 25. Let f be defined by $f(x, y) = \dfrac{x + y}{x^2 + y^2}$ for all (x, y) in $\mathscr{R}^2$, $(x, y) \neq (0, 0)$. We shall show in a very simple fashion that $\lim_{(x,y)\to(0,0)} f(x, y)$

does not exist. Let $\mathcal{L} = \left\{ t\begin{pmatrix} 1 \\ 1 \end{pmatrix} \;\middle|\; t \text{ real} \right\}$. If $X \in \mathcal{L}$ and $X \neq (0,0)$, then

$X = \begin{pmatrix} t_0 \\ t_0 \end{pmatrix}$ and $f(X) = \dfrac{2t_0}{2t_0^2} = \dfrac{1}{t_0}$. Now when t_0 is close to 0, $|f(x)|$ is large. The argument used in Example 23 can be used here to show that f does not approach a limit as X approaches $(0, 0)$. ∎

The second technique involves showing that the values of f approach two different limits as X approaches X_0 along two different paths. If this occurs, then Theorem 3.11 assures us that f does not approach a limit as X approaches X_0.

Example 26. Let f be defined by

$$f(x, y) = \frac{x + y^2}{y + x^2} \text{ for } (x, y) \text{ in } \mathcal{R}^2, y \neq -x^2.$$

Consider the two following lines in $\mathcal{R}^2$:

$$(1) \quad X(t) = t \begin{pmatrix} 1 \\ 1 \end{pmatrix}, t \text{ real,}$$

$$(2) \quad Y(t) = t \begin{pmatrix} 1 \\ 2 \end{pmatrix}, t \text{ real.}$$

If $t \neq 0$, $X(t) \in dom f$ and

$$f(X(t)) = f(t, t) = \frac{t + t^2}{t + t^2} = 1.$$

Obviously, as t gets close to 0, $f(X(t))$ gets close to 1, or $\lim\limits_{t \to 0} f(X(t)) = 1$. Also, for $t \neq 0$, $Y(t) \in dom f$ and

$$f(Y(t)) = f(t, 2t) = \frac{t + 4t^2}{2t + t^2}$$
$$= \frac{1 + 4t}{2 + t}.$$

In this case $\lim\limits_{t \to 0} f(Y(t)) = \frac{1}{2}$. Thus, we see that no matter what value of ρ we pick, there exist X_1 and X_2 in $\mathcal{B}(\tilde{O}_2; \rho)$ such that fX_1 is close to 1 and fX_2 is close to 1/2. Therefore, by Theorem 3.11 we know that $\lim\limits_{X \to O_2} fX$ does not exist. ∎

EXERCISES

Using Theorems 3.12 to 3.15, compute the following limits and justify your results:

1. $\lim\limits_{t \to 4} \left(\dfrac{t + 1}{t^2 - 1} \right).$

2. $\lim\limits_{t \to 3} \left(\dfrac{x^2 + 1}{x^2 - 4} \atop 5 - 2x \right).$

3. $\displaystyle \lim_{(x,y)\to(1,2)} (x^2 + 2y^2)$.

4. $\displaystyle \lim_{(x,y)\to(2,0)} \begin{pmatrix} x + 3y \\ y^2 - 2x \end{pmatrix}$.

5. $\displaystyle \lim_{(x,y)\to(-1,2)} \begin{pmatrix} \dfrac{x-1}{x^2+y^2} \\ x^3 y \end{pmatrix}$.

6. $\displaystyle \lim_{(x,y)\to(1,-1)} \dfrac{x^2 - y^2}{x + y}$.

7. $\displaystyle \lim_{(x,y)\to(0,0)} \begin{pmatrix} x + y - 2 \\ x^2 - xy \\ y + 2 \end{pmatrix}$.

8. $\displaystyle \lim_{(x,y,z)\to(0,1,-1)} (x^3 + xy + 3z)$.

9. $\displaystyle \lim_{(x,y)\to(2,3)} \dfrac{4x^2 - 9y^2}{2x + 3y}$.

10. $\displaystyle \lim_{t\to3} \begin{pmatrix} \dfrac{t^2 - 9}{t - 3} \\ 4t - 1 \end{pmatrix}$.

Show that the following functions do not approach a limit as X approaches X_0 (hint: study Examples 24 to 26):

11. $f(x) = \dfrac{1}{x - 2}$ for $x \neq 2$, $x_0 = 2$.

12. $f(x) = \begin{cases} 1 & \text{for } x \geq 0, \ x_0 = 0. \\ -1 & \text{for } x < 0 \end{cases}$

13. $f(x,y) = \dfrac{x - y}{x^2 + y^2}$ for $(x, y) \neq (0, 0)$, $X_0 = (0, 0)$.

14. $f(x,y) = \dfrac{x^2}{y^2}$ for $(x, y) \in \mathcal{R}^2$, $y \neq 0$, $X_0 = (0, 0)$.

ANSWERS

2. $\begin{pmatrix} 10 \\ 5 \\ -1 \end{pmatrix}$. 4. $\begin{pmatrix} 2 \\ -4 \end{pmatrix}$. 6. 2. 8. -3. 10. $\begin{pmatrix} 6 \\ 11 \end{pmatrix}$.

MISCELLANEOUS EXERCISES

Compute the following limits:

1. $\displaystyle \lim_{t\to2} \begin{pmatrix} \dfrac{t^2 - t}{2t} \end{pmatrix}$.

2. $\displaystyle \lim_{(x,y)\to(4,-3)} (x^3 - y^2)$.

3. $\displaystyle \lim_{(x,y)\to(2,-1)} \dfrac{4x^2 - y^2}{2x + y}$.

4. $\displaystyle \lim_{(x,y)\to(0,1)} \begin{pmatrix} x + y - 2 \\ x^2 - 3y \\ 4 \end{pmatrix}$.

5. $\displaystyle \lim_{(x,y)\to(0,0)} \begin{pmatrix} \dfrac{x^2 - y^2}{x + y} \\ x^3 - 3 \end{pmatrix}$.

6. $\displaystyle \lim_{(x,y,z)\to(-1,2,-1)} (x^2 z + z^2 y - 3yx)$.

7. $\displaystyle \lim_{(x,y,z)\to(0,2,0)} \begin{pmatrix} \dfrac{x + zy - 3}{xy} \end{pmatrix}$.

8. $\displaystyle \lim_{t\to4} \begin{pmatrix} \dfrac{t^3 - 1}{t^2 - 16} \\ t - 4 \\ 2t + 5 \end{pmatrix}$.

9. $\displaystyle \lim_{(x,y)\to(3,2)} \dfrac{1}{x - y}$.

10. $\displaystyle \lim_{t\to1} \dfrac{t^3 - 1}{t - 1}$.

Determine which of the following limits exist. If the limit exists, determine its value; otherwise, prove that the limit does not exist. Note that in each case we are considering a function $\dfrac{f}{g}$ such that $\lim\limits_{X \to X_0} fX = 0$ and $\lim\limits_{X \to X_0} gX = 0$.

11. $\lim\limits_{x \to 1} \dfrac{x^3 + 2x^2 - 7x + 4}{x - 1}$.

12. $\lim\limits_{x \to 3} \dfrac{x - 3}{x^2 - 6x + 9}$.

13. $\lim\limits_{x \to 2} \dfrac{x^3 - 6x^2 + 12x - 8}{x^2 - 4x + 4}$.

14. $\lim\limits_{(x,y) \to (1,-1)} \dfrac{x^2 - y^2}{x + y}$.

15. $\lim\limits_{(x,y) \to (0,0)} \dfrac{xy}{x^2 - y^2}$.

16. $\lim\limits_{(x,y) \to (1,3)} \dfrac{(x - 1)^2 - (y - 3)^2}{x + y - 4}$.

17. $\lim\limits_{(x,y) \to (0,0)} \dfrac{x^2 + y^2}{x + y}$.

18. $\lim\limits_{t \to 0} \dfrac{t^2 + t}{2t}$.

ANSWERS

2. 55.   **4.** $\begin{pmatrix} -1 \\ -3 \\ 4 \end{pmatrix}$.   **6.** 7.   **8.** $\begin{pmatrix} 63 \\ 8 \\ 13 \end{pmatrix}$.   **10.** 3.   **12.** Does not exist.   **14.** 2.

16. 0.   **18.** 1/2.

4

CONTINUITY OF FUNCTIONS; INFINITE LIMITS

INTRODUCTION

In Chapter 3 we concerned ourselves with vector functions and with the problem of limits of vector functions. If X_0 is a cluster point of the domain of a vector function f, we said that f approached a limit as X approached X_0 when the values of f were all close to some element L of the range space for all X in a small deleted open ball about X_0. The most orderly situation in this context is the one where X_0 is an element of the domain of f, $\lim_{X \to X_0} fX$ exists, and $\lim_{X \to X_0} fX = fX_0$. In this situation (Definition 4.1) we shall say that f is continuous at X_0.

In Chapter 3 we also saw several examples of functions that do not approach a limit as X approaches X_0, but that do behave in a reasonably orderly fashion. For example, $f(x) = \dfrac{1}{x}$ for $x > 0$ does not approach a limit as x approaches 0, but the behavior of f is consistent in that the values of f increase as x gets close to 0. In the latter part of this chapter we shall formally discuss similar situations.

193

1. CONTINUOUS FUNCTIONS

DEFINITION 4.1 Let $f:\mathcal{R}^n \rightarrow \mathcal{R}^m$ and let $X_0 \in \mathcal{R}^n$. f is continuous at X_0 if and only if
    i. $X_0 \in dom\, f$,
    ii. fX approaches a limit as X approaches X_0,
and iii. $\lim\limits_{X \to X_0} fX = fX_0$.

If f is continuous at each point of $dom\, f$, we say that f is continuous.

Thus, there are three ways in which a function f can fail to be continuous at X_0: (i) $X_0 \notin dom\, f$, (ii) fX does not approach a limit as X approaches X_0, or (iii) $\lim\limits_{X \to X_0} fX$ exists but does not equal fX_0.

Example 1. Consider the function f defined by $f(x) = x^2$ for x in $[1, 2)$. We have previously shown that $\lim\limits_{x \to 2} f(x) = 4$. However, in this case, since 2 is not in the domain of f, f is not continuous at 2. The extended function $\bar{f}$, defined by $\bar{f}(x) = x^2$ for x in $[1, 2]$, is continuous at $x = 2$. ∎

Example 2. Let f be defined by

$$f(x) = \begin{cases} 1/x & \text{for } x \neq 0 \\ 0 & \text{for } x = 0. \end{cases}$$

By Example 24, page 189, we know that fX does not approach a limit as x approaches 0. Therefore, f is not continuous at $x_0 = 0$. ∎

Example 3. Let f be defined by

$$f(x) = \begin{cases} x^2 & \text{for } x \text{ in } [1, 2) \\ 0 & \text{for } x = 2. \end{cases}$$

In this case $2 \in dom\, f$, $\lim\limits_{x \to 2} f(x) = 4$, but $f(2) = 0$, which is unequal to 4. Condition iii of Definition 4.1 is not satisfied and f is not continuous at $x_0 = 2$. ∎

The following three theorems are corollaries of Theorem 3.12 or Theorem 3.15. The proofs are quite straightforward and are called for as exercises at the end of this section.

THEOREM 4.2 Let $f:\mathcal{R}^n \rightarrow \mathcal{R}^m$ and $g:\mathcal{R}^n \rightarrow \mathcal{R}^m$. Let c be a constant and $X_0 \in \mathcal{R}^n$. Assume that f and g are continuous at X_0 and that X_0 is a cluster point of $(dom\ f) \cap (dom\ g)$. Then

    i. $f + g$ is continuous at X_0,

    ii. cf is continuous at X_0,

    iii. fg is continuous at X_0.

and iv. if $m = 1$ and $gX_0 \neq 0$, f/g is continuous at X_0.

THEOREM 4.3 Let $f:\mathcal{R} \to \mathcal{R}$, defined by $f(x) = a_0x^n + a_1x^{n-1} + \cdots + a_n$, for all real x, where $a_0, a_1, \ldots, a_n$ are constants. If $x_0 \in \mathcal{R}$, f is continuous at x_0.

THEOREM 4.4 Let $f:\mathcal{R}^n \to \mathcal{R}^m$ and let $X_0 \in \mathcal{R}^n$. If $f_1, \ldots, f_m$ are the coordinate functions of f, then f is continuous at X_0 if and only if f_k is continuous at X_0 for each k, $k = 1, \ldots, m$.

Example 4. In view of Theorem 4.3 we know that if f is defined by $f(x) = x^3 - 4x^2 + 2$ for x in $\mathcal{R}$, then f is continuous at each real x_0. Moreover, if g is defined by $g(x) = x^2$ for x in $\mathcal{R}$, g is continuous at each real x_0. It follows from Theorem 4.2iv that f/g is continuous for all nonzero x. Note that $\left(\dfrac{f}{g}\right)(x) = $ ~~x³−4x²+2~~ $(x-4) + \dfrac{2}{x^2}$. ∎

Example 5. We shall now consider a function which is *not* a polynomial and show that it is continuous at a certain point. Let $f:\mathcal{R} \to \mathcal{R}$, defined by

$$f(x) = \begin{cases} x \sin \dfrac{1}{x} & \text{for } x \neq 0 \\ 0 & \text{for } x = 0. \end{cases}$$

Let $x_0 = 0$. Then

    i. $x_0 \in dom\ f$.

Moreover, $|f(x) - 0| = \left| x \sin \dfrac{1}{x} \right| \leq |x|$, since $|\sin y| \leq 1$ for all real y. It follows easily from Definition 3.7 that

    ii. $\lim_{x \to 0} f(x) = 0$.

    iii. From ii and the rule defining f we know that $\lim_{x \to 0} f(x) = f(0)$.

Therefore, f satisfies the three conditions of Definition 4.1 and is continuous at $x_0 = 0$. ∎

Example 6. Now consider the following function f, $f : \mathcal{R}^2 \to \mathcal{R}^2$:

$$f(x, y) = \begin{pmatrix} x^2 - 2y \\ y^2 + 4x \end{pmatrix}.$$

If $X_0 = (1, 2)$, we have

  i. $X_0 \in dom\, f$

  ii. $\displaystyle\lim_{(x,y) \to (1,2)} f(x, y) = \begin{pmatrix} \displaystyle\lim_{(x,y) \to (1,2)} (x^2 - 2y) \\ \displaystyle\lim_{(x,y) \to (1,2)} (y^2 + 4x) \end{pmatrix}$ by Theorem 3.15

$$= \begin{pmatrix} -3 \\ 8 \end{pmatrix}.$$

  iii. Finally, $fX_0 = \begin{pmatrix} -3 \\ 8 \end{pmatrix} = \displaystyle\lim_{(x,y) \to (1,2)} f(x, y)$. Thus, by Definition 4.1 f is continuous at X_0. ∎

Example 7. Let f be defined by

$$f(x, y) = \begin{cases} \dfrac{x - y}{x^2 + y^2} & \text{for } (x, y) \in \mathcal{R}^2 \text{ and } (x, y) \neq (0, 0) \\ 0 & \text{for } (x, y) = (0, 0). \end{cases}$$

If f were continuous at $X_0 = (0, 0)$, then $\displaystyle\lim_{X \to (0,0)} f(X)$ would equal 0 by Definition 4.1. However, we can show that this limit does not exist by using the same technique that we used in Example 25, Chapter 3. Therefore, f is not continuous at $X_0 = (0, 0)$. ∎

EXERCISES

Show that the following functions are continuous at X_0:

1. $\begin{cases} f(x) = x^2 + 5x \\ x_0 = -1. \end{cases}$

2. $\begin{cases} f(x) = \begin{pmatrix} x + 4x^2 \\ 3 \end{pmatrix} \\ x_0 = 4. \end{cases}$

3. $\begin{cases} f(x, y) = x^2 + y^2 \\ (x_0, y_0) = (1, -2). \end{cases}$

4. $\begin{cases} f(t) = \begin{pmatrix} t^2 \\ t^3 \\ 1 \end{pmatrix} \\ t_0 = 3. \end{cases}$

5. $\begin{cases} f(x) = \begin{cases} x \cos (1/x), & x \neq 0 \\ 0, & x = 0 \end{cases} \\ x_0 = 0. \end{cases}$
   (Hint: see Example 5).

Show that the following functions are not continuous at X_0:

6. $\begin{cases} f(x) = x + 5 \text{ for } x \text{ in } (1, 3) \\ x_0 = 4. \end{cases}$

7. $\begin{cases} f(x) = \begin{cases} 1 + x \text{ for } x \geq 2 \\ 2 - x \text{ for } x < 2 \end{cases} \\ x_0 = 2. \end{cases}$

8. $\begin{cases} f(x) = \begin{cases} 1 + x \text{ for } x \neq -1 \\ 2 \qquad\quad \text{ for } x = -1 \end{cases} \\ x_0 = -1. \end{cases}$ 9. $\begin{cases} f(x) = \begin{cases} \dfrac{1}{2x} \text{ for } x \neq 0 \\ 3 \text{ for } x = 0 \end{cases} \\ x_0 = 0. \end{cases}$

10. $\begin{cases} f(x,y) = \begin{cases} \dfrac{2x + 1}{x^2 + y^2} \text{ for } (x, y) \neq (0, 0) \\ 0 \qquad\quad \text{ for } (x, y) = (0, 0). \end{cases} \\ X_0 = (0, 0) \end{cases}$

Determine which of the following functions are continuous at X_0. If the function is not continuous, explain why:

11. $f(x) = \begin{cases} \dfrac{x^2 - 9}{x - 3} \text{ for } x \neq 3, \\ 6 \qquad\quad \text{ for } x = 3 \end{cases} \quad x_0 = 3.$

12. $\begin{cases} f(x,y) = \begin{cases} \dfrac{x^2 - y^2}{x - y} \text{ for } (x, y) \in R^2,\, x \neq y \\ 0 \qquad\quad \text{ for } x = y. \end{cases} \\ X_0 = (0, 0) \end{cases}$

13. Define f by $f(X) = \|X\|$ for all X in R^n. Prove that f is continuous at each X_0 in R^n.

14. Define g by $g(X, Y) = X \cdot Y$ for all X, Y in R^n. Prove that g is continuous at (X_0, Y_0).

15. Prove Theorem 4.2.   16. Prove Theorem 4.3.

17. Prove Theorem 4.4.

2. IMPORTANT PROPERTIES OF CONTINUOUS FUNCTIONS

In Section 1 of this chapter we considered the algebraic properties of continuous functions. Recall that there is another important way of combining functions, that of composition. The following theorem tells us that $g \circ f$ is continuous under certain conditions on f and g:

THEOREM 4.5 Let $f : R^n \to R^m$ and $g : R^m \to R^p$. Let $X_0 \in dom\, f$ and $Y_0 = fX_0$. If f is continuous at X_0, g is continuous at Y_0, and X_0 is a cluster point of $dom\, (g \circ f)$, then $g \circ f$ is continuous at X_0.

Proof. It follows easily from the hypotheses of the theorem that $X_0 \in dom\, (g \circ f)$. We shall now prove that $\lim_{X \to X_0} (g \circ f)(X)$ exists and equals $(g \circ f)X_0$. Let $r > 0$.

i. $\begin{cases} \text{Since } g \text{ is continuous at } Y_0, \\ \text{there exists a } \rho > 0 \text{ such} \\ \text{that } gY \in \mathcal{B}(gY_0; r) \text{ when-} \\ \text{ever } Y \in \mathcal{B}(Y_0; \rho) \cap dom\, g. \end{cases}$

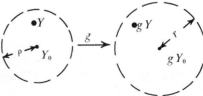

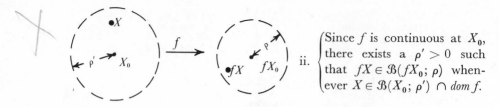

ii. $\begin{cases} \text{Since } f \text{ is continuous at } X_0, \\ \text{there exists a } \rho' > 0 \text{ such} \\ \text{that } fX \in \mathcal{B}(fX_0;\, \rho) \text{ when-} \\ \text{ever } X \in \mathcal{B}(X_0;\, \rho') \cap dom\, f. \end{cases}$

By hypothesis we have $Y_0 = fX_0$. For each $X \in dom\, g \circ f$, let $Y = fX$. Consider the following diagram:

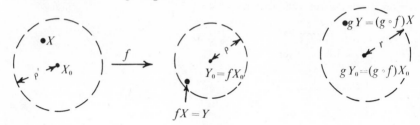

This diagram is justified by combining (i) and (ii) as follows: Let $X \in \mathcal{B}(X_0;\, \rho') \cap dom\, g \circ f$. By (i) we have $Y = fX \in \mathcal{B}(fX_0;\, \rho)$ and by (ii) we have $gY = (g \circ f)X \in \mathcal{B}((g \circ f)X_0;\, r)$. Thus, by Definition 3.7, p. 173,

$$\lim_{X \to X_0} (g \circ f)X = (g \circ f)X_0,$$

and by Definition 4.1, p. 194, $g \circ f$ is continuous at X_0. ∎

Example 8. Let f and g be defined as follows:

$$f(x) = x^2 + 1 \quad \text{for } x \text{ in } \mathcal{R}$$
$$g(y) = \sin y \quad \text{for } y \text{ in } \mathcal{R}.$$

By Theorem 4.3 we know that f is continuous at each real x_0. Let us assume that g is continuous for each real y_0 (see Section 3). Then by Theorem 4.5, $g \circ f$, which is defined by

$$(g \circ f)(x) = g(f(x))$$
$$= g(x^2 + 1)$$
$$= \sin (x^2 + 1),$$

is continuous at each real x_0. ∎

The following three theorems are of unusual importance in the study of continuous functions. The proofs of these theorems are relatively sophisticated and will not be given here (see Appendix IV).

A simple intuitive notion can be associated with the graph of certain continuous functions, that is, if f is defined and continuous at each point of an interval, a rectangle, or any other "connected" set, then the graph of f is "connected," or unbroken. Theorem 4.6 states this fact more precisely for the one dimensional case.

THEOREM 4.6 (Intermediate Value Theorem.) Let $f : \mathcal{R} \to \mathcal{R}$ and let $[a, b] \subset \operatorname{dom} f$. Assume that f is continuous at each point of $[a, b]$. If $f(a) < f(b)$, and c is a number such that $f(a) < c < f(b)$, then there exists an x_0 such that

i. $a < x_0 < b$

and

ii. $f(x_0) = c$.

Figure 4.1 illustrates a situation as described by Theorem 4.6. Notice that if there exists a number c, $f(a) < c < f(b)$, for which there is no x_0 such that $a < x_0 < b$ and $f(x_0) = c$, there must be a break in the graph of f.

It should be noted that for Theorem 4.6 to apply, f must be continuous and $[a, b]$ must be contained in the domain of f. Consider the following examples:

Let f be defined by

$$f(x) = \begin{cases} x \text{ for } x \text{ in } [0, 1) \\ x + 1 \text{ for } x \text{ in } [1, 2]. \end{cases}$$

Then $\operatorname{dom} f = [0, 2]$, but f is not continuous at $x_0 = 1$. Figure 4.2 is a sketch of the graph of f. If c is any number such that $1 < c < 2$, then $f(a) < c < f(b)$; but there exists no x_0 such that $a < x_0 < b$ and $f(x_0) = c$.

On the other hand let g be defined by $g(x) = x$ for x in $[0, 1] \cup [2, 3]$. In this case g is continuous at each point of its domain, but $[0, 3]$ is not

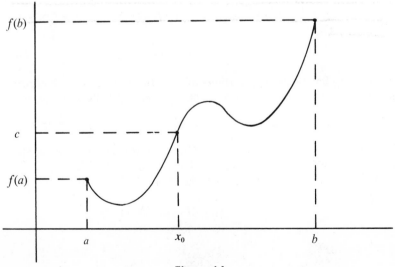

Figure 4.1

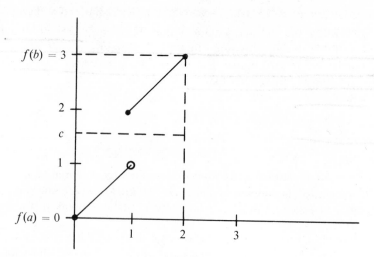

Figure 4.2

contained in *dom g*. If we choose $a = 0$, $b = 3$, and $c = 3/2$, there is no x_0 such that $0 < x_0 < 3$ and $g(x_0) = c$.

The reader can easily verify the following variation of Theorem 4.6: If $f : \mathcal{R} \to \mathcal{R}$, $[a, b] \subset dom\ f$, f is continuous at each point of $[a, b]$ and $f(a) > f(b)$, then whenever c is a real number such that $f(a) > c > f(b)$, there exists an x_0 such that (i) $a < x_0 < b$ and (ii) $f(x_0) = c$.

THEOREM 4.7 Let $f : \mathcal{R}^n \to \mathcal{R}^m$ and let $\overline{\mathcal{B}}(X_0; r) \subset dom\ f$. If f is continuous at each point of $\overline{\mathcal{B}}(X_0; r)$, there exists an $M > 0$ such that $\|fX\| \leq M$ for all X in $\overline{\mathcal{B}}(X_0; r)$. (In this case we say that f is *bounded* on $\overline{\mathcal{B}}(X_0; r)$.)

In considering the implications of this theorem, let us first restrict the theorem to the case in which $f : \mathcal{R} \to \mathcal{R}$. The closed balls in $\mathcal{R}$ are of course the closed intervals $[a, b]$: $[a, b] = \overline{\mathcal{B}}\left(\dfrac{a + b}{2}; \dfrac{b - a}{2}\right)$. In this context Theorem 4.6 tells us that when f is continuous at each point of $[a, b]$, there exists a real number M such that $-M \leq f(x) \leq M$ for all x in $[a, b]$.

Let us now consider the value of the hypotheses. Consider the following examples:

$$f(x) = \begin{cases} 1/x & \text{for } x \in (0, 1] \\ 0 & \text{for } x = 0 \end{cases}$$

and

$$g(x) = \frac{1}{x} \quad \text{for } x \in (0, 1].$$

The function f is not continuous at $x_0 = 0$ and also is not bounded on $[0, 1]$. The domain of g, $(0, 1]$, is not closed and g is not bounded on $(0, 1]$. Thus, in each case one of the hypotheses of the theorem is not satisfied and the conclusion is not satisfied. In Exercise 11 we ask the reader to find an example of a function that satisfies the conclusion of Theorem 4.7, but not all the hypotheses.

THEOREM 4.8 Let $f: \mathcal{R}^n \to \mathcal{R}$ and let $\bar{\mathcal{B}}(X_0; r) \subset dom\ f$. If f is continuous at each point of $\bar{\mathcal{B}}(X_0; r)$, there exist points X_1 and X_2 in $\bar{\mathcal{B}}(X_0; r)$ such that $f(X_1) \leq f(X) \leq f(X_2)$ for each X in $\bar{\mathcal{B}}(X_0; r)$. $f(X_1)$ is called the *minimum value* of f on $\bar{\mathcal{B}}(X_0; r)$. $f(X_2)$ is called the *maximum value* of f on $\bar{\mathcal{B}}(X_0; r)$.

Notice that the hypotheses of Theorem 4.8 are the same as those of Theorem 4.7 for case $m = 1$. However, the conclusion of Theorem 4.8 is stronger, for it says that not only is f bounded on $\bar{\mathcal{B}}(X_0; r)$, but it actually assumes a greatest and a least value on $\bar{\mathcal{B}}(X_0; r)$.

Example 9. Let f be defined by $f(x) = x^2$ for x in $[-1, 2]$. It is easily seen that for x in $[-1, 2]$, $0 \leq f(x) \leq 4$. Thus, 0 is a candidate for the minimum value of f and 4 is a candidate for the maximum value of f. If we choose $x_1 = 0$ and $x_2 = 2$, $f(x_1) = 0$ and $f(x_2) = 4$, so that $f(x_1)$ and $f(x_2)$ are indeed the minimum and maximum values of f, respectively, on $[-1, 2]$. ∎

EXERCISES WOW

Using the theorems in Sections 1 and 2, show that the following functions are continuous at X_0. You may assume the following:

i. The sine and cosine functions are continuous at each real x.

ii. The square root function is continuous for each non-negative x.

1. $f(x) = \sin(2x + 1)$, $x_0 = -\frac{1}{2}$ 2. $f(t) = \sqrt{1 + t^2}$, $t_0 = -4$.

3. $f(x) = \dfrac{\sin(x^2 + \pi)}{\cos(x^2 - \pi)}$, $x_0 = 0$.

4. $f(x) = \sqrt{2 + \sin(x^2)}$, $x_0 = \sqrt{\pi/2}$.

5. $f(x, y) = \begin{cases} \dfrac{x(x^2 - y^2)}{x^2 + y^2}, & (x, y) \neq (0, 0), \ X_0 = (0, 0) \\ 0, & (x, y) = (0, 0) \end{cases}$

$\left(\text{Hint:}\ \left| \dfrac{x^2 - y^2}{x^2 + y^2} \right| \leq 1 \text{ for all } (x, y). \right)$

At which points do the following functions fail to be continuous?

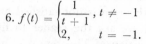

6. $f(t) = \begin{cases} \dfrac{1}{t+1}, & t \neq -1 \\ 2, & t = -1. \end{cases}$

7. $f(x, y) = \dfrac{x+y}{1-x^2}.$

8. $f(x, y) = \left(\dfrac{3x - 2y + 1}{7xy} \right).$

9. $f(x, y) = \dfrac{x-y}{x^2 - y^2}.$

10. $f(t) = \dfrac{1}{t^2 - 6t + 5}.$

11. Find an example of a function that satisfies the conclusion of Theorem 4.7 but does not satisfy one of the hypotheses.

12. Using Theorem 4.6, prove the following: Let $f : \mathcal{R} \to \mathcal{R}$ and let $[a, b] \subset$ *dom* f. Assume that f is continuous at each point of $[a, b]$ and $f(a) > f(b)$. If c is a number such that $f(a) > c > f(b)$, there exists an x_0 such that (i) $a < x_0 < b$ and (ii) $f(x_0) = c$ (Hint: apply Theorem 4.6 to $(-f)$).

ANSWERS

6. $t = -1.$ **8.** Everywhere continuous.   **10.** $t = 5, t = 1.$

3.   CONTINUITY OF THE SINE AND COSINE FUNCTIONS

So far in our discussion and examples of limits and continuous functions, we have restricted ourselves almost entirely to polynomials. The trigonometric functions are of vast importance both in pure and in applied mathematics. In view of the fact that it is extremely difficult to define the sine and cosine functions rigorously without the use of integrals, our basic proofs in this section will be largely geometric in nature.

LEMMA 4.9

i. $\lim\limits_{x \to 0} \sin x = 0.$

ii. $\lim\limits_{x \to 0} \cos x = 1.$

Geometric Motivation.   Since we are concerned with values of x close to 0, we may assume that $-\dfrac{\pi}{2} < x < \dfrac{\pi}{2}$. Moreover, since $\sin(-x) = -\sin x$ and $\cos(-x) = \cos x$, we may assume that $0 < x < \dfrac{\pi}{2}$.

Consider a right triangle, as shown in Figure 4.3, with vertices Q, R, and T such that the measure of $\angle RQT$ is x radians, $0 < x < \dfrac{\pi}{2}$. Assume that $\angle RTQ$ is a right angle and that $\|Q - R\| = 1$. Then $\sin x = \|T - R\|$

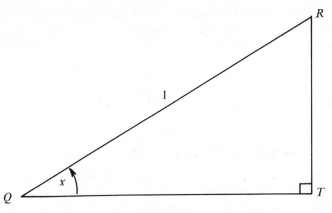

Figure 4.3

and $\cos x = \|Q - T\|$. As x approaches 0, we see that $\|T - R\|$ approaches 0 and $\|Q - T\|$ approaches 1. Consequently,

$$\lim_{x \to 0} \sin x = 0 \text{ and } \lim_{x \to 0} \cos x = 1. \quad \blacksquare$$

THEOREM 4.10 Let x_0 be any real number;

i. $\lim\limits_{x \to x_0} \sin x = \sin x_0$,

ii. $\lim\limits_{x \to x_0} \cos x = \cos x_0$.

Consequently, the sine and cosine functions are continuous at each real number x_0.

Proof. By Theorem 3.9 (p. 177) we know that $\lim\limits_{x \to x_0} \sin x$ exists if and only if $\lim\limits_{h \to 0} \sin (x_0 + h)$ exists, and if these limits exist, they are equal. By the well-known addition formula,

$$\sin (x_0 + h) = \sin x_0 \cos h + \cos x_0 \sin h.$$

Since, by the preceding lemma, $\lim\limits_{h \to 0} \cos h = 1$ and $\lim\limits_{h \to 0} \sin h = 0$, we have

$$\lim_{h \to 0} \sin (x_0 + h) =^1 \sin x_0 \lim_{h \to 0} \cos h + \cos x_0 \lim_{h \to 0} \sin h$$
$$= \sin x_0(1) + \cos x_0(0)$$
$$= \sin x_0.$$

(1) By Theorem 3.12 (p. 182).

To prove equation ii we could proceed in a similar fashion. However, if we note that $\cos x = \sin (\pi/2 - x)$ for all real x, we have

$$\lim_{x \to x_0} \cos x = \lim_{x \to x_0} \sin \left(\frac{\pi}{2} - x \right)$$

$$=^1 \sin \left(\frac{\pi}{2} - x_0 \right)$$

$$= \cos x_0.$$

(1) By Theorems 4.5 (p. 197) and 4.10i. ∎

The results of the preceding theorem should seem natural to the reader familiar with the sine and cosine functions. We shall leave it to the reader to verify that the tangent and secant functions are continuous at all real numbers x, $x \neq \dfrac{(2n + 1)\pi}{2}$, n an integer, and the cotangent and cosecant functions are continuous at all real numbers x, $x \neq n\pi$, n an integer. This follows easily from Theorem 3.12iv (p. 182) and Theorem 4.10.

The reader has already encountered the situation in which $\lim\limits_{x \to x_0} \dfrac{f(x)}{g(x)}$ exists, but $\lim\limits_{x \to x_0} f(x) = \lim\limits_{x \to x_0} g(x) = 0$ (see Example 13, p. 167). The following theorem is another example of such a situation:

THEOREM 4.11

$$\lim_{x \to 0} \frac{\sin x}{x} = 1.$$

Geometric Motivation. Since we are concerned with values of x close to 0, we may assume that $-\dfrac{\pi}{2} < x < \dfrac{\pi}{2}$. Moreover, $\dfrac{\sin (-x)}{(-x)} = \dfrac{-\sin x}{-x} = \dfrac{\sin x}{x}$ and we may assume that $0 < x < \pi/2$.

Consider an angle of x radians, $0 < x < \pi/2$, with vertex at Q (see Figure 4.4). Let P and R represent, respectively, the intersection of a circle of radius 1 with the initial and terminal sides of this angle. Drop a perpendicular from R to QP and let T be the point of intersection. Construct a circle of radius $\| Q - T \|$ about Q and let S be the intersection of this circle with QR. Note that

(4.1)   area (sector TQS) < area (triangle TQR) < area (sector PQR).

Recall that the area of a circular sector is $\frac{1}{2}r^2\theta$, where r is the radius of the circle and θ is the angle of the sector, given in radians. The area of a right triangle is $\frac{1}{2}ab$, where a is the altitude and b is the base of the triangle.

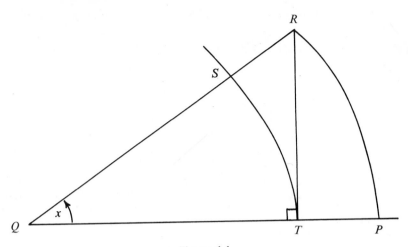

Figure 4.4

Note that
$$\cos x = \| T - Q \| \quad \text{and} \quad \sin x = \| R - T \|.$$

Then

(4.2) $\begin{cases} \text{area (sector } TQS) = \frac{1}{2} \| T - Q \|^2 x = \frac{1}{2} x (\cos x)^2, \\ \text{area (triangle } TQR) = \frac{1}{2} \| T - Q \| \, \| R - T \| = \frac{1}{2} \cos x \sin x, \text{ and} \\ \text{area (sector } PQR) = \frac{1}{2} \| P - Q \|^2 x = \frac{1}{2} (1)^2 x = \frac{1}{2} x. \end{cases}$

Combining (4.1) and (4.2) we have

(4.3) $\qquad\qquad \frac{1}{2} x (\cos x)^2 < \frac{1}{2} \cos x \sin x < \frac{1}{2} x.$

Dividing (4.3) through by $\frac{1}{2} x (\cos x)$, we have

$$\cos x < \frac{\sin x}{x} < \frac{1}{\cos x}.$$

By Lemma 4.9, $\lim_{x \to 0} \cos x = 1$. From this fact and Theorem 3.12iv (p. 182),

we also have $\lim_{x \to 0} \dfrac{1}{\cos x} = 1.$ Therefore, $\lim_{x \to 0} \dfrac{\sin x}{x} = 1$ and the argument

is complete. ∎

Example 10. We shall use Theorem 4.11 to compute the following

limit:

$$\lim_{x \to 0} \frac{\sin (2x)}{x} = \lim_{x \to 0} \frac{2 \sin (2x)}{(2x)}$$

$$=^1 2 \lim_{x \to 0} \frac{\sin (2x)}{(2x)}$$

$$=^2 2(1)$$

$$= 2.$$

(1) By Theorem 3.12ii (p. 182).
(2) By Theorem 4.5 (p. 197), and Theorem 4.11.

An alternative method of determining this limit is as follows:

$$\lim_{x \to 0} \frac{\sin (2x)}{x} = \lim_{x \to 0} \frac{2(\sin x)(\cos x)}{x}$$

$$=^3 2 \lim_{x \to 0} \frac{\sin x}{x} \lim_{x \to 0} \cos x$$

$$=^4 2(1)(1)$$

$$= 2.$$

(3) By Theorem 3.12ii and iii (p. 182).
(4) By Theorem 4.11 and Lemma 4.9. ■

EXERCISES

Determine whether the following limits exist. If the limit exists, find the value of the limit; if the limit does not exist, state the reason:

1. $\displaystyle\lim_{x \to 0} \frac{\tan x}{x}$.

2. $\displaystyle\lim_{x \to 0} \tan x$.

3. $\displaystyle\lim_{x \to \pi/2} \tan x$.

4. $\displaystyle\lim_{x \to 0} \frac{\sin^2 x}{x^2}$.

5. $\displaystyle\lim_{x \to 0} \frac{\sin^2 x}{x^3}$.

6. $\displaystyle\lim_{x \to 0} x \csc x$.

7. $\displaystyle\lim_{x \to \pi/2} \frac{\cos x}{\dfrac{\pi}{2} - x}$.

8. $\displaystyle\lim_{x \to \pi/2} \cot x$.

9. $\displaystyle\lim_{x \to 0} x \sin \frac{1}{x}$.

10. $\displaystyle\lim_{(x,y) \to (0,0)} \frac{\sin x}{\cos y}$.

11. $\displaystyle\lim_{(x,y) \to (0,0)} \frac{\cos x}{\sin y}$.

12. $\displaystyle\lim_{x \to 0} \frac{\sin 3x}{x}$.

13. $\displaystyle\lim_{x \to 0} \cos \left(\frac{\pi}{4} - x\right)$.

14. $\displaystyle\lim_{(x,y) \to (\pi/2,0)} \left(\frac{\sin\left(\dfrac{\pi}{2} - x\right)}{\left(\dfrac{\pi}{2} - x\right)} \atop \sin x \cos y \right)$.

15. $\displaystyle\lim_{t \to 0} \left(\begin{matrix} t^2 \sin t \\ \dfrac{\tan t}{\cos t} \\ t \end{matrix} \right)$.

ANSWERS

1. 1. **3.** Does not exist. **6.** 1. **9.** 0. **11.** Does not exist. **12.** 3.
14. $\begin{pmatrix} 1 \\ 1 \end{pmatrix}$.

4. INFINITE LIMITS

In this section we shall concern ourselves with limits of functions, such that the functional values increase without bound. Consider the function f defined by $f(x) = \dfrac{1}{x^2}$, $x \neq 0$. We know that as x approaches 0, $f(x)$ is positive and increases without bound. Consider now the function g defined by $g(x) = \dfrac{1}{x}$, $x \neq 0$. This function does not behave as nicely as f; for if x approaches 0, $g(x)$ may be either positive or negative, depending upon whether x is positive or negative. In either case $|g(x)|$ increases without bound. Let us formalize this situation by the following definition:

DEFINITION 4.12 Let $f : \mathcal{R} \to \mathcal{R}$. Let x_0 be a cluster point of $(dom\, f)$.

i. We say that f *diverges to* $+\infty$ *as x approaches* x_0 if to each positive number M there corresponds a $\rho > 0$ such that $f(x) > M$ whenever $x \in \mathcal{B}(\tilde{x}_0; \rho) \cap (dom\, f)$. We denote this by

$$\lim_{x \to x_0} f(x) = +\infty$$

ii. We say that f *diverges to* $-\infty$ *as x approaches* x_0 if to each negative number N there corresponds a $\rho > 0$ such that $f(x) < N$ whenever $x \in \mathcal{B}(\tilde{x}_0; \rho) \cap (dom\, f)$. We denote this by

$$\lim_{x \to x_0} f(x) = -\infty.$$

Do not fall into the trap of thinking of $+\infty$ and $-\infty$ as real numbers, for they are not real numbers. These symbols are mathematical shorthand for exactly the situations described in Definition 4.12i and ii. When we write $\lim\limits_{x \to x_0} f(x) = +\infty$, we mean that for values of x close to x_0, $f(x)$ is positive and large and that there exists no real number M such that $f(x) \leq M$ for all values of x close to x_0. We can describe the notation $\lim\limits_{x \to x_0} f(x) = -\infty$ in a similar fashion.

Example 11. Let $f(x) = \tan x$ for x in $(0, \pi/2)$. Let $M > 0$. Since $\lim\limits_{x \to \pi/2} \sin x = 1$ and $\lim\limits_{x \to \pi/2} \cos x = 0$, there exists a $\rho > 0$ such that whenever $x \in B\left(\dfrac{\tilde{\pi}}{2}; \rho\right) \cap dom\, f$,

$$\sin x > \frac{1}{2} \quad \text{and} \quad 0 < \cos x < \frac{1}{2M}.$$

Therefore,

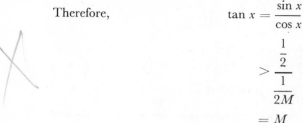

$$\tan x = \frac{\sin x}{\cos x}$$

$$> \frac{\dfrac{1}{2}}{\dfrac{1}{2M}}$$

$$= M$$

whenever $x \in B\left(\dfrac{\pi}{2}; \rho\right) \cap dom\, f$. By Definition 4.12, $\lim\limits_{x \to \pi/2} f(x) = +\infty$. ∎

Notice that in the preceding example we restricted the domain of f to $\left(0, \dfrac{\pi}{2}\right)$, where the tangent function is positive. If instead we had used the domain $\left(\dfrac{\pi}{2}, \pi\right)$, where the tangent function is negative, we would have found that $\lim\limits_{x \to \pi/2} f(x) = -\infty$. What could we say about the function defined by $f(x) = \tan x$ for x in $\left(0, \dfrac{\pi}{2}\right) \cup \left(\dfrac{\pi}{2}, \pi\right)$? It should be obvious from the preceding examples and discussion that f approaches neither $+\infty$ nor $-\infty$ as x approaches $\dfrac{\pi}{2}$, for f is positive to the left of $\dfrac{\pi}{2}$ and negative to the right of $\dfrac{\pi}{2}$. To describe this sort of situation we make the following definition:

DEFINITION 4.13 Let $f: \mathcal{R} \to \mathcal{R}$ and let x_0 be a cluster point of $dom\, f$. Let $L \in \mathcal{R}$. Let f_1 and f_2 be defined by

$$f_1(x) = f(x) \text{ for } x \text{ in } (dom\, f) \cap (-\infty, x_0), \text{ and}$$
$$f_2(x) = f(x) \text{ for } x \text{ in } (dom\, f) \cap (x_0, +\infty).$$

i. *f approaches L as x approaches x_0 from the left* if and only if $\lim\limits_{x \to x_0} f_1(x) = L$. We denote this by

$$\lim\limits_{x \to x_0^-} f(x) = L.$$

ii. *f approaches L as x approaches x_0 from the right* if and only if $\lim\limits_{x \to x_0} f_2(x) = L$. We denote this by

$$\lim\limits_{x \to x_0^+} f(x) = L.$$

iii. *f diverges to $+\infty$ $(-\infty)$ as x approaches x_0 from the left* if and only if $\lim\limits_{x \to x_0} f_1(x) = +\infty(-\infty)$. We denote this by

$$\lim\limits_{x \to x_0^-} f(x) = +\infty(-\infty).$$

iv. *f diverges to* $+\infty(-\infty)$ *as x approaches* x_0 *from the right* if and only if $\lim\limits_{x \to x_0} f_2(x) = +\infty(-\infty)$. We denote this by

$$\lim_{x \to x_0^+} f(x) = +\infty(-\infty).$$

As we have seen, the tangent function has the property that $\lim\limits_{x \to \pi/2^-} f(x) = +\infty$ and $\lim\limits_{x \to \pi/2^+} f(x) = -\infty$, but $\lim\limits_{x \to \pi/2} f(x)$ does not exist either in the sense of Definition 3.7 or Definition 4.12. The function g, defined by $g(x) = \dfrac{-1}{x}$, $x \neq 0$, has the same property at $x_0 = 0$.

The function f defined by $f(x) = \begin{cases} 1 & \text{for } x \geq 0 \\ -1 & \text{for } x < 0 \end{cases}$ has the property that $\lim\limits_{x \to 0^-} f(x) = -1$ and $\lim\limits_{x \to 0^+} f(x) = +1$, but $\lim\limits_{x \to 0} f(x)$ does not exist either as a real number or in the sense of Definition 4.12. It should be obvious that in this case as well as others consideration of the left and right hand limits will yield information about the behavior of f whether or not $\lim\limits_{x \to x_0} f(x)$ exists. In Exercise 18 the reader is asked to prove that $\lim\limits_{x \to x_0} f(x)$ exists if and only if $\lim\limits_{x \to x_0^+} f(x)$ and $\lim\limits_{x \to x_0^-} f(x)$ exist and are equal.

It must be emphasized that we have extended the interpretation of the notation "$\lim\limits_{x \to x_0} f(x)$" by means of Definition 4.12. The reader must adjust his thinking to allow the possibility that "$\lim\limits_{x \to x_0} f(x)$ exists" means either that f diverges to $+\infty$, diverges to $-\infty$, or approaches L as x approaches x_0, where L is some real number. Prior to our consideration of Definition 4.12, there were only two possibilities: (i) $\lim\limits_{x \to x_0} f(x) = L$, where L is some real number or (ii) $\lim\limits_{x \to x_0} f(x)$ does not exist. Now there are four possibilities: (i) $\lim\limits_{x \to x_0} f(x) = L$, where L is some real number, (ii) $\lim\limits_{x \to x_0} f(x) = +\infty$, (iii) $\lim\limits_{x \to x_0} f(x) = -\infty$, or (iv) $\lim\limits_{x \to x_0} f(x)$ does not exist either in the sense of Definition 3.7 or that of Definition 4.12.

Example 12

(a) Let f be defined by $f(x) = \dfrac{1}{(x-1)^2}$ for $x \neq 0$. We shall see that it is quite straightforward to show that $\lim\limits_{x \to 1} f(x) = +\infty$. We let M be any positive number and note that $\dfrac{1}{(x-1)^2} > M$ if and only if $\dfrac{1}{M} > (x-1)^2$, or $|x-1| < \dfrac{1}{\sqrt{M}}$. We let $\rho = \dfrac{1}{\sqrt{M}}$. Then $f(x) > M$ whenever

$x \in \mathcal{B}(\tilde{1}; \rho) \cap dom\, f$, and we see that $\lim\limits_{x \to 1} f(x) = +\infty$, according to Definition 4.12.

(b) Let g be defined by $g(x) = \dfrac{1}{x - 1}$ for $x \neq 1$ (notice that this function is closely related to the function f in **(a)**). In this case, we shall show that $\lim\limits_{x \to 1^+} g(x) = +\infty$, $\lim\limits_{x \to 1^-} g(x) = -\infty$, and hence that $\lim\limits_{x \to 1} g(x)$ does not exist. As in Definition 4.13, we let $g_1(x) = g(x)$ for x in $dom\, f \cap (-\infty, 1)$. If N is any negative number and $x \in dom\, g_1$, $g_1(x) = \dfrac{1}{x - 1} < N$ if and only if $\dfrac{1}{N} < x - 1$ or $|x - 1| < -\dfrac{1}{N}$. Letting $\rho = -\dfrac{1}{N}$, we see that $g_1(x) < N$ whenever $|x - 1| < \rho$ and $x \in dom\, g_1$ and that $\lim\limits_{x \to 1^-} g(x) = -\infty$. We argue in a similar fashion to show that $\lim\limits_{x \to 1^+} g(x) = +\infty$. ■

EXERCISES

Evaluate the following limits whenever they exist. Justify your answer if the limit does not exist:

1. $\lim\limits_{x \to \pi/2^+} \sec x$.

2. $\lim\limits_{x \to 0} \dfrac{\sin x}{x}$.

3. $\lim\limits_{x \to 1} \dfrac{1}{x^2 - 1}$.

4. $\lim\limits_{x \to 1^+} \dfrac{1}{x^2 - 1}$.

5. $\lim\limits_{x \to 2} \dfrac{x}{(x - 2)^2}$.

6. $\lim\limits_{x \to 3} \dfrac{1}{(x - 3)^3}$.

7. $\lim\limits_{x \to 3^-} \dfrac{1}{(x - 3)^3}$.

8. $\lim\limits_{x \to 4} \dfrac{x^2 - 16}{x - 4}$.

9. $\lim\limits_{x \to 4} \dfrac{x^2 - 16}{(x - 4)^2}$.

10. $\lim\limits_{x \to \pi} \cot x$.

11. $\lim\limits_{t \to 2} \dfrac{t^2 - 5t + 6}{t - 2}$.

12. $\lim\limits_{h \to 0} \dfrac{(x + h)^3 - x^3}{h}$.

13. $\lim\limits_{x \to \pi/2^+} \tan x$.

14. $\lim\limits_{x \to 0^+} \csc x$.

15. Prove that if $\lim\limits_{x \to x_0^+} f(x) = +\infty$ and $\lim\limits_{x \to x_0^+} g(x) = +\infty$, then

$$\lim\limits_{x \to x_0^+} (f(x) + g(x)) = +\infty.$$

16. Find an example of two functions f and g such that $\lim\limits_{x \to x_0} f(x) = +\infty$, $\lim\limits_{x \to x_0} g(x) = -\infty$, and $\lim\limits_{x \to x_0} (f(x) + g(x)) = 1$.

17. Find an example of two functions f and g such that $\lim\limits_{x \to x_0} f(x) = +\infty$, $\lim\limits_{x \to x_0} g(x) = 0$, and $\lim\limits_{x \to x_0} (fg)(x) = 1$.

18. i. Prove that $\lim\limits_{x \to x_0} f(x) = L$ if and only if $\lim\limits_{x \to x_0^+} f(x) = \lim\limits_{x \to x_0^-} f(x) = L$.

    ii. Prove that $\lim\limits_{x \to x_0} f(x) = +\infty$ if and only if

$$\lim\limits_{x \to x_0^+} f(x) = \lim\limits_{x \to x_0^-} f(x) = +\infty.$$

19. Prove that if $\lim\limits_{x \to x_0^+} f(x) = L$ and $\lim\limits_{x \to x_0^+} g(x) = K$, then $\lim\limits_{x \to x_0^+} (f + g)(x) = L + K$.

20. Prove that if $\lim\limits_{x \to x_0^+} f(x) = L$ and $\lim\limits_{x \to x_0^+} g(x) = +\infty$, then

$$\lim\limits_{x \to x_0^+} (f(x) + g(x)) = +\infty.$$

ANSWERS

1. $-\infty$. **3.** Does not exist. **5.** $+\infty$. **7.** $-\infty$. **9.** Does not exist. **11.** -1. **12.** $3x^2$. **14.** $+\infty$. **16.** $f(x) = x$, $g(x) = 1 - x$ is one such function.

5. MORE ABOUT INFINITE LIMITS

We now come to a third type of limit, one in which the domain variable increases without limit. First let us consider several examples. The function f defined by $f(x) = x^2$ has the property that when x gets large, so does $f(x)$. On the other hand the function g, defined by $g(x) = \dfrac{1}{x}$ for $x \neq 0$, has the property that when x is large, $g(x)$ is near zero. There are, of course, other possibilities. For example, let h be defined by $h(x) = \sin x$. For $x_1 = n\pi$, n a positive integer, $h(x_1) = 0$; for $x_2 = \dfrac{(4n + 1)\pi}{2}$, $h(x_2) = 1$. Thus, there exist arbitrarily large numbers x_1 and x_2 such that $h(x_1) = 0$ and $h(x_2) = 1$. We feel intuitively that h cannot approach a limit as x increases without bound. The following definition will clarify these ideas:

DEFINITION 4.14 Let $f: \mathcal{R} \to \mathcal{R}$ and let L be a real number. Assume that for each real number M, $(M, +\infty) \cap (dom\, f) \neq \varnothing$.

i. *f approaches L as x approaches* $+\infty$ *if and only if*

for each $r > 0$

there exists an $M > 0$ such that

$|f(x) - L| < r$

whenever $x \in (M, +\infty) \cap (dom\, f)$.

We denote this situation by

$$\lim_{x \to +\infty} f(x) = L.$$

ii. *f diverges to* $+\infty$ *as x approaches* $+\infty$ *if and only if*

for each $N > 0$

there exists a $M > 0$ such that

$f(x) > N$

whenever $x \in (M, +\infty) \cap (dom\, f)$.

We denote this situation by

$$\lim_{x \to +\infty} f(x) = +\infty$$

Obviously, similar definitions can be made for

$$\lim_{x \to -\infty} f(x) = L, \qquad \lim_{x \to -\infty} f(x) = -\infty,$$

$$\lim_{x \to +\infty} f(x) = -\infty, \quad \text{and} \quad \lim_{x \to -\infty} f(x) = +\infty.$$

The reader should at this point be able to formulate such definitions.

There are several techniques for evaluating infinite limits with which the reader should become familiar. One of these techniques is that of substitution. The following theorem will justify one type of substitution:

THEOREM 4.15 Let $f: \mathcal{R} \to \mathcal{R}$.

i. $\lim\limits_{x \to +\infty} f(x)$ exists if and only if $\lim\limits_{t \to 0^+} f(1/t)$ exists; in this case the two limits are equal.

ii. $\lim\limits_{x \to -\infty} f(x)$ exists if and only if $\lim\limits_{t \to 0^-} f\left(\dfrac{1}{t}\right)$ exists; in this case the two limits are equal.

Proof of i. This theorem covers the three cases in which $\lim\limits_{x \to +\infty} f(x)$ exists and is equal to a real number L, $+\infty$, or $-\infty$. We shall prove the theorem only for the finite case.

Assume that $\lim\limits_{x \to +\infty} f(x) = L$, L a real number. Let $r > 0$. We wish to find a $\rho > 0$ such that $\left| f\left(\dfrac{1}{t}\right) - L \right| < r$ whenever $t > 0$, $\dfrac{1}{t} \in dom\, f$, and $t \in \mathcal{B}(\tilde{0}; \rho)$. Since $\lim\limits_{x \to +\infty} f(x) = L$, we know by Definition 4.14 that there exists an $M > 0$ such that

$$(4.4) \quad |f(x) - L| < r \quad \text{whenever} \quad x > M \quad \text{and} \quad x \in dom\, f.$$

Now if $x \in (M, +\infty)$, then $x > M$ and $0 < \dfrac{1}{x} < \dfrac{1}{M}$. Letting $t = \dfrac{1}{x}$ and $\rho = \dfrac{1}{M}$, we can restate (4.4) as follows:

$$\left| f\left(\dfrac{1}{t}\right) - L \right| < r \quad \text{whenever} \quad 0 < t < \rho \quad \text{and} \quad \dfrac{1}{t} \in dom\, f.$$

Therefore, $\lim\limits_{t \to 0^+} f\left(\dfrac{1}{t}\right) = L$, by Definition 4.136.

This argument can be reversed to show that if $\lim\limits_{t \to 0^+} f(1/t) = L$, then $\lim\limits_{x \to +\infty} f(x) = L$. Similar arguments can be used for the two remaining cases in (i) and the three cases in (ii). ∎

Example 13. Let $f(x) = \dfrac{x^2 + 1}{2x^2 + 3}$. We wish to determine whether $\lim\limits_{x \to +\infty} f(x)$ exists. We shall use Theorem 4.15 for this purpose. Let $t = \dfrac{1}{x}$, $x \neq 0$. We have then

$$f\left(\frac{1}{t}\right) = \frac{\left(\dfrac{1}{t}\right)^2 + 1}{2\left(\dfrac{1}{t}\right)^2 + 3}$$

$$= \frac{1 + t^2}{2 + 3t^2}$$

By Theorem 3.12 (p. 182),

$$\lim_{t \to 0^+} f\left(\frac{1}{t}\right) = \frac{\lim\limits_{t \to 0^+} (1 + t^2)}{\lim\limits_{t \to 0^+} (2 + 3t^2)}$$

$$= \frac{1}{2}.$$

Therefore, by Theorem 4.15,

$$\lim_{x \to +\infty} \frac{x^2 + 1}{2x^2 + 3} = \frac{1}{2}. \quad ∎$$

Example 14. Let $f(x) = \dfrac{x-1}{x^2+1}$. Again we wish to determine whether $\lim\limits_{x \to +\infty} f(x)$ exists. Using the same method as in Example 13, we have

$$\lim_{x \to +\infty} f(x) = \lim_{t \to 0^+} f\left(\frac{1}{t}\right)$$

$$= \lim_{t \to 0^+} \frac{\left(\frac{1}{t}\right) - 1}{\left(\frac{1}{t}\right)^2 + 1}$$

$$= \lim_{t \to 0^+} \frac{t - t^2}{1 + t^2}$$

$$= \frac{0}{1}$$

$$= 0. \quad \blacksquare$$

Example 15. Now let f be defined by $f(x) = \dfrac{x^2}{x+1}$ for $x \neq -1$. Proceeding as above we shall evaluate $\lim\limits_{x \to +\infty} f(x)$:

$$\lim_{x \to +\infty} f(x) = \lim_{t \to 0^+} f\left(\frac{1}{t}\right)$$

$$= \lim_{t \to 0^+} \frac{\left(\frac{1}{t}\right)^2}{\left(\frac{1}{t}\right) + 1}$$

$$= \lim_{t \to 0^+} \frac{1}{t + t^2}:$$

Note that $\dfrac{1}{t+t^2} > \dfrac{1}{2t}$ whenever $0 < t < 1$. We know that $\lim\limits_{t \to 0^+} \dfrac{1}{2t} = +\infty.$

It is easy to see that Definition 4.13ii is satisfied and $\lim\limits_{t \to 0^+} \dfrac{1}{t+t^2} = +\infty.$ Therefore, $\lim\limits_{x \to +\infty} f(x) = +\infty. \quad \blacksquare$

EXERCISES

Compute the following limits if they exist:

1. $\lim\limits_{x \to +\infty} \dfrac{x}{x+1}.$

2. $\lim\limits_{x \to -\infty} \dfrac{x^2}{2x^2 - 1}.$

3. $\lim\limits_{x \to +\infty} x \sin x.$

4. $\lim\limits_{x \to +\infty} \dfrac{\sin x}{x}.$

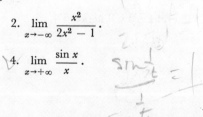

5. $\lim\limits_{x \to +\infty} \dfrac{x}{x^2 - 1}$.

6. $\lim\limits_{x \to -\infty} \dfrac{x^3}{1 - 3x^3}$.

7. $\lim\limits_{x \to +\infty} \dfrac{3x + x^2 - 1}{7x^2 - 2}$.

8. $\lim\limits_{x \to -\infty} \dfrac{x^3}{1 + x^2}$.

9. $\lim\limits_{x \to +\infty} 3x^2 - x$.

10. $\lim\limits_{x \to +\infty} \dfrac{x^4 + 1}{1 - 3x^2}$.

11. $\lim\limits_{x \to +\infty} (x \sin x)^2$.

12. $\lim\limits_{x \to -\infty} (x - 1)^2$.

13. Using Definitions 4.12ii and 4.14ii as a guide, define $\lim\limits_{x \to -\infty} f(x) = +\infty$.

14. Prove that if $\lim\limits_{x \to -\infty} f(x) = +\infty$ and $\lim\limits_{x \to -\infty} g(x) = +\infty$, then $\lim\limits_{x \to -\infty}$
$[f(x) + g(x)] = +\infty$.

15. Give an example of two functions, f and g, such that $\lim\limits_{x \to -\infty} f(x) = +\infty$,
$\lim\limits_{x \to -\infty} g(x) = +\infty$, and $\lim\limits_{x \to -\infty} \dfrac{f(x)}{g(x)} = +2$.

ANSWERS

2. $1/2$. **3.** Does not exist. **6.** $-1/3$. **8.** $-\infty$. **10.** $-\infty$. **12.** $+\infty$.

MISCELLANEOUS EXERCISES

Compute the following limits, if they exist:

1. $\lim\limits_{x \to 0} \dfrac{\tan^2 x}{x}$.

2. $\lim\limits_{x \to 2} \dfrac{-1}{x^2 - 4}$.

3. $\lim\limits_{x \to 3} \dfrac{2x^2}{(x - 3)^4}$.

4. $\lim\limits_{x \to 0} \dfrac{\tan x}{x^2}$.

5. $\lim\limits_{x \to 2^+} \dfrac{5}{x^2 - 4}$.

6. $\lim\limits_{x \to +\infty} \dfrac{x^2}{2x^2 - 1}$.

7. $\lim\limits_{x \to -\infty} \dfrac{1 - 3x^2}{x^2 - 2x}$.

8. $\lim\limits_{x \to 0} x \cos \left(\dfrac{1}{x^2}\right)$.

9. $\lim\limits_{x \to 2^-} \dfrac{-1}{2 - x}$.

10. $\lim\limits_{x \to -\infty} \dfrac{2x}{4x^2 + 1}$.

11. $\lim\limits_{x \to -\infty} \dfrac{-1}{x}$.

12. $\lim\limits_{x \to 0} \dfrac{\tan 2x}{x}$.

13. $\lim\limits_{x \to 0^-} \csc x$.

14. $\lim\limits_{x \to +\infty} (x - \sqrt{x^2 - 4})$.

15. $\lim\limits_{x \to +\infty} \dfrac{\sqrt{x} - \sqrt{x + 1}}{\sqrt{x} + \sqrt{x + 1}}$.

ANSWERS

1. 0. **3.** $+\infty$. **5.** $+\infty$. **7.** -3. **9.** $-\infty$. **11.** 0. **12.** 2. **14.** 0.

5

DERIVATIVES

INTRODUCTION

Let f be a function with both domain and range contained in $\mathfrak{R}$. In the preceding chapters we discussed in some detail what was meant by "f approaches a limit L as x approaches t." We decided that this means that $f(x)$ gets close to L whenever x gets close to t. Thus, we can think of f as being approximated in a neighborhood of t by the function g, defined by $g(x) = L$ for all x.

Figure 5.1 is a sketch of a function continuous in a neighborhood of t. Therefore $f(t) = \lim\limits_{x \to t} f(x) = L$. Let g be defined by $g(x) = L$ for all x. Look again at the sketch and see how much better an approximation of f we would have if the graph of g "slanted in the same direction as the graph of f." It is obvious, after our discussion of continuous functions in the preceding chapter, that every continuous function can be nicely approximated in a neighborhood of t by the constant function $g(x) = f(t)$. The question then arises as to whether this approximation can be improved. Perhaps we can find a function h such that (1) $h(t) = f(t)$, (2) the graph of h

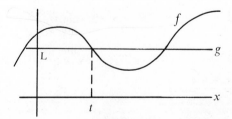

Figure 5.1

is a straight line K, and (3) the slope of K is the same as the "slope of the graph of f at t." Obviously we must say what we mean by the "slope of the graph of f at t" before we can discuss the problem in the usual mathematical fashion; heretofore we have discussed the slopes of lines but of no other graphs.

1. AFFINE APPROXIMATIONS

DEFINITION 5.1 Let $L:\mathcal{R} \to \mathcal{R}$ and $A:\mathcal{R} \to \mathcal{R}$.

i. L is a *linear transformation* if and only if there exists a constant m such that $L(x) = mx$ for all x in $\mathcal{R}$.

ii. A is an *affine transformation* if and only if there exist constants m and k such that $A(x) = mx + k$ for all x in $\mathcal{R}$.

We notice immediately that the graph of a linear transformation L is a line, passing through the origin, with slope m. The graph of an affine transformation is a line with slope m and y-intercept k. It is easily seen that if L is a linear transformation then

i. $L(x + y) = L(x) + L(y)$ for all x and y in $\mathcal{R}$,

and ii. $L(rx) = rL(x)$ for all r and x in $\mathcal{R}$.

Indeed, these two conditions are the usual definition for a general linear transformation.

If $f:\mathcal{R} \to \mathcal{R}$ and t is a real number in the interior of the domain of f, we wish to determine conditions under which there exists an affine transformation A such that

i. $A(t) = f(t)$

and ii. the slope of the secant line through $(t, f(t))$ and $(x, f(x))$ is close to the slope of the graph of A, whenever x is close to t.

Figure 5.2 should be used by the reader as a geometric guide in the following development:

We note first that the slope of the secant line through $(t, f(t))$ and $(x, f(x))$ is $\dfrac{f(x) - f(t)}{x - t}$. Moreover, the slope of the graph of A is

$$\frac{A(x) - A(t)}{x - t}.$$

If we require that $A(t) = f(t)$, we find that the difference of these two slopes is

$$\frac{f(x) - f(t)}{x - t} - \frac{A(x) - A(t)}{x - t} = \frac{f(x) - A(x)}{x - t}.$$

If we also require that these two slopes be close to one another when x is

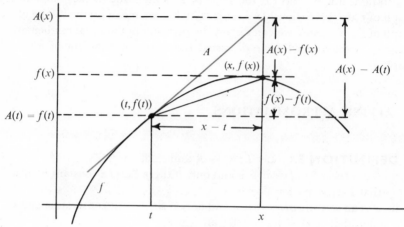

Figure 5.2

close to t, we see from the previous equation that $\dfrac{f(x) - A(x)}{x - t}$ must be small when x is close to t. Let us now consider the following theorem:

THEOREM 5.2 Let $f: \mathcal{R} \to \mathcal{R}$ and let t be in the interior of the domain of f. Then

i. there exists an affine transformation A such that

$$A(t) = f(t)$$

and

$$\lim_{x \to t} \frac{f(x) - A(x)}{x - t} = 0$$

if and only if

ii. $\displaystyle\lim_{x \to t} \frac{f(x) - f(t)}{x - t}$ exists and is finite.

Moreover, in this case $A(x) = m(x - t) + f(t)$, where

$$m = \lim_{x \to t} \frac{f(x) - f(t)}{x - t} = \frac{A(x) - A(t)}{x - t} \text{ for } x \neq t.$$

Proof. Let A be any affine transformation such that $A(t) = f(t)$. As we noted above, for $x \neq t$,

$$(5.1) \quad \frac{f(x) - f(t)}{x - t} - \frac{A(x) - A(t)}{x - t} = \frac{f(x) - A(x)}{x - t}.$$

Since A is an affine transformation, there exist constants m and k such that $A(x) = mx + k$. Since $A(t) = f(t)$, we have $f(t) = mt + k$ or $k = f(t) - mt$. Therefore,

$$(5.2) \quad A(x) = mx + (f(t) - mt) = m(x - t) + f(t).$$

Moreover, for $x \neq t$,

$$(5.3) \quad \frac{A(x) - A(t)}{x - t} = \frac{m(x - t) + f(t) - f(t)}{x - t} = m.$$

Combining (5.1) and (5.3), we have

$$\frac{f(x) - f(t)}{x - t} - m = \frac{f(x) - A(x)}{x - t}.$$

Therefore, $\lim\limits_{x \to t} \dfrac{f(x) - A(x)}{x - t} = 0$ if and only if $\lim\limits_{x \to t} \dfrac{f(x) - f(t)}{x - t}$ exists and is equal to m. This last observation, combined with (5.2) and (5.3) gives us the final conclusion of the theorem. ∎

We are now ready to make a formal definition of an affine approximation of a function about a point:

DEFINITION 5.3 Let $f: \mathcal{R} \to \mathcal{R}$. Let t be in the interior of the domain of f. Then A is the *affine approximation of f about t* if and only if A is an affine transformation such that

    i. $A(t) = f(t)$

and

    ii. $\lim\limits_{x \to t} \dfrac{f(x) - A(x)}{x - t} = 0.$

A second glance at Theorem 5.2 is quite rewarding at this point. This theorem not only tells us how to determine whether for a given function f and point t an affine approximation exists, but it also tells us how to find the affine approximation when there is one. This will become even more apparent to the reader after he has considered the following examples:

Example 1. Let $f(x) = 3x + 4$ and $t = 2$. Then for $x \neq 2$,

$$\frac{f(x) - f(2)}{x - 2} = \frac{3x + 4 - 10}{x - 2} = \frac{3(x - 2)}{x - 2} = 3.$$

Therefore, $\lim\limits_{x \to 2} \dfrac{f(x) - f(2)}{x - 2} = 3$ and

$$A(x) = 3(x - 2) + f(2) = 3x + 4.$$

Since f is itself affine, it is not at all surprising that it is its own affine approximation. ∎

Example 2. Let $f(x) = x^2$. Let us find the affine approximation of f at $t = 2$. For $x \neq 2$, $\dfrac{f(x) - f(2)}{x - 2} = \dfrac{x^2 - 4}{x - 2} = \dfrac{(x - 2)(x + 2)}{x - 2} = x + 2.$

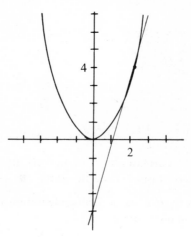

Figure 5.3

Therefore, $\lim\limits_{x \to 2} \dfrac{f(x) - f(2)}{x - 2} = \lim\limits_{x \to 2} (x + 2) = 4$, and

$$A(x) = 4(x - 2) + 4 = 4x - 4.$$

Figure 5.3 is a sketch of the graphs of f and A. ∎

Example 3. Let $f(x) = x^3 + 3$ and $t = 1$.

For $x \neq 1$, $\dfrac{f(x) - f(1)}{x - 1} = \dfrac{x^3 - 1}{x - 1} = \dfrac{(x - 1)(x^2 + x + 1)}{x - 1} = x^2 + x + 1.$

Thus, $\lim\limits_{x \to 1} \dfrac{f(x) - f(1)}{x - 1} = \lim\limits_{x \to 1} (x^2 + x + 1) = 3$. Therefore, $A(x) = 3(x - 1)$
$+ f(1) = 3x + 1$. ∎

EXERCISES

In Exercises 1 to 8

(a) Sketch the graphs of the given functions and affine approximations.
Use a single diagram for each problem and take particular note of the
point $(t, f(t))$.

(b) In each case show that $\lim\limits_{x \to t} \dfrac{f(x) - A(x)}{x - t} = 0$:

1. $f(x) = 1 + x^2$
   $A(x) = 4x - 3$
   $t = 2$.

2. $f(x) = x^2 + 2x$
   $A(x) = -2x - 4$
   $t = -2$.

3. $f(x) = 1 - 2x^2$
   $A(x) = 1$
   $t = 0$.

4. $f(x) = x^3 + 1$
   $A(x) = 3x - 1$
   $t = 1$.

5. $f(x) = x^3 + x^2$
   $A(x) = x + 1$
   $t = -1.$

6. $f(x) = \dfrac{1}{x}, \; x \neq 0$

   $A(x) = \dfrac{-(x-6)}{9}$

   $t = 3.$

7. $f(x) = \sqrt{x}, \; x \geq 0$
   $A(x) = \frac{1}{2}(x + 1)$
   $t = 1.$

8. $f(x) = \dfrac{1}{x^2 + 1}$

   $A(x) = \dfrac{x + 2}{2}$

   $t = -1.$

In Exercises 9 to 18 find the affine approximation of f about t for the following choices of f and t:

9. $f(x) = 2x + 3$
   $t = 3.$

10. $f(x) = x^2 + 2$
    $t = 2.$

11. $f(x) = x - x^2$
    $t = -1.$

12. $f(x) = \dfrac{1}{x^2}$

    $t = -2.$

13. $f(x) = k, \; k$ a constant
    $t = 24.$

14. $f(x) = 2x^3 + x^2$
    $t = 1.$

15. $f(x) = 8 - x^3$
    $t = 1.$

16. $f(x) = \dfrac{2}{x}$

    $t = -1.$

17. $f(x) = \dfrac{1}{x^2}, \; x \neq 0$

    $t = 2.$

18. $f(x) = \sqrt{x} + 1$
    $t = 4.$

ANSWERS

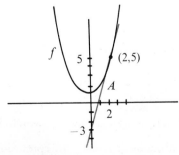

1a.

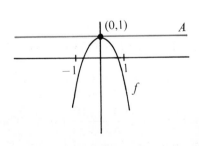

3a.

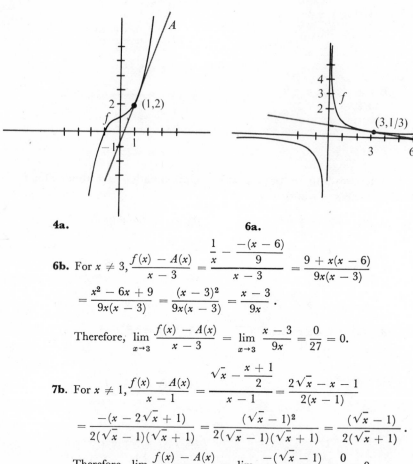

4a.   **6a.**

6b. For $x \neq 3, \dfrac{f(x) - A(x)}{x - 3} = \dfrac{\dfrac{1}{x} - \dfrac{-(x - 6)}{9}}{x - 3} = \dfrac{9 + x(x - 6)}{9x(x - 3)}$

$$= \dfrac{x^2 - 6x + 9}{9x(x - 3)} = \dfrac{(x - 3)^2}{9x(x - 3)} = \dfrac{x - 3}{9x} \,.$$

Therefore, $\displaystyle\lim_{x \to 3} \dfrac{f(x) - A(x)}{x - 3} = \lim_{x \to 3} \dfrac{x - 3}{9x} = \dfrac{0}{27} = 0.$

7b. For $x \neq 1, \dfrac{f(x) - A(x)}{x - 1} = \dfrac{\sqrt{x} - \dfrac{x + 1}{2}}{x - 1} = \dfrac{2\sqrt{x} - x - 1}{2(x - 1)}$

$$= \dfrac{-(x - 2\sqrt{x} + 1)}{2(\sqrt{x} - 1)(\sqrt{x} + 1)} = \dfrac{(\sqrt{x} - 1)^2}{2(\sqrt{x} - 1)(\sqrt{x} + 1)} = \dfrac{(\sqrt{x} - 1)}{2(\sqrt{x} + 1)} \,.$$

Therefore, $\displaystyle\lim_{x \to 1} \dfrac{f(x) - A(x)}{x - 1} = \lim_{x \to 1} \dfrac{-(\sqrt{x} - 1)}{2(\sqrt{x} + 1)} = \dfrac{0}{4} = 0.$

10. $A(x) = 4x - 2.$ **12.** $A(x) = \dfrac{x + 3}{4} \,.$ **14.** $A(x) = 8x - 5.$

16. $A(x) = -2x - 4.$ **18.** $A(x) = \dfrac{x + 8}{4} \,.$

2. THE DERIVATIVE AND THE DIFFERENTIAL

Let $f:\mathcal{R} \to \mathcal{R}$ and let t be in the interior of the domain of f. Theorem 5.2 and Definition 5.3 tell us that f has an affine approximation at t if and only if $\displaystyle\lim_{x \to t} \dfrac{f(x) - f(t)}{x - t}$ exists. This limit has a special name, which is given in the following definition. Keep in mind the geometric interpretation of this limit as the slope of the affine approximation of f at t:

DEFINITION 5.4 (Derivative.) Let $f: \mathcal{R} \to \mathcal{R}$.

i. If f has an affine approximation about t, we say that f is *differentiable at* t and we define the *derivative of* f *at* t, denoted $f'(t)$, by

$$f'(t) = \lim_{x \to t} \frac{f(x) - f(t)}{x - t}.$$

ii. The number $f'(t)$ is also called the *slope* of the graph of f at $(t, f(t))$.

If f has an affine approximation about t, we shall in what follows denote this function by $A_t f$. We noted in Theorem 5.2 that $A_t f(x) = m(x - t) + f(t)$, where $m = \lim\limits_{x \to t} \dfrac{f(x) - f(t)}{x - t}$. Therefore, by Definition 5.4,

(5.4)
$$A_t f(x) = f'(t)(x - t) + f(t).$$

We know that the graph of $A_t f$ is a line, and thus, from Equation 5.4, $f'(t)$ is the slope of this line. A re-examination of Figure 5.2 should justify in the reader's mind our calling $f'(t)$ the slope of the graph of f at $(t, f(t))$.

DEFINITION 5.5 Let $f: \mathcal{R} \to \mathcal{R}$. If f is differentiable at t, then $d_t f$ is the function defined by

$$d_t f(x) = f'(t)x \text{ for all real } x.$$

$d_t f$ is called the *differential of* f *at* t.

We immediately see that $d_t f$ is a linear transformation (Definition 5.1). Moreover,

$$\begin{aligned}
A_t f(x) &= f'(t)(x - t) + f(t) \\
&= f'(t)x - f'(t)t + f(t) \\
&= d_t f(x) + [f(t) - d_t f(t)].
\end{aligned}$$

Noting that $[f(t) - d_t f(t)]$ is a constant, we see that the graphs of $A_t f$ and $d_t f$ are parallel lines.

Example 4. Let $f(x) = x^2 + 1$. We shall prove that f is differentiable and find the affine approximation and differential of f at an arbitrary point t. Note that for $x \neq t$,

$$\frac{f(x) - f(t)}{x - t} = \frac{x^2 - t^2}{x - t} = \frac{(x - t)(x + t)}{x - t} = x + t.$$

Therefore, $\lim\limits_{x \to t} \dfrac{f(x) - f(t)}{x - t} = \lim\limits_{x \to t} (x + t) = 2t$. Then, by Definition 5.4,

$f'(t) = 2t$ for all real t. Moreover,

$$A_t f(x) = 2t(x - t) + (t^2 + 1)$$

and $$d_t f(x) = 2tx.$$

Figure 5.4 is a sketch of f, $A_t f$, and $d_t f$ for $t = 2$.

$$A_2 f(x) = 4(x - 2) + 5 = 4x - 3$$
$$d_2 f(x) = 4x. \quad \blacksquare$$

Example 5. Let $f(x) = x^{1/3}$ for all real x. Notice that f is continuous at each point in $\mathcal{R}$. We shall show that f is *differentiable everywhere except at* $t = 0$. Thus, we have an example of a function *continuous* at a point, but *not differentiable* there. For $x \neq t$,

$$\frac{f(x) - f(t)}{x - t} = \frac{x^{1/3} - t^{1/3}}{x - t} = \frac{x^{1/3} - t^{1/3}}{(x^{1/3} - t^{1/3})(x^{2/3} + x^{1/3}t^{1/3} + t^{2/3})}$$

$$= \frac{1}{x^{2/3} + x^{1/3}t^{1/3} + t^{2/3}}.$$

For $t = 0$, $\dfrac{f(x) - f(t)}{x - t} = \dfrac{1}{x^{2/3}}$, which does not approach a finite limit as x approaches 0.

But, for $t \neq 0$, $\lim\limits_{x \to t} \dfrac{f(x) - f(t)}{x - t} = \dfrac{1}{3t^{2/3}}. \quad \blacksquare$

The preceding example shows us that differentiability does not necessarily follow from continuity. One should immediately ask if perhaps a logical

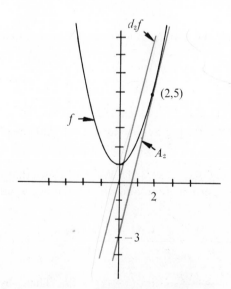

Figure 5.4

implication results when one reverses the conditions, and indeed it does, as we see from the following:

THEOREM 5.6 Let $f: \mathcal{R} \to \mathcal{R}$. If f is differentiable at t, then f is continuous at t.

Proof. Note that for $x \neq t$

$$f(x) - f(t) = \frac{[f(x) - f(t)](x - t)}{x - t}.$$

Since f is differentiable at t, $\lim\limits_{x \to t} \dfrac{f(x) - f(t)}{x - t}$ exists, and in any case

$\lim\limits_{x \to t} (x - t) = 0$. Therefore,

$$\lim_{x \to t} [f(x) - f(t)] = \lim_{x \to t} \frac{f(x) - f(t)}{x - t} \cdot \lim_{x \to t} (x - t) \text{ by Theorem 3.12 (p. 182),}$$

$$= \left[\lim_{x \to t} \frac{f(x) - f(t)}{x - t} \right] \cdot 0$$

$$= 0.$$

Consequently, $f(t) = \lim\limits_{x \to t} f(x)$ and f is continuous at t by Definition 4.1 (p. 194). ∎

EXERCISES

In Exercises 1 to 12
(a) Use Definition 5.4 to find $f'(t)$ where t is an arbitrary point of the domain of f.
(b) Find the affine approximation $A_t f$ and the differential $d_t f$, where $t = 1$. In each case sketch the graph of f, $A_t f$, and $d_t f$:

1. $f(x) = 2x$.

2. $f(x) = 3 - 5x$.

3. $f(x) = x^2$.

4. $f(x) = x^2 - 2x$,

5. $f(x) = 4$.

6. $f(x) = \dfrac{1}{x}$, $x \neq 0$.

7. $f(x) = \dfrac{1}{x^2}$, $x \neq 0$.

8. $f(x) = 3x^2 + x - 6$.

9. $f(x) = \dfrac{x}{x + 1}$.

10. $f(x) = \sqrt{x}$, $x > 0$.

11. $f(x) = x^{3/2}$, $x > 0$.   12. $f(x) = \dfrac{x^2 + 1}{x - 2}$, $x \neq 2$.

13. Let $f(x) = |x|$. Show that f is not differentiable at $t = 0$.

14. Let $f(x) = \begin{cases} x \sin \dfrac{1}{x} & \text{for } x \neq 0 \\ 0 & \text{for } x = 0. \end{cases}$ Show that f is not differentiable at $t = 0$.

15. Let $f(x) = \begin{cases} x^2 \sin \dfrac{1}{x} & \text{for } x \neq 0 \\ 0 & \text{for } x = 0. \end{cases}$ Show that f is differentiable at $t = 0$

and that $f'(0) = 0$.

ANSWERS

1a. $f'(t) = 2$.   **1b.** $A_1 f(x) = 2x$ **3a.** $f'(t) = 2t$.   **3b.** $A_1 f(x) = 2x - 1$
$\qquad\qquad\qquad\quad d_1 f(x) = 2x$. $\qquad\qquad\qquad\qquad\qquad d_1 f(x) = 2x$.

5a. $f'(t) = 0$.   **5b.** $A_1 f(x) = 4$ **7a.** $f'(t) = \dfrac{-2}{t^3}$.
$\qquad\qquad\qquad\quad d_1 f(x) = 0$.

7b. $A_1 f(x) = -2x + 3$ **9a.** $f'(t) = \dfrac{1}{(t + 1)^2}$.   **9b.** $A_1 f(x) = \frac{1}{4}(x + 1)$
$\quad\; d_1 f(x) = -2x$. $\qquad\qquad\qquad\qquad\qquad\qquad\qquad d_1 f(x) = \frac{1}{4}x$.

10a. $f'(t) = \dfrac{1}{2\sqrt{t}}$.   **11b.** $A_1 f(x) = \frac{3}{2}x - \frac{1}{2}$
$\qquad\qquad\qquad\qquad\qquad\quad d_1 f(x) = \frac{3}{2}x$.

3. THE SUM AND PRODUCT RULES FOR DIFFERENTIATION

As we saw in the preceding section, it is a relatively lengthy process to find the derivatives of even simple functions when we restrict ourselves to the definition. In the next three sections we shall develop some general rules for *differentiation* (the process of finding derivatives). One of our main objectives in this section will be that of learning to differentiate polynomials.

LEMMA 5.7

i. Let $f : \mathcal{R} \to \mathcal{R}$ be defined by $f(x) = c$, where c is a constant. Then $f' = 0$.

ii. Let $g : \mathcal{R} \to \mathcal{R}$ be defined by $g(x) = x$. Then $g' = 1$.

Proof

i. For $x \neq t$, $\dfrac{f(x) - f(t)}{x - t} = \dfrac{c - c}{x - t} = 0$.

Therefore, $f'(t) = \lim\limits_{x \to t} \dfrac{f(x) - f(t)}{x - t} = 0$ for all t.

ii. For $x \neq t$, $\dfrac{g(x) - g(t)}{x - t} = \dfrac{x - t}{x - t} = 1$.

Therefore, $g'(t) = \lim\limits_{x \to t} \dfrac{g(x) - g(t)}{x - t} = 1$ for all t. ∎

THEOREM 5.8 Let $u: \Re \to \Re$, $v: \Re \to \Re$, and let c be a constant. Let S be the set of all t, such that u and v are differentiable at t. Then $u + v$, uv, and cv are differentiable on S, and on S we have the following formulae:

  i. (Sum Rule.) $(u + v)' = u' + v'$.
  ii. (Product Rule.) $(uv)' = u'v + uv'$.
  iii. (Scalar Rule.) $(cu)' = cu'$.

$$\left(\frac{u}{v}\right) = \frac{u'v - uv'}{v^2}$$

Proof. For each t in S, u and v are differentiable by hypothesis, so that

$$u'(t) = \lim_{x \to t} \frac{u(x) - u(t)}{x - t} \quad \text{and} \quad v'(t) = \lim_{x \to t} \frac{v(x) - v(t)}{x - t}.$$

i. For $x \neq t$, $\dfrac{(u + v)(x) - (u + v)(t)}{x - t} = \dfrac{u(x) + v(x) - [u(t) + v(t)]}{x - t}$

$$= \frac{u(x) - u(t)}{x - t} + \frac{v(x) - v(t)}{x - t}.$$

Therefore, by Theorem 3.12 (p. 182), $\lim\limits_{x \to t} \dfrac{(u + v)(x) - (u + v)(t)}{x - t}$

exists and is equal to $\lim\limits_{x \to t} \dfrac{u(x) - u(t)}{x - t} + \lim\limits_{x \to t} \dfrac{v(x) - v(t)}{x - t}$. By definition then, $(u + v)'(t) = u'(t) + v'(t)$ for each t in S, so that $(u + v)' = u' + v'$ on S.

ii. Again let $x \neq t$. In the series of equations that follow we shall merely be performing some algebraic manipulations. The key step is the second equation in which we add and subtract $u(t)v(x)$.

$$\frac{(uv)(x) - (uv)(t)}{x - t} = \frac{u(x)v(x) - u(t)v(t)}{x - t}$$

$$= \frac{u(x)v(x) - u(t)v(x) + u(t)v(x) - u(t)v(t)}{x - t}$$

$$= \left(\frac{u(x) - u(t)}{x - t}\right)v(x) + u(t)\left(\frac{v(x) - v(t)}{x - t}\right).$$

By Theorem 5.6, we know that $\lim_{x \to t} v(x) = v(t)$. Note also that $u(t)$ is a constant. Therefore, using the familiar sum and product theorems on limits, we have

$$\lim_{x \to t} \frac{(uv)(x) - (uv)(t)}{x - t} = \lim_{x \to t} \frac{u(x) - u(t)}{x - t} \lim_{x \to t} v(x) + u(t) \lim_{x \to t} \frac{v(x) - v(t)}{x - t}$$

$$= u'(t)v(t) + u(t)v'(t).$$

Thus, on S we have the desired formula: $(uv)' = u'v + uv'$.

iii. This, in fact, is a simple corollary of (ii) and Lemma 5.7. Let v be defined by $v(x) = c$. Then $v'(t) = 0$ for all t and

$$(cu)' = (vu)' = v'u + vu' = 0 \cdot u + cu' = cu'. \quad \blacksquare$$

Before proceeding to find the formula for differentiating $f(x) = x^n$, where n is a positive integer, let us collect some of the facts we have and consider some examples. In Lemma 5.7 we showed that $f'(x) = 0$ if $f(x) = 1$ and $f'(x) = 1$ if $f(x) = x$. In Exercise 3 of the preceding section you showed that $f'(x) = 2x$ if $f(x) = x^2$.

Example 6. Let $f(x) = x^3$. Let $u(x) = x^2$ and $v(x) = x$, then $f = uv$. Therefore, $f' = u'v + uv'$, or $f'(x) = (2x)(x) + (x^2)(1) = 3x^2$. $\quad \blacksquare$

Example 7. Let $f(x) = x^4$. Let $u(x) = x^3$ and $v(x) = x$, then $f = uv$. Example 6 tells us that $u'(x) = 3x^2$. Therefore, since $f' = u'v + uv'$ and $v'(x) = 1$,

$$f'(x) = (3x^2)(x) + (x^3)(1) = 4x^3. \quad \blacksquare$$

Not only do the two preceding examples give us a hint as to the general formula for differentiating $f(x) = x^n$, but they also indicate a simple method of proof.

THEOREM 5.9 Let $f: \mathfrak{R} \to \mathfrak{R}$ be defined by $f(x) = x^n$, where n is a positive integer. Then $f'(x) = nx^{n-1}$ for all real x.

*Proof.** We shall proceed by induction. The case when $n = 1$ was proved in Lemma 5.7. Assume now that $u'(x) = kx^{k-1}$ where $u(x) = x^k$ and k is a positive integer. Let $f(x) = x^{k+1}$ and $v(x) = x$. Then $f = uv$ and $f' = u'v + uv'$. Consequently, $f'(x) = (kx^{k-1})(x) + (x^k)(1) = (k + 1)x^k$ and the induction is complete. $\quad \blacksquare$

Example 8. Let $f(x) = x^5 + 4x^3 - 2x$.
Then $f'(x) = (x^5)' + (4x^3)' - (2x)'$ by Theorem 5.8i (Sum Rule).

$$= (x^5)' + 4(x^3)' - 2(x)' \quad \text{by Theorem 5.8iii (Scalar Rule).}$$

$$= 5x^4 + 12x^2 - 2. \quad \text{by Theorem 5.9.} \quad \blacksquare$$

---

* See Appendix I for discussion of proof by induction.

Example 9. Let $f(x) = (x^2 + 2x - 3)(x^3 - 4x^2 + 1)$. To differ-entiate this function we could, of course, multiply it out and differentiate term by term. However, notice that this is a product of two polynomials. Therefore, using the product rule of Theorem 5.8ii we have

$$f'(x) = (2x + 2)(x^3 - 4x^2 + 1) + (x^2 + 2x - 3)(3x^2 - 8x).$$ ∎

EXERCISES

In Exercises 1 to 10 use Theorems 5.8 and 5.9 to differentiate the following:

1. $f(x) = 3x^2$.

2. $f(x) = x^2 - 2x + 1$.

3. $f(x) = 4x^3$.

4. $f(x) = x - 2x^3$.

5. $f(x) = 2x^5 - 4x + 1$.

6. $f(x) = x^7 + 2$.

7. $f(x) = \frac{1}{3}x^6 + \frac{1}{2}x^2 - 2x$.

8. $f(x) = (x^2 + 1)^2$.

9. $f(x) = x^{12} - 2x^{10}$.

10. $f(x) = -x - 1$.

In Exercises 11 to 15 differentiate f without multiplying first. After differentiating, simplify the answer algebraically:

11. $f(x) = (x^2 + 1)(x^2 - 1)$.

12. $f(x) = x^3(2x - 1)$.

13. $f(x) = (x^3 - 2x)(x^2 + 2)$.

14. $f(x) = x\sqrt{x}$ (use Exercise 10 of Section 2).

15. $f(x) = (x + 1)(x - 2)(x + 3)$.

16. Let $f = uvw$, where u, v, and w are differentiable. Using Theorem 5.8ii, find a formula for f' involving u, v, w, u', v', and w'.

17. Prove Theorem 5.8iii directly from Definition 5.4.

ANSWERS

1. $f'(x) = 6x$. **3.** $f'(x) = 12x^2$. **5.** $f'(x) = 10x^4 - 4$.
7. $f'(x) = 2x^5 + x - 2$. **9.** $f'(x) = 12x^{11} - 20x^9$. **11.** $f'(x) = 4x^3$.
13. $f'(x) = 5x^4 - 4$. **15.** $f'(x) = 3x^2 + 4x - 5$.

4. THE CHAIN RULE

We should like now to extend our knowledge of differentiation to a larger class of functions. Theorems 5.8 and 5.9 enable us to differentiate polynomials of any degree. However, consider the following function: $f(x) = (x^2 + 3x + 1)^{17}$. This function is indeed a polynomial of degree 34. With only our present theorems we would be forced to multiply it out in order to differentiate it, and that would be a formidable task. Note though that $f = g \circ h$, where $h(x) = x^2 + 3x + 1$ and $g(y) = y^{17}$. Both h and g

are polynomials which we can differentiate. The "chain rule" will give us a method for differentiating composite functions, which in this case would relieve us of an onerous algebraic manipulation.

THEOREM 5.10 (The Chain Rule.) Let $u: \mathcal{R} \to \mathcal{R}$ and $v: \mathcal{R} \to \mathcal{R}$. Let S be the set of points s such that v is differentiable at s and u is differentiable at $v(s)$. Then $u \circ v$ is differentiable on S and

$$(u \circ v)' = (u' \circ v)(v').$$

Before proving the theorem, let us consider some examples:

Example 10. Let $f(x) = (x^2 + 3x + 1)^{17}$. Then $f(x) = (u \circ v)(x)$ where $v(x) = x^2 + 3x + 1$ and $u(y) = y^{17}$. By Theorem 5.10, $f'(x) = (u' \circ v)(x)v'(x)$. Now $v'(x) = 2x + 3$ and $u'(y) = 17y^{16}$. Therefore, $(u' \circ v)(x) = 17(x^2 + 3x + 1)^{16}$. Finally, $f'(x) = 17(x^2 + 3x + 1)^{16}(2x + 3)$. ■

Example 11. Let $f(x) = \sqrt{3x^2 + 1}$. Let $v(x) = 3x^2 + 1$ and $u(y) = \sqrt{y}$ for $y > 0$. Then $f = u \circ v$ and

$$f'(x) = (u' \circ v)(x)v'(x)$$

$$= \frac{1}{2\sqrt{v(x)}}(6x) = \frac{3x}{\sqrt{3x^2 + 1}}. \quad ■$$

Proof of Theorem 5.10. We shall give the proof of the chain rule for the special case in which $v(x) \neq v(t)$ for all x in some deleted open ball $\mathcal{B}(t; \rho)$ about t. Only such values of x need be considered in evaluating the limits involved in calculating $(u \circ v)'(t)$. Thus,

$$(5.5) \quad (u \circ v)'(t) =^1 \lim_{x \to t} \frac{(u \circ v)(x) - (u \circ v)(t)}{x - t}$$

$$=^2 \lim_{x \to t} \frac{u(v(x)) - u(v(t))}{x - t}$$

$$=^3 \lim_{x \to t} \frac{u(v(x)) - u(v(t))}{v(x) - v(t)} \cdot \frac{v(x) - v(t)}{x - t}$$

$$=^4 \lim_{x \to t} \frac{u(v(x)) - u(v(t))}{v(x) - v(t)} \cdot \lim_{x \to t} \frac{v(x) - v(t)}{x - t}$$

$$=^5 \left(\lim_{x \to t} \frac{u(y) - u(s)}{y - s} \right) v'(t), \text{ where } y = v(x) \text{ and } s = v(t).$$

(1) Definition 5.4 applied to $(u \circ v)$ to obtain $(u \circ v)'$.
(2) Definition of $(u \circ v)(t) = u(v(t))$.
(3) Multiplying and dividing by $v(x) - v(t)$ which is nonzero for x in $\mathcal{B}(t; \rho)$.
(4) Product theorem for limits, Theorem 3.12iii (p. 182).
(5) Setting $y = v(x)$, $s = v(t)$, and applying Definition 5.4 to obtain $v'(t)$.

To complete the proof, we must establish that the last limit above is equal to $(u' \circ v)(t)$. To accomplish this we introduce the function $U: \mathfrak{R} \to \mathfrak{R}$ given by

$$(5.6) \qquad \begin{cases} \text{(a)} \ \ U(y) = \dfrac{u(y) - u(s)}{y - s} \ \text{for } y \neq s \\[2mm] \text{(b)} \ \ U(s) = u'(s) = u'(v(t)). \end{cases}$$

According to the definition of $u'(s)$ and the hypothesis that u is differentiable at $s = v(t)$, we then have

$$\lim_{y \to s} U(y) = \lim_{y \to s} \frac{u(y) - u(s)}{y - s} = u'(s) = U(s).$$

Hence, U is continuous at $s = v(t)$. Moreover, v is continuous at t since v is differentiable at t. Consequently, we may use Theorem 4.5 (which assures us that the composition of continuous functions is continuous) to conclude that

$$(5.7) \qquad \begin{aligned} \lim_{x \to t} \frac{u(y) - u(s)}{y - s} &=^1 \lim_{x \to t} U(y) \\ &=^2 \lim_{x \to t} U(v(x)) \\ &=^3 U(v(t)) \\ &=^4 u'(v(t)) \\ &=^5 (u' \circ v)(t). \end{aligned}$$

(1) From Equation 5.6a.
(2) Setting $y = v(x)$.
(3) Applying Theorem 4.5 (p. 197).
(4) From Equation 5.6b.
(5) Definition of $u' \circ v$.

The theorem follows from Equations 5.5 and 5.7. Again we emphasize that the above proof is valid only with the restriction that $v(x) \neq v(t)$ for all x in some deleted open ball $\mathcal{B}(t; \rho)$ about t. The theorem, however, is true without restriction. The proof for the more general situation is given for a still larger class of functions in Chapter 16. ∎

EXERCISES

In Exercises 1 to 10 use the Chain Rule, and other theorems when necessary, to find $f'(x)$.

1. $f(x) = (x + 1)^2$.

2. $f(x) = (x^3 + 1)^2$.

3. $f(x) = 2\sqrt{2x^3 + 1}$.

4. $f(x) = \sqrt{x^2 + \sqrt{x}}$.

5. $f(x) = (\sqrt{2x^2 + 1} + 1)^2$.

6. $f(x) = (4x^3 + 3x^2 - 5x)^5$.

7. $f(x) = (x - 1)^{12}(x + 2)^{17}$.   8. $f(x) = (x^4 + 1)^{16}$.

9. $f(x) = [3x^2 + (x - 5)^5]^2$.   10. $f(x) = [(4x + 1)(x^5 + 2)^7]^2$.

In Exercises 11 to 20 find the affine approximation A_t of f about t:

11. $f(x) = 2\sqrt{2x^3 + 1}$, $t = 0$.   12. $f(x) = (x + 1)^2$, $t = 1$.

13. $f(x) = (x^2 - 1)^2$, $t = 2$.   14. $f(x) = x^2 + 3x - 1$, $t = 0$.

15. $f(x) = x^n$, n a non-negative integer, $t = a$.

16. $f(x) = (x^3 - 3x)^6$, $t = -1$.   17. $f(x) = x^5(1 - x^2)^3$, $t = 2$.

18. $f(x) = 5$.   19. $f(x) = (x^3 + 1)^2$, $t = -1$.

20. $f(x) = \sqrt{1 + \sqrt{x}}$, $t = 4$.

ANSWERS

3. $\dfrac{6x^2}{\sqrt{2x^3 + 1}}$.   4. $\dfrac{4x\sqrt{x} + 1}{4\sqrt{x}\sqrt{x^2 + \sqrt{x}}}$.   6. $5(4x^3 + 3x^2 - 5x)^4 \cdot$

$(12x^2 + 6x - 5)$.   7. $(x - 1)^{11}(x + 2)^{16}(29x + 7)$.

10. $2(4x + 1)(x^5 + 2)^{13}(144x^5 + 35x^4 + 8)$.   11. $A_t f(x) = 2$.

13. $A_t f(x) = 24x - 39$.   15. $A_t f(x) = na^{n-1}x + (1 - n)a^n$.

19. $A_t f(x) = 0$.

5. DIFFERENTIATION OF RATIONAL POWERS AND QUOTIENTS

In previous sections we have learned how to differentiate x^n, where n is a non-negative integer. Recall that if $f(x) = x^n$, then $f'(x) = nx^{n-1}$ for $n > 0$. In this section we consider the extension of this rule to all real powers of x.

THEOREM 5.11   Let $f(x) = x^{-1} = 1/x$ for $x \neq 0$.
Then $f'(x) = -1/x^2$ for $x \neq 0$.

Proof.   Let $t \neq 0$ and $x \neq t, 0$.   Then

$$\frac{f(x) - f(t)}{x - t} = \frac{(1/x) - (1/t)}{x - t} = \frac{\left(\dfrac{t - x}{xt}\right)}{x - t} = \frac{-1}{xt}.$$

Therefore, $f'(t) = \lim\limits_{x \to t} \dfrac{-1}{xt} = \dfrac{-1}{t^2}$.   ∎

Notice that Theorem 5.11 tells us that if $f(x) = x^n$, $n = -1$, then $f'(x) = nx^{n-1}$. The following lemma is a special case of the "quotient rule," which we shall see later in this section. It will allow us very simply to differentiate functions with negative integral powers.

LEMMA 5.12 Let $v: \mathcal{R} \to \mathcal{R}$. Let S be the set of all t such that v is differentiable at t and $v(t) \neq 0$. Then $1/v$ is differentiable on S and

$$(1/v)' = -v'/v^2.$$

Proof. Let $u(y) = y^{-1}$ for $y \neq 0$. Let $f(x) = 1/v(x)$ for $v(x) \neq 0$. Then $f(x) = u(v(x)) = (u \circ v)(x)$. By hypothesis and Theorem 5.10 (the Chain Rule), f is differentiable, and for x in S, $f'(x) = (u' \circ v)(x)v'(x)$. By Theorem 5.11, $u'(y) = -1/y^2$. Therefore, $u' \circ v(x) = -1/[v(x)]^2$. Finally,

$$f'(x) = \frac{v'(x)}{(v(x))^2}. \quad \blacksquare$$

THEOREM 5.13
Let $f(x) = x^n$ where n is a negative integer and $x \neq 0$.
Then for $x \neq 0, f'(x) = nx^{n-1}$.

Proof. Let $m = -n$. Then m is a positive integer and $f(x) = 1/x^m$. Let $v(x) = x^m$. By Lemma 5.12,

$$f'(x) = \frac{-v'(x)}{v(x)^2} = \frac{-mx^{m-1}}{x^{2m}} = -mx^{-m-1} = nx^{n-1}. \quad \blacksquare$$

The following theorem will be proved in its full generality in Chapter 10. We state it here without proof.

THEOREM 5.14 Let $f(x) = x^r$, where r is a real number.
Then $f'(x) = rx^{r-1}$ for all x in the domain of f.

We have already verified the cases when n is a nonzero integer, $n = 1/2$, and $n = 3/2$ (these last two occurring as exercises in previous sections). If $n = 0, f(x) = 1$ and $f'(x) = 0$, which we will interpret as $f'(x) = 0(x^{-1})$. At present we are primarily concerned with the application of the theorem to rational powers.

Example 12. Let $f(x) = x^{4/3}$.
Then by Theorem 5.14, $f'(x) = (4/3)x^{(4/3-1)} = (4/3)x^{1/3}$. $\blacksquare$

Example 13. Let $f(x) = (x^2 + 1)^{-2}$.

Then by Theorem 5.14 and the Chain Rule (Theorem 5.10),

$$f'(x) = (-2)(x^2 + 1)^{-3}(2x) = \frac{-4x}{(x^2 + 1)^3} \cdot \quad \blacksquare$$

The last theorem in this section follows easily as a corollary of Theorem 5.8 (Product Rule) and Lemma 5.12. For obvious reasons it is called the "quotient rule," and though it is a simple corollary of our previous work, it is an important and useful tool.

THEOREM 5.15 (Quotient Rule.) Let $u: \mathcal{R} \to \mathcal{R}$ and $v: \mathcal{R} \to \mathcal{R}$. Let S be the set of all t such that u and v are differentiable at t and $v(t) \neq 0$. Then u/v is differentiable on S and

$$\left(\frac{u}{v}\right)' = \frac{u'v - uv'}{v^2} \cdot$$

den. ≠ 0

Proof. Let $f(x) = \dfrac{u(x)}{v(x)}$ for x in S. Then $f = u \cdot \dfrac{1}{v}$ on S. Lemma 5.12 tells us that $1/v$ is differentiable at each x in S and $(1/v)'(x) = \dfrac{-v'(x)}{[v(x)]^2} \cdot$ Therefore, using the Product Rule of Theorem 5.8, we have

$$f'(x) = u'(x)\frac{1}{v(x)} + u(x)(1/v)'(x)$$

$$= \frac{u'(x)}{v(x)} - u(x)\frac{v'(x)}{[v(x)]^2} = \frac{u'(x)v(x) - u(x)v'(x)}{[v(x)]^2} \cdot \quad \blacksquare$$

Example 14. Let $f(x) = \dfrac{x}{x^2 + 1}$. We wish to find $f'(x)$. Let $u(x) = x$ and $v(x) = x^2 + 1$. Then $u'(x) = 1$ and $v'(x) = 2x$. Therefore, $f'(x) = \dfrac{u'(x)v(x) - u(x)v'(x)}{v(x)^2} = \dfrac{(1)(x^2 + 1) - (x)(2x)}{(x^2 + 1)^2} = \dfrac{1 - x^2}{(x^2 + 1)^2} \cdot \quad \blacksquare$

Example 15. Let $f(x) = \dfrac{1 - x^{1/2}}{1 + x^{1/2}}$. Let $u(x) = 1 - x^{1/2}$ and $v(x) = 1 + x^{1/2}$. Then $u'(x) = (-1/2)x^{-1/2}$ and $v'(x) = (1/2)x^{-1/2}$. Therefore, by Theorem 5.15,

$$f'(x) = \frac{(-1/2)x^{-1/2}(1 + x^{1/2}) - (1/2)(x^{-1/2})(1 - x^{1/2})}{(1 + x^{1/2})^2}$$

$$= \frac{-x^{-1/2}}{(1 + x^{1/2})^2} = \frac{-1}{x^{1/2}(1 + x^{1/2})^2} \cdot \quad \blacksquare$$

Example 16. Let $f(x) = \left(\dfrac{3x^2 + 2x + 1}{x - 1}\right)^{3/2}$. This function, of course, is the composite of a power and a quotient. Therefore, we differentiate it in two steps. Let $g(x) = \dfrac{3x^2 + 2x + 1}{x - 1}$ and let $h(y) = y^{3/2}$. Then $f = h \circ g$, and by Theorem 5.10 (the Chain Rule), $f'(x) = (h' \circ g)(x)g'(x)$. By Theorem 5.14, $h'(y) = (3/2)y^{1/2}$, so that $h' \circ g(x) = (3/2)\left(\dfrac{3x^2 + 2x + 1}{x - 1}\right)^{1/2}$. Since g is a quotient, we have by Theorem 5.15,

$$g'(x) = \frac{(6x + 2)(x - 1) - (3x^2 + 2x + 1)(1)}{(x - 1)^2} = \frac{3x^2 - 6x - 3}{(x - 1)^2}.$$

Finally, $f'(x) = (9/2)\dfrac{(3x^2 + 2x + 1)^{1/2}(x^2 - 2x - 1)}{(x - 1)^{5/2}}.$ ∎

EXERCISES

Differentiate the following functions:

1. $f(x) = x^2 + x^{3/2} - 2x^{1/2}$.

2. $f(x) = x^{4/7} - x^{7/4}$.

3. $f(x) = (1 + x^2)^{1/2}$.

4. $f(x) = x(1 + x)^{2/3}$.

5. $f(x) = \dfrac{x^2 - 1}{x}$.

6. $f(x) = \dfrac{(x + 1)(x - 1)}{x^2 - 1}$.

7. $f(x) = \left(\dfrac{x - 1}{x + 1}\right)^{1/2}$.

8. $f(x) = \dfrac{(2x + 1)^3}{x^2}$.

9. $g(t) = \dfrac{t}{\sqrt{a^2 - t^2}}$.

10. $h(s) = \dfrac{1}{(3 - 4s)^5}$.

11. $k(x) = \left(\dfrac{x}{1 + x}\right)^{1/2}$.

12. $f(x) = [1 + (x^2 - 1)^3]^{3/2}$.

13. $f(x) = \dfrac{1}{(x^2 - 3)^5}$.

14. $f(x) = \dfrac{(x^2 - 1)^4(x + 5)^6}{(x + 1)}$.

15. $s(t) = t^{-5} - 3t^{-2} + 1$.

16. $g(y) = 7y(1 + y^2)^{3/2} - (1 + y^2)^{5/2}$.

17. $f(x) = \dfrac{x^{-4}}{1 - x^{-3}}$.

18. $f(x) = x^{\sqrt{2}}$.

19. $f(x) = \sqrt{1 + x^{2\pi}}$.

20. $s(x) = x\dfrac{x^{\sqrt{2}}}{\sqrt{1 + x^{2\pi}}}$.

ANSWERS

3. $f'(x) = x/(1 + x^2)^{1/2}$. **4.** $f'(x) = \dfrac{5x + 3}{3(1 + x)^{1/3}}$. **6.** $f'(x) = 0$.

7. $f'(x) = \dfrac{1}{(x + 1)^{3/2}(x - 1)^{1/2}}$. **8.** $f'(x) = 8 - 6x^{-2} - 2x^{-3}$.

9. $g'(t) = \dfrac{a^2}{(a^2 - t^2)^{3/2}}$. **10.** $h'(s) = \dfrac{20}{(3 - 4s)^6}$.

11. $k'(x) = \dfrac{x^{-1/2}(1 + x)^{-3/2}}{2}$. **12.** $f'(x) = 9x(x^2 - 1)^2\sqrt{1 + (x^2 - 1)^3}$.

15. $s'(t) = -5t^{-6} + 6t^{-3}$. **16.** $g'(y) = (7 - 5y)(1 + y^2)^{3/2} +$

$21y^2(1 + y^2)^{1/2}$. **18.** $f'(x) = \sqrt{2}x^{\sqrt{2}-1}$. **19.** $f'(x) = \dfrac{\pi x^{2\pi-1}}{\sqrt{1 + x^{2\pi}}}$.

6. DERIVATIVES OF TRIGONOMETRIC FUNCTIONS

Up to this point we have restricted ourselves to a discussion of the differentiation of functions that are powers of x and composites of such functions. Such functions, of course, form a large class. One of the most important sets of functions not included in this class is the set of trigonometric functions.

Before developing the differentiation formulae for trigonometric functions, we say just a word about notation. In previous courses you have perhaps had the point emphasized that such notation as "sin/cos" is not clear and that sin x/cos x must be used. However, since we have introduced the notation f/g for the quotient of two functions, sin/cos simply means that function whose value at x is sin x/cos x, whenever this quotient is defined (that is, whenever cos $x \neq 0$).

THEOREM 5.16

    i. $(\sin)' = \cos$.   ii. $(\cos)' = -\sin$.

    iii. $(\tan)' = \sec^2$.   iv. $(\cot)' = -(\csc^2)$.

    v. $(\sec)' = (\sec)(\tan)$.   vi. $(\csc)' = -(\csc)(\cot)$.

Proof of i. To find $(\sin)'$ we shall use the trigonometric identity

$$\sin x - \sin t = 2 \sin\left(\frac{x - t}{2}\right) \cos\left(\frac{x + t}{2}\right),$$

the proof of which is left as an exercise. Keeping this identity in mind, we have, for $x \neq t$,

$$\frac{\sin x - \sin t}{x - t} = \frac{2 \sin\left(\dfrac{x - t}{2}\right) \cos\left(\dfrac{x + t}{2}\right)}{x - t} = \frac{\sin \dfrac{x - t}{2}}{\dfrac{x - t}{2}} \cos\left(\frac{x + t}{2}\right).$$

By Theorems 3.12 and 4.11, we know that $\lim\limits_{x\to t} \dfrac{\sin\dfrac{x-t}{2}}{\dfrac{x-t}{2}} = 1$.   Moreover,

since cos is a continuous function, we have $\lim\limits_{x\to t} \cos\dfrac{x+t}{2} = \cos\dfrac{2t}{2} =$

cos t. Therefore, $(\sin)'(t) = \lim\limits_{x\to t} \dfrac{\sin x - \sin t}{x-t} = (1)(\cos t) = \cos t$.

The proof of (ii) follows easily from (i) and the chain rule for differentiation. Recall that $\cos x = \sin\left(\dfrac{\pi}{2} - x\right)$ and $\sin x = \cos\left(\dfrac{\pi}{2} - x\right)$ for all real x. Therefore, using the first of these two identities and the Chain Rule

$$(\cos)'(x) = \left[(\sin)'\left(\frac{\pi}{2} - x\right)\right]\left(\frac{\pi}{2} - x\right)'$$

$$= \cos\left(\frac{\pi}{2} - x\right)(-1)$$

$$= -\sin x \text{ (by the second identity above).}$$

The proofs of (iii) to (vi) will be left as exercises. Note that each of the other four trigonometric functions is a combination of sin and cos—either as a product or a quotient—so that the careful reader should be able to handle them without undue difficulties. ∎

Example 17.   Let $f(x) = 3\sin(2x)$.

Then
$$f'(x) = 3(\sin 2x)'$$
$$= 3(\cos 2x)(2x)'$$
$$= 3(\cos 2x)(2)$$
$$= 6\cos 2x. \quad ∎$$

Example 18.   Let $f(x) = \sqrt{\tan x}$.

Then
$$f'(x) = (1/2)(\tan x)^{-1/2}(\sec^2 x)$$

$$= \frac{\sec^2 x}{2\sqrt{\tan x}}. \quad ∎$$

Example 19.   Let $f(x) = \cos(\cos x^3)$.

Then
$$f'(x) = -\sin(\cos x^3)(\cos x^3)'$$
$$= -\sin(\cos x^3)(-\sin x^3)(x^3)'$$
$$= -\sin(\cos x^3)(-\sin x^3)(3x^2)$$
$$= 3x^2(\sin x^3)\sin(\cos x^3). \quad ∎$$

EXERCISES

In Exercises 1 to 15 differentiate the function given:

1. $f(x) = 3 \cos 4x$.

2. $g(x) = 6 \sec (1/2)x$.

3. $h(\theta) = 2 \sin (\pi\theta)$.

4. $f(x) = \sqrt{1 - 2 \sin x}$

5. $f(x) = x^2 \cot^2 2x$.

6. $s(t) = (\sin t + \cos t)^2$.

7. $f(x) = 4x \tan \sqrt{x}$.

8. $f(\theta) = \dfrac{1 + \cos 2\theta}{\sin 2\theta}$.

9. $h(x) = \sin (x \tan x)$.

10. $h(x) = x \sin (x \tan x)$.

11. $f(x) = \sin \sqrt{1 + x^2}$.

12. $f(x) = \dfrac{\cos (2x + 3)}{1 - x^3}$.

13. $f(x) = x^3 \sin (1 + \sqrt{x \cos x})$.

14. $h(y) = \dfrac{1}{1 + \cos y}$.

15. $s(\theta) = \sec^2 \theta - \tan^2 \theta$.

16. Prove Theorem 5.16iii to vi.

17. Prove the trigonometric identity

$$\sin x - \sin t = 2 \sin \frac{x - t}{2} \cos \frac{x + t}{2}$$

$\left(\text{Hint: use the identities for } \sin (a + b) \text{ and } \sin (b - a), \text{ where } a = \dfrac{x - t}{2}\right.$

$\left.\text{and } b = \dfrac{x + t}{2}\right)$.

ANSWERS

2. $g'(x) = 3 \sec (1/2x) \tan (1/2x)$.   4. $f'(x) = \dfrac{-\cos x}{\sqrt{1 - 2 \sin x}}$.

6. $s'(t) = \cos 2t$.   9. $h'(x) = (\tan x + x \sec^2 x) \cos (x \tan x)$.

10. $h'(x) = \sin (x \tan x) + x(\tan x + x \sec^2 x) \cos (x \tan x)$.

12. $f'(x) = \dfrac{3x^2 \cos (2x + 3) - 2(1 - x^3) \sin (2x + 3)}{(1 - x^3)^2}$.

14. $h'(y) = \dfrac{\sin y}{(1 + \cos y)^2}$.

7. HIGHER ORDER DERIVATIVES AND OTHER NOTATIONS FOR THE DERIVATIVE

If a function f is defined on a subset S of $\mathcal{R}$, its derivative function f' will be defined on a (possibly empty) set S', where $S' \subset S \subset \mathcal{R}$. Thus, f' is a function that has a derivative function also. Hence,

DEFINITION 5.17 Let $f: \mathfrak{R} \to \mathfrak{R}$. The derivative function f' of f will be called the *first derivative function of f*. The derivative of f' will be called the *second derivative function of f* and denoted by f'' or $f^{(2)}$. In general the derivative of the *n*th derivative function of f will be called the $(n+1)$st *derivative function of f* and denoted by $f^{(n+1)}$.

Example 20. Let $f(x) = x^4 + 2x^2 - x + 1$.

then
$$f'(x) = 4x^3 + 4x - 1$$
$$f''(x) = 12x^2 + 4$$
$$f'''(x) = 24x$$
$$f^{(4)} = 24.$$

It is easy to see then that $f^{(5)}$, $f^{(6)}$, and so forth, will be the identically zero function, since $f^{(4)}$ is a constant function. In fact, if we are given any polynomial whatsoever, it is easy to see that when we differentiate it a sufficient number of times, we will eventually arrive at the zero function. ∎

Example 21. Let $g = \cos$. Then $g' = -\sin$
$$g'' = -\cos$$
$$g''' = \sin$$
$$g^{(4)} = \cos.$$

Therefore, we know how to continue, since we have returned to our original function, and we already know how its first four derivatives behave. ∎

Now we come to the question of other notations for the derivative. Unfortunately, there are quite a few different notations for this concept. For the most part we shall use the f' notation that we have introduced. However, for certain purposes some of the others will be convenient and will be used. Furthermore, it is desirable that the reader be familiar with alternative notations that are found in other books and that are necessary to communicate with people who prefer them.

Suppose f is a differentiable function. Then $\dfrac{df}{dx}$, $\dfrac{d}{dx} f$, and Df are alternative notations for f', and $\dfrac{d^2f}{dx^2}$, $\dfrac{d^2}{dx^2} f$, and D^2f are alternatives for f''. Moreover, if $y = f(x)$, then $\dfrac{dy}{dx}$, y', and $D_x y$ are alternatives for $f'(x)$, just as $\dfrac{d^2y}{dx^2}$, y'', and $D_x^2 y$ are for $f''(x)$. Thus, the following sets of expressions are different ways of denoting exactly the same thing:

$$\begin{cases} f(x) = 2x^2 + 1 \\ f'(x) = 4x \\ f''(x) = 4 \end{cases} \qquad \begin{cases} f(x) = 2x^2 + 1 \\ (Df)(x) = 4x \\ (D^2f)(x) = 4 \end{cases} \qquad \begin{cases} f(x) = 2x^2 + 1 \\ \dfrac{df(x)}{dx} = 4x \\ \dfrac{d^2f(x)}{dx^2} = 4 \end{cases}$$

$$\begin{cases} f(x) = 2x^2 + 1 \\ \dfrac{d}{dx}f(x) = 4x \\ \dfrac{d^2}{dx^2}f(x) = 4 \end{cases} \qquad \begin{cases} y = 2x^2 + 1 \\ y' = 4x \\ y'' = 4 \end{cases} \qquad \begin{cases} y = 2x^2 + 1 \\ \dfrac{dy}{dx} = 4x \\ \dfrac{d^2y}{dx^2} = 4 \end{cases}$$

$$\begin{cases} y = 2x^2 + 1 \\ D_x y = 4x \\ D_x^2 y = 4 \end{cases} \qquad \begin{cases} f(x) = 2x^2 + 1 \\ f'(x) = (2x^2 + 1)' = 4x \\ f''(x) = (4x)' = 4. \end{cases}$$

EXERCISES *look at*

In Exercises 1 to 10 find f' and f'':

1. $f(x) = 4x^3 + 3$.   2. $f(x) = (2x^2 + 1)^3$.

3. $f(x) = (x^2 + 3)\sin x$.   4. $f(x) = \sqrt{\sin 5x}$.

5. $f(x) = (1/6)x^3 + 4x^{3/2} + 2x$.   6. $f(x) = \sqrt{6 - x^2}$.

7. $f(x) = \dfrac{x}{x - 1}$.   8. $f(x) = 4x\sqrt{x^2 + 1}$.

9. $f(x) = \cos^3 x$.   10. $f(x) = \sec x \tan x$.

11. Let $f(x) = \sin 3x$. Show that $D^2f + 9f = 0$.

12. Find a function f such that $Df = 3$.

13. Find a function f such that $D^2f = 4$.

14. Find a function f such that $D^2f = Df$.

15. Find a nonzero function f such that $D^2f + 16f = 0$ (Hint: see Exercise 11).

ANSWERS

2. $f''(x) = 12(2x^2 + 1)(10x^2 + 1)$.

4. $f''(x) = \dfrac{-25}{4}[(\sin 5x)^{-3/2}\cos^2 5x + 2(\sin 5x)^{1/2}]$.

6. $f''(x) = \dfrac{-6}{(6 - x^2)^{3/2}}$.   8. $f''(x) = \dfrac{8x^3 + 12x}{(x^2 + 1)^{3/2}}$.

10. $f''(x) = \sec x \tan x[5\sec^2 x + \tan^2 x]$.

MISCELLANEOUS EXERCISES

In Exercises 1 to 12 differentiate the function given:

1. $f(t) = (t^4 + 3t^2 - 1)^{5/2}$.
2. $g(x) = x^3(x + 1)^2$.
3. $f(x) = (1 + x)^2/x$.
4. $h(t) = 4\sqrt{3 + t}$.
5. $y = \tan 3x$.
6. $y = x^2 \sin^3 x$.
7. $f(\theta) = \sec^2 \theta - \tan^2 \theta$.
8. $f(x) = \sin (\cos x)$.
9. $g(x) = 2 \cos^2 \left(\dfrac{x}{3}\right)$.
10. $y = \tan (x \cos x)$.
11. $f(t) = \dfrac{\tan 2t}{1 - \cot 2t}$.
12. $k(x) = \sqrt{x + \sqrt{x + \sqrt{x}}}$.

In Exercises 13 to 17 find the affine approximation of f about t:

13. $f(x) = \sin x$, $t = \pi/4$.
14. $f(x) = \cos (\cos x)$, $t = \pi$.
15. $f(x) = \sqrt{2 - x^2}$, $t = 1$.
16. $f(x) = x^{5/2}$, $t = 2$.
17. $f(x) = \dfrac{x^2}{x^3 - 1}$, $t = 2$.

ANSWERS

2. $g'(x) = x^2(x + 1)(5x + 3)$. **4.** $\dfrac{dh}{dt} = \dfrac{2}{\sqrt{3 + t}}$. **5.** $\dfrac{dy}{dx} = 3 \sec^2 3x$.

7. $Df = 0$. **8.** $f'(x) = -(\sin x) \cos (\cos x)$.

10. $y' = (\cos x - x \sin x) \sec^2 (x \cos x)$.

13. $A_{\pi/4}f(x) = \dfrac{1}{\sqrt{2}} [x - (\pi/4) + 1]$. **14.** $A_\pi f(x) = \cos (-1)$.

15. $A_1 f(x) = -x + 2$. **16.** $A_2 f(x) = \sqrt{2}(5x - 6)$.

17. $A_2 f(x) = \dfrac{-20}{49} x + \dfrac{68}{49}$.

8. FUNCTIONS FROM ℛ INTO ℛ^m

At this point we have at our fingertips all the material necessary for developing a differential calculus for functions from ℛ into ℛ^m. As you will see in this section, this constitutes only a minor generalization of our development of differential calculus of functions from ℛ into ℛ. As in Section 1, we shall begin with a definition of linear and affine transformations:

DEFINITION 5.18 Let $L: ℛ \to ℛ^m$ and $A: ℛ \to ℛ^m$.

i. L is a *linear transformation* if and only if there exists an element M of $ℛ^m$ such that $L(x) = xM$ for all x in $ℛ$.

ii. A is an *affine transformation* if and only if there exist elements M and K in $ℛ^m$ such that $A(x) = xM + K$ for all x in $ℛ$.

Example 22. Let $A(x) = x \begin{pmatrix} 1 \\ 2 \end{pmatrix} + \begin{pmatrix} -3 \\ 1 \end{pmatrix}$ and $L(x) = x \begin{pmatrix} 1 \\ 2 \end{pmatrix}$. We see that A is an affine transformation from $\mathcal{R}$ into $\mathcal{R}^2$ and L is a linear transformation from $\mathcal{R}$ into $\mathcal{R}^2$. ∎

The reader will be asked to show that whenever A is an affine transformation from $\mathcal{R}$ into $\mathcal{R}^n$, then each of the coordinate functions A_i, $i = 1$, $\ldots$, n, of A is an affine transformation from $\mathcal{R}$ into $\mathcal{R}$ (Definition 5.1). ⋅

Without an attempt at geometrical motivation at this point we shall proceed with a development of affine approximations and derivatives of functions from $\mathcal{R}$ into $\mathcal{R}^n$. The reader should compare the following definition with Definition 5.3:

DEFINITION 5.19 Let $f : \mathcal{R} \to \mathcal{R}^m$ and let t be in the interior of the domain of f. Then $A_t f$ is the *affine approximation of f about t* if and only if $A_t f$ is an affine transformation such that

　　i. $A_t f(t) = f(t)$,

and

　　ii. $\lim_{x \to t} \dfrac{f(x) - A_t f(x)}{x - t} = O_m.$

Example 23. Let $f(x) = \begin{pmatrix} x \\ x^2 \end{pmatrix}$, $A(x) = x \begin{pmatrix} 1 \\ 2 \end{pmatrix} + \begin{pmatrix} 0 \\ -1 \end{pmatrix}$, and $t = 1$. Note the following:

i. $f(1) = \begin{pmatrix} 1 \\ 1 \end{pmatrix}$ and $A(1) = \begin{pmatrix} 1 \\ 2 \end{pmatrix} + \begin{pmatrix} 0 \\ -1 \end{pmatrix} = \begin{pmatrix} 1 \\ 1 \end{pmatrix}$.

ii. for $x \neq 1$,

$$\frac{f(x) - A(x)}{x - 1} = \frac{\begin{pmatrix} x \\ x^2 \end{pmatrix} - \left[x \begin{pmatrix} 1 \\ 2 \end{pmatrix} + \begin{pmatrix} 0 \\ -1 \end{pmatrix} \right]}{x - 1}$$

$$= \frac{\begin{pmatrix} 0 \\ x^2 - 2x + 1 \end{pmatrix}}{x - 1} = \frac{\begin{pmatrix} 0 \\ (x - 1)^2 \end{pmatrix}}{x - 1} = \begin{pmatrix} 0 \\ (x - 1) \end{pmatrix}.$$

Therefore,

$$\left\| \frac{f(x) - A(x)}{x - 1} - O_2 \right\| =^1 \left\| \begin{pmatrix} 0 \\ (x - 1) \end{pmatrix} \right\|$$

$$=^2 \sqrt{0^2 + (x - 1)^2}$$

$$=^3 |x - 1|.$$

(1) Substitution by previous calculation.
(2) Definition of length.
(3) Definition of absolute value.

Consequently, when $|x - 1| < r$, $\left\| \dfrac{f(x) - A(x)}{x - 1} - O_2 \right\| < r$, and (by

Definition 3.7), we have $\lim\limits_{x \to t} \dfrac{f(x) - A(x)}{x - 1} = O_2$. Applying Definition 5.19,

we see that A is the affine approximation of f about $t = 1$. ■

At this point we shall rely on our previous development of differential calculus of functions from $\mathcal{R}$ into $\mathcal{R}$ to give us a theorem analogous to Theorem 5.2.

THEOREM 5.20 Let $f: \mathcal{R} \to \mathcal{R}^m$ and let t be in the interior of the domain of f. f has an affine approximation about t if and only if each coordinate function f_i, $i = 1, \ldots, m$, is differentiable at t. Moreover, in this case, the affine approximation of f at t is defined by

$$A_t f(x) = (x - t) \begin{pmatrix} f_1'(t) \\ f_2'(t) \\ \cdot \\ \cdot \\ \cdot \\ f_m'(t) \end{pmatrix} + f(t).$$

Proof. Consider the following implication chain:
$A_t f$ is the affine approximation of f about t.

$\Leftrightarrow^1 \begin{cases} \quad \text{i. } A_t f(t) = f(t) \\ \text{and ii. } \lim\limits_{x \to t} \dfrac{f(x) - A_t f(x)}{x - t} = O_m \end{cases}$

$\Leftrightarrow^2 \begin{cases} \quad \text{i. } (A_t f)_i(t) = f_i(t), \, i = 1, \ldots, m \\ \text{and ii. } \lim\limits_{x \to t} \dfrac{f_i(x) - (A_t f)_i(x)}{x - t} = 0, \, i = 1, \ldots, m \end{cases}$

$\Leftrightarrow^3 \lim\limits_{x \to t} \dfrac{f_i(x) - f_i(t)}{x - t}$ exists and is finite for $i = 1, \ldots, m$

$\Leftrightarrow^4 f_i$, $i = 1, \ldots, m$ is differentiable at t.

(1) Definition 5.19.
(2) Definition 1.6 (p. 9) and Theorem 3.15.
(3) Theorem 5.2, p. 218.
(4) Definition 5.4, p. 223.

This implication chain tells us that f has an affine approximation at t if and only if each coordinate function of f is differentiable at t.

Assume that f_i is differentiable at t for each $i = 1, \ldots, m$ and let A be the affine transformation defined by

$$A(x) = (x - t)\begin{pmatrix} f_1'(t) \\ f_2'(t) \\ \cdot \\ \cdot \\ \cdot \\ f_m'(t) \end{pmatrix} + f(t).$$

We note that $A(t) = f(t)$ and

$$\lim_{x \to t} \frac{f(x) - A(x)}{x - t} =^1 \lim_{x \to t} \left(\frac{f(x) - \left[(x - t)\begin{pmatrix} f_1'(t) \\ f_2'(t) \\ \cdot \\ \cdot \\ \cdot \\ f_m'(t) \end{pmatrix} + f(t) \right]}{x - t} \right)$$

$$=^2 \lim_{x \to t} \left[\frac{f(x) - f(t)}{x - t} - \begin{pmatrix} f_1'(t) \\ f_2'(t) \\ \cdot \\ \cdot \\ \cdot \\ f_m'(t) \end{pmatrix} \right]$$

$$=^3 \lim_{x \to t} \begin{pmatrix} \dfrac{f_1(x) - f_1(t)}{x - t} - f_1'(t) \\ \cdot \\ \cdot \\ \cdot \\ \dfrac{f_m'(x) - f_m(t)}{x - t} - f_m'(t) \end{pmatrix}$$

$$=^4 \begin{pmatrix} \lim\limits_{x \to t} \dfrac{f_1(x) - f_1(t)}{x - t} - f_1'(t) \\ \cdot \\ \cdot \\ \cdot \\ \lim\limits_{x \to t} \dfrac{f_m(x) - f_m(t)}{x - t} - f_m'(t) \end{pmatrix}$$

$$=^5 O_m.$$

(1) Substitution for A.
(2) Algebraic rearrangement.
(3) Substitution of coordinates.
(4) Theorem 3.15.
(5) Definition 5.4 applied to $f_1, \ldots, f_m$.

Therefore, A is the affine approximation of f about t (Definition 5.19). ∎

Example 24. Let $f(x) = \begin{pmatrix} x^2 + 1 \\ 3x - 4 \\ 4x^3 \end{pmatrix}$ and $t = 2$. We shall find the

affine approximation of f about t. Consider the following coordinate functions:

$$f_1(x) = x^2 + 1, \qquad f_2(x) = 3x - 4, \quad \text{and} \quad f_3(x) = 4x^3.$$

The corresponding derivative functions are

$$f_1'(x) = 2x, \qquad f_2'(x) = 3, \quad \text{and} \qquad f_3'(x) = 12x^2,$$

and their values at $t = 2$ are

$$f_1'(2) = 4, \qquad f_2'(2) = 3, \quad \text{and} \qquad f_3'(2) = 48.$$

Therefore, by Theorem 5.20 the affine approximation of f about $t = 2$ is given by

$$A_2 f(x) = (x - 2)\begin{pmatrix} 4 \\ 3 \\ 48 \end{pmatrix} + f(2)$$

$$= (x - 2)\begin{pmatrix} 4 \\ 3 \\ 48 \end{pmatrix} + \begin{pmatrix} 5 \\ 2 \\ 32 \end{pmatrix}$$

$$= \begin{pmatrix} 4x - 3 \\ 3x - 4 \\ 48x - 64 \end{pmatrix}. \quad \blacksquare$$

Our last definition in this section and chapter gives the reader more terminology and notation for the differential calculus of functions from $\mathcal{R}$ into $\mathcal{R}^m$. This definition should be compared with Definitions 5.4 and 5.5:

DEFINITION 5.21 Let $f : \mathcal{R} \to \mathcal{R}^m$. Let S be the set of all points t in the domain of f at which f has an affine approximation. Let f' be the function defined by

$$f'(t) = \begin{pmatrix} f_1'(t) \\ \cdot \\ \cdot \\ \cdot \\ f_m'(t) \end{pmatrix} = \lim_{x \to t} \frac{f(x) - f(t)}{x - t} \quad \text{for all } t \text{ in } S.$$

Then f' is called the *derivative function of* f. We call $f'(t)$ the *derivative of* f *at* t. If f has a derivative at t, we say that f is *differentiable at* t. If f has a derivative at each point in the domain of f, we say that f is *differentiable*.

Let $t \in S$. Define $d_t f$ by

$$d_t f(x) = x f'(t) \quad \text{for all real } x.$$

$d_t f$ is called the *differential of* f *at* t.

The reader should notice that $d_t f$ is a linear transformation (Definition 5.18).

Example 25. Let $f(x) = \begin{pmatrix} \sqrt{1-x^2} \\ 3x^3 + x - 1 \end{pmatrix}$. We shall find the derivative function of f, the affine approximation $A_{1/2}f$ of f about $t = 1/2$, and the differential $d_{1/2}f$ of f at $t = 1/2$:

By Definition 5.21, we see that

$$f'(t) = \begin{pmatrix} \dfrac{-t}{\sqrt{1-t^2}} \\ 9t^2 + 1 \end{pmatrix} \text{ for } -1 < t < 1.$$

Therefore, by Theorem 5.20 and Definition 5.21,

$$A_{1/2}f(x) = (x - 1/2)f'(1/2) + f(1/2)$$

$$= (x - \tfrac{1}{2})\begin{pmatrix} \dfrac{-1/2}{\sqrt{1-\frac{1}{4}}} \\ 9(\tfrac{1}{4}) + 1 \end{pmatrix} + \begin{pmatrix} \sqrt{1 - \tfrac{1}{4}} \\ 3(\tfrac{1}{8}) + (\tfrac{1}{2}) - 1 \end{pmatrix}$$

$$= x\begin{pmatrix} \dfrac{-1}{\sqrt{3}} \\ 13/4 \end{pmatrix} + \begin{pmatrix} \dfrac{2}{\sqrt{3}} \\ -7/4 \end{pmatrix},$$

and $\quad d_{1/2}f(x) = xf'(1/2)$

$$= x\begin{pmatrix} \dfrac{-1}{\sqrt{3}} \\ 13/4 \end{pmatrix}. \quad \blacksquare$$

Example 26. Let $f(x) = \begin{pmatrix} 2\sin x \\ \sin x - \cos x \\ \cos 2x \end{pmatrix}$. This time we shall use Theorem 5.20 and Definition 5.21 to find f', $A_0 f$, and $d_0 f$:

i. $f'(t) = \begin{pmatrix} 2\cos t \\ \cos t + \sin t \\ -2\sin 2t \end{pmatrix}$.

ii. $A_0(x) = (x - 0)f'(0) + f(0)$

$$= x\begin{pmatrix} 2 \\ 1 \\ 0 \end{pmatrix} + \begin{pmatrix} 0 \\ -1 \\ 1 \end{pmatrix}.$$

iii. $d_0 f(x) = xf'(0)$

$$= x\begin{pmatrix} 2 \\ 1 \\ 0 \end{pmatrix}. \quad \blacksquare$$

We hope that the reader has recognized the relation between the derivative function of f, the affine approximation to f about t, and the differential of f at t:

$$A_t f(x) = (x - t)f'(t) + f(t)$$
$$= xf'(t) - tf'(t) + f(t)$$
$$= d_t f(x) + [f(t) - tf'(t)].$$

Compare this equation to the one found in the discussion following Definition 5.5, p. 223.

EXERCISES

Find the derivative function for each of Exercises 1 to 10:

1. $f(x) = \left(\dfrac{1}{x + 2} \right).$

2. $f(x) = \left(\dfrac{\sin x}{x^3 - 3x + 4} \right).$

3. $f(x) = \left(\dfrac{(x + 1)^4 (x - 3)^3}{x^2 + 3} \right).$
$(1 + x^2)^5$

4. $f(x) = \left(\dfrac{1}{x + 1} \atop \sin (3x^2) \right).$

5. $f(x) = \left(\dfrac{\sqrt{7x^3 - 2x}}{x^{2/3} - 1} \right).$

6. $f(x) = \begin{pmatrix} x \\ x \\ x \end{pmatrix}.$

7. $f(x) = \left(\dfrac{\tan x}{\sec x} \right).$

8. $f(x) = \begin{pmatrix} \sin x \\ x \\ \csc x \end{pmatrix}.$

9. $f(x) = \left(\dfrac{\sin (\cos x^2)}{2} \right).$

10. $f(x) = \begin{pmatrix} 1 + x^2 - x \\ 7x^3 - 13 \\ x + 2 \\ 4x - 6x^2 \end{pmatrix}.$

In Exercises 11 to 20 find the differential of f at t and the affine approximation of f about t:

11. $f(x) = \left(\dfrac{4x^2 - 2x}{3x^3 + x} \right), t = 1.$

12. $f(x) = \left(\dfrac{x^3 + 2x^2 - x}{5x^4 - x} \right), t = 0.$

13. $f(x) = \left(\dfrac{2 \sin x}{\cos^2 x} \right), t = \pi/2.$

14. $f(x) = \begin{pmatrix} x^3 \\ \sqrt{x} \\ x^2 - 1 \end{pmatrix}, t = 4.$

15. $f(x) = \begin{pmatrix} x\sqrt{1 - x^2} \\ 1 \\ x \end{pmatrix}, t = \tfrac{1}{2}.$

16. $f(x) = \left(\dfrac{\tan x}{\sec x \tan x} \right), t = \pi/4.$

17. $f(x) = \dfrac{(1 - x)^4 (2 + x)^3}{(3 - x)}, t = 2.$

18. $f(x) = \left(\dfrac{x^{2/3} - 4}{(1 - x)^3} \right), t = 8.$

19. $f(x) = \left(\dfrac{\sin (x^2 - 1)}{3x^2 + 4x} \right), t = 0.$

20. $f(x) = \begin{pmatrix} x^{-1/2} \\ 1 - x^{1/2} \\ x^3 + 2x \\ 7 \end{pmatrix}, t = 9.$

21. Let A be an affine approximation from $\mathcal{R}$ into $\mathcal{R}^m$. Prove that each coordinate function A_i, $i = 1, \ldots, m$, of A is an affine approximation from $\mathcal{R}$ into $\mathcal{R}$.

22. Let $f: \mathcal{R} \to \mathcal{R}^m$. Prove that whenever f is differentiable at t (Definition 5.21), then f is continuous at t (Definition 4.1) (Hint: see the proof of Theorem 5.6, p. 225).

ANSWERS

2. $f'(t) = \begin{pmatrix} \cos t \\ 3t^2 - 3 \end{pmatrix}$. **4.** $f'(t) = \begin{pmatrix} \dfrac{-1}{(t+1)^2} \\ 6t \cos (3t^2) \end{pmatrix}$. **6.** $f'(t) = \begin{pmatrix} 1 \\ 1 \\ 1 \end{pmatrix}$.

8. $f'(t) = \begin{pmatrix} \dfrac{t \cos t - \sin t}{t^2} \\ -\csc t \cot t \end{pmatrix}$. **10.** $f'(t) = \begin{pmatrix} 2t - 1 \\ 21t^2 \\ 1 \\ 4 - 12t \end{pmatrix}$.

12. $\begin{cases} d_0 f(x) = x \begin{pmatrix} -1 \\ -1 \end{pmatrix} \\ A_0 f(x) = x \begin{pmatrix} -1 \\ -1 \end{pmatrix}. \end{cases}$
14. $\begin{cases} d_4 f(x) = x \begin{pmatrix} 48 \\ \frac{1}{4} \\ 8 \end{pmatrix} \\ A_4 f(x) = x \begin{pmatrix} 48 \\ \frac{1}{4} \\ 8 \end{pmatrix} + \begin{pmatrix} -128 \\ 1 \\ -17 \end{pmatrix}. \end{cases}$

16. $\begin{cases} d_{\pi/4} f(x) = x \begin{pmatrix} 2 \\ 3\sqrt{2} \end{pmatrix} \\ A_{\pi/4} f(x) = x \begin{pmatrix} 2 \\ 3\sqrt{2} \end{pmatrix} + \begin{pmatrix} 1 - \pi/2 \\ \sqrt{2} - \dfrac{3\pi\sqrt{2}}{4} \end{pmatrix}. \end{cases}$

18. $\begin{cases} d_8 f(x) = x \begin{pmatrix} \frac{1}{3} \\ -147 \end{pmatrix} \\ A_8(x) = x \begin{pmatrix} \frac{1}{3} \\ -147 \end{pmatrix} + \begin{pmatrix} -8/3 \\ 833 \end{pmatrix}. \end{cases}$

20. $\begin{cases} d_9 f(x) = x \begin{pmatrix} -\frac{1}{54} \\ -1/6 \\ 245 \\ 0 \end{pmatrix} \\ A_9 f(x) = x \begin{pmatrix} -\frac{1}{54} \\ -1/6 \\ 245 \\ 0 \end{pmatrix} + \begin{pmatrix} 1/2 \\ -\frac{1}{2} \\ -1458 \\ 7 \end{pmatrix}. \end{cases}$

6

APPLICATIONS OF THE DERIVATIVE

1. MAXIMA AND MINIMA OF FUNCTIONS DEFINED ON A CLOSED AND BOUNDED INTERVAL

One of the principal applications of the derivative is its use in finding "maxima" and "minima" of certain functions. Consider the graph sketched in Figure 6.1. Affine approximations are drawn in at x_1, x_2, and x_3. It seems that these are constant functions at the "peaks" and "valleys" of f. This is indeed often the case, but we shall need to be careful. To determine the points where such peaks and valleys occur we shall develop precise terminology and several important theorems.

DEFINITION 6.1 Let $f: \mathcal{R} \rightarrow \mathcal{R}$, $x_0 \in \mathcal{R}$. We say f is *increasing at* x_0 if there are real numbers a and b such that $a < x_0 < b$ and such that

$$\begin{cases} f(x) \leq f(x_0) \text{ whenever } a < x < x_0, \text{ and} \\ f(x) \geq f(x_0) \text{ whenever } x_0 < x < b. \end{cases}$$

If

$$\begin{cases} f(x) < f(x_0) \text{ whenever } a < x < x_0, \text{ and} \\ f(x) > f(x_0) \text{ whenever } x_0 < x < b, \end{cases}$$

we say f is *strictly increasing at* x_0. Functions *decreasing at a point* and *strictly decreasing at a point* are defined in an analogous way.

Intuitively we feel that if f is increasing at each point of an interval containing x_0, then as we move from left to right across x_0 the graph of f is

249

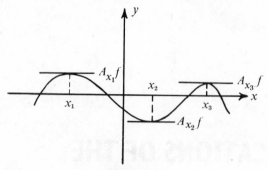

Figure 6.1

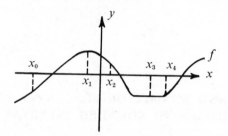

Figure 6.2

rising (or at least not falling). Consider the function f illustrated in Figure 6.2. This function is strictly increasing at x_0, neither increasing nor decreasing at x_1, strictly decreasing at x_2, both increasing and decreasing at x_3, and increasing (but not strictly increasing) at x_4.

Now look at Figure 6.3. It appears that the slopes of the affine approximations should tell us something about whether or not f is increasing at a point, and conversely the fact that f is increasing or decreasing should tell us something about the slope. This is indeed the case as we shall see in the following two theorems:

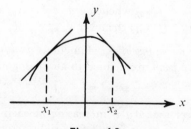

Figure 6.3

THEOREM 6.2 Let $f: \mathcal{R} \to \mathcal{R}$ have a derivative f' at x_0.
i. If f is increasing at x_0, then $f'(x_0) \geq 0$.
ii. If f is decreasing at x_0, then $f'(x_0) \leq 0$.

Proof. We use the following implication chain to prove *i*:

f is increasing at x_0

$\Leftrightarrow^1$ there exist elements a and b of $\mathcal{R}$ such that $a < x_0 < b$

$$\text{and} \begin{cases} f(x) \leq f(x_0) \text{ whenever } a < x < x_0 \\ f(x) \geq f(x_0) \text{ whenever } x_0 < x < b \end{cases}$$

$\Leftrightarrow^2$ for each x such that $a < x < b$ and $x \neq x_0$ we have

$$\frac{f(x) - f(x_0)}{x - x_0} \geq 0$$

$\Rightarrow^3 \lim\limits_{x \to x_0} \dfrac{f(x) - f(x_0)}{x - x_0} \geq 0$

$\Rightarrow^4 f'(x_0) \geq 0$.

(1) Definition 6.1.

(2) If $x > x_0$, we have $f(x) \geq f(x_0)$ and $x - x_0 > 0$; hence, $\dfrac{f(x) - f(x_0)}{x - x_0} \geq 0$.

A similar argument holds if $x < x_0$.

(3) See Theorem 3.10i (p. 180).

(4) Definition of derivative (p. 223).

Statement ii is proved in a similar manner. ∎

THEOREM 6.3 Let $f: \mathcal{R} \to \mathcal{R}$ be differentiable at x_0.
i. If $f'(x_0) > 0$, then f is strictly increasing at x_0.
ii. If $f'(x_0) < 0$, then f is strictly decreasing at x_0.

Proof. We prove only the first part since the proof of the second part is similar. We use the implication chain:

$f'(x_0) > 0$

$\Leftrightarrow^1 \lim\limits_{x \to x_0} \dfrac{f(x) - f(x_0)}{x - x_0} > 0$

$\Rightarrow^2$ for some a and b in $\mathcal{R}$ such that $a < x_0 < b$, we have $\dfrac{f(x) - f(x_0)}{x - x_0} > 0$

whenever $a < x < b$, $x \neq x_0$

$\Leftrightarrow^3 \begin{cases} f(x) - f(x_0) < 0 \text{ whenever } a < x < x_0 \\ f(x) - f(x_0) > 0 \text{ whenever } x_0 < x < b \end{cases}$

$\Leftrightarrow \begin{cases} f(x) < f(x_0) \text{ whenever } a < x < x_0 \\ f(x) > f(x_0) \text{ whenever } x_0 < x < b \end{cases}$

$\Leftrightarrow^4 f$ is increasing at x_0.

(1) Definition of $f'(x_0)$ (Definition 5.4, p. 223).

(2) Definition of limit (Definition 3.7, p. 173).

(3) If $x < x_0$, we have $x - x_0 < 0$; hence, $f(x) - f(x_0)$ must be negative in order that $\dfrac{f(x) - f(x_0)}{x - x_0} > 0$. Similarly for $x > x_0$.

(4) Definition 6.1. ∎

NOTE. The function f given by $f(x) = x^3$ is strictly increasing at $x = 0$, although $f'(0) = 0$. Thus, it is not necessary that $f'(x_0) > 0$ in order that $f(x)$ be strictly increasing at x_0. Study the statements of Theorems 6.2 and 6.3 carefully in the light of this example.

We now wish to give precise terminology for what we called, in the opening paragraph of this section, "peaks" and "valleys" of a function.

DEFINITION 6.4 Let $f : \mathcal{R} \to \mathcal{R}$, $x_0 \in dom\, f$. We say f has a maximum at x_0, $(x_0, f(x_0))$ is a maximum point of f, and $f(x_0)$ is a maximum value of f if $f(x) \le f(x_0)$ for every x in $dom\, f$. We say f has a relative maximum at x_0, $(x_0, f(x_0))$ is a relative maximum point of f, and $f(x_0)$ is a relative maximum value of f if there exist a, b such that $a < x_0 < b$ and such that $f(x) \le f(x_0)$ whenever $x \in dom\, f \cap (a, b)$. We define minimum and relative minimum in an analogous way. Any relative maximum or relative minimum value of f is said to be an extreme value or extremum of f.

Notice that every maximum is a relative maximum and every minimum is a relative minimum but that the converse is not true. The sketch in Figure 6.4 of a function f from $[0, 7]$ into $\mathcal{R}$ illustrates the aforementioned concepts. In the figure

$f(0)$ is a relative maximum value but not a maximum value.

$f(1)$ is a relative maximum value but not a maximum value.

$f(2)$ is a maximum value and a relative maximum value.

$f(3)$ is a minimum value and a relative minimum value.

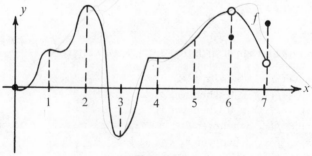

Figure 6.4

$f(4)$ is a relative maximum value and a relative minimum value (read the definition carefully).

$f(5)$ is neither a relative maximum value nor a relative minimum value.

$f(6)$ is a relative minimum value but not a minimum value.

$f(7)$ is a relative maximum value but not a maximum value.

Quite often it is desirable to locate relative maxima and minima for functions defined on an interval. The following theorem will be of considerable help to us in solving such problems:

THEOREM 6.5 Let $f: \mathcal{R} \rightarrow \mathcal{R}$ be defined on an interval (possibly infinite) and let x_0 be an element of this interval. If f has an extremum at x_0, at least one of the following must be true:

i. f is discontinuous at x_0;
ii. x_0 is an endpoint of the interval;
iii. $f'(x_0)$ is undefined;
iv. $f'(x_0) = 0$.

Proof. Suppose $f(x_0)$ is an extremum and none of (i), (ii), or (iii) holds. We shall show that (iv) must hold and this will complete the proof. Since x_0 is not an endpoint and $f(x_0)$ is an extremum, f cannot be either strictly increasing or strictly decreasing at x_0. But f' is defined at x_0, so either $f'(x_0) < 0, f'(x_0) > 0$, or $f'(x_0) = 0$. The first two cannot hold because of Theorem 6.3; hence, $f'(x_0) = 0$. ∎

If a and b are real numbers and f is defined and continuous on $[a, b]$, Theorem 6.5 gives us a procedure that, in most cases, leads to the relative maxima and minima of f on $[a, b]$. We note first that because of Theorem 4.8, p. 201, these extrema exist. Hence, we examine the function for the points satisfying one of the four conditions of Theorem 6.5 (since f is continuous, no points satisfy condition (i)). If there are only a finite number of such points, we evaluate the function at each of them and then choose the largest and smallest values found. These must of necessity be the maximum

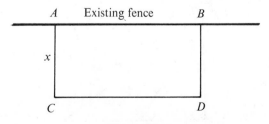

Figure 6.5

and minimum values of f. We outline the procedure as follows:

Procedure 6.6. For finding maximum and minimum values of a function f defined and continuous on a closed interval $[a, b]$.

  i. Differentiate f.
 ii. Note all points at which f' is undefined.
iii. Determine all values of x in (a, b) at which $f'(x) = 0$. These values of x together with the values noted in step (ii) and the endpoints a and b are called the *critical* points for f.
 iv. If (as will usually be the case) only a finite number of critical points have been noted, evaluate f at each of these points.
  v. Choose the largest and smallest of the values determined in step (iv). These are the desired maximum and minimum values. (Note that there may be more than one point in $[a, b]$ at which f takes on its maximum or minimum value.)

Example 1. A farmer wishes to build a rectangular corral next to an existing fence (see Figure 6.5). He has 100 ft. of wire. What dimensions will provide the largest area?

Let x be the length of AC. Then BD will also have length x, so that the length of CD will be $100 - 2x$. The area, $A(x)$, is then given by $A(x) = x(100 - 2x) = -2x^2 + 100x$. Obviously x can be no less than 0 and no more than 50, so that A is defined on $[0, 50]$. We wish to find the maximum value of A so we follow Procedure 6.6.

  i. $A'(x) = -4x + 100$
 ii. A' is everywhere defined so we note no points.
iii. Solve $A'(x) = -4x + 100 = 0$ to obtain $x = 25$.
 iv. Our critical points are at the end points and at $x = 25$. We find $A(0) = A(50) = 0$ and $A(25) = 1,250$.
  v. The largest value of the three we have found is 1,250 sq. ft. Hence, this is the maximum area, and the best dimensions are 25 ft. $\times$ 50 ft. ∎

Example 2. A man is in a rowboat at C, 6 miles from shore (see Figure 6.6). He wishes to reach a point A on the shore 10 miles away. He

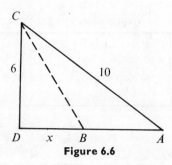

Figure 6.6

can row 2 m.p.h. and walk 4 m.p.h. Where should he land in order to reach A in the least time?

SOLUTION. Let x be the distance from D to B where D is the closest point on land and B is the landing point. Then since $\triangle ADC$ is a right triangle, we easily find the distance from A to D to be 8 miles. The time T required is a function of x. It is obvious from physical considerations that if there is a minimum value of T we must choose B to be at D or at A or between A and D; hence, we may consider $T(x)$ to be defined on $[0, 8]$. The time to go from C to B is the distance $\sqrt{x^2 + 6^2}$ divided by the rate 2, and the time from B to A is the distance $8 - x$ divided by the rate 4. Hence the total time from C to A is given by

$$T(x) = \tfrac{1}{2}\sqrt{x^2 + 36} + \tfrac{1}{4}(8 - x), \qquad 0 \le x \le 8.$$

T is then the function we wish to minimize. T is continuous on $[0, 8]$ so we follow Procedure 6.6.

i. $T'(x) = \dfrac{x}{2\sqrt{x^2 + 36}} - \dfrac{1}{4}.$

ii. T' is everywhere defined so we note no points.

iii. Solve $\dfrac{x}{2\sqrt{x^2 + 36}} - \dfrac{1}{4} = 0$ to obtain $x = 2\sqrt{3}$.

iv. We have found only one critical point other than the end points, 0 and 8. We evaluate T at each of these points and obtain $T(0) = 5$, $T(2\sqrt{3}) = 4.6$ (approx.), and $T(8) = 5$.

v. The smallest value of the three found in step iv is 4.6 (approx.). Hence, this is the minimum time possible and the solution is $x = 2\sqrt{3}$. ∎

EXERCISES

Locate the maximum and minimum points for each of the functions defined below:

1. $f(x) = x^2 + x + 1$, $-1 \le x \le 1$.

2. $f(x) = x^2 - x + 3$, $-1 \le x \le 1$.

3. $g(x) = \dfrac{1}{x} + \dfrac{1}{x^2}$, $-3 \le x \le -1$.

4. $h(x) = \dfrac{1}{x^2 + 1}$, $-1 \le x \le 1$.

5. $g(\theta) = \sin\theta$, $0 \le \theta \le \dfrac{\pi}{2}$.

6. $h(\theta) = \cos\theta$, $\dfrac{\pi}{4} \le \theta \le \dfrac{3\pi}{4}$.

7. Find two numbers whose sum is 12 and whose product is maximum.

8. Find two positive numbers whose sum is two and such that the sum of the second and the square of the first is maximum.

9. A box (without top) is to be made from a piece of sheet metal 20 in. × $12\frac{1}{2}$ in. by cutting squares from the corners and folding up the sides. What is the largest volume possible?

10. A rectangular building is to be constructed on a right triangular lot where the legs of the triangle are 50 and 100 ft. in length. If one corner of the building is to be at the vertex of the right angle, find the dimensions that will give the largest area.

11. Find the most economical proportions for a covered box of fixed volume whose base is a rectangle with one side three times as long as the other.

12. Two posts, one 10 ft. high and the other 15 ft. high, stand 30 ft. apart. They are to be stayed by wires attached to a single stake at the ground level, the wires running to the top of the posts. Where should the stake be placed to use the least amount of wire?

13. A man runs out of gasoline at point B (see Figure 6.7). There is a filling station at point D. He can walk 3 m.p.h. on the road and 2 m.p.h. in the woods. What is the shortest possible time in which he can reach the station?

ANSWERS

1. Max at $\begin{pmatrix}1\\3\end{pmatrix}$, min at $\begin{pmatrix}-\frac{1}{2}\\ \frac{3}{4}\end{pmatrix}$. **3.** Max at $\begin{pmatrix}-1\\0\end{pmatrix}$, min at $\begin{pmatrix}-2\\-\frac{1}{4}\end{pmatrix}$.

5. Max at $\begin{pmatrix}\frac{\pi}{2}\\1\end{pmatrix}$, min at $\begin{pmatrix}0\\0\end{pmatrix}$. **7.** Both numbers 6. **9.** 281.25 in³.

11. Altitude $= \frac{3}{2} \times$ shorter side of base. **13.** 1 hour. $\left(\text{If you got } \dfrac{4 + \sqrt{5}}{6}\right.$

hours you fell into a trap. Go back and read Procedure 6.6 *carefully.*$\Big)$

2. THE FIRST DERIVATIVE TEST

The procedure that we developed in the previous section for finding maxima and minima applied only to functions defined on a closed and

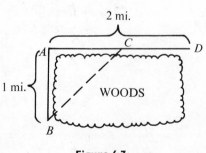

Figure 6.7

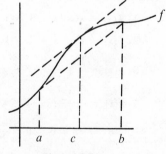

Figure 6.8

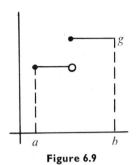

Figure 6.9

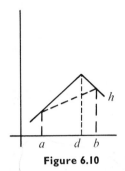

Figure 6.10

bounded interval. We may wish to examine functions defined on other subsets of $\mathfrak{R}$, for example those defined for all $\mathfrak{R}$. Furthermore, while the first four steps of the procedure can be used to find critical points of such a function, we have as yet found no way to learn whether a given point of *graph f* is a *relative* maximum point or *relative* minimum point. The first and second derivative tests described in the following pages often prove useful for this purpose.

Consider now the function *f* indicated in Figure 6.8. It seems intuitively evident, if *f* is continuous on $[a, b]$ and differentiable on (a, b), that there must be at least one point *c* such that $a < c < b$ and the tangent line at *c* is parallel to the line joining the points $\begin{pmatrix} a \\ f(a) \end{pmatrix}$ and $\begin{pmatrix} b \\ f(b) \end{pmatrix}$. A glance at Figure 6.9 shows why we assume continuity, since for the function *f* indicated there is clearly no such point between *a* and *b*. Figure 6.10 illustrates a situation where a function *h* is continuous on $[a, b]$ and there is no such *c*. In this case there is one point, *d*, where *h* is not differentiable.

The next theorem summarizes the previous discussion (refer to Appendix IV for proof).

THEOREM 6.7 (Mean Value Theorem for Derivatives.) Let $f : \mathfrak{R} \to \mathfrak{R}$ such that *f* is continuous on $[a, b]$, $a < b$, and let f' exist on (a, b). Then there is at least one real number *c* such that

i. $a < c < b$ and

ii. $\dfrac{f(b) - f(a)}{b - a} = f'(c).$

Consider now Figure 6.11. For points close to but less than x_0, *f* is increasing, and for points close to but greater than x_0, *f* is decreasing. This seems to indicate that *f* has a relative maximum at x_0. Here again we must be careful. Look at x_1. Both these conditions hold at x_1, but *f* has a relative *minimum* and not a relative maximum at x_1. This situation came about by

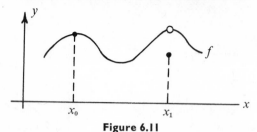

Figure 6.11

letting f be discontinuous at x_1. If we rule out this possibility we can often determine whether or not f has a relative maximum (or relative minimum) at an interior point of its domain by examining its derivative at nearby points. We state this precisely in the following theorem:

THEOREM 6.8 (First Derivative Test for Interior Points.) Let $f:\mathcal{R} \to \mathcal{R}$ be continuous at x_0. Let a and b be real numbers such that $a < x_0 < b$. Then

  i. if $\begin{cases} f'(x) \geq 0 \text{ for } a < x < x_0 \\ f'(x) \leq 0 \text{ for } x_0 < x < b, \end{cases}$

  f has a relative maximum at x_0.

  ii. if $\begin{cases} f'(x) \leq 0 \text{ for } a < x < x_0 \\ f'(x) \geq 0 \text{ for } x_0 < x < b, \end{cases}$

  f has a relative minimum at x_0.

  iii. If $f'(x) \geq 0(f'(x) > 0)$ for $x \in (a, x_0) \cup (x_0, b)$ then f is increasing (strictly increasing) at x_0.

  iv. If $f'(x) \leq 0(f'(x) < 0)$ for $x \in (a, x_0) \cup (x_0, b)$ then f is decreasing (strictly decreasing) at x_0.

(Note that $f'(x_0)$ need not be defined.)

We shall prove the somewhat weaker theorem that results from replacing $\leq$ by $<$ and $\geq$ by $>$ wherever these symbols occur in (i) and (ii). The stronger theorem can be proved with a little more effort and will be left as an exercise for the reader (see Exercise 11), as will (iii) and (iv) (see Exercise 12).

Proof of Weaker Theorem. Suppose $f'(x) > 0$ for $a < x < x_0$ and $f'(x) < 0$ for $x_0 < x < b$. Choose a_1 and b_1 such that $a < a_1 < x_0$ and $x_0 < b_1 < b$. Then f is continuous on $[a_1, b_1]$, since it is differentiable at every point other than x_0 and is continuous at x_0 by hypothesis. Hence, by Theorem 4.8 (p. 201), there is an x_1 in $[a_1, b_1]$ such that $f(x) \leq f(x_1)$ for every x in $[a_1, b_1]$ (that is, $f(x_1)$ is a maximum value of f in $[a_1, b_1]$). But if $a_1 \leq x < x_0$, x cannot be x_1, because $f'(x) > 0$, which implies that f is strictly increasing at x by Theorem 6.3. Similarly x cannot be x_1 if $x_0 < x \leq b_1$. Thus $x_1 = x_0$ is the point at which f has a maximum.

The proof of (ii) (weaker theorem) is similar. ■

We can now use Theorem 6.5 in combination with the first derivative test to find all extreme values of a function defined on an open interval.

> **Procedure 6.9.** Finding extreme values of a function f defined and continuous on an open interval (a, b) (bounded or not).
>   i. Differentiate f.
>   ii. Note points at which f' is undefined.
>   iii. Set $f'(x) = 0$ and solve, noting all solutions.
>   iv. If we have noted only a finite number of points in steps ii and iii, determine the values of $f'(x)$ for x near each point.
>   v. Apply the first derivative test to each of these points.

(Note: It is possible for the first derivative test to be not applicable at a point that gives an extremum. This occurs only for rather pathological functions.)

In most cases arising in practice the domain of a function can be divided into a finite number of open intervals and closed and bounded intervals in such a manner that by using Procedures 6.6 and 6.9 we can determine all extreme values.

Example 3. We wish to find extrema for the function f such that $f(x) = 2x^3 + 3x^2 - 12x + 4$, x real. Following Procedure 6.9:
  i. $f'(x) = 6x^2 + 6x - 12 = 6(x + 2)(x - 1)$.
  ii. f' is everywhere defined.
  iii. Set $f'(x) = 6(x + 2)(x - 1) = 0$. The solutions are $x = -2$, $x = 1$.
  iv. $f'(x)$ is positive for $x < -2$, since $(x + 2)$ and $(x - 1)$ are both negative;
  negative for $-2 < x < 1$, since $(x + 2)$ is positive and $(x - 1)$ is negative;
  positive for $x > 1$, since both $(x + 2)$ and $(x - 1)$ are positive.
  v. Using (iv) we can now apply the first derivative test at -2 and 1. We find -2 to be a relative maximum point since $f'(x) \geq 0$ for $a < x < -2$, where a is any number less than -2, and $f'(x) \leq 0$ for $-2 < x < 1$. In a similar way we find 1 to be a relative minimum point. ■

If we wish to sketch the graph of the function in the previous example, we now have some information to go on. We know that -2 is a relative maximum point of f and 1 is a relative minimum point of f, and that f is increasing in $(-\infty, -2)$, decreasing in $(-2, 1)$, and increasing in $(1, \infty)$. We may *guess* that the sketch will look something like that of Figure 6.12 (as is indeed the case). However, from our present information we have no way of knowing that it does not look more like Figure 6.13. In the following sections we shall develop means of obtaining more information on the shape of the graph.

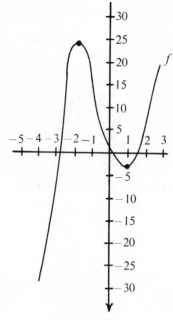

Figure 6.12

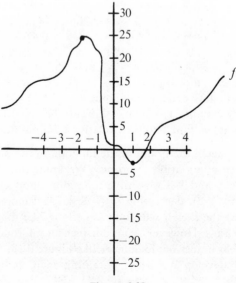

Figure 6.13

EXERCISES

In Exercises 1 to 10 locate all relative extreme points for the functions defined below:

1. $f(x) = x^2 + 2x + 1$.
2. $f(x) = -2x^2 + 3$.
3. $f(x) = \cos x$.
4. $f(x) = x^3 + 5x^2 + 1$.
5. $f(x) = -x^3 - 4x^2 - 4x + 2$.
6. $f(x) = 4x^3 + 4x^2 + x - 1$.
7. $f(x) = \sin^2 x$.
8. $f(x) = \sin x \cos x$.
9. $f(x) = \dfrac{x}{x^2 + 1}$.
10. $f(x) = \dfrac{x^2 + 1}{x}$, $x \neq 0$.

11. Use the Mean Value Theorem to prove part (i) of Theorem 6.8 as stated (Hint: suppose the hypothesis of part (i) holds and $a < x < x_0$. Supply reasons for the following implication chain:

$a < x < x_0 \Rightarrow$ there is a point e such that $x < e < x_0$ and

$$f'(e) = \frac{f(x_0) - f(x)}{x_0 - x}$$

$$\Rightarrow f(x_0) - f(x) = f'(e)(x_0 - x)$$
$$\Rightarrow f(x_0) - f(x) \geq 0$$
$$\Rightarrow f(x_0) \geq f(x).$$

Also take care of the case where $x_0 < x < b$.)

12. Use the Mean Value Theorem to prove Theorem 6.8iii and iv.

ANSWERS

1. $\begin{pmatrix} -1 \\ 0 \end{pmatrix}$ is a rel. min. point. 4. $\begin{pmatrix} 0 \\ 1 \end{pmatrix}$ is a rel. min. point. $\begin{pmatrix} \dfrac{-10}{3} \\ \dfrac{527}{27} \end{pmatrix}$ is a rel.

max. point. 6. $\begin{pmatrix} \dfrac{-1}{6} \\ \dfrac{-29}{27} \end{pmatrix}$ is a rel. min. point. $\begin{pmatrix} \dfrac{-1}{2} \\ -1 \end{pmatrix}$ is a rel. max. point.

7. $\begin{pmatrix} n\pi \\ 0 \end{pmatrix}$, n an integer, is rel. min. point. $\begin{pmatrix} \dfrac{2n+1}{2}\pi \\ 1 \end{pmatrix}$, n an integer, is a rel.

max. point. 9. $\begin{pmatrix} 1 \\ 1 \\ 2 \end{pmatrix}$ is a rel. max. point. $\begin{pmatrix} -1 \\ -1 \\ 2 \end{pmatrix}$ is a rel. min. point.

3. CONCAVITY AND THE SECOND DERIVATIVE TEST

In this section we shall develop the concept of concavity, which will be useful in curve sketching. The second derivative test is another convenient means for testing points that may be extrema.

Consider the function f illustrated in Figure 6.14. We notice that for points x close to a, the function is above the affine approximation at a. In such a case we shall say that f is "concave upward at a." Similarly for

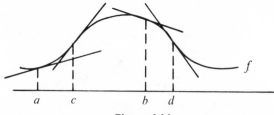

Figure 6.14

points close to b, f is below its affine approximation at b. We say that f is "concave downward at b." For points near c, but to the left of c, f is above its affine approximation at c; for points near c, but to the right of c, f is below its affine approximation at c. A point such as $(c, f(c))$ is said to be an "inflection point" of f. We shall also want to call points such as $(d, f(d))$ inflection points.

DEFINITION 6.10 Let $f:\mathcal{R} \rightarrow \mathcal{R}$, $t \in dom f$ and suppose f is differentiable at t and continuous in an interval about t. Let $A_t f$ be the affine approximation to f about t. If there exists an $r > 0$ such that for x in $\mathcal{B}(\bar{t}; r) \cap dom f$ we have $f(x) > A_t f(x)$, we say f is *concave upward at* t. If there exists an $r > 0$ such that for x in $\mathcal{B}(\bar{t}; r) \cap dom f$ we have $f(x) < A_t f(x)$, we say f is *concave downward at* t.

If, for some $r > 0$, we have

i. $f(x) > A_t f(x)$ if $x \in \mathcal{B}(t; r)$, $x > t$

and

$f(x) < A_t f(x)$ if $x \in \mathcal{B}(t; r)$, $x < t$

or   ii. $f(x) < A_t f(x)$ if $x \in \mathcal{B}(t; r)$, $x > t$

and

$f(x) > A_t f(x)$ if $x \in \mathcal{B}(t; r)$, $x < t$,

we say $(t, f(t))$ is an *inflection point of* f.

Look now at Figure 6.15 where we have sketched a function f and the related functions f', f'', and $A_t f$. The fact that f'' is positive at each point tells us that f' is increasing at each point (see Theorem 6.3). This in turn tells us that f is "turning upward" as we move from left to right. But $A_t f$ has constant slope $f'(t)$ and hence does not "turn upward" as we move from left to right. It seems reasonable then, since f and $A_t f$ have the same slope at t, to suppose that f stays above $A_t f$ to the right of t. This is indeed the case as is shown in part (i) of the following theorem. Parts (ii), (iii), and (iv) may be interpreted in a similar way:

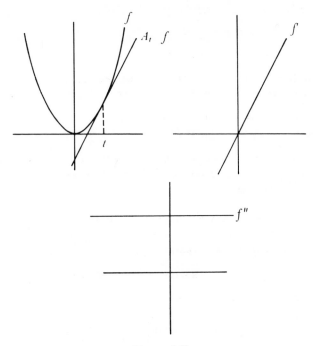

Figure 6.15

THEOREM 6.11   Let $f: \mathfrak{R} \to \mathfrak{R}$, $t \in dom\, f$, and let $f''(t)$ exist. Let $a < t < b$.

  i. If $f''(x) > 0$ for each $x \in (t, b)$, then $f(x) > A_t f(x)$ for each $x \in (t, b)$.

  ii. If $f''(x) < 0$ for each $x \in (t, b)$, then $f(x) < A_t f(x)$ for each $x \in (t, b)$.

  iii. If $f''(x) > 0$ for each $x \in (a, t)$, then $f(x) > A_t f(x)$ for each $x \in (a, t)$.

  iv. If $f''(x) < 0$ for each $x \in (a, t)$, then $f(x) < A_t f(x)$ for each $x \in (a, t)$.

We shall prove (i) by the method of indirect argument:

Proof of i.   Suppose $f''(x) > 0$ for $x \in (t, b)$ and that $f(c) \le A_t f(c)$ for some c such that $t < c < b$. We shall have proved our theorem if we can reach a contradiction. Since f'' is defined in $[t, b)$, we have both f' and f continuous in $[t, b)$. We now have the implication chain:

(a) $f(c) \le A_t f(c)$ and⎫
(b) $t < c < b$ ⎬

    $\Rightarrow^1$ there exists d such that $t < d < c$ and

$$f'(d) = \frac{f(c) - f(t)}{c - t} \le^2 \frac{A_t f(c) - f(t)}{c - t} =^3 f'(t)$$

$\Rightarrow^4$ there exists e such that $t < e < d$ and

$$f''(e) = \frac{f'(d) - f'(t)}{d - t} \le^5 0.$$

(1) From (b) and Theorem 6.7, (Mean Value Theorem).
(2) From (a) and the fact that $c - t$ is positive.
(3) Equation 5.4 (p. 223).
(4) Theorem 6.7, (Mean Value Theorem).
(5) Since $d - t > 0$ and we have shown $f'(d) \le f'(t)$.

This contradicts our assumption that $f''(x) > 0$ if $x \in (t, b)$. Thus, $f(x) > A_t f(x)$ in (t, b). The proofs for (ii), (iii), and (iv) are similar and will be left to the exercises. ∎

We can sketch a function more accurately if we know its inflection points and its concavity at each point of its domain. The following theorem gives us a means of finding this information:

THEOREM 6.12 Let $f : \Re \to \Re$, $t \in \Re$, such that f'' is continuous at t.
  i. If $f''(t) > 0$, then f is concave upward at t.
  ii. If $f''(t) < 0$, then f is concave downward at t.
  iii. If $(t, f(t))$ is an inflection point of f, then $f''(t) = 0$.
  iv. If $f''(t) = 0$ and $f''(x) < 0$ for $a < x < t$, $f''(x) > 0$ for $t < x < b$, $(t, f(t))$ is an inflection point of f.
  v. If $f''(t) = 0$ and $f''(x) > 0$ for $a < x < t$, $f''(x) < 0$ for $t < x < b$, $(t, f(t))$ is an inflection point of f.

Proof

  i. Since f'' is continuous at t, it is positive in some (a, b) where $a < b < t$. Hence (by Theorem 6.11i and iii), f is concave upward at t (see Definition 6.10).
  ii. Similar to (i).
  iii. If $f''(t) > 0$, f is concave upward by (i), and if $f''(t) < 0$, f is concave downward at t by (ii). But both of these are impossible if $(t, f(t))$ is an inflection point of f (read the definitions carefully). Hence $f''(t)$ must be zero.
  iv. Follows immediately from Theorem 6.11i and iv and Definition 6.10i.
  v. Follows immediately from Theorem 6.11ii and iii and Definition 6.10ii. ∎

If $f'(t) = 0$, $A_t f(x) = f'(t)(x - t) + f(t) = f(t)$, and $A_t f$ is a constant function. Hence, if f is concave upward at t, it seems evident that $(t, f(t))$ is a relative minimum point of f, and that if f is concave downward at t,

$(t, f(t))$ is a relative maximum point of f. Such is indeed the case as the following theorem states:

THEOREM 6.13 (The Second Derivative Test.) Let $f: \mathscr{R} \to \mathscr{R}$. If $f'(t) = 0$ and f'' is continuous at t, then
  i. $(t, f(t))$ is a relative maximum point of f if $f''(t) < 0$, and
  ii. $(t, f(t))$ is a relative minimum point of f if $f''(t) > 0$.

Proof of i. By Theorem 6.12i f is concave downward at t so $f(x) > A_t f(x) = f(t)$, a constant, for each x in some $B(\tilde{t}; r)$. Thus, by Definition 6.4 f is a relative maximum point of f. The proof for (ii) is similar and will be left to the exercises. ∎

We now have a second method for determining whether a point at which $f'(t) = 0$ is a relative maximum or a relative minimum. If the second derivative $f''(t)$ exists and is continuous, $(t, f(t))$ will be a relative maximum if $f''(t)$ is negative and will be a relative minimum if $f''(t)$ is positive.

Example 4. We wish to find maxima, minima, and inflection points for the function $h(x) = x^3 - 3x - 2$.

We compute the first and second derivatives and find the set of zeros of the first derivative. We then test each point of the set of zeros of f' by the second derivative test:

$$h'(x) = 3x^2 - 3$$
$$h''(x) = 6x.$$

Setting $h'(x) = 3x^2 - 3 = 0$, we find that the critical points are at ± 1. Then $h''(1) = 6$ and $h''(-1) = -6$, so $(1, h(1))$ is a relative minimum point and $(-1, h(-1))$ a relative maximum point. To test for inflection points, we set $h''(x) = 6x = 0$, or $x = 0$. Hence, if there is an inflection point it is at $(0, h(0))$ (Theorem 6.12iii). We note that if $x > 0$, then $h''(x) > 0$, and if $x < 0$, then $h''(x) < 0$.

Thus, by Theorem 6.12iv, $(0, h(0))$ is indeed an inflection point. The function h is sketched in Figure 6.16. We shall discuss curve sketching in the next section. ∎

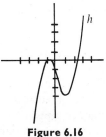

Figure 6.16

EXERCISES

Find all relative extrema and inflection points for the following functions. Specify the intervals in which the functions are concave upward and those in which they are concave downward:

1. $f(x) = x^2 + x - 2$.   2. $f(x) = 4x - x^2 - 3$.

3. $f(x) = \tan x$.   4. $f(x) = \sec x$.

5. $f(x) = 2x^3 + 9x^2 + 12x + 6$.   6. $f(x) = 2x^3 - 15x^2 - 36x + 4$.

7. $f(x) = 3x^4 + 8x^3 + 6x^2 - 48x - 30$.

8. $f(x) = 3x^4 + 16x^3 + 6x^2 - 72x + 16$.

9. $f(x) = \cos^2 x - \sin^2 x$.   10. $f(x) = \sin^2\left(\dfrac{x}{2}\right)$.

11. Prove parts ii), iii) and iv) of Theorem 6.11.

12. Prove Theorem 6.13ii).

ANSWERS

1. Rel. min. at $(-\frac{1}{2}, f(-\frac{1}{2}))$; concave upward in $(-\infty, \infty)$; no inflection points. **3.** No rel. extrema; inflection at $(n\pi, 0)$; n an integer; concave downward in $\left(n\pi - \dfrac{\pi}{2}, n\pi\right)$; concave upward in $\left(n\pi, n\pi + \dfrac{\pi}{2}\right)$.

5. Rel. min. at $(-1, f(-1))$; rel. max. at $(-2, f(-2))$; concave upward for $x > \dfrac{-3}{2}$; concave downward for $x < \dfrac{-3}{2}$; inflection at $\left(\dfrac{-3}{2}, f\left(\dfrac{-3}{2}\right)\right)$.

7. Rel. min. at $(1, f(1))$; concave upward in $(-\infty, -1)$, $(-\frac{1}{3}, \infty)$; concave downward in $(-1, -\frac{1}{3})$; inflection points at $(-1, f(-1))$ and $(-\frac{1}{3}, f(-\frac{1}{3}))$.

9. Rel. max. at $(n\pi, f(n\pi))$; rel. min. at $\left(\dfrac{(2n+1)\pi}{2}, f\left(\dfrac{2n+1}{2}\pi\right)\right)$; inflection at $\left(\dfrac{(2n+1)\pi}{4}, f\left(\dfrac{2n+1}{4}\pi\right)\right)$; concave upward in $\left(\dfrac{(4n+1)\pi}{4}, \dfrac{(4n+3)\pi}{4}\right)$; concave downward in $\left(\dfrac{(4n-1)\pi}{4}, \dfrac{(4n+1)\pi}{4}\right)$.

4. CURVE SKETCHING

We now have at our command methods that are quite useful in curve sketching. We shall develop our techniques principally through the use of examples.

Example 5. Sketch the set of all points (x, y) such that $y = x^2 - x + 1$. We can write $y = f(x) = x^2 - x + 1$. Following Procedure 6.9, we find that f has a relative minimum at $x = \frac{1}{2}$. We calculate $f''(x) = 2$ and note that this is positive for all x. Hence, by Theorem 6.12 there are no inflection points and f is concave upward everywhere. Thus, by plotting a few points, including the relative minimum point, we can make a

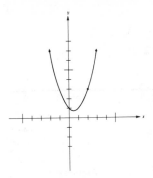

Figure 6.17

fairly accurate sketch (see Figure 6.17).

x	$f(x)$
-2	7
-1	3
0	1
$\frac{1}{2}$	$\frac{3}{4}$ (rel. min.)
1	1
2	3
3	7

∎

Example 6. We wish to sketch the sine curve, $f(x) = \sin x$. Knowing that the sine function is periodic and has period 2π, we shall sketch the graph of f for the interval $\left(\dfrac{-\pi}{4}, \dfrac{9\pi}{4}\right)$. (We choose our interval a little larger than the interval $[0, 2\pi]$ in which we are interested in order to take proper care of the end points, 0 and 2π.) To find possible extrema we find $f'(x) = \cos x$ and set $\cos x = 0$. The only solutions in the given interval are $\dfrac{\pi}{2}$ and $\dfrac{3\pi}{2}$. We apply the second derivative test to the these points. We find $f''(x) = -\sin x$, so $f''\left(\dfrac{\pi}{2}\right) = -\sin \dfrac{\pi}{2} = -1 < 0$ and $f''\left(\dfrac{3\pi}{2}\right) = -\sin \dfrac{3\pi}{2} = 1 > 0$. Hence, f has a relative maximum at $\dfrac{\pi}{2}$ and a relative minimum at $\dfrac{3\pi}{2}$. Finding the function values at these two points we begin our sketch as in Figure 6.18.

To find possible inflection points we find $f''(x) = -\sin x$ and set $-\sin x = 0$ to obtain $x = 0$, $x = \pi$, and $x = 2\pi$ as the solutions. We find $-\sin x > 0$ for x in $\left(\dfrac{-\pi}{4}, 0\right) \cup (\pi, 2\pi)$; $-\sin x < 0$ for x in $(0, \pi) \cup \left(2\pi, \dfrac{9\pi}{4}\right)$. Thus, by Theorem 6.12, the points $(0, 0)$, $(\pi, 0)$, and $(2\pi, 0)$ are

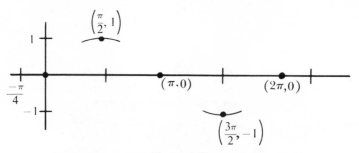

Figure 6.18

inflection points and f is concave upward for x in $\left(\dfrac{-\pi}{4}, 0\right) \cup (\pi, 2\pi)$ and

concave downward for x in $(0, \pi) \cup \left(2\pi, \dfrac{9\pi}{4}\right)$. We plot the inflection

points (see Figure 6.18). With the information we now have, we can fill in the remainder of the sketch with a fair degree of accuracy if we are careful to keep the proper concavity at each point (see Figure 6.19). ∎

Example 7. We wish to sketch the set of points (x, y) in $\mathfrak{R}^2$ such that $y = x^4 + 4x^3 - 2$. We set $y = f(x) = x^4 + 4x^3 - 2$. Then $f'(x) = 4x^3 + 12x^2 = 4x^2(x + 3)$. The only solutions to $f'(x) = 0$ are then $x = 0$ and -3. We find $f''(x) = 12x^2 + 24x = 12x(x + 2)$. We apply the second derivative test to f at 0 and at -3 to obtain $f''(0) = 0$, $f''(-3) = 36$. This test tells us that f has a relative minimum at -3 but gives no information about f at 0. Hence, we examine the first derivative and find that $f'(x) > 0$ for x in $(-3, 0) \cup (0, \infty)$. Thus, by Theorem 6.8iii, f is strictly increasing at 0 and so does not have an extremum there.

Returning to the second derivative function, we find that

$$f''(x) = 0 \text{ if } x = 0 \text{ or } x = -2,$$
$$f''(x) > 0 \text{ if } x < -2 \text{ or } x > 0,$$
$$f''(x) < 0 \text{ if } -2 < x < 0.$$

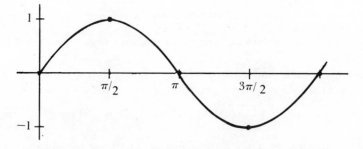

Figure 6.19

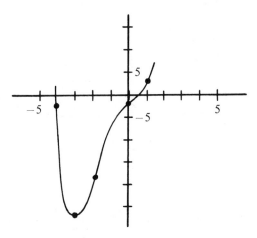

Figure 6.20

Thus, $(0, f(0))$ and $(-2, f(-2))$ are inflection points of f; f is concave upward for x in $(-\infty, -2) \cup (0, \infty)$ and concave downward for x in $(-2, 0)$. We compute the function values at 0, -2, and -3 and thus determine the relative minimum point and the inflection points of f. It is then helpful to compute a few more function values. These, together with the information on concavity that we have found, give us the sketch shown in Figure 6.20. (Note that we have used different scales on the x and y axes.)

x	$f(x)$
-4	-2
-3	-29 (rel. min. point)
-2	-18 (inflection point)
0	-2 (inflection point)
1	3 ∎

Example 8. We wish to sketch the set of points (x, y) in $\mathfrak{R}^2$ such that $x^2y^2 - 4x^2 + y^2 = 0$. Solving for y we find $y = \dfrac{\pm 2x}{\sqrt{x^2 + 1}}$. We shall obtain our sketch by setting $f(x) = \dfrac{2x}{\sqrt{x^2 + 1}}$ and then sketching the graphs of f and $-f$ on the same pair of axes. We obtain

$$f(x) = \frac{2x}{\sqrt{x^2 + 1}} = \frac{2x}{(x^2 + 1)^{1/2}}$$

$$f'(x) = \frac{2}{(x^2 + 1)^{3/2}}$$

$$f''(x) = \frac{-6x}{(x^2 + 1)^{5/2}}.$$

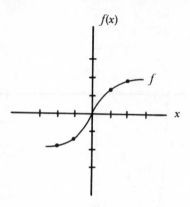

Figure 6.21

We note that $f'(x)$ is everywhere positive so there are no relative extrema and f is strictly increasing everywhere by Theorem 6.3. The only solution to $f''(x) = 0$ is $x = 0$; hence, $(0, f(0))$ is the only possible inflection point. We note that $f''(x)$ is positive if x is negative and negative if f is positive; hence, by Theorem 6.12, f is concave upward for x in $(-\infty, 0)$ and concave downward for x in $(0, \infty)$, and at $x = 0$ f has an inflection point. By plotting just a few more points we are now able to make a fairly accurate sketch (see Figure 6.21).

x	$f(x)$
-2	$\dfrac{-4}{\sqrt{5}} = 1.8$ (approx.)
-1	$-\sqrt{2} = 1.4$ (approx.)
0	0 (inflection point)
1	$\sqrt{2} = 1.4$ (approx.)
2	$\dfrac{4}{\sqrt{5}} = 1.8$ (approx.)

From this sketch we are led to suspect that perhaps f is *bounded*, that is, there is a real number M such that $f(x) \leq M$ for every x. To check this we set $f(x) = a = \dfrac{2x}{(x^2 + 1)^{1/2}}$ and solve for x to obtain $x = \dfrac{\pm a}{\sqrt{4 - a^2}}$. These equations have solutions if and only if $|a| < 2$. Hence, since f is strictly increasing we conclude that $\lim\limits_{x \to \infty} f(x) = 2$ (the reader will check the details in Exercise 17). Similarly $\lim\limits_{x \to -\infty} f(x) = -2$. This enables us to improve our sketch of the graph of f. By adding the graph of $(-f)$ to the set of points sketched we complete the figure (see Figure 6.22). ∎

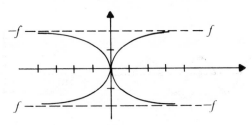

Figure 6.22

EXERCISES

As Exercises 1 to 10 sketch the graphs of the functions in Exercises 1 to 10 at the end of Section 3. Show all relative extreme points and inflection points.

In Exercises 11 to 16 sketch the sets of points (x, y) that satisfy each of the following equations. Show all relative extreme points and inflection points:

11. $x^2y^2 - 4x^2 - y^2 = 0.$ 12. $x^2y^2 - 4x^2 + y^2 = 0.$

13. $(y + 1)^3 = x - 1$ (Hint: express x as a function of y).

14. $y^4 + 4y^3 - 2 - x = 0.$ 15. $x^3 + y^2 = 0.$

16. $x^3 + y^2 = 1.$

17. Let $f: \mathcal{R} \to \mathcal{R}$ such that f is increasing on $(-\infty, \infty)$ and such that the equation $f(x) = a$ has a solution if and only if $|a| < M$. Show that $\lim_{x \to \infty} f(x) = M.$

ANSWERS

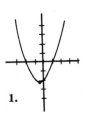

1.

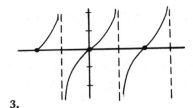

3.

5.

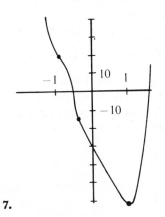

7.

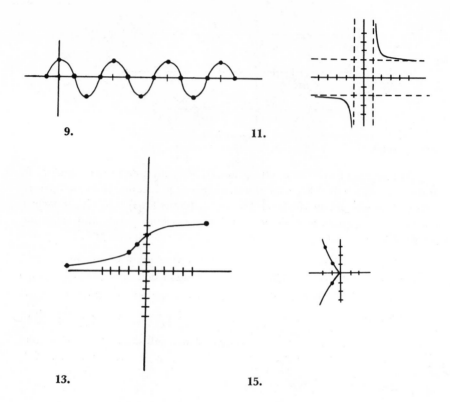

9.   11.

13.   15.

5. VELOCITY AND ACCELERATION

Consider now an automobile moving down a straight road from A to B, where A and B are two miles apart. Suppose that the position, in miles from A, t minutes from starting time, is given by $p(t) = \frac{3}{2}t^2 - \frac{1}{2}t^3$. Figure 6.23 is a sketch of the graph of p, where the time t is plotted on the horizontal axis and the distance, $p(t)$, from A to the automobile, is plotted on the vertical axis.

For the given two minutes the *average speed* of our automobile has been $\dfrac{p(2) - p(0)}{2 - 0} = \dfrac{2 - 0}{2 - 0} = 1$ mile/min. or 60 m.p.h. Can we conclude that at time $t_0 = 1$ the driver was not breaking the 65 m.p.h. speed limit? To decide this we must give a precise meaning to the statement "the car was traveling at x miles per hour at time $t_0 = 1$." In other words, just what is the speedometer reading at $t_0 = 1$ supposed to mean? We should like it to

t	$p(t)$
0	0
$\frac{1}{2}$	$\frac{5}{16}$
1	1
$\frac{3}{2}$	$\frac{27}{16}$
2	2

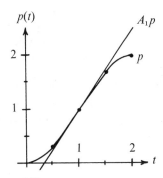

Figure 6.23

mean that if the reading had remained constant at x m.p.h., or $\dfrac{x}{60} = k$
miles/min., until some time $t > 1$, then the car would have traveled $k(t - 1)$
miles between t_0 and t. If this had been the case then the position at time
$t > 1$ would have been given by $A(t) = p(1) + k(t - 1)$. But this is an
affine function whose graph passes through the point $\begin{pmatrix} 1 \\ 1 \end{pmatrix} = \begin{pmatrix} t_0 \\ p(t_0) \end{pmatrix}$. Now
of all possible such affine functions the one that seems to best approximate
the behavior of our automobile at $t_0 = 1$ is the affine approximation $A_1 p$.
Accordingly we set $A_1 p(t) = A(t)$ or $p'(t)(t - 1) + p(1) = p(1) + k(t - 1)$
from which we deduce $p'(t) = k$. Hence, we make the following definition:

DEFINITION 6.14 Let an object Q be moving in a straight line
path in such a way that its position p is a differentiable function of the
time t. The *instantaneous velocity of Q at time t* is defined to be $p'(t)$. The
speed of Q at time t is defined to be $|p'(t)|$.

(Note: We shall often refer to the "instantaneous velocity" simply as the
"velocity.")

Thus, the velocity of our car at $t = 1$ was 90 m.p.h. in the direction of B,
since $p'(t) = 3t - \frac{3}{2}t^2$ and $p'(1) = \frac{3}{2}$ miles/min. We wonder now whether
our automobile has at some point reached an even higher velocity. Hence,
we find the maximum value of $p'(t)$ on $[0, 2]$ by setting $p''(t) = 3 - 3t = 0$
to obtain $t = 1$. Thus, the car reached its highest velocity at $t = 1$.

We now sketch the graph of the instantaneous velocity of the car as a
function of the time between 0 and 2 (see Figure 6.24). We note that the

velocity is increasing in the interval $(0, 1)$ and decreasing in $(1, 2)$. In $(0, 1)$ the velocity increases from 0 to $\frac{3}{2}$ so we might say that our velocity increased at $\frac{3}{2}$ miles/min. Again this is only an average increase. For reasons similar to those for "instantaneous velocity" we make the following definition:

DEFINITION 6.15 Let an object Q be moving in a straight line path in such a way that its position p is a twice differentiable function of the time t. The *instantaneous acceleration* of Q at time t is defined to be $p''(t)$.

(Note: We shall sometimes simply say "acceleration" when we mean instantaneous acceleration.)

Acceleration, then, is the rate at which the velocity is changing. If the acceleration is positive, the velocity is increasing and if it is negative, the velocity is decreasing. The acceleration of the automobile of our previous discussion is given by $p''(t) = 3 - 3t$. The graph of this function is sketched in Figure 6.24b. This sketch tells us that the automobile was speeding up before $t = 1$ and slowing down thereafter.

Example 9. Sir Isaac Newton discovered that if air resistance is neglected, the height above the starting point of an object thrown vertically upward from a position near the surface of the earth with initial velocity K is given approximately by $p(t) = Kt - 16t^2$, where t is the time in seconds from the beginning. Suppose now that a ball is thrown downward from a cliff at a velocity of 50 ft./sec. If the cliff is 200 ft. high, with what velocity will the ball strike the ground? What will be the acceleration at this time?

Since the ball is thrown *downward*, its initial *upward* velocity K is -50 ft./sec. Thus, its position at time t is given by $p(t) = -50t - 16t^2$. To find the time of fall, we let $p(t) = -200$ (since the final position is 200 ft. *below* the initial position) and solve for t to obtain $t = 2.3$ (approx.). The velocity is given by $p'(t) = -50 - 32t$. Thus, the final velocity is $p'(2.3) = -123.6$ ft./sec. (approx.). The negative sign indicates that the motion is downward.

The acceleration at time t is given by $p''(t) = -32$. Hence, the acceleration is -32 ft./sec.^{2} at all times, including the one in question. (For this reason the number $g = 32$ is called the *constant acceleration of gravity*.) ∎

Notice that the velocity of an object moving in a straight line may be negative, whereas its speed can never be negative. Therefore, the velocity of an object tells us not only how fast it is moving but the direction in which it is moving. We wish now to generalize this idea.

Suppose that an object Q is moving in a coordinate plane in such a way that its position at time t is given by $p(t) = \begin{pmatrix} x \\ y \end{pmatrix} = \begin{pmatrix} -t^2 + 4t \\ t^3 - 5t^2 + 6t \end{pmatrix}$,

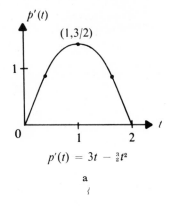

$$p'(t) = 3t - \tfrac{3}{2}t^2$$

a

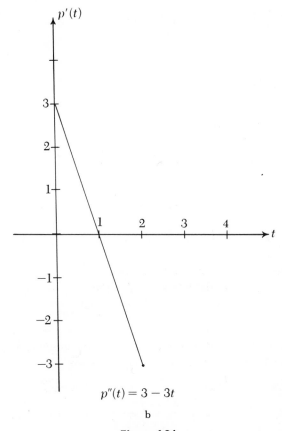

$$p''(t) = 3 - 3t$$

b

Figure 6.24

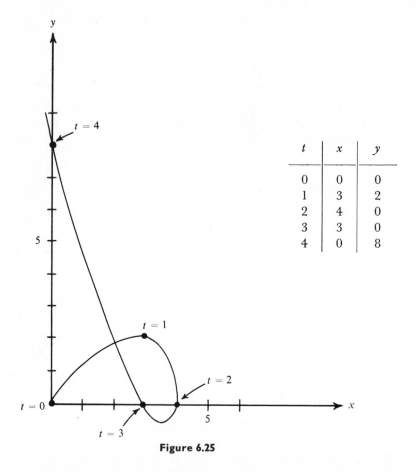

t	x	y
0	0	0
1	3	2
2	4	0
3	3	0
4	0	8

Figure 6.25

$0 \leq t \leq 4$. A sketch of the path of the object is given in Figure 6.25.

We wish to describe mathematically the state of motion of the object at time t_0. We begin with the following definition:

DEFINITION 6.16 Let an object P be moving in a coordinate space in such a way that its position is given by the function $p(t)$, where t is the time measured from some fixed instant. Then if $t_0 < t_1$, $\dfrac{p(t_1) - p(t_0)}{t_1 - t_0}$ will be called the *average velocity of P between t_0 and t_1*. The non-negative real number $\left\| \dfrac{p(t_1) - p(t_0)}{t_1 - t_0} \right\|$ will be called the *average speed of P between t_0 and t_1*.

Thus, the average velocity between $t = 1$ and $t = 4$ of the object Q of our example is $\dfrac{p(4) - p(1)}{4 - 1} = \begin{pmatrix} -1 \\ 2 \end{pmatrix}$, and the average speed of Q between $t = 1$ and $t = 4$ is $\left\| \begin{pmatrix} -1 \\ 2 \end{pmatrix} \right\| = \sqrt{5}$. Notice that the average velocity of Q is a two-dimensional vector whereas the average speed is a non-negative real number. If we are given only the position of Q at $t = 1$ and the average speed of Q from $t = 1$ to $t = 4$, we know only that Q is somewhere on the circle of radius $(4 - 1)\sqrt{5} = 3\sqrt{5}$ and center $\begin{pmatrix} 3 \\ 2 \end{pmatrix}$. But if we are given the average velocity vector between $t = 1$ and $t = 4$ and the position at $t = 1$, we know the exact position of Q at $t = 4$.

However, neither the average velocity between $t = 1$ and $t = 4$ nor the average speed between $t = 1$ and $t = 4$ portrays accurately the state of motion of Q at the time $t = 1$. For if Q had proceeded from its position at $t = 1$ at the average velocity $\begin{pmatrix} -1 \\ 2 \end{pmatrix}$, then at time $t = 2$ it would have been at $\begin{pmatrix} 2 \\ 4 \end{pmatrix}$, whereas in actuality it is at $\begin{pmatrix} 4 \\ 0 \end{pmatrix}$. Taking the average velocity between $t = 1$ and $t = 2$ gives a better, but still not adequate, idea of the behavior of Q at $t = 1$. Indeed, the shorter the period of time over which the average velocity is taken, the better our description of the state of motion at the given instant. It seems reasonable then to describe the *instantaneous velocity* at t_0 as $\lim\limits_{t \to t_0} \dfrac{p(t) - p(t_0)}{t - t_0}$, if this limit exists. But this is just $p'(t_0)$. Hence, we make the following definition:

DEFINITION 6.17 Let an object Q be moving in a coordinate space. If its position at time t is given by $p(t)$, where $p: \Re \to \Re^n$ and $p'(t_0)$ is defined, then $p'(t_0)$ will be called the *instantaneous velocity of Q at time t_0* and $\|p'(t_0)\|$ will be called the *instantaneous speed of Q at time t_0*.

(Note: We shall usually say simply "speed" for "instantaneous speed" and "velocity" for "instantaneous velocity.")

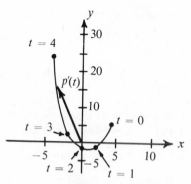

Figure 6.26

In our example the instantaneous velocity of Q at time t is $p'(t) = \begin{pmatrix} -2t + 4 \\ 3t^2 - 10t + 6 \end{pmatrix}$, and at $t = 1$ the instantaneous velocity is $p'(1) = \begin{pmatrix} 2 \\ -1 \end{pmatrix}$. The speed at $t = 1$ is $\left\| \begin{pmatrix} 2 \\ -1 \end{pmatrix} \right\| = \sqrt{5}$.

We now evaluate $p'(t)$ at $t = 0$, 1, 2, 3, and 4 and plot these points on a pair of axes (see Figure 6.26). We join these points with a smooth curve and the resulting sketch gives us a picture of how the velocity of Q is changing with time. The length of the vector $p'(t)$ indicates the speed of Q at time t,

t	x	y
0	4	6
1	2	−1
2	0	−2
3	−2	3
4	−4	14

and the direction of the vector indicates the direction in which Q is moving at time t. For example, at $t = 2$ we can see that Q is moving in the direction of the negative y-axis at a speed of 2 units per unit of time. Compare this information with the path of Q in Figure 6.25. Also compare the sketches of Figures 6.25 and 6.26 at $t = 0$, 1, 3, and 4.

Now $p'(t)$ is also a function of the time t, that is, the velocity of Q is varying with the time. A change of velocity is called an *acceleration*. We now make the following definition which is analogous to the first part of Definition 6.16:

DEFINITION 6.18 Let an object Q be moving in a coordinate space. If the velocity of Q at time t_0 is V_0 and its velocity at time t_1 is V_1, then $\dfrac{V_1 - V_0}{t_1 - t_0}$ will be called the *average acceleration of Q between t_0 and t_1.*

The average acceleration over a period of time is not an adequate description of what is happening at a given instant, as an examination of Figure 6.26 shows. For instance, the average acceleration (or change of the velocity vector per unit of time) between $t = 0$ and $t = 4$ is $\begin{pmatrix} -2 \\ 2 \end{pmatrix}$. If the velocity vector had changed from its value at $t = 0$ in accordance with the average acceleration, then at time $t = 1$ it would have been $\begin{pmatrix} 4 \\ 6 \end{pmatrix} + \begin{pmatrix} -2 \\ 2 \end{pmatrix} = \begin{pmatrix} 2 \\ 8 \end{pmatrix}$, whereas actually it is $\begin{pmatrix} 2 \\ -1 \end{pmatrix}$. Hence, we make the following definition which should be compared with Definitions 6.15 and 6.17:

DEFINITION 6.19 Let an object P be moving in a coordinate space. If its position at time t is given by $p(t)$, where $p : \Re \to \Re^n$, and p'' is defined at t_0, $p''(t_0)$ will be called the *instantaneous acceleration of P at time t_0*, or simply the *acceleration* of P at time t_0.

Returning to our example, we see that the instantaneous acceleration at time t is given by $p''(t) = \begin{pmatrix} -2 \\ 6t - 10 \end{pmatrix}$ and at $t = 1$ the acceleration is $p''(1) = \begin{pmatrix} -2 \\ -4 \end{pmatrix}$. The acceleration of P is sketched in Figure 6.27. It should be instructive to the reader to compare Figures 6.26 and 6.27, noting that at each t, $p''(t)$ gives us the rate at which $p'(t)$ is changing.

Example 10. A particle is moving in 3-space according to the function $p(t) = \begin{pmatrix} \cos t \\ \sin t \\ -16t^2 \end{pmatrix}$. What is its velocity and acceleration at $t = 0$?

We compute $p'(t) = \begin{pmatrix} -\sin t \\ \cos t \\ -32t \end{pmatrix}$; hence, $p'(0) = \begin{pmatrix} 0 \\ 1 \\ 0 \end{pmatrix}$. The acceleration is given by $p''(t) = \begin{pmatrix} -\cos t \\ -\sin t \\ -32 \end{pmatrix}$, so $p''(0) = \begin{pmatrix} -1 \\ 0 \\ -32 \end{pmatrix}$. ∎

t	x	y
0	-2	-10
1	-2	-4
2	-2	2
3	-2	8
4	-2	14

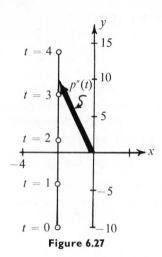

Figure 6.27

EXERCISES

1. Suppose an object is dropped from a tower (then its initial velocity is zero).
   (a) Find its position at the end of 1 second.
   (b) Find its velocity at the end of 1 second.
   (c) Find its acceleration at the end of 1 second.

2. A ball is thrown vertically upward from ground level with an initial velocity of 64 ft./sec.
   (a) How high will it go?
   (b) With what velocity will it strike the ground?

In Exercises 3 to 5 an object is moving so that its position at time t is given by $p(t)$. Find its velocity v and acceleration a, at time t_0.

3. $p(t) = t^2 + \sqrt{t}, t_0 = 2.$

4. $p(t) = \sin t + \cos t, t_0 = \pi.$

5. $p(t) = \dfrac{t^2}{1 + t^2}, t_0 = 1.$

6. A stone is dropped from a hovering helicopter. A second stone is thrown downward after the first has fallen 100 ft. With what velocity must the second stone be thrown downward in order to catch up with the first one 10 sec. after the second stone is thrown?

7. Gabby Street once caught a baseball dropped from the Washington Monument, which is 555 ft. high. What was the velocity of the ball when it was caught?

8. Two particles, P_1 and P_2, start at the same time, $t = 0$, from the same point and move along the same line. The first moves so that its position is given by $p_1(t) = t^2 + 2t$ and the second so that its position is given by $p_2(t) = t^3 + 2t^2$. Find the velocities and accelerations at the time when the particle that begins more slowly catches the other.

In Exercises 9 to 11 a particle is moving according to the function p. What is its velocity, speed, and acceleration at time t? (Notice particularly

Exercise 9; the particle is moving around a circle of radius a at a uniform speed, yet its velocity and acceleration are varying!)

9. $p(t) = \begin{pmatrix} a\cos t \\ a\sin t \end{pmatrix}$.

10. $p(t) = \begin{pmatrix} \sin t \\ \cos t \\ t \end{pmatrix}$.

11. $p(t) = \begin{pmatrix} t \\ t^2 \\ t^3 \end{pmatrix}$.

> "I shot an arrow into the air.
> It fell to earth, I know not where."

If the ground is level and if Longfellow gives us the initial velocity, we can tell him "where." For, neglecting air resistance, the position of a projectile given initial velocity $\begin{pmatrix} m \\ n \\ k \end{pmatrix}$ is given by $p(t) = \begin{pmatrix} mt \\ nt \\ kt - 16t^2 \end{pmatrix}$, where the third coordinate gives the distance above ground. In the following Exercises 12 to 17 let us assume that the initial velocity was $\begin{pmatrix} 200 \\ 0 \\ 80 \end{pmatrix}$ ft./sec.

12. How long did the arrow remain in the air?

13. Where did it land?

14. What was its velocity at the end of (a) 1 sec? (b) 4 sec?

15. What was its speed at the end of (a) 1 sec? (b) 4 sec?

16. At what time during its flight was the speed least?

17. What was the acceleration at the end of (a) 1 sec? (b) 2 sec? (c) t sec, $0 \le t \le 5$?

ANSWERS

4. $v = -1, a = 1.$ **5.** $v = \dfrac{1}{2}, a = \dfrac{-1}{2}.$ **6.** 90 ft./sec.

8. $p_1'(1) = 4, p_2'(1) = 7.$ **10.** $p''(t) = \begin{pmatrix} -\sin t \\ -\cos t \\ 0 \end{pmatrix}, \|p'(t)\| = \sqrt{2}.$
      $p_1''(1) = 2, p_2''(1) = 10.$

13. $\begin{pmatrix} 1000 \\ 0 \\ 0 \end{pmatrix}.$ **14.** (a) $\begin{pmatrix} 200 \\ 0 \\ 48 \end{pmatrix}$, (b) $\begin{pmatrix} 200 \\ 0 \\ -48 \end{pmatrix}.$ **16.** $t = 2.5.$

6. RELATED RATES

Often one physical quantity is related to another in such a way that we can express the first as a function of the second. In such a case we say that the first is *functionally* related to the second. For example, in the previous section, we discussed the position of objects as a function of the time. In this case we thought of the derivative of the function as the velocity or rate of change of position with respect to time. It is frequently useful, therefore, to think of the derivative as the rate of change of one physical quantity with respect to a second.

It may be that one physical quantity is functionally related to a second that is in turn functionally related to a third. In such a case the composite function relates the first to the third. We may be given the first functional relation and the rate of change of the second quantity with respect to the third and wish to compute the rate of change of the first quantity with respect to the third. Such situations are called *related rates problems*. Consider the following problem:

Example II. A ladder 13 ft. long is being pulled away from a vertical wall at 10 ft./sec at the moment the foot of the ladder is 5 ft. from the wall. How fast is the top falling?

We formalize the preceding procedure as follows:

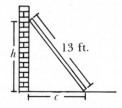

SOLUTION. Let h be the height of the top of the ladder from the ground and let c be the distance of the foot of the ladder from the wall. Then h can be expressed as a function of c by $h(c) = \sqrt{13^2 - c^2}$. But c is a function of the time t, say $c = f(t)$, even though we do not know the function f. Therefore, the height of the top at any given time is given by $h(c) = h(f(t)) = (h \circ f)(t)$. Thus, we have expressed the height of the ladder as a function of the time t, and the velocity of the top is given by $(h \circ f)'(t)$. We seem to have too little information to answer our question since we know neither the function f nor the particular time, say t_0. However, on closer examination we find that if we apply the chain rule we have all the information we need. We have $(h \circ f)'(t_0) = [(h' \circ f) \cdot f'](t_0) = h'(f(t_0)) \cdot f'(t_0)$. But $f(t_0)$, the position of the foot of the ladder at the given time, is 5 ft., and $f'(t_0)$, the velocity of the foot of the ladder, is 10 ft./sec. Hence,

$$(h \circ f)'(t_0) = h'(5)10. \text{ Now } h'(c) = \frac{-c}{\sqrt{13^2 - c^2}}, \text{ so } (h \circ f)'(t) = 10h'(5) =$$

$\dfrac{-50}{12}$ ft./sec. The negative sign indicates that the height is decreasing. ∎

THEOREM 6.20 Let X be a variable quantity (real number or n-vector) related to a real valued quantity t by the following composition:

$$X = f(Y), \qquad Y = g(t).$$

Let $Y_0 = g(t_0)$ and $g'(t_0)$ be specified. Then the rate of change of X with respect to t for the same value t_0 is

$$f'(g(t_0))g'(t_0) = f'(Y_0)g'(t_0).$$

(Note: t will usually, but not always, be *time*.)

Example 12. If the radius of a sphere is increasing at a rate of 1 in./min at the time t_0 when the sphere is 6 in. in diameter, how fast is the volume increasing?

SOLUTION. Recall that the volume of a sphere is given by $V(r) = \frac{4}{3}\pi r^3$, where r is the radius. Then $V'(r) = 4\pi r^2$. We also have $r(t_0) = 3$ in. and $r'(t_0) = 1$ in./min. Hence, the rate of increase of volume with respect to time is $V'(r(t_0)) \cdot r'(t_0) = V'(3) \cdot 1 = 4\pi(3)^2 = 36\pi$ cu. in./min. ∎

Example 13. A wheel 2 ft. in diameter is rolling along a road at 10 ft./sec. Compute the velocity and speed of a point P on the rim at the time P is 1 ft. from the surface of the road and directly behind the center of the wheel.

SOLUTION. Choose a coordinate system in such a way that the positive x-axis is on the road and in the direction of motion of the wheel, the y-axis is vertical, and the origin is at the bottom of the wheel (see Figure 6.28a). Suppose that the point P is at the origin. As the wheel moves along the x-axis, let θ be the distance the center has moved (see Figure 6.28b). The coordinates of P are given by $f(\theta) = \begin{pmatrix} x \\ y \end{pmatrix} = \begin{pmatrix} \theta - \sin\theta \\ 1 - \cos\theta \end{pmatrix}$. Now θ is a function of the time t and at the time t_0 in question, $\theta(t_0) = \dfrac{\pi}{2}$, $\theta'(t_0) = 10$ ft/sec.

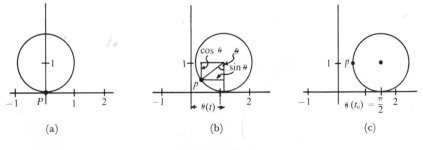

(a)   (b)   (c)

Figure 6.28

Hence, the velocity of P is $(f \circ \theta)'(t_0) = f'(\theta(t_0)) \cdot \theta'(t_0) = f'\left(\dfrac{\pi}{2}\right) \cdot 10 =$

$\begin{pmatrix} 1 - \cos \dfrac{\pi}{2} \\ \sin \dfrac{\pi}{2} \end{pmatrix} \cdot 10 = \begin{pmatrix} 10 \ \text{ft./sec} \\ 10 \ \text{ft./sec} \end{pmatrix}$. The speed is $\left\| \begin{matrix} 10 \\ 10 \end{matrix} \right\| = \sqrt{200} =$

$10\sqrt{2}$ ft./sec. ∎

EXERCISES

1. A man 6 ft. tall is walking away from a lamp post 10 ft. tall at a rate of 4 ft./sec. At what rate is the end of his shadow moving when he is 20 ft. from the post? K ft. from the post?

2. A man on a bridge 12 ft. high observes a boat passing under the bridge at a speed of 13 ft./sec. At what rate is the distance from the man to the boat increasing when the boat is 5 ft. past the bridge?

3. A conical tank with the vertex downward is 10 ft. tall and the diameter of the top is 6 ft. If water is being drained from the tank at 6 cu. ft./min., how fast is the level of the water falling when the depth of the water is 5 ft.?

4. A spherical toy balloon has a picture of Pinocchio on it. If Pinocchio's nose is 1 in. long when the diameter of the balloon is 4 in., and if the volume is increasing at 16 cu. in./sec., how fast is the nose growing? (The volume of a sphere is given by $V(r) = \frac{4}{3}\pi r^3$, where r is the radius.)

5. A particle is moving in a plane according to the function $h(t) = \begin{pmatrix} t^{\frac{1}{3}} \\ \dfrac{t+2}{3} \end{pmatrix}$.

   Find the points of the plane at which $h'(t) = \begin{pmatrix} k \\ 9k \end{pmatrix}$ for some k.

6. A ladder 15 ft. in length is leaning against a fence 8 ft. high. If the foot of the ladder is being pulled away from the fence at 10 ft./sec., at the time

Figure 6.29

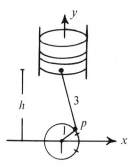

Figure 6.30

it is 6 ft. from the fence what is the velocity of the top of the ladder? (Let the positive direction of the first coordinate be toward the fence from the foot of the ladder.)

7. Water is being poured into a conical tank of height 12 in. and radius 2 in. at the rate of 3 cu. in./sec. (see Figure 6.29.) At what rate is the area of the water surface increasing when the volume is 2 cu. in.? (The volume of a cone is given by $V(r, h) = \frac{1}{3}\pi r^2 h$, where r is the radius and h is the height.)

8. The crankshaft, piston, and piston rod of a small gasoline engine are as represented in Figure 6.30, where the radius of the crank is 1, the length of the connecting rod is 3, and h is the height of the wrist pin above the center of the crankshaft. We consider the center of the crankshaft to be the origin and the horizontal line through the plane of rotation to be the x-axis. The y-axis is considered to be vertical. We shall compute the velocity of pivot P when $h = 3$ and the piston is moving downward at 36 in./sec.
   (a) Write x as a function of h by first writing h as a function of x and then solving for x.
   (b) Write y as a function of h by first writing h as a function of y and then solving for y.
   (c) Write $\binom{x}{y}$ as a function $f(h) = \binom{x}{y}$ by combining (a) and (b).
   (d) Compute $f'(3)$.
   (e) Compute $(f \circ h)'(t_0)$.

ANSWERS

1. 10 ft./sec. in both cases. **3.** $\dfrac{8}{3\pi}$ ft./min. **4.** $\dfrac{1}{2\pi}$ in./sec. **5.** $\begin{pmatrix} 3 \\ 29 \\ \frac{3}{3} \end{pmatrix}$ and

$\begin{pmatrix} -3 \\ -25 \\ 3 \end{pmatrix}$. **6.** $\begin{pmatrix} \frac{-2}{5} \text{ ft./sec.} \\ \frac{-36}{5} \text{ ft./sec.} \end{pmatrix}$. **8.** $\begin{pmatrix} \frac{-36}{35} + 2\sqrt{35} \text{ in./sec.} \\ -34 \qquad \text{in./sec.} \end{pmatrix}$.

PART III

INTEGRAL CALCULUS OF $\mathcal{R} \to \mathcal{R}^n$ FUNCTIONS

7

GEOMETRY OF VECTOR FUNCTIONS

INTRODUCTION

In Chapter 6 we used the derivative to sketch geometric representations of functions from $\mathcal{R}$ into $\mathcal{R}$. These are often fairly easy to represent since the figures are two-dimensional. But if our functions are from $\mathcal{R}^n$ into $\mathcal{R}^m$, it is sometimes hard to represent them in the usual manner. For example, if $f:\mathcal{R}^2 \to \mathcal{R}^2$, the "graph" is a subset of $\mathcal{R}^4$, and for most people it is very difficult to draw four-dimensional sketches. However, we shall find other ways of making sketches to help us to visualize the behavior of such functions.

1. FUNCTION-LOCI

If a subset S of a vector space is in some way determined by a vector function, S is called a *function-locus*. More precise uses of this term will be given in definitions that follow. The three types of function-loci which we shall discuss are the *graph*, the *range*, and the *K-level set*.

DEFINITION 7.1 (Explicitly Defined Function-Locus.) Let $f:\mathcal{R}^n \to \mathcal{R}^m$. The *graph* of f is the set of vectors in $\mathcal{R}^{n+m}$ denoted and defined by

$$\text{graph } f = \left\{ \begin{pmatrix} X \\ fX \end{pmatrix} \,\middle|\, X \in dom f \right\}.$$

The graph of f is called the *function-locus explicitly defined by f*.

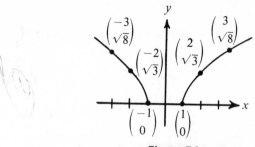

Figure 7.1

The graph of a function is perhaps the type of function-locus most familiar to the reader. However, many readers are accustomed to thinking of the graph of a function as a sketch. It is *not* a sketch but a set of elements of $\mathcal{R}^{n+m}$; a sketch of a function-locus is a pictorial representation of the set of points described.

Example Ia. Let $f(x) = \sqrt{x^2 - 1}$. Then f is a function from $\mathcal{R}^1$ into $\mathcal{R}^1$, and the function-locus explicitly defined by f is the subset of $\mathcal{R}^{n+m} = \mathcal{R}^2$, consisting of all points $\begin{pmatrix} x \\ f(x) \end{pmatrix} = \begin{pmatrix} x \\ \sqrt{x^2 - 1} \end{pmatrix}$ such that the second coordinate is defined.

To sketch this graph, we first determine *dom f*. This is easily seen to be $\{x \mid x \in R, |x| \geq 1\}$. Hence, using methods described in Chapter 6 we obtain Figure 7.1 as the sketch of *graph f*. ∎

DEFINITION 7.2 (Implicitly Defined Function-Locus.) Let $f: \mathcal{R}^n \to \mathcal{R}^m$ and let $K \in \mathcal{R}^m$. The *K-level set of f* is the set of vectors in $\mathcal{R}^n$ denoted and defined by

$$K\text{-lev} f = \{X \mid fX = K, X \in dom f\}.$$

A *K*-level set of f is called a *function-locus implicitly defined by f.*

(Note: *K-lev f* may well be the empty set as will be seen in Example 1b.)

Example Ib. We continue to examine the function f defined in Example 1a, namely, $f(x) = \sqrt{x^2 - 1}$. This time we seek several function-loci implicitly defined by f. These will be subsets of $\mathcal{R}^1$ since $n = 1$. To obtain 0-*lev f*, we set $f(x) = \sqrt{x^2 - 1} = 0$ and solve for x to obtain $x = \pm 1$. To obtain 1-*lev f*, we set $\sqrt{x^2 - 1} = 1$ and solve to obtain $x = \pm\sqrt{2}$; hence, $1\text{-}lev = \{-\sqrt{2}, \sqrt{2}\}$. Similarly, we obtain $2\text{-}lev f = \{-\sqrt{5}, \sqrt{5}\}$ and $(-1)\text{-}lev f = \emptyset$. Our sketches of 0-*lev f*, 1-*lev f*, and 2-*lev f* are shown in Figure 7.2. ∎

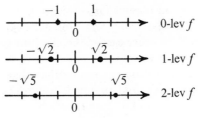

Figure 7.2

DEFINITION 7.3 (Parametrically Defined Function-Locus.) Let $f: \mathcal{R}^n \to \mathcal{R}^m$. The *range of f* is the set of vectors in $\mathcal{R}^m$ denoted and defined by

$$ran\, f = \{Y \mid Y = fX \text{ for some } X \text{ in } dom\, f\}.$$

This is often abbreviated

$$ran\, f = \{fX \mid X \in dom\, f\}.$$

The range of f is called the *function-locus parametrically defined by f*.

Example Ic. Again we examine the function $f(x) = \sqrt{x^2 - 1}$ of Example 1a, this time looking for the function-locus parametrically defined by f. Thus, we wish to determine those elements y in $\mathcal{R}$ such that $f(x) = \sqrt{x^2 - 1} = y$ for some x in $\mathcal{R}$, that is, those values of y for which a solution to this equation exists. This is clearly $\{y \mid y \in R, y \geq 0\}$. We sketch *ran f* in Figure 7.3. ■

NOTE: Function-loci are often considered without expressly mentioning the function involved. For example, if we write $y = 4x + 17$, we are simply substituting the symbol "y" for the symbol "$f(x)$" and understand that our explicitly defined function-locus is *graph f*. If we write $x^2 + y^2 = 1$, we have in mind a function g such that $g\begin{pmatrix} x \\ y \end{pmatrix} = x^2 + y^2$ and understand that our implicitly defined function-locus is 1-*lev g*. A parametric representation is often written in a form similar to $x = 2t^2$, $y = 4t$, or $\begin{pmatrix} x \\ y \end{pmatrix} = \begin{pmatrix} 2t^2 \\ 4t \end{pmatrix}$. The reader will learn by experience to determine from context the mode of description intended. The reader may well observe that the sketch of *graph f* (Figure 7.1) conveys more information than the sketch of *ran f* (Figure 7.3) and the sketches of the K-level sets (Figure 7.2) combined. Hence, he may ask, "Why bother with these other function-loci?"

Figure 7.3

There are many good reasons for considering them, of which we shall mention a few. One is that many sets cannot be described explicitly, for example a circle in $\mathcal{R}^2$. In other cases a sketch of the graph may be difficult to construct, as in those instances where $n + m \geq 4$, but still a great deal of information may be obtained by examining the range and the K-level sets. In still other cases implicitly and parametrically defined function-loci are useful in sketching the graph of a given function.

EXERCISES

For each of the functions defined below, sketch the graph, the range, and the K-level set. Indicate empty sets where necessary:

1. $f(x) = x$, $K = 1$.
2. $f(x) = \sqrt{x - 1}$, $K = 3$.

3. $f(x) = \sin x$, $K = 1$.
4. $f(x) = \cos x$, $K = 2$.

5. $f(x) = \sqrt{4 - x^2}$, $K = 0$.
6. $f(x) = \dfrac{1}{x}$, $K = 1$.

7. $f(x) = x^2 + 1$, $K = 5$.

Sketch in $\mathcal{R}^2$ the function-loci determined by the following equations. Indicate empty sets where necessary:

8. $x + y = 1$.
9. $y - x^2 = 0$.

10. $y + \dfrac{1}{x} - 1 = 0$.
11. $x^2 + x = 0$.

12. $x^3 + 2x^2 - y + 1 = 0$.

Sketch the function-loci defined parametrically by the following equations:

13. $X(t) = \begin{pmatrix} 1 \\ 2 \end{pmatrix} + t \begin{pmatrix} 1 \\ 1 \end{pmatrix}$, $t \in \mathcal{R}$.

14. $X\begin{pmatrix} s \\ t \end{pmatrix} = \begin{pmatrix} 1 \\ 1 \\ 1 \end{pmatrix} + s \begin{pmatrix} 1 \\ 0 \\ 0 \end{pmatrix} + t \begin{pmatrix} 0 \\ 1 \\ 1 \end{pmatrix}$, $\begin{pmatrix} s \\ t \end{pmatrix} \in \mathcal{R}^2$.

15. $f(t) = 2 \begin{pmatrix} \sin t \\ \cos t \end{pmatrix}$, $t \in \mathcal{R}$.

16. $x = t + 1$, $y = t$, $t \in \mathcal{R}$.

17. $x = 2t$, $y = t^2 - 1$, $t \in \mathcal{R}$ (Hint: express y as a function of x).

ANSWERS

1.

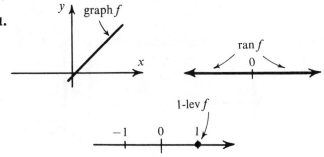

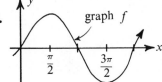

3.

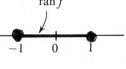

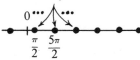

5.

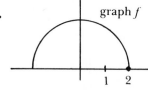

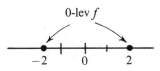

7.

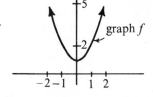

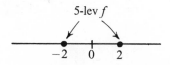

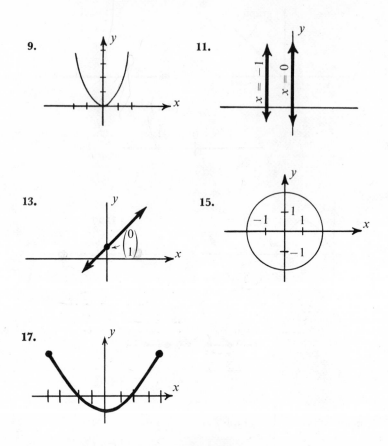

2. CONICS: THE ELLIPSE AND THE CIRCLE

We consider now the problems of expressing certain given sets as function-loci and of recognizing sets so expressed. We have already seen in Chapters 1 and 2 how lines and planes may be expressed implicitly, explicitly, and parametrically. Three other important types of sets are parabolas, hyperbolas, and ellipses, all of which are subsets of a plane. These subsets are collectively called the *conic sections* or *conics*. We shall regard the conics as subsets of $\mathcal{R}^2$ and at first consider these subsets to be in positions related to O_2 in ways which make our functions simple. Later we shall learn to "translate" these subsets to other parts of the plane.

The Ellipse

Consider Figure 7.4. Think of X_0 and X_1 as being nails in a flat surface and of placing a loop of string loosely around X_0 and X_1. Now pull the

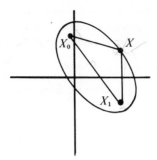

Figure 7.4

string taut with the point of a pencil, and move the pencil on the surface while keeping the string taut. The sketch we obtain in this manner will be that of an *ellipse*.

In order to define this figure mathematically, we note that if X denotes the location of the point of the pencil the sum of the distances from X to X_0 and from X to X_1 is always the same, namely, the length of the loop minus $\|X_1 - X_0\|$, and that this sum is greater than the distance from X_0 to X_1. Thus, in our definition we require that $\|X - X_0\| + \|X - X_1\|$ be a constant, which we choose to call $2a$, and that $2a > \|X_0 - X_1\|$. We now make our formal definition:

DEFINITION 7.4 (The Ellipse.) Let X_0 and X_1 be points in $\Re^2$ and let a be a real number such that $a > 0$, $\|X_0 - X_1\| < 2a$. The set

$$\mathscr{E} = \{X \mid \|X - X_1\| + \|X - X_0\| = 2a\}$$

is said to be an *ellipse*. The points X_0 and X_1 are called the *foci* of $\mathscr{E}$ and $\dfrac{X_0 + X_1}{2}$ is called the *center* of $\mathscr{E}$.

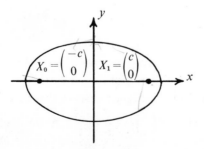

Figure 7.5

We wish now to find an implicit representation for an ellipse $\mathcal{E}$. In so doing we shall use an implication chain. This particular chain should be read completely through in one direction and then in the other.

We simplify our problem by setting $X_1 = \begin{pmatrix} c \\ 0 \end{pmatrix}$, $c \geq 0$, and $X_0 = -X_1$ (see Figure 7.5). Let $X = \begin{pmatrix} x \\ y \end{pmatrix}$ denote a general point on $\mathcal{E}$ and consider the implication scheme:

$$X = \begin{pmatrix} x \\ y \end{pmatrix} \in \mathcal{E}$$

$\Leftrightarrow^1 \|X - X_1\| + \|X - X_0\| = 2a, \quad 0 \leq c < a$

$\Leftrightarrow \left\| \begin{pmatrix} x \\ y \end{pmatrix} - \begin{pmatrix} c \\ 0 \end{pmatrix} \right\| + \left\| \begin{pmatrix} x \\ y \end{pmatrix} - \begin{pmatrix} -c \\ 0 \end{pmatrix} \right\| = 2a$

$\Leftrightarrow^2 \sqrt{(x-c)^2 + y^2} + \sqrt{(x+c)^2 + y^2} = 2a$

$\Leftrightarrow \sqrt{(x-c)^2 + y^2} = 2a - \sqrt{(x+c)^2 + y^2}$

$\Leftrightarrow^3 (x-c)^2 + y^2 = 4a^2 - 4a\sqrt{(x+c)^2 + y^2} + (x+c)^2 + y^2$

$\Leftrightarrow 4a\sqrt{(x+c)^2 + y^2} = 4a^2 - (x-c)^2 + (x+c)^2$

$\qquad\qquad\qquad\qquad = 4a^2 - (x^2 - 2cx + c^2) + (x^2 + 2cx + c^2)$

$\qquad\qquad\qquad\qquad = 4a^2 + 4cx$

$\Leftrightarrow a\sqrt{(x+c)^2 + y^2} = a^2 + cx$

$\Leftrightarrow^4 a^2((x+c)^2 + y^2) = a^4 + 2a^2cx + c^2x^2$

$\Leftrightarrow a^2x^2 + 2a^2cx + a^2c^2 + a^2y^2 = a^4 + 2a^2cx + c^2x^2$

$\Leftrightarrow a^2x^2 - c^2x^2 + a^2y^2 = a^4 - a^2c^2$

$\Leftrightarrow x^2(a^2 - c^2) + a^2y^2 = a^2(a^2 - c^2)$

$\Leftrightarrow^5 x^2b^2 + a^2y^2 = a^2b^2, \quad \text{where} \quad (a^2 - c^2) = b^2, \quad b > 0, \quad c \geq 0$

$\Leftrightarrow \dfrac{x^2}{a^2} + \dfrac{y^2}{b^2} = 1, \quad 0 < b \leq a.$

(1) Our definition requires $\|X_0 - X_1\| < 2a$, hence $0 \leq 2c = \|X_0 - X_1\| < 2a$, so $0 \leq c < a$.

(2) Definition 2.21 (Length) (p. 133).

(3) $\Rightarrow$ Squaring.
    $\Leftarrow$ See Exercise 20.

(4) $\Rightarrow$ Squaring.
    $\Leftarrow$ See Exercise 19.

(5) $\Rightarrow$ This definition of b^2 is possible since $a > c > 0 \Rightarrow (a^2 - c^2) > 0$.
    $\Leftarrow$ This definition of c^2 is possible since $a \geq b > 0 \Rightarrow a^2 - b^2 > 0$.

From the preceding implication scheme we see that $1\text{-}lev\,f$ gives us an implicit representation of $\mathcal{E}$, where $f\begin{pmatrix} x \\ y \end{pmatrix} = \dfrac{x^2}{a^2} + \dfrac{y^2}{b^2}$. Accordingly we represent $\mathcal{E}$ by the equation $\dfrac{x^2}{a^2} + \dfrac{y^2}{b^2} = 1$.

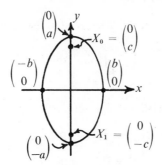

Figure 7.6

If we set first y and then x equal to zero, we find that $\mathcal{E}$ intersects the axes at $\pm\begin{pmatrix} a \\ 0 \end{pmatrix}$ and $\pm\begin{pmatrix} 0 \\ b \end{pmatrix}$. These points are called, respectively, the x and y *intercepts* of $\mathcal{E}$.

The equation $\dfrac{x^2}{a^2} + \dfrac{y^2}{b^2} = 1$ is referred to as the *standard form of the equation of the ellipse with center at O_2 and foci on the x-axis*. $\mathcal{E}$ may be defined implicitly by other equations such as $x^2 b^2 + a^2 y^2 = a^2 b^2$, but the standard form makes the task of sketching $\mathcal{E}$ simpler.

It is easy to see that if $X_0 = \begin{pmatrix} 0 \\ -c \end{pmatrix}$, $X_1 = \begin{pmatrix} 0 \\ c \end{pmatrix}$ we can interchange all coordinates and the preceding argument gives us $\mathcal{E} = 1$-*lev f*, where $f\begin{pmatrix} x \\ y \end{pmatrix} = \dfrac{x^2}{b^2} + \dfrac{y^2}{a^2}$ (see Figure 7.6). Notice that we have required $b^2 = (a^2 - c^2)$, so $b \le a$. Hence, if we see an equation such as $\dfrac{x^2}{4} + \dfrac{y^2}{9} = 1$, we identify a^2 with the larger denominator and thus determine that the foci are on the y-axis. Of course we can similarly identify ellipses with foci on the x-axis (Fig. 7.7).

Example 2. We wish to sketch the ellipse $\mathcal{E}$ defined implicitly by $\dfrac{x^2}{4} + \dfrac{y^2}{2} = 1$. We have $a^2 = 4$, $b^2 = 2$, and $c^2 = a^2 - b^2 = 2$, hence,

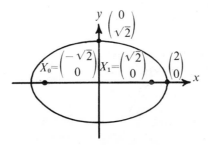

Figure 7.7

$a = 2$, $b = \sqrt{2}$, and $c = \sqrt{2}$. The x-intercepts are $\pm \begin{pmatrix} a \\ 0 \end{pmatrix} = \pm \begin{pmatrix} 2 \\ 0 \end{pmatrix}$, the y-intercepts are $\pm \begin{pmatrix} 0 \\ b \end{pmatrix} = \pm \begin{pmatrix} 0 \\ \sqrt{2} \end{pmatrix}$, and the foci are $\pm \begin{pmatrix} \sqrt{2} \\ 0 \end{pmatrix}$. We sketch $\mathcal{E}$ in Figure 7.7. ∎

Example 3. We wish to represent implicitly an ellipse with center at O_2, foci at $\begin{pmatrix} 0 \\ \sqrt{2} \end{pmatrix}$ and $\begin{pmatrix} 0 \\ -\sqrt{2} \end{pmatrix}$, and passing through the point $\begin{pmatrix} 1 \\ \sqrt{2} \end{pmatrix}$. We have $c = \sqrt{2}$, so $a^2 - c^2 = b^2$ implies $a^2 = b^2 + 2$. Letting $x = 1$, $y = \sqrt{2}$, and $a^2 = b^2 + 2$ in the standard equation, we obtain the implication chain

$$\frac{1}{b^2} + \frac{2}{b^2 + 2} = 1 \Rightarrow b^4 - b^2 - 2 = 0$$
$$\Rightarrow (b^2 - 2)(b^2 + 1) = 0$$
$$\Rightarrow^1 b^2 = 2$$
$$\Rightarrow a^2 = 4.$$

(1) Since $b^2 = -1$ gives no real solution for b.

Thus, $\dfrac{x^2}{2} + \dfrac{y^2}{4} = 1$ represents the desired ellipse implicitly. ∎

We come now to the problem of representing an ellipse $\mathcal{E}$ explicitly. We consider the standard form equation $\dfrac{x^2}{a^2} + \dfrac{y^2}{b^2} = 1$. If we solve this for y, we obtain $y = \pm \dfrac{b}{a} \sqrt{a^2 - x^2}$. Thus, for each choice of x we have two distinct values of y and hence an explicit representation is impossible. This should also be evident from the sketch (see Figure 7.5). However, if we set $f(x) = \dfrac{b}{a} \sqrt{a^2 - x^2}$, we have $\mathcal{E} = graph\, f \cup graph\, (-f)$. Thus, $\mathcal{E}$ is expressed as a union of two explicitly defined function-loci. For simplicity we shall say that $\mathcal{E}$ is represented explicitly by $graph\, f \cup graph\, (-f)$.

If we consider the ellipse $\mathcal{E} = \left\{ \begin{pmatrix} x \\ y \end{pmatrix} \middle| \frac{x^2}{b^2} + \frac{y^2}{a^2} = 1 \right\}$, we obtain $y = \pm \dfrac{a}{b} \sqrt{b^2 - x^2}$, thus, $f(x) = \dfrac{a}{b} \sqrt{b^2 - x^2}$ is such that $\mathcal{E} = graph\, f \cup graph\, (-f)$.

Example 4. We wish to sketch $graph\, f$, where $f(x) = \sqrt{12 - 3x^2}$. We let $y = f(x)$ and square to obtain

$$y^2 = 12 - 3x^2$$
$$3x^2 + y^2 = 12$$

and

$$\frac{x^2}{(2)^2} + \frac{y^2}{(2\sqrt{3})^2} = 1.$$

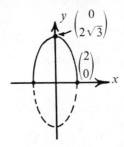

Figure 7.8

This is the standard form of an ellipse. If we remember that y can have only non-negative values, we see that *graph f* is that part of the ellipse in Figure 7.8 which is not below the x-axis. ∎

To obtain a parametric representation of an ellipse with foci on the x-axis, we consider the standard equation $\dfrac{x^2}{a^2} + \dfrac{y^2}{b^2} = \left(\dfrac{x}{a}\right)^2 + \left(\dfrac{y}{b}\right)^2 = 1$ and note its resemblance to the trigonometric identity $\cos^2 t + \sin^2 t = 1$. This suggests that if we set $\cos t = \dfrac{x}{a}$, $\sin t = \dfrac{y}{b}$, or $f(t) = \begin{pmatrix} x \\ y \end{pmatrix} = \begin{pmatrix} a \cos t \\ b \sin t \end{pmatrix}$, $0 \le t < 2\pi$, we shall obtain the desired function. Checking this, we find the following scheme:

$$\begin{pmatrix} x \\ y \end{pmatrix} \in ran\, f \Rightarrow \begin{pmatrix} x \\ y \end{pmatrix} = \begin{pmatrix} a \cos t \\ b \sin t \end{pmatrix}$$

$$\Rightarrow \frac{x^2}{a^2} + \frac{y^2}{b^2} = \frac{a^2 \cos^2 t}{a^2} + \frac{b^2 \sin^2 t}{b^2}$$

$$= \cos^2 t + \sin^2 t = 1$$

$$\Rightarrow \begin{pmatrix} x \\ y \end{pmatrix} \in \mathcal{E},$$

which shows that $ran\, f \subset \mathcal{E}$.

Conversely, suppose $\begin{pmatrix} x \\ y \end{pmatrix} \in \mathcal{E}$. We consider first the case where $x > 0$, $y > 0$. Since $0 \le \dfrac{x}{a} \le 1$ (see Exercise 20a), there is a real number t such that $\cos t = \dfrac{x}{a}$ and $0 \le t \le \dfrac{\pi}{2}$. For this choice of t we have $\sin t = \sqrt{1 - \cos^2 t} = \sqrt{1 - \dfrac{x^2}{a^2}} = \sqrt{\dfrac{y^2}{b^2}} = \dfrac{y}{b}$. Thus, $y = b \sin t$, and t was chosen so that $x = a \cos t$. In a similar way we can find a real number t, $0 \le t < 2\pi$, such that $\begin{pmatrix} x \\ y \end{pmatrix} = \begin{pmatrix} a \cos t \\ b \sin t \end{pmatrix}$ for $\begin{pmatrix} x \\ y \end{pmatrix}$ in any quadrant. Thus, we have the scheme

$$\begin{pmatrix} x \\ y \end{pmatrix} \in \mathcal{E} \Rightarrow \begin{pmatrix} x \\ y \end{pmatrix} = \begin{pmatrix} a \cos t \\ b \sin t \end{pmatrix}, \text{ for some } t, \quad 0 \le t < 2\pi$$

$$\Rightarrow \begin{pmatrix} x \\ y \end{pmatrix} \in ran\, f,$$

so $\mathcal{E} \subset ran\, f$. We therefore conclude that $ran\, f = \mathcal{E}$. Thus, $f\colon [0, 2\pi) \to \mathcal{R}^2$ and gives us a parametric representation for E. Other parametric representations are possible, but this is a useful one.

Example 5. We seek a parametric representation for the ellipse $\mathcal{E}$ (of Example 3), which is given implicitly by $\dfrac{x^2}{2} + \dfrac{y^2}{4} = 1$. We set $\dfrac{x^2}{2} = \cos^2 t, \dfrac{y^2}{4} = \sin^2 t$, to obtain $\mathcal{E} = \left\{ \begin{pmatrix} \sqrt{2}\,\cos t \\ 2\,\sin t \end{pmatrix} \middle|\ 0 \le t < 2\pi \right\}$. ∎

The Circle

DEFINITION 7.5 Let $X_0 \in \mathcal{R}^2$, $r > 0$. Then the set

$$\mathcal{E} = \{ X \mid \|X - X_0\| = r \}$$

is said to be a *circle* with *radius r* and *center* X_0.

We notice that the circle is merely a special case of the ellipse. We have only to let $X_1 = X_0$ and $a = r$ in Definition 7.4 and the definitions become the same. (Since in this case the foci and the center are identical, we do not mention the foci in our definition of the circle.) If our circle is centered at O_2, we have $c = 0$ and $a = b = r$, so we obtain $x^2 + y^2 = r^2$ as an equation of our circle. This equation is known as the *standard equation of the circle with center at O_2 and radius r*. It is easy to check that the parametric representation is given by $\left\{ \begin{pmatrix} r\,\cos t \\ r\,\sin t \end{pmatrix} \middle|\ 0 \le t < 2\pi \right\}$ and that $\mathcal{E} = graph\, f \cup graph\ (-f)$, where $f(x) = \sqrt{r^2 - x^2}$.

We summarize below the results of this section:

THEOREM 7.6 An ellipse with center O_2, foci $\pm \begin{pmatrix} c \\ 0 \end{pmatrix}$, x-intercepts $\pm \begin{pmatrix} a \\ 0 \end{pmatrix}$, y-intercepts $\pm \begin{pmatrix} 0 \\ b \end{pmatrix}$, where $a^2 = b^2 + c^2$, and $0 \le c < a$, $b > 0$, is given

i. explicitly by $graph\, f \cup graph\ (-f)$, where

$$f(x) = \frac{b}{a}\sqrt{a^2 - x^2};$$

ii. implicitly by 1-*lev g*, where

$$g\begin{pmatrix} x \\ y \end{pmatrix} = \frac{x^2}{a^2} + \frac{y^2}{b^2};$$

iii. parametrically by $\operatorname{ran} h$, where

$$h(t) = \begin{pmatrix} a \cos t \\ b \sin t \end{pmatrix}, \quad 0 \leq t < 2\pi.$$

An ellipse with center O_2, foci $\pm \begin{pmatrix} 0 \\ c \end{pmatrix}$, x-intercepts $\pm \begin{pmatrix} b \\ 0 \end{pmatrix}$, y-intercepts $\pm \begin{pmatrix} 0 \\ a \end{pmatrix}$, where $a^2 = b^2 + c^2$ and $0 \leq c < a$, $b > 0$, is given explicitly, implicitly, and parametrically by the equations obtained by exchanging a and b in the corresponding equations above.

Explicit, implicit, and parametric representations of a circle with center O_2 and radius r are obtained by letting $a = b = r$ in the above.

EXERCISES

Sketch the ellipses expressed implicitly by the following equations. Show the foci:

1. $\dfrac{x^2}{4} + \dfrac{y^2}{1} = 1$.

2. $\dfrac{x^2}{1} + \dfrac{y^2}{4} = 1$.

3. $2x^2 + 3y^2 = 6$.

4. $4x^2 + y^2 = 16$.

5. $(x - 1)(x + 1) + y^2 = 3$.

6 to 10. Give explicit and parametric representations for the ellipses in Exercises 1 to 5.

Sketch the ellipses represented explicitly or parametrically by

11. $f(x) = \sqrt{1 - \dfrac{x^2}{9}}$.

12. $f(x) = 6\sqrt{\dfrac{25 - x^2}{25}}$.

13. $\begin{pmatrix} x \\ y \end{pmatrix} = \begin{pmatrix} 3 \cos t \\ \sin t \end{pmatrix}, \ 0 \leq t < 2\pi$.

14. $\begin{pmatrix} x \\ y \end{pmatrix} = \begin{pmatrix} 4 \cos t \\ \sqrt{3} \sin t \end{pmatrix}, \ 0 \leq t < 2\pi$.

15. $Q(s) = \begin{pmatrix} \cos s \\ 4 \sin s \end{pmatrix}, \ 0 \leq s < 2\pi$.

Find implicit representations for ellipses with center at the origin and

16. foci at $\pm \begin{pmatrix} 1 \\ 0 \end{pmatrix}$ and including the point $\begin{pmatrix} 1 \\ -3 \\ 2 \end{pmatrix}$.

17. including the points $\begin{pmatrix} 1 \\ 2 \end{pmatrix}$ and $\begin{pmatrix} 2 \\ 1 \end{pmatrix}$.

18. including the points $\begin{pmatrix} 0 \\ 2 \end{pmatrix}$ and $\begin{pmatrix} 1 \\ 1 \end{pmatrix}$.

19. If $\dfrac{x^2}{a^2} + \dfrac{y^2}{b^2} = 1$, where $a \geq b > 0$, and if $c = \sqrt{a^2 - b^2}$, show that $a^2 + cx \geq 0$ (Hint: use the fact that $c < a$ to show that $a^2 + cx < 0 \Rightarrow \dfrac{x^2}{a^2} > 1$, which is impossible).

20. (a) Show that if $\dfrac{x^2}{a^2} + \dfrac{y^2}{b^2} = 1$, where $0 < b \le a$, then $a^2 \ge x^2$.

   (b) If $a\sqrt{(x+c)^2 + y^2} = a^2 + cx$ and $a^2 > x^2$, $a > c > 0$, show that

$$2a - \sqrt{(x+c)^2 + y^2} \ge 0.$$

ANSWERS

1.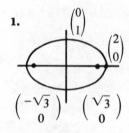

3.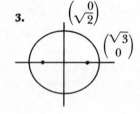

5.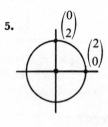

6. Explicit: *graph* $f \cup$ *graph* $(-f)$ where $f(x) = \frac{1}{2}\sqrt{4 - x^2}$, parametric: $\begin{pmatrix} 2\cos t \\ \sin t \end{pmatrix}$, $0 \le t < 2\pi$. **8.** Explicit: *graph* $f \cup$ *graph* $(-f)$ where

$f(x) = \sqrt{\dfrac{6 - 2x^2}{3}}$, parametric: $f(t) = \begin{pmatrix} \sqrt{3}\cos t \\ \sqrt{2}\sin t \end{pmatrix}$, $0 \le t < 2\pi$.

11.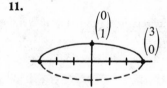

13.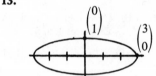

17. $x^2 + y^2 = 5$.

3. CONICS: THE HYPERBOLA

   Another conic section which is quite closely related to the ellipse is the *hyperbola*. The reader will notice that the following definition differs from

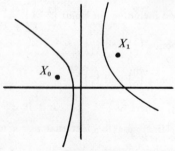

Figure 7.9

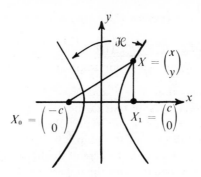

Figure 7.10

Definition 7.4 (p. 295) only in the fact that we require $\big|\,\|X - X_0\| - \|X - X_1\|\,\big| = 2a$ instead of $\|X - X_0\| + \|X - X_1\| = 2a$. This results in a sketch similar to that of Figure 7.9.

> **DEFINITION 7.7** Let X_0 and X_1 be points in $\mathcal{R}^2$ and a a real number, where $\|X_1 - X_0\| > 2a$. The set $a > 0$
>
> $$\mathcal{H} = \{X \mid \big|\,\|X - X_0\| - \|X - X_1\|\,\big| = 2a\}$$
>
> is said to be a *hyperbola*. The points X_0 and X_1 are called the *foci* of $\mathcal{H}$ and $\dfrac{X_0 + X_1}{2}$ is called the *center* of $\mathcal{H}$.

As in the case of the ellipse, we simplify the problem of obtaining an implicit representation of a hyperbola $\mathcal{H}$ by setting $X_1 = \begin{pmatrix} c \\ 0 \end{pmatrix}$, $c > 0$, $X_0 = -X_1$ (see Figure 7.10). By calculations quite similar to those used for the ellipse, we find $\dfrac{x^2}{a^2} - \dfrac{y^2}{b^2} = 1$, where b is chosen so that $b > 0$ and $b^2 = c^2 - a^2$, to be an implicit representation of $\mathcal{H}$. This equation is

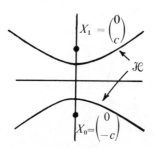

Figure 7.11

called the *standard form equation of the hyperbola with foci on the x-axis and center at O_2.* The x-intercepts are at $\pm \begin{pmatrix} a \\ 0 \end{pmatrix}$. There are no y-intercepts.

If $X_1 = \begin{pmatrix} 0 \\ c \end{pmatrix}$, $X_0 = -X_1$, our implicit representation is $\dfrac{y^2}{a^2} - \dfrac{x^2}{b^2} = 1$, where b is so chosen that $b > 0$ and $b^2 = c^2 - a^2$. Such a hyperbola is sketched in Figure 7.11. Note that in the case of the ellipse, we have $b^2 = a^2 - c^2$, whereas in the case of the hyperbola, $b^2 = c^2 - a^2$. To remember this, recall that in the case of the ellipse $0 < c < a$ (that is, the foci lie between the intercepts), and in the case of the hyperbola $0 < a < c$. Notice also that in the general equation for the hyperbola, b^2 appears in the negative term and may be larger than, smaller than, or equal to a^2, whereas in that of the ellipse b^2 is the smaller of the two denominators.

If the reader finds it difficult to distinguish among the formulae $\pm \dfrac{x^2}{a^2} \pm \dfrac{y^2}{b^2} = 1$, he can, in any given case, fix one of the variables at zero and solve for the other. In this way he can find the points at which the figure intersects the axes and then easily determine the general form of the sketch.

Example 6. Suppose we are given the equation $\dfrac{x^2}{4} - \dfrac{y^2}{9} = 1$. We recognize this as representing either an ellipse or a hyperbola with center at O_2. If we set $y = 0$, we obtain $x = \pm 2$, so the figure intersects the x-axis. On the other hand, if we set $x = 0$, we get $y^2 = -9$, and there is no real number y satisfying this equation. Thus, our figure does not intersect the y-axis. We conclude that the implicitly defined function-locus is a hyperbola intersecting the x-axis at ± 2. Plotting a few more points, we obtain a rough sketch as in Figure 7.12. To find the foci we note that $c^2 = a^2 + b^2 = 13$, so $c = \pm \sqrt{13}$. ∎

We shall find ways to improve our sketching technique after we develop a parametric representation for the hyperbola.

In most cases it is impossible to express a hyperbola as the graph of a single function, but a hyperbola of the form $\dfrac{x^2}{a^2} - \dfrac{y^2}{b^2} = 1$ can be represented

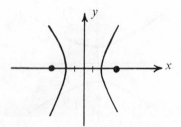

Figure 7.12

as *graph* $f \cup graph$ $(-f)$, where $f(x) = \dfrac{b}{a}\sqrt{x^2 - a^2}$. A hyperbola of the form $\dfrac{y^2}{a^2} - \dfrac{x^2}{b^2} = 1$ can be written as *graph* $f \cup graph$ $(-f)$, where $f(x) = \dfrac{a}{b}\sqrt{x^2 + b^2}$. As in the case of an ellipse, we shall say that $\mathcal{H}$ is represented explicitly by *graph* $f \cup graph$ $(-f)$.

In seeking a parametric representation of the hyperbola $\mathcal{H}$ represented implicitly by $\dfrac{x^2}{a^2} - \dfrac{y^2}{b^2} = 1$, we note the resemblance of the equation to the trigonometric identity $\sec^2 t - \tan^2 t = 1$. This suggests that we set $\dfrac{x}{a} = \sec t, \dfrac{y}{b} = \tan t, 0 \le t < 2\pi, t \notin \left\{\dfrac{\pi}{2}, \dfrac{3\pi}{2}\right\}$ and solve for x and y. A check similiar to that for the ellipse in Section 2 shows us that

$$f(t) = \begin{pmatrix} x \\ y \end{pmatrix} = \begin{pmatrix} a\sec t \\ b\tan t \end{pmatrix}, \quad 0 \le t < 2\pi, \quad t \notin \left\{\dfrac{\pi}{2}, \dfrac{3\pi}{2}\right\}$$

does indeed represent $\mathcal{H}$ parametrically. In the case where $\mathcal{H}$ is given by $\dfrac{y^2}{a^2} - \dfrac{x^2}{b^2} = 1$, a parametric representation is

$$f(t) = \begin{pmatrix} b\tan t \\ a\sec t \end{pmatrix}, \quad 0 \le t < 2\pi, \quad t \notin \left\{\dfrac{\pi}{2}, \dfrac{3\pi}{2}\right\}.$$

Example 7. We wish to find an implicit representation for the function-locus determined parametrically by $h(t) = \begin{pmatrix} x \\ y \end{pmatrix} = \begin{pmatrix} 4\sec t \\ 3\tan t \end{pmatrix}$, $0 \le t < 2\pi, t \notin \left\{\dfrac{\pi}{2}, \dfrac{3\pi}{2}\right\}$. We find $x^2 = 16\sec^2 t, y^2 = 9\tan^2 t$ or $\sec^2 t = \dfrac{x^2}{16}$, and $\tan^2 t = \dfrac{y^2}{9}$. Substituting these equalities in the identity $\sec^2 t - \tan^2 t = 1$, we obtain $\dfrac{x^2}{16} - \dfrac{y^2}{9} = 1$ as our solution. ∎

Suppose $\mathcal{H}$ is a hyperbola such that $\mathcal{H} = ran\ h$, where $h(t) = \begin{pmatrix} x \\ y \end{pmatrix} = \begin{pmatrix} a\sec t \\ b\tan t \end{pmatrix}$, $0 \le t < 2\pi, t \notin \left\{\dfrac{\pi}{2}, \dfrac{3\pi}{2}\right\}$. Suppose $X = \begin{pmatrix} x \\ y \end{pmatrix} \in \mathcal{H}$, such that $x \ge 0, y \ge 0$. In Figure 7.13, we are considering only points of $\mathcal{H}$ in the

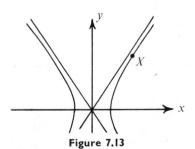

Figure 7.13

first quadrant. Since a and b are positive, we must have $\sec t \geq 0$, $\tan t \geq 0$, which implies $0 \leq t < \dfrac{\pi}{2}$. However, the line through X and O_2 has slope $m = \dfrac{y}{x}$, and as $t \to \pi/2^-$ we observe that $x \to +\infty$. Moreover,

$$\lim_{t \to \pi/2^-} \frac{y}{x} = \lim_{t \to \pi/2^-} \frac{b \tan t}{a \sec t}$$

$$= \lim_{t \to \pi/2^-} \frac{b}{a} \sin t$$

$$\overset{1}{=} \frac{b}{a}.$$

(1) See Theorem 4.10 (p. 203).

This suggests that as we move to the right along the x-axis, our function-locus comes very close to the line defined implicitly by $\dfrac{y}{x} = \dfrac{b}{a}$, or $y = \dfrac{b}{a} x$ (see Exercise 12). $\left(\text{Note that since } \sin t < 1 \text{ for every } t \text{ in } \left(0, \dfrac{\pi}{2}\right), \dfrac{y}{x} < \dfrac{b}{a} \text{ for every such } t. \text{ Therefore, } \begin{pmatrix} x \\ y \end{pmatrix} \text{ is always below the line } y = \dfrac{b}{a} x.\right)$

A similar examination of the behavior of $\mathscr{H}$ in the third quadrant reveals that as we go far to the left our sketch of the hyperbola approaches the same line $y = \dfrac{b}{a} x$. We say that this line is an *asymptote* of the hyperbola. Similarly, we find $y = \dfrac{-b}{a} x$ to be an asymptote in the second and fourth quadrants. These asymptotes can be quite helpful in completing our sketches with a greater degree of accuracy.

In the case of a hyperbola with foci on the y-axis, we find by arguments similar to the above that the asymptotes are given by $y = \pm \dfrac{a}{b} x$.

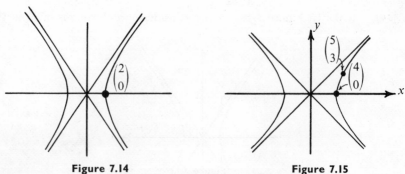

Figure 7.14   Figure 7.15

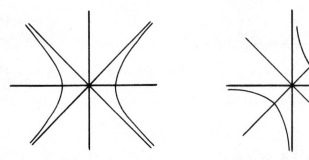

Figure 7.16   Figure 7.17

Example 8. We wish to complete, with greater accuracy, the sketch we started in Example 6. We note that the given equation $\dfrac{x^2}{4} - \dfrac{y^2}{9} = 1$ is in standard form, so $a^2 = 4$, $b^2 = 9$. Hence, the asymptotes are given by $y = \pm \dfrac{b}{a} x = \pm \dfrac{3}{2} x$. With these additional aids we sketch the hyperbola as shown in Figure 7.14. ∎

Example 9. We wish to find an implicit representation and sketch a hyperbola with center at the origin, foci on the x-axis, passing through the point $\begin{pmatrix} 5 \\ 3 \end{pmatrix}$ and having asymptotes $y = \pm x$. From the equations of the asymptotes we note that $a = b$. Hence, $\dfrac{5^2}{a^2} - \dfrac{3^2}{a^2} = 1$ and $a = b = 4$. Thus, an implicit representation is $\dfrac{x^2}{16} - \dfrac{y^2}{16} = 1$ (see Figure 7.15). ∎

We can obtain other hyperbolas centered at O_2 by "rotating" the ones we have discussed. We shall not concern ourselves with this except in one special case. Let $\dfrac{x^2}{a^2} - \dfrac{y^2}{b^2} = 1$ be the equation of a hyperbola and require that $a = b$ (see Figure 7.16). It is clear from the sketch that if we consider the asymptotes to be the coordinate axes, we obtain a hyperbola resembling that of Figure 7.17. We shall not prove the fact, but such a hyperbola is

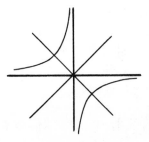

Figure 7.18

given explicitly by $f(x) = \dfrac{a}{x}$, $a > 0$, $x \neq 0$. If we rotate in the opposite direction, we obtain Figure 7.18, which is given explicitly by $f(x) = \dfrac{-a}{x}$, $a > 0$, $x \neq 0$. Details may be found in a number of texts on analytic geometry.

We summarize the results of this section in the following theorem:

THEOREM 7.8 A hyperbola with center at O_2, foci at $\pm\begin{pmatrix} c \\ 0 \end{pmatrix}$, x-intercepts at $\pm\begin{pmatrix} a \\ 0 \end{pmatrix}$, asymptotes given by $y = \pm\dfrac{b}{a}x$, where $c^2 = a^2 + b^2$ and $0 < a < c$, $b > 0$, is given

  i. explicitly by *graph f* $\cup$ *graph* $(-f)$, where

$$f(x) = \frac{b}{a}\sqrt{x^2 - a^2};$$

  ii. implicitly by 1-*lev g*, where

$$g\begin{pmatrix} x \\ y \end{pmatrix} = \frac{x^2}{a^2} - \frac{y^2}{b^2};$$

  iii. parametrically by *ran h*, where

$$h(t) = \begin{pmatrix} a\sec t \\ b\tan t \end{pmatrix}, \quad 0 \leq t < 2\pi, \quad t \notin \left\{ \frac{\pi}{2}, \frac{3\pi}{2} \right\}.$$

A hyperbola with center at O_2, foci at $\pm\begin{pmatrix} 0 \\ c \end{pmatrix}$, y-intercepts $\pm\begin{pmatrix} 0 \\ a \end{pmatrix}$, and asymptotes given by $y = \pm\dfrac{a}{b}x$, where $a^2 = b^2 + c^2$ and $0 < a < c$, $b > 0$, is given

$$C^2 = a^2 + b^2$$

  i. explicitly by *graph f* $\cup$ *graph* $(-f)$, where

$$f(x) = \frac{a}{b}\sqrt{b^2 + x^2};$$

$$\frac{y^2}{a^2} = 1 + \frac{x^2}{b^2}$$

$$y^2 = a^2 + \frac{a^2 x^2}{b^2}$$

  ii. implicitly by 1-*lev g*, where

$$g\begin{pmatrix} x \\ y \end{pmatrix} = \frac{y^2}{a^2} - \frac{x^2}{b^2};$$

$$y^2 = \frac{a^2 b^2 + a^2 x}{b^2}$$

$$y = \frac{a}{b}\sqrt{b^2 + y^2}$$

  iii. parametrically by *ran h*, where

$$h(t) = \begin{pmatrix} b\tan t \\ a\sec t \end{pmatrix}, \quad 0 \leq t < 2\pi, \quad t \notin \left\{ \frac{\pi}{2}, \frac{3\pi}{2} \right\}.$$

EXERCISES

Find foci and asymptotes and sketch the following hyperbolas:

1. $\dfrac{y^2}{4} - \dfrac{x^2}{9} = 1.$

2. $\dfrac{x^2}{25} - \dfrac{y^2}{16} = 1.$

3. $3x^2 - 2y^2 = 6.$

4. $2y^2 - 3x^2 = 6.$

5. $f(t) = \begin{pmatrix} 2\tan t \\ \sec t \end{pmatrix},\ 0 \le t < 2\pi,\ t \notin \left(\dfrac{\pi}{2}, \dfrac{3\pi}{2}\right).$

6. $\begin{pmatrix} x \\ y \end{pmatrix} = \begin{pmatrix} 2\sec t \\ 2\tan t \end{pmatrix},\ 0 \le t < 2\pi,\ t \notin \left(\dfrac{\pi}{2}, \dfrac{3\pi}{2}\right).$

Find implicit (standard form) and parametric representations for the hyperbolas meeting the following conditions. In each case the center is to be at O_2 and the foci are to be one of the axes:

7. Foci $\begin{pmatrix} \pm 4 \\ 0 \end{pmatrix}$, asymptotes, $y = \pm x.$

8. Foci $\begin{pmatrix} 0 \\ \pm 1 \end{pmatrix}$, asymptotes, $y = \pm 2x.$

9. Containing the points $\begin{pmatrix} \sqrt 2 \\ 0 \end{pmatrix}$ and $\begin{pmatrix} 2 \\ 2 \end{pmatrix}.$

10. Containing the points $\begin{pmatrix} 0 \\ \pm 3 \end{pmatrix}$ and having asymptotes $y = \dfrac{\pm\sqrt 3}{3}\, x.$

11. Show that an implicit representation of a hyperbola with center at the origin and foci at $\pm \begin{pmatrix} c \\ 0 \end{pmatrix}$ is given by $\dfrac{x^2}{a^2} - \dfrac{y^2}{b^2} = 1,$ where $b^2 = c^2 - a^2$ (Hint: study the implicit representation for the ellipse as shown on page 296).

12. Show that as x becomes large the subset of the hyperbola $\dfrac{x^2}{a^2} - \dfrac{y^2}{b^2} = 1,$ given explicitly by $y_1 = \dfrac{b}{a}\sqrt{x^2 - a^2},$ comes arbitrarily close to the line given by $y_2 = \dfrac{b}{a}\, x.$

ANSWERS

1. Foci $\pm \begin{pmatrix} 0 \\ \sqrt{13} \end{pmatrix}$, asymptotes $y = \pm\frac{2}{3}x.$ 3. Foci $\pm \begin{pmatrix} \sqrt 5 \\ 0 \end{pmatrix}$, asymptotes $y = \pm \dfrac{\sqrt 3}{\sqrt 2}\, x.$ 5. Foci $\pm \begin{pmatrix} 0 \\ \sqrt 5 \end{pmatrix}$, asymptotes $y = \pm \dfrac{x}{2}.$

7. $\dfrac{x^2}{8} - \dfrac{y^2}{8} = 1,\ \begin{pmatrix} \sqrt 8\,\sec t \\ \sqrt 8\,\tan t \end{pmatrix},\ 0 \le t < 2\pi,\ t \notin \left(\dfrac{\pi}{2}, \dfrac{3\pi}{2}\right).$

9. $\dfrac{x^2}{2} - \dfrac{y^2}{4} = 1,\ \begin{pmatrix} \sqrt 2\,\sec t \\ 2\tan t \end{pmatrix},\ 0 \le t < 2\pi,\ t \notin \left(\dfrac{\pi}{2}, \dfrac{3\pi}{2}\right).$

4. CONICS: THE PARABOLA

DEFINITION 7.9 Let $\mathcal{L}$ be a line in $\mathcal{R}^2$ and $X_0 \in \mathcal{R}^2$ such that $X_0 \notin L$. The set

$$\mathcal{F} = \{X \mid X \in \mathcal{R}^2 \quad \text{and} \quad d(X, \mathcal{L}) = \|X - X_0\|\}$$

is called a *parabola* with *focus* X_0 and *directrix* $\mathcal{L}$. If M is the point of $\mathcal{L}$ which is nearest to X_0, the point $\dfrac{M + X_0}{2}$ is called the *vertex* of $\mathcal{F}$.

Consider Figure 7.19. The curve $\mathcal{F}$ represents a parabola. The focus is X_0 and $\mathcal{L}$ is the directrix. The vertex is at V. Note that if X is any point of P, the distance from X to $\mathcal{L}$ is the same as the distance from X to X_0.

In order to simplify the problem of representing a parabola analytically, we shall set $X_0 = \begin{pmatrix} 0 \\ a \end{pmatrix}$, where $a \neq 0$, and let $\mathcal{L}$ be the line explicitly defined by $y = -a$ (see Figure 7.20). If we let $X = \begin{pmatrix} x \\ y \end{pmatrix}$ and $Q = \begin{pmatrix} x \\ -a \end{pmatrix}$, we have $d(X, \mathcal{L}) = \|X - Q\|$. Therefore,

$$X \in \mathcal{F} \Longleftrightarrow^1 \|X - Q\| = \|X - X_0\|$$

$$\Longleftrightarrow \left\| \begin{pmatrix} x \\ y \end{pmatrix} - \begin{pmatrix} x \\ -a \end{pmatrix} \right\| = \left\| \begin{pmatrix} x \\ y \end{pmatrix} - \begin{pmatrix} 0 \\ a \end{pmatrix} \right\|$$

$$\Longleftrightarrow^2 \sqrt{(x - x)^2 + (y - (-a))^2} = \sqrt{x^2 + (y - a)^2}$$

$$\Longleftrightarrow^3 y^2 + 2ay + a^2 = x^2 + (y^2 - 2ay + a^2)$$

$$\Longleftrightarrow x^2 = 4ay$$

$$(7.1) \qquad\qquad \Longleftrightarrow y = \frac{x^2}{4a}.$$

(1) By definition of parabola.

(2) Definition of distance.

(3) $\Rightarrow$ Squaring both sides and then multiplying out the terms in parentheses. $\Leftarrow$ Since both members are sums of squares and are therefore non-negative and hence have square roots.

Thus, an explicit representation of $\mathcal{F}$ is given by $f(x) = \dfrac{x^2}{4a}$. Since $f''(x) = \dfrac{2}{4a}$, the graph is concave upward if $a > 0$ and concave downward if $a < 0$.

To obtain an implicit representation, we have merely to set $g\begin{pmatrix} x \\ y \end{pmatrix} = x^2 - 4ay$ and then $0\text{-}lev\ g = \mathcal{F}$. For a parametric representation, we set

$$h(x) = \begin{pmatrix} x \\ y \end{pmatrix} = \begin{pmatrix} x \\ \dfrac{x^2}{4a} \end{pmatrix}.$$

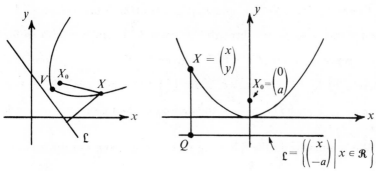

Figure 7.19   **Figure 7.20**

If we choose $X_0 = \begin{pmatrix} b \\ 0 \end{pmatrix}$ and $\mathcal{L} = \left\{ \begin{pmatrix} -b \\ y \end{pmatrix} \middle| y \in \mathcal{R} \right\}$, it is clear that an

argument similar to the above will give $x = \dfrac{y^2}{4b}$. Since this equation can be

obtained from Equation 7.1 by exchanging first and second coordinates and replacing a by b, the sketch will appear as in Figure 7.21, with the opening to the right if $b > 0$ and to the left if $b < 0$. $\mathcal{F}$ cannot be represented explicitly, but for simplicity we shall say that $\mathcal{F}$ is represented explicitly by *graph $f \cup$ graph $(-f)$* where $f(x) = 2\sqrt{bx}$. An implicit representation is

given by $g\begin{pmatrix} x \\ y \end{pmatrix} = y^2 - 4bx = 0$, and $h(y) = \begin{pmatrix} x \\ y \end{pmatrix} = \begin{pmatrix} \dfrac{y^2}{4b} \\ y \end{pmatrix}$ is a parametric

representation.

Example 10. We wish to represent analytically the parabola whose

focus is at $X_0 = \begin{pmatrix} 2 \\ 0 \end{pmatrix}$ and having directrix $\mathcal{L} = \left\{ \begin{pmatrix} -2 \\ y \end{pmatrix} \middle| y \in \mathcal{R} \right\}$. An

implicit representation is given by $g\begin{pmatrix} x \\ y \end{pmatrix} = y^2 - 4bx = y^2 - 8x = 0$, and

$\begin{pmatrix} x \\ y \end{pmatrix} = \begin{pmatrix} \dfrac{y^2}{8} \\ y \end{pmatrix}$ is a parametric representation. ∎

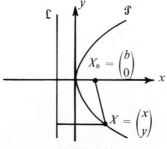

Figure 7.21

Example 11. We are told that a parabola $\mathcal{P}$ has vertex $\begin{pmatrix} 0 \\ 0 \end{pmatrix}$, focus on the y-axis, and passes through the point $\begin{pmatrix} 2 \\ 1 \end{pmatrix}$, and we wish to find an explicit representation of $\mathcal{P}$. This is given by $f(x) = \dfrac{x^2}{4a}$, so our problem is merely to find a. Since we know $\begin{pmatrix} 2 \\ 1 \end{pmatrix} \in \mathcal{P}$, we have $f(2) = 1 = \dfrac{2^2}{4a}$; thus, $a = 1$. Hence $f(x) = \dfrac{x^2}{4}$ is the desired explicit representation of $\mathcal{P}$. ∎

We summarize in the following theorem the results of this section:

THEOREM 7.10 A parabola with vertex at O_2, focus at $\begin{pmatrix} 0 \\ a \end{pmatrix}$, and directrix $\mathcal{L}$ defined by $y = -a$, where $a \neq 0$ is given

i. explicitly by *graph f*, where

$$f(x) = \frac{x^2}{4a};$$

ii. implicitly by *0-lev g*, where

$$g\begin{pmatrix} x \\ y \end{pmatrix} = x^2 - 4ay;$$

iii. parametrically by *ran h*, where

$$h(t) = \begin{pmatrix} x \\ y \end{pmatrix} = \begin{pmatrix} t \\ \dfrac{t^2}{4a} \end{pmatrix}.$$

A parabola with vertex at O_2, focus at $\begin{pmatrix} b \\ 0 \end{pmatrix}$, $b \neq 0$, and directrix defined by $\mathcal{L} = \left\{ \begin{pmatrix} -b \\ y \end{pmatrix} \,\middle|\, y \in \mathcal{R} \right\}$ is given

i. explicitly by *graph f* ∪ *graph* $(-f)$, where

$$f(x) = 2\sqrt{bx};$$

ii. implicitly by *0-lev g*, where

$$g\begin{pmatrix} x \\ y \end{pmatrix} = y^2 - 4bx;$$

iii. parametrically by *ran h*, where

$$h(t) = \begin{pmatrix} x \\ y \end{pmatrix} = \begin{pmatrix} \dfrac{t^2}{4b} \\ t \end{pmatrix}.$$

EXERCISES

Find implicit, parametric, and explicit representations for the following parabolas:

1. Focus at $\begin{pmatrix} 0 \\ -2 \end{pmatrix}$, directrix $\left\{ \begin{pmatrix} x \\ y \end{pmatrix} \middle| y = 2 \right\}$.

2. Focus at $\begin{pmatrix} \frac{1}{2} \\ 0 \end{pmatrix}$, directrix $\left\{ \begin{pmatrix} x \\ y \end{pmatrix} \middle| x = -\frac{1}{2} \right\}$.

3. Focus at $\begin{pmatrix} -2 \\ 0 \end{pmatrix}$, directrix $\left\{ \begin{pmatrix} x \\ y \end{pmatrix} \middle| x = 2 \right\}$.

4. Focus at $\begin{pmatrix} 0 \\ 1 \end{pmatrix}$, directrix $\left\{ \begin{pmatrix} x \\ y \end{pmatrix} \middle| y = -1 \right\}$.

5. Vertex at O_2 and passing through the points $\begin{pmatrix} -1 \\ 4 \end{pmatrix}, \begin{pmatrix} -16 \\ -16 \end{pmatrix}$.

6. Vertex at O_2, focus on the x-axis and passing through the point $\begin{pmatrix} -1 \\ -2 \end{pmatrix}$.

Sketch the conics determined by the following equations:

7. $y^2 - 4x = 0$.   8. $x^2 + y^2 = 4$.

9. $4x^2 - 3y^2 = 12$. *hyp*   10. $\begin{pmatrix} x \\ y \end{pmatrix} = \begin{pmatrix} 4 \sec t \\ \tan t \end{pmatrix}$. *hyp*

11. $25x^2 + 100 = 4y^2$.   12. $f(x) = \begin{pmatrix} x \\ 4x^2 \end{pmatrix}$.

13. $x^2 = \dfrac{12 - 3y^2}{4}$.   14. $\begin{pmatrix} x \\ y \end{pmatrix} = \begin{pmatrix} 25 \cos t \\ 16 \sin t \end{pmatrix}$. *ellipse*

ANSWERS

1. Explicit, $f(x) = \dfrac{-x^2}{8}$.   **3.** Implicit, $x + \dfrac{y^2}{8} = 0$.

5. Implicit $x + \dfrac{y^2}{16} = 0$.

7.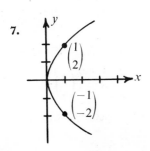

9.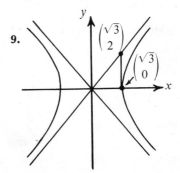

11.

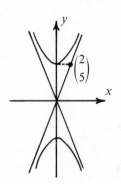

13.

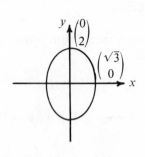

5. SET TRANSLATIONS

Hitherto we have discussed function loci that are quite closely related to the zero vector. For example our ellipses and hyperbolas have all had center O_2. We wish now to think of more general sets. We can often describe such a set by describing a similar one that is related to zero, as in the previous sections, and then "moving" the described set to a new position. In order to be more specific we shall need the following definition:

DEFINITION 7.11   (Translation of a Set.)   Let $S \subset \mathcal{R}^n$ and let $X_0 \in \mathcal{R}^n$. The *translation of S by X_0* is the set denoted and defined by

$$S + X_0 = \{X + X_0 \mid X \in S\}.$$

Example 12a.   Suppose we wish to describe explicitly, implicitly, and parametrically the parabola $\mathcal{S}$ with focus $\begin{pmatrix} 2 \\ 3 \end{pmatrix}$ and directrix $\mathcal{L} = \left\{ \begin{pmatrix} 0 \\ 1 \end{pmatrix} + t \begin{pmatrix} 1 \\ 0 \end{pmatrix} \middle| t \in \mathcal{R} \right\}$ (see Figure 7.22). It is geometrically evident that $\mathcal{S}$ is the translation by $B = \begin{pmatrix} 2 \\ 2 \end{pmatrix}$ of the parabola $\mathcal{S}'$ with focus at $\begin{pmatrix} 0 \\ 1 \end{pmatrix}$ and directrix $\mathcal{L}' = \left\{ \begin{pmatrix} 0 \\ -1 \end{pmatrix} + t \begin{pmatrix} 1 \\ 0 \end{pmatrix} \middle| t \in \mathcal{R} \right\}$. Now, by Section 7.3, $\mathcal{S}'$ is represented explicitly by $f(x) = \dfrac{x^2}{4}$, implicitly by 0-*lev* g, where $g \begin{pmatrix} x \\ y \end{pmatrix} = y - \dfrac{x^2}{4}$, and parametrically by $h(t) = \begin{pmatrix} t \\ \dfrac{t^2}{4} \end{pmatrix}$. The next three theorems will show us how to effect the desired translation.   ∎

Consider the function f whose graph is depicted in Figure 7.23. Suppose we wish to construct a function F whose graph will be the translated set of

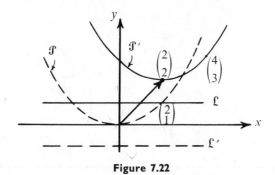

Figure 7.22

points $graph\, f + \begin{pmatrix} x_0 \\ y_0 \end{pmatrix}$. We first "move" our set to the position of $graph\, f_1$ by defining $f_1(x) = f(x - x_0)$. Thus, to plot $f_1(d)$, where $d = a + x_0$, we find $f(d - x_0) = f(a)$ and plot the point which is at the same height above the x-axis as $\begin{pmatrix} a \\ f(a) \end{pmatrix}$ but is directly above d. We do likewise for each point of the interval $[d, e]$, where $e = b + x_0$. Thus, $graph\, f_1$ will be an exact copy of $graph\, f$, but moved to the right by the distance x_0. Now to "move" $graph\, f_1$ to the desired position, we simply add $\begin{pmatrix} 0 \\ y_0 \end{pmatrix}$ to each point of $graph\, f_1$. We do this by defining $F(x) = f(x - x_0) + y_0$.

It will perhaps be helpful to the reader to keep Figure 7.23 in mind while reading the following theorem and its proof. If we let $n = m = 1$ in the theorem, we shall have a situation quite similar to the one we have pictured.

THEOREM 7.12 (Translation of a Graph.) Let $f : \mathcal{R}^n \to \mathcal{R}^m$, $X_0 \in \mathcal{R}^n$, $Y_0 \in \mathcal{R}^m$, and $B = \begin{pmatrix} X_0 \\ Y_0 \end{pmatrix}$. If F is defined by

$$FX = f(X - X_0) + Y_0,$$

then

$$graph\, F = graph\, f + B.$$

Proof. $graph\, F =^1 \left\{ \begin{pmatrix} X \\ FX \end{pmatrix} \,\middle|\, X \in dom\, F \right\}$ *pick X_0*

$=^2 \left\{ \begin{pmatrix} (X - X_0) + X_0 \\ f(X - X_0) + Y_0 \end{pmatrix} \,\middle|\, (X - X_0) \in dom\, f \right\}$

$=^3 \left\{ \begin{pmatrix} (X - X_0) \\ f(X - X_0) \end{pmatrix} \,\middle|\, (X - X_0) \in dom\, f \right\} + \begin{pmatrix} X_0 \\ Y_0 \end{pmatrix}$

$=^4 \left\{ \begin{pmatrix} W \\ fW \end{pmatrix} \,\middle|\, W \in dom\, f \right\} + B$

$=^1 graph\, f + B.$

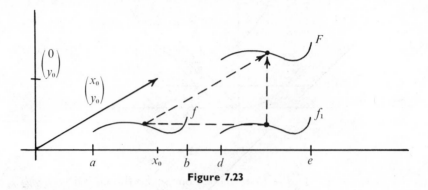

Figure 7.23

(1) Definition 7.1 (Explicitly Defined Function-Locus).
(2) Adding and subtracting X_0 in first coordinate; definition of F in second.
(3) Definition 7.11 (Translation of a Set).
(4) Defining $W = X - X_0$. ∎

(Note: If a function-locus is the union of two graphs, say *graph f* ∪ *graph* $(-f)$, we must be careful to translate each graph separately (see Example 13 and note especially the explicit representation of the translated hyperbola).)

Example 12b. An explicit representation of $\mathfrak{I} = \mathfrak{I}' + B$ (Example 12a) is given by $F(x) = f(x - 2) + 2 = \dfrac{x^2 - 4x + 12}{4}$. ∎

Consider now the function g whose graph is pictured in Figure 7.24. *W-lev g* is, by definition, the set of points x such that $g(x) = W$. In this case it is just $\{c, d, e\}$. Then *W-lev g* $+ x_0$ (see Definition 7.11) is the set $\{c + x_0 = c_1, d + x_0 = d_1, e + x_0 = e_1\}$ as pictured. If we define $G(x) = g(x - x_0)$, we can see that *graph G* will be *graph g* $+ \begin{pmatrix} x_0 \\ 0 \end{pmatrix}$. From the sketch it is clear that *W-lev G = W-lev g* $+ x_0$. If we set $m = n = 1$ in the following theorem, we have a situation similar to the one we have sketched. It will perhaps be helpful to the reader to keep this in mind in reading the theorem and its proof.

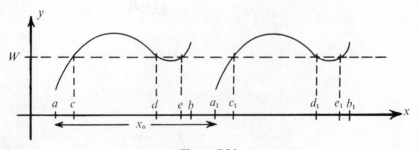

Figure 7.24

THEOREM 7.l3 (Translation of a Level Set.) Let $g:\mathcal{R}^n \to \mathcal{R}^m$, $W \in \mathcal{R}^m$, $X_0 \in \mathcal{R}^n$. If G is defined by

$$GX = g(X - X_0),$$

then

$$W\text{-}lev\ G = W\text{-}lev\ g + X_0.$$

Proof.
$$\begin{aligned}
W\text{-}lev\ g + X_0 &=^1 \{X \mid gX = W\} + X_0 \\
&=^2 \{X + X_0 \mid gX = W\} \\
&=^3 \{X + X_0 \mid g((X + X_0) - X_0) = W\} \\
&=^4 \{Y \mid g(Y - X_0) = W\} \\
&=^5 \{Y \mid GY = W\} \\
&=^1 W\text{-}lev\ G.
\end{aligned}$$

(1) Definition 7.2 (Implicitly Defined Function-Locus).
(2) Definition 7.11 (Translation of a Set).
(3) Adding and subtracting X_0.
(4) Substituting Y for $X + X_0$.
(5) Definition of G in statement of theorem. ■

Example l2c. An implicit representation of $\mathcal{I} = \mathcal{I}' + X_0$ of Example 12a is given by 0-*lev* G, where

$$\begin{aligned}
G\begin{pmatrix} x \\ y \end{pmatrix} &= g\left(\begin{pmatrix} x \\ y \end{pmatrix} - \begin{pmatrix} 2 \\ 2 \end{pmatrix}\right) \\
&= (y - 2) - \frac{(x - 2)^2}{4} \\
&= \frac{4y - x^2 + 4x - 12}{4}. \quad ■
\end{aligned}$$

In reading the following theorem and its proof, the reader should construct for himself a sketch similar to those used for the previous two theorems.

THEOREM 7.l4 (Translation of the Range of a Function.) Let $h:\mathcal{R}^n \to \mathcal{R}^m$, $Y_0 \in \mathcal{R}^m$. If H is defined by

$$HX = hX + Y_0, \quad \text{for all } X \text{ in } dom\ h,$$

then

$$ran\ H = ran\ h + Y_0.$$

Proof.
$$\begin{aligned}
ran\ h + Y_0 &=^1 \{hX \mid X \in dom\ h\} + Y_0 \\
&=^2 \{hX + Y_0 \mid X \in dom\ h\} \\
&=^3 \{HX \mid X \in dom\ h\} \\
&=^4 \{HX \mid X \in dom\ H\} \\
&= ran\ H.
\end{aligned}$$

(1) Definition 7.3 (Parametrically Defined Function-Locus).
(2) Definition 7.11 (Translation of a Set).
(3) Definition of H in statement of theorem.
(4) Since $dom\ h = dom\ H$ by definition. ∎

Example I2d. A parametric representation of $\mathfrak{I} = \mathfrak{I}' + X_0$ of Example 12a is given by

$$H(t) = h(t) + X_0$$

$$= \begin{pmatrix} t \\ \dfrac{t^2}{4} \end{pmatrix} + \begin{pmatrix} 2 \\ 2 \end{pmatrix}$$

$$= \begin{pmatrix} t + 2 \\ \dfrac{t^2}{4} + 2 \end{pmatrix}. \quad ∎$$

Example I3. A hyperbola $\mathfrak{K}$, centered at the origin, is given implicitly by

$$g\begin{pmatrix} x \\ y \end{pmatrix} = \frac{x^2}{9} - \frac{y^2}{4} = 1.$$

We wish to translate this hyperbola so that its center will be at $X_0 = \begin{pmatrix} x_0 \\ y_0 \end{pmatrix} = \begin{pmatrix} -1 \\ -2 \end{pmatrix}$ and to give implicit, explicit, and parametric representations for the translated set $\mathfrak{K} + X_0$.

First it is necessary to find explicit and parametric representations for $\mathfrak{K}$. We find
$\mathfrak{K} = graph\ f \cup graph\ (-f)$, where

$$f(x) = \frac{2\sqrt{x^2 - 9}}{3}, \quad \text{and}$$

$\mathfrak{K} = ran\ h$, where

$$h(t) = \begin{pmatrix} x \\ y \end{pmatrix} = \begin{pmatrix} 3 \sec t \\ 2 \tan t \end{pmatrix}, \quad 0 \le t < 2\pi, \quad t \notin \left\{ \frac{\pi}{2}, \frac{3\pi}{2} \right\}.$$

Then by Theorem 7.13, $\mathfrak{K} + X_0 = 1\text{-}lev\ G$, where

$$G\begin{pmatrix} x \\ y \end{pmatrix} = g\left(\begin{pmatrix} x \\ y \end{pmatrix} - \begin{pmatrix} -1 \\ -2 \end{pmatrix} \right)$$

$$= g\begin{pmatrix} x + 1 \\ y + 2 \end{pmatrix}$$

$$= \frac{(x + 1)^2}{9} - \frac{(y + 2)^2}{4}.$$

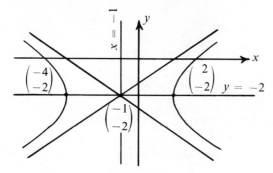

Figure 7.25

By Theorem 7.12, $\mathcal{H} + X_0 = graph\ F_1 \cup graph\ F_2$, where

$$F_1(x) = f(x - x_0) + y_0$$
$$= f(x + 1) - 2$$
$$= \frac{-6 + 2\sqrt{x^2 + 2x - 8}}{3}, \quad \text{and}$$

$$F_2(x) = (-f)(x - x_0) + y_0$$
$$= \frac{-6 - 2\sqrt{x^2 + 2x - 8}}{3}.$$

By Theorem 7.14, $\mathcal{H} + X_0 = ran\ H$, where

$$H(t) = h(t) + X_0$$
$$= \begin{pmatrix} 3 \sec t - 1 \\ 2 \tan t - 2 \end{pmatrix}, \quad 0 \le t < 2\pi, \quad t \notin \left\{\frac{\pi}{2}, \frac{3\pi}{2}\right\}.$$

To sketch the translated set we consider the lines $x = -1, y = -2$ to be the axes for the translated figure, and then sketch $\mathcal{H}$ with these axes by using the techniques of Section 3 (see Figure 7.25). ∎

EXERCISES

In Exercises 1 to 8 translate by X_0 the conics represented by the given equations. Give implicit, explicit, and parametric representations and sketch:

1. $x^2 + y^2 = 4, X_0 = \begin{pmatrix} 2 \\ -1 \end{pmatrix}$.

2. $x^2 + y = 0, X_0 = \begin{pmatrix} 1 \\ 1 \end{pmatrix}$.

3. $f(t) = \begin{pmatrix} 2 \cos t \\ \sin t \end{pmatrix}, 0 \le t < 2\pi, X_0 = \begin{pmatrix} 0 \\ 4 \end{pmatrix}$.

4. $4y^2 - 9x^2 = 36, X_0 = \begin{pmatrix} 1 \\ 2 \end{pmatrix}$.

5. f graph $\cup$ graph $(-f)$ where

$$f(x) = \tfrac{1}{3}\sqrt{36 - x^2}, X_0 = \begin{pmatrix} -1 \\ -1 \end{pmatrix}.$$

6. $y^2 = 4x$, $X_0 = \begin{pmatrix} 2 \\ 3 \end{pmatrix}$.   7. $\dfrac{x^2}{4} + \dfrac{y^2}{2} = 1$, $X_0 = \begin{pmatrix} 3 \\ 4 \end{pmatrix}$.

8. $y = 2\sqrt{-2x}$.

Give implicit, explicit, and parametric representations for the following sets in $\mathcal{R}^2$:

9. A parabola with focus $\begin{pmatrix} -2 \\ 2 \end{pmatrix}$ and directrix $y = 0$.

10. A circle of radius 1 centered at $\begin{pmatrix} 2 \\ 2 \end{pmatrix}$.

11. An ellipse with foci at $\begin{pmatrix} 0 \\ 0 \end{pmatrix}$, $\begin{pmatrix} 2 \\ 0 \end{pmatrix}$, and including the point $\begin{pmatrix} 3 \\ 0 \end{pmatrix}$.

12. A parabola with focus $\begin{pmatrix} 4 \\ 3 \end{pmatrix}$ and directrix $x = 1$.

13. A hyperbola with foci $\begin{pmatrix} -2 + \sqrt{5} \\ -2 \end{pmatrix}$, $\begin{pmatrix} -2 - \sqrt{5} \\ -2 \end{pmatrix}$ and including the point $\begin{pmatrix} \sqrt{5} - 2 \\ \frac{1}{2} \end{pmatrix}$.

14. An ellipse including the points $\begin{pmatrix} 0 \\ 0 \end{pmatrix}$, $\begin{pmatrix} -1 \\ 2 \end{pmatrix}$, $\begin{pmatrix} -1 \\ -2 \end{pmatrix}$, $\begin{pmatrix} -2 \\ 0 \end{pmatrix}$.

ANSWERS

1. Implicit: $x^2 - 4x + y^2 + 2y = -1$;

parametric: $\begin{pmatrix} x \\ y \end{pmatrix} = \begin{pmatrix} 2\cos t + 2 \\ 2\sin t - 1 \end{pmatrix}$, $0 \le t < 2\pi$.

3. Implicit: $\dfrac{x^2}{4} + (y - 4)^2 = 1$; explicit: $F_1(x) = \dfrac{8 + \sqrt{4 - x^2}}{2}$,

$F_2(x) = \dfrac{8 - \sqrt{4 - x^2}}{2}$; parametric: $\begin{pmatrix} 2\cos t \\ \sin t + 4 \end{pmatrix}$, $0 \le t < 2\pi$.

5. Implicit: $\dfrac{(x + 1)^2}{36} + \dfrac{(y + 1)^2}{4} = 1$;

explicit: $F_1(x) = \dfrac{-3 + \sqrt{35 - x^2 - 2x}}{3}$,

$F_2(x) = \dfrac{-3 - \sqrt{35 - x^2 - 2x}}{3}$;

parametric: $\begin{pmatrix} 6\cos t - 1 \\ 2\sin t - 1 \end{pmatrix}$, $0 \le t < 2\pi$.

7. Implicit: $\dfrac{(x - 3)^2}{4} + \dfrac{(y - 4)^2}{2} = 1$.   **9.** Parametric: $\begin{pmatrix} t - 2 \\ \frac{t^2}{4} + 1 \end{pmatrix}$.

11. Explicit: $F_1(x) = \dfrac{\sqrt{3}}{2} \sqrt{2x - x^2} + 3$, $F_2(x) = \dfrac{-\sqrt{3}}{2} \sqrt{2x - x^2} + 3$.

13. Implicit: $x^2 + 4x - 4y^2 - 16y = 16$.

6. SET TRANSLATIONS (CONTINUED)

We wish now to learn to recognize the various representations of translated conics. As an example, we consider the implicit representation of an ellipse. If an ellipse $\mathcal{E}$ is a translation of $\mathcal{E}' = 1\text{-}lev \left(\dfrac{x^2}{a^2} + \dfrac{y^2}{b^2} \right)$ by $X_0 = \begin{pmatrix} x_0 \\ y_0 \end{pmatrix}$, we can write $\mathcal{E}' = 1\text{-}lev\ h$, where $h\begin{pmatrix} x \\ y \end{pmatrix} = \dfrac{x^2}{a^2} + \dfrac{y^2}{b^2}$. Hence, by Theorem 7.13 we have $\mathcal{E} = 1\text{-}lev\ H$, where

$$H\begin{pmatrix} x \\ y \end{pmatrix} = h\left(\begin{pmatrix} x \\ y \end{pmatrix} - \begin{pmatrix} x_0 \\ y_0 \end{pmatrix} \right)$$

$$= h\begin{pmatrix} x - x_0 \\ y - y_0 \end{pmatrix}$$

$$= \frac{(x - x_0)^2}{a^2} + \frac{(y - y_0)^2}{b^2}.$$

Thus, an equation for $\mathcal{E}$ is

(7.2)
$$\frac{(x - x_0)^2}{a^2} + \frac{(y - y_0)^2}{b^2} = 1$$

translated

(Note: Since we are working with ellipses in general, we have no way of knowing whether a is larger than b or not. Thus, the roles of a and b may be reversed from those of Theorem 7.6. This is true in the following example.)

Example 14. We wish to sketch the function-locus determined implicitly by

$$\frac{(x - 2)^2}{4} + \frac{(y + 4)^2}{9} = 1.$$

We recognize this as being of the form of Equation 7.2, where $x_0 = 2$, $y_0 = -4$, $a = 2$, and $b = 3$. We can make the sketch of our translated ellipse by regarding the lines $x = 2$ and $y = -4$ as axes and proceeding as in Section 2 (see Figure 7.26). ∎

Consider now the translation by $B = \begin{pmatrix} x_0 \\ y_0 \end{pmatrix}$ of a hyperbola $\mathcal{H}$ given parametrically by $h(t) = \begin{pmatrix} b \tan t \\ a \sec t \end{pmatrix}$, $0 \le t < 2\pi$, $t \notin \left\{ \dfrac{\pi}{2}, \dfrac{3\pi}{2} \right\}$. By Theorem 7.14 $\mathcal{H} + B$ is given parametrically by

$$H(t) = h(t) + B$$

$$= \begin{pmatrix} b \tan t \\ a \sec t \end{pmatrix} + \begin{pmatrix} x_0 \\ y_0 \end{pmatrix}, \quad 0 \le t < 2\pi, \quad t \notin \left\{ \frac{\pi}{2}, \frac{3\pi}{2} \right\}$$

$$= \begin{pmatrix} x_0 + b \tan t \\ y_0 + a \sec t \end{pmatrix}, \quad 0 \le t < 2\pi, \quad t \notin \left\{ \frac{\pi}{2}, \frac{3\pi}{2} \right\}.$$

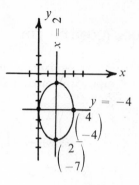

Figure 7.26

Example 15. Sketch the hyperbola determined parametrically by

$$\binom{x}{y} = \binom{1 + 2 \sec t}{2 + \tan t}, \quad 0 \le t < 2\pi, \quad t \notin \left\{\frac{\pi}{2}, \frac{3\pi}{2}\right\}.$$

We recognize this as the translation by $\binom{1}{2}$ of the hyperbola determined by

$\binom{x}{y} = \binom{2 \sec t}{\tan t}, \; 0 \le t < 2\pi, \; t \notin \left\{\frac{\pi}{2}, \frac{3\pi}{2}\right\}.$ Hence, we sketch this second hyperbola by considering the lines $x = 1$ and $y = 2$ as our axes (see Figure 7.27). ■

We turn now to the translation by $X_0 = \binom{x_0}{y_0}$ of the parabola given explicitly by *graph f* $\cup$ *graph* $(-f)$, where $f(x) = 2\sqrt{ax}$. By Theorem 7.12, graph $f + X_0$ is given by graph F_1, where

$$F_1(x) = f(x - x_0) + y_0$$

$$= 2\sqrt{a(x - x_0)} + y_0$$

$$= y_0 + \sqrt{4a(x - x_0)}.$$

Similarly graph $(-f) + X_0$ is given by graph F_2, where

$$F_2(x) = y_0 - \sqrt{4a(x - x_0)}.$$

Hence, the translated parabola is given by *graph* F_1 $\cup$ *graph* F_2.

Example 16. Sketch the function-locus defined explicitly by $y = 2 - \sqrt{4x + 8}$. Putting this equation in the form of F_2 above, we obtain

$$F_2(x) = 2 - \sqrt{4[x - (-2)]}$$

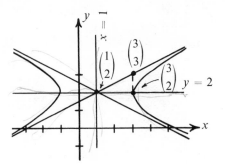

Figure 7.27

Thus, our function-locus is the translation by $\begin{pmatrix} x_0 \\ y_0 \end{pmatrix} = \begin{pmatrix} -2 \\ 2 \end{pmatrix}$ of the lower half of the parabola with focus at $\begin{pmatrix} a \\ 0 \end{pmatrix} = \begin{pmatrix} 1 \\ 0 \end{pmatrix}$, and directrix the line $x = -1$ (see Figure 7.28). ∎

Various representations of translated conics are given in the following table. Some of these have been derived; others are left as exercises. In all cases a, b, c, and r are positive and d and e are nonzero.

A conic will frequently be encountered in the implicit form

$$ax^2 + bx + cy^2 + dy + e = 0.$$

We shall wish to convert such a form to one of the forms in Table 7.1 in order to identify properly the function-locus defined. The theorem on p. 326 gives us conditions under which the equation can be converted to the form indicated.

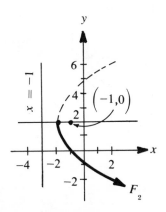

Figure 7.28

Table 7.1.   Conics

Representation of Conic Translated by $X_0 = \binom{x_0}{y_0}$

	Explicit	Implicit	Parametric
i. Parabola with vertex $\binom{0}{0}$ focus $\binom{d}{0}$	(a) $F_1(x) = y_0 + 2\sqrt{d(x-x_0)}$ $F_2(x) = y_0 - 2\sqrt{d(x-x_0)}$	(b) $(y-y_0)^2 - 4d(x-x_0) = 0$	(c) $\binom{x}{y} = \binom{x_0 + \dfrac{t^2}{4d}}{y_0 + t}$
ii. Parabola with vertex $\binom{0}{0}$ focus $\binom{0}{e}$	(a) $F(x) = \dfrac{(x-x_0)^2}{4e} + y_0$	(b) $(x-x_0)^2 - 4e(y-y_0) = 0$	(c) $\binom{x}{y} = \binom{x_0 + t}{y_0 + \dfrac{t^2}{4e}}$
iii. Ellipse with center $\binom{0}{0}$ x-intercepts $\binom{\pm a}{0}$ y-intercepts $\binom{0}{\pm b}$	(a) $F_1(x) = y_0 + b\sqrt{1 - \dfrac{(x-x_0)^2}{a^2}}$ $F_2(x) = y_0 - b\sqrt{1 - \dfrac{(x-x_0)^2}{a^2}}$	(b) $\dfrac{(x-x_0)^2}{a^2} + \dfrac{(y-y_0)^2}{b^2} = 1$	(c) $\binom{x}{y} = \binom{x_0 + a\cos t}{y_0 + b\sin t}$ $0 \le t < 2\pi$

	(a)	(b)	(c)
iv. Circle with center $\begin{pmatrix} 0 \\ 0 \end{pmatrix}$ radius r	$F_1(x) = y_0 + \sqrt{r^2 - (x - x_0)^2}$ $F_2(x) = y_0 - \sqrt{r^2 - (x - x_0)^2}$	$(x - x_0)^2 + (y - y_0)^2 = r^2$	$\begin{pmatrix} x \\ y \end{pmatrix} = \begin{pmatrix} x_0 + r\cos t \\ y_0 + r\sin t \end{pmatrix}$ $0 \le t < 2\pi$
v. Hyperbola with center $\begin{pmatrix} 0 \\ 0 \end{pmatrix}$ foci $\begin{pmatrix} \pm c \\ 0 \end{pmatrix}$ x-intercepts $\begin{pmatrix} \pm a \\ 0 \end{pmatrix}$	$F_1(x) = y_0 + b\sqrt{\dfrac{(x - x_0)^2}{a^2} - 1}$ $F_2(x) = y_0 - b\sqrt{\dfrac{(x - x_0)^2}{a^2} - 1}$ $b = \sqrt{c^2 - a^2}$	$\dfrac{(x - x_0)^2}{a^2} - \dfrac{(y - y_0)^2}{b^2} = 1$	$\begin{pmatrix} x \\ y \end{pmatrix} = \begin{pmatrix} x_0 + a\sec t \\ y_0 + b\tan t \end{pmatrix}$ $0 \le t < 2\pi$ $t \notin \left\{\dfrac{\pi}{2}, \dfrac{3\pi}{2}\right\}$
vi. Hyperbola with center $\begin{pmatrix} 0 \\ 0 \end{pmatrix}$ foci $\begin{pmatrix} 0 \\ \pm c \end{pmatrix}$ y-intercepts $\begin{pmatrix} 0 \\ \pm a \end{pmatrix}$	$F_1(x) = y_0 + a\sqrt{\dfrac{(x - x_0)^2}{b^2} + 1}$ $F_2(x) = y_0 - a\sqrt{\dfrac{(x - x_0)^2}{b^2} + 1}$ $b = \sqrt{c^2 - a^2}$	$\dfrac{(y - y_0)^2}{a^2} - \dfrac{(x - x_0)^2}{b^2} = 1$	$\begin{pmatrix} x \\ y \end{pmatrix} = \begin{pmatrix} x_0 + b\tan t \\ y_0 + a\sec t \end{pmatrix}$ $0 \le t < 2\pi$ $t \notin \left\{\dfrac{\pi}{2}, \dfrac{3\pi}{2}\right\}$

THEOREM 7.15 The function-locus implicitly defined by the equation

(7.3) $ax^2 + bx + cy^2 + dy + e = 0$

is a translation of the conic indicated in the left column if and only if the conditions stated in the center column hold. Such an equation will frequently appear in the form shown in the right column.

Parabola with vertex $\begin{pmatrix} 0 \\ 0 \end{pmatrix}$ and focus on the x-axis.	$a = 0, c \neq 0.$	$cy^2 + dy + bx = -e.$
Parabola with vertex $\begin{pmatrix} 0 \\ 0 \end{pmatrix}$ and focus on the y-axis.	$a \neq 0, c = 0.$	$ax^2 + bx + dy = -e.$
Ellipse with center $\begin{pmatrix} 0 \\ 0 \end{pmatrix}$ and foci on the x-axis.	$c > a > 0,$ $\dfrac{b^2}{4a} + \dfrac{d^2}{4c} > e.$	$ax^2 + bx + cy^2 + dy = e.$
Ellipse with center $\begin{pmatrix} 0 \\ 0 \end{pmatrix}$ and foci on the y-axis.	$a > c > 0,$ $\dfrac{b^2}{4a} + \dfrac{d^2}{4c} > e.$	$ax^2 + bx + cy^2 + dy = e.$
Circle with center $\begin{pmatrix} 0 \\ 0 \end{pmatrix}.$	$a = c > 0,$ $\dfrac{b^2}{4a} + \dfrac{d^2}{4c} > e.$	$ax^2 + bx + ay^2 + dy = e.$
Hyperbola with center $\begin{pmatrix} 0 \\ 0 \end{pmatrix}$ and foci on the x-axis.	$a > 0, c < 0,$ $\dfrac{b^2}{4a} + \dfrac{d^2}{4c} > e.$	$ax^2 + bx + cy^2 + dy = e.$
Hyperbola with center $\begin{pmatrix} 0 \\ 0 \end{pmatrix}$ and foci on the y-axis.	$a < 0, c > 0,$ $\dfrac{b^2}{4a} + \dfrac{d^2}{4c} > e.$	$ax^2 + bx + cy^2 + dy = e.$

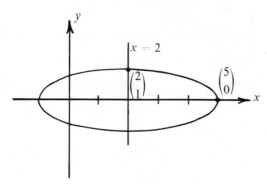

Figure 7.29

The proof will be left to the exercises. Note that, in the cases of the ellipse and circle, $\dfrac{b^2}{4a} + \dfrac{d^2}{4c} > e$ whenever $e < 0$, $a > 0$, and $c > 0$.

Example 17. We wish to sketch the function-locus S defined implicitly by $x^2 + 9y^2 - 4x - 5 = 0$. We note that this is of the form of Equation 7.3 and that $a = 1$, $c = 9$, and $e = -5$. Hence, by the theorem, S is a translation of an ellipse with center $\begin{pmatrix} 0 \\ 0 \end{pmatrix}$ and foci on the x-axis. We write our equation in the form of (iii-b) of Table 7.1 by proceeding as follows:

$$x^2 + 9y^2 - 4x - 5 = 0$$
$$\Rightarrow^1 (x - 2)^2 + 9y^2 = 9$$
$$\Rightarrow^2 \frac{(x - 2)^2}{3^2} + \frac{(y - 0)^2}{1^2} = 1.$$

(1) Completing squares.
(2) Dividing through by 9.

Thus, $x_0 = 2$, $y_0 = 0$, $a = 3$, and $b = 1$. We make our sketch by using the lines $x = 2$ and $y = 0$ as our axes (see Figure 7.29). ∎

EXERCISES

Sketch the conics defined by the following equations. Identify each as an ellipse, circle, hyperbola, or parabola:

1. $y^2 = 4(x - 2)$.

2. $f(t) = \begin{pmatrix} 3 + 2\cos t \\ -1 + 2\sin t \end{pmatrix}$, $0 \le t < 2\pi$.

3. $(x - 1)^2 - (y - 2)^2 = 4$.

4. $25x^2 + 100x - 4y^2 + 8y + 196 = 0$.

5. $4x^2 + 9y^2 + 8x - 36y + 4 = 0$.

6. $f(t) = \begin{pmatrix} t + 1 \\ t^2 - 4 \end{pmatrix}, 0 \le t < 2$.

7. $f(x) = \dfrac{x^2 + 4x - 8}{8}$.

8. $f(t) = \begin{pmatrix} 1 + \tan t \\ 4 + 2 \sec t \end{pmatrix}, 0 \le t < 2\pi, t \notin \left[\dfrac{\pi}{2}, \dfrac{3\pi}{2}\right]$.

9. $y^2 - 4y - 8x = -28$.

10. $y = 2 + 3 \sqrt{\dfrac{(x - 1)^2}{4} - 1}$.

Find implicit and parametric representations for the following conics:

11. Circle of radius 2 with center $\begin{pmatrix} 4 \\ 3 \end{pmatrix}$.

12. Ellipse with foci $\begin{pmatrix} 2 \\ 3 \end{pmatrix}$, $\begin{pmatrix} 4 \\ 3 \end{pmatrix}$ and including the point $\begin{pmatrix} 6 \\ 3 \end{pmatrix}$.

13. Hyperbola with asymptotes $y = x + 2$, $y = -x - 4$ and one focus $\begin{pmatrix} -2 \\ -1 \end{pmatrix}$.

14. Parabola with directrix $\left\{ \begin{pmatrix} 1 \\ 2 \end{pmatrix} + t \begin{pmatrix} 1 \\ 0 \end{pmatrix} t \in \mathcal{R} \right\}$ and focus $\begin{pmatrix} 3 \\ 0 \end{pmatrix}$.

Find implicit representations for the following conics:

15. Hyperbola with the axes as asymptotes and including the point $\begin{pmatrix} 1 \\ 2 \end{pmatrix}$.

16. Hyperbola with the lines $x = 3$, $y = 2$ as asymptotes and including the point $\begin{pmatrix} 4 \\ 1 \end{pmatrix}$.

In Exercises 17 to 20 derive the indicated formula of Table 7.1:

17. i-b.   18. iii-a.

19. iii-c.   20. v-c.

ANSWERS

1. Parabola.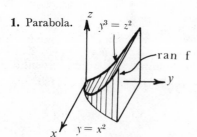

3. Hyperbola.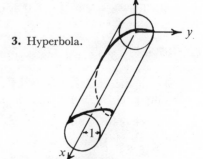

5. Ellipse.   **7.** Parabola.   **9.** Parabola.

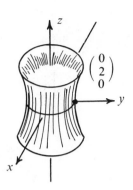

13. $(x + 3)^2 - (y + 1)^2 = \frac{1}{2}$.   **15.** $y = \dfrac{2}{x}$.

7. FUNCTION-LOCI IN $\mathscr{R}^3$

Frequently in sketching function-loci that are subsets of $\mathscr{R}^3$ it is helpful to study related function-loci in $\mathscr{R}^2$. We shall find level sets, projections, and sections of a given function-locus to be especially helpful.

Example 18. We wish to sketch the function-locus defined para-metrically by $f(t) = \begin{pmatrix} t \\ t^2 \\ t^2 \end{pmatrix}$, $t \geq 0$. We can visualize the figure better if we look at its "shadow" or projection on the coordinate planes. Hence, we first consider the set of points having the same first and second coordinates as those in *ran f*, but whose third coordinate is zero, that is,

$$\left\{ \begin{pmatrix} x_1 \\ x_2 \\ x_3 \end{pmatrix} = \begin{pmatrix} t \\ t^2 \\ 0 \end{pmatrix}, \quad t \geq 0 \right\}.$$

We shall call this set the *projection of ran f* on the 12-plane and denote it P_{12} *ran f*. Substituting x_1 for t in the second coordinate equation, we obtain $x_2 = x_1^2$, which we recognize as an explicit representation of a parabola in the 12-plane (see Figure 7.30a).

Similarly we look at $\left\{ \begin{pmatrix} x_1 \\ x_2 \\ x_3 \end{pmatrix} = \begin{pmatrix} 0 \\ t^2 \\ t^2 \end{pmatrix}, t \geq 0 \right\}$, which we call the projection of *ran f* on the 23-plane, denoted P_{23} *ran f*. We note that this is simply the half-line $x_2 = x_3$ (see Figure 7.30b). We deduce that if $\begin{pmatrix} x_1 \\ x_2 \\ x_3 \end{pmatrix} \in$ *ran f*, then it must be both on the surface indicated in Figure 7.30c and on the plane

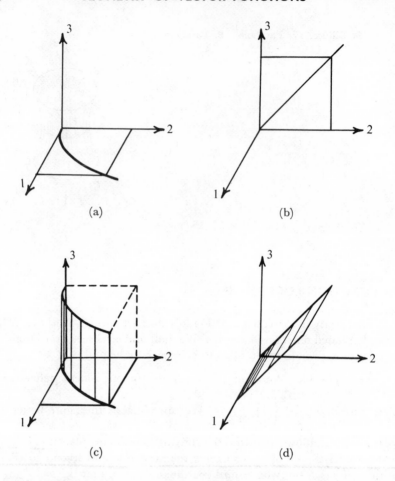

(a)

(b)

(c)

(d)

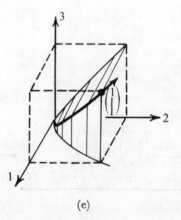

(e)

Figure 7.30

indicated in Figure 7.30d. (Note: A surface such as that indicated in Figure 7.30c is often called a *cylinder*.)

We can now see easily that *ran f*, which is the intersection of the surface shown in Figures 7.30c and d, is a curve through the origin. We can indicate its general form by showing a nonzero point in it, say $\begin{pmatrix} 1 \\ 1 \\ 1 \end{pmatrix}$, and sketching a curve through the origin and this point (see Figure 7.30c). ∎

We now make a general definition:

DEFINITION 7.16 Let $S \subset \mathcal{R}^3$. The set

$$\left\{ \begin{pmatrix} x_1 \\ x_2 \\ 0 \end{pmatrix} \middle| \text{ there exists } x_3 \text{ such that } \begin{pmatrix} x_1 \\ x_2 \\ x_3 \end{pmatrix} \in S \right\}$$

will be called the *projection of S on the 12-plane* (read one-two plane) of $\mathcal{R}^3$ and denoted $P_{12}S$. We define the projections of S on the 13-plane and on the 23-plane in an analogous way.

Example 19. Sometimes we can locate a function-locus in $\mathcal{R}^n$ by "slicing" it and looking at the "slice." Suppose we wish to sketch the function-locus $\mathcal{L}$ in $\mathcal{R}^3$ defined implicitly by

$$\frac{x_1^2}{4} + \frac{x_2^2}{9} + \frac{x_3^2}{1} = 1.$$

It may be that some points of $\mathcal{L}$ are on the 13-plane, so we "slice" the figure along this plane by looking at those points of $\mathcal{L}$ whose third coordinate is zero. More formally, we refer to the *section of $\mathcal{L}$ determined by the requirement that $x_3 = 0$*, denoted $S_{x_3=0}(\mathcal{L})$. Thus,

$$S_{x_3=0}(\mathcal{L}) = \left\{ \begin{pmatrix} x_1 \\ x_2 \\ x_3 \end{pmatrix} \middle| \frac{x_1^2}{4} + \frac{x_2^2}{9} = 1, \quad x_3 = 0 \right\},$$

which is clearly an ellipse in the 12-plane (see Figure 7.31a).

(Note carefully that $S_{x_3=0}(\mathcal{L})$ is a set of points *in* $\mathcal{L}$ whose third coordinate is zero, whereas the projection of $\mathcal{L}$ on the 12-plane is the set of points $\begin{pmatrix} x_1 \\ x_2 \\ 0 \end{pmatrix}$ such that $\begin{pmatrix} x_1 \\ x_2 \\ x_3 \end{pmatrix} \in \mathcal{L}$ for *some* x_3. Thus, $S_{x_3=0}(\mathcal{L}) \subset P_{12}\mathcal{L}$, but in general the two sets are not equal.)

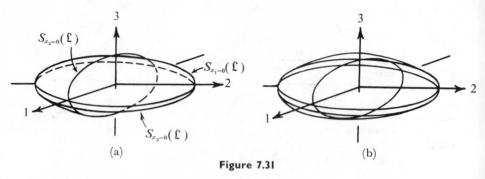

Figure 7.31

In a similar way we find $S_{x_1=0}(\mathcal{L})$ and $S_{x_2=0}(\mathcal{L})$. These are ellipses in the 23- and 13-planes, respectively, which we sketch. Note that if $|K| > 1$, then $S_{x_3=K}(\mathcal{L}) = \varnothing$, so the figure is confined between the planes $x_3 = -1$ and $x_3 = 1$. Similarly it is confined between the planes $x_1 = -2$ and $x_1 = 2$ and between the planes $x_2 = -3$ and $x_2 = 3$. Now if x_3 is a constant K such that $-1 < K < 1$, we obtain $\dfrac{x_1^2}{4} + \dfrac{x_2^2}{9} = 1 - K$, which is an ellipse in the plane $x_3 = K$. We obtain similar results when we consider sections parallel to the 13-plane and to the 23-plane. These results are sufficient for the completion of our sketch (see Figure 7.31b). The figure whose sketch we have completed is called an *ellipsoid*. ∎

We make a general definition:

DEFINITION 7.17   Let $\mathcal{B} \subset \mathcal{R}^3$ and let T be a rule that determines a subset $\mathcal{A}$ of $\mathcal{R}^3$. Then $\mathcal{B} \cap \mathcal{A}$ will be called the *section of $\mathcal{B}$ determined by T* and will be denoted $S_T\mathcal{B}$. We usually omit specific mention of $\mathcal{A}$.

The level set (see Definition 7.2, p. 290), which is just a special kind of section, is often useful in sketching explicitly defined function-loci.

Example 20.   Let S be the function-locus determined explicitly by $f\begin{pmatrix} x_1 \\ x_2 \end{pmatrix} = x_1^2 + x_2^2$ and note that f is everywhere defined. Hence, we may think of *graph f* as an infinite sheet spread over the 12-plane and then moved upward or downward at each point in accordance with the function values. We can get some idea of the figure by examining the level sets. (We may think of these as being similar to contour lines on a map.) First observe that if $K < 0$, then $K\text{-}lev\, f = \varnothing$. Thus, the sketch will be entirely on and above the 12-plane. To find the intersection with the 12-plane, we examine $S_{x_1=0}\, graph\, f$, which we find to be $\{O_3\}$. If $K > 0$, we find $K\text{-}lev\, f$ to be the circle implicitly defined by $x_1^2 + x_2^2 = K$ (see Figure 7.32). Thus, as "higher and higher" level sets are considered, we obtain larger and larger circles about the 3-axis. We need a somewhat better picture of how the

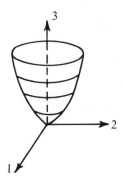

Figure 7.32

circles increase, so we consider $S_{x_1=0}$ *graph f.* We obtain $f\begin{pmatrix} 0 \\ x_2 \end{pmatrix} = x_2^2$, which
we may consider as a function of one variable. This gives us a parabola in the.
23-plane, which each of our circles must intersect. Thus, we are able to
complete our sketch (see Figure 7.32). We may think of this figure as the
set of points generated by rotating the parabola $f(x_2) = x_2^2$ about the 3-axis.
Such a set is called a *paraboloid of revolution.* ■

Example 21. We wish to sketch the function-locus defined parametrically by

$$g\begin{pmatrix} u \\ v \end{pmatrix} = \begin{pmatrix} x \\ y \\ z \end{pmatrix} = \begin{pmatrix} \dfrac{\cos u}{\sin v} \\ \dfrac{\sin u}{\sin v} \\ \cos v \end{pmatrix}, \quad 0 \le u \le 2\pi, \quad 0 < v \le \dfrac{\pi}{2}.$$

The domain of g is the rectangle sketched in Figure 7.33a, b, and c. The
range will, of course, be a subset of $\mathcal{R}^3$. We shall find the range by fixing
first one and then another of the variables u and v, and then letting the other
vary over its domain. First consider the set of all points $g\begin{pmatrix} u \\ \pi \\ 2 \end{pmatrix}$, that is, see
where the line segment between $\begin{pmatrix} 0 \\ \pi \\ 2 \end{pmatrix}$ and $\begin{pmatrix} 2\pi \\ \pi \\ 2 \end{pmatrix}$ goes under the function g
(see Figure 7.33a). We have

$$g\begin{pmatrix} u \\ \dfrac{\pi}{2} \end{pmatrix} = \begin{pmatrix} x \\ y \\ z \end{pmatrix} = \begin{pmatrix} \dfrac{\cos u}{\sin \dfrac{\pi}{2}} \\ \dfrac{\sin u}{\sin \dfrac{\pi}{2}} \\ \cos \dfrac{\pi}{2} \end{pmatrix} = \begin{pmatrix} \cos u \\ \sin u \\ 0 \end{pmatrix}, \quad 0 \le u \le 2\pi.$$

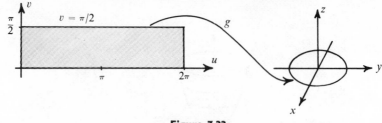

Figure 7.33a

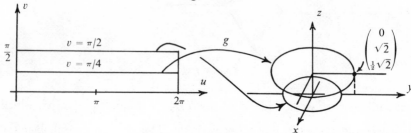

Figure 7.33b

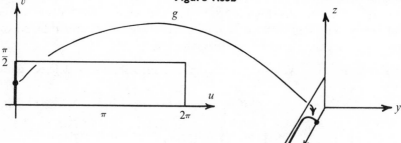

Figure 7.33c

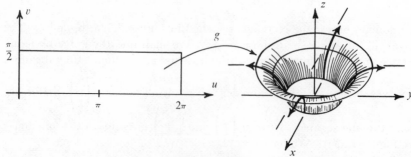

Figure 7.33d

This is recognizable as a circle of radius one and center O_3 in the xy plane. Similarly, if we fix v at any value of its range and let u vary from zero to 2π, we get a circle in the plane $z = \cos v$, with center at $\begin{pmatrix} 0 \\ 0 \\ \cos v \end{pmatrix}$ and radius $\dfrac{1}{\sin v}$. The situation for $v = \dfrac{\pi}{4}$ is illustrated in Figure 7.33b.

Now by fixing u at 0 and letting v vary over its domain, we shall find one point on each of the circles. This will completely determine the figure. For this choice of u we find

$$g\begin{pmatrix}0\\v\end{pmatrix} = \begin{pmatrix}x\\y\\z\end{pmatrix} = \begin{pmatrix}\dfrac{1}{\sin v}\\0\\\cos v\end{pmatrix}, \quad 0 < v \le \frac{\pi}{2}.$$

For $v = \dfrac{\pi}{2}$ we find $g\begin{pmatrix}0\\\frac{\pi}{2}\end{pmatrix} = \begin{pmatrix}1\\0\\0\end{pmatrix}$, and as we let v approach zero we find that $x = \dfrac{1}{\sin v}$ increases toward $+\infty$ and $z = \cos v$ approaches 1. Thus, the line segment between $\begin{pmatrix}0\\0\end{pmatrix}$ and $\begin{pmatrix}0\\\frac{\pi}{2}\end{pmatrix}$, $\begin{pmatrix}0\\0\end{pmatrix}$ excluded, maps into a curve

something like that shown in Figure 7.33c.

Since each of the circles previously determined must intersect this curve, we conclude that the function-locus looks something like the sketch of Figure 7.33d (of course it extends infinitely in the directions of the xy-plane). ∎

EXERCISES

Sketch the function-loci defined parametrically by the following functions:

1. $f(t) = \begin{pmatrix}t\\t^2\\t^3\end{pmatrix}, 0 \le t \le 1.$ 2. $f(t) = \begin{pmatrix}t\\t\\t^2+1\end{pmatrix}, 0 \le t \le 1.$

3. $\begin{pmatrix}x\\y\\z\end{pmatrix} = \begin{pmatrix}t\\\cos t\\\sin t\end{pmatrix}, t \ge 0.$ 4. $\begin{pmatrix}x\\y\\z\end{pmatrix} = \begin{pmatrix}t\cos t\\t\sin t\\t\end{pmatrix}, 0 \le t.$

5. $f(t) = \begin{pmatrix}t\\t^2\\\sin t + 1\end{pmatrix}, -\frac{\pi}{2} < t < \frac{\pi}{2}.$

6. $g\begin{pmatrix}u\\v\end{pmatrix} = \begin{pmatrix}\cos u \sin v\\\sin u \sin v\\\cos v\end{pmatrix}, 0 \le u \le 2\pi, -\frac{\pi}{2} \le v < \frac{\pi}{2}.$

Sketch the function-loci in $\mathfrak{R}^3$ defined implicitly by the following functions:

7. $x_1^2 + x_2^2 = 1$ (a circular cylinder).

8. $\dfrac{x_1^2}{4} + \dfrac{x_2^2}{4} + \dfrac{x_3^2}{9} = 1$ (an ellipsoid of revolution).

9. $\dfrac{x^2}{4} + \dfrac{y^2}{4} - \dfrac{z^2}{9} = 1$ (a hyperboloid of revolution of one sheet).

10. $9x_1^2 + 4x_2^2 = 36(1 + x_3^2)$ (a hyperboloid of one sheet).

11. $\dfrac{x^2}{9} - \dfrac{y^2}{4} - \dfrac{z^2}{4} = 1$ (a hyperboloid of revolution of two sheets).

12. $9x_1^2 - 18x_1 + 9x_2^2 - 18x_2 - 4x_3^2 = 14$ (a hyperboloid of revolution of one sheet).

13. $\dfrac{x^2}{2} + \dfrac{z^2}{3} = y$ (an elliptic paraboloid).

14. $x + \dfrac{y^2}{4} = \dfrac{z^2}{2}$ (a hyperbolic paraboloid).

Sketch the function-loci defined explicitly by the following functions:

15. $f\begin{pmatrix} x \\ y \end{pmatrix} = x^2 + 2y^2.$

16. $f\begin{pmatrix} x \\ y \end{pmatrix} = \sin y.$

17. $z = y + \cos x.$

18. $z = x^2 - \dfrac{y^2}{3}.$

ANSWERS

1.

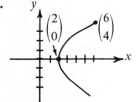

3.

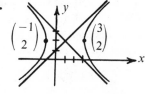

5.

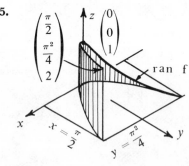

7.

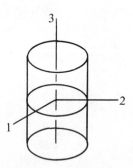

9.   11.

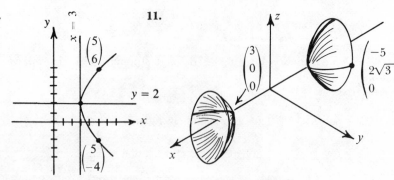

13.

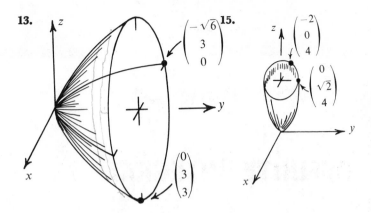

15.

17.

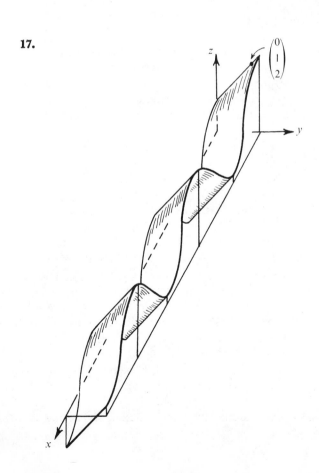

8

THE DEFINITE INTEGRAL

1. THE AREA PROBLEM

The concept of the length of a line segment is easily defined in mathematical terms. To the person who possesses knowledge of the real numbers, the mathematical description of this concept seems natural and acceptable.

A second concept to which one is introduced at an early age in his education is that of area. It is accepted that the area of that region in the plane bounded by a rectangle is the product of the lengths of two adjacent sides of the rectangle. Similar rules are given to compute the area bounded by a triangle, a trapezoid, or a circle.

Unfortunately, it is not easy to verbalize what the word *area* means. Consider the region $\mathcal{R} = \{(x, y) \mid 0 \leq x \leq 1, 0 \leq y \leq x^2\}$, illustrated in Figure 8.1a. Is it a reasonable question to ask what the *area* of $\mathcal{R}$ is? Let us assume that this is a reasonable question. Let $A(\mathcal{R})$ denote the area of this region. By what criteria shall we determine $A(\mathcal{R})$? Since $\mathcal{R}$ is contained in the rectangle $[0, 1] \times [0, 1]$ (see Figure 8.1b), which has area 1, it is reasonable to assume that $A(\mathcal{R}) \leq 1$. Similarly, since $[\frac{1}{2}, 1] \times [0, \frac{1}{4}] \subset \mathcal{R}$ (see Figure 8.1c) and has area $\frac{1}{8}$, we conclude that $\frac{1}{8} \leq A(\mathcal{R})$. These assumptions should be intuitively acceptable, for if area is a measure of the size of regions in the plane, it should have the following property: If $\mathcal{R} \subset \mathcal{S}$, then area $\mathcal{R} \leq$ area $\mathcal{S}$.

Let us assume now (1) that we have a function A (called an area function) which is defined on a certain class $\mathcal{C}$ of bounded subsets of the plane, (2) that this class of subsets contains all rectangles of the form $[a, b] \times [c, d]$, and (3) that $A([a, b] \times [c, d]) = A((a, b) \times (c, d)) = (b - a)(d - c)$. We shall assume that our function A has the following properties:

i. $A(\mathcal{R}) \geq 0$ if $\mathcal{R} \in \mathcal{C}$.

338

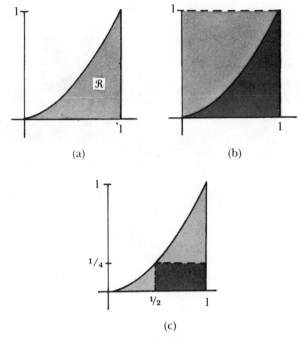

(a)   (b)

(c)

Figure 8.1

ii. $\phi \in \mathfrak{C}$ and $A(\phi) = 0$.

iii. $A(\mathfrak{R}) \leq A(\mathfrak{S})$ if $\mathfrak{R} \subset \mathfrak{S}$, $\mathfrak{R} \in \mathfrak{C}$, and $\mathfrak{S} \in \mathfrak{C}$.

iv. $A(\mathfrak{R} \cup \mathfrak{S}) = A(\mathfrak{R}) + A(\mathfrak{S})$ if $\mathfrak{R} \cap \mathfrak{S} = \phi$, $\mathfrak{R} \in \mathfrak{C}$, and $\mathfrak{S} \in \mathfrak{C}$.

v. $A(\mathfrak{R} \cup \mathfrak{S}) = A(\mathfrak{R}) + A(\mathfrak{S}) - A(\mathfrak{R} \cap \mathfrak{S})$ if $\mathfrak{R} \in \mathfrak{C}$, and $\mathfrak{S} \in \mathfrak{C}$.

Property (i) merely states that the area of a region $\mathfrak{R}$ is either a positive number or zero, and property (ii) asserts that the area of the empty set is indeed zero. Property (iii) may be thought of as an assertion that the area of part of a region is not greater than the area of the whole region. Property (iv) asserts that the sum of the areas of two non-overlapping regions is the area of the union of those two regions. Notice that in view of property (ii), property (iv) is a special case of property (v). Figure 8.2 illustrates the plausibility of property (v) in case $\mathfrak{R}$ and $\mathfrak{S}$ are rectangles.

Using only the five properties listed above we can compute certain areas as follows:

Example 1. Let $\mathfrak{R} = ([0, 1] \times [0, 2]) \cup ((1, 2] \times [0, 1])$. Figure 8.3 is a sketch of the region $\mathfrak{R}$. The area of $\mathfrak{R}$ is (by property (iv)) the sum of the areas of $\mathfrak{R}_1 = [0, 1] \times [0, 2]$ and $\mathfrak{R}_2 = (1, 2] \times [0, 1]$. We compute $A(\mathfrak{R}_1) = (1 - 0)(2 - 0) = 2$, and $A(\mathfrak{R}_2) = (2 - 1)(1 - 0) = 1$, and thus $A(\mathfrak{R}) = 2 + 1 = 3$. ∎

In the future we shall assume that a straight line in the plane has area 0. This is plausible if we think of a straight line as a region bounded by a

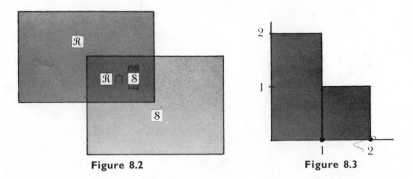

Figure 8.2   Figure 8.3

degenerate rectangle (one side being a point). Thus, in Example 1 we could have written $\mathcal{R} = \mathcal{R}_1 \cup \mathcal{R}_2$, where $\mathcal{R}_1 = [0, 1] \times [0, 2]$ and $\mathcal{R}_2 = [1, 2] \times [0, 1]$. Since $\mathcal{R}_1 \cap \mathcal{R}_2 = [1, 1] \times [0, 1]$, $\mathcal{R}_1 \cap \mathcal{R}_2$ is a line segment and $A(\mathcal{R}_1 \cap \mathcal{R}_2) = 0$. Thus, by property (v), $A(\mathcal{R}) = A(\mathcal{R}_1) + A(\mathcal{R}_2) - A(\mathcal{R}_1 \cap \mathcal{R}_2) = A(\mathcal{R}_1) + A(\mathcal{R}_2)$. For convenience, in future examples we shall ignore $A(\mathcal{R}_1 \cap \mathcal{R}_2)$ in our computations if $\mathcal{R}_1 \cap \mathcal{R}_2$ is a line segment.

For many regions properties (i) to (v) listed previously for an area function are not adequate. They do not tell us how to compute the area of the region $\mathcal{R}$ sketched in Figure 8.1. However, we did conclude that if $\mathcal{R}$ has an area, $\frac{1}{8} \leq A(\mathcal{R}) \leq 1$. In the following example we shall improve these estimates:

Example 2.   Let $\mathcal{R} = \{(x, y) \mid 0 \leq x \leq 1, 0 \leq y \leq x^2\}$.

(**a**) Divide the interval $[0, 1]$ into three subintervals of equal length. Now let

$$\mathcal{R}_3 = \left(\left[0, \frac{1}{3} \right] \times \left[0, \left(\frac{1}{3} \right)^2 \right] \right) \cup \left(\left[\frac{1}{3}, \frac{2}{3} \right] \times \left[0, \left(\frac{2}{3} \right)^2 \right] \right)$$

$$\cup \left(\left[\frac{2}{3}, \frac{3}{3} \right] \times \left[0, \left(\frac{3}{3} \right)^2 \right] \right)$$

and

$$\mathcal{S}_3 = \left(\left[0, \frac{1}{3} \right] \times \left[0, \left(\frac{0}{3} \right)^2 \right] \right) \cup \left(\left[\frac{1}{3}, \frac{2}{3} \right] \times \left[0, \left(\frac{1}{3} \right)^2 \right] \right)$$

$$\cup \left(\left[\frac{2}{3}, \frac{3}{3} \right] \times \left[0, \left(\frac{2}{3} \right)^2 \right] \right).$$

The regions $\mathcal{R}_3$ and $\mathcal{R}$ are sketched in Figure 8.4a; $\mathcal{S}_3$ and $\mathcal{R}$ are sketched in Figure 8.4b. Notice that $\mathcal{S}_3 \subset \mathcal{R} \subset \mathcal{R}_3$. Moreover,

$$A(\mathcal{R}_3) = \frac{1}{3}\left(\frac{1^2}{3^2}\right) + \frac{1}{3}\left(\frac{2^2}{3^2}\right) + \frac{1}{3}\left(\frac{3^2}{3^2}\right) = \frac{1}{3^3}[1^2 + 2^2 + 3^2]$$

and

$$A(\mathcal{S}_3) = \frac{1}{3}\left(\frac{0^2}{3^2}\right) + \frac{1}{3}\left(\frac{1^2}{3^2}\right) + \frac{1}{3}\left(\frac{2^2}{3^2}\right) = \frac{1}{3^3}[0^2 + 1^2 + 3^2].$$

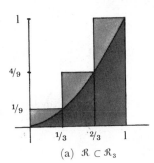

(a) $\mathcal{R} \subset \mathcal{R}_3$

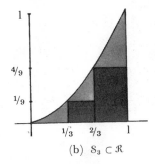

(b) $\mathcal{S}_3 \subset \mathcal{R}$

Figure 8.4

Thus, if $A(\mathcal{R})$ exists,

$$\frac{1}{3^3}\,[0^2 + 1^2 + 2^2] \leq A(\mathcal{R}) \leq \frac{1}{3^3}\,[1^2 + 2^2 + 3^2]$$

or

$$\frac{5}{27} \leq A(\mathcal{R}) \leq \frac{14}{27}.$$

If we consider the difference of the upper and lower bounds on $A(\mathcal{R})$ as a measure of whether one approximation is better than another, this estimate on $A(\mathcal{R})$, with a difference of $\frac{9}{27} = \frac{1}{3}$, definitely improves our first estimate in which we had $\frac{1}{8} \leq A(\mathcal{R}) \leq 1$ for a difference of $\frac{7}{8}$.

(**b**) Now divide $[0, 1]$ into four subintervals of equal length and let

$$\mathcal{R}_4 = \bigcup_{k=1}^{4}\left[\frac{k-1}{4}, \frac{k}{4}\right] \times \left[0, \left(\frac{k}{4}\right)^2\right]$$

and

$$\mathcal{S}_4 = \bigcup_{k=1}^{4}\left[\frac{k-1}{4}, \frac{k}{4}\right] \times \left[0, \left(\frac{k-1}{4}\right)^2\right]$$

(see Figures 8.5a and 8.5b). In this case

$$A(\mathcal{R}_4) = \frac{1}{4^3}\,[1^2 + 2^2 + 3^2 + 4^2]$$

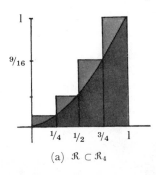

(a) $\mathcal{R} \subset \mathcal{R}_4$

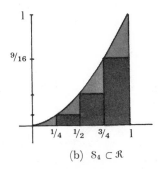

(b) $\mathcal{S}_4 \subset \mathcal{R}$

Figure 8.5

and

$$A(S_4) = \frac{1}{4^3}[0^2 + 1^2 + 2^2 + 3^2].$$

Thus,

$$\frac{1}{4^3}[0^2 + 1^2 + 2^2 + 3^2] \le A(\mathcal{R}) \le \frac{1}{4^3}[1^2 + 2^2 + 3^2 + 4^2]$$

or

$$\frac{14}{64} \le A(\mathcal{R}) \le \frac{30}{64}.$$

Again we have improved our estimate of $A(\mathcal{R})$.

(c) We now generalize our procedure; Divide $[0, 1]$ into n subintervals of equal length and let

$$\mathcal{R}_n = \bigcup_{k=1}^{n}\left[\frac{k-1}{n}, \frac{k}{n}\right] \times \left[0, \left(\frac{k}{n}\right)^2\right]$$

and

$$S_n = \bigcup_{k=1}^{n}\left[\frac{k-1}{n}, \frac{k}{n}\right] \times \left[0, \left(\frac{k-1}{n}\right)^2\right].$$

We compute $A(\mathcal{R}_n)$ and $A(S_n)$ as follows:

$$A(\mathcal{R}_n) = \sum_{k=1}^{n} \frac{k-(k-1)}{n}\left(\frac{k}{n}\right)^2$$

$$= \frac{1}{n^3}\sum_{k=1}^{n} k^2$$

$$=^1 \frac{1}{n^3}\frac{n(n+1)(2n+1)}{6}$$

$$= \frac{1}{6}\left(1 + \frac{1}{n}\right)\left(2 + \frac{1}{n}\right);$$

$$A(S_n) = \sum_{k=1}^{n} \frac{k-(k-1)}{n}\left(\frac{k-1}{n}\right)^2$$

$$= \frac{1}{n^3}\sum_{k=1}^{n} (k-1)^2$$

$$=^2 \frac{1}{n^3}\frac{(n-1)(n)(2n-1)}{6}$$

$$= \frac{1}{6}\left(1 - \frac{1}{n}\right)\left(2 - \frac{1}{n}\right).$$

(1) Exercise 3a.
(2) Exercise 3b.

Note that $S_n \subset \mathcal{R} \subset \mathcal{R}_n$, $A(S_n) \le A(\mathcal{R}) \le A(\mathcal{R}_n)$, and that $\lim_{n\to\infty} A(S_n) = \lim_{n\to\infty} A(\mathcal{R}_n) = \frac{1}{3}$. We conclude then that $A(\mathcal{R}) = \frac{1}{3}$. ∎

In the preceding example we assumed that the area of the region $\mathcal{R}$ under consideration existed and concluded that it must be equal to $1/3$ on the basis of property (iii). In a later section of this chapter we shall define the area of certain regions by this sort of approximation, and consequently we add to our list of properties of area the following:

vi. If $\mathcal{R}$ is a bounded region of the plane possessing area $A(\mathcal{R})$, and, for each n, $\mathcal{R}_n$ and $\mathcal{S}_n$ are each the union of a finite number of regions bounded by rectangles, such that $\mathcal{S}_n \subset \mathcal{R} \subset \mathcal{R}_n$ and $\lim\limits_{n\to\infty} A(\mathcal{S}_n) = \lim\limits_{n\to\infty} A(\mathcal{R}_n)$, then $A(\mathcal{R}) = \lim\limits_{n\to\infty} A(\mathcal{R}_n) = \lim\limits_{n\to\infty} A(\mathcal{S}_n)$.

Notice that in property (vi) we do not insist that $\mathcal{R}_n = \bigcup\limits_{j=1}^{n} \mathcal{R}_n^j$ where each of the $\mathcal{R}_n^j$ has the same width (as we did in Example 2). In Example 2 we subdivided $[0, 1]$ into subintervals of equal length $\dfrac{1}{n}$ in order to simplify the resulting computations. The very problems posed by these computations lead us to a formalization and generalization of the procedure used in Example 2. The remainder of this chapter will be devoted to these considerations.

EXERCISES

1. Prove that $\sum\limits_{k=1}^{n} k = 1 + 2 + \cdots + n = \dfrac{n(n + 1)}{2}$ by showing that

   (a) $\sum\limits_{k=1}^{1} k = \dfrac{1(1 + 1)}{2}$

   and

   (b) if $\sum\limits_{k=1}^{n} k = \dfrac{n(n + 1)}{2}$ for any positive integer n, then $\sum\limits_{k=1}^{n+1} k = \dfrac{(n + 1)(n + 2)}{2}$.

2. Let $\mathcal{R} = \{(x,y) \mid 0 \leq x \leq 1, 0 \leq y \leq 3x\}$.

   Let $\mathcal{R}_n = \bigcup\limits_{k=1}^{n} \left[\dfrac{k - 1}{n}, \dfrac{k}{n}\right] \times \left[0, \dfrac{3k}{n}\right]$ and

   $\mathcal{S}_n = \bigcup\limits_{k=1}^{n} \left[\dfrac{k - 1}{n}, \dfrac{k}{n}\right] \times \left[0, \dfrac{3(k - 1)}{n}\right]$.

   (a) Compute $A(\mathcal{R}_n)$.
   (b) Compute $A(\mathcal{S}_n)$.
   (c) Show that $A(\mathcal{R}) = \lim\limits_{n\to\infty} A(\mathcal{R}_n) = \lim\limits_{n\to\infty} A(\mathcal{S}_n)$.

3. (a) Prove that $\sum\limits_{k=1}^{n} k^2 = \dfrac{n(n + 1)(2n + 1)}{6}$ for each positive integer n

   (Hint: either use induction or the relation

   $$\sum_{k=1}^{n} (k + 1)^3 = \sum_{k=1}^{n} k^3 + 3\sum_{k=1}^{n} k^2 + 3\sum_{k=1}^{n} k + \sum_{k=1}^{n} 1.$$

   (b) Prove that $\sum\limits_{k=1}^{n} (k - 1)^2 = \dfrac{(n - 1)(n)(2n - 1)}{6}$ for each positive integer n.

2. STEP FUNCTIONS

In the preceding section we considered the problem of approximating the area of a certain region by considering the area of a union of rectangles, either contained in or containing the given region. Our first step in this approximation was that of dividing an interval $[a, b]$ into subintervals. This simple idea is formalized as follows:

> **DEFINITION 8.1** A *partition* of a closed interval $[a, b]$ is a finite set $\mathcal{S} = \{x_0, x_1, \ldots, x_n\}$ of real numbers such that
>
> $$a = x_0 < x_1 < \cdots < x_n = b.$$
>
> The interval $(x_{j-1}, x_j), j = 1, \ldots, n$ is called the jth *open subinterval* of the partition $\mathcal{S}$ of $[a, b]$. If $\mathcal{S}$ and $\mathcal{Q}$ are partitions of $[a, b]$ and $\mathcal{S} \subset \mathcal{Q}$, then $\mathcal{Q}$ is said to be a *refinement of* $\mathcal{S}$.

Example 3. Let $[a, b] = [0, 4]$.

(a) $\mathcal{S} = \{0, 4\}$ is a partition of $[0, 4]$ with $x_0 = 0$ and $x_1 = 4$.

(b) $\mathcal{Q} = \{0, 1, 2, 3, 4\}$ is a partition of $[0, 4]$ with $x_0 = 0$, $x_1 = 1$, $x_2 = 2$, $x_3 = 3$, and $x_4 = 4$. Furthermore, $\mathcal{Q}$ is a refinement of $\mathcal{S}$ since $\{0, 4\} \subset \{0, 1, 2, 3, 4\}$.

(c) $\mathcal{R} = \{0, 1/2, 3, 4\}$ is a partition of $[a, b]$ with $x_0 = 0$, $x_1 = 1/2$, $x_2 = 3$, and $x_3 = 4$. Furthermore, $\mathcal{R}$ is a refinement of $\mathcal{S}$, but neither $\mathcal{Q}$ nor $\mathcal{R}$ is a refinement of the other.

The open subintervals of $\mathcal{S}$ are $(0, 4)$; the open subintervals of $\mathcal{Q}$ are $(0, 1)$, $(1, 2)$, $(2, 3)$, and $(3, 4)$; and the open subintervals of $\mathcal{R}$ are $(0, 1/2)$, $(1/2, 3)$, and $(3, 4)$. ∎

We continue now to generalize the approximation method we used in Example 2 of Section 1. We approximated the area in question by the area of a region which is the union of a finite number of regions bounded by rectangles. Such a region is bounded to the left and to the right by the vertical lines $x = a$ and $x = b$, (in Example 2 $a = 0$ and $b = 1$) and above and below by the graph of a very simple function which is constant on the open subintervals of some partition of the interval $[a, b]$.

Consider now the partition of $[0, 1]$, $\mathcal{S} = \{0, \frac{1}{4}, \frac{1}{2}, 1\}$. If we define $f : [0, 1] \to \mathcal{R}$ by

$$f(x) = \begin{cases} 1/2 & \text{for } 0 \leq x < \frac{1}{4} \\ 1 & \text{for } \frac{1}{4} < x < \frac{1}{2} \\ -1/2 & \text{for } \frac{1}{2} < x \leq 1 \\ -1 & \text{for } x = \frac{1}{4}, \frac{1}{2}, \end{cases}$$

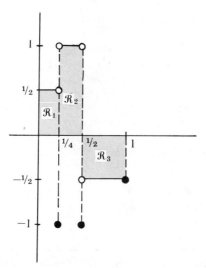

Figure 8.6

we have a function which is constant on the open subintervals of $\mathfrak{I}$. Figure 8.6 is a sketch of the graph of f. With the exception of the points $(\frac{1}{4}, -1)$ and $(\frac{1}{2}, -1)$ the graph of f resembles a set of stair steps. Moreover, the region lying between the graph of f and the corresponding points of the x-axis is the union of three rectangles, $\mathfrak{R}_1$, $\mathfrak{R}_2$, and $\mathfrak{R}_3$.

Without much effort the reader can verify the fact that if the region bounded by the graph of a function f ($f:[a, b] \to \mathfrak{R}$) and the x-axis is the union of a finite number of rectangles, there exists a partition $\mathfrak{I}$ of $[a, b]$ such that f is constant on the open subintervals of $\mathfrak{I}$. We formalize this situation as follows:

DEFINITION 8.2 Let $s:[a, b] \to \mathfrak{R}$. Then s is a *step function on* $[a, b]$ if and only if there exists a partition $\mathfrak{I} = \{x_0, \ldots, x_n\}$ of $[a, b]$ for which s is constant on each of the open subintervals of $\mathfrak{I}$. That is, there exist constants $s_1, \ldots, s_n$ such that

$$s(x) = s_j \quad \text{for} \quad x_{j-1} < x < x_j.$$

Example 4. Let $\mathfrak{I} = \{0, 2, 4\}$ be a partition of $[0, 4]$.
(**a**) If $s:[0, 4] \to \mathfrak{R}$ which is defined by

$$s(x) = 2 \quad \text{for all } x \text{ in } [0, 4],$$

then s is a step function on $[0, 4]$ constant on the open subintervals of $\mathfrak{I}$.

(**b**) If $s:[0, 4] \to \mathcal{R}$ which is defined by

$$s(x) = \begin{cases} 1 & \text{for} \quad 0 \leq x < 2 \\ -1 & \text{for} \quad x = 2 \\ 2 & \text{for} \quad 2 < x \leq 4, \end{cases}$$

then s is a step function on $[0, 4]$ constant on the open subintervals of $\mathcal{S}$.

(**c**) If $s:[0, 4] \to \mathcal{R}$ which is defined by

$$s(x) = \begin{cases} 3 & \text{for} \quad 0 \leq x \leq 3 \\ 5 & \text{for} \quad 3 < x \leq 4, \end{cases}$$

then s is a step function on $[0, 4]$ constant on the open subintervals of the partitions $\mathcal{Q} = \{0, 3, 4\}$ and $\mathcal{R} = \{0, 1/2, 3, 4\}$, but *not* constant on the open subintervals of $\mathcal{S}$. ∎

Reconsider the step function s, defined in Example 4b, on the partition $\mathcal{S} = \{0, 2, 4\}$ of $[0, 4]$. The graph of s is sketched in Figure 8.7a. Notice that if we refine $\mathcal{S}$ by further subdividing $[0, 4]$, as in Figure 8.7b, s is constant on the open subintervals of the new partition and $\mathcal{R}_1$ and $\mathcal{R}_2$ are merely subdivided into smaller rectangles, the sum of whose areas is equal to the sum of the areas of $\mathcal{R}_1$ and $\mathcal{R}_2$. Lemma 8.3, which we present without proof, is a more formal statement of this simple situation:

LEMMA 8.3   Let $s:[a, b] \to \mathcal{R}$ be a step function that is constant on the open subintervals of a partition $\mathcal{S} = \{x_0, \ldots, x_n\}$. If $\mathcal{Q} = \{y_0, \ldots, y_m\}$ is any refinement of $\mathcal{S}$, then s is constant on the open subintervals of $\mathcal{Q}$ and

(8.1)
$$\sum_{j=1}^{n} s_j(x_j - x_{j-1}) = \sum_{k=1}^{m} s'_k(y_k - y_{k-1}),$$

where $s(x) = s_j$ for $x_{j-1} < x < x_j$ and $s(x) = s'_k$ for $y_{k-1} < x < y_k$.

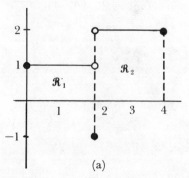

(a)

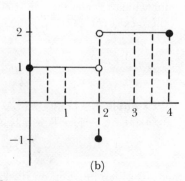

(b)

Figure 8.7

The simple nature of step functions leads us readily to the following theorem:

THEOREM 8.4 Let s and t be step functions on $[a, b]$ and $r \in \mathcal{R}$.
i. (Additive Property) $s + t$ is a step function on $[a, b]$,
and ii. (Homogeneous Property) rs is a step function on $[a, b]$.

The reader familiar with vector space terminology will recognize that Theorem 8.4 asserts that the set of step functions on $[a, b]$ is a vector subspace of the set of all real-valued functions on $[a, b]$.

Proof of i. If s is constant on the open subintervals of a partition $\mathcal{S}$ of $[a, b]$ and t is constant on the open subintervals of a partition $\mathcal{Q}$ of $[a, b]$, then each of s and t is constant on the open subintervals of $\mathcal{S} \cup \mathcal{Q}$ (by Definition 8.1 and Lemma 8.3). Thus $s + t$ is constant on the open subintervals of the partition $\mathcal{S} \cup \mathcal{Q}$ and by Definition 8.2 is a step function on $[a, b]$. The proof of (ii) is called for in Exercise 10. ∎

Example 5. Let $s(x) = \begin{cases} 1 & \text{for} & -1 \le x < 0 \\ -1 & \text{for} & 0 \le x \le 1 \end{cases}$

and $t(x) = \begin{cases} 2 & \text{for} & -1 \le x < \frac{1}{2} \\ 0 & \text{for} & x = \frac{1}{2} \\ 1 & \text{for} & \frac{1}{2} < x \le 1. \end{cases}$

Note that s is a step function on $[-1, 1]$, constant on the open subintervals of $\mathcal{S} = \{-1, 0, 1\}$, and t is a step function on $[-1, 1]$, constant on the open subintervals of $\mathcal{Q} = \{-1, 1/2, 1\}$. We let $\mathcal{R} = \mathcal{S} \cup \mathcal{Q} = \{-1, 0, 1/2, 1\}$ and note that

$$(s + t)(x) = \begin{cases} 3 & \text{for} & -1 \le x < 0 \\ 1 & \text{for} & 0 \le x < 1/2 \\ -1 & \text{for} & x = 1/2 \\ 0 & \text{for} & \frac{1}{2} < x \le 1. \end{cases}$$

Thus, $s + t$ is a step function on $[-1, 1]$ which is constant on the open subintervals of $\mathcal{R}$. ∎

EXERCISES

In Exercises 1 to 4 a partition $\mathcal{S}$ of an interval $[a, b]$ is given:
(a) Describe three distinct refinements of $\mathcal{S}$.
(b) Describe two refinements of $\mathcal{S}$, neither of which is a refinement of the other.
(c) Describe a partition of $[a, b]$ which is not a refinement of $\mathcal{S}$.

  1. $\mathcal{S} = \{0, 1, 2, 5\}$; $[a, b] = [0, 5]$.

2. $\mathcal{I} = \{-1, 0, 2\};\ [a, b] = [-1, 2]$.

3. $\mathcal{I} = \{-8, 0, 4, 6\};\ [a, b] = [-8, 6]$.

4. $\mathcal{I} = \{-1, 1\};\ [a, b] = [-1, 1]$. *there is none*

In Exercises 5 to 9 a step function is defined on an interval $[a, b]$. In each case compute $\sum_{j=1}^{n} s_j(x_j - x_{j-1})$, where $\{x_0, \ldots, x_n\}$ is a partition such that s is constant on the open subintervals (x_{j-1}, x_j) and s_j is the value of s on (x_{j-1}, x_j):

5. $s(x) = \begin{cases} -1 & \text{for} \quad 0 \le x < 1 \\ 2 & \text{for} \quad 1 \le x \le 2. \end{cases}$

6. $s(x) = \begin{cases} -1 & \text{for} \quad 0 \le x < 1 \\ 2 & \text{for} \quad 1 \le x < 3 \\ -3 & \text{for} \quad 3 \le x \le 5. \end{cases}$

7. $s(x) = \begin{cases} -2 & \text{for} \quad -1 \le x < 0 \\ 5 & \text{for} \quad x = 0 \\ 1/2 & \text{for} \quad 0 < x \le 2. \end{cases}$

8. $s(x) = \begin{cases} 1 & \text{for} \quad -2 \le x \le 0 \\ -1 & \text{for} \quad 0 < x \le 2. \end{cases}$

9. $s(x) = \begin{cases} 1 & \text{for} \quad 0 \le x < 1 \\ 2 & \text{for} \quad 1 \le x < 2 \\ 5 & \text{for} \quad 2 \le x \le 3. \end{cases}$

10. Prove Theorem 8.4ii.

ANSWERS

1. (a) and (b): $\{0, 1, 2, 3, 5\}$ and $\{0, 1, 2, 4, 5\}$ are refinements of $\mathcal{I}$, but neither is a refinement of the other. $\{0, 3, 5\}$ is a partition of $[0, 5]$ which is not a refinement of $\mathcal{I}$. **3.** (a) and (b): $\{-8, 0, 2, 4, 6\}$ and $\{-8, -3, 0, 4, 5, 6\}$ are refinements of $\mathcal{I}$, but neither is a refinement of the other. $\{-8, 6\}$ is a partition of $[-8, 6]$ which is not a refinement of $\mathcal{I}$.

5. Let $x_0 = 0, x_1 = 1, x_2 = 2, s_1 = -1$, and $s_2 = 2$. Then $\sum_{j=1}^{2} s_j(x_j - x_{j-1}) = -1(1 - 0) + 2(2 - 1) = 1$. **7.** Let $x_0 = -1, x_1 = 0, x_2 = 2, s_1 = -2$, and $s_2 = 1/2$. Then $\sum_{j=1}^{2} s_j(x_j - x_{j-1}) = -2(0 - (-1)) + \frac{1}{2}(2 - 0) = -1$.

9. Let $x_0 = 0, x_1 = 1, x_2 = 2, x_3 = 3, s_1 = 1, s_2 = 2$, and $s_3 = 5$. Then $\sum_{j=1}^{3} s_j(x_j - x_{j-1}) = 1(1 - 0) + 2(2 - 1) + 5(3 - 2) = 8$.

3. THE INTEGRAL OF A STEP FUNCTION

Assume now that $\mathcal{I} = \{x_0, x_1, \ldots, x_n\}$ and $\mathcal{Q} = \{y_0, \ldots, y_m\}$ are each partitions of $[a, b]$ and that s is a step function on $[a, b]$ which is constant on each of the open subintervals of $\mathcal{I}$ and of $\mathcal{Q}$:

$$s(x) = s_j \quad \text{for} \quad x_{j-1} < x < x_j, \quad j = 1, \ldots, n$$

and
$$s(x) = s'_k \quad \text{for} \quad y_{k-1} < x < y_k, \quad k = 1, \ldots, m.$$

If $\mathcal{P}$ is a refinement of $\mathcal{Q}$ or $\mathcal{Q}$ a refinement of $\mathcal{P}$, we know (by Lemma 8.3) that $\sum_{j=1}^{n} s_j(x_j - x_{j-1}) = \sum_{k=1}^{m} s'_k(y_k - y_{k-1})$. If $s(x) \geq 0$ for all x in $[a, b]$, we can interpret each of these sums as the area of the region which is bounded on the left by $x = a$, and the right by $x = b$, below by the x-axis and above by the graph of s (see Figure 8.8).

However, even if neither $\mathcal{P}$ nor $\mathcal{Q}$ is a refinement of the other, it is not difficult to show that these two sums are equal. The proof is as follows: Let $\mathcal{R} = \mathcal{P} \cup \mathcal{Q}$. Notice that $\mathcal{R}$ is a partition of $[a, b]$ and is a refinement of each of $\mathcal{P}$ and $\mathcal{Q}$ (Definition 8.1, p. 344). Let $\mathcal{R} = \{z_0, \ldots, z_p\}$, $a = z_0 < z_1 < \cdots < z_p = b$. By Lemma 8.3 there exist constants s''_i such that $s(x) = s''_i$ for $z_{i-1} < x < z_i$. Moreover two applications of Equation 8.1 tell us that
$$\sum_{j=1}^{n} s_j(x_j - x_{j-1}) = \sum_{i=1}^{p} s''_i(z_i - z_{i-1})$$
and
$$\sum_{k=1}^{m} s'_k(y_k - y_{k-1}) = \sum_{i=1}^{p} s''_i(z_i - z_{i-1}).$$

A combination of these last two equations yields Equation 8.2 of the following lemma:

LEMMA 8.5 Let $s:[a, b] \to \mathcal{R}$ be a step function that is constant on the open subintervals of partitions $\mathcal{P} = \{x_0, \ldots, x_n\}$ and $\mathcal{Q} = \{y_0, \ldots, y_m\}$ of $[a, b]$. If $s(x) = s_j$ for $x_{j-1} < x < x_j$ and $s(x) = s'_k$ for $y_{k-1} < x < y_k$, then

$$(8.2) \qquad \sum_{j=1}^{n} s_j(x_j - x_{j-1}) = \sum_{k=1}^{m} s'_k(y_k - y_{k-1}).$$

In Definition 8.6 we give a special name to the sums appearing in Equation 8.2.

DEFINITION 8.6 Let $s:[a, b] \to \mathcal{R}$ be a step function on $[a, b]$:
$$s(x) = s_j \quad \text{for} \quad x_{j-1} < x < x_j$$
where $\{x_0, x_1, \ldots, x_n\}$ is a partition of $[a, b]$. The *integral of s over* $[a, b]$ is denoted by $\int_a^b s$ and is defined by

$$(8.3) \qquad \int_a^b s = \sum_{j=1}^{n} s_j(x_j - x_{j-1}).$$

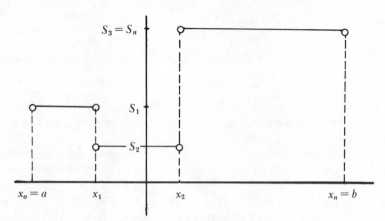

Figure 8.8

Notice that by virtue of Lemma 8.5 the integral of a step function s is well-defined; that is, it is independent of the particular partition used to describe the function s. Moreover, the value of $\int_a^b s$ depends only upon the values of s on the open subintervals of $\{x_0, x_1, \ldots, x_n\}$ and is independent of the value of s at the points $x_j, j = 0, 1, \ldots, n$.

Example 6. Let $s:[-1, 2] \to \mathfrak{R}$ be defined by

$$s(x) = \begin{cases} -3. & \text{for} \quad -1 \leq x < 0 \\ 2 & \text{for} \quad 0 \leq x \leq 3/2 \\ 1 & \text{for} \quad 3/2 < x \leq 2. \end{cases}$$

According to Definition 8.6,

$$\int_{-1}^{2} s = -3[0 - (-1)] + 2[\tfrac{3}{2} - 0] + 1[2 - \tfrac{3}{2}]$$
$$= \tfrac{1}{2}. \quad \blacksquare$$

The next two theorems list the basic properties of the integral of a step function. In view of the simple nature of step functions, each property should seem natural to the reader.

THEOREM 8.7 (Properties of the Integral of a Step Function.) Let s and t be step functions on $[a, b]$ and $r \in \mathfrak{R}$.
i. (Additive Property.) $\int_a^b (s + t) = \int_a^b s + \int_a^b t,$
ii. (Homogeneous Property.) $\int_a^b (rs) = r \int_a^b s,$
iii. (Order-Preserving Property.) If $s(x) \leq t(x)$ for all x in $[a, b]$, then $\int_a^b s \leq \int_a^b t.$
iv. If $a < c < b, \int_a^c s + \int_c^b s = \int_a^b s.$

The reader familiar with functions on a vector space will recognize that Theorem 8.7i and ii assert that $\int_a^b$ is a linear transformation or linear functional from the vector space of all step functions on $[a, b]$ into the vector space $\mathcal{R}$.

Proof of i. By Theorem 8.4 we know that $s + t$ is a step function on $[a, b]$. We may assume that s and t are each constant on the open subintervals of a partition $\mathcal{P} = \{x_0, x_1, \ldots, x_n\}$ of $[a, b]$ (see Exercise 1). Thus, we let $s(x) = s_j$ and $t(x) = t_j$ for $x_{j-1} < x < x_j$. We then have $(s + t)(x) = s_j + t_j$ for $x_{j-1} < x < x_j$ and

$$\int_a^b (s + t) =^1 \sum_{j=1}^n (s_j + t_j)(x_j - x_{j-1})$$

$$=^2 \sum_{j=1}^n s_j(x_j - x_{j-1}) + \sum_{j=1}^n t_j(x_j - x_{j-1})$$

$$=^1 \int_a^b s + \int_a^b t.$$

(1) Definition 8.6 (p. 349).
(2) Properties of real numbers.

The proofs of (ii), (iii), and (iv) are called for in Exercises 2 to 4. ■

THEOREM 8.8 Let s be a step function on $[a, b]$. Then $|s|$ is a step function on $[a, b]$ and

$$\left| \int_a^b s \right| \leq \int_a^b |s|.$$

Proof. The fact that $|s|$ is also a step function on $[a, b]$ follows readily from Definition 8.2 (p. 345) and the definition of absolute value. We note also that $-s$ (by Theorem 8.4ii) is a step function on $[a, b]$ and that for all x in $[a, b]$

$$- |s|(x) \leq s(x) \leq |s|(x).$$

Thus, by Theorem 8.7ii and iii we have

$$- \int_a^b |s| \leq \int_a^b s \leq \int_a^b |s|$$

or

$$\left| \int_a^b s \right| \leq \int_a^b |s|. \quad ■$$

Example 7. Let $s:[-1, 2] \to \mathcal{R}$ be the step function considered in Example 6 (p. 350). Then

$$|s|(x) = \begin{cases} 3 & \text{for } -1 \leq x < 0 \\ 2 & \text{for } 0 \leq x \leq 3/2 \\ 1 & \text{for } 3/2 < x \leq 2. \end{cases}$$

and

$$\int_{-1}^{2} |s| = 3[0 - (-1)] + 2[3/2 - 0] + 1[2 - 3/2]$$
$$= \frac{13}{2}.$$

In Example 6 we found that $\int_{-1}^{2} s = \frac{1}{2}$, which has an absolute value obviously less than $\int_{-1}^{2} |s|$. ∎

EXERCISES

1. Let s and t be step functions on $[a, b]$. Show that there exists a partition $\mathcal{P}$ of $[a, b]$ such that each of s and t is constant on the open subintervals of $\mathcal{P}$.

2. Prove Theorem 8.7ii.

3. Prove Theorem 8.7iii.

4. Prove Theorem 8.7iv.

5. Let s be a step function on $[a, b]$ and $c \in \mathcal{R}$. Define $t(x) = s(x + c)$ for x in $[a - c, b - c]$.
   (a) Prove that t is a step function on $[a - c, b - c]$.
   (b) Prove that $\int_{a-c}^{b-c} t = \int_{a}^{b} s$.

6. Let s be a step function on $[a, b]$ and $c > 0$. Define $t(x) = s(cx)$ for x in $\left[\dfrac{a}{c}, \dfrac{b}{c} \right]$.

   (a) Prove that t is a step function on $\left[\dfrac{a}{c}, \dfrac{b}{c} \right]$.

   (b) Prove that $c \int_{a/c}^{b/c} t = \int_{a}^{b} s$.

4. THE UPPER AND LOWER INTEGRALS OF A BOUNDED FUNCTION

Now that we have investigated various properties of what we have called step functions, we return to our original problem of approximating the area of the region bounded by the graph of a function, the x-axis, and the vertical lines $x = a$ and $x = b$. We are restricting ourselves to functions whose graphs lie entirely within a region bounded by a rectangle. Such functions are described as follows:

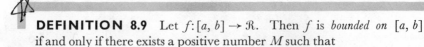

DEFINITION 8.9 Let $f : [a, b] \to \mathcal{R}$. Then f is *bounded on* $[a, b]$ if and only if there exists a positive number M such that

$$-M \leq f(x) \leq M \quad \text{for all } x \text{ in } [a, b].$$

Notice that if f is bounded on $[a, b]$, $-M \le f(x) \le M$, there exist step functions l and u such that $l(x) \le f(x) \le u(x)$ for all x in $[a, b]$, namely, $l(x) = -M$ and $u(x) = M$ for all x in $[a, b]$. We use the notation l and u to call to mind the words "lower" and "upper." Indeed if l and u are any step functions such that $l(x) \le f(x) \le u(x)$ for all x in $[a, b]$, we shall refer to l as a *lower step function* of f and u as an *upper step function* of f.

In Example 2 (p. 340) we constructed lower and upper step functions, l_n and u_n (although we did not use this terminology in the example), for the function f defined by $f(x) = x^2$, $0 \le x \le 1$. These step functions had the property that $\lim_{n \to \infty} \int_0^1 u_n = \lim_{n \to \infty} \int_0^1 l_n$ (check this). In other words we could approximate the size of the region $\mathcal{R} = \{(x, y) \mid 0 \le x \le 1, 0 \le y \le x^2\}$ both above and below by using the integrals of step functions and arrive at the same result.

In general the results of such an approximation may not be so orderly. However, if f is a bounded function on $[a, b]$, we can approximate the size of the region $\mathcal{R} = \{(x, y) \mid a \le x \le b, 0 \le y \le f(x) \text{ or } f(x) \le y \le 0\}$ both above and below by the integrals of step functions. Definition 8.10 describes the approximation with which we are concerned.

Let us recall the meaning of the terminology and notation used in Definition 8.10. An *upper bound* of a set S of real numbers is a real number r such that $s < r$ for all s in S. The *least upper bound of S* is the smallest upper bound of S. The *Least Upper Bound Axiom* (sometimes called the Completeness Axiom or the Continuity Axiom) of the real numbers states that if S is a nonempty set of real numbers which has an upper bound, then S has a least upper bound, denoted by lub S. Similar definitions are given for a lower bound and the greatest lower bound of a set S of real numbers. The greatest lower bound of S is denoted by glb S.

DEFINITION 8.10   Let $f : [a, b] \to \mathcal{R}$ be bounded on $[a, b]$. Let

$$\mathcal{L}_f = \{l \mid l \text{ is a lower step function for } f\}$$

and

$$\mathcal{U}_f = \{u \mid u \text{ is an upper step function for } f\}.$$

The *lower integral of f on $[a, b]$* is denoted by $\int_{\underline{a}}^b f$ and is defined by

$$\int_{\underline{a}}^b f = \mathrm{lub} \left\{ \int_a^b l \, \middle| \, l \in L_f \right\}.$$

The *upper integral of f on $[a, b]$* is denoted by $\overline{\int}_a^b f$ and defined by

$$\overline{\int}_a^b f = \mathrm{glb} \left\{ \int_a^b u \, \middle| \, u \in U_f \right\}.$$

In Figure 8.9 there is sketched the graph of a bounded, non-negative function f on the interval $[-1, 2]$. The graphs of two step functions u and l,

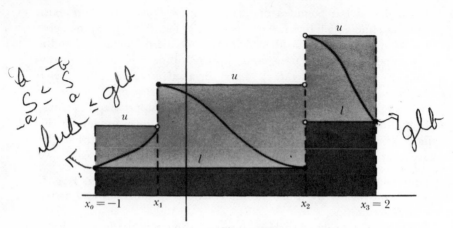

Figure 8.9

$l(x) \leq f(x) \leq u(x)$, are also sketched. It is easily seen that by taking a finer partition of $[-1, 2]$ one can construct other step functions u and l, the graphs of which lie closer to that of f. However, since f is not a step function on $[-1, 2]$, it is easy to see that there does not exist an element u of $\mathfrak{U}_f$ or an element l of $\mathfrak{L}_f$ such that $\int_{-1}^{2} u = \int_{-1}^{2} l$. Is it possible then that $\overline{\int_{-1}^{2}} f = \underline{\int_{-1}^{2}} f$ for a function f such as that sketched in Figure 8.9? This question we shall not answer immediately. However, from practical considerations it does seem reasonable that $\underline{\int_a^b} f \leq \overline{\int_a^b} f$, for any function f which is bounded on an interval $[a, b]$. We state this result formally in Theorem 8.11:

THEOREM 8.11   If $f: [a, b] \to \mathfrak{R}$ is bounded on $[a, b]$, then

(8.4)
$$\underline{\int_a^b} f \leq \overline{\int_a^b} f.$$

Proof.   If $u \in \mathfrak{U}_f$ and $l \in \mathfrak{L}_f$, then by Definition 8.10 $l(x) \leq u(x)$ for all x in $[a, b]$. Thus, by Theorem 8.7iii,

$$\int_a^b l \leq \int_a^b u \quad \text{for all } l \text{ in } \mathfrak{L}_f \text{ and } u \text{ in } \mathfrak{U}_f.$$

Therefore, if $u \in \mathfrak{U}_f$, $\int_a^b u$ is an upper bound for the set of real numbers $\{\int_a^b l \mid l \in \mathfrak{L}_f\}$, and consequently,

(8.5)
$$\text{lub} \left\{ \int_a^b l \,\middle|\, l \in \mathfrak{L}_f \right\} \leq \int_a^b u.$$

However, Inequality 8.5 tells us that lub $\{\int_a^b l \mid l \in \mathcal{L}_f\}$ is a lower bound for the set of real numbers $\{\int_a^b u \mid u \in \mathcal{U}_f\}$. Therefore,

$$(8.6) \qquad \text{lub}\left\{\int_a^b l \,\middle|\, l \in \mathcal{L}_f\right\} \le \text{glb}\left\{\int_a^b u \,\middle|\, u \in \mathcal{U}_f\right\}.$$

Finally, by Definition 8.10, Inequality 8.6 is the same as Inequality 8.4 and the proof is complete. ■

In Exercise 3 the reader is asked to prove that if f and g are bounded on $[a, b]$ and $f(x) \le g(x)$ for all x in $[a, b]$, then $\underline{\int_a^b} f \le \underline{\int_a^b} g$ and $\overline{\int_a^b} f \le \overline{\int_a^b} g$. This property and those listed in Theorem 8.12 will be crucial in the development which follows:

THEOREM 8.12 Let $f:[a, b] \to \mathcal{R}$ and $g:[a, b] \to \mathcal{R}$ be bounded on $[a, b]$, and let $r \in \mathcal{R}$.

  i. $\overline{\int_a^b}(f + g) \le \overline{\int_a^b} f + \overline{\int_a^b} g$
  ii. $\underline{\int_a^b}(f + g) \ge \underline{\int_a^b} f + \underline{\int_a^b} g$
  iii. If $r \ge 0$, $\underline{\int_a^b}(rf) = r\underline{\int_a^b} f$ and $\overline{\int_a^b}(rf) = r\overline{\int_a^b} f$
  iv. If $r < 0$, $\underline{\int_a^b} rf = r\overline{\int_a^b} f$ and $\overline{\int_a^b}(rf) = r\underline{\int_a^b} f$.

Proof of i. Let $u_1 \in \mathcal{U}_f$ and $u_2 \in \mathcal{U}_g$. It is immediate from Definition 8.10 (p. 353) and Theorem 8.4i (p. 347) that $u_1 + u_2 \in \mathcal{U}_{f+g}$. Therefore,

$$(8.7) \qquad \begin{cases} \overline{\int_a^b}(f + g) =^1 \text{glb}\left\{\int_a^b u \,\middle|\, u \in \mathcal{U}_{f+g}\right\} \\ \qquad\qquad \le^2 \int_a^b (u_1 + u_2) \\ \qquad\qquad =^3 \int_a^b u_1 + \int_a^b u_2. \end{cases}$$

(1) Definition 8.10.
(2) Definition of glb and the fact that $u_1 + u_2 \in \mathcal{U}_{f+g}$.
(3) Theorem 8.7i (p. 354).

From Inequality 8.7 it follows that for each u_1 in $\mathcal{U}_f$ and each u_2 in $\mathcal{U}_g$

$$(8.8) \qquad \overline{\int_a^b}(f + g) - \int_a^b u_2 \le \int_a^b u_1.$$

Thus, $\overline{\int_a^b}(f + g) - \int_a^b u_2$ is a lower bound for the set $\{\int_a^b u_1 \mid u_1 \in \mathcal{U}_f\}$, and hence

$$(8.9) \qquad \begin{cases} \overline{\int_a^b}(f + g) - \int_a^b u_2 \le^4 \text{glb}\left\{\int_a^b u_1 \,\middle|\, u_1 \in \mathcal{U}_f\right\} \\ \qquad\qquad =^5 \overline{\int_a^b} f. \end{cases}$$

(4) Definition of glb and Inequality 8.8.

(5) Definition 8.10.

It now follows from Inequality 8.9 that $\overline{\int_a^b} (f + g) - \overline{\int_a^b} f$ is a lower bound for the set $\{\int_a^b u_2 \mid u_2 \in \mathcal{U}_g\}$, and thus

$$\overline{\int_a^b} (f + g) - \overline{\int_a^b} f \le \text{glb} \left\{ \int_a^b u_2 \,\middle|\, u_2 \in \mathcal{U}_g \right\}$$
$$= \overline{\int_a^b} g,$$

or

$$\overline{\int_a^b} (f + g) \le \overline{\int_a^b} f + \overline{\int_a^b} g,$$

as claimed.

Proof of iv. Let $r < 0$ and $u \in \mathcal{U}_f$.

(8.10)
$$\begin{cases}
u \in U_f \Rightarrow^1 f(x) \le u(x) & \text{for all } x \text{ in } [a, b] \\[4pt]
\Rightarrow^2 ru(x) \le rf(x) & \text{for all } x \text{ in } [a, b] \\[4pt]
\Rightarrow^3 \int_a^b (ru) \le \underline{\int_a^b} (rf) \\[4pt]
\Rightarrow^4 r \int_a^b u \le \underline{\int_a^b} (rf) \\[4pt]
\Rightarrow^2 \frac{1}{r} \underline{\int_a^b} (rf) \le \int_a^b u & \text{for all } u \text{ in } \mathcal{U}_f.
\end{cases}$$

(1) Definition of $\mathcal{U}_f$ (p. 353).

(2) Since $r < 0$.

(3) Definition 8.10 (ru is a lower step function of rf).

(4) Theorem 8.7 ($r \int_a^b u = \int_a^b ru$).

The implication chain (8.10) tells us that $\frac{1}{r} \underline{\int_a^b} (rf) \le \int_a^b u$ for all u in U_f. Therefore, by definition of the greater lowest bound and Definition 8.10,

$$\frac{1}{r} \underline{\int_a^b} (rf) \le \text{glb} \left\{ \int_a^b u \,\middle|\, u \in \mathcal{U}_f \right\} = \overline{\int_a^b} f.$$

Thus, since $r < 0$, we have $\underline{\int_a^b} rf \ge r \overline{\int_a^b} f$. In a similar fashion we could show, by starting with any step function l in $\mathcal{L}_{rf}$, that $\underline{\int_a^b} (rf) \le r \overline{\int_a^b} f$ and thereby arrive at the desired equality: $r \overline{\int_a^b} f = \underline{\int_a^b} (rf)$.

The second equality in (iv) follows easily from the first, for if $r < 0$,

$$r \underline{\int_a^b} f = r \underline{\int_a^b} \frac{1}{r} (rf)$$
$$= r \left(\frac{1}{r}\right) \overline{\int_a^b} (rf)$$
$$= \overline{\int_a^b} (rf). \qquad \blacksquare$$

The reader is asked to prove parts (ii) and (iii) of Theorem 8.12 in Exercises 1 and 2 which follow:

EXERCISES

1. Prove Theorem 8.12ii.

2. Prove Theorem 8.12iii.

3. Let f and g be bounded functions on $[a, b]$ such that $f(x) \leq g(x)$ for all x in $[a, b]$. Show that $\overline{\int}_a^b f \leq \overline{\int}_a^b g$ and $\underline{\int}_a^b f \leq \underline{\int}_a^b g$.

4. Let s be a step function on $[a, b]$. Show that $\int_a^b s = \underline{\int}_a^b s = \overline{\int}_a^b s$.

5. Let f and g be bounded functions on $[a, b]$ and $r \in \mathcal{R}$. Show that
   (a) lub $\{f(x) + g(x) \mid x \in [a, b]\}$
       $\leq$ lub $\{f(x) \mid x \in [a, b]\}$ + lub $\{g(x) \mid x \in [a, b]\}$;
   (b) glb $\{f(x) + g(x) \mid x \in [a, b]\}$
       $\geq$ glb $\{f(x) \mid x \in [a, b]\}$ + glb $\{g(x) \mid x \in [a, b]\}$;
   (c) if $r \geq 0$, lub $\{rf(x) \mid x \in [a, b]\}$ = r lub $\{f(x) \mid x \in [a, b]\}$;
   (d) if $r < 0$, lub $\{rf(x) \mid x \in [a, b]\}$ = r glb $\{f(x) \mid x \in [a, b]\}$.

ANSWERS

1. (Outline.) Let $l_1 \in \mathcal{L}_f$ and $l_2 \in \mathcal{L}_g$. Then $l_1 + l_2 \in \mathcal{L}_{f+g}$ and $\underline{\int}_a^b (f + g) \geq$ $\int_a^b l_1 + \int_a^b l_2$. Use an argument similar to that in the proof of Theorem 8.12i to show that $\underline{\int}_a^b (f + g) \geq \underline{\int}_a^b f + \underline{\int}_a^b g$. **3.** Assume $f(x) \leq g(x)$ and let $u \in \mathcal{U}_g$. Then $u \in \mathcal{U}_f$, and by Definition 8.10, $\overline{\int}_a^b f \leq \int_a^b u$. Since this is true for each u in $\mathcal{U}_g$, $\overline{\int}_a^b f \leq$ glb $\{\int_a^b u \mid u \in \mathcal{U}_g\} = \overline{\int}_a^b g$.

5. INTEGRABLE BOUNDED FUNCTIONS

In the preceding section we investigated several properties of the upper and lower integrals of bounded functions. The work of Section 1 showed that $\underline{\int}_0^1 f = \overline{\int}_0^1 f$ where $f(x) = x^2$, and in Exercise 4 of Section 4 the reader was asked to show that every step function has this same property: $\int_a^b s = \overline{\int}_a^b s$. Any function which has this property will be called *integrable* (Definition 8.13). We now ask the following two questions: (1) Does there exist a bounded function which is *not* integrable and (2) what classes of functions other than step functions are necessarily integrable? Before answering either of these questions, we formally define the term *integrable*.

DEFINITION 8.13 Let $f: [a, b] \to \mathcal{R}$ be bounded on $[a, b]$. Then *f is integrable on* $[a, b]$ if and only if the upper and lower integrals of f on $[a, b]$ have the same value ($\overline{\int}_a^b f = \underline{\int}_a^b f$). Moreover, if f is integrable on $[a, b]$, the *integral of f on* $[a, b]$ is denoted by $\int_a^b f$ and defined to be the common value of the upper and lower integrals of f on $[a, b]$:

$$\int_a^b f = \overline{\int}_a^b f = \underline{\int}_a^b f.$$

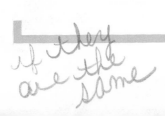
if they are the same

We now answer, by example, the first question posed prior to Definition 8.13:

Example 8.   Let $f:[0, 1] \to \mathcal{R}$ be defined by

$$f(x) = \begin{cases} 1, & \text{if } x \text{ is rational} \\ 0, & \text{if } x \text{ is irrational.} \end{cases}$$

We shall show that $\underline{\int}_a^b f = 0$ and $\overline{\int}_a^b f = 1$, and thus that f is not integrable on $[0, 1]$.

In Exercise 3 the reader is asked to show that if x and y are real numbers with $x < y$, then there exists a rational number r such that $x < r < y$. Now consider any element u of $\mathcal{U}_f : u(x) = u_j$ on the jth open subinterval (x_{j-1}, x_j) of a partition $\{x_0, x_1, \ldots, x_n\}$ of $[0, 1]$. By the previous statement we know that there exists a rational number r_j in (x_{j-1}, x_j). Since $u_j = u(r_j) \geq f(r_j) = 1$, we have $u_j \geq 1$ for $j = 1, \ldots, n$, and $\int_0^1 u = \sum_{j=1}^n u_j(x_j - x_{j-1}) \geq \sum_{j=1}^n (1)(x_j - x_{j-1}) = 1$. Therefore, $\overline{\int}_0^1 f \geq 1$. Moreover, $\int_0^1 u_0 = 1$ where $u_0(x) = 1$ for all x in $[0, 1]$. Since $u_0 \in \mathcal{U}_f$, we conclude that $\overline{\int}_0^1 f = 1$.

In Exercise 4 the reader is asked to show that if x and y are real numbers, $x < y$, then there exists an irrational number r such that $x < r < y$. We then argue in a fashion similar to that used in the preceding paragraph that $\underline{\int}_0^1 f = 0$.   ∎

Thus, we have answered the first question: There does exist a bounded function which is not integrable. Before attempting to describe classes of functions (other than step functions) which are integrable, we shall investigate certain properties of the set of all integrable functions on $[a, b]$ and the function $\int_a^b$.

THEOREM 8.14   (Properties of the Integral.) Let $f:[a, b] \to \mathcal{R}$ and $g:[a, b] \to \mathcal{R}$ be integrable on $[a, b]$ and let $r \in \mathcal{R}$.

i. (Additive Property.) $f + g$ is integrable on $[a, b]$ and

$$\int_a^b (f + g) = \int_a^b f + \int_a^b g.$$

ii. (Homogeneous Property.) rf is integrable on $[a, b]$ and

$$\int_a^b rf = r \int_a^b f.$$

(In other words the set of all functions integrable on $[a, b]$ is a vector subspace of the set of all real-valued functions on $[a, b]$, and $\int_a^b$ is a linear transformation from this vector subspace into the reals.)

Proof.   Note first that $f + g$ and rf are bounded, since f and g have this property. To prove that a bounded function is integrable we must show that its upper and lower integrals have the same value. To prove part (i) we proceed as follows:

$$\overline{\int_a^b}(f+g) \le^1 \overline{\int_a^b} f + \overline{\int_a^b} g$$

$$=^2 \int_a^b f + \int_a^b g$$

$$=^2 \underline{\int_a^b} f + \underline{\int_a^b} g$$

$$\le^1 \underline{\int_a^b}(f+g)$$

$$\le^3 \overline{\int_a^b}(f+g).$$

(1) Theorem 8.12.
(2) By Definition 8.13, since f and g are integrable.
(3) Theorem 8.11.

The extreme members of this chain of inequalities are equal, and thus *all* of the terms are equal. In particular we have

$$\overline{\int_a^b}(f+g) = \int_a^b f + \int_a^b g = \underline{\int_a^b}(f+g),$$

which proves simultaneously that $f+g$ is integrable (since the upper and lower integrals are equal) and that $\int_a^b (f+g) = \int_a^b f + \int_a^b g$ (since, by definition, $\int_a^b (f+g)$ is the common value of the upper and lower integrals of $(f+g)$).

The proof of (ii) is left to the reader (see Exercise 5). ∎

Not only is $\int_a^b$ a linear transformation on the set of all integrable functions on $[a, b]$, but it is order-preserving in the sense that if $f(x) \le g(x)$ for all x in $[a, b]$, then $\int_a^b f \le \int_a^b g$. This is not at all surprising in view of Theorem 8.7iii (which states the same property for step functions) and in view of the definition of integrability.

What may be surprising is that there do exist functions $f:[a, b] \to \Re$ such that f is not integrable, but $|f|$ is integrable. The two functions, f and $|f|$, are intimately related, but, as we discovered in our study of differential calculus, behave differently. Recall that $f(x) = x$ is differentiable everywhere, *but* that $|f|(x) = |x|$ is *not* differentiable at the origin. In Example 9 we have a function f which is not integrable but is such that $|f|$ *is* integrable.

Example 9. Let $f:[0, 1] \to \Re$ be defined by

$$f(x) = \begin{cases} 1 & \text{if } x \text{ is rational} \\ -1 & \text{if } x \text{ is irrational.} \end{cases}$$

Note that $|f|(x) = 1$ for all x in $[0, 1]$. Hence, $|f|$ is a step function and

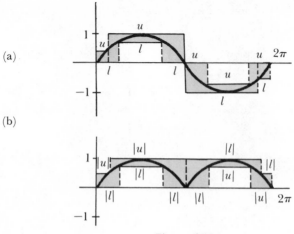

Figure 8.10

$\int_0^1 f = 1$. However, by an argument similar to that used in Example 8, we could show that $\overline{\int}_0^1 f = 1$ and $\underline{\int}_0^1 f = -1$, and hence that f is not integrable. ■

Let us now consider the function f, defined by $f(x) = \sin x, 0 \leq x \leq 2\pi$. We assume that f is integrable (this will be justified in Section 7). Figures 8.10a and 8.10b show, respectively, the graphs of f, an element u of $\mathfrak{U}_f$, an element l of $\mathfrak{L}_f$, and the corresponding graphs of $|f|$, $|u|$, and $|l|$. The shaded areas in the two graphs show the regions between the graphs of u and l and the graphs of $|u|$ and $|l|$. Notice that the areas of these two regions are the same. Thus, if $\int_0^{2\pi} u$ is close to $\int_0^{2\pi} l$, the same holds true for $\int_0^{2\pi} |u|$ and $\int_0^{2\pi} |l|$. Thus, we see a motivation for the proposition that if f is integrable on $[a, b]$, $|f|$ is also integrable on $[a, b]$. This theorem we shall not prove in the text but shall leave as an exercise for the reader.

THEOREM 8.15 (Order Properties of the Integral.) Let $f:[a, b] \rightarrow \mathfrak{R}$ and $g:[a, b] \rightarrow \mathfrak{R}$ be integrable on $[a, b]$:
i. if $f(x) \leq g(x)$ for all x in $[a, b]$, then $\int_a^b f \leq \int_a^b g$;
ii. $|f|$ is integrable on $[a, b]$, and $|\int_a^b f| \leq \int_a^b |f|$.

Proof of i. Let $l \in \mathfrak{L}_f$ and $f(x) \leq g(x)$ for all x in $[a, b]$.

$$l \in \mathfrak{L}_f \quad \text{and} \quad f(x) \leq g(x) \Rightarrow^1 l \in \mathfrak{L}_g$$

$$\Rightarrow^2 \underline{\int}_a^b l \leq \underline{\int}_a^b g$$

$$\Rightarrow^3 \int_a^b l \leq \int_a^b g.$$

(1) Definition of $\mathcal{L}_f$ and $\mathcal{L}_g$.
(2) Definition 8.10 (p. 353).
(3) Definition 8.13 (p. 357) and the hypothesis that g is integrable.

The above implication chain shows that $\int_a^b g$ is an upper bound for the set $\{\int_a^b l \mid l \in \mathcal{L}_f\}$. Thus, by Definitions 8.13 and 8.10, we have

$$\int_a^b f = \int_{\underline{a}}^b f$$
$$= \text{lub} \left\{ \int_{\underline{a}}^b l \,\Big|\, l \in \mathcal{L}_f \right\}$$
$$\leq \int_a^b g,$$

and we have proven the desired result.

Proof of ii. The proof that $|f|$ is integrable is called for in Exercises 6 to 9. Assuming that $|f|$ is integrable whenever f is integrable, we have, by Theorem 8.15i and Theorem 8.14ii,

$$-\int_a^b |f| \leq \int_a^b f \leq \int_a^b |f|.$$

This last series of inequalities is equivalent to

$$\left| \int_a^b f \right| \leq \int_a^b |f|. \quad \blacksquare$$

EXERCISES

1. If $x > 0$, prove that there is a positive integer n such that $\dfrac{1}{n} < x$.

2. If x is a real number, prove that there is a (unique) integer n such that $x \leq n < x + 1$.

3. Prove that if $x < y$, there exists a rational number r such that $x < r < y$.

4. Prove that if $x < y$, there exists an irrational number r' such that $x < r' < y$ (Hint: By Exercise 3 there exists a rational number r such that $x < r < y$. Add a sufficiently small irrational number to r to find r').

5. Prove Theorem 8.14ii (Hint: Prove the theorem first for $r < 0$ by using Theorem 8.12iv).

For each $g : [a, b] \to \mathfrak{R}$, define g^+ and g^- as follows:

$$g^+(x) = \begin{cases} 0 & \text{if } g(x) < 0 \\ g(x) & \text{if } g(x) \geq 0 \end{cases}$$
$$g^-(x) = \begin{cases} g(x) & \text{if } g(x) < 0 \\ 0 & \text{if } g(x) \geq 0. \end{cases}$$

Let $f:[a, b] \to \mathcal{R}$ be integrable on $[a, b]$.

6. Show that $|f| = f^+ - f^-$.

7. Show that if $u \in \mathfrak{U}_f$ and $l \in \mathfrak{L}_f$, $u^+ - l^- \in \mathfrak{U}_{|f|}$ and $l^+ - u^- \in \mathfrak{L}_{|f|}$.

8. Show that $\int_a^b u - \int_a^b l = \int_a^b (u^+ - l^-) - \int_a^b (l^+ - u^-)$.

9. Show that $|f|$ is integrable on $[a, b]$.

ANSWERS

1. Assume that $\frac{1}{n} \geq x$ for all positive integers n. Then $n \leq \frac{1}{x}$ for all positive integers n, which results in a contradiction since the positive integers are not bounded above. **3.** Assume that $x < y$. By Exercise 1 there exists a positive integer q such that $\frac{1}{q} < y - x$. Then $qx + 1 < qy$. By Exercise 2 there exists an integer p such that $qx \leq p < qx + 1$. Show that $x \leq \frac{p}{q} < y$. If $x = \frac{p}{q}$, find a rational number r such that $\frac{p}{q} < r < y$.

6. FURTHER PROPERTIES OF INTEGRABLE FUNCTIONS

At this point the reader should realize that a function f is integrable on the interval $[a, b]$ if and only if there exist step functions u in $\mathfrak{U}_f$ and l in $\mathfrak{L}_f$ whose integrals have nearly the same value. Theorem 8.16 formalizes this intuitive notion of integrability.

THEOREM 8.16 (Riemann Condition for Integrability.) Let $f:[a, b] \to \mathcal{R}$ be bounded on $[a, b]$. Then f is integrable on $[a, b]$ if and only if the following condition holds: for each $r > 0$ there is an l_r in $\mathfrak{L}_f$ and a u_r in $\mathfrak{U}_f$ such that

$$\int_a^b u_r - \int_a^b l_r < r.$$

Proof. Assume first that f is integrable on $[a, b]$ and let $r > 0$.

$$f \text{ integrable on } [a, b] \Rightarrow^1 \overline{\int_a^b} f = \underline{\int_a^b} f$$

$$\Rightarrow^2 \text{glb} \left\{ \int_a^b u \,\middle|\, u \in \mathfrak{U}_f \right\} = \text{lub} \left\{ \int_a^b l \,\middle|\, l \in \mathfrak{L}_f \right\}$$

$$\Rightarrow^3 \text{ there exists a } u_r \text{ in } \mathfrak{U}_f \text{ such that}$$

$$\int_a^b u_r - r < \text{lub} \left\{ \int_a^b l \,\middle|\, l \in \mathfrak{L}_f \right\}.$$

$\Rightarrow^4$ there exists an l_r in $\mathcal{L}_f$ such that

$$\int_a^b u_r - r < \int_a^b l_r$$

$$\Rightarrow^5 \int_a^b u_r - \int_a^b l_r < r.$$

(1) Definition 8.13 (p. 357).
(2) Definition 8.10 (p. 353).
(3) Definition of greatest lower bound.
(4) Definition of least upper bound.
(5) Property of real numbers.

The above implication chain shows that if f is integrable and $r > 0$, there exist u_r in $\mathcal{U}_f$ and l_r in $\mathcal{L}_f$ such that $\int_a^b u_r - \int_a^b l_r < r$.

Now assume that f is not integrable on $[a, b]$. Let $r_0 = \bar{\int}_a^b f - \underline{\int}_a^b f$. By the definition of integrability and Theorem 8.11, we have $r_0 > 0$. Moreover, for each l in $\mathcal{L}_f$ and u in $\mathcal{U}_f$ we have

$$r_0 + \int_a^b l \leq^6 r_0 + \text{lub}\left\{ \int_a^b l \,\middle|\, l \in \mathcal{L}_f \right\}$$

$$=^7 r_0 + \underline{\int}_a^b f$$

$$=^8 \bar{\int}_a^b f$$

$$= \text{glb}\left\{ \int_a^b u \,\middle|\, u \in \mathcal{U}_f \right\}$$

$$\leq^9 \int_a^b u.$$

(6) Definition of upper bound.
(7) Definition 8.10.
(8) Since $r_0 = \bar{\int}_a^b f - \underline{\int}_a^b f$.
(9) Definition of lower bound.

Thus, if f is not integrable, $\int_a^b u - \int_a^b l \geq r_0$ for all u in $\mathcal{U}_f$ and all l in $\mathcal{L}_f$, and hence the Riemann Condition does not hold for $r = r_0$. ∎

Assume now that f is integrable on $[a, b]$ and that $[c, d] \subset [a, b]$. The following question immediately arises: Is f integrable on $[c, d]$ also? We answer this question rather easily by using the Riemann Condition. Let $r > 0$, $u_r \in \mathcal{U}_f$, and $l_r \in \mathcal{L}_f$ such that $\int_a^b u_r - \int_a^b l_r < r$. Refer now to Figure 8.11 and assume that $a < c < d < b$. We shall denote the restriction of u_r to $[c, d]$ by u_r and the restriction of l_r to $[c, d]$ by l_r. It is immediate both from the sketch in Figure 8.11 and from the fact that $u_r(x) \geq l_r(x)$ for all x

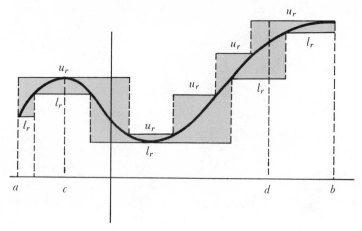

Figure 8.11

in $[a, b]$ that

$$\int_c^d u_r - \int_c^d l_r \leq \left[\int_a^c u_r - \int_a^c l_r\right] + \left[\int_c^d u_r - \int_c^d l_r\right] + \left[\int_d^b u_r - \int_d^b l_r\right]$$

$$=^1 \int_a^b u_r - \int_a^b l_r$$

$$<^2 r.$$

(1) Theorem 8.7iv.
(2) Previous choice of u_r and l_r.

Thus, for each $r > 0$ we can find a u_r in $\mathfrak{U}_f$ and an l_r in $\mathfrak{L}_f$ such that $\int_c^d u_r - \int_c^d l_r < r$. By the Riemann Condition (Theorem 8.16) we have thus shown f to be integrable on $[c, d]$.

This result is now stated as a theorem:

THEOREM 8.17 If $f:[a, b] \to \mathfrak{R}$ is integrable on $[a, b]$, then f is integrable on every closed subinterval of $[a, b]$.

We now turn to the problem of the function f which is integrable on two adjacent closed intervals $[a, c]$ and $[c, b]$. The reader shall be asked in Exercise 1 to show that in this case f is also integrable on $[a, b]$ and moreover that

(8.11) $$\int_a^b f = \int_a^c f + \int_c^b f.$$

This property corresponds to that stated for step functions in Theorem 8.7iv and is the "if" part of the following theorem:

THEOREM 8.18 Let $f:[a, b] \to \mathfrak{R}$ and let $a < c < b$. Then f is integrable on $[a, b]$ if and only if f is integrable on $[a, c]$ and on $[c, b]$. Moreover, in this case

(8.11)
$$\int_a^b f = \int_a^c f + \int_c^b f.$$

Proof. We shall use the Riemann Condition to prove the necessity or "only if" portion of this theorem. Assume that f is integrable on $[a, b]$. By Theorem 8.17 we know that f is integrable on each of $[a, c]$ and $[c, b]$. We must still show the validity of Equation 8.11. Let $r > 0$ and let u_r in $\mathfrak{U}_f$ and l_r in $\mathfrak{L}_f$ be step functions on $[a, b]$ such that

(8.12)
$$\int_a^b u_r - \int_a^b l_r < r.$$

We then have

$$\int_a^b f \leq^1 \int_a^b u_r$$
$$<^2 \int_a^b l_r + r$$
$$=^3 \int_a^c l_r + \int_c^b l_r + r$$
$$\leq^1 \int_a^c f + \int_c^b f + r.$$

(1) Definitions 8.10 and 8.13.
(2) Inequality 8.12.
(3) Theorem 8.7iv, (p. 350).

Thus $-r \leq \int_a^c f + \int_c^b f - \int_a^b f$.

In Exercise 2 the reader is asked to show that

$$\int_a^c f + \int_c^b f - \int_a^b f < r.$$

Thus, for each $r > 0$ we have

(8.13)
$$\left| \int_a^c f + \int_c^b f - \int_a^b f \right| < r.$$

If Equation 8.11 does not hold and we define $r_0 = |\int_a^c f + \int_c^b f - \int_a^b f|$, then $r_0 > 0$ and Inequality 8.13 does not hold for $r = r_0$. Since we have shown Inequality 8.13 to hold for all $r > 0$, Equation 8.11 must hold. ∎

For many reasons, to be covered in the following chapters, it is convenient to extend our definition of the integral function, $\int_a^b$, to the cases in which $a = b$ and $a > b$. This is easily done by Definition 8.19:

DEFINITION 8.19

i. If $a \in dom\, f$, f is integrable on $[a, a]$ and $\int_a^a f = 0$.

ii. If f is integrable on $[a, b]$, we define

$$\int_b^a f = -\int_a^b f.$$

In this context when we say that $\int_a^b f$ exists we mean that f is integrable on $[a, b]$ if $a < b$, or f is integrable on $[b, a]$ if $b < a$.

With this extension of the definition of the integral we can restate Theorem 8.18 as follows: If any two of the integrals $\int_a^b f$, $\int_a^c f$, and $\int_c^b f$ exist, then the third integral also exists, and $\int_a^b f = \int_a^c f + \int_c^b f$. In this restatement we need not be concerned with the order of the real numbers a, b, and c. We can also restate Theorem 8.15ii as follows. If a and b are real numbers and $\int_a^b f$ exists, then $\int_a^b |f|$ exists and $|\int_a^b f| \le |\int_a^b |f||$.

EXERCISES

1. Let f be integrable on each of the intervals $[a, c]$ and $[c, b]$. Prove
   (a) f is integrable on $[a, b]$,
   (b) $\int_a^b f = \int_a^c f + \int_c^b f$.

2. Let f be integrable on $[a, b]$, $a < c < b$, and $r > 0$. Show that $\int_a^c f + \int_c^b f < \int_a^b f + r$ and thus that $\int_a^c f + \int_c^b f - \int_a^b f < r$.

ANSWERS

1. (Outline.) (a) Let $r > 0$. Find step functions u_1 and l_1 on $[a, c]$ such that $u_1 \in \mathfrak{U}_f$, $l_1 \in \mathfrak{L}_f$, and $\int_a^c u_1 - \int_a^c l_1 < r/2$. Similarly, find step functions u_2 and l_2 on $[c, b]$ such that $u_2 \in \mathfrak{U}_f$, $l_2 \in \mathfrak{L}_f$, and $\int_c^b u_2 - \int_c^b l_2 < r/2$. From u_1, u_2, l_1, and l_2 construct step functions u_3 and l_3 on $[a, b]$ such that $u_3 \in \mathfrak{U}_f$, $l_3 \in \mathfrak{L}_f$, and $\int_a^b u_3 - \int_a^b l_3 < r$. Apply Theorem 8.16.

7. INTEGRABILITY OF SPECIAL CLASSES OF FUNCTIONS

In the introductory paragraph of Section 5 we asked the following question: Do there exist classes of functions other than step functions which are integrable? This section will be devoted to showing that two classes of functions are, by their very nature, also integrable.

DEFINITION 8.20
Let $f: [a, b] \to \mathfrak{R}$. We say that f is *monotone on* $[a, b]$ if and only if f is increasing on $[a, b]$ or f is decreasing on $[a, b]$ (Definition 6.1, p. 249).

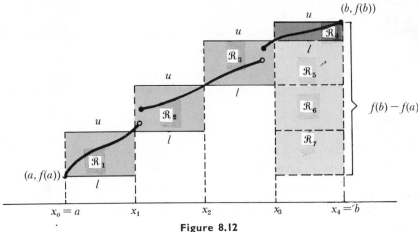

Figure 8.12

Consider a function $f : [a, b] \to \mathfrak{R}$ which is increasing on $[a, b]$. Let $\{x_0, \ldots, x_n\}$ be any partition of $[a, b]$ for which the open subintervals are all of the same length: Let $u \in \mathfrak{U}_f$ and $l \in \mathfrak{L}_f$ such that $u_j = f(x_j)$ and $l_j = f(x_{j-1})$ for $j = 1, \ldots, n$.[*] Figure 8.12 is a sketch of such a situation. Since, in the figure, $\mathfrak{R}_3$ and $\mathfrak{R}_5$ are rectangles of the same height and width, they have the same area. The same is true for $\mathfrak{R}_2$ and $\mathfrak{R}_6$, and for $\mathfrak{R}_1$ and $\mathfrak{R}_7$. Thus, the area bounded above by u and below by l is the same as the area of $\mathfrak{R}_4 \cup \mathfrak{R}_5 \cup \mathfrak{R}_6 \cup \mathfrak{R}_7$. The union of these four rectangles is a rectangle of height $f(b) - f(a)$ and width $x_4 - x_3$, and thus of area $[f(b) - f(a)] \times (x_4 - x_3)$. That is, $\int_a^b u - \int_a^b l = [f(b) - f(a)](x_4 - x_3)$ for the step functions sketched in Figure 8.12. If we took a similar partition, consisting of $n + 1$ elements of $[a, b]$, we would have $\int_a^b u - \int_a^b l = [f(b) - f(a)] \times (x_n - x_{n-1})$. Obviously we need only make $x_n - x_{n-1}$ small in order to make $\int_a^b u - \int_a^b l$ small (this is what is called for by the Riemann Condition). This geometric motivation leads us to the following theorem and proof:

founded

THEOREM 8.21 If $f : [a, b] \to \mathfrak{R}$ is monotone on $[a, b]$, then f is integrable on $[a, b]$.

Proof. We shall assume that f is increasing on $[a, b]$. Let $r > 0$. Let $\mathfrak{F} = \{x_0, x_1, \ldots, x_n\}$ be any partition of $[a, b]$ for which $x_j - x_{j-1} < \dfrac{r}{f(b) - f(a)}$. We define the step functions l_r and u_r on the open subintervals of $\mathfrak{F}$ as follows:

$$l_r(x) = f(x_{j-1}) \quad \text{for } x \text{ in } (x_{j-1}, x_j), \quad j = 1, \ldots, n$$
$$u_r(x) = f(x_j) \quad \text{for } x \text{ in } (x_{j-1}, x_j), \quad j = 1, \ldots, n.$$

Note that since f is increasing on $[a, b]$, $l_r(x) \leq f(x) \leq u_r(x)$ for all x in $[a, b]$. Moreover,

$$\int_a^b u_r - \int_a^b l_r =^1 \sum_{j=1}^n f(x_j)(x_j - x_{j-1}) - \sum_{j=1}^n f(x_{j-1})(x_j - x_{j-1})$$

$$= \sum_{j=1}^n [f(x_j) - f(x_{j-1})](x_j - x_{j-1})$$

$$<^2 \frac{r}{f(b) - f(a)} \sum_{j=1}^n [f(x_j) - f(x_{j-1})]$$

$$=^3 \frac{r}{f(b) - f(a)} [f(b) - f(a)]$$

$$= r.$$

(1) Definition of u_r, l_r, and the integral of a step function.

(2) Since $x_j - x_{j-1} < \dfrac{r}{f(b) - f(a)}$.

(3) $\displaystyle\sum_{j=1}^n [f(x_j) - f(x_{j-1})] = f(x_n) - f(x_0) = f(b) - f(a)$.

Thus, the Riemann Condition is satisfied and f is integrable on $[a, b]$. ∎

By means of Theorem 8.18, we can generalize Theorem 8.21. First let us consider a particular example:

Example 10. Let $f: [0, 2\pi] \to \mathcal{R}$ be defined by $f(x) = \sin x$ (see Figure 8.13). The function f is not monotone on $[0, 2\pi]$. However, f is increasing on each of $\left[0, \dfrac{\pi}{2}\right]$ and $\left[\dfrac{3\pi}{2}, 2\pi\right]$ and decreasing on $\left[\dfrac{\pi}{2}, \dfrac{3\pi}{2}\right]$. Thus, by Theorem 8.21 f is integrable on $\left[0, \dfrac{\pi}{2}\right]$, $\left[\dfrac{\pi}{2}, \dfrac{3\pi}{2}\right]$, and $\left[\dfrac{3\pi}{2}, 2\pi\right]$; and consequently, by Theorem 8.18, f is integrable on $[0, 2\pi]$. Unfortunately, we do not yet have a technique by which we can evaluate $\int_a^b f$. ∎

We shall say that a function f is *piecewise monotone* on an interval $[a, b]$ if there exists a partition $\mathcal{G} = \{x_0, x_1, \ldots, x_n\}$ of $[a, b]$ such that f is monotone on each of the *open* subintervals of $\mathcal{G}$. It would seem that we could immediately conclude (using Theorems 8.18 and 8.21) that f is integrable on $[a, b]$.

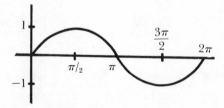

Figure 8.13

However, remember that we have defined integrability only for bounded functions. The function $f:[-1, 1] \rightarrow \mathcal{R}$, defined by $f(x) = \dfrac{1}{x^2}$ for $x \neq 0$ and $f(0) = 0$, is, by our definition, piecewise monotone. It is increasing on $(-1, 0)$ and decreasing on $(0, 1)$, but it is not bounded and hence not integrable.

In Exercise 2 the reader is asked to prove the following generalization of Theorem 8.21: If $f:[a, b] \rightarrow \mathcal{R}$ is bounded and is monotone on the open interval (a, b), then f is integrable on $[a, b]$. By virtue of this theorem, if f is bounded and monotone on each of the open subintervals of a partition $\{x_0, x_1, \ldots, x_n\}$ of $[a, b]$, then f is integrable on each of the closed subintervals $[x_{j-1}, x_j], j = 1, \ldots, n$. Thus, by Theorem 8.18 f is integrable on $[a, b]$. This proves the following theorem:

THEOREM 8.22 If $f:[a, b] \rightarrow \mathcal{R}$ is piecewise monotone and bounded, then f is integrable on $[a, b]$.

This theorem enables us to conclude (1) that every polynomial is integrable on each closed interval $[a, b]$, (2) that both the sine and cosine functions are integrable on each closed interval $[a, b]$, and (3) that the square root function is integrable on any finite subinterval of $[0, \infty)$, and so on. All of these functions are examples of *continuous*, piecewise, monotone functions. Unfortunately, Theorem 8.22 cannot be used to prove that every continuous function is integrable. The function f, defined by

$$f(x) = \begin{cases} x \sin \dfrac{1}{x} & \text{for } x \text{ in } [0, 1] \\ 0 & \text{for } x = 0, \end{cases}$$

is continuous on $[0, 1]$ and bounded, but is *not* piecewise monotone (Why?).

It is reassuring, though, that every continuous function on $[a, b]$ is integrable. We shall only suggest what can lead to a proof of this assertion. Choose a partition $\mathcal{P} = \{x_0, x_1, \ldots, x_n\}$ of $[a, b]$ and define u and l as follows:

$$u(x) = M_j = \text{lub } \{f(x) \mid x \in [x_{j-1}, x_j]\} \quad \text{for} \quad x_{j-1} < x < x_j$$
$$l(x) = m_j = \text{glb } \{f(x) \mid x \in [x_{j-1}, x_j]\} \quad \text{for} \quad x_{j-1} < x < x_j.$$

That u and l are well defined on the open subintervals of $\mathcal{P}$ depends upon the fact that a continuous function is bounded on a closed interval. We then have

$$\int_a^b u - \int_a^b l = \sum_{j=1}^n M_j(x_j - x_{j-1}) - \sum_{j=1}^n m_j(x_j - x_{j-1})$$
$$= \sum_{j=1}^n (M_j - m_j)(x_j - x_{j-1}).$$

If for each $r > 0$ we can find a partition $\{x_0, \ldots, x_{n_r}\}$ of $[a, b]$ with the subintervals sufficiently small that $M_j - m_j < \dfrac{r}{b - a}$ for $j = 1, \ldots, n_r$, then we have

$$\int_a^b u - \int_a^b l < \frac{r}{b - a} \sum_{j=1}^{n_r} (x_j - x_{j-1})$$

$$= \frac{r}{b - a} (b - a) = r,$$

and the Riemann Condition is satisfied.

It seems quite reasonable to suppose that if f is continuous, we can partition $[a, b]$ in the way called for in the preceding paragraph. Note that this calls for the following: Given an $r > 0$, find a $\rho > 0$ such that if $|x_j - x_{j-1}| < \rho, j = 1, \ldots, n_r$, then $|f(x) - f(y)| < \dfrac{r}{b - a}$ for all x and y in $[x_{j-1}, x_j]$. This is a requirement that looks *very* much like the definition of continuity. However, this requirement is stronger, for it calls for one ρ to work for all pairs x and y: $x, y \in [x_{j-1}, x_j]$.

To further illustrate the difficulties involved consider the function f defined by $f(x) = \dfrac{1}{x}$ for x in $(0, 1]$. This function is continuous but not bounded. Let $0 < r < 1$. Assume that there exists a partition $\{x_0, \ldots, x_{n_r}\}$ of $(0, 1]$ such that $|f(x) - f(y)| < r$ whenever x and y are in the same open subinterval of this partition. Let k be a positive integer such that $\dfrac{1}{k} < x_1$. Then $x = \dfrac{1}{k}$ and $y = \dfrac{1}{k+1}$ are both in (x_0, x_1). Therefore, $|f(x) - f(y)| = \left| \dfrac{1}{x} - \dfrac{1}{y} \right| = (k + 1) - k = 1 > r$, contradicting our assumption that $|f(x) - f(y)| < r$ for all x and y in (x_0, x_1). Thus, there exists no such partition of $[0, 1]$.

We note two things about the preceding example. The domain of f is not a closed interval and f cannot be extended to a continuous function on $[0, 1]$. We now state without proof the following theorem:

If $f : [a, b] \to \mathcal{R}$ is continuous and $r > 0$, there exists a $\rho > 0$ such that

$$|f(x) - f(y)| < r \text{ whenever } |x - y| < \rho, \; x, y \in [a, b].$$

With this assertion we now have:

THEOREM 8.23 If $f : [a, b] \to \mathcal{R}$ is continuous, then f is integrable on $[a, b]$.

EXERCISES

1. Let $f:[a, b] \rightarrow \mathcal{R}$ and $g:[a, b] \rightarrow \mathcal{R}$ such that $f(x) = g(x)$, except possibly at a finite number of points. Prove that if g is integrable on $[a, b]$, (a) f is also integrable on $[a, b]$ and (b) $\int_a^b f = \int_a^b g$.

2. Let $f:[a, b] \rightarrow \mathcal{R}$ be bounded on $[a, b]$ and monotone on (a, b). Prove that f is integrable on $[a, b]$ (Hint: Find a function $g:[a, b] \rightarrow \mathcal{R}$ which is monotone on $[a, b]$ and such that $f(x) = g(x)$ for all x in (a, b). Then apply Exercise 1).

8.   THE AREA PROBLEM REVISITED

We began this chapter with a discussion of the problems involved in defining the term "area." In Example 2 (p. 340) we examined the region $\mathcal{R}$ of the plane bounded by the x-axis and the graph of the function $f(x) = x^2$, $0 \le x \le 1$. In an intuitive fashion we decided that the "area" of this region is $1/3$. In order to arrive at this number we approximated the region $\mathcal{R}$ by a union of rectangles containing $\mathcal{R}$ and by a union of rectangles contained in $\mathcal{R}$, and we computed the areas of these two unions. This, of course, corresponds to considering two step functions u and l, $l \le f \le u$, and computing $\int_0^1 u$ and $\int_0^1 l$. In effect, in Example 2 we showed that $\overline{\int_0^1} f = \underline{\int_0^1} f = 1/3$. This motivates the following definition of area:

DEFINITION 8.24   Let f be integrable on $[a, b]$ and $f(x) \ge 0$ for all x in $[a, b]$. The *area* of the region $\mathcal{R}$ bounded below by the x-axis and above by the graph of f is defined to be $A(\mathcal{R}) = \int_b^a f$.

Example 11.   Let $f(x) = \sin x$, $0 \le x \le \pi$. Since f is continuous, it follows that f is integrable. Moreover, $f(x) \ge 0$ for $0 \le x \le \pi$. Thus, by Definition 8.24

$$A(\mathcal{R}_1) = \int_0^\pi \sin,$$

where $\mathcal{R}_1 = \{(x, y) \mid 0 \le x \le \pi, 0 \le y \le \sin x\}$.   ∎

Let us now consider the function g defined by $g(x) = \sin x$, $0 \le x \le 2\pi$. Figure 8.14 is a sketch of the region $\mathcal{R}$ bounded by the graph of g and the x-axis. Recall that $\int_0^{2\pi} \sin = \int_0^\pi \sin + \int_\pi^{2\pi} \sin$. Moreover, $\sin x \le 0$ for $\pi \le x \le 2\pi$ and $\int_\pi^{2\pi} \sin \le 0$. It is easy to show that $\int_\pi^{2\pi} \sin < 0$. Therefore, $\int_0^{2\pi} \sin < \int_0^\pi \sin$. By Example 11, $\int_0^\pi \sin = A(\mathcal{R}_1)$. We see immediately that it would be unsuitable to define the area $A(\mathcal{R}) = \int_0^{2\pi} \sin$, for then we would have $\mathcal{R}_1 \subset \mathcal{R}$ and $A(\mathcal{R}_1) > A(\mathcal{R})$. This situation is taken care of by the following definition:

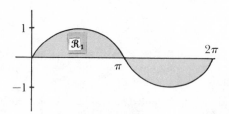

Figure 8.14

DEFINITION 8.25 Let f be integrable on $[a, b]$. Let $\mathcal{R}$ be the region bounded by the graph of f and the x-axis. The *area of* $\mathcal{R}$ is denoted by $A(\mathcal{R})$ and defined by

$$A(\mathcal{R}) = \int_a^b |f|.$$

Thus, for the function $g(x) = \sin x$, $0 \le x \le 2\pi$, $A(\mathcal{R}) = \int_0^{2\pi} |\sin| = \int_0^\pi \sin + \int_\pi^{2\pi} (-\sin)$, since $|\sin x| = \sin x$ for $0 \le x \le \pi$ and $|\sin x| = -\sin x$ for $\pi \le x \le 2\pi$.

Example 12. Let $f(x) = x^3$, $-1 \le x \le 2$, and let $\mathcal{R}$ be the region bounded by the graph of f and the x-axis. Notice that $f(x) \ge 0$ for $0 \le x \le 2$ and $f(x) \le 0$ for $-1 \le x \le 0$. Thus, by Definition 8.25

$$A(\mathcal{R}) = \int_{-1}^2 |x^3|$$

$$= \int_{-1}^0 -x^3 + \int_0^2 x^3. \quad \blacksquare$$

Definitions 8.24 and 8.25 deal only with the areas of certain regions at least partially bounded by the x-axis. Consider now the region

$$\mathcal{R} = \{(x, y) \mid 0 \le x \le 1, x^2 \le y \le x^3\},$$

sketched in Figure 8.15a. Let $\mathcal{R}_1 = \{(x, y) \mid 0 \le x \le 1, 0 \le y \le x^2\}$ and $\mathcal{R}_2 = \{(x, y) \mid 0 \le x \le 1, 0 \le y \le x^3\}$ (see Figures 8.15b and c). Notice that $\mathcal{R}_2 \cup \mathcal{R} = \mathcal{R}_1$. We know that $A(\mathcal{R}_1) = \int_0^1 x^2$ and $A(\mathcal{R}_2) = \int_0^1 x^3$. It seems plausible to assume that $A(\mathcal{R}_1) - A(\mathcal{R}_2) = \int_0^1 (x^2 - x^3)$. Indeed this is how we shall define the area of this region $\mathcal{R}$.

DEFINITION 8.26 Let f and g be integrable on $[a, b]$ and let $\mathcal{R}$ be the region bounded by the graphs of f and g. The *area of* $\mathcal{R}$, denoted by $A(\mathcal{R})$, is defined by

$$A(\mathcal{R}) = \int_a^b |f - g|.$$

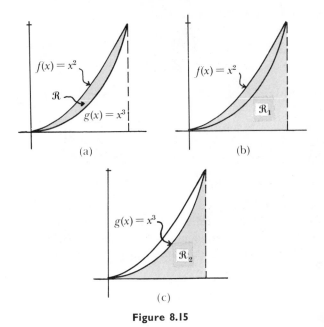

Figure 8.15

To the reader who has followed the discussion, the necessity for the absolute value signs in Definition 8.26 should be obvious. If it is not obvious, consider carefully the following example:

Example 13. Let $f(x) = x^3$ and $g(x) = x$ for $-1 \leq x \leq 1.5$. We wish to find the area of the region $\mathcal{R}$ (sketched in Figure 8.16), bounded by

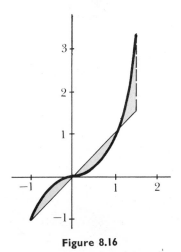

Figure 8.16

the graphs of f and g. Notice that $f(x) \leq g(x)$ for $0 \leq x \leq 1$ and $g(x) \leq f(x)$ for $-1 \leq x \leq 0$ and $1 \leq x \leq 1.5$. Therefore,

$$|f(x) - g(x)| = \begin{cases} x^3 - x & \text{for } x \text{ in } [-1, 0] \cup [1, 1.5] \\ x - x^3 & \text{for } x \text{ in } [0, 1], \end{cases}$$

and consequently,

$$A(\mathcal{R}) = \int_{-1}^{1.5} |x^3 - x|$$

$$= \int_{-1}^{0} (x^3 - x) + \int_{0}^{1} (x - x^3) + \int_{1}^{1.5} (x^3 - x). \quad \blacksquare$$

The reader will have noted that we have not computed the numerical values of the areas considered in Examples 11 to 13. To do so now would involve the type of algebraic computations used in the first section. Early in the next chapter we shall have at our disposal a more effective technique to compute these numerical values.

EXERCISES

Indicate the integral(s) giving the area of the region $\mathcal{R}$ bounded by the graph of f and the x-axis. Sketch the region $\mathcal{R}$:

1. $f(x) = x + 1, 0 \leq x \leq 1$.

2. $f(x) = x + 1, -2 \leq x \leq 0$.

3. $f(x) = \sin x, -\pi \leq x \leq \pi$.

4. $f(x) = x^3 - 1, -1 \leq x \leq 0$.

5. $f(x) = x^3 - 1, 0 \leq x \leq 2$.

Find the integral(s) giving the area of the region $\mathcal{R}$ bounded by the graphs of f and g. Sketch the region $\mathcal{R}$:

6. $\begin{cases} f(x) = 4 \\ g(x) = x^2 \end{cases}, -2 \leq x \leq 2$.

7. $\begin{cases} f(x) = 4 \\ g(x) = x^2 \end{cases}, 0 \leq x \leq 3$.

8. $\begin{cases} f(x) = x^3 \\ g(x) = x^2 \end{cases}, -1 \leq x \leq 1$.

9. $\begin{cases} f(x) = \sin x \\ g(x) = \cos x \end{cases}, 0 \leq x \leq \pi$.

10. $\begin{cases} f(x) = \sqrt{x} \\ g(x) = x \end{cases}, 0 \leq x \leq 4$.

ANSWERS

2. $A(\mathcal{R}) = \int_{-2}^{-1} -(x + 1) + \int_{-1}^{0} (x + 1)$.   4. $A(\mathcal{R}) = \int_{-1}^{0} -(x^3 - 1)$.

6. $A(\mathcal{R}) = \int_{-2}^{2} (4 - x^2)$.   8. $A(\mathcal{R}) = \int_{-1}^{1} (x^2 - x^3)$.

10. $A(\mathcal{R}) = \int_{0}^{1} (\sqrt{x} - x) + \int_{1}^{4} (x - \sqrt{x})$.

2.

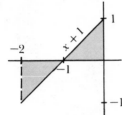

4.

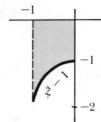

6.

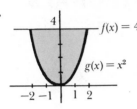

8.

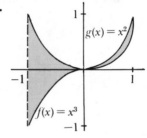

10.

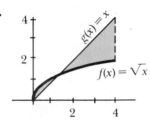

9

EVALUATION OF INTEGRALS AND AREAS

INTRODUCTION

In Chapter 8 we investigated in some detail various properties of the class of integrable functions. We discovered that every monotone and every continuous function is integrable. Moreover we found that the integral itself is

i. order-preserving: if $f \leq g$ then $\int_a^b f \leq \int_a^b g$,

and

ii. linear: $\begin{cases} \int_a^b (f + g) = \int_a^b f + \int_a^b g \\ \int_a^b (rf) = r(\int_a^b f). \end{cases}$

In this and subsequent chapters we shall concern ourselves with the practical problem of *how* to evaluate the integral of a given function f. In principle the definition of the integrable function tells us how to perform this task. We approximate the given function above or below by step functions and then compute the upper or lower integral, respectively. However, the reader will recall that rather tedious algebraic computations were involved in the very first application of this definition, to determine $\int_0^1 x^2$. Attempts to evaluate integrals of more complex functions by direct application of the definition involve even more difficult algebraic manipulations. The Fundamental Theorems of Calculus, presented in the next two sections, will (for certain functions) very much simplify the actual evaluation of integrals.

1. FIRST FUNDAMENTAL THEOREM OF CALCULUS

Let $f:[a, b] \to \Re$ be integrable on $[a, b]$. We know from Theorem 8.17 (p. 364) that f is also integrable on $[a, x]$ for each x in $[a, b]$. Consequently, $\int_a^x f$ is also a real-valued function on $[a, b]$. Let $F(x) = \int_a^x f$. If $f \geq 0, F(x)$ is by Definition 8.24 (p. 371) the area of the region below the graph of f and above the x-axis in the interval $[a, x]$ (see Figure 9.1).

Now let $x, x_0 \in (a, b), x_0 < x$. Since $F(x) - F(x_0) = \int_{x_0}^x f, F(x) - F(x_0)$ is the area of the region below the graph of f and above the x-axis in the interval $[x_0, x]$ (see Figure 9.2a). Consider now the rectangle $\Re$ of base $[x_0, x]$ and with height $f(x_0)$. Figure 9.2a and b seems to indicate that when x is close to x_0, $F(x) - F(x_0)$ is closely approximated by the area of this rectangle, $f(x_0)(x - x_0)$. $(A = F(x) - F(x_0) \approx f(x_0)(x - x_0).)$ If this is the case, then $\dfrac{F(x) - F(x_0)}{x - x_0}$ is close to $f(x_0)$ when x is close to x_0. $\left(\dfrac{F(x) - F(x_0)}{x - x_0} \approx f(x_0).\right)$ The quotient $\dfrac{F(x) - F(x_0)}{x - x_0}$ should bring to mind the definition of the derivative:

$$F'(x_0) = \lim_{x \to x_0} \frac{F(x) - F(x_0)}{x - x_0}.$$

It seems plausible to expect that $F'(x_0) = f(x_0)$. The following example will indicate why f must be continuous at x_0 to draw this conclusion:

Example 1. Let $f:[0, 2] \to \Re$ be defined by

$$f(x) = \begin{cases} 1 & \text{for } 0 \leq x \leq 1 \\ 2 & \text{for } 1 < x \leq 2. \end{cases}$$

Let $F(x) = \int_0^x f$ for $0 \leq x \leq 2$. We shall compute $F(x)$.

(a) For $0 \leq x \leq 1, f(x) = 1$ and hence

$$F(x) = \int_0^x f = 1(x - 0) = x.$$

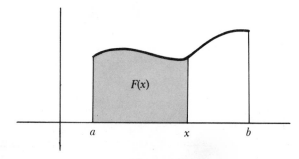

Figure 9.1

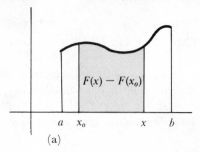

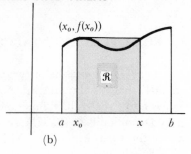

(a)   (b)

Figure 9.2

(b) For $1 < x \leq 2, f(x) = 2$ and

$$F(x) = \int_0^x f$$
$$= \int_0^1 f + \int_1^x f$$
$$= 1 + 2(x - 1)$$
$$= 2x - 1.$$

Thus,

$$F(x) = \begin{cases} x & \text{for} \quad 0 \leq x \leq 1 \\ 2x - 1 & \text{for} \quad 1 < x \leq 2. \end{cases}$$

See Figure 9.3a and b for sketches of the graphs of f and F, respectively.

We notice that $F'(x) = 1 = f(x)$ for $0 < x < 1$ and $F'(x) = 2 = f(x)$ for $1 < x < 2$. However, since $\lim_{x \to 1^-} \dfrac{F(x) - F(1)}{x - 1} = 1$ and $\lim_{x \to 1^+} \dfrac{F(x) - F(1)}{x - 1} = 2$, $\lim_{x \to 1} \dfrac{F(x) - F(1)}{x - 1}$ does not exist, and F is not differentiable at $x_0 = 1$.

We notice further that for $x > 1$, $(x - 1)f(1)$ is not a good approximation of the area $F(x) - F(1)$, because f is discontinuous at $x_0 = 1$ and the value $f(1)$ is too small. ∎

We now state and prove one of the most important theorems in calculus. Because of its importance it is called the Fundamental Theorem of Calculus or the First Fundamental Theorem of Calculus. (In the next section we shall state and prove a theorem that we shall call the Second Fundamental

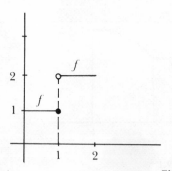

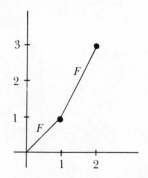

Figure 9.3

Theorem of Calculus. Some authors refer to this second theorem as the Fundamental Theorem of Calculus.)

THEOREM 9.1 (First Fundamental Theorem of Calculus.) Let $f : [a, b] \to \mathcal{R}$ be integrable and let $F : [a, b] \to \mathcal{R}$ be defined by $F(x) = \int_a^x f$. If f is continuous at x_0, $x_0 \in (a, b)$, then F is differentiable at x_0 and $F'(x_0) = f(x_0)$.

Proof. Assume that f is continuous at x_0, $x_0 \in (a, b)$. We wish to show that

$$(9.1) \qquad \lim_{x \to x_0} \frac{F(x) - F(x_0)}{x - x_0} = f(x_0).$$

This limit is by definition $F'(x_0)$. Assume that $x \neq x_0$ and consider the following chain of equalities:

$$(9.2) \quad \begin{cases} \dfrac{F(x) - F(x_0)}{x - x_0} - f(x_0) =^1 \dfrac{1}{x - x_0}\left[\left(\int_a^x f - \int_a^{x_0} f\right) - (x - x_0)f(x_0)\right] \\[2mm] =^2 \dfrac{1}{x - x_0}\left[\int_a^x f + \int_{x_0}^a f - \int_{x_0}^x f(x_0)\right] \\[2mm] =^3 \dfrac{1}{x - x_0}\left[\int_{x_0}^x f - \int_{x_0}^x f(x_0)\right] \\[2mm] =^4 \dfrac{1}{x - x_0}\left[\int_{x_0}^x [f - f(x_0)]\right]. \end{cases}$$

(1) Definition of F and algebra.
(2) Definition 8.19ii ($\int_{x_0}^a f = -\int_a^{x_0} f$) and definition of the integral of the step function with constant value $f(x_0)$.
(3) Theorem 8.18.
(4) Theorem 8.14.

Let $r > 0$. In order to establish that Equation 9.1 holds we need to show that there exists a $\rho > 0$ such that $\left| \dfrac{F(x) - F(x_0)}{x - x_0} - f(x_0) \right| < r$ whenever $|x - x_0| < \rho$. Since f is continuous at x_0, there exists a $\rho > 0$ such that $|f(x) - f(x_0)| < r$ whenever $|x - x_0| < \rho$. This ρ is the one we seek, for then

$$\left| \frac{F(x) - F(x_0)}{x - x_0} - f(x_0) \right| =^5 \left| \frac{1}{x - x_0} \int_{x_0}^x [f - f(x_0)] \right|$$

$$\leq^6 \frac{1}{|x - x_0|} \left| \int_{x_0}^x |f - f(x_0)| \right|$$

$$<^7 \frac{1}{|x - x_0|} \left| \int_{x_0}^x r \right|$$

$$=^8 \frac{1}{|x - x_0|} r |x - x_0|$$

$$= r.$$

Thus, we see that $\lim\limits_{x \to x_0} \dfrac{F(x) - F(x_0)}{x - x_0} = f(x_0)$ at each point x_0 at which f is continuous.

(5) Taking absolute values of extreme members of Equation 9.2.
(6) Theorem 8.15ii.
(7) Theorem 8.15i and assuming $|x - x_0| < \rho$. Hence $|f(t) - f(x_0)| < r$ for all t between x and x_0.
(8) Definition of the integral of the step function with constant value r. ∎

Example 2. Let $f:[0, 2] \to \mathfrak{R}$ be defined by $f(x) = x^3$ and let $F(x) = \int_0^x f$ for $0 \leq x \leq 2$. By Theorem 9.1 we have for each x in $(0, 2)$, $F'(x) = f(x) = x^3$. ∎

Example 3. Let $f:[-\pi, \pi] \to \mathfrak{R}$ be defined by $f(x) = \sin x$ and let $G(x) = \int_x^\pi f$ for $-\pi \leq x \leq \pi$. Theorem 9.1 does not tell us that $G'(x) = f(x)$ since x appears here as the lower limit of the integral and x appears as the upper limit of the integral in Theorem 9.1. However, using Theorem 8.18 it is easy to show that $G(x) = \int_x^\pi f = \int_{-\pi}^\pi f - \int_{-\pi}^x f$. Therefore, if we let $F(x) = \int_{-\pi}^x f$, we have (1) $G(x) = F(\pi) - F(x)$ and (2) $F'(x) = f(x)$ for $-\pi < x < \pi$. Thus, $G'(x) = [F(\pi) - F(x)]' = -F'(x) = -f(x) = -\sin x$ for each x in $(-\pi, \pi)$. ∎

Consider now an arbitrary integrable function f on $[a, b]$ and let $F(x) = \int_a^x f, x \in [a, b]$. We know that if f is continuous at x_0, F is differentiable at x_0. Furthermore, in Example 1 we discovered a function f for which F is not differentiable at a certain point. However, F is continuous at that point.

At first glance it may seem surprising that F is continuous even if f is discontinuous. However, if we think of $F(x) - F(x_0)$ as the area under the graph of f and above the line segment $[x_0, x]$ and recall that f is bounded above by some constant function M, we see that $F(x) - F(x_0) \leq M(x - x_0)$. This last quantity's being small when x is close to x_0 assures us that $F(x)$ is close to $F(x_0)$ (see Figure 9.4).

The formal proof given below of the continuity of F is an immediate generalization of this argument.

THEOREM 9.2 (Continuity of the Integral.) Let $f:[a, b] \to \mathfrak{R}$ be integrable and let $F:[a, b] \to \mathfrak{R}$ be defined by $F(x) = \int_a^x f$. The function F is continuous at each point of $[a, b]$.

Proof. Let $r > 0$ and $x_0 \in [a, b]$. We wish to find a $\rho > 0$ such that $|F(x) - F(x_0)| < r$ whenever $|x - x_0| < \rho, x \in [a, b]$. Since f is integrable, it is bounded and there exists a positive number M such that $|f(x)| \leq M$

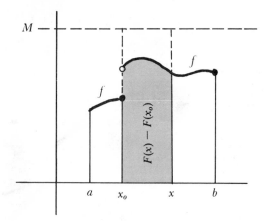

Figure 9.4

for all x in $[a, b]$. We then have

$$|F(x) - F(x_0)| =^1 \left| \int_a^x f - \int_a^{x_0} f \right|$$

$$=^2 \left| \int_a^x f + \int_{x_0}^a f \right|$$

$$=^3 \left| \int_{x_0}^x f \right|$$

$$\leq^4 \left| \int_{x_0}^x |f| \right|$$

$$\leq^5 \left| \int_{x_0}^x M \right|$$

$$=^6 M |x - x_0|.$$

Thus, if we choose $\rho = \dfrac{r}{M}$ and restrict x such that $|x - x_0| < \rho$ we have

$|F(x) - F(x_0)| < \dfrac{r}{M} M = r$, proving that F is continuous at x_0.

 (1) Definition of F.
 (2) Definition 8.19ii (p. 366).
 (3) Theorem 8.18.
 (4) Theorem 8.15ii.
 (5) Theorem 8.15i.
 (6) Definition of integral of step function with constant value M. ∎

EXERCISES

 For the functions f in Exercises 1 to 3 compute $F(x) = \int_0^x f$ for each x in the domain of f (see Example 1, p. 377):

 1. $f(x) = \begin{cases} 1 & \text{for} & 0 \leq x \leq 1 \\ -1 & \text{for} & 1 < x \leq 2. \end{cases}$

2. $f(x) = \begin{cases} 2 & \text{for} & 0 \le x \le 1 \\ -3 & \text{for} & 1 < x \le 3' \\ 1 & \text{for} & 3 < x \le 4. \end{cases}$

3. $f(x) = \begin{cases} 1 & \text{for} & -1 \le x \le 0 \\ -1 & \text{for} & 0 < x \le 1. \end{cases}$

4. Sketch the graphs of f and $F(x) = \int_0^x f$, where f is the function defined in Exercise 1.

5. Sketch the graphs of f and $F(x) = \int_0^x f$, where f is the function defined in Exercise 2.

6. Sketch the graphs of f and $F(x) = \int_0^x f$, where f is the function defined in Exercise 3.

In Exercises 7 to 12 find $F'(x_0)$ for x_0 in (a, b):

7. $F(x) = \int_1^x (t^{3/2})$, $x \in [1, 3]$.

8. $F(x) = \int_2^x \frac{1}{t}$, $x \in [2, 4]$.

9. $F(x) = \int_{-\pi}^x \sin(t^2)$, $x \in [-\pi, \pi]$.

10. $F(x) = \int_x^1 \sqrt{1 + t^2}$, $x \in [-1, 1]$.

11. $F(x) = \int_0^x (\sin t)$, $x \in [-\pi, \pi]$.

12. $F(x) = \int_a^x \frac{1}{t^2 + 4}$, $x \in [a, b]$.

In Exercises 13 to 16 determine the points at which F is defined and the points at which F is differentiable. In this latter case determine the derivative of F:

13. $F(x) = \int_0^x f$, $f(t) = \begin{cases} t^2 & \text{for} & 0 \le t \le 1 \\ t & \text{for} & t > 1. \end{cases}$

14. $F(x) = \int_0^x f$, $f(t) = \begin{cases} \sin t & \text{for} & 0 \le t \le \frac{\pi}{2} \\ \cos t & \text{for} & \frac{\pi}{2} < t \le \pi. \end{cases}$

15. $F(x) = \int_0^x f$, $f(t) = \begin{cases} 1 & \text{for} & 0 \le t \le 1 \\ t^2 & \text{for} & 1 < t \le 2 \\ t^3 & \text{for} & 2 < t \le 5. \end{cases}$

16. $F(x) = \int_0^x f$, $f(t) = \begin{cases} t^2 & \text{for} & -1 \le t \le 0 \\ 1 + t & \text{for} & 0 < t \le 2 \\ 3t & \text{for} & 2 < t \le 4. \end{cases}$

ANSWERS

1. $F(x) = \begin{cases} x & \text{for} & 0 \le x \le 1 \\ 2 - x & \text{for} & 1 \le x \le 2. \end{cases}$ 3. $F(x) = \begin{cases} x & \text{for} & -1 \le x \le 0 \\ -x & \text{for} & 0 \le x \le 1 \end{cases}$

5.

wrong answer

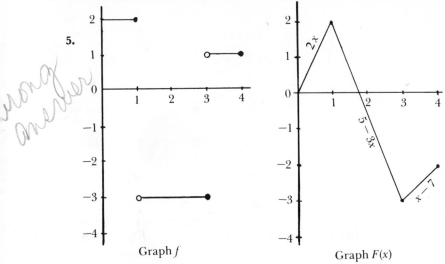

Graph f

Graph $F(x)$

7. $F'(x_0) = x_0^{3/2}$. **9.** $F'(x_0) = \sin(x_0^2)$. **11.** $F'(x_0) = \sin x_0$.
13. F is defined on $[0, \infty)$. F is differentiable at each point of $(0, \infty)$ since f is continuous at each of these points. $F'(x) = x^2$ for $0 < x < 1$, and $F'(x) =$ ✗ for $x \geq 1$. **15.** F is defined on $[0, 5]$ and is differentiable on $(0, 2) \cup (2, 5)$. $F'(x) = 1$ for $0 < x \leq 1$, $F'(x) = x^2$ for $1 < x < 2$, and $F'(x) = x^3$ for $2 < x < 5$.

2. THE SECOND FUNDAMENTAL THEOREM OF CALCULUS

Let us now summarize the results of Theorems 9.1 and 9.2 for the special case in which f is a continuous function on $[a, b]$ and $F(x) = \int_a^x f$ for x in $[a, b]$. By Theorem 9.2 F is continuous on the closed interval $[a, b]$; by Theorem 9.1 F is differentiable on the open interval (a, b) with $F'(x) = f(x)$.

We ask the following question: Do there exist functions G, other than F, which have basically the same characteristics? That is, do there exist functions G, continuous on $[a, b]$, differentiable on (a, b), with $G'(x) = f(x)$ and $G \neq F$? If we let $G(x) = F(x) + 5$, $x \in [a, b]$, we answer this question affirmatively. Indeed, if for all x and some real number C, $G(x) = F(x) + C$, we have $G'(x) = F'(x)$. Moreover, if G is continuous on $[a, b]$, differentiable on (a, b), and $G'(x) = F'(x)$ on (a, b), the slope of the graphs of G and F are the same at each point of (a, b). A glance at Figure 9.5 suggests that $G = F + C$ for some constant function C.

The aforementioned ideas provide the content of the following theorem:

THEOREM 9.3 Let $F: [a, b] \to \mathcal{R}$ and $G: [a, b] \to \mathcal{R}$ be differentiable on the open interval (a, b) and continuous on the closed interval $[a, b]$. Then $G'(x) = F'(x)$ for each x in (a, b) if and only if there exists a real number C such that $G(x) = F(x) + C$ for all x in $[a, b]$.

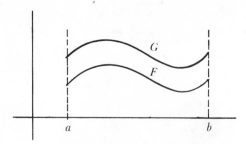

Figure 9.5

Proof. Assume that $G'(x) = F'(x)$ for each x in (a, b).

Let $H = G - F$. Note that H is continuous on $[a, b]$ and differentiable on (a, b), and that $H'(x) = 0$ for all x in (a, b). Let $x_1 \in (a, b]$. By the Mean Value Theorem (Appendix IV, p. 888) there exists an x_2 in (a, x_1) such that $H(x_1) - H(a) = H'(x_2)(x_1 - a)$. Since $H'(x) = 0$ on (a, b), we have $H(x_1) - H(a) = 0$ for any x_1 in $(a, b]$. Thus, $H(x_1) = H(a)$ for all x_1 in $[a, b]$ or, equivalently, $H(x) = H(a)$ for all x in $[a, b]$. Setting $C = H(a)$ and recalling that $H = G - F$, we have $G(x) - F(x) = C$ for all x in $[a, b]$.

The proof of the converse is trivial and is left to the reader. ∎

We now apply Theorem 9.3 to the special case in which $F(x) = \int_c^x f$, with f continuous on $[a, b]$, $c \in [a, b]$.

THEOREM 9.4   (Second Fundamental Theorem of Calculus.) Let $f : [a, b] \to \mathcal{R}$ be continuous on $[a, b]$ and $F : [a, b] \to \mathcal{R}$ be continuous on $[a, b]$ and differentiable on (a, b). If $F'(x) = f(x)$ for all x in (a, b), then for each x in $[a, b]$ and each c in $[a, b]$

(9.3) $$\int_c^x f = F(x) - F(c).$$

Proof. Let $F_1(x) = \int_a^x f$ for each x in $[a, b]$. By Theorem 9.2 F_1 is continuous on $[a, b]$, and by Theorem 9.1 F_1 is differentiable on (a, b) with $F_1'(x) = f(x)$. Thus, by the hypothesis of the theorem, $F_1'(x) = F'(x)$ for all x in (a, b). By Theorem 9.3 there exists a real number C such that $F(x) = F_1(x) + C$ for all x in $[a, b]$. We therefore have

(9.4) $$F(x) =^1 \int_a^x f + C.$$

Substituting $x = a$ in Equation 9.4, we have

$$F(a) = \int_a^a f + C =^2 C,$$

and therefore, for each x in $[a, b]$

(9.5) $$\int_a^x f =^3 F(x) - F(a).$$

Finally,

$$\int_c^x f =^4 \int_a^x f - \int_a^c f$$
$$=^5 [F(x) - F(a)] - [F(c) - F(a)]$$
$$= F(x) - F(c).$$

(1) Definition of F_1.
(2) Definition 8.19 (p. 366) ($\int_a^a f = 0$).
(3) Solving Equation 9.4 for $\int_a^x f$ and substituting $C = F(a)$.
(4) Theorem 8.18 (p. 365).
(5) Applying Equation 9.5 to $\int_a^x f$ and $\int_a^c f$. ∎

One great value of Theorem 9.4 is that it allows us to evaluate the integrals of many functions. Indeed it tells us that if f is continuous on $[a, b]$ and we can find a function G such that G is continuous on $[a, b]$ and $G'(x) = f(x)$ on (a, b), then $\int_a^b f = G(b) - G(a)$. This will enable us to evaluate most of the areas considered at the end of Chapter 8.

Example 4. Let $f(t) = t^2$, $G_1(t) = \dfrac{t^3}{3}$, and $G_2(t) = \dfrac{t^3}{3} + 5$. Since $G_1' = f$ and $G_2' = f$, we have by Theorem 9.4

(a) $$\int_1^2 t^2 =^1 G_1(2) - G_1(1)$$
$$= \frac{2^3}{3} - \frac{1^3}{3} = \frac{7}{3},$$

and

(b) $$\int_1^2 t^2 =^1 G_2(2) - G_2(1)$$
$$= \left(\frac{2^3}{3} + 5\right) - \left(\frac{1^3}{3} + 5\right)$$
$$= \frac{7}{3}.$$

(1) By Theorem 9.4. ∎

In Example 4 we found two functions G_1 and G_2 for which $G_1' = G_2' = f$. In the future we shall search for only one such function and in general we shall search for the simplest one.

Example 5. Since $(\sin)' = \cos$, we have by Equation 9.3 for all real x

$$\int_0^x \cos = \sin x - \sin 0$$
$$= \sin x.$$

We now turn our attention to the problem of evaluating certain areas.

∎

Example 6. Find the area of the region $\mathcal{R}$ bounded by the x-axis and the graph of $f(x) = \sin x$, $0 \le x \le 2\pi$. By Definition 8.25 (p. 372) $A(\mathcal{R}) = \int_0^{2\pi} |f|$. Since $f(x) \ge 0$ for $0 \le x \le \pi$ and $f(x) \le 0$ for $\pi \le x \le 2\pi$, we have

$$A(\mathcal{R}) = \int_0^\pi f + \int_\pi^{2\pi} (-f).$$

If we let $F(x) = -\cos x$, then $F'(x) = f(x)$, and hence by Theorem 9.4

$$\int_0^\pi f = F(\pi) - F(0) = -\cos \pi - (-\cos 0) = 2,$$

and

$$\int_\pi^{2\pi} (-f) = -\int_\pi^{2\pi} f$$
$$= -[F(2\pi) - F(\pi)] = -[-\cos 2\pi - (-\cos \pi)] = 2.$$

Therefore, $A(\mathcal{R}) = 4$. ∎

Example 7. Find the area of the region $\mathcal{R}$ specified in Example 13, Chapter 8 (p. 373). In Example 13 we found that for $h(x) = x^3 - x$

$$A(\mathcal{R}) = \int_{-1}^0 h + \int_0^1 -h + \int_1^{1.5} h.$$

The function H defined by $H(x) = \dfrac{x^4}{4} - \dfrac{x^2}{2}$ is such that $H'(x) = x^3 - x$ for all x. Therefore, by Theorem 9.4,

$$A(\mathcal{R}) = \int_{-1}^0 h - \int_0^1 h + \int_1^{1.5} h$$
$$= [H(0) - H(-1)] - [H(1) - H(0)] + [H(1.5) - H(1)]$$
$$= 2H(0) - H(-1) - 2H(1) + H(1.5)$$
$$= 2(0) - (-\tfrac{1}{4}) - 2(-\tfrac{1}{4}) + \tfrac{9}{64}$$
$$= \tfrac{57}{64}.$$

Notice again the importance of dividing $\mathcal{R}$ into subregions in which $x^3 - x \ge 0$ and $x^3 - x \le 0$. If we evaluate $\int_{-1}^{1.5} (x^3 - x) = \left(\dfrac{(1.5)^4}{4} - \dfrac{(1.5)^2}{2} \right) - \left(\dfrac{(-1)^4}{4} - \dfrac{(-1)^2}{2} \right) = \dfrac{25}{64}$, we arrive at a number strictly less than the area just computed. ∎

EXERCISES

For each of the following functions f, find a function G such that $G' = f$. Using Theorem 9.4, evaluate the specified integral:

1. $\begin{cases} f(t) = t \\ \int_{-1}^x f. \end{cases}$

2. $\begin{cases} f(t) = 4 - t \\ \int_{-1}^1 f. \end{cases}$

3. $\begin{cases} f(t) = t^4 - 3t^2 \\ \int_0^1 f. \end{cases}$

4. $\begin{cases} f(t) = \sin 2t \\ \int_0^{\pi/2} f. \end{cases}$

5. $\begin{cases} f(t) = \frac{3}{2}t^{1/2} \\ \int_0^4 f. \end{cases}$

6. $\begin{cases} f(t) = 2\cos t \\ \int_{\pi/2}^{\pi/2} f. \end{cases}$

7. $\begin{cases} f(t) = t^{-1/2} \\ \int_1^4 f. \end{cases}$

8. $\begin{cases} f(t) = t\cos(t^2) \\ \int_0^1 f. \end{cases}$

9. $\begin{cases} f(t) = t(1 - t^2)^{1/2} \\ \int_1^x f, \ -1 \le x \le 1. \end{cases}$

10. $\begin{cases} f(t) = \sec^2 t \\ \int_0^{\pi/4} f. \end{cases}$

In Exercises 11 to 20 evaluate the area of the region specified in Exercises 1 to 10, Section 8, Chapter 8:

11. Exercise 1.   12. Exercise 2.   13. Exercise 3.

14. Exercise 4.   15. Exercise 5.   16. Exercise 6.

17. Exercise 7.   18. Exercise 8.   19. Exercise 9.

20. Exercise 10.

ANSWERS

1. $\begin{cases} G(x) = \dfrac{x^2}{2} \\ \int_{-1}^x f = \dfrac{x^2}{2} - \dfrac{1}{2}. \end{cases}$
3. $\begin{cases} G(x) = \dfrac{x^5}{5} - x^3 \\ \int_0^1 f = -\dfrac{4}{5}. \end{cases}$
5. $\begin{cases} G(x) = x^{3/2} \\ \int_0^4 f = 8. \end{cases}$

7. $\begin{cases} G(x) = 2x^{1/2}. \\ \int_1^4 f = 2. \end{cases}$
9. $\begin{cases} F(x) = -\dfrac{(1 - x^2)^{3/2}}{3} \\ \int_1^x f = -\dfrac{(1 - x^2)^{3/2}}{3}. \end{cases}$

11. $A(R) = \int_0^1 (x + 1) = \frac{1}{2} + 1 - (0 + 0) = \frac{3}{2}.$ 13. $A(R) = -\int_{-\pi}^0 \sin +$ $\int_0^\pi \sin = 4.$ 15. $A(R) = -\int_0^1 (x^3 - 1) + \int_1^2 (x^3 - 1) = \frac{7}{2}.$

17. $A(R) = \int_0^2 (4 - x^2) + \int_2^3 (x^2 - 4) = \frac{23}{3}.$

19. $A(R) = \int_0^{\pi/4} (\cos x - \sin x) + \int_{\pi/4}^\pi (\sin x - \cos x) = 2\sqrt{2}.$

3. PRIMITIVES OF FUNCTIONS

We discovered in the preceding section that a convenient method for evaluating $\int_a^x f$ was to find a function F such that $F' = f$. In Definition 9.5 we give a name to such a function F.

DEFINITION 9.5 Let $f : [a, b] \to \mathcal{R}$ and $F : [a, b] \to \mathcal{R}$. If F is continuous on $[a, b]$ and differentiable on (a, b) with $F'(x) = f(x)$, we say that F is *primitive* of f.

It follows from Theorem 9.3 that if F and G are primitives of f and $\text{dom} f = [a, b]$, then there exists a constant function C such that $G = F + C$ on $[a, b]$. Had we not required that the domain of f be an interval, the preceding statement would not necessarily be true. For example, let $f(x) = 1$ for x in $[0, 1] \cup [2, 3]$. Then if $G_1(x) = x$ for x in $[0, 1] \cup [2, 3]$, and $G_2(x) = x$ for x in $[0, 1]$ and $G_2(x) = x + 1$ for x in $[2, 3]$, each of G_1 and G_2 is such that $G_1' = G_2' = f$, but there is no constant function C such that $G_2(x) = G_1(x) + C$ for all x in $[0, 1] \cup [2, 3]$. Therefore, we have restricted our attention to functions f defined on intervals.

We now restate the Second Fundamental Theorem of Calculus in terms of primitives:

THEOREM 9.6 Let f be continuous on $[a, b]$. If F is a primitive of f on $[a, b]$, then for c and d in $[a, b]$

$$\int_c^d f = F(d) - F(c).$$

At this point we introduce the notation $[F]_c^d = F(d) - F(c)$. From Theorem 9.6 it follows immediately that if F and G are both primitives of f, then

$$\int_c^d f = [F]_c^d = [G]_c^d.$$

Since Theorem 9.6 applies to any primitive of f, it is convenient to have a notation for this set.

The set of all primitives of f (on some given interval) will be denoted by $\int f$.

This set is frequently called the *indefinite integral* of f. We have already noted that if F is a primitive of f on $[a, b]$, then any other primitive of f differs from F by at most a constant function.

We shall let $[F]$ denote the set $\{F + C \mid C$ a constant function on $[a, b]\}$.

Thus, if F is a primitive of f on $[a, b]$, $\int f$ and $[F]$ stand for exactly the same sets, namely, the set of all primitives of f on $[a, b]$.

DEFINITION 9.7 Let $f : [a, b] \to \mathcal{R}$. The set of all primitives of f is called the *primitive set* of f and is denoted by $\int f$. If F is a primitive of f, $[F]$ also denotes the set of all primitives of f and is called the *primitive set containing* F.

Let us make the following observations concerning the notation in Definition 9.7. We have $\int f = [F] = \{G \mid G$ is continuous on $[a, b]$ and

$G'(x) = f(x)$ on (a, b)}. This set, the primitive set of f, is a set of functions any two of which differ by at most a constant function on $[a, b]$. That is, $G \in \int f$ and $H \in \int f$ only if $G(x) = H(x) + C$ for some real number C and all x in $[a, b]$.

We emphasize that the two symbols $\int f$ and $\int_c^d f$ stand for two different type objects: $\int f$ is the set of all primitive functions of f and $\int_c^d f$ is a real number. Furthermore, if F is a primitive of f, $[F]$ is the set of all functions differing from F by at most a constant function, whereas $[F]_c^d = F(d) - F(c)$. The basic connection between the various symbols lies in Theorem 9.6, which states that if $\int f = [F]$, then $\int_c^d f = [F]_c^d$. The following example illustrates this application of Theorem 9.6.

Example 8. Let us evaluate $\int_1^2 6x^5$. The function F defined by $F(x) = x^6$ is a primitive of f on any interval $[a, b]$. Thus, the primitive set relation

(a)
$$\int 6x^5 = [x^6]$$
$$= \{G \mid G(x) = x^6 + C, C \text{ a real number}\}$$

gives rise to the following evaluation:

(b)
$$\int_1^2 6x^5 = [x^6]_1^2$$
$$= 2^6 - 1^6$$
$$= 63. \quad \blacksquare$$

Table 9.1 lists both primitive sets and the integrals of several familiar functions. The justifications for these formulae are called for in the exercises. The formulae are understood to be valid on intervals where the functions are continuous.

Table 9.1   Primitives and Integrals of Several Functions.

Primitives	Integrals
i. $\int x^r = \left[\dfrac{x^{r+1}}{r + 1}\right] \quad (r \neq -1).$	$\int_a^b x^r = \dfrac{b^{r+1} - a^{r+1}}{r + 1} \quad (r \neq -1).$
ii. $\int \sin x = [-\cos x].$	$\int_a^b \sin x = -(\cos b - \cos a).$
iii. $\int \cos x = [\sin x].$	$\int_a^b \cos x = \sin b - \sin a.$
iv. $\int \sec^2 x = [\tan x].$	$\int_a^b \sec^2 x = \tan b - \tan a.$
v. $\int \sec x \tan x = [\sec x].$	$\int_a^b \sec x \tan x = \sec b - \sec a.$
vi. $\int \csc^2 x = [-\cot x].$	$\int_a^b \csc^2 x = -(\cot b - \cot a).$
vii. $\int \csc x \cot x = [-\csc x].$	$\int_a^b \csc x \cot x = -(\csc b - \csc a).$

We complete our basic discussion of primitive sets by introducing an addition between the sets and a scalar multiplication.

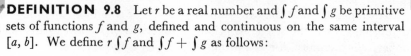

DEFINITION 9.8 Let r be a real number and $\int f$ and $\int g$ be primitive sets of functions f and g, defined and continuous on the same interval $[a, b]$. We define $r \int f$ and $\int f + \int g$ as follows:

$$\text{i. } r \int f = \int rf,$$

$$\text{ii. } \int f + \int g = \int (f + g).$$

Definition 9.8 is meaningful in view of the fact that $\int f_1 = \int f_2$ if and only if $f_1 = f_2$ (why?). Thus, $\int f_1 = \int f_2 \Rightarrow f_1 = f_2 \Rightarrow rf_1 = rf_2 \Rightarrow \int rf_1 = \int rf_2$. Similarly, $\int f_1 = \int f_2$ and $\int g_1 = \int g_2 \Rightarrow f_1 = f_2$ and $g_1 = g_2 \Rightarrow f_1 + g_1 = f_2 + g_2 \Rightarrow \int (f_1 + g_1) = \int (f_2 + g_2)$.

We note one particular case of Definition 9.8i. For $r = 0$, we have $0 \cdot \int f = \int (0 \cdot f) = \{C \mid C \text{ is a constant function on } [a, b]\}$.

In the exercises the reader will be asked to prove several properties of this addition and scalar multiplication. One of the most important of these properties is the cancellation law: If $\int f + \int g = \int f + \int h$, then $\int g = \int h$.

Definition 9.8 leads us to the following primitive set relations:

(9.6)
$$\begin{cases} \text{i. } r[F] = [rF] \\ \text{ii. } [F] + [G] = [F + G]. \end{cases}$$

Example 9. Let $f(x) = x^{-3} - 3x^{2/3} + 7$. We then have

$$\int f = \int (x^{-3} - 3x^{2/3} + 7)$$

$$=^1 \int x^{-3} - 3 \int x^{2/3} + \int 7$$

$$=^2 \left[\frac{x^{-2}}{-2} \right] - 3 \left[\frac{x^{5/3}}{5/3} \right] + [7x]$$

$$=^3 \left[-\frac{x^{-2}}{2} - \frac{9}{5} x^{5/3} + 7x \right].$$

(1) Definition 9.8.
(2) Table 9.1i.
(3) Equations 9.6. ∎

Example 10. Let $f(x) = x^{1/2} \sin (x^{3/2})$. This function is more complicated than any of those listed in Table 9.1. We shall, nonetheless, find the primitive set of this function—partially by inspection and partially by use of Table 9.1. We seek a function F such that $F' = f$. We note that

in the process of differentiation the sine function comes from ($-$cosine).
Thus, we attempt to arrive at a primitive by considering $F_1(x) = -\cos(x^{3/2})$.
We differentiate F_1: $F_1'(x) = \frac{3}{2}x^{1/2}\sin(x^{3/2})$. $F_1' \neq f$, but we note that the
function $F = \frac{2}{3}F_1$ is such that $F' = f$. Therefore,

$$\int x^{1/2}\sin(x^{3/2}) = [-\tfrac{2}{3}\cos(x^{3/2})].$$

As a double check we now differentiate:

$$(-\tfrac{2}{3}\cos(x^{3/2}))' = -\tfrac{2}{3}(-\sin(x^{3/2}))(\tfrac{3}{2}x^{1/2}) = x^{1/2}\sin(x^{3/2}). \quad \blacksquare$$

The reader is asked in several of the exercises to find primitive sets of
composite functions by this sort of inspection. The last two sections of this
chapter will be devoted to developing more elegant techniques for determining primitive sets of such functions.

We emphasize now the check which is available to us after having
found a prospective primitive: *Differentiate the primitive* and determine
whether the original function is obtained.

EXERCISES

In Exercises 1 to 7 justify the Equations (i) to (vii) of Table 9.1.
In the following find the primitive sets, evaluate the integrals, or state
why the integral does not exist, as appropriate:

8. $\int 2\cos x.$

9. $\int (\csc^2 x + \sec^2 x).$

10. $\int_{-1}^{1}(x^2 + 1).$

11. $\int_{-1}^{1} x^{-2}.$

12. $\int 4\csc^2 x.$

13. $\int \tan^2 x$ (Hint: use a trigonometric identity).

14. $\int \dfrac{\sin x}{\cos^2 x}.$

15. $\int_{0}^{\pi} \dfrac{\cos x}{\sin^2 x}.$

16. $\int_{\pi/4}^{3\pi/4} \dfrac{\cos x}{\sin^2 x}.$

17. $\int x^{4/5}.$

18. $\int (x^{-1/2} + 2).$

19. $\int_{0}^{1} x(1 + x)^2.$

20. $\int (3x^2 - 7x^{-4}).$

21. $\int \cos(2x).$

22. $\int \sec^2(3x).$

23. $\int 2x(1 + x^2)^{3/2}.$

24. $\int 2\sin x \cos x$ (Hint: use a trigonometric identity).

25. $\int x^2 \cos (x^3)$.

Verify the following formulae by differentiation:

take set of the pum + take derivative

26. $\int x \sin x = [-x \cos x + \sin x]$.

27. $\int \cos^2 x = [\frac{1}{2}(x + \sin x \cos x)]$.

28. $\int \dfrac{x}{\sqrt{1 - x}} = [-\frac{2}{3}(2 + x) \sqrt{1 - x}]$.

29. $\int \dfrac{1}{\sqrt{x}\sqrt{1 + \sqrt{x}}} = [4\sqrt{1 + \sqrt{x}}]$.

30. $\int x \sin x^2 \cos (\cos x^2) = [-\frac{1}{2} \sin (\cos x^2)]$.

Show that the following are valid:

31. $\int f + \int g = \int g + \int f$.

32. $\left(\int f + \int g\right) + \int h = \int f + \left(\int g + \int h\right)$.

33. If $\int f + \int h = \int g + \int h$, then $\int f = \int g$.

ANSWERS

1. For $r \neq -1$, $\left(\dfrac{x^{r+1}}{r + 1}\right)' = (r + 1) \dfrac{x^r}{r + 1} = x^r$. Thus, by Definition 9.7,

$\int x^r = \left[\dfrac{x^{r+1}}{r + 1}\right]$ and by Theorem 9.6, $\displaystyle\int_a^b x^r = \dfrac{b^{r+1} - a^{r+1}}{r + 1}$, provided

that $[a, b]$ is an interval on which x^r is continuous. **9.** $[\tan x - \cot x]$.
11. Undefined, because x^{-2} is not defined at $x = 0$. **13.** $[\tan x - x]$.
15. Undefined. **17.** $[\frac{5}{9}x^{9/5}]$. **19.** $\frac{17}{12}$. **21.** $[\frac{1}{2} \sin 2x]$. **23.** $[\frac{2}{5}(1 + x^2)^{5/2}]$.
25. $[\frac{1}{3} \sin (x^3)]$. **33.** Assume $\int f + \int h = \int g + \int h$. Then by Definition
9.8ii, $\int (f + h) = \int (g + h)$. Show that $f + h = g + h$ and consequently
$f = g$, and thus $[f] = [g]$.

4. FINDING PRIMITIVES BY SUBSTITUTION

The first technique for finding primitives to be discussed here is that of substitution. Although this is a technique, it demands on the part of the reader an element of discernment, for he must learn to recognize combinations of functions that are applicable. The basic principle upon which substitution depends is the Chain Rule of Differentiation, as will be seen by the proof of Theorem 9.9 and the subsequent discussion.

THEOREM 9.9 Let g be differentiable and let f and F be such that

$$\int f = [F];$$

then

(9.7)
$$\int (f \circ g)g' = [F \circ g].$$

Alternately, setting $u = g(x)$ and $u' = g'(x)$ we have

(9.8)
$$\int f(u)u' = [F(u)] = [F(g(x))].$$

Proof. Let $h = F \circ g$. To show the validity of Equation 9.7 we need only show that $h' = (f \circ g)g'$. From the Chain Rule for Differentiation (Theorem 5.10, p. 230), we have $h' = (F \circ g)' = (F' \circ g)g'$. Since $\int f = [F]$, F is a primitive of f and we have $F' = f$ (see Definition 9.7, p. 388, and Definition 9.5, p. 387). Equation 9.7 follows at once. Equation 9.8 is nothing more than a restatement of Equation 9.7 with the variable x appearing. ■

Although Equation 9.7 is a more immediate application of the Chain Rule, Equation 9.8 lends itself more readily to the development of a method for finding primitive sets. The general procedure is outlined in Method 9.10.

METHOD 9.10 (Primitive Sets by Substitution.) The determination of primitive sets of $\int h(x)$ by the method of substitution involves the following five steps:

i. Choose a substitution function $u = g(x)$.

ii. Determine $u' = g'(x)$ by differentiation of $g(x)$.

iii. Substitute from steps (i) and (ii) to rewrite $h(x)$ as

$$h(x) = f(u)u',$$

using any legitimate algebraic technique available.

iv. If f has a primitive F, set up the expression

$$\int h(x) = \int f(u) \cdot u' = [F(u)].$$

v. Substitute from step (i) to conclude that

$$\int h(x) = [F(g(x))].$$

Of course, if any of steps (i) to (iv) are not possible, the method fails. Consider the following applications of Method 9.10:

Example 11. We wish to find the following primitive sets.

(a) $$\int (x+1)^{10}.$$

　i. Choose $u = x + 1$, since $(x+1)^{10}$ is composite.
　ii. $u' = 1$.
　iii. $h(x) = (x+1)^{10} = u^{10} \cdot u'$, $f(u) = u^{10}$.
　iv. $\int (x+1)^{10} = \int u^{10} \cdot u'$

$$= \left[\frac{u^{11}}{11} \right] \text{ (by Table 9.1)}.$$

　v. $\int (x+1)^{10} = \left[\dfrac{(x+1)^{11}}{11} \right]$ (by Theorem 9.9).

(b) $$\int 3x^2 \cos x^3.$$

　i. Choose $u = x^3$, since $\cos x^3$ is composite.
　ii. $u' = 3x^2$.
　iii. $h(x) = 3x^2 \cos x^3 = (\cos u) \cdot u'$, $f(u) = \cos u$.
　iv. $\int 3x^2 \cos x^3 = \int (\cos u) u'$
　　　　$= [\sin u]$ (by Table 9.1).
　v. $\int 3x^2 \cos x^3 = [\sin x^3]$ (by Theorem 9.9). ■

The next examples will illustrate the need for the algebraic manipulations called for in part (iii) of Method 9.10.

Example 12. We wish to find the primitive set $\int x(1-x^2)^{5/2}$.
　i. Choose $u = 1 - x^2$, since $(1-x^2)^{5/2}$ is composite.
　ii. $u' = -2x$ (and $x = -\tfrac{1}{2}u'$).
　iii. $h(x) = x(1-x^2)^{5/2}$

$$= xu^{5/2}$$

$$= -\tfrac{1}{2}u^{5/2} \cdot u' \text{ (by part (ii))};$$

$f(u) = -\tfrac{1}{2}u^{5/2}$.
　iv. $\int x(1-x^2)^{5/2} = \int -\tfrac{1}{2}u^{5/2} \cdot u'$
　　　　　　$= -\tfrac{1}{2} \int u^{5/2} \cdot u'$ (by Definition 9.8i)

$$= -\frac{1}{2}\left[\frac{u^{7/2}}{7/2} \right] \text{(by Table 9.1)}$$

　　　　　　$= [-\tfrac{1}{7}u^{7/2}]$ (by Equation 9.6i).
　v. $\int x(1-x^2)^{5/2} = [-\tfrac{1}{7}(1-x^2)^{7/2}]$ (by Theorem 9.9). ■

Example 13. Consider $\int \cos x \cos(\sin x)$.
　i. Choose $u = \sin x$, since $\cos(\sin x)$ is composite.

ii. $u' = \cos x$.

iii. $h(x) = \cos x \cos (\sin x)$

$\qquad = (\cos u) \cdot u'$ (by parts (i) and (ii)).

iv. $\int \cos x \cos (\sin x) = \int \cos u \cdot u'$

$\qquad\qquad\qquad\qquad = [\sin u]$ (by Table 9.1).

v. $\int \cos x \cos (\sin x) = [\sin (\cos x)]$. ∎

Example 14. Consider $\int \dfrac{1+x}{\sqrt{1-x}}$.

$$x = 1 - u$$
$$1 + x = 2 - u$$

i. Choose $u = 1 - x$ and note that $1 + x = 2 - u$.

ii. $u' = -1$.

iii. $h(x) = \dfrac{1+x}{\sqrt{1-x}} = \dfrac{2-u}{u^{1/2}} (-u') = (u^{1/2} - 2u^{-1/2})u'$.

iv. $\displaystyle\int \dfrac{1+x}{\sqrt{1-x}} = \int (u^{1/2} - 2u^{-1/2})u'$

$\qquad\qquad\qquad = [\tfrac{2}{3}u^{3/2} - 4u^{1/2}]$.

v. $\displaystyle\int \dfrac{1+x}{\sqrt{1-x}} = [\tfrac{2}{3}(1-x)^{3/2} - 4(1-x)^{1/2}]$. ∎

The reader will have noticed that in both Example 12 and Example 14 we used Method 9.10 with certain algebraic manipulations to reduce $\int h(x)$ to the form $\int f(u)u'$. In each case we could find the primitive set of $f(u)u'$ by Theorem 9.9 and Table 9.1. Particular note should be taken of the algebraic manipulations involved in our solution of Example 14. Several of the exercises will call for the same type manipulation.

At this point the reader may well wonder why we chose the particular substitutions $u = 1 - x^2$ in Example 12 and $u = 1 - x$ in Example 14. We did so because of the composite nature of $(1 - x^2)^{5/2}$ and $(1 - x)^{-1/2}$. Each of these functions was simplified by the substitution. This is, of course, the sort of substitution for which the reader should search—namely, a substitution which will simplify the function in question.

We now turn to the problem of using the Method of Substitution in the integral. Theorem 9.11 states the theory involved and Method 9.12 describes the procedure.

THEOREM 9.11 (Substitution in the Integral.) Let g be continuous on $[a, b]$ and differentiable on (a, b) with $g([a, b]) \subset [c, d]$. Let $\int f = [F]$ on $[c, d]$. Then

(9.9) $$\int_a^b (f \circ g)g' = \int_{g(a)}^{g(b)} f = [F]_{g(a)}^{g(b)}.$$

Proof. By Theorem 9.9, $\int (f \circ g)g' = [F \circ g]$. Therefore, by Theorem 9.6 (Second Fundamental Theorem) we have

$$\int_a^b (f \circ g)g' = [F \circ g]_b^a = F(g(b)) - F(g(a)) = [F]_{g(a)}^{g(b)}.$$

A direct application of Theorem 9.6 gives

$$\int_{g(a)}^{g(b)} f = [F]_{g(a)}^{g(b)}.$$

Thus, each of the equalities of Equation 9.9 is valid. ∎

The following procedure simplifies the application of Theorem 9.11:

METHOD 9.12 (Integration by Substitution.) To evaluate $\int_a^b h(x)$ by the method of substitution, we carry out the following operations:

i. Choose a substitution function $u = g(x)$ and rewrite the function $h(x)$ in terms of u and u' as in Method 9.10:

$$h(x) = f(u) \cdot u'.$$

ii. Compute $u_a = g(a)$ and $u_b = g(b)$.

iii. Evaluate $\int_a^b h(x) = \int_{x=a}^{x=b} f(u)u' = \int_{u_a}^{u_b} f(u) = [F(u)]_{u_a}^{u_b} = F(u_b) - F(u_a)$.

Note: The appearance of u' in the second member of step (iii) serves to indicate that u is there regarded as a function $u = g(x)$ and that the limits used are limits on x. The absence of the u' in the third member indicates the application of Theorem 9.11 in which the limits used are u limits and u is considered to be the variable.

We illustrate the preceding method by the following examples:

Example 15. We wish to compute $\int_1^2 (1 + 2x)^5$.

i. Let $u = g(x) = 1 + 2x$. Then $u' = 2$ or $\frac{1}{2}u' = 1$. Thus, $(1 + 2x)^5 = u^5(\frac{1}{2}u')$.

ii. $u_1 = g(1) = 1 + 2(1) = 3$ and $u_2 = g(2) = 1 + 2(2) = 5$.

iii. $\int_1^2 (1 + 2x)^5 = \int_{x=1}^{x=2} \frac{1}{2}u^5 u'$

$$= \int_3^5 \frac{1}{2}u^5$$

$$= \frac{1}{2}\left[\frac{u^6}{6}\right]_3^5$$

$$= \frac{1}{2}\frac{5^6 - 3^6}{6}$$

$$= \frac{3724}{3}. \blacksquare$$

Example 16. We now compute $\int_0^{\pi/6} (\cos 3x)(\sin 3x)^4$.

$$\int_0^{\pi/6} (\cos 3x)(\sin 3x)^4 =^1 \int_{x=0}^{x=\pi/6} u^4 (\tfrac{1}{3}u')$$

$$=^2 \frac{1}{3} \int_0^1 u^4$$

$$=^3 \frac{1}{3} \left[\frac{u^5}{5} \right]_0^1$$

$$=^4 \frac{1}{3} \frac{1^5 - 0^5}{5}$$

$$= \tfrac{1}{15}.$$

(1) $u = g(x) = \sin 3x$
    $u' = 3\cos 3x$
    $\cos 3x = \tfrac{1}{3}u'$
(2) $g(0) = 0;\ g(\pi/6) = 1.$
(3) Table 9.1.
(4) $[F]_a^b = F(b) - F(a).$ ∎

Our final example of the Method of Substitution will be more complex than those previously encountered.

Example 17. Consider $\int x^3 \sec^2 (x^4) \tan^{2/3} (x^4)$. We first use the substitution $u = x^4$ and then $v = \tan u$.

$$\int x^3 \sec^2 (x^4) \tan^{2/3}(x^4) =^1 \int \sec^2 (u) \tan^{2/3} (u)(\tfrac{1}{4}u')$$

$$=\frac{1}{4} \int \sec^2 (u) \tan^{2/3} (u)u'$$

$$=^2 \frac{1}{4} \int v^{2/3}v'$$

$$=^3 \frac{1}{4} \left[\frac{v^{5/3}}{5/3} \right]$$

$$= \left[\tfrac{3}{20} v^{5/3} \right]$$

$$=^4 \left[\tfrac{3}{20} (\tan u)^{5/3} \right]$$

$$=^5 \left[\tfrac{3}{20} (\tan x^4)^{5/3} \right]$$

(1) $u = x^4$
    $u' = 4x^3$
    $x^3 = \tfrac{1}{4}u'.$
(2) $v = \tan u$
    $v' = \sec^2 u\ u'.$
(3) Table 2.1.
(4) $v = \tan u.$
(5) $u = x^4.$

We have in this solution used two substitutions to arrive at our answer. The single substitution $u = \tan (x^4)$ would have resulted more rapidly in the same answer. ∎

EXERCISES

In Exercises 1 to 20 use the Method of Substitution to find the specified primitive sets:

1. $\int \frac{3}{2}\sqrt{1+x}$. (Let $u = 1 + x$.) 2. $\int (1 - 2x)^{3/2}$. (Let $u = 1 - 2x$.)

3. $\int \frac{1}{\sqrt{1+4x}}$. (Let $u = 1 + 4x$.) 4. $\int x\sqrt{1-x}$. (Let $u = 1 - x$.)

5. $\int \frac{x^2}{\sqrt{1-x}}$. 6. $\int x \cos (x^2)$.

7. $\int x^3 \sin (x^4)$. 8. $\int \sin x \cos x$.

9. $\int x \sin (x^2) \cos (x^2)$. 10. $\int \frac{\sec^2 (x^{1/2})}{x^{1/2}}$.

11. $\int (2x^2 - 2)(4x^3 - 12x - 3)^5$. 12. $\int (7x + 1)^{10}$.

13. $\int \frac{\sin \sqrt{1-x}}{\sqrt{1-x}}$. 14. $\int x \sin (3 + x^2) \cos^{-2} (3 + x^2)$.

15. $\int \sin x \sin (\cos x)$. 16. $\int \frac{\sec \sqrt{x} \tan \sqrt{x}}{\sqrt{x}}$.

17. $\int \sqrt{\tan x} \sec^2 x$. 18. $\int x^2 \tan^2 (x^3)$.

19. $\int (3x - 1)^{1/2} \sin ((3x - 1)^{3/2})$. 20. $\int \frac{x \sin^2 \sqrt{x^2-1} \cos \sqrt{x^2-1}}{\sqrt{x^2-1}}$.

In Exercises 21 to 30 use the Method of Integration by Substitution to evaluate the specified definite integrals:

21. $\int_1^2 (1 - x)^7$. 22. $\int_0^{\sqrt{\pi}} 2x \sin (x^2)$.

23. $\int_0^{1/3} \frac{x}{\sqrt{2-3x}}$. 24. $\int_0^{\pi/2} \cos x \sqrt{4 - \sin x}$.

25. $\int_0^3 \frac{\cos \sqrt{x+1}}{\sqrt{x+1}}$. 26. $\int_\pi^{2\pi} (\cos^4 x)(\sin x)$.

27. $\int_0^{(\pi/4)^2} \frac{\sec^2 \sqrt{x} \tan \sqrt{x}}{\sqrt{x}}$. 28. $\int_{-1}^1 \frac{x}{(4 + x^2)^3}$.

29. $\int_3^8 \frac{1-x}{\sqrt{1+x}}$. 30. $\int_{-\pi}^\pi \sin x \cos x(1 + \sin^2 x)$.

31. Show that for each real number r, $\int_a^b f(x) = \int_{a+r}^{b+r} f(t - r)$.

32. Show that for each real number $r \neq 0$, $\int_a^b f(x) = \frac{1}{r} \int_{ra}^{rb} f(rx)$.

ANSWERS

1. $[(1 + x)^{3/2}]$. **3.** $[\frac{1}{2}(1 + 4x)^{1/2}]$.

5. $[-\frac{2}{15}\sqrt{1 - x}(8 + 4x + 3x^2)](u = 1 - x)$. **7.** $\left[-\dfrac{\cos(x^4)}{4}\right](u = x^4)$.

9. $\left[\dfrac{\sin^2(x^2)}{4}\right](u = x^2 \text{ and } v = \sin u; \text{ or } u = \sin(x^2)$.

11. $\left[\dfrac{(4x^3 - 12x - 3)^6}{36}\right]$. **13.** $[2\cos\sqrt{1 - x}]$. **15.** $[\cos(\cos x)]$.

17. $\left[\dfrac{2(\tan x)^{3/2}}{3}\right](u = \tan x)$. **19.** $\left[\dfrac{-2\cos(3x - 1)^{3/2}}{9}\right]$.

21. $-1/8$ $(u = g(x) = 1 - x)$. **23.** $\dfrac{8\sqrt{2} - 10}{27}$ $(u = g(x) = 2 - 3x)$.

25. $2(\sin 2 - \sin 1)(u = g(x) = \sqrt{x + 1})$. **27.** 1. **29.** $-26/3$.

5. FINDING PRIMITIVES BY PARTS

In Section 4 we found that the Chain Rule of Differentiation led to the method of finding primitives by substitution. The product rule of differentiation gives rise to a useful technique called the *Method of Parts*. If f and g are differentiable, we know that $(fg)' = f'g + gf'$. This leads us to a corresponding formula for primitive sets, which is given in the following theorem.

THEOREM 9.13 If f and g are differentiable, then

(9.10) $$\int fg' = [fg] - \int f'g,$$

(where $[fg] - \int f'g$ means $[fg] + \int -(f'g)$).

Proof. By the product rule for differentiation, $(fg)' = f'g + fg'$. Therefore, $fg' = (fg)' - f'g$, and by our definition of primitive sets and addition of primitive sets, $\int fg' = \int ((fg)' - f'g) = \int (fg)' + \int -(f'g) = [fg] + \int -(f'g)$. If we interpret $[fg] - \int f'g$ to be $[fg] + \int -(f'g)$, we have Equation 9.10. ∎

As in the case with the Method of Substitution, we wish to use the Method of Parts to simplify the problem of finding a primitive set. Thus, if we wish to find the primitive set of a function h by parts, we attempt to express h as a product: $h = fg'$, such that g' is a function for which we know the primitive set and such that $\int f'g$ is (hopefully) more readily determined than $\int fg'$. We shall illustrate this technique in the following examples:

Example 18.

(a) Consider $\int x \sin x$. We note that the function $x \sin x$ is the product of two functions, x and $\sin x$. Since we wish to simplify this function, we choose

$$f(x) = x \quad \text{and} \quad g'(x) = \sin x.$$

Choosing $g(x) = -\cos x$, we then have

$$f'(x) = 1 \quad \text{and} \quad g(x) = -\cos x.$$

Thus, by Equation 9.10,

$$\int x \sin x = [-x \cos x] - \int -\cos x,$$

and since $-\int -\cos x = -[-\sin x] = [\sin x]$, we finally have

(9.11)
$$\int x \sin x = [-x \cos x] + [\sin x]$$
$$= [-x \cos x + \sin x].$$

(b) We now use the Method of Parts to determine $\int x^2 \cos x$. We let

$$f(x) = x^2 \quad \text{and} \quad g'(x) = \cos x.$$

Choosing $g(x) = \sin x$, we have

$$f'(x) = 2x \quad \text{and} \quad g(x) = \sin x.$$

Thus, by Equation 9.10,

(9.12)
$$\int x^2 \cos x = [x^2 \sin x] - \int 2x \sin x.$$

In Equation 9.11 we determined $\int x \sin x$.

Substituting Equation 9.11 in Equation 9.12 we have

$$\int x^2 \cos x = [x^2 \sin x] - 2[-x \cos x + \sin x]$$
$$= [x^2 \sin x + 2x \cos x - 2 \sin x].$$

The reader should take note that to determine the primitive set $\int x^2 \cos x$ we used the Methods of Parts twice. ■

Example 19. We use the Method of Parts to find the primitive set of $h(x) = \dfrac{1 + x}{\sqrt{1 - x}}$ (see Example 14, p. 395).

Let

$$f(x) = 1 + x \quad \text{and} \quad g'(x) = (1 - x)^{-1/2}.$$

Choosing $g(x) = -2(1 - x)^{1/2}$, we have

$$f'(x) = 1, \quad \text{and} \quad g(x) = -2(1 - x)^{1/2},$$

and thus,

$$\int \frac{1+x}{\sqrt{1-x}} = [-2(1+x)(1-x)^{1/2}] - \int -2(1-x)^{1/2}$$

$$= [-2(1+x)(1-x)^{1/2} - \tfrac{4}{3}(1-x)^{3/2}].$$

This answer looks quite different from the one given by Example 14. The reader should check that the two answers differ by at most a constant function C and that each is a primitive of $h(x) = \dfrac{1+x}{\sqrt{1-x}}$. ∎

The preceding example and Example 14 demonstrate that some primitive sets can be found by either the Method of Substitution or the Method of Parts. They further demonstrate that two primitives belonging to the same primitive set can differ radically in algebraic form.

The next example illustrates a technique, combining the Method of Parts and algebra. Several of the exercises will call for this same technique.

Example 20. Consider $\int \sin 2x \cos 3x$. We obviously have a choice for the division by parts. We choose

$$f(x) = \sin 2x \quad \text{and} \quad g'(x) = \cos 3x,$$
$$f'(x) = 2 \cos 2x \quad \text{and} \quad g(x) = \tfrac{1}{3} \sin 3x.$$

Thus,

(9.13) $$\int \sin 2x \cos 3x = [\tfrac{1}{3} \sin 2x \sin 3x] - \tfrac{2}{3} \int \cos 2x \sin 3x.$$

This new primitive set is similar to the original. Again we use the Method of Parts. Note that the chosen order which follows does not reverse our original efforts, whereas the opposite choice will.

Let

$$f(x) = \cos 2x \quad \text{and} \quad g'(x) = \sin 3x,$$
$$f'(x) = -2 \sin 2x \quad \text{and} \quad g(x) = -\tfrac{1}{3} \cos 3x.$$

(9.14) $$\int \cos 2x \sin 3x = [-\tfrac{1}{3} \cos 2x \cos 3x] - \tfrac{2}{3} \int \sin 2x \cos 3x.$$

Substituting Equation 9.14 in Equation 9.13, we have

(9.15) $$\int \sin 2x \cos 3x$$

$$= [\tfrac{1}{3} \sin 2x \sin 3x] - \tfrac{2}{3}\left([-\tfrac{1}{3} \cos 2x \cos 3x] - \tfrac{2}{3} \int \sin 2x \cos 3x\right)$$

$$= \tfrac{1}{9}[3 \sin 2x \sin 3x + 2 \cos 2x \cos 3x] + \tfrac{4}{9} \int \sin 2x \cos 3x.$$

This last primitive set is the same as the original. Subtracting $\tfrac{4}{9} \int \sin 2x \cos 3x$ from both members of Equation 9.15, we have

$$\tfrac{5}{9} \int \sin 2x \cos 3x = \tfrac{1}{9}[3 \sin 2x \sin 3x + 2 \cos 2x \cos 3x].$$

Multiplying by $\frac{9}{5}$, we finally arrive at

$$\int \sin 2x \cos 3x = \tfrac{1}{5}[3 \sin 2x \sin 3x + 2 \cos 2x \cos 3x]$$

$$= [\tfrac{1}{5}(3 \sin 2x \sin 3x + 2 \cos 2x \cos 3x)]. \quad \blacksquare$$

Nothing is gained by stating a special theorem for using the Method of Parts to evaluate integrals. In order to evaluate $\int_a^b f'g$ we need only find a primitive F of $f'g$ and evaluate $[F]_a^b$. We illustrate this technique in the next example.

Example 21.   We shall evaluate $\int_{-\pi/2}^{\pi/2} \cos^2$ by parts.   First we shall determine $\int \cos^2$.   Let $f(x) = \cos x$ and $g'(x) = \cos x$.   Then $f'(x) = -\sin x$ and we choose $g(x) = \sin x$.   Thus,

$$\int \cos^2 x =^1 [\cos x \sin x] - \int -\sin^2 x$$

$$=^2 [\cos x \sin x] + \int \sin^2 x$$

$$=^3 [\cos x \sin x] + \int (1 - \cos^2 x)$$

$$=^4 [\cos x \sin x] + \int 1 - \int \cos^2 x$$

$$=^5 [\cos x \sin x] + [x] - \int \cos^2 x.$$

Adding $\int \cos^2 x$ to extreme members of this equation, we have

$$2\int \cos^2 x = [\cos x \sin x + x]$$

and

$$\int \cos^2 x = \tfrac{1}{2}[\cos x \sin x + x].$$

Therefore,

$$\int_{-\pi/2}^{\pi/2} \cos^2 x = \tfrac{1}{2}[\cos x \sin x + x]_{-\pi/2}^{\pi/2}$$

$$= \tfrac{1}{2}(\cos \pi/2 \sin \pi/2 + \pi/2 - \cos(-\pi/2) \sin(-\pi/2))$$

$$- (-\pi/2))$$

$$= \frac{\pi}{2}.$$

(1) Theorem 9.13.
(2) Definition 9.8, p. 390.
(3) Trigonometric substitution: $\sin^2 x = 1 - \cos^2 x$.
(4) Definition 9.8.
(5) Table 9.1i $(r = 0)$, p. 389. $\blacksquare$

EXERCISES

In Exercises 1 to 15 use the Method of Parts to find the indicated primitive sets:

1. $\int x(1 + x)^{10}$.

2. $\int (1 + 2x)\sqrt{1 + x}$.

3. $\int x \cos x$.

4. $\int x \cos 3x$.

5. $\int x \sin 2x$.

6. $\int x^2 \cos x$ (use Exercise 3).

7. $\int x^3 \cos x$ (use Exercise 6).

8. $\int \sin x \cos 4x$ (study Example 20).

9. $\int \cos 2x \cos 3x$.

10. $\int \sin x \cos 4x$.

11. $\int \sin 5x \cos 5x$.

12. $\int x \sin x \cos x$.

13. $\int \sin^2 x$ (study Example 21).

14. $\int \sin^3 x$ (use Exercise 13).

15. $\int \sin^4 x$.

In Exercises 16 to 20 use the Method of Parts to evaluate the given integral:

16. $\int_{-\pi}^{\pi} x \cos x$.

17. $\int_0^1 x^2(1 - x)^3$.

18. $\int_0^{\pi} \sin^2 x$.

19. $\int_0^3 (1 - x) \sqrt{1 + x}$.

20. $\int_{-1}^{1} (1 - x^2) \sin x$.

In Exercises 21 to 30 use either the Method of Parts and/or the Method of Substitution to find the indicated primitive sets:

21. $\int x^2(1 + 2x)^5$.

22. $\int x \sin (1 - x)$.

23. $\int \sin (2x) \cos (2x)$.

24. $\int x^3 \sin (x^2)$.

25. $\int x^2 \sin (x^3)$.

26. $\int x^4(1 + x^5)^5$.

27. $\int \tan^2 x$.

28. $\int \cot^2 x$.

29. $\int \sin^2 x \cos^2 x$.

30. $x \dfrac{\sin \sqrt{1 + x} \cos \sqrt{1 + x}}{\sqrt{1 + x}}$.

ANSWERS

1. $\left[\dfrac{x(1 + x)^{11}}{11} - \dfrac{(1 + x)^{12}}{132}\right]$.   3. $[x \sin x + \cos x]$.

5. $[-\frac{1}{2}x \cos 2x + \frac{1}{4} \sin 2x]$.   7. $[x^3 \sin x + 3x^2 \cos x - 6x \sin x - 6 \cos x]$.

9. $[\frac{1}{5}(3 \cos 2x \sin 3x - 2 \sin 2x \cos 3x)]$.   11. $[\frac{1}{10}(\sin^2 5x - \cos^2 5x)]$.

13. $[\frac{1}{2}(x - \sin x \cos x)]$.   15. $[-\frac{1}{4} \sin^3 x \cos x + \frac{3}{8}x - \frac{3}{8} \sin x \cos x]$.

17. $\frac{1}{60}$.   19. $-\frac{46}{15}$.   21. $[\frac{1}{12}(x^2(1 + 2x)^6 - \frac{1}{7}x(1 + 2x)^7 + \frac{1}{112}(1 + 2x)^8)]$.

23. $[-\frac{1}{8} \cos 4x]$ or $[\frac{1}{4} \sin^2 2x]$.   25. $[-\frac{1}{3} \cos (x^3)]$.   27. $[\tan x - x]$.

29. $[\frac{1}{8}(x - \sin x \cos^3 x + \sin^3 x \cos x)]$.

10

SPECIAL FUNCTIONS

INTRODUCTION

In this chapter we shall introduce certain special functions which are of great importance in mathematics. These functions are the logarithmic and exponential functions, the inverse trigonometric functions, and the hyperbolic functions.

The first section concerns itself with the inverse function theorem for $\mathcal{R} \to \mathcal{R}$ functions. This theorem assures the existence of differentiable inverse functions for a broad class of functions and will be used in discussing the exponential and inverse trigonometric functions.

1. THE INVERSE FUNCTION THEOREM

If we are given a function f from a set $\mathcal{A}$ into a set $\mathcal{B}$ and given an element x of $\mathcal{A}$, the element $f(x)$ of $\mathcal{B}$ is uniquely determined. On the other hand, if we are given an element y in $\mathcal{B}$, it is not always possible to determine a unique x such that $f(x) = y$. There are two conditions which might prevent such a determination: either (1) there is *no* element x of $\mathcal{A}$ such that $f(x) = y$, or (2) there is *more than one* element x of $\mathcal{A}$ such that $f(x) = y$.

It is convenient to introduce special terms to describe those functions for which one or the other of conditions (1) and (2) does not apply.

DEFINITION 10.1 Let f be a function with $dom\, f = \mathcal{A}$ and $ran\, f \subset \mathcal{B}$.

   i. If for each y in $\mathcal{B}$ there is an x such that $f(x) = y$, we say that f is a function from $\mathcal{A}$ *onto* $\mathcal{B}$ and that f is *onto*.

404

ii. If for each y in $\mathcal{B}$, there is no more than one x such that $f(x) = y$, we say f is a *one-to-one function* from $\mathcal{A}$ *into* $\mathcal{B}$.

iii. If for each y in $\mathcal{B}$ there is exactly one x such that $f(x) = y$, we say f is *one-to-one* from $\mathcal{A}$ *onto* $\mathcal{B}$, and that f is *one-to-one and onto*.

Example 1.   (*a*) Let $f(x) = 2x$ for every x in $\mathcal{R}$, then f is one-to-one from $\mathcal{R}$ onto $\mathcal{R}$, since for each y in $\mathcal{R}$, $w = \dfrac{y}{2}$ is a unique element of $\mathcal{R}$ such that $f(w) = y$.

(**b**) Let $f(x) = x^2$ for every x in $\mathcal{R}$; then f is neither one-to-one nor onto when considered as a function from $\mathcal{R}$ to $\mathcal{R}$. It is not one-to-one since $f(1) = 1$ and $f(-1) = 1$. It is not onto since there is no real number x such that $f(x) = x^2 = -1$. However, if we consider f to be a function from $\mathcal{R}$ to the non-negative reals, it is an *onto* function; and if we consider f to be a function from the non-negative reals to the non-negative reals, it is one-to-one and onto.   ■

Let $f : \mathcal{A} \to \mathcal{B}$ which is one-to-one and onto. For each y in $\mathcal{B}$ we can determine a unique x such that $f(x) = y$. In such a case, we shall write $x = f^{-1}(y)$ and say that x is equal to the *f*-inverse of y. More precisely we introduce the following terminology:

DEFINITION 10.2   Let f be a one-to-one function from $\mathcal{A}$ onto $\mathcal{B}$. The *inverse function* of f is denoted by f^{-1} and defined by

$$f^{-1} = \{(y, x) \mid (x, y) \in f\}.$$

Note: $f(f^{-1})(y)) = y$ and $f^{-1}(f(x)) = x$ for each y in $\mathcal{B}$ and each x in $\mathcal{A}$.

Example 2.   Let $f(x) = 2x + 1$. Then f is one-to-one from $\mathcal{R}$ onto $\mathcal{R}$. We can write

$$f = \{(x, y) \mid y = 2x + 1\}.$$

Hence,

$$f^{-1} = \{(y, x) \mid y = 2x + 1\}$$

$$= \left\{(y, x) \mid x = \frac{y - 1}{2}\right\}$$

$$= \left\{(x, y) \mid y = \frac{x - 1}{2}\right\},$$

where in the last step we have interchanged the coordinates y and x to obtain the more familiar arrangement (x, y) in the ordered pair description of f^{-1}. We sketch *graph f* and *graph* f^{-1} along with the line $y = x$ in Figure 10.1.   ■

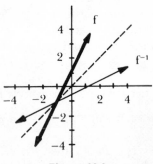

Figure 10.1

Example 3.   Let $f(x) = x^2$. If *dom f* is taken to be all of $\mathcal{R}$, *f* is neither one-to-one nor onto, and therefore does not have an inverse. However, if we take the domain of *f* to be the interval $[0, \infty)$, *f* is one-to-one from $[0, \infty)$ onto $[0, \infty)$. We can write $f = \{(x, y) \mid x \geq 0 \text{ and } y = x^2\}$. Hence,

$$f^{-1} = \{(y, x) \mid x \geq 0 \text{ and } y = x^2\}$$
$$= \{(y, x) \mid x = \sqrt{y}, y \geq 0\}$$
$$= \{(x, y) \mid y = \sqrt{x}, x \geq 0\},$$

where in the last step we have reversed *y* and *x* as we did in Example 2. We sketch *graph f* and *graph f^{-1}* in Figure 10.2. ■

Observe from Figures 10.1 and 10.2 that the sketch of *graph f^{-1}* is the mirror image of the sketch of *graph f* through the line $y = x$. This is true of every function from $\mathcal{R}$ into $\mathcal{R}$ which has an inverse.

We note from Definition 10.1 that a function *f*, defined on some subset of $\mathcal{R}$, has an inverse on *ran f* if and only if *f* is one-to-one. We are particularly interested in the existence and continuity of inverses of continuous functions. The following theorem will be useful in this respect:

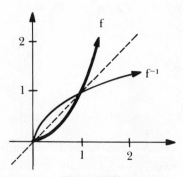

Figure 10.2

THEOREM 10.3 Let $f:\mathcal{R} \to \mathcal{R}$ be continuous on (a, b), where either a or b may be infinite. Define

$$c = \begin{cases} \text{glb } \{f(x) \mid x \in (a, b)\}, \text{ if this set is bounded below,} \\ -\infty, \text{ if } \{f(x) \mid x \in (a, b)\} \text{ is not bounded below.} \end{cases}$$

$$d = \begin{cases} \text{lub } \{f(x) \mid x \in (a, b)\}, \text{ if this set is bounded above,} \\ \infty, \text{ if } \{f(x) \mid x \in (a, b)\} \text{ is not bounded above.} \end{cases}$$

i. If f is strictly increasing (Definition 6.1, p. 249), then f^{-1} is defined, continuous, and strictly increasing on (c, d).

ii. If f is strictly decreasing (Definition 6.1, p. 349), then f^{-1} is defined, continuous, and strictly decreasing on (c, d).

Proof. We prove only case (i). Since f is strictly increasing, it is one-to-one. To show that f is onto (c, d), let $y \in (c, d)$. By definition of c and d we can choose c', d' in $ran f$ such that $c' < y < d'$. Hence, by the Intermediate Value Theorem 4.6 (p. 199), there is an x in (a, b) such that $f(x) = y$. Thus, f is one-to-one and onto, so f^{-1} is defined.

We now show that f^{-1} is strictly increasing. Let $a' < b'$ and assume, contrary to expectation, that $f^{-1}(a') \geq f^{-1}(b')$. Since f is strictly increasing, we have $f(f^{-1}(a')) \geq f(f^{-1}(b'))$. But by the note after Definition 10.2, this implies $a' \geq b'$, a contradiction. Therefore, $f^{-1}(a') < f^{-1}(b')$, and f^{-1} is strictly increasing.

To show that f^{-1} is continuous, let $x \in (a, b)$, $r > 0$. We may assume r is chosen so that $x - r$ and $x + r$ are elements of (a, b) (see Figure 10.3). Let $r' = \min \{f(x) - f(x - r), f(x + r) - f(x)\}$. Then it is easy to see that $f^{-1}(y) \in \mathcal{B}(x, r)$ for each y in $\mathcal{B}(f(x), r')$. ∎

Example 4. Let $f(x) = x^2$. Then f is strictly decreasing on $(-\infty, 0]$ and strictly increasing on $[0, \infty)$. We have seen in Example 3 that if we

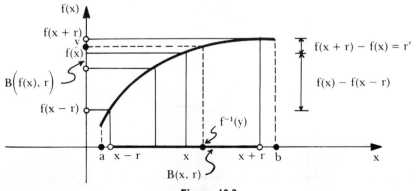

Figure 10.3

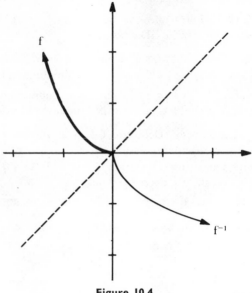

Figure 10.4

consider f to be defined on $[0, \infty)$, $f^{-1}(x) = \sqrt{x}$ is the inverse function of f. If, on the other hand, we consider f to be defined on $(-\infty, 0]$, $f^{-1}(x) = -\sqrt{x}$. We sketch f and its inverse with this last restriction in Figure 10.4. The reader should compare this figure with Figure 10.2. ∎

It is often important to know whether the inverse of a function is differentiable, and if so, to know the derivative of the inverse function. With this in mind, look at Figure 10.5. It seems, in the case of the function f which we have sketched, that *graph* $A_{f(x_0)}f^{-1}$ is the mirror image of *graph* $A_{x_0}f$ through the line $y = x$. If this is indeed the case and if $A_{x_0}f = mx + b$,

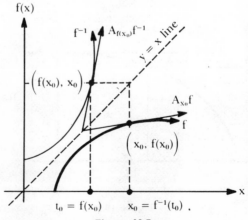

Figure 10.5

then, since $A_{x_0}f = \{(x, y) \mid y = mx + b\}$, we have $A_{f(x_0)}f^{-1} = \left\{(x, y) \mid y = \frac{1}{m}(x - b)\right\}$ (check this!). Thus, the slope of $A_{f(x_0)}f^{-1}$ is $1/m$. This suggests that if we wish to find the derivative of f^{-1} at a point t_0, we first find a point x_0 such that $t_0 = f(x_0)$, then find $f'(x_0)$ and take the reciprocal of this number. Thus, if our suppositions are correct, we have

$$(f^{-1})'(t_0) = \frac{1}{f'(f^{-1}(t_0))} = \frac{1}{f' \circ f^{-1}(t_0)}.$$

We state this in the following theorem. We consider only the case where f is strictly increasing. The case where f is strictly decreasing is similar.

THEOREM 10.4 (Inverse Function Theorem.) Let $f: \mathcal{R} \to \mathcal{R}$ be continuous, differentiable with non-zero derivative, and strictly increasing on (a, b), where a or b may be infinite. Then f^{-1} is continuous, differentiable, and strictly increasing on $ran\, f$. Furthermore,

$$(f^{-1})' = \frac{1}{f' \circ f^{-1}}.$$

Proof. The fact that f^{-1} is continuous and strictly increasing was proved in Theorem 10.3. To prove that f^{-1} is differentiable at the point y_0 in $f(a, b)$, where $f(x_0) = y_0$, we shall compute the derivative assuming that $f'(x_0) \neq 0$.

$$(f^{-1})'(y_0) =^1 \lim_{y \to y_0} \frac{f^{-1}(y) - f^{-1}(y_0)}{y - y_0}$$

$$=^2 \lim_{x \to x_0} \frac{f^{-1}(f(x)) - f^{-1}(f(x_0))}{f(x) - f(x_0)}$$

$$=^3 \lim_{x \to x_0} \frac{x - x_0}{f(x) - f(x_0)}$$

$$=^4 \lim_{x \to x_0} \frac{1}{\left(\dfrac{f(x) - f(x_0)}{x - x_0}\right)}$$

$$=^5 \frac{1}{\lim_{x \to x_0} \dfrac{f(x) - f(x_0)}{x - x_0}}$$

$$=^1 \frac{1}{f'(x_0)}$$

$$=^6 \frac{1}{f'(f^{-1}(y_0))}$$

$$=^7 \frac{1}{f' \circ f^{-1}(y_0)}.$$

(1) Definition 5.4 (p. 223) (Derivative).
(2) Since f has an inverse, we may set $y = f(x)$ for some x and also replace y_0 by $f(x_0)$. Moreover, since f is strictly increasing and continuous at y_0, we know that $y = f(x)$ approaches $y_0 = f(x_0)$ as x approaches x_0.
(3) By the note after Definition 10.2, $f^{-1}(f(x)) = x$ and $f^{-1}(f(x_0)) = x_0$.
(4) Elementary properties of real numbers.
(5) Theorem 3.12iv (p. 182) (Limit of a Quotient).
(6) Since $x_0 = f^{-1}(y_0)$.
(7) Definition 3.3 (p. 154) (Composition of Functions). ■

Example 5. Let $f(x) = x^2$. We then have $f'(t) = 2t$, $f^{-1}(x) = \sqrt{x}$ on $[0, \infty)$, and consequently,

$$(f^{-1})'(x) = \frac{1}{f'(f^{-1}(x))} = \frac{1}{f'(\sqrt{x})} = \frac{1}{2\sqrt{x}}. \quad ■$$

EXERCISES

For each of the following, determine an interval I of definition such that the prescribed function has an inverse which is also a function. Sketch the function and its inverse on the same set of axes:

1. $f(x) = x^2 - 2$.

2. $f(x) = x^3$.

3. $f(x) = |x|$.

4. $f(x) = |x| - 1$.

5. $f(x) = x^3 - 1$.

6. $f(x) = x^3 + 2x^2 - 1$.

7. $f(x) = x^2 - 2x + 1$.

8. $f(x) = \sqrt{x^2 - 1}$.

9. $f(x) = |x^2 - 1|$.

10. $f(x) = 1/x$, $x \neq 0$.

For each of the following functions f, determine $(f^{-1})'(t)$ by applying the Inverse Function Theorem:

11. $f(x) = x^2 + 1$.

12. $f(x) = 1/(2x)$, $x \neq 0$.

13. $f(x) = \sqrt{x - 1}$.

14. $f(x) = \dfrac{1}{\sqrt{x}}$.

15. $f(x) = (x^2 - 1)^{3/2}$.

16. $f(x) = (x^{3/2} - 1)^2$.

17. $f(x) = x^2 + x - 2$.

18. $f(x) = x^2 - 2x + 1$.

ANSWERS

1.

3.   **5.**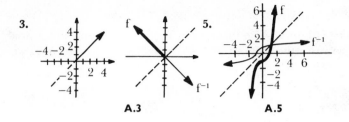

A.3   A.5

7.

9.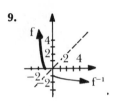

11. $\dfrac{1}{2\sqrt{t-1}}$.

13. $2t$. **15.** $\dfrac{1}{3t^{1/3}(t^{2/3}+1)^{1/2}}$. **17.** $\dfrac{\pm 1}{\sqrt{9+4t}}$.

2. THE LOGARITHMIC FUNCTION

In Chapter 9 we were able to find primitive sets for all functions of the form $f(x) = x^n$, where n is an integer, with the single exception of $f(x) = x^{-1} = \dfrac{1}{x}$. For $x > 0$, the function $f(x) = \dfrac{1}{x}$ is continuous; hence, by Theorem 9.1 (p. 380) we can find a primitive for this function, namely, $\int_a^x \dfrac{1}{t}$, where $a > 0$. For the sake of simplicity we choose $a = 1$ and make the following definition:

DEFINITION 10.5 Let $\ln: (0, \infty) \to \Re$ be defined by $\ln x = \displaystyle\int_1^x \dfrac{1}{t}$.
We call ln the *logarithmic function* and $\ln x$ the *logarithm of x*.

In accordance with custom, we have written "$\ln x$" instead of "$\ln (x)$." Many writers refer to $\ln x$ as the *natural logarithm of x*. Note that ln is defined only for $x > 0$.

THEOREM 10.6 The function ln is continuous and differentiable on $(0, \infty)$. Furthermore,

i. $\ln' x = \dfrac{1}{x}, x > 0$ and

ii. $\displaystyle\int \dfrac{1}{x} = [\ln x], x > 0.$

The theorem follows immediately from the definitions and Theorems 8.23 (p. 370) and 9.1 (p. 380). ∎

Example 6. Let $f(x) = \ln (x^2 + \cos x)$, for those values of x such that $x^2 + \cos x > 0$. We have

$$f'(x) =^1 \frac{1}{x^2 + \cos x} (x^2 + \cos x)'$$

$$=^2 \frac{2x - \sin x}{x^2 + \cos x}.$$

(1) Theorem 10.6i and the Chain Rule.
(2) $(x^2 + \cos x)' = 2x - \sin x$. ∎

Note that Theorem 10.6 implies that ln is strictly increasing. Furthermore, the fact that $\ln'' x = \dfrac{-1}{x^2} < 0$ tells us that ln is concave downward at x for all $x > 0$.

If $x \geq 1$, we may think of $\ln x$ as the area $\displaystyle\int_1^x \dfrac{1}{t}$ under the curve $y = \dfrac{1}{t}$ between 1 and x (see Figure 10.6a). If $0 < x < 1$, we may think of $\ln x$ as the negative of the area under this curve and between x and 1 (see Figure 10.6b) since $\displaystyle\int_1^x = - \int_x^1.$

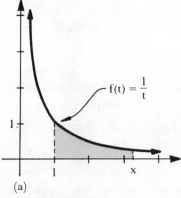

(a)

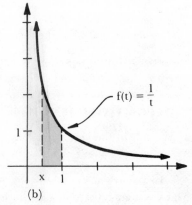

(b)

Figure 10.6

Certain properties of the logarithmic function become immediately obvious. We state these along with some previous observations, in the following theorem:

THEOREM 10.7 The logarithmic function ln has the following properties:
  i. $\ln 1 = 0$;
  ii. If $0 < x < 1$, $\ln x < 0$;
  iii. If $x > 1$, $\ln x > 0$;
  iv. ln is strictly increasing;
  v. ln is defined only for positive x;
  vi ln is concave downward in $(0, \infty)$;
  vii ln is one-to-one.

The facts we have accumulated permit us to make a rough sketch of *graph* ln. We do this in Figure 10.7.

The logarithmic function has other significant properties, some of which we state in the following theorem.

THEOREM 10.8 Let x and y be positive real numbers and r any real number. Then

  i. $\ln xy = \ln x + \ln y$;

  ii. $\ln \dfrac{1}{x} = - \ln x$;

  iii. $\ln \dfrac{x}{y} = \ln x - \ln y$;

  iv. $\ln x^r = r \ln x$.

NOTE. Using only elementary algebra we can give meaning to the expression x^r only when r is a rational number and our present proof will be restricted to this case. In Section 5 we give meaning to the general case in which r is an arbitrary real number and shall extend the validity of Theorem 10.8iv to include this case. Until then we require r to be rational for proofs of theorems, but permit r to be a real number in examples and exercises.

Proof of Theorem 10.8
  i. Let $p > 0$. Then for each $x > 0$,

$$\ln' px =^1 \frac{1}{px} p =^2 \frac{1}{x} =^3 \ln' x$$

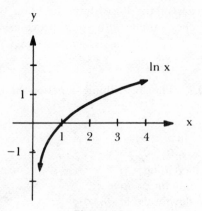

Figure 10.7

(10.1) $\Rightarrow^4 \ln px = \ln x + k$ for all $x > 0$ and for some fixed real number k

$\qquad \Rightarrow^5 \ln p \cdot 1 = \ln 1 + k = 0 + k$

$\qquad\qquad = k;$

(10.2) $\Rightarrow^2 \ln p = k$

$\qquad \Rightarrow^6 \ln px = \ln x + \ln p$

$\qquad \Rightarrow^7 \ln x \cdot y = \ln x + \ln y$

ii. $\ln x + \ln \dfrac{1}{x} =^8 \ln \left(x \cdot \dfrac{1}{x}\right) =^2 \ln 1 =^9 0 \quad \Rightarrow^2 \ln \dfrac{1}{x} = -\ln x;$

iii. $\ln \dfrac{x}{y} =^2 \ln \left(x \dfrac{1}{y}\right) =^8 \ln x + \ln \dfrac{1}{y} =^{10} \ln x - \ln y$

iv. If the exponent $r > 0$, r an integer

$$\qquad\qquad (r \text{ factors})$$

$$\ln x^r =^{11} \ln (x \cdot x \cdots x)$$

$$\qquad\qquad (r \text{ terms})$$

$$=^8 \ln x + \ln x + \cdots + \ln x$$

$$=^2 r \ln x.$$

If $r < 0$, then $-r > 0$ and we may write

$$\ln x^r =^2 \ln \frac{1}{x^{-r}}$$

$$=^{10} -\ln x^{-r}$$

$$=^{12} -(-r \ln x)$$

$$=^2 r \ln x.$$

If $r = 0$, it is immediate that $\ln x^r = r \ln x$.

Finally, let $r = \dfrac{n}{m}$, where n and m are integers and $m \neq 0$. Then, if we set $y = x^{\frac{n}{m}}$, we have

$$y = x^{\frac{n}{m}} \Leftrightarrow^{13} y^m = x^n$$

$$\Leftrightarrow \ln (y^m) = \ln (x^n)$$

$$\Leftrightarrow^{14} m \ln y = n \ln x$$

$$\Leftrightarrow^2 \ln y = \frac{n}{m} \ln x$$

$$\Leftrightarrow^{15} \ln x^{\frac{n}{m}} = \frac{n}{m} \ln x$$

$$\Leftrightarrow^{16} \ln x^r = r \ln x$$

(1) Theorem 10.6 and the Chain Rule.
(2) Elementary properties of real numbers.
(3) Theorem 10.6.
(4) Theorem 9.3 (p. 383).
(5) Letting $x = 1$ in Equation 10.1.
(6) Substituting $\ln p$ for k in Equation 10.1, by Equation 10.2.
(7) Letting $p = y$ since p is an arbitrary positive number.
(8) By Theorem 10.8i.
(9) Theorem 10.7i.
(10) By Theorem 10.8ii.
(11) Definition of x^r where r is a positive integer.
(12) Theorem 10.8iv for the positive exponent $-r$.
(13) Definition of rational exponents.
(14) By the preceding proof of iv for integers.
(15) Substituting $x^{\frac{n}{m}}$ for y.
(16) Substituting r for $\dfrac{n}{m}$. ∎

Theorem 10.8 can often be used to convert the logarithm of a complicated algebraic expression into a sum of simpler expressions.

Example 7. (a) $\ln \left(\dfrac{x^2 \cos x}{\sin x} \right) =^1 \ln (x^2 \cos x) - \ln (\sin x)$

$$=^2 \ln x^2 + \ln (\cos x) - \ln (\sin x)$$

$$=^3 2 \ln x + \ln (\cos x) - \ln (\sin x).$$

(This is true, of course, only for those values of x such that $\sin x > 0$ and $\cos x > 0$.)

(1) Theorem 10.8iii.
(2) Theorem 10.8i.
(3) Theorem 10.8iv. ∎

(b) $\ln \dfrac{(x^2 - 4)^{2/3}(x + 1)}{(2x^3 - 3)^{3/2}} =^1 \ln (x^2 - 4)^{2/3}(x + 1) - \ln (2x^3 - 3)^{3/2}$

$$=^2 \ln (x^2 - 4)^{2/3} + \ln (x + 1) - \ln (2x^3 - 3)^{3/2}$$

$$=^3 \tfrac{2}{3} \ln (x^2 - 4) + \ln (x + 1) - \tfrac{2}{3} \ln (2x^3 - 3).$$

(1) Theorem 10.8iii.
(2) Theorem 10.8i.
(3) Theorem 10.8iv. ∎

Theorem 10.8 also gives us a valuable differentiation technique known as "logarithmic differentiation." This will be illustrated in the following example:

Example 8. Differentiate $f(x) = \dfrac{(x^2 - 1)^{2/3}(x^3 + 1)}{x^3 + 2}$. We compose

ln with each side of the equation obtaining $\ln \circ f(x) = \ln \dfrac{(x^2 - 1)^{2/3}(x^3 + 1)}{x^3 + 2}$

$\Rightarrow^1 \ln \circ f(x) = \tfrac{2}{3} \ln (x^2 - 1) + \ln (x^3 + 1) - \ln (x^3 + 2)$

$\Rightarrow^2 \dfrac{1}{f(x)} \cdot f'(x) = \dfrac{4x}{3(x^2 - 1)} + \dfrac{3x^2}{x^3 + 1} - \dfrac{3x^2}{x^3 + 2}$

$\Rightarrow^3 f'(x) = f(x)\left(\dfrac{4x}{3(x^2 - 1)} + \dfrac{3x^2}{x^3 + 1} - \dfrac{3x^2}{x^3 + 2} \right)$

$=^4 \dfrac{(x^2 - 1)^{2/3}(x^3 + 1)}{x^3 + 2}\left(\dfrac{4x}{3(x^2 - 1)} + \dfrac{3x^2}{x^3 + 1} - \dfrac{3x^2}{x^3 + 2} \right).$

(1) Theorem 10.8 using the technique of Example 7.
(2) Differentiating by means of the Chain Rule.
(3) Solving for $f'(x)$.
(4) Substitution. ∎

LEMMA 10.9 Let $M \in \mathcal{R}$. There exist real numbers $r, s > 0$ such that $\ln s < M < \ln r$.

Proof. $g(x) = \tfrac{1}{2}$ is an underapproximating step function to $f(x) = \dfrac{1}{x}$ on $[1, 2]$ (see Figure 10.8). Hence, $\ln 2 = \int_1^2 \dfrac{1}{x} > \int_1^2 \dfrac{1}{2} = \dfrac{1}{2}$. Consequently

$$\ln 2^n = n \ln 2 > \tfrac{1}{2}n.$$

Thus, if we let $r = 2^n$ and choose n sufficiently large, we have $\ln r = \ln 2^n > M$.

Now choose $\dfrac{1}{s} > 0$ such that $\ln \dfrac{1}{s} > -M$. This is possible by the

previous calculation. Then $\ln s = \ln \dfrac{1}{\left(\dfrac{1}{s}\right)} = -\ln \dfrac{1}{s} < M$. ∎

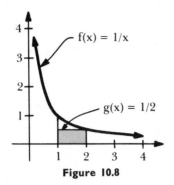

Figure 10.8

THEOREM 10.10 The function ln is onto $\mathcal{R}$, that is $ran\ \ln = (-\infty, \infty)$.

Proof. Let $y \in \mathcal{R}$. Choose $s > 0$ and $r > 0$ such that $\ln s < y < \ln r$ (Lemma 10.9). Then by Theorem 4.6 (p. 199) (Intermediate Value Theorem) there exists $x > 0$ such that $\ln x = y$. ∎

We summarize below the results of this section regarding the logarithmic function:

THE LOGARITHMIC FUNCTION. ln is *one-to-one* function from the set of positive reals *onto* $\mathcal{R}$ defined by

$$\ln x = \int_{t=1}^{x} \frac{1}{t}.$$

It has the following properties:

i. ln is continuous and differentiable on $(0, \infty)$ and $\ln' x = \dfrac{1}{x}$.

ii. $\displaystyle\int \frac{1}{x} = [\ln x], \ x > 0.$

iii. $\ln 1 = 0$.

iv. $\ln x < 0$ if and only if $0 < x < 1$.

v. $\ln x > 0$ if and only if $x > 1$.

vi. ln is strictly increasing.

If x and y are positive real numbers and r is a real number.

vii. $\ln xy = \ln x + \ln y$.

viii. $\ln \dfrac{1}{x} = - \ln x$.

ix. $\ln \dfrac{x}{y} = \ln x - \ln y$.

x. $\ln x^r = r \ln x$.

EXERCISES

In Exercises 1 to 10 use Theorem 10.8 to convert the given expression into a sum of simpler expressions:

1. $\ln \dfrac{(x^2 + 2)(2x - 1)}{x^2}$.

2. $\ln \dfrac{(x - 2x^2)}{(x^2 - 3x + 2)}$.

3. $\ln [(x^2 - 1)(x + 2)]^2$.

4. $\ln (x^2 + 2x + 1)^2$.

5. $\ln \sqrt{(\sin^2 x \cos x)}$.

6. $\ln (1 + \tan^2 x)$.

7. $\ln (x^{x+1})$.

8. $\ln (\sin^x \cos^{x+1} x)$.

9. $\ln \sqrt{x^2(x^2 + 1)}$.

10. $\ln \cot^2 x \csc^2 x$.

Differentiate the following functions using the method of logarithmic differentiation:

11. $f(x) = \dfrac{(x^2 + 1)(x - 1)}{2x^2 + 1}$.

12. $f(x) = \dfrac{x^3 - x^2 - x + 1}{2x^3 - 2x^2 + x - 1}$.

13. $f(x) = x^x$.

14. $f(x) = x^{\cos^2 x}$.

15. $f(x) = 2^{x^2 - 1}$ (remember ln 2 is a constant).

16. $f(x) = 3^{x^2}$.

17. $f(x) = x^{(x^x)}$ (refer to Exercise 13).

18. $f(x) = 10^{\sqrt{1-x^2}}$.

19. $f(x) = 2^{x^2 \ln x}$.

20. $f(x) = x^{\ln x}$.

ANSWERS

1. $\ln (x^2 + 2) + \ln (2x - 1) - 2 \ln x$.   **3.** $2 \ln (x^2 - 1) + 2 \ln (x + 2)$.
5. $\ln |\sin x| + \frac{1}{2} \ln \cos x$, for $\cos x > 0$ and $\sin x \neq 0$.   **7.** $x \ln x + \ln x$.
9. $\ln |x| + \frac{1}{2} \ln (x^2 + 1)$.   **11.** $f(x) \left(\dfrac{2x}{x^2 + 1} + \dfrac{1}{x - 1} - \dfrac{4x}{2x^2 + 1} \right)$.
13. $x^x + x^x \ln x$.   **15.** $2^{x^2} x \ln 2$.   **17.** $x^{(x^x)} x^x \left(\ln x + \ln^2 x + \dfrac{1}{x} \right)$.
19. $2^{x^2 \ln x} \ln 2 (2x \ln x + x)$.

3. THE LOGARITHMIC FUNCTION *(Continued)*

One frequently needs to find primitives for functions of the form $\dfrac{u'}{u}$.
For example, $-\tan = \dfrac{-\sin}{\cos}$ is of this form. If it happens that $u(x) > 0$ so
that $\ln \circ u$ is defined we have by the Chain Rule and Theorem 10.6, p. 412,

$$(\ln \circ u)'(x) = (\ln' \circ u)(x) \cdot u'(x)$$

$$= \dfrac{u'}{u}(x).$$

If $u(x) < 0$, we can write

$$(\ln \circ |u|)'(x) = (\ln (-u))'(x)$$

$$= \left(\frac{1}{-u} (-u)' \right) (x)$$

$$= \frac{u'}{u} (x).$$

Hence, the following theorem holds:

THEOREM 10.11   If u is a differentiable function of x, then for those values of x for which $u(x) \neq 0$

i. $(\ln |u|)' = \dfrac{u'}{u}$.

ii. $\displaystyle\int \frac{u'}{u} = [\ln |u|].$

iii. $\displaystyle\int_a^b \frac{u'}{u} = [\ln |u|]_a^b.$

Example 9

$$(\ln |x^2 - 2|)' =^1 \frac{1}{x^2 - 2} 2x$$

$$=^2 \frac{2x}{x^2 - 2}, \quad x \neq \pm\sqrt{2},$$

and

$$\int \frac{2x}{x^2 - 2} =^3 \int \frac{u'}{u} \qquad \text{(a) Let } u = x^2 - 2.$$
$$\qquad\qquad\qquad\qquad\qquad u' = 2x.$$

$$=^4 [\ln |u|]$$

$$=^3 [\ln |x^2 - 2|], \quad x \neq \pm\sqrt{2}.$$

(1) Theorem 10.11i and the Chain Rule.
(2) Elementary properties of real numbers.
(3) See substitution (a) above.
(4) Theorem 10.11ii.   ■

It is important to note the restriction that $u(x) \neq 0$ in Theorem 10.11. Indeed, the notation $\int_1^2 \dfrac{4x}{x^2 - 2}$ is not meaningful since the function $\dfrac{4x}{x^2 - 2}$ is not bounded on the interval $[1, 2]$.

Example 10. $\displaystyle\int_2^3 \frac{4x}{x^2-2} =^1 \int_2^3 \frac{2u'}{u}$ (a) Let $u = x^2 - 2$

$$u' = 2x.$$

$$=^2 2\int_2^3 \frac{u'}{u}$$

$$=^3 2[\ln |u|]_{x=2}^{x=3}$$

$$=^1 2[\ln |x^2 - 2|]_2^3$$

$$= 2(\ln 7 - \ln 2).$$

(1) See substitution (a) above.
(2) Theorem 8.14 (Linear Properties of the Integral).
(3) Theorem 10.11iii. ∎

Theorem 10.11 gives us a tool with which we can find several more formulae for finding primitive sets. We collect some of these in the following theorem:

THEOREM 10.12   If u is a differentiable function of x, then, on intervals for which the functions in question are defined,
  i. $\int (\tan u)u' = [-\ln |\cos u|]$;
 ii. $\int (\sec u)u' = [\ln |\sec u + \tan u|]$;
iii. $\int (\cot u)u' = [\ln |\sin u|]$;
 iv. $\int (\csc u)u' = [-\ln |\csc u + \cot u|]$.

We prove Theorem 10.12i and ii. Proofs of (iii) and (iv) are similar and are left to the exercises.

Proof of i

$$(-\ln |\cos u|)' =^1 \frac{-1}{\cos u} (-\sin u)u'$$

$$=^2 \frac{(\sin u)u'}{\cos u}$$

$$=^3 (\tan u)u'$$

$$\Rightarrow^4 \int (\tan u)u' = [-\ln |\cos u|].$$

Proof of ii

$$\int (\sec u)u' =^5 \int \frac{(\sec u + \tan u)\sec u}{\sec u + \tan u} u'$$

$$=^2 \int \frac{(\sec^2 u)u' + (\sec u \tan u)u'}{\sec u + \tan u}$$

$$=^6 \int \frac{w'}{w}$$ (a) $w = \sec u + \tan u$

$$w' = (\sec u \tan u)u' + (\sec^2 u)u'.$$

$$=^7 [\ln |w|]$$

$$=^6 [\ln |\sec u + \tan u|].$$

(1) Differentiating using Theorem 10.11i and the Chain Rule.
(2) Elementary properties of real numbers.
(3) Trigonometric Identity.
(4) Definition of primitive set.
(5) Multiplying numerator and denominator by $\sec u + \tan u$.
(6) See substitution (a) above.
(7) Theorem 10.11i. ∎

Example 11

$$\int x \tan x^2 =^1 \int \frac{1}{2} (\tan u)u' \qquad \text{(a) Let } u = x^2$$
$$u' = 2x.$$

$$=^2 \frac{1}{2} \int (\tan u)u'$$

$$=^3 \tfrac{1}{2}[-\ln|\cos u|]$$

$$=^1 [-\tfrac{1}{2}\ln|\cos x^2|].$$

(1) See substitution (a) above.
(2) Definition 9.8 (p. 390).
(3) Theorem 10.12i. ∎

Example 12

$$\int x^3 \csc x^2 \cot x^2 \qquad\qquad \text{(a) Let } u = x^2$$
$$u' = 2x.$$

$$=^1 \int \frac{1}{2} (u \csc u \cot u)u'$$

$$\text{(b) Let } f = u, \quad g' = (\csc u \cot u)u'$$
$$f' = u', \quad g = -\csc u.$$

$$=^2 \tfrac{1}{2}([-u \csc u] + \int (\csc u)u')$$

$$=^3 \tfrac{1}{2}([-u \csc u] + [-\ln|\csc u + \cot u|])$$

$$=^4 \tfrac{1}{2}[-u \csc u - \ln|\csc u + \cot u|]$$

$$=^1 -\tfrac{1}{2}[x^2 \csc x^2 + \ln|\csc x^2 + \cot x^2|].$$

(1) See substitution (a) above.
(2) See substitution (b) above.
(3) Theorem 10.12iv.
(4) Equation 9.6 (p. 390). ∎

EXERCISES

Differentiate the following functions:

1. $f(x) = \ln|2x^2|$.

2. $f(x) = \ln|\sin x^2|$.

3. $f(x) = x \ln x$.

4. $f(x) = x \ln|\cos x|$.

5. $f(x) = \ln|2x^2 - 2x + 1|$.

6. $f(x) = \dfrac{\ln x}{\tan x}$.

7. $f(x) = (\ln x^2)^2$.   8. $f(x) = \sin(\ln x)$.

9. $f(x) = \sin x \ln x$.   10. $f(x) = \sqrt{\ln x}$.

Find the following primitive sets:

11. $\int \dfrac{x}{x^2 + 1}$.   12. $\int \dfrac{1}{x - 1}$.

13. $\int \dfrac{1}{2x + 3}$.   14. $\int \dfrac{2x}{x^2 - 1}$.

15. $\int \dfrac{2x^2 + 1}{x - 1}$ (Hint: first perform the division).

16. $\int \dfrac{3x^3 - 2x}{x^2 + 1}$.   17. $\int \dfrac{\sin x - 1}{\cos x + x}$.

18. $\int \dfrac{\cot \sqrt{x}}{\sqrt{x}}$.   19. $\int \dfrac{x \csc \sqrt{x^2 + 1}}{\sqrt{x^2 + 1}}$.

20. $\int \dfrac{x^2 + x + 1}{x^2 + 1}$.   21. $\int \dfrac{x^2 + 3x + 1}{x^2 + x}$.

22. $\int \dfrac{x \cos x + \sin x}{x \sin x}$.   23. $\int \dfrac{1}{\sin x \cos x}$.

24. $\int \cos x \sec(\sin x)$.   25. $\int \dfrac{x^2 - 1}{x^2 \cot\left(x + \dfrac{1}{x}\right)}$.

26. $\int \ln |x|$.   27. $\int \sin \theta \ln |\cos \theta|$.

28. $\int x \sec^2 x$.   29. Prove Theorem 10.12iii.

30. Prove Theorem 10.12iv.

ANSWERS

1. $\dfrac{2}{x}$.   **3.** $1 + \ln x$.   **5.** $\dfrac{2(2x - 1)}{2x^2 - 2x + 1}$.   **7.** $\dfrac{4 \ln x^2}{x}$.

9. $\cos x \ln x + \dfrac{\sin x}{x}$.   **11.** $[\frac{1}{2} \ln |x^2 + 1|]$.   **13.** $[\frac{1}{2} \ln |2x + 3|]$.

15. $[x^2 + 2x + 3 \ln |x - 1|]$.   **17.** $[-\ln |\cos x + x|]$.

19. $[-\ln |\csc \sqrt{x^2 + 1} + \cot \sqrt{x^2 + 1}|]$.   **21.** $[x + \ln |x^2 + x|]$.

23. $[\ln |\tan x|]$.   **25.** $\left[-\ln \left| \cos \left(x + \dfrac{1}{x}\right) \right| \right]$.

27. $[\cos \theta - \cos \theta \ln |\cos \theta|]$.

4. THE EXPONENTIAL FUNCTION

We have shown that the logarithmic function ln is a continuous, strictly increasing, differentiable function from $(0, \infty)$ *onto* $\mathcal{R}$ whose derivative is non-zero. Hence, by Theorem 10.4 (Inverse Function Theorem) it has a continuous, strictly increasing, differentiable inverse from $\mathfrak{R}$ *onto* $(0, \infty)$. In this section we shall investigate this inverse function.

DEFINITION 10.13   The function $\ln^{-1}:\mathfrak{R} \to (0, \infty)$ which is the inverse of ln, is called the *exponential function* and denoted by exp.

That is, $\exp = \ln^{-1}$. The following are different ways of saying the same thing:

(10.3) $$\ln = \exp^{-1};$$

(10.4) $$\ln \circ \exp (x) = x, \text{ for each } x \in \mathfrak{R};$$

(10.5) $$\exp \circ \ln (y) = y, \text{ for each } y \in (0, \infty);$$

(10.6) $$\ln x = y \text{ if and only if } \exp y = x.$$

To obtain a sketch of the graph of the exponential function, we reverse the coordinates of each point of the graph of ln in Figure 10.7. That is, if $(x, y) \in graph$ ln we plot the point (y, x). The resulting set of points is the graph of the exponential function (see Figure 10.9).

THEOREM 10.14   The exponential function, exp, is continuous, differentiable, and strictly increasing on $\mathfrak{R}$. Furthermore,

$$\exp' = \exp.$$

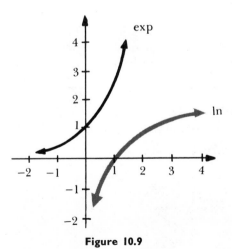

Figure 10.9

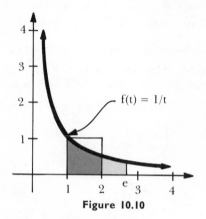

Figure 10.10

Proof. Since $\exp = \ln^{-1}$ we may use Theorem 10.4 (Inverse Function Theorem) and the fact that ln is continuous, differentiable with non-zero derivative, and strictly increasing on $(0, \infty)$ to show that exp is continuous, differentiable, and strictly increasing on $\mathcal{R}$. Again, by Theorem 10.4,

$$\exp' = \frac{1}{\ln' \circ \exp} = \frac{1}{1/\exp} = \exp. \quad \blacksquare$$

Example 13. Let $f(x) = \exp(\cos x) = \exp \circ \cos(x)$. Then

$$\begin{aligned} f' &= (\exp \circ \cos)' \\ &=^1 (\exp' \circ \cos) \cos' \\ &=^2 (\exp \circ \cos)(-\sin) \\ &= -\sin \cdot (\exp \circ \cos). \end{aligned}$$

(1) Chain Rule.
(2) Theorem 10.14 and the fact that $\cos' = -\sin$.
(3) Commuting. $\blacksquare$

The following theorem should be evident from Figure 10.10. The proofs are easy and will be left to the exercises.

THEOREM 10.15   Let $y \in \mathcal{R}$.

   i. $0 < \exp y < 1$ if and only if $y < 0$.

   ii. $\exp y > 1$ if and only if $y > 0$.

   iii. $\exp 0 = 1$.

   iv. $\lim\limits_{x \to \infty} \exp x = \infty$.

   v. $\lim\limits_{x \to -\infty} \exp x = 0$.

The real number exp 1 is of sufficient interest to justify denoting it by a special symbol.

DEFINITION 10.16 The real number exp 1 is denoted by e (in honor of the mathematician Euler).

To find the value of e, we note that $e = \exp 1 = \ln^{-1} 1$. That is, e is that real number such that $\ln e = \int_1^e \frac{1}{t} = 1$. Thus, we are seeking the real number e such that the area under the graph of the function $\frac{1}{t}$ between 1 and e is 1 (see Figure 10.10). Consider the rectangle indicated in Figure 10.10 having base $[1, 2]$ and height 1. The area of this rectangle is 1. From the figure it is clear that $\int_1^2 1/t < 1$. Consequently, we must have $e > 2$ in order that $\int_1^e 1/t = 1$. On the other hand, $\ln 4 = \ln 2^2 = 2 \ln 2 = 2\int_1^2 \frac{1}{t} > 2 \cdot \frac{1}{2} = 1$, so $e < 4$. It can be shown that e is an irrational number. To five decimal places it is 2.71828 (see page 526).

One of the important properties of e is stated in the following theorem:

THEOREM 10.17 Let r be any rational number; then

$$\exp r = e^r.$$

Proof. Let $y = e^r$. By Definition 10.16 we then have $y = (\exp 1)^r$. Thus

$$\ln y = \ln ((\exp 1)^r)$$

$$=^1 r \ln (\exp 1)$$

$$=^2 r \cdot 1$$

$$= r.$$

From Equation 10.6 we then have

$$\exp r = y =^3 e^r.$$

    (1) Theorem 10.8iv (proved for rational r).
    (2) Since ln and exp are inverse functions (Equation 10.4).
    (3) By the previous definition of y. ∎

The preceding theorem motivates the following definition:

DEFINITION 10.18 For all real numbers x

$$e^x = \exp x.$$

Equations 10.4, 10.5, and 10.6 can now be restated as follows:

(10.7) $\ln e^x = x$, for each x in $\mathcal{R}$.

(10.8) $e^{\ln y} = y$, for each y in $(0, \infty)$.

(10.9) $\ln x = y$ if and only if $x = e^y$.

The following theorem lists useful properties of the exponential function:

THEOREM 10.19 Let x and y be real numbers. Then

i. $e^{x+y} = e^x e^y$.

ii. $e^{-x} = \dfrac{1}{e^x}$.

iii. $e^{x-y} = \dfrac{e^x}{e^y}$.

Proof. We prove (i) and assign (ii) and (iii) to the exercises. Let $u = e^x, v = e^y$. Then, by Equation 10.9, $x = \ln u, y = \ln v$ and

$$\begin{aligned} e^{x+y} &=^1 e^{\ln u + \ln v} \\ &=^2 e^{\ln (u \cdot v)} \\ &=^3 u \cdot v \\ &=^1 e^x \cdot e^y. \end{aligned}$$

(1) Substitution for x and y.
(2) Theorem 10.8i.
(3) Equation 10.8. ■

From the differentiation formula $\exp' = \exp$ and Definition 10.19, we obtain the following theorem:

THEOREM 10.20 If $u: \mathcal{R} \to \mathcal{R}$ is differentiable, then
    i. $(e^u)' = e^u u'$,
and ii. $\int e^u u' = [e^u]$.

Example 14. Let $f(x) = e^{x^2-1} = \exp (x^2 - 1)$.
Then

$$\begin{aligned} f'(x) &= \exp' (x^2 - 1) \cdot 2x \\ &= 2x e^{x^2-1}. \quad \blacksquare \end{aligned}$$

Example 15.   Let $f(x) = \sin e^{x^2}$.

We have

$$f'(x) = (\cos e^{x^2} e^{x^2}) \cdot 2x$$
$$= 2xe^{x^2} \cos e^{x^2}. \quad \blacksquare$$

Example 16.   Find the primitive set of $e^{2e^x + x}$.

$$\int e^{2e^x + x} \overset{1}{=} \int e^{2e^x} e^x$$

$$\overset{2}{=} \int \frac{1}{2} e^u u' \qquad \text{(a) Let } u = 2e^x, \ u' = 2e^x.$$

$$\overset{3}{=} \frac{1}{2} \int e^u u'$$

$$\overset{4}{=} \tfrac{1}{2} [e^u]$$

$$\overset{2}{=} [\tfrac{1}{2} e^{2e^x}].$$

(1) Theorem 10.16i.
(2) See substitution (a) above.
(3) Definition 9.8, (p. 390).
(4) Theorem 10.20. $\blacksquare$

Example 17

$$\int e^x \cos x$$

$$\overset{1}{=} [e^x \sin x] - \int e^x \sin x$$

(a) Let $f(x) = e^x, \quad g'(x) = \cos x;$
   $f'(x) = e^x, \quad g(x) = \sin x.$

(b) Let $f(x) = e^x, \quad g'(x) = \sin x;$
   $f'(x) = e^x, \quad g(x) = -\cos x.$

$$\overset{2}{=} [e^x \sin x] - \left([-e^x \cos x] - \int -e^x \cos x \right)$$

$$\overset{3}{\Rightarrow} 2 \int e^x \cos x = [e^x \sin x + e^x \cos x]$$

$$\Rightarrow \int e^x \cos x = [\tfrac{1}{2} e^x \sin x + \tfrac{1}{2} e^x \cos x].$$

(1) See substitution (a) above.
(2) See substitution (b) above.
(3) Algebraic manipulations. $\blacksquare$

EXERCISES

Differentiate:

1. $f(x) = e^{x^2}$.

2. $f(x) = e^{3x-1}$.

3. $f(x) = e^{x + \ln x}$.

4. $f(x) = e^{x \ln x}$.

5. $f(x) = \ln(1 + e^{-x})$

6. $f(x) = e^{-x^2}$.

7. $f(x) = e^{1/\ln x}$.

8. $f(x) = e^{1/x}$.

9. $f(x) = \dfrac{e^x - e^{-x}}{2}$.

10. $f(x) = \dfrac{e^x + e^{-x}}{2}$.

11. $f(x) = e^x \ln x$.

12. $f(x) = \dfrac{e^x}{\ln x}$.

13. $f(x) = e^{x \sin x}$.

14. $f(x) = \sin x e^{x \sin x}$.

15. $f(x) = x e^{x^2}$.

16. $f(x) = e^{x^2} \ln x$.

17. $f(x) = e^{\sqrt{x-1}}$.

18. $f(x) = \ln(e^x + 1)$.

19. $f(x) = \cos(e^x \ln x)$.

20. $f(x) = e^{\cos(\ln x)}$.

Find primitives for the following functions:

21. $e^{\cos x} \sin x$.

22. $4x e^{x^2}$.

23. $\dfrac{e^{\ln x}}{x}$.

24. $x e^{x^2 + 2x} + e^{x^2 + 2x}$.

25. $e^{x \ln x} \ln x + e^{x \ln x}$.

26. $e^{(e^x) + x}$.

27. $e^x(x + 1)$.

28. $e^x\left(\dfrac{1}{x} + \ln x\right)$.

29. $e^x(\cos x - \sin x)$.

30. $e^x((1 + x)\cos x - x \sin x)$.

31. $\dfrac{e^{\sqrt{x-1}}}{\sqrt{x-1}}$.

32. $x e^{(x^2-1)^{3/2}}(x^2 - 1)^{1/2}$.

33. $e^x(x^2 + 4x + 2)$.

34. $e^x(x^3 + 3x^2 + 1)$.

35. $e^x \cos x$.

36. Prove Theorem 10.15i.

37. Prove Theorem 10.15ii.

38. Prove Theorem 10.15iii.

39. Prove Theorem 10.15iv.

40. Prove Theorem 10.15v.

41. Prove Theorem 10.16ii.

42. Prove Theorem 10.16iii.

ANSWERS

1. $2x e^{x^2}$. 3. $e^x(1 + x)$. 5. $\dfrac{-1}{e^x + 1}$. 7. $\dfrac{-e^{(1/\ln x)}}{x(\ln x)^2}$. 9. $\dfrac{e^x + e^{-x}}{2}$.

11. $e^x \ln x + \dfrac{e^x}{x}$. 13. $e^{x \sin x}(x \cos x + \sin x)$. 15. $e^{x^2}(2x^2 + 1)$.

17. $\dfrac{e^{\sqrt{x-1}}}{2\sqrt{x-1}}$. 19. $-e^x \sin(e^x \ln x)\left(\ln x + \dfrac{1}{x}\right)$. 21. $-e^{\cos x}$. 23. x.

25. $e^{x \ln x}$. 27. $x e^x$. 29. $e^x \cos x$. 31. $2e^{\sqrt{x-1}}$. 33. $e^x(x^2 + 2x)$.

5. THE EXPONENTIAL FUNCTION (*Continued*)

In this section we shall take care of some unfinished business. We shall give a careful definition for a^x where a and x are real numbers and $a > 0$. We shall then prove that $(a^x)' = a^x \ln a$ and consider the resulting formulas for finding primitive sets.

If $a > 0$ and r is any rational number, we have

$$a^r = e^{\ln a^r}, \quad \text{by Equation 10.8}$$
$$= e^{r \ln a}, \quad \text{by Theorem 10.8iv}$$

We use this equation as motivation for the following definition:

DEFINITION 10.21 If $a > 0$ and r is any real number, we define

$$a^r = e^{r \ln a}.$$

Note: The above definition applies only if $a > 0$. For $a < 0$ the usual definition of a^r, for integral values of r, is all that we shall use.

With the preceding definition we can now prove Theorem 10.8iv for arbitrary real numbers r, as follows:

$$x^r =^1 e^{r \ln x}$$
$$\Rightarrow \ln x^r = \ln \left(e^{r \ln x} \right)$$
$$=^2 r \ln x.$$

(1) Definition 10.21.
(2) Equation 10.8. ■

Using Definition 10.21 we extend the familiar laws of exponents to include arbitrary real exponents. We also give other important properties of exponential functions.

THEOREM 10.22 Let $a, b, r, s \in \mathfrak{R}$ and let $a, b > 0$.

  i. $a^{r+s} = a^r a^s$.

  ii. $a^{-r} = \dfrac{1}{a^r}$.

  iii. $a^{r-s} = \dfrac{a^r}{a^s}$.

  iv. $1^r = 1$.

  v. $a^0 = 1$.

  vi. $a^r > 0$.

  vii. $(a^r)^s = a^{rs}$.

  viii. If $a > 1$, $\lim\limits_{x \to \infty} a^x = \infty$.

  ix. If $a > 1$, $\lim\limits_{x \to -\infty} a^x = 0$.

  x. If $0 < a < 1$, $\lim\limits_{x \to \infty} a^x = 0$.

  xi. If $0 < a < 1$, $\lim\limits_{x \to -\infty} a^x = \infty$.

  xii. If $a > 1$, then a^r is a strictly increasing, continuous function of r on $(-\infty, \infty)$.

  xiii. If $a < 1$, a^r is a strictly decreasing, continuous function of r on $(-\infty, \infty)$.

We shall prove (i), (iv), (vii), (viii), and (xii). The remaining proofs are called for in the exercises.

Proof of i

$$a^{r+s} =^1 e^{(r+s)\ln a}$$
$$=^2 e^{r \ln a + s \ln a}$$
$$=^3 e^{r \ln a} e^{s \ln a}$$
$$=^1 a^r a^s.$$

Proof of iv

$$1^r =^1 e^{r \ln 1}$$
$$=^4 e^0$$
$$=^5 1.$$

Proof of vii

$$(a^r)^s =^1 e^{s \ln a^r}$$
$$=^6 e^{s(r \ln a)}$$
$$=^2 e^{rs \ln a}$$
$$=^1 a^{rs}.$$

Proof of viii

$$\lim_{x \to \infty} a^x =^1 \lim_{x \to \infty} e^{x \ln a}$$
$$=^7 \lim_{x \ln a \to \infty} e^{x \ln a}$$
$$=^8 \infty.$$

Proof of xii

Since $a > 1$, $\ln a > 0$ by Theorem 10.7iii, hence, $f(x) = x \ln a$ is a continuous strictly increasing function of x. By Theorem 10.14, exp is continuous and strictly increasing. By Theorem 4.5 (p. 197), $\exp \circ f$ is continuous and it is obvious that the composition of two strictly increasing functions is strictly increasing. Thus, $a^x = e^{x \ln a} = \exp \circ f(x)$ is continuous and strictly increasing on $(-\infty, \infty)$.

    (1) Definition 10.21.
    (2) Properties of real numbers.
    (3) Theorem 10.19i (p. 426).
    (4) Theorem 10.7i (p. 413).
    (5) Theorem 10.15iii.
    (6) Theorem 10.8iv (p. 413).
    (7) Since $a > 1$, $\ln a > 0$, so $x \ln a \to \infty$ as $x \to \infty$.
    (8) Theorem 10.15iv. ∎

Now we present the long delayed proof for the differentiation formula for the power function $f(x) = x^r$.

THEOREM 10.23   Let $f(x) = x^r$ where $x > 0$ and r is any real number. Then $f'(x) = rx^{r-1}$.

Note: This is Theorem 5.14 (p. 233) which was proved only for the case in which r is an integer.

Proof. $f(x) = x^r =^1 e^{r \ln x}$

$$\Rightarrow f'(x) =^2 e^{r \ln x} \cdot r \cdot \frac{1}{x}$$

$$=^1 rx^r \cdot \frac{1}{x}$$

$$=^3 rx^r \cdot x^{-1}$$

$$=^4 rx^{r-1}.$$

(1) Definition 10.21.
(2) Chain Rule.
(3) Definition of x^{-1}.
(4) Theorem 10.22i. ∎

THEOREM 10.24

i. Let $f(x) = a^x$, $a > 0$. Then

$$f'(x) = a^x \ln a.$$

ii. For $a > 0$, u a differentiable function of x,

$$\int a^u u' = \left[\frac{a^u}{\ln a} \right].$$

Proof. $f(x) = a^x = e^{x \ln a}$

$$\Rightarrow f'(x) = e^{x \ln a} \cdot \ln a$$

$$= a^x \ln a.$$

The second statement can easily be proved by differentiating $\dfrac{a^u}{\ln a}$. ∎

Example 18

$$(2^{3x})' =^1 ((2^3)^x)'$$
$$= (8^x)'$$
$$=^2 8^x \ln 8.$$

(1) Theorem 10.22vii.
(2) Theorem 10.24i. ∎

Example 19

$$(2^{\sin x})' = 2^{\sin x} \cdot \cos x \cdot \ln 2. ∎$$

Example 20

$$\int 4^{x^2+x}(2x + 1)$$

(a) Let $u = x^2 + x$, $u' = 2x + 1$.

$$=^1 \int 4^u u'$$

$$=^2 \left[\frac{4^u}{\ln 4}\right]$$

$$=^1 \left[\frac{4^{x^2+x}}{\ln 4}\right].$$

(1) See substitution (a) above.
(2) Theorem 10.24. ■

The function given by $f(x) = a^x$ is known as the *exponential function base a* and is denoted $\exp_a x$. For $a > 0$, $a \neq 1$, such functions are strictly monotonic, continuous, and one-to-one from $\mathcal{R}$ onto $(0, \infty)$. The inverses then exist and map $(0, \infty)$ into $\mathcal{R}$. The inverse of $\exp_a$ is called the log- arithmic function to the base a and denoted $\log_a$. Of course $\log_e = \ln$. The only logarithmic function other than ln which is widely used is $\log_{10}$, frequently called the common logarithm and used in trigonometry.

EXERCISES

Differentiate the following:

1. $f(x) = 3^{x^2}$.

2. $f(x) = 5^{\sqrt{x}}$.

3. $f(x) = 10^{e^x}$.

4. $f(x) = 4^{\ln x}$.

5. $f(x) = (\sqrt{e})^{x^2}$.

6. $f(x) = 4^{\cos x}$.

7. $f(x) = 4^{\ln x^2}$.

8. $f(x) = 9^{(\ln x+1)}$.

9. $f(x) = (\ln 4)^{\ln x}$.

10. $f(x) = 2^{(\cos x+2)}$.

Find a member of each of the following primitive sets:

11. $\int 4^x$.

12. $\int \frac{3^{\sqrt{x}}}{\sqrt{x}}$.

13. $\int x\, 2^{x^2}$.

14. $\int 3^{2x}$.

15. $\int 4^{x^2+x}(2x + 1)$.

16. $\int 3^{x^2+2x}(x + 1)$.

17. $\int 4^x(1 + x \ln 4)$.

18. $\int 4^x \cos 4^x$.

19. $\int 4^x(\ln 4 \sin x + \cos x)$.

20. $\int 4^x\left(\frac{1}{x} + \ln 4 \ln x\right)$.

21. Prove Theorem 10.22v.

22. Prove Theorem 10.22iv (Hint: use Exercise 21).

23. Prove Theorem 10.22iii.   24. Prove Theorem 10.22vi.

25. Prove Theorem 10.22ix.   26. Prove Theorem 10.22x.

27. Prove Theorem 10.22xi.   28. Prove Theorem 10.22xiii.

ANSWERS

1. $2x \ln 3 \cdot 3^{x^2}$. **3.** $10^{e^x} e^x \ln 10$. **5.** $x(\sqrt{e})^{x^2}$. **7.** $\dfrac{16^{\ln x} \ln 16}{x}$.

9. $\dfrac{\ln 4^{\ln x} \ln (\ln 4)}{x}$. **11.** $\dfrac{4^x}{\ln 4}$. **13.** $\dfrac{2^{x^2}}{2 \ln 2}$. **15.** $\dfrac{4^{x^2+x}}{\ln 4}$. **17.** $x \, 4^x$.

19. $4^x \sin x$.

6. THE INVERSE TRIGONOMETRIC FUNCTIONS

In order that a function have an inverse it must be one-to-one and onto. None of the six basic trigonometric functions have inverses since all of them are periodic, and hence, not one-to-one. However, we can restrict the domains of these in such a way that the resulting functions do have inverses. We consider only sin, cos, tan, and cot, restricting these functions to sets on which they are continuous, differentiable with non-zero derivative, and strictly monotonic; that is, sets on which they meet the conditions of Theorem 10.4 (Inverse Function Theorem). We restrict

$$\sin \text{ to } \left[-\frac{\pi}{2}, \frac{\pi}{2} \right],$$

$$\cos \text{ to } [0, \pi],$$

$$\tan \text{ to } \left(-\frac{\pi}{2}, \frac{\pi}{2} \right),$$

and

$$\cot \text{ to } (0, \pi).$$

We graph the restricted functions in Figures 10.11 to 10.14.

For convenience, we shall refer to the inverses of these *restricted* functions as the inverses of the original functions.

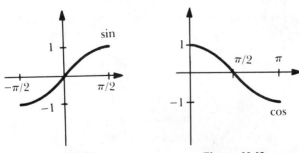

Figure 10.11   Figure 10.12

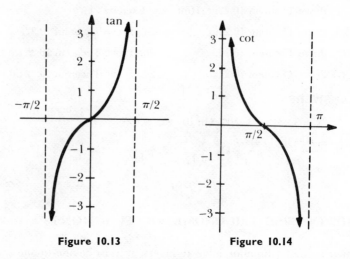

Figure 10.13   **Figure 10.14**

DEFINITION 10.25

i. The inverse of the restriction of the sine function to $\left[-\dfrac{\pi}{2}, \dfrac{\pi}{2}\right]$ will be called the *arc sine function* and denoted arc sin.

ii. The inverse of the restriction of the cosine function to $[0, \pi]$ will be called the *arc cosine function* and denoted arc cos.

iii. The inverse of the restriction of the tangent function to $\left(-\dfrac{\pi}{2}, \dfrac{\pi}{2}\right)$ will be called the *arc tangent function* and denoted arc tan.

iv. The inverse of the restriction of the cotangent function to $(0, \pi)$ will be called the *arc cotangent function* and denoted arc cot.

NOTE. In line with the conventional notation for functional inverses which we have been using, it would seem natural to denote the inverse functions by $\sin^{-1}$, $\cos^{-1}$, and so forth, and some authors do this. However, we choose not to do so in order to avoid confusion. The symbol $\sin^{-1} x$ could be interpreted to mean $\dfrac{1}{\sin x}$, which is certainly not the inverse *function* of the sine function.

We obtain sketches of the graphs of the inverse trigonometric functions by sketching the mirror images of these functions in the line $y = x$, as shown in Figures 10.15 to 10.18.

Note that the domains of the inverse trigonometric functions are given by

$$dom \ \text{arc sin} \ = [-1, 1],$$
$$dom \ \text{arc cos} \ = [-1, 1],$$
$$dom \ \text{arc tan} \ = (-\infty, \infty),$$

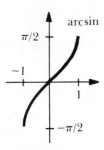

Figure 10.15

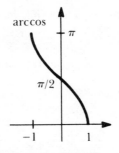

Figure 10.16

and

$$dom \text{ arc cot} = (-\infty, \infty).$$

We now use Theorem 10.4 (Inverse Function Theorem) to find the derivative functions for the inverse trigonometric functions.

THEOREM 10.26 For x in the domain of the respective functions, we have

    i. $(\text{arc sin})'(x) = \dfrac{1}{\sqrt{1 - x^2}}$,

    ii. $(\text{arc cos})'(x) = \dfrac{-1}{\sqrt{1 - x^2}}$,

    iii. $(\text{arc tan})'(x) = \dfrac{1}{1 + x^2}$,

and

    iv. $(\text{arc cot})'(x) = \dfrac{-1}{1 + x^2}$.

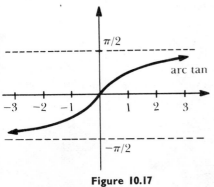

Figure 10.17

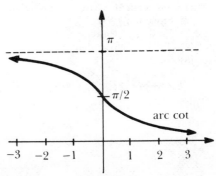

Figure 10.18

We prove (i) and (iii), leaving (ii) and (iv) to the exercises.

Proof of i

$$(\text{arc sin})'(x) =^1 \frac{1}{\sin' \circ \text{arc sin } x}$$

$$=^2 \frac{1}{\cos (\text{arc sin } x)}$$

$$=^3 \frac{1}{\cos y}$$

$$=^4 \frac{1}{\sqrt{1 - \sin^2 y}}$$

$$=^5 \frac{1}{\sqrt{1 - x^2}}.$$

(1) Theorem 10.4 (Inverse Function Theorem).
(2) Since $\sin' = \cos$.
(3) Letting $y = \text{arc sin } x$.
(4) Since $\cos^2 = 1 - \sin^2$ and cos is positive on the range of arc sin.
(5) Since $\sin y = x$. ∎

Proof of iii

$$(\text{arc tan})'(x) =^1 \frac{1}{\tan' \circ \text{arc tan } x}$$

$$=^2 \frac{1}{\sec^2 (\text{arc tan } x)}$$

$$=^3 \frac{1}{\sec^2 y}$$

$$=^4 \frac{1}{1 + \tan^2 y}$$

$$=^5 \frac{1}{1 + x^2}.$$

(1) Theorem 10.4 (Inverse Function Theorem).
(2) Since $\tan' = \sec^2$.
(3) Letting $y = \text{arc tan } x$.
(4) Trigonometric identity.
(5) Since $x = \tan y$. ∎

Example 21. Let $f(x) = \text{arc tan } (x^2 + 1)$. Then, using the Chain Rule,

$$f'(x) = \frac{1}{1 + (x^2 + 1)^2} \cdot 2x$$

$$= \frac{2x}{x^4 + 2x^2 + 2}. \blacksquare$$

Since $(\text{arc sin})' = -(\text{arc cos})'$ and $(\text{arc tan})' = -(\text{arc cot})'$, we shall need only two of the corresponding primitive set formulas.

THEOREM 10.27 If u is a differentiable function,

i. $\displaystyle\int \frac{u'}{\sqrt{1-u^2}} = [(\text{arc sin}) \circ u], \ |u| < 1$

and ii. $\displaystyle\int \frac{u'}{1+u^2} = [(\text{arc tan}) \circ u].$

Example 22

$\displaystyle\int \frac{\sin x}{1 + \cos^2 x} =^1 \int \frac{-u'}{1+u^2}$ (a) Let $u = \cos x, \ u' = -\sin x.$

$=^2 -[\text{arc tan } u]$

$=^1 -[\text{arc tan } (\cos x)].$

(1) See substitution (a) above.
(2) Theorem 10.27. ■

Example 23

$\displaystyle\int x^2(1-x^2)^{-3/2}$ (a) $f(x) = x, \ g'(x) = x(1-x^2)^{-3/2}$
   $\qquad\qquad f'(x) = 1, \ \ g(x) = (1-x^2)^{-1/2}$

$=^1 [x(1-x^2)^{-1/2}] - \displaystyle\int (1-x^2)^{-1/2}$

$=^2 [x(1-x^2)^{-1/2} - \text{arc sin } x]$

$=^3 [x(1-x^2)^{-1/2} - \text{arc sin } x].$

(1) See substitution (a) above.
(2) Theorem 10.27i.
(3) Equation 9.6ii (p. 390). ■

Example 24

$\displaystyle\int \frac{(\text{arc tan } x) \sin (\text{arc tan } x)}{1 + x^2}$ (a) Let $u = \text{arc tan } x$
   $\qquad\qquad\qquad u' = \dfrac{1}{1+x^2}.$

$=^1 \displaystyle\int u(\sin u)u'$

$=^2 [-u \cos u] + \displaystyle\int (\cos u)u'$ (b) $f(x) = u(x), \ g'(x) = \sin u(x)u'(x)$
   $\qquad\qquad\qquad\qquad f'(x) = u'(x), \ g(x) = -\cos u(x).$

$=^3 [-u \cos u + \sin u]$

$=^1 [-(\text{arc tan } x) \cos (\text{arc tan } x)] + [\sin (\text{arc tan } x)].$

(1) See substitution (a) above.
(2) See substitution (b) above.
(3) Table 9.1 (p. 389) and Equation 9.6ii (p. 390). ■

EXERCISES

Differentiate the following:

1. $f(x) = \arc \sin x^2$.   2. $f(x) = \arc \cos \sqrt{x}$.

3. $f(x) = x \arc \cot x$.   4. $f(x) = \dfrac{\arc \tan x}{x}$.

5. $f(x) = (\arc \sin x)^3$.   6. $f(x) = (\arc \sin \sqrt{x})^2$.

7. $f(x) = \cot(\arc \tan x)$.   8. $f(x) = \arc \tan(\cot x)$.

9. $f(x) = \arc \cot \sqrt{x^2 + 1}$.   10. $f(x) = \sqrt{\arc \cos x}$.

Find the primitive sets called for in Exercises 11 to 25:

11. $\displaystyle\int \frac{1}{\sqrt{1 - 4x^2}}$.   12. $\displaystyle\int \frac{1}{1 + 4x^2}$.

13. $\displaystyle\int \frac{x + 1}{1 + (x + 1)^4}$.   14. $\displaystyle\int \frac{x^2}{\sqrt{1 - x^6}}$.

15. $\displaystyle\int \frac{x + 1}{\sqrt{1 - x^2}}$.   16. $\displaystyle\int \frac{x^3 + x}{1 + x^4}$.

17. $\displaystyle\int \frac{1}{x + x(\ln x)^2}$.   18. $\displaystyle\int \frac{\sin x + x \cos x}{1 + x^2 \sin^2 x}$.

19. $\displaystyle\int \frac{\sqrt{1 - (\ln x)^2}}{x - x(\ln x)^2}$.   20. $\displaystyle\int \frac{1}{e^x + e^{-x}}$.

21. $\displaystyle\int \frac{1}{\sqrt{1 - x^2} \arc \cos x}$.   22. $\displaystyle\int x^5(1 - x^4)^{-3/2}$.

23. $\displaystyle\int \frac{(\ln x)^2}{x(1 + (\ln x)^2)^2}$.   24. $\displaystyle\int \frac{x^8}{(1 - x^6)^{3/2}}$.

25. Evaluate: $\displaystyle\int_0^1 \frac{\sqrt{x}}{(1 + x)^2}$.   26. Prove Theorem 10.26ii.

27. Prove Theorem 10.26iv.

ANSWERS

1. $\dfrac{2x}{\sqrt{1 - x^4}}$.   3. $\arc \cot x - \dfrac{x}{1 + x^2}$.   5. $\dfrac{3(\arc \sin x)^2}{\sqrt{1 - x^2}}$.

7. $\dfrac{-\csc^2(\arc \tan x)}{1 + x^2}$.   9. $\dfrac{-x}{(x^2 + 2)\sqrt{x^2 + 1}}$.   11. $[\frac{1}{2} \arc \sin 2x]$.

13. $[\frac{1}{2} \arc \tan (x + 1)^2]$.   15. $[(\arc \sin x) - \sqrt{1 - x^2}]$.

17. $[\arc \tan (\ln |x|)]$.   19. $[\arc \sin (\ln |x|)]$.   21. $[-\ln |\arc \cos x|]$.

23. $\frac{1}{2}\left[\arc \tan \ln x - \dfrac{\ln x}{1 + (\ln x)^2}\right]$.   25. $\dfrac{\pi - 2}{4}$.

7. THE HYPERBOLIC FUNCTIONS

Certain combinations of the exponential functions occur so frequently that they have been given special names. The names chosen are due to their relationship to the hyperbola and some remarkable resemblances between the properties of these functions and those of the trigonometric functions.

DEFINITION 10.28 The following functions, collectively referred to as the hyperbolic functions, are defined for all x in $\mathcal{R}$, except as noted:

i. Hyperbolic sine: $\sinh x = \dfrac{e^x - e^{-x}}{2}$.

ii. Hyperbolic cosine: $\cosh x = \dfrac{e^x + e^{-x}}{2}$.

iii. Hyperbolic tangent: $\tanh x = \dfrac{\sinh x}{\cosh x} = \dfrac{e^x - e^{-x}}{e^x + e^{-x}}$.

iv. Hyperbolic cosecant: $\operatorname{csch} x = \dfrac{1}{\sinh x} = \dfrac{2}{e^x - e^{-x}}$, $x \neq 0$.

v. Hyperbolic secant: $\operatorname{sech} x = \dfrac{1}{\cosh x} = \dfrac{2}{e^x + e^{-x}}$.

vi. Hyperbolic cotangent: $\coth x = \dfrac{1}{\tanh x} = \dfrac{e^x + e^{-x}}{e^x - e^{-x}}$, $x \neq 0$.

A number of identities for the hyperbolic functions bear strong resemblances to identities for the trigonometric functions. We give three of these in the following theorem. Others are left to the exercises.

THEOREM 10.29
i. $\cosh^2 - \sinh^2 = 1$.
ii. $\operatorname{sech}^2 + \tanh^2 = 1$.
iii. $\sinh (x + y) = \sinh x \cosh y + \cosh x \sinh y$.

Proof

i. $\cosh^2 x - \sinh^2 x =^1 \left(\dfrac{e^x + e^{-x}}{2}\right)^2 - \left(\dfrac{e^x - e^{-x}}{2}\right)^2$.

$=^2 \dfrac{e^{2x} + 2e^x e^{-x} + e^{-2x}}{4} - \dfrac{e^{2x} - 2e^x e^{-x} + e^{-2x}}{4}$

$=^3 \dfrac{4e^x e^{-x}}{4} =^3 e^x e^{-x} =^4 1$.

ii. $\operatorname{sech}^2 x + \tanh^2 x =^5 \dfrac{1}{\cosh^2 x} + \dfrac{\sinh^2 x}{\cosh^2 x}$

$$=^3 \dfrac{1 + \sinh^2 x}{\cosh^2 x} =^6 \dfrac{\cosh^2 x}{\cosh^2 x} =^3 1.$$

iii. $\sinh x \cosh y + \cosh x \sinh y$

$$=^1 \dfrac{e^x - e^{-x}}{2} \dfrac{e^y + e^{-y}}{2} + \dfrac{e^x + e^{-x}}{2} \dfrac{e^y - e^{-y}}{2}$$

$$=^2 \dfrac{e^{x+y} + e^{x-y} - e^{-x+y} - e^{-x-y}}{4} + \dfrac{e^{x+y} - e^{x-y} + e^{-x+y} - e^{-x-y}}{4}$$

$$=^3 \dfrac{e^{x+y} - e^{-x-y}}{2} =^1 \sinh (x + y)$$

(1) Definition 10.28.
(2) Properties of real numbers and Theorem 10.22i.
(3) Properties of real numbers.
(4) Theorem 10.22i and v.
(5) Definition 10.28.
(6) By (i) of the above proof. ∎

Note the resemblance of the implicit representation of a hyperbola,

$$x^2 - y^2 = 1,$$

to the identity

$$\cosh^2 t - \sinh^2 t = 1.$$

If we let $x = \cosh t$, $y = \sinh t$, we obtain a parametric representation for the right half of the hyperbola $x^2 - y^2 = 1$. (We obtain only this half of the hyperbola since $\cosh t$ is never negative.) This, along with the identities in Theorem 10.29, seems to be reason enough to name these functions as we have. The following theorem points out other resemblances between the trigonometric and hyperbolic functions. Be sure to note the differences as well as the similarities.

THEOREM 10.30

  i. $\sinh' = \cosh$.

  ii. $\cosh' = \sinh$.

  iii. $\tanh' = \operatorname{sech}^2$.

  iv. $\coth' = -\operatorname{csch}^2$.

  v. $\operatorname{sech}' = -\operatorname{sech} \tanh$.

  vi. $\operatorname{csch}' = -\operatorname{csch} \coth$.

We prove (i) and (ii). Proofs of the remaining statements are left to the exercises.

Proof of i

$$\sinh x =^1 \frac{e^x - e^{-x}}{2}$$

$$\sinh' x =^2 \frac{e^x - e^{-x}(-1)}{2}$$

$$=^3 \frac{e^x + e^{-x}}{2}$$

$$=^1 \cosh x.$$

Proof of ii

$$\cosh x =^1 \frac{e^x + e^{-x}}{2}$$

$$\cosh' x =^2 \frac{e^x - e^{-x}}{2}$$

$$=^1 \sinh x.$$

(1) Definition 10.28.
(2) Theorem 10.14 and Definition 10.18, p. 246.
(3) Properties of real numbers. ∎

Example 25. $(\sin(\cosh x))' = \cos(\cosh x) \cdot \sinh x.$ ∎

We seek now information which will permit us to sketch the graph of sinh, cosh, and tanh. It is left to the reader to prove successively the following:
(a) sinh is an odd function, that is, $\sinh(-x) = -\sinh x$.
(b) cosh is an even function, that is, $\cosh(-x) = \cosh x$.
(c) tanh is an odd function.
(d) $\cosh x > 0$ for every $x \in R$.
(e) $\cosh 0 = 1$.
(f) $\sinh x > 0$ if $x > 0$.
(g) $\sinh x < 0$ if $x < 0$.
(h) $\sinh 0 = 0$.
(i) cosh is strictly decreasing on $(-\infty, 0)$ and strictly increasing on $(0, \infty)$.
(j) cosh is concave upward everywhere.
(k) cosh has a minimum at 0.
(l) cosh is unbounded above.

By plotting a few points and using (b), (d), (e), (i), (j), (k), and the fact that cosh is everywhere defined, we can now sketch graph cosh in Figure 10.19.

It will be left to the exercises for the reader to sketch the graphs of the remaining functions.

Remark: Clearly sinh, cosh, and tanh are everywhere continuous.

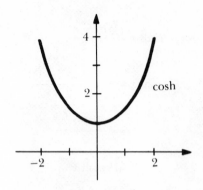

Figure 10.19

EXERCISES

1. Prove (iii) to (vi) of Theorem 10.30.

2. Differentiate:
   (a) $\sinh (x^2 + 1)$.   (b) $\cosh (\ln x)$.   (c) $e^{\tanh x}$.

   (d) $\operatorname{sech}^3 x$.   (e) $\operatorname{sech} x \tan x$.   (f) $\dfrac{\cosh x}{\ln x}$.

3. Give primitive set formulas corresponding to each differentiation formula in Theorem 10.30.

4. Find primitives for the following functions:
   (a) $x \operatorname{sech}^2 x^2$.
   (b) $\dfrac{\sinh (\ln x)}{x}$.
   (c) $\sinh x \cos x - \cosh x \sin x$.
   (d) $\operatorname{sech} x$.

5. Prove statements (a) to (l) which follow Example 25.

6. Sketch the graph of sinh.

7. Sketch the graph of tanh.

ANSWERS

2. (a) $2x \cosh (x^2 + 1)$.   (c) $e^{\tanh x} \operatorname{sech}^2 x$.
   (e) $\operatorname{sech} x \sec^2 x + \operatorname{sech} x \tanh x \tan x$.
4. (a) $\frac{1}{2} \tanh x^2$.   (c) $\cosh x \cos x$.

11

ADDITIONAL TECHNIQUES FOR FINDING PRIMITIVES

This chapter is devoted to developing further techniques for finding primitives. In Section 1, trigonometric substitutions are investigated. In Section 2, additional examples are given in which both the method of substitution and the method of parts are invoked. In Sections 3 and 4, we study a new technique, called the Method of Partial Fractions, which will enable us to find primitives of functions such as $f(x) = \dfrac{1}{(x-2)(x+3)}$ and $f(x) = \dfrac{1}{(x-1)(x^2+4)}$.

1. SPECIAL TECHNIQUES INVOLVING TRIGONOMETRIC IDENTITIES

The trigonometric identities

$$(11.1) \qquad \sin^2 \theta = \frac{1 - \cos 2\theta}{2},$$

$$(11.2) \qquad \cos^2 \theta = \frac{1 + \cos 2\theta}{2},$$

$$(11.3) \qquad \tan^2 \theta = \sec^2 \theta - 1,$$

(11.4) $$\cot^2 \theta = \csc^2 \theta - 1,$$

(11.5) $$\sin^2 \theta = 1 - \cos^2 \theta,$$

and

(11.6) $$\sin 2\theta = 2 \sin \theta \cos \theta$$

can often be used to find members of primitive sets. We shall illustrate their usefulness by means of examples.

We first consider primitive sets of the form $\int \sin^n x$ and $\int \cos^n x$ where n is a positive integer:

Example 1

$$\int \sin^2 x =^1 \int \frac{1 - \cos 2x}{2}$$

$$=^2 \tfrac{1}{2}\left(\int 1 - \tfrac{1}{2}\int 2 \cos 2x\right)$$

$$=^3 \tfrac{1}{2}[x] - \tfrac{1}{4}[\sin 2x]$$

$$=^4 \left[\tfrac{1}{2}x - \tfrac{1}{4}\sin 2x\right].$$

(1) Equation 11.1.
(2) Definition 9.8i (p. 390).
(3) Table 9.1 and the method of substitution ($u = 2x$).
(4) Definition 9.8i and ii. ∎

Example 2

$$\int \sin^3 x =^1 \int (1 - \cos^2 x) \sin x$$

$$=^2 \int \sin x - \int \cos^2 x \sin x$$

$$=^3 \left[-\cos x + \frac{\cos^3 x}{3}\right].$$

(1) Equation 11.5.
(2) Definition 9.8ii (p.390).
(3) Table 9.1 and the method of substitution ($u = \cos x$). ∎

In Chapter 9, we used the Method of Parts to find the primitive sets in Examples 1 and 2. In fact we could use the result of Example 1 together with the Method of Parts to find the primitive set in Example 2. However, this technique is more involved than that exhibited in Example 2.

Combinations of the techniques used in the previous examples can be used to find any primitive set of the form $\int \sin^n x$ or $\int \cos^n x$, where n is a positive integer. We further illustrate this in the next example.

Example 3

$$\int \cos^6 x = \int (\cos^2 x)^3$$

$$= \int \left(\frac{1 + \cos 2x}{2}\right)^3$$

$$= \frac{1}{8} \int (1 + 3 \cos 2x + 3 \cos^2 2x + \cos^3 2x)$$

$$= \frac{1}{8}\left(\int 1 + 3 \int \cos 2x + 3 \int \cos^2 2x + \int \cos^3 2x\right)$$

$$= \frac{1}{8}\left(\int 1 + 3 \int \cos 2x + 3 \int \frac{1 + \cos 4x}{2}\right.$$

$$\left. + \int (1 - \sin^2 2x) \cos 2x\right)$$

$$= \frac{1}{8}\left(\int 1 + 3 \int \cos 2x + \int \frac{3}{2} + \frac{3}{2} \int \cos 4x\right.$$

$$\left. + \int \cos 2x - \int \sin^2 2x \cos 2x\right)$$

$$= \frac{1}{8}\left(\int \frac{5}{2} + 4 \int \cos 2x + \frac{3}{2} \int \cos 4x - \int \sin^2 2x \cos 2x\right)$$

$$= \left[\frac{1}{8}\left(\frac{5}{2} x + 2 \sin 2x + \frac{3}{8} \sin 4x - \frac{1}{6} \sin^3 2x\right)\right]$$

$$= \left[\frac{5}{16} x + \frac{1}{4} \sin 2x + \frac{3}{64} \sin 4x - \frac{1}{48} \sin^3 2x\right]. \quad \blacksquare$$

Primitive sets of the form $\int \tan^n x$, where n is a positive integer, can be found by using Equation 11.3, and sets of the form $\int \cot^n x$, n a positive integer, can be found by using Equation 11.4. We illustrate this statement in the following examples:

Example 4

$$\int \tan^2 x =^1 \int (\sec^2 x - 1)$$

$$=^2 \int \sec^2 x - \int 1$$

$$=^3 [\tan x - x].$$

(1) Equation 11.3.
(2) Definition 9.8 (p. 390) (Addition of Primitive Sets).
(3) Table 9.1 (p. 389) (Primitives and Integrals).   ∎

Example 5

$$\int \cot^5 x =^1 \int (\csc^2 x - 1) \cot^3 x$$

$$=^2 \int \cot^3 x \csc^2 x - \int \cot^3 x$$

$$=^1 \int \cot^3 x \csc^2 x - \int \cot x(\csc^2 x - 1)$$

$$=^2 \int \cot^3 x \csc^2 x - \int \cot x \csc^2 x + \int \cot x$$

$$=^3 \left[\frac{-\cot^4 x}{4} + \frac{\cot^2 x}{2} + \ln |\sin x| \right].$$

(1) Equation 11.4.
(2) Definition 9.8 (Addition of Primitive Sets).
(3) Table 9.1 and substitution $(u = \cot x)$ for the first two primitive sets. Theorem 10.12iii (p. 420) for the last. ∎

In Chapter 9 we found the substitution of a variable u for the function $g(x)$ to be a powerful method for finding primitive sets of functions in the form $(f' \circ g)g'(x)$. The reverse process can also be useful at times, that is, we substitute a function of θ for the variable x. This process is justified by the following theorem, which is analogous to Theorem 9.9 (p. 393).

THEOREM II.I   Let f and g be differentiable and g be invertible.

(11.7) $$\int (f' \circ g^{-1})(g^{-1})' = [f \circ g^{-1}].$$

Moreover, if $\theta(x) = g^{-1}(x)$, then $\theta'(x) = (g^{-1})'(x)$ and Equation 11.7 becomes

$$\int (f' \circ g^{-1})(g^{-1})' = \int f'(\theta)\theta'$$

$$= [f(\theta)]$$

$$= [f \circ g^{-1}].$$

The proof of Theorem 11.1 is a direct application of Theorem 9.9 (p. 393) and Theorem 10.4 (p. 409).

Members of primitive sets for functions containing expressions of the form $(a^2 - x^2)^r$, $(a^2 + x^2)^r$, or $(x^2 - a^2)^r$, where r is a real number, can often be found by substituting for x a trigonometric function of θ and using Theorem 11.1. The particular substitution required depends upon the expression involved. If the function contains a factor of the form $(a^2 - x^2)^r$

and $a^2 \geq x^2$, we attempt the substitution $x = a \sin \theta$. Then $(a^2 - x^2)^r = (a^2 - a^2 \sin^2 \theta)^r = a^{2r}(1 - \sin^2 \theta)^r = a^{2r} \cos^{2r} \theta$. This substitution is suggested by the trigonometric identity $1 - \sin^2 \theta = \cos^2 \theta$.

Similarly, for a function involving an expression of the form $(a^2 + x^2)^r$, we substitute $x = a \tan \theta$. This substitution is suggested by the identity $1 + \tan^2 \theta = \sec^2 \theta$. Finally, for an expression of the form $(x^2 - a^2)$ where $x \geq a^2$, we substitute $x = a \sec \theta$, a substitution suggested by the identity $\sec^2 \theta - 1 = \tan^2 \theta$. With each of these substitutions we obtain a simpler expression.

Note carefully that in applying Theorem 11.1 by substituting $\sin \theta$ for x, we are assuming that θ is a function of x—that is, we are assuming that $\theta(x) = \arcsin x$ is well defined. This calls for a restriction on x, namely, $-1 \leq x \leq 1$. Similar restrictions are called for if we substitute $\cos \theta$, $\sec \theta$, or $\tan \theta$ for x.

Example 6

(a) Let us find the primitive set $\int \dfrac{1}{\sqrt{1 - x^2}}$ by trigonometric substitution. If we substitute $x = \sin \theta$ and note that $\theta(x) = \arcsin x$ and $\theta'(x) = \dfrac{1}{\sqrt{1 - x^2}}$, we have, by Theorem 11.1,

$$\int \frac{1}{\sqrt{1 - x^2}} = \int \theta'(x)$$

$$= [\theta(x)]$$

$$= [\arcsin x].$$

This agrees with Theorem 10.27 (p. 437).

(b) Consider now $\int \dfrac{x}{a^2 - x^2}$, where $x^2 < a^2$ and $a > 0$. This time we substitute $x = a \sin \theta$, $\theta(x) = \arcsin \dfrac{x}{a}$, and $\theta'(x) = \dfrac{1/a}{\sqrt{1 - x^2/a^2}} = \dfrac{1}{\sqrt{a^2 - x^2}}$.

Invoking Theorem 11.1, we have

$$\int \frac{x}{a^2 - x^2} = \int a \sin \theta \cdot \theta'$$

$$=^1 [-a \cos \theta]$$

$$=^2 \left[-a\sqrt{1 - \frac{x^2}{a^2}} \right]$$

$$= [-\sqrt{a^2 - x^2}].$$

(1) Table 9.1ii.

(2) If $x = a \sin \theta$, then $\cos^2 \theta = 1 - \sin^2 \theta = 1 - \dfrac{x^2}{a^2}$. ∎

Each of the functions considered in Example 6 was of a type previously studied. As noted in Example 6a, $\displaystyle\int \frac{1}{\sqrt{1 - x^2}} = [\arcsin x]$. In Example 6b we could have used the familiar substitution $u = a^2 - x^2$ and arrived at the same primitive set. The next example will illustrate more positively the value of trigonometric substitution.

Example 7

(**a**) Consider $\displaystyle\int \sqrt{4 - x^2}$. The reader is encouraged to attempt to arrive at a solution using the substitution $u = 4 - x^2$. We shall use the substitutions $x = 2 \sin \theta$, $\theta = \arcsin \left(\dfrac{x}{2}\right)$, and $\theta' = \dfrac{1}{2\sqrt{1 - x^2/4}} = \dfrac{1}{\sqrt{4 - x^2}}$.

We then note that

$$\int \sqrt{4 - x^2} = \int \frac{4 - x^2}{\sqrt{4 - x^2}}$$

$$=^1 \int (4 - 4 \sin^2 \theta)\theta'$$

$$=^2 \int 2(1 + \cos 2\theta)\theta'$$

$$=^3 \left[2\left(\theta + \frac{\sin 2\theta}{2}\right)\right]$$

$$=^4 [2\theta + 2 \sin \theta \cos \theta]$$

$$=^5 \left[2 \arcsin \left(\frac{x}{2}\right) + \frac{x\sqrt{4 - x^2}}{2}\right].$$

(1) Using the indicated substitutions and Theorem 11.1.

(2) Since $4 - 4 \sin^2 \theta = 4(1 - \sin^2 \theta) = 4 \cos^2 \theta = 4\dfrac{1 + \cos 2\theta}{2} = 2(1 + \cos 2\theta)$.

(3) Table 9.1.

(4) Since $\sin 2\theta = 2 \sin \theta \cos \theta$.

(5) Resubstituting $\theta = \arcsin \left(\dfrac{x}{2}\right)$, $\sin \theta = \dfrac{x}{2}$, and $\cos \theta = \sqrt{1 - \sin^2 \theta} = \sqrt{1 - \dfrac{x^2}{4}} = \dfrac{\sqrt{4 - x^2}}{2}$.

(**b**) Now consider $\displaystyle\int \frac{x^2}{\sqrt{a^2 - x^2}}$ for $x^2 < a^2$ and $a > 0$. Again we use the substitution $x = a \sin \theta$, $\theta = \arcsin \dfrac{x}{a}$, and $\theta' = \dfrac{1/a}{\sqrt{1 - x^2/a^2}} = \dfrac{1}{\sqrt{a^2 - x^2}}$. Thus,

$$\int \frac{x^2}{\sqrt{a^2 - x^2}} =^1 \int a^2 \sin^2 \theta \cdot \theta'$$

$$=^2 a^2[\tfrac{1}{2}\theta - \tfrac{1}{4}\sin 2\theta]$$

$$=^3 a^2[\tfrac{1}{2}\theta - \tfrac{1}{2}\sin \theta \cos \theta]$$

$$=^4 a^2\left[\frac{1}{2}\arcsin \frac{x}{a} - \frac{1}{2}\frac{x\sqrt{a^2 - x^2}}{a^2}\right]$$

$$= \left[\frac{1}{2}a^2 \arcsin \frac{x}{a} - \frac{x}{2}\sqrt{a^2 - x^2}\right].$$

(1) Using the indicated substitutions and Theorem 11.1.
(2) By Example 1 (p. 444).
(3) Since $\sin 2\theta = 2 \sin \theta \cos \theta$.
(4) Resubstituting $\theta = \arcsin \dfrac{x}{a}$, $\sin \theta = \dfrac{x}{a}$, and $\cos \theta = \sqrt{1 - \sin^2 \theta} =$

$$\sqrt{1 - \frac{x^2}{a^2}} = \frac{\sqrt{a^2 - x^2}}{a}. \quad \blacksquare$$

Thus far we have concentrated on functions involving expressions of the form $(a^2 - x^2)^r$, $a^2 \geq x^2$. Let us now investigate two examples of functions involving expressions of the form $(a^2 + x^2)^r$.

Example 8

(a) Consider $\displaystyle\int \frac{1}{(x^2 + 1)^{3/2}}$. We use the suggested substitutions $x = \tan \theta$, $\theta = \arctan x$, and $\theta' = \dfrac{1}{1 + x^2}$ where $x > 0$. Thus,

$$\int \frac{1}{(1 + x^2)^{3/2}} = \int \frac{1}{\sqrt{1 + x^2}(1 + x^2)}$$

$$=^1 \int \frac{\theta'}{\sqrt{1 + \tan^2 \theta}}$$

$$=^2 \int \cos \theta \cdot \theta'$$

$$=^3 [\sin \theta]$$

$$=^4 \left[\frac{x}{\sqrt{1 + x^2}}\right].$$

(1) Using the indicated substitutions and Theorem 11.1.

(2) Since $\dfrac{1}{\sqrt{1 + \tan^2 \theta}} = \dfrac{1}{\sec \theta} = \cos \theta$.

(3) Table 9.1.

(4) Note first that $\cos^2 \theta = \dfrac{1}{\sec^2 \theta} = \dfrac{1}{1 + \tan^2 \theta} = \dfrac{1}{1 + x^2}$. Then $\sin \theta =$

$\sqrt{1 - \cos^2 \theta} = \sqrt{\dfrac{x^2}{1 + x^2}} = \dfrac{x}{\sqrt{1 + x^2}}$.

(b) Consider $\displaystyle\int \dfrac{1}{(x^2 - 2x + 5)^{3/2}}$. If we note that $x^2 - 2x + 5 = (x - 1)^2 + 4$, we see that we are concerned with finding the primitive set of $\dfrac{1}{(u^2 + 4)^{3/2}}$, where $u = x - 1$. This suggests the substitution $u = 2 \tan \theta$ or $x - 1 = 2 \tan \theta$, $\theta = \arctan \dfrac{x - 1}{2}$, and

$$\theta' = \frac{\frac{1}{2}}{1 + \dfrac{(x - 1)^2}{4}} = \frac{2}{4 + (x - 1)^2}.$$

Therefore,

$$\int \frac{1}{(x^2 - 2x + 5)^{3/2}} = \frac{1}{2} \int \frac{1}{\sqrt{4 + (x - 1)^2}} \cdot \frac{2}{(4 + (x - 1)^2)}$$

$$= \frac{1}{2} \int \frac{1}{\sqrt{4 + 4 \tan^2 \theta}} \cdot \theta'$$

$$= \frac{1}{4} \int \frac{1}{\sec \theta} \cdot \theta'$$

$$= \frac{1}{4} [\sin \theta]$$

$$= \frac{1}{4} \left[\frac{x - 1}{\sqrt{x^2 - 2x + 5}} \right].$$

The reader is encouraged to justify each step in this solution. ∎

In attempting the exercises which follow the reader should remember to make use of the familiar trigonometric identities, exhibited in Equations 11.1 to 11.6, and to also use the following substitutions: $x = a \sin \theta$ in $(a^2 - x^2)^r$, $x = a \tan \theta$ in $(a^2 + x^2)^r$, and $x = a \sec \theta$ in $(x^2 - a^2)^r$.

EXERCISES

Find the following primitive sets:

1. $\int \cos^2 2x.$

2. $\int \sin^4 x.$

3. $\int \cot^3 x.$

4. $\int \tan^3 3x.$

5. $\int \cos^5 x.$

6. $\int \sin^6 x.$

7. $\int \dfrac{\sqrt{9 - x^2}}{x}.$

8. $\int \dfrac{\sqrt{x^2 - 4}}{x^2}.$

9. $\int \dfrac{1}{(a^2 - x^2)^{3/2}}.$

10. $\int \dfrac{x^2}{(x^2 + 9)^{3/2}}.$

11. $\int \dfrac{x^2}{(4 - x^2)^{5/2}}.$

12. $\int \dfrac{x}{\sqrt{4 - x^2}}.$

13. $\int \tan^2 \theta \sec \theta.$

14. $\int x^2 \sqrt{1 - x^2}.$

ANSWERS

1. $\left[\dfrac{x}{2} + \dfrac{1}{8} \sin 4x \right].$ 3. $\left[\dfrac{-\cot^2 x}{2} - \ln |\sin x| \right].$

5. $\left[\sin x - \dfrac{2}{3} \sin^3 x + \dfrac{\sin^5 x}{5} \right].$ 7. $\left[\sqrt{9 - x^2} - 3 \ln \left| \dfrac{3}{x} + \dfrac{\sqrt{9 - x^2}}{x} \right| \right].$

9. $\left[\dfrac{x}{a^2 \sqrt{a^2 - x^2}} \right].$ 11. $\left[\dfrac{x^3}{12(4 - x^2)^{3/2}} \right].$

13. $[\tfrac{1}{2}(\sec \theta \tan \theta - \ln |\sec \theta + \tan \theta|)].$

2. COMBINED METHODS OF SUBSTITUTION AND PARTS

In Chapter 9, we saw several examples of primitive sets which could be determined by using both the Method of Substitution and the Method of Parts (see the exercises at the end of Chapter 9). In Chapter 9 the exponential, logarithmic, and inverse trigonometric functions had not yet been introduced. Therefore, we devote this section to the aforementioned technique as applied to these special functions.

Example 9. Consider $\int e^{2x} \sin e^x$. The obvious substitution called for is $u = e^x$. We then have

$$\int e^{2x} \sin e^x = \int e^x (\sin e^x) e^x$$

$$=^1 \int u(\sin u) u'$$

$$=^2 [-u \cos u] + \int (\cos u) u'$$

$$=^3 [-u \cos u] + [\sin u]$$

$$=^4 [\sin e^x - e^x \cos e^x].$$

(1) Using the substitution $u(x) = e^x = u'(x)$ and Theorem 9.9.
(2) Using the Method of Parts with

$$f(u) = u, \qquad g'(u) = (\sin u) u',$$

$$f'(u) = u', \quad \text{and} \quad g(u) = -\cos u.$$

(3) Table 9.1.
(4) Addition of primitive sets and resubstitution of $u = e^x$. ∎

Example 10. Consider

$$\int \frac{\sqrt{1 - x^2} \arcsin x + x}{\sqrt{1 - x^2}}.$$

We attempt the substitution $u = \arcsin x$. Thus, we have $u' = \dfrac{1}{\sqrt{1 - x^2}}$, $\sin u = x$, and $\cos u = \sqrt{1 - x^2}$. Therefore,

$$\int \frac{\sqrt{1 - x^2} \arcsin x + x}{\sqrt{1 - x^2}} =^1 \int ((\cos u) u + \sin u) u'$$

$$=^2 \int u \cos u u' + \int \sin u u'$$

$$=^3 [u \sin u + \cos u] - [\cos u]$$

$$=^4 [x \arcsin x].$$

(1) Using the indicated substitutions.
(2) Addition of primitive sets.
(3) Method of Parts on the first primitive set and Table 9.1.
(4) Addition of primitive sets and resubstituting $u = \arcsin x$ and $\sin x = u$. ∎

The next example will be abbreviated. The reader is encouraged to determine the rationale for each step.

Example II. $\displaystyle\int \frac{(\ln x^2)\cos(\ln x^2)}{x} = \frac{1}{2}\int u(\cos u)u'$

$$= \frac{1}{2}\left([u\sin u] - \int \sin uu'\right)$$

$$= \frac{1}{2}\left([u\sin u + \cos u]\right)$$

$$= \frac{1}{2}[(\ln x^2)\sin(\ln x^2) + \cos(\ln x^2)].$$

∎

EXERCISES

1. $\displaystyle\int x\sin\sqrt{x^2-1}.$ 2. $\displaystyle\int \sin\theta\cos\theta(1+\sin\theta)^{10}.$

3. $\displaystyle\int \sqrt{x}\sin\sqrt{x}.$ 4. $\displaystyle\int \cos x(\sin^3 x)e^{\sin^2 x}.$

5. $\displaystyle\int \sin x(\cos x)4^{\cos x}.$ 6. $\displaystyle\int 3^{\tan x}\sec^2 x\tan x.$

7. $\displaystyle\int \frac{(\ln x^2)\sec^2(\ln x^2)\tan(\ln x^2)}{x}.$

8. $\displaystyle\int \sqrt{1-x^2}\arcsin\sqrt{1-x^2}$ (Hint: let $x=\cos\theta$).

9. $\displaystyle\int ye^{\sqrt{1-y^2}}.$ 10. $\displaystyle\int \sin(\cos x) - x\sin x\cos(\cos x).$

ANSWERS

1. $[\sin\sqrt{x^2-1} - \sqrt{x^2-1}\cos\sqrt{x^2-1}].$

3. $[4\sqrt{x}\sin\sqrt{x} + 4\cos\sqrt{x} - 2x\cos\sqrt{x}].$ **5.** $\left[\dfrac{4^{\cos x}}{(\ln 4)^2} - \dfrac{4^{\cos x}\cos x}{\ln 4}\right].$

7. $\frac{1}{4}[(\ln x^2)\sec^2(\ln x^2) - \tan(\ln x^2)].$ **9.** $[e^{\sqrt{1-y^2}}(1-\sqrt{1-y^2})].$

3. THE METHOD OF PARTIAL FRACTIONS

We recall that a (real) *polynomial of degree n* is a function of the form $a_nx^n + a_{n-1}x^{n-1} + \cdots + a_1x + a_0$, where n is a non-negative integer, $a_0, \ldots, a_n$ are real numbers, and $a_n \neq 0$. a_n is called the *leading* coefficient of the polynomial and a_j, $j = 0, \ldots, n$ is called the *coefficient of the term of degree j*. Polynomials of degree 1 are frequently called *linear polynomials* and those of degree 2 are called quadratic polynomials. The function p, defined by $p(x) = 0$ for all x, is called the zero polynomial and is *not* assigned a degree. If the leading coefficient a_n equals 1, the polynomial is called *monic*. A polynomial Q is said to be a *factor* of a polynomial P if and only if there exists a polynomial R such that $P = QR$. A quadratic polynomial is called *irreducible* if and only if it has no linear factors (otherwise it is called *reducible*).

A function of the form $\dfrac{P}{Q}$, where P and Q are polynomials, is called a *rational function*. If P and Q have no common factors, then $\dfrac{P}{Q}$ is said to be in *reduced form*. In this and the following section we shall develop a technique by which we can find primitives of rational functions. With this technique we shall be able to find primitives of such functions as $f(x) = \dfrac{1}{(x-2)(x+3)}$ and $g(x) = \dfrac{x+4}{(x+1)^2(x^2+x+2)^3}$. The basis for this technique is algebraic.

Before starting the development of this technique, let us consider simple examples of polynomials and rational functions.

Example 12

a. $2x$ and $4x + 1$ are polynomials of degree 1, that is, linear polynomials.

b. $4x^2$, $3x^2 - 1$, and $x^2 + x + 1$ are polynomials of degree 2—quadratic polynomials. Since $4x^2 = (4x)(x)$ and $4x^2 - 1 = (2x - 1)(2x + 1)$, these two polynomials are reducible. The reader is asked to check that $x^2 + x + 1$ has no linear factors and hence is irreducible.

c. $4x^3 + 1$ is a polynomial of degree 3 whereas $0 \cdot x^3 + x^2 + 3$ is a polynomial of degree 2.

d. $\dfrac{x+1}{x-4}$ and $\dfrac{x^2+3}{x^3+2x^2-1}$ are rational functions. ∎

Let us return to the problem of determining primitives of rational functions. Let $R = \dfrac{P}{Q}$ be a rational function; then, by division, we can rewrite

$$R(x) = \frac{P(x)}{Q(x)} = a_n x^n + a_{n-1} x^{n-1} + \cdots + a_1 x + a_0 + \frac{P_1(x)}{Q(x)},$$

where P_1 is a polynomial of degree less than that of Q. Thus, the problem of finding primitives of rational functions reduces to the case in which the degree of the numerator is less than that of the denominator.

Example 13

a. $\dfrac{x^2 - 2}{x + 1} = x - 1 - \dfrac{1}{x + 1}$, and

b. $\dfrac{x^2 + 3x - 1}{x^2 + x + 2} = 1 + \dfrac{2x - 3}{x^2 + x + 2}$. ∎

The following theorems of algebra are basic in our development but will be stated without proof.

THEOREM 11.2 Let P be a polynomial of degree n.
i. If P is monic and $n > 0$, then P can be uniquely factored as the product of not more than n monic linear and monic irreducible quadratic polynomials.
ii. If $P(a) = 0$, there exists a unique polynomial Q such that $P(x) = (x - a)Q(x)$ for all real x.

In the following theorem the expression $i!$ stands for the product $i(i - 1) \cdots 2 \cdot 1$. For example $3! = 3 \cdot 2 \cdot 1$ and $6! = 6 \cdot 5 \cdot 4 \cdot 3 \cdot 2 \cdot 1$. The superscript (k) denotes the k-th derivative. For example $f^{(2)}$ denotes the second derivative of f.

THEOREM 11.3 Let $\dfrac{P}{Q}$ be a rational function of the form

$$\frac{P(x)}{Q(x)} = \frac{P(x)}{(x - a)^n Q_1(x)},$$

where the degree of P is less than the degree of Q and $Q_1(a) \neq 0$. There exists unique real numbers $A_0, \ldots, A_{n-1}$ and a unique polynomial P_n of degree less than that of Q_1 such that

(11.8) $\dfrac{P(x)}{(x - a)^n Q_1(x)} = \dfrac{A_0}{(x - a)^n} + \dfrac{A_1}{(x - a)^{n-1}} + \cdots$

$$+ \frac{A_{n-1}}{x - a} + \frac{P_n(x)}{Q_1(x)}.$$

Moreover, the numbers $A_i (i = 0, 1, \ldots, n - 1)$ are given by

$$A_0 = \frac{P(a)}{Q_1(a)},$$

(11.9) $A_i = \dfrac{1}{i!} \left(\dfrac{P}{Q_1} \right)^{(i)} (a), \ (i = 1, \ldots, n - 1).$

Proof

i. We first show that there is a number A_0 and polynomial P_1 such that

(11.10) $$\frac{P(x)}{(x-a)^n Q_1(x)} = \frac{A_0}{(x-a)^n} + \frac{P_1(x)}{(x-a)^{n-1} Q_1(x)}.$$

To this end, consider the following implication chain:

$$\text{(11.10) is valid} \Leftrightarrow^1 P(x) = A_0 Q_1(x) + P_1(x)(x-a)$$

(11.11) $$\Leftrightarrow P(x) - A_0 Q_1(x) = P_1(x)(x-a)$$

$$\Leftrightarrow^2 P(a) - A_0 Q_1(a) = 0$$

$$\Leftrightarrow^3 A_0 = \frac{P(a)}{Q_1(a)}.$$

(1) ($\Rightarrow$) Adding fractions in right member of Equation 11.10 and equating numerators of left and right members.
  ($\Leftarrow$) Division by $(x-a)^n Q_1(x)$ and simplification.
(2) .($\Rightarrow$) Evaluation at a.
  ($\Leftarrow$) Theorem 11.2ii with P replaced by $P - A_0 Q_1$.
(3) ($\Rightarrow$) Solving for A_0 since, by hypothesis, $Q_1(a) \neq 0$.
  ($\Leftarrow$) Multiplying by $Q_1(a)$ and subtracting $A_0 Q_1(a)$. ∎

Thus, we see that there is exactly one number A_0 and, from Equation 11.11, one polynomial

$$P_1(x) = \frac{P(x) - A_0 Q_1(x)}{(x-a)} \quad \text{such that Equation 11.10 is valid.}$$

ii. We now apply the argument given in part i to the rational function $\dfrac{P_1(x)}{(x-a)^{n-1} Q_1(x)}$ to obtain a real number A_1 and polynomial P_2 such that

(11.12) $$\frac{P_1(x)}{(x-a)^{n-1} Q_1(x)} = \frac{A_1}{(x-a)^{n-1}} + \frac{P_2(x)}{(x-a)^{n-2} Q_1(x)}.$$

Substitution from (11.12) into (11.10) then gives

$$\frac{P(x)}{(x-a)^n Q_1(x)} = \frac{A_0}{(x-a)^n} + \frac{A_1}{(x-a)^{n-1}} + \frac{P_2(x)}{(x-a)^{n-2} Q_1(x)}.$$

iii. Applying the argument of parts (i) and (ii) n times, we obtain real numbers $A_0, A_1, \ldots, A_{n-1}$ and a polynomial P_n such that Equation 11.8 holds.

iv. To show that the numbers $A_0, A_1, \ldots, A_{n-1}$ are given by Equation 11.9, we multiply both members of Equation 11.8 by $(x-a)^n$ to obtain

$$\frac{P(x)}{Q_1(x)} = A_0 + A_1(x-a) + \cdots + A_{n-1}(x-a)^{n-1} + \frac{P_n(x)(x-a)^n}{Q_1(x)}.$$

Evaluation at a gives $\dfrac{P(a)}{Q_1(a)} = A_0$. Computation of the first derivative of $\dfrac{P}{Q_1}$ (using the quotient and product rules on the last term) followed by evaluation at a gives

$$\left(\frac{P}{Q_1}\right)'(a) = A_1.$$

Computation and evaluation of successive derivatives verify the remainder of Equations 11.9. The uniqueness of the A's and thus the uniqueness of the polynomial

$$P_n(x) = \frac{P(x)}{(x-a)^n Q_1(x)} - \frac{A_0 Q_1(x)}{(x-a)^n} + \cdots + \frac{A_{n-1} Q_1(x)}{x-a}$$

is thereby established.

v. We still must show that the degree of P_n is less than the degree of Q_1: $\deg P_n < \deg Q_1$. Multiplication of both members of Equation 11.8 by $(x-a)^n Q_1(x)$ leads to

(11.13) $(x-a)^n P_n(x)$

$$= P(x) - [A_0 + A_1(x-a) + \cdots + A_{n-1}(x-a)^{n-1}] Q_1(x).$$

Let $L.M.$ denote the left member of Equation 11.13 and $R.M.$ denote the right member. Since we assumed that

$$\deg P < \deg Q = \deg [(x-a)^n Q_1(x)] = n + \deg Q_1,$$

we find that each term of $R.M.$ has degree $< n + \deg Q_1$. Thus,

$$\deg R.M. < n + \deg Q_1.$$

But $\deg L.M. = \deg [(x-a)^n P_n(x)] = n + \deg P_n$.

Hence, $n + \deg P_n = \deg L.M. = \deg R.M. < n + \deg Q_1$.

Therefore, $\deg P_n < \deg Q_1$. ∎

Equation 11.8 is called a *partial fraction decomposition* of the rational function $\dfrac{P}{Q}$. We now consider the value of a partial fraction decomposition in finding primitives of a rational function $\dfrac{P}{Q}$. Assume that $\dfrac{P}{Q}$ is in reduced form, that $Q(x) = (x-a)^n Q_1(x)$, and that $Q_1(a) \neq 0$. We may then apply Theorem 11.3 to rewrite $\dfrac{P}{Q}$ in the form given in Equation 11.8, that is, as a sum of terms of the form $\dfrac{A_j}{(x-a)^j}$ and a last term $\dfrac{P_n}{Q_1}$. Application of the formula

$$\int \frac{1}{(x-a)^j} = \begin{cases} [\ln|x-a|], & \text{if } j = 1 \\ \left[\dfrac{-1}{(j-1)(x-a)^{j-1}}\right], & \text{if } j \neq 1 \end{cases}$$

yields the primitive sets for all but the last term: $\dfrac{P_n}{Q_1}$. If Q_1 has other linear factors we may reapply Theorem 11.3 to the rational function $\dfrac{P_n}{Q_1}$ to further resolve the problem.

When the denominator of $\dfrac{P}{Q}$ is a product of distinct linear factors the following consequence of Theorem 11.3 provides a very efficient means for effecting the partial fraction decomposition for $\dfrac{P}{Q}$. The proof (by induction on the number n of linear factors) is left to the reader.

THEOREM 11.4 Let $\dfrac{P}{Q}$ be a rational function in reduced form such that Q is a product of distinct linear factors:

(11.14) $$\frac{P(x)}{Q(x)} = \frac{P(x)}{(x - r_1)(x - r_2) \cdots (x - r_n)}.$$

Then

(11.15) $$\frac{P(x)}{Q(x)} = \frac{A_1}{x - r_1} + \frac{A_2}{x - r_2} + \cdots + \frac{A_n}{x - r_n},$$

where

(11.16) $$A_j = \frac{P(r_j)}{(r_j - r_1) \cdots (r_j - r_{j-1})(r_j - r_{j+1}) \cdots (r_j - r_n)}.$$

The content of Theorem 11.4 is frequently called the *cover-up method* of determining the partial fraction decomposition of (11.15). For according to (11.16) the coefficient A_j can be obtained by striking the j-th factor from the denominator of (11.14) and substituting r_j for x in what remains. That is, we "cover up" the factor $(x - r_j)$ and evaluate the resulting rational function at r_j.

It should be obvious that we can easily find primitives of the rational functions P/Q described in Theorem 11.4. Each of the terms in the right member of Equation 11.15 has a primitive of the form $A_j \ln |x - r_j|$.

The first examples we consider shall be applications of Theorem 11.4. It is convenient to symbolize the evaluation of a function $R(x)$ at a by the symbol $R(x)|_{x=a}$.

Example 14

(a) Consider the rational function $\dfrac{1}{x^2 - 1} = \dfrac{1}{(x - 1)(x + 1)}$. By Theorem 11.4,

$$\frac{1}{x^2 - 1} = \frac{1}{(x - 1)(x + 1)}$$

$$= \frac{A}{x - 1} + \frac{B}{x + 1},$$

where, by the cover-up method,

$$A = \frac{1}{x+1}\bigg|_{x=1} = \frac{1}{2}$$

and

$$B = \frac{1}{x-1}\bigg|_{x=-1} = -\frac{1}{2}.$$

Thus,

$$\frac{1}{x^2-1} = \frac{\frac{1}{2}}{x-1} - \frac{\frac{1}{2}}{x+1}$$

and

$$\int \frac{1}{x^2-1} = \frac{1}{2}\int \frac{1}{x-1} - \frac{1}{2}\int \frac{1}{x+1}$$

$$= \frac{1}{2}\left[\ln|x-1|\right] - \frac{1}{2}\left[\ln|x+1|\right]$$

$$= \left[\ln\sqrt{\left|\frac{x-1}{x+1}\right|}\right].$$

(**b**) Consider $\dfrac{2x+3}{x^2-1} = \dfrac{2x+3}{(x-1)(x+1)}$. Applying Theorem 11.4,

we have

$$\frac{2x+3}{x^2-1} = \frac{2x+3}{(x+1)(x-1)}$$

$$= \frac{A}{x-1} + \frac{B}{x+1},$$

where

$$A = \frac{2x+3}{x+1}\bigg|_{x=1} = \frac{5}{2}$$

and

$$B = \frac{2x+3}{x-1}\bigg|_{x=-1} = -\frac{1}{2}.$$

Therefore,

$$\frac{2x+3}{x^2-1} = \frac{\frac{5}{2}}{x-1} - \frac{\frac{1}{2}}{x+1}$$

and

$$\int \frac{2x+3}{x^2-1} = \frac{5}{2}\int \frac{1}{x-1} - \frac{1}{2}\int \frac{1}{x+1}$$

$$= \frac{5}{2}\left[\ln|x-1|\right] - \frac{1}{2}\left[\ln|x+1|\right]$$

$$= \left[\ln\sqrt{\left|\frac{(x-1)^5}{x+1}\right|}\right].$$

(c) Consider $\dfrac{x^2 + x}{(x-1)(x-2)(x+3)}$. By Theorem 11.4 we can write

$$\frac{x^2 + x}{(x-1)(x-2)(x+3)} = \frac{A}{x-1} + \frac{B}{x-2} + \frac{C}{x+3},$$

where

$$A = \frac{x^2 + x}{(x-2)(x+3)}\bigg|_{x=1} = -\frac{2}{4} = -\frac{1}{2},$$

$$B = \frac{x^2 + x}{(x-1)(x+3)}\bigg|_{x=2} = \frac{6}{5},$$

and

$$C = \frac{x^2 + x}{(x-1)(x-2)}\bigg|_{x=-3} = \frac{6}{20} = \frac{3}{10}.$$

Therefore,

$$\int \frac{x^2 + x}{(x-1)(x-2)(x+3)} = -\frac{1}{2}\int \frac{1}{x-1} + \frac{6}{5}\int \frac{1}{x-2} + \frac{3}{10}\int \frac{1}{x+3}$$

$$= \left[-\frac{1}{2}\ln|x-1| + \frac{6}{5}\ln|x-2| + \frac{3}{10}\ln|x+3| \right]. \quad \blacksquare$$

Now let us consider applications of Theorem 11.3 to rational functions in which the denominator has multiple linear factors.

Example 15. We wish to find a primitive for the rational function $\dfrac{x^2}{(x+1)^2(x+2)}$. By Theorem 11.3, with $P(x) = x^2$ and $Q_1(x) = x + 2$, we can write

$$\frac{x^2}{(x+1)^2(x+2)} = \frac{A}{(x+1)^2} + \frac{B}{(x+1)} + \frac{P_2(x)}{x+2}.$$

where

$$A = \frac{x^2}{x+2}\bigg|_{x=-1} = 1$$

and

$$B = \left(\frac{x^2}{x+2}\right)'\bigg|_{x=-1} = \frac{x^2 + 4x}{(x+2)^2}\bigg|_{x=-1} = -3,$$

and P_2 is a polynomial of degree less than 1—hence, a constant, C. Therefore,

$$\frac{x^2}{(x+1)^2(x+2)} = \frac{1}{(x+1)^2} - \frac{3}{x+1} + \frac{C}{x+2}.$$

To evaluate C we need only substitute a real number $x \neq -1$ or -2.

Choosing $x = 0$, we arrive at $0 = 1 - 3 + \dfrac{C}{2}$, or $C = 4$. Thus,

$$\frac{x^2}{(x + 1)^2(x + 2)} = \frac{1}{(x + 1)^2} - \frac{3}{x + 1} + \frac{4}{x + 2},$$

and consequently,

$$\int \frac{x^2}{(x + 1)^2(x + 2)} = \int \frac{1}{(x + 1)^2} - 3\int \frac{1}{x + 1} + 4\int \frac{1}{x + 2}$$

$$= \left[\frac{-1}{x + 1} - 3 \ln |x + 1| + 4 \ln |x + 2|\right]. \quad \blacksquare$$

Example 16. Consider the rational function $\dfrac{1}{(x - 1)^3(x + 1)^2}$.

Applying Theorem 11.3 with $P(x) = 1$ and $Q_1(x) = (x + 1)^2$, we find that

$$\frac{1}{(x - 1)^3(x + 1)^2} = \frac{A}{(x - 1)^3} + \frac{B}{(x - 1)^2} + \frac{C}{(x - 1)} + \frac{Dx + E}{(x + 1)^2},$$

where

$$A = \frac{P(x)}{Q_1(x)}\bigg|_{x=1} = \frac{1}{(x + 1)^2}\bigg|_{x=1} = \frac{1}{4},$$

$$B = \left(\frac{P}{Q_1}\right)'(x)\bigg|_{x=1} = \left(\frac{1}{(x + 1)^2}\right)'\bigg|_{x=1} = \frac{-2}{(x + 1)^3}\bigg|_{x=1} = -\frac{1}{4}$$

and

$$C = \frac{1}{2!}\left(\frac{P}{Q_1}\right)''(x)\bigg|_{x=1} = \frac{1}{2!}\left(\frac{1}{(x + 1)^2}\right)''\bigg|_{x=1} = \frac{1}{2!}\frac{6}{(x + 1)^4}\bigg|_{x=1} = \frac{3}{16}.$$

Therefore,

$$\frac{1}{(x - 1)^3(x + 1)^2} = \frac{\frac{1}{4}}{(x - 1)^3} - \frac{\frac{1}{4}}{(x - 1)^2} + \frac{\frac{3}{16}}{x - 1} + \frac{Dx + E}{(x + 1)^2}.$$

Substituting any two real numbers not equal to 1 or -1 and solving the resulting equations, we find that $D = -\frac{3}{16}$ and $E = -\frac{5}{16}$. Consequently,

$$\frac{1}{(x - 1)^3(x + 1)^2} = \frac{\frac{1}{4}}{(x - 1)^3} - \frac{\frac{1}{4}}{(x - 1)^2} + \frac{\frac{3}{16}}{x - 1} + \frac{-\frac{3}{16}x - \frac{5}{16}}{(x + 1)^2}$$

and

$$\int \frac{1}{(x - 1)^3(x + 1)^2} = \frac{1}{4}\int \frac{1}{(x - 1)^3} - \frac{1}{4}\int \frac{1}{(x - 1)^2}$$

$$+ \frac{3}{16}\int \frac{1}{x - 1} - \frac{1}{16}\int \frac{3x + 5}{(x + 1)^2}.$$

The first three primitive sets in the right member of this last equation can be determined either by the power formula or by the logarithm formula. The last primitive set can be determined by the substitution $u = x + 1$. $\quad \blacksquare$

The determination of primitive sets of rational functions P/Q is most efficiently accomplished by the following procedure:

i. Put P/Q in reduced form by elimination of common factors of P and Q.

ii. If $\deg P \geq \deg Q$, divide Q into P to obtain

$$\frac{P(x)}{Q(x)} = p(x) + \frac{P_0(x)}{Q(x)},$$

where $p(x)$ is a polynomial and $\deg P_0 < \deg Q$.

iii. Apply the partial fraction decomposition to $\dfrac{P_0}{Q}$.

iv. Find primitive sets of each term obtained.

EXERCISES

Find the following primitive sets

1. $\displaystyle\int \frac{x + 5}{x^2 - x - 6}$.

2. $\displaystyle\int \frac{3x + 7}{x^2 - 2x - 3}$.

3. $\displaystyle\int \frac{1}{x^2 + x}$.

4. $\displaystyle\int \frac{x}{x - 3}$.

5. $\displaystyle\int \frac{2x^2 + 5x - 4}{x^3 + x^2 - 2x}$.

6. $\displaystyle\int \frac{x^2 + 1}{(x - 1)(x - 4)}$.

7. $\displaystyle\int \frac{2x^2 + 5x + 5}{x(x + 5)(x + 2)}$.

8. $\displaystyle\int \frac{x}{x^3 + 3x - 4}$.

9. $\displaystyle\int \frac{x + 1}{2(x^4 + 6x^3 + 12x^2 + 8x)}$.

10. $\displaystyle\int \frac{x^3 + x^2 + x + 1}{x^4 - 2x^2 + 1}$.

11. $\displaystyle\int \frac{2x^3 + 1}{x^3(x - 1)}$.

12. $\displaystyle\int \frac{15x^4}{x^{10} - 4x^5 - 5}$ (Hint: $y = x^5$).

13. $\displaystyle\int \frac{(2 \sin \theta - 1) \cos \theta}{\sin^2 \theta + 3 \sin \theta - 10}$.

14. $\displaystyle\int \frac{\sin \theta(\cos \theta + 1)}{(\cos \theta - 2)(\cos \theta - 1)}$.

15. $\displaystyle\int \frac{1}{x\sqrt{a^2 + x^2}}$.

16. $\displaystyle\int \frac{1}{e^x - 1}$.

17. $\displaystyle\int \frac{e^x}{e^{2x} - 4}$.

18. $\displaystyle\int \frac{2e^{3x} + 1}{e^{2x}(e^x - 1)}$.

19. $\displaystyle\int \frac{4}{e^x + 4}$.

20. $\displaystyle\int \frac{1}{\sqrt{1 - e^{2x}}}$ (Hint: let $e^x = \sin \theta$.)

ANSWERS

1. $\left[\frac{8}{5}\ln|x-3| - \frac{3}{5}\ln|x+2|\right]$.   3. $\left[\ln\left|\frac{x}{x+1}\right|\right]$.

5. $\left[\ln\left|\frac{x^2(x-1)}{x+2}\right|\right]$.   7. $\left[\ln\left|\frac{\sqrt{|x|}(x+5)^2}{\sqrt{|x+2|}}\right|\right]$.

9. $\frac{1}{2}\left[\ln\left|\left(\frac{x}{x+2}\right)^{1/8}\right| + \frac{x+1}{4(x+2)^2}\right]$.   11. $\left[\frac{2x+1}{2x^2} + \ln\left|\frac{(x-1)^3}{x}\right|\right]$.

13. $\left[\frac{11}{7}\ln|\sin\theta+5| + \frac{3}{7}\ln|\sin\theta-2|\right]$.   15. $\left[-\frac{1}{a}\ln\left|\frac{\sqrt{a^2+x^2}+a}{x}\right|\right]$.

17. $\left[\frac{1}{4}\ln\left|\frac{e^x-2}{e^x+2}\right|\right]$.   19. $[x - \ln|e^x + 4|]$.

4. THE METHOD OF PARTIAL FRACTIONS *(Continued)*

In the examples of the preceding section we considered only rational functions such that the denominator could be factored completely into linear factors. Theorem 11.3 enables us to do more, as the next example shows:

Example 17.   Consider $\int \frac{-x^2+x+1}{x(x^2+1)^2}$. By Theorem 11.3 we have

(11.16)
$$\frac{-x^2+x+1}{x(x^2+1)^2} = \frac{A}{x} + \frac{P_1(x)}{(x^2+1)^2},$$

where

$$A = \frac{-x^2+x+1}{(x^2+1)^2}\bigg|_{x=0} = 1$$

and P_1 is a polynomial of degree less than 4. Setting $A = 1$ in Equation 11.16, we solve for $P_1(x)$ and find that

$$P_1(x) = -x^3 - 3x + 1.$$

Consequently,

$$\frac{-x^2+x+1}{x(x^2+1)^2} = \frac{1}{x} + \frac{-x^3-3x+1}{(x^2+1)^2} .$$

Since $\int \frac{1}{x} = [\ln|x|]$, we need only determine $\int \frac{-x^3-3x+1}{(x^2+1)^2}$. This may be accomplished by the substitution $x = \tan\theta$ which results in

$$\int \frac{-x^3-3x+1}{(x^2+1)^2} = \frac{1}{2}\left[\arctan x - \ln(x^2+1) + \frac{x-2x^2}{x^2+1}\right]. \quad ∎$$

The weakness of Theorem 11.3 lies in the fact that it does not enable us to decompose a rational function which has two distinct irreducible quadratic factors in the denominator. For example, we cannot use Theorem 11.3 to decompose $\dfrac{1}{(x^2 + 1)(x^2 + 4)}$ in order to find its primitive set.

To handle rational functions where the denominator has more than one irreducible quadratic factor, we make use of the following theorem. The interested reader is referred to Theorem 12.8d of *Calculus and Analytic Geometry* by Britton, Kreigh, and Rutland (Freeman Publishing Company).

THEOREM II.5 Let $\dfrac{P}{Q}$ be a rational function in reduced form such that the degree of P is less than that of Q. Assume that $Q(x) = (x^2 + ax + b)^n Q_1(x)$, where $x^2 + ax + b$ is an irreducible quadratic polynomial which is not a factor of Q_1. There exist unique real numbers $A_1, \ldots, A_n, B_1, \ldots, B_n$ such that

$$(11.17) \quad \frac{P(x)}{Q(x)} = \frac{P(x)}{(x^2 + ax + b)^n Q_1(x)} = \frac{A_1 x + B_1}{(x^2 + ax + b)^n}$$
$$+ \cdots + \frac{A_n x + B_n}{(x^2 + ax + b)} + \frac{P_1(x)}{Q_1(x)},$$

where the degree of P_1 is less than that of Q_1.

It can be shown that every polynomial can be factored completely into a product of linear and/or irreducible quadratic factors. Thus the denominator of each rational function can be so factored. Repeated applications of Theorems 11.3 and 11.5 then provide us with a technique for finding primitive sets for rational functions. We illustrate this technique by way of several examples.

Example 18. Find the primitive set $\displaystyle\int \frac{x^4 - x^2 - 3x + 1}{(x + 1)(x^2 + 1)^2}$. By Theorem 11.3 we may write

$$\frac{x^4 - x^2 - 3x + 1}{(x + 1)(x^2 + 1)^2} = \frac{A}{x + 1} + \frac{P(x)}{(x^2 + 1)^2}.$$

Application of Theorem 11.5 to the last term then gives

$$(11.18) \quad \frac{x^4 - x^2 - 3x + 1}{(x + 1)(x^2 + 1)^2} = \frac{A}{x + 1} + \frac{Bx + C}{(x^2 + 1)^2} + \frac{Dx + E}{x^2 + 1}$$

where A, B, C, D and E are real number coefficients which need to be determined. We may use the method of Theorem 11.3 to evaluate A. However, we introduce the following method which uses a single technique to determine all of the coefficients. Multiply both members of Equation 11.18 by $(x + 1)(x^2 + 1)^2$ to obtain

(11.19) $x^4 - x^2 - 3x + 1 = A(x^2 + 1)^2 + (Bx + C)(x + 1) +$
$$(Dx + E)(x + 1)(x^2 + 1).$$

By substitution of five different values for x into Equation 11.19, we obtain five equations for the five coefficients. We note that the substitution $x = -1$ will give an equation involving A alone, since the linear factor $(x + 1)$ of the last two terms is then zero. Choosing the four other values to be $x = 0$, $1, 2, -2$, we obtain

$$x = -1: \quad 4 = 4A$$
$$x = 0: \quad 1 = A + C + E$$
$$x = 1: \quad -2 = 4A + 2B + 2C + 4D + 4E$$
$$x = 2: \quad 7 = 25A + 6B + 3C + 30D + 15E$$
$$x = -2: \quad 19 = 25A + 2B - C + 10D - 5E.$$

Solving this system of equations, we find that

$$A = 1, \quad B = -3, \quad C = D = E = 0.$$

Thus, making these substitutions in Equation 11.18 we find

$$\int \frac{x^4 - x^2 - 3x + 1}{(x + 1)(x^2 + 1)^2} = \int \left(\frac{1}{x + 1} + \frac{-3x}{(x^2 + 1)^2} \right)$$
$$=^1 \int \frac{1}{x + 1} - \frac{3}{2} \int \frac{2x}{(x^2 + 1)^2}$$
$$=^2 [\ln |x + 1| + \tfrac{3}{2}(x^2 + 1)^{-1}]$$

(1) Putting the second term in the form $\dfrac{u'}{u^2}$, where $u = x^2 + 1$.

(2) Using the fact $\displaystyle\int \frac{u'}{u^2} = \left[\frac{u^{-1}}{-1} \right] = [-u^{-1}]$. ∎

Example 19. Evaluate $\displaystyle\int \frac{x^2 + 3x + 1}{(x^2 + 2x + 2)^2}$. The denominator of the fraction has only one repeated quadratic factor. From Theorem 11.5 we seek coefficients A, B, C, D such that

(11.20) $$\frac{x^2 + 3x + 1}{(x^2 + 2x + 2)^2} = \frac{Ax + B}{(x^2 + 2x + 2)^2} + \frac{Cx + D}{(x^2 + 2x + 2)}.$$

Multiplying both members of this equation by the denominator on the left we obtain

$$x^2 + 3x + 1 = Ax + B + (Cx + D)(x^2 + 2x + 2).$$

In this example we illustrate another method for determining the coefficients. We expand the terms on the right and collect coefficients of the various powers of x:

$$x^2 + 3x + 1 = Cx^3 + (2C + D)x^2 + (A + 2C + 2D)x + (B + 2D).$$

Equating coefficients of like powers of x we obtain a system of four equations:

Equating coefficients of	we obtain
x^3	$0 = C$
x^2	$1 = 2C + D$
x	$3 = A + 2C + 2D$
x^0	$1 = B + 2D.$

Solving this system we find that $A = 1$, $B = -1$, $C' = 0$, $D = 1$. Hence, from Equation 11.20, we have

$$\int \frac{x^2 + 3x + 1}{(x^2 + 2x + 2)^2} =^1 \int \frac{x - 1}{(x^2 + 2x + 2)^2} + \int \frac{1}{x^2 + 2x + 2}$$

$$=^2 \int \frac{\frac{1}{2}(2x + 2) - 2}{(x^2 + 2x + 2)^2} + \int \frac{1}{x^2 + 2x + 2}$$

$$=^3 \frac{1}{2} \int \frac{2x + 2}{(x^2 + 2x + 2)^2} - 2 \int \frac{1}{(x^2 + 2x + 2)^2}$$

$$+ \int \frac{1}{x^2 + 2x + 2}$$

$$=^4 \frac{1}{2}[-(x^2 + 2x + 2)^{-1}] - 2 \int \frac{1}{((x + 1)^2 + 1)^2}$$

$$+ \int \frac{1}{(x + 1)^2 + 1}$$

$$=^5 \left[\frac{-1}{2(x^2 + 2x + 2)} \right] - \left[\frac{x + 1}{x^2 + 2x + 2} \right.$$

$$+ \arctan(x + 1) \Big] + [\arctan (x + 1)]$$

(1) Substituting the values of A, B, C, D into the partial fraction decomposition given by Equation 11.20.

(2) Setting $u = x^2 + 2x + 2$ we find $u' = 2x + 2$. Therefore

$$x - 1 = (\tfrac{1}{2}u' - 1) - 1$$

$$= \tfrac{1}{2}(2x + 2) - 2$$

(3) Separating the first term into two parts where the first part has the form

$$\int \frac{u'}{u^2} .$$

(4) Finding the primitive set for the first term and completing the square in the denominators of the second and third terms.

(5) Using the substitution $x + 1 = \tan\theta$, $x' = \sec^2\theta\theta'$ and the technique of Example 8b to determine the primitive sets for the last two terms. For example:

$$\int \frac{1}{((x+1)^2 + 1)^2} = \int \frac{\sec^2\theta\theta'}{\sec^4\theta} = \int \cos^2\theta\theta'$$

$$= [\tfrac{1}{2}(\theta + \sin\theta\cos\theta)]$$

$$= \left[\tfrac{1}{2}\left(\arctan(x+1) + \frac{x+1}{x^2 + 2x + 2} \right) \right].$$

It is well to note that the single substitution $x + 1 = \tan\theta$ made at the beginning also provides a successful solution to this problem. The reader may wish to explore this possibility. ■

Example 20. We shall find the primitive set $\int \dfrac{12x^3 + 30x}{(x^2 + 1)^2(x^2 + 4)^2}$.

By Theorem (11.5)

$$\frac{12x^3 + 30x}{(x^2 + 1)^2(x^2 + 4)^2} = \frac{Ax + B}{(x^2 + 1)^2} + \frac{Cx + D}{(x^2 + 1)} + \frac{Ex + F}{(x^2 + 4)^2} + \frac{Gx + H}{x^2 + 4}.$$

By either of the methods of the preceding two examples we find that all coefficients are equal to zero except $A = 2$ and $E = -2$. Therefore

$$\int \frac{12x^3 + 30x}{(x^2 + 1)^2(x^2 + 4)^2} = \int \frac{2x}{(x^2 + 1)^2} - \int \frac{2x}{(x^2 + 4)^2}$$

$$= \left[\frac{-1}{x^2 + 1} + \frac{1}{x^2 + 4} \right]. \quad ■$$

Consider the partial fraction decomposition of the rational function $R(x) = \dfrac{1}{(x - 1)^3(x^2 + 1)^2}$. By Theorems 11.3 and 11.5 we can write

$$R(x) = \frac{A_1}{(x - 1)^3} + \frac{A_2}{(x - 1)^2} + \frac{A_3}{x - 1} + \frac{A_4 x + B_4}{(x^2 + 1)^2} + \frac{A_5 + B_5}{x^2 + 1}.$$

We may choose to evaluate the coefficients A_1, A_2, and A_3 by the method of Theorem 11.3 and substitute four values for x to determine the four coefficients A_4, A_5, B_4, and B_5 algebraically. On the other hand, we may substitute seven values for x and solve for all seven algebraically. This same choice will be possible in many problems. Which technique is the "more efficient" will depend upon the nature of the individual problem. With practice the reader will gain experience in which approach is likely to be best.

EXERCISES

1. $\int \dfrac{16x}{(x^2 + 1)(x^2 + 9)}$.

2. $\int \dfrac{x^2 - 2x^3 - 2x + 4}{(x^2 + 1)(x^2 + 4)}$.

3. $\int \dfrac{x^3 - 24x^2 + 25x - 24}{(x^2 + 1)^2(x^2 + 25)}$.

4. $\int \dfrac{1}{(x^2 + 2)(x^2 + 1)}$.

5. $\int \dfrac{2x + 7}{(x^2 + 4x + 8)(x + 1)^2}$.

6. $\int \dfrac{2x + 1}{x^4 + 3x^2 + 2}$.

7. $\int \dfrac{1}{(x^2 + 3)^2(x^2 + 1)}$.

8. $\int \dfrac{x^2 + 2}{(x^2 + 3)^2(x^2 + 1)}$.

9. $\int \dfrac{5 - x^2}{(x - 1)(x + 2)^2(x^2 + 1)^2}$.

10. $\int \dfrac{x}{(x + 1)^2(x^2 + 1)^2(x^2 + 4)^2}$.

ANSWERS

1. $\left[\ln\left(\dfrac{x^2 + 1}{x^2 + 9}\right)\right]$.   2. $[\arctan x - \ln(x^2 + 4)]$.

3. $\left[-\dfrac{1}{2(x^2 + 1)} - \arctan x + \dfrac{1}{5}\arctan \dfrac{x}{5}\right]$.

4. $\left[\arctan x - \dfrac{\sqrt{2}}{2}\arctan \dfrac{x}{\sqrt{2}}\right]$.   5. $\left[-\dfrac{1}{2}\arctan\left(\dfrac{x + 2}{2}\right) - \dfrac{1}{x + 1}\right]$.

6. $\left[\ln\dfrac{x^2 + 1}{x^2 + 2} + \arctan x - \dfrac{\sqrt{2}}{2}\arctan \dfrac{x}{\sqrt{2}}\right]$.

7. $\left[\dfrac{1}{4}\arctan x - \dfrac{1}{3\sqrt{3}}\arctan \dfrac{x}{\sqrt{3}} - \dfrac{x}{12(x^2 + 3)}\right]$.

9. $\left[\dfrac{1}{9}\ln\left|\dfrac{x - 1}{x + 2}\right| + \dfrac{3}{9(x + 2)} + \dfrac{1}{2(x^2 + 1)}\right]$.

5.   TABLES OF PRIMITIVES;   TRADITIONAL NOTATION

The material in this and the preceding chapters discloses that the evaluation of an integral of a given function f on some interval requires the determination of a primitive set for f. If one has frequent need to evaluate integrals, a Table of Primitives is most helpful. Several such tables, more commonly called Tables of Integrals, are available: Pierce, Foster: *A Short Table of Integrals*, Gill and Company. Dwight, H. B.: *Tables of Integrals and other Mathematical Data*, The Macmillan Co. Burington, R. S.: *Handbook of Mathematical Tables and Formulas*, McGraw-Hill. Selby, S. M.: *CRC Handbook of Tables for Mathematics*, Chemical Rubber Co. Appendix VI (p.891) gives a very brief Table of Primitives for use with this book. We give several illustrations of its use here:

Example 21.   Evaluate $\int_1^2 \dfrac{x^2}{\sqrt{x+1}}$.   Each of the formulas of Appendix VI is of the form $\int f(u)u'$. It is necessary, therefore, to put the above problem in this form, which we can do by setting $u = x$ and computing $u' = 1$. The primitive set we seek is then

$$\int \frac{x^2}{\sqrt{x+1}} = \int \frac{u^2 u'}{\sqrt{u+1}} .$$

A search of the table in Appendix VI reveals that Formula 21 is applicable. This formula is

$$\int \frac{u^2 u'}{\sqrt{au+b}} = \left[\frac{2(3a^2u^2 - 4abu + 8b^2)}{15a^3} \sqrt{au+b} \right].$$

Setting $u = x$, $u' = 1$, and $a = b = 1$ gives

$$\int_1^2 \frac{x^2}{\sqrt{x+1}} = \left[\frac{2(3x^2 - 4x + 8)}{15} \sqrt{x+1} \right]_1^2 .$$

The final evaluation is left to the reader.   ■

Example 22.   Evaluate $\int_3^5 x^2(2 - x)^{-1/2}$. Setting $u = x$ and computing $u' = 1$, we find that we may rewrite the problem as follows:

$$\int_3^5 x^2(2 - x)^{-1/2} = \int_3^5 \frac{u^2 u'}{\sqrt{2 - u}} .$$

We then note that the integrand is defined only when $u > 2$ and that the interval of integration falls within this domain of definition. Again, a search of Appendix VI shows that Formula 21 is applicable with $u = x$, $a = -1$, and $b = 2$. Thus,

$$\int_3^5 x^2(2 - x)^{-1/2} = \left[\frac{2(3x^2 + 8x + 32)}{-15} \sqrt{2 - x} \right]_3^5 .   ■$$

Example 23.   Evaluate $\int_2^4 \dfrac{1}{x \ln x}$.   We put this in appropriate form by setting $u = x$, $u' = 1$ to obtain

$$\int_2^4 \frac{1}{x \ln x} = \int_{x=2}^{x=4} \frac{u'}{u \ln u} .$$

Applying Formula 100 of Appendix VI with $u = x$ and $a = 1$, we find

$$\int_2^4 \frac{1}{x \ln x} = [\ln |\ln x|]_2^4 .   ■$$

Example 24.   Evaluate $\displaystyle\int_2^7 \frac{(\ln x)^2 \cos(\ln x)}{x}$.   First note that the function is defined, continuous, and hence integrable on the interval of integration. To put the integrand in a form which will be found in Appendix VI, we set $u = \ln x$ and compute

$$u' = \frac{1}{x}.$$

Thus,

$$\int_2^7 \frac{(\ln x)^2 \cos (\ln x)}{x} = \int_{x=2}^{7} u^2(\cos u)u'$$

$$=^1 [2u \cos u - 2 \sin u + u^2 \sin u]_{x=2}^7$$

$$=^2 [2(\ln x) \cos (\ln x) - 2 \sin (\ln x) + (\ln x)^2 \sin (\ln x)]_2^7.$$

(1) Applying Formula 82, Appendix VI, with $a = 1$.
(2) Resubstituting from $u = \ln x$.

In practice it is sometimes necessary to find primitive sets which are not immediately reducible to one of the forms given in the table at hand. In such cases it is often helpful to use the techniques presented in preceding sections to obtain a different form. The previous example is one instance of this situation; we now present another:

Example 25.

$$\int \tan^4 x(\tan^2 x - a^2)^{-1/2} \sec^2 x =^1 \int u^4(u^2 - a^2)^{-1/2}u'$$

$$=^2 [u^3(u^2 - a^2)^{1/2}] - 3\int u^2(u^2 - a^2)^{1/2}u'.$$

(1) Setting $u = \tan x$, computing $u' = \sec^2 x$ and substituting.
(2) Using method of parts with

$$f(u) = u^3 \qquad g'(u) = (u^2 - a^2)^{-1/2}uu'$$

$$f'(u) = 3u^2 \qquad g(u) = \sqrt{u^2 - a^2}.$$

$g(u)$ is a primitive of $g'(u)$ and may be found by using Formula 31 of Appendix VI. ∎

The last primitive set above is given by Formula 32 of Appendix VI. Application of this formula and resubstitution from $u = \tan x$ yields

$$[(\tfrac{1}{4} \tan^3 x + \tfrac{3}{8}a^2 \tan x)\sqrt{\tan^2 x - a^2} + \tfrac{3}{8}a^4 \ln |\tan x + \sqrt{\tan^2 x - a^2}|].$$

Example 26.   Compute the area A of the set $\mathcal{C}$ bounded by a circle of radius r. Consider the circle to be centered at $(0, 0)$. $\mathcal{C}$ may then be regarded as the area between the graphs of the functions f and g given by

(11.21) $f(x) = \sqrt{r^2 - x^2}$ and $g(x) = -\sqrt{r^2 - x^2}$, $-r \leq x \leq r$.

Thus,

$$A(\mathcal{C}) =^1 \int_{-r}^{r} |f(x) - g(x)|$$

$$=^2 2\int_{-r}^{r} \sqrt{r^2 - x^2}$$

$$=^3 2 \cdot \tfrac{1}{2}\left[x\sqrt{r^2 - x^2} + r^2 \arcsin \frac{x}{r}\right]_{-r}^{r}$$

$$=^4 r^2(\arcsin 1 - \arcsin(-1))$$

$$= r^2\left(\frac{\pi}{2} - \left(-\frac{\pi}{2}\right)\right) = \pi r^2.$$

(1) Definition 8.26 (Area Between Two Curves).
(2) Substitution from Equation 11.21.
(3) Formula 29, Appendix VI.
(4) Evaluation at upper and lower limits. ∎

The traditional notation used for definite integrals and primitive sets in tables of integrals (and, indeed, in other texts) differs from that developed herein. It is well that the reader be aware of the two alternatives summarized as follows:

	Notation Herein	*Traditional Notation*
Definite Integral:	$\int_{a}^{b} f$ or $\int_{a}^{b} f(x)$	$\int_{a}^{b} f(x)\, dx$
Primitive Sets:	$\int f = [F]$	
or		$\int f(x)\, dx = F(x) + C$
	$\int f(x) = [F(x)]$	

The traditional notation is widely adopted for historical reasons. The reader will observe that this notation always employs the symbol "*dx*" (or "*du*," or some other "*dee*" symbol) and calls for specification of the function f by its functional value $f(x)$ (or $f(u)$, or some other similar symbol). This points up one of the reasons why we have not used this more conventional symbolism, namely, the integral $\int_{a}^{b} f$ is completely determined by the function itself, not by the variable, say x, used to write the recipe for f. A second reason for not employing the dx notation is the tendency of this notation to cause students and others to ascribe to the symbol dx a real, albeit mystic, mathematical content of its own rather than regarding it is a mnemonic device for recalling various properties of the integral.

The traditional notation for primitive sets is $F(x) + C$. The function F denotes one primitive for f, and $F(x) + C$ symbolizes the generation of all other primitives by the addition of a constant C (the so-called "arbitrary constant"). Many students soon come to regard C as a fixed (even though arbitrary) constant and $\int f(x)\, dx = F(x) + C$ to represent a single primitive for f instead of the set of all primitives. With this understanding we have, for $k = 0$,

$$\int kf(x)\, dx = \int 0\, dx = 0 + C_1 = C_1$$

and

$$k\left(\int f(x)\, dx\right) = k(F(x) + C) = 0(Fx + C_2) = 0.$$

Thus, if $C_1 \neq 0$ we have $\int kf(x)\, dx \neq k(\int f(x)\, dx)$. Compare this result with Definition 9.8, p. 390.

A further objection to the $F(x) + C$ symbolism lies in the fact that two different primitives of one and the same function f do not always differ by a constant. For example,

$$F_1(x) = \ln |x|, \quad x \neq 0$$

and

$$F_2(x) = \begin{cases} \ln |x| + 1, & x < 0 \\ \ln |x| + 2, & x > 0 \end{cases}$$

are both primitives for $f(x) = 1/x$. However, $F_1 - F_2$ is not a constant function. Such a condition is possible whenever the domain of definition for f is separated by one or more points at which f is not defined. The assertion that "two primitives of a single function differ by a constant" is valid only if the domain of definition of all functions involved is an interval.

Many of the longer tables of integrals omit the C from their formulas and there will be little difficulty with the notation for primitive sets when using such tables. One need only regard the right member as a single element of the primitive set. For example, Formula 101 of Burington's Tables is

$$\int \frac{x}{ax^2 + c}\, dx = \frac{1}{2a} \log (ax^2 + c).$$

This formula is equivalent to Formula 26 of Appendix VI:

$$\int \frac{uu'}{au^2 + c} = \left[\frac{1}{2a} \ln |au^2 + c|\right].$$

The presence of u' in each of the formulas of Appendix VI corresponds to the presence of the dx in the traditional notation.

EXERCISES

Use the Table of Primitives of Appendix VI to find primitive sets for each of the following:

1. $\int \dfrac{x}{(2 + 3x)^2}$

2. $\int x \sinh x.$

3. $\int x^3 \ln x.$

4. $\int x^2 \sin x.$

5. $\int \sin 2x \cos 3x.$

6. $\int \dfrac{\sin x \cos x}{1 - \cos x}.$

7. $\int \dfrac{\cot x}{\sqrt{4 - \sin^2 x}}$ (let $u(x) = \sin x$).

8. $\int \dfrac{(x^4 - 1)^{1/2}}{x}$ (let $u(x) = x^2$).

9. $\int \dfrac{\sqrt{2 \ln x - 1}}{x \ln x}.$

10. $\int \dfrac{\sec^2 x}{2 \tan x - \tan^2 x}.$

11. $\int e^{4x} \cos e^x.$

12. $\int \dfrac{x^3 \ln x^2}{\sqrt{1 + x^4}}.$

13. $\int \dfrac{\sqrt{e^{2x} - 4}}{e^x}.$

ANSWERS

1. $\dfrac{1}{9} \left[\dfrac{2}{2 + 3x} + \ln |2 + 3x| \right].$ 3. $\left[x^4 \left(\dfrac{\ln x}{4} - \dfrac{1}{16} \right) \right].$

5. $\left[\dfrac{\cos x}{2} - \dfrac{\cos 5x}{10} \right].$ 7. $\left[-\dfrac{1}{2} \ln \left| \dfrac{2 + (4 - \sin^2 x)^{1/2}}{\sin x} \right| \right].$

9. $2[\sqrt{-1 + 2 \ln x} - \arctan \sqrt{-1 + 2 \ln x}\,].$

11. $[(e^{3x} - 6e^x) \sin e^x + (3e^{2x} - 6) \cos e^x].$

13. $\left[\ln |e^x + \sqrt{e^{2x} - 4}| - \dfrac{\sqrt{e^{2x} - 4}}{e^x} \right].$

12

APPLICATIONS OF THE INTEGRAL

The reader will recall that the problem of definition and computation of area of plane regions lead to the definition of the integral. The computation of plane areas is only one of many applications of the integral. Its uses in solving the problems of physics, engineering, economics, and many other fields are numerous. In this chapter, we shall study a few of these.

1. VOLUMES

We characterize the concept of the volume of a solid in a manner similar to that in which the concept of area was specified in Chapter 8. We assume

i. that the volume function V is defined on a certain class C of bounded subsets of $\mathcal{R}^3$,

ii. that the class C contains all rectangular parallelepipeds $[a, b] \times [c, d] \times [e, f]$ (see Figure 12.1),

and

iii. that V has the following properties:

(1) $V([a, b] \times [c, d] \times [e, f]) = V((a, b) \times (c, d) \times (e, f))$
$$= (b - a)(d - c)(f - e).$$

(2) $V(R) \geq 0$ if $R \in C$.

(3) $\emptyset \in C$ and $V(\emptyset) = 0$.

(4) $V(R) \leq V(S)$ if $R \subset S$, $R \in C$, $S \in C$.

(5) $V(R \cup S) = V(R) + V(S)$, if $R \cap S = \emptyset$, $R \in C$ and $S \in C$.

(6) $V(R \cup S) = V(R) + V(S) - V(R \cap S)$ if $R \in C$ and $S \in C$.

The reader should be able to supply the reasons why each of the above properties is desirable. Reference to Chapter 8 (p. 338) may be helpful.

474

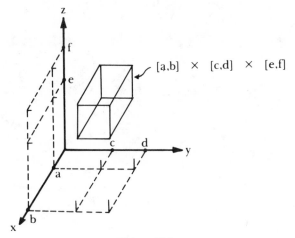

Figure 12.1

A more complete development of the volume concept is given in Chapter 18. For our purposes here we shall make one additional assumption concerning the volume of a right cylinder. Let $\mathcal{B}$ be a set of points in the xy-plane of $\mathcal{R}^3$. By a right cylinder of height h and base $\mathcal{B}$ we shall mean the set $\mathcal{K}$ given by

$$\mathcal{K} = \{(x, y, z) \mid (x, y, 0) \in \mathcal{B} \text{ and } 0 \le z \le h\}.$$

We shall assume that

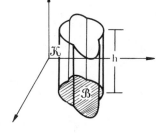

(7) $V(\mathcal{K}) = h \cdot$ area of $\mathcal{B}$.

This assumption is consistent with the treatment given in Chapter 18.

Let $0 \le a < b$ and consider the function $f : [a, b] \to \mathcal{R}$, the graph of which is sketched in Figure 12.2a. If the area under f is rotated about the y-axis, as illustrated in Figure 12.2b, the figure generated is called a solid of revolution. (We may also obtain a solid of revolution by rotating a plane figure about any other axis.) We wish to define and compute the volume of such a solid.

Let l be a step function from $[a, b]$ into $\mathcal{R}$ such that $l(x) \le f(x)$ for each x. By assumption 4 above, the volume of revolution generated by the area under l is less than or equal to that generated by the area under f (see Figure 12.3). Let $P = \{x_0, \ldots, x_n\}$ be a partition of $[a, b]$ such that l is constant with value l_i, on the ith open subinterval of P. Then the area under l and between x_{i-1} and x_i generates a right circular shell whose volume, in accordance with assumption 7, is $\pi x_i^2 l_i - \pi x_{i-1}^2 l_i = \pi l_i(x_i^2 - x_{i-1}^2)$. The volume generated by the area under l is then the sum of the volumes of

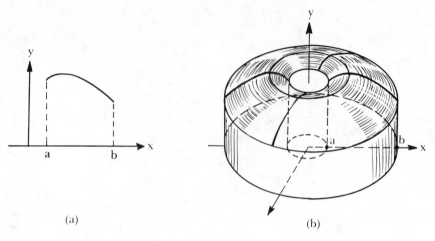

(a)   (b)

Figure 12.2

these shells and is given by

$$V_l = \sum_{i=1}^{n} \pi l_i (x_i^2 - x_{i-1}^2)$$

(12.1)
$$=^1 \sum_{i=1}^{n} \pi l_i \int_{x_{i-1}}^{x_i} 2t$$

$$= 2\pi \sum_{i=1}^{n} \int_{x_{i-1}}^{x_i} l_i t.$$

(1) Theorem 9.4 (p. 384) (Second Fundamental Theorem).

Similarly, if u, a step function which is constant on the open subintervals of P, has value u_i on the ith subinterval and is such that $u(x) \geq f(x)$ for each

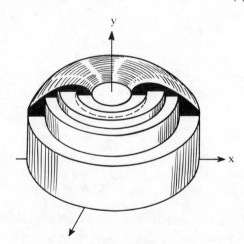

Figure 12.3

x, the volume generated by the area under u is given by

$$(12.2) \qquad V_u = \sum_{i=1}^{n} \pi u_i (x_i^2 - x_{i-1}^2) = 2\pi \sum_{i=1}^{n} \int_{x_{i-1}}^{x_i} u_i t$$

and is greater than or equal to that generated by the area under f.

Observe now that if $tf(t)$ is integrable on $[a, b]$, then

$$V_l =^1 2\pi \sum_{i=1}^{n} \int_{x_{i-1}}^{x_i} l_i t$$

$$\leq^2 2\pi \sum_{i=1}^{n} \int_{x_{i-1}}^{x_i} tf(t)$$

$$\leq^2 2\pi \sum_{i=1}^{n} \int_{x_{i-1}}^{x_i} tu$$

$$=^3 V_u.$$

(1) Equation 12.1.
(2) Theorem 8.15i (p. 360) (Order Properties of the Integral).
(3) Equation 12.2.

We denote the middle expression in the above chain by V. We then have

$$V = 2\pi \sum_{i-1}^{n} \int_{x_{i-1}}^{x_i} tf(t) = 2\pi \int_{a}^{b} tf(t),$$

where we have used the interval additivity of the integral (Theorem 8.18, p. 365). The preceding argument shows that

$$V_l \leq V \leq V_u.$$

for each pair of step functions such that $l(x) \leq f(x) \leq u(x)$. (We may assume that both l and u are constant on the open subintervals of the *same* partition P. Why?) We shall show in Theorem 12.1 that V is the *only* real number having this property. Accepting this result, we define the *volume generated by the area under the graph of f* to be

$$(12.3) \qquad V = 2\pi \int_{a}^{b} tf(t).$$

Example I. Let $f(x) = 1 - x^2, 0 \leq x \leq 1$. We rotate the area under the graph of f about the y-axis to obtain the paraboloid sketched in Figure 12.4. We wish to find the volume of the resulting solid. By Equation 12.3 we have

$$V = 2\pi \int_{0}^{1} xf(x)$$

$$= 2\pi \int_{0}^{1} x(1 - x^2)$$

$$= 2\pi \left[\frac{x^2}{2} - \frac{x^4}{4} \right]_{0}^{1}$$

$$= \frac{\pi}{2}. \quad \blacksquare$$

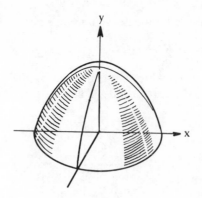

Figure 12.4

In the following theorem we establish the previous assertion concerning the uniqueness of the volume formula for solids of revolution. The theorem is stated in a more general setting which is applicable to problems to be considered in later sections. The application of the theorem to the preceding discussion is obtained by replacing F_l, F_u, and F by V_l, V_u and V, in addition to selecting $g(x) = 2\pi x^2$.

THEOREM 12.1 Let f and g be $\mathcal{R} \to \mathcal{R}$ functions such that f is integrable on $[a, b]$, g is increasing and differentiable in $[a, b]$ and g' is continuous on $[a, b]$. Let $\mathcal{S} = \{x_0, \ldots, x_n\}$ be a partition of $[a, b]$ and let l and u be lower and upper step functions for f on $[a, b]$, constant on the open subintervals of $\mathcal{S}$:

$$l(x) = l_i \text{ and } u(x) = u_i \text{ for } x \text{ in } (x_{i-1}, x_i).$$

For each choice of $\mathcal{S}$, l and u define F_l and F_u by

$$F_l = \sum_{i=1}^{n} l_i(g(x_i) - g(x_{i-1}))$$

and

$$F_u = \sum_{i=1}^{n} u_i(g(x_i) - g(x_{i-1})).$$

If fg' is integrable on $[a, b]$, then there is exactly one number F such that

(12.4) $F_l \leq F \leq F_u$ for *every* choice of $\mathcal{S}$, l, and u.

In fact,

(12.5) $$F = \int_a^b fg'.$$

Proof. We first show that there is but one number F satisfying Equation 12.4 for every choice of $\mathfrak{S}$, l, and u by demonstrating that $F_u - F_l$ is positive and can be made smaller than any real number r by appropriate choices of $\mathfrak{S}$, l, and u. For each choice of $\mathfrak{S}$, l, and u let F_u and F_l be as described in the theorem. $F_u - F_l$ is clearly non-negative since $l \leq f \leq u$ and g is increasing. Moreover,

$$F_u - F_l =^1 \sum_{i=1}^{n} u_i(g(x_i) - g(x_{i-1})) - \sum_{i=1}^{n} l_i(g(x_i) - g(x_{i-1}))$$

$$=^2 \sum_{i=1}^{n} (u_i - l_i)\, g'(a_i)(x_i - x_{i-1}), \text{ for some } a_i \text{ in } (x_{i-1}, x_i)$$

$$\leq^3 M \sum_{i=1}^{n} (u_i - l_i)(x_i - x_{i-1}), \text{ for some } M$$

$$=^4 M \int_a^b (u - l) = M\left(\int_a^b u - \int_a^b l\right).$$

(1) Definition of F_u and F_l.
(2) Mean Value Theorem: $g(x_i) - g(x_{i-1}) = g'(a_i)(x_i - x_{i-1})$, where $a_i \in (x_{i-1}, x_i)$.
(3) g' continuous on $[a, b]$ $\Rightarrow$ g' bounded on $[a, b]$ $\Rightarrow g'(x) \leq M$, for some M.
(4) Definition of the integral of a step function.

The integrability of f allows us to apply Theorem 8.16 (Riemann Condition for Integrability) to find functions l and u such that

$$\int_a^b u - \int_a^b l < \frac{r}{M}.$$

The preceding two inequalities assure us that there are functions l and u such that

$$F_u - F_l < M\frac{r}{M} = r,$$

and that the F of Equation 12.4 is unique. (For $\mathfrak{S}$ we take $\mathfrak{S}_u \cup \mathfrak{S}_l$ where $\mathfrak{S}_u$ and $\mathfrak{S}_l$ are the partitions of $[a, b]$ used to define u and l.)

To complete our proof we need only show that $\int_a^b fg'$ lies between F_l and F_u for all $\mathfrak{S}$, l, and u. Let F_l be any function defined in accordance with the theorem. Then

(12.6)
$$F_l =^1 \sum_{i=1}^{n} l_i(g(x_i) - g(x_{i-1}))$$

$$=^2 \sum_{i=1}^{n} l_i \int_{x_{i-1}}^{x_i} g'(x) =^3 \sum_{i=1}^{n} \int_{x_{i-1}}^{x_i} lg'(x)$$

$$=^4 \int_a^b lg'(x)$$

$$\leq^5 \int_a^b f(x)g'(x) = \int_a^b fg'$$

(1) Definition of F_l.
(2) Using the continuity of g' and Theorem 9.4 (Second Fundamental Theorem of Calculus).
(3) Since $l(x) = l_i$ is constant on the interval $[x_{i-1}, x_i]$ of integration.
(4) Using Theorem 8.18 (Interval Additivity of Integrals).
(5) Theorem 8.15 (Order Properties of the Integral) and the fact that $l \leq f$ on $[a, b]$.

Similarly, we can show that

(12.7)
$$F_u \geq \int_a^b fg'.$$

Combining Inequalities 12.6 and 12.7, we conclude that

$$F_l \leq \int_a^b fg' \leq F_u,$$

and hence that

$$F = \int_a^b fg'. \quad \blacksquare$$

As an illustration of another use of the above theorem for finding the volume of a figure, let $\mathcal{F}$ be a figure the cross-sectional area of which can be expressed as an integrable function $A(x)$ (see Figure 12.5). If l is a lower step function to A, the volume of a cylinder of base area l_i and length $x_i - x_{i-1}$ is given by $l_i(x_i - x_{i-1})$. Then $V_l = \sum_{i=1}^n l_i(x_i - x_{i-1})$ is an underapproximation to the volume of $\mathcal{F}$. Similarly, if u is an upper step function for A, $V_u = \sum_{i=1}^n u_i(x_i - x_{i-1})$ is an overapproximation to the volume of $\mathcal{F}$. We apply Theorem 12.1, letting $A = f$ and $g(x) = x$, and define the volume of $\mathcal{F}$ to be

(12.8)
$$V = \int_a^b g'f = \int_a^b A.$$

Example 2. We wish to find the volume of a conoid (see Figure 12.6) whose base is the closed ball given by $x^2 + y^2 \leq r^2$, whose vertex is the line through the z-axis parallel to the x-axis of length $2r$ and whose height is h. The area of a cross-section parallel to the yz-plane is $A(x) = h\sqrt{r^2 - x^2}$. Then, by Equation 12.8, the volume is given by

$$V = \int_{-r}^r h\sqrt{r^2 - x^2} = \tfrac{1}{2}\pi r^2 h. \quad \blacksquare$$

If the area under the graph of a function $f: [a, b] \to \mathcal{R}$ is rotated about the x-axis (see Figure 12.7), the area of a section parallel to the yz-plane is given by $\pi(f(x)^2)$. Then, by Equation 12.8, the volume is given by

(12.9)
$$V = \int_a^b A(x) = \pi \int_a^b (f(x))^2.$$

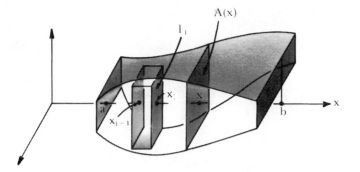

Figure 12.5

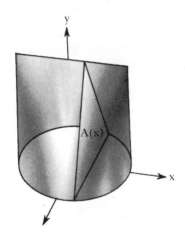

Figure 12.6

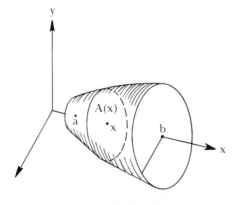

Figure 12.7

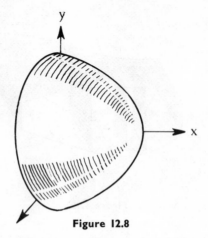

Figure 12.8

Example 3. The area under the graph of f where $f(x) = \sqrt{1-x}$, $0 \le x \le 1$, is rotated about the x-axis (see Figure 12.8 and note that this figure is similar to Figure 12.4). By Equation 12.9, the volume is given by

$$V = \pi \int_a^b (f(x))^2 = \pi \int_0^1 (1-x)$$

$$= \frac{\pi}{2}. \quad \blacksquare$$

This is the same as the volume we obtained for the congruent figure of Example 1.

Example 4. A spherical water tank has a radius of 5 ft. How much water is in the tank when its depth is 3 ft?

We may consider the tank to be centered at the origin and generated by revolving the area under the curve $y = \sqrt{25 - x^2}$ about the x-axis (see Figure 12.9). The volume of the water is then given by

$$\pi \int_2^5 (\sqrt{25 - x^2})^2 = \pi \left[25x - \frac{x^3}{3} \right]_2^5$$

$$= 36\pi. \quad \blacksquare$$

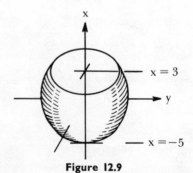

Figure 12.9

EXERCISES

In Exercises 1 to 8 compute the volume generated by rotating the areas beneath the given functions and between the two endpoints about the y-axis. Sketch the figures:

1. $f(x) = x$, $[0, 1]$.

2. $f(x) = kx$, $[0, a]$.

3. $f(x) = \cos x$, $\left[0, \dfrac{\pi}{2}\right]$.

4. $f(x) = \sin x$, $[0, \pi]$.

5. $f(x) = e^x$, $[0, 1]$.

6. $f(x) = \ln x$, $[1, 2]$.

7. $f(x) = \dfrac{1}{x+1}$, $[0, 1]$.

8. $f(x) = \dfrac{x}{x+1}$, $[0, 1]$.

9 to 16. Compute the volumes generated by rotating the areas in Exercises 1 to 8 about the x-axis. Sketch the figures.

17. Find the volume of the ellipsoid formed by rotating $\left\{ (x, y) \,\middle|\, \dfrac{x^2}{4} + y^2 \le 1 \right\}$ about the x-axis.

18. If a tank in the form of the ellipsoid of Exercise 17 is tipped so that the x-axis is vertical and filled with water to a height of 1, what is the volume of the water?

19. A cylindrical hole 4 in. in diameter is drilled through the center of a sphere 6 in. in diameter. Find the volume of the remaining part of the sphere. Hint: Rotate the shaded part of the figure about the y-axis.

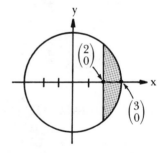

20. The radius of a solid sphere is r. A cylindrical hole is to be drilled through the sphere with the axis of the hole passing through the center of the sphere. What must the radius of the hole be in order to remove half the volume of the sphere.

21. The base of a pyramid is a rectangle with sides a and b and height h. The vertex of the pyramid is above one corner of the base. Compute its volume.

22. A figure has the base $x^2 + y^2 = 25$. Its sections parallel to the yz-plane are squares. Sketch the figure and compute its volume.

23. Find the volume of an elliptical conoid whose base is the ellipse $\dfrac{x^2}{4} + \dfrac{y^2}{9} = 1$, whose altitude is 8, and whose vertex is parallel to and above the x-axis.

24. A wedge is cut from a log 6 in. in diameter by making a cut halfway through the log at a right angle to the axis and making a second cut at an angle of 45 degrees to the axis meeting the first at the center of the log. Find the volume of the wedge.

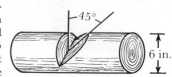

25. Two pipes, each having 2 in. inside diameter, meet each other right angles. Find the volume of their intersection.

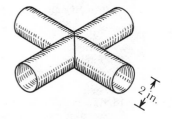

ANSWERS

1. $\dfrac{2\pi}{3}$. 3. $\pi^2 - 2\pi$. 5. 2π. 7. $2\pi(1 - \ln 2)$. 9. $\dfrac{\pi}{3}$. 11. $\dfrac{\pi^2}{4}$.

13. $\dfrac{\pi(e^2 - 1)}{2}$. 15. $\dfrac{\pi}{2}$. 17. $\dfrac{8\pi}{3}$. 19. $\dfrac{10\sqrt{5}}{3}\pi$. 21. $\dfrac{abh}{3}$. 23. 24π.

25. $\dfrac{16}{3}$ in³.

2. INTEGRALS OF $\mathcal{R} \to \mathcal{R}^n$ VECTOR FUNCTIONS

In Section 8 of Chapter 5 we defined the derivative function of f, where $f : R \to R^m$, by the formula

$$f'(t) = \begin{pmatrix} f'_1(t) \\ \cdot \\ \cdot \\ \cdot \\ f'_m(t) \end{pmatrix}.$$

This formula is valid, of course, only when each coordinate function f_k of f is differentiable at t. In this section we shall define the integral of an $\mathcal{R} \to \mathcal{R}^m$ function in an analogous fashion and shall discuss several properties of this integral.

DEFINITION 12.2 Let $f:\mathcal{R} \to \mathcal{R}^m$. f is *integrable* on $[a, b]$ if and only if each of the coordinate functions $f_1, \ldots, f_m$ is integrable on $[a, b]$. In this case, the *integral of f* on $[a, b]$, denoted by $\int_a^b f$, is defined by

$$\int_a^b f = \begin{pmatrix} \int_a^b f_1 \\ \vdots \\ \int_a^b f_m \end{pmatrix}.$$

Example 5. Let $f(x) = \begin{pmatrix} x^2 \\ \sin x \end{pmatrix}$, $0 \leq x \leq \dfrac{\pi}{2}$. By Definition 12.2, f is integrable on $\left[0, \dfrac{\pi}{2}\right]$ and

$$\int_0^{\pi/2} f = \begin{pmatrix} \int_0^{\pi/2} x^2 \\ \int_0^{\pi/2} \sin x \end{pmatrix}$$

$$= \begin{pmatrix} \left[\dfrac{x^3}{3}\right]_0^{\pi/2} \\ [-\cos x]_0^{\pi/2} \end{pmatrix}$$

$$= \begin{pmatrix} \dfrac{\pi^3}{24} \\ 1 \end{pmatrix}. \quad \blacksquare$$

Many theorems can be proven concerning integrable $\mathcal{R} \to \mathcal{R}^m$ functions. The reader is urged to formulate and prove theorems other than those discussed in this section.

THEOREM 12.3 Let $f:\mathcal{R} \to \mathcal{R}^m$ and $g:\mathcal{R} \to \mathcal{R}^m$. Let r and s be real numbers and C be a vector in $\mathcal{R}^m$. If f and g are integrable on $[a, b]$, then

i. the linear combination $rf + sg$ is integrable on $[a, b]$ and

$$\int_a^b (rf + sg) = r \int_a^b f + s \int_a^b g;$$

ii. the inner product $C \cdot f$ is integrable on $[a, b]$ and

$$\int_a^b C \cdot f = C \cdot \int_a^b f.$$

Proof. A proof of i is called for in Exercise 5 at the end of the section. In order to prove ii, we let $C = (c_1, \ldots, c_m)$. Since $f_1, \ldots, f_m$ are integrable $\mathcal{R} \to \mathcal{R}$ functions on $[a, b]$, $C \cdot f = \sum_{i=1}^{m} c_i f_i$ is also integrable on $[a, b]$. Furthermore,

$$\int_a^b C \cdot f =^1 \int_a^b \sum_{i=1}^{m} c_i f_i$$

$$=^2 \sum_{i=1}^{m} c_i \int_a^b f_i$$

$$=^3 C \cdot \int_a^b f.$$

(1) Substituting $\sum_{i=1}^{m} c_i f_i$ for $C \cdot f$ (Definition of inner product).
(2) Theorem 8.14 (Linear Properties of the Integral).
(3) Definition 12.2 and definition of inner product. ∎

Consider now a function $f: \mathcal{R} \to \mathcal{R}^m$ which is integrable on $[a, b]$. Define $F(x) = \int_a^x f$ for x in $[a, b]$. It follows immediately from Definition 12.2 and Theorem 9.1 (p. 379) that if f is continuous at a point x_0 of (a, b), then F is differentiable at x_0 and $F'(x_0) = f(x_0)$. Thus the First Fundamental Theorem of Calculus holds for $\mathcal{R} \to \mathcal{R}^m$ functions. It is obvious that the Second Fundamental Theorem of Calculus (see Theorem 9.4, p. 384) can be stated and proven for $\mathcal{R} \to \mathcal{R}^m$ functions.

Recall that if f is an $\mathcal{R} \to \mathcal{R}$ function which is integrable on $[a, b]$, then $|f|$ is also integrable on $[a, b]$ and $|\int_a^b f| \leq \int_a^b |f|$. The next theorem is the analogous statement for $\mathcal{R} \to \mathcal{R}^m$ functions.

THEOREM 12.4   Let $f: [a, b] \to \mathcal{R}^m$. If f and $\|f\|$ are integrable on $[a, b]$, then

$$\left\| \int_a^b f \right\| \leq \int_a^b \|f\|.$$

Proof. Let $C = \int_a^b f$ and note that, by Definition 12.2, C is an element of $\mathcal{R}^m$. Therefore,

$$\|C\|^2 =^1 C \cdot C$$

$$=^2 C \cdot \int_a^b f$$

$$=^3 \int_a^b C \cdot f$$

$$\leq^4 \int_a^b \|C\| \, \|f\|$$

$$=^5 \|C\| \int_a^b \|f\|.$$

(1) Theorem 2.22 (p. 134).
(2) Substituting $C = \int_a^b f$.

(3) Theorem 12.3ii.

(4) Theorem 2.23 (p. 135) (Cauchy-Schwarz Inequality) and Theorem 8.15 (p. 360) (Order Properties of the Integral).

(5) Theorem 12.3i.

The above chain of inequalities tells us that $\|C\|^2 \leq \|C\| \int_a^b \|f\|$. Thus, since $\|C\|$ and $\int_a^b \|f\|$ are non-negative, $\|C\| \leq \int_a^b \|f\|$. Recalling that $C = \int_a^b f$, we have $\|\int_a^b f\| \leq \int_a^b \|f\|$. ∎

EXERCISES

In Exercises 1 and 2 evaluate $\int_a^b f$:

1. $f(x) = \begin{pmatrix} x \\ \sqrt{x+1} \\ \sqrt{\dfrac{x}{x}} \end{pmatrix}$, $[a, b] = [0, 2]$.

2. $f(x) = \begin{pmatrix} x \sin x \\ \cos 2x \end{pmatrix}$, $[a, b] = \left[-\dfrac{\pi}{2}, \dfrac{\pi}{2} \right]$.

3. For the function given in Exercise 1, compute $\|\int_a^b f\|$ and $\int_a^b \|f\|$.

4. For the function given in Exercise 2 compute $\|\int_a^b f\|$ and $\int_a^b \|f\|$.

5. Prove Theorem 12.3i.

6. A particle P moves in the plane. At time t its velocity vector is given by $v(t)$ and its position at time t is $s(t)$. From Chapter 6 we know that $s'(t) = v(t)$.

   i. Show that at any time T
   $$s(T) = \int_0^T v(t) + s(0).$$

   ii. If $v(t) = \begin{pmatrix} t^2 \\ t \end{pmatrix}$ and $s(0) = \begin{pmatrix} 1 \\ 1 \end{pmatrix}$, find $s(2)$.

ANSWERS

1. $\begin{pmatrix} \dfrac{6\sqrt{3}-2}{3} \\ \dfrac{4\sqrt{2}}{3} \end{pmatrix}$.   3. $\|\int_a^b f\| = \sqrt{20 - \dfrac{8\sqrt{3}}{3}} \leq \int_a^b \|f\| = 4$.

3. PATH LENGTH

Consider the polygonal path in $\mathcal{R}^m$ sketched in Figure 12.10. If a particle moves from A to C along the line segments as indicated by the arrows, the distance traversed is $\|B - A\| + \|C - B\|$. If upon reaching C the particle returns to B and then returns to C again, the total distance traversed

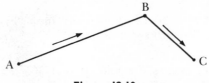

Figure 12.10

is $\|B - A\| + 3\|C - B\|$. In both cases the set of points traversed in $\mathcal{R}^m$ is the same, but the distance travelled is greater in the second case. We shall make a distinction between these two paths, although the set of points involved in the two paths are the same.

A continuous function $p:[a, b] \to \mathcal{R}^m$ is called a *path* in $\mathcal{R}^m$. By *the path $p[a, b]$ from $p(a)$ to $p(b)$* we shall mean not only the range of p but also position of the point $p(t)$ for each t in $[a, b]$. It is helpful to think of t as time and $p(t)$ the position of a particle at time t.

Example 6. Consider the two paths in $\mathcal{R}^2$, defined by

$$p_1(t) = \begin{pmatrix} \cos t \\ \sin t \end{pmatrix} \text{ for } t \text{ in } [0, 2\pi]$$

and

$$p_2(t) = \begin{pmatrix} \cos 2t \\ \sin 2t \end{pmatrix} \text{ for } t \text{ in } [0, 2\pi].$$

The range of p_1 is $\left\{ \begin{pmatrix} x \\ y \end{pmatrix} \middle| x^2 + y^2 = 1 \right\}$. This set is also the range of p_2. Figure 12.11a and b illustrates the difference between the two paths. If a particle follows path $p_1[0, 2\pi]$ it will traverse the unit circle once, starting and ending at the point $\begin{pmatrix} 1 \\ 0 \end{pmatrix}$. If a particle follows path $p_2[0, 2\pi]$,

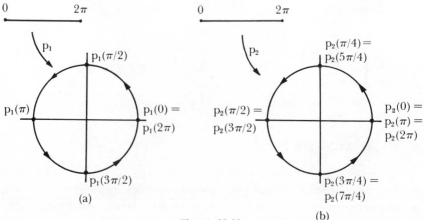

Figure 12.11

it will traverse the unit circle twice, passing through the point $\begin{pmatrix} 1 \\ 0 \end{pmatrix}$ at $t = 0$, $t = \pi$, and $t = 2\pi$. Furthermore, it will pass through every other point of the unit circle exactly twice. ∎

The problem of defining the length of an arbitrary path $\mathcal{C}$ in $\mathcal{R}^m$ is more complicated than that of defining the length of a polygonal path in $\mathcal{R}^m$. We seek a length function L which has the following properties:

1. If $\mathcal{C}$ is a path in $\mathcal{R}^m$ with endpoints A and B, then the length of $\mathcal{C}$ (if it is defined) is greater than or equal to $\|B - A\|$, the distance from A to B.
2. If a path $\mathcal{C}$ is divided into two paths $\mathcal{C}_1$ and $\mathcal{C}_2$ such that the terminal point of $\mathcal{C}_1$ is the initial point of $\mathcal{C}_2$, then the length of C is the sum of the length of $\mathcal{C}_1$ and the length of $\mathcal{C}_2$.

As the reader will see in the following discussion, these assumptions together with the definition of the length of a polygonal path will enable us to find the length of certain paths. Indeed, we shall use the length of certain polygonal paths to approximate the length of a given path $\mathcal{C}$.

Consider the paths $\mathcal{C}$, defined parametrically by $p : [0, 4] \to \mathcal{R}^m$ and sketched in Figure 12.12a. Assume that the length of $\mathcal{C}$ can be determined. We partition $[0, 4]$ by $\mathcal{P}_1 = \{0, 1, 2, 4\}$ and let $\mathcal{A}_1$ be the polygonal path connecting $p(0)$ to $p(1), p(1)$ to $p(2)$, and $p(2)$ to $p(4)$. $\mathcal{C}$ and $\mathcal{A}_1$ are sketched in Figure 12.12b. The length of $\mathcal{A}_1$ is $\|p(1) - p(0)\| + \|p(2) - p(1)\| + \|p(4) - p(2)\|$. By our basic assumptions about length, the length of $\mathcal{C}$ is greater than or equal to the length of $\mathcal{A}_1$. Now let $P_2 = \{0, 1, 2, 3, 4\}$ and let $\mathcal{A}_2$ be the polygonal path from $p(0)$ to $p(1)$, $p(1)$ to $p(2)$, $p(2)$ to $p(3)$, and $p(3)$ to $p(4)$.

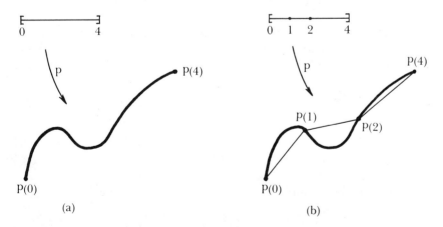

(a)   (b)

Figure 12.12

Note now that the length of $\mathcal{A}_1$

$$=^1 \|p(1) - p(0)\| + \|p(2) - p(1)\| + \|p(4) - p(2)\|$$
$$\leq^2 \|p(1) - p(0)\| + \|p(2) - p(1)\| + \|p(3) - p(2)\| + \|p(4) - p(3)\|$$
$$=^1 \text{ the length of } \mathcal{A}_2$$
$$\leq^3 \text{ the length of } \mathcal{C}.$$

(1) Definition of length of a polygonal path.
(2) Theorem 2.24 (p. 136) (the Triangle Inequality).
(3) By our basic assumptions.

Obviously, if we choose finer partitions of $[0, 4]$ and compute the length of the corresponding polygonal paths, we obtain better approximations for the length of C. This is our motivation for the following definition:

DEFINITION 12.5 Let $p : [a, b] \to \mathcal{R}^m$ be continuous.
i. For each partition $\mathcal{P} = \{x_0, \ldots, x_n\}$ of $[a, b]$, let

$$\Pi(\mathcal{P}) = \sum_{i=1}^{n} \|p(x_i) - p(x_{i-1})\|.$$

ii. If $\{\Pi(\mathcal{P}) \mid \mathcal{P} \text{ is a partition of } [a, b]\}$ has an upper bound, define $L_p[a, b] = \text{lub } \{\Pi(\mathcal{P}) \mid \mathcal{P} \text{ is a partition of } [a, b]\}$. In this case $L_p[a, b]$ is called the *length of the path* $p[a, b]$ and $p[a, b]$ is called a *rectifiable path*.
iii. If $\{\Pi(\mathcal{P}) \mid \mathcal{P} \text{ is a partition of } [a, b]\}$ does not have an upper bound, we say that $p[a, b]$ is *not rectifiable*.

For each $\mathcal{P}$ we have $\Pi(\mathcal{P}) \geq 0$. Hence $L_p[a, b] \geq 0$ if $p[a, b]$ is a rectifiable path. There are paths which are not rectifiable. The reader is encouraged to find one.

We shall now prove that our second basic assumption holds with this definition of length. In the proof of this theorem we shall need the fact that if $\mathcal{P}$ is a partition of $[a, b]$ and $c \in (a, b)$ then $\Pi(\mathcal{P}) \leq \Pi(\mathcal{P} \cup \{c\})$. This should be intuitively obvious. It follows from the fact that if $x_{i-1} \leq c \leq x_i$, then $\|p(x_i) - p(x_{i-1})\| \leq \|p(x_i) - p(c)\| + \|p(c) - p(x_{i-1})\|$ by the Triangle Inequality.

THEOREM 12.6 (Additivity of Path Length.) Let $p : [a, b] \to \mathcal{R}^m$ be continuous and let $c \in [a, b]$. The path $p[a, b]$ is rectifiable if and only if both $p[a, c]$ and $p[c, b]$ are rectifiable. In this case

$$L_p[a, b] = L_p[a, c] + L_p[c, b].$$

Proof. Assume that $p[a, b]$ is rectifiable. Let P_1 and P_2 be partitions of $[a, c]$ and $[c, b]$, respectively. It follows immediately from Definition 12.5 that $\Pi(\mathfrak{F}_1) \leq L_p[a, b]$ and $\Pi(\mathfrak{F}_2) \leq L_p[a, b]$. Consequently, a second application of Definition 12.5 shows that $p[a, c]$ and $p[c, b]$ are rectifiable paths and that

$$(12.10) \qquad L_p[a, c] + L_p[c, b] \leq L_p[a, b].$$

If strict inequality holds in Inequality 12.10, then by Definition 12.5 there exists a partition $\mathfrak{F}$ of $[a, b]$ such that

$$(12.11) \qquad L_p[a, c] + L_p[c, b] < \Pi(\mathfrak{F}).$$

Let $\mathfrak{F}_0 = \mathfrak{F} \cup \{c\}$, $\mathfrak{F}_1 = \mathfrak{F}_0 \cap [a, c]$, and $\mathfrak{F}_2 = \mathfrak{F}_0 \cap [c, b]$. $\mathfrak{F}_1$ is a partition of $[a, c]$ and $\mathfrak{F}_2$ is a partition of $[c, b]$. Furthermore, $\Pi(\mathfrak{F}) \leq \Pi(\mathfrak{F}_0)$ and $\Pi(\mathfrak{F}_1) + \Pi(\mathfrak{F}_2) = \Pi(\mathfrak{F}_0)$ (check these facts). Therefore,

$$(12.12) \qquad \begin{aligned} L_p[a, c] + L_p(c, b] &<^1 \Pi(\mathfrak{F}) \\ &\leq^2 \Pi(\mathfrak{F}_0) \\ &=^3 \Pi(\mathfrak{F}_1) + \Pi(\mathfrak{F}_2) \\ &\leq^4 L_p[a, c] + L_p[c, b]. \end{aligned}$$

(1) Inequality 12.11.
(2) By Definition 12.5i, since $\mathfrak{F}_0 = \mathfrak{F} \cup \{c\}$.
(3) Since $\mathfrak{F}_0 = \mathfrak{F}_1 \cup \mathfrak{F}_2$ and $\mathfrak{F}_1 \cap \mathfrak{F}_2 = \{c\}$.
(4) Definition 12.5ii.

The Inequality Chain 12.12 contains an obvious contradiction. Therefore, equality must hold in Equation 12.10 if $p[a, b]$ is rectifiable.

The remainder of the proof consists of showing that $p[a, b]$ is rectifiable if $p[a, c]$ and $p[c, b]$ are each rectifiable. This proof is called for in Exercise 15. ∎

A straightforward result of Theorem 12.6 is that if $p[a, b]$ is a rectifiable path and $x \in (a, b]$, then $p[a, x]$ is a rectifiable arc. Thus, the following definition is meaningful:

DEFINITION 12.7 (The Path Length Function Λ_p.) Let $p : [a, b] \rightarrow \mathcal{R}^m$. If $p[a, b]$ is a rectifiable path, we define the *path length function* Λ_p on $p[a, b]$ as follows:

$$\Lambda_p(x) = \begin{cases} L_p[a, x] & \text{for } a < x \leq b \\ 0 & \text{for } x = a. \end{cases}$$

In later sections of this chapter we shall make use of the path length function. One of the most basic and intuitively obvious properties of this function is that it is increasing. This follows easily from Theorem 12.6: If

$a \leq x \leq y \leq b$, then $L_p[a, x] + L_p[x, y] = L_p[a, y]$. Consequently,

$$L_p[a, x] \leq L_p[a, y]$$

if $x \leq y$. This proves Theorem 12.8.

THEOREM 12.8 Let $p:[a, b] \to \mathcal{R}^m$. If $p[a, b]$ is a rectifiable path and Λ_p is the path length function of $f[a, b]$, then Λ_p is an increasing function.

We have not yet attempted to find the length of any path other than a polygonal path. If we attempted to find the length of even a simple path such as a semi-circle by Definition 12.5 we would encounter algebraic difficulties similar to those which we encountered in Chapter 8 where the integral was first defined. The next theorem obviates these difficulties for a certain class of paths:

THEOREM 12.9 Let $p:\mathcal{R} \to \mathcal{R}^m$. If p is differentiable on $[a, b]$ and p' is continuous on $[a, b]$, then
 i. $p[a, b]$ is a rectifiable path in $\mathcal{R}^m$.
 ii. Λ_p is differentiable on $[a, b]$ and $\Lambda_p' = \|p'\|$.
 iii. $\Lambda_p(b) - \Lambda_p(a) = L_p[a, b] = \int_a^b \|p'\|$.

Proof. If $\mathcal{I} = \{x_0, \ldots, x_n\}$ is a partition of $[a, b]$, we have

(12.13)
$$\Pi(\mathcal{I}) =^1 \sum_{i=1}^{n} \|p(x_i) - p(x_{i-1})\|$$

$$=^2 \sum_{i=1}^{n} \left\| \int_{x_{i-1}}^{x_i} p' \right\|$$

$$\leq^3 \sum_{i=1}^{n} \int_{x_{i-1}}^{x_i} \|p'\|$$

$$=^4 \int_a^b \|p'\|.$$

(1) Definition 12.5 (p. 490).
(2) Theorem 9.4 (p. 384) (Second Fundamental Theorem).
(3) Theorem 12.4 (p. 486). (In Exercise 16 the reader is asked to show that $\|p'\|$ is integrable on $[a, b]$.)
(4) Theorem 8.18 (p. 365).

Thus, by Inequality Chain 12.13, $\int_a^b \|p'\|$ is an upper bound for $\{\Pi(\mathcal{I}) \mid \mathcal{I}$ is a partition on $[a, b]\}$. By Definition 12.5, then $p[a, b]$ is a rectifiable path and

$$L_p[a, b] \leq \int_a^b \|p'\|.$$

If we can show that $\Lambda_p'(x) = \|p'(x)\|$ for x in (a, b) then we shall have

(12.14)
$$L_p[a, b] =^1 \Lambda_p(b) - \Lambda_p(a)$$

$$=^2 \int_a^b \|p'\|.$$

(1) By Definition 12.7 (of Λ_p).
(2) Theorem 9.4 (p. 384) (Second Fundamental Theorem).

The Equality Chain 12.14 is what we need to establish to complete the proof of the theorem. Thus we need only prove that $\Lambda_p'(x) = \|p'(x)\|$ for all x in (a, b).

Assume first that $h > 0$ and let $t \in (a, b)$. Then

$$\left\| \frac{p(t + h) - p(t)}{h} \right\| \leq^1 \frac{1}{h} L_p[t, t + h] =^2 \frac{\Lambda_p(t + h) - \Lambda_p(t)}{h}$$

$$\leq^3 \frac{1}{h} \int_t^{t+h} \|p'\|.$$

(1) Since $\mathfrak{F} = \{t, t + h\}$ is a partition of $[t, t + h]$.
(2) Definition 12.7 and Theorem 12.6.
(3) Equation 12.14, applied with $t = a$ and $t + h = b$.

A similar argument can be used to show that if $h < 0$,

(12.15)
$$\left\| \frac{p(t) + h) - p(t)}{h} \right\| \leq \frac{\Lambda_p(t + h) - \Lambda_p(t)}{h} \leq \frac{1}{h} \int_t^{t+h} \|p'\|.$$

Thus Inequality Chain 12.15 holds for all $h \neq 0$. Consider the extreme members of Inequality 12.15. Since p is differentiable,

$$\lim_{h \to 0} \left\| \frac{p(t + h) - p(t)}{h} \right\| = \|p'(t)\|$$

and since $\|p'\|$ is continuous, $\lim_{h \to 0} \int_t^{t+h} \|p'\| = \|p'(t)\|$, by the Second Fundamental Theorem. Therefore it follows from Inequality 12.15 that Λ_p is differentiable and $\Lambda_p'(t) = \|p'(t)\|$ for each t in (a, b). This fact completes the proof of the theorem. ∎

Example 7. Find the circumference of a circle of radius r. If the circle has center at O_2 it can be defined parametrically by

$$p(t) = \begin{pmatrix} r \cos t \\ r \sin t \end{pmatrix}, \qquad 0 \leq t \leq 2\pi.$$

Theorem 12.9 tells us that

$$L_p[0, 2\pi] = \int_0^{2\pi} \|p'\|.$$

We note that

$$p'(t) = \begin{pmatrix} -r \sin t \\ r \cos t \end{pmatrix}$$

and
$$\|p'(t)\| = \sqrt{r^2 \sin^2 t + r^2 \cos^2 t} = r.$$
Therefore,
$$L_p[0, 2\pi] = \int_0^{2\pi} \|p'(t)\|$$
$$= \int_0^{2\pi} r$$
$$= 2\pi r. \quad \blacksquare$$

In this book we have not given a rigorous mathematical definition of the sine and cosine functions. Therefore, Example 7 does not prove that the circumference of a circle of radius r is $2\pi r$. It merely illustrates an application of Theorem 12.9 and shows that our results are consistent with the definition which we do give for the sine and cosine functions.

Example 8. Find the length of the path defined by

$$p(t) = \begin{pmatrix} t \\ \dfrac{2t^{3/2}}{3} \end{pmatrix}, \qquad 0 \le t \le 3.$$

By Theorem 12.9,

$$L_p[0, 3] = \int_0^3 \|p'\|$$
$$= \int_0^3 \left\| \begin{pmatrix} 1 \\ t^{1/2} \end{pmatrix} \right\|$$
$$= \int_0^3 \sqrt{1 + t}$$
$$= [\tfrac{2}{3}(1 + t)^{3/2}]_0^3$$
$$= \tfrac{14}{3}. \quad \blacksquare$$

The path considered in Example 8 can be described explicitly by $f(x) = \tfrac{2}{3}x^{3/2}$, $0 \le x \le 3$. Note that $\|p'\| = \sqrt{1 + (f')^2}$. This is a special case of the next Corollary:

COROLLARY 12.10 Let $f:[a, b] \to \mathcal{R}$ be differentiable on $[a, b]$ and assume that f' is continuous on $[a, b]$. The length of the graph of f between a and x is given by

$$\Lambda_p(x) = \int_a^x \sqrt{1 + (f')^2},$$

where p is the path defined by $p(x) = \begin{pmatrix} x \\ f(x) \end{pmatrix}$ for $a \le x \le b$.

Proof. Let $p(x) = \begin{pmatrix} x \\ f(x) \end{pmatrix}$ for x in $[a, b]$. Since f is differentiable and f' is continuous on $[a, b]$, we have p differentiable and p' continuous on $[a, b]$. Therefore, by Theorem 12.9,

$$\Lambda_p(x) = L_p[a, x]$$

$$= \int_a^x \|p'\|$$

$$= \int_a^x \sqrt{1 + (f')^2}. \quad \blacksquare$$

In applying either Theorem 12.8 or Corollary 12.10 one should check carefully to see that the hypotheses of differentiability and continuity are satisfied. For example $f(x) = \sqrt{1 - x^2}$, $-1 \leq x \leq 1$, defines explicitly the upper half of the unit circle. However, f is not differentiable at either $x = 1$ or $x = -1$.

EXERCISES

Find the lengths of the paths defined parametrically by the following functions:

1. $p(x) = \begin{pmatrix} x \\ 2x \end{pmatrix}$, $0 \leq x \leq 1$.

2. $p(x) = \begin{pmatrix} x \\ x \\ x \end{pmatrix}$, $2 \leq x \leq 3$.

3. $p(x) = \begin{pmatrix} x^2 \\ x^2 \\ x^2 \\ x^2 \end{pmatrix}$, $-1 \leq x \leq 1$.

4. $p(x) = \begin{pmatrix} \cos x \\ \sin x \\ \cos x \\ \sin x \end{pmatrix}$, $0 \leq x \leq \pi$.

5. $p(x) = \begin{pmatrix} x \\ x^{3/2} \end{pmatrix}$, $0 \leq x \leq \frac{4}{9}$.

6. $p(x) = \begin{pmatrix} x \\ \ln \cos x \end{pmatrix}$, $0 \leq x \leq \frac{\pi}{3}$.

7. $p(t) = \begin{pmatrix} t \\ \sin t \\ \cos t \end{pmatrix}$, $0 \leq t \leq 2$.

8. $p(t) = \begin{pmatrix} t^2 \\ \frac{2}{3}t^3 - \frac{1}{2}t \end{pmatrix}$, $0 \leq t \leq 2$.

9. $p(t) = \begin{pmatrix} 4\sqrt{2}t^3 \\ 3t^4 \\ 6t^2 \end{pmatrix}$, $-1 \leq t \leq 1$.

10. $p(x) = \begin{pmatrix} x \\ \ln |\cos x| \end{pmatrix}$, $0 \leq x \leq \frac{\pi}{3}$.

11. $p(x) = 2 \begin{pmatrix} x/2 \\ e^{x/4} + e^{-x/4} \end{pmatrix}$, $0 \leq x \leq 4$. 12. $p(t) = \begin{pmatrix} t \\ 2e^{t/2} \end{pmatrix}$, $0 \leq t \leq 2$.

13. $p(x) = \begin{pmatrix} x \\ \ln x \end{pmatrix}$, $\frac{1}{2} \leq x \leq 2$.

14. Try to find the length of the path defined by $p(x) = \begin{pmatrix} x \\ 2x^{1/2} \end{pmatrix}$, $0 \leq x \leq 1$, using Theorem 12.9. You will run into difficulty. Tell why.

15. Let $p:[a, b] \to \mathcal{R}^m$ and $c \in (a, b)$. Show that if $p[a, b]$ and $p[c, b]$ are rectifiable paths, then $p[a, b]$ is a rectifiable path.

16. Let $p:[a, b] \to \mathcal{R}$ be such that p' is continuous on $[a, b]$. Show that p' is integrable on $[a, b]$.

ANSWERS

1. $\sqrt{5}$. **3.** 4. **5.** $\frac{8}{27}[2\sqrt{2} - 1]$. **7.** $2\sqrt{2}$. **9.** 18. **11.** $\dfrac{2(e^2 - 1)}{e}$.

13. $\dfrac{\sqrt{5}}{2} + \ln \dfrac{4 + 2\sqrt{5}}{1 + \sqrt{5}}$.

4. SURFACES OF REVOLUTION

Let $f:[a, b] \to \mathcal{R}$ and assume that $f(x) \geq 0$ for all x in $[a, b]$. If the graph of f is rotated about the x-axis (as in Figure 12.13), a *surface of revolution* is generated. In this section we shall define the area of such a surface.

The simplest surface of revolution is that generated by a constant function $g(x) = k$ on $[a, b]$ (see Figure 12.14a). In this case we *define* the surface area to be

$$(12.16) \qquad\qquad S = 2\pi k(b - a).$$

This definition agrees with our definition of the area of a rectangle, for we can think of the surface of a right circular cylinder of radius k and height $b - a$ as a rectangle of length $2\pi k$ and width $b - a$.

We return now to the more general situation. Let $f:[a, b] \to \mathcal{R}$. Assume that f is differentiable on $[a, b]$ and that f' is continuous on $[a, b]$. With these assumptions the path length function Λ of the graph of f is defined on $[a, b]$.

By Theorem 12.9 and Corollary 12.10, Λ is differentiable and

$$(12.17) \qquad\qquad \Lambda'(x) = \sqrt{1 + f'(x)^2} \quad \text{for } x \text{ in } [a, b].$$

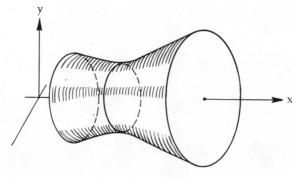

Figure 12.13

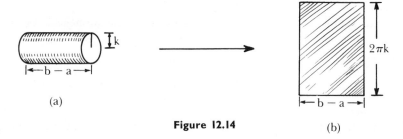

Figure 12.14

(a)

(b)

Let $\mathfrak{F} = \{x_0, \ldots, x_n\}$ be a partition of $[a, b]$. Let l and u be lower and upper step functions for f on $[a, b]$ with $l(x) = l_i$ and $u(x) = u_i$ for x in (x_{i-1}, x_i). Let us assume that the area S of the surface generated by rotating the graph of f about the x-axis can be defined. We shall show below that it is reasonable to require S to be such that

$$(12.18) \quad \sum_{i=1}^{n} 2\pi l_i (\Lambda(x_i) - \Lambda(x_{i-1})) \leq S \leq \sum_{i=1}^{n} 2\pi u_i (\Lambda(x_i) - \Lambda(x_{i-1}))$$

for all $\mathfrak{F}$, l, and u. It follows from Theorem 12.1 that $S = \int_a^b 2\pi f(x) \Lambda'(x)$. Thus, using 12.17, we shall *define* the surface area S generated by rotating the graph of f about the x-axis by

$$(12.19) \qquad S = 2\pi \int_a^b f(x) \sqrt{1 + f'(x)^2}.$$

We now attempt to motivate Inequality 12.18. Consider the cylinder of radius l_i and height $\Lambda(x_i) - \Lambda(x_{i-1})$, sketched in Figure 12.15a. By Equation 12.16 the surface area generated is $(2\pi l_i)(\Lambda(x_i) - \Lambda(x_{i-1}))$. Recall that $l_i \leq f(x)$ on (x_{i-1}, x_i) and that $\Lambda(x_i) - \Lambda(x_{i-1})$ is the length of the graph of f from $f(x_{i-1})$ to $f(x_i)$. In Figure 12.15b we have sketched the surface generated by f on $[x_{i-1}, x_i]$ together with the cylinder shown in Figure 12.15a.

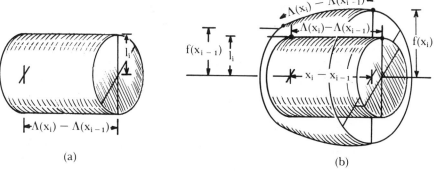

(a)

(b)

Figure 12.15

If we think of distorting the surface of the right circular cylinder of radius l_i and height $\Lambda(x_i) - \Lambda(x_{i-1})$ to match the surface generated by f on (x_{i-1}, x_i), we see that we shall have to stretch the surface of the right circular cylinder. Thus we require that area S_i of the surface generated by f on $[x_{i-1}, x_i]$ to be such that

(12.20) $$2\pi l_i(\Lambda(x_i) - \Lambda(x_{i-1})) \le S_i.$$

If we assume that the area S generated by f on $[a, b]$ is the sum of the areas generated by f on $[x_{i-1}, x_i]$, $i = 1, \ldots, n$, then we have motivated the left-hand inequality in Inequality 12.18. The right hand inequality is motivated in a similar fashion.

Example 9. Find the area generated by rotating the graph of $f(x) = \dfrac{x^3}{3}$ on $[0, 1]$ about the x-axis. By our definition and Equation 12.19 this area is

$$S = 2\pi \int_0^1 f(x)\sqrt{1 + f'(x)^2}$$

$$= 2\pi \int_0^1 \frac{x^3}{3} \sqrt{1 + x^4}$$

$$= \frac{2\pi}{3} \left[\tfrac{1}{6}(1 + x^4)^{3/2}\right]_0^1$$

$$= \frac{\pi}{9} (2\sqrt{2} - 1). \quad \blacksquare$$

The reader should keep in mind that to apply Formula 12.19 the function f must be differentiable on the interval $[a, b]$ in question. The next example illustrates the problem involved if f is not differentiable on all of $[a, b]$.

Example 10. Let $f(x) = \sqrt{r^2 - x^2}$ for x in $[-r, r]$. If the graph of f is rotated about the x-axis a sphere of radius r is generated. We cannot apply Formula 12.19 to find the surface area of the whole sphere since f is not differentiable at $x = r$ or at $x = -r$. We handle this problem as follows: Let h be a real number such that $0 < h < r$. The area S_h generated by f on $[-r, +h, r - h]$ is

(12.21) $$S_h = 2\pi \int_{-r+h}^{r-h} f(x)\sqrt{1 + f'(x)^2}$$

$$= 2\pi \int_{-r+h}^{r-h} \sqrt{r^2 - x^2} \sqrt{1 + \frac{x^2}{r^2 - x^2}}$$

$$= 2\pi \int_{-r+h}^{r-h} r$$

$$= 4\pi r(r - h).$$

Since Equation 12.21 is valid for each $h > 0$, we define the area of the sphere to be S, where $S = \lim_{h \to 0} S_h = 4\pi r^2$. ∎

Assume now that $f:[a, b] \to \mathfrak{R}$, that f is differentiable, and that f has an inverse on $[a, b]$. If we rotate the graph of f about the y-axis, the surface generated is the same as that generated by rotating the graph of f^{-1} about the x-axis. In particular, if f is *increasing*, the area is

$$(12.22a) \qquad S = 2\pi \int_{f(a)}^{f(b)} f^{-1}(x)\sqrt{1 + (f^{-1})'(x)^2},$$

and if f is *decreasing*, the area is

$$(12.22b) \qquad S = 2\pi \int_{f(b)}^{f(a)} f^{-1}(x)\sqrt{1 + (f^{-1})'(x)^2}.$$

Example 11. Consider the surface generated by rotating the graph of $f(x) = x^{1/3}$ on $[1, 8]$ about the y-axis. The function f is increasing on $[1, 8]$ and $f^{-1}(x) = x^3$ on $[1, 8]$. Thus, the area generated is

$$S = 2\pi \int_{f(1)}^{f(8)} f^{-1}(x)\sqrt{1 + (f^{-1})'(x)^2}$$

$$= 2\pi \int_{1}^{2} x^3\sqrt{1 + 9x^4}$$

$$= 2\pi[\tfrac{1}{54}(1 + 9x^4)^{3/2}]_1^2$$

$$= \frac{\pi}{27}[(145)^{3/2} - (10)^{3/2}]. \quad ∎$$

EXERCISES

In Exercises 1 to 4 compute the area generated by rotating the graph of f about the x-axis:

1. $f(x) = \tfrac{3}{4}x$ for x in $[0, 2]$. 2. $f(x) = e^x$ for x in $[0, 1]$.

3. $f(x) = x^2$ for x in $[0, 2]$. 4. $f(x) = 2\sqrt{x}$ for x in $[0, 3]$.

In Exercises 5 to 8 compute the area generated by rotating the graph of f about the y-axis:

5. $f(x) = x^{1/3}$ for x in $[0, 1]$. 6. $f(x) = x^2$ for x in $[0, 2]$.

7. $f(x) = \ln x$ for x in $[1, 2]$. 8. $f(x) = \sqrt{x}$ for x in $[1, 4]$.

9. A right circular cone has height h and base radius r. Find the area of the surface between heights a and b.

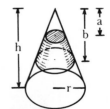

10. The parabolic reflector of a searchlight is 12 in. in diameter and 4 in deep. What is its area?

11. A spherical water tank of diameter 8 ft is filled to a depth of 2 ft. Find the area of the submerged surface of the tank.

12. Find the area of the ellipsoid given implicitly by

$$\frac{x^2}{4} + y^2 + z^2 = 1.$$

13. If the circle defined implicitly by $x^2 + (y - R)^2 = r^2$ is rotated about the x-axis, the resulting figure is known as a "torus." Find the area of its surface.

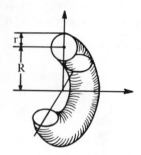

ANSWERS

1. $\frac{15}{4}\pi$. **3.** $\pi(\frac{1}{4}\sqrt{17^3} - \frac{1}{8}\sqrt{17} - \frac{1}{32}\ln(4 + \sqrt{17}))$. **5.** $\frac{\pi}{27}[10^{3/2} - 1]$.

7. $\pi\left(2\sqrt{5} - \sqrt{2} + \ln\frac{2 + \sqrt{5}}{1 + \sqrt{2}}\right)$. **9.** $\frac{\pi r}{h}\sqrt{1 + \frac{r^2}{h^2}}(a^2 - b^2 + 2h(b - a))$.

11. 16π ft². **13.** $4\pi^2 Rr$.

5. FORCES ON A SUBMERGED SURFACE

Consider a tank filled with a fluid of homogeneous density δ. The *pressure* exerted by the fluid at depth t is found experimentally to be δt. (For example, if the density is 50 lbs/cu ft, the pressure at a depth of 10 ft is 500 lbs/sq ft.) If a plate of area A is submerged t_0 units in such a way that the face of the plate is parallel to the surface of the fluid, as in Figure 12.16, the *force* exerted on one face of the plate by the liquid is the product of the area of the plate and the pressure at depth t_0: $F = A(\delta t_0)$. In this situation

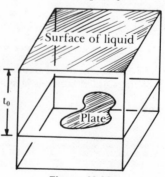

Figure 12.16

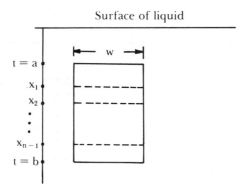

Surface of liquid

Figure 12.17

the pressure of the liquid δt_0 is the same at each point of the surface of the plate.

If a plate is submerged and held perpendicular to the surface of the liquid, the force exerted on the plate is not so simply determined, for the pressure at different points on the plate varies with the depth of the liquid. We seek a reasonable definition of the total force F on one side of a plate so situated. We consider first the case in which the plate is rectangular and one edge of the plate is parallel to the surface of the liquid. Assume that the width of the plate is w, the top edge is at depth $t = a$, the bottom edge is at depth $t = b$, and the density of the fluid is δ (see Figure 12.17). Let $\mathcal{P} = \{x_0, \ldots, x_n\}$ be a partition of $[a, b]$ and consider the resulting partitioning of the plate into rectangles. The pressure at depth $t = x_{k-1}$ is by definition δx_{k-1}, and the pressure at depth $t = x_k$ is δx_k. Since the pressure increases with the depth, we shall assume that the resulting force also increases with depth. We further assume that the total force exerted on two nonoverlapping plates is the sum of the forces exerted on the two plates. Thus it is reasonable to expect that the force exerted on the strip of width and edge (x_{k-1}, x_k) is greater than $w(x_k - x_{k-1})(\delta x_{k-1})$ and less than $w(x_k - x_{k-1})(\delta x_k)$, and that the total force F is such that

$$(12.23) \qquad \sum_{i=1}^{n} w(x_k - x_{k-1})(\delta x_{k-1}) \leq F \leq \sum_{i=1}^{n} w(x_k - x_{k-1})(\delta x_k).$$

Applying Theorem 12.1 with g given by $g(t) = \delta t$ and f given by $f(t) = wt$, we see that $\int_a^b w\, \delta t$ is the only number lying between all such pairs of upper and lower estimates of the force. With this motivation we *define* the force F on the rectangular plate to be

$$(12.24) \qquad F = \delta \int_a^b wt = \delta w \int_a^b t.$$

Consider now a vertical plate suspended in a liquid of density δ and let $w(t)$ be the width of the plate at depth t, $a \leq t \leq b$ (see Figure 12.18a and b). Let $\mathcal{P} = \{x_0, \ldots, x_n\}$ be a partition of $[a, b]$ and let l and u be lower and

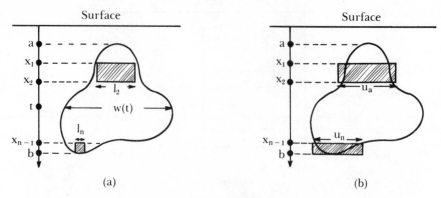

Figure 12.18

upper step functions of w on $\mathcal{S}$: $l(t) = l_i$ and $u(t) = u_i$ for t in (x_{i-1}, x_i). It follows from our preceding discussion and Equation 12.24 that

$$(12.25) \qquad F_l = \sum_{i=1}^{n} l_i \, \delta \int_{x_{i-1}}^{x_i} t = \sum_{i=1}^{n} l_i \, \delta(x_i^2 - x_{i-1}^2)$$

and

$$(12.26) \qquad F_u = \sum_{i=1}^{n} u_i \, \delta \int_{x_{i-1}}^{x_i} t = \sum_{i=1}^{n} u_i \, \delta(x_i^2 - x_{i-1}^2)$$

are respectively lower and upper estimates for the total force exerted on the plate. If we apply Theorem 12.1 with $g(t) = \dfrac{t^2}{2}$ and $f(t) = w(t)$, we know that $\delta \int_a^b tw(t)$ is the only number lying between all possible sums appearing in Equations 12.25 and 12.26. Thus we define the total force F on the plate to be

$$(12.27) \qquad F = \delta \int_a^b tw(t),$$

where δ is the density of the fluid and $w(t)$ is the width of the plate at depth t.

Example 12. A dam with a trapezoidal face is 40 ft wide at the bottom, 60 ft wide at the top, and 10 ft high. When the lake is full, what is the total force on the face of the dam?

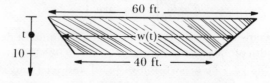

SOLUTION. Since the face of the dam is trapezoidal, the width at depth t can be shown to be $w(t) = 60 - 2t$. Thus the total force on the dam is, by Equation 12.27,

$$F = \delta \int_0^{10} t(60 - 2t)$$

$$= \delta[30t^2 - \tfrac{2}{3}t^3]_0^{10}$$

$$\approx (2334)\delta.$$

The density δ of water is approximately 62.4 lbs/cu ft, so $F \approx (62.4)(2334)$ lbs. ■

Example 13. A circular plate of radius 5 ft is suspended vertically in a fluid of density δ so that the top edge is 3 ft below the surface of the liquid. What is the total force on the face of the plate?

SOLUTION. Consider Figure 12.19 in which we have sketched the plate with its center 8 ft below the surface of the liquid. Before computing the force F, we must determine the width $w(t)$ of the plate at depth t. A glance at Figure 12.19 assures us that $w(t) = 2x(t)$ where $x(t)^2 + (8 - t)^2 = 25$. Thus,

$$w(t) = 2\sqrt{25 - (8 - t)^2}$$

and

$$F = \delta \int_3^{13} 2t\sqrt{25 - (\delta - t)^2}$$

$$= 200\pi\delta. \quad ■$$

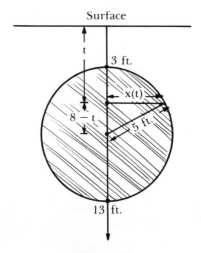

Surface

t

3 ft.

x(t)

8 − t

5 ft.

13 ft.

Figure 12.19

EXERCISES

In Exercises 1 to 5 assume that P is a plate submerged vertically in a liquid of density δ. Determine the total force exerted on the plate by the liquid.

1. P is rectangular; the top edge is 2 ft below and parallel to the surface of the liquid; the width of the plate is 5 ft and the bottom edge is 8 ft below the surface of the liquid.

2. P is rectangular; the top edge is at the surface of the liquid, the plate is 10 ft wide and the bottom edge is 9 ft below the surface of the liquid.

3. P is triangular and is sketched in Figure 12.20a.

4. P is triangular. A side of length 8 ft is parallel to and 10 ft below the surface of the liquid. The vertex opposite this side is 4 ft below the surface and the other two sides are equal in length.

5. P is square and each side is of length $\sqrt{2}$. One diagonal of the square is vertical. The top corner of P is 6 ft below the surface of the liquid.

In Exercises 6 to 10 assume that P is a vertical dam. Find the total force exerted on P by the water behind the dam.

6. The dam is semicircular, with the top edge 10 ft wide. The water comes to the top edge.

7. In Problem 6 the surface of the water is 3 ft below the top edge of the dam.

8. The width of the dam is given by $w(t) = 100 - t^2$ (where t denotes the depth) and the water comes to the top edge.

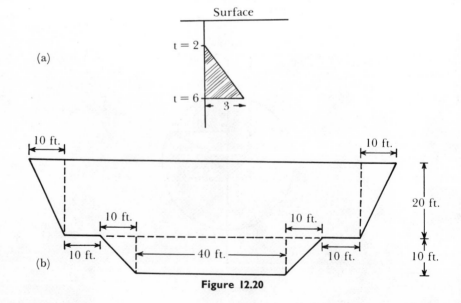

Figure 12.20

9. In Problem 8 the surface of the water is 4 ft below the top edge of the dam.

10. The dam is sketched in Figure 12.20b and the surface of the water is at the top edge of the dam.

ANSWERS

1. $F = \delta \int_2^8 5t = 150\delta$.   **3.** $F = \delta \int_2^6 t \cdot \frac{3}{4}(t - 2) = 28\delta$.

5. $F = \delta \int_6^7 t \cdot 2(t - 6) + \delta \int_7^8 t \cdot 2(8 - t) = 14\delta$.

7. $F = \delta \int_3^5 t \cdot 2\sqrt{25 - t^2} = \frac{128}{3}\delta$.   **9.** $F = \delta \int_4^{10} t(100 - t^2) = 1764\delta$.

6. WORK

If a constant force F moves an object P along a straight line path and if the force F is applied in the direction of the path, the work done is *defined* to be the force F multiplied by the distance P has moved: Work = Force × Distance. For example, if a person stands squarely behind a table and pushes the table 3 ft with a force of 2 lbs, the work done is 6 ft-lbs.

Assume now that an object P is to be moved along a path $r[a, b]$ in $\mathcal{R}^m$ from $r(a)$ to $r(b)$. Assume further that this object P is to be moved from $r(a)$ to $r(b)$ by a force F which has constant magnitude F_0 but has the same direction at time t as the path. (For example, one can push a chair in a circle, applying exactly 2 lbs of force at every moment, but constantly changing direction to keep the chair moving in the circle.) If $r[a, b]$ is a rectifiable path in R^m, we *define* the work done in applying F to move P from $r(a)$ to $r(b)$ to be F_0 times the length of $r[a, b]$. Let us now see why this definition is reasonable. Since $r[a, b]$ is rectifiable, its length can be closely approximated by the length of a polygonal path from $r(a)$ to $r(b)$ (see Figure 12.21). If the force F were applied along this polygonal path, the work done would be F_0 times the length of the polygonal path. We assume here that the work done in moving P along the union of straightline paths is the result obtained by computing the work done on each path and then adding.)

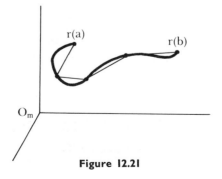

r(a)

r(b)

O_m

Figure 12.21

Let us continue to consider the problem discussed in the preceding paragraph and assume further that the function r is differentiable and $\|r'(t)\| > 0$. Let Λ be the arc length function on $r[a, b]$. Recall that $\Lambda(b) - \Lambda(a) = \int_a^b \Lambda'(t)$ is the length of the path $r[a, b]$. Thus the work done in moving P from $r(a)$ to $r(b)$ by exerting F is (by our definition)

$$(12.28) \qquad W = F_0[\Lambda(b) - \Lambda(a)]$$

$$= \int_a^b F_0 \Lambda'(t).$$

The general problem of defining the work done in applying a force to move an object along a path is more complex than the two cases previously considered. In the general case the force involved may vary in magnitude or may differ in direction from the path. We shall in the remainder of this section assume that $r : [a, b] \to \mathcal{R}^m$ defines a rectifiable path in $\mathcal{R}^m$, that r is differentiable and $\|r'(t)\| > 0$ for t in $[a, b]$. We further assume that $F : [a, b] \to \mathcal{R}^m$ is a force function in $\mathcal{R}^m$ and that $F \cdot r'$ is integrable on $[a, b]$.

If F is applied to move an object P along the path $r[a, b]$, the effective force involved at time t is the component of $F(t)$ in the direction of $r(t)$; this component is given by $(F(t) \cdot T) T$, where $T = \dfrac{r'(t)}{\|r'(t)\|}$ is the unit vector in the direction of $r(t)$ (see Figure 12.22 and refer to Section 5, Chapter 6). The *coefficient* (signed magnitude) of $F(t)$ in the direction of $r(t)$ is given by

$$(12.29) \qquad f(t) = F(t) \cdot T = F(t) \cdot \frac{r'(t)}{\|r'(t)\|}.$$

Let $\mathcal{Q} = \{x_0, \ldots, x_n\}$ be a partition of $[a, b]$ and let l and u be lower and upper step functions of f, respectively, such that

$$l(t) = l_i \quad \text{for } t \text{ in } (x_{i-1}, x_i),$$

$$u(t) = u_i \quad \text{for } t \text{ in } (x_{i-1}, x_i),$$

and

$$l(t) \leq f(t) \leq u(t) \quad \text{for } t \text{ in } [a, b].$$

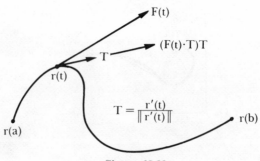

Figure 12.22

Let F_l and F_u be the force functions defined by

$$F_l(t) = l_i \frac{r'(t)}{\|r'(t)\|} \quad \text{for } t \text{ in } (x_{i-1}, x_i)$$

and

$$F_u(t) = u_i \frac{r'(t)}{\|r'(t)\|} \quad \text{for } t \text{ in } (x_{i-1}, x_i).$$

Each of F_l and F_u has constant magnitude on (x_{i-1}, x_i) and the same direction as $r(t)$ at time t. Thus by Equation 12.28 the work done in moving P from $r(a)$ to $r(b)$ along $r[a, b]$ with F_l is defined by

$$(12.30) \qquad W_l = \sum_{i=1}^{n} l_i[\Lambda(x_i) - \Lambda(x_{i-1})] = \sum_{i=1}^{n} \int_{x_{i-1}}^{x_i} l_i \Lambda'(t)$$

and the work done by applying F_u is

$$(12.31) \qquad W_u = \sum_{i=1}^{n} u_i[\Lambda(x_i) - \Lambda(x_{i-1})] = \sum_{i=1}^{n} \int_{x_{i-1}}^{x_i} u_i \Lambda'(t).$$

Since l is a lower step function and u an upper step function for f, and Λ is an increasing function,

$$W_l = \sum_{i=0}^{n-1} l_i[\Lambda(x_{i+1}) - \Lambda(x_i)] \le \sum_{i=0}^{n-1} u_i[\Lambda(x_{i+1}) - \Lambda(x_i)] = W_u.$$

Theorem 12.1 assures us that there is exactly one number, namely $\int_a^b f(t) \Lambda'(t)$, lying between all such upper and lower estimates. Thus we define the work W done in applying F to move a particle P along $r[a, b]$ from $r(a)$ to $r(b)$ to be

$$W = \int_a^b f(t) \Lambda'(t).$$

It will be more convenient to have a formula in which F and r appear, rather than f and Λ. Recall that $\Lambda'(t) = \|r'(t)\|$ and $f(t) = F(t) \cdot \dfrac{r'(t)}{\|r'(t)\|}$.

Therefore, $\displaystyle\int_a^b f(t) \Lambda'(t) = \int F(t) \cdot \frac{r'(t)}{\|r'(t)\|} \|r'(t)\| = \int_a^b F(t) \cdot r'(t).$ Thus the work W done in applying F to move P along $r[a, b]$ from $r(a)$ to $r(b)$ is given by

$$(12.32) \qquad W = \int_a^b F \cdot r'.$$

Example 14. The force required to stretch a coil spring is directly proportional to the amount of elongation. A given spring requires a force of 4 lbs to be stretched 1 ft from its free position. How much work is done in stretching it 5 ft?

SOLUTION. Let $r(t) = t$, $0 \le t \le 5$, represent the straightline path of the moving end of the spring. By hypothesis the force $F(t)$ required to stretch the coil t units is proportional to t. Thus $F(t) = kt$ for some constant t. Further we are given that $F(1) = 4$. Therefore, $k = 4$ and $F(t) = 4t$. To compute the work done in stretching the spring 5 ft. we apply Equation 12.32 and find that

$$W = \int_0^5 F \cdot r'$$

$$= \int_0^5 (4t) \cdot (t)'$$

$$= \int_0^5 4t$$

$$= [2t^2]_0^5$$

$$= 50 \text{ ft.-lbs.} \quad \blacksquare$$

Example 14 illustrates a situation in which the force varies in magnitude but has the same direction as the path. The next example illustrates a situation in which the force is constant in magnitude but varies in direction from that of the path.

Example 15. Assume that a plane travels from point A to point B (see Figure 12.23) along the path defined by $r(t) = \begin{pmatrix} t \\ t^2 \end{pmatrix}$, $0 \le t \le 2$ miles. At each moment of the flight the wind is blowing due north and exerting a force of 1000 lbs. upon the plane. The force then can be written $F(t) = 1000 \begin{pmatrix} 0 \\ 1 \end{pmatrix}$. The work W done by F is

$$W = \int_0^2 F \cdot r'$$

$$= \int_0^2 1000 \begin{pmatrix} 0 \\ 1 \end{pmatrix} \cdot \begin{pmatrix} 1 \\ 2t \end{pmatrix}$$

$$= \int_0^2 1000(2t)$$

$$= 4000 \text{ mile-lbs.} \quad \blacksquare$$

Notice that in Example 15, if the plane were flying from B to A (rather than from A to B) the work done by the force of the wind would be negative.

Example 16. A spring of length 5 in. requires a force of 4 lbs. to be stretched 1 in. Assume that the attached end of the spring is at $\begin{pmatrix} 0 \\ -5 \end{pmatrix}$ and when the spring is not under tension the free end is at $\begin{pmatrix} 0 \\ 0 \end{pmatrix}$ (see Figure 12.24). We wish to determine the amount of work done when the free end of the spring is moved from $\begin{pmatrix} 0 \\ 0 \end{pmatrix}$ to $\begin{pmatrix} 5 \\ 0 \end{pmatrix}$ along the path $r(t) = \begin{pmatrix} t \\ 0 \end{pmatrix}$, $0 \le t \le 5$ in.

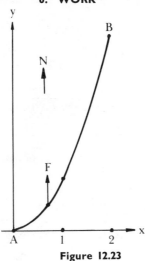

Figure 12.23

SOLUTION.   Recall that the force to stretch a string is proportional to the amount of elongation and in this case we know the constant of proportionality is 4. If the free end of the spring has moved to the point $\begin{pmatrix} t \\ 0 \end{pmatrix}$, the amount of elongation is the distance from $\begin{pmatrix} 0 \\ -5 \end{pmatrix}$ to $\begin{pmatrix} t \\ 0 \end{pmatrix}$ minus the original length of the spring: $\sqrt{t^2 + 25} - 5$. Let $F(t)$ be the force required to stretch the spring from rest to $\begin{pmatrix} t \\ 0 \end{pmatrix}$ in the direction from $\begin{pmatrix} 0 \\ -5 \end{pmatrix}$ to $\begin{pmatrix} t \\ 0 \end{pmatrix}$. This direction is $\begin{pmatrix} t \\ 0 \end{pmatrix} - \begin{pmatrix} 0 \\ -5 \end{pmatrix} = \begin{pmatrix} t \\ 5 \end{pmatrix}$. The unit vector in this direction is $\dfrac{1}{\sqrt{t^2 + 25}} \begin{pmatrix} t \\ 5 \end{pmatrix}$, and the amount of elongation is $\sqrt{t^2 + 25} - 5$. Therefore,

$$F(t) = 4\left(\frac{\sqrt{t^2 + 25} - 5}{\sqrt{t^2 + 25}}\right)\begin{pmatrix} t \\ 5 \end{pmatrix} = 4\left(1 - \frac{5}{\sqrt{t^2 + 25}}\right)\begin{pmatrix} t \\ 5 \end{pmatrix}.$$

Since $r = \begin{pmatrix} t \\ 0 \end{pmatrix}$, we have

$$r' = \begin{pmatrix} 1 \\ 0 \end{pmatrix}.$$

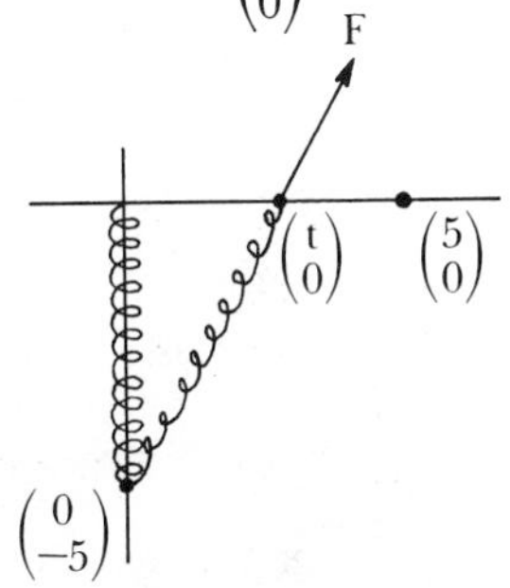

Figure 12.24

Thus the work done is

$$W = \int_0^5 F \cdot r'$$

$$= \int_0^5 4t\left(1 - \frac{5}{\sqrt{t^2 + 25}}\right)$$

$$= [2t^2 - 20\sqrt{t^2 + 25}]_0^5$$

$$= (150 - 100\sqrt{2}) \text{ in.-lbs.} \quad \blacksquare$$

Assume now that $F: \mathcal{R}^m \to \mathcal{R}^m$ is defined at each point of a rectifiable path $\mathcal{C}$ in $\mathcal{R}^m$. If Q is a point on $\mathcal{C}$, $F(Q)$ can be considered as a force acting on a particle P, located at Q. If $r: [a, b] \to \mathcal{R}^m$ gives a parametric representation of $\mathcal{C}$ and is differentiable on $[a, b]$, we define the work W done in moving a particle P from $r(a)$ to $r(b)$ along $r[a, b]$ by

$$W = \int_a^b F(r(t)) \cdot r'(t).$$

This definition coincides with our previous definition, since

$$F(r(t)): [a, b] \to \mathcal{R}^m.$$

Example 17. Let $F: \mathcal{R}^2 \to \mathcal{R}^2$ be defined by

$$F\begin{pmatrix} x \\ y \end{pmatrix} = \begin{pmatrix} x^2 \\ -xy \end{pmatrix}.$$

Let $r(t) = \begin{pmatrix} \cos t \\ \sin t \end{pmatrix}$, $0 \le t \le \pi$. Note that r gives a parametric representation of the path from $\begin{pmatrix} 1 \\ 0 \end{pmatrix}$ to $\begin{pmatrix} -1 \\ 0 \end{pmatrix}$ counterclockwise on the circle of radius 1 with center at $\begin{pmatrix} 0 \\ 0 \end{pmatrix}$. In this case

$$F(r(t)) = F\begin{pmatrix} \cos t \\ \sin t \end{pmatrix} = \begin{pmatrix} \cos^2 t \\ -\cos t \sin t \end{pmatrix}$$

and

$$r'(t) = \begin{pmatrix} -\sin t \\ \cos t \end{pmatrix}.$$

Thus the work done in moving a particle P from $r(0)$ to $r(\pi)$ is

$$W = \int_0^\pi \begin{pmatrix} \cos^2 t \\ -\cos t \sin t \end{pmatrix} \cdot \begin{pmatrix} -\sin t \\ \cos t \end{pmatrix}$$

$$= \int_0^\pi -\sin t \cos^2 t - \cos^2 t \sin t$$

$$= -\tfrac{4}{3}. \quad \blacksquare$$

Suppose now that in Example 17 we choose an arbitrary differentiable parameterization r of the path described, that is, choose a function

$$r:[a, b] \to \mathcal{R}^2 \text{ such that } r(t) = \begin{pmatrix} f(t) \\ g(t) \end{pmatrix}$$

and

$$r(a) = \begin{pmatrix} f(a) \\ g(a) \end{pmatrix} = \begin{pmatrix} 1 \\ 0 \end{pmatrix},$$

$$r(b) = \begin{pmatrix} f(b) \\ g(b) \end{pmatrix} = \begin{pmatrix} -1 \\ 0 \end{pmatrix},$$

and $f(t)^2 + g(t)^2 = 1$ for $a \le t \le b$. If we compute $\int_a^b F(r(t)) \cdot r'(t)$ we have

$$\int_a^b F(r(t)) \cdot r'(t) = \int_a^b \begin{pmatrix} f(t)^2 \\ -f(t)g(t) \end{pmatrix} \cdot \begin{pmatrix} f'(t) \\ g'(t) \end{pmatrix}$$

$$= \int_a^b [f(t)^2 f'(t) - f(t)g(t)g'(t)]$$

$$=^1 \int_a^b f(t)^2 f'(t) - f(t)(-f(t)f'(t))$$

$$= \left[\frac{2f(t)^3}{3} \right]_a^b =^2 -\frac{4}{3}.$$

(1) Since $f(t)^2 + g(t)^2 = 1$, $2f(t)f'(t) + 2g(t)g'(t) = 0$ and thus $g(t)g'(t) = -f(t)f'(t)$ for all t in $[a, b]$.

(2) Since $f(b) = -1$ and $f(a) = 1$.

Thus we see that in Example 17 we could have chosen any differentiable parameterization of the path in question and arrived at the same answer. The reader should note that the parameterization $r(t) = \begin{pmatrix} -t \\ \sqrt{1 - t^2} \end{pmatrix} - 1$,

$\le t \le 1$, of this path is *not* admissible since it is not differentiable at $t = 1$ or $t = -1$.

In the example just considered we discovered that the work done in moving a particle P along the given path by exerting F was independent of the parameterization r chosen to represent the path. Even under restrictive conditions on the path it is not easy to prove that this is generally the case. In Exercise 11 the reader is asked to prove the following theorem: Let $r_1:[a, b] \to \mathcal{R}^m$ and $r_2:[c, d] \to \mathcal{R}^m$ be differentiable parametric representations of the same path in $\mathcal{R}^m$. If there exists a strictly increasing function $h:[c, d] \to [a, b]$ such that $h(c) = a$, $h(d) = b$, and $r_2(t) = r_1(h(t))$ for $c \le t \le d$, then

$$\int_a^b F(r_1) \cdot r_1' = \int_c^d F(r_2) \cdot r_2'$$

for all F such that $F(r_1) \cdot r_1'$ and $F(r_2) \cdot r_2'$ are integrable. Thus, if two parameterizations are related in this fashion, the work done in moving a particle along the path $r_1[a, b]$ is the same as that done in moving the particle along $r_2[c, d]$.

EXERCISES

In Exercises 1 to 4 compute the work done in applying the force F to move a particle P from $r(a)$ to $r(b)$ along the path $r[a, b]$:

1. $F(t) = \begin{pmatrix} t \\ 2t - 1 \end{pmatrix}$; $r(t) = t\begin{pmatrix} 1 \\ 1 \end{pmatrix}$, $0 \le t \le 1$.

2. $F(t) = \begin{pmatrix} t^2 \\ (1 - t)^2 \end{pmatrix}$; $r(t) = \begin{pmatrix} 1 \\ 1 \end{pmatrix} + t\begin{pmatrix} -1 \\ -1 \end{pmatrix}$, $0 \le t \le 1$.

3. $F(t) = \begin{pmatrix} t \\ \cos t \\ \sin t \end{pmatrix}$; $r(t) = \begin{pmatrix} t^2 \\ t \\ t^2 \end{pmatrix}$, $-\dfrac{\pi}{2} \le t \le \dfrac{\pi}{2}$.

4. $F(t) = \begin{pmatrix} e^t \\ e^{-t} \end{pmatrix}$; $r(t) = \begin{pmatrix} t \\ t^2 \end{pmatrix}$; $-1 \le t \le 1$.

In Exercises 5 to 10 compute the work done in applying the force F to move a particle P from A to B along the given path $\mathcal{C}$:

5. $F\begin{pmatrix} x \\ y \end{pmatrix} = \begin{pmatrix} x - y \\ x + y \end{pmatrix}$; $\mathcal{C}$ is the straight line path from $\begin{pmatrix} 2 \\ 1 \end{pmatrix}$ to $\begin{pmatrix} -3 \\ 0 \end{pmatrix}$.

6. $F\begin{pmatrix} x \\ y \end{pmatrix} = \begin{pmatrix} x^2 - 2xy \\ y^2 + 2xy \end{pmatrix}$; $\mathcal{C}$ is the path from $\begin{pmatrix} -1 \\ 1 \end{pmatrix}$ to $\begin{pmatrix} 1 \\ 1 \end{pmatrix}$, defined by $y = x^2$.

7. $F\begin{pmatrix} x \\ y \end{pmatrix} = \begin{pmatrix} y \\ x \end{pmatrix}$; $\mathcal{C}$ is the path from $\begin{pmatrix} 1 \\ 0 \end{pmatrix}$ to $\begin{pmatrix} -1 \\ 0 \end{pmatrix}$ along the circle of radius 1 with center at $\begin{pmatrix} 0 \\ 0 \end{pmatrix}$. $\left(\text{Hint: choose the parameterization } r(t) = \begin{pmatrix} \cos t \\ \sin t \end{pmatrix}, 0 \le t \le \pi.\right)$

8. In Exercise 7 choose an arbitrary differentiable parameterization of the path $r(t) = \begin{pmatrix} f(t) \\ g(t) \end{pmatrix}$, $a \le t \le b$, $r(a) = \begin{pmatrix} 1 \\ 0 \end{pmatrix}$, $r(b) = \begin{pmatrix} -1 \\ 0 \end{pmatrix}$, $f(t)^2 + g(t)^2 = 1$.

9. $F\begin{pmatrix} x \\ y \end{pmatrix} = \begin{pmatrix} y^2 \\ x^2 \end{pmatrix}$; $\mathcal{C}$ is the polygonal path obtained by moving in a straight line from $\begin{pmatrix} 1 \\ 0 \end{pmatrix}$ to $\begin{pmatrix} 1 \\ 1 \end{pmatrix}$ and then in a straight line from $\begin{pmatrix} 1 \\ 1 \end{pmatrix}$ to $\begin{pmatrix} 0 \\ 1 \end{pmatrix}$.

10. $F\begin{pmatrix} x \\ y \end{pmatrix} = \begin{pmatrix} 2x + y \\ y - x \end{pmatrix}$; $\mathcal{C}$ is the path obtained by moving counterclockwise around the square with vertices $\begin{pmatrix} 1 \\ 1 \end{pmatrix}$, $\begin{pmatrix} -1 \\ 1 \end{pmatrix}$, $\begin{pmatrix} -1 \\ -1 \end{pmatrix}$, $\begin{pmatrix} 1 \\ -1 \end{pmatrix}$.

11. Let $r:[a, b] \to \mathcal{R}^m$ and $r_2:[a, b] \to \mathcal{R}^m$ be differentiable parameterizations of the same path in $\mathcal{R}^m$. Assume that there exists a strictly increasing function $h:[c, d] \to [a, b]$ such that $h(c) = a$, $h(d) = b$ and $r_2(t) = r_1(h(t))$ for all t in $[c, d]$. Prove that $\int_a^b F(r_1) \cdot r_1' = \int_c^d F(r_2) \cdot r_2'$ for all $F:\mathcal{R}^m \to \mathcal{R}^m$ such that $F(r_1) \cdot r_1'$ and $F(r_2) \cdot r_2'$ are integrable.

12. Two particles attract each other with a force inversely proportional to the square of the distance between them, that is, $\|F\| = \dfrac{k}{d^2}$, where F is the force, d is the distance, and k is a constant. If one is located 5 units from the other, find the work done in moving them 5 additional units apart.

13. A cylindrical tank measures 10 ft in diameter and is 8 ft deep. Find the work done in pumping all the water out over the top of the tank. Assume the density of water to be 62.4 lb/ft³. Hint: Imagine a piston pushing the water up from the bottom of the tank.

14. A cable 100 ft long weighing 1 lb/ft of length is suspended from the top of a building. Find the work done in winding it onto a drum at the top of the building.

15. A cable weighing 2 lb/ft is used to lift coal from a mine 1,500 ft below the surface. The container for the coal weighs 100 lbs. How much work is done in lifting 1,000 lbs of coal to the surface?

16. A ball is rolled 10 ft across a level concrete court against a wind which is exerting a constant force of 1 lb at an angle of 45° to the direction of motion. Neglecting rolling friction, find the work done.

17. By Coulomb's law the attraction or repulsion between particles varies inversely as the square of the distance between them, that is, $\|F\| = \dfrac{k}{d^2}$ where F is the force vector, d is the distance, and k is a constant. A particle P_1 is located in the coordinate plane at $(0, 2)$ and a second particle P_2 is moved along the x-axis from $(-1, 0)$ to $(1, 0)$. Find the work done.

18. In Exercise 17 let the particle move from $(-1, 0)$ to $(1, 0)$ along the path given parametrically by $r(t) = (\cos(\pi - t), \sin(\pi - t)), 0 \le t \le \pi$. Compute the work done.

19. A force field is said to be "conservative" if the amount of work done to move a particle from a point X to a point Y is independent of the path along which it is moved. Show that the force field described in Exercise 17 is conservative.

20. In Example 15 let the spring be first stretched to its final length and then rotated to its final position. Compute the work done.

21. The force, in pounds, with which the earth attracts a body of mass w located x miles away from the center of the earth is given approximately by

$$\|F\| = \frac{16,000,000w}{x^2}.$$

Find the work in moving a 1 lb mass from the surface of the earth to a point 1,000 miles above the surface. (Assume the diameter of the earth to be 8,000 miles.)

22. Boyle's law states that the product of pressure P and volume V of a given mass of gas at constant temperature is constant, that is $PV = K$, where K is a constant. Using this formula compute the work done by a piston, 3 in. in diameter, in moving from a position 6 in. below the head of a cylinder to a position 3 in. below the head. (Actually the work would

have to be done very slowly for the temperature to remain approximately constant.)

23. A circular cistern 6 ft in diameter and 10 ft deep is filled with water to a depth of 8 ft. Find the work done in emptying the cistern if the water is discharged 4 ft above the surface of the ground. (Assume the density of water to be d.)

24. A spherical tank 4 ft in radius is to be filled to a depth of 3 ft with oil, weighing 50 lb/ft³, from a large reservoir 2 ft below the bottom of the tank. Find the work to be done.

25. A circular cistern 12 ft in diameter and 8 ft deep is filled with water to a depth of 3 ft. The water is to be pumped into an upright circular tank whose bottom is at ground level and whose diameter is 8 ft. Find the work to be done.

26. A cylindrical tank with a hemispherical bottom has a radius of 3 ft. It is to be filled to a depth of 5 feet with water from a lake 7 feet below the bottom of the tank. How much work is to be done?

27. A tank in the form of a right circular cylinder 2 ft in radius and 10 ft long is lying with its axis parallel to the ground. It is filled to a depth of 1 ft with oil weighing 50 lb/ft³. The oil is to be pumped to a height of 12 ft from the bottom of the tank and discharged into a vat (from the top). Find the work to be done.

ANSWERS

1. $W = \int_0^1 \left(2t - \dfrac{t}{1}\right) \cdot \left(\dfrac{t}{t}\right)' = \frac{1}{2}$. **3.** $W = \int_{-\pi/2}^{\pi/2} \begin{pmatrix} t \\ \cos t \\ \sin t \end{pmatrix} \cdot \begin{pmatrix} t^2 \\ t \\ t^2 \end{pmatrix}' = 6 + \dfrac{\pi^3}{6}$.

5. $r(t) = \begin{pmatrix} 2 \\ 1 \end{pmatrix} + t\begin{pmatrix} -5 \\ -1 \end{pmatrix}; \ 0 \leq t \leq 1$, is a parameterization of C.

$$W = \int_0^1 F(r(t)) \cdot r'(t) = \int_0^1 \begin{pmatrix} 1 - 4t \\ 3 - 6t \end{pmatrix} \cdot \begin{pmatrix} -5 \\ -1 \end{pmatrix} = 5.$$

7. $W = \int_0^\pi \begin{pmatrix} \sin t \\ \cos t \end{pmatrix} \cdot \begin{pmatrix} \cos t \\ \sin t \end{pmatrix}' = 0.$ **9.** Let $r_1(t) = \begin{pmatrix} 1 \\ 0 \end{pmatrix} + t\begin{pmatrix} 1 \\ 1 \end{pmatrix}, 0 \leq t \leq 1$

and $r_2(t) = \begin{pmatrix} 1 \\ 1 \end{pmatrix} + t\begin{pmatrix} -1 \\ 0 \end{pmatrix}, 0 \leq t \leq 1.$ $W = \int_0^1 F(r_1(t)) \cdot r_1'(t) +$ $\int_0^1 F(r_2(t)) \cdot r_2'(t) = 1 + (-1) = 0.$ **11.** Hint: Use the substitution $u = h(t)$ in $\int_a^b F(r_1(t)) \cdot r_1'(t).$ **13.** 49,120π ft.-lb. **15.** 3,900,000 ft.-lb. **17.** 0. **21.** 800 mi.-lb. **23.** 720πd ft.-lb. **25.** 912πd ft.-lb. **27.** $\left(\dfrac{20,000\pi}{3} - 4,000\sqrt{3}\right)$ ft.-lb.

13

AN INTRODUCTION TO TAYLOR POLYNOMIALS AND INFINITE SERIES

In this chapter we present the Taylor polynomials of a function f as a generalization of the affine approximation of f. This leads to the problem of the convergence of infinite series. Only a few tests for convergence are given in this chapter. A more complete development will be given in Chapter 21.

1. TAYLOR POLYNOMIALS

Let $f: \mathcal{R} \to \mathcal{R}$. If f is differentiable at a, the affine approximation of f about a is the polynomial P_1 defined by

$$P_1(x) = f'(a)(x - a) + f(a).$$

If $f'(a) \neq 0$, then P_1 is a polynomial of degree 1. If $f'(a) = 0$, then P_1 is a constant function. This polynomial P_1 has the property that

$$P_1(a) = f(a) \quad \text{and} \quad P_1'(a) = f'(a).$$

(In this and the subsequent discussion the reader should recall that the first derivative of f gives us the slope of the graph of f, the second derivative of f gives us the concavity of f, the third derivative gives us the rate of change of concavity of f, and so on.) Thus the affine approximation P_1 of f about a tells us the value of f at a and the slope of the graph of f at a. However, since $P_1'' = 0$, P_1 does not necessarily give us any information about the concavity of f.

Consider the function f, defined by $f(x) = x^3 - 3x$ and sketched in Figure 13.1a. Using the techniques developed in Chapter 6, we find that $x = -1$ is a relative maximum point, $x = 1$ is a relative minimum point, f is concave downward on $(-\infty, 0)$, and concave upward on $(0, \infty)$. The affine approximation P_1 of f about $x = 1$ is given by

$$(13.1) \qquad P_1(x) = f'(1)(x - 1) + f(1) = -2.$$

The graphs of f and P_1 are sketched in Figure 13.1b. Certainly a quadratic function having a parabolic graph provides a better approximation for f about $x = 1$ than does P_1. We seek to add a quadratic expression to P_1 to

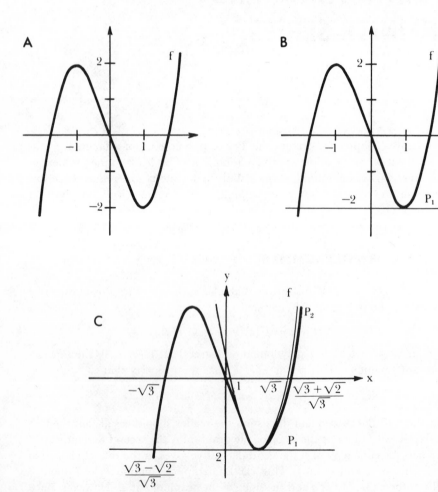

Figure 13.1. Affine and quadratic approximations to $f(x) = x^3 - 3x$. (a) Graph: $f(x) = x^3 - 3x$. (b) Graph: $f(x) = x^3 - 3x$ $P_1(x) = -2$. (c) Graph: $f(x) = x^3 - 3x$ $P_1(x) = -2$ $P_2(x) = 3(x - 1)^2 - 2$.

obtain a new function P_2 such that P_2 has the same basic relation to f as does P_1:

(13.2)
$$\begin{cases} P_2(1) = f(1) = -2 \\ P_2'(1) = f'(1) = 0; \end{cases}$$

but such that P_2 and f have the same concavity at $x = 1$:

(13.3)
$$P_2''(1) = f''(1) = 6.$$

Consider the function P_2 defined by

(13.4)
$$P_2(x) = P_1(x) + \frac{f''(1)}{2}(x-1)^2$$
$$=^1 f(1) + f'(1)(x-1) + \frac{f''(1)}{2}(x-1)^2$$
$$=^2 3(x-1)^2 - 2.$$

(1) P_1 is defined in Equation 13.1.
(2) Substituting values for $f(1), f'(1)$, and $f''(1)$.

Note that P_2 has the properties called for in Equations 13.2 and 13.3. The graphs of f, P_1, and P_2 appear in Figure 13.1c. Although P_2 and f differ at every point except $x = 1$, it is obvious from Figure 13.1c that P_2 is a better approximating function for f about $x = 1$ than is P_1. Indeed,

$$|f(x) - P_2(x)| \le |f(x) - P_1(x)|$$

for all x in $[-\frac{1}{2}, \infty)$.

Consider now the third derivatives of f and P_2 at $x = 1$. Since $f'''(1) = 6$ and $P_3'''(1) = 0$, P_2 is not an approximating function which can tell us the rate of change of concavity of f. We shall now augment P_2 in such a way that the new function will satisfy the same relations as P_2 (given in Equations 13.2 and 13.3) and will have the same third derivative at $x = 1$ as does f:

(13.5)
$$P_3'''(1) = f'''(1) = 6.$$

We do this by defining

$$P_3(x) = P_2(x) + \frac{f'''(1)}{3!}(x-1)^3.$$
$$=^1 f(1) + f'(1)(x-1) + \frac{f''(1)}{2!}(x-1)^2 + \frac{f'''(1)}{3!}(x-1)^3$$
$$=^2 -2 + 3(x-1)^2 + (x-1)^3.$$

(1) P_2 is defined in Equation 13.4.
(2) Substituting for $f(1), f'(1), f''(1)$, and $f'''(1)$.

(Recall that $n! = n(n-1) \cdots 2 \cdot 1$ if n is a positive integer and $0! = 1$.) The function P_3 does satisfy Equations 13.2, 13.3, and 13.5, and consequently at $x = 1$, P_3 and f have the same functional value, the same slope, the same concavity, and the same rate of change of concavity. Indeed, $P_3(x) = f(x)$

for all values of x (Check this!). The following theorem assures us that P_3 is the only polynomial of degree 3 which satisfies Equations 13.2, 13.3, and 13.4, and further that every polynomial f can be expanded in terms of the derivatives $f^{(k)}(a)$ and powers of $(x - a)$.

LEMMA 13.1 Let $P: \mathcal{R} \to \mathcal{R}$ be a polynomial of degree n and let $a \in \mathcal{R}$. Then

i. $P(x) = P(a) + \dfrac{P'(a)}{1!}(x - a) + \cdots + \dfrac{P^{(n)}(a)}{n!}(x - a)^n$,

and

ii. if Q is a polynomial of degree n such that $Q^{(k)}(a) = P^{(k)}(a)$ for $k = 0, 1, \ldots, n$, then $Q = P$.

Proof

i. Since P is a polynomial of degree n, there exist real numbers $a_0, \ldots, a_n$ such that

$$P(x) = a_0 + a_1 x + \cdots + a_n x^n (a_n \neq 0).$$

By means of elementary algebra (see Exercise 11) P can be rewritten as

$$P(x) = b_0 + b_1(x - a) + \cdots + b_n(x - a)^n,$$

where $b_0, \ldots, b_n$ are real numbers and $b_n = a_n$. In Exercise 12 the reader is asked to show that $b_k = \dfrac{P^{(k)}(a)}{k!}$ and thus part (i) is proven.

ii. Let P and Q be polynomials of degree n. From part (i) we can write

$$(13.6) \quad \begin{cases} P(x) = P(a) + \dfrac{P'(a)}{1!}(x - a) + \cdots + \dfrac{P^{(n)}(a)}{n!}(x - a)^n \\ \text{and} \\ Q(x) = Q(a) + \dfrac{Q'(a)}{1!}(x - a) + \cdots + \dfrac{Q^{(n)}(a)}{n!}(x - a)^n. \end{cases}$$

Thus, if $Q^{(k)}(a) = P^{(k)}(a)$ for $k = 0, 1, \ldots, n$. Equation 13.6 yields exactly the same expression for each of $P(x)$ and $Q(x)$. Thus, $P = Q$. ∎

Example 1. Let $f(x) = x^4 - 4x^3 + x - 3$ and $a = 2$.

By Lemma 13.1, we have

$$f(x) = f(2) + \frac{f'(2)}{1!}(x - 2) + \frac{f''(2)}{2!}(x - 2)^2 + \frac{f'''(2)}{3!}(x - 2)^3$$

$$+ \frac{f^{iv}(2)}{4!}(x - 2)^4$$

$$= -17 - \frac{15}{1}(x - 2) + \frac{0}{2}(x - 2)^2 + \frac{24}{6}(x - 2)^3 + \frac{24}{24}(x - 2)^4$$

$$= -17 - 15(x - 2) + 4(x - 3)^3 + (x - 2)^4. \quad ∎$$

Assume now that $f: \mathcal{R} \to \mathcal{R}$ is a function which is n times differentiable at a. We seek a polynomial P such that $P^{(k)}(a) = f^{(k)}(a)$ for $k = 0, 1, \ldots, n$. (Here we interpret $f^{(0)}(a) = f(a)$ as we have previously.) The form occurring in Lemma 13.1i leads us to consider the approximating polynomial

$$P_n(x) = f(a) + \frac{f'(a)}{1!}(x - a) + \cdots + \frac{f^{(n)}(a)}{n!}(x - a)^n.$$

The following theorem asserts that this polynomial and no other polynomial (of degree not greater than n) is the one we seek.

THEOREM 13.2 (Existence and Uniqueness of the nth Taylor Polynomial.) Let $f: \mathcal{R} \to \mathcal{R}$ be n times differentiable at a. There exists a unique polynomial P of degree not greater than n such that

(13.7) $P^{(k)}(a) = f^{(k)}(a)$ for $k = 0, 1, \ldots, n$.

This polynomial P is defined by

$$(13.8) \quad P(x) = f(a) + f'(a)(x - a) + \frac{f''(a)}{2!}(x - a)^2 + \cdots$$

$$+ \frac{f^{(n)}(a)}{n!}(x - a)^n$$

$$= \sum_{k=0}^{n} \frac{f^{(k)}(a)}{k!}(x - a)^k.$$

Proof. In Exercise 13 the reader is asked to show that the polynomial P defined by Equation 13.8 satisfies the properties called for in Equation 13.7. Thus, P is one polynomial satisfying Equation 13.7. If Q is a second polynomial satisfying Equation 13.7, then $Q^{(k)}(a) = f^{(k)}(a) = P^{(k)}(a)$ for $k = 0, 1, \ldots, n$. Thus, by Lemma 13.1, if the degree of Q is not greater than n, $Q = P$. ∎

In the remainder of this chapter we shall be using the approximating polynomials defined by Equation 13.8. These are called Taylor approximating polynomials, in honor of Brook Taylor, an English mathematician.

DEFINITION 13.3 Let $f: \mathcal{R} \to \mathcal{R}$ be n times differentiable at $x = a$. The nth *Taylor approximating polynomial of* f *about* a is denoted by $_nT_a f$ and defined by

$$(13.9) \qquad\qquad _nT_a f(x) = \sum_{k=0}^{n} \frac{f^{(k)}(a)}{k!}(x - a)^k.$$

The reader should keep in mind that the first Taylor polynomial of f about a is what we have previously called the affine approximation of f about a: $_1T_a f(x) = f(a) + f'(a)(x - a)$. A second fact to keep in mind is that the nth Taylor polynomial of f about a is equal to f whenever f is a polynomial of degree n. This was proven in Lemma 13.1.

Example 2. Let $f(x) = e^x$, $x \in \mathcal{R}$. Since $f^{(n)}(x) = e^x$ for each n, $f^{(n)}(a) = e^a$ for all a. In particular, $f^{(n)}(0) = 1$ and $f^{(n)}(1) = e$. Therefore,

(**a**) the nth Taylor polynomial of f about 0 is

$$_nT_0 f(x) = f(0) + f'(0)(x - 0) + \frac{f''(0)}{2!}(x - 0)^2 + \cdots + \frac{f^{(n)}(0)}{n!}(x - 0)^n$$

$$= 1 + x + \frac{x^2}{2!} + \cdots + \frac{x^n}{n!}.$$

In particular,

$$_0T_0 f(x) = 1$$

$$_1T_0 f(x) = 1 + x$$

$$_2T_0 f(x) = 1 + x + \frac{x^2}{2!}$$

and

$$_3T_0 f(x) = 1 + x + \frac{x^2}{2!} + \frac{x^3}{3!}.$$

(**b**) The nth Taylor polynomial of f about 1 is given by

$$_nT_1 f(x) = f(1) + f'(1)(x - 1) + \frac{f''(1)}{2!}(x - 1)^2 + \cdots + \frac{f^{(n)}(1)}{n!}(x - 1)^n$$

$$= e\left(1 + (x - 1) + \frac{(x - 1)^2}{2!} + \cdots + \frac{(x - 1)^n}{n!}\right).$$

In particular,

$$_0T_1 f(x) = e$$

$$_1T_1 f(x) = e(1 + (x - 1))$$

$$_2T_1 f(x) = e\left(1 + (x - 1) + \frac{(x - 1)^2}{2!}\right)$$

$$_3T_1 f(x) = e\left(1 + \frac{(x - 1)^2}{2!} + \frac{(x - 1)^3}{3!}\right). \quad \blacksquare$$

EXERCISES

In Exercises 1 to 8 find the Taylor approximating polynomials of f about a: $_kT_af$, $k = 0, 1, 2, 3, 4$:

1. $f(x) = x^6 - 3x^2 + 4$, $a = 0$.   2. $f(x) = x^{1/2}$, $a = 1$.
3. $f(x) = \sin x$, $a = 0$.   4. $f(x) = \sin x$, $a = \pi/2$.
5. $f(x) = \cos x$, $a = 0$.   6. $f(x) = e^{2x}$, $a = 1$.
7. $f(x) = \ln x$, $a = 1$.   8. $f(x) = \arcsin x$, $a = 0$.

9. Let $f(x) = \sin x$. Show that the nth Taylor polynomials of f about 0 are given by

$$_{(2k-1)}T_0f(x) = x - \frac{x^3}{3!} + \cdots + \frac{(-1)^{(k-1)}x^{2k-1}}{(2k-1)!} \quad \text{for} \quad k = 1, 2, \ldots$$

$$_{2k}T_0f(x) = {}_{(2k-1)}T_0f(x) \quad \text{for} \quad k = 1, 2, \ldots.$$

10. Let $f(x) = \cos x$. Show that the nth Taylor polynomials of f about 0 are given by

$$_{2k}T_0f(x) = 1 - \frac{x^2}{2!} + \cdots + \frac{(-1)^{k-1}x^{2k}}{(2k)!} \quad \text{for} \quad k = 0, 1, \ldots$$

$$_{2k+1}T_0f(x) = {}_{2k}T_0f(x) \quad \text{for} \quad k = 0, 1, \ldots.$$

11. Let $P(x) = 2x^3 - 3x^2 - 4x + 1$. By use of algebra find real numbers b_0, b_1, b_2, b_3 such that

$$P(x) = b_0 + b_1(x - 1) + b_2(x - 1)^2 + b_3(x - 1)^3.$$

12. Let P be a polynomial defined by

$$P(x) = b_0 + b_1(x - a) + \cdots + b_n(x - a)^n.$$

Show that $b_k = \dfrac{P^{(k)}(a)}{k!}$ for $k = 0, 1, \ldots, n$.

13. Let f be n times differentiable at a. Define P by

$$P(x) = f(a) + f'(a)(x - a) + \frac{f''(a)}{2!}(x - a)^2 + \cdots + \frac{f^{(n)}(a)}{n!}(x - a)^n.$$

Show that $P^{(k)}(a) = f^{(k)}(a)$ for $k = 0, 1, \ldots, n$.

ANSWERS

1. $_4T_0f(x) = 4 - 3x^2$.   3. $_4T_0f(x) = x - \dfrac{x^3}{3!}$.   5. $_4T_0f(x) = 1 - \dfrac{x^2}{2!} + \dfrac{x^4}{4!}$.

7. $_4T_1f(x) = (x - 1) - \dfrac{(x - 1)^2}{2} + \dfrac{(x - 1)^3}{3} - \dfrac{(x - 1)^4}{4}$.

9. Let $f(x) = \sin x$. Then $f^{(2p)}(0) = 0$ for $p = 0, 1, \ldots$; $f^{(2p-1)}(0) = (-1)^{p-1}$ for $p = 1, 2, \ldots$. Therefore,

$$_{(2k-1)}T_0f(x) = \sum_{j=0}^{2k-1} \frac{f^{(j)}(0)}{j!}x^j$$

$$= x - \frac{x^3}{3!} + \cdots + \frac{(-1)^{k-1}}{(2k-1)!}x^{2k-1},$$

and $_{2k}T_0f(x) = {}_{(2k-1)}T_0f(x) + \dfrac{f^{(2k)}(0)}{(2k)!}x^{2k} = {}_{2k-1}T_0f(x)$.

11. $P(x) = [2(x^3 - 3x^2 + 3x - 1) + 6x^2 - 6x + 2] - 3x^2 - 4x + 1$

$\qquad = 2(x - 1)^3 + 3x^2 - 10x + 3$

$\qquad = 2(x - 1)^3 + [3(x^2 - 2x + 1) + 6x - 3] - 10 + 3$

$\qquad = 2(x - 1)^3 + 3(x - 1)^2 - 4x$

$\qquad = 2(x - 1)^3 + 3(x - 1)^2 - 4(x - 1) - 4.$

2. THE INTEGRAL FORM FOR THE REMAINDER

In the preceding section we noted that a polynomial function f of degree n coincides with its nth Taylor polynomial: $f = {}_n T_a f$. The polynomials are obviously the only functions having this property.

Consider now an arbitrary $\mathcal{R} \to \mathcal{R}$ function f, which is n times differentiable at $x = a$, and its nth Taylor polynomial, ${}_n T_a f$. We know that ${}_n T_a f(a) = f(a)$; however, we do not know that ${}_n T_a f(x) = f(x)$ if $x \neq a$. Let ${}_1 T_a f$ be the first Taylor polynomial for f. If we define the function ${}_1 R_a f$ by

$${}_1 R_a f(x) = f(x) - {}_1 T_a f(x),$$

we have

$${}_1 R_a f(x) = f(x) - (f(a) + f'(a)(x - a))$$

$$= (f(x) - f(a)) - f'(a)(x - a)$$

$$= \int_a^x f'(t) - f'(a)(\dot{x} - a)$$

$$= \int_a^x f'(t) - \int_a^x f'(a).$$

Thus, the error made if ${}_1 T_a f(x)$ is used to approximate $f(x)$ is given by

(13.10) $$\qquad {}_1 R_a f(x) = \int_a^x (f'(t) - f'(a)).$$

If the second derivative of f is defined and continuous on $[a, x]$, we may write this "remainder" term in another manner:

(13.11) $\begin{cases} {}_1 R_a f(x) =^1 \displaystyle\int_a^x (f'(t) - f'(a)) \\[2mm] \qquad =^2 [(f'(t) - f'(a))(t - x)]_a^x - \displaystyle\int_a^x f''(t)(t - x) \\[2mm] \qquad = \displaystyle\int_a^x f''(t)(x - t). \end{cases}$

(1) Equation 13.10.
(2) Method of Parts, differentiating $f'(t) - f'(a)$ and using $t - x$ as a primitive for the constant 1.

Thus, if f'' is defined and continuous on $[a, x]$, Equation 13.11 gives us two formulae for computing the difference ${}_1 R_a f(x)$. The following theorem gives us the general formula for computing the difference

$$_nR_af(x) = f(x) - {}_nT_af(x).$$

Thus, we shall be able to determine how well $_nT_af$ approximates f.

THEOREM 13.4 (Integral Remainder Theorem.)

Let $a, x \in (c, d)$. Assume that f is an $\mathfrak{R} \to \mathfrak{R}$ function, defined on (c, d), and that $f^{(n)}$ exists and is continuous on $[a, x]$. Then

$$(13.12) \quad {}_nR_af(x) = f(x) - {}_nT_af(x)$$

$$= \frac{1}{(n-1)!} \int_a^x (f^{(n)}(t) - f^{(n)}(a))(x-t)^{n-1}.$$

Furthermore, if $f^{(n+1)}$ exists and is continuous on $[a, x]$,

$$(13.13) \quad {}_nR_af(x) = \frac{1}{n!} \int_a^x f^{(n+1)}(t)(x-t)^n.$$

Proof. For $k = 1, \ldots, n$ consider the formula

$$(13.14) \quad {}_kR_af(x) = f(x) - {}_kT_af(x)$$

$$= \frac{1}{(k-1)!} \int_a^x (f^{(k)}(t) - f^{(k)}(a))(x-t)^{k-1}.$$

In the discussion preceding the statement of the theorem, we proved that Equation 13.14 is valid for $k = 1$. (See Equation 13.10). We shall show that if Equation 13.14 holds for any $k < n$, it also holds for $k + 1$. Thus, by induction, Equation 13.14 is valid for $k = n$, and consequently Equation 13.12 is valid.

Assume now that Equation 13.14 holds for some $k < n$. In Exercise 4 the reader is asked to use the Method of Parts to show that

$$(13.15) \quad \frac{1}{(k-1)!} \int_a^x (f^{(k)}(t) - f^{(k)}(a))(x-t)^{k-1}$$

$$= \frac{f^{(k+1)}(a)(x-a)^{k+1}}{(k+1)!} + \frac{1}{k!} \int_a^x (f^{(k+1)}(t) - f^{(k+1)}(a))(x-t)^k.$$

Substituting the expression in Equation 13.15 for the integral term in 13.14, we have

$$(13.16) \quad {}_kR_af(x) = \frac{f^{(k+1)}(a)}{(k+1)!}(x-a)^{k+1}$$

$$+ \frac{1}{k!} \int_a^x (f^{(k+1)}(t) - f^{(k+1)}(a))(x-t)^k.$$

Therefore,

$$
\begin{aligned}
_{(k+1)}R_a f(x) &=^1 f(x) - {}_{(k+1)}T_a f(x) \\
&=^2 f(x) - \left({}_k T_a f(x) + \frac{f^{(k+1)}(a)}{(k+1)!}(x-a)^{k+1} \right) \\
&=^1 {}_k R_a f(x) - \frac{f^{(k+1)}(a)}{(k+1)!}(x-a)^{k+1} \\
&=^3 \frac{1}{k!} \int_a^x (f^{(k+1)}(t) - f^{(k+1)}(a))(x-t)^k.
\end{aligned}
$$

(1) Definition of $_{(k+1)}R_a f$ and $_k R_a f$.
(2) Definition 13.3 of $_k T_a f$ and $_{(k+1)}T_a f$.
(3) Equation 13.16.

This proves the first statement of the theorem.

We must now show that Equation 13.13 is valid if $f^{(n+1)}$ is continuous on $[a, x]$. We know that Equation 13.12 holds under the hypothesis. An application of the Method of Parts (see Exercise 5) shows that

$$
\frac{1}{(n-1)!} \int_a^x (f^{(n)}(t) - f^{(n)}(a))(x-t)^{n-1} = \frac{1}{n!} \int_a^x f^{(n+1)}(t) \ (x-t)^n.
$$

If we substitute this expression in Equation 13.12 we have 13.13. ∎

The following example will show—for at least one case—why the nth Taylor approximation of f about a closely approximates the function f for x close to a.

Example 3. Let $f: \mathcal{R} \to \mathcal{R}$ be defined by $f(x) = e^x$. In Example 2 we discovered that the nth Taylor approximation of f about 0 is

(13.17)
$$
_n T_0 f(x) = 1 + x + \frac{x^2}{2!} + \cdots + \frac{x^n}{n!}.
$$

Thus, by Theorem 13.4, Equation 13.13, we have

(13.18)
$$
\begin{cases}
_n R_0 f(x) = e^x - \left(1 + x + \frac{x^2}{2!} + \cdots + \frac{x^n}{n!} \right) \\
\qquad = \frac{1}{n!} \int_0^x e^t (x-t)^n,
\end{cases}
$$

since $f^{(n+1)}(t) = e^t$ for any positive integer n. If we consider only values of

x such that $|x| \leq 1$, we have the following for each positive integer n:

(13.19)
$$\begin{aligned}
|_n R_0 f(x)| &=^1 \left| \frac{1}{n!} \int_0^x e^t (x - t)^n \right| \\
&\leq^2 \frac{1}{n!} \left| \int_0^x |e^t| \, |x - t|^n \right| \\
&\leq^3 \frac{1}{n!} \left| \int_0^x e \, |x - t|^n \right| \\
&=^4 \frac{e}{n!} \left[\frac{-|x - t|^{n+1}}{n + 1} \right]_0^x \\
&=^5 \frac{e}{(n + 1)!} \, |x|^{n+1} \\
&\leq^6 \frac{e}{(n + 1)!} \, .
\end{aligned}$$

(1) Equation 13.18.
(2) Theorem 8.15ii (p. 360).
(3) Theorem 8.15i (p. 360) since $e^t \leq e$ if $|t| \leq 1$.
(4) $\int (x - t)^n = \left[\dfrac{-(x - t)^{n+1}}{n + 1} \right]$.
(5) Evaluation of the primitive.
(6) Since $|x| \leq 1$.

The chain of inequalities (13.19) shows us that when n is large, $|e^x - {}_n T_0 f(x)|$ is small for all x, $|x| \leq 1$. Indeed we note the following (using the fact that $e < 4$ (see page 425)):

(a) for $n = 1$,
$$|e^x - {}_1 T_0 f(x)| = |e^x - (1 + x)|$$
$$\leq \frac{e}{2!} < \frac{4}{2} \quad \text{for } x \text{ in } [-1, 1];$$

(b) for $n = 2$,
$$|e^x - {}_2 T_0 f(x)| = \left| e^x - \left(1 + x + \frac{x^2}{2!} \right) \right|$$
$$\leq \frac{e}{3!} < \frac{4}{6} \quad \text{for } x \text{ in } [-1, 1];$$

(c) for $n = 3$,
$$|e^x - {}_3 T_0 f(x)| = \left| e^x - \left(1 + x + \frac{x^2}{2!} + \frac{x^3}{3!} \right) \right|$$
$$\leq \frac{e}{4!} < \frac{4}{24} \quad \text{for } x \text{ in } [-1, 1].$$

Indeed, if we wish to force $|e^x - {}_nT_0f(x)| < \dfrac{1}{1000}$ for all x in $[-1, 1]$, we

can choose $n = 6$ $\left(\text{Check that } \dfrac{e}{7!} < \dfrac{1}{1000}\right)$. Thus, in this example if n is

large, the values of f are closely approximated by those of ${}_nT_0f$ for x in $[-1, 1]$.   ■

In the sections which follow we develop techniques to show that certain functions (such as the exponential function) can be closely approximated by the Taylor polynomials for all values of x.

Example 4.   We shall use the results of Example 3 to evaluate the number e to five decimal places. In Example 3 we showed that

$$|e^x - {}_nT_0f(x)| \le \frac{e}{(n+1)!}$$

for each positive integer n and each x in $[-1, 1]$. We know that $e \le 4$. Thus, we have

$$|e^x - {}_nT_0f(x)| \le \frac{4}{(n+1)!} \quad \text{for } x \text{ in } [-1, 1].$$

In particular, for $x = 1$

$$|e - {}_nT_0f(1)| \le \frac{4}{(n+1)!}$$

or

(13.20) $${}_nT_0f(1) - \frac{4}{(n+1)!} \le e \le {}_nT_0f(1) + \frac{4}{(n+1)!}.$$

We use these last inequalities to approximate the value of e. From Inequality 13.20 and Table 13.1, we see that e lies between 2.7182803 and 2.7182827. Thus, to five decimal places we have

$$e = 2.71828.   ■$$

Table 13.1. Evaluation of e to Five Decimal Places Using Taylor Polynomials.

		Approximations for e		
n	$\dfrac{4}{(n+1)!}$	**Lower Bound** ${}_nT_0f(1) - \dfrac{4}{(n+1)!}$	**Taylor Approximation** ${}_nT_0f(1)$	**Upper Bound** ${}_nT_0f(1) + \dfrac{4}{(n+1)!}$
1	2	0	2.0	4
2	0.6667	1.83332	2.5000	3.16667
3	0.16667	2.50000	2.66667	2.83334
4	0.03334	2.67500	2.70833	2.74167
5	0.00556	2.71111	2.71667	2.72223
6	0.00080	2.71742	2.71822	2.71903
7	0.00010	2.71813	2.71824	2.71834
8	0.00001	2.71826	2.71827	2.71828
9	0.0000011	2.7182803	2.7182815	2.7182827

EXERCISES

1. Let $f(x) = \sin x$. Show that
$$|_nR_0f(x)| \leq \frac{|x|^{n+1}}{(n+1)!}$$
for each positive integer n and each real x.
2. Use the result in Exercise 1 to find the 4 decimal place approximation of $\sin 1$ (Study Example 4).
3. Let $f(x) = \sin x$. Show that
$$|_nR_1f(x)| \leq \frac{|x-1|^{n+1}}{(n+1)!}$$
for each positive integer n and each real x.
4. Use the Method of Parts to demonstrate the validity of Equation 13.15.
5. Show that if $f^{(n+1)}$ is defined and continuous on $[a, x]$, then
$$\frac{1}{(n-1)!}\int_a^x (f^{(n)}(t) - f^{(n)}(a))(x-t)^{n-1} = \frac{1}{n!}\int_a^x f^{(n+1)}(t)(x-t)^n.$$

3.   LIMITS OF SEQUENCES

In the preceding section we discovered that an $(n+1)$ times differentiable function f differs from its nth Taylor approximation $_nT_af$ by $_nR_af(x) = \frac{1}{n!}\int_a^x f^{(n+1)}(t)(x-t)^n$. This last expression is a real number. In Example 3 we demonstrated for one choice of f that if n is large, the remainder $_nR_af$ is small, and consequently $_nT_af$ is a good approximation for f. In this section we present the preliminaries which will enable us to strengthen these results.

DEFINITION 13.5   A *sequence* of real numbers is a function s from the set of positive integers into the set of real numbers $\mathcal{R}$.

Example 5

(a) The function $s:\mathcal{R} \to \mathcal{R}$, defined by $s(n) = n^2$ for each positive integer n, is a sequence of real numbers.

(b) The function $s:\mathcal{R} \to \mathcal{R}$, defined by $s(n) = \frac{1}{n+1}$ for each positive integer n, is a sequence of real numbers.

(c) The function $s:\mathcal{R} \to \mathcal{R}$, defined by $s(n) = (-1)^n$ for each positive integer n, is a sequence of real numbers.

We note that in (a), $s(n)$ is large if n is large; in (b), $s(n)$ is close to 0 if n is large; and in (c), $s(n)$ alternates between 1 and -1. ∎

For convenience we shall use the following notation with sequences. If s is a sequence, we shall speak of the sequence $\{s(n)\}$ or the sequence $\{s(n)\}_{n=1}^{\infty}$. For example, $\left\{\dfrac{1}{n}\right\}$ denotes the sequence s defined by $s(n) = \dfrac{1}{n}$.

The preceding example suggests that certain sequences behave in an "orderly" fashion and others do not. The following definition describes four possible ways in which a sequence s may behave. The reader should refer to Definition 4.14 (p. 212) for similar definitions.

DEFINITION 13.6 Let $s: \mathcal{R} \to \mathcal{R}$ be a sequence of real numbers. Let $L \in \mathcal{R}$.

i. $\{s(n)\}$ *converges to* L if and only if for each $r > 0$, there exists an integer N_r such that $|s(n) - L| < r$ for all $n \geq N_r$. In this case we write

$$\lim_{n \to \infty} s(n) = L.$$

ii. $\{s(n)\}$ *diverges to* $+\infty$ if and only if for each $M > 0$ there exists an integer N_M sych that $s(n) > M$ for all $n \geq N_M$. In this case we write

$$\lim_{n \to \infty} s(n) = +\infty.$$

iii. $\{s(n)\}$ *diverges to* $-\infty$ if and only if for each $M < 0$ there exists an integer N_M such that $s(n) < M$ for all $n \geq N_M$. In this case we write

$$\lim_{n \to \infty} s(n) = -\infty.$$

iv. $\{s(n)\}$ *diverges and oscillates* if and only if none of statements (i), (ii), or (iii) is true.

We shall frequently say that $\{s(n)\}$ *converges* or $\{s(n)\}$ *diverges*. In the first case we mean that $\{s(n)\}$ converges to some real number L, and in the second case we mean that $\{s(n)\}$ either diverges to $+\infty$, diverges to $-\infty$, or diverges and oscillates.

Example 6. Let us reconsider the sequences given in Example 5. In a straightforward fashion the following results can be demonstrated:

(**a**) If $s(n) = n^2$, then $\lim_{n \to \infty} s(n) = +\infty$.

(**b**) If $s(n) = \dfrac{1}{n+1}$, then $\lim_{n \to \infty} s(n) = 0$.

(**c**) If $s(n) = (-1)^n$, then $\{s(n)\}$ diverges and oscillates. An example of a sequence which diverges to $-\infty$ is given by $s(n) = 2 - n$ for each positive integer n. ∎

Many theorems concerning the limits of sequences can be determined by techniques similar to those used in Chapter 4. The most obvious results are given in the following theorem. The proofs are called for in the exercises at the end of the section.

THEOREM 13.7 Let s and t be sequences of real numbers. Let $\lim_{n \to \infty} s(n) = L$ and $\lim_{n \to \infty} t(n) = K$, where L and K are real numbers. The following are true:

i. $\lim_{n \to \infty} (s(n) + t(n)) = L + K$;

ii. $\lim_{n \to \infty} rs(n) = rL$ for each real r;

iii. $\lim_{n \to \infty} s(n)t(n) = LK$;

iv. $\lim_{n \to \infty} \dfrac{s(n)}{t(n)} = \dfrac{L}{K}$, if $K \neq 0$ and $t(n) \neq 0$ for each n.

The following theorem will be particularly useful in the development which follows. The reader is urged to state and prove its analogue for decreasing sequences of real numbers.

THEOREM 13.8 Let s be an increasing sequence of real numbers, that is, $s(n) \leq s(n + 1)$ for each positive integer n.

i. If the set $\{s(n) \mid n \text{ is an integer}\}$ has an upper bound, then

$$\lim_{n \to \infty} s(n) = L$$

where $L = \text{lub } \{s(n) \mid n \text{ is an integer}\}$.

ii. If the set $\{s(n) \mid n \text{ is an integer}\}$ does not have an upper bound, then

$$\lim_{n \to \infty} s(n) = +\infty.$$

Proof

i. Assume that $\{s(n) \mid n \text{ is an integer}\}$ has an upper bound and let $L = \text{lub}$ $\{s(n) \mid n \text{ is an integer}\}$. Let $r > 0$ and note that $L - r$ is not an upper bound for this set. Thus, there exists an integer N such that $L - r < s(N)$. Since $s(n) \geq s(N)$ for all $n \geq N$, we have

$$L - r < s(n) \quad \text{for all } n \geq N,$$

and since L is an upper bound for $\{s(n) \mid n \text{ is an integer}\}$, we have

$$s(n) \leq L \quad \text{for all } n.$$

These two inequalities show that $|s(n) - L| < r$ for all $n \geq N$, and thus, by Definition 13.6i,

$$\lim_{n \to \infty} s(n) = L.$$

ii. Assume that $\{s(n) \mid n \text{ is an integer}\}$ has no upper bound. Let $M > 0$. Since M is not an upper bound for $\{s(n) \mid n \text{ is an integer}\}$, there exists an integer N such that $s(N) > M$. Therefore, if $n \geq N$, $s(n) \geq s(N) > M$, and by Definition 13.6ii

$$\lim_{n \to \infty} s(n) = +\infty. \quad \blacksquare$$

Example 7

(**a**) The sequence $\left\{\dfrac{n}{n+1}\right\}$ is increasing (check this). Furthermore, $\dfrac{n}{n+1} \leq 1$ for each positive integer n. Thus, by Theorem 13.8, $\left\{\dfrac{n}{n+1}\right\}$ converges. Moreover, we note that

$$\lim_{n \to \infty} \frac{n}{n+1} =^1 \lim_{n \to \infty} \frac{1}{1 + \dfrac{1}{n}}$$

$$=^2 1.$$

(1) Algebraic manipulation.
(2) By Theorem 13.7iv and the fact that $\lim\limits_{n \to \infty} \dfrac{1}{n} = 0$.

(**b**) The sequence $\left\{\dfrac{n^2}{n+1}\right\}$ is increasing (check this). Furthermore, if M is any real number and $n - 1 > M$, then $\dfrac{n^2}{n+1} > M$. Thus, by Theorem 13.8,

$$\lim_{n \to \infty} \frac{n^2}{n+1} = +\infty. \quad \blacksquare$$

EXERCISES

In Exercises 1 to 10 determine whether the sequence converges to a finite limit, diverges to $+\infty$, diverges to $-\infty$, or diverges and oscillates.

1. $\left\{\dfrac{1}{n+2}\right\}$.

2. $\{(-2)^n\}$.

3. $\left\{\dfrac{n^2}{2n^2+1}\right\}$.

4. $\left\{\dfrac{n}{n^2+1}\right\}$.

5. $\{(-1)^n n\}$.

6. $\left\{\dfrac{n^2-1}{n+1}\right\}$.

7. $\left\{\sin \dfrac{n\pi}{2}\right\}$.

8. $\left\{n \sin \dfrac{1}{n}\right\}$ $\left(\text{Hint: recall that } \lim\limits_{x \to 0} \dfrac{\sin x}{x} = 1\right)$.

9. $\{\sqrt{n+1} - \sqrt{n}\}$.

10. $\left\{\dfrac{2^n}{n!}\right\}$.

11. Prove Theorem 13.7i.   12. Prove Theorem 13.7ii.

13. Show that if $\lim_{n \to \infty} s(n) = +\infty$ and $\lim_{n \to \infty} t(n) = +\infty$, then $\lim_{n \to \infty} (s(n) + t(n)) = +\infty$.

14. Show that if $\{s(n)\}$ is bounded (that is, $|s(n)| \leq M$ for all n) and $\lim_{n \to \infty} t(n) = +\infty$, then $\lim_{n \to \infty} \dfrac{s(n)}{t(n)} = 0$.

ANSWERS

1. 0. **3.** $\frac{1}{2}$. **5.** Diverges and oscillates.

7. Diverges and oscillates $\left(\sin \dfrac{n\pi}{2} = 1, 0, \text{ or } -1 \right)$.

9. $\lim_{n \to \infty} (\sqrt{n+1} - \sqrt{n}) = \lim_{n \to \infty} \dfrac{1}{\sqrt{n+1} + \sqrt{n}} = 0$.

11. Let $r > 0$. Choose N_1 and N_2 such that

$$|s(n) - L| < \frac{r}{2} \quad \text{for} \quad n \geq N_1$$

and

$$|t(n) - K| < \frac{r}{2} \quad \text{for} \quad n \geq N_2.$$

Let N_3 be the larger of N_1 and N_2 and note that
$$|s(n) + t(n) - (L + K)| \leq |s(n) - L| + |t(n) - K|$$
$$< r \quad \text{for} \quad n \geq N_3.$$

13. Use an argument similar to that in Exercise 11, but applying Definition 13.6ii.

4. SERIES

Our primary concern in the remainder of this chapter will be with sequences of a very special form. These special sequences are called series. Before defining the term *series*, we consider the following example.

Example 8. Let t be the sequence defined by $t(n) = 1/n$. We form a new sequence S as follows:

$$S(1) = t(1) = 1$$
$$S(2) = t(1) + t(2) = 1 + \tfrac{1}{2}$$

.

.

.

$$S(n) = \sum_{k=1}^{n} t(k) = \sum_{k=1}^{n} \frac{1}{k}$$
$$= 1 + \cdots + \frac{1}{n}.$$

We shall call the sequence S a series.

It follows easily from Definition 13.6 that $\lim_{n \to \infty} t(n) = 0$. The behavior of the sequence S is not so easily determined. Since $S(n + 1) \geq S(n)$ for each value of n, it follows from Theorem 13.8 that either $\lim_{n \to \infty} S(n)$ is a finite number or $\lim_{n \to \infty} S(n) = +\infty$. The reader is asked to prove the latter in Exercise 11. ∎

Refer to Example 8 while reading the following definition.

DEFINITION 13.9 Let t be a sequence of real numbers. The sequence S, defined by

$$(13.21) \qquad S(n) = \sum_{k=1}^{n} t(k) \quad \text{for each positive integer } n,$$

is called the nth *partial sum sequence for* t. A sequence S is called a *series* when it is regarded as the nth partial sum sequence of a sequence t. The number $t(n)$ is called the nth *term of series* S. The sum $S(n)$ is called the nth *partial sum of the series* S.

Thus, in Example 8, $1 + \frac{1}{2}$ is the second partial sum of S, and $\frac{1}{3}$ is the third term of S.

The problem of determining whether a given series has a limit and of determining the value of that limit is a complicated one. In this and subsequent sections we shall only consider special cases. A more detailed study of series is found in Chapter 21.

One particular type of series lends itself easily to analysis. These series are called *geometric series* and are the subject of the following theorem.

THEOREM 13.10 (Geometric Series.) Let x be a real number and let S be the series defined for each positive integer by

$$S(n) = \sum_{k=1}^{n} x^{k-1}.$$

i. If $|x| < 1$, $\lim_{n \to \infty} S(n) = \dfrac{1}{1 - x}$.

ii. If $x \geq 1$, $\lim_{n \to \infty} S(n) = +\infty$.

iii. If $x \leq -1$, $\{S(n)\}$ oscillates and diverges.

Proof. If $x = 1$, it is obvious that $S(n) = \sum_{k=1}^{n} 1 = n$ diverges to $+\infty$.

We demonstrate the validity of the remainder of the theorem by algebraic techniques assuming that $x \neq 1$:

(13.22)

$$S(n) = \sum_{k=1}^{n} x^{k-1} \Rightarrow^1 xS(n) = \sum_{k=1}^{n} x^k$$

$$\Rightarrow^2 (1 - x)S(n) = S(n) - xS(n)$$

$$= \sum_{k=1}^{n} x^{k-1} - \sum_{k=1}^{n} x^k$$

$$=^3 1 - x^n$$

$$\Rightarrow^4 S(n) = \frac{1 - x^n}{1 - x} .$$

(1) Multiplying $S(n)$ by x.

(2) Subtracting $xS(n)$ from $S(n)$.

(3) $\sum_{k=1}^{n} x^{k-1} = 1 + x + \ldots + x^{n-1}$ and $\sum_{k=1}^{n} x^k = x + \ldots + x^{n-1} + x^n$.

Thus $x, \ldots, x^{n-1}$ are common to both sums and cancel when we subtract.
(4) Dividing the preceding expression by $1 - x$.

From the implication chain (13.22) we have

(13.23) $$S(n) = \frac{1 - x^n}{1 - x}, \quad \text{for} \quad x \neq 1.$$

Thus,

i. if $|x| < 1$, then $\lim\limits_{n \to \infty} x^{n+1} = 0$, and consequently $\lim\limits_{n \to \infty} S(n) = \frac{1}{1 - x}$;

ii. if $x > 1$, then $\lim\limits_{n \to \infty} x^n = +\infty$ and $\lim\limits_{n \to \infty} S(n) = \lim\limits_{n \to \infty} \frac{x^n - 1}{x - 1} = +\infty$;

iii. if $x \leq -1$, x^n oscillates and diverges. Thus $\{S(n)\}$ oscillates and diverges. ∎

Example 9. Using Theorem 13.10 we shall determine the limit of several series.

(a) Let $x = 1/2$ and

$$S(n) = \sum_{k=1}^{n} x^{k-1} = 1 + \tfrac{1}{2} + \tfrac{1}{4} + \cdots + (\tfrac{1}{2})^{n-1}.$$

Since $|\tfrac{1}{2}| < 1$, we have

$$\lim_{n \to \infty} S(n) = \frac{1}{1 - \tfrac{1}{2}} = 2.$$

(b) Let $x = -\tfrac{1}{3}$ and

$$S(n) = \sum_{k=1}^{n} x^{k-1} = 1 - \tfrac{1}{3} + \cdots + (-\tfrac{1}{3})^{n-1}.$$

Thus $\lim\limits_{n \to \infty} S(n) = \dfrac{1}{1 - (-\frac{1}{3})} = \dfrac{3}{4}$.

(**c**) Consider the series defined by

$$S(n) = \sum_{k=1}^{n} \frac{2^k}{3^k}.$$

This series is not in the form called for in Theorem 13.10. However,

$$S(n) = \tfrac{2}{3} \sum_{k=1}^{n} \left(\tfrac{2}{3}\right)^{k-1}$$

and $\lim\limits_{n \to \infty} \sum\limits_{k=1}^{n} \left(\tfrac{2}{3}\right)^{k-1} = \dfrac{1}{1 - \frac{2}{3}} = 3$ by Theorem 13.10. Therefore,

$\lim\limits_{n \to \infty} S(n) = \tfrac{2}{3} \cdot 3 = 2.$ ■

The following theorem will be useful not only in the subsequent development but also in showing that certain series do not converge:

THEOREM 13.11 Let S be the series of nth partial sums of the sequence t:

$$S(n) = \sum_{k=1}^{n} t(k).$$

If $\lim\limits_{n \to \infty} S(n)$ exists and is finite, then $\lim\limits_{k \to \infty} t(k) = 0$.

Proof. Assume that $\lim\limits_{n \to \infty} S(n) = L$. By Definition 13.6 (p. 528), if $r > 0$, there exists a number N_r such that

$$|S(n) - L| < \frac{r}{2} \quad \text{for all} \quad n \geq N_r.$$

Let us choose m so that $m - 1 > N_r$ and $m > N_r$. We then have

(13.24) $\qquad |S(m) - L| < \dfrac{r}{2} \quad \text{and} \quad |S(m - 1) - L| < \dfrac{r}{2}.$

Noting that $S(m) - S(m - 1) = t(m)$ (check this), we have

$$
\begin{aligned}
|t(m)| &= |S(m) - S(m - 1)| \\
&\leq^1 |S(m) - L| + |L - S(m - 1)| \\
&<^2 \frac{r}{2} + \frac{r}{2} \\
&= r.
\end{aligned}
$$

(1) Triangle Inequality
(2) Inequality 13.24.

Thus, by Definition 13.6, $\lim\limits_{m \to \infty} t(m) = 0$. This is the same as saying that $\lim\limits_{k \to \infty} t(k) = 0$. ∎

The reader is cautioned to observe that the converse of Theorem 13.11 is *not* valid. For example, the series $\left\{ \sum\limits_{k=1}^{n} \dfrac{1}{k} \right\}$ diverges to $+\infty$ (see Exercise 11), whereas $\lim\limits_{k \to \infty} \dfrac{1}{k} = 0$.

From Theorem 13.11 it follows that if t is a sequence which *does not* converge to 0, then the series $\left\{ \sum\limits_{k=1}^{n} t(k) \right\}$ *does not* converge. We apply this in the following example:

Example 10

(**a**) Consider the series $\left\{ \sum\limits_{k=1}^{n} \dfrac{k}{k+1} \right\}$. Since $\lim\limits_{n \to \infty} \dfrac{k}{k+1} = 1 \neq 0$,

$$\left\{ \sum_{k=1}^{n} \dfrac{k}{k+1} \right\} \text{ does not converge.}$$

In this case we note that $S(n+1) \geq S(n)$, where $S(n) = \sum\limits_{k=1}^{n} \dfrac{k}{k+1}$, and thus, by Theorem 13.8, $\lim\limits_{n \to \infty} \sum\limits_{k=1}^{n} \dfrac{k}{k+1} = +\infty$.

(**b**) Consider the series $\left\{ \sum\limits_{k=1}^{n} \sin\left(\dfrac{k\pi}{2}\right) \right\}$. The values of the sequence $t(k) = \sin\left(\dfrac{k\pi}{2}\right)$ alternate between 1, 0, and -1. Thus $\{t(k)\}$ oscillates and diverges, and consequently, $\left\{ \sum\limits_{k=1}^{n} \sin\left(\dfrac{k\pi}{2}\right) \right\}$ diverges.

In subsequent sections the following theorem will be invaluable. We defer the proof of this theorem to Chapter 21.

THEOREM 13.12 Let t be a sequence of real numbers. If the series $\left\{ \sum\limits_{k=1}^{n} |t(k)| \right\}$ converges, then $\left\{ \sum\limits_{k=1}^{n} t(k) \right\}$ also converges (the converse is *not* true).

EXERCISES

In Exercises 1 to 6 apply Theorem 13.10 to show that the given geometric series diverges or to show that it converges and to find its limit:

1. $\left\{ \sum\limits_{k=1}^{n} (\frac{3}{4})^{k-1} \right\}$.

2. $\left\{ \sum\limits_{k=1}^{n} \frac{(-1)^k 3^{k-1}}{4^k} \right\}$.

3. $\left\{ \sum\limits_{k=1}^{n} (-2)^{k-1} \right\}$.

4. $\left\{ \sum\limits_{k=1}^{n} (\frac{1}{2})^{k+1} \right\}$.

5. $\left\{ \sum\limits_{k=1}^{n} (\frac{7}{8})^{k+1} \right\}$.

6. $\left\{ \sum\limits_{k=1}^{n} (-1)^k \right\}$.

In Exercises 7 to 10 apply Theorem 13.11 to show that the series diverges:

7. $\left\{ \sum\limits_{k=1}^{n} \frac{2k}{k+1} \right\}$.

8. $\left\{ \sum\limits_{k=1}^{n} \left(1 + \frac{1}{2^k} \right) \right\}$.

9. $\left\{ \sum\limits_{k=1}^{n} \cos (k\pi) \right\}$.

10. $\left\{ \sum\limits_{k=1}^{n} (-1)^k \left(1 + \frac{1}{k} \right) \right\}$.

11. Let S be the series $\left\{ \sum\limits_{k=1}^{n} \frac{1}{k} \right\}$. Show that $\lim\limits_{n \to \infty} \sum\limits_{k=1}^{n} \frac{1}{k} = +\infty$.

12. Assume that the series $\left\{ \sum\limits_{k=1}^{n} \frac{1}{k^2} \right\}$ converges. Show that the series $\left\{ \sum\limits_{k=1}^{n} \frac{1}{k^3} \right\}$ converges. Hint: Use Theorem 13.8.

13. Let t be a sequence of real numbers such that $|t(k)| \le \frac{1}{2^{k-1}}$ for each k. Show that the series $\left\{ \sum\limits_{k=1}^{n} t(k) \right\}$ converges.

ANSWERS

1. $\dfrac{1}{1 - 3/4} = 4$. **3.** Oscillates and diverges since $-2 \le -1$.

5. $\lim\limits_{n \to \infty} \sum\limits_{k=1}^{n} (\frac{7}{8})^{k+1} = \lim \, (\frac{7}{8})^2 \sum\limits_{k=1}^{n} (\frac{7}{8})^{k-1}$

$$= (\tfrac{7}{8})^2 \frac{1}{1 - \frac{7}{8}} = \frac{49}{8}.$$

7. $\lim\limits_{k \to \infty} \dfrac{2k}{k+1} = 2$. Therefore, $\left\{ \sum\limits_{k=1}^{n} \dfrac{2k}{k+1} \right\}$ diverges.

9. $\{\cos (k\pi)\}$ oscillates between $+1$ and -1, and therefore, $\left\{ \sum\limits_{k=1}^{n} \cos (k\pi) \right\}$ diverges.

11. Let $S(n) = \sum\limits_{k=1}^{n} \dfrac{1}{k}$. Show that (i) $S(1) = 1$, (ii) $S(2) = 1 + \frac{1}{2}$, (iii) $S(4) \ge 1 + \frac{1}{2} + \frac{1}{2}$, (iv) $S(8) \ge 1 + \frac{1}{2} + \frac{1}{2} + \frac{1}{2}$, (v) $S(16) \ge 1 + \frac{1}{2} + \frac{1}{2} + \frac{1}{2} + \frac{1}{2}$. Now use an inductive argument to show that

$$\lim\limits_{n \to \infty} \sum\limits_{k=1}^{n} \frac{1}{k} = +\infty.$$

13. Since $|t(k)| \leq \dfrac{1}{2^{k-1}}$ for each k,

$$\sum_{k=1}^{n} |t(k)| \leq \sum_{k=1}^{n} (\tfrac{1}{2})^{k-1} = \frac{1}{1 - \frac{1}{2}} = 2.$$

Thus $\left\{ \displaystyle\sum_{k=1}^{n} |t(k)| \right\}$ is an increasing sequence which is bounded above and (by Theorem 13.8) converges. Thus, by Theorem 13.12, $\left\{ \displaystyle\sum_{k=1}^{n} t(k) \right\}$ also converges.

5. CONVERGENCE OF SERIES OF POSITIVE TERMS

It follows from Theorem 13.8 that if $S(n) = \displaystyle\sum_{k=1}^{n} t(n)$ and $t(n) \geq 0$; then either $\lim_{n \to \infty} S(n)$ is a real number or it is $+\infty$. Thus, series of positive terms are simpler than arbitrary sequences, since the latter may converge, diverge to $+\infty$, diverge to $-\infty$, or diverge and oscillate. In this section we shall develop three tests which will enable us to show that certain series of positive terms converge.

THEOREM 13.13 (The Comparison Test.) Let r and t be sequences of non-negative real numbers such that $r(k) \leq t(k)$ for each positive integer k.

i. If the series $\left\{ \displaystyle\sum_{k=1}^{n} t(k) \right\}$ converges, then the series $\left\{ \displaystyle\sum_{k=1}^{n} r(k) \right\}$ converges.

ii. If $\left\{ \displaystyle\sum_{k=1}^{n} r(k) \right\}$ diverges to $+\infty$, then the series $\left\{ \displaystyle\sum_{k=1}^{n} t(k) \right\}$ diverges to $+\infty$.

Proof. Since $\left\{ \displaystyle\sum_{k=1}^{n} r(k) \right\}$ and $\left\{ \displaystyle\sum_{k=1}^{n} t(k) \right\}$ are series of non-negative terms, each of them either converges to a real number or diverges to $+\infty$. Consequently, assertion (ii) is the contrapositive of (and is logically equivalent to) statement (i). Therefore, we need only prove statement (i).

Assume that $\left\{ \displaystyle\sum_{k=1}^{n} t(k) \right\}$ converges to a real number. Then, by Theorem 13.8, there exists a member M such that $\displaystyle\sum_{k=1}^{n} t(k) \leq M$ for all positive integers n. Since $r(k) \leq t(k)$ for each k, we also have $\displaystyle\sum_{k=1}^{n} r(k) \leq M$. Thus $\left\{ \displaystyle\sum_{k=1}^{n} r(k) \right\}$ is an increasing sequence which has an upper bound. Another application of Theorem 13.8 assures us that $\left\{ \displaystyle\sum_{k=1}^{n} r(k) \right\}$ converges to a real number. ∎

Note. The Comparison Test is applicable if we know that $r(k) \leq t(k)$ only for each positive integer k greater than some fixed integer K. This extension will be treated in Chapter 21 but will be used in the proof of Theorem 13.14.

Example 11. We know that $\left\{\sum_{k=1}^{n} x^{k-1}\right\}$ converges for $0 \leq x < 1$. We shall use this fact and the Comparison Test to show that two other series converge.

(**a**) Consider $\left\{\sum_{k=1}^{n} \frac{1}{k!}\right\}$. Note that $\frac{1}{k!} \leq \frac{1}{2^{k-1}}$ for each positive integer k.

Therefore, by the Comparison Test, $\lim_{n \to \infty} \sum_{k=1}^{n} \frac{1}{k!}$ is a real number. In

fact, from Examples 3 and 4 the reader should be able to show that

$$e = \lim_{n \to \infty} \sum_{k=1}^{n} \frac{1}{k!}.$$

(**b**) Consider $\left\{\sum_{k=1}^{n} \frac{1}{k^k}\right\}$. Note that $\frac{1}{k^k} \leq \frac{1}{2^{k-1}}$ for $k \geq 1$.

Therefore, by the Comparison Test, $\lim_{n \to \infty} \sum_{k=1}^{n} \frac{1}{k^k}$ is a real number. ∎

Two of the most useful applications of the Comparison Test involve the geometric series and are found in the two following theorems.

THEOREM 13.14 (The Root Test.) Let t be a sequence of non-negative real numbers such that $\lim_{k \to \infty} t(k)^{1/k} = L$, where $L \in \mathcal{R}$ or $L = +\infty$.

i. If $L < 1$, $\lim_{n \to \infty} \sum_{k=0}^{n} t(k)$ exists and is finite.

ii. If $L > 1$, $\lim_{n \to \infty} \sum_{k=0}^{n} t(k) = +\infty$.

iii. If $L = 1$, the series $\left\{\sum_{k=0}^{n} t(k)\right\}$ may converge or may diverge to $+\infty$.

Proof

i. Assume $L < 1$ and let c be such that $L < c < 1$. Since $\lim_{k \to \infty} t(k)^{1/k} = L < c$, there exists a positive integer K such that $t(k)^{1/k} < c$ for all $k \geq K$. Therefore, $t(k) < c^k$ for each $k \geq K$. Since $0 < c < 1$, $\left\{\sum_{k=1}^{n} c^k\right\}$

converges and thus by the Comparison Test we know that $\left\{\sum_{k=1}^{n} t(k)\right\}$ converges. (See Note following Theorem 13.13.)

ii. A similar argument to that given in (i) can be used.

iii. To show the validity of statement (iii), we must exhibit a sequence s such that $\lim_{k \to \infty} s(k)^{1/k} = 1$ and $\left\{ \sum_{k=1}^{n} s(k) \right\}$ converges, and a sequence t such that $\lim_{k \to \infty} t(k)^{1/k} = 1$ and $\left\{ \sum_{k=1}^{n} t(n) \right\}$ diverges to $+\infty$. For the first example we choose $s(k) = \dfrac{1}{k^2}$. We shall develop techniques in Chapter 21 to show that $\lim_{k \to \infty} \left(\dfrac{1}{k^2} \right)^{1/k} = 1$ and that $\left\{ \sum_{k=1}^{n} \dfrac{1}{k^2} \right\}$ converges. For the second example we need only choose $t(k) = 1$. Then $\lim_{k \to \infty} 1^{1/k} = 1$ and obviously $\left\{ \sum_{k=1}^{n} 1 \right\} = \{n\}$ diverges to $+\infty$. ∎

Example 12. The Root Test lends itself to the following types of problems:

(a) Let $t(k) = \dfrac{1}{k^k}$. Then $\lim_{k \to \infty} t(k)^{1/k} = \lim_{k \to \infty} \dfrac{1}{k} = 0$, and consequently, $\left\{ \sum_{k=1}^{n} \dfrac{1}{k^k} \right\}$ converges.

(b) Let $t(k) = (\ln k)^k$. Then $\lim_{k \to \infty} t(k)^{1/k} = \lim_{k \to \infty} \ln k = +\infty$ and $\left\{ \sum_{k=1}^{n} (\ln k)^k \right\}$ diverges to $+\infty$. ∎

THEOREM 13.15 (The Ratio Test.) Let t be a sequence of non-negative real numbers such that $\lim_{k \to \infty} \dfrac{t(k + 1)}{t(k)} = L$, where $L \in R$ or $L = +\infty$.

i. If $L < 1$, $\lim_{n \to \infty} \sum_{k=1}^{n} t(k)$ exists and is finite;

ii. If $L > 1$, $\lim_{n \to \infty} \sum_{k=1}^{n} t(k) = +\infty$;

iii. If $L = 1$, the series $\left\{ \sum_{k=1}^{n} t(k) \right\}$ may converge or may diverge to $+\infty$.

The proof of this theorem is similar to the proof of the Root Test and will be given in Chapter 21. (See Theorem 21.25, p.826.)

Example 13. The Ratio Test is useful in determining the behavior of series of the following types:

(a) Consider $\left\{ \sum_{k=1}^{n} \dfrac{1}{k!} \right\}$ (see Example 11).

Let $t(k) = \dfrac{1}{k!}$. Then

$$\lim_{k \to \infty} \frac{t(k+1)}{t(k)} = \lim_{k \to \infty} \frac{\dfrac{1}{(k+1)!}}{\dfrac{1}{k!}} = \lim_{k \to \infty} \frac{1}{k+1} = 0.$$

Therefore, $\left\{ \displaystyle\sum_{k=1}^{n} \frac{1}{k!} \right\}$ converges to a real number.

(b) Consider $\left\{ \displaystyle\sum_{k=1}^{n} \frac{k}{2^k} \right\}$. Let $t(k) = \dfrac{k}{2^k}$. Then

$$\lim_{k \to \infty} \frac{t(k+1)}{t(k)} = \lim_{k \to \infty} \frac{\dfrac{k+1}{2^{k+1}}}{\dfrac{k}{2^k}}$$

$$= \lim_{k \to \infty} \frac{1}{2} \frac{k+1}{k}$$

$$= \frac{1}{2} \lim_{k \to \infty} \left(1 + \frac{1}{k} \right)$$

$$= \frac{1}{2}.$$

Thus, by the Ratio Test, $\left\{ \displaystyle\sum_{k=1}^{n} \frac{k}{2^k} \right\}$ converges. ∎

With the techniques at our disposal the reader will discover that the Ratio Test is more valuable than the Root Test. This is true because we are not able to evaluate limits such as $\lim_{k \to \infty} (k)^{1/k}$ or $\lim_{k \to \infty} (k!)^{1/k}$.

EXERCISES

Use the Comparison Test, the Root Test, and/or the Ratio Test to show that the following series converge or diverge to $+\infty$:

1. $\left\{ \displaystyle\sum_{k=1}^{n} \frac{k!}{2^k} \right\}$.

2. $\left\{ \displaystyle\sum_{k=1}^{n} \frac{k!}{(k+2)!} \right\}$.

3. $\left\{ \displaystyle\sum_{k=1}^{n} \frac{2^k}{k^k} \right\}$.

4. $\left\{ \displaystyle\sum_{k=1}^{n} \frac{1}{10^k} \right\}$.

5. $\left\{ \displaystyle\sum_{k=1}^{n} \frac{k+1}{2^k} \right\}$.

6. $\left\{ \displaystyle\sum_{k=1}^{n} \frac{3^k k!}{k^k} \right\}$.

7. $\left\{ \displaystyle\sum_{k=1}^{n} \frac{k}{e^{k^2}} \right\}$.

8. $\left\{ \displaystyle\sum_{k=1}^{n} \frac{k^2}{2^{k^2}} \right\}$.

9. $\left\{ \sum\limits_{k=1}^{n} \dfrac{k\,|\cos k|}{2^k} \right\}.$

10. $\left\{ \sum\limits_{k=1}^{n} \dfrac{|\sin k|}{5^k} \right\}.$

11. Use Examples 3, 4 and 11i to show that $e = \lim\limits_{n \to \infty} \sum\limits_{k=1}^{n} \dfrac{1}{k!}.$

ANSWERS

1. Let $t(k) = \dfrac{k!}{2^k}$. Show that $\lim\limits_{k \to \infty} \dfrac{t(k+1)}{t(k)} = \lim\limits_{k \to \infty} \dfrac{k+1}{2} = +\infty$, and thus that the series diverges to $+\infty$. **3.** Let $t(k) = \dfrac{2^k}{k^k}$. Show that $\lim\limits_{k \to \infty} t(k)^{1/k} = 0$ and thus that the series converges. **5.** Let $t(k) = \dfrac{k+1}{2^k}$. Show that $\lim\limits_{k \to \infty} \dfrac{t(k+1)}{t(k)} = \dfrac{1}{2}$ and thus that the series converges. **7.** Let $t(k) = \dfrac{k}{e^{k^2}}$. Show that $\lim\limits_{k \to \infty} \dfrac{t(k+1)}{t(k)} = 0$, and thus that the series converges. **9.** Let $t(k) = \dfrac{k\,|\cos k|}{2^k}$ and $r(k) = \dfrac{k}{2^k}$. Show (i) that $t(k) \le r(k)$ and (ii) that $\lim\limits_{k \to \infty} \dfrac{r(k+1)}{r(k)} = \dfrac{1}{2}$. Thus, by the Ratio Test, $\left\{ \sum\limits_{k=1}^{n} r(k) \right\}$ converges and by the Comparison Test $\left\{ \sum\limits_{k=1}^{n} t(k) \right\}$ converges.

6. TAYLOR APPROXIMATIONS (continued)

We started this chapter with a discussion of Taylor approximations for a function f. Each of these approximating polynomials is of the form $\sum\limits_{k=0}^{n} \dfrac{f^k(a)}{k!}(x-a)^k$. Thus the nth Taylor polynomial is an nth partial sum of a sequence. The techniques discussed in the preceding sections will enable us to determine (in certain cases) for what values of x the nth Taylor polynomial is a "good" approximation of $f(x)$. The following theorem is one example of what we consider a "good" approximation of f by the Taylor polynomials; in this theorem we shall consider a function f which has derivatives of every order n. For convenience we shall say that f is *infinitely differentiable* on (c, d) if and only if $f^{(n)}(t)$ exists for each t in (c,d) and for each positive integer n.

THEOREM 13.16 (Convergence of Taylor Polynomials.) Let $f:\mathcal{R} \to \mathcal{R}$ be infinitely differentiable on $(a - r, a + r)$ for some real number a and some $r > 0$. If there exists a real number M such that $|f^{(n)}(t)| \le M$ for each positive integer n and all t in $(a - r, a + r)$, then $f(x) = \lim\limits_{n \to \infty} {}_n T_a f(x)$ for each x in $(a - r, a + r)$.

Proof. Under the hypotheses of this theorem it follows from Theorem 13.4, p. 523, that

(13.25)
$$f(x) - {}_nT_af(x) = \frac{1}{n!}\int_a^x f^{(n+1)}(t)(x-t)^n$$

for each x in $(a - r, a + r)$ and for each positive integer n.

(13.26)
$$
\begin{cases}
|f(x) - {}_nT_af(x)| =^1 \frac{1}{n!}\left|\int_a^x f^{(n+1)}(t)(x-t)^n\right| \\[2mm]
\qquad\qquad \leq^2 \frac{1}{n!}\left|\int_a^x M\,|x-t|^n\right| \\[2mm]
\qquad\qquad =^3 \frac{M}{n!}\left[\frac{-\,|x-t|^{n+1}}{(n+1)}\right]_a^x \\[2mm]
\qquad\qquad =^4 \frac{M}{(n+1)!}\,|x-a|^{n+1}.
\end{cases}
$$

(1) Taking absolute values in Equation 13.25.
(2) Since $|f^{(n+1)}(t)| \leq M$ for all n and t, we can apply Theorem 8.15 (p. 360).
(3) Finding the primitive of $|x - t|^n$ for each x and n.
(4) Evaluating the primitive at $t = x$ and $t = a$.

In Exercise 11 the reader is asked to prove that $\lim\limits_{n\to\infty} \dfrac{b^n}{n!} = 0$ for any real number b. Therefore, $\lim\limits_{n\to\infty} \dfrac{M\,|x-a|^{n+1}}{(n+1)!} = 0$ for any real numbers x and a. It now follows from Inequality Chain 13.26 that $\lim\limits_{n\to\infty} {}_nT_af(x) = f(x)$ for each x in $(a - r, a + r)$. ■

Example 14. Let $f(x) = e^x$. We shall show that

(13.27)
$$e^x = \lim_{n\to\infty} {}_nT_0f(x) = \lim_{n\to\infty} \sum_{k=0}^n \frac{x^k}{k!} \text{ for all real } x.$$

We have previously determined (Example 2, p. 520) the nth Taylor approximation of f about 0 to be

$$
{}_nT_0f(x) = \sum_{k=0}^n \frac{x^k}{k!}.
$$

Let x be any real number and r be a positive real number such that $x \in (-r, r)$. Since $f^{(n)}(x) = e^x$ for all n and f is an increasing function, we have $|f^{(n)}(x)| \leq e^r$ for all x in $(-r, r)$ and all n. We can now apply Theorem 13.16 to conclude that Equation 13.27 is valid. ■

Note that we can substitute any value of x in Equation 13.27 and determine e^x as the limit of the corresponding series. For example, substituting $x = 1$ we have

(13.28)
$$
e = \lim_{n\to\infty} \sum_{k=0}^n \frac{1}{k!}.
$$

> **DEFINITION 13.17** Let $f: \mathcal{R} \to \mathcal{R}$ be infinitely differentiable at a and let x be such that
>
> (13.29) $\qquad f(x) = \lim_{n \to \infty} {}_nT_af(x) = \lim_{n \to \infty} \sum_{k=0}^{n} \frac{f^{(k)}(a)}{k!}(x-a)^k.$
>
> In this case $\left\{ \sum_{k=0}^{n} \frac{f^{(k)}(a)}{k!}(x-a)^k \right\}$ is called the *Taylor series representation for $f(x)$ about a.*

Thus, in Example 14 we showed that $\left\{ \sum_{k=0}^{n} \frac{x^k}{k!} \right\}$ is the Taylor series representation for e^x about 0.

In the exercises the reader is asked to show that the Taylor series representations for $\sin x$ and $\cos x$ about 0 are

(13.30) $\qquad\qquad \sin x = \lim_{n \to \infty} \sum_{k=0}^{n} \frac{(-1)^k x^{2k+1}}{(2k+1)!}$

and

(13.31) $\qquad\qquad \cos x = \lim_{n \to \infty} \sum_{k=0}^{n} \frac{(-1)^k x^{2k}}{(2k)!}.$

The hypotheses of Theorem 13.16 are quite restrictive. Many functions are infinitely differentiable, but not such that $|f^{(n)}(t)| \leq M$ for all n and t. For example, let $f(x) = e^{2x}$. Then $f^{(n)}(0) = 2^n$. The following example will illustrate a technique which sometimes allows us to avoid this difficulty.

Example 15

(*a*) Let $f(x) = e^{2x}$. Substituting $2x$ for x in Equation 13.27, we have

$$e^{2x} = \lim_{n \to \infty} \sum_{k=0}^{n} \frac{(2x)^k}{k!}$$
$$= \lim_{n \to \infty} \sum_{k=0}^{n} \frac{2^k x^k}{k}.$$

We shall now verify that $\left\{ \sum_{k=0}^{n} \frac{2^k x^k}{k!} \right\}$ is the Taylor series representation for e^{2x} about 0, and thus, by the last equality chain, we have $f(x) = \lim_{n \to \infty} {}_nT_0f(x)$ for all real x. The verification consists merely of noting that $f^{(k)}(x) = 2^k e^{2x}$ for all x and k, and consequently, that $f^{(k)}(0) = 2^k$ for each k and ${}_nT_0f(x) = \sum_{k=0}^{n} \frac{2^k x^k}{k!}$ for each n.

(*b*) Let $f(x) = e^{x^2}$. Substituting x^2 for x in Equation 13.27, we have

$$e^{x^2} = \lim_{n \to \infty} \sum_{k=0}^{n} \frac{x^{2k}}{k!}.$$

It is not at all obvious that this is the Taylor series expansion of f about 0.

It can be shown that $_{2n}T_0 f(x) = {}_{(2n+1)}T_0 f(x) = \sum\limits_{k=0}^{n} \dfrac{x^{2k}}{k!}$ and thus that

$$e^{x^2} = \lim_{n \to \infty} {}_{2n}T_0 f(x) = \lim_{p \to \infty} {}_{p}T_0 f(x)$$

for all values of x. ∎

EXERCISES

1. Show that

$$\sin x = \lim_{n \to \infty} \sum_{k=0}^{n} \frac{(-1)^k x^{2k+1}}{(2k+1)!} \quad \text{for all } x \text{ in } \mathcal{R}$$

and that this is the Taylor series representation of $\sin x$ about $a = 0$.

2. Show that

$$\cos x = \lim_{n \to \infty} \sum_{k=0}^{n} \frac{(-1)^k x^{2k}}{(2k)!} \quad \text{for all } x \text{ in } \mathcal{R}$$

and that this is the Taylor series representation of $\cos x$ about $a = 0$.

3. Let $f(x) = \dfrac{1}{1-x}$. Show that $\lim\limits_{n \to \infty} \sum\limits_{k=0}^{n} x^k$ is the Taylor series representation of $f(x)$ about 0 for $-1 < x < 1$.

In Exercises 4 to 7 use Equations 13.27, 13.30, and 13.31 to justify the given expansion:

4. $e^{-x} = \lim\limits_{n \to \infty} \sum\limits_{k=0}^{n} \dfrac{(-1)^k x^k}{k!}$ for all x.

5. $\sin(x^2) = \lim\limits_{n \to \infty} \sum\limits_{k=0}^{n} \dfrac{(-1)^k x^{4k+2}}{(2k+1)!}$ for all x.

6. $\cos(3x) = \lim\limits_{n \to \infty} \sum\limits_{k=0}^{n} \dfrac{(-1)^k 9^k x^{2k}}{(2k)!}$ for all x.

7. $\sin(1-x) = \lim\limits_{n \to \infty} \sum\limits_{k=0}^{n} \dfrac{(-1)^k (1-x)^{2k+1}}{(2k+1)!}$ for all x.

In Exercises 8 to 10 show that the given equations are valid:

8. $\lim\limits_{n \to \infty} \sum\limits_{k=0}^{n} \dfrac{2^k}{k!} = e^2$.

9. $\lim\limits_{n \to \infty} \sum\limits_{k=0}^{n} \dfrac{(-1)^k}{(2k!)} = \cos 1$.

10. $\lim\limits_{n \to \infty} \sum\limits_{k=0}^{n} \dfrac{(-1)^k \pi^{2k+1}}{(2k+1)!} = 0$.

11. Let b be a real number. Show that $\lim\limits_{n \to \infty} \dfrac{b^n}{n!} = 0$.

ANSWERS

1. Let $f(x) = \sin x$. Note that $|f^{(n)}(x)| \leq 1$ for all x and all n, and apply Theorem 13.16 to show that

$$f(x) = \lim_{p \to \infty} {}_p\mathsf{T}_0 f(x) \quad \text{for all real } x.$$

Note that ${}_{2n}\mathsf{T}_0 f(x) = {}_{(2n+1)}\mathsf{T}_0 f(x) = \sum_{k=0}^{n} \frac{(-1)^k x^{2k+1}}{(2k+1)!}$ and conclude that

$$\lim_{p \to \infty} {}_p\mathsf{T}_0 f(x) = \lim_{n \to \infty} {}_{(2n+1)}\mathsf{T}_0 f(x) = \lim_{n \to \infty} \sum_{k=0}^{n} \frac{(-1)^k x^{2k+1}}{(2k+1)!}.$$

3. By Theorem 13.10, $\dfrac{1}{1-x} = \lim_{n \to \infty} \sum_{k=0}^{n} x^k$ for $-1 < x < 1$. Show that

${}_n\mathsf{T}_0 f(x) = \sum_{k=0}^{n} x^k$. **5.** Substitute x^2 for x in Equation 13.30. **7.** Substitute $1 - x$ for x in Equation 13.30. **9.** Substitute 1 for x in Equation 13.31. **11.** Consider the series $\left\{ \sum_{k=1}^{n} \dfrac{|b|^k}{k!} \right\}$. Use the Ratio Test to show that this series converges and Theorem 13.11 to show that $\lim_{k \to \infty} \dfrac{|b|^k}{k!} = 0$. Show that this implies that $\lim_{k \to \infty} \dfrac{b^k}{k!} = 0$.

7. POWER SERIES

In Section 6 we showed that in certain cases the nth Taylor approximations of f about a converged to $f(x)$ for all x in some open interval centered at a. Each of these Taylor polynomials is of the form

$$\sum_{k=0}^{n} t(k)(x - a)^k,$$

where t is a sequence of real numbers. We shall now consider the problem from a different viewpoint.

Assume that t is a sequence of real numbers, that a is a real number, and that $\{S(n, x)\}$ is the sequence defined by

$$S(n, x) = \sum_{k=0}^{n} t(k)(x - a)^k$$

for each non-negative integer n and each real x. The sequence $\{S(n, x)\}$ is called a *power series*. It may be that the only value of x for which $\{S(n, x)\}$ converges is $x = a$. (Note that $S(n, a) = \sum_{k=0}^{n} t(k)(a - a)^k = t(0)$ for each n and thus $\lim_{n \to \infty} S(n, a) = t(0)$.) A second possibility is that $\{S(n, x)\}$ converges for all real x, and the remaining possibility is that $\{S(n, x)\}$ converges for some $x \neq a$ but not for all x. We define $S(x) = \lim_{n \to \infty} S(n, x)$ for the values of x for which this limit exists and is finite and note that S is an $\mathcal{R} \to \mathcal{R}$ function. $S(x)$ is called the *sum* of the sequence $\{S(n, x)\}$. We can now ask the questions: For what values of x is $S(x)$ defined and what are the properties of the function S?

The largest interval I centered at a such that $\lim\limits_{n \to \infty} S(n, x)$ exists and is finite for all x in I is called the *interval of convergence* of S. The radius of I is called the *radius of convergence* of S. It is possible that $r = 0$, $r = +\infty$, or $0 < r < +\infty$.

In Chapter 21 we shall show that each power series does not converge outside its interval of convergence. The following theorem will give us a technique by which we can determine the interval and radius of convergence for certain power series:

THEOREM 13.18   (The Root and Ratio Test.) Let t be a sequence of real numbers and $\{S(n, x)\}$ be the power series defined by

$$S(n, x) = \sum_{k=0}^{n} t(k)(x - a)^k.$$

If $\lim\limits_{k \to \infty} |t(k)|^{1/k} = L$ or $\lim\limits_{k \to \infty} \left| \dfrac{t(k + 1)}{t(k)} \right| = L$ (where $L \in \mathcal{R}$ or $L = +\infty$), then

  i. $S(x)$ is defined for all x such that

$$L\,|x - a| < 1;$$

  ii. $S(x)$ is not defined for any x such that

$$L\,|x - a| > 1;$$

  iii. if $L\,|x - a| = 1$, $S(x)$ may or may not be defined.

Proof i. Apply Theorems 13.14 and 13.15 to the series with k^{th} term

$$|t(k)(x - a)^k|$$

and recall $\sum\limits_{k=0}^{n} t(k)(x - a)^k$ converges if $\sum\limits_{k=0}^{n} |t(k)(x - a)^k|$ converges.

ii. Consider the case in which $\lim\limits_{k \to 0} |t(k)|^{1/k} = L$. By Theorem 13.11, if $\left\{ \sum\limits_{k=0}^{n} t(k)(x - a)^k \right\}$ converges, then $\lim\limits_{k \to \infty} t(k)(x - a)^k = 0$. Thus, for k sufficiently large we have $|t(k)(x - a)^k| < 1$ or $|t(k)|^{1/k}\,|x - a| < 1$. Therefore, if $L = \lim\limits_{k \to \infty} |t(k)|^{1/k}$, we have $L\,|x - a| \le 1$. We have shown that $L\,|x - a| \le 1$ if $\left\{ \sum\limits_{k=0}^{n} t(k)(x - a)^k \right\}$ converges. Therefore, if $L\,|x - a| > 1$, $\left\{ \sum\limits_{k=0}^{n} t(k)(x - a)^k \right\}$ must diverge.

iii. The proof of (iii) is deferred to Chapter 21. ■

Example 16

(**a**) Let $S(n, x) = \sum\limits_{k=0}^{n} 2^k (x - 3)^k$.

Applying the Root Test we find that $\lim\limits_{k \to \infty} |2^k|^{1/k} = 2$. Thus $S(x)$ exists if $|x - 3| < 1/2$ and does not exist if $|x - 3| > 1/2$. If $x - 3 = \frac{1}{2}$,

$$\sum_{k=0}^{n} 2^k (x - 3)^k = \sum_{k=0}^{n} 1 = n + 1,$$

which diverges to $+\infty$. If $x - 3 = -\frac{1}{2}$, $\sum\limits_{k=0}^{n} 2^k (x - 3)^k = \sum\limits_{k=0}^{n} (-1)^k$, which oscillates and diverges. Thus $S(x)$ exists if and only if $|x - 3| < 1/2$. Consequently, $(3 - \frac{1}{2}, 3 + \frac{1}{2}) = (\frac{5}{2}, \frac{7}{2})$ is the *interval of convergence* of S, and $r = 1/2$ is the *radius of convergence* of S.

(**b**) Let $S(n, x) = \sum\limits_{k=0}^{n} \dfrac{(x + 1)^k}{k}$.

Applying the Ratio Test we find that $\lim\limits_{k \to \infty} \dfrac{\dfrac{1}{k + 1}}{\dfrac{1}{k}} = \lim\limits_{k \to \infty} \dfrac{k}{k + 1} = 1$.

Thus $S(x)$ exists if $|x + 1| < 1$ and does not exist if $|x + 1| > 1$. We do not currently have the techniques to test the existence of $S(x)$ if $x + 1 = 1$ or $x + 1 = -1$. However, we do know that the radius of convergence is $r = 1$ and the interval of convergence is $(-2, 0)$, $(-2, 0]$, $[-2, 0)$, or $[-2, 0]$.

(**c**) Let $S(n, x) = \sum\limits_{k=0}^{n} k^k x^k$.

Applying the Root Test we have

$$\lim_{k \to \infty} |k^k|^{1/k} = \lim_{k \to \infty} k = +\infty.$$

Thus $S(x)$ exists only if $x = 0$.

(**d**) Let $S(n, x) = \sum\limits_{k=1}^{n} \dfrac{x^k}{k!}$.

Applying the Ratio Test we have

$$\lim_{k \to \infty} \dfrac{\dfrac{1}{(k + 1)!}}{\dfrac{1}{k!}} = \lim_{k \to \infty} \dfrac{1}{k + 1} = 0.$$

Therefore, $S(x)$ exists for all real x, the radius of convergence of S is $+\infty$, and the interval of convergence is $(-\infty, \infty)$. ∎

EXERCISES

Find interval and radius of convergence of the following power series:

1. $\left\{ \sum\limits_{k=1}^{n} \dfrac{(x-1)^k}{2^k} \right\}.$

2. $\left\{ \sum\limits_{k=0}^{n} 3^k(x-2)^k \right\}.$

3. $\left\{ \sum\limits_{k=0}^{n} k(x+4)^k \right\}.$

4. $\left\{ \sum\limits_{k=0}^{n} \dfrac{x^k}{k} \right\}.$

5. $\left\{ \sum\limits_{k=0}^{n} k!\,(x+2)^k \right\}.$

6. $\left\{ \sum\limits_{k=0}^{n} e^k(x-3)^k \right\}.$

7. $\left\{ \sum\limits_{k=0}^{n} \dfrac{(x-3)^k}{e^k} \right\}.$

8. $\left\{ \sum\limits_{k=0}^{n} \dfrac{x^k}{k^{1/2}} \right\}$

9. $\left\{ \sum\limits_{k=0}^{n} \dfrac{2^k(x-1)^k}{k!} \right\}.$

10. $\left\{ \sum\limits_{k=0}^{n} \dfrac{k!\,(x-1)^k}{2^k} \right\}.$

ANSWERS

1. Using the Root Test, $\lim\limits_{k\to\infty} \left| \dfrac{1}{2^k} \right|^{1/k} = \tfrac{1}{2}$. Thus the series converges for $|x-1| < 2$ and diverges for $|x-1| > 2$. The radius of convergence is 2 and the interval of convergence is $(-1, 3)$ (check the divergence at $x = -1$ and $x = 3$). **3.** Using the Ratio Test, $\lim\limits_{k\to\infty} \dfrac{(k+1)}{k} = 1$. Thus the series converges for $|x+4| < 1$ and diverges for $|x+4| > 1$. The radius of convergence is 1 and the interval of convergence is $(-5, -3)$ (check the divergence at $x = -5$ and $x = -3$). **5.** The series diverges except at $x = -2$. **7.** The radius of convergence is e; the interval of convergence is $(3-e, 3+e)$ (check the divergence at $x = 3-e$ and $x = 3+e$). **9.** The series converges for all real x.

8. PROPERTIES OF POWER SERIES

In this section we catalogue several properties of power series. Some of the proofs follow immediately from the definitions. Others are more difficult and will be deferred to Chapter 21.

The first theorem gives the algebraic properties of power series and follows immediately from the Definitions 13.6, 13.9, and Theorem 13.7.

THEOREM 13.19 Let c_1, c_2, and a be real numbers, t_1 and t_2 sequences of real numbers. If

$$S_1(x) = \lim_{n\to\infty} \sum_{k=0}^{n} t_1(k)(x-a)^k \text{ for } x \text{ in } (a-r, a+r)$$

and

$$S_2(x) = \lim_{n \to \infty} \sum_{k=0}^{n} t_2(k)(x-a)^k \text{ for } x \text{ in } (a-r, a+r),$$

then

$$(c_1 S_1 + c_2 S_2)(x) = \lim_{n \to \infty} \sum_{k=0}^{n} (c_1 t_1(k) + c_2 t_2(k))(x-a)^k$$

for x in $(a-r, a+r)$.

Thus we see that convergent power series can be combined term by term as though they were polynomials.

Example 17. We shall use Theorem 13.19 to find a power series representation of $3e^x - e^{2x}$ about $a = 0$. By Equation 13.27 we have

$$e^x = \lim_{n \to \infty} \sum_{k=0}^{n} \frac{x^k}{k!} \text{ for all } x$$

and

$$e^{2x} = \lim_{n \to \infty} \sum_{k=0}^{n} \frac{2^k x^k}{k!} \text{ for all } x.$$

Therefore, by Theorem 13.19,

$$3e^x - e^{2x} = \lim_{n \to \infty} \sum_{k=0}^{n} \frac{(3 - 2^k)x^k}{k!} \text{ for all } x. \quad \blacksquare$$

The next theorem is one of the most useful theorems concerning power series. We defer the proof of this theorem to Chapter 21.

THEOREM 13.20 Let t be a sequence of real numbers and

$$S(x) = \lim_{n \to \infty} \sum_{k=0}^{n} t(k)(x-a)^k \text{ for } x \text{ in } (a-r, a+r).$$

Then

i. (Differentiation) S is differentiable on $(a-r, a+r)$ and

$$S'(x) = \lim_{n \to \infty} \sum_{k=1}^{n} k t(k)(x-a)^{k-1} \text{ for } x \text{ in } (a-r, a+r).$$

ii. (Integration) S is integrable on $(a-r, a+r)$ and

$$\int_{a}^{x} S = \lim_{n \to \infty} \sum_{k=0}^{n} \frac{t(k)(x-a)^{k+1}}{k+1} \text{ for all } x \text{ in } (a-r, a+r).$$

This theorem tells us that power series can be integrated and differentiated term by term as can polynomials.

Example 18.   In Theorem 13.10 we proved that

(13.32) $$\frac{1}{1-x} = \lim_{n \to \infty} \sum_{k=0}^{n} x^k \text{ for all } x \text{ in } (-1, 1).$$

(**a**) If we differentiate this power series, we discover that for all x in $(-1, 1)$,

$$\frac{1}{(1-x)^2} =^1 \lim_{n \to \infty} \sum_{k=1}^{n} k x^{k-1}$$

$$=^2 \lim_{n \to \infty} \sum_{j=0}^{n} (j+1) x^j.$$

(1) Theorem 13.20i.
(2) Substituting $j = k - 1$.

(**b**) If we integrate the power series (13.32) we arrive at

$$\int_0^x \frac{1}{1-t} = \lim_{n \to \infty} \sum_{k=0}^{n} \frac{x^{k+1}}{k+1}$$

$$= \lim_{n \to \infty} \sum_{j=1}^{n} \frac{x^j}{j}.$$

Since $\int_0^x \frac{1}{1-t} = [-\ln(1-t)]_0^x = -\ln(1-x)$, we have $-\ln(1-x) =$

$\lim_{n \to \infty} \sum_{j=1}^{n} \frac{x^j}{j}$ for x in $(-1, 1)$. ∎

The last theorem in this section will demonstrate a basic connection between power series, convergent on an interval $(a - r, a + r)$, and Taylor series representations. Indeed, it asserts that if $f(x) = \lim_{n \to \infty} \sum_{k=0}^{n} t(k)(x - a)^k$ is convergent on $(a - r, a + r)$, then f is infinitely differentiable on $(a - r, a + r)$ and $\left\{ \sum_{k=0}^{n} t(k)(x - a)^k \right\}$ is the Taylor series representation of f on $(a - r, a + r)$.

THEOREM 13.21   (Uniqueness of a Power Representation.)   Let $r > 0$, t be a sequence of real numbers, and $f : \mathcal{R} \to \mathcal{R}$ such that

$$f(x) = \lim_{n \to \infty} \sum_{k=0}^{n} t(k)(x - a)^k \text{ for all } x \text{ in } (a - r, a + r).$$

Then
i. f is infinitely differentiable on $(a - r, a + r)$
and
ii. $t(k) = \dfrac{f^{(k)}(a)}{k!}$.

Consequently, $\left\{ \sum\limits_{k=0}^{n} t(k)(x-a)^k \right\}$ is the Taylor series representation of $f(x)$ about a.

Proof. We shall prove (i) and (ii) simultaneously. We have previously noted that $f(a) = t(0)$. By Theorem 13.20,

$$f'(x) = \lim_{n \to \infty} \sum_{k=1}^{n} kt(k)(x-a)^{k-1} \text{ for } x \text{ in } (a-r, a+r).$$

Thus $f'(a) = t(1)$. By successive differentiations (assured by Theorem 13.20) we have

$$f^{(j)}(x) = \lim_{n \to \infty} \sum_{k=j}^{n} k(k-1) \cdots (k-j+1)t(k)(x-a)^{k-j}.$$

Thus $f^{(j)}(a) = j!\, t(j)$, or equivalently, $t(j) = \dfrac{f^{(j)}(a)}{j!}$. Finally,

$$\sum_{k=0}^{n} t(k)(x-a)^k \overset{1}{=} \sum_{k=0}^{n} \frac{f^{(k)}(a)}{k!}(x-a)^k$$
$$\overset{2}{=} {}_n\mathsf{T}_a f(x).$$

(1) Since $t(k) = \dfrac{f^{(k)}(a)}{k!}$ for each k.

(2) Definition of ${}_n\mathsf{T}_a f$.

Example 19. In Example 15 we showed that

(13.33) $$e^{x^2} = \lim_{n \to \infty} \sum_{k=0}^{n} \frac{x^{2k}}{k!} \quad \text{for all } x,$$

but we did not show directly that this is the Taylor series representation of e^{x^2}. We can do this easily now. We know that $\left\{ \sum\limits_{k=0}^{n} \dfrac{x^{2k}}{k!} \right\}$ converges for all x to $f(x) = e^{x^2}$. By Theorem 13.21 we are assured that Equation 13.33 gives the Taylor series representation of e^{x^2} about 0. Examination of Equation 13.33 then discloses that the coefficients of each odd power term x^{2k+1} is zero, whereas the coefficient of the even power x^{2k} is $1/(k!)$. Accordingly, we use Theorem 13.21 to write

$$t(2k) = \frac{f^{(2k)}(0)}{(2k)!} = 1/(k!),$$

$$t(2k+1) = \frac{f^{(2k+1)}(0)}{(2k+1)!} = 0, \quad k = 0, 1, 2, \ldots .$$

Thus we have,

$$D^{(2k)} e^{x^2}\Big|_{x=0} = f^{(2k)}(0) = (2k)!/k!,$$

and

$$D^{(2k+1)} e^{x^2}\Big|_{x=0} = f^{(2k+1)}(0) = 0, \quad k = 0, 1, 2, \ldots. \quad \blacksquare$$

Example 20. Consider the function $f(x) = \dfrac{1}{1+x^2}$.

We wish to determine the values of successive derivatives of f. Substituting $-x^2$ for x in Equation 13.32, we arrive at

$$\frac{1}{1+x^2} = \lim_{n \to \infty} \sum_{k=0}^{n} (-1)^k x^{2k} \quad \text{for} \quad -1 < x < 1.$$

Thus, by Theorem 13.21ii,

$$D^{(2k+1)} \left(\frac{1}{1+x^2} \right)\Big|_{x=0} = 0 \quad \text{for} \quad k = 0, 1, \ldots,$$

and

$$D^{(2k)} \left(\frac{1}{1+x^2} \right)\Big|_{x=0} = (2k)! \, (-1)^k \quad \text{for} \quad k = 0, 1, \ldots. \quad \blacksquare$$

In the two preceding examples we used the uniqueness of the Taylor representation of $f(x)$ about a to evaluate the successive derivatives of f at a. In the next example we shall use the Taylor series representation of $f(x)$ to evaluate an integral, unattainable by the techniques of Chapters 8 to 11.

Example 21. Consider the integral $\int_0^x e^{t^2}$.

By Equation 13.33

$$e^{x^2} = \lim_{n \to \infty} \sum_{k=0}^{n} \frac{x^{2k}}{k!} \quad \text{for all } x.$$

Thus, by Theorem 13.20ii,

$$\int_0^x e^{t^2} = \lim_{n \to \infty} \sum_{k=0}^{n} \frac{x^{2k+1}}{(2k+1)k!}.$$

The last expression will allow us to numerically approximate $\int_0^x e^{t^2}$ for any fixed value of x. For example, if $x = 1$, the sequence $\left\{ \sum_{k=0}^{n} \dfrac{1}{(2k+1)k!} \right\}$ approximates the number $\int_0^1 e^{t^2}$. $\quad \blacksquare$

EXERCISES

Use the Taylor series representations of $\dfrac{1}{1-x}$, $\sin x$, $\cos x$, or e^x about 0 to determine (a) the Taylor series representation of $f(x)$ about 0 and (b) the values of all derivatives of f at 0:

1. $f(x) = \dfrac{1}{1 + 9x^2}$.

2. $f(x) = \dfrac{x}{1 + x^2}$.

3. $f(x) = \arctan x$.

4. $f(x) = x \sin x^2$.

5. $f(x) = \sin x \cos x$.

6. $f(x) = x^2 e^{x^3}$.

Use the Taylor series representations of $\dfrac{1}{1 - x}$, $\sin x$, $\cos x$, or e^x about 0 to evaluate the indicated integrals:

7. $\displaystyle\int_0^2 e^{-x^3}$.

8. $\displaystyle\int_0^{1/2} \arctan x$.

9. $\displaystyle\int_0^3 x^2 \sin x^2$.

10. $\displaystyle\int_0^1 \cos(x^4)$.

ANSWERS

1. $\dfrac{1}{1 + 9x^2} - \displaystyle\lim_{n \to \infty} \sum_{k=0}^{n} (-9)^k x^{2k}$ for $-\tfrac{1}{3} < x < \tfrac{1}{3}$

$D^{(2k+1)}f\big|_{x=0} = 0 \quad$ for $\quad k = 0, 1, \ldots$

$D^{(2k)}f\big|_{x=0} = (-9)^k (2k)! \quad$ for $\quad k = 0, 1, \ldots$.

3. $\arctan x = \displaystyle\int_0^x \dfrac{1}{1 + t^2} = \lim_{n \to \infty} \sum_{k=0}^{n} \dfrac{(-1)^k x^{2k+1}}{(2k + 1)}$ for $-1 < x < 1$

$D^{(2k+1)}f\big|_{x=0} = (-1)^k (2k)! \quad$ for $\quad k = 0, 1, \ldots$

$D^{(2k)}f\big|_{x=0} = 0 \quad$ for $\quad k = 0, 1, \ldots$.

5. $\sin x \cos x = \tfrac{1}{2} \sin 2x = \displaystyle\lim_{n \to \infty} \sum_{k=0}^{n} \dfrac{(-1)^k 4^k x^{2k+1}}{(2k + 1)!}$

$D^{(2k+1)}f\big|_{x=0} = (-4)^k \quad$ for $\quad k = 0, 1, \ldots$

$D^{(2k)}f\big|_{x=0} = 0 \quad$ for $\quad k = 0, 1, \ldots$.

7. $\displaystyle\int_0^2 e^{-x^3} = \lim_{n \to \infty} \sum_{k=0}^{n} \dfrac{(-1)^k 2^{(3k+1)}}{(3k + 1)k!}$.

9. $\displaystyle\int_0^3 x^2 \sin x^2 = \lim_{n \to \infty} \sum_{k=0}^{n} \dfrac{(-1)^k 3^{(4k+5)}}{(4k + 5)(2k + 1)!}$.

PART IV

DIFFERENTIAL CALCULUS OF $\mathcal{R}^n \to \mathcal{R}^m$ FUNCTIONS

14

LINEAR SUBSPACES AND
LINEAR TRANSFORMATIONS

INTRODUCTION

In this chapter we shall return to the study of $\mathcal{R}^n$ which we began in Chapters 1 and 2. In those chapters we emphasized the algebraic properties of $\mathcal{R}^n$, as well as the geometry of lines and planes.

In this chapter we shall investigate various other facets of the structure of $\mathcal{R}^n$. Many of the concepts presented, as will be seen in Section 5, can be generalized and put in the more sophisticated setting of an abstract vector space. Many of the proofs presented are valid for any vector space.

One of the most important concepts studied in this chapter is that of dimension. Every reader should have some intuitive feeling for this concept. He should think of a line as being one-dimensional and a plane as being two-dimensional. This feeling will motivate the mathematical definition of dimension to be given in Section 7 of this chapter.

1. LINEAR INDEPENDENCE

In Chapter 1 the concept of parallelism of two vectors was defined and discussed. We said that vectors A and B were parallel if one of them was a scalar multiple of the other. Another manner of describing this situation is to say that one of the vectors is *dependent* on the other. We also defined the term plane, a subset $\mathcal{M}$ of the form $\{A + tB + sC \mid t \text{ and } s \text{ scalar}\}$, where B and C are nonparallel vectors. Thus we see that every element X_0 of $\mathcal{M}$ is *dependent* upon A, B, and C, in the sense there exist scalars t_0 and s_0 such that

$X_0 = A + t_0 B + s_0 C$. The following two definitions describe mathematical ideas which generalize the situations we have discussed in this paragraph and which generalize Definitions 1.28, p. 43, and 1.29, p. 44.

DEFINITION 14.1 (Linear Combination.) Let A, $X_1, \ldots, X_k$ $(k \geq 1)$ be elements of $\mathcal{R}^n$. A is a *linear combination* of the elements of $\{X_1, \ldots, X_k\}$ if and only if there exist scalars $a_1, \ldots, a_k$ such that

$$A = \sum_{i=1}^{k} a_i X_i.$$

For example, $(-1, 8, 2)$ is a linear combination of the elements of $\{(1, 2, 4), (-1, 3, -1)\}$ since $(-1, 8, 2) = (1)(1, 2, 4) + (2)(-1, 3, -1)$.

DEFINITION 14.2 (Linear Dependence and Independence.) Let $\mathcal{S}$ be a subset of $\mathcal{R}^n$. $\mathcal{S}$ is *linearly dependent* if and only if $\mathcal{S}$ is nonempty and

    i. if $\mathcal{S}$ has only one element A, then $A = O_n$,

    ii. if $\mathcal{S}$ has two or more elements, then at least one element of $\mathcal{S}$ is a linear combination of other elements of $\mathcal{S}$.

A subset $\mathcal{S}$ of $\mathcal{R}^n$ is *linearly independent* if and only if $\mathcal{S}$ is not linearly dependent.

Let us consider some implications of Definition 14.2. Since a necessary condition for a set $\mathcal{S}$ to be linearly dependent is that $\mathcal{S}$ be nonempty, we know that the empty set is a linearly independent set. Moreover, a singleton set $\{A\}$, where $A \in \mathcal{R}^n$ and $A \neq O_n$, is a linearly independent subset of $\mathcal{R}^n$. Finally, any subset $\mathcal{S}$ of $\mathcal{R}^n$, consisting of two or more elements, is linearly independent if and only if no element of $\mathcal{S}$ is a linear combination of other elements of $\mathcal{S}$.

Notice that the only difference between Definition 1.28 and Definition 14.1 is that we now allow ourselves to talk about the linear combinations of a singleton set $\{X_1\}$. Any such combination is merely a scalar multiple of X_1. Definition 14.2 generalizes Definition 1.28 in allowing us to talk about a singleton set $\{A\}$ as being linearly independent or linearly dependent.

In view of the fact that it is tedious to apply the definition of linear dependence, we shall next develop a theorem which is equivalent to the definition for the case in which we have a finite nonempty set $\mathcal{S}$.

THEOREM 14.3 (Criterion for Dependence.) Let $\mathcal{S} = \{X_1, \ldots, X_k\}$, where $\mathcal{S}$ is a nonempty subset of $\mathcal{R}^n$. $\mathcal{S}$ is linearly dependent if and only if there exist scalars $a_1, \ldots, a_k$, at least one of which is nonzero, such that

$$a_1 X_1 + \cdots + a_k X_k = O_n.$$

Proof. The theorem is obviously true for the case $k = 1$. Therefore, we assume that $k \geq 2$. We give the proof in two parts. Part (i) for the "only if" and part (ii) for the "if."

i. First let us assume that S is linearly dependent. We shall find scalars $a_1, \ldots, a_k$, at least one of which is nonzero, such that $a_1 X_1 + \cdots + a_k X_k = O_n$.

By Definition 14.2, there exists an integer i, $1 \leq i \leq k$, such that X_i is a linear combination of other elements of S. By Definition 14.1 there exist scalars $a_1, \ldots, a_{i-1}, a_{i+1}, \ldots, a_k$ such that

$$X_i = a_1 X_1 + \cdots + a_{i-1} X_{i-1} + a_{i+1} X_{i+1} + \cdots + a_k X_k.$$

Subtracting X_i from both members of this equation, we have

$$O_n = a_1 X_1 + \cdots + a_{i-1} X_{i-1} - X_i + a_{i+1} X_{i+1} + \cdots + a_k X_k,$$

which is just what we seek if we set $a_i = -1$. (Note: a_i is thus nonzero.)

ii. Now assume that there exist scalars $b_1, \ldots, b_k$ such that $b_i \neq 0$ for some integer i and

(14.1) $$b_1 X_1 + \cdots + b_k X_k = O_n.$$

We shall show that X_i is a linear combination of the elements of $\{X_1, \ldots, X_{i-1}, X_{i+1}, \ldots, X_k\}$. Thus by Definition 14.2, $\{X_1, \ldots, X_k\}$ is linearly dependent.

By subtracting $b_i X_i$ from both members of Equation 14.1 and multiplying both members by $-\dfrac{1}{b_i}$, we have

$$X_i = \left(-\frac{b_1}{b_i}\right) X_1 + \cdots + \left(-\frac{b_{i-1}}{b_i}\right) X_{i-1}$$
$$+ \left(-\frac{b_{i+1}}{b_i}\right) X_{i+1} + \cdots + \left(-\frac{b_k}{b_i}\right) X_k.$$

Thus X_i is a linear combination of the other elements of S. ∎

Notice that if $\{X_1, \ldots, X_k\}$ is a nonempty subset of $\mathcal{R}^n$, it is always possible to find scalars $a_1, \ldots, a_k$ such that

(14.2) $$a_1 X_1 + \cdots + a_k X_k = O_n,$$

by choosing $a_i = 0$ for all i, $i = 1, \ldots, k$. The force of Theorem 14.3 lies in the fact that when $\{X_1, \ldots, X_k\}$ is linearly dependent, there exist scalars $a_1, \ldots, a_k$, at least one of which is nonzero, for which Equation 14.2 is true.

COROLLARY 14.4 (Criterion for Independence.) Let S be a nonempty subset of $\mathcal{R}^n$, $S = \{X_1, \ldots, X_k\}$. S is linearly independent if and only if the only choice of scalars $a_1, \ldots, a_k$ for which $a_1 X_1 + \cdots + a_k X_k = O_n$ is the trivial choice, that is, $a_1 = \cdots = a_k = 0$.

Proof. We prove the corollary by the following chain of implications: S is linearly independent

$\Leftrightarrow^1$ S is not linearly dependent

$\Leftrightarrow^2$ there exist no scalars $a_1, \ldots, a_k$, at least one of which is nonzero, such that $a_1 X_1 + \cdots + a_k X_k = O_n$

$\Leftrightarrow^3$ the only choice of scalars $a_1, \ldots, a_k$ for which $a_1 X_1 + \cdots + a_k X_k = O_n$ is the trivial choice: $a_1 = \cdots = a_k = 0$.

    (1)   Definition 14.2.
    (2)   Theorem 14.3.
    (3)   Logical equivalence. ∎

With Theorem 14.3 and Corollary 14.4 we are able to develop the following method for determining whether a subset $\{X_1, \ldots, X_k\}$ of $\mathcal{R}^n$ is linearly independent or linearly dependent. We assume that $a_1, \ldots, a_k$ are scalars such that

$$a_1 X_1 + \cdots + a_k X_k = O_n.$$

If we can *prove* that each of these scalars *must* be zero ($a_i = 0$, $i = 1, \ldots, k$), then we know that $\{X_1, \ldots, X_k\}$ is linearly independent. On the other hand if we can find a set of scalars $a_1, \ldots, a_k$ such that at least one a_i is nonzero we know that $\{X_1, \ldots, X_k\}$ is linearly dependent.

Example I. We wish to determine whether the set $\{(2, -4, 16), (3, 6, -12), (4, -4, 20)\}$ is linearly independent. To do this we assume that a, b, and c are scalars such that

$$(14.3) \qquad a(2, -4, 16) + b(3, 6, -12) + c(4, -4, 20) = (0, 0, 0).$$

Using the definitions of scalar multiplication, addition, and equality of vectors, we find that Equation 14.3 is equivalent to the following system of equations:

$$(14.4) \qquad\qquad 2a + 3b + 4c = 0,$$

$$(14.5) \qquad\qquad -4a + 6b - 4c = 0,$$

$$(14.6) \qquad\qquad 16a - 12b + 20c = 0.$$

If we multiply Equation 14.4 by 4 and add this to Equation 14.6, we have

$$(14.7) \qquad\qquad 24a + 36c = 0.$$

If we multiply Equation 14.5 by 2 and add this to Equation 14.6, we have

(14.8) $$8a + 12c = 0.$$

We notice that Equations 14.7 and 14.8 have the same set of solutions, namely any scalars a and c such that $a = -(3/2)c$. Substituting $a = -(3/2)c$ in Equation 14.4, 14.5, or 14.6 and solving for b, we find that $b = -\frac{1}{3}c$. Let $c = 6$. Then $a = -\frac{3}{2}(6) = -9$ and $b = -\frac{1}{3}(6) = -2$. We substitute these values for a, b, and c in Equation 14.3 and find that they satisfy Equation 14.3: that is,

$$-9(2, -4, 16) - 2(3, 6, -12) + 6(4, -4, 20) = (0, 0, 0).$$

Therefore, by Theorem 14.3 we know that this set of vectors is linearly dependent. ∎

The reader should notice that the particular values of a, b, and c, which we found in order to satisfy Equation 14.3 in the preceding example, are not the only values satisfying Equation 14.3. Once we know that $a = -(3/2)c$ and $b = -\frac{1}{3}c$, we may choose any nonzero value for c and evaluate a and b accordingly. For example, $c = 1$, $a = -3/2$, and $b = -1/3$ satisfy Equation 14.3.

Example 2. We shall show, using Theorem 14.3, that the set $\{(1, 0, 0),$ $(0, 1, 0), (0, 0, 1)\}$ is linearly independent. Assume that a, b, and c are scalars such that

$$a(1, 0, 0) + b(0, 1, 0) + c(0, 0, 1) = (0, 0, 0).$$

This vector equation is equivalent to the following system of linear equations:

$$\begin{cases} a = 0, \\ b = 0, \\ c = 0. \end{cases}$$

Therefore, since all three scalars *must* equal zero, we know that the set $\{(1, 0, 0), (0, 1, 0), (0, 0, 1)\}$ is linearly independent. ∎

There are many theorems which may be proven about linearly independent or linearly dependent sets of vectors in $\mathcal{R}^n$. The following two theorems will be of use to us in succeeding sections of this chapter.

THEOREM 14.5 If S is a linearly independent set of vectors in $\mathcal{R}^n$, then every subset of S is linearly independent.

Proof. Let S' be a subset of the linearly independent set S. If $S' = \varnothing$, then by Definition 14.2, S' is linearly independent. We shall assume then that S' is linearly dependent and strive to reach a contradiction. If S' is linearly dependent, then there exist a vector A in S' and a subset $\{X_1, \ldots, X_k\}$ of S' such that A is a linear combination of the elements of $\{X_1, \ldots, X_k\}$ and $A \notin \{X_1, \ldots, X_k\}$. But since A is also an element of S and $\{X_1, \ldots, X_k\} \subset S$, this implies (by Definition 14.2) that S is linearly dependent. Our hypothesis being that S is linearly independent, our assumption that S' is linearly dependent must be false. ∎

The following corollary is logically equivalent to Theorem 14.5, and the proof will be left as an exercise for the reader.

COROLLARY 14.6 If S is a subset of $\mathcal{R}^n$ and S' is a linearly dependent subset of S, then S is linearly dependent.

THEOREM 14.7 Let S be a linearly independent set of vectors in $\mathcal{R}^n$, $S = \{X_1, \ldots, X_k\}$, and let A be a linear combination of elements of S, $A \neq O_n$. There exists an element X_i of S such that

i. the set $S' = \{X_1, \ldots, X_{i-1}, A, X_{i+1}, \ldots, X_k\}$ is linearly independent,

ii. every linear combination of the elements of S is a linear combination of the elements of S', and

iii. every linear combination of the elements of S' is a linear combination of the elements of S.

Proof. Since A is a linear combination of the elements of S (by Definition 14.1), there exist scalars $a_1, \ldots, a_k$ such that $A = \sum_{j=1}^{k} a_j X_j$. Since $A \neq O_n$, at least one of these scalars is nonzero: call it a_i. We shall show that the corresponding X_i is the one we seek.

We shall prove (i) by an indirect argument. If $\{X_1, \ldots, X_{i-1}, A, X_{i+1}, \ldots, X_k\}$ is linearly dependent, then there exist scalars $b_1, \ldots, b_k$, at least one of which is nonzero, such that

$$(14.9) \quad b_1 X_1 + \cdots + b_{i-1} X_{i-1} + b_i A + b_{i+1} X_{i+1} + \cdots + b_k X_k = O_n.$$

If $b_i = 0$, then by Theorem 14.3 $\{X_1, \ldots, X_{i-1}, X_{i+1}, \ldots, X_k\}$ is linearly dependent, which by Corollary 14.6 contradicts our hypothesis that S is linearly independent. If $b_i \neq 0$, then substituting $A = \sum_{j=1}^{k} a_j X_j$ in Equation 14.9, we have

$$(b_1 + b_i a_1) X_1 + \cdots + (b_{i-1} + b_i a_{i-1}) X_{i-1} + b_i a_i X_i$$
$$+ (b_{i+1} + b_i a_{i+1}) X_{i+1} + \cdots + (b_k + b_i a_k) X_k = O_n.$$

However, since $b_i a_i \neq 0$, Theorem 14.3 implies that $\{X_1, \ldots, X_k\}$ is linearly dependent, which is again a contradiction of our hypothesis. This leads us to the desired conclusion: namely, that $S' = \{X_1, \ldots, X_{i-1}, A, X_{i+1}, \ldots, X_k\}$ is linearly independent.

Conclusion (ii) of the theorem follows from the fact that in any linear combination of the elements of S we may substitute for X_i:

$$X_i = -\frac{1}{a_i}(a_1 X_1 + \cdots + a_{i-1} X_{i-1} + a_{i+1} X_{i+1} + \cdots + a_k X_k - A).$$

We argue in a similar fashion for conclusion (iii). ■

We paraphrase Theorem 14.7 as follows: whenever $\{X_1, \ldots, X_k\}$ is a linearly independent set of vectors and A is a nonzero linear combination of the elements of $\{X_1, \ldots, X_k\}$, then A may be substituted for some element of the set; the set resulting from this substitution is also linearly independent, and one can form exactly the same set of linear combinations with the original set of vectors and the new set of vectors.

We now offer a word of caution. Theorem 14.7 does *not* say that we can substitute A for *any* X_i and obtain the desired result. However, our proof shows us that when $A = \sum\limits_{j=1}^{k} a_j X_j$, we may substitute A for any X_j for which $a_j \neq 0$.

EXERCISES

In each of the following, determine which of the vectors X are linear combinations of the set S.

1. $S = \{(1, 1), (2, 2)\}$,
   (a) $X = (1/2, 1/2)$,
   (b) $X = (0, 0)$,
   (c) $X = (-1, 1)$.

2. $S = \{(1, 0, 1), (2, 3, 0)\}$,
   (a) $X = (-1, -3, 1)$,
   (b) $X = (0, 0, 2)$,
   (c) $X = (6, 3, 4)$.

3. $S = \{(1, 0, 0), (0, 1, 0), (0, 0, 1)\}$,
   (a) $X = (11, 15, -4)$,
   (b) $X = (5, -1, 0)$,
   (c) $X = (x, y, z)$.

Determine which of the following sets are linearly independent and which are linearly dependent:

4. $\{(1, 2), (-2, -4)\}$.

5. $\{(0, 0), (3, 2)\}$.

6. $\{(1, 2), (-1, 3)\}$.

7. $\{(1, 0, 0), (1, 2, 1), (0, 1, 0)\}$.

8. $\{(-1, 2), (0, 1), (3, 5)\}$.

9. $\{(1, 2, 1), (3, -1, 5), (-5, 4, -11)\}$.

10. Prove that a subset $\{A, B\}$ consisting of two elements of $\mathcal{R}^n$ is linearly dependent if and only if A is parallel to B.

11. Give any example of a subset $\{A, B, C\}$ of $\mathcal{R}^3$ which is linearly dependent but such that no pair of vectors is parallel.

12. Prove Corollary 14.6.

ANSWERS

1. (a), (b).   **2.** (a), (c).   **3.** All.   **4.** Dependent.   **6.** Independent.
8. Dependent.

2. LINEAR SUBSPACES

There are certain subsets of $\mathcal{R}^n$ which behave algebraically in the same fashion as $\mathcal{R}^m$, for some $m \leq n$. For example, we frequently think of the xy-plane in $\mathcal{R}^3$ as being a copy of $\mathcal{R}^2$; or we think of the x-axis or the y-axis in $\mathcal{R}^2$ as a copy of $\mathcal{R}^1$. Perhaps there are other subsets of $\mathcal{R}^2$ and $\mathcal{R}^3$ which satisfy all the properties of a vector space (see Theorem 1.14, p. 16). Two of the very basic properties of $\mathcal{R}^n$ are that $X + Y \in \mathcal{R}^n$ and $rX \in \mathcal{R}^n$ whenever X and Y are elements of $\mathcal{R}^n$ and r is a scalar. The following definition gives a name to any subset of $\mathcal{R}^n$ which satisfies these two properties:

DEFINITION 14.8   (Linear Subspace.) Let M be a subset of $\mathcal{R}^n$. $\mathcal{M}$ is a *linear (vector) subspace of* $\mathcal{R}^n$ if and only if
   i. $\mathcal{M}$ is nonempty,
   ii. $X + Y \in \mathcal{M}$ whenever $X \in \mathcal{M}$ and $Y \in \mathcal{M}$, and
   iii. $rX \in \mathcal{M}$ whenever $X \in \mathcal{M}$ and $r \in \mathcal{R}$.

The use of the word linear in this context is reasonable, for if $\mathcal{M}$ is a linear subspace of $\mathcal{R}^n$ and X and Y are distinct elements of $\mathcal{M}$, it is easy to verify that the line passing through X and Y is entirely contained in $\mathcal{M}$.

Example 3.   Let $\mathcal{M} = \{(x, y) \mid (x, y) \in \mathcal{R}^2 \text{ and } y = 2x\}$. Thus $\mathcal{M}$ is the set of all elements of $\mathcal{R}^2$ for which the second component is twice the first. We shall show that $\mathcal{M}$ is a linear subspace of $\mathcal{R}^2$.
   i. $\mathcal{M} \neq \varnothing$ since $(0, 0) \in \mathcal{M}$.
   ii. Let (a, b) and (c, d) be elements of $\mathcal{M}$. We shall show that $(a, b) + (c, d) \in \mathcal{M}$. Since $(a, b) \in \mathcal{M}$, $b = 2a$ and since $(c, d) \in \mathcal{M}$, $d = 2c$. Thus $b + d = 2a + 2c$ and we have $(a + c, b + d) \in \mathcal{M}$. Since $(a + c, b + d) = (a, b) + (c, d)$, we know that $(a, b) + (c, d) \in \mathcal{M}$.
   iii. Let $(a, b) \in \mathcal{M}$ and $r \in \mathcal{R}$. We shall show that $r(a, b) \in \mathcal{M}$ by showing that $rb = 2(ra)$. Since $(a, b) \in \mathcal{M}$, $b = 2a$. Therefore, $rb = r(2a) = 2(ra)$, as desired.

Therefore $\mathcal{M}$ satisfies the three conditions of Definition 14.8 and is a linear subspace of $\mathcal{R}^2$. ∎

In view of the rather weak conditions imposed upon a set by Definition 14.8 in order that it be a linear subspace, one is justified in being surprised by the following theorem. However, remember that a linear subspace of $\mathcal{R}^n$ *is* a subset of $\mathcal{R}^n$ and that $\mathcal{R}^n$ itself has a powerful algebraic structure.

THEOREM 14.9 Let $\mathcal{M}$ be a nonempty subset of $\mathcal{R}^n$. $\mathcal{M}$ is a linear subspace of $\mathcal{R}^n$ if and only if the elements of $\mathcal{M}$ satisfy all of the vector space properties. (See Theorem 1.14, p. 16.)

Proof. Certainly if the elements of $\mathcal{M}$ satisfy all of these properties, then $\mathcal{M}$ satisfies Definition 14.8 and hence is a linear subspace.

Assume now that $\mathcal{M}$ is a linear subspace of $\mathcal{R}^n$. Since $\mathcal{M}$ is nonempty, there exists a vector X in $\mathcal{M}$. By property (iii) of Definition 14.8, $(-1)X \in \mathcal{M}$ and since $0(X) = O_n$ and $0(X) \in \mathcal{M}$, we have $O_n \in \mathcal{M}$. Therefore, $\mathcal{M}$ contains the zero vector of $\mathcal{R}^n$ and the additive inverse of every element of $\mathcal{M}$. The other properties listed in Theorem 1.14, p. 16, are satisfied by the elements of $\mathcal{M}$ because these are also elements of $\mathcal{R}^n$. ∎

Example 4. We shall show that the set $\mathcal{M} = \{(x, y, z) \mid z = x + y + 1\}$ is not a linear subspace of $\mathcal{R}^3$. To demonstrate the general method, we shall start by proving that this set does not satisfy one of the properties of Definition 14.8. Let $(a, b, c) \in \mathcal{M}$ and $(d, e, f) \in \mathcal{M}$. Then

$$(14.10) \qquad c = a + b + 1$$

and

$$(14.11) \qquad f = d + e + 1.$$

In order that $(a, b, c) + (d, e, f) = (a + d, b + e, c + f)$ be an element of $\mathcal{M}$, it is necessary that $(c + f) = (a + d) + (b + e) + 1$. But, in view of Equations 14.10 and 14.11 $(c + f) = (a + d) + (b + e) + 2$. Therefore, $(a, b, c) + (d, e, f) \notin \mathcal{M}$. ∎

Another very easy way by which we may see that the set $\mathcal{M}$ in the preceding example is not a linear subspace is to observe that $(0, 0, 0) \notin \mathcal{M}$ (for if $(x, y, z) \in \mathcal{M}$ and $x = y = 0$, then $z = 1$). The following example will show that it is possible for a set $\mathcal{M}$ not to be a linear subspace even though $O_n \in \mathcal{M}$.

Example 5. Let $\mathcal{M} = \{(x, y) \mid y = x^2\}$. Note that $(0, 0) \in \mathcal{M}$. If $(a, b) \in \mathcal{M}$ and $(c, d) \in \mathcal{M}$, then $b = a^2$ and $d = c^2$. In order that $(a, b) + (c, d) = (a + c, b + d)$ be an element of $\mathcal{M}$, it is necessary that $b + d = (a + c)^2$. However, it is very easy to find examples of scalars a, b, c, and d such that $b = a^2$, $d = c^2$, but $b + d \neq (a + c)^2$, for example, $a = b = c = d = 1$. ∎

EXERCISES

Determine which of the following are linear subspaces.

1. $\{t(1, 0) \mid t \text{ scalar}\}$.

2. $\{(x, y) \mid 2x + 3y = 0\}$.

3. $\{(x, y, z) \mid x^2 + y = 3z\}$.

4. $\{(x, y) \mid y \geq 0\}$.

5. $\{(x, y, z) \mid 2x - 3y + 4z = 0\}$.

6. $\{(x, y) \mid x = 0\}$.

7. $\{r(1, 2, -1) + s(1, 0, 0) \mid r$ and s scalar$\}$.

8. $\{(x, y) \mid (x, y) \cdot (-1, 3) = 0\}$.

9. $\{(x, y, z) \mid (x, y, z) \cdot (-1, 3, 2) = 4\}$.

10. $\{(0, 0)\}$.

11. Let $\mathcal{M}$ and $\mathcal{N}$ be linear subspaces of $\mathcal{R}^n$. Prove that $\mathcal{M} \cap \mathcal{N}$ is a linear subspace of $\mathcal{R}^n$. (Query: How do you know that $\mathcal{M} \cap \mathcal{N}$ is nonempty?)

12. Let $\mathcal{M}$ and $\mathcal{N}$ be linear subspaces of $\mathcal{R}^n$. Prove that $\mathcal{M} \cup \mathcal{N}$ is a linear subspace of $\mathcal{R}^n$ if and only if $\mathcal{M} \subset \mathcal{N}$ or $\mathcal{N} \subset \mathcal{M}$. (Hint: Let $X \in \mathcal{M} \setminus \mathcal{N}$ and $Y \in \mathcal{N} \setminus \mathcal{M}$. Where is $X + Y$?)

13. Let $\mathcal{M} = \{O_n\}$. Prove from Definition 4.8 that $\mathcal{M}$ is a linear subspace of $\mathcal{R}^n$.

14. Show that the only linear subspaces of $\mathcal{R}^3$ are $\{O_3\}$, lines passing through the origin, planes passing through the origin, and $\mathcal{R}^3$ itself. (You may assume that any subset of $\mathcal{R}^3$ containing 4 or more vectors is linearly dependent.)

15. Determine the linear subspaces of $\mathcal{R}^2$.

16. Give an example of a subset $\mathcal{M}$ of $\mathcal{R}^2$ such that $X + Y \in \mathcal{M}$ whenever $X \in \mathcal{M}$ and $Y \in \mathcal{M}$, but $rX \notin \mathcal{M}$ for some X in $\mathcal{M}$ and r in $\mathcal{R}$.

17. Give an example of a subset of $\mathcal{R}^2$ such that $rX \in \mathcal{M}$ whenever $X \in \mathcal{M}$ and $r \in \mathcal{R}$, but $X + Y \notin \mathcal{M}$ for some X in $\mathcal{M}$ and Y in $\mathcal{M}$.

ANSWERS

2. Yes. **4.** No. **6.** Yes. **8.** Yes. **10.** Yes.

3. BASES OF LINEAR SUBSPACES

Let us consider an arbitrary linear subspace $\mathcal{M}$ of $\mathcal{R}^3$. There are several possibilities (i) $\mathcal{M} = \{O_n\}$, (ii) $\mathcal{M}$ contains a nonzero vector X_1 and every other element of $\mathcal{M}$ is a linear combination of X_1, (iii) $\mathcal{M}$ contains two linearly independent vectors X_1 and X_2 and every other element of $\mathcal{M}$ is a linear combination of X_1 and X_2, or (iv) $\mathcal{M}$ contains a set of three linearly independent vectors X_1, X_2, and X_3. In case (ii) we see that $\mathcal{M}$ is a line passing through O_n, with direction vector X_1. In case (iii) $\mathcal{M}$ is a plane, passing through O_3, with direction set $\{X_1, X_2\}$. In case (iv) $\mathcal{M} = \mathcal{R}^3$ (we will discover why this is true in Theorem 14.12). The fact that these are all of the possible descriptions of linear subspaces of $\mathcal{R}^3$ follows from Theorem 14.13 below.

Notice that in the preceding paragraph we "constructed $\mathcal{M}$" by considering how many elements could occur in a linearly independent subset of $\mathcal{M}$ and forming all possible linear combinations of the elements of this linearly independent set. We generalize this "construction process" as follows:

DEFINITION 14.10 (Basis of a Linear Subspace.) A *basis* for a linear subspace $\mathcal{M}$ of $\mathcal{R}^n$ is a set of vectors $\mathcal{S}$ such that

  i. $\mathcal{S} \subset \mathcal{M}$,
  ii. $\mathcal{S}$ is linearly independent, and
  iii. every element of $\mathcal{M}$ is a linear combination of the elements of $\mathcal{S}$.

We should note at this point that the subspace $\{O_n\}$ of $\mathcal{R}^n$ does not have a basis. If $\{O_n\}$ did have a basis $\mathcal{S}$, then by (i) of Definition 14.10, $\mathcal{S} \subset \{O_n\}$. Therefore, either $\mathcal{S} = \varnothing$ or $\mathcal{S} = \{O_n\}$. But $\{O_n\}$ does not satisfy (ii) and $\varnothing$ does not satisfy (iii).

Let us note another point related to the definition of a basis. Let $\{X_1, \ldots, X_k\}$ be a basis for a linear subspace $\mathcal{M}$ of $\mathcal{R}^n$ and let A be a nonzero linear combination of elements of $\{X_1, \ldots, X_k\}$. Thanks to Theorem 14.7, we may substitute A for a certain element X_i of $\{X_1, \ldots, X_k\}$ and the resulting set is also a basis for $\mathcal{M}$. Thus, if a linear subspace $\mathcal{M}$ has a basis, consisting of a finite number of vectors, there exist other bases consisting of the same number of vectors.

Certainly the simplest example for a basis of $\mathcal{R}^n$ itself is what is called the *natural basis* for $\mathcal{R}^n$. This is the set $\{E_1, \ldots, E_n\}$ such that each coordinate of E_j is zero except the j-th coordinate which is equal to one. For example, in $\mathcal{R}^3$ the set $\{(1, 0, 0), (0, 1, 0), (0, 0, 1)\}$ is called the natural basis. Note that this set is linearly independent and that if $(x, y, z) \in \mathcal{R}^3$, then

$$(x, y, z) = x(1, 0, 0) + y(0, 1, 0) + z(0, 0, 1).$$

Before we can fully appreciate the concept of a basis for a linear subspace of $\mathcal{R}^n$, there are several questions which should be answered: 1. Does every linear subspace of $\mathcal{R}^n$, except $\{O_n\}$, have a basis? 2. Can a linear subspace of $\mathcal{R}^n$ have two bases with different numbers of elements? 3. If $\mathcal{M}$ and $\mathcal{N}$ are linear subspaces and $\mathcal{M} \subset \mathcal{N}$, can $\mathcal{M}$ have a basis with more elements than a given basis of $\mathcal{N}$? The following lemma will put us on the road to an answer for each of these questions.

LEMMA 14.11   Let $\{X_1, \ldots, X_k\}$ be a basis for a linear subspace $\mathcal{M}$ of $\mathcal{R}^n$ and let $A \in \mathcal{M}$. If for some $p < k$, the set $\{A, X_1, \ldots, X_p\}$ is linearly independent, then there exists an integer i, $p < i \le k$, such that $\{X_1, \ldots, X_p, \ldots, X_{i-1}, A, X_{i+1}, \ldots, X_k\}$ is a basis for $\mathcal{M}$.

Proof.   Since $A \in \mathcal{M}$ and $\{X_1, \ldots, X_k\}$ is a basis for $\mathcal{M}$, there exist scalars $a_1, \ldots, a_k$ such that $A = \sum_{j=1}^{k} a_j X_j$. If each of the scalars a_j, $j = p+1, \ldots, k$, is zero, then $A = \sum_{j=1}^{p} a_j X_j$, which contradicts the fact that $\{A, X_1, \ldots, X_p\}$ is a linearly independent set. Therefore, since $A \ne O_n$ (by Theorem 14.5, p. 561), there exists a_i, $i > p$, such that $a_i \ne 0$. Using an argument identical to that used in the proof of Theorem 14.7 we can show that $\{X_1, \ldots, X_p, \ldots, X_{i-1}, A, X_{i+1}, \ldots, X_k\}$ is linearly independent and has exactly the same set of linear combinations as does $\{X_1, \ldots, X_k\}$.   ∎

THEOREM 14.12   Let $\{X_1, \ldots, X_k\}$ be a basis consisting of k vectors, for a linear subspace $\mathcal{M}$ of $\mathcal{R}^n$, and let $\{Y_1, \ldots, Y_k\}$ be a linearly independent subset consisting of k vectors of $\mathcal{M}$. Then $\{Y_1, \ldots, Y_k\}$ is also a basis for $\mathcal{M}$.

Proof.   Since $Y_1 \ne O_n$ and Y_1 is a linear combination of elements of $\{X_1, \ldots, X_k\}$, then by Theorem 14.7, we may substitute Y_1 for one of the X_i and the resulting set will also be a basis for $\mathcal{M}$. For simplicity of argument we assume that we can substitute Y_1 for X_1. (Otherwise we rename the elements of $\{X_1, \ldots, X_k\}$ so that we substitute Y_1 for the first element.)

The rest of the argument proceeds by induction. Assume that we have successfully replaced $X_1, \ldots, X_p$ by $Y_1, \ldots, Y_p$, where $p < k$, and obtained a basis for $\mathcal{M}$, $\{Y_1, \ldots, Y_p, X_{p+1}, \ldots, X_k\}$. Letting $A = Y_{p+1}$, by Lemma 14.11 we may replace one of the X_i, $p + 1 \le i \le k$, by A and the resulting set will still be a basis for $\mathcal{M}$. Again, either we can assume that we replace X_{p+1} with Y_{p+1} or we can reorder $\{X_{p+1}, \ldots, X_k\}$ so that we may replace the first one with Y_{p+1}. Finally, since $\{X_1, \ldots, X_k\}$ is a finite set, we shall obtain the basis $\{Y_1, \ldots, Y_k\}$ for $\mathcal{M}$ at the k-th step.   ∎

One of the important results of this theorem is that, whenever $\{Y_1, \ldots, Y_n\}$ is a linearly independent subset of $\mathcal{R}^n$, $\{Y_1, \ldots, Y_n\}$ is a basis for $\mathcal{R}^n$. This follows from Theorem 14.12 and the fact that the natural basis for $\mathcal{R}^n$ has n elements. This, of course, will much simplify the problem of determining whether a given set of vectors in $\mathcal{R}^n$ is indeed a basis for $\mathcal{R}^n$. For example, $\{(1, 0, 2), (1, -1, 0), (0, 0, 1)\}$ is a basis for $\mathcal{R}^3$ since it is linearly independent, consists of *three* vectors, and since we know that $\{(1, 0, 0),$

$(0, 1, 0)$, $(0, 0, 1)\}$ is a set of *three* vectors forming a basis for $\mathscr{R}^3$. More generally, we have the following answer to the second and third questions posed in the paragraph preceding Lemma 14.11.

THEOREM 14.13 If a linear subspace $\mathscr{M}$ of $\mathscr{R}^n$ has a basis consisting of k elements, then every basis of $\mathscr{M}$ has exactly k elements. Moreover, any subset of $\mathscr{M}$ containing more than k elements is linearly dependent.

Proof. We shall prove the last assertion first. Let $\{X_1, \ldots, X_k\}$ be a basis for $\mathscr{M}$ and let S be a subset of $\mathscr{M}$ containing at least $k + 1$ elements, $Z_1, \ldots, Z_{k+1}$. If $\{Z_1, \ldots, Z_k\}$ is linearly dependent, then by Corollary 14.6, S is linearly dependent. If $\{Z_1, \ldots, Z_k\}$ is linearly independent, then by Theorem 14.12 $\{Z_1, \ldots, Z_k\}$ is a basis for $\mathscr{M}$. Therefore, Z_{k+1} is a linear combination of elements of $\{Z_1, \ldots, Z_k\}$; hence $\{Z_1, \ldots, Z_{k+1}\}$ is linearly dependent and, as a consequence, so also is S. We see then that in either case S must be linearly dependent.

Now let $\{Y_1, \ldots, Y_p\}$ be a second basis for $\mathscr{M}$. By the preceding we see that it is impossible for p to be greater than k, for in that case $\{Y_1, \ldots, Y_p\}$ would be linearly dependent. Therefore, $p \leq k$. Now consider the set $\{X_1, \ldots, X_p\}$. This set is linearly independent since it is a subset of the basis $\{X_1, \ldots, X_k\}$. Moreover, Theorem 14.12 tells us that this set, $\{X_1, \ldots, X_p\}$, is also a basis for $\mathscr{M}$. Thus, if $p < k$, X_{p+1} is a linear combination of $\{X_1, \ldots, X_p\}$, contradicting our assumption that $\{X_1, \ldots, X_k\}$ is a basis for $\mathscr{M}$. Therefore, we also have $p \geq k$. Since we have $h \leq k$ and $p \geq k$ we must have $p = k$. ∎

Our first question in the paragraph preceding Lemma 14.11 will be answered in Section 7 by Lemma 14.20. We shall find that whenever $\mathscr{M}$ is a linear subspace of $\mathscr{R}^n$, $\mathscr{M} \neq \{O_n\}$, $\mathscr{M}$ has a basis consisting of k elements, where $k \leq n$.

The reader will benefit by thinking of a basis $\{X_1, \ldots, X_k\}$ of a linear subspace $\mathscr{M}$ of $\mathscr{R}^n$ as a set of building blocks for $\mathscr{M}$. Every element of $\mathscr{M}$ can be obtained by taking an appropriate linear combination of elements of $\{X_1, \ldots, X_k\}$. Moreover, if we omit one of the elements of this basis and take the set of all possible linear combinations of the remaining basis elements, this set of combinations will be a proper subset of $\mathscr{M}$. In particular, we cannot write X_i as a linear combination of $\{X_1, \ldots, X_{i-1}, X_{i+1}, \ldots, X_k\}$. Thus we may regard a basis for $\mathscr{M}$ as a minimal set of elements which will generate $\mathscr{M}$ by taking all possible linear combinations.

The only examples of bases we have discussed so far have been the natural basis for $\mathscr{R}^n$ and one example of another basis for $\mathscr{R}^3$. Let us, therefore, consider other examples.

Example 6. Let $\mathcal{M} = \{r(1, 3, -2) + s(1, 0, 1) \mid r \text{ and } s \text{ scalar}\}$. It is easily checked that $\mathcal{M}$ is a linear subspace of $\mathcal{R}^3$. Note the following:

   i. $\{(1, 3, -2), (1, 0, 1)\} \subset \mathcal{M}$,

  ii. $\{(1, 3, -2), (1, 0, 1)\}$ is linearly independent, and

 iii. every element of $\mathcal{M}$ is obviously a linear combination of the elements of $\{(1, 3, -2), (1, 0, 1)\}$.

Therefore, $\{(1, 3, -2), (1, 0, 1)\}$ is by definition a basis for $\mathcal{M}$. ■

Example 7. Let $\mathcal{M} = \{(x, y, z) \mid z = 2x + 3y\}$. Using the techniques developed in Chapter 2, we rewrite this plane in the parametric form:

$$\mathcal{M} = \{t(0, 1, 3) + s(1, 0, 2) \mid t \text{ and } s \text{ scalar}\}.$$

As in the preceding example, we observe that $\{(0, 1, 3), (1, 0, 2)\}$ is a basis for $\mathcal{M}$. ■

[handwritten: $N = 2x + 3y - z = 0$]

[handwritten: $N = \begin{pmatrix} 2 \\ 3 \\ -1 \end{pmatrix} \quad \begin{pmatrix} -3 \\ 3 \\ 0 \end{pmatrix} \begin{pmatrix} 1 \\ 0 \\ -2 \end{pmatrix} \begin{pmatrix} 0 \\ 1 \\ 3 \end{pmatrix}$]

EXERCISES

Find two different bases for the linear subspace:

1. In exercise 1, Section 2.   2. In exercise 2, Section 2.

3. In exercise 5, Section 2.   4. In exercise 6, Section 2.

5. In exercise 7, Section 2.   6. In exercise 8, Section 2.

7. Let $\{A + tB \mid t \in \mathcal{R}, B \neq O_n\}$ be a line in $\mathcal{R}^n$. Prove that this line is a linear subspace of $\mathcal{R}^n$ if and only if it contains O_n.

8. Let $\mathcal{M}$ be a linear subspace of $\mathcal{R}^n$. Assume that $\{X_1, \ldots, X_k\}$ is a basis for $\mathcal{M}$ and prove that $k \leq n$. (Hint: Assume $k > n$ and use Theorem 14.13 to arrive at a contradiction.)

ANSWERS

2. $\{(3, -2)\}$.   **3.** $\{(3, 2, 0), (0, 4, 3)\}$.   **6.** $\{(3, 1)\}$.

4. LINEAR TRANSFORMATIONS

We shall digress from our study of the structure of linear subspaces. In this section we shall introduce a new type of function, called a linear transformation. The use of the word linear here is reasonable, for these functions are the ones which take the linear subspaces of one vector space into the linear subspaces of a second vector space. More particularly, they take a line in the domain space into a line or a point (which may be considered to be a degenerate line) in the range space.

Before defining precisely what we mean by a linear transformation, let us consider the function f from $\mathcal{R}^2$ into $\mathcal{R}$, defined by $f\begin{pmatrix} x \\ y \end{pmatrix} = x + y$. We note that if $\begin{pmatrix} a \\ b \end{pmatrix} \in \mathcal{R}^2$ and $\begin{pmatrix} c \\ d \end{pmatrix} \in \mathcal{R}^2$, then

$$f\left[\begin{pmatrix} a \\ b \end{pmatrix} + \begin{pmatrix} c \\ d \end{pmatrix} \right] = f\begin{pmatrix} a+c \\ b+d \end{pmatrix} = (a+c) + (b+d) = (a+b) + (c+d).$$

However, $f\begin{pmatrix} a \\ b \end{pmatrix} = a + b$ and $f\begin{pmatrix} c \\ d \end{pmatrix} = c + d$, so that $f\begin{pmatrix} a \\ b \end{pmatrix} + f\begin{pmatrix} c \\ d \end{pmatrix} = f\left[\begin{pmatrix} a \\ b \end{pmatrix} + \begin{pmatrix} c \\ d \end{pmatrix} \right]$. Similarly, we could show that $f\left[r\begin{pmatrix} a \\ b \end{pmatrix} \right] = rf\begin{pmatrix} a \\ b \end{pmatrix}$ whenever $\begin{pmatrix} a \\ b \end{pmatrix} \in \mathcal{R}^2$ and $r \in \mathcal{R}$. These two properties are exactly those we shall demand for a "linear transformation":

DEFINITION 14.14 (Linear Transformation.) Let $T:\mathcal{R}^n \to \mathcal{R}^m$, dom $T = \mathcal{R}^n$. T is a *linear transformation* if and only if
  i. $T(X + Y) = TX + TY$ whenever $X \in \mathcal{R}^n$ and $Y \in \mathcal{R}^n$ and
  ii. $T(rX) = rTX$ whenever $X \in \mathcal{R}^n$ and $r \in \mathcal{R}$.
Property (i) is called the *additive property* and property (ii) is called the *homogeneous property* of T.

The reader should note that the additive and homogeneous properties are independent of one another. There do exist examples of functions (too complicated for inclusion here) which satisfy the additive property but not the homogeneous property. On the other hand it is quite simple to find a function which satisfies the homogeneous property but not the additive property. We give one such example. Define $T:\mathcal{R}^2 \to \mathcal{R}$ by

$$T(x, y) = \begin{cases} 0 \text{ if } x \neq 0 \text{ and } y \neq 0, \\ y \text{ if } x = 0, \\ x \text{ if } y = 0. \end{cases}$$

Note that $T(1, 0) = 1$ and $T(0, 1) = 1$, but that $T[(1, 0) + (0, 1)] = T(1, 1) = 0$. Hence $T(1, 0) + T(0, 1) \neq T[(1, 0) + (0, 1)]$. We leave it to the reader to check that the homogeneous property is indeed satisfied for this function.

Example 8. Let $T:\mathcal{R}^2 \to \mathcal{R}^2$ be defined by $T\begin{pmatrix} x \\ y \end{pmatrix} = \begin{pmatrix} x+2y \\ 3x-y \end{pmatrix}$. We shall show that T satisfies both properties of Definition 14.14 Let $X = \begin{pmatrix} a \\ b \end{pmatrix}$,

$Y = \begin{pmatrix} c \\ d \end{pmatrix}$, and $r \in \mathcal{R}$. Then

i. $T(X + Y) =^1 T\begin{pmatrix} a + c \\ b + d \end{pmatrix}$

$=^2 \begin{pmatrix} (a + c) + 2(b + d) \\ 3(a + c) - (b + d) \end{pmatrix}$

$=^3 \begin{pmatrix} (a + 2b) + (c + 2d) \\ (3a - b) + (3c - d) \end{pmatrix}$

$=^4 \begin{pmatrix} a + 2b \\ 3a - b \end{pmatrix} + \begin{pmatrix} c + 2d \\ 3c - d \end{pmatrix} =^2 TX + TY$

(1) Addition of X and Y.
(2) Definition of T.
(3) Properties of real numbers.
(4) Addition of vectors.

ii. $T(rX) =^6 T\begin{pmatrix} ra \\ rb \end{pmatrix}$

$=^7 \begin{pmatrix} ra + 2(rb) \\ 3(ra) - (rb) \end{pmatrix}$

$=^8 \begin{pmatrix} r(a + 2b) \\ r(3a - b) \end{pmatrix}$

$=^9 r\begin{pmatrix} a + 2b \\ 3a - b \end{pmatrix} =^7 rTX$

(6) Scalar multiplication of X by r.
(7) Definition of T.
(8) Properties of real numbers.
(9) Scalar multiplication of vectors.

Thus we have shown (i) $T(X + Y) = TX + TY$ whenever $X \in \mathcal{R}^2$ and $Y \in \mathcal{R}^2$ and (ii) $T(rX) = rTX$ whenever $X \in \mathcal{R}^2$ and $r \in \mathcal{R}$. Consequently, T is a linear transformation. ■

Linear transformations have many "nice" properties, as we shall discover in the next few sections. The following theorem lists three of the simplest properties common to all linear transformations.

THEOREM 14.15 Let T be a linear transformation from $\mathcal{R}^n$ into $\mathcal{R}^m$. Then

i. $T(O_n) = O_m$,
ii. $T(-X) = -T(X)$ whenever $X \in \mathcal{R}^n$, and
iii. $T(X - Y) = T(X) - T(Y)$ whenever $X \in \mathcal{R}^n$ and $Y \in \mathcal{R}^n$.

Proof of i. Recall that $0 \cdot X = O_n$ whenever $X \in \mathcal{R}^n$. Thus

$$T(O_n) = T(0X)$$
$$= 0(T\ X) \quad \text{by the homogeneous property of } T$$
$$= O_m \quad \text{since } 0(Y) = O_m \text{ whenever } Y \in \mathcal{R}^m.$$

Proof of ii. Recall that $-X = (-1)X$ whenever $X \in \mathcal{R}^n$. Therefore,

$$T(-X) = T[(-1)X]$$
$$= (-1)TX \quad \text{by the homogeneous property of } T$$
$$= -TX \quad \text{since } (-1)Y = -Y \text{ whenever } Y \in \mathcal{R}^m.$$

Proof of iii. $T(X - Y) = T[X + (-Y)]$ by the definition of subtraction

$$= TX + T(-Y) \quad \text{by the additive property of } T$$
$$= TX + [-T(Y)] \quad \text{by ii.}$$
$$= TX - TY \quad \text{by the definition of subtraction.} \quad \blacksquare$$

Example 9. We shall use the preceding theorem to prove that the following function T, mapping $\mathcal{R}^2$ into $\mathcal{R}$, is not linear: $T\begin{pmatrix} x \\ y \end{pmatrix} = xy + 1$ for all $\begin{pmatrix} x \\ y \end{pmatrix}$ in $\mathcal{R}^2$. By Theorem 14.15, we know that $T\begin{pmatrix} 0 \\ 0 \end{pmatrix} = 0$ *if* T is linear. However, for this function $T\begin{pmatrix} 0 \\ 0 \end{pmatrix} = 1$ and consequently T is not linear. $\blacksquare$

The reader should at this point take careful note of the fact that there do exist *nonlinear* functions T from $\mathcal{R}^n$ into $\mathcal{R}^m$ such that $T(O_n) = O_m$. For example, define $T\begin{pmatrix} x \\ y \end{pmatrix} = xy$ for all $\begin{pmatrix} x \\ y \end{pmatrix}$ in $\mathcal{R}^2$. In this case $T(O_2) = O_1$. However, $T\begin{pmatrix} 1 \\ 1 \end{pmatrix} = 1$ and $T\begin{pmatrix} 2 \\ 2 \end{pmatrix} = 4$, so that $2T\begin{pmatrix} 1 \\ 1 \end{pmatrix} \neq T\left[2\begin{pmatrix} 1 \\ 1 \end{pmatrix} \right]$.

Besides the simple algebraic properties, linear transformations possess other strong properties.

omit

LEMMA 14.16 Let $T : \mathcal{R}^n \to \mathcal{R}^m$. If T is a linear transformation, there exists a constant k such that $\| TX \| \leq k \| X \|$ for each X in $\mathcal{R}^n$.

Proof. Let $\{E_1, \ldots, E_n\}$ be the natural basis for $\mathcal{R}^n$ and $X = (x_1, \ldots, x_n)$. We then have the implication chain: $X = \sum_{i=1}^{n} x_i E_i$

$$\Rightarrow TX =^1 T\left(\sum_{i=1}^{n} x_i E_i\right) =^2 \sum_{i=1}^{n} T(x_i E_i) =^3 \sum_{i=1}^{n} x_i TE_i$$

$$\Rightarrow \|TX\| = \left\|\sum_{i=1}^{n} x_i TE_i\right\| \leq^4 \sum_{i=1}^{n} \|x_i TE_i\| =^5 \sum_{i=1}^{n} |x_i| \, \|TE_i\|$$

$$\Rightarrow \|TX\| \leq \sum_{i=1}^{n} |x_i| \, k_0 =^6 k_0 \sum_{i=1}^{n} |x_i|, \quad \text{where } k_0 \text{ is an upper}$$

bound for $\{\|TE_i\| \mid i = 1, \ldots, n\}$,

$$\Rightarrow \|TX\| \leq^7 k_0 n \, \|X\| =^8 k \, \|X\|.$$

(1) Substitution.
(2) Definition 14.14, additive property.
(3) Definition 14.14, homogeneous property.
(4) Theorem 2.24iv, (p. 136) (Triangle Inequality).
(5) Theorem 2.24iii, (p. 136) (Positive Homogeneous Property).
(6) Distributive property of real numbers.
(7) See Exercise 14.
(8) We define k to be equal to $k_0 n$. ∎

From this lemma, the continuity of a linear transformation T from $\mathcal{R}^n$ into $\mathcal{R}^m$ follows easily. Let $X_0 \in \mathcal{R}^n$ and let $r > 0$. Let k be a positive number such that $\|TX\| \leq k \|X\|$ for all X in $\mathcal{R}^n$. If we choose $\rho = \dfrac{r}{k}$, then whenever $\|X - X_0\| < \rho$,

$$\|TX - TX_0\| =^1 \|T(X - X_0)\|$$
$$\leq^2 k \, \|X - X_0\|$$
$$<^3 k\rho$$
$$=^4 r.$$

(1) Theorem 14.15.
(2) Lemma 14.16.
(3) Since $\|X - X_0\| < \rho$.
(4) Since $\rho = \dfrac{r}{k}$.

Thus if $r > 0$, there exists a $\rho > 0$ such that $\|TX - TX_0\| < r$ whenever $\|X - X_0\| < \rho$. We have now proved:

THEOREM 14.17 (Continuity of a Linear Transformation.) Every linear transformation from $\mathcal{R}^n$ into $\mathcal{R}^m$ is continuous at each point of $\mathcal{R}^n$.

We could have, at the outset, defined our linear transformations in a more general fashion by requiring only that the domain of T be some linear subspace of $\mathcal{R}^n$ and that T still satisfy properties (i) and (ii) of Definition 14.14. However, this more general definition is more sophisticated and, indeed, is unnecessary at this point. Exercises 9 and 10 in this section indicate why this more general definition would yield the same results as those produced by Definition 14.14.

EXERCISES

Prove either that T is a linear transformation or that T is not a linear transformation.

1. $T\begin{pmatrix} x \\ y \end{pmatrix} = 2x - 3y$ for all $\begin{pmatrix} x \\ y \end{pmatrix}$ in $\mathcal{R}^2$.

2. $T\begin{pmatrix} x \\ y \end{pmatrix} = x + y - 3$ for all $\begin{pmatrix} x \\ y \end{pmatrix}$ in $\mathcal{R}^2$.

3. $T(x) = \begin{pmatrix} 2x \\ 0 \\ -x \end{pmatrix}$ for all x in $\mathcal{R}$.

4. $T(x) = \begin{pmatrix} x^2 \\ 3x \end{pmatrix}$ for all x in $\mathcal{R}$.

5. $T\begin{pmatrix} x \\ y \end{pmatrix} = \begin{pmatrix} xy \\ y+1 \end{pmatrix}$ for all $\begin{pmatrix} x \\ y \end{pmatrix}$ in $\mathcal{R}^2$.

6. $T\begin{pmatrix} x \\ y \end{pmatrix} = \begin{pmatrix} x+y \\ x-2y \\ 3y \end{pmatrix}$ for all $\begin{pmatrix} x \\ y \end{pmatrix}$ in $\mathcal{R}^2$.

7. $T\begin{pmatrix} x \\ y \\ z \end{pmatrix} = x - yz$ for all $\begin{pmatrix} x \\ y \\ z \end{pmatrix}$ in $\mathcal{R}^3$.

8. $T\begin{pmatrix} x \\ y \\ z \end{pmatrix} = \begin{pmatrix} y \\ x+z \end{pmatrix}$ for all $\begin{pmatrix} x \\ y \\ z \end{pmatrix}$ in $\mathcal{R}^3$.

Let T be a linear transformation from $\mathcal{R}^n$ into $\mathcal{R}^m$.

9. Prove that if $\mathcal{M}$ is a linear subspace of $\mathcal{R}^n$, then $T(\mathcal{M})$ is a linear subspace of $\mathcal{R}^m$, where

$$T(\mathcal{M}) = \{TX \mid X \in \mathcal{M}\} = \{Y \mid Y = TX, X \in \mathcal{M}\}.$$

(The "image" of a linear subspace under a linear transformation is also a linear subspace.)

10. Prove that if $\mathcal{N}$ is a linear subspace of $\mathcal{R}^m$, then $T^{-1}(\mathcal{N})$ is a linear subspace of $\mathcal{R}^n$, where $T^{-1}(\mathcal{N}) = \{X \mid X \in \mathcal{R}^n$ and $TX \in \mathcal{N}\}$. (The "inverse image" of a linear subspace under a linear transformation is also a linear subspace.)

11. Let Y_0 be a fixed element of $\mathcal{R}^n$. For each X in $\mathcal{R}^n$, define $TX = X \cdot Y_0$. Prove that T is a linear transformation from $\mathcal{R}^n$ into $\mathcal{R}$.

12. For each X in $\mathcal{R}^n$, define T by $TX = 5X$. Prove that T is a linear transformation from $\mathcal{R}^n$ into $\mathcal{R}^n$.

13. Let $T\begin{pmatrix} x \\ y \end{pmatrix} = x + y$ for all $\begin{pmatrix} x \\ y \end{pmatrix}$ in $\mathcal{R}^2$. Define $K(T) = \left\{ \begin{pmatrix} x \\ y \end{pmatrix} \middle| T\begin{pmatrix} x \\ y \end{pmatrix} = 0 \right\}$.

   (a) Show that $K(T)$ is a linear subspace of $\mathcal{R}^2$.

   (b) Show that $K(T) = \left\{ t\begin{pmatrix} 1 \\ -1 \end{pmatrix} \middle| t \text{ scalar} \right\}$.

14. Let $X = (x_1, x_2, \ldots, x_n)$ be an element of $\mathcal{R}^n$. Show that $\sum_{i=1}^{n} |x_i| \le n \|X\|$. (Hint: Use the fact that $|x_i| = \sqrt{x_i^2} \le \sqrt{x_1^2 + x_2^2 + \cdots + x_n^2} = \|X\|$, for each coordinate x_i.)

ANSWERS

2. No.   **3.** Yes.   **6.** Yes.   **8.** Yes.

5. EXAMPLES OF LINEAR SUBSPACES AND LINEAR TRANSFORMATIONS

This section contains several examples of linear subspaces and transformations not previously mentioned.

We shall in this section discuss a more sophisticated approach to linear subspaces and linear transformations than that presented in Sections 2, 3, and 4. There exist sets of elements $\mathcal{V}$, other than $\mathcal{R}^n$, in which there is defined addition and scalar multiplication with the following properties:

1. $X + Y \in \mathcal{V}$ whenever $X \in \mathcal{V}$ and $Y \in \mathcal{V}$.
2. $X + Y = Y + X$ whenever $X \in \mathcal{V}$ and $Y \in \mathcal{V}$.
3. $X + (Y + Z) = (X + Y) + Z$ whenever $X, Y, Z \in \mathcal{V}$.
4. There exists an element $O_\mathcal{V}$ in $\mathcal{V}$ such that $X + O_\mathcal{V} = O_\mathcal{V} + X = X$ for each X in $\mathcal{V}$.
5. If $X \in \mathcal{V}$, there exists an element $-X$ in $\mathcal{V}$ such that $X + (-X) = (-X) + X = O_\mathcal{V}$.
6. $r(X) \in \mathcal{V}$ whenever $X \in \mathcal{V}$ and $r \in \mathcal{R}$.
7. $1(X) = X$ whenever $X \in \mathcal{V}$.
8. $r(sX) = (rs)X$ whenever $X \in \mathcal{V}$ and $r, s \in \mathcal{R}$.
9. $r(X + Y) = rX + rY$ whenever $X, Y \in \mathcal{V}$ and $r \in \mathcal{R}$.
10. $(r + s)(X) = rX + sX$ whenever $X \in \mathcal{V}$ and $r, s \in \mathcal{R}$.

We call any such set a *vector space*. These properties are, of course, exactly the vector properties listed for $\mathcal{R}^n$, Theorem 1.14, p. 16.

If $\mathcal{V}$ is a vector space and if $\mathcal{M}$ is a nonempty subset of $\mathcal{V}$, then we say that $\mathcal{M}$ is a *linear subspace* of $\mathcal{V}$ if and only if $X + Y \in \mathcal{M}$ and $rX \in \mathcal{M}$ whenever $X \in \mathcal{M}$, $Y \in \mathcal{M}$, and $r \in \mathcal{R}$. If we repeat the proof of Theorem 1.9 in this more general context, we find that each linear subspace of $\mathcal{V}$ is itself a vector space: That is, it satisfies all ten listed properties.

We can also define linear combination, linear dependence, and linear independence for sets of elements of a vector space $\mathcal{V}$, as we did for $\mathcal{R}^n$. To define a basis for a linear subspace $\mathcal{M}$ of $\mathcal{V}$ we repeat Definition 14.10 in this more general context. Moreover, the following theorems can be proved in this more abstract setting. The same proofs we used before apply now.

1. Theorem 14.3 (Criterion for Dependence), p. 558.
2. Corollary 14.4 (Criterion for Independence), p. 560.
3. Theorem 14.5, p. 561.
4. Corollary 14.6, p. 562.
5. Lemma 14.11, p. 568.
6. Theorem 14.12, p. 568.
7. Theorem 14.13, p. 569.

Example 10. Let $\mathcal{V} = \{f \mid f : \mathcal{R}^n \to \mathcal{R}^m, \, dom\, f = \mathcal{A}\}$. Definition 3.3i, p. 154, defines an addition on $\mathcal{V}$ and Definition 3.3iv, p. 154, defines a scalar multiplication on $\mathcal{V}$. It is easily shown that this addition and scalar multiplication satisfy all the ten properties listed for a vector space.

An example of one linear subspace of $\mathcal{V}$ is the set $\mathcal{M}$, where

$$\mathcal{M} = \{f \mid f \in \mathcal{V} \quad \text{and} \quad f(X_0) = O_m\}, \, X_0 \in \mathcal{A}.$$

The reader should check the three properties to show that $\mathcal{M}$ is indeed a linear subspace of $\mathcal{V}$. (Why is $\mathcal{M}$ nonempty?) ∎

Once we have defined a general vector space, it is natural to ask whether it is possible to define a linear transformation from a vector space $\mathcal{V}$ into a vector space $\mathcal{W}$. A quick look at Definition 14.14, p. 571, assures us that we can do this: Let $\mathcal{V}$ and $\mathcal{W}$ be vector spaces, $T : \mathcal{V} \to \mathcal{W}, \, dom\, T = \mathcal{V}$. We say that T is a linear transformation if and only if $T(X + Y) = TX + TY$ and $T(rX) = rTX$ whenever $X \in \mathcal{V}$, $Y \in \mathcal{V}$, and $r \in \mathcal{R}$. Theorem 14.15 ($T(O_n) = O_m$, and so forth) is true in this context, but Lemma 14.16 and Theorem 14.17 (Continuity) are not true in general. Indeed, the concept of continuity is not defined in the most general case.

Example 11. Let $\mathcal{C} = \{f \mid f : (a, b) \to \mathcal{R}, f \text{ continuous on } (a, b)\}$. Let $\mathcal{D} = \{f \mid f : (a, b) \to \mathcal{R}, f \text{ differentiable on } (a, b), \text{ and } f' \text{ continuous on } (a, b)\}$. Theorem 4.2 assures us that $\mathcal{C}$ is a vector space and Theorem 5.8 assures us that $\mathcal{D}$ is a vector space. Can you show that $\mathcal{D}$ is a linear subspace of $\mathcal{C}$? We define a function T from $\mathcal{D}$ into $\mathcal{C}$ as follows: $Tf = f'$. Since $(f + g)' = f' + g'$ and $(rf)' = rf'$ whenever f and g are differentiable and r is a scalar, we have $T(f + g) = Tf + Tg$ and $T(rf) = rTf$ whenever $f \in \mathcal{D}$, $g \in \mathcal{D}$, and $r \in \mathcal{R}$. Therefore T is a linear transformation from $\mathcal{D}$ into $\mathcal{C}$. That is, differentiation may be regarded as a linear transformation from the space of all functions with continuous derivatives to the space of all continuous functions. ∎

More generally, if $\mathcal{F}$ is the set of all real valued-functions on (a, b) and $\mathcal{F}_0$ is the set of all differentiable functions on (a, b), the function $T : \mathcal{F}_0 \to \mathcal{F}$, defined by $Tf = f'$, is a linear transformation. The reader should verify that both $\mathcal{F}$ and $\mathcal{F}_0$ are indeed vector spaces.

EXERCISES

Let $\mathcal{V}$ be the set of all real polynomials of degree less than or equal to two (every element of $\mathcal{V}$ is a function f, $f: \mathcal{R} \to \mathcal{R}$, such that there exist real numbers a_0, a_1, a_2 for which $f(x) = a_0 x^2 + a_1 x + a_2$ for all x in $\mathcal{R}$).

1. Show that $\mathcal{V}$ is a vector space.

2. Find a linear subspace $\mathcal{M}$ of $\mathcal{V}$, $\mathcal{M} \neq \mathcal{V}$ and $\mathcal{M} \neq O_{\mathcal{V}}$.

3. For f in $\mathcal{V}$, define $Df = g$, where $g(x) = 2a_0 x + a_1$ when $f(x) = a_0 x^2 + a_1 x + a_2$. Show that D is a linear transformation from $\mathcal{V}$ into $\mathcal{V}$.

4. Let D be defined as in Exercise 3. Which elements of $\mathcal{V}$ are in the following sets?

$$\mathcal{K} = \{f \mid f \in \mathcal{V}, Df = O_{\mathcal{V}}\}.$$
$$\mathcal{R} = \{Df \mid f \in \mathcal{V}\}.$$

5. Find a basis for $\mathcal{V}$.

6. Define $T: \mathcal{V} \to \mathcal{R}^3$ as follows: if $f(x) = a_0 x^2 + a_1 x + a_2$, $Tf = (a_0, a_1, a_2)$. Prove that T is a linear transformation.

7. Let T be defined as in Exercise 6. Prove that $Tf = Tg$ if and only if $f = g$.

ANSWERS

4. $\mathcal{K} = \{f \mid f \text{ is constant}\}$, $\mathcal{R} = \{g \mid g(x) = b_1 x + b_2 \text{ for all } x \text{ in } \mathcal{R}\}$.
5. $\{f_0, f_1, f_2\}$, where $f_0(x) = 1, f_1(x) = x$, and $f_2(x) = x^2$ for all x in $\mathcal{R}$.

6. LINEAR SUBSPACES ASSOCIATED WITH A GIVEN LINEAR TRANSFORMATION

Let us once again consider Exercises 9 and 10 of Section 4. They are restated here as follows: If $\mathcal{M}$ is a linear subspace of the domain of a linear transformation T, then the image of $\mathcal{M}$ under T is a linear subspace of the range space of T and if $\mathcal{N}$ is a linear subspace of the range space of T, then the inverse image of $\mathcal{N}$ under T is a linear subspace of the domain of T. The following definition describes three sets, each of which is associated with a given linear transformation. Theorem 14.19 states that each of these is a linear subspace.

DEFINITION 14.18 Let T be a linear transformation from $\mathcal{R}^n$ into $\mathcal{R}^m$.

i. The *kernel of* T, denoted by *ker* T, is the subset of $\mathcal{R}^n$

$$ker\ T = \{X \mid X \in \mathcal{R}^n \text{ and } T(X) = O_m\}.$$

ii. The *range of T*, denoted by *ran T*, is the subset of $\mathcal{R}^m$

$$ran\ T = \{TX \mid X \in \mathcal{R}^n\}.$$

iii. The *graph of T*, denoted by *graph T*, is the subset of $\mathcal{R}^{n+m}$

$$graph\ T = \left\{ \begin{pmatrix} X \\ TX \end{pmatrix} \mid X \in \mathcal{R}^n \right\}.$$

We have already defined these concepts in Chapter 7, but for emphasis we have repeated the definitions in the context of linear transformations. Notice that the kernel of T is a subset of $\mathcal{R}^n$ and the range of T is a subset of $\mathcal{R}^m$. For convenience, we consider the graph of T to be a subset of $\mathcal{R}^{n+m}$. If $X \in \mathcal{R}^n$ and $TX \in \mathcal{R}^m$, X is an n-tuple of real numbers and TX is an m-tuple of real numbers, so that the ordered pair $\begin{pmatrix} X \\ TX \end{pmatrix}$ may be thought of as an $(n + m)$-tuple of real numbers. Notice that if X_1 and X_2 are in $\mathcal{R}^n$ and r is a real number, then from the definition of vector addition and scalar multiplication in $\mathcal{R}^{n+m}$

$$\begin{pmatrix} X_1 \\ TX_1 \end{pmatrix} + \begin{pmatrix} X_2 \\ TX_2 \end{pmatrix} = \begin{pmatrix} X_1 + X_2 \\ TX_1 + TX_2 \end{pmatrix}$$

and

$$r \begin{pmatrix} X_1 \\ TX_1 \end{pmatrix} = \begin{pmatrix} rX_1 \\ rTX_1 \end{pmatrix}.$$

THEOREM 14.19 Let T be a linear transformation from $\mathcal{R}^n$ into $\mathcal{R}^m$.

   i. *ker T* is a linear subspace of $\mathcal{R}^n$,
  ii. *ran T* is a linear subspace of $\mathcal{R}^m$, and
 iii. *graph T* is a linear subspace of $\mathcal{R}^{n+m}$.

Proof of i. In Theorem 14.15 we saw that $T(O_n) = O_m$. Therefore, by definition of *ker T*, $O_n \in ker\ T$ and thus *ker T* is a nonempty subset of $\mathcal{R}^n$. We wish to show that *ker T* satisfies the rest of Definition 14.8; that is, we wish to show that $X + Y \in ker\ T$ and $rX \in ker\ T$ whenever $X \in ker\ T$, $Y \in ker\ T$, and $r \in \mathcal{R}$. We note that an element Z of $\mathcal{R}^n$ is in *ker T* if and only if $TZ = O_m$. Therefore, we wish to show that $T(X + Y) = O_m$ and $T(rX) = O_m$ whenever $TX = O_m$, $TY = O_m$, and $r \in \mathcal{R}$. This follows easily from the additive and homogeneous properties of T:

$$T(X + Y) = TX + TY = O_m + O_m = O_m$$

and

$$T(rX) = rTX = rO_m = O_m.$$

Proof of ii. Since $\mathcal{R}^n$ is nonempty and T is a function, certainly *ran T* is a nonempty subset of $\mathcal{R}^m$. This time we wish to show that $Z + W \in ran\ T$

and $rZ \in ran\ T$ whenever $Z \in ran\ T$, $W \in ran\ T$, and $r \in \mathcal{R}$. However, if $Z \in ran\ T$ and $W \in ran\ T$, then by the definition of $ran\ T$, there exist elements X and Y of $\mathcal{R}^n$ such that $Z = TX$ and $W = TY$. Using the additive and homogeneous properties of T, we see that

$$Z + W = TX + TY = T(X + Y)$$

and

$$rZ = rTX = T(rX).$$

Therefore $Z + W \in ran\ T$ and $rZ \in ran\ T$; it follows that $ran\ T$ is a linear subspace of $\mathcal{R}^m$.

Proof of iii. This set is obviously nonempty (Why?). Let $Z \in graph\ T$, $W \in graph\ T$, and $r \in \mathcal{R}$. By the definition of $graph\ T$, there exist vectors X and Y in $\mathcal{R}^n$ such that

$$Z = \begin{pmatrix} X \\ TX \end{pmatrix} \quad \text{and} \quad W = \begin{pmatrix} Y \\ TY \end{pmatrix}.$$

Therefore,

$$Z + W = \begin{pmatrix} X \\ TX \end{pmatrix} + \begin{pmatrix} Y \\ TY \end{pmatrix} =^1 \begin{pmatrix} X + Y \\ TX + TY \end{pmatrix}$$

$$=^2 \begin{pmatrix} X + Y \\ T(X + Y) \end{pmatrix}$$

$$=^3 \begin{pmatrix} X_1 \\ TX_1 \end{pmatrix}, \text{ where } X_1 = X + Y,$$

and

$$rZ = r \begin{pmatrix} X \\ TX \end{pmatrix} =^4 \begin{pmatrix} rX \\ rTX \end{pmatrix}$$

$$=^5 \begin{pmatrix} rX \\ T(rX) \end{pmatrix}$$

$$= \begin{pmatrix} X_2 \\ TX_2 \end{pmatrix}, \text{ where } X_2 = rX.$$

(1) Addition of vectors.
(2) Additive property of T.
(3) Substituting X_1 for $X + Y$.
(4) Scalar multiplication in $\mathcal{R}^{n+m}$.
(5) Homogeneous property of T.

Finally, since $Z + W = \begin{pmatrix} X_1 \\ TX_1 \end{pmatrix}$ and $rZ = \begin{pmatrix} X_2 \\ TX_2 \end{pmatrix}$, where X_1 and X_2 are elements of $\mathcal{R}^n$, we have $Z + W \in graph\ T$ and $rZ \in graph\ T$. Thus, $graph\ T$ is a linear subspace of $\mathcal{R}^{n+m}$. ∎

Example 12. Let T be the linear transformation from $\mathcal{R}^3$ into $\mathcal{R}^2$ defined by

$$T \begin{pmatrix} x \\ y \\ z \end{pmatrix} = \begin{pmatrix} x - y \\ z + x \end{pmatrix}.$$

We wish to find out exactly what *ker T*, *ran T*, and *graph T* are.

i. $ker\ T =^1 \left\{ \begin{pmatrix} x \\ y \\ z \end{pmatrix} \middle| T\begin{pmatrix} x \\ y \\ z \end{pmatrix} = \begin{pmatrix} 0 \\ 0 \end{pmatrix} \right\}$

$=^2 \left\{ \begin{pmatrix} x \\ y \\ z \end{pmatrix} \middle| \begin{pmatrix} x - y \\ z + x \end{pmatrix} = \begin{pmatrix} 0 \\ 0 \end{pmatrix} \right\}$

$=^3 \left\{ \begin{pmatrix} x \\ y \\ z \end{pmatrix} \middle| x - y = 0 \text{ and } z + x = 0 \right\}$

$=^4 \left\{ \begin{pmatrix} x \\ y \\ z \end{pmatrix} \middle| y = x \text{ and } z = -x \right\}$

$=^5 \left\{ \begin{pmatrix} x \\ x \\ -x \end{pmatrix} \middle| x \in \mathcal{R} \right\}$

$=^6 \left\{ x\begin{pmatrix} 1 \\ 1 \\ -1 \end{pmatrix} \middle| x \in \mathcal{R} \right\}.$

(1) Definition of *ker T*.

(2) Since $T\begin{pmatrix} x \\ y \\ z \end{pmatrix} = \begin{pmatrix} x - y \\ z + x \end{pmatrix}.$

(3) Equality of vectors.
(4) Algebra.
(5) Substituting for *y* and *z* in terms of *x*.
(6) Scalar multiplication.

Thus we see that *ker T* in this case is the line in $\mathcal{R}^3$ passing through the origin, with direction vector $\begin{pmatrix} 1 \\ 1 \\ -1 \end{pmatrix}.$

ii. $ran\ T =^1 \left\{ T\begin{pmatrix} x \\ y \\ z \end{pmatrix} \middle| \begin{pmatrix} x \\ y \\ z \end{pmatrix} \in \mathcal{R}^3 \right\}$

$=^2 \left\{ \begin{pmatrix} x - y \\ z + x \end{pmatrix} \middle| \begin{pmatrix} x \\ y \\ z \end{pmatrix} \in \mathcal{R}^3 \right\}$

$=^3 \left\{ x\begin{pmatrix} 1 \\ 1 \end{pmatrix} + y\begin{pmatrix} -1 \\ 0 \end{pmatrix} + z\begin{pmatrix} 0 \\ 1 \end{pmatrix} \middle| x, y, z \in \mathcal{R} \right\}.$

(1) Definition of *ran T*.
(2) Definition of *T*.
(3) Scalar multiplication and addition in $\mathcal{R}^2$.

Note the following:

    a. *ran T* is a linear subspace of $\mathcal{R}^2$ (Theorem 14.19);

    b. $\left\{ \begin{pmatrix} -1 \\ 0 \end{pmatrix}, \begin{pmatrix} 0 \\ 1 \end{pmatrix} \right\}$ is a basis for $\mathcal{R}^2$; and

    c. $\left\{ \begin{pmatrix} -1 \\ 0 \end{pmatrix}, \begin{pmatrix} 0 \\ 1 \end{pmatrix} \right\} \subset$ *ran T*. Therefore, we know that *ran T* = $\mathcal{R}^2$, since
every element of $\mathcal{R}^2$ is an element of *ran T*.

    iii. *graph* $T =^1 \left\{ \begin{pmatrix} X \\ TX \end{pmatrix} \;\middle|\; X \in \mathcal{R}^3 \right\}$

$$=^2 \left\{ \begin{pmatrix} x \\ y \\ z \\ x - y \\ z + x \end{pmatrix} \;\middle|\; \begin{pmatrix} x \\ y \\ z \end{pmatrix} \in \mathcal{R}^3 \right\}$$

$$=^3 \left\{ x\begin{pmatrix} 1 \\ 0 \\ 0 \\ 1 \\ 1 \end{pmatrix} + y\begin{pmatrix} 0 \\ 1 \\ 0 \\ -1 \\ 0 \end{pmatrix} + z\begin{pmatrix} 0 \\ 0 \\ 1 \\ 0 \\ 1 \end{pmatrix} \;\middle|\; x, y, z \in \mathcal{R} \right\}.$$

(1) Definition of *graph T*.
(2) Substitution for *X* and *TX*.
(3) Scalar multiplication and addition in $\mathcal{R}^5$.

We note that the set S given by

$$\left\{ \begin{pmatrix} 1 \\ 0 \\ 0 \\ 1 \\ 1 \end{pmatrix}, \begin{pmatrix} 0 \\ 1 \\ 0 \\ -1 \\ 0 \end{pmatrix}, \begin{pmatrix} 0 \\ 0 \\ 1 \\ 0 \\ 1 \end{pmatrix} \right\}$$

is a linearly independent subset of $\mathcal{R}^5$ (Why?) and that every element in
graph T is a linear combination of elements of S. Therefore, *graph T* is the
linear subspace of $\mathcal{R}^5$ that has S as a basis. ∎

Before proceeding to the next example, we shall make some observations about Example 12. The domain of *T*, $\mathcal{R}^3$, has a basis consisting of
three elements, as does the graph of *T*. The kernel of *T* has a basis consisting
of one element and the range has a basis consisting of two elements. Do you
observe any relation involving the number of elements in these bases? Can
you generalize your observation?

Example 13. Let T be the linear transformation from $\mathcal{R}^2$ into $\mathcal{R}^3$, defined by

$$T\begin{pmatrix} x \\ y \end{pmatrix} = \begin{pmatrix} x + 2y \\ y - x \\ 2x + 3y \end{pmatrix}.$$

We shall describe *ker T* and *ran T*. The justifications for the various steps are the same as those in Example 12.

i. $ker\ T = \left\{ \begin{pmatrix} x \\ y \end{pmatrix} \middle| T\begin{pmatrix} x \\ y \end{pmatrix} = \begin{pmatrix} 0 \\ 0 \\ 0 \end{pmatrix} \right\}$

$= \left\{ \begin{pmatrix} x \\ y \end{pmatrix} \middle| \begin{pmatrix} x + 2y \\ y - x \\ 2x + 3y \end{pmatrix} = \begin{pmatrix} 0 \\ 0 \\ 0 \end{pmatrix} \right\}$

$= \left\{ \begin{pmatrix} x \\ y \end{pmatrix} \middle| x + 2y = 0, y - x = 0, 2x + 3y = 0 \right\}.$

In this case we note that the only pair (x, y) satisfying the equations $x + 2y = 0$, $y - x = 0$, and $2x + 3y = 0$ is the pair $(0, 0)$. Therefore, *ker T* $= \{O_2\}$.

ii. $ran\ T = \left\{ T\begin{pmatrix} x \\ y \end{pmatrix} \middle| \begin{pmatrix} x \\ y \end{pmatrix} \in \mathcal{R}^2 \right\}$

$= \left\{ \begin{pmatrix} x + 2y \\ y - x \\ 2x + 3y \end{pmatrix} \middle| \begin{pmatrix} x \\ y \end{pmatrix} \in \mathcal{R}^2 \right\}$

$= \left\{ x\begin{pmatrix} 1 \\ -1 \\ 2 \end{pmatrix} + y\begin{pmatrix} 2 \\ 1 \\ 3 \end{pmatrix} \middle| x, y \in \mathcal{R} \right\}.$

Since $\left\{ \begin{pmatrix} 1 \\ -1 \\ 2 \end{pmatrix}, \begin{pmatrix} 2 \\ 1 \\ 3 \end{pmatrix} \right\}$ is a linearly independent set of vectors in $\mathcal{R}^3$, we see that *ran T* is the plane, passing through the origin and containing points $\begin{pmatrix} 1 \\ -1 \\ 2 \end{pmatrix}$ and $\begin{pmatrix} 2 \\ 1 \\ 3 \end{pmatrix}$. ∎

The reader should take note of the fact that in both Example 12 and Example 13 we have endeavored to express *ker T* and *ran T* in the simplest possible fashion. That is, when *ker T* $\neq \{O_n\}$ and *ran T* $\neq \{O_m\}$, we have expressed these sets as linear combinations of basis elements. There are, as we found in Section 3, many different ways in which this can be done. Therefore, the reader, while working the exercises at the end of this section, should keep in mind that one set may have two quite different looking descriptions.

For example:

a. $\left\{ t\begin{pmatrix} 1 \\ 2 \end{pmatrix} \,\middle|\, t \in \mathcal{R} \right\} = \left\{ x\begin{pmatrix} 1/2 \\ 1 \end{pmatrix} \,\middle|\, x \in \mathcal{R} \right\}$,

b. $\left\{ x\begin{pmatrix} 1 \\ -2 \end{pmatrix} + y\begin{pmatrix} 3 \\ 1 \end{pmatrix} \,\middle|\, x, y \in \mathcal{R} \right\} = \mathcal{R}^2$,

c. $\left\{ x\begin{pmatrix} 1 \\ 3 \\ -1 \end{pmatrix} + y\begin{pmatrix} 2 \\ 0 \\ 4 \end{pmatrix} \,\middle|\, x, y \in \mathcal{R} \right\} = \left\{ r\begin{pmatrix} 3 \\ 3 \\ 3 \end{pmatrix} + s\begin{pmatrix} -1 \\ 3 \\ -5 \end{pmatrix} \,\middle|\, r, s \in \mathcal{R} \right\}$,

d. $\left\{ x\begin{pmatrix} 1 \\ 0 \\ 1 \end{pmatrix} + y\begin{pmatrix} 2 \\ -1 \\ 3 \end{pmatrix} + z\begin{pmatrix} 3 \\ -1 \\ 4 \end{pmatrix} \,\middle|\, x, y, z \in \mathcal{R} \right\}$

$= \left\{ r\begin{pmatrix} 1 \\ 0 \\ 1 \end{pmatrix} + s\begin{pmatrix} 2 \\ -1 \\ 3 \end{pmatrix} \,\middle|\, r, s \in \mathcal{R} \right\}$.

To elaborate on the procedure to follow in finding *ker T*, *ran T*, and *graph T* for a given linear transformation let us consider one further example.

Example 14.   Let $T:\mathcal{R}^3 \to \mathcal{R}^3$ be defined by

$$T\begin{pmatrix} x \\ y \\ z \end{pmatrix} = \begin{pmatrix} 2x - y \\ z \\ 2x - y \end{pmatrix}.$$

Applying our various definitions, we find that

i. $ker\,T = \left\{ \begin{pmatrix} x \\ y \\ z \end{pmatrix} \,\middle|\, T\begin{pmatrix} x \\ y \\ z \end{pmatrix} = \begin{pmatrix} 2x - y \\ z \\ 2x - y \end{pmatrix} = \begin{pmatrix} 0 \\ 0 \\ 0 \end{pmatrix} \right\}$

$= \left\{ \begin{pmatrix} x \\ y \\ z \end{pmatrix} \,\middle|\, 2x - y = 0,\, z = 0,\, x, y, z \in \mathcal{R} \right\}$

$= \left\{ \begin{pmatrix} x \\ 2x \\ 0 \end{pmatrix} \,\middle|\, x \in \mathcal{R} \right\}$

$= \left\{ x\begin{pmatrix} 1 \\ 2 \\ 0 \end{pmatrix} \,\middle|\, x \in \mathcal{R} \right\}$.

Note that we might have arrived also at

$$\ker T = \left\{ y \begin{pmatrix} 1/2 \\ 1 \\ 0 \end{pmatrix} \;\middle|\; y \in \mathcal{R} \right\}.$$

ii. $\operatorname{ran} T = \left\{ T\begin{pmatrix} x \\ y \\ z \end{pmatrix} \;\middle|\; x, y, z \in \mathcal{R} \right\}$

$$= \left\{ \begin{pmatrix} 2x - y \\ z \\ 2x - y \end{pmatrix} \;\middle|\; x, y, z \in \mathcal{R} \right\}.$$

$$= \left\{ x\begin{pmatrix} 2 \\ 0 \\ 2 \end{pmatrix} + y\begin{pmatrix} -1 \\ 0 \\ -1 \end{pmatrix} + z\begin{pmatrix} 0 \\ 1 \\ 0 \end{pmatrix} \;\middle|\; x, y, z \in \mathcal{R} \right\}.$$

At this point, we should ask ourselves if this is a satisfactory description of $\operatorname{ran} T$. If we note that $\begin{pmatrix} 2 \\ 0 \\ 2 \end{pmatrix}$ and $\begin{pmatrix} -1 \\ 0 \\ -1 \end{pmatrix}$ are both multiples of $\begin{pmatrix} 1 \\ 0 \\ 1 \end{pmatrix}$, we realize

that every vector of the form $x\begin{pmatrix} 2 \\ 0 \\ 2 \end{pmatrix} + y\begin{pmatrix} -1 \\ 0 \\ -1 \end{pmatrix}$ can be written in the form

$r\begin{pmatrix} 1 \\ 0 \\ 1 \end{pmatrix}$, where $r = 2x - y$. Thus a simpler expression for $\operatorname{ran} T$ and one that

expresses $\operatorname{ran} T$ in terms of basis elements is given by

$$\operatorname{ran} T = \left\{ r\begin{pmatrix} 1 \\ 0 \\ 1 \end{pmatrix} + z\begin{pmatrix} 0 \\ 1 \\ 0 \end{pmatrix} \;\middle|\; r, z \in \mathcal{R} \right\}.$$

iii $\operatorname{graph} T = \left\{ \begin{pmatrix} X \\ TX \end{pmatrix} \;\middle|\; X \in \mathcal{R}^3 \right\}$

$$= \left\{ \begin{pmatrix} x \\ y \\ z \\ 2x - y \\ z \\ 2x - y \end{pmatrix} \;\middle|\; x, y, z \in \mathcal{R} \right\}$$

$$= \left\{ x\begin{pmatrix} 1 \\ 0 \\ 0 \\ 2 \\ 0 \\ 2 \end{pmatrix} + y\begin{pmatrix} 0 \\ 1 \\ 0 \\ -1 \\ 0 \\ -1 \end{pmatrix} + z\begin{pmatrix} 0 \\ 0 \\ 1 \\ 0 \\ 1 \\ 0 \end{pmatrix} \;\middle|\; x, y, z \in \mathcal{R} \right\}.$$

Since $\left\{ \begin{pmatrix} 1 \\ 0 \\ 0 \\ 2 \\ 0 \\ 2 \end{pmatrix}, \begin{pmatrix} 0 \\ 1 \\ 0 \\ -1 \\ 0 \\ -1 \end{pmatrix}, \begin{pmatrix} 0 \\ 0 \\ 1 \\ 0 \\ 1 \\ 0 \end{pmatrix} \right\}$ is a set of three linearly independent

elements of *graph T*, we have expressed *graph T* in the simplest possible fashion, for we have expressed *graph T* as the set of all linear combinations of a basis. ∎

EXERCISES

Express *ker T*, *ran T*, and *graph T* in terms of basis elements for each set:

1. $T\begin{pmatrix} x \\ y \end{pmatrix} = \begin{pmatrix} x - y \\ x - y \end{pmatrix}.$

2. $T\begin{pmatrix} x \\ y \end{pmatrix} = \begin{pmatrix} 2x - y \\ x + y \end{pmatrix}.$

3. $T\begin{pmatrix} x \\ y \end{pmatrix} = 2x - y.$

4. $T\begin{pmatrix} x \\ y \end{pmatrix} = 3x.$

5. $T\begin{pmatrix} x \\ y \end{pmatrix} = \begin{pmatrix} x + 2y \\ 2x - 3y \\ x \end{pmatrix}.$

6. $T\begin{pmatrix} x \\ y \end{pmatrix} = \begin{pmatrix} 0 \\ 0 \\ 0 \end{pmatrix}.$

7. $T\begin{pmatrix} x \\ y \\ z \end{pmatrix} = \begin{pmatrix} x - 2z \\ y + z \end{pmatrix}.$

8. $T\begin{pmatrix} x \\ y \\ z \end{pmatrix} = \begin{pmatrix} y \\ z \end{pmatrix}.$

9. $T(x) = \begin{pmatrix} 2x \\ -x \end{pmatrix}.$

10. $T\begin{pmatrix} x \\ y \\ z \end{pmatrix} = \begin{pmatrix} x + y \\ 2x - 3y \\ z + y \end{pmatrix}.$

11. $T\begin{pmatrix} x \\ y \\ z \end{pmatrix} = \begin{pmatrix} y \\ 0 \\ x + 2z \end{pmatrix}.$

12. $T\begin{pmatrix} x \\ y \\ z \end{pmatrix} = \begin{pmatrix} 0 \\ x - y \\ x - y \end{pmatrix}.$

13. $T\begin{pmatrix} x \\ y \\ z \end{pmatrix} = 2x - y.$ (Compare this with Exercise 3.)

14. $T\begin{pmatrix} x \\ y \\ z \\ w \end{pmatrix} = \begin{pmatrix} 2x + 3y \\ 0 \\ z - w \\ 2x + 3y \end{pmatrix}.$

15. Let $T\begin{pmatrix} x \\ y \end{pmatrix} = 2y + 3x.$ Show that $\left\{ \begin{pmatrix} x \\ y \end{pmatrix} \,\middle|\, T\begin{pmatrix} x \\ y \end{pmatrix} = 1 \right\}$ is a line in $\mathcal{R}^2$ with slope $-3/2$ and *y*-intercept $1/2$.

ANSWERS

1. $ker\ T = \left\{ x\begin{pmatrix} 1 \\ 1 \end{pmatrix} \,\middle|\, x \in \mathcal{R} \right\}$, $ran\ T = \left\{ r\begin{pmatrix} 1 \\ 1 \end{pmatrix} \,\middle|\, r \in \mathcal{R} \right\}.$

2. $graph\ T = \left\{ x\begin{pmatrix} 1 \\ 0 \\ 2 \\ 1 \end{pmatrix} + y\begin{pmatrix} 0 \\ 1 \\ -1 \\ 1 \end{pmatrix} \,\middle|\, x, y \in \mathcal{R} \right\}.$

3. $ker\ T = \left\{ x\begin{pmatrix} 1 \\ 2 \end{pmatrix} \,\middle|\, x \in \mathcal{R} \right\}$, $ran\ T = \mathcal{R}.$

4. graph $T = \left\{ x \begin{pmatrix} 1 \\ 0 \\ 3 \end{pmatrix} + y \begin{pmatrix} 0 \\ 1 \\ 0 \end{pmatrix} \;\middle|\; x, y \in \mathcal{R} \right\}.$

5. ker $T = \{O_2\}$, ran $T = \left\{ \begin{pmatrix} 1 \\ 2 \\ 1 \end{pmatrix} + y \begin{pmatrix} 2 \\ -3 \\ 0 \end{pmatrix} \;\middle|\; x, y \in \mathcal{R} \right\}.$

6. graph $T = \left\{ x \begin{pmatrix} 1 \\ 0 \\ 0 \\ 0 \\ 0 \end{pmatrix} + y \begin{pmatrix} 0 \\ 1 \\ 0 \\ 0 \\ 0 \end{pmatrix} \;\middle|\; x, y \in \mathcal{R} \right\}.$

7. ker $T = \left\{ z \begin{pmatrix} 2 \\ -1 \\ 1 \end{pmatrix} \;\middle|\; z \in \mathcal{R} \right\}$, ran $T = \mathcal{R}^2.$

9. ker $T = \{O_1\}$, ran $T = \left\{ x \begin{pmatrix} 2 \\ -1 \end{pmatrix} \;\middle|\; x \in \mathcal{R} \right\}$, graph $T = \left\{ x \begin{pmatrix} 1 \\ 2 \\ -1 \end{pmatrix} \;\middle|\; x \in \mathcal{R} \right\}.$

11. ker $T = \left\{ z \begin{pmatrix} -2 \\ 0 \\ 1 \end{pmatrix} \;\middle|\; z \in \mathcal{R} \right\}$, ran $T = \left\{ y \begin{pmatrix} 1 \\ 0 \\ 0 \end{pmatrix} + r \begin{pmatrix} 0 \\ 0 \\ 1 \end{pmatrix} \;\middle|\; y, r \in \mathcal{R} \right\}.$

12. graph $T = \left\{ x \begin{pmatrix} 1 \\ 0 \\ 0 \\ 0 \\ 1 \\ 1 \end{pmatrix} + y \begin{pmatrix} 0 \\ 1 \\ 0 \\ 0 \\ -1 \\ -1 \end{pmatrix} + z \begin{pmatrix} 0 \\ 0 \\ 1 \\ 0 \\ 0 \\ 0 \end{pmatrix} \;\middle|\; x, y, z \in \mathcal{R} \right\}.$

13. ker $T = \left\{ x \begin{pmatrix} 1 \\ 2 \\ 0 \end{pmatrix} + z \begin{pmatrix} 0 \\ 0 \\ 1 \end{pmatrix} \;\middle|\; x, z \in \mathcal{R} \right\}$, ran $T = \mathcal{R}$,

graph $T = \left\{ x \begin{pmatrix} 1 \\ 0 \\ 0 \\ 2 \end{pmatrix} + y \begin{pmatrix} 0 \\ 1 \\ 0 \\ -1 \end{pmatrix} + z \begin{pmatrix} 0 \\ 0 \\ 1 \\ 0 \end{pmatrix} \;\middle|\; x, y, z \in \mathcal{R} \right\}.$

7. THE DIMENSION OF LINEAR SUBSPACES

We have yet to show that when $\mathcal{M}$ is a linear subspace of $\mathcal{R}^n$, $\mathcal{M} \neq \{O_n\}$, $\mathcal{M}$ has a basis. We know that $\mathcal{R}^n$ has a basis consisting of n elements—the natural basis—and we have exhibited bases for certain given linear subspaces of $\mathcal{R}^2$ and $\mathcal{R}^3$. The following lemma will answer the more general question:

LEMMA 14.20 Let $\mathcal{M}$ be a linear subspace of $\mathcal{R}^n$, $\mathcal{M} \neq \{O_n\}$. There exists a basis $\{X_1, \ldots, X_k\}$ for $\mathcal{M}$ and $k \leq n$. Moreover, if $\{X_1, \ldots, X_k\}$ is a basis for $\mathcal{M}$ and $k < n$, there exist vectors $X_{k+1}, \ldots,$

[handwritten: not including M]

X_n in $\mathcal{R}^n \backslash \mathcal{M}$ such that $\{X_1, \ldots, X_k\} \cup \{X_{k+1}, \ldots, X_n\}$ is a basis for $\mathcal{R}^n$.

Proof. Let $\mathcal{M}$ satisfy the hypothesis of the lemma. There exists a vector X_1 in $\mathcal{M}$, $X_1 \neq O_n$ since $\mathcal{M} \neq \{O_n\}$. Either every element of $\mathcal{M}$ is a scalar multiple of X_1 and $\{X_1\}$ is a basis for $\mathcal{M}$, or there exists a vector X_2 in $\mathcal{M}$ such that $\{X_1, X_2\}$ is linearly independent. In the first case we have a basis for $\mathcal{M}$ and in the second case, either every element of $\mathcal{M}$ is a linear combination of elements of $\{X_1, X_2\}$ or there exists an element X_3 of $\mathcal{M}$ such that $\{X_1, X_2, X_3\}$ is linearly independent.

We proceed inductively: assume that we have found a linearly independent subset $\{X_1, \ldots, X_k\}$ in $\mathcal{M}$ for some $k \geq 1$. Then either every element of $\mathcal{M}$ is a linear combination of elements of $\{X_1, \ldots, X_k\}$ and $\{X_1, \ldots, X_k\}$ is a basis for $\mathcal{M}$ or there exists an element X_{k+1} in $\mathcal{M}$ such that $\{X_1, \ldots, X_k, X_{k+1}\}$ is linearly independent. Two possibilities arise. Either (i) there is a value of $k < n$ such that $\{X_1, \ldots, X_k\}$ is a basis for $\mathcal{M}$ and we are done, or (ii) we shall find a set of n linearly independent vectors $\{X_1, \ldots, X_n\}$ in $\mathcal{M}$. In this latter case, we note by Theorem 14.12, p. 568, that $\{X_1, \ldots, X_n\}$ is a basis for $\mathcal{R}^n$ and thus $\mathcal{R}^n \subset \mathcal{M}$. However, since $\mathcal{M} \subset \mathcal{R}^n$, we may conclude that $\{X_1, \ldots, X_n\}$ is a basis for $\mathcal{M}$. Thus in any case, there is a basis $\{X_1, \ldots, X_k\}$ for $\mathcal{M}$, with $k \leq n$.

Assume now that $\{X_1, \ldots, X_k\}$ is a basis for $\mathcal{M}$ and $k < n$. If every element of $\mathcal{R}^n$ is a linear combination of elements of $\{X_1, \ldots, X_k\}$, then $\{X_1, \ldots, X_k\}$ is a basis for $\mathcal{R}^n$, which contradicts Theorem 14.13 and our assumption that $k < n$. Therefore, any element X of $\mathcal{R}^n \backslash \mathcal{M}$ is such that $\{X_1, \ldots, X_k, X\}$ is linearly independent. We choose one such vector X_{k+1}. We repeat this construction process as we did in the first part of the proof and find elements $\{X_{k+1}, \ldots, X_n\}$ such that $\{X_1, \ldots, X_k\} \cup \{X_{k+1}, \ldots, X_n\}$ is a basis for $\mathcal{R}^n$. ∎

Before we define what we mean by the dimension of a linear subspace $\mathcal{M}$ of $\mathcal{R}^n$, let us reiterate two facts. First, if $\mathcal{M} \neq \{O_n\}$, then by Lemma 14.20 $\mathcal{M}$ has a basis, consisting of k elements. Second, by Theorem 14.13, p. 569, every basis of $\mathcal{M}$ consists of exactly k elements. These two facts assure us that the following definition is meaningful.

DEFINITION 14.21 (Dimension.) Let $\mathcal{M}$ be a linear subspace of $\mathcal{R}^n$.

   i. If $\mathcal{M} = \{O_n\}$, we define the *dimension of* $\mathcal{M}$, denoted by *dim* $\mathcal{M}$, to be the real number 0.

   ii. If $\mathcal{M} \neq \{O_n\}$ and $\mathcal{M}$ has a basis consisting of k elements, we define the *dimension of* $\mathcal{M}$, denoted by *dim* $\mathcal{M}$, to be the number k.

This definition of dimension should coincide with our intuitive feeling for this word. Note that if $\mathcal{M}$ is a line in $\mathcal{R}^n$ which passes through the origin, $\mathcal{M} = \{tU: t \text{ scalar}, U \neq O_n\}$, then $\{U\}$ is a basis for $\mathcal{M}$ and then according to Definition 14.21 $dim \, \mathcal{M} = 1$. Similarly, if $\mathcal{M}$ is a plane in $\mathcal{R}^n$, $\mathcal{M} = \{rU + sV \mid r \text{ and } s \text{ scalar}, U \text{ and } V \text{ nonparallel}\}$, then $\{U, V\}$ is a basis for $\mathcal{M}$ and $\mathcal{M}$ has dimension 2, according to Definition 14.21. Also, $\mathcal{R}^3$ has

dimension 3, since $\left\{ \begin{pmatrix} 1 \\ 0 \\ 0 \end{pmatrix}, \begin{pmatrix} 0 \\ 1 \\ 0 \end{pmatrix}, \begin{pmatrix} 0 \\ 0 \\ 1 \end{pmatrix} \right\}$ is a basis for $\mathcal{R}^3$. Thus we see that this

definition of dimension coincides with our intuitive feeling for this concept.

Note that all Definition 14.21 does is give a *name* to the number of elements in the bases of a given linear subspace of $\mathcal{R}^n$, with the exception of the case when $\mathcal{M} = \{O_n\}$. Since $\{O_n\}$ is a set consisting of just one point, we still find Definition 14.21 reasonable.

Example 15. In Example 7 of Section 3, p. 570, we considered the linear subspace $\mathcal{M}$, where $\mathcal{M} = \{(x, y, z) \mid (x, y, z) \in \mathcal{R}^3 \text{ and } z = 2x + 3y\}$. We discovered that $\{(0, 1, 3), (1, 0, 2)\}$ was a basis for $\mathcal{M}$. Therefore, $dim \, \mathcal{M} = 2$. ∎

In Example 12 of Section 6, p. 580, we encountered a linear transformation T from $\mathcal{R}^3$ into $\mathcal{R}^2$ such that $dim(dom \, T) = 3$, $dim(graph \, T) = 3$, $dim \, (ker \, T) = 1$, and $dim \, (ran \, T) = 2$. We suggested, at that time, that there is a relation involving these numbers and that this relationship can be generalized. The following two theorems complete the program.

THEOREM 14.22 (Dimension Theorem for Linear Transformations.) Let T be a linear transformation from $\mathcal{R}^n$ into $\mathcal{R}^m$. If $dim \, (ker \, T) = k$ and $dim \, (ran \, T) = r$, then $k + r = n$.

Proof. We shall not prove the theorem in the cases in which $dim \, (ker \, T) = 0$ or $dim \, (ker \, T) = n$. These will be left as exercises for the reader.

We shall assume then that $dim \, (ker \, T) = k$ and that $0 < k < n$. By Lemma 14.20, we can choose $X_1, \ldots, X_k$ in $ker \, T$ and $X_{k+1}, \ldots, X_n$ in $\mathcal{R}^n \backslash ker \, T$ such that $\{X_1, \ldots, X_k\}$ is a basis for $ker \, T$ and $\{X_1, \ldots, X_k\} \cup \{X_{k+1}, \ldots, X_n\}$ is a basis for $\mathcal{R}^n$. We assert that $\{TX_{k+1}, \ldots, TX_n\}$ is a basis for $ran \, T$, so that $dim \, (ran \, T) = n - k$.

To prove this assertion we first establish that $\{TX_{k+1}, \ldots, TX_n\}$ is a linearly independent set, a fact which follows from Corollary 14.4, p. 560, and the following implication scheme. Let $b_{k+1}, \ldots, b_n$ be scalars.

$$\sum_{i=k+1}^{n} b_i TX_i = O_m \Rightarrow^1 \sum_{i=k+1}^{n} T(b_i X_i) = O_m$$

$$\Rightarrow^2 T\left(\sum_{i=k+1}^{n} b_i X_i\right) = O_m$$

$$\Rightarrow^3 \sum_{i=k+1}^{n} b_i X_i \in ker\ T$$

$\Rightarrow^4$ there exist scalars $c_1, \ldots, c_k$ such that

$$\sum_{i=k+1}^{n} b_i X_i = \sum_{i=1}^{k} c_i X_i$$

$$\Rightarrow \sum_{i=k+1}^{n} b_i X_i = \sum_{i=1}^{k} (-b_i) X_i, \text{ setting } c_i = -b_i$$

$$\Rightarrow^5 \sum_{i=1}^{n} b_i X_i = O_n$$

$$\Rightarrow^6 b_1 = \cdots = b_n = 0$$

$$\Rightarrow^7 b_{k+1} = \cdots = b_n = 0.$$

(1) Definition 14.14 (homogeneous property).
(2) Definition 14.14 (additive property).
(3) Definition 14.18 (kernel of T).
(4) Definition 14.10 since $\{X_1, \ldots, X_k\}$ is a basis for *ker T*.
(5) Algebraic property of R^n.
(6) Corollary 14.4, p. 560, since $\{X_1, \ldots, X_n\}$ is linearly independent.
(7) Special case of preceding line.

The second step in proving that $\{TX_{k+1}, \ldots, TX_n\}$ is a basis for *ran T* is to show that every element of *ran T* is a linear combination of $\{TX_{k+1}, \ldots, TX_n\}$. This we shall do by the following implication scheme.

$$Y \in ran\ T \Rightarrow^1 \text{ there exists an } X \text{ in } \mathcal{R}^n \text{ such that } Y = TX$$

$$\Rightarrow^2 Y = TX = \sum_{i=1}^{n} a_i TX_i$$

$$\Rightarrow^3 Y = \sum_{i=k+1}^{n} a_i TX_i.$$

(1) Definition 14.18, p. 578.

(2) By Definition 14.10, since $\{X_1, \ldots, X_n\}$ is a basis for $\mathcal{R}^n$ there exist scalars $a_1, \ldots, a_n$ such that $X = \sum_{i=1}^{n} a_i X_i$. Substitute this expression for X and use the additive and homogeneous properties of T.

(3) Since $\{X_1, \ldots, X_k\} \subset ker\ T$ and by Definition 14.18, $T\ X_i = O_m$ for $i = 1, \ldots, k$.

Thus by the two preceding arguments we have shown that $\{TX_{k+1}, \ldots, TX_n\}$ is a basis for *ran T*. Therefore, $r = n - k$ and $k + r = n$, as desired. ∎

COROLLARY 14.23 (Dimension of *graph T.*) Let T be a linear transformation from $\mathcal{R}^n$ into $\mathcal{R}^m$. Then $dim\ (graph\ T) = n$.

Proof. For each X in $\mathcal{R}^n$ define $SX = (X, TX)$. It will be left as an exercise for the reader to show that S is a linear transformation from $\mathcal{R}^n$ into $\mathcal{R}^{n+m}$. Notice that $dom\ S = \mathcal{R}^n$ and $ran\ S = graph\ T$. Therefore, by Theorem 14.22, $dim\ (ker\ S) + dim\ (graph\ T) = n$. If we can show that $dim\ (ker\ S) = 0$, then we have $dim\ (graph\ T) = n$, as desired.

Let $X \in ker\ S$. Then $SX = (X, TX) = O_{n+m}$. But $(X, TX) = O_{n+m}$ if and only if $X = O_n$. Therefore, $ker\ S = \{O_n\}$ and $dim\ (ker\ S) = 0$ by Definition 14.21. ∎

Theorem 14.22 can be of particular help in determining the range of certain linear transformations. Suppose that T is a linear transformation from $\mathcal{R}^3$ into $\mathcal{R}^2$ and that we have already determined that $dim\ (ker\ T) = 1$. Then by this theorem we know that $dim\ (ran\ T) = 3 - dim\ (ker\ T) = 2$. The only two-dimensional subspace of $\mathcal{R}^2$ is $\mathcal{R}^2$ itself, so that $ran\ T = \mathcal{R}^2$. On the other hand, if we had known that $dim\ (ker\ T) = 2$, then we would have known that $dim\ (ran\ T) = 1$, or that $ran\ T$ was necessarily a line, passing through the origin, in $\mathcal{R}^2$.

Example 16. Let T be the linear transformation from $\mathcal{R}^4$ into $\mathcal{R}^4$ defined by

$$T\begin{pmatrix} x \\ y \\ z \\ w \end{pmatrix} = \begin{pmatrix} x - y \\ z + w \\ x - y \\ 0 \end{pmatrix}.$$

We find that

$$ker\ T = \left\{ x\begin{pmatrix} 1 \\ 1 \\ 0 \\ 0 \end{pmatrix} + z\begin{pmatrix} 0 \\ 0 \\ 1 \\ -1 \end{pmatrix} \,\middle|\, x, z \in \mathcal{R} \right\}.$$

Therefore, $dim\ (ker\ T) = 2$. By Theorem 14.22, $dim\ (ran\ T) = 2$, also, so that the range of T is a two-dimensional plane, passing through the origin, in $\mathcal{R}^4$. More specifically

$$ran\ T = \left\{ r\begin{pmatrix} 1 \\ 0 \\ 1 \\ 0 \end{pmatrix} + s\begin{pmatrix} 0 \\ 1 \\ 0 \\ 0 \end{pmatrix} \,\middle|\, r, s \in \mathcal{R} \right\}.$$

By Theorem 14.23 we know that *graph T* is a four-dimensional subspace of $\mathcal{R}^8$. ∎

EXERCISES

Find the dimension of the following linear subspaces.

1. $\mathcal{M} = \left\{ \begin{pmatrix} x \\ y \end{pmatrix} \middle| x = -3y \right\}$.

2. $\mathcal{M} = \left\{ \begin{pmatrix} x \\ y \\ z \end{pmatrix} \middle| z - 2x = y \right\}$.

3. $\mathcal{M} = \left\{ \begin{pmatrix} x \\ y \end{pmatrix} \middle| x + y = 0,\ 2x - 3y = 0 \right\}$.

4. $\mathcal{M} = \left\{ \begin{pmatrix} z + w \\ x \\ z + w \end{pmatrix} \middle| x, z, w \in \mathcal{R} \right\}$.

5. $\mathcal{M} = \left\{ r \begin{pmatrix} 1 \\ 0 \\ 1 \end{pmatrix} + s \begin{pmatrix} 2 \\ 0 \\ 2 \end{pmatrix} \middle| r, s \in \mathcal{R} \right\}$.

For each of the following linear transformations, (a) find *dim (ker T)*, *dim (ran T)*, and *dim (graph T)*, and (b) express *ran T* in terms of basis elements.

6. $T \begin{pmatrix} x \\ y \end{pmatrix} = \begin{pmatrix} x + 2y \\ x + 2y \end{pmatrix}$.

7. $T \begin{pmatrix} x \\ y \end{pmatrix} = \begin{pmatrix} 3x - 4y \\ x + y \end{pmatrix}$.

8. $T \begin{pmatrix} x \\ y \\ z \end{pmatrix} = \begin{pmatrix} x \\ z - y \end{pmatrix}$.

9. $T \begin{pmatrix} x \\ y \end{pmatrix} = \begin{pmatrix} y \\ 0 \\ x + y \end{pmatrix}$.

10. $T \begin{pmatrix} x \\ y \\ z \end{pmatrix} = \begin{pmatrix} 2x + y \\ z - y \\ 2x + z \end{pmatrix}$.

11. $T \begin{pmatrix} x \\ y \\ z \\ w \end{pmatrix} = \begin{pmatrix} x + y - w \\ z \\ y - 2w \end{pmatrix}$.

12. $T \begin{pmatrix} x \\ y \\ z \\ w \end{pmatrix} = \begin{pmatrix} z - w \\ x + 2y \\ 0 \\ z - w \end{pmatrix}$.

Let *T* be a linear transformation from $\mathcal{R}^n$ into $\mathcal{R}^m$.

13. Prove that if *dim (ker T)* = 0, then *dim (ran T)* = *n*.

14. Prove that if *dim (ker T)* = *n*, then *dim (ran T)* = 0.

15. Define *S* by *SX* = (*X, TX*) for each *X* in $\mathcal{R}^n$. Prove that *S* is a linear transformation from $\mathcal{R}^n$ into $\mathcal{R}^{n+m}$.

ANSWERS

1. *dim* $\mathcal{M}$ = 1. 3. *dim* $\mathcal{M}$ = 0. 5. *dim* $\mathcal{M}$ = 1.

6. *dim (ker T)* = 1, *dim (ran T)* = 1, *dim (graph T)* = 2.

9. *dim (ker T)* = 0, *dim (ran T)* = 2, *dim (graph T)* = 2.

11. *dim (ker T)* = 1, *dim (ran T)* = 3, *dim (graph T)* = 4.

15

MATRICES

1. VECTOR PROPERTIES OF MATRICES

At this point the reader should be thoroughly familiar with the vector properties in $\mathcal{R}^n$, as well as the inner product and length functions. We hope that he is as at ease manipulating n-vectors as he is manipulating real numbers. Thus we shall in this section leave many of the proofs to the reader.

Consider now the three vectors $(1, 3, 4)$, $(-1, 2, 5)$, and $(1, 3, 4, -1, 2, 5)$. Immediately we see that the last vector is closely related to the first two and can be thought of as a "pairing" of the first two vectors. Now consider the symbol $\begin{pmatrix} 1 & 3 & 4 \\ -1 & 2 & 5 \end{pmatrix}$. It also can be thought of as a pairing of the two vectors $(1, 3, 4)$ and $(-1, 2, 5)$. This sort of rectangular array of numbers has a special name: *matrix*, the plural of which is *matrices*.

DEFINITION 15.1 ($m \times n$ Matrices.) An *$m \times n$ matrix* is a rectangular array of real numbers with m (horizontal) rows and n (vertical) columns. If A is an $m \times n$ matrix, we denote the number in the i-th row and j-th column by a_{ij} and we call this number the *ij-th coordinate of A*.

Two $m \times n$ matrices A and B are *equal* if and only if $a_{ij} = b_{ij}$ for $i = 1, \ldots, m$ and $j = 1, \ldots, n$.

The set of all $m \times n$ matrices will be denoted by $\mathcal{R}_{mn}$.

The reader should recognize immediately that an $m \times 1$ matrix is an m-vector written vertically and a $1 \times n$ matrix is an n-vector written horizontally. For example, $\begin{pmatrix} 1 \\ 2 \end{pmatrix}$ is a 2×1 matrix as well as a 2-vector and

(0 1 3) is a 1 × 3 matrix as well as a 3-vector. The symbol (0 1 3) stands for the ordered triple of numbers (0, 1, 3). We shall use the two notations interchangeably.

Example 1. Let $A = \begin{pmatrix} 2 & 3 & 0 \\ 5 & -4 & 1 \end{pmatrix}$ and $B = \begin{pmatrix} 6 & 5 \\ 1 & 0 \\ 2 & 4 \end{pmatrix}$. A is a 2 × 3

matrix and B is a 3 × 2 matrix. What is a_{12}? Referring to Definition 15.1, we see that a_{12} is the number in the first row and second column: $a_{12} = 3$. Similarly, $a_{21} = 5$ and $a_{23} = 1$. We also see that $b_{22} = 0$, $b_{31} = 2$, and $b_{32} = 4$. ■

Let us emphasize at this point the conventions we are using. When we consider an $m \times n$ matrix and denote it by A, we denote the element of the i-th row and j-th column by a_{ij}. If you are given such a matrix and asked to find a_{34}, look down to the 3rd row and across to the 4th column (from the extreme left of the array) and you will find the desired number.

Example 2. We wish to find real numbers $x, y,$ and z such that

$$\begin{pmatrix} (x+y) & (x-z) \\ (y-z) & 3 \end{pmatrix} = \begin{pmatrix} 3 & -1 \\ 4 & 3 \end{pmatrix}.$$

Applying the criterion for equality of matrices given in Definition 15.1, we see that

$$\begin{cases} x + y = 3, \\ x - z = -1, \\ y - z = 4, \\ \quad\ 3 = 3. \end{cases}$$

It is a simple process then to show that $x = -1, y = 4,$ and $z = 0$. ■

In the case of vectors, we defined addition only when two vectors had the same number of coordinates; in the case of matrices we shall insist not only that two matrices have the same number of coordinates but also that they have the same number of columns and the same number of rows. We shall not attempt to add a 2 × 3 matrix to a 3 × 2 matrix. Our definition of multiplying a matrix by a real number is the exact parallel of the definition for vectors.

DEFINITION 15.2 Let A and B be $m \times n$ matrices and $c \in \mathcal{R}$.
  i. (Matrix Addition.) $A + B$ is the $m \times n$ matrix with $a_{ij} + b_{ij}$ in the i-th row and j-th column.
  ii. (Multiplication by a Scalar.) cA is the $m \times n$ matrix with ca_{ij} in the i-th row and j-th column.

Example 3. Using Definition 15.2, we simplify the following:

i. $\begin{pmatrix} 1 & 2 \\ -3 & 1 \end{pmatrix} + \begin{pmatrix} 2 & -1 \\ 5 & 6 \end{pmatrix} = \begin{pmatrix} 3 & 1 \\ 2 & 7 \end{pmatrix}$

ii. $4\begin{pmatrix} 5 & 0 \\ -1 & 2 \end{pmatrix} = \begin{pmatrix} 20 & 0 \\ -4 & 8 \end{pmatrix}$

iii. $2\begin{pmatrix} 1 & 3 \\ 0 & 0 \\ 5 & 2 \end{pmatrix} + 3\begin{pmatrix} -1 & 1 \\ 4 & 2 \\ 1 & 0 \end{pmatrix} = \begin{pmatrix} 2 & 6 \\ 0 & 0 \\ 10 & 4 \end{pmatrix} + \begin{pmatrix} -3 & 3 \\ 12 & 6 \\ 3 & 0 \end{pmatrix}$

$$= \begin{pmatrix} -1 & 9 \\ 12 & 6 \\ 13 & 4 \end{pmatrix}. \quad \blacksquare$$

These examples should leave the reader with the feeling that "nothing is new" and that "matrices are just overgrown vectors." And indeed, the reader is correct. Consider the following theorem:

THEOREM 15.3 (Properties of Matrix Addition and Multiplication by a Scalar.) The elements of $\mathcal{R}_{mn}$, together with the operations of addition and multiplication by a scalar (Definition 15.2), satisfy the vector space properties (given in Theorem 1.14, p. 16). Whenever A, B, and C are elements of $\mathcal{R}_{mn}$ and r, $s \in \mathcal{R}$,
  i. $A + B = B + A$.
  ii. $A + (B + C) = (A + B) + C$.
  iii. There exists an element $O_{m \times n}$ of $\mathcal{R}_{mn}$ such that $A + O_{m \times n} = A$ for all A in $\mathcal{R}_{mn}$. ($O_{m \times n}$ is the $m \times n$ matrix with each component equal to zero.)
  iv. There exists an element $-A$ of $\mathcal{R}_{mn}$ such that $(-A) + A = O_{m \times n}$. ($-A$ is the $m \times n$ matrix with ij-th component equal to the additive inverse of the ij-th component of A.)
  v. $1 \cdot A = A$.
  vi. $r(sA) = (rs)A$.
  vii. $(r + s)A = rA + sA$.
  viii. $r(A + B) = rA + rB$.

In view of the marked similarity between vector and matrix additive properties we do not give a formal definition of matrix subtraction. We merely note that it is the same as subtraction of vectors: $A - B = A + (-B) = A + (-1)B$ whenever $A, B \in \mathcal{R}_{mn}$.

EXERCISES

(a) Each of the following is an $m \times n$ matrix. In each case specify the values of m and n.

(b) For each case give the values of a_{11}, a_{21}, a_{23}, and a_{31}.

1. $A = \begin{pmatrix} 1 & 2 \\ -3 & 4 \end{pmatrix}$.

2. $A = \begin{pmatrix} 3 & 0 & 1 \\ 4 & 1 & 7 \end{pmatrix}$.

3. $A = \begin{pmatrix} 2 & 1 \\ 0 & 5 \\ -7 & 0 \end{pmatrix}$.

4. $A = \begin{pmatrix} 3 & 0 & 7 & 5 \\ 1 & 2 & 0 & -1 \end{pmatrix}$.

5. $A = (1 \quad 2 \quad 0 \quad 7)$.

6. $A = \begin{pmatrix} 1 & 1 & 1 & 1 \\ 0 & 1 & 3 & 2 \\ 1 & 2 & 1 & 0 \end{pmatrix}$.

Simplify the following expressions:

7. $\begin{pmatrix} 1 & 2 \\ 3 & -1 \end{pmatrix} + \begin{pmatrix} 4 & 7 \\ -3 & 1 \end{pmatrix}$

8. $5\begin{pmatrix} 2 & 1 \\ 3 & -1 \\ 0 & 4 \end{pmatrix}$.

9. $2\begin{pmatrix} 3 & 2 & 1 \\ 0 & 5 & 9 \end{pmatrix} + 3\begin{pmatrix} -1 & 2 & 0 \\ 7 & 3 & -6 \end{pmatrix}$.

10. $\left[\begin{pmatrix} 1 & 2 & 3 \\ 4 & 5 & 6 \end{pmatrix} + \begin{pmatrix} 2 & 4 & 9 \\ 3 & 1 & 0 \end{pmatrix} \right] + \begin{pmatrix} -2 & -4 & -9 \\ -3 & -1 & 0 \end{pmatrix}$.

11. $3\begin{pmatrix} 1 & 0 \\ -5 & 4 \\ 7 & 2 \end{pmatrix} + (-3)\begin{pmatrix} 1 & 0 \\ -5 & 4 \\ 7 & 2 \end{pmatrix}$.

12. $\begin{pmatrix} x & y \\ z & 2w \end{pmatrix} + \begin{pmatrix} 3y & x \\ x + w & z \end{pmatrix}$.

Find the additive inverse of each of the following:

13. $A = \begin{pmatrix} 1 & 5 \\ -7 & 4 \end{pmatrix}$.

14. $A = \begin{pmatrix} 0 & 3 & -1 \\ -4 & 2 & -5 \end{pmatrix}$.

15. $A = \begin{pmatrix} 4 & -2 & 11 & -7 \\ 0 & 0 & -1 & 4 \end{pmatrix}$

16. $A = \begin{pmatrix} 3 & 3 & -3 \\ -1 & 2 & 4 \\ 7 & -2 & -15 \end{pmatrix}$.

Prove the following:

17. Theorem 15.3ii.
18. Theorem 15.3iv.
19. Theorem 15.3viii.

ANSWERS

1. (a) $m = 2, n = 2$; (b) $a_{11} = 1, a_{21} = -3$, a_{23} does not exist, a_{31} does not exist. 4. (a) $m = 2, n = 4$; (b) $a_{11} = 3, a_{21} = 1, a_{23} = 0$, a_{31} does not exist. 9. $\begin{pmatrix} 3 & 10 & 2 \\ 21 & 19 & 0 \end{pmatrix}$. 11. $O_{3\times2}$. 14. $-A = \begin{pmatrix} 0 & -3 & 1 \\ 4 & -2 & 5 \end{pmatrix}$.

2. MATRIX MULTIPLICATION

One of the important operations in vector algebra is that of the inner product, a product whose values are always scalars. We shall in this section use this inner product to develop a product between certain matrices. We note the following difference between the inner product for vectors and the matrix product for matrices: the inner product of two n-vectors is a scalar; the matrix product of two $n \times n$ matrices is again an $n \times n$ matrix.

DEFINITION 15.4 (Matrix Multiplication.) Let $A \in \mathcal{R}_{mn}$ and $B \in \mathcal{R}_{np}$. The *matrix product AB* is the $m \times p$ matrix with ij-th coordinate equal to the inner product of the i-th row of A and the j-th column of B for $i = 1, \ldots, m$ and $j = 1, \ldots, p$; i.e.,

if $C = AB$, $C_{ij} = \sum_{k=1}^{n} a_{ik}b_{kj}$ for $i = 1, \ldots, m$ and $j = 1, \ldots, p$.

The last summation in Definition 15.4 may be interpreted as follows: The i-th row of A is the n-vector $(a_{i1}, a_{i2}, \ldots, a_{in})$ and the j-th column of B is the n-vector $(b_{1j}, b_{2j}, \ldots, b_{nj})$ and $\sum_{k=1}^{n} a_{ik}b_{kj}$ is the inner product of these two vectors.

Example 4. Let $A = \begin{pmatrix} 1 & -1 \\ 2 & 3 \end{pmatrix}$ and $B = \begin{pmatrix} 2 & 4 \\ 7 & -1 \end{pmatrix}$. Using Definition 15.4 we shall compute the 2×2 matrix AB.

$$AB = \begin{pmatrix} 1 & -1 \\ 2 & 3 \end{pmatrix}\begin{pmatrix} 2 & 4 \\ 7 & -1 \end{pmatrix}$$

$$= \begin{pmatrix} (1, & -1) \cdot \begin{pmatrix} 2 \\ 7 \end{pmatrix} & (1, & -1) \cdot \begin{pmatrix} 4 \\ -1 \end{pmatrix} \\ (2, & 3) \cdot \begin{pmatrix} 2 \\ 7 \end{pmatrix} & (2, & 3) \cdot \begin{pmatrix} 4 \\ -1 \end{pmatrix} \end{pmatrix}$$

$$= \begin{pmatrix} -5 & 5 \\ 25 & 5 \end{pmatrix}.$$

In this case we can also compute BA:

$$BA = \begin{pmatrix} 2 & 4 \\ 7 & -1 \end{pmatrix} \begin{pmatrix} 1 & -1 \\ 2 & 3 \end{pmatrix}$$

$$= \begin{pmatrix} (2,\ 4) \cdot \begin{pmatrix} 1 \\ 2 \end{pmatrix} & (2,\ 4) \cdot \begin{pmatrix} -1 \\ 3 \end{pmatrix} \\ (7,\ -1) \cdot \begin{pmatrix} 1 \\ 2 \end{pmatrix} & (7,\ -1) \cdot \begin{pmatrix} -1 \\ 3 \end{pmatrix} \end{pmatrix}$$

$$= \begin{pmatrix} 10 & 10 \\ 5 & -10 \end{pmatrix}.$$

We note that $AB \neq BA$ and that matrix multiplication is not a commutative operation. ■

Example 5. Let $A = \begin{pmatrix} 3 & -1 \\ 0 & 4 \\ 7 & -1 \end{pmatrix}$ and $B = \begin{pmatrix} 7 & 5 & -3 & 0 \\ 2 & 4 & 1 & 4 \end{pmatrix}$. Since A is a 3×2 matrix and B is a 2×4 matrix, then by Definition 15.4, AB is the 3×4 matrix given by

$$AB = \begin{pmatrix} (3\ -1) \cdot \begin{pmatrix} 7 \\ 2 \end{pmatrix} & (3\ -1) \cdot \begin{pmatrix} 5 \\ 4 \end{pmatrix} & (3\ -1) \cdot \begin{pmatrix} -3 \\ 1 \end{pmatrix} & (3\ -1) \cdot \begin{pmatrix} 0 \\ 4 \end{pmatrix} \\ (0) \cdot \begin{pmatrix} 7 \\ 2 \end{pmatrix} & (0) \cdot \begin{pmatrix} 5 \\ 4 \end{pmatrix} & (0) \cdot \begin{pmatrix} -3 \\ 1 \end{pmatrix} & (0) \cdot \begin{pmatrix} 0 \\ 4 \end{pmatrix} \\ (7\ -1) \cdot \begin{pmatrix} 7 \\ 2 \end{pmatrix} & (7\ -1) \cdot \begin{pmatrix} 5 \\ 4 \end{pmatrix} & (7\ -1) \cdot \begin{pmatrix} -3 \\ 1 \end{pmatrix} & (7\ -1) \cdot \begin{pmatrix} 0 \\ 4 \end{pmatrix} \end{pmatrix}$$

$$= \begin{pmatrix} 19 & 11 & -10 & -4 \\ 8 & 16 & 4 & 16 \\ 47 & 31 & -22 & -4 \end{pmatrix}$$

In this example we note that it is impossible to find the product BA since the rows of B are 4-vectors and the columns of A are 3-vectors. Note also that we omitted the commas in the horizontal vectors $(3\ -1)$, $(0\ 4)$, and $(7\ -1)$. ■

In spite of the fact that matrix multiplication is not defined for all pairs of matrices, it is an orderly operation. The following theorem gives us certain relations among the operations of matrix multiplication, matrix addition, and multiplication of matrices by scalars.

THEOREM 15.5 (Properties of Matrix Multiplication.) Let A, $B \in \mathcal{R}_{mn}$; $C, D \in \mathcal{R}_{np}$; and $E \in \mathcal{R}_{pq}$ and let $r \in \mathcal{R}$.

    i. (Associative Law.) $A(CE) = (AC)E$.

    ii. (Distributive Laws.) $A(C + D) = AC + AD$,

                               $(A + B)C = AC + BC$.

    iii. (Mixed Associative Law.) $r(AC) = (rA)C = A(rC)$.

Proof. To avoid tedium we shall only prove (i). We note that all the products are defined and that each of the products $A(CE)$ and $(AC)E$ is an $m \times q$ matrix (Check this!).

Let $F = CE$. Then $f_{kj} = \sum_{s=1}^{p} c_{ks} \ell_{sj}$.

Therefore, if $G = A(CE)$, we have for $i = 1, \ldots, m$ and $j = 1, \ldots, q$

$$g_{ij} =^1 \sum_{k=1}^{n} a_{ik} f_{kj}$$

$$=^2 \sum_{k=1}^{n} a_{ik} \left[\sum_{s=1}^{p} c_{ks} \ell_{sj} \right]$$

$$=^3 \sum_{k=1}^{n} \left[\sum_{s=1}^{p} a_{ik} c_{ks} \ell_{sj} \right]$$

$$=^4 \sum_{s=1}^{p} \left[\sum_{k=1}^{n} a_{ik} c_{ks} \right] \ell_{sj}$$

$$=^1 \sum_{s=1}^{p} h_{is} \ell_{sj}, \quad \text{where } H = AC,$$

$$=^1 t_{ij}, \quad \text{where } T = (AC)E,$$

(1) Definition 15.4 (Matrix Multiplication).
(2) Substitution for f_{kj}.
(3) Distributive property of reals.
(4) Properties of real numbers needed to rearrange order of addition. (See Exercise 14.)

Thus we have shown that $g_{ij} = t_{ij}$ for $i = 1, \ldots, m$ and $j = 1, \ldots, r$, where $G = A(CE)$ and $T = (AC)E$.

The proofs of (ii) and (iii) will be left as exercises for the reader. ∎

To emphasize the fact that matrix multiplication is not completely analogous to multiplication of real numbers, we shall consider the following example:

Example 6. Let $A = \begin{pmatrix} 0 & 1 \\ 0 & 2 \end{pmatrix}$ and $B = \begin{pmatrix} 0 & 3 \\ 0 & 0 \end{pmatrix}$. We have

$$AB = \begin{pmatrix} (0, 1) \cdot \begin{pmatrix} 0 \\ 0 \end{pmatrix} & (0, 1) \cdot \begin{pmatrix} 3 \\ 0 \end{pmatrix} \\ (0, 2) \cdot \begin{pmatrix} 0 \\ 0 \end{pmatrix} & (0, 2) \cdot \begin{pmatrix} 3 \\ 0 \end{pmatrix} \end{pmatrix}$$

$$= \begin{pmatrix} 0 & 0 \\ 0 & 0 \end{pmatrix}.$$

Thus we have an example of two matrices A, B such that $A \neq O_{2 \times 2}$, $B \neq O_{2 \times 2}$, but $AB = O_{2 \times 2}$.

Moreover, we can observe that

$$B^2 = BB = \begin{pmatrix} 0 & 0 \\ 0 & 0 \end{pmatrix},$$

whereas $B \neq O_{2\times 2}$. ∎

An $n \times n$ matrix A, $A \neq O_{n\times n}$, is called a *proper divisor* of zero if there exists a second nonzero $n \times n$ matrix B such that $AB = O_{n\times n}$ or $BA = O_{n\times n}$. The matrices in Example 6 are therefore proper divisors of zero. We know that in the real number system there exist no proper divisors of zero: if $a, b \in \mathcal{R}$, $a \neq 0$ and $b \neq 0$, then $ab \neq 0$.

Assume now that A and B are $m \times n$ matrices and C is a nonzero $n \times p$ matrix and that $AC = BC$. Can we say then that $A = B$? In general, the answer to this question is no. (See Exercise 15.) The following theorem will give us two equivalent criteria for determining when $A = B$.

THEOREM 15.6 (Matrix Cancellation Law.) Let $A \in \mathcal{R}_{mn}$, $B \in \mathcal{R}_{mn}$. The following are equivalent:
  i. $A = B$.
  ii. $AX = BX$ for all X in $\mathcal{R}^n$.
  iii. $AX_j = BX_j$, $j = 1, \ldots, n$, where $\{X_1, \ldots, X_n\}$ is some basis for $\mathcal{R}^n$.

NOTE. In this theorem we have identified $\mathcal{R}_{n\times 1}$ and $\mathcal{R}^n$: That is, when we write AX, $A \in \mathcal{R}_{mn}$, $X \in \mathcal{R}^n$, we consider X to be an $n \times 1$ matrix.

Proof. We shall prove the theorem by proving that (iii) $\Rightarrow$ (ii), (ii) $\Rightarrow$ (i) and (i) $\Rightarrow$ (iii). Thus if one of the three is true, all are true.

To prove that (iii) implies (ii) we let $\{X_1, \ldots, X_n\}$ be a basis for $\mathcal{R}^n$ and assume that $AX_i = BX_i$, $i = 1, \ldots, n$. Let $X \in \mathcal{R}^n$. By Definition 14.10, p. 567, there exist scalars such that

(15.1)
$$X = \sum_{j=1}^{n} c_i X_i.$$

Therefore,

$$AX =^1 A\left(\sum_{i=1}^{n} c_i X_i \right)$$

$$=^2 \sum_{i=1}^{n} c_i A X_i$$

$$=^3 \sum_{i=1}^{n} c_i B X_i$$

$$=^2 B\left(\sum_{i=1}^{n} c_i X_i \right)$$

$$=^1 BX.$$

(1) By Equation 15.1.
(2) Theorem 15.5ii and iii.
(3) By hypothesis, $AX_i = BX_i$ for $i = 1, \ldots, n$.

Thus we have $AX = BX$ whenever $X \in \mathcal{R}^n$.

Now let us prove that (ii) $\Rightarrow$ (i). We assume that $AX = BX$ for all X in $\mathcal{R}^n$. In particular $AE_j = BE_j$, where $\{E_1, \ldots, E_n\}$ is the natural basis in $\mathcal{R}^n$. Therefore by Definition 15.4 (Matrix Multiplication)

$$\begin{pmatrix} a_{1j} \\ a_{2j} \\ \cdot \\ \cdot \\ \cdot \\ a_{mj} \end{pmatrix} = AE_j = BE_j = \begin{pmatrix} b_{1j} \\ b_{2j} \\ \cdot \\ \cdot \\ \cdot \\ b_{mj} \end{pmatrix} \quad \text{for } j = 1, \ldots, m.$$

Thus we have $a_{ij} = b_{ij}$ for $i = 1, \ldots, m$ and $j = 1, \ldots, n$. By the definition of equality of matrices in Definition 15.1, we have $A = B$.

The fact that (i) $\Rightarrow$ (iii) follows immediately from the definitions of matrix equality (Definition 15.1) and of matrix multiplication (Definition 15.4). ■

EXERCISES

Perform the following matrix operations.

1. $\begin{pmatrix} -1 & 3 \\ 0 & 2 \end{pmatrix} \begin{pmatrix} 3 & 4 \\ -1 & 5 \end{pmatrix}$.

2. $\begin{pmatrix} 0 & 3 \\ -4 & 5 \end{pmatrix} \begin{pmatrix} 1 & 3 & 2 \\ 0 & 7 & 1 \end{pmatrix}$.

3. $\begin{pmatrix} 5 & 1 \\ 2 & 4 \\ 7 & -1 \end{pmatrix} \begin{pmatrix} 1 & 1 & 7 \\ 3 & 3 & -1 \end{pmatrix}$.

4. $\begin{pmatrix} 3 & 3 & -1 \\ 4 & 1 & 0 \\ 7 & -1 & 1 \end{pmatrix} \begin{pmatrix} 2 & 1 \\ 4 & 2 \\ 2 & -1 \end{pmatrix}$.

5. $\begin{pmatrix} 7 & 7 \\ 0 & 0 \\ 1 & 1 \end{pmatrix} \begin{pmatrix} 0 & 5 & 0 & 1 \\ 1 & 0 & 1 & 0 \end{pmatrix}$.

6. $\begin{pmatrix} 2 & 1 & 3 \\ 1 & -1 & 1 \end{pmatrix} \left[\begin{pmatrix} 3 & 0 & 7 \\ 2 & 0 & -1 \\ 0 & 0 & 2 \end{pmatrix} \begin{pmatrix} 1 \\ 2 \\ 3 \end{pmatrix} \right]$.

7. $\begin{pmatrix} 3 & 1 & 0 & 5 \\ 7 & 2 & -1 & 1 \\ 3 & 4 & 5 & 1 \end{pmatrix} \begin{pmatrix} 0 \\ 1 \\ 1 \\ 0 \end{pmatrix}$.

8. $(1 \ 3 \ 0 \ 4) \begin{pmatrix} 2 & 3 \\ -1 & 1 \\ 4 & 2 \\ 1 & 0 \end{pmatrix}$. $+3$ 6

9. Find a 2×2 matrix I_2 such that $AI_2 = I_2A = A$ for all 2×2 matrices A.

10. Find a 2×2 matrix A such that

$$A\begin{pmatrix} 3 & 1 \\ 5 & 2 \end{pmatrix} = \begin{pmatrix} 3 & 1 \\ 5 & 2 \end{pmatrix}A = \begin{pmatrix} 1 & 0 \\ 0 & 1 \end{pmatrix}.$$

11. Show that there exists no 2×2 matrix A such that

$$A\begin{pmatrix} 2 & 1 \\ 4 & 2 \end{pmatrix} = \begin{pmatrix} 2 & 1 \\ 4 & 2 \end{pmatrix}A = \begin{pmatrix} 1 & 0 \\ 0 & 1 \end{pmatrix}.$$

12. Prove Theorem 15.5ii.
13. Prove Theorem 15.5iii.
14. Let A, C, and E be 2×2 matrices. Show that

$$\sum_{k=1}^{2}\left[\sum_{s=1}^{2} a_{1k}c_{ks}e_{s2}\right] = \sum_{s=1}^{2}\left[\sum_{k=1}^{2} a_{1k}c_{ks}\right]e_{s2}$$

by writing out the sums involved and using properties of the real numbers.
15. Find an example of 2×2 matrices, A, B, and C such that $AC = BC$, $C \neq 0_{2\times 2}$, and $A \neq B$.

ANSWERS

2. $\begin{pmatrix} 0 & 21 & 3 \\ -4 & 23 & -3 \end{pmatrix}$. **4.** $\begin{pmatrix} 16 & 10 \\ 12 & 6 \\ 12 & 4 \end{pmatrix}$. **6.** $\begin{pmatrix} 65 \\ 31 \end{pmatrix}$. **8.** $(3 \quad 6)$.

3. MATRIX REPRESENTATION OF A LINEAR TRANSFORMATION

Let $A \in \mathcal{R}_{mn}$, $X \in \mathcal{R}^n$, $Y \in \mathcal{R}^n$, and $c \in \mathcal{R}$. We note the following facts:

1. $AX \in \mathcal{R}^m$ (Definition 15.4).
2. $A(X + Y) = AX + AY$ (Theorem 15.5ii).
3. $A(cX) = c(AX)$ (Theorem 15.5iii).

Therefore, we can consider A to be a linear transformation from $\mathcal{R}^n$ into $\mathcal{R}^m$ (Definition 14.14, p. 571).

In true mathematical style we ask if the converse of the preceding statement is true: if T is a linear transformation from $\mathcal{R}^n$ into $\mathcal{R}^m$, does there exist an element A of $\mathcal{R}_{mn}$ such that $TX = AX$ for all X in $\mathcal{R}^n$? Before answering the question in general let us consider the following example:

$$T\begin{pmatrix} x \\ y \end{pmatrix} = \begin{pmatrix} 3x - 2y \\ 4x + y \end{pmatrix}, \quad \begin{pmatrix} x \\ y \end{pmatrix} \in \mathcal{R}^2.$$

Using the definition of matrix multiplication, we see that

$$T\begin{pmatrix} x \\ y \end{pmatrix} = \begin{pmatrix} 3 & -2 \\ 4 & 1 \end{pmatrix}\begin{pmatrix} x \\ y \end{pmatrix} \text{ for all } \begin{pmatrix} x \\ y \end{pmatrix} \text{ in } \mathcal{R}^2.$$

Consequently, the preceding comments assure us that T is linear. To find the matrix $A = \begin{pmatrix} 3 & -2 \\ 4 & 1 \end{pmatrix}$ such that $TX = AX$, we needed only to read off the coefficients of x and y in the expression for T. We note further that $T\begin{pmatrix} 1 \\ 0 \end{pmatrix} = \begin{pmatrix} 3 \\ 4 \end{pmatrix}$ and $T\begin{pmatrix} 0 \\ 1 \end{pmatrix} = \begin{pmatrix} -2 \\ 1 \end{pmatrix}$; that is, the columns of the matrix A are obtained by evaluating T at the vectors of the natural basis. The general situation is exactly the same, as is seen in the following theorem.

THEOREM 15.7 Let T be a linear transformation from $\mathcal{R}^n$ into $\mathcal{R}^m$. There exists a unique $m \times n$ matrix, denoted by $[T]$, such that $TX = [T]X$ for all X in $\mathcal{R}^n$.

The matrix $[T]$ is called the *natural matrix representation of* T and is found as follows:

Let $\{E_1, \ldots, E_n\}$ be the natural basis for $\mathcal{R}^n$ and let the j-th column of $[T]$ be equal to $T(E_j)$ for $j = 1, \ldots, n$.

Proof. We form the matrix $[T]$ as described in the latter part of the theorem: the j-th column of $[T]$ is equal to $T(E_j)$, $j = 1, \ldots, n$, where $\{E_1, \ldots, E_n\}$ is the natural basis for $\mathcal{R}^n$. We note then that $[T]E_j = TE_j$ by Definition 15.4 (Matrix Multiplication). It follows from the linearity of T, the properties of matrix multiplication, and the fact that $\{E_1, \ldots, E_n\}$ is a basis for $\mathcal{R}^n$ that $[T]X = TX$ for all X in $\mathcal{R}^n$.

The fact that there is only one matrix $[T]$ such that $[T]X = TX$ for all X in $\mathcal{R}^n$ follows immediately from Theorem 15.6. ∎

Example 7. (a) Let $T\begin{pmatrix} x \\ y \\ z \end{pmatrix} = \begin{pmatrix} 2x - 3y + z \\ y - x + 4z \end{pmatrix}$. Using the definition of matrix multiplication, we have

$$T\begin{pmatrix} x \\ y \\ z \end{pmatrix} = \begin{pmatrix} 2 & -3 & 1 \\ -1 & 1 & 4 \end{pmatrix}\begin{pmatrix} x \\ y \\ z \end{pmatrix}.$$

for all $\begin{pmatrix} x \\ y \\ z \end{pmatrix}$ in $\mathcal{R}^3$. Therefore

$$[T] = \begin{pmatrix} 2 & -3 & 1 \\ -1 & 1 & 4 \end{pmatrix}.$$

We note also that

$$T\begin{pmatrix} 1 \\ 0 \\ 0 \end{pmatrix} = \begin{pmatrix} 2 \\ -1 \end{pmatrix}, \quad T\begin{pmatrix} 0 \\ 1 \\ 0 \end{pmatrix} = \begin{pmatrix} -3 \\ 1 \end{pmatrix}, \quad \text{and } T\begin{pmatrix} 0 \\ 0 \\ 1 \end{pmatrix} = \begin{pmatrix} 1 \\ 4 \end{pmatrix}.$$

(b) Let $T\begin{pmatrix} x \\ y \\ z \end{pmatrix} = z - 4x + 3y$. This time we see that

$$T\begin{pmatrix} x \\ y \\ z \end{pmatrix} = (-4 \quad 3 \quad 1) \cdot \begin{pmatrix} x \\ y \\ z \end{pmatrix}$$

and $[T] = (-4 \quad 3 \quad 1)$.

(c) Let $T\begin{pmatrix} x \\ y \end{pmatrix} = \begin{pmatrix} 2xy \\ 3x + 4y \end{pmatrix}$. No matter how hard we try, we cannot find a matrix A such that $TX = AX$ for all X in $\mathcal{R}^2$. We are suspicious of the linearity of T. If we check, we find that $T\begin{pmatrix} 2 \\ 2 \end{pmatrix} \neq 2T\begin{pmatrix} 1 \\ 1 \end{pmatrix}$ and that T is not linear. ∎

The preceding theorem and examples show us two things:

1. How to find the natural matrix of a linear transformation.
2. If T is a linear transformation and if we know the values of T at each element of the natural basis $\{E_1, \ldots, E_n\}$, then we can find the values of T at any element X of $\mathcal{R}^n$.

If this last assertion is not clear, consider the next examples.

Example 8. Assume that T is a linear transformation from $\mathcal{R}^2$ into $\mathcal{R}^3$ and that $T\begin{pmatrix} 1 \\ 0 \end{pmatrix} = \begin{pmatrix} 3 \\ -4 \\ 0 \end{pmatrix}$, $T\begin{pmatrix} 0 \\ 1 \end{pmatrix} = \begin{pmatrix} 5 \\ 7 \\ -1 \end{pmatrix}$. We wish to find the general expression for $T\begin{pmatrix} x \\ y \end{pmatrix}$, for any $\begin{pmatrix} x \\ y \end{pmatrix}$ in $\mathcal{R}^2$ and we proceed as follows: by Theorem 15.7

$$[T] = \begin{pmatrix} 3 & 5 \\ -4 & 7 \\ 0 & -1 \end{pmatrix}$$

and

$$T\begin{pmatrix} x \\ y \end{pmatrix} = [T]\begin{pmatrix} x \\ y \end{pmatrix}$$

$$= \begin{pmatrix} 3 & 5 \\ -4 & 7 \\ 0 & -1 \end{pmatrix}\begin{pmatrix} x \\ y \end{pmatrix}$$

$$= \begin{pmatrix} 3x + 5y \\ -4x + 7y \\ -y \end{pmatrix} \text{ for all } \begin{pmatrix} x \\ y \end{pmatrix} \text{ in } \mathcal{R}^2. \quad ∎$$

Indeed, with a little thought we see that we can handle a more general situation. For example, suppose that T is a linear transformation from $\mathcal{R}^2$ into $\mathcal{R}^2$ and that $T\begin{pmatrix} 2 \\ 0 \end{pmatrix} = \begin{pmatrix} 4 \\ -2 \end{pmatrix}$ and $T\begin{pmatrix} 0 \\ 3 \end{pmatrix} = \begin{pmatrix} 6 \\ 9 \end{pmatrix}$. Since T is linear, we know that $T\begin{pmatrix} 1 \\ 0 \end{pmatrix} = \begin{pmatrix} 2 \\ -1 \end{pmatrix}$ and $T\begin{pmatrix} 0 \\ 1 \end{pmatrix} = \begin{pmatrix} 2 \\ 3 \end{pmatrix}$. If we proceed as in Example 8, we find that $T\begin{pmatrix} x \\ y \end{pmatrix} = \begin{pmatrix} 2x + 2y \\ -x + 3y \end{pmatrix}$ for all $\begin{pmatrix} x \\ y \end{pmatrix}$ in $\mathcal{R}^2$.

Another example of this type of reasoning is one in which T is a linear transformation from $\mathcal{R}^2$ into $\mathcal{R}^2$, $T\begin{pmatrix} 1 \\ 0 \end{pmatrix} = \begin{pmatrix} 5 \\ 2 \end{pmatrix}$ and $T\begin{pmatrix} 1 \\ 1 \end{pmatrix} = \begin{pmatrix} 4 \\ 3 \end{pmatrix}$. To find the general expression for $T\begin{pmatrix} x \\ y \end{pmatrix}$ we find the value of $T\begin{pmatrix} 0 \\ 1 \end{pmatrix}$. This is simply calculated:

$$T\begin{pmatrix} 0 \\ 1 \end{pmatrix} = T\left[\begin{pmatrix} 1 \\ 1 \end{pmatrix} - \begin{pmatrix} 1 \\ 0 \end{pmatrix} \right]$$

$$= T\begin{pmatrix} 1 \\ 1 \end{pmatrix} - T\begin{pmatrix} 1 \\ 0 \end{pmatrix} = \begin{pmatrix} 4 \\ 3 \end{pmatrix} - \begin{pmatrix} 5 \\ 2 \end{pmatrix} = \begin{pmatrix} -1 \\ 1 \end{pmatrix}.$$

Therefore, $[T] = \begin{pmatrix} 5 & -1 \\ 2 & 1 \end{pmatrix}$ and $T\begin{pmatrix} x \\ y \end{pmatrix} = \begin{pmatrix} 5x - y \\ 2x + y \end{pmatrix}$ for all $\begin{pmatrix} x \\ y \end{pmatrix}$ in $\mathcal{R}^2$.

Theorem 15.8 generalizes the situations discussed in the last two paragraphs.

THEOREM 15.8 Let $\{X_1, \ldots, X_n\}$ be a basis for $\mathcal{R}^n$ and let $B_1, \ldots, B_n$ be elements of $\mathcal{R}^m$. There exists a unique linear transformation T from $\mathcal{R}^n$ into $\mathcal{R}^m$ such that $TX_i = B_i$ for $i = 1, \ldots, n$.

Proof. Let $X \in \mathcal{R}^n$ and let $\{X_1, \ldots, X_n\}$ be a basis for $\mathcal{R}^n$. There exist unique scalars $c_1, \ldots, c_n$ such that $X = \sum_{i=1}^{n} c_i X_i$. If there is a linear transformation $T : \mathcal{R}^n \to \mathcal{R}^m$ such that $TX_i = B_i$, $i = 1, \ldots, n$, then TX is uniquely determined by

$$TX = T\left(\sum_{i=1}^{n} c_i X_i \right)$$

$$= \sum_{i=1}^{n} c_i TX_i = \sum_{i=1}^{n} c_i B_i,$$

or

(15.2) $$TX = \sum_{i=1}^{n} c_i B_i, \quad \text{when} \quad X = \sum_{i=1}^{n} c_i B_i.$$

To complete our proof we need to show that Equation (15.2) indeed defines a linear transformation.

Let X and Y be elements of $\mathcal{R}^n$, $X = \sum_{i=1}^{n} c_i X_i$ and $Y = \sum_{i=1}^{n} d_i X_i$. Let $r \in \mathcal{R}$. We then have

(15.3) $$X + Y = \sum_{i=1}^{n} (c_i + d_i) X_i$$

and

(15.4) $$rX = \sum_{i=1}^{n} (rc_i) X_i.$$

Using Equation (15.3), we have

$$T(X + Y) =^1 \sum_{i=1}^{n} (c_i + d_i)B_i$$

$$=^2 \sum_{i=1}^{n} c_i B_i + \sum_{i=1}^{n} d_i B_i$$

$$=^1 TX + TY.$$

Using Equation (15.4), we have

$$T(rX) =^1 \sum_{i=1}^{n} (rc_i)B_i$$

$$=^2 r\left(\sum_{i=1}^{n} c_i B_i\right)$$

$$=^1 rTX.$$

(1) By definition of T in Equation 15.2.
(2) Vector properties in $\mathcal{R}^m$.

Thus we see that $T(X + Y) = TX + TY$ and $T(rX) = rTX$ for all X, Y in $\mathcal{R}^n$ and r in $\mathcal{R}$. By Definition 14.14, T is a linear transformation. ■

Example 9. Assume that T is a linear transformation from $\mathcal{R}^3$ into $\mathcal{R}^3$ and that

$$T\begin{pmatrix}2\\1\\0\end{pmatrix} = \begin{pmatrix}0\\-1\\3\end{pmatrix}, \quad T\begin{pmatrix}1\\-1\\1\end{pmatrix} = \begin{pmatrix}7\\2\\1\end{pmatrix}, \quad T\begin{pmatrix}0\\4\\1\end{pmatrix} = \begin{pmatrix}-15\\-3\\5\end{pmatrix}.$$

We note first that $\left\{\begin{pmatrix}2\\1\\0\end{pmatrix}, \begin{pmatrix}1\\-1\\1\end{pmatrix}, \begin{pmatrix}0\\4\\1\end{pmatrix}\right\}$ is a basis for $\mathcal{R}^3$. Therefore, by Theorem 15.8, we know that there exists exactly one linear transformation from $\mathcal{R}^3$ into $\mathcal{R}^3$ with these specified values. We wish to find the general formula of $T\begin{pmatrix}x\\y\\z\end{pmatrix}$ for all $\begin{pmatrix}x\\y\\z\end{pmatrix}$ in $\mathcal{R}^3$. This we can do if we can find the values $T\begin{pmatrix}1\\0\\0\end{pmatrix}$, $T\begin{pmatrix}0\\1\\0\end{pmatrix}$, and $T\begin{pmatrix}0\\0\\1\end{pmatrix}$.

One can easily show that

(15.5)
$$\begin{pmatrix}1\\0\\0\end{pmatrix} = \frac{5}{11}\begin{pmatrix}2\\1\\0\end{pmatrix} + \frac{1}{11}\begin{pmatrix}1\\-1\\1\end{pmatrix} - \frac{1}{11}\begin{pmatrix}0\\4\\1\end{pmatrix}$$

(15.6)
$$\begin{pmatrix}0\\1\\0\end{pmatrix} = \frac{1}{11}\begin{pmatrix}2\\1\\0\end{pmatrix} - \frac{2}{11}\begin{pmatrix}1\\-1\\1\end{pmatrix} + \frac{2}{11}\begin{pmatrix}0\\4\\1\end{pmatrix}$$

(15.7)
$$\begin{pmatrix}0\\0\\1\end{pmatrix} = -\frac{4}{11}\begin{pmatrix}2\\1\\0\end{pmatrix} + \frac{8}{11}\begin{pmatrix}1\\-1\\1\end{pmatrix} + \frac{3}{11}\begin{pmatrix}0\\4\\1\end{pmatrix}.$$

Therefore, since T is linear,

$$T\begin{pmatrix}1\\0\\0\end{pmatrix} = \frac{5}{11}\begin{pmatrix}0\\-1\\3\end{pmatrix} + \frac{1}{11}\begin{pmatrix}7\\2\\1\end{pmatrix} - \frac{1}{11}\begin{pmatrix}-15\\-3\\5\end{pmatrix} = \begin{pmatrix}2\\0\\1\end{pmatrix}$$

$$T\begin{pmatrix}0\\1\\0\end{pmatrix} = -\frac{1}{11}\begin{pmatrix}0\\-1\\3\end{pmatrix} - \frac{2}{11}\begin{pmatrix}7\\2\\1\end{pmatrix} + \frac{2}{11}\begin{pmatrix}-15\\-3\\5\end{pmatrix} = \begin{pmatrix}-4\\-1\\1\end{pmatrix}$$

$$T\begin{pmatrix}0\\0\\1\end{pmatrix} = -\frac{4}{11}\begin{pmatrix}0\\-1\\3\end{pmatrix} + \frac{8}{11}\begin{pmatrix}7\\2\\1\end{pmatrix} + \frac{3}{11}\begin{pmatrix}-15\\-3\\5\end{pmatrix} = \begin{pmatrix}1\\1\\1\end{pmatrix}.$$

Hence by Theorem 15.7, for each $\begin{pmatrix}x\\y\\z\end{pmatrix}$ in $\mathcal{R}^3$,

$$T\begin{pmatrix}x\\y\\z\end{pmatrix} = [T]\begin{pmatrix}x\\y\\z\end{pmatrix} = \begin{pmatrix}2 & -4 & 1\\0 & -1 & 1\\1 & 1 & 1\end{pmatrix}\begin{pmatrix}x\\y\\z\end{pmatrix}$$

$$= \begin{pmatrix}2x - 4y + z\\z - y\\x + y + z\end{pmatrix}. \quad \blacksquare$$

The last theorem in this section gives us a vital connection between multiplication of matrices and composition of linear transformations.

THEOREM 15.9 Let $T:\mathcal{R}^n \to \mathcal{R}^m$, $S:\mathcal{R}^m \to \mathcal{R}^p$, T and S linear. Then $S \circ T$ is linear and $[S \circ T] = [S][T]$.

Proof. In Exercise 15, the student will be asked to prove that $S \circ T$ is linear.

To prove that $[S \circ T] = [S][T]$, it suffices to show that $[S \circ T]X = [S][T]X$ for all X in $\mathcal{R}^n$ (Theorem 15.6, p. 600, Matrix Cancellation Law). This follows simply, for if $X \in \mathcal{R}^n$,

$$[S \circ T]X =^1 (S \circ T)X$$

$$=^2 S(TX)$$

$$=^1 S([T]X)$$

$$=^1 [S][T]X.$$

(1) Theorem 15.7. (Note that $(S \circ T)$ is the composite of S and T and $[S \circ T]$ is the natural matrix of $(S \circ T)$.)
(2) Definition of composition. $\blacksquare$

Example 10. Let $T\begin{pmatrix} x \\ y \end{pmatrix} = \begin{pmatrix} 2x - 3y \\ y - 4x \end{pmatrix}$ and $S\begin{pmatrix} x \\ y \end{pmatrix} = \begin{pmatrix} y \\ x + 2y \end{pmatrix}$. We note

that $[T] = \begin{pmatrix} 2 & -3 \\ -4 & 1 \end{pmatrix}$, $[S] = \begin{pmatrix} 0 & 1 \\ 1 & 2 \end{pmatrix}$, and compute $S \circ T$ as follows:

$$(S \circ T)\begin{pmatrix} x \\ y \end{pmatrix} = [S \circ T]\begin{pmatrix} x \\ y \end{pmatrix}$$

$$= [S][T]\begin{pmatrix} x \\ y \end{pmatrix} \text{ by Theorem 15.9.}$$

$$= \begin{pmatrix} 0 & 1 \\ 1 & 2 \end{pmatrix}\begin{pmatrix} 2 & -3 \\ -4 & 1 \end{pmatrix}\begin{pmatrix} x \\ y \end{pmatrix}$$

$$= \begin{pmatrix} -4x + y \\ -6x - y \end{pmatrix}.$$

The student should check by direct computation that this is the correct formula for $S \circ T$. ■

EXERCISES

Find the natural matrix $[T]$ for each of the following linear transformations:

1. $T\begin{pmatrix} x \\ y \end{pmatrix} = \begin{pmatrix} x - 5y \\ y \end{pmatrix}$.

2. $T\begin{pmatrix} x \\ y \end{pmatrix} = 3x - y$.

3. $T\begin{pmatrix} x \\ y \\ z \end{pmatrix} = \begin{pmatrix} 4z - x - y \\ x + 3z \end{pmatrix}$.

4. $T\begin{pmatrix} x \\ y \end{pmatrix} = \begin{pmatrix} x + y \\ x - y \\ 3x + 2y \end{pmatrix}$.

5. $T(x) = \begin{pmatrix} 2x \\ -x \\ 0 \end{pmatrix}$.

6. $T\begin{pmatrix} x \\ y \\ z \end{pmatrix} = \begin{pmatrix} 7z - y \\ x \\ y + z - x \end{pmatrix}$.

7. $T\begin{pmatrix} x \\ y \\ z \\ w \end{pmatrix} = \begin{pmatrix} 3x - 4y + z \\ z + w \\ x - y \\ 2z + 3y \end{pmatrix}$

8. $T\begin{pmatrix} x \\ y \\ z \\ w \end{pmatrix} = 4w + 2y - x + z$.

Assume that each of the following functions, T, is a linear transformation. Find $[T]$ and the formula for TX, $X \in dom\ T$.

9. $\begin{cases} T\begin{pmatrix} 1 \\ 0 \end{pmatrix} = \begin{pmatrix} 2 \\ 4 \end{pmatrix} \\ T\begin{pmatrix} 0 \\ 1 \end{pmatrix} = \begin{pmatrix} -1 \\ 3 \end{pmatrix}. \end{cases}$

10. $\begin{cases} T\begin{pmatrix} 1 \\ 0 \\ 0 \end{pmatrix} = \begin{pmatrix} 2 \\ 5 \end{pmatrix} \\ T\begin{pmatrix} 0 \\ 1 \\ 0 \end{pmatrix} = \begin{pmatrix} -1 \\ 1 \end{pmatrix} \\ T\begin{pmatrix} 0 \\ 0 \\ 1 \end{pmatrix} = \begin{pmatrix} 0 \\ 4 \end{pmatrix}. \end{cases}$

11. $\begin{cases} T\begin{pmatrix}3\\0\end{pmatrix} = \begin{pmatrix}12\\15\end{pmatrix} \\ T\begin{pmatrix}0\\5\end{pmatrix} = \begin{pmatrix}5\\0\end{pmatrix}. \end{cases}$

12. $\begin{cases} T\begin{pmatrix}1\\1\end{pmatrix} = \begin{pmatrix}2\\3\\6\end{pmatrix} \\ T\begin{pmatrix}1\\-1\end{pmatrix} = \begin{pmatrix}4\\-1\\-2\end{pmatrix}. \end{cases}$

13. $\begin{cases} T\begin{pmatrix}1\\1\\0\end{pmatrix} = \begin{pmatrix}-1\\1\\0\end{pmatrix} \\ T\begin{pmatrix}1\\0\\1\end{pmatrix} = \begin{pmatrix}7\\1\\0\end{pmatrix} \\ T\begin{pmatrix}0\\1\\1\end{pmatrix} = \begin{pmatrix}6\\0\\2\end{pmatrix}. \end{cases}$

14. Let $T\begin{pmatrix}x\\y\\z\end{pmatrix} = \begin{pmatrix}2x + 4y\\y - 3x\end{pmatrix}$ and $S\begin{pmatrix}x\\y\end{pmatrix} = y + 5x$. Compute the formula for

$(S \circ T)\begin{pmatrix}x\\y\\z\end{pmatrix}$ in two different ways.

15. Let $T:\mathcal{R}^n \rightarrow \mathcal{R}^m$, $S:\mathcal{R}^m \rightarrow \mathcal{R}^q$, T and S linear. Prove that $S \circ T$ is a linear transformation from $\mathcal{R}^n$ into $\mathcal{R}^q$.

16. Let $T:\mathcal{R}^n \rightarrow \mathcal{R}^m$, $S:\mathcal{R}^n \rightarrow \mathcal{R}^m$, T and S linear. Let $c \in \mathcal{R}$.
Prove (i) $T + S$ is linear and $[T + S] = [T] + [S]$
(ii) cT is linear and $[cT] = c[T]$.

17. Let $T:\mathcal{R}^n \rightarrow \mathcal{R}^m$, T linear. Let $\{E_1, \ldots, E_n\}$ be the natural basis for $\mathcal{R}^n$ and $Y_1, \ldots, Y_n$ the columns of $[T]$. Show that $TE_i = Y_i$, $i = 1, \ldots, n$.

ANSWERS

2. $[T] = (3 \quad -1)$. 4. $[T] = \begin{pmatrix} 1 & 1 \\ 1 & -1 \\ 3 & 2 \end{pmatrix}$.

7. $[T] = \begin{pmatrix} 3 & -4 & 1 & 0 \\ 0 & 0 & 1 & 1 \\ 1 & -1 & 0 & 0 \\ 0 & 3 & 2 & 0 \end{pmatrix}$ 10. $\begin{cases} [T] = \begin{pmatrix} 2 & -1 & 0 \\ 5 & 1 & 4 \end{pmatrix} \\ T\begin{pmatrix}x\\y\\z\end{pmatrix} = \begin{pmatrix} 2x - y \\ 5x + y + 4z \end{pmatrix}. \end{cases}$

12. $\begin{cases} [T] = \begin{pmatrix} 3 & -1 \\ 1 & 2 \\ 2 & 4 \end{pmatrix} \\ T\begin{pmatrix}x\\y\end{pmatrix} = \begin{pmatrix} 3x - y \\ x + 2y \\ 2x + 4y \end{pmatrix}. \end{cases}$

4. INVERTIBLE LINEAR TRANSFORMATIONS AND MATRICES

In the preceding section, we discovered a close connection between $m \times n$ matrices and linear transformations from $\mathcal{R}^n$ into $\mathcal{R}^m$. Indeed we found that every linear transformation from $\mathcal{R}^n$ into $\mathcal{R}^m$ can be represented by an $m \times n$ matrix, and conversely, every $m \times n$ matrix represents a linear transformation from $\mathcal{R}^n$ into $\mathcal{R}^m$. Moreover, in Theorem 15.9, we found that $[S \circ T] = [S][T]$ and in Exercise 16 that $[S + T] = [S] + [T]$ and $[rT] = r[T]$. In the present section we shall discuss special types of $m \times n$ matrices and linear transformations from $\mathcal{R}^n$ into $\mathcal{R}^n$, and shall discuss the connection between these two.

DEFINITION 15.10 Let T be a linear transformation from $\mathcal{R}^n$ into $\mathcal{R}^n$. T is *invertible* if and only if there exists a function $S: \mathcal{R}^n \to \mathcal{R}^n$ such that $(S \circ T)X = (T \circ S)X = X$ for all X in $\mathcal{R}^n$. In this case we call S the *inverse of* T and write $S = T^{-1}$.

Example 11. Let $T: \mathcal{R}^2 \to \mathcal{R}^2$ be defined by $T\begin{pmatrix} x \\ y \end{pmatrix} = \begin{pmatrix} 2x + y \\ 3x + 2y \end{pmatrix}$ and

$S: \mathcal{R}^2 \to \mathcal{R}^2$ be defined by $S\begin{pmatrix} x \\ y \end{pmatrix} = \begin{pmatrix} 2x - y \\ -3x + 2y \end{pmatrix}$. We compute $(S \circ T)\begin{pmatrix} x \\ y \end{pmatrix}$

and find that

$$(S \circ T)\begin{pmatrix} x \\ y \end{pmatrix} = [S][T]\begin{pmatrix} x \\ y \end{pmatrix}$$

$$= \begin{pmatrix} 2 & -1 \\ -3 & 2 \end{pmatrix}\begin{pmatrix} 2 & 1 \\ 3 & 2 \end{pmatrix}\begin{pmatrix} x \\ y \end{pmatrix}$$

$$= \begin{pmatrix} 1 & 0 \\ 0 & 1 \end{pmatrix}\begin{pmatrix} x \\ y \end{pmatrix} = \begin{pmatrix} x \\ y \end{pmatrix}.$$

Similarly, $(T \circ S)\begin{pmatrix} x \\ y \end{pmatrix} = \begin{pmatrix} x \\ y \end{pmatrix}$ for all $\begin{pmatrix} x \\ y \end{pmatrix} \in \mathcal{R}^2$. By Definition 15.10, T is invertible and $S = T^{-1}$. ∎

We call attention to the relation between $[S]$ and $[T]$ in Example 11 by noting that

$$[S][T] = [T][S] = \begin{pmatrix} 1 & 0 \\ 0 & 1 \end{pmatrix}.$$

In Exercise 9 of Section 2 (p. 601) the reader is asked to find a 2×2 matrix I_2 such that $AI_2 = I_2A = A$ for all A in $\mathcal{R}_{22}$. The solution to the problem is

$I_2 = \begin{pmatrix} 1 & 0 \\ 0 & 1 \end{pmatrix}$. Moreover, this matrix has the property that $I_2 \begin{pmatrix} x \\ y \end{pmatrix} = \begin{pmatrix} x \\ y \end{pmatrix}$ for all $\begin{pmatrix} x \\ y \end{pmatrix}$ in $\mathcal{R}^2$ and hence is the matrix representation of the identity function $I : IX = X,\ X \in \mathcal{R}^2$.

DEFINITION 15.11 (Identity Matrix.) Let I_n be the $n \times n$ matrix with the ij-th coordinates equal 1 for $i = j$ and equal 0 for $i \neq j$. I_n is called the *identity matrix* of $\mathcal{R}_{nn}$.

In Exercise 10 the student will be asked to prove that $I_n A = A I_n = A$ for all A in $\mathcal{R}_{nn}$ and that $I_n X = X$ for all X in $\mathcal{R}^n$.

The next definition is the matrix analogue of Definition 15.10.

DEFINITION 15.12 (Inverse of a Matrix.) Let $A \in \mathcal{R}_{nn}$. A is *invertible* if and only if there exists an element B of $\mathcal{R}_{nn}$ such that $AB = BA = I_n$.

We call B the *inverse* of A and write $B = A^{-1}$.

In Exercise 11 the student will be asked to show that if A is an invertible $n \times n$ matrix, and if $AB = BA = I_n$ and $AC = CA = I_n$, then $B = C$. Hence, the inverse of an invertible matrix is unique.

Now that we have defined the use of the term invertible for linear transformations and for matrices, can we relate the two uses? Example 11 gives us a strong hint as to the formulation of the next theorem.

THEOREM 15.13 Let T be a linear transformation from $\mathcal{R}^n$ into $\mathcal{R}^n$.

    i. If T is invertible, T^{-1} is linear.

    ii. T is invertible if and only if $[T]$ is invertible, in which case $[T]^{-1} = [T^{-1}]$.

Proof of i. Assume that T is invertible and let $S = T^{-1}$. By Definition 15.10, $(T \circ S)X = X$ for all X in $\mathcal{R}^n$. Therefore, if $X \in \mathcal{R}^n$, $X = TY$, where $Y = SX$ and hence $X \in ran\ T$. This shows us that $ran\ T = \mathcal{R}^n$.

Now let X and Y be elements of $\mathcal{R}^n$. Since $ran\ T = \mathcal{R}^n$, there exist Z and W in $\mathcal{R}^n$ such that

(15.8) $$X = TZ \quad \text{and} \quad Y = TW.$$

Consequently,

(15.9) $X + Y = T(Z + W)$ since T is linear.

Therefore,

$$S(X + Y) =^1 (S \circ T)(Z + W)$$
$$=^2 Z + W$$
$$=^2 (S \circ T)Z + (S \circ T)W$$
$$=^3 SX + SY.$$

(1) By equation 15.9.
(2) Since $S = T^{-1}$ (Definition 15.10).
(3) By Equation 15.8.

We see that $S(X + Y) = SX + SY$ for all X, Y in $\mathcal{R}^n$. In an analogous fashion, we can show that $S(cX) = cSX$ for all X in $\mathcal{R}^n$ and c in $\mathcal{R}$. Thus S is linear (Definition 14.14, p. 571).

Proof of ii. Consider the following implication chain: T is invertible and $S = T^{-1}$

$$\Leftrightarrow^1 (T \circ S)X = (S \circ T)X = X, X \in \mathcal{R}^n$$
$$\Leftrightarrow^2 [T][S]X = [S][T]X = I_n X, X \in \mathcal{R}^n$$
$$\Leftrightarrow^3 [T][S] = [S][T] = I_n$$
$$\Leftrightarrow^4 [T] \text{ is invertible and } [T]^{-1} = [S].$$

(1) Definition 15.10 (Invertible Linear Transformation).
(2) Theorems 15.9, p. 607, and 15.7, p. 603, and Exercise 10.
(3) Theorem 15.6, p. 600.
(4) Definition 15.12 and Exercise 11.

This implication chain establishes statement (ii) of the theorem. ∎

Now that we have given the connection between invertible linear transformations and invertible matrices, we face two other questions:

1. Given a linear transformation from $\mathcal{R}^n$ into $\mathcal{R}^n$ or an $n \times n$ matrix, how can one tell if it is invertible (other than by finding the inverse)?

2. If it is invertible, how can one find the inverse transformation or matrix? Theorem 15.14 will answer the first question.

THEOREM 15.14 Let T be a linear transformation from $\mathcal{R}^n$ into $\mathcal{R}^n$. The following are equivalent:

    i. T in invertible,
    ii. $\ker T = \{O_n\}$,
    iii. $\operatorname{ran} T = \mathcal{R}^n$.

Proof. We shall prove the following sequence of implications: (i) $\Rightarrow$ (ii), (ii) $\Leftrightarrow$ (iii) and (iii) $\Rightarrow$ (i). Thus any one of the conditions implies the other two.

To show that (i) $\Rightarrow$ (ii), assume that T is invertible and $X \in ker\ T$. Then

$$X = (T^{-1} \circ T)X \quad \text{by Definition 15.10}$$

$$= T^{-1}(TX)$$

$$= T^{-1}O_n \quad \text{since } TX = O_n$$

$$= O_n \quad \text{by Theorems 15.13, p. 611, and 14.15, p. 572.}$$

Thus the only element of $ker\ T$ is O_n.

By Theorem 14.22, p. 589, we know that $dim\ (ran\ T) = n - dim\ (ker\ T)$. Therefore,

$$ker\ T = \{O_n\} \Leftrightarrow^1 dim\ (ker\ T) = 0$$

$$\Leftrightarrow^2 dim\ (ran\ T) = n$$

$$\Leftrightarrow^3 ran\ T = \mathcal{R}^n.$$

(1) Definition 14.21 (Dimension), p. 588.
(2) Theorem 14.22 (Dimension Theorem), p. 589.
(3) Theorem 14.13, p. 569.

This proves that (ii) and (iii) are equivalent.

To show that (iii) $\Rightarrow$ (i), assume that $ran\ T = \mathcal{R}^n$. If $Y \in \mathcal{R}^n$, there exists X in $\mathcal{R}^n$ such that $Y = TX$. If $X_1 \in \mathcal{R}^n$ and $Y = TX_1$ also, then $TX = TX_1$ or $T(X - X_1) = O_n$. Therefore, $X - X_1 \in ker\ T$. However, since (ii) and (iii) are equivalent, $ker\ T = \{O_n\}$, so that $X_1 = X$. Thus for each Y in $\mathcal{R}^n$, there exists a *unique* X in $\mathcal{R}^n$ such that $Y = TX$.

We are now ready to define the inverse, S, of T and do so as follows: if $Y \in \mathcal{R}^n$, define $S(Y) = X$ if and only if $Y = TX$. Since $ran\ T = \mathcal{R}^n$, $dom\ T^{-1} = \mathcal{R}^n$ and since $TX_1 = TX_2$ if and only if $X_1 = X_2$, we know that T^{-1} is a function. Moreover, we have

$$(S \circ T)X = S(TX)$$

$$= SY, \quad \text{where } Y = TX,$$

$$= X \quad \text{for all } X \text{ in } \mathcal{R}^n$$

and

$$(T \circ S)Y = T(SY)$$

$$= TX, \quad \text{where } X = SY,$$

$$= Y \quad \text{for all } Y \text{ in } \mathcal{R}^n.$$

Therefore, S satisfies the definition (15.10) of the inverse of T. ∎

Example 12. Let us use this theorem to show that the matrix $[T] = \begin{pmatrix} 1 & 3 & 2 \\ 4 & 0 & 0 \\ 7 & 0 & 1 \end{pmatrix}$ is invertible. To do this we define $T: \mathcal{R}^3 \to \mathcal{R}^3$ by

$$T\begin{pmatrix} x \\ y \\ z \end{pmatrix} = \begin{pmatrix} 1 & 3 & 2 \\ 4 & 0 & 0 \\ 7 & 0 & 1 \end{pmatrix}\begin{pmatrix} x \\ y \\ z \end{pmatrix}$$

$$= \begin{pmatrix} x + 3y + 2z \\ 4x \\ 7x + z \end{pmatrix}.$$

By Theorem 15.13, $[T]$ is invertible if and only if T is invertible, and by Theorem 15.14, T is invertible if and only if $ker\ T = \{O_3\}$. Therefore, if we show that $ker\ T = \{O_3\}$, we shall know that both $[T]$ and T are invertible. We proceed to do this:

$$ker\ T = \{X \mid X \in \mathcal{R}^3,\ TX = O_3\}$$

$$= \left\{ \begin{pmatrix} x \\ y \\ z \end{pmatrix} \middle| \begin{pmatrix} x + 3y + 2z \\ 4x \\ 7x + z \end{pmatrix} = \begin{pmatrix} 0 \\ 0 \\ 0 \end{pmatrix} \right\}$$

$$= \left\{ \begin{pmatrix} x \\ y \\ z \end{pmatrix} \middle| x + 3y + 2z = 0,\ 4x = 0,\ \text{and}\ 7x + z = 0 \right\}$$

$$=^1 \left\{ \begin{pmatrix} 0 \\ 0 \\ 0 \end{pmatrix} \right\}$$

$$= O_3.$$

(1) Solving the system of linear equations. ∎

In the discussion just prior to Theorem 15.14, we also asked the question: How do we find the inverse of a given linear transformation or matrix? Let us again consider the matrix given in Example 12.

Example 13. Let $[T] = \begin{pmatrix} 1 & 3 & 2 \\ 4 & 0 & 0 \\ 7 & 0 & 1 \end{pmatrix}$. From our work in Example 12, we know that $[T]$ is invertible. Let us use Definition 15.12, p. 611, to find $[T]^{-1}$. Let

$$[T]^{-1} = \begin{pmatrix} a & b & c \\ d & e & f \\ g & h & k \end{pmatrix}.$$

Since $[T]^{-1}[T] = I_3$, we have

$$\begin{pmatrix} a & b & c \\ d & e & f \\ g & h & k \end{pmatrix}\begin{pmatrix} 1 & 3 & 2 \\ 4 & 0 & 0 \\ 7 & 0 & 1 \end{pmatrix} = \begin{pmatrix} 1 & 0 & 0 \\ 0 & 1 & 0 \\ 0 & 0 & 1 \end{pmatrix},$$

or

$$\begin{pmatrix} (a + 4b + 7c) & 3a & (2a + c) \\ (d + 4e + 7f) & 3d & (2d + f) \\ (g + 4h + 7k) & 3g & (2g + k) \end{pmatrix} = \begin{pmatrix} 1 & 0 & 0 \\ 0 & 1 & 0 \\ 0 & 0 & 1 \end{pmatrix}.$$

Using the definition of equality of matrices, we find that

$$\begin{cases} a + 4b + 7c = 1 \\ 3a \qquad\qquad = 0 \\ 2a + \qquad c = 0 \end{cases}$$

$$\begin{cases} d + 4e + 7f = 0 \\ 3d \qquad\qquad = 1 \\ 2d + \qquad f = 0 \end{cases}$$

$$\begin{cases} g + 4h + 7k = 0 \\ 3g \qquad\qquad = 0 \\ 2g + \qquad k = 1. \end{cases}$$

Solving these three systems of equations, we have

$$a = 0, \qquad b = \tfrac{1}{4}, \qquad c = 0,$$
$$d = \tfrac{1}{3}, \qquad e = \tfrac{13}{12}, \qquad f = -\tfrac{2}{3},$$
$$g = 0, \qquad h = -\tfrac{7}{4}, \qquad k = 1.$$

Therefore,

$$[T]^{-1} = \begin{pmatrix} 0 & 1/4 & 0 \\ 1/3 & 13/12 & -2/3 \\ 0 & -7/4 & 1 \end{pmatrix}.$$

That student should check that $[T]^{-1}[T] = [T][T]^{-1} = I_3$. ∎

The actual calculations involved in finding $[T]^{-1}$ in Example 13 were not too involved but we note that they could have been worse. If $[T]$ had been a 4×4 matrix we would have been forced to solve 4 sets of equations in 4 unknowns each.

The last two sections of this chapter will be devoted to using matrices to solve systems of linear equations. One of our objectives in the last section will be to develop a reasonably graceful method of inverting (finding inverses of) matrices. In the exercises of this section we shall develop a special method for inverting 2×2 matrices.

EXERCISES

Using Theorem 15.14 determine which of the following linear transformations are invertible:

1. $T(x) = 3x.$

2. $T\begin{pmatrix} x \\ y \end{pmatrix} = \begin{pmatrix} 2x - y \\ -4x + 2y \end{pmatrix}.$

3. $T\begin{pmatrix} x \\ y \end{pmatrix} = \begin{pmatrix} 3x - 4y \\ 2y + x \end{pmatrix}.$

4. $T\begin{pmatrix} x \\ y \\ z \end{pmatrix} = \begin{pmatrix} 2x - y \\ x + z \\ 2x - y \end{pmatrix}.$

5. $T\begin{pmatrix} x \\ y \\ z \end{pmatrix} = \begin{pmatrix} x + y + z \\ x - y + z \\ x + y - z \end{pmatrix}.$

Using Theorem 15.14 and Theorem 15.13, determine which of the following matrices are invertible.

6. $\begin{pmatrix} 2 & 3 \\ -4 & -6 \end{pmatrix}.$

7. $\begin{pmatrix} 2 & 4 \\ 1 & 3 \end{pmatrix}.$

8. $\begin{pmatrix} 3 & 1 & 4 \\ 0 & 1 & 0 \\ 1 & -1 & 1 \end{pmatrix}.$

9. $\begin{pmatrix} 2 & 1 & 4 \\ 4 & 2 & 8 \\ 3 & 0 & 1 \end{pmatrix}.$

10. Let I_n be the $n \times n$ identity matrix. Show that $AI_n = I_n A = A$ for all A in $\mathcal{R}_{nn}$ and that $I_n X = X$ for all X in $\mathcal{R}^n$.

11. Let A, B, and C be $n \times n$ matrices. Show that if $AB = BA = I_n$ and $AC = CA = I_n$, then $B = C$.

12. Prove that if A is an invertible $n \times n$ matrix, then A^{-1} is invertible and $(A^{-1})^{-1} = A$.

13. Prove that if A and B are invertible $n \times n$ matrices, then AB is invertible and $(AB)^{-1} = B^{-1}A^{-1}$.

14. Let $A = \begin{pmatrix} a & b \\ c & d \end{pmatrix}$. Prove that when $\begin{vmatrix} a & b \\ c & d \end{vmatrix} = ad - bc = 0$, then A is not invertible.

15. Let $A = \begin{pmatrix} a & b \\ c & d \end{pmatrix}$. Prove that when $ad - bc \neq 0$, A is invertible and

$$A^{-1} = \frac{1}{ad - bc}\begin{pmatrix} d & -b \\ -c & a \end{pmatrix}.$$

For the following matrices use Exercises 14 and 15 to either prove that A is not invertible or prove that A is invertible and find A^{-1}.

16. $A = \begin{pmatrix} 4 & 1 \\ 11 & 3 \end{pmatrix}.$

17. $A = \begin{pmatrix} 1 & 2 \\ 2 & 4 \end{pmatrix}.$

18. $A = \begin{pmatrix} 5 & 2 \\ 4 & 2 \end{pmatrix}.$

19. $A = \begin{pmatrix} -1 & 2 \\ -1 & 3 \end{pmatrix}.$

20. $A = \begin{pmatrix} 5 & 4 \\ 4 & 5 \end{pmatrix}.$

21. Let $T:\mathcal{R}^n \rightarrow \mathcal{R}^n$, T linear. Let $Y_1, \ldots, Y_n$ be the columns of $[T]$. Prove that T is invertible if and only if $\{Y_1, \ldots, Y_n\}$ is a linearly independent set. (Hint: Use Theorem 15.14 and Exercise 17, Section 3.)

ANSWERS

2. Not invertible. **5.** Invertible. **7.** Invertible. **9.** Not invertible.

16. $A^{-1} = \begin{pmatrix} 3 & -1 \\ -11 & 4 \end{pmatrix}$. **18.** $A^{-1} = \begin{pmatrix} 1 & -1 \\ -2 & 5/2 \end{pmatrix}$.

20. $A^{-1} = \begin{pmatrix} 5/9 & -4/9 \\ -4/9 & 5/9 \end{pmatrix}$

5. AFFINE TRANSFORMATIONS AND k-PLANES

In Chapter 5, we defined the concepts of affine and linear transformations from $\mathcal{R}$ to $\mathcal{R}^m$. (See Definition 5.18, p. 241.) The reader will recall that the following are equivalent:

1. A is an affine transformation from $\mathcal{R}$ into $\mathcal{R}^m$.
2. There exist elements M and Y_0 of $\mathcal{R}^m$ such that

$$A(x) = xM + Y_0 \text{ for all } x \text{ in } \mathcal{R}.$$

3. There exists a linear transformation L from $\mathcal{R}$ into $\mathcal{R}^m$ and an element Y_0 of $\mathcal{R}^m$ such that

$$A(x) = L(x) + Y_0 \text{ for all } x \text{ in } \mathcal{R}.$$

In view of the fact that we have studied linear transformations in detail, we shall generalize our concept of an affine transformation by generalizing 3.

DEFINITION 15.15 (Affine Transformation $\mathcal{R}^n \rightarrow \mathcal{R}^m$.) Let $A:\mathcal{R}^n \rightarrow \mathcal{R}^m$. A is an *affine transformation* if and only if there exists a linear transformation $L:\mathcal{R}^n \rightarrow \mathcal{R}^m$ and an element Y_0 of $\mathcal{R}^m$ such that

$$AX = LX + Y_0 \text{ for all } X \text{ in } \mathcal{R}^n.$$

Example 14. Let $A:\mathcal{R}^2 \rightarrow \mathcal{R}^2$ be defined by

$$A\begin{pmatrix} x \\ y \end{pmatrix} = \begin{pmatrix} 3x - 2y + 1 \\ -4x + y - 2 \end{pmatrix} \text{ for all } \begin{pmatrix} x \\ y \end{pmatrix} \text{ in } \mathcal{R}^2.$$

Then

$$A\begin{pmatrix} x \\ y \end{pmatrix} = \begin{pmatrix} 3x - 2y \\ -4x + y \end{pmatrix} + \begin{pmatrix} 1 \\ -2 \end{pmatrix}$$

$$= \begin{pmatrix} 3 & -2 \\ -4 & 1 \end{pmatrix}\begin{pmatrix} x \\ y \end{pmatrix} + \begin{pmatrix} 1 \\ -2 \end{pmatrix}.$$

If we define $L\begin{pmatrix} x \\ y \end{pmatrix} = \begin{pmatrix} 3 & -2 \\ -4 & 1 \end{pmatrix} \begin{pmatrix} x \\ y \end{pmatrix}$ and $Y_0 = \begin{pmatrix} 1 \\ -2 \end{pmatrix}$, we see that L is linear and $AX = LX + Y_0$ for all X in $\mathcal{R}^2$. Thus A is an affine transformation from $\mathcal{R}^2$ into $\mathcal{R}^2$ (Definition 15.15). ∎

Assume now that A is an affine transformation from $\mathcal{R}^n$ into $\mathcal{R}^m$, that L_1 and L_2 are linear transformations, and that Y_1 and Y_2 are elements of $\mathcal{R}^m$ such that

$$(15.10) \qquad\qquad AX = L_1 X + Y_1 \text{ for all } X \text{ in } \mathcal{R}^n$$

and

$$(15.11) \qquad\qquad AX = L_2 X + Y_2 \text{ for all } X \text{ in } \mathcal{R}^n.$$

Combining Equations 15.10 and 15.11, we have

$$(15.12) \qquad\qquad L_1 X + Y_1 = L_2 X + Y_2 \text{ for all } X \text{ in } \mathcal{R}^n.$$

In particular, Equation 15.12 holds for $X = O_n$. Moreover, by Theorem 14.15i, p. 572, we know that $L_1 O_n = L_2 O_n = O_m$. Consequently,

$$(15.13) \qquad\qquad\qquad Y_1 = Y_2.$$

Combining Equations 15.12 and 15.13, we have

$$(15.14) \qquad\qquad L_1 X = L_2 X \text{ for all } X \text{ in } \mathcal{R}^n.$$

However, this implies that $L_1 = L_2$ by definition of equality of functions. Thus we see that the linear transformation L and the element Y_0 of $\mathcal{R}^m$, defining A, are unique. We state this result formally as:

THEOREM 15.16   If $A : \mathcal{R}^n \to \mathcal{R}^m$ is an affine transformation, there is exactly one linear transformation L from $\mathcal{R}^n$ into $\mathcal{R}^m$ and exactly one element Y_0 of $\mathcal{R}^m$ such that

$$AX = LX + Y_0 \text{ for all } X \text{ in } \mathcal{R}^n.$$

In Chapter 14, we discovered an intimate relation between linear transformations from $\mathcal{R}^n$ into $\mathcal{R}^m$ and linear subspaces of $\mathcal{R}^n$ and $\mathcal{R}^m$. In Theorem 14.19, p. 579, we showed that the kernel of a linear transformation is a linear subspace of $\mathcal{R}^n$, the range is a linear subspace of $\mathcal{R}^m$, and the graph is a linear subspace of $\mathcal{R}^{n+m}$. What shall we call the function-loci defined implicitly, parametrically, and explicitly by an affine transformation? Since an affine transformation is a "translate" of a linear transformation, let us now consider translates of linear subspaces and attempt to relate them to affine transformations.

DEFINITION 15.17 A subset $\mathcal{S}$ of $\mathcal{R}^n$ is a *k-plane*, $0 \leq k \leq n$, if and only if there exists a k-dimensional linear subspace $\mathcal{M}$ of $\mathcal{R}^n$ and an element X_0 of $\mathcal{R}^n$ such that $\mathcal{S} = \mathcal{M} + X_0$.

Example 15. Let $\mathcal{S} = \left\{ \begin{pmatrix} 3x - 2y + 1 \\ -4x + y - 2 \end{pmatrix} \middle| \begin{pmatrix} x \\ y \end{pmatrix} \in \mathcal{R}^2 \right\}$. We note that

$\mathcal{S} = ran\ L + \begin{pmatrix} 1 \\ -2 \end{pmatrix}$, where L is the linear transformation appearing in Example 14. Since $ran\ L$ is a linear subspace of $\mathcal{R}^2$, we know that $\mathcal{S}$ is a k-plane of $\mathcal{R}^2$. It is easy to check that $dim\ (ran\ L) = 2$, so that $\mathcal{S}$ is a 2-plane of $\mathcal{R}^2$. ■

The reader will realize at this point that a 1-plane of $\mathcal{R}^n$ is a line and a 2-plane of $\mathcal{R}^n$ is what we have previously called a plane. Moreover, note that a 0-plane of $\mathcal{R}^n$ is merely a point of $\mathcal{R}^n$.

Without further discourse let us state the theorem giving the connection between affine transformations and k-planes.

THEOREM 15.18 Let $\mathcal{S}$ be a nonempty subset of $\mathcal{R}^n$.

　　i. There exists an integer k, $0 \leq k \leq n$, such that $\mathcal{S}$ is a k-plane if and only if $\mathcal{S}$ can be expressed parametrically by an affine transformation A.

　　ii. There exists an integer k, $0 \leq k \leq n$, such that $\mathcal{S}$ is a k-plane if and only if $\mathcal{S}$ can be expressed implicitly by an affine transformation A.

　　iii. If $\mathcal{S}$ can be expressed explicitly by an affine transformation A, then $\mathcal{S}$ is a k-plane for some integer k, $0 \leq k \leq n$.

Proof of i. Assume that $\mathcal{S}$ is a k-plane of $\mathcal{R}^n$. Then $\mathcal{S} = \mathcal{M} + X_0$, where $X_0 \in \mathcal{R}^n$ and $\mathcal{M}$ is a k-dimensional linear subspace of $\mathcal{R}^n$. Let $\{X_1, \ldots, X_k\}$ be a basis for $\mathcal{M}$ and $\{E_1, \ldots, E_k\}$ be the natural basis for $\mathcal{R}^k$. By Theorem 15.8, there exists a unique linear transformation $L: \mathcal{R}^k \to \mathcal{R}^n$ such that $LE_i = X_i$, $i = 1, \ldots, k$. In Exercise 16, the reader will be asked to show that $ran\ L = \mathcal{M}$.

Now define $A: \mathcal{R}^k \to \mathcal{R}^n$ by $AX = LX + X_0$. We have

$$
\begin{aligned}
ran\ A &=^1 \{LX + X_0 \mid X \in \mathcal{R}^k\} \\
&=^2 \{LX \mid X \in \mathcal{R}^k\} + X_0 \\
&=^1 ran\ L + X_0 \\
&=^3 \mathcal{M} + X_0 \\
&=^4 \mathcal{S}.
\end{aligned}
$$

(1) Definition of range.
(2) Definition 7.11, p. 314, translation of a set.
(3) Exercise 16.
(4) Since $\mathfrak{T} = \mathcal{M} + X_0$.

Thus we have shown that when $\mathfrak{T}$ is a k-plane of $\mathcal{R}^n$, there exists an affine transformation A such that $\mathfrak{T} = ran\ A$ (that is, A defines $\mathfrak{T}$ parametrically).

In Exercise 17, we shall ask the reader to show that the range of any affine transformation is a k-plane of $\mathcal{R}^n$ for some value of k and n.

The proofs of (ii) and (iii) will also be left as exercises for the student. (See Exercises 18, 19, and 20.) ∎

The reader should certainly wonder why Theorem 15.18iii is not reversible. Do there exist k-planes in $\mathcal{R}^n$ that cannot be expressed explicitly? The answer to this question was actually given in Chapter 7. The y-axis in $\mathcal{R}^2$ is certainly a 1-plane but equally certainly cannot be expressed as the graph of a function.

Example 16. In Chapters 1 and 2 we learned how to express a plane in $\mathcal{R}^3$ parametrically, implicitly, and explicitly. Consider the plane $\mathfrak{T}$ passing through $\begin{pmatrix} 1 \\ 0 \\ 0 \end{pmatrix}$, $\begin{pmatrix} 0 \\ 1 \\ 0 \end{pmatrix}$, and $\begin{pmatrix} 0 \\ 0 \\ 1 \end{pmatrix}$. We wish to find the affine transformations which define $\mathfrak{T}$ in these three fashions.

1. $\mathfrak{T} = \left\{ \begin{pmatrix} 1 \\ 0 \\ 0 \end{pmatrix} + r\begin{pmatrix} -1 \\ 1 \\ 0 \end{pmatrix} + s\begin{pmatrix} -1 \\ 0 \\ 1 \end{pmatrix} \;\middle|\; r,\, s \in \mathcal{R} \right\}$ by Theorem 1.45, p. 70.

If we define $A_1 : \mathcal{R}^2 \to \mathcal{R}^3$ by

$$A_1\begin{pmatrix} x \\ y \end{pmatrix} = \begin{pmatrix} 1 \\ 0 \\ 0 \end{pmatrix} + x\begin{pmatrix} -1 \\ 1 \\ 0 \end{pmatrix} + y\begin{pmatrix} -1 \\ 0 \\ 1 \end{pmatrix}, \quad \begin{pmatrix} x \\ y \end{pmatrix} \in \mathcal{R}^2,$$

we see that A_1 is affine, $ran\ A_1 = \mathfrak{T}$, and A_1 defines $\mathfrak{T}$ parametrically.

2. $\mathfrak{T} = \left\{ \begin{pmatrix} x \\ y \\ z \end{pmatrix} \;\middle|\; \left[\begin{pmatrix} x \\ y \\ z \end{pmatrix} - \begin{pmatrix} 1 \\ 0 \\ 0 \end{pmatrix} \right] \cdot \begin{pmatrix} 1 \\ 1 \\ 1 \end{pmatrix} = 0 \right\}$ by Theorem 2.19i, p. 115.

If we define $A_2 : \mathcal{R}^3 \to \mathcal{R}^1$ by

$$A_2\begin{pmatrix} x \\ y \\ z \end{pmatrix} = \left[\begin{pmatrix} x \\ y \\ z \end{pmatrix} - \begin{pmatrix} 1 \\ 0 \\ 0 \end{pmatrix} \right] \cdot \begin{pmatrix} 1 \\ 1 \\ 1 \end{pmatrix}, \quad \begin{pmatrix} x \\ y \\ z \end{pmatrix} \in \mathcal{R}^3,$$

we can show that A_2 is affine, $ker\ A_2 = 0\text{-}lev\ A_2 = \mathfrak{T}$, and hence A_2 defines $\mathfrak{T}$ implicitly.

3. $\mathcal{S} = \left\{ \begin{pmatrix} x \\ y \\ z \end{pmatrix} \middle| x + y + z = 1 \right\}$ by Theorem 2.19ii, p. 115.

If we define $A_3 : \mathcal{R}^2 \to \mathcal{R}^1$ by

$$A_3 \begin{pmatrix} x \\ y \end{pmatrix} = 1 - x - y, \quad \begin{pmatrix} x \\ y \end{pmatrix} \in \mathcal{R}^2,$$

we can show that A_3 is affine, *graph* $A_3 = \mathcal{S}$, and hence A_3 defines $\mathcal{S}$ explicitly. ∎

Example 17. Let us now consider the problem of finding the affine transformations defining a given line $\mathcal{S}$ (1-plane) of $\mathcal{R}^3$ parametrically, implicitly, and explicitly.

Let $\mathcal{S}$ be the line passing through $\begin{pmatrix} 1 \\ 0 \\ 0 \end{pmatrix}$ and $\begin{pmatrix} 0 \\ 1 \\ 0 \end{pmatrix}$.

1. $\mathcal{S} = \left\{ \begin{pmatrix} 1 \\ 0 \\ 0 \end{pmatrix} + t \begin{pmatrix} -1 \\ 1 \\ 0 \end{pmatrix} \middle| t \in \mathcal{R} \right\}$ by Theorem 1.25, p. 36.

If we define $A_1 : \mathcal{R} \to \mathcal{R}^3$ by $A_1(t) = \begin{pmatrix} 1 \\ 0 \\ 0 \end{pmatrix} + t \begin{pmatrix} -1 \\ 1 \\ 0 \end{pmatrix}$, we see that A_1 is an affine transformation defining $\mathcal{S}$ parametrically (*ran* $A_1 = \mathcal{S}$).

2. To find the implicit representation of $\mathcal{S}$ we choose two independent vectors, each of which is perpendicular to the direction vector of $\mathcal{S}$; namely, we choose $\begin{pmatrix} 1 \\ 1 \\ 0 \end{pmatrix}$ and $\begin{pmatrix} 0 \\ 0 \\ 1 \end{pmatrix}$ and we set

$$\mathcal{M} = \left\{ \begin{pmatrix} x \\ y \\ z \end{pmatrix} \middle| \left[\begin{pmatrix} x \\ y \\ z \end{pmatrix} - \begin{pmatrix} 1 \\ 0 \\ 0 \end{pmatrix} \right] \cdot \begin{pmatrix} 1 \\ 1 \\ 0 \end{pmatrix} = 0 \text{ and} \right.$$

$$\left. \left[\begin{pmatrix} x \\ y \\ z \end{pmatrix} - \begin{pmatrix} 1 \\ 0 \\ 0 \end{pmatrix} \right] \cdot \begin{pmatrix} 0 \\ 0 \\ 1 \end{pmatrix} = 0 \right\}.$$

We see immediately that $\mathcal{S} \subset \mathcal{M}$. However, $\mathcal{M}$ is the intersection of two nonparallel planes in $\mathcal{R}^3$ and hence by Theorem 1.50, p. 79, $\mathcal{M} = \mathcal{S}$. It only remains to find an affine transformation A_2 such that $\mathcal{S} = \mathcal{M} = ker\ A_2$. With little effort we think of

$$A_2 \begin{pmatrix} x \\ y \\ z \end{pmatrix} = \begin{pmatrix} 1 & 1 & 0 \\ 0 & 0 & 1 \end{pmatrix} \cdot \left[\begin{pmatrix} x \\ y \\ z \end{pmatrix} - \begin{pmatrix} 1 \\ 0 \\ 0 \end{pmatrix} \right].$$

Then

$$ker\ A_2 = \left\{ \begin{pmatrix} x \\ y \\ z \end{pmatrix} \middle| (1 \quad 1 \quad 0) \cdot \left[\begin{pmatrix} x \\ y \\ z \end{pmatrix} - \begin{pmatrix} 1 \\ 0 \\ 0 \end{pmatrix} \right] = 0 \right.$$

$$\left. \text{and } (0 \quad 0 \quad 1) \cdot \left[\begin{pmatrix} x \\ y \\ z \end{pmatrix} - \begin{pmatrix} 1 \\ 0 \\ 0 \end{pmatrix} \right] = 0 \right\}$$

$$= \mathfrak{S},$$

so that A_2 defines $\mathfrak{S}$ implicitly.

3. To find the explicit representation of $\mathfrak{S}$, we make use of the implicit representation:

$$\mathfrak{S} = \left\{ \begin{pmatrix} x \\ y \\ z \end{pmatrix} \middle| \begin{pmatrix} 1 & 1 & 0 \\ 0 & 0 & 1 \end{pmatrix} \left[\begin{pmatrix} x \\ y \\ z \end{pmatrix} - \begin{pmatrix} 1 \\ 0 \\ 0 \end{pmatrix} \right] = \begin{pmatrix} 0 \\ 0 \end{pmatrix} \right\}$$

$$= \left\{ \begin{pmatrix} x \\ y \\ z \end{pmatrix} \middle| x + y = 1 \text{ and } z = 0 \right\}$$

$$= \left\{ \begin{pmatrix} x \\ 1 - x \\ 0 \end{pmatrix} \middle| x \in \mathcal{R} \right\}$$

$$= graph\ A_3, \qquad \text{where } A_3(x) = \begin{pmatrix} 1 - x \\ 0 \end{pmatrix}, \quad x \in \mathcal{R}.$$

Thus $\mathfrak{S}$ is represented explicitly by A_3. ■

Careful study should be made of the examples in this section before attempting the exercises.

Before ending this section let us discuss briefly one more problem. If $\mathfrak{S}$ is a k-plane of $\mathcal{R}^n$ and $X_1 \in \mathcal{R}^n$, how do we determine whether $X_1 \in \mathfrak{S}$? The reader has already had experience with this problem for 1-planes and 2-planes. The problem reduces to determining whether a system of n equations in k unknowns has a solution. We met a similar problem in the preceding section when we were faced with finding the inverse of a given $n \times n$ matrix. Accordingly, the last sections of this chapter will be devoted to using matrices to solve systems of linear equations.

EXERCISES

Show that the following functions are affine transformations:

1. $A(x) = 2x + 3$.

2. $A(x) = \begin{pmatrix} 2x - 1 \\ x - 4 \end{pmatrix}$.

3. $A\begin{pmatrix} x \\ y \end{pmatrix} = \begin{pmatrix} 2x + y - 1 \\ 3x - 2 \\ y - 1 \end{pmatrix}$.

4. $A\begin{pmatrix} x \\ y \end{pmatrix} = (2 \quad 1) \cdot \left[\begin{pmatrix} x \\ y \end{pmatrix} - \begin{pmatrix} 3 \\ -2 \end{pmatrix} \right]$.

5. $A\begin{pmatrix} x \\ y \\ z \end{pmatrix} = \begin{pmatrix} 3 & 2 & 1 \\ -1 & 0 & 2 \end{pmatrix} \cdot \begin{pmatrix} x - 2 \\ y + 1 \\ z - 4 \end{pmatrix}$.

Show that the following sets are k-planes of $\mathcal{R}^n$. Determine the value of k in each case.

6. $\left\{ \begin{pmatrix} 1 \\ 0 \\ 1 \end{pmatrix} + r \begin{pmatrix} -1 \\ 2 \\ 1 \end{pmatrix} + s \begin{pmatrix} 3 \\ 0 \\ 1 \end{pmatrix} \;\middle|\; r, s \in \mathcal{R} \right\}.$

7. $\left\{ \begin{pmatrix} 2 \\ 0 \\ 0 \end{pmatrix} + r \begin{pmatrix} -1 \\ 2 \\ 1 \end{pmatrix} + s \begin{pmatrix} 2 \\ -4 \\ -2 \end{pmatrix} \;\middle|\; r, s \in \mathcal{R} \right\}.$

8. $\left\{ \begin{pmatrix} x \\ y \\ z \end{pmatrix} \;\middle|\; \left[\begin{pmatrix} x \\ y \\ z \end{pmatrix} - \begin{pmatrix} 1 \\ 0 \\ 0 \end{pmatrix} \right] \cdot \begin{pmatrix} 0 \\ 2 \\ 0 \end{pmatrix} = 0 \right\}.$

9. $\left\{ \begin{pmatrix} 0 \\ 1 \\ 0 \\ 0 \end{pmatrix} + r \begin{pmatrix} 0 \\ 2 \\ 1 \\ 0 \end{pmatrix} + s \begin{pmatrix} 1 \\ 0 \\ 0 \\ 0 \end{pmatrix} + t \begin{pmatrix} 0 \\ 0 \\ 0 \\ 1 \end{pmatrix} \;\middle|\; r, s, t \in \mathcal{R} \right\}.$

10. $\left\{ \begin{pmatrix} 3 \\ 2 \\ 1 \\ 0 \end{pmatrix} + t \begin{pmatrix} 7 \\ -1 \\ 1 \\ 5 \end{pmatrix} \;\middle|\; t \in \mathcal{R} \right\}.$

Express the following k-planes parametrically, implicitly, and explicitly (where possible) by means of affine transformations.

11. The line $\mathcal{L}$ in $\mathcal{R}^2$ through $\begin{pmatrix} 1 \\ 2 \end{pmatrix}$ and $\begin{pmatrix} 2 \\ 3 \end{pmatrix}$.

12. The plane $\mathcal{S}$ in $\mathcal{R}^3$ through $\begin{pmatrix} 1 \\ 1 \\ 1 \end{pmatrix}$, $\begin{pmatrix} 1 \\ 2 \\ 2 \end{pmatrix}$, and $\begin{pmatrix} 0 \\ 0 \\ 1 \end{pmatrix}$.

13. The line $\mathcal{L}$ in $\mathcal{R}^3$ through $\begin{pmatrix} 1 \\ 1 \\ 1 \end{pmatrix}$ and $\begin{pmatrix} 3 \\ 3 \\ 3 \end{pmatrix}$.

14. The plane $\mathcal{S}$ in $\mathcal{R}^3$ through $\begin{pmatrix} 1 \\ 2 \\ 2 \end{pmatrix}$, $\begin{pmatrix} 2 \\ 2 \\ 2 \end{pmatrix}$, and $\begin{pmatrix} -1 \\ -1 \\ -1 \end{pmatrix}$.

15. The line $\mathcal{L}$ in $\mathcal{R}^2$ through $\begin{pmatrix} 3 \\ 2 \end{pmatrix}$ and $\begin{pmatrix} 3 \\ -2 \end{pmatrix}$.

16. Let $\{X_1, \ldots, X_k\}$ be a basis for a k-dimensional linear subspace $\mathcal{M}$ of $\mathcal{R}^n$. Let $\{E_1, \ldots, E_k\}$ be the natural basis for $\mathcal{R}^k$. Let $L:\mathcal{R}^k \to \mathcal{R}^n$ be the unique linear transformation such that $LE_i = X_i$, $i = 1, \ldots, k$. Prove that $ran\ L = \mathcal{M}$.

17. Let $A:\mathcal{R}^p \to \mathcal{R}^n$ be an affine transformation. Prove that $ran\ A$ is a k-plane of $\mathcal{R}^n$, for some value of k.

18. Let $A:\mathcal{R}^n \to \mathcal{R}^m$ be an affine transformation. Prove that each X_0-lev A is a k-plane of $\mathcal{R}^n$, for some k. Moreover, X-lev A is a k-plane if and only if some $m - k$ columns of $[L]$ are linearly independent, where L is the linear transformation defining A.

19. Let $\mathcal{S}$ be a k-plane of $\mathcal{R}^n$. Prove that $\mathcal{S}$ can be represented implicitly by some affine transformation.

20. Let $A:\mathcal{R}^k \to \mathcal{R}^{n-k}$ be an affine transformation. Prove that *graph A* is a k-plane of $\mathcal{R}^n$.

ANSWERS

2. Let $L(x) = x\begin{pmatrix}2\\1\end{pmatrix}$, $Y_0 = \begin{pmatrix}-1\\-4\end{pmatrix}$.

4. Let $L\begin{pmatrix}x\\y\end{pmatrix} = (2, 1) \cdot \begin{pmatrix}x\\y\end{pmatrix} = 2x + y$, $Y_0 = (2, 1) \cdot \begin{pmatrix}-3\\2\end{pmatrix} = -4$.

6. $k = 2$.   **8.** $k = 2$.   **9.** $k = 3$.

11. $\mathcal{S}$ defined parametrically by $A_1(t) = \begin{pmatrix}1\\2\end{pmatrix} + t\begin{pmatrix}1\\1\end{pmatrix}$, implicitly as 0-lev A_2,

where $A_2\begin{pmatrix}x\\y\end{pmatrix} = \left[\begin{pmatrix}x\\y\end{pmatrix} - \begin{pmatrix}1\\2\end{pmatrix}\right] \cdot \begin{pmatrix}1\\-1\end{pmatrix}$, explicitly by $A_3(x) = x + 1$.

13. $\mathcal{S}$ defined parametrically by $A_1(t) = t\begin{pmatrix}1\\1\\1\end{pmatrix}$, implicitly as 0-lev A_2,

where $A_2\begin{pmatrix}x\\y\\z\end{pmatrix} = \begin{pmatrix}1 & -1 & 0\\1 & 1 & -2\end{pmatrix}\begin{pmatrix}x\\y\\z\end{pmatrix}$, explicitly by $A_3(x) = \begin{pmatrix}x\\x\end{pmatrix}$.

17. Hint: $AX = LX + Y_0$ for some linear transformation $L:\mathcal{R}^p \to \mathcal{R}^n$ and some element Y_0 of $\mathcal{R}^n$. Show that *ran A* is a k-plane if and only if *ran L* is a k-dimensional subspace of $\mathcal{R}^n$. To do this use Theorem 7.11, p. 314.

19. Hint: $\mathcal{S} = \mathcal{M} + X_0$ for some $X_0 \in \mathcal{R}^n$ and some k-dimensional linear subspace $\mathcal{M}$ of $\mathcal{R}^n$.

    (i) If $k = n$, show that $\mathcal{M} = $ 0-lev L, where $LX = X \cdot O_n$.

    (ii) If $k < n$, let $\mathcal{M}^\perp = \{X \mid X \perp Y \text{ for all } Y \text{ in } \mathcal{M}\}$.

Show that $\mathcal{M}^\perp$ is a linear subspace of $\mathcal{R}^n$ of dimension $n - k$ and let $\{X_{n-k+1}, \ldots, X_n\}$ be a basis for $\mathcal{M}^\perp$. Show that $\mathcal{M} = O_{n-k}$-lev L, where

L is defined by $LX = \begin{pmatrix}X \cdot X_{n-k+1}\\X \cdot X_{n-k+2}\\\cdot\\\cdot\\\cdot\\X \cdot X_n\end{pmatrix}$ for X in $\mathcal{R}^n$.

6. MATRICES AND SYSTEMS OF LINEAR EQUATIONS

Many times during the course of this book we have been faced with solving systems of linear equations: in determining whether a point was on a given line or plane, in determining the line of intersection of two planes, in finding the inverse of a given matrix.

We shall in this section develop the "*Method of Matrix Reduction*" in order to solve these types of problems.

Consider the following system of m equations in n unknowns:

(15.15)

$$\begin{cases} a_{11}x_1 + \cdots + a_{1n}x_n = b_1 \\ a_{21}x_1 + \cdots + a_{2n}x_n = b_2 \\ \\ \\ \\ a_{r1}x_1 + \cdots + a_{rn}x_n = b_r \\ \\ \\ \\ a_{m1}x_1 + \cdots + a_{m1}x_n = b_m \end{cases}$$

In this system the a_{ij} and the b_i, $i = 1, \ldots, m$, $j = 1, \ldots, n$, are fixed real numbers. An n-vector $(x_1, \ldots, x_n)$ is called a *solution of this system of equations* if and only if the numbers $x_1, \ldots, x_n$ satisfy each of the m equations

in Equation 15.15. Notice that $X = \begin{pmatrix} x_1 \\ \cdot \\ \cdot \\ \cdot \\ x_n \end{pmatrix}$ is a solution of Equation 15.15 if

and only if $A_1 X = B$, where A_1 is the $m \times n$ matrix whose ij-th coordinate

is a_{ij} and $B = \begin{pmatrix} b_1 \\ \cdot \\ \cdot \\ \cdot \\ b_m \end{pmatrix}$. Thus the n-vector X is a *solution of the matrix equation*

$A_1 X = B$ if and only if X is also a solution of Equation 15.15. In this context A_1 is called the *matrix of coefficients* of Equation 15.15. The *set of solutions* or *solution set* of Equation 15.15 is the set of all X's satisfying $A_1 X = B$. If two systems of linear equations have exactly the same set of solutions, we say that the two *systems are equivalent.*

Permissible Operations on Systems of Linear Equations and Matrices

I. ADDITION OF ROWS

Consider a second system of equations which is identical with Equation 15.15 except that the r-th equation is replaced by the sum of the r-th and s-th equations:

(15.16)

$$\begin{cases} a_{11}x_1 + \cdots + a_{1n}x_n = b_1 \\ a_{21}x_1 + \cdots + a_{2n}x_n = b_2 \\ \\ \\ \\ (a_{r1} + a_{s1})x_1 + \cdots + (a_{rn} + a_{sn})x_n = b_r + b_s \\ \\ \\ \\ a_{m1}x_1 + \cdots + a_{mn}x_n = b_m \end{cases}$$

The reader should be able to show that $X = (x_1, \ldots, x_n)$ is a solution of Equation 15.15 if and only if X is a solution of Equation 15.16. (See Exercise 14.)

Let A_1 be the matrix of coefficients of Equation 15.15 and A_2 the matrix of coefficients of Equation 15.16 and let $B_2 = (b_1, \ldots, b_{r-1}, b_r + b_s, b_{r+1}, \ldots, b_n)$. From the preceding we see that X is a solution of $A_1 X = B$ if and only if X is a solution of $A_2 X = B_2$.

Thus in searching for a solution for Equation 15.15 or for the matrix equation $A_1 X = B$, we may change the system of equations or the matrix equation by substituting for a given row the sum of that row with any other row, and we may be assured that we shall arrive at the same set of solutions. The substitution of the sum of the r-th and s-th equations for the r-th equation in the system of linear equations is called *The Rule of Addition of Row Equations*. This rule permits us to form a new system of linear equations with exactly the same set of solutions as those of the original system. The new system is equivalent to the original system.

Example 18a. Consider the system of equations

(15.17)
$$\begin{cases} x + 3y - 2z = 4 \\ -x + y + z = 2 \end{cases}$$

and its related matrix equation

(15.18)
$$\begin{pmatrix} 1 & 3 & -2 \\ -1 & 1 & 1 \end{pmatrix} \begin{pmatrix} x \\ y \\ z \end{pmatrix} = \begin{pmatrix} 4 \\ 2 \end{pmatrix}.$$

By the rule for Addition of Rows the solutions of Equations 15.17 and 15.18 are the same as the solutions for

(15.19)
$$\begin{cases} x + 3y - 2z = 4 \\ 0 + 4y - z = 6 \end{cases}$$

and

(15.20)
$$\begin{pmatrix} 1 & 3 & -2 \\ 0 & 4 & -1 \end{pmatrix} \begin{pmatrix} x \\ y \\ z \end{pmatrix} = \begin{pmatrix} 4 \\ 6 \end{pmatrix},$$

which we obtain by substituting the sum of the two rows in Equation 15.17 for the second row in Equation 15.17. The student should be quite familiar with this particular step toward the solution of a system of linear equations. ∎

II. MULTIPLICATION OF A ROW BY A NONZERO REAL NUMBER

Consider now a third system of equations which is identical with Equation 15.15 except that both members of the r-th equation have been multiplied by some nonzero number c:

(15.21)
$$\begin{cases} a_{11}x_1 + \cdots + a_{1n}x_n = b_1 \\ a_{21}x_1 + \cdots + a_{2n}x_n = b_2 \\ \quad\vdots \\ ca_{r1}x_1 + \cdots + ca_{rn}x_n = cb_r \\ \quad\vdots \\ a_{m1}x_1 + \cdots + a_{mn}x_n = b_m. \end{cases}$$

Since $x_1, \ldots, x_1$ satisfy one of these equations if and only if they satisfy the other, $X = (x_1, \ldots, x_n)$ is a solution of Equation 15.15 if and only if it is a solution of Equation 15.21. Moreover, if A_3 is the matrix of coefficients for Equation 15.21 and $B_3 = (b_1, \ldots, b_{r-1}, cb_r, b_{r+1}, \ldots, b_m)$, the solutions of $A_1X = B$ and $A_3X = B_3$ are exactly the same.

Thus in searching for a solution for Equation 15.15 or for the matrix equation $A_1X = B$, we may change the system of equations or the matrix equation by substituting for a given row a nonzero multiple of that row and retain exactly the same set of solutions. The substitution of a nonzero multiple of an equation for the equation in a system of linear equations is called *The Rule of Scalar Multiplication of Row Equations*.

Example 18b. In Example 18a we found that (x, y, z) is a solution of

(15.18)
$$\begin{pmatrix} 1 & 3 & -2 \\ -1 & 1 & 1 \end{pmatrix} \begin{pmatrix} x \\ y \\ z \end{pmatrix} = \begin{pmatrix} 4 \\ 2 \end{pmatrix}$$

if and only if (x, y, z) is a solution of

(15.20)
$$\begin{pmatrix} 1 & 3 & -2 \\ 0 & 4 & -1 \end{pmatrix} \begin{pmatrix} x \\ y \\ z \end{pmatrix} = \begin{pmatrix} 4 \\ 6 \end{pmatrix}.$$

Let us multiply the second row of Equation 15.20 by $-3/4$ and substitute the result for the second row. We arrive at

(15.22)
$$\begin{pmatrix} 1 & 3 & -2 \\ 0 & -3 & 3/4 \end{pmatrix} \begin{pmatrix} x \\ y \\ z \end{pmatrix} = \begin{pmatrix} 4 \\ -9/2 \end{pmatrix}.$$

Combining the results in I and II we see that the solutions of Equation 15.22 are exactly the same as those for Equation 15.18.

Now let us replace the first row in Equation 15.22 by the sum of the two rows to obtain

(15.23)
$$\begin{pmatrix} 1 & 0 & -5/4 \\ 0 & -3 & 3/4 \end{pmatrix} \begin{pmatrix} x \\ y \\ z \end{pmatrix} = \begin{pmatrix} -1/2 \\ -9/2 \end{pmatrix}.$$

Finally, let us replace the second row of Equation 15.23 by $-1/3$ times that row to obtain

$$(15.24) \qquad \begin{pmatrix} 1 & 0 & -5/4 \\ 0 & 1 & -1/4 \end{pmatrix} \begin{pmatrix} x \\ y \\ z \end{pmatrix} = \begin{pmatrix} -1/2 \\ 3/2 \end{pmatrix}.$$

The solutions of Equations 15.23 and 15.24 are still exactly the solutions of Equation 15.18. Thus (x, y, z) satisfies

$$\begin{cases} x + 3y - 2z = 4 \\ -x + y + z = 2 \end{cases} \Leftrightarrow \begin{pmatrix} 1 & 0 & -5/4 \\ 0 & 1 & -1/4 \end{pmatrix} \begin{pmatrix} x \\ y \\ z \end{pmatrix} = \begin{pmatrix} -1/2 \\ 3/2 \end{pmatrix}$$

$$\Leftrightarrow \begin{pmatrix} x & -5/4z \\ y & -1/4z \end{pmatrix} = \begin{pmatrix} -1/2 \\ 3/2 \end{pmatrix}$$

$$\Leftrightarrow \begin{cases} x = \frac{5}{4}z - \frac{1}{2} \\ y = \frac{1}{4}z + \frac{3}{2} \end{cases}$$

$$\Leftrightarrow \begin{pmatrix} x \\ y \\ z \end{pmatrix} = \begin{pmatrix} \frac{5}{4}z - \frac{1}{2} \\ \frac{1}{4}z + \frac{3}{2} \\ z \end{pmatrix}$$

$$\Leftrightarrow \begin{pmatrix} x \\ y \\ z \end{pmatrix} = z\begin{pmatrix} \frac{5}{4} \\ \frac{1}{4} \\ 1 \end{pmatrix} + \begin{pmatrix} -1/2 \\ 3/2 \\ 0 \end{pmatrix}.$$

Thus the set of solutions of Equation 15.17 is a line in $\mathcal{R}^3$ passing through $\begin{pmatrix} -1/2 \\ 3/2 \\ 0 \end{pmatrix}$ with direction vector $\begin{pmatrix} 5/4 \\ 1/4 \\ 1 \end{pmatrix}$. ∎

The third permissible operation used in solving a system of linear equations or a matrix equation is the simplest of the three:

III. INTERCHANGE OF TWO ROWS

Let us consider a system of equations identical with Equation 15.15 except that two rows have been interchanged. The corresponding matrix equations obviously have the same sets of solutions. This procedure is called *The Rule of Interchange of Row Equations*.

Let us reiterate the three permissible operations in solving a system of linear equations or a matrix equation.

PERMISSIBLE OPERATIONS 15.19

I. We may substitute for a given row the sum (or difference) of that row with any other row.

II. We may substitute for a given row the multiple of that row by any nonzero scalar.

III. We may interchange any two rows.

The set of solutions for the new system is identical with that for the original system.

The student should notice how simple the permissible operations are when applied to matrices. One is required to write down a solution vector X only once in the matrix equations, as opposed to writing each of the components m times in the system of linear equations.

The purpose in considering the three permissible operations is to change the original matrix equation to an equivalent matrix equation from which the set of solutions can be obtained easily. Suppose that we start with a matrix equation $AX = B$ and arrive at the equivalent equation

$$\begin{pmatrix} 1 & 0 & 0 & 0 \\ 0 & 1 & 0 & 0 \\ 0 & 0 & 1 & 2 \end{pmatrix} \begin{pmatrix} x \\ y \\ z \\ w \end{pmatrix} = \begin{pmatrix} 2 \\ -3 \\ 4 \end{pmatrix}.$$

Then we immediately read off the solutions to be all vectors $\begin{pmatrix} x \\ y \\ z \\ w \end{pmatrix}$ such that

$x = 2$, $y = -3$, and $z + 2w = 4$, or all vectors of the form $\begin{pmatrix} 2 \\ -3 \\ 4 - 2w \\ w \end{pmatrix}.$

DEFINITION 15.20   Let $A \in \mathcal{R}_{mn}$. A is in reduced form if and only if

i. The first nonzero component of each row is 1, and

ii. if $a_{ij} = 1$ is the first nonzero component of the i-th row, then $a_{kj} = 0$ for all $k \neq i$ and $a_{k'j'} = 0$ for all $k' > i$ and $j' \leq j$ (that is, all other components of the j-th column are zero and all components "southwest" of a_{ij} are zero).

Notice that the matrix appearing in Equation 15.24 of Example 18b is in reduced form. When the matrix appearing in a matrix equation is in reduced form, the solutions are most easily obtained. Moreover, every element of $\mathcal{R}_{mn}$ can be changed into a matrix in reduced form by means of a finite number of permissible operations. (This can be proved by induction.)

Example 19.   (a) $\begin{pmatrix} 1 & 0 & 0 \\ 0 & 1 & 0 \\ 0 & 0 & 1 \end{pmatrix}$ is in reduced form.

(b) $\begin{pmatrix} 0 & 1 & 0 \\ 1 & 0 & 0 \\ 0 & 0 & 1 \end{pmatrix}$ is not in reduced form, for $a_{12} = 1$ is the first nonzero element of the first row, but a_{21} is southwest of a_{12} and $a_{21} \neq 0$.

(c) $\begin{pmatrix} 1 & 0 & 0 \\ 0 & 1 & 1 \\ 0 & 0 & 1 \end{pmatrix}$ is not in reduced form because $a_{33} = 1$

is the first nonzero element of the third row, but $a_{23} \neq 0$.

(d) $\begin{pmatrix} 1 & 0 & 0 & 0 & 3 \\ 0 & 1 & 4 & 0 & 2 \\ 0 & 0 & 0 & 1 & 0 \end{pmatrix}$ is in reduced form. ∎

Notice that in Example 19b one can form a reduced matrix by interchanging the first and second row. In Example 19c one can form a reduced matrix by substituting for the second row the difference of the second and third rows.

Henceforth we shall use the method of reduction of matrices by the three permissible operations in order to solve systems of linear equations and matrix equations.

Example 20. We wish to find the set of solutions of the following system of linear equations:

$$x + 2y + 7z + 4w = 1$$
$$2x - 3y \qquad + 2w = -6$$
$$x + y + 5z + 3w = 0.$$

Thus we are looking for all $X = \begin{pmatrix} x \\ y \\ z \\ w \end{pmatrix}$, satisfying

(15.25) $\begin{pmatrix} 1 & 2 & 7 & 4 \\ 2 & -3 & 0 & 2 \\ 1 & 1 & 5 & 3 \end{pmatrix} X = \begin{pmatrix} 1 \\ -6 \\ 0 \end{pmatrix}.$

We shall use the three permissible operations on this matrix equation to reduce the matrix of coefficients.

1. Since $a_{11} = 1$, we start by finding an equivalent system with $a_{21} = a_{31} = 0$. We substitute for the second row the difference of the two times the first row and the second row. At the same time we substitute for the third row the difference of the first and third rows.

$\begin{pmatrix} 1 & 2 & 7 & 4 \\ 0 & 7 & 14 & 6 \\ 0 & 1 & 2 & 1 \end{pmatrix} X = \begin{pmatrix} 1 \\ 8 \\ 1 \end{pmatrix}.$

2. We divide the second row by 7, in order to have the first nonzero element of the second row equal to 1.

$$\begin{pmatrix} 1 & 2 & 7 & 4 \\ 0 & 1 & 2 & 6/7 \\ 0 & 1 & 2 & 1 \end{pmatrix} X = \begin{pmatrix} 1 \\ 8/7 \\ 1 \end{pmatrix}.$$

3. Since the second row now "starts" with 1, we wish to have zeros elsewhere in that column and we substitute for the first row the difference of the first row and two times the second row. We substitute for the third row the difference of the second and third rows.

$$\begin{pmatrix} 1 & 0 & 3 & 16/7 \\ 0 & 1 & 2 & 6/7 \\ 0 & 0 & 0 & -1/7 \end{pmatrix} X = \begin{pmatrix} -9/7 \\ 8/7 \\ 1/7 \end{pmatrix}$$

4. Since the first nonzero element of the third row is $-1/7$, we want to have zeros elsewhere in the fourth column and we substitute for the first row the sum of the first row and 16 times the third row. We substitute for the second row the sum of the second row and 6 times the third row.

$$\begin{pmatrix} 1 & 0 & 3 & 0 \\ 0 & 1 & 2 & 0 \\ 0 & 0 & 0 & -1/7 \end{pmatrix} X = \begin{pmatrix} 1 \\ 2 \\ 1/7 \end{pmatrix}$$

5. We substitute for the third row -7 times the third row.

(15.26)
$$\begin{pmatrix} 1 & 0 & 3 & 0 \\ 0 & 1 & 2 & 0 \\ 0 & 0 & 0 & 1 \end{pmatrix} \begin{pmatrix} x \\ y \\ z \\ w \end{pmatrix} = \begin{pmatrix} 1 \\ 2 \\ -1 \end{pmatrix}.$$

The matrix of coefficients is now in reduced form and the solutions of Equation 15.26, which are the same as those of Equation 15.25, are the set of

all $\begin{pmatrix} x \\ y \\ z \\ w \end{pmatrix}$ such that $x + 3z = 1$, $y + 2z = 2$, and $w = -1$. Since x and y

can both be expressed in terms of z, we write

$$\begin{pmatrix} x \\ y \\ z \\ w \end{pmatrix} = \begin{pmatrix} 1 - 3z \\ 2 - 2z \\ z \\ -1 \end{pmatrix} = \begin{pmatrix} 1 \\ 2 \\ 0 \\ -1 \end{pmatrix} + z \begin{pmatrix} -3 \\ -2 \\ 1 \\ 0 \end{pmatrix}.$$

Thus the set of solutions of Equation 15.25 is the line in $\mathcal{R}^4$, passing

through $\begin{pmatrix} 1 \\ 2 \\ 0 \\ -1 \end{pmatrix}$ with direction vector $\begin{pmatrix} -3 \\ -2 \\ 1 \\ 0 \end{pmatrix}$. ∎

EXERCISES

Which of the following matrices are in reduced form?

1. $\begin{pmatrix} 1 & 0 & 1 \\ 0 & 1 & 0 \end{pmatrix}$.

2. $\begin{pmatrix} 1 & 0 & 1 \\ 0 & 0 & 1 \end{pmatrix}$.

3. $\begin{pmatrix} 0 & 1 & 0 \\ 0 & 0 & 1 \end{pmatrix}$.

4. $\begin{pmatrix} 1 & 2 & 0 & 0 \\ 0 & 0 & 1 & 0 \\ 0 & 0 & 0 & 1 \end{pmatrix}$.

5. $\begin{pmatrix} 1 & 2 & 0 & 1 \\ 0 & 0 & 1 & 0 \\ 0 & 0 & 1 & 0 \end{pmatrix}$.

6. $\begin{pmatrix} 0 & 1 & 0 & 0 \\ 1 & 0 & 0 & 0 \\ 0 & 0 & 1 & 0 \\ 0 & 0 & 0 & 1 \end{pmatrix}$.

Use the method of reduction of matrices to find the set of solutions of the following systems of linear equations.

7. $\begin{cases} x + y = 1 \\ x - y = 3. \end{cases}$

8. $\begin{cases} 2x + 3y = 6 \\ 3x - 4y = 12. \end{cases}$

9. $\begin{cases} x + y + z = 1 \\ 2x - y + 3z = 4 \\ 6x - 5y - z = 0. \end{cases}$

10. $\begin{cases} x + y + z = 1 \\ 2x - y + 3z = 4. \end{cases}$

11. $\begin{cases} 2x - y + 4z - 3w = 0 \\ x + y + z + w = 10 \\ 3x - z + w = 4 \\ 3y + 2z - 2w = 4. \end{cases}$

12. $\begin{cases} x + y + 2z - w = 33 \\ 2x + y = 27 \\ y + 2z = 23. \end{cases}$

13. $\begin{cases} 2x - y + 4z - 3w = 0 \\ x + y + z + w = 10. \end{cases}$

14. Show that $(x_1, \ldots, x_n)$ is a solution of

$$\begin{cases} a_{j1}x_1 + \cdots + a_{jn}x_n = b_j \\ a_{k1}x_1 + \cdots + a_{kn}x_n = b_k \end{cases}$$

if and only if $(x_1, \ldots, x_n)$ is a solution of

$$\begin{cases} (a_{j1} + a_{k1})x_1 + \cdots + (a_{jn} + a_{kn})x_n = b_j + b_k \\ a_{k1}x_1 + \cdots + a_{kn}x_n = b_k. \end{cases}$$

ANSWERS

1. Yes. **3.** Yes. **5.** No. **8.** $\begin{pmatrix} x \\ y \end{pmatrix} = \begin{pmatrix} \frac{60}{17} \\ -\frac{6}{17} \end{pmatrix}$.

10. $\left\{ \begin{pmatrix} x \\ y \\ z \end{pmatrix} \middle| x = -\frac{4}{3}z + \frac{5}{3}, y = \frac{1}{3}z - \frac{2}{3} \right\} = \left\{ z\begin{pmatrix} -4/3 \\ 1/3 \\ 1 \end{pmatrix} + \begin{pmatrix} 5/3 \\ -2/3 \\ 0 \end{pmatrix} \middle| z \in \mathcal{R} \right\}$.

12. $\left\{ \begin{pmatrix} x \\ y \\ z \\ w \end{pmatrix} \middle| x = 10 + w, y = 7 - 2w, z = 8 + w \right\}$

$= \left\{ w\begin{pmatrix} 1 \\ -2 \\ 1 \\ 1 \end{pmatrix} + \begin{pmatrix} 10 \\ 7 \\ 8 \\ 0 \end{pmatrix} \middle| w \in \mathcal{R} \right\}$.

7. INVERSION OF MATRICES

After studying Section 6 and working the exercises, the reader may be left with the (erroneous) impression that no matter what system of linear equations appears, he can find a solution. This impression is erroneous for a very simple reason: there exist systems of linear equations and matrix equations for which there are no solutions. This in no way invalidates the procedure we developed in Section 6. Indeed, we see that our method does exactly what we anticipated; it tells us the set of solutions of a system of linear equations, whether that set is empty or nonempty.

For example, consider the system

$$(15.27) \qquad \begin{cases} x + y = 1 \\ x + y = 2. \end{cases}$$

The set of all (x, y) satisfying $x + y = 1$ is a line in $\mathcal{R}^2$ with slope -1 and y-intercept 1. The set of all (x, y), satisfying $x + y = 2$, is a line with slope -1 and y-intercept 2. Thus these two sets represent distinct parallel lines and have no point in common.

Let us consider now the related matrix equation

$$(15.28) \qquad \begin{pmatrix} 1 & 1 \\ 1 & 1 \end{pmatrix} \begin{pmatrix} x \\ y \end{pmatrix} = \begin{pmatrix} 1 \\ 2 \end{pmatrix}$$

which is equivalent to Equation 2.27. If we replace the second row by the difference of the two rows, we obtain

$$(15.29) \qquad \begin{pmatrix} 1 & 1 \\ 0 & 0 \end{pmatrix} \begin{pmatrix} x \\ y \end{pmatrix} = \begin{pmatrix} 1 \\ 1 \end{pmatrix}.$$

There are obviously no values of x and y satisfying Equation 15.29, for if there were, then we would have $x + y = 1$ and $(0)x + (0)y = 1$. The problem, of course, lies in the fact that the second row of the matrix is identically zero and the second component of the vector on the right is nonzero.

From these observations, we derive the following rule. Let $AX = B$ be a matrix equation. Let $A'X = B'$ be the matrix equation, obtained from $AX = B$, in which A' is in reduced form. If the i-th row of A' is identically zero and the i-th component of B' is nonzero, there is no solution to the equation $AX = B$.

Example 21. Let A_1 be the affine transformation defined by

$$A_1 \begin{pmatrix} x \\ y \end{pmatrix} = \begin{pmatrix} x + y - 3 \\ x - y + 1 \\ x + 3y + 2 \end{pmatrix}.$$

We wish to determine whether $O_3 \in ran\ A_1$. The following implication chain will give us our answer:

$$O_3 \in ran\ A_1 \Leftrightarrow^1 \text{There exists } \begin{pmatrix} x \\ y \end{pmatrix} \text{ such that } A_1 \begin{pmatrix} x \\ y \end{pmatrix} = \begin{pmatrix} 0 \\ 0 \\ 0 \end{pmatrix}$$

$$\Leftrightarrow \begin{pmatrix} x + y - 3 \\ x - y + 1 \\ x + 3y + 2 \end{pmatrix} = \begin{pmatrix} 0 \\ 0 \\ 0 \end{pmatrix}$$

$$\Leftrightarrow^2 \begin{pmatrix} 1 & 1 \\ 1 & -1 \\ 1 & 3 \end{pmatrix} \begin{pmatrix} x \\ y \end{pmatrix} = \begin{pmatrix} 3 \\ -1 \\ -2 \end{pmatrix}$$

$$\Leftrightarrow^3 \begin{pmatrix} 1 & 1 \\ 0 & 2 \\ 0 & -2 \end{pmatrix} \begin{pmatrix} x \\ y \end{pmatrix} = \begin{pmatrix} 3 \\ 4 \\ 5 \end{pmatrix}$$

$$\Leftrightarrow^3 \begin{pmatrix} 1 & 1 \\ 0 & 2 \\ 0 & 0 \end{pmatrix} \begin{pmatrix} x \\ y \end{pmatrix} = \begin{pmatrix} 3 \\ 4 \\ 9 \end{pmatrix}.$$

(1) Definition of range.
(2) Matrix multiplication and equality of matrices.
(3) Addition and subtraction of rows.

Since there is no $\begin{pmatrix} x \\ y \end{pmatrix}$ satisfying this last matrix equation, $O_3 \notin ran\ A_1$. ∎

We are now ready to apply our method of reduction of matrices to the problem of finding the inverse of a given $n \times n$ matrix, or alternatively showing that this matrix has no inverse. We state the following theorem:

THEOREM 15.21 Let $A \in \mathcal{R}_{nn}$. Let A' be a matrix obtained by reducing A.
　　i. A is invertible if and only if $A' = I_n$.
　　ii. A is not invertible if and only if A' has an identically zero row.
　　iii. If A is invertible, A^{-1} can be found by performing on I_n the permissible operations used in reducing A to $A' = I_n$.

Proof. We shall prove (i) and (ii) simultaneously.
Let T and T' be the linear transformations defined by

$$TX = AX \quad \text{and} \quad T'X = A'X \quad \text{for all } X \text{ in } \mathcal{R}^n.$$

Consider the following implication chain:

$$A \text{ is invertible} \Leftrightarrow^1 T \text{ is invertible}$$

$$\Leftrightarrow^2 ker\, T = \{O_n\}$$

$$\Leftrightarrow^3 ker\, T' = \{O_n\}$$

$$\Leftrightarrow^2 T' \text{ is invertible}$$

$$\Leftrightarrow^1 A' \text{ is invertible.}$$

(1) Theorem 15.13, p. 611.
(2) Theorem 15.14, p. 612.
(3) Exercise 14.

To prove (i) and (ii), it suffices to show
    (iv) A' is invertible if and only if $A' = I_n$ (or equivalently, A' is not invertible if and only if $A' \neq I_n$) and
    (v) A' is not invertible if and only if A' has an identically zero row.
We note that by Definition 15.20, the only way a reduced $n \times n$ matrix can fail to be I_n is for that matrix to have an identically zero row. Thus, $A' \neq I_n$ if and only if A' has an identically zero row and we see that statements (iv) and (v) are equivalent.

    If $A' = I_n$, then A' is certainly invertible. (What is the inverse of I_n?)

    Assume that $A' \neq I_n$ and has the j-th row identically zero. Consequently, for each X in $\mathcal{R}^n$, the j-th component of $T'X = A'X$ is zero and $ran\, T' \neq \mathcal{R}^n$ since $E_j \notin ran\, T'$. By Theorem 15.14 T' is not invertible and by Theorem 15.13 A' is not invertible. Therefore, A' is not invertible if $A' \neq I_n$, and we have completed the proofs of (iv) and (v).

    Proof of iii. Now assume that A is invertible and let $\{E_1, \ldots, E_n\}$ be the natural basis for $\mathcal{R}^n$. Let $X_i = A^{-1}E_i$ for $i = 1, \ldots, n$. We have

$$AX_i = E_i, \quad i = 1, \ldots, n$$

and

(15.30) $$\qquad A'X_i = E_i', \quad i = 1, \ldots, n,$$

where E_i' is obtained by performing on E_i the same operations performed in reducing A to $A' = I_n$. Since $X_i = A^{-1}E_i$ by definition, it follows from Equation 15.30 that $X_i = E_i' = A^{-1}E_i$, so that E_i' is the i-th column of A^{-1}. This proves the last assertion of the theorem. ∎

    Theorem 15.21 gives us a relatively simple method by which we can determine whether an $n \times n$ matrix A is invertible and find its inverse when it is invertible. This method is as follows:

    1. Place A and I_n side by side.
    2. Perform the necessary operations to reduce A and at the same time perform exactly the same operations on I_n.

3. If A reduces to I_n, A is invertible and the inverse of A has been obtained by performing the given operations on I_n.

4. If A reduces to a matrix with an identically zero row, A is not invertible.

Example 22. Let $A = \begin{pmatrix} 2 & 1 & 0 \\ 1 & -3 & 5 \\ 3 & 0 & 2 \end{pmatrix}$. We wish to find A^{-1} or show

that A^{-1} does not exist. We use the method just described:

$$\begin{pmatrix} 2 & 1 & 0 \\ 1 & -3 & 5 \\ 3 & 0 & 2 \end{pmatrix} \qquad \begin{pmatrix} 1 & 0 & 0 \\ 0 & 1 & 0 \\ 0 & 0 & 1 \end{pmatrix}$$

Replace row 1 by the difference of row 1 and 2 times row 2. Replace row 3 by the difference of three times row 2 and row 3.

$$\begin{pmatrix} 0 & 7 & -10 \\ 1 & -3 & 5 \\ 0 & -9 & 13 \end{pmatrix} \qquad \begin{pmatrix} 1 & -2 & 0 \\ 0 & 1 & 0 \\ 0 & 3 & -1 \end{pmatrix}$$

Interchange rows 1 and 2.

$$\begin{pmatrix} 1 & -3 & 5 \\ 0 & 7 & -10 \\ 0 & -9 & 13 \end{pmatrix} \qquad \begin{pmatrix} 0 & 1 & 0 \\ 1 & -2 & 0 \\ 0 & 3 & -1 \end{pmatrix}$$

Replace row 1 by the sum of row 1 and $\frac{3}{7}$ times row 2. Replace row 3 by the sum of row 3 and $\frac{9}{7}$ times row 2.

$$\begin{pmatrix} 1 & 0 & 5/7 \\ 0 & 7 & -10 \\ 0 & 0 & 1/7 \end{pmatrix} \qquad \begin{pmatrix} 3/7 & 1/7 & 0 \\ 1 & -2 & 0 \\ 9/7 & 3/7 & -1 \end{pmatrix}$$

Replace row 1 by the difference of row 1 and 5 times row 3. Replace row 2 by the sum of row 2 and 70 times row 3.

$$\begin{pmatrix} 1 & 0 & 0 \\ 0 & 7 & 0 \\ 0 & 0 & 1/7 \end{pmatrix} \qquad \begin{pmatrix} -6 & -2 & 5 \\ 91 & 28 & -70 \\ 9/7 & 3/7 & -1 \end{pmatrix}$$

Replace row 2 by 1/7 times row 2. Replace row 3 by 7 times row 3.

$$\begin{pmatrix} 1 & 0 & 0 \\ 0 & 1 & 0 \\ 0 & 0 & 1 \end{pmatrix} \qquad \begin{pmatrix} -6 & -2 & 5 \\ 13 & 4 & -10 \\ 9 & 3 & -7 \end{pmatrix}$$

Therefore A^{-1} exists and

$$A^{-1} = \begin{pmatrix} -6 & -2 & 5 \\ 13 & 4 & -10 \\ 9 & 3 & -7 \end{pmatrix}.$$

The student should check to see that indeed we have computed A^{-1} correctly; that is, $AA^{-1} = A^{-1}A = I_n$. ■

Example 23. Let $A = \begin{pmatrix} 1 & 3 & -2 \\ 4 & 2 & 1 \\ 5 & 5 & -1 \end{pmatrix}$. Again we wish to find A^{-1} or show that A^{-1} does not exist.

$$\begin{pmatrix} 1 & 3 & -2 \\ 4 & 2 & 1 \\ 5 & 5 & -1 \end{pmatrix} \qquad \begin{pmatrix} 1 & 0 & 0 \\ 0 & 1 & 0 \\ 0 & 0 & 1 \end{pmatrix}$$

Replace row 2 by the difference of row 2 and 4 times row 1. Replace row 3 by the difference of row 3 and 5 times row 1.

$$\begin{pmatrix} 1 & 3 & -2 \\ 0 & -10 & 9 \\ 0 & -10 & 9 \end{pmatrix} \qquad \begin{pmatrix} 1 & 0 & 0 \\ -4 & 1 & 0 \\ -5 & 0 & 1 \end{pmatrix}$$

Replace row 3 by the difference of row 2 and row 3.

$$\begin{pmatrix} 1 & 3 & -2 \\ 0 & -10 & 9 \\ 0 & 0 & 0 \end{pmatrix} \qquad \begin{pmatrix} 1 & 0 & 0 \\ -4 & 1 & 0 \\ 1 & 1 & -1 \end{pmatrix}$$

Although this last matrix is not in reduced form, we can stop, for it is evident that A can be reduced to a matrix with an identically zero row and hence is not invertible. ■

EXERCISES

Using the method of reduction of matrices, find the set of all solutions of the following systems of equations. Indicate empty sets where they arise.

1. $\begin{cases} 2x - 3y = 1 \\ -4x + 6y = -2. \end{cases}$

2. $\begin{cases} 2x - 3y = 1 \\ -4x + 6y = -3. \end{cases}$

3. $\begin{cases} 2x - 3y = 1 \\ x + 4y = 2. \end{cases}$

4. $\begin{cases} 5x - 3y + z = 2 \\ x + 7y - 5z = 0 \\ 3x - y + 4z = 13. \end{cases}$

5. $\begin{cases} 5x - 3y + z = 2 \\ x + 7y - 5z = 0 \\ 4x - 10y + 6z = 2. \end{cases}$

6. $\begin{cases} 5x + 3y + 7z = 2 \\ x + 7y - 5z = 0 \\ x + y + z = 0. \end{cases}$

Using Theorem 15.21, either find the inverse of the matrix or show that the matrix is not invertible.

7. $\begin{pmatrix} 3 & 1 \\ 11 & 4 \end{pmatrix}$.

8. $\begin{pmatrix} 2 & 3 & 1 \\ -1 & 4 & 2 \\ 1 & 7 & 3 \end{pmatrix}$.

9. $\begin{pmatrix} 1 & 3 & 2 \\ 0 & 4 & 11 \\ 0 & 1 & 3 \end{pmatrix}$.

10. $\begin{pmatrix} 2 & 3 & 2 \\ 0 & 4 & 2 \\ 1 & 2 & 1 \end{pmatrix}$.

11. $\begin{pmatrix} 3 & 2 & 5 \\ 4 & -1 & 3 \\ 0 & 5 & 5 \end{pmatrix}$.

12. $\begin{pmatrix} 1 & 0 & 0 & 0 \\ 2 & 1 & 3 & 2 \\ 3 & 0 & 4 & 11 \\ 4 & 0 & 1 & 3 \end{pmatrix}$.

13. Using the method of reduction of matrices find the set defined implicitly by $A_1 X = \begin{pmatrix} 3 \\ 2 \end{pmatrix}$, where A_1 is defined by

$$A_1 \begin{pmatrix} x \\ y \\ z \end{pmatrix} = \begin{pmatrix} 3x + 2y - 2 \\ x - y + 2z \end{pmatrix}, \qquad \begin{pmatrix} x \\ y \\ z \end{pmatrix} \in \mathcal{R}^3.$$

14. Let $A \in \mathcal{R}_{nn}$ and let A' be a matrix obtained by reducing A. Let T and T' be defined by $TX = AX$ and $T'X = A'X$ for all X in $\mathcal{R}^n$. Prove that $\ker T = \ker T'$.

ANSWERS

1. $\left\{ x \begin{pmatrix} 1 \\ 2/3 \end{pmatrix} + \begin{pmatrix} 0 \\ -1/3 \end{pmatrix} \Big| x \in \mathcal{R} \right\}$.   2. $\varnothing$.   5. $\left\{ z \begin{pmatrix} \frac{4}{19} \\ \frac{13}{19} \\ 1 \end{pmatrix} + \begin{pmatrix} \frac{7}{19} \\ -\frac{1}{19} \\ 0 \end{pmatrix} \Big| z \in \mathcal{R} \right\}$.

6. $\varnothing$.   8. Not invertible.   9. $A^{-1} = \begin{pmatrix} 1 & -7 & 25 \\ 0 & 3 & -11 \\ 0 & -1 & 4 \end{pmatrix}$.

12. $A^{-1} = \begin{pmatrix} 1 & 0 & 0 & 0 \\ -81 & 1 & -7 & 25 \\ 35 & 0 & 3 & -11 \\ -13 & 0 & -1 & 4 \end{pmatrix}$.

16

DIFFERENTIAL CALCULUS
OF $\mathcal{R}^n \to \mathcal{R}^m$ FUNCTIONS

INTRODUCTION

In Chapters 3, 4, and 5 we were introduced to one of the basic viewpoints of much of mathematics. This is the viewpoint often referred to as *functional approximation*. Given a function $f: \mathcal{R}^n \to \mathcal{R}^m$ we may find that, for a specific application, an approximation to f by a simpler function g will disclose all the information about f that we need. Moreover, the function g would be easier to deal with.

This viewpoint immediately raises two questions. First, what set of "simple functions" shall be selected as the set of approximating functions? Second, what set of criteria for "goodness of approximation" shall be selected for determining which of the approximating functions is the *best* approximation to the given function f? There is no single answer to either of these questions. In fact, the study of functional approximation is enriched by the variety of choices which one may make for the set of approximating functions and for the criteria of goodness of approximation.

In Chapter 5 we selected the set of affine $\mathcal{R} \to \mathcal{R}$ functions as our approximating functions and set about to determine the criteria for defining an affine approximation to an $\mathcal{R} \to \mathcal{R}$ function about the point t in *dom f*. This approach led us to introduce the concepts of the derivative and the differential and the operation of differentiation for $\mathcal{R} \to \mathcal{R}$ functions.

In the preceding chapters we became acquainted with various vector functions from $\mathcal{R}^n$ into $\mathcal{R}^m$, and we certainly must regard the affine transformation as the simplest kind of $\mathcal{R}^n \to \mathcal{R}^m$ function, other than the strictly constant function. We shall then in the present chapter seek to define the concept of an affine approximation to an $\mathcal{R}^n \to \mathcal{R}^m$ function and shall be led to generalizations of the $\mathcal{R} \to \mathcal{R}$ concepts of derivative, differential, and differentiation.

1. AFFINE APPROXIMATIONS OF $\mathcal{R}^2 \to \mathcal{R}$ FUNCTIONS

Let $f:\mathcal{R}^2 \to \mathcal{R}$, and let T be an interior point of $dom\,f$. We seek an affine transformation A which will serve as a suitable approximation to f for points X in $dom\,f$ which are close to T. To determine the criteria by which we single out a "best" affine approximation, we proceed in a manner similar to that used in Chapter 5 for $\mathcal{R} \to \mathcal{R}$ functions. That is, we shall consider the graphs of the function f and an approximating affine function A.

Figure 16.1a is a sketch of the graphs of a continuous $\mathcal{R}^2 \to \mathcal{R}$ function f and its constant approximation K at the point T, where $KX = fT$. Figure 16.1b illustrates the graphs of f and an affine function A which, from the

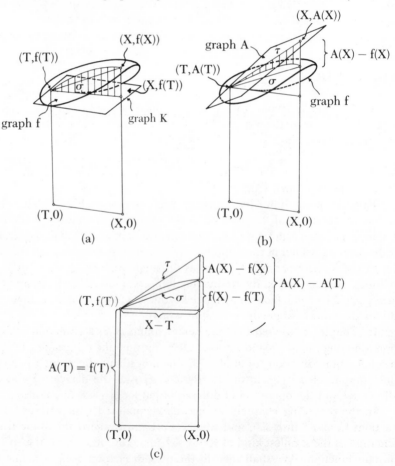

Figure 16.1 Affine approximations to $\mathcal{R}^2 \to \mathcal{R}$ functions.

figure, would seem to be a better approximation to the function f for points close to T. For, in Figure 16.1a the vertical lines representing the difference between fX and KX are noticeably longer than the corresponding lines representing the difference $fX - AX$ in Figure (16.1b). Figure (16.1c) is the sketch of the section of the graphs of Figure 16.1b by the plane containing the points $(T, 0)$, (T, fT) and the arbitrary point $(X, 0)$, for X in $dom\,f$. The line segment τ lies in the planar graph of A (see Figure 16.1b) and joins (T, AT) to (X, AX); the line segment σ (see Figure 16.1a) is the secant line of $graph\,f$ joining the points (T, fT) and (X, fX). (Of course the lines τ and σ as well as the planar section will vary with different choices of X.)

Let us define the slope of the lines σ and τ as the slope of these lines in the planar section indicated in Figure 16.1c. Thus

$$\text{slope of } \tau = \frac{AX - AT}{\|X - T\|}, \quad \text{slope of } \sigma = \frac{fX - fT}{\|X - T\|},$$

where $\|X - T\| = \|(X, 0) - (T, 0)\|$ and thus denotes the distance between $(X, 0)$ and $(T, 0)$. The graphs of A and f are said to slope in the same direction at T if the slope of the secant line σ approaches the slope of the line τ as X approaches T, or, in other words, if the difference between the slope of σ and the slope of τ approaches 0 as X approaches T:

$$(16.1) \quad \lim_{X \to T} (\text{slope } \sigma - \text{slope } \tau) = \lim_{X \to T}\left[\frac{fX - fT}{\|X - T\|} - \frac{AX - AT}{\|X - T\|}\right] = 0.$$

Equation 16.1 thus appears as a reasonable criterion for an approximating affine function. Implicit in the discussion is a second requirement, namely, that $AT = fT$. If we impose this (reasonable) restriction the Relation 16.1 takes on a simpler form. For if in Equation 16.1 we substitute fT for AT and carry out the subtraction, we obtain

$$\lim_{X \to T}\left[\frac{fX - AX}{\|X - T\|}\right] = 0.$$

The foregoing considerations suggest a criteria for an affine approximation to a given function. In the definition which follows we denote an affine approximation by the symbol $A_T f$. This symbol is intended to emphasize the fact that a given affine approximation depends not only on the function f but also upon the point T about which the approximation is "best."

DEFINITION 16.1 Let $f: \mathcal{R}^2 \to \mathcal{R}$ and let T be an interior point of $dom\,f$. An affine function $A_T f: \mathcal{R}^2 \to \mathcal{R}$ is called *an affine approximation of f about T* if and only if

    i. $fT = (A_T f)\,T$, and

    ii. $\displaystyle\lim_{X \to T} \frac{fX - (A_T f)X}{\|X - T\|} = 0.$

Let $A_T f$ be an affine approximation about T of an $\mathcal{R}^2 \to \mathcal{R}$ function f. Associated with $A_T f$ is a unique linear transformation $L : \mathcal{R}^2 \to \mathcal{R}$ such that

(16.2) $$(A_T f)X = LX + y_0,$$

where y_0 is a constant (Theorem 15.16, p. 618). Condition (i) of Definition 16.1 then generates the implication chain:

$$fT = (A_T f)T \Rightarrow^1 fT = LT + y_0$$
$$\Rightarrow^2 y_0 = fT - LT$$
$$\Rightarrow^3 (A_T f)X = LX + fT - LT$$

(1) Substituting for $(A_T f)T$ from Equation 16.2 (for $X = T$).
(2) Solving for y_0 by subtracting LT.
(3) Substituting for y_0 in Equation 16.2.

Employing the linearity of L in the last member of this chain we obtain

(16.3) $$(A_T f)X = fT + L(X - T)$$

and hence

(16.4) $$L(X - T) = (A_T f)X - fT.$$

Thus, if we regard $(A_T f)X$ as approximating fX, so that $(A_T f)X - fT$ approximates $fX - fT$, Equation 16.4 states that $L(X - T)$ is an approximation to the difference $fX - fT$. Therefore, to emphasize the usefulness of L in obtaining an approximation to the *difference* $fX - fT$, we shall refer to L as the *differential of f at T associated with $A_T f$* and shall henceforth denote it by a special symbol now to be introduced.

DEFINITION 16.2 Let $f : \mathcal{R}^2 \to \mathcal{R}$ and let T be an interior point of *dom f* about which f has an affine approximation $A_T f$. Then,

i. f is said to be *differentiable at T*, and
ii. the linear function associated with $A_T f$ is called a *differential of f about T* and shall be denoted by $d_T f$.

If f is differentiable at every point of a set $\mathcal{S}$, we say that f is *differentiable on $\mathcal{S}$*.
If f is differentiable on *dom f*, we say merely that f is *differentiable*.

The result contained in Equation 16.3 is useful enough to be stated as a theorem. We change the notation somewhat by using the preceding definition to write $d_T f$ instead of L:

THEOREM 16.3 Let $f : \mathcal{R}^2 \to \mathcal{R}$ and let $A_T f$ be an affine approximation of f about T with $d_T f$ as differential. Then $A_T f$ is given by

$$(A_T f)X = fT + (d_T f)(X - T).$$

At this point we do not know if any $\mathcal{R}^2 \to \mathcal{R}$ function f will have an affine approximation in the sense we have defined it. Moreover, we have no method for determining what particular affine function is an affine approximation about a point T in *dom f* even if we knew that one does exist. Finally, we must face the logical possibility that a given $\mathcal{R}^2 \to \mathcal{R}$ function f may have more than one affine approximation about T, and, consequently, more than one differential about T. Some of the uncertainties just cited will be removed by the next theorem in which we shall devise a method for determining an affine approximation, assuming that it exists, and shall show that for each point T, there is at most one such approximation. In an example following the theorem, we shall show that some $\mathcal{R}^2 \to \mathcal{R}$ functions are differentiable at least at some points.

Before stating and proving the result just promised, we wish to recall to the reader's mind that a differential of an $\mathcal{R}^2 \to \mathcal{R}$ function about a point T is a linear transformation from $\mathcal{R}^2 \to \mathcal{R}$ and as such will have a matrix representation (Theorem 15.7, p. 603), say

$$(16.5) \qquad [d_T f] = [a_1 \quad a_2].$$

If we introduce the natural basis vectors into $\mathcal{R}^2$, that is, the vectors

$$E_1 = \begin{pmatrix} 1 \\ 0 \end{pmatrix}, \qquad E_2 = \begin{pmatrix} 0 \\ 1 \end{pmatrix},$$

we obtain

$$(16.6) \qquad (d_T f)E_1 = [a_1 \quad a_2]\begin{pmatrix} 1 \\ 0 \end{pmatrix} = a_1,$$

$$(d_T f)E_2 = [a_1 \quad a_2]\begin{pmatrix} 0 \\ 1 \end{pmatrix} = a_2.$$

We are now ready to consider the problem of the determination of affine approximations.

THEOREM 16.4 Let $f: \mathcal{R}^2 \to \mathcal{R}$ have an affine approximation $A_T f$ about $T = \begin{pmatrix} t_1 \\ t_2 \end{pmatrix}$. Then the limits

$$(16.7) \qquad (\partial_1 f) T = \lim_{h \to 0} \frac{f(T + hE_1) - fT}{h},$$

$$(\partial_2 f) T = \lim_{h \to 0} \frac{f(T + hE_2) - fT}{h}$$

exist. Moreover, $A_T f$ is uniquely determined and the matrix of the (unique) differential is

$$(16.8) \qquad [d_T f] = [(\partial_1 f) T \quad (\partial_2 f) T].$$

Proof.   Since f is assumed to have an affine approximation about T, T must be an interior point of $dom\,f$ (Definition 16.1, p. 641). Thus, for suitably small values of h, the points on the lines

(16.9)

$$\mathcal{L}_1 : X(h) = T + hE_1 = \begin{pmatrix} t_1 \\ t_2 \end{pmatrix} + h\begin{pmatrix} 1 \\ 0 \end{pmatrix} = \begin{pmatrix} t_1 + h \\ t_2 \end{pmatrix}$$

$$\mathcal{L}_2 : X(h) = T + hE_2 = \begin{pmatrix} t_1 \\ t_2 \end{pmatrix} + h\begin{pmatrix} 0 \\ 1 \end{pmatrix} = \begin{pmatrix} t_1 \\ t_2 + h \end{pmatrix}$$

are in $dom\,f$. $\mathcal{L}_1$ is clearly the line through T parallel to the 1-coordinate axis, since $E_1 = \begin{pmatrix} 1 \\ 0 \end{pmatrix}$ is a direction vector for $\mathcal{L}_1$ (see Figure 16.2); similarly, $\mathcal{L}_2$ is the line through T parallel to the 2-coordinate axis.

From Theorem 16.3 we have, for X in $\mathcal{L}_1 \cap dom\,f$,

(16.10)
$$(A_T f)X = fT + (d_T f)(X - T)$$

$$= fT + (d_T f)(T + hE_1 - T)$$

$$= fT + (d_T f)(hE_1)$$

$$= fT + h(d_T f)E_1, \quad \text{since } d_T f \text{ is linear.}$$

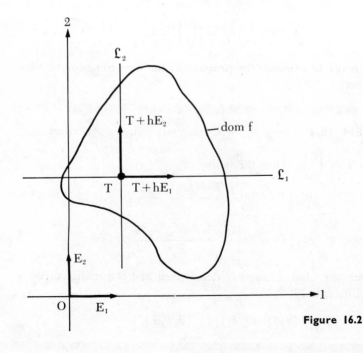

Figure 16.2

Condition (ii) of Definition 16.1, p. 641, then initiates the implication chain:

$$\lim_{X \to T} \frac{fX - (A_T f)X}{\|X - T\|} =^1 0 \Rightarrow^2 \lim_{X \to T} \frac{fX - fT - h(d_T f)E_1}{\|X - T\|} = 0$$

$$\Rightarrow^3 \lim_{h \to 0} \frac{f(T_1 + hE_1) - fT - h(d_T f)E_1}{|h|} = 0$$

$$\Rightarrow^4 \lim_{h \to 0} \frac{(f(T + hE_1) - fT) - h(d_T f)E_1}{h} = 0$$

$$\Rightarrow^5 \lim_{h \to 0} \left[\frac{f(T + hE_1) - fT}{h} - (d_T f)E_1 \right] = 0$$

$$\Rightarrow^6 \lim_{h \to 0} \frac{f(T + hE_1) - fT}{h} \text{ exists and is equal to } (d_T f)E_1.$$

(1) Definition 16.1.
(2) Substituting from Equation 16.10.
(3) Using the fact that $X = T + hE_1$ and hence $\|X - T\| = \|T + hE_1 - T\| = \|hE_1\| = |h| \, \|E_1\| = |h|$ and the fact that $\lim_{X \to T}$ may be written as $\lim_{h \to 0}$, Theorem 3.9, p. 177.
(4) Using the fact that $\lim_{h \to 0} \frac{g(h)}{|h|} = 0$ if and only if $\lim_{h \to 0} \frac{g(h)}{h} = 0$, Theorem 3.10, p. 180.
(5) Splitting into two terms and dividing out the h in the second.
(6) Employing Exercise 18, p. 186.

Thus we have shown that the first limit in Equations 16.7 exists and, in view of Equation 16.6, that the first entry in the matrix of Equation 16.5 is

$$a_1 \equiv (d_T f)E_1 = \lim_{h \to 0} \frac{f(T + hE_1) - fT}{h}.$$

If we consider points $X = T + hE_2$ on $\mathfrak{L}_2$ we can, by a similar line of reasoning, show that the second limit in Equations 16.7 exists and that the second entry in the matrix of Equation 16.4 is

$$a_2 \equiv (d_T f)E_2 = \lim_{h \to 0} \frac{f(T + hE_2) - fT}{h}.$$

If we now introduce the symbolism $(\partial_1 f)T$ and $(\partial_2 f)T$ for these limits, we find that the matrix of the differential *must* be that given in Equation 16.8. Hence the differential $d_T f$ and, consequently, the affine approximation $A_T f$ of f about T is *unique*.

Example Ia. Let $f\begin{pmatrix} x \\ y \end{pmatrix} = 3x + 5y + 4$ and $T = \begin{pmatrix} 1 \\ 2 \end{pmatrix}$. We first deter-

mine whether the limits $(\partial_1 f)T$ and $(\partial_2 f)T$ exist. We use the fact that

$\begin{pmatrix} 1 \\ 2 \end{pmatrix} = 3 + 10 + 4 = 17$.

$$(\partial_1 f)T = \lim_{h \to 0} \frac{f(T + hE_1) - fT}{h} = \lim_{h \to 0} \frac{f\begin{pmatrix} 1 + h \\ 2 \end{pmatrix} - f\begin{pmatrix} 1 \\ 2 \end{pmatrix}}{h}$$

$$= \lim_{h \to 0} \frac{3(1 + h) + 5 \cdot 2 + 4 - 17}{h} = \lim_{h \to 0} \frac{3h}{h} = 3,$$

$$(\partial_2 f)T = \lim_{h \to 0} \frac{f(T + hE_2) - fT}{h} = \lim_{h \to 0} \frac{f\begin{pmatrix} 1 \\ 2 + h \end{pmatrix} - f\begin{pmatrix} 1 \\ 2 \end{pmatrix}}{h}$$

$$= \lim_{h \to 0} \frac{3 + 5(2 + h) + 4 - 17}{h} = \lim_{h \to 0} \frac{5h}{h} = 5.$$

Thus, in view of Equation 16.8, if f has an affine approximation at $T = \begin{pmatrix} 1 \\ 2 \end{pmatrix}$,

the matrix of the differential is $[3 \quad 5]$ and from Theorem 16.3, p. 642, the
affine approximation would be

$$(16.11) \qquad (A_T f)X = (A_T f)\begin{pmatrix} x \\ y \end{pmatrix} = fT + (d_T f)(X - T)$$

$$= 17 + [3 \quad 5]\begin{pmatrix} x - 1 \\ y - 2 \end{pmatrix}$$

$$= 3x + 5y + 4. \quad \blacksquare$$

At this point we emphasize that Theorem 16.4 does *not* assure us that
the affine function $A_T f$ just obtained is the affine approximation of f about T.
The theorem does tell us that the $A_T f$ of Equation 16.11 is the only affine
function that we need to test. If the conditions of Definition 16.1 are met, f
has an affine approximation at $T = \begin{pmatrix} 1 \\ 2 \end{pmatrix}$ and we have found it! If the
conditions of this definition are not met, f has no affine approximation at T.
Let us now formalize the observations just made.

THEOREM 16.5 Let $f : \mathcal{R}^2 \to \mathcal{R}$ and let T be a point of *dom f* at
which the entries of the matrix

$$J_T f = [(\partial_1 f)T \quad (\partial_2 f)T]$$

are defined. Then f has an affine approximation at $T = \begin{pmatrix} t_1 \\ t_2 \end{pmatrix}$ if and
only if the function $A_T f$ given by

$$(A_T f)X = (A_T f)\begin{pmatrix} x_1 \\ x_2 \end{pmatrix} = fT + [(\partial_1 f)T \quad (\partial_2 f)T]\begin{pmatrix} x_1 - t_1 \\ x_2 - t_2 \end{pmatrix}$$

is such that

$$\lim_{X \to T} \frac{fX - (A_T f)X}{\|X - T\|} = 0.$$

Moreover, $A_T f$ is the affine approximation and J is the matrix of the differential:

$$[d_T f] = J.$$

The formal proof of this theorem is straight-forward and is given as Exercise 16.

Example 1b. We return to the function of Example 1a, namely $f\binom{x}{y} =$ $3x + 5y + 4$. We showed that if f has an affine approximation at $T = \binom{1}{2}$ it is the function

$$(A_T f)\binom{x}{y} = 3x + 5y + 4,$$

that is $(A_T f)X = fX$. It is therefore indeed evident that $A_T f$ satisfies the conditions of Definition 16.1 or, equivalently, the condition of Theorem 16.5. That is,

$$\lim_{X \to T} \frac{fX - (A_T f)X}{\|X - T\|} = \lim_{X \to T} \frac{fX - fX}{\|X - T\|} = \lim_{X \to T} 0 = 0. \quad \blacksquare$$

This example illustrates the expected but satisfying result contained in the following theorem, the proof of which is called for in Exercise 17.

THEOREM 16.6 Let A be an affine function from $\mathcal{R}^2 \to \mathcal{R}$, and let L be its associated linear function. Then A is differentiable and is its own affine approximation about any point in $\mathcal{R}^2$. Moreover, L is the differential of A at each point of $\mathcal{R}^2$.

Let $f: \mathcal{R}^2 \to \mathcal{R}$ and let T be an interior point of $dom\ f$. Theorems 16.4 and 16.5 have led us to the following four-step method of testing whether f has an affine approximation about T, or not.

   i. Determine the elements of the matrix $J = [(\partial_1 f)T \quad (\partial_2 f)T] = [m \quad n]$, if they exist.

   ii. Construct the affine test function

$$(A_T f)\binom{x}{y} = fT + [m \quad n]\binom{x - t_1}{y - t_2}.$$

   iii. Verify that $\lim_{X \to T} \dfrac{fX - (A_T f)X}{\|X - T\|} = 0.$

iv. Conclusion: (a) If the elements of J are not defined or if the limit of (iii) is not verifiable then f has *no* affine approximation about T. (b) If all elements of J are defined and the limit in (iii) is verified then the function $A_T f$ constructed in (ii) is the affine approximation to f about T and the differential of f at T has the matrix

$$[d_T f] = [m \quad n].$$

Example 2. Let us apply the outlined four-step procedure to the function $f\begin{pmatrix} x \\ y \end{pmatrix} = x^2 + y^2 + 4$ with $T = \begin{pmatrix} 0 \\ 1 \end{pmatrix}$.

i. We note first that $fT = f\begin{pmatrix} 0 \\ 1 \end{pmatrix} = 5$. Consequently,

$$\frac{f(T + hE_1) - fT}{h} = \frac{f\begin{pmatrix} h \\ 1 \end{pmatrix} - f\begin{pmatrix} 0 \\ 1 \end{pmatrix}}{h} = \frac{h^2 + 1 + 4 - 5}{h} = h,$$

and

$$\frac{f(T + hE_2) - fT}{h} = \frac{f\begin{pmatrix} 0 \\ 1 + h \end{pmatrix} - f\begin{pmatrix} 0 \\ 1 \end{pmatrix}}{h}$$

$$= \frac{0 + (1 + h)^2 + 4 - 5}{h} = \frac{2h + h^2}{h} = 2 + h.$$

Therefore

$$(\partial_1 f) T = \lim_{h \to 0} \frac{f(t + hE_1) - fT}{h} = \lim_{h \to 0} h = 0,$$

$$(\partial_2 f) T = \lim_{h \to 0} \frac{f(t + hE_2) - fT}{h} = \lim_{h \to 0} 2 + h = 2,$$

and

$$J = [0 \quad 2].$$

ii. The affine test function is therefore

$$(A_T f)\begin{pmatrix} x \\ y \end{pmatrix} = f\begin{pmatrix} 0 \\ 1 \end{pmatrix} + [0 \quad 2]\begin{pmatrix} x - 0 \\ y - 1 \end{pmatrix}$$

$$= 5 + 0(x - 0) + 2(y - 1)$$

$$= 5 + 2y - 2$$

$$= 2y + 3.$$

iii. Noting that $\|X - T\| = \left\| \begin{pmatrix} x - 0 \\ y - 1 \end{pmatrix} \right\| = \sqrt{x^2 + (y - 1)^2}$

and that

$$\frac{fX - (A_T f)X}{\|X - T\|} = \frac{x^2 + y^2 + 4 - (2y + 3)}{\sqrt{x^2 + (y - 1)^2}}$$

$$= \frac{x^2 + y^2 - 2y + 1}{\sqrt{x^2 + (y - 1)^2}} = \frac{x^2 + (y - 1)^2}{\sqrt{x^2 + (y - 1)^2}}$$

$$= \sqrt{x^2 + (y - 1)^2},$$

we find

$$\lim_{X \to T} \frac{fX - (A_T f)X}{\|X - T\|} = \lim_{\binom{x}{y} \to \binom{0}{1}} \sqrt{x^2 + (y - 1)^2}$$

$$= 0.$$

iv. Conclusion: $(A_T f)\binom{x}{y} = 2y + 3$ is the affine approximation of f about $(0, 1)$.

EXERCISES

In Exercises 1 to 5 sketch in a single figure the graphs of f and its affine approximation $A_T f$ about the point T, noting the position of the point T with respect to *graph f* and *graph $A_T f$.*

1. $f\binom{x}{y} = x + y + 2$

    $A_T f\binom{x}{y} = x + y + 2$

    $T = \binom{0}{0}.$

2. $f\binom{x}{y} = x^2 + y^2$

    $A_T f\binom{x}{y} = 2x + 2y - 2$

    $T = \binom{1}{1}.$

3. $f\binom{x}{y} = x^2 + y$

    $A_T f\binom{x}{y} = 2x + y - 1$

    $T = \binom{1}{1}.$

4. $f\binom{x}{y} = \cos^2 x$

    $A_T f\binom{x}{y} = 1$

    $T = \binom{0}{1}.$

5. $f\binom{x}{y} = x \sin y$

    $A_T f\binom{x}{y} = 0$

    $T = \binom{0}{\pi}.$

In Exercises 6 to 10 show that the function $A_T f$ is indeed the affine approximation to the given function f about the given point T by using Definition 16.1.

6. For f, $A_T f$, T of Exercise 1.   7. For f, $A_T f$, T of Exercise 2.

8. For f, $A_T f$, T of Exercise 3.   9. For f, $A_T f$, T of Exercise 4.

10. For f, $A_T f$, T of Exercise 5.

In Exercises 11 to 15 use the four-step method outlined in the text to show that f has an affine approximation about T.

11. $f\begin{pmatrix} x \\ y \end{pmatrix} = x - y - 4$

$$T = \begin{pmatrix} 1 \\ 1 \end{pmatrix}.$$

12. $f\begin{pmatrix} x \\ y \end{pmatrix} = x^2 - 2y^2 + 1$

$$T = \begin{pmatrix} 0 \\ 1 \end{pmatrix}.$$

13. $f\begin{pmatrix} x \\ y \end{pmatrix} = x^2 - 2y + 1$

$$T = \begin{pmatrix} 1 \\ 0 \end{pmatrix}.$$

14. $f\begin{pmatrix} x \\ y \end{pmatrix} = \sin^2 x$

$$T = \begin{pmatrix} 0 \\ 1 \end{pmatrix}.$$

15. $f\begin{pmatrix} x \\ y \end{pmatrix} = x \tan y$

$$T = \begin{pmatrix} 0 \\ \pi \end{pmatrix}.$$

16. Prove Theorem 16.5, p. 646.

17. Prove Theorem 16.6, p. 647.

ANSWERS

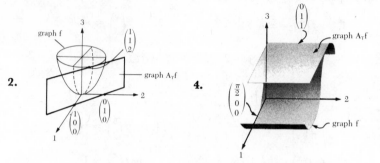

2.

4.

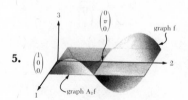

5.

6. See Examples 1a and 1b.

7a. $fT = f\begin{pmatrix} 1 \\ 1 \end{pmatrix} = 2, \; A_T f\begin{pmatrix} 1 \\ 1 \end{pmatrix} = 2, \; A_T f\begin{pmatrix} 1 \\ 1 \end{pmatrix} = f\begin{pmatrix} 1 \\ 1 \end{pmatrix}.$

7b.
$$\frac{fX - (A_T f)X}{\|X - T\|} = \frac{(x^2 + y^2) - (2x + 2y - 2)}{\sqrt{(x-1)^2 + (y-1)^2}}$$

$$= \frac{(x^2 - 2x + 1) + (y^2 - 2y + 1)}{\sqrt{(x-1)^2 + (y-1)^2}}$$

$$= \sqrt{(x-1)^2 + (y-1)^2},$$

$$\lim_{X \to T} \frac{fX - AX}{\|X - T\|} = \lim_{\binom{x}{y} \to \binom{1}{1}} \sqrt{(x-1)^2 + (y-1)^2} = 0.$$

9a. $fT = f\begin{pmatrix}0\\1\end{pmatrix} = \cos 0 = 1 = A_T f\begin{pmatrix}0\\1\end{pmatrix}.$

9b. $\left| \dfrac{fX - (A_T f)X}{\|X - T\|} \right| = \left| \dfrac{\cos^2 x - 1}{\sqrt{x^2 + (y-1)^2}} \right| = \left| \dfrac{\sin^2 x}{\sqrt{x^2 + (y-1)^2}} \right|$

$$\leq \frac{\sin^2 x}{|x|} = \frac{\sin x}{|x|} \sin x.$$

Therefore, $\displaystyle\lim_{X \to T} \left| \dfrac{fX - (A_T f)X}{\|X - T\|} \right| \leq \lim_{x \to 0} \left(\dfrac{\sin x}{|x|} \sin x \right) = 1 \cdot 0 = 0,$

$$\lim_{X \to T} \frac{fX - (A_T f)X}{\|X - T\|} = 0.$$

(Notice the difficulty encountered in establishing the limit for the simple function of this exercise. A later theorem will alleviate this part of the test for certain kinds of functions.)

11. $A_T f\begin{pmatrix}x\\y\end{pmatrix} = x - y - 4.$

12. (i) $m = (\partial_1 f)T = \displaystyle\lim_{h \to 0} \dfrac{f\begin{pmatrix}h\\1\end{pmatrix} - f\begin{pmatrix}0\\1\end{pmatrix}}{h}$

$$= \lim_{h \to 0} \frac{(h^2 - 2 + 1) - (-1)}{h} = 0,$$

$$n = (\partial_2 f)T = \lim_{h \to 0} \frac{f\begin{pmatrix}0\\1+h\end{pmatrix} - f\begin{pmatrix}0\\1\end{pmatrix}}{h}$$

$$= \lim_{h \to 0} \frac{[-2(1+h)^2 + 1] - (-1)}{h} = -4;$$

(ii) test function is therefore

$$A_T f\begin{pmatrix}x\\y\end{pmatrix} = [m \;\; n]\begin{pmatrix}x - 0\\y - 1\end{pmatrix} + f\begin{pmatrix}0\\1\end{pmatrix} = -4y + 3;$$

(iii) $\left| \dfrac{fX - (A_T f)X}{\|X - T\|} \right| = \left| \dfrac{x^2 - 2y^2 + 1 - (-4y + 3)}{\sqrt{x^2 + (y-1)^2}} \right|$

$$= \left| \frac{x^2 - 2(y^2 - 2y + 1)}{\sqrt{x^2 + (y-1)^2}} \right|$$

$$\leq \frac{2|x^2 + (y-1)^2|}{\sqrt{x^2 + (y-1)^2}}$$

$$= 2\sqrt{x^2 + (y-1)^2}.$$

Therefore,

$$\lim_{X \to T} \left| \frac{fX - (A_T f)X}{\|X - T\|} \right| \leq 2 \lim_{\binom{x}{y} \to \binom{0}{1}} \sqrt{x^2 + (y-1)^2} = 0$$

$$\lim_{X \to T} \frac{fX - (A_T f)X}{\|X - T\|} = 0.$$

(iv) Therefore, f has an affine approximation about $\binom{0}{1}$ and it is the function in (ii).

15. $A_T f \binom{x}{y} = 0.$

2. PARTIAL DIFFERENTIATION OF $\mathcal{R}^2 \to \mathcal{R}$ FUNCTIONS

From Theorem 16.4, p. 643, we know that an $\mathcal{R}^2 \to \mathcal{R}$ function will have an affine approximation about T only if each of the two limits

$$(\partial_1 f)\, T = \lim_{h \to 0} \frac{f(T + hE_1) - fT}{h}, \quad (\partial_2 f)\, T = \lim \frac{f(T + hE_2) - fT}{h}$$

do, in fact, exist. The observant reader has already noticed that these limits closely resemble the limit

$$g'(t) = \lim_{h \to 0} \frac{g(t + h) - g(t)}{h} = \lim_{x \to t} \frac{g(x) - g(t)}{x - t}$$

which defines the derivative of an $\mathcal{R} \to \mathcal{R}$ function g at the point t (Chapter 5, p. 223). Indeed, if we consider g to be the $\mathcal{R} \to \mathcal{R}$ function defined by

$$g(x) = f\binom{x}{t_2},$$

we may write

$$f(T + hE_1) = f\left[\binom{t_1}{t_2} + h\binom{1}{0}\right] = f\binom{t_1 + h}{t_2} = g(t_1 + h),$$

$$fT = f\binom{t_1}{t_2} = g(t_1),$$

and thus

$$(\partial_1 f)\, T = \lim_{h \to 0} \frac{f(T + hE_1) - fT}{h} = \lim_{h \to 0} \frac{g(t_1 + h) - g(t_1)}{h} = g'(t_1).$$

Since g is a restriction of f to the set $\left\{ \binom{x_1}{t_2} \,\middle|\, \binom{x_1}{t_2} \in dom\, f \right\}$, $g'(t_1)$ and hence $(\partial_1 f)\, T$ is the rate at which f is increasing with respect to a unit increase in the first $\mathcal{R}^2$-coordinate, the second domain-space coordinate being kept fixed at its value t_2 at the point $T = \binom{t_1}{t_2}$. Accordingly, we shall refer to $(\partial_1 f)\, T$ as the *partial derivative of f at T with respect to the first $\mathcal{R}^2$-coordinate.*

By introducing the function k given by $k(x_2) = f\begin{pmatrix} t_1 \\ x_2 \end{pmatrix}$, the reader can easily show that $(\partial_2 f) T = k'(t_2)$. Thus $(\partial_2 f) T$ is the rate at which f is increasing with respect to a unit increase in the second $\mathcal{R}^2$-coordinate, the first coordinate being held fixed at its value t_1 at the point $T = \begin{pmatrix} t_1 \\ t_2 \end{pmatrix}$. $(\partial_2 f) T$ shall be called the *partial derivative of f at T with respect to the second $\mathcal{R}^2$-coordinate*.

The function f may possess partial derivatives with respect to one (or both) coordinates at many points in *dom f*, and this thereby suggests consideration of a partial derivative function (or functions) on a subset of *dom f*. We shall formalize this and the preceding observations.

DEFINITION 16.7   Let $f : \mathcal{R}^2 \to \mathcal{R}$

i. Let $\mathcal{C}_1$ be the set of all points T in *dom f* at which the function $\partial_1 f$, given by

$$(\partial_1 f) T = \lim_{h \to 0} \frac{f(T + hE_1) - fT}{h},$$

is defined. Then $\partial_1 f$ is called the *first coordinate (partial) derivative function of f*; for each T in $\mathcal{C}_1$, $(\partial_1 f) T$ is called the *(partial) derivative of f at T with respect to the first $\mathcal{R}^2$-coordinate*. Finally, f is said to be *differentiable on $\mathcal{C}_1$ with respect to the first $\mathcal{R}^2$-coordinate*.

ii. Let $\mathcal{C}_2$ be the set of all points T in *dom f* at which the function $\partial_2 f$, given by

$$(\partial_2 f) T = \lim_{h \to 0} \frac{f(T + hE_2) - fT}{h},$$

is defined. Then $\partial_2 f$ is called the *second coordinate (partial) derivative function of f*; for each T in $\mathcal{C}_2$, $(\partial_2 f) T$ is called the *(partial) derivative of f at T with respect to the second $\mathcal{R}_2$-coordinate*. Finally, f is said to be *differentiable on $\mathcal{C}_2$ with respect to the second $\mathcal{R}^2$-coordinate*.

Figure 16.3a represents the graph of a function $f : \mathcal{R}^2 \to \mathcal{R}$ for which both $\partial_1 f$ and $\partial_2 f$ are defined at the point $T = \begin{pmatrix} t_1 \\ t_2 \end{pmatrix}$. Included in the same figure are the implicitly defined planes

$$\mathcal{S}_1 : p(x_1, x_2, x_3) \equiv x_2 - t_2 = 0,$$
$$\mathcal{S}_2 : q(x_1, x_2, x_3) \equiv x_1 - t_1 = 0.$$

The section of *graph f* by the plane $\mathcal{S}_1$ is the curve $\mathcal{G}$ described parametrically by

$$\begin{pmatrix} x \\ y \\ z \end{pmatrix} = \begin{pmatrix} t \\ t_2 \\ f\begin{pmatrix} t \\ t_2 \end{pmatrix} \end{pmatrix}$$

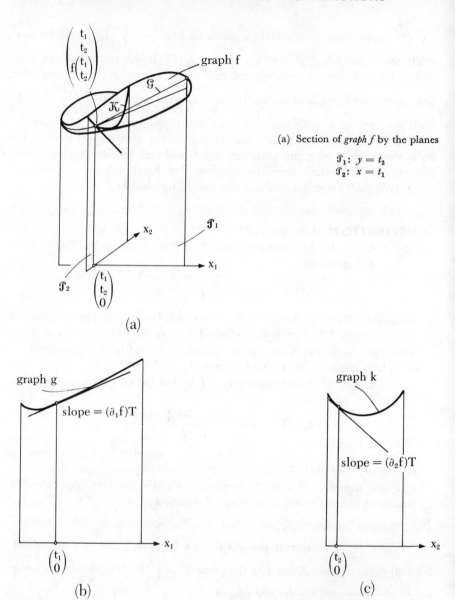

(a) Section of *graph f* by the planes

$$\mathcal{P}_1: \quad y = t_2$$
$$\mathcal{P}_2: \quad x = t_1$$

(a)

graph g

slope $= (\partial_1 f)T$

(b)

graph k

slope $= (\partial_2 f)T$

(c)

(b) Graph of $g(x_1) = f\begin{pmatrix} x_1 \\ t_2 \end{pmatrix}$

(P_1 section of *graph f*).

(c) Graph of $k(x_2) = f\begin{pmatrix} t_1 \\ x_1 \end{pmatrix}$

(P_2 section of *graph f*).

Figure 16.3 Geometrical interpretation of partial derivatives.

Figure 16.3b is a sketch of *graph g*, where $g(t) = f\begin{pmatrix} t \\ t_2 \end{pmatrix}$ and is geometrically congruent to the planar section of $\mathcal{G}$ of *graph f*. Now, as seen early in this section, $(\partial_1 f) T = g'(t_1)$ and thus represents the slope of *graph g* at $(t_1, g(t_1))$; that is $(\partial_1 f) T$ is the slope of the tangent to *graph g* at $(t_1, g(t_1))$. Thus $(\partial_1 f) T$ may be regarded as the slope in the plane $\mathcal{G}_1$ of the tangent line to $\mathcal{G}$ at $\begin{pmatrix} T \\ fT \end{pmatrix}$.

Similarly $(\partial_2 f) T$ may be regarded as the slope in $\mathcal{G}_2$ of the tangent line to K (the $\mathcal{G}_2$-section of *graph f*), as suggested by Figure 16.3a and c. Of course this same geometrical viewpoint of $(\partial_1 f) T$ and $(\partial_2 f) T$ may be obtained by referring to Figure 16.1 and the considerations which gave rise to the concept of affine approximation.

In what follows it will be convenient to use an alternate notation for partial derivatives of $\mathcal{R}^2 \to \mathcal{R}$ functions. To this end we shall henceforth extend the use of the coordinate function notation to the degenerate case of $\mathcal{R}^n \to \mathcal{R}$ functions, in which case there is only one coordinate function (see Definition 3.14, p. 187). Thus, for the $\mathcal{R}^2 \to \mathcal{R}$ function

$$f\begin{pmatrix} x \\ y \end{pmatrix} = x^2 + xy$$

we shall also write

$$f\begin{pmatrix} x \\ y \end{pmatrix} = f_1\begin{pmatrix} x \\ y \end{pmatrix} = x^2 + xy$$

and for

$$g\begin{pmatrix} x \\ y \end{pmatrix} = \sin(x + \sqrt{y})$$

we shall also write

$$g\begin{pmatrix} x \\ y \end{pmatrix} = g_1\begin{pmatrix} x \\ y \end{pmatrix} = \sin(x + \sqrt{y}),$$

and so forth. The utility of this notation will become apparent in what follows.

Using the preceding coordinate notation, we note that if f is an $\mathcal{R}^2 \to \mathcal{R}$ function, then $\partial_1 f = \partial_1 f_1$ and $\partial_2 f = \partial_2 f_1$. However, it is more convenient to use the alternate notation $f_{1,1}$ for $\partial_1 f$ and $f_{1,2}$ for $\partial_2 f$. Thus

$$f_{1,1} T \equiv (\partial_1 f) T \quad \text{and} \quad f_{1,2} \equiv (\partial_2 f) T.$$

In this notation, the matrix J of Theorem 16.5 may be written

$$J = [f_{1,1} T \quad f_{1,2} T].$$

Thus the subscripts may be regarded as row and column indices for the entries in the matrix J. We shall find that there are other advantages to this alternate notation. The reason for the comma in the subscript will become apparent later in this section.

The next three examples shall employ the notation just introduced.

Example 3.   Let $f\begin{pmatrix} x \\ y \end{pmatrix} = f_1\begin{pmatrix} x \\ y \end{pmatrix} = 3xy + 4y^2 + 16x^2y + 3$. To compute $f_{1,1}\begin{pmatrix} t_1 \\ t_2 \end{pmatrix}$ we proceed as follows.

 i. Set the second coordinate equal to t_2 to obtain the function:

$$g(x) = 3xt_2 + 4t_2^2 + 16x^2t_2 + 3.$$

 ii. Differentiate g with respect to x (remember t_2 is a constant)

$$g'(x) = 3t_2 + 32xt_2.$$

 iii. Evaluate $g'(t_1)$

$$f_{1,1}\begin{pmatrix} t_1 \\ t_2 \end{pmatrix} = g'(t_1) = 3t_2 + 32t_1t_2.$$

Similarly, for $f_{1,2}\begin{pmatrix} t_1 \\ t_2 \end{pmatrix}$ we write

 i. $h(y) = 3t_1y + 4y^2 + 16t_1^2y + 3.$

 ii. $h'(y) = 3t_1 + 8y + 16t_1^2.$

 iii. $f_{1,2}\begin{pmatrix} t_1 \\ t_2 \end{pmatrix} = h'(t_2) = 3t_1 + 8t_2 + 16t_1^2.$ ∎

Most frequently, if we define a function f by specifying its value at a general point $\begin{pmatrix} x \\ y \end{pmatrix}$ in $\mathcal{R}^2$, we prefer to have the partial derivative functions also described by their values at the general point $\begin{pmatrix} x \\ y \end{pmatrix}$ and not at $\begin{pmatrix} t_1 \\ t_2 \end{pmatrix}$. Such a description for the preceding example is readily obtained by replacing $\begin{pmatrix} t_1 \\ t_2 \end{pmatrix}$ by $\begin{pmatrix} x \\ y \end{pmatrix}$ in the final results. However, we may arrive at the same result by a shortening of the whole procedure according to the following rule:

RULE FOR PARTIAL DIFFERENTIATION   Let $f : \mathcal{R}^2 \to \mathcal{R}$ be given by a value specification $f\begin{pmatrix} x \\ y \end{pmatrix} = f_1\begin{pmatrix} x \\ y \end{pmatrix}$.

 i. To compute $f_{1,1}\begin{pmatrix} x \\ y \end{pmatrix}$, treat the second coordinate as though it were fixed and differentiate $f_1\begin{pmatrix} x \\ y \end{pmatrix}$ as though it were an $\mathcal{R} \to \mathcal{R}$ function given by specification of its value at the point x given by the first coordinate.

 ii. To compute $f_{1,2}\begin{pmatrix} x \\ y \end{pmatrix}$, treat the first coordinate, x, as though it were fixed and differentiate $f_1\begin{pmatrix} x \\ y \end{pmatrix}$ as though it were an $\mathcal{R} \to \mathcal{R}$ function given by specification of its value at the point y given by the second coordinate.

Example 4.   We use the same function as in Example 3: $f\begin{pmatrix} x \\ y \end{pmatrix} = f_1\begin{pmatrix} x \\ y \end{pmatrix} = 3xy + 4y^2 + 16x^2y + 3$. To compute $f_{1,1}\begin{pmatrix} x \\ y \end{pmatrix}$ regard y as fixed.

As an aid to our thinking we may rewrite the expression for $f_1\begin{pmatrix} x \\ y \end{pmatrix}$ under-lining all symbols which are to be treated as constants:

$$f_1\begin{pmatrix} x \\ y \end{pmatrix} = 3x\underline{y} + 4\underline{y}^2 + 16x^2\underline{y} + \underline{3}.$$

We then differentiate as though working with an $\mathcal{R} \to \mathcal{R}$ function given in terms of x:

$$f_{1,1}\begin{pmatrix} x \\ y \end{pmatrix} = 3\underline{y} + 0 + 16 \cdot 2x\underline{y} + 0$$

or, removing the underlining,

$$f_{1,1}\begin{pmatrix} x \\ y \end{pmatrix} = 3y + 32xy.$$

To compute $f_{1,2}\begin{pmatrix} x \\ y \end{pmatrix}$, we write

$$f_1\begin{pmatrix} x \\ y \end{pmatrix} = 3\underline{x}y + 4y^2 + 16\underline{x}^2y + 3,$$

$$f_{1,2}\begin{pmatrix} x \\ y \end{pmatrix} = 3\underline{x} \cdot 1 + 4 \cdot 2y + 16\underline{x}^2 \cdot 1 + 0$$

$$= 3x + 8y + 16x^2.$$

Of course, it is not necessary to rewrite f_1 with either x or y underlined if we merely *remember* which coordinate is held fixed.   ■

Example 5.   Let $f: \mathcal{R}^2 \to \mathcal{R}$ be given by

$$f\begin{pmatrix} x \\ y \end{pmatrix} = f_1\begin{pmatrix} x \\ y \end{pmatrix} = \sqrt{x^2 + y} + \frac{y}{x}.$$

(a) To compute $f_{1,1}\begin{pmatrix} x \\ y \end{pmatrix}$, we use the underlining method.

i. $f_1\begin{pmatrix} x \\ y \end{pmatrix} = \sqrt{x^2 + \underline{y}} + \underline{y}/x$

ii. $f_{1,1}\begin{pmatrix} x \\ y \end{pmatrix} = \frac{1}{2}(x^2 + \underline{y})^{-1/2}2x + \underline{y}(-1)x^{-2}$

$$f_{1,1}\begin{pmatrix} x \\ y \end{pmatrix} = \frac{x}{\sqrt{x^2 + y}} - \frac{y}{x^2}.$$

(b) To compute $f_{1,2}\begin{pmatrix} x \\ y \end{pmatrix}$, we shall merely *remember* that x is to be regarded as fixed and compute $f_{1,2}\begin{pmatrix} x \\ y \end{pmatrix}$ immediately.

$$f_{1,2}\begin{pmatrix} x \\ y \end{pmatrix} = \frac{1}{2}(x^2 + y)^{-\frac{1}{2}} \cdot 1 + \frac{1}{x}.   ■$$

HIGHER ORDER PARTIAL DERIVATIVES The partial derivative functions $f_{1,1}$ and $f_{1,2}$ of a given $\mathcal{R}^2 \to \mathcal{R}$ function $f = f_1$ may also possess partial derivatives at some points in their domains. Such derivatives of $f_{1,1}$ and $f_{1,2}$ are called second-order partial derivatives of f. The second order partial derivative functions may, in turn, also possess derivatives which are referred to as third order partial derivatives of f or f_1, and so on.

Example 6a. Let $f\begin{pmatrix} x \\ y \end{pmatrix} = f_1\begin{pmatrix} x \\ y \end{pmatrix} = 2x^2y + y^2x$. Then $f_{1,1}$ is given by

$$f_{1,1}\begin{pmatrix} x \\ y \end{pmatrix} = 4xy + y^2$$

and has (a) a first coordinate partial derivative function $f_{1,11}$ given by

$$f_{1,11}\begin{pmatrix} x \\ y \end{pmatrix} = 4y$$

and (b) a second coordinate partial derivative function $f_{1,12}$ given by

$$f_{1,12}\begin{pmatrix} x \\ y \end{pmatrix} = 4x + 2y.$$

On the other hand $f_{1,2}$ is given by

$$f_{1,2}\begin{pmatrix} x \\ y \end{pmatrix} = 2x^2 + 2xy$$

and has (c) a first coordinate partial derivative function $f_{1,21}$ given by

$$f_{1,21}\begin{pmatrix} x \\ y \end{pmatrix} = 4x + 2y$$

and (d) a second coordinate partial derivative function $f_{1,22}$ given by

$$f_{1,22}\begin{pmatrix} x \\ y \end{pmatrix} = 2x. \quad \blacksquare$$

Take note of the notation in this example. The first subscript indicates the coordinate function being differentiated. Since we are dealing with an $\mathcal{R}^2 \to \mathcal{R}$ function, there is only one coordinate function and the first subscript is always a one. The subscripts following the comma indicate which derivatives are to be taken and in which order. Thus, $f_{1,12}$ denotes the second order derivative (of the first coordinate function) obtained by differentiating first with respect to the first coordinate and the resulting function with respect to the second coordinate. On the other hand $f_{1,21}$ denotes the second order derivative function obtained by differentiating in the opposite order. The order of differentiation is important; the fact that the two functions $f_{1,12}$ and $f_{1,21}$ are identical is a result of the very special nature of $f = f_1$ itself. The following theorem will shed some light on this situation.

THEOREM 16.8 Let $f:\mathcal{R}^2 \to \mathcal{R}$ and let $f_{1,1}$ and $f_{1,2}$ be defined on some set S. If

    i. one of the second partial derivatives $f_{1,12}$ or $f_{1,21}$ is defined in some neighborhood of the point $\begin{pmatrix} x_0 \\ y_0 \end{pmatrix} \in S$ and

    ii. is continuous at $\begin{pmatrix} x_0 \\ y_0 \end{pmatrix}$,

then

    iii. the other' of the two second partial derivative functions is defined at least at $\begin{pmatrix} x_0 \\ y_0 \end{pmatrix}$ and

    iv. $f_{1,12}\begin{pmatrix} x_0 \\ y_0 \end{pmatrix} = f_{1,21}\begin{pmatrix} x_0 \\ y_0 \end{pmatrix}$.

We shall not prove the theorem but encourage the interested student to investigate the proof elsewhere.*

The function of Example 6a clearly satisfies the hypothesis of the theorem.

Example 6b. In Example 6a we showed that the $\mathcal{R}^2 \to \mathcal{R}$ function f given by $f\begin{pmatrix} x \\ y \end{pmatrix} = f_1\begin{pmatrix} x \\ y \end{pmatrix} = 2x^2y + y^2x$ has two first partial derivatives

$$f_{1,1}\begin{pmatrix} x \\ y \end{pmatrix} = 4xy + y^2 \quad \text{and} \quad f_{1,2}\begin{pmatrix} x \\ y \end{pmatrix} = 2x^2 + 2yx$$

and four second derivatives

$$f_{1,11}\begin{pmatrix} x \\ y \end{pmatrix} = 4y, \quad f_{1,12}\begin{pmatrix} x \\ y \end{pmatrix} = f_{1,21}\begin{pmatrix} x \\ y \end{pmatrix} = 4x + 2y, \quad f_{1,22}\begin{pmatrix} x \\ y \end{pmatrix} = 2x.$$

The eight third derivatives are readily obtainable, but because of Theorem 16.8 not all will be different. In fact, we have

$$f_{1,111}\begin{pmatrix} x \\ y \end{pmatrix} = 0, \quad f_{1,211}\begin{pmatrix} x \\ y \end{pmatrix} = f_{1,112}\begin{pmatrix} x \\ y \end{pmatrix} = f_{1,121}\begin{pmatrix} x \\ y \end{pmatrix} = 4,$$

$$f_{1,212}\begin{pmatrix} x \\ y \end{pmatrix} = f_{1,221}\begin{pmatrix} x \\ y \end{pmatrix} = f_{1,122}\begin{pmatrix} x \\ y \end{pmatrix} = 2, \quad f_{1,222}\begin{pmatrix} x \\ y \end{pmatrix} = 0.$$

It is apparent that all derivatives of order four or higher are zero. ■

Again take note of the symbolism. After the comma, there is one subscript for each differentiation, each subscript denotes the coordinate with respect to which differentiation is performed, and the order—reading from left to right—indicates the order in which the various differentiations are to

---

* Apostol, *Mathematical Analysis*, Addison-Wesley, 1957, page 121.

be carried out. Thus $f_{1,221}$ is the third-order derivative function obtained by differentiation of f_1 with respect to the second coordinate to obtain $f_{1,2}$, the differentiation of $f_{1,2}$ with respect to the second coordinate to obtain $f_{1,22}$ and finally differentiation of $f_{1,22}$ with respect to the first coordinate to obtain $f_{1,221}$.

There are, of course, alternate notations for higher order derivatives. For example,

$$f_{1,211}, f_{yxx}, \frac{\partial^3 f}{\partial x\, \partial x\, \partial y}, \frac{\partial^3 z}{\partial x\, \partial x\, \partial y},$$

where $z = f\begin{pmatrix} x \\ y \end{pmatrix}$, are symbols for the same third order derivative function.

The reader is asked to observe that when the "round dee" notation ∂ is employed that the order of differentiation is given by reading the lower portion of the symbol from *right to left*, while the "subscript-on-f" notation is read in the usual order—left to right. Still other systems of notation are employed, notably those using the upper case D. Whether the symbol is to be read from right to left or from left to right must be checked with each book in which it is used.

In the next section we shall define the affine approximation of $\mathcal{R}^n \to \mathcal{R}$ functions. The concept of partial derivative functions of such functions will play a role in the study of $\mathcal{R}^n \to \mathcal{R}$ affine approximations in much the same manner as in the case of $\mathcal{R}^2 \to \mathcal{R}$ functions. We, therefore, generalize the notion of a partial derivative.

DEFINITION 16.9   Let $f: \mathcal{R}^n \to \mathcal{R}$.

Let $\mathcal{C}_i$ be the set of all points $T = \begin{pmatrix} t_1 \\ \cdot \\ \cdot \\ \cdot \\ t_n \end{pmatrix}$ in *dom f* at which the function $f_{1,i}$, given by

$$f_{1,i}T = \lim_{h \to 0} \frac{f(T + hE_i) - fT}{h}$$

is defined. Then $f_{1,i}$ is called the *i-th coordinate derivative function of* f; for $T = \begin{pmatrix} t_1 \\ \cdot \\ \cdot \\ \cdot \\ t_n \end{pmatrix} \in \mathcal{C}_i$, $f_{1,i}T$ is called the *i-th coordinate derivative of* f and f is said to be differentiable with respect to the *i*-th $\mathcal{R}^n$ coordinate at T.

This generalization is such that the derivative function $f_{1,i}$ may be obtained according to the following rule:

GENERALIZED RULE FOR PARTIAL DIFFERENTIA-TION   Let $f:\mathcal{R}^n \to \mathcal{R}$ be given by a value specification

$$f\begin{pmatrix} x_1 \\ \cdot \\ \cdot \\ \cdot \\ x_n \end{pmatrix} = f_1\begin{pmatrix} x_1 \\ \cdot \\ \cdot \\ \cdot \\ x_n \end{pmatrix}. \text{ To compute } f_{1,i}\begin{pmatrix} x_1 \\ \cdot \\ \cdot \\ \cdot \\ x_n \end{pmatrix}, \text{ treat all coordinates other}$$

than x_i as though they were constants and differentiate $f_1\begin{pmatrix} x_1 \\ \cdot \\ \cdot \\ \cdot \\ x_n \end{pmatrix}$ as

though it were an $\mathcal{R} \to \mathcal{R}$ function given by specification of its value at x_i.

Example 7.   Let $f:\mathcal{R}^3 \to \mathcal{R}$ be given by $f\begin{pmatrix} x_1 \\ x_2 \\ x_3 \end{pmatrix} = f_1\begin{pmatrix} x_1 \\ x_2 \\ x_3 \end{pmatrix} = 3x_1^2 + 2x_1x_3 + 4x_2x_3^2$. To compute $f_{1,1}\begin{pmatrix} x_1 \\ x_2 \\ x_3 \end{pmatrix}$ we regard x_2 and x_3 as constants and differentiate f_1 as an $\mathcal{R} \to \mathcal{R}$ function with x_1 as the domain coordinate. We shall first rewrite $f_1\begin{pmatrix} x_1 \\ x_2 \\ x_3 \end{pmatrix}$ underlining all symbols to be regarded as constant; differentiate and remove the underlining:

$$f_1\begin{pmatrix} x_1 \\ x_2 \\ x_3 \end{pmatrix} = 3x_1^2 + 2x_1\underline{x_3} + 4\underline{x_2x_3^2},$$

$$f_{1,1}\begin{pmatrix} x_1 \\ x_2 \\ x_3 \end{pmatrix} = 6x_1 + 2\underline{x_3} + 0,$$

$$f_{1,1}\begin{pmatrix} x_1 \\ x_2 \\ x_3 \end{pmatrix} = 6x_1 + 2x_3.$$

The reader should be able to verify that

$$f_{1,2}\begin{pmatrix} x_1 \\ x_2 \\ x_3 \end{pmatrix} = 4x_3^2 \quad \text{and} \quad f_{1,3}\begin{pmatrix} x_1 \\ x_2 \\ x_3 \end{pmatrix} = 2x_1 + 8x_2x_3.$$

This function will also possess second and higher order derivatives.

Thus, the nine second-order derivatives are

$$f_{1,11}\begin{pmatrix} x_1 \\ x_2 \\ x_3 \end{pmatrix} = 6, \quad f_{1,21}\begin{pmatrix} x_1 \\ x_2 \\ x_3 \end{pmatrix} = f_{1,12}\begin{pmatrix} x_1 \\ x_2 \\ x_3 \end{pmatrix} = 0, \quad f_{1,31}\begin{pmatrix} x_1 \\ x_2 \\ x_3 \end{pmatrix} = f_{1,13}\begin{pmatrix} x_1 \\ x_2 \\ x_3 \end{pmatrix} = 2,$$

$$f_{1,22}\begin{pmatrix} x_1 \\ x_2 \\ x_3 \end{pmatrix} = 0, \quad f_{1,32}\begin{pmatrix} x_1 \\ x_2 \\ x_3 \end{pmatrix} = f_{1,23}\begin{pmatrix} x_1 \\ x_2 \\ x_3 \end{pmatrix} = 8x_3, \quad f_{1,33}\begin{pmatrix} x_1 \\ x_2 \\ x_3 \end{pmatrix} = 8x_2.$$

Only three of the twenty-seven third-order derivative functions are not the zero function, namely

$$f_{1,332}\begin{pmatrix} x_1 \\ x_2 \\ x_3 \end{pmatrix} = f_{1,323}\begin{pmatrix} x_1 \\ x_2 \\ x_3 \end{pmatrix} = f_{1,233}\begin{pmatrix} x_1 \\ x_2 \\ x_3 \end{pmatrix} = 8.$$

All higher order derivatives are of course zero.

In this example you will again note that not all higher order derivative functions are different. This is due, in part, to the validity of the generalization of Theorem 16.8. ∎

THEOREM 16.10 (Equality of Second Order Derivatives.) Let $f:\mathcal{R}^n \to \mathcal{R}$ and let $f_{1,i}$, and $f_{1,j}$ be defined on some region $\mathcal{C}$. If

   i. one of the second order derivatives $f_{1,ij}$ or $f_{1,ji}$ is defined on some neighborhood of a point T in $\mathcal{C}$ and

   ii. is continuous at T,

then

   iii. the other of the two second order derivatives is defined at T and

   iv. $f_{1,ij}T = f_{1,ji}T$.

The alternate symbolism for partial derivatives of $\mathcal{R}^2 \to \mathcal{R}$ functions readily generalizes to the $\mathcal{R}^n \to \mathcal{R}$ case as the following example shows.

Example 8. Let $f:\mathcal{R}^4 \to \mathcal{R}$ be given by $z = f\begin{pmatrix} x_1 \\ x_2 \\ x_3 \\ x_4 \end{pmatrix} = f_1\begin{pmatrix} x_1 \\ x_2 \\ x_3 \\ x_4 \end{pmatrix} = x_1 x_2^2 x_3 + 4x_2 x_4^3$. Then

$$f_{1,2}(X) = \frac{\partial f}{\partial x_2} X = \frac{\partial z}{\partial x_2} = 2x_1 x_2 x_3 + 4x_4^3,$$

$$f_{1,24}(X) = \frac{\partial^2 f}{\partial x_4 \partial x_2} X = \frac{\partial^2 z}{\partial x_4 \partial x_2} = 12x_4^2,$$

$$f_{1,241}(X) = \frac{\partial^3 f}{\partial x_1 \partial x_4 \partial x_2} X = \frac{\partial^3 z}{\partial x_1 \partial x_4 \partial x_2} = 0,$$

$$f_{1,244}(X) = \frac{\partial^3 f}{\partial x_4^2 \partial x_2} X = \frac{\partial^3 z}{\partial x_4^2 \partial x_2} = 24x_4. \quad \blacksquare$$

EXERCISES

Find $f_{1,1}$ and $f_{1,2}$ for each of the following functions f:

1. $f\binom{x}{y} = x^2 + 2y$.

2. $f\binom{x}{y} = 2x - 3y^2 + 1$.

3. $f\binom{x}{y} = \cos x + y$.

4. $f\binom{x}{y} = 2 \sin y + x$.

5. $f\binom{x}{y} = x \cos y$.

6. $f\binom{x}{y} = x^2 - \sqrt{xy}$.

7. $f\binom{x}{y} = x^2 \cos (xy)$.

8. $f\binom{x}{y} = \sqrt{\dfrac{x}{y}}$.

Find $f_{1,11}, f_{1,12}, f_{1,21}$, and $f_{1,22}$ for each of the following functions f:

9. $f\binom{x}{y} = 2x^2 - y$.

10. $f\binom{x}{y} = 2x^2y - y^2x$.

11. $f\binom{x}{y} = \cos x \sin y$.

Find the eight third-order partial derivatives for the following functions:

12. $f\binom{x}{y} = y \tan x$.

13. $f\binom{x}{y} = \sin (x^2y)$.

Find the indicated partial derivatives.

14. $f\begin{pmatrix} x_1 \\ x_2 \\ x_3 \end{pmatrix} = x_1^2 - x_2 \cos x_3, f_{1,13}, f_{123}, f_{1,321}.$

15. $f\begin{pmatrix} x_1 \\ x_2 \\ x_3 \end{pmatrix} = x_1^2 \sin (x_2x_3), f_{1,13}, f_{1,31}, f_{1,131}, f_{1,313}.$

ANSWERS

1. $f_{1,1}\binom{x}{y} = 2x, f_{1,2}\binom{x}{y} = 2.$

3. $f_{1,1}\binom{x}{y} = -\sin x, f_{1,2}\binom{x}{y} = 1.$

5. $f_{1,1}\binom{x}{y} = \cos y, f_{1,2}\binom{x}{y} = -x \sin y.$

7. $f_{1,1}\binom{x}{y} = 2x \cos xy - x^2y \sin xy, f_{1,2}\binom{x}{y} = -x^3 \sin xy.$

9. $f_{1,11}\binom{x}{y} = 4, f_{1,12}\binom{x}{y} = f_{21}\binom{x}{y} = f_{22}\binom{x}{y} = 0.$

11. $f_{1,11}\binom{x}{y} = -\cos x \sin y, f_{1,12}\binom{x}{y} = -\sin x \cos y,$

$f_{1,21}\binom{x}{y} = -\sin x \cos y, f_{1,22}\binom{x}{y} = -\cos x \sin y.$

12. $f_{1,111}\begin{pmatrix} x \\ y \end{pmatrix} = [2\sec^2 x \tan^2 x + \sec^4 x](2y),$

$f_{1,211}\begin{pmatrix} x \\ y \end{pmatrix} = f_{1,112}\begin{pmatrix} x \\ y \end{pmatrix} = f_{1,121}\begin{pmatrix} x \\ y \end{pmatrix} = 2\sec^2 x \tan x,$

$f_{1,122}\begin{pmatrix} x \\ y \end{pmatrix} = f_{1,212}\begin{pmatrix} x \\ y \end{pmatrix} = f_{1,221}\begin{pmatrix} x \\ y \end{pmatrix} = f_{1,222}\begin{pmatrix} x \\ y \end{pmatrix} = 0.$

15. $f_{1,313} = -2x_1 x_2^2 \sin(x_2 x_3).$

3. AFFINE APPROXIMATIONS OF $\mathcal{R}^n \to \mathcal{R}^m$ FUNCTIONS

We shall define an affine approximation of an $\mathcal{R}^n \to \mathcal{R}^m$ function by generalizing Definition 16.1 ($\mathcal{R}^2 \to \mathcal{R}$ Affine Approximations), p. 641.

DEFINITION 16.11 ($\mathcal{R}^n \to \mathcal{R}^m$ Affine Approximations.) Let $f:\mathcal{R}^n \to \mathcal{R}^m$ and let T be an interior point of $dom\,f$. An affine function $A_T f:\mathcal{R}^n \to \mathcal{R}^m$ is called an affine approximation of f about T if and only if

   i. $fT = (A_T f)\,T$ and

   ii. $\displaystyle\lim_{X \to T} \frac{fX - (A_T f)X}{\|X - T\|} = O_m.$

The reader will recall that Definition 16.1 was motivated by geometrical considerations of the graphs of the functions involved. A similar interpretation is possible here as well, but, because of the difficulty in depicting the graph of an $\mathcal{R}^n \to \mathcal{R}^m$ function when $n + m > 3$, such an interpretation is not easily visualized. Therefore, we shall give a different interpretation. It should be noted that this interpretation is also valid, useful, and instructive for the special cases of $\mathcal{R} \to \mathcal{R}$ and $\mathcal{R}^2 \to \mathcal{R}$ affine approximations considered previously.

To this end, suppose that for a given function f we have selected an affine function A which we shall use to estimate f at points X close to T. We introduce the function E given by

$$EX = fX - AX,$$

and refer to E as the *error function associated with A*. For different choices of A we will have different associated error functions. If we are careful to select A so that condition (i) of Definition 16.11 is satisfied, we find that

$$ET = fT - AT = O_m;$$

that is, the error associated with A at T is zero—a reasonable requirement for a function which is to approximate f about T. Now, certainly, we cannot expect to find an A such that $EX = O_m$ for *all* X close to T—this could occur only if f, when restricted to some neighborhood of T, were itself an affine function. In fact, as we consider points X at greater and greater distances $d = \|X - T\|$ from T, we would expect the magnitude $\|EX\|$ of the error to grow (unless f is rather special or we look only at special points X). Condition (ii) of Definition 16.11 thus considers the error-to-distance ratio

$$\frac{EX}{d} = \frac{fX - AX}{\|X - T\|}.$$

This ratio defines what might reasonably be termed the error *per unit distance* and regarded as a rough measure of the distance rate of growth of the error as we proceed from the point T, where the error is O_m, to the point X, where the error is EX. Thus if A satisfies (ii) of Definition 16.11, we are assured that the distance rate of growth of error is close to zero for points X close to T—again, a reasonable requirement. With this point of view, an affine approximation $A_T f$ of f about T is an affine function such that the associated error function $EX = fX - (A_T f)X$ is zero at T and has at X a distance rate of growth close to zero for X close to T.

Appealing to Theorem 15.16, p. 618, we note that there is associated with each affine approximation $A_T f$ a unique linear transformation which we denote by $d_T f$ and which is such that

$$(A_T f)X = (d_T f)X + Y_0,$$

where Y_0 is a fixed vector easily shown to be given by $Y_0 = fT - (d_T f)T$. (See Exercise 13.) This observation sets the stage for the following analogues of Definition 16.2 and Theorem 16.3.

DEFINITION 16.12 Let $f:\mathcal{R}^n \to \mathcal{R}^m$, and let T be an interior point of *dom f* about which f has an affine approximation $A_T f$. Then
  i. f is said to be differentiable at T, and
  ii. the linear function $d_T f$ is called the *differential of f about T*.
If f is differentiable at every point of a set S, we say that f is *differentiable on* S.
If f is differentiable on *dom f*, we say merely that f is differentiable.

THEOREM 16.13 Let $f:\mathcal{R}^n \to \mathcal{R}^m$, and let $A_T f$ be an affine approximation of f about T with $d_T f$ as differential. The affine approximation $A_T f$ may be expressed by

(16.12) $$(A_T f)X = (d_T f)(X - T) + fT.$$

From Equation 16.12 we find that

$$(d_T f)(X - T) = (A_T f)X - fT.$$

Regarding $(A_T f)X - fT$ as an approximation to $fX - fT$, the last equation shows that the *differential* $d_T f$ evaluated at $X - T$ approximates the *difference* $fX - fT$. The use of the term "differential" is therefore natural.

We begin the problem of actually determining affine approximations of general $\mathcal{R}^n \rightarrow \mathcal{R}^m$ functions by considering the case of an $\mathcal{R}^n \rightarrow \mathcal{R}$ function f. The result is a straightforward generalization of Theorem 16.4.

THEOREM 16.14 Let $f: \mathcal{R}^n \rightarrow \mathcal{R}$ have an affine approximation $A_T f$ about T, then each of the partial derivatives

$$f_{1,j}T = \lim_{h \to 0} \frac{f(T + hE_j) - fT}{h}, \qquad j = 1, \ldots, n,$$

exists. Moreover, $A_T f$ is uniquely determined and the matrix of the (unique) differential is

$$[d_T f] = [f_{1,1}T \quad f_{1,2}T \quad \cdots \quad f_{1,n}T].$$

The proof is similar to that of Theorem 16.4 and is left to the reader (Exercise 14).

Using Theorem 16.14 it is easy to show that an affine approximation of an $\mathcal{R}^n \rightarrow \mathcal{R}^m$ function f is uniquely given in terms of the affine approximations of the coordinate functions of f.

THEOREM 16.15 Let $f: \mathcal{R}^n \rightarrow \mathcal{R}^m$ and let $f_1, \ldots, f_i, \ldots, f_m$ denote the $\mathcal{R}^n \rightarrow \mathcal{R}$ coordinate functions of f. f has an affine approximation $A_T f$ about T if and only if each of the coordinate functions f_i has an affine approximation $A_T(f_i)$ about T. Moreover,

$$(16.13) \qquad A_T f = \begin{pmatrix} A_T(f_1) \\ \cdot \\ \cdot \\ \cdot \\ A_T(f_i) \\ \cdot \\ \cdot \\ \cdot \\ A_T(f_m) \end{pmatrix}$$

and $A_T f$ is unique.

Proof. Let $f : \mathcal{R}^n \to \mathcal{R}^m$ and let f have an affine approximation $A_T f$ about the point T. We then have

$$O_m =^1 \lim_{X \to T} \frac{fX - (A_T f)X}{\|X - T\|}$$

$$=^2 \lim_{X \to T} \frac{1}{\|X - T\|} \begin{pmatrix} f_1 X - (A_T f)_1 X \\ \cdot \\ \cdot \\ f_i X - (A_T f)_i X \\ \cdot \\ \cdot \\ f_m X - (A_T f)_m X \end{pmatrix}$$

$$=^3 \begin{pmatrix} \lim_{X \to T} \dfrac{f_i X - (A_T f)_1 X}{\|X - T\|} \\ \cdot \\ \cdot \\ \lim_{X \to T} \dfrac{f_i X - (A_T f)_i X}{\|X - T\|} \\ \cdot \\ \cdot \\ \lim_{X \to T} \dfrac{f_m X - (A_T f)_m X}{\|X - T\|} \end{pmatrix}.$$

(1) Using the definition of an affine approximation.
(2) Introducing the coordinate functions for f and for $A_T f$.
(3) Employing the limit theorem for coordinate functions, Theorem 3.15, p. 187.

By a simple comparison of the first and last members of the preceding equality chain, we conclude that each coordinate of the last member is zero:

$$(16.14) \qquad \lim_{X \to T} \frac{f_i X - (A_T f)_i X}{\|X - T\|} = 0, \qquad i = 1, 2, \ldots, m.$$

Since each of the coordinate functions $(A_T f)_i$ is an *affine* $\mathcal{R}^n \to \mathcal{R}$ function (see Exercise 15), Equation 16.14 assures us that $f_i X$ has an affine approximation about T and that $(A_T f)_i$ is the approximating function, that is the i-th coordinate function of $A_T f$ is the affine approximation $A_T(f_i)$ of f_i:

$$(A_T f)_i = A_T(f_i).$$

Equation 16.13 of the theorem readily follows. The uniqueness of $A_T f$ follows from that equation and the uniqueness of the $A_T(f_i)$ which is assured by Theorem 16.14.

We leave it to the reader to complete the proof by showing that if each of the coordinate functions f_i has an affine approximation then f will have an affine approximation. (See Exercise 16.) ∎

From Theorems 16.14 and 16.15 the following theorem is obtained.

THEOREM 16.16 (Differentials for $\mathcal{R}^n \to \mathcal{R}^m$ Functions.) Let $f: \mathcal{R}^n \to \mathcal{R}^m$ have an affine approximation $A_T f$ about T. Then each of the partial derivatives

$$f_{i,j} = \lim_{h \to 0} \frac{f_i(T + hE_j) - f_i T}{h} \qquad (i = 1, \ldots, m; \, j = 1, \ldots, n)$$

exists. Moreover, $A_T f$ is uniquely determined and the matrix of the (unique) differential is

$$(16.15) \quad [d_T f] = \begin{pmatrix} [d_T(f_1)] \\ \cdot \\ \cdot \\ \cdot \\ [d_T(f_i)] \\ \cdot \\ \cdot \\ \cdot \\ [d_T(f_m)] \end{pmatrix} = \begin{pmatrix} f_{1,1}T & \cdots & f_{1,n}T \\ \cdot & & \cdot \\ \cdot & & \cdot \\ f_{i,1}T & \cdots & f_{i,n}T \\ \cdot & & \cdot \\ \cdot & & \cdot \\ f_{m,1}T & \cdots & f_{m,n}T \end{pmatrix}$$

Proof. Let $d_T f$ denote the differential of f at T and let $d_T(f_i)$ denote the differential of f_i at T. We then have

$$(16.16) \qquad (A_T f)X = (d_T f)(X - T) + fT$$

and

$$(16.17) \qquad [A_T(f_i)]X = [d_T(f_i)](X - T) + f_i T.$$

Substitution from these two equations into Equation 16.13 of Theorem 16.15 gives, after simplification,

$$(d_T f)(X - T) = \begin{pmatrix} [d_T(f_1)] \\ \cdot \\ \cdot \\ [d_T(f_i)] \\ \cdot \\ \cdot \\ [d_T(f_m)] \end{pmatrix} (X - T).$$

From this relation the first of Equations 16.15 follows at once. The second of Equations 16.15 then follows from the application of Theorem 16.14 to determine $[d_T f_1], \ldots, [d_T f_i], \ldots, [d_T f_n]$. Theorem 16.14 also assures the existence of each of the partial derivatives $f_{i,j}$. ∎

The usefulness of the "comma" notation to denote partial derivatives is readily apparent in the preceding theorem. The symbol $f_{i,j}$ denotes the partial derivative of the i-th coordinate function of f with respect to the j-th coordinate. On the other hand the subscripts i, j also designate the row (i) and column (j) in which $f_{i,j}X$ appears in the matrix $[d_T f]$.

The matrix appearing as the last member of Equation 16.15 may be defined even though f is not differentiable. This matrix is given a special name.

DEFINITION 16.17   Let $f: \mathfrak{R}^n \to \mathfrak{R}^m$ such that each of the partial derivatives $f_{i,j}T$ of f at T is defined. The matrix of partial derivatives is called the *Jacobian matrix* and is denoted by $J_T f$:

$$J_T f = \begin{pmatrix} f_{1,1}T & \cdots & f_{1,n}T \\ & & \\ & & \\ f_{i,1}T & \cdots & f_{i,n}T \\ & & \\ & & \\ f_{m,1}T & \cdots & f_{m,n}T \end{pmatrix}$$

Example 9.   Consider the function given by

$$f(x, y, z) = \begin{pmatrix} x^2 + yz \\ \sin x \\ \cos xz + y \end{pmatrix} = \begin{pmatrix} f_1 X \\ f_2 X \\ f_3 X \end{pmatrix}.$$

If f has an affine approximation at $T = \begin{pmatrix} \pi \\ 1 \\ 0 \end{pmatrix}$ it is the function

$$(A_T f)X = (d_T f)(X - T) + f(T)$$

$$= (d_T f)\begin{pmatrix} x - \pi \\ y - 1 \\ z - 0 \end{pmatrix} + \begin{pmatrix} \pi^2 \\ 0 \\ 2 \end{pmatrix},$$

where

$$[d_T f] = J_T f = \begin{pmatrix} f_{1,1}T & f_{1,2}T & f_{1,3}T \\ f_{2,1}T & f_{2,2}T & f_{2,3}T \\ f_{3,1}T & f_{3,2}T & f_{3,3}T \end{pmatrix}$$

$$= {}^1 \begin{pmatrix} 2x & z & y \\ \cos x & 0 & 0 \\ -z\sin xz & 1 & -x\sin xz \end{pmatrix}_{\begin{pmatrix}\pi\\1\\0\end{pmatrix}}$$

$$= \begin{pmatrix} 2\pi & 0 & 1 \\ -1 & 0 & 0 \\ 0 & 1 & 0 \end{pmatrix},$$

hence

$$A_T f \begin{pmatrix} x \\ y \\ z \end{pmatrix} = \begin{pmatrix} 2\pi & 0 & 1 \\ -1 & 0 & 0 \\ 0 & 1 & 0 \end{pmatrix} \begin{pmatrix} x-\pi \\ y-1 \\ z-0 \end{pmatrix} + \begin{pmatrix} \pi^2 \\ 0 \\ 2 \end{pmatrix} = \begin{pmatrix} 2\pi(x-\pi)+z+\pi^2 \\ -x+\pi \\ y-1+2 \end{pmatrix}$$

$$= \begin{pmatrix} 2\pi x + z - \pi^2 \\ \pi - x \\ y+1 \end{pmatrix}$$

(1) The vector subscript $\begin{pmatrix}\pi\\1\\0\end{pmatrix}$ denotes that the entries in the matrix are to be evaluated by setting $x=\pi, y=1$, and $z=0$.

We have not shown that the affine function $A_T f$ which we have just determined is actually the affine approximation to f about the point T. We only know that *if* f has an affine approximation at all, it must be the $A_T f$ we have constructed. However, the present problem is just a particular case of a general situation in which the existence of an affine approximation is assured. The following theorem assures the existence of an affine approximation in this case and in a large class of other cases. ∎

THEOREM 16.18 Let $f: \mathcal{R}^n \to \mathcal{R}^m$. If the partial derivatives $f_{i,j}$ are continuous in an open ball about the point T, then f is differentiable at T, and $[d_T f] = J_T f$.

In other words, the affine function given by

$$AX = \begin{pmatrix} f_{1,1}T & \cdots & f_{1,n}T \\ \vdots & & \vdots \\ \vdots & & \vdots \\ f_{m,1}T & \cdots & f_{m,n}T \end{pmatrix} [X - T] + fT$$

is the affine approximation to f about T if the functions in the Jacobian matrix,

$$Jf = \begin{pmatrix} f_{1,1} & \cdots & f_{1,n} \\ \cdot & & \cdot \\ \cdot & & \cdot \\ \cdot & & \cdot \\ f_{m,1} & \cdots & f_{m,n} \end{pmatrix},$$

are continuous in an open ball about T. This fact will be useful in working the exercises. The proof of Theorem 16.18 may be found in Crowell and Williamson, *Calculus of Vector Functions*, Prentice-Hall, 1962, p. 204.

Example 10. Consider the function $f: \mathcal{R}^2 \rightarrow \mathcal{R}^2$ given by

$$f(x, y) = \begin{pmatrix} x^2 + 2xy \\ x \\ y \end{pmatrix}$$

and the point $T = \begin{pmatrix} 1 \\ 1 \end{pmatrix}$. The Jacobian matrix of f is

$$J_X f = \begin{pmatrix} 2x + 2y & 2x \\ \dfrac{1}{y} & \dfrac{-x}{y^2} \end{pmatrix}.$$

The entries of this matrix are clearly continuous in an open ball about the point $\begin{pmatrix} 1 \\ 1 \end{pmatrix}$. According to Theorem 16.18, f is differentiable at T and

$$[d_T f] = J_T f = \begin{pmatrix} 2x + 2y & 2x \\ \dfrac{1}{y} & -\dfrac{x}{y^2} \end{pmatrix}_{\binom{1}{1}}$$

$$= \begin{pmatrix} 4 & 2 \\ 1 & -1 \end{pmatrix}.$$

Thus

$$(A_{\binom{1}{1}} f)X = [d_{\binom{1}{1}} f](X - T) + fT$$

$$= \begin{pmatrix} 4 & 2 \\ 1 & -1 \end{pmatrix} \begin{pmatrix} x - 1 \\ y - 1 \end{pmatrix} + f\begin{pmatrix} 1 \\ 1 \end{pmatrix}$$

$$= \begin{pmatrix} 4x + 2y - 3 \\ x - y + 1 \end{pmatrix}. \quad \blacksquare$$

Example 11. Consider the function $f: \mathcal{R}^2 \rightarrow \mathcal{R}$ given by

$$f(x, y) = \begin{cases} 1, & \text{if } x = 0 \text{ or } y = 0, \\ 0, & \text{if } x \neq 0 \text{ and } y \neq 0. \end{cases}$$

Thus f has value 0 for all points in $\mathcal{R}^2$ except those on the x- or y-axis, and the value of f there is 1. The graph of f is indicated in Figure 16.4. Now f has

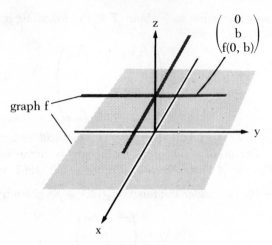

Figure 16.4

a single coordinate function $f_1 = f$. Moreover

$$f_{1,1}\begin{pmatrix}0\\0\end{pmatrix} =^1 \lim_{h\to 0} \frac{f\left[\begin{pmatrix}0\\0\end{pmatrix} + h\begin{pmatrix}1\\0\end{pmatrix}\right] - f\begin{pmatrix}0\\0\end{pmatrix}}{h}$$

$$= \lim_{h\to 0} \frac{f\begin{pmatrix}h\\0\end{pmatrix} - f\begin{pmatrix}0\\0\end{pmatrix}}{h}$$

$$=^2 \lim_{h\to 0} \frac{1-1}{h}$$

$$= 0.$$

(1) Definition 16.9.
(2) From definition of f.

Similarly, one can show that

$$f_{1,2}\begin{pmatrix}0\\0\end{pmatrix} = \lim_{h\to 0} \frac{f\left[\begin{pmatrix}0\\0\end{pmatrix} + h\begin{pmatrix}0\\1\end{pmatrix}\right] - f\begin{pmatrix}0\\0\end{pmatrix}}{h} = 0.$$

Thus the Jacobian matrix of f at $\begin{pmatrix}0\\0\end{pmatrix}$ is

$$J_{\begin{pmatrix}0\\0\end{pmatrix}}f = \left(f_{1,1}\begin{pmatrix}0\\0\end{pmatrix} \quad f_{1,2}\begin{pmatrix}0\\0\end{pmatrix}\right)$$

$$= (0 \quad 0).$$

Now, if f has an affine approximation about $T = \begin{pmatrix} 0 \\ 0 \end{pmatrix}$, it is given by

$$AX = (J_T f)(X - T) + fT$$

$$= (0 \quad 0)\begin{pmatrix} x - 0 \\ y - 0 \end{pmatrix} + f\begin{pmatrix} 0 \\ 0 \end{pmatrix}$$

$$= f\begin{pmatrix} 0 \\ 0 \end{pmatrix}$$

$$= 1.$$

But

$$\frac{fX - AX}{\left\| X - \begin{pmatrix} 0 \\ 0 \end{pmatrix} \right\|} = \frac{fX - 1}{\|X\|} = \frac{0 - 1}{\|X\|} = \frac{-1}{\|X\|}$$

for all points in every open ball about $\begin{pmatrix} 0 \\ 0 \end{pmatrix}$ except those points on either the

x- or y-axis. Thus as X approaches $\begin{pmatrix} 0 \\ 0 \end{pmatrix}$ it is possible for $\dfrac{fX - AX}{\left\| X - \begin{pmatrix} 0 \\ 0 \end{pmatrix} \right\|}$ to

increase without bound. Hence

$$\lim_{X \to \binom{0}{0}} \frac{fX - AX}{\left\| X - \begin{pmatrix} 0 \\ 0 \end{pmatrix} \right\|} \quad \text{does not exist}$$

and f is not differentiable at $T = \begin{pmatrix} 0 \\ 0 \end{pmatrix}$.

We note that the Jacobian matrix of f is not continuous in an open ball about $\begin{pmatrix} 0 \\ 0 \end{pmatrix}$ since $f_{1,1}$ is not defined at any point on the y-axis except at $\begin{pmatrix} 0 \\ 0 \end{pmatrix}$.

In particular, for any $\begin{pmatrix} 0 \\ b \end{pmatrix}$ point on the y-axis with $b \neq 0$ we have

$$f_{1,1}\begin{pmatrix} 0 \\ b \end{pmatrix} = \lim_{h \to 0} \frac{f\left[\begin{pmatrix} 0 \\ b \end{pmatrix} - h\begin{pmatrix} 1 \\ 0 \end{pmatrix}\right] - f\begin{pmatrix} 0 \\ b \end{pmatrix}}{h}$$

$$= \lim_{h \to 0} \frac{f\begin{pmatrix} h \\ b \end{pmatrix} - f\begin{pmatrix} 0 \\ b \end{pmatrix}}{h}$$

$$= \lim_{h \to 0} \frac{0 - 1}{h}$$

$$= \lim_{h \to 0} \frac{-1}{h}$$

This limit does not exist. ∎

From the preceding example we see that mere existence of the Jacobian matrix $J_T f$ at the point T does not assure us of the existence of an affine approximation to f about T. The following procedure should be used to determine whether a function $f : \mathcal{R}^n \to \mathcal{R}^m$ has an affine approximation about the point T.

    i. Determine the elements of the matrix

$$J_X f = \begin{pmatrix} f_{1,1}X & \cdots & f_{1,n}X \\ \cdot & & \cdot \\ \cdot & & \cdot \\ \cdot & & \cdot \\ f_{m,1}X & \cdots & f_{m,n}X \end{pmatrix}.$$

    ii. Construct the affine test function

$$(16.18) \qquad (A_T f)X = (J_T f)(X - T) + fT$$

and verify, if possible, that $\displaystyle \lim_{X \to T} \frac{fX - (A_T f)X}{\|X - T\|} = O_m,$

or

    ii . show that each entry in $J_X f$ is continuous in an open ball about T.

    iii. Conclusion: (a) If the limit of (ii) is established or if each entry in $J_X f$ is continuous in an open ball about T, then the function $(A_T f)$ given by Equation 16.18 is the affine approximation to f about T and

$$[d_T f] = J_T f.$$

(b) If any element of $J_X f$ is not defined at T or if the limit of (ii) is not valid then f has *no* affine approximation about T.

EXERCISES

In Exercises 1 to 5 determine the Jacobian matrix $J_X f$ for the prescribed f and evaluate $J_T f$ at the given point T.

1. $f(x, y) = \begin{pmatrix} 2x + y \\ x^2 - y^2 \end{pmatrix}$, $T = \begin{pmatrix} 2 \\ 1 \end{pmatrix}$.

.2. $f(x, y) = \begin{pmatrix} \sqrt{xy} \\ \sin(x - y) \end{pmatrix}$, $T = \begin{pmatrix} \pi \\ \pi \end{pmatrix}$.

3. $f(u, v) = \begin{pmatrix} u \cos v \\ u \sin v \\ v \end{pmatrix}$ $T = \begin{pmatrix} 1 \\ \pi \\ \frac{\pi}{2} \end{pmatrix}$.

4. $f(r, \theta, \phi) = \begin{pmatrix} r \cos \theta \sin \phi \\ r \sin \theta \sin \phi \\ r \cos \phi \end{pmatrix}$, $T = \begin{pmatrix} 1 \\ \pi \\ 0 \end{pmatrix}$.

5. $f(x, y, z) = \begin{pmatrix} xy + yz \\ yz + y^2 \\ xyz \end{pmatrix}$, $T = \begin{pmatrix} 1 \\ 1 \\ 1 \end{pmatrix}$.

In Exercises 6 to 10 determine the affine approximation to f about the given T.

6. f and T of Exercise 1.   7. f and T of Exercise 2.

8. f and T of Exercise 3.   9. f and T of Exercise 4.

10. f and T of Exercise 5.

In Exercises 11 to 12 determine those points at which the given function is *not* differentiable and justify your reply.

11. $f(x, y) = \begin{pmatrix} \sqrt[3]{xy} \\ x^2 + y^2 \end{pmatrix}$.

12. $g(u, v) = \begin{pmatrix} 3u - v \\ \sqrt{u^2 - v} \end{pmatrix}$.

13. Let $f : \mathcal{R}^n \to \mathcal{R}^m$ have an affine approximation $A_T f$ about T. Show that if $(A_T f)X = (d_T f)X + Y_0$ then $Y_0 = fT - (d_T f)T$.

14. Prove Theorem 16.14, p. 666.

15. Let $A : \mathcal{R}^n \to \mathcal{R}^m$ be an affine transformation whose coordinate functions are denoted by A_i. Show that each A_i is an $\mathcal{R}^n \to \mathcal{R}$ affine transformation.

16. Let $f : \mathcal{R}^n \to \mathcal{R}^m$ be such that each coordinate function f_i has an affine approximation $A_T(f_i)$ about T. Show that the $\mathcal{R}^n \to \mathcal{R}^m$ function

$$A_T f = \begin{pmatrix} A_T(f_1) \\ \vdots \\ A_T(f_m) \end{pmatrix}$$

is the affine approximation to f about T.

ANSWERS

1. $J_X f = \begin{pmatrix} 2 & 1 \\ 2x & -2y \end{pmatrix}$, $J_{\binom{2}{1}} f = \begin{pmatrix} 2 & 1 \\ 4 & -2 \end{pmatrix}$.

3. $J_X f = \begin{pmatrix} \cos v & -u \sin v \\ \sin v & u \cos v \\ 0 & 1 \end{pmatrix}$, $J_T f = \begin{pmatrix} 0 & -1 \\ 1 & 0 \\ 0 & 1 \end{pmatrix}$.

5. $J_X f = \begin{pmatrix} y & x + z & y \\ 0 & z + 2y & y \\ yz & xz & xy \end{pmatrix}$, $J_T f = \begin{pmatrix} 1 & 2 & 1 \\ 0 & 3 & 1 \\ 1 & 1 & 1 \end{pmatrix}$.

7. $[A_{\binom{\pi}{\pi}} f] \begin{pmatrix} x \\ y \end{pmatrix} = \begin{pmatrix} (x + y)/2 \\ x - y \end{pmatrix}$.   **9.** $[A_{\binom{1}{\frac{\pi}{0}}} f] \begin{pmatrix} r \\ \theta \\ \phi \end{pmatrix} = \begin{pmatrix} -\phi \\ 0 \\ r - 2 \end{pmatrix}$.

11. Jacobian matrix is not defined if $x = 0$ or if $y = 0$ but is defined and continuous elsewhere. Hence f is differentiable at all points except those for which $x = 0$ or $y = 0$.

4. THE SPECIAL CASE OF $\mathcal{R} \to \mathcal{R}^m$ FUNCTIONS

The purpose of this brief section is to emphasize the fact that the definitions of affine approximations and differentials of $\mathcal{R} \to \mathcal{R}$ and $\mathcal{R} \to \mathcal{R}^m$

functions given in Chapter 5 (Definitions 5.3, 5.5, 5.19, and 5.21) are special cases of the concepts discussed in the preceding section.

Let $f: \mathcal{R} \to \mathcal{R}$ and let t be an interior point of $dom\, f$. The reader should observe that Definition 16.11 with $n = m = 1$, X replaced by x, and T replaced by (t) reduces to Definition 5.3. Moreover the matrix $[d_t f]$ of the differential of f given in Theorem 16.14 is

$$[d_t f] = [f_{1,1}(t)].$$

But $f_{1,1}(t) = f'(t)$ so that

(16.19) $$[d_t f] = [f'(t)]$$

in agreement with Definition 5.5.

Let $f: \mathcal{R} \to \mathcal{R}^m$. The reader should be able to show that Definitions 5.19 and 5.21 are consistent with the content of Definition 16.11 and Theorem 16.16.

5. CONTINUITY AND DIFFERENTIABILITY

In Chapter 5, p. 225, it was shown that if an $\mathcal{R} \to \mathcal{R}$ function f is differentiable at t then f is continuous at t. This same property is shared by all differentiable $\mathcal{R}^n \to \mathcal{R}^m$ functions. The proof given in Chapter 5 readily generalizes to the general case, and we leave it to the reader to show this. We shall give a proof which goes back to first principles simply because one of the intermediate results is needed in the next section.

THEOREM 16.19   (Continuity of Differentiable Functions.) Let $f: \mathcal{R}^n \to \mathcal{R}^m$ and let f have an affine approximation about the point T. Then f is continuous at T.

Proof. Let $A_T f$ denote the affine approximation of f at T. To show that f is continuous at T we must show that

$$\lim_{X \to T} (fX - fT) = O_m$$

or, equivalently, for each real number r there is a deleted open ball $\mathcal{B}(\tilde{T};\ \rho)$ of radius ρ about T such that

(16.20) $$X \in \mathcal{B}(\tilde{T};\ \rho) \Rightarrow \|fX - fT\| < r.$$

Since f has an affine approximation at T, we know that

$$\lim_{X \to T} \frac{fX - (A_T f)X}{\|X - T\|} = O_m.$$

This means that for any real number r' there is a deleted open ball $\mathcal{B}(\check{T};\ \rho')$ of radius ρ' about T such that

$$(16.21) \qquad X \in \mathcal{B}(\check{T};\ \rho') \Rightarrow \frac{\|fX - (A_Tf)X\|}{\|X - T\|} < r'.$$

For any $\rho < \rho'$ we then have

(16.22)

$$X \in \mathcal{B}(\check{T};\ \rho) \Rightarrow X \in B(\check{T};\ \rho')$$

$$\Rightarrow^1 \frac{\|fX - (A_Tf)X\|}{\|X - T\|} < r'$$

$$\Rightarrow^2 \|fX - (A_Tf)X\| < r'\,\|X - T\|$$

$$\Rightarrow^3 \|(fX - fT) - [d_Tf](X - T)\| < r'\,\|X - T\|$$

$$\Rightarrow^4 |\|fX - fT\| - \|[d_Tf](X - T)\|| < r'\,\|X - T\|$$

$$\Rightarrow^5 \|fX - fT\| < r'\,\|X - T\| + \|[d_Tf](X - T)\|$$

$$\left.\begin{array}{l}\Rightarrow^6 \|fX - fT\| < r'\,\|X - T\| + k\,\|X - T\| \\[4pt] \Rightarrow \|fX - fT\| < (r' + k)\,\|X - T\| \\[4pt] \Rightarrow^7 \|fX - fT\| < (r' + k)\rho \end{array}\right\} \begin{array}{l}\text{for some}\\ \text{real num-}\\ \text{ber } k > 0.\end{array}$$

(1) From Equation 16.21.
(2) Multiplying by $\|X - T\|$.
(3) Substituting for $(A_Tf)X$ in terms of $[d_Tf]$.
(4) Using the fact that $|\|A\| - \|B\|| \le \|A - B\|$, with $A = fX - fT$ and $B = [d_Tf](X - T)$ (see Exercise 1).
(5) Using the fact that $|a - b| < c \Rightarrow a < b + c$, with $a = \|fX - fT\|$ and $b = \|[d_Tf](X - T)\|$ (see Exercise 2).
(6) Using Theorem 14.16, p. 573, and the linearity of d_Tf.
(7) $X \in \mathcal{B}(\check{T};\ \rho) \Rightarrow \|X - T\| < \rho$.

Equation 16.20 follows at once from the preceding implication chain if we further restrict ρ so that $(r' + k)\rho < r$, that is, so that $\rho < \dfrac{r}{r' + k}$ and $\rho < \rho'$.

EXERCISES

1. Let A and B be two vectors in $\mathcal{R}^m$. Show that $|\|A\| - \|B\|| \le \|A - B\|$

2. Let a, b, and c be real numbers. Show that if $|a - b| < c$ then $a < b + c$.

HINTS

1. Apply the Triangle Property to the right members of $\|A\| = \|(A - B) + B\|$ and $\|B\| = \|(B - A) + A\|$ and then "solve" for $\|A\| - \|B\|$.
2. Express the relation $|a - b| < c$ without absolute value signs.

6. THE GENERAL CHAIN RULE

In this section we determine the relationship between the differential of a composite function $g \circ f$ and the differentials of the component functions f and g.

We begin by reinterpreting the chain rule for the differentiation of $\mathcal{R} \to \mathcal{R}$ functions. Let $f : \mathcal{R} \to \mathcal{R}$ and $g : \mathcal{R} \to \mathcal{R}$ such that f is differentiable at t and g is differentiable at $s = f(t)$. According to Theorem 5.10, p. 230, we know that the composite function $g \circ f : \mathcal{R} \to \mathcal{R}$ is differentiable at t and that

$$(16.23) \qquad\qquad (g \circ f)' = (g' \circ f)f'.$$

From the discussion of Section 4, in particular, from Equation 16.19, we know that the matrix of the differential $d_t f$ is given by

$$(16.24) \qquad\qquad [d_t f] = [f'(t)].$$

The matrix for $d_s g$ and $d_t(g \circ f)$ may be obtained from Equation 16.24 by replacing f and t by g and s and then by $g \circ f$ and t.

$$(16.25) \qquad\qquad [d_s g] = [g'(s)],$$

$$(16.26) \qquad\qquad [d_t(g \circ f)] = [(g \circ f)'(t)].$$

Applying the chain rule in Equation 16.23 to the second member of Equation 16.26 initiates the following chain:

$$
\begin{aligned}
[d_t(g \circ f)] &=^1 [((g' \circ f) \cdot f')(t)] \\
&=^2 [(g' \circ f)(t) \cdot f'(t)] \\
&=^3 [g'(f(t)) \cdot f'(t)] \\
&=^4 [g'(f(t))][f'(t)] \\
&=^5 [d_{f(t)}g][d_t f].
\end{aligned}
$$

(1) Application of chain rule to $(g \circ f)'$.
(2) Definition of product of the two functions $(g' \circ f)$ and f'.
(3) Definition of $g' \circ f$.
(4) Properties of 1×1 matrices.
(5) Using Equations 16.24 and 16.25 with $s = f(t)$.

Thus the chain rule (Equation 16.23) has the matrix equivalent

$$(16.27) \qquad\qquad [d_t(g \circ f)] = [d_{f(t)}g][d_t f].$$

The preceding considerations assure the validity of Equation 16.27 only if f and g are both $\mathcal{R} \to \mathcal{R}$ functions. We shall refer to Equation 16.27 as the $\mathcal{R} \to \mathcal{R} \to \mathcal{R}$ chain rule. The form of this equation readily suggests a generalization to the case in which f and g are arbitrary vector functions.

Indeed, the following $\mathcal{R}^n \to \mathcal{R}^m \to \mathcal{R}^p$ chain rule is suggested:

(16.28) $[d_T(g \circ f)] = [d_{fT}g][d_Tf].$

Happily, this general chain rule is valid. Before developing the details, we consider several examples.

Example 12.   (An $\mathcal{R} \to \mathcal{R} \to \mathcal{R}^2$ case.) Let f and g be given by

$$f(x) = \sin x \qquad g(y) = \begin{pmatrix} g_1(y) \\ g_2(y) \end{pmatrix} = \begin{pmatrix} y \\ y^2 \end{pmatrix}.$$

From Equation 16.24 we know that

$$[d_t f] = [f'(t)] = [\cos t].$$

For an $\mathcal{R} \to \mathcal{R}^2$ function g Equation 16.25 becomes

$$[d_s g] = \begin{pmatrix} g_1'(s) \\ g_2'(s) \end{pmatrix} = \begin{pmatrix} 1 \\ 2s \end{pmatrix}.$$

Thus, assuming the validity of the $\mathcal{R} \to \mathcal{R} \to \mathcal{R}^2$ chain rule given by Equation 16.28, we have

$$[d_t(g \circ f)] = [d_{f(t)}g][d_t f]$$
$$= \begin{pmatrix} 1 \\ 2f(t) \end{pmatrix} (\cos t)$$
$$= \begin{pmatrix} \cos t \\ 2 \sin t \cos t \end{pmatrix}.$$

As a verification of this result let us form the composite function and then differentiate. Thus

$$(g \circ f)(x) = g(f(x)) = \begin{pmatrix} f(x) \\ [f(x)]^2 \end{pmatrix} = \begin{pmatrix} \sin x \\ \sin^2 x \end{pmatrix}.$$

Hence

$$[d_t(g \circ f)] = [(g \circ f)'(t)]$$
$$= \begin{pmatrix} \cos t \\ 2 \sin t \cos t \end{pmatrix},$$

a result in complete agreement with the preceding.

Example 13.   (An $\mathcal{R}^2 \to \mathcal{R}^2 \to \mathcal{R}^3$ case.) Let f and g be given by

$$f(x_1, x_2) = \begin{pmatrix} \sin x_1 \\ x_2^2 + x_1^{-2} \end{pmatrix}, \qquad g(y_1, y_2) = \begin{pmatrix} y_1^2 + y_2 \\ \cos y_2 \\ y_1^3 \end{pmatrix}.$$

Using the methods of the preceding section, we have

(16.29) $[d_{\binom{t_1}{t_2}} f] = \begin{pmatrix} f_{1,1}\binom{t_1}{t_2} & f_{1,2}\binom{t_1}{t_2} \\ f_{2,1}\binom{t_1}{t_2} & f_{2,2}\binom{t_1}{t_2} \end{pmatrix} = \begin{pmatrix} \cos t_1 & 0 \\ -2t_1^{-3} & 2t_2 \end{pmatrix},$

$$(16.30) \quad [d_{\binom{s_1}{s_2}} \ g] = \begin{pmatrix} g_{1,1}\binom{s_1}{s_2} & g_{1,2}\binom{s_1}{s_2} \\ g_{2,1}\binom{s_1}{s_2} & g_{2,2}\binom{s_1}{s_2} \\ g_{3,1}\binom{s_1}{s_2} & g_{3,2}\binom{s_1}{s_2} \end{pmatrix} = \begin{pmatrix} 2s_1 & 1 \\ 0 & -\sin s_2 \\ 3s_1^2 & 0 \end{pmatrix}.$$

Assuming the validity of Equation 16.28 and using Equations 16.29 and 16.30 with

$$S = \binom{s_1}{s_2} = f(t_1, t_2) = \binom{\sin t_1}{t_2^2 + t^{-2}_{\ 1}},$$

we find

$$[d_{\binom{t_1}{t_2}} g \circ f] = [d_{f(t_1, t_2)} \ g][d_{\binom{t_1}{t_2}} f]$$

$$= \begin{pmatrix} 2(\sin t_1 \cos t_1 - t_1^{-3}) & 2t_2 \\ 2t_1^{-3} \sin (t_2^2 + t_1^{-2}) & -2t_2 \sin (t_2^2 + t_1^{-2}) \\ 3 \sin^2 t_1 \cos t_1 & 0 \end{pmatrix}.$$

The unbelieving reader might wish to verify this result by forming the composite function and then differentiating.

We now state and prove the general chain rule.

THEOREM 16.20 ($\mathcal{R}^n \to \mathcal{R}^m \to \mathcal{R}^p$ Chain Rule.) Let $f: \mathcal{R}^n \to \mathcal{R}^m$ and $g: \mathcal{R}^m \to \mathcal{R}^p$. Let $\mathcal{C}$ be the set of points T such that f is differentiable at T and g is differentiable at fT. The composite function $g \circ f$ is differentiable on $\mathcal{C}$. Moreover, for each T in $\mathcal{C}$, we have

$$(16.31) \qquad [d_T(g \circ f)] = [d_{fT} g][d_T f].$$

Proof. Let $T \in \mathcal{C}$ and $S = fT$. The affine approximation to f about T is given by

$$(A_T f)X = fT + [d_T f](X - T).$$

According to Definition 16.11, p. 664, we then have

$$(16.32) \quad \lim_{X \to T} \frac{fX - (A_T f)X}{\|X - T\|} = \lim_{X \to T} \frac{fX - fT - [d_T f](X - T)}{\|X - T\|} = O_m.$$

Similarly, since g is differentiable at $S = fT$, we have

$$(16.33) \qquad \lim_{Y \to S} \frac{gY - gS - [d_S g](Y - S)}{Y - S} = O_p,$$

where $d_S g$ denotes the differential of g at S.

For the purposes of our proof, we shall introduce the function $G: \mathcal{R}^m \to \mathcal{R}^p$ given by

(16.34)
$$GY = \begin{cases} \text{(a)} \quad \dfrac{gY - gS - [d_S g](Y - S)}{\|Y - S\|}, & \text{if } Y \neq S \\ \text{(b)} \quad O_p, & \text{if } Y = S. \end{cases}$$

Equation 16.33 may then be written as

(16.35)
$$\lim_{Y \to S} GY = O_p = GS.$$

Thus G is continuous at S. We note that from Equation 16.34a, we may write, if $Y \neq S$,

$$\|Y - S\| \, GY = gY - gS - [d_S g](Y - S).$$

But this equation is also valid when $Y = S$. (Check this using Equation 16.34b.) Thus for all $Y \in dom\ g$ we have

(16.36)
$$gY - gS = \|Y - S\| GY + [d_S g](Y - S).$$

To establish our theorem, we need only show that

(16.37)
$$AX = (g \circ f)T + [d_{fT} g][d_T f](X - T)$$

is the affine approximation of $g \circ f$ about T. We therefore consider the following:

$$\lim_{X \to T} \frac{(g \circ f)X - AX}{\|X - T\|} = \lim_{X \to T} \frac{(g \circ f)X - \{(g \circ f)T + [d_{fT} g][d_T f](X - T)\}}{\|X - T\|}$$

$$=^1 \lim_{X \to T} \frac{g(fX) - g(fT) - [d_{fT} g][d_T f](X - T)}{\|X - T\|}$$

$$=^2 \lim_{X \to T} \frac{\|fX - fT\| G(fX) + [d_{fT} g](fX - fT) - [d_{fT} g][d_T f](X - T)}{\|X - T\|}$$

$$=^3 \lim_{X \to T} G(fX) \frac{\|fX - fT\|}{\|X - T\|} + \lim_{X \to T} \left\{ d_{fT} g \frac{(fX - fT) - [d_T f](X - T)}{\|X - T\|} \right\}$$

$$=^4 \lim_{X \to T} \left\{ G(fX) \frac{\|fX - fT\|}{\|X - T\|} \right\} + [d_{fT} g] \left\{ \lim_{X \to T} \frac{fX - fT - [d_T f](X - T)}{\|X - T\|} \right\}$$

$$=^5 \lim_{X \to T} \left\{ (G \circ f)X \frac{\|fX - fT\|}{\|X - T\|} \right\}$$

$$=^6 O_p.$$

(1) Definition of $g \circ f$.

(2) Using Equation 16.36 with Y and S replaced by fX and fT, respectively.

(3) Limit theorem for sums, and linearity of $d_{fT}g$.

(4) Using the continuity of $d_{fT}g$, Theorem 14.17, p. 574.

(5) Using Equation 16.32 and the fact that $[d_{fT}g]O_m = O_p$.

(6) Since $\lim_{X \to T} (G \circ f)X = O_p$ and since from Equation 16.22 with $r' = k$, we have

$$h(X) = \frac{\|fX - fT\|}{\|X - T\|} < 2k \text{ for some positive integer } k, \text{ provided } X \text{ is suffi-}$$

ciently close to T. (See Exercise 12.)

We have therefore shown that the affine function given by Equation 16.37 is the affine approximation about T. Equation 16.31 of the theorem follows at once. ∎

EXERCISES

In Exercises 1 to 4 find $d_t w$ where $w = g \circ f$.

1. $\begin{pmatrix} x \\ y \end{pmatrix} = f(t) = \begin{pmatrix} t^2 \\ \sin t \end{pmatrix}$, $g\begin{pmatrix} x \\ y \end{pmatrix} = x^2 + y^2$.

2. $\begin{pmatrix} x \\ y \\ z \end{pmatrix} = f(t) = \begin{pmatrix} \dfrac{1}{t} \\ e^t \\ \ln t \end{pmatrix}$, $g\begin{pmatrix} x \\ y \\ z \end{pmatrix} = x \sin y + z \sin x$.

3. f any differentiable $\mathcal{R} \to \mathcal{R}^2$ function, g any differentiable $\mathcal{R}^2 \to \mathcal{R}$ function.

4. $\begin{pmatrix} x \\ y \end{pmatrix} = f(t) = \begin{pmatrix} \sqrt{t} \\ t^4 \end{pmatrix}$, $g\begin{pmatrix} x \\ y \end{pmatrix} = x^2 y$.

In Exercises 5 to 8 find the matrix $[d_T w]$ of the differential of $w = g \circ f$.

5. $\begin{pmatrix} x \\ y \end{pmatrix} = f\begin{pmatrix} r \\ \theta \end{pmatrix} = \begin{pmatrix} r \cos \theta \\ r \sin \theta \end{pmatrix}$, $g\begin{pmatrix} x \\ y \end{pmatrix} = x^2 + y^2$.

6. $\begin{pmatrix} x \\ y \end{pmatrix} = f\begin{pmatrix} u \\ v \end{pmatrix} = \begin{pmatrix} u \ln v \\ u \ln u \end{pmatrix}$, $g\begin{pmatrix} x \\ y \end{pmatrix} = xe^y + ye^x$.

7. $\begin{pmatrix} x \\ y \\ z \end{pmatrix} = f\begin{pmatrix} R \\ \varphi \\ \theta \end{pmatrix} = \begin{pmatrix} R \cos \theta \sin \varphi \\ R \sin \theta \sin \varphi \\ R \cos \theta \end{pmatrix}$, $g\begin{pmatrix} x \\ y \\ z \end{pmatrix} = x^2 + y^2$.

8. $\begin{pmatrix} x \\ y \\ z \end{pmatrix} = f\begin{pmatrix} r \\ s \\ t \end{pmatrix} = \begin{pmatrix} r \cos s \\ r \sin t \\ st \end{pmatrix}$, $g\begin{pmatrix} x \\ y \\ z \end{pmatrix} = xy + xz + yz$.

9. Let $F\begin{pmatrix} x \\ t \end{pmatrix} = k\begin{pmatrix} u \\ v \end{pmatrix}$ where $\begin{pmatrix} u \\ v \end{pmatrix} = h(x, t) = \begin{pmatrix} x + ct \\ x - ct \end{pmatrix}$.

Show that

$$\partial_1 F \Big|_{\binom{x}{t}} = (\partial_1 k + \partial_2 k)\Big|_{\binom{x+ct}{x-ct}}$$

and

$$\partial_2 F \Big|_{\binom{x}{t}} = (c\,\partial_1 k - c\,\partial_2 k)\Big|_{\binom{x+ct}{x-ct}},$$

or, in a frequently used alternate notation,

$$\frac{\partial F}{\partial x}\Big|_{\binom{x}{t}} = \left(\frac{\partial k}{\partial u} + \frac{\partial k}{\partial v}\right)\Big|_{\binom{x+ct}{x-ct}}$$

and

$$\frac{\partial F}{\partial t}\Big|_{\binom{x}{t}} = \left(c\,\frac{\partial k}{\partial u} - c\,\frac{\partial k}{\partial v}\right)\Big|_{\binom{x+ct}{x-ct}}.$$

10. Let F be an $\mathcal{R}^2 \to \mathcal{R}$ funtion given by

$$F(x, t) = f(x + ct) + g(x - ct)$$

where f and g are $\mathcal{R}^2 \to \mathcal{R}$ functions. Show that F satisfies the wave equation:

$$\frac{\partial^2 F}{\partial x^2} = c^2\,\frac{\partial^2 F}{\partial t^2}.$$

[Hint: Select $K: \mathcal{R}^2 \to \mathcal{R}$ such that $k\binom{u}{v} = f(u) + g(v)$ and compute $\dfrac{\partial F}{\partial x}, \dfrac{\partial F}{\partial t}, \dfrac{\partial^2 F}{\partial x^2},$ and $\dfrac{\partial^2 F}{\partial t^2}$ by use of Exercise 9.]

11. Let $z = f(x, y)$ and $\binom{x}{y} = g(r, \theta) = \binom{r\cos\theta}{r\sin\theta}$. Define $Z = (f \circ g)(r, \theta)$. Show that

$$\left(\frac{\partial Z}{\partial r}\right)^2 + \frac{1}{r^2}\left(\frac{\partial Z}{\partial \theta}\right)^2 = \left(\frac{\partial z}{\partial x}\right)^2 + \left(\frac{\partial z}{\partial y}\right)^2.$$

(Hint: Use the chain rule to compute $\begin{pmatrix} \dfrac{\partial Z}{\partial r} \\[2mm] \dfrac{\partial Z}{\partial \theta} \end{pmatrix} = d_{\binom{r}{\theta}} f$

in terms of $\dfrac{\partial z}{\partial x}$ and $\dfrac{\partial z}{\partial y}$. By substitution show that the left member of the given equation reduces to the right member.)

12. Let $H: \mathcal{R}^n \to \mathcal{R}^p$ be such that $\lim\limits_{X \to T} HX = O_p$ and let $h: \mathcal{R}^n \to \mathcal{R}$ be such that $\|hX\| \le m$ for some real number m and for every X in some open ball about T. Show that $\lim\limits_{X \to T} Hx \cdot hx = O_p$.

ANSWERS

1. $4t^3 + \sin 2t$.

3. $\partial_1 g\binom{u}{v} \cdot f_1' + \partial_2 g\binom{u}{v} \cdot f_2'$, where $\binom{u}{v} = f(t) = \binom{f_1(t)}{f_2(t)}$.

5. $\left[\dfrac{\partial w}{\partial r} \quad \dfrac{\partial w}{\partial \theta}\right] = [2r \quad 0]$.

7. $\left[\dfrac{\partial w}{\partial R} \quad \dfrac{\partial w}{\partial \varphi} \quad \dfrac{\partial w}{\partial \theta}\right] = [2R\sin^2\varphi \quad 2R^2\sin\varphi\cos\varphi \quad 0]$.

17

TANGENT LINES AND PLANES; DIRECTIONAL DERIVATIVE

1. TANGENT LINES AND PLANES TO EXPLICITLY DEFINED FUNCTION-LOCI

In elementary geometry a line is said to be tangent to a circle if the line and the circle have exactly one point in common. Similarly, a plane is said to be tangent to a sphere if the plane and the sphere have exactly one point in common. It is natural to try to extend these ideas to other function-loci in $\mathcal{R}^n$. We would like, for example, to say that the line $\mathcal{L}$ pictured in Figure 17.1 is tangent to the set C at P and that the plane $\mathcal{M}$ in Figure 17.2 is tangent to the set $\mathcal{D}$ at Q. However, $\mathcal{L}$ and C have more than one point in common, as do $\mathcal{M}$ and $\mathcal{D}$. Hence our definition will of necessity be more general than the ones just mentioned.

We consider the lines and planes that we shall define to be tangent to explicitly defined function-loci, restricting ourselves at first to $\mathcal{R} \to \mathcal{R}^m$ functions. The figures suggest that a tangent line exists at a point where the function is differentiable and is the function-locus defined explicitly by the affine approximation to f at that point.

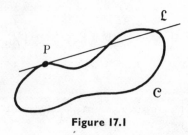

Figure 17.1

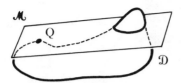

Figure 17.2

DEFINITION 17.1 (Tangent Line to an Explicitly Defined Function-Locus.) Let $f : \mathcal{R} \to \mathcal{R}^m$, and let f be differentiable at x_0. We say that *graph* $A_{x_0}f$ is the *f-tangent line* to *graph* f at $(x_0, f(x_0))$. A direction vector for *graph* $A_{x_0}f$ will be called an *f-tangent vector* to *graph* f at $(x_0, f(x_0))$. If f is differentiable at x_0, *graph* f is said to be *one-dimensional* at $(x_0, f(x_0))$. If *graph* f is one-dimensional at each of its points, we say that *graph* f is a *one-dimensional surface.*

NOTE. The reason for the terminology "*f*-tangent" instead of just "tangent" will become apparent in Section 2. When no confusion is likely, we shall simply say "tangent."

The reader is asked to verify in Exercise 15 that a tangent line, as just defined, intersects a circle in exactly one point.

Example I. Let $f(x) = \cos x$, $x_0 = \dfrac{\pi}{4}$. The tangent line to *graph* f at x_0 is given by *graph* $A_{x_0}f$, where $A_{x_0}f(x) = -\left(\dfrac{1}{\sqrt{2}}\right)x + \dfrac{(\pi + 4)}{4\sqrt{2}}$ (see Figure 17.3). Using the methods of Chapter 1, Section 7, we find that $\left(1, \dfrac{-1}{\sqrt{2}}\right)$ is a direction vector for *graph* $A_{x_0}f$ and thus is a tangent vector to *graph* f. ∎

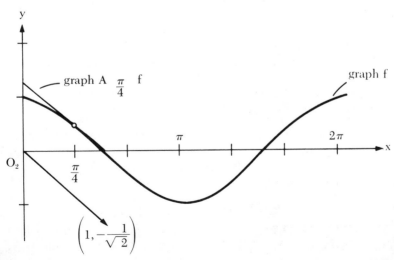

Figure 17.3

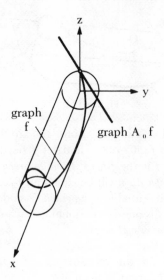

Figure 17.4

Example 2. Let $f(x) = \begin{pmatrix} y \\ z \end{pmatrix} = \begin{pmatrix} \sin x \\ \cos x \end{pmatrix}$, $x_0 = 0$. We give the sketch of *graph f* in Figure 17.4.

The reader should show that $A_0 f(x) = x\begin{pmatrix} 1 \\ 0 \end{pmatrix} + \begin{pmatrix} 0 \\ 1 \end{pmatrix}$.

Then

$$graph\ A_0 f(x) = \left\{ \begin{pmatrix} x \\ A_0 f(x) \end{pmatrix} \middle| x \in R \right\}$$

$$= \left\{ \begin{pmatrix} x \\ x\begin{pmatrix} 1 \\ 0 \end{pmatrix} + \begin{pmatrix} 0 \\ 1 \end{pmatrix} \end{pmatrix} \middle| x \in R \right\}$$

$$= \left\{ x\begin{pmatrix} 1 \\ 1 \\ 0 \end{pmatrix} + \begin{pmatrix} 0 \\ 0 \\ 1 \end{pmatrix} \middle| x \in R \right\}.$$

This set is the tangent line to *graph f* at x_0 and $(1, 1, 0)$ is a tangent vector. ∎

The geometrical motivation, used in Chapter 16, Section 1, for the definition of $A_{X_0} f$, where $f : \mathcal{R}^2 \to \mathcal{R}$, suggests also that we call *graph $A_{X_0} f$* the tangent to *graph f* at X_0. The following definition is a generalization of this idea. Notice that this definition parallels Definition 17.1 and that by this definition a tangent line is a tangent 1-plane.

DEFINITION 17.2 (*f*-tangent *m*-plane to an Explicitly Defined Functions-Locus.) Let $f : \mathcal{R}^m \to \mathcal{R}^n$ and let f be differentiable at X_0. We say that *graph $A_{X0} f$* is the *f-tangent m-plane to graph f at* $(X_0, f X_0)$.

A direction vector for *graph* $A_{X_0}f$ will be called an *f-tangent vector to graph f at* (X_0, fX_0). If f is differentiable at X_0, *graph f* is said to be *m*-dimensional at (X_0, fX_0). *If graph f is m-dimensional at each of its points, we say that graph f is an m-dimensional surface.*

NOTE.   We shall often refer to an *f*-tangent 2-plane simply as a tangent plane.

Example 3.   Let $f\begin{pmatrix} x \\ y \end{pmatrix} = x^2 + 2y^2$. We find

$$A_{\begin{pmatrix} x_0 \\ y_0 \end{pmatrix}} f \begin{pmatrix} x \\ y \end{pmatrix} = 2x_0(x - x_0) + 4y_0(y - y_0) + f\begin{pmatrix} x_0 \\ y_0 \end{pmatrix}.$$

Since f is differentiable at each $\begin{pmatrix} x_0 \\ y_0 \end{pmatrix}$ in *dom f*, *graph f* is a 2-dimensional surface. If $X_0 = \begin{pmatrix} 1 \\ 0 \end{pmatrix}$, we find that the tangent plane at $\begin{pmatrix} X_0 \\ fX_0 \end{pmatrix} = \begin{pmatrix} 1 \\ 0 \\ 1 \end{pmatrix}$ is *graph* $A_{X_0}f$, where $A_{X_0}f\begin{pmatrix} x \\ y \end{pmatrix} = 2x - 1$. The graph of f and its tangent plane at $\begin{pmatrix} 1 \\ 0 \\ 1 \end{pmatrix}$ are sketched in Figure 17.5. ∎

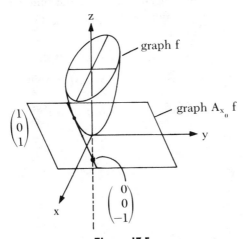

Figure 17.5

Example 4.   Let $f\begin{pmatrix} x \\ y \\ z \end{pmatrix} = x^2 + y^2 + z^2$, $X_0 = \begin{pmatrix} 1 \\ 1 \\ 1 \end{pmatrix}$. Then, by the methods of Chapter 16, $A_{X_0}f\begin{pmatrix} x \\ y \\ z \end{pmatrix} = 2x + 2y + 2z - 3$.

and

$$\text{graph } A_{X_0}f = \left\{ \left(\begin{pmatrix} x \\ y \\ z \\ A_0 f \begin{pmatrix} x \\ y \\ z \end{pmatrix} \end{pmatrix} \right) \middle| x, y, z \in \mathcal{R} \right\}$$

$$= \left\{ \left(\begin{pmatrix} x \\ y \\ z \\ 2x + 2y + 2z - 3 \end{pmatrix} \right) \middle| x, y, z \in \mathcal{R} \right\}$$

$$= \left\{ x \begin{pmatrix} 1 \\ 0 \\ 0 \\ 2 \end{pmatrix} + y \begin{pmatrix} 0 \\ 1 \\ 0 \\ 2 \end{pmatrix} + z \begin{pmatrix} 0 \\ 0 \\ 1 \\ 2 \end{pmatrix} + \begin{pmatrix} 0 \\ 0 \\ 0 \\ -3 \end{pmatrix} \middle| x, y, z \in \mathcal{R} \right\}.$$

According to Definition 17.2, this set is the tangent 3-plane to *graph f* at $(1, 1, 1, 3)$. ∎

EXERCISES

Find tangent lines or tangent planes to the function-loci defined explicitly by the following functions at the points x_0 (or X_0). Sketch all two- and three-dimensional figures.

1. $f(x) = x^2$, $x_0 = 1$.

2. $f(x) = x^3$, $x_0 = 0$.

3. $f(x) = \sin^2 x$, $x_0 = \dfrac{3\pi}{2}$.

4. $f(x) = \sqrt{\sin x}$, $x_0 = \dfrac{\pi}{4}$.

5. $f(x) = \begin{pmatrix} \cos x \\ \sin x \end{pmatrix}$, $x_0 = \dfrac{\pi}{2}$.

6. $f(x) = \begin{pmatrix} \cos x \\ x \end{pmatrix}$, $x_0 = 0$.

7. $f(x) = \begin{pmatrix} x \cos x \\ x \sin x \end{pmatrix}$, $x_0 = 0$.

8. $f(x) = \begin{pmatrix} x \\ x^2 \end{pmatrix}$, $x_0 = 1$.

9. $f\begin{pmatrix} x \\ y \end{pmatrix} = x^2 + y^2$, $X_0 = \begin{pmatrix} 1 \\ 1 \end{pmatrix}$.

10. $f\begin{pmatrix} x \\ y \end{pmatrix} = 4 - x^2 - y^2$, $X_0 = \begin{pmatrix} 1 \\ 1 \end{pmatrix}$.

11. $f\begin{pmatrix} x \\ y \end{pmatrix} = \sqrt{4 - x^2 - y^2}$, $X_0 = \begin{pmatrix} 1 \\ 1 \end{pmatrix}$.

12. $f\begin{pmatrix} x \\ y \end{pmatrix} = x^2 + y$, $X_0 = \begin{pmatrix} 0 \\ 1 \end{pmatrix}$.

13. $f\begin{pmatrix} x \\ y \\ z \end{pmatrix} = \cos x + \sin y + z$, $X_0 = \begin{pmatrix} \frac{\pi}{2} \\ \frac{\pi}{2} \\ 1 \end{pmatrix}$.

14. $f\begin{pmatrix} x \\ y \\ z \end{pmatrix} = x^2 + y - 2z + 1)$, $X_0 = \begin{pmatrix} 1 \\ 1 \\ 1 \end{pmatrix}$.

15. Let a circle $\mathcal{C}$ be defined explicitly as *graph f* $\cup$ *graph* $(-f)$, where $f(x) = \sqrt{a^2 - x^2}$, and let x_0 be an interior point of *dom f*. Show that $\mathcal{C} \cap$ *graph* $A_{x_0}f$ contains exactly one point. (Hint: Show that $A_{x_0}f(x) = f(x)$ has a unique solution and that $A_{x_0}f(x) = (-f)(x)$ has no solution.)

ANSWERS

1. $\left\{ x\binom{1}{2} + \binom{0}{-1} \,\middle|\, x \in \mathcal{R} \right\}.$ **3.** $\left\{ x\binom{1}{0} + \binom{0}{1} \,\middle|\, x \in \mathcal{R} \right\}.$

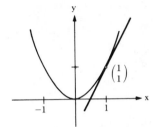   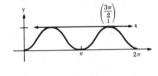

5. $\left\{ \begin{pmatrix} 1 \\ -1 \\ 0 \end{pmatrix} x + \begin{pmatrix} 0 \\ \frac{\pi}{2} \\ 1 \end{pmatrix} \,\middle|\, x \in \mathcal{R} \right\}$ **7.** $\left\{ \begin{pmatrix} 1 \\ 1 \\ 0 \end{pmatrix} x \,\middle|\, x \in \mathcal{R} \right\}$

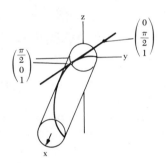   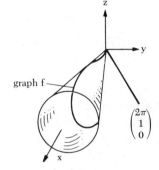

9. $\left\{ \begin{pmatrix} 1 \\ 0 \\ 2 \end{pmatrix} x + \begin{pmatrix} 0 \\ 1 \\ 2 \end{pmatrix} y + \begin{pmatrix} 0 \\ 0 \\ -2 \end{pmatrix} \,\middle|\, x, y \in \mathcal{R} \right\}$

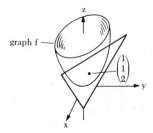

11. $\left\{ \left(\begin{pmatrix} 1 \\ 0 \\ -1 \\ \sqrt{2} \end{pmatrix} x + \begin{pmatrix} 0 \\ 1 \\ -1 \\ \sqrt{2} \end{pmatrix} y + \begin{pmatrix} 0 \\ 0 \\ 2\sqrt{2} \end{pmatrix} \right) \Big| \, x, y \in \mathcal{R} \right\}$

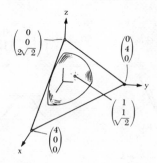

13. $\left\{ \left(\begin{pmatrix} 1 \\ 0 \\ 0 \\ -1 \end{pmatrix} x + \begin{pmatrix} 0 \\ 1 \\ 0 \\ 0 \end{pmatrix} y + \begin{pmatrix} 0 \\ 0 \\ 1 \\ 1 \end{pmatrix} z + \begin{pmatrix} 0 \\ 0 \\ 0 \\ 1 + \dfrac{\pi}{2} \end{pmatrix} \right) \Big| \, x, y, z \in \mathcal{R} \right\}$

2. TANGENT LINES AND PLANES TO PARAMETRICALLY DEFINED FUNCTION-LOCI

In the previous section we found that if $f : \mathcal{R}^n \to \mathcal{R}^m$ and $X_0 \in \mathcal{R}^n$, then *graph* $A_{X_0} f$ is a satisfactory set to be called the tangent to *graph* f at $(X_0, f X_0)$. This suggests that we define *ran* $A_{X_0} f$ to be the tangent plane to *ran* f at $f X_0$. Two questions need to be answered: is this definition consistent with the usual geometric definition of a tangent to a circle and is it consistent with Definition 17.2 of the previous section? The following example suggests that the answers are in the affirmative.

Example 5. Consider the parametric representation for the circle $\mathcal{C}$ given by $f(t) = \begin{pmatrix} x \\ y \end{pmatrix} = \begin{pmatrix} \cos t \\ \sin t \end{pmatrix}$, $0 \le t < 2\pi$, $t_0 = \dfrac{\pi}{4}$. By the methods of Chapter 16,

$$A_{t_0} f(s) = s \begin{pmatrix} \dfrac{-1}{\sqrt{2}} \\ \dfrac{1}{\sqrt{2}} \end{pmatrix} + \begin{pmatrix} \dfrac{4 + \pi}{4\sqrt{2}} \\ \dfrac{4 - \pi}{4\sqrt{2}} \end{pmatrix}$$

and

$$ran\ A_{t_0}f = \left\{ s\begin{pmatrix} \dfrac{-1}{\sqrt{2}} \\ 1 \\ \dfrac{1}{\sqrt{2}} \end{pmatrix} + \begin{pmatrix} \dfrac{4+\pi}{4\sqrt{2}} \\ \dfrac{4-\pi}{4\sqrt{2}} \end{pmatrix} \Bigg| s \in \mathcal{R} \right\}$$

$$=^1 \left\{ s\begin{pmatrix} 1 \\ -1 \end{pmatrix} + \begin{pmatrix} \dfrac{1}{\sqrt{2}} \\ 1 \\ \dfrac{1}{\sqrt{2}} \end{pmatrix},\ s \in \mathcal{R} \right\}.$$

(1) Theorem 1.23, p. 34 (since $\begin{pmatrix} \dfrac{1}{\sqrt{2}} \\ 1 \\ \dfrac{1}{\sqrt{2}} \end{pmatrix} = f(t_0) \in ran\ A_{t_0}f$).

We leave it to the reader to show, in Exercise 15, that $ran\ A_{t_0}f \cap ran f = \begin{pmatrix} \dfrac{1}{\sqrt{2}} \\ 1 \\ \dfrac{1}{\sqrt{2}} \end{pmatrix}$. ■

Now the circle $\mathcal{C}$ is given explicitly by $graph\ g \cup graph\ (-g)$, where $g(x) = \sqrt{1 - x^2}$. If we let $x_0 = \dfrac{1}{\sqrt{2}}$ then $graph\ A_{x_0}g = ran\ A_{y_0}f$ (see Exercise 16). Thus, in this case, the proposed definition of a tangent line to a parametrically defined function-locus yields the same tangent to $\mathcal{C}$ as Definition 17.2.

We make the formal definition:

DEFINITION 17.3 (Tangent n-plane to a Parametrically-Defined Function-Locus.) Let $f: \mathcal{R}^n \to \mathcal{R}^m$. Let f have an affine approximation $A_{X_0}f$ about X_0 such that $ran\ A_{X_0}f$ is an n-plane. *The f-tangent n-plane to ran f at fX$_0$ is ran A$_{X_0}$f.* A direction vector for $ran\ A_{X_0}f$ is called an *f-tangent vector to ran f at fX$_0$*. If a set S in $\mathcal{R}^n$ can be described parametrically by a function f with the preceding properties, S is said to be *n-dimensional at fX$_0$*. If S is n-dimensional at each of its points, S is said to be *n-dimensional*.

NOTE. An f-tangent 1-plane is referred to as an f-tangent line.

Theorem 17.4, which follows, shows that this definition is consistent with Definition 17.2. In the following example we shall see reasons for the requirement that $ran\ A_{X_0}f$ be n-dimensional and for the terminology "f-tangent" instead of "tangent."

Example 6. Let $f(t) = \begin{pmatrix} t^3 \\ t^3 \end{pmatrix}$, $t \in \mathcal{R}$, $t_0 = 0$. Thus $f: \mathcal{R}^1 \to \mathcal{R}^2$ and $n = 1, m = 2$. Now $A_{t_0}f(x) = \begin{pmatrix} 0 \\ 0 \end{pmatrix}$ for all x; hence $ran\ A_{t_0}f$ is 0-dimensional, not 1-dimensional. Thus, according to our definition $ran\ f$ has no f-tangent line at $f(0) = \begin{pmatrix} 0 \\ 0 \end{pmatrix}$, although the set of points, $ran\ f$, is a line $\mathcal{L}$ through the origin, and we feel intuitively that a line is tangent to itself at every point. Had the definition not required that an f-tangent to $ran\ f$ be 1-dimensional, we would have found ourselves referring to $\{O_2\}$ as a tangent line. On the other hand, if we let $g(t) = \begin{pmatrix} t \\ t \end{pmatrix}$, $t \in \mathcal{R}$, then $A_0g(t) = t\begin{pmatrix} 1 \\ 1 \end{pmatrix}$ and since this defines a one-dimensional space, $ran\ g$ does have a g-tangent line at $g(0) = (0)$.

Now $ran\ g$ and $ran\ f$ are the same set of points $\mathcal{L}$. Thus we see that the f-tangent to $\mathcal{L} = ran\ f$ at one of its points depends on the function f used to define it as well as on the set of points $\mathcal{L}$. ∎

Example 7. Let $f(t) = \begin{pmatrix} t^2 - 1 \\ t^3 - 2t^2 - t + 2 \end{pmatrix}$, $t \in \mathcal{R}$. The f-tangent

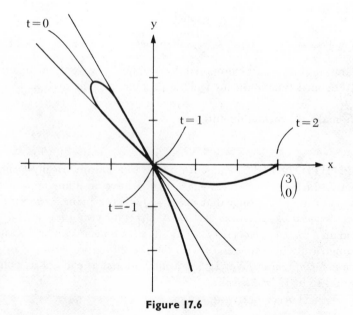

Figure 17.6

line to $ran\ f$ at $f(1) = O_2$ is given by $\mathcal{L}_0 = \left\{ t\begin{pmatrix} 2 \\ -2 \end{pmatrix} \ \middle|\ t \in \mathcal{R} \right\}$. On the other

hand the f-tangent to $ran\ f$ at $f(-1) = O_2$ is given by $\mathcal{L}_1 = \left\{ t\begin{pmatrix} -2 \\ 6 \end{pmatrix} \ \middle|\ t \in \mathcal{R} \right\}$.
We sketch $ran\ f$, along with its f-tangents at $f(0)$ and $f(1)$, in Figure 17.6. ∎

From Example 7, we see that the f-tangent to $ran\ f$ at fX_0 depends upon X_0 as well as upon f. It is even possible for a function-locus to have an f-tangent line at fX_0 but not at fX_1, even though $fX_0 = fX_1$.

THEOREM 17.4 Let $S \subset \mathcal{R}^m$, defined parametrically by a function $f : \mathcal{R}^n \to \mathcal{R}^m$ and explicitly by $g : \mathcal{R}^n \to \mathcal{R}^{m-n}$. Let $P \in S$. An f-tangent n-plane to S at P and the g-tangent n-plane to S at P coincide, providing they both exist.

Proof. We give a proof only for the case in which $n = 1$, $m = 2$. We

note first that for any Q in S we can write $Q = f(t) = \begin{pmatrix} f_1(t) \\ f_2(t) \end{pmatrix}$ for some t in

$\mathcal{R}$, since $S = ran\ f$. Similarly since $S = graph\ g$, we can write $Q = \begin{pmatrix} x \\ g(x) \end{pmatrix}$

for some x in $\mathcal{R}$. Hence $Q = \begin{pmatrix} f_1(t) \\ f_2(t) \end{pmatrix} = \begin{pmatrix} x \\ g(x) \end{pmatrix}$, and $f_2(t) = g(x) = g(f_1(t)) = g \circ f_1(t)$. Thus

(17.1) $f_2(t) = g \circ f_1(t)$ for every t in $dom\ f$.

Now

$$A_{t_0} f(t) =^1 \begin{pmatrix} f_1'(t_0) \\ f_2'(t_0) \end{pmatrix} (t - t_0) + \begin{pmatrix} f_1(t_0) \\ f_2(t_0) \end{pmatrix}$$

$$=^2 \begin{pmatrix} f_1'(t_0) \\ (g \circ f_1)'(t_0) \end{pmatrix} (t - t_0) + \begin{pmatrix} f_1(t_0) \\ g \circ f_1(t_0) \end{pmatrix}$$

$$=^3 \begin{pmatrix} f_1'(t_0) \\ g' \circ f_1(t_0) f_1'(t_0) \end{pmatrix} (t - t_0) + \begin{pmatrix} f_1(t_0) \\ g \circ f_1(t_0) \end{pmatrix}$$

$$=^4 f_1'(t_0) \begin{pmatrix} 1 \\ g' \circ f_1(t_0) \end{pmatrix} (t - t_0) + \begin{pmatrix} f_1(t_0) \\ g \circ f_1(t_0) \end{pmatrix}.$$

(1) Theorem 5.20, p. 243.

(2) By Equation 17.1.

(3) Theorem 5.10 (Chain Rule), p. 230.

(4) Factoring out $f_1'(t_0)$. Note that in view of Equation 17.1 and Definition 17.3 we have $f_1'(t_0) \neq 0$.

Thus $ran\ A_{t_0} f$ is a line in $\mathcal{R}^2$ with direction vector $\begin{pmatrix} 1 \\ g' \circ f_1(t_0) \end{pmatrix}$ and passing
through the point $\begin{pmatrix} f_1(t_0) \\ g \circ f_1(t_0) \end{pmatrix}$.

On the other hand

$$A_{f_1(t_0)}g(x) =^1 g'(f_1(t_0))(x - f_1(t_0)) + g(f_1(t_0))$$

and

$$\text{graph } A_{f_1(t_0)}g =^2 \left\{ \left(\begin{matrix} x \\ g' \circ f_1(t_0)(x - f_1(t_0)) + g \circ f_1(t_0) \end{matrix} \right) \,\middle|\, x \in \mathcal{R} \right\}$$

$$=^3 \left\{ x \left(\begin{matrix} 1 \\ g' \circ f_1(t_0) \end{matrix} \right) - \left(\begin{matrix} 0 \\ g' \circ f_1(t_0)f_1(t_0) - g \circ f_1(t_0) \end{matrix} \right) \,\middle|\, x \in \mathcal{R} \right\}.$$

(1) Equation 5.4, p. 223.

(2) Definition 7.1, p. 289 and definition of composition of functions.

(3) Algebraic manipulations.

Consequently *graph* $A_{f_1(t_0)}g$ is also a line in $\mathcal{R}^2$ with direction vector $\left(\begin{matrix} 1 \\ g \circ f_1(t_0) \end{matrix} \right)$.
It too passes through $\left(\begin{matrix} f_1(t_0) \\ g \circ f_1(t_0) \end{matrix} \right)$; hence *graph* $A_{f_1(t_0)}g = ran\ A_{t_0}f$ and our
proof is complete. ∎

The following theorem gives an alternate way of describing the *f*-tangent *n*-plane to a parametrically-defined function locus and will be used to simplify calculations. Proof will be left to Exercise 17.

THEOREM 17.5 Let $f: \mathcal{R}^n \to \mathcal{R}^m$, $X_0 \in dom\ f$. If *ran* f has an *f*-tangent *n*-plane at fX_0, it is $\{[d_{X_0}f]X + fX_0 \mid X \in \mathcal{R}^n\}$.

Example 7. Let $f(t) = \left(\begin{matrix} x \\ y \\ z \end{matrix} \right) = \left(\begin{matrix} t \cos t \\ t \sin t \\ t \end{matrix} \right)$, $t \geq 0$, $t_0 = \dfrac{\pi}{2}$. The parametrically defined function-locus is pictured in Figure 17.7.

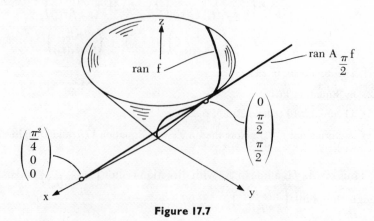

Figure 17.7

Using Theorem 17.5, we find the f-tangent line at $f\left(\dfrac{\pi}{2}\right)$ to be given by

$$X(t) = t\begin{pmatrix} \cos t - t \sin t \\ \sin t + t \cos t \\ 1 \end{pmatrix}_{\pi/2} + f\left(\frac{\pi}{2}\right)$$

$$= t\begin{pmatrix} \dfrac{-\pi}{2} \\ 1 \\ 1 \end{pmatrix} + \begin{pmatrix} 0 \\ \dfrac{\pi}{2} \\ \dfrac{\pi}{2} \end{pmatrix}.$$

Moreover, $\begin{pmatrix} \dfrac{-\pi}{2} \\ 1 \\ 1 \end{pmatrix}$ is a tangent vector to $ran\ f$ at $\begin{pmatrix} 0 \\ \dfrac{\pi}{2} \\ \dfrac{\pi}{2} \end{pmatrix}$. ■

Example 8. Let

$$g\begin{pmatrix} u \\ v \end{pmatrix} = \begin{pmatrix} x \\ y \\ z \end{pmatrix} = \begin{pmatrix} \cos u \\ \sin v \\ \sin u \\ \sin v \\ \cos v \end{pmatrix},\ 0 \le u \le 2\pi,\ 0 < v \le \frac{\pi}{2},\ X_0 = \begin{pmatrix} \pi \\ \pi \\ 4 \end{pmatrix}.$$

We have sketched $ran\ g$ in Example 21 of Chapter 7, p. 333. According to Theorem 17.5, the g-tangent plane to $ran\ g$ at gX_0 is given parametrically by

$$[d_{X_0}g]X + gX_0 = \begin{bmatrix} -\sin u & \dfrac{-\cos u \cos v}{\sin^2 v} \\ \sin v & \\ \cos u & \dfrac{-\sin u \cos v}{\sin^2 v} \\ \sin v & \\ 0 & -\sin v \end{bmatrix}_{\left(\frac{\pi}{4}\right)}\begin{pmatrix} u \\ v \end{pmatrix} + g\begin{pmatrix} \pi \\ \pi \\ 4 \end{pmatrix}$$

$$= \begin{bmatrix} 0 & \sqrt{2} \\ -\sqrt{2} & 0 \\ 0 & \dfrac{-1}{\sqrt{2}} \end{bmatrix}\begin{pmatrix} u \\ v \end{pmatrix} + \begin{pmatrix} -\sqrt{2} \\ 0 \\ \dfrac{1}{\sqrt{2}} \end{pmatrix}.$$

We sketch $ran\ g$ and its tangent plane at gX_0 in Figure 17.8. ■

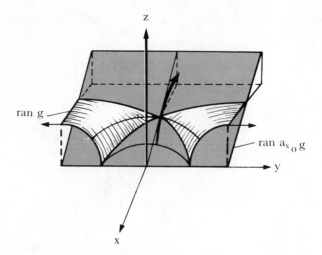

Figure 17.8

EXERCISES

Find f-tangent lines or f-tangent planes to the following parametrically defined function-loci at the points indicated. If the function-locus is a subset of $\mathcal{R}^2$ or $\mathcal{R}^3$, sketch the figure.

1. $f(t) = \begin{pmatrix} \cos t \\ \sin t \end{pmatrix}$, $0 \le t < 2\pi$, $t_0 = \dfrac{5\pi}{4}$.

2. $f(t) = \begin{pmatrix} \sin t \\ \cos t \end{pmatrix}$, $0 \le t < 2\pi$, $t_0 = \dfrac{7\pi}{4}$.

3. $f(t) = \begin{pmatrix} 2\cos t \\ \sin t \end{pmatrix}$, $0 \le t < 2\pi$, $t_0 = \dfrac{3\pi}{4}$.

4. $f(t) = \begin{pmatrix} \sec t \\ \tan t \end{pmatrix}$, $0 \le t < 2\pi$, $t \notin \left\{ \dfrac{\pi}{2}, \dfrac{3\pi}{2} \right\}$, $t_0 = \pi$.

5. $f(t) = \begin{pmatrix} \cos^2 t \\ t \end{pmatrix}$, $t_0 = \dfrac{3\pi}{4}$.

6. $f(t) = \begin{pmatrix} \sec^2 t \\ t \end{pmatrix}$, $t \notin \left\{ \dfrac{\pi}{2} + n\pi,\ n \text{ an integer} \right\}$, $t_0 = 0$.

7. $f(t) = \begin{pmatrix} \sin t \\ \sin t \\ \sin t \end{pmatrix}$, $t \in \mathcal{R}$, $t_0 = 0$.

8. $f(t) = \begin{pmatrix} \sin t \\ \sin t \\ t \end{pmatrix}$, $t \in \mathcal{R}$, $t_0 = \dfrac{\pi}{2}$.

9. $f\begin{pmatrix} s \\ t \end{pmatrix} = \begin{pmatrix} s \\ t \\ \sqrt{s^2 + t^2} \end{pmatrix}$, $\begin{pmatrix} s_0 \\ t_0 \end{pmatrix} = \begin{pmatrix} 1 \\ 1 \end{pmatrix}$.

10. $f\begin{pmatrix} s \\ t \end{pmatrix} = \begin{pmatrix} s \\ 2t \\ \sqrt{s^2 + t^2} \end{pmatrix}$, $\begin{pmatrix} s_0 \\ t_0 \end{pmatrix} = \begin{pmatrix} 1 \\ 1 \end{pmatrix}$.

11. $f\begin{pmatrix} s \\ t \end{pmatrix} = \begin{pmatrix} s \cos t \\ s \sin t \\ t \end{pmatrix}$, $\begin{cases} 0 \le s < 4 \\ 0 \le t \le \pi, \end{cases}$ $\begin{pmatrix} s_0 \\ t_0 \end{pmatrix} = \begin{pmatrix} 2 \\ \dfrac{\pi}{2} \end{pmatrix}$.

12. $f\begin{pmatrix} u \\ v \end{pmatrix} = \begin{pmatrix} \cos u \\ \sin v \\ \sin u \\ \sin v \\ \sin v \end{pmatrix}$, $\begin{cases} 0 \le u \le 2\pi \\ 0 < v \le \dfrac{\pi}{2}, \end{cases}$ $X_0 = \begin{pmatrix} \dfrac{\pi}{4} \\ \dfrac{\pi}{4} \end{pmatrix}$.

13. $f\begin{pmatrix} u \\ v \end{pmatrix} = \begin{pmatrix} \cos u \\ \sin u \\ \cos v \\ \sin v \end{pmatrix}$, $X_0 = \begin{pmatrix} 0 \\ \dfrac{\pi}{2} \end{pmatrix}$.

14. $f\begin{pmatrix} u \\ v \\ w \end{pmatrix} = \begin{pmatrix} \cos u \cos w \\ \sin u \\ u^2 + v^2 \\ w \end{pmatrix}$, $X_0 = \begin{pmatrix} 0 \\ 0 \\ \dfrac{\pi}{2} \end{pmatrix}$.

15. In Example 5 show that $ran\ A_{t_0} f \cap ran\ f = \left\{ \begin{pmatrix} \dfrac{1}{\sqrt{2}} \\ \dfrac{1}{\sqrt{2}} \end{pmatrix} \right\}$.

16. In Example 5 show that $ran\ A_{t_0} f = graph\ A_{x_0} g$.

17. Prove Theorem 17.5.

ANSWERS

1. $\left\{ t\begin{pmatrix} 1 \\ -1 \end{pmatrix} + \begin{pmatrix} -\dfrac{1}{\sqrt{2}} \\ -\dfrac{1}{\sqrt{2}} \end{pmatrix} \,\middle|\, t \in \mathcal{R} \right\}$.   3. $\left\{ t\begin{pmatrix} -2 \\ -1 \end{pmatrix} + \begin{pmatrix} -\dfrac{2}{\sqrt{2}} \\ \dfrac{1}{\sqrt{2}} \end{pmatrix} \,\middle|\, t \in \mathcal{R} \right\}$.

5. $\left\{ t\begin{pmatrix} 1 \\ 1 \end{pmatrix} + \begin{pmatrix} \frac{1}{2} \\ \frac{3\pi}{4} \end{pmatrix} \,\middle|\, t \in \mathcal{R} \right\}.$

7. $\left\{ t\begin{pmatrix} 1 \\ 1 \\ 1 \end{pmatrix} \,\middle|\, t \in \mathcal{R} \right\}.$

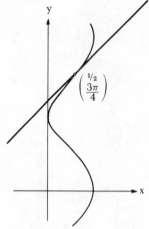

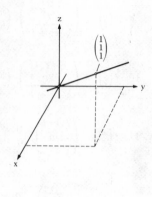

9. $\left\{ s\begin{pmatrix} 1 \\ 0 \\ \frac{1}{\sqrt{2}} \end{pmatrix} + t\begin{pmatrix} 0 \\ 1 \\ \frac{1}{\sqrt{2}} \end{pmatrix} + \begin{pmatrix} 1 \\ 1 \\ \sqrt{2} \end{pmatrix} \,\middle|\, s, t \in \mathcal{R} \right\}.$

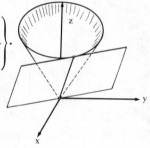

11. $\left\{ s\begin{pmatrix} 0 \\ 1 \\ 0 \end{pmatrix} + t\begin{pmatrix} -2 \\ 0 \\ 1 \end{pmatrix} + \begin{pmatrix} 0 \\ 2 \\ \frac{\pi}{2} \end{pmatrix} \,\middle|\, s, t \in \mathcal{R} \right\}.$

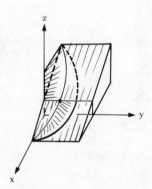

13. $\left\{ u\begin{pmatrix} 0 \\ 1 \\ 0 \\ 0 \end{pmatrix} + v\begin{pmatrix} 0 \\ 0 \\ -1 \\ 0 \end{pmatrix} + \begin{pmatrix} 1 \\ 0 \\ 0 \\ 1 \end{pmatrix} \middle| u, v \in \mathcal{R} \right\}.$

3. TANGENT LINES AND PLANES TO IMPLICITLY DEFINED FUNCTION-LOCI

Proceeding as in the last two sections, we make our definition of a tangent plane to an implicitly defined function-locus in terms of the affine approximation to the defining function.

DEFINITION 17.6 (Tangent n-plane to an Implicitly Defined Function Locus.) Let $f : \mathcal{R}^{n+m} \to \mathcal{R}^m$, $K \in \text{ran} f$, and $X_0 \in K\text{-lev} f$. Let f have an affine approximation $A_{X_0} f$ about X_0 such that $K\text{-lev} A_{X_0} f$ is an n-plane. The f-tangent n-plane to $K\text{-lev} f$ at X_0 is $K\text{-lev} A_{X_0} f$. A direction vector for $K\text{-lev} A_{X_0} f$ is called an f-tangent vector to $K\text{-lev} f$ at X_0. If a set S in $\mathcal{R}^{n+m}$ can be described implicitly by a function f with the preceding properties, S is said to be n-dimensional at X_0. If S is n-dimensional at each of its points, S is said to be an n-dimensional surface.

Notice that by Exercise 18, Sec. 15.5, p. 624, $K\text{-lev} A_{X_0} f$ is an n-plane if and only if some m columns of $[d_{X_0} f]$ are linearly independent.

Example 9. We wish to find the f-tangent line to the unit circle $\mathcal{C}$ defined implicitly by

$$f(x, y) = x^2 + y^2 = 1$$

at $X_0 = \left(\dfrac{1}{\sqrt{2}}, \dfrac{1}{\sqrt{2}} \right)$. The affine approximation is given by

$$A_{X_0} f(x, y) = [2x \quad 2y]_{\left(\frac{1}{\sqrt{2}}, \frac{1}{\sqrt{2}}\right)} \begin{pmatrix} x - \dfrac{1}{\sqrt{2}} \\ y - \dfrac{1}{\sqrt{2}} \end{pmatrix} + 1$$

$$= \sqrt{2}\, x + \sqrt{2}\, y - 1.$$

Setting this equal to one, we obtain

$$x + y = \sqrt{2}$$

as an implicit representation of the f-tangent line.

Furthermore, it is clear that $(1, 1)$ is a normal vector for this line and hence that $(1, -1)$ is a direction vector for the f-tangent line and thus an f-tangent vector to $\mathcal{C}$. ∎

Example 10. Let $g(x, y) = (x^2 + y^2 - 1)^2$. We find that *0-lev g* is the unit circle $\mathcal{C}$ of Example 9. However, a *g*-tangent line to $\mathcal{C}$ is not definable for any point of $\mathcal{C}$. This is a consequence of the fact that $[d_{X_0}g] = [0, 0]$ for any X_0 in *0-lev g* (see Exercise 9 and the note following Definition 17.6). ∎

From Examples 9 and 10, we see that the existence of an *f*-tangent line to an implicitly defined function-locus depends not only on the set of points in the function locus, but also on the defining function *f*. This is why we refer to the "*f*-tangent *n*-plane" rather than just the "tangent *n*-plane." Frequently when a function-locus is defined implicitly and the function *f* is not specifically mentioned, we shall refer to the *f*-tangent *n*-plane merely as the tangent *n*-plane to the function-locus; but we shall always understand that its existence depends on the defining function.

At this point we check the consistency of our definition with those previously given.

THEOREM 17.7 Let $\mathcal{S}$ be a subset of $\mathcal{R}^{n+m}$, defined implicitly by a function $f: \mathcal{R}^{n+m} \to \mathcal{R}^m$ and explicitly by $g: \mathcal{R}^n \to \mathcal{R}^m$. Let $P \in \mathcal{S}$. The *f*-tangent *n*-plane to $\mathcal{S}$ at P and the *g*-tangent *n*-plane to $\mathcal{S}$ at P coincide, provided they both exist.

Proof. We consider only the case where $n = m = 1$. Let $P = (x_0, y_0)$ and let k be such that $\mathcal{S} = k\text{-lev} f$. Let $\mathcal{C}_f$ be the *f*-tangent line to $\mathcal{S}$ at P and $\mathcal{C}_g$ be the *g*-tangent line to $\mathcal{S}$ at P. By Definition 17.6

$$\mathcal{C}_f = k\text{-lev } A_{(x_0, y_0)} f$$

$$= \left\{ \begin{pmatrix} x \\ y \end{pmatrix} \Bigg| A_{(x_0, y_0)} f \begin{pmatrix} x \\ y \end{pmatrix} = k, \begin{pmatrix} x \\ y \end{pmatrix} \in \mathcal{R} \right\}$$

$$= \left\{ \begin{pmatrix} x \\ y \end{pmatrix} \Bigg| [f_{1,1}(x_0, y_0) \quad f_{1,2}(x_0, y_0)] \begin{pmatrix} x - x_0 \\ y - y_0 \end{pmatrix} = O_2, \begin{pmatrix} x \\ y \end{pmatrix} \in \mathcal{R}^2 \right\}.$$

Therefore $\mathcal{C}_f$ is the line through (x_0, y_0) with normal

$$N = [f_{1,1}(x_0, y_0) \quad f_{1,2}(x_0, y_0)].$$

On the other hand, by Definition 17.2,

$$\mathcal{C}_g = \text{graph } A_{x_0} g$$

$$= \left\{ \begin{pmatrix} x \\ g'(x_0)(x - x_0) + g(x_0) \end{pmatrix} \Bigg| x \in \mathcal{R} \right\}$$

$$= \left\{ \begin{pmatrix} x - x_0 + x_0 \\ g'(x_0)(x - x_0) + y_0 \end{pmatrix} \Bigg| x \in \mathcal{R} \right\}$$

$$= \left\{ (x - x_0) \begin{pmatrix} 1 \\ g'(x_0) \end{pmatrix} + \begin{pmatrix} x_0 \\ y_0 \end{pmatrix} \Bigg| x \in \mathcal{R} \right\}$$

$$= \left\{ t \begin{pmatrix} 1 \\ g'(x_0) \end{pmatrix} + \begin{pmatrix} x_0 \\ y_0 \end{pmatrix} \Bigg| t \in \mathcal{R} \right\},$$

where $t = x - x_0$. We see that $\mathfrak{C}_g$ is the line passing through (x_0, y_0) with direction vector $D = \begin{pmatrix} 1 \\ g'(x_0) \end{pmatrix}$. Thus, $\mathfrak{C}_f = \mathfrak{C}_g$ if

$$[f_{1,1}(x_0, y_0) \quad f_{1,2}(x_0, y_0)]\begin{pmatrix} 1 \\ g'(x_0) \end{pmatrix} = 0.$$

To show that this last equality holds we define

$$G(x) = (x, g(x))$$

and

$$h(x) = f \circ G(x)$$
$$= f(x, g(x)).$$

Since $G(x) = \begin{pmatrix} x \\ g(x) \end{pmatrix}$ is a point of *graph* $g = k$-lev f, we have $h(x) = f(G(x)) = k$, for every $x \in dom\ h$. Thus h is a constant $\mathcal{R} \to \mathcal{R}$ function. Consequently

$$0 =^1 h'(x_0)$$
$$=^2 (f \circ G)'(x_0)$$
$$=^3 [d_{G(x_0)}f][d_{x_0}G]$$
$$=^4 \left[f_{1,1}\begin{pmatrix} x_0 \\ y_0 \end{pmatrix} \quad f_{1,2}\begin{pmatrix} x_0 \\ y_0 \end{pmatrix} \right] \cdot [1\ \ g'(x_0)]$$
$$=^5 N \cdot D.$$

(1) h is a constant $\mathcal{R} \to \mathcal{R}$ function.

(2) Definition of h.

(3) Theorem 16.20 (Chain Rule), p. 680.

(4) Evaluating $G(x_0)$ and $G'(x_0)$.

(5) Definitions of N and D. ∎

The preceding theorem assures us that, if a function-locus is given both explicitly and implicitly, the tangent planes at a given point will be the same, if both exist; hence Definitions 17.1 and 17.6 are consistent. The next theorem will assure us of the consistency of Definitions 17.3 and 17.6.

THEOREM 17.8 Let S be a subset of $\mathcal{R}^{n+m}$, defined implicitly by $f: \mathcal{R}^{n+m} \to \mathcal{R}^m$ and parametrically by $g: \mathcal{R}^n \to \mathcal{R}^{n+m}$. Let $P \in S$. The f-tangent n-plane to S at P and the g-tangent n-plane to S at P coincide, provided they both exist.

Proof. We consider only the case where $m = n = 1$. Let $\mathfrak{C}_f$ and $\mathfrak{C}_g$ be respectively the f-tangent and g-tangent planes to S at P. We can write

$P = g(t_0) = \begin{pmatrix} g_1(t_0) \\ g_2(t_0) \end{pmatrix}$ for some $t_0 \in \Re$. For some $k \in \Re$, $S = k\text{-}lev\, f$ and $\mathscr{C}_f = k\text{-}lev\, A_P f = k\text{-}lev\, A_{g(t_0)} f$. Thus $\mathscr{C}_f$ is given implicitly by

$$A_{g(t_0)} f \begin{pmatrix} x \\ y \end{pmatrix} = [f_{1,1}(g(t_0))\ \ f_{1,2}(g(t_0))] \begin{pmatrix} x - g_1(t_0) \\ y - g_2(t_0) \end{pmatrix} + f(g(t_0)) = k.$$

Hence $\mathscr{C}_f$ has the normal $(f_{1,1}(g(t_0)), f_{1,2}(g(t_0))$ and includes the point $P = \begin{pmatrix} g_1(t_0) \\ g_2(t_0) \end{pmatrix}$.

On the other hand $\mathscr{C}_g$ is given parametrically by

$$A_{t_0}(g(t)) = g'(t_0)(t - t_0) + g(t_0)$$

$$= \begin{pmatrix} g_1(t_0) \\ g_2(t_0) \end{pmatrix}(t - t_0) + g(t_0).$$

This is a parametric equation of a line in $\Re^2$ including $g(t_0) = P$ and having the direction vector $\begin{pmatrix} g_1(t_0) \\ g_2(t_0) \end{pmatrix}$. Our proof will be complete if we can show that the normal $N = (f_{1,1}(g(t_0)), f_{1,2}(g(t_0)))$ to $\mathscr{C}_f$ is perpendicular to the direction vector $D = (g_1(t_0), g_2(t_0))$ for $\mathscr{C}_g$. This will be left to the reader in Exercise 11. ∎

Example 11. A satellite is traveling in an elliptical polar orbit given by

$$\frac{x^2}{a^2} + \frac{y^2}{b^2} = 1.$$

We wish to find the direction of its motion at the time it is moving northward over the forty-fifth parallel; that is, we wish to find a tangent vector to its orbit at the point P shown in Figure 17.9.

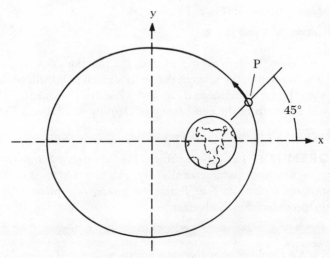

Figure 17.9

Setting $x = y$ in the given equation, we find that $P = \left(\dfrac{ab}{\sqrt{a^2 + b^2}},\right.$
$\left.\dfrac{ab}{\sqrt{a^2 + b^2}}\right)$. Using this value for P we find that the tangent line is given by
$b^2x + a^2y = ab\sqrt{a^2 + b^2}$. This equation represents a line with normal
(b^2, a^2) and direction vector $(-a^2, b^2)$. A tangent vector to a one-dimensional surface (or curve) may be in either of two opposite directions. Since
the given equation tells us nothing of the direction of motion, we must
decide from physical considerations whether or not this vector is in the
proper direction. It is, if the direction of motion is as indicated in Figure
17.9. ∎

Example 12. We wish to find the plane which is tangent at $X_0 =$
$(1, 1, 2)$ to the function-locus given implicitly by

$$f(x, y, z) = x^2 + y^2 + z = 4.$$

We have

$$A_{X_0}f(x, y, z) = [2x \quad 2y \quad 1]_{(1,1,2)}\begin{pmatrix} x - 1 \\ y - 1 \\ z - 2 \end{pmatrix} + 4 = 4$$

as the implicit representation of the tangent plane. This reduces to

$$2x + 2y + z = 6.$$

Notice that $[d_X f] = (2x \quad 2y \quad 1) \neq [0, 0, 0]$ for any choice of X on the
function-locus. Hence our function-locus is a two-dimensional surface.
The surface (a paraboloid of revolution) and its tangent plane at $(1, 1, 2)$
are sketched in Figure 17.10. ∎

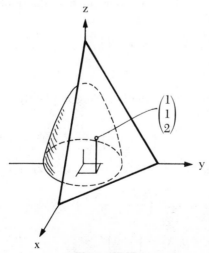

Figure 17.10

Example 13. Consider the function-locus defined implicitly by $2\sqrt{x^2+y^2} - x^2 - y^2 + z = 1$, $x^2+y^2 < 2$. The sketch is given in Figure 17.11. (Details are left to Exercise 10.)

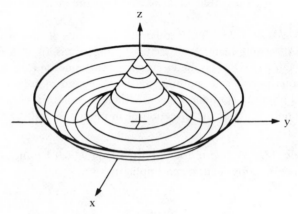

Figure 17.11

Observe that there seems to be a "sharp peak" at $(0, 0, 1)$. This leads us to suspect that perhaps there will not be a tangent plane at this point. The Jacobian matrix is

$$\left[\frac{2x}{\sqrt{x^2+y^2}} - 2x, \frac{2y}{\sqrt{x^2+y^2}} - 2y, 1 \right]_{X_0}.$$

As we suspected, this is not defined at $X_0 = (0, 0, 1)$. However, the Jacobian matrix is defined and unequal to $[0, 0, 0]$ at every other point of the function-locus. Hence our function-locus is not a two-dimensional surface; but if we delete $(0, 0, 1)$ from the domain of our function, the resulting function-locus is a two-dimensional surface. ■

EXERCISES

Find tangent surfaces to the following implicitly defined function loci at the points X_0. Sketch.

1. $x^2 + 2y^2 = 2$, $X_0 = \left(-1, \dfrac{-1}{\sqrt{2}} \right)$. 2. $x^2 - y^2 = 1$, $X_0 = (1, 0)$.

3. $x^2 + y = 4$, $X_0 = (-1, 3)$. 4. $2x + 2y - y^2 = 3$, $X_0 = (3, -1)$.

5. $y - \sin x = 1$, $X_0 = \left(\dfrac{3\pi}{4}, \dfrac{2+\sqrt{2}}{2} \right)$.

6. $y - 2\cos x = 0$, $X_0 = \left(\dfrac{\pi}{2}, 0 \right)$.

7. $x^2 + y^2 + z^2 = 1$, $X_0 = \left(\dfrac{1}{\sqrt{2}}, 0, \dfrac{1}{\sqrt{2}} \right)$.

8. $x^2 + y^2 + z = 4$, $X_0 = (1, 1, 2)$.

9. $(x^2 + y^2 - 1)^2 = 0$, $x_0^2 + y_0^2 = 1$.

10. Supply the details for Example 13.

11. Complete the proof of Theorem 17.8 (Hint: Observe that $(f \circ g)(t) = k$ for every t; hence $(f \circ g)' = 0$. Write out this equation using the chain rule.

ANSWERS

1. $x + \sqrt{2}y = -2$.

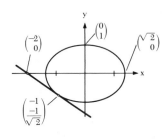

3. $-2x + y = 5$.

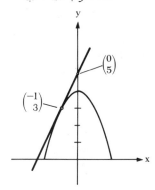

5. $y + \dfrac{1}{\sqrt{2}} x = \dfrac{3\pi + 4\sqrt{2} + 4}{4\sqrt{2}}$

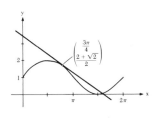

7. $x + z = \sqrt{2}$.

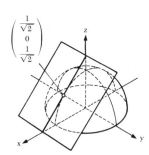

4. THE DIRECTIONAL DERIVATIVE AND THE GRADIENT

Let $f : \mathcal{R}^2 \to \mathcal{R}$, $X_0 = (x_0, y_0) \in dom\, f$, $X = (x, y)$. We may think of $f_{1,1}X_0$, the partial derivative of f with respect to x at X_0 (Definition 16.7, p. 653) as the rate of change of $f(x, y_0)$ as we move away from X_0 in the positive direction of the x-axis (see Figure 17.12). Similarly we may think of $f_{1,2}X_0$ as the rate of change of $f(x_0, y)$ as we move away from X_0 in the positive direction of the y-axis (see Figure 17.13).

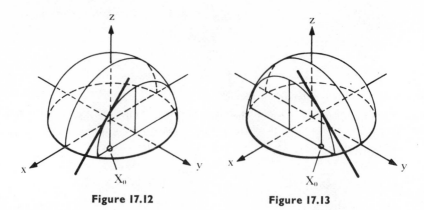

Figure 17.12   Figure 17.13

It may be that we wish to find the rate of change of $f(x, y)$ as we move away from X_0 in an arbitrary direction such as that illustrated in Figure 17.14. In this section we shall deal with this problem.

We have not been very careful in our use of the word "direction," although we have an intuitive feeling for its meaning. It should be obvious that there is a one-to-one correspondence between what we think of as "directions" in $\mathcal{R}^2$ and the points of the unit circle with center at O_2. Hence in order to give a precise meaning to the word, we shall identify the "directions" in $\mathcal{R}^2$ and the points of the unit circle. Similarly, we identify the "directions" in $\mathcal{R}^n$ with the points of the unit n-sphere.

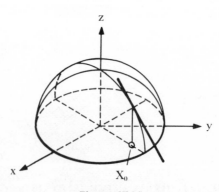

Figure 17.14

DEFINITION 17.9 (Direction, Direction of a Vector.) A *direction* in $\mathcal{R}^n$ is a unit vector in $\mathcal{R}^n$. If $Y \neq O_n$ is a vector in $\mathcal{R}^n$, the *direction of Y* is $\dfrac{Y}{\|Y\|}$.

In $\mathcal{R}$ there are only two directions, -1 and 1, but in higher dimensions there are infinitely many.

We generalize the idea of the partial derivative as follows:

DEFINITION 17.10   (Directional Derivative.) Let $f: \mathcal{R}^n \to \mathcal{R}$ and let U be a direction in $\mathcal{R}^n$. The *derivative of f in the direction U*, denoted $\dfrac{\partial f}{\partial U}$, is the vector function defined by

$$\frac{\partial f}{\partial U} X = \lim_{t \to 0} \frac{f(X + tU) - fX}{t},$$

whenever this limit exists.

If $n = 2$ and $U = E_1 = (1, 0)$ we have

$$\frac{\partial f}{\partial E_1} X = \lim_{t \to 0} \frac{f(X + tE_1) - fX}{t}$$
$$= (\partial_1 f) X,$$

by Definition 16.7, p. 653. A similar statement holds when $f: \mathcal{R}^n \to \mathcal{R}$ and U is any natural basis vector.

Example 14.   Let $f(x, y) = x^2 + y^2$, $X_0 = (1, 1)$, $U = \left(\dfrac{1}{\sqrt{2}}, \dfrac{1}{\sqrt{2}} \right)$. Find $\dfrac{\partial f}{\partial U} X_0$. The graph of f was sketched in Example 19 of Chapter 7, p. 331. We reproduce it in Figure 17.15, adding $S_{x=y}\mathcal{R}^3$ and the line in $S_{x=y}\mathcal{R}^3$ which is tangent to $S_{x=y}$ *graph f* at $(1, 1, g(1, 1))$. We may think of $S_{x=y}\mathcal{R}^3$ as a copy of $\mathcal{R}^2$. It is the slope of the line $\mathfrak{L}$ in this copy of $\mathcal{R}^2$ that we

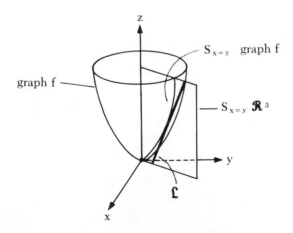

Figure 17.15

wish to determine. By Definition 17.7,

$$\frac{\partial f}{\partial U}\begin{pmatrix}1\\1\end{pmatrix} = \lim_{t\to 0}\frac{f\left(\begin{pmatrix}1\\1\end{pmatrix} + t\begin{pmatrix}\dfrac{1}{\sqrt{2}}\\ \dfrac{1}{\sqrt{2}}\end{pmatrix}\right) - f\begin{pmatrix}1\\1\end{pmatrix}}{t}$$

$$= \lim_{t\to 0}\frac{\left(1 + \dfrac{t}{\sqrt{2}}\right)^2 + \left(1 + \dfrac{t}{\sqrt{2}}\right)^2 - 2}{t}$$

$$= \lim_{t\to 0}\frac{t^2 + \dfrac{4t}{\sqrt{2}}}{t}$$

$$= \lim_{t\to 0} t + \frac{4}{\sqrt{2}} = \frac{4}{\sqrt{2}} = 2\sqrt{2}. \quad\blacksquare$$

The following theorem will permit us to evaluate directional derivatives by means of techniques previously developed.

THEOREM 17.11 Let $f:\mathcal{R}^n \to \mathcal{R}$, $X \in \mathcal{R}^n$, U a unit vector in $\mathcal{R}^n$. Let f be differentiable at X; then $\dfrac{\partial f}{\partial U}X$ exists and is given by

$$\frac{\partial f}{\partial U} X = [d_X f]U.$$

Proof. Define $g:\mathcal{R} \to \mathcal{R}^n$ such that

$$g(t) = X + tU,$$

and assume, for the moment, that $\dfrac{\partial f}{\partial U}$ exists. Then

$$\frac{\partial f}{\partial U} X =^1 \lim_{t\to 0}\frac{f(X + tU) - fX}{t}$$

$$=^2 \lim_{t\to 0}\frac{f(g(t)) - f(g(0))}{t}$$

$$=^3 \lim_{t\to 0}\frac{f \circ g(t) - f \circ g(0)}{t}$$

$$=^4 (f \circ)'(0)$$

$$=^5 [d_0(f \circ g)]$$

$$=^6 [dg_{(0)}f][d_0 g]$$

$$=^7 [d_X f]U.$$

(1) Definition 17.10.

(2) Definition of g.

(3) Definition of $f \circ g$.

(4) Definition 5.4, p. 223. Note that $f \circ g$ is a function $\mathcal{R} \to \mathcal{R}$ and substitute t for 0, x for t, f for $f \circ g$.

(5) Theorem 16.16, p. 668, letting $m = n = 1$.

(6) Theorem 16.20, (Chain Rule), p. 680.

(7) Substituting X for $g(0)$ and evaluating $[d_0 g]$.

Since f is given to be differentiable at X, the existence of $[d_X f]U$ is assured. We note that the preceding equalities are reversible, so $\dfrac{\partial f}{\partial U} X$ exists. $\blacksquare$

With each differentiable function $f : \mathcal{R}^n \to \mathcal{R}$ and each X in $\mathcal{R}^n$ we now associate the constant vector $[d_X f]$. If we are given a unit vector U in $\mathcal{R}^n$, we need only compute $[d_X f]U$ to obtain the derivative of f at X in the direction of U. The vector $[d_X f]$ is of sufficient importance to give it a special name.

DEFINITION 17.12 (The Gradient.) Let $f : \mathcal{R}^n \to \mathcal{R}$, $X \in \mathcal{R}^n$. Then $[d_X f]$ is called the *gradient of f at X* and denoted $\nabla f X$ and read "del f at X."

In Example 14, $\nabla f X_0 = (2, 2)$.

Example 15. Let $f(x, y) = x \tan y$, $X_0 = (x_0, y_0) = \left(1, \dfrac{\pi}{4}\right)$, $Y = (1, 2)$. We compute the derivative of f in the direction of Y. The direction of Y is

$$U = \frac{Y}{\|Y\|} = \left(\frac{1}{\sqrt{5}}, \frac{2}{\sqrt{5}}\right). \quad \text{Hence}$$

$$\frac{\partial f}{\partial U} X = \nabla f X \cdot U$$

$$= (\tan y_0, \, x_0 \sec^2 y_0) \cdot \left(\frac{1}{\sqrt{5}}, \frac{2}{\sqrt{5}}\right)$$

$$= (1, 2) \cdot \left(\frac{1}{\sqrt{5}}, \frac{2}{\sqrt{5}}\right) = \sqrt{5}. \quad \blacksquare$$

With the notation introduced in Definition 17.12 we restate Theorem 17.11 as follows:

THEOREM 17.13 Let $f: \mathcal{R}^n \to \mathcal{R}$, $X \in \mathcal{R}^n$, U a unit vector in $\mathcal{R}^n$. Let f be differentiable at X; then $\dfrac{\partial f}{\partial U} X$ exists and is given by

$$\frac{\partial f}{\partial U} X = (\nabla f X) \cdot U.$$

Sometimes we wish to know the direction for which the directional derivative at a point X attains a maximum. For this purpose the following theorem, which is an easy consequence of the Cauchy-Schwartz Inequality (Theorem 2.23, p. 135) and Theorem 17.13, is useful. The proof is called for in Exercise 9.

THEOREM 17.14 The direction U for which $\dfrac{\partial f}{\partial U} X$ is a maximum is $\dfrac{\nabla f X}{\|\nabla f X\|}$. Furthermore, for this choice of U we have $\dfrac{\partial f}{\partial U} X = \|\nabla f X\|$.

We may think of the graph of a function $\mathcal{R}^2 \to \mathcal{R}$ as a mountain (see Figure 17.16) and the point (X, fX) as the position of a mountain climber on its side. Then by the theorem, the direction of $\nabla f X$ is the direction in which he must travel if he wishes his ascent to be as steep as possible. If we interpret the level sets of f as the contour lines on the map of our mountain (see Figure 17.17), it seems natural to suppose that the direction of steepest ascent is normal to the contour line which passes through X. This is indeed the case as we shall see in Exercise 10.

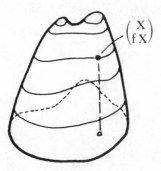

Figure 17.16

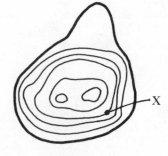

Figure 17.17

EXERCISES

For the following functions, find the directional derivative in the direction U at the point X_0, directly from the definition as in Example 14. Then compute the gradient at X_0 and find the directional derivative using Theorem 17.11.

1. $f\begin{pmatrix} x \\ y \end{pmatrix} = x^2 + y^2$, $X_0 = \begin{pmatrix} 1 \\ 1 \end{pmatrix}$, $U = \begin{pmatrix} \frac{1}{2} \\ \frac{\sqrt{3}}{2} \end{pmatrix}$.

2. $f\begin{pmatrix} x \\ y \end{pmatrix} = \sin x + \sin y$, $X_0 = \begin{pmatrix} 0 \\ 0 \end{pmatrix}$, $U = \begin{pmatrix} \frac{1}{\sqrt{2}} \\ \frac{-1}{\sqrt{2}} \end{pmatrix}$.

3. $f\begin{pmatrix} x \\ y \end{pmatrix} = xy$, $X_0 = \begin{pmatrix} 1 \\ -1 \end{pmatrix}$, $U = \begin{pmatrix} \frac{1}{\sqrt{2}} \\ \frac{-1}{\sqrt{2}} \end{pmatrix}$.

4. $f\begin{pmatrix} x \\ y \end{pmatrix} = x$, $y \neq 0$, $X_0 = \begin{pmatrix} 1 \\ 1 \end{pmatrix}$, $U = \begin{pmatrix} \frac{-1}{\sqrt{2}} \\ \frac{1}{\sqrt{2}} \end{pmatrix}$.

For the following functions find the gradient at X_0 and use this to (a) compute the derivative of f in the direction of Y, (b) determine the direction U in which the directional derivative at x_0 is a maximum and give the value of the directional derivative in this direction.

5. $f\begin{pmatrix} x \\ y \end{pmatrix} = \sin x \cos y$, $X_0 = \begin{pmatrix} \pi \\ \pi \end{pmatrix}$, $Y = \begin{pmatrix} 1 \\ 2 \end{pmatrix}$.

6. $f\begin{pmatrix} x \\ y \end{pmatrix} = x \sin y$, $X_0 = \begin{pmatrix} 1 \\ \pi \end{pmatrix}$, $Y = \begin{pmatrix} 2 \\ 3 \end{pmatrix}$.

7. $f\begin{pmatrix} x \\ y \\ z \end{pmatrix} = xyz$, $X_0 = \begin{pmatrix} 1 \\ 1 \\ 1 \end{pmatrix}$, $Y = \begin{pmatrix} 1 \\ 1 \\ 0 \end{pmatrix}$.

8. $f\begin{pmatrix} x \\ y \\ z \end{pmatrix} = xy - xz$, $X_0 = \begin{pmatrix} 0 \\ 0 \\ 0 \end{pmatrix}$, $Y = \begin{pmatrix} 1 \\ 1 \\ 0 \end{pmatrix}$.

9. Prove Theorem 17.13.

10. Prove that if $f : R^2 \to R$, $K \in dom\, f$, $X_0 \in K\text{-}lev\, f$, $\nabla f X_0$ is normal to the tangent line to $K\text{-}lev\, f$ at X_0.

ANSWERS

1. $1 + \sqrt{3}$. **3.** $-\sqrt{2}$.

5. (a) $\dfrac{\partial f}{\partial y} x_0 = \dfrac{1}{\sqrt{5}}$, (b) $U = \begin{pmatrix} 1 \\ 0 \end{pmatrix}$, $\dfrac{\partial f}{\partial U} x_0 = 1$.

7. (a) $\dfrac{\partial f}{\partial y} x_0 = \sqrt{2}$, (b) $U = \dfrac{1}{\sqrt{3}} \begin{pmatrix} 1 \\ 1 \\ 1 \end{pmatrix}$, $\dfrac{\partial f}{\partial U} x_0 = \sqrt{3}$.

PART V

INTEGRAL CALCULUS OF $\mathcal{R}^n \to \mathcal{R}$ FUNCTIONS

18

INTEGRATION OF $\mathcal{R}^n \to \mathcal{R}$ FUNCTIONS

1. THE VOLUME PROBLEM

In Chapter 12 the volume problem was anticipated and the volume concept characterized as follows:

We assume,

(i) that the volume function V is defined on a certain class $\mathcal{C}$ of bounded subsets of $\mathcal{R}^3$,

(ii) that the class $\mathcal{C}$ contains all rectangular parallelepipeds $[a, b] \times [c, d] \times [e, f]$. (See Figure 18.1.)

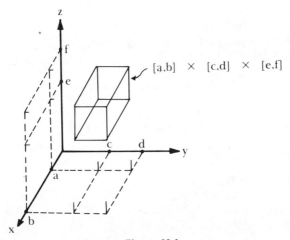

$[a,b] \times [c,d] \times [e,f]$

Figure 18.1

(iii) that V has the following properties:

(1) $V([a, b] \times [c, d] \times [e, f]) = V((a, b) \times (c, d) \times (e, f))$
$$= (b - a)(d - c)(f - e).$$

(2) $V(R) \geq 0$ if $R \in \mathcal{C}$.

(3) $\varnothing \in \mathcal{C}$ and $V(\varnothing) = 0$.

(4) $V(R) \leq V(S)$, if $R \subset S$, $R \in \mathcal{C}$, $S \in \mathcal{C}$.

(5) $V(R \cup S) = V(R) + V(S)$, if $R \cap S = \varnothing$, $R \in \mathcal{C}$ and $S \in \mathcal{C}$.

(6) $V(R \cup S) = V(R) + V(S) - V(R \cap S)$, if $R \in \mathcal{C}$ and $S \in \mathcal{C}$.

In the next section we begin the generalization of the integral of an $\mathcal{R} \to \mathcal{R}$ function to $\mathcal{R}^2 \to \mathcal{R}$ functions. This generalization will provide us with a volume function just as the $\mathcal{R} \to \mathcal{R}$ integral provided us with an area function.

2. STEP FUNCTIONS IN $\mathcal{R}^2$

The development of the $\mathcal{R} \to \mathcal{R}$ integral previously given involves the concept of a step function which, in turn, involves the idea of the partition of an interval of $\mathcal{R}$. Following this pattern, we define a partition for the $\mathcal{R}^2$ analog of an interval. $\mathcal{R}^2$ is the Cartesian product of $\mathcal{R}$ with itself: $\mathcal{R}^2 = \mathcal{R} \times \mathcal{R}$. This suggests that as the $\mathcal{R}^2$ analog of an interval we take the Cartesian product of two intervals, that is, a rectangle. For each rectangle $R = [a, b] \times [c, d]$ of $\mathcal{R}^2$ we define a partition to be the Cartesian product of two partitions, one for $[a, b]$ and one for $[c, d]$. We state this more precisely as:

DEFINITION 18.1 A *partition* $\mathcal{P}$ of a closed rectangle $R = [a, b] \times [c, d]$ is the Cartesian product of a partition $\mathcal{P}_1 = \{x_0, x_1, \ldots, x_m\}$ for $[a, b]$ and a partition $\mathcal{P}_2 = \{y_0, y_1, \ldots, y_n\}$ for $[c, d]$:

$$\mathcal{P} = \mathcal{P}_1 \times \mathcal{P}_2 = \{(x_0, y_0), (x_0, y_1), \ldots, (x_m, y_n)\}$$

The open rectangle $R = (x_{j-1}, x_j) \times (y_{k-1}, y_k)$, $(j = 1, \ldots, m; k = 1, \ldots, n)$, is called the $(j, k)^{\text{th}}$ *open subrectangle* of R and is denoted by R_{jk}.

Example 1. Let $R = [1, 4] \times [1, 2]$.
(a) $\mathcal{P}_1 = \{1, 3, 4\}$ is a partition of $[1, 4]$, $\mathcal{P}_2 = \{1, 2\}$ is a partition of $[1, 2]$, and $\mathcal{P} = \mathcal{P}_1 \times \mathcal{P}_2 = \{(1, 1), (1, 2), (3, 1), (3, 2), (4, 1), (4, 2)\}$ is the corresponding partition of R. (See Figure 18.2.) Points corresponding to $\mathcal{P}_1$ and $\mathcal{P}_2$ are indicated as well as the points of $\mathcal{P}$ itself. The open subrectangles are R_{11} and R_{21}.

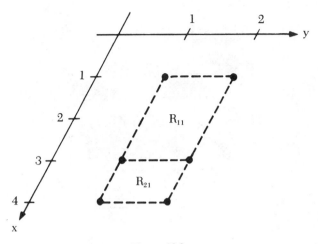

Figure 18.2

(b) With $\mathfrak{I}_1 = \{1, 3/2, 3, 4\}$ and $\mathfrak{I}_2 = \{1, 3/2, 2\}$, we have the situation depicted in Figure 18.3 for $\mathfrak{I} = \mathfrak{I}_1 \times \mathfrak{I}_2$. The partition $\mathfrak{I}$ may be thought of as separating R into a matrix of subrectangles R_{jk} where j denotes the column and k denotes the row in which R_{jk} is located. ■

We are now prepared to define an $\mathcal{R}^2 \to \mathcal{R}$ step function.

DEFINITION 18.2 Let $s : R \to \mathcal{R}$, where $R = [a, b] \times [c, d]$. We say *s is a step function on R* if and only if there is a partition $\mathfrak{I}$ of R for which s is constant on each of the open subrectangles R_{jk} of $\mathfrak{I}$. That is, there exist constants $s_{11}, \ldots, s_{mn}$ such that

$$s(x, y) = s_{jk} \text{ whenever } (x, y) \in R_{jk}.$$

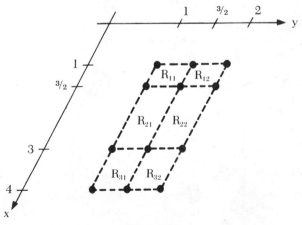

Figure 18.3

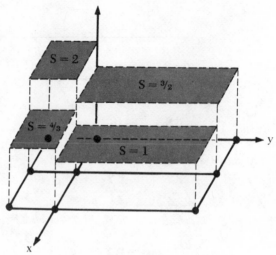

Figure 18.4

Example 2. Let $R = [0, 2] \times [-1, 3]$ be partitioned as shown in Figure 18.4. The shaded portion represents the graph of a step function s on R. The value of s at a point on the side of any of the open subrectangles is of no interest, except that it be defined. ∎

The following properties of step functions follow easily from the definition.

THEOREM 18.3 Let s and t be step functions on $R = [a, b] \times [c, d]$ and let $r \in \mathcal{R}$.
  (i) (Additive Property) $s + t$ is a step function on R.
  (ii) (Homogeneous Property) rs is a step function on R.

This result assures us that the set of step functions on a rectangle R is a subspace of the vector space of all $R \to \mathcal{R}$ functions.

3. THE INTEGRAL OF AN $\mathcal{R}^2 \to \mathcal{R}$ STEP FUNCTION

Consider the step function s on $R = [0, 2] \times [-1, 3]$ given by

$$s(x, y) = \begin{cases} 2, & \text{if } (x, y) \text{ is in } R_{11} = (0, 1) \times (-1, 0) \\ 3/2, & \text{if } (x, y) \text{ is in } R_{12} = (0, 1) \times (0, 3) \\ 4/3, & \text{if } (x, y) \text{ is in } R_{21} = (1, 2) \times (-1, 0) \\ 1, & \text{if } (x, y) \text{ is in } R_{22} = (1, 2) \times (0, 3) \\ 0, & \text{otherwise.} \end{cases}$$

This is the step function of Example 2, and its graph is given in Figure 18.4. The areas of the open subrectangles of the indicated partition of R are

$$A_{11} = 1, \quad A_{12} = 3, \quad A_{21} = 1 \quad \text{and} \quad A_{23} = 3.$$

We define the integral of s over R to be the sum of the volumes of the parallelepipeds bounded by the xy-plane and the graph of s.

$$\int_R s = s_{11}A_{11} + s_{12}A_{12} + s_{21}A_{21} + s_{22}A_{22}$$
$$= 2 \cdot 1 + (3/2) \cdot 3 + (4/3) \cdot 1 + 1 \cdot 3$$
$$= 65/6.$$

This illustration should clarify the following general definition.

DEFINITION 18.4 (Integral of an $\mathcal{R}^2 \to \mathcal{R}$ Step Function.) Let $s : R \to \mathcal{R}$ be a step function on $R = [a, b] \times [c, d]$, where $s(x, y) = s_{jk}$ for (x, y) in R_{jk} and $\{R_{jk} \mid j = 1, \ldots, m; k = 1, \ldots, n\}$ is a set of open subrectangles of R on which s is constant. The *integral of s over R* is denoted by $\int_R s$ and is defined by

(18.1)
$$\int_R s = \sum_{j=1, k=1}^{m, n} s_{jk}A_{jk},$$

where A_{jk} is the area of R_{jk}: $A_{jk} = (x_j - x_{j-1})(y_k - y_{k-1})$.

Certain important properties follow from the preceding definition in a manner similar to those for the $\mathcal{R} \to \mathcal{R}$ integral. They are collected in the following theorem (with proof left to the reader).

THEOREM 18.5 (Properties of the Integral of an $\mathcal{R}^2 \to \mathcal{R}$ Step Function.) Let s and t be step functions on $R = [a, b] \times [c, d]$ and $r \in \mathcal{R}$. Let $\mathring{R}$ denote the open rectangle $(a, b) \times (c, d)$,

(i) (Additive Property) $\quad \int_R (s + t) = \int_R s + \int_R t$
(ii) (Homogeneous Property) $\quad \int_R rs = r \int_R s$
(iii) (Order-Preserving Property)
   if $sX \le tX$ for all $X \in R$, then $\quad \int_R s \le \int_R t$.
(iv) If $R = R_1 \cup R_2$ and $\mathring{R}_1 \cap \mathring{R}_2 = \varnothing$, then

$$\int_{R_1} s + \int_{R_2} s = \int_{R_1 \cup R_2} s = \int_R s$$

(v) $|s|$ is a step function on R and

$$\left| \int_R s \right| \le \int_R |s|.$$

The reader should note that Properties (i) and (ii) are the defining properties of linear transformation. Thus the operation of integration over R which we symbolize by $\int_R$ is a linear transformation from the vector space of step functions on R into the vector space of real numbers:

$$\int_R : \{\text{step functions on } R\} \to \mathcal{R}.$$

4. THE UPPER AND LOWER INTEGRALS OF A BOUNDED $\mathcal{R}^2 \to \mathcal{R}$ FUNCTION

The integral of an $\mathcal{R}^2 \to \mathcal{R}$ function f over a rectangle shall be defined in terms of the integrals of upper and lower step functions for f. We therefore require f to admit upper and lower step functions; that is, we require f to be bounded. We now give the sequence of ideas leading to the concept of the upper and the lower integrals of an $\mathcal{R}^2 \to \mathcal{R}$ function.

DEFINITION 18.6 Let $f : R \to \mathcal{R}$ where $R = [a, b] \times [c, d]$.

(i) (Boundedness) f *is bounded on* R if and only if there is a positive number M such that

$$-M \leq f(x, y) \leq M \text{ for all } (x, y) \text{ in } R.$$

(ii) (Upper Step Function for f) A step function u defined on R is called an *upper step function for* f if and only if

$$u(x, y) \geq f(x, y) \text{ for each } (x, y) \text{ in } R.$$

(iii) (Lower Step Function for f) A step function l defined on R is called a *lower step function for* f if and only if

$$l(x, y) \leq f(x, y) \text{ for each } (x, y) \text{ in } R.$$

If f is an $\mathcal{R}^2 \to \mathcal{R}$ function which is bounded on the rectangle R, then there will be at least one lower step function for f and at least one upper step function for f, namely, the l and the u given by

$$l(x, y) = -M \qquad \text{and} \qquad u(x, y) = M \qquad \text{for each } (x, y) \text{ in } R.$$

This fact gives meaning to the following definition.

DEFINITION 18.7 (Upper and Lower Integrals of Bounded $\mathcal{R}^2 \to \mathcal{R}$ Functions.) Let $f : R \to \mathcal{R}$ be bounded on the rectangle $R = [a, b] \times [c, d]$.

(i) The *lower integral of f on R* is denoted by $\underline{\int}_R f$ and defined by

$$\underline{\int}_R f = \mathrm{lub}\left\{ \int_R l \,\middle|\, l \text{ is a lower step function for } f \text{ on } R \right\}.$$

(ii) The *upper integral of f on R* is denoted by $\overline{\int}_R f$ and defined by

$$\overline{\int}_R f = \mathrm{glb}\left\{ \int_R u \,\middle|\, u \text{ is an upper step function for } f \text{ on } R \right\}.$$

For each bounded function f the upper and lower integrals are defined and possess the properties listed in the next theorem. Again we shall omit the proof since it is a straightforward generalization of the proofs of the corresponding properties given in Chapter 8 for the $\mathcal{R} \to \mathcal{R}$ integral.

THEOREM 18.8 Let $f : R \to \mathcal{R}$ be bounded on $R = [a, b] \times [c, d]$ and let $r \in \mathcal{R}$; then

(i) $\underline{\int} f \le \overline{\int} f.$

(ii) $\overline{\int}(f + g) \le \overline{\int} f + \overline{\int} g.$

(iii) $\underline{\int}(f + g) \ge \underline{\int} f + \underline{\int} g.$

(iv) $\left. \begin{array}{l} \underline{\int}(rf) = r\underline{\int} f \\ \overline{\int}(rf) = r\overline{\int} f \end{array} \right\}$ if $r \ge 0.$

(v) $\left. \begin{array}{l} \underline{\int}(rf) = r\overline{\int} f \\ \overline{\int}(rf) = r\underline{\int} f \end{array} \right\}$ if $r < 0.$

5.   INTEGRABLE BOUNDED $\mathcal{R}^2 \to \mathcal{R}$ FUNCTIONS

Every $\mathcal{R}^2 \to \mathcal{R}$ function which is bounded on a rectangle $R = [a, b] \times [c, d]$ will possess both an upper and a lower integral on R. For many $\mathcal{R}^2 \to \mathcal{R}$ functions the upper and the lower integrals over R will be unequal. Functions for which the upper and lower integrals are equal form a special class of functions which are said to be integrable on R.

DEFINITION 18.9 Let $f : R \to \mathcal{R}$ be bounded on $R = [a, b] \times [c, d]$. We say that f is *integrable on R* if and only if the upper and lower integrals of f on R have the same value ($\underline{\int}_R f = \overline{\int}_R f$). Moreover, the common value of the upper and lower integrals of f on R is defined to be the *integral of f on R* and is denoted by $\int_R f$:

$$\int_R f = \int_{\underline{R}} f = \int_R^{-} f.$$

The next example shows that not all bounded functions on a rectangle R are integrable.

Example 3. Let $f:[0, 1] \times [0, 1] \to \mathfrak{R}$ be defined by

$$f(x, y) = \begin{cases} 1, & \text{if } x \text{ is rational.} \\ 0, & \text{if } x \text{ is irrational.} \end{cases}$$

Since every subrectangle of $R = [0, 1] \times [0, 1]$ will contain points for which x is rational and points for which x is irrational, it follows that every lower step function l for f is such that

$$l(x, y) \leq 0,$$

and every upper step function u for f is such that

$$u(x, y) \geq 1.$$

Thus,

$$\int_{\underline{R}} f = \text{glb } \{l \mid l \text{ is a lower step function for } f\} \leq \int_R 0 = 0$$

and

$$\int_R^{-} f = \text{lub } \{u \mid u \text{ is an upper step function for } f\} \geq \int_R 1 = 1.$$

Consequently, $\int_{\underline{R}} f \leq 0 < 1 \leq \int_R^{-} f$ and we must have $\int_{\underline{R}} f \neq \int_R^{-} f$; that is, f is not integrable on R. ∎

Definition 18.4, page 719, defines the basic concept of the integral $\int_R s$ of a step function s over a rectangle R. Definition 18.9 applied to s is a *logically different* definition for an integral of a step function s. We leave it to the reader to establish that the numbers arrived at are identical:

(18.2) $$\int_R s = \int_R s$$

<small>(using Definition 18.4)   (using Definition 18.9)</small>

The use of the former definition is easier than the latter, and is the preferred technique to be used for evaluation of the integral of a step function.

Unfortunately, Definition 18.9 is the only one which is applicable to functions f which are not step functions. Application of this definition is, at best, tedious. Therefore, in the next section we develop a technique which, for a wide class of functions, enables one to use the integration techniques previously developed for integrating $\mathfrak{R} \to \mathfrak{R}$ functions over an interval.

Various properties of the integral of an $\mathfrak{R}^2 \to \mathfrak{R}$ function over a rectangle R follow as a consequence of Definition 18.9 and the properties of integrals of $\mathfrak{R}^2 \to \mathfrak{R}$ step functions, given in Theorem 18.5, page 719. We list some of these in the following theorem.

THEOREM 18.10   (Properties of the Integrals of $\mathfrak{R}^2 \to \mathfrak{R}$ Functions.) Let f and g be integrable on the rectangle $R = [a, b] \times [c, d]$ and let $r \in \mathfrak{R}$. Let $\mathring{R}$ again denote the open rectangle $(a, b) \times (c, d)$.

(i) (Additive Property) $f + g$ is integrable on R and

$$\int_R (f + g) = \int_R f + \int_R g$$

(ii) (Homogeneous Property) rf is integrable on R and

$$\int_R rf = r \int_R f.$$

(iii) (Order Property) If $fX \le gX$ on R, then

$$\int_R f \le \int_R g.$$

(iv) (Additivity of Domain) If $R = R_1 \cup R_2$ and $\mathring{R}_1 \cap \mathring{R}_2 = \emptyset$, then f is integrable on R_1 and R_2 and

$$\int_{R_1} f + \int_{R_2} f = \int_{R_1 \cup R_2} f = \int_R f.$$

(v) $|f|$ is integrable on R and

$$\left| \int_R f \right| \le \int_R |f|.$$

It is these properties which will enable us to use the integral of $\mathfrak{R}^2 \to \mathfrak{R}$ functions to define a volume function.

EXERCISES

1. Prove the assertion of Equation 18.2, p. 722.

In Exercises 2 to 6, use Definition 18.9 and Theorem 18.5 to establish the indicated properties of Theorem 18.10.

2. Establish property (i), Theorem 18.10.

3. Establish property (ii), Theorem 18.10.

4. Establish property (iii), Theorem 18.10.

5. Establish property (iv), Theorem 18.10.

6. Establish property (v), Theorem 18.10.

6. THE INTEGRAL OF $\mathcal{R}^2 \to \mathcal{R}$ FUNCTIONS AS A REPEATED INTEGRAL

We restrict our attention to $\mathcal{R}^2 \to \mathcal{R}$ functions and $\mathcal{R} \to \mathcal{R}$ functions. We shall refer to the integral of $\mathcal{R}^2 \to \mathcal{R}$ functions given in Definition 18.9 as the $\mathcal{R}^2$-integral. The integral of an $\mathcal{R} \to \mathcal{R}$ function given in Chapter 8 will be referred to as the $\mathcal{R}^1$-integral.

Consider an $\mathcal{R}^2 \to \mathcal{R}$ step function $s : R \to \mathcal{R}$ defined on the rectangle $R = [a, b] \times [c, d]$. The $\mathcal{R}^2$-integral of s over R may be evaluated by means of two $\mathcal{R}^1$-integrals.

THEOREM 18.11 Let $s : R \to \mathcal{R}$ be an $\mathcal{R}^2 \to \mathcal{R}$ step function defined on the rectangle $R = [a, b] \times [c, d]$. Then

$$(18.3) \qquad \int_R s = \int_{x=a}^{b} \left(\int_{y=c}^{d} s(x, y) \right)$$

$$= \int_{y=c}^{d} \left(\int_{x=a}^{b} s(x, y) \right).$$

NOTE. The meaning of the symbolism is made clear in the proof and again in the text following Example 5.

Proof.

(i) We first consider a step function which is constant over the entire rectangle R:

$$s(x, y) = s_0 \quad \text{for } (x, y) \text{ in } R.$$

Then

$$\int_R s =^1 \text{area of } R \cdot s_0 = (b - a) \cdot (d - c) \cdot s_0$$

$$=^2 \int_{x=a}^{b} (d - c) \cdot s_0$$

$$=^3 \int_{x=a}^{b} \left(\int_{y=c}^{d} s_0 \right)$$

$$=^4 \int_{x=a}^{b} \left(\int_{y=c}^{d} s(x, y) \right)$$

(1) Definition of integral over R.
(2) Since $(d - c) \cdot s_0$ is a constant, we have $\int_{x=a}^{b} (d - c) \cdot s_0 = [(d - c) \cdot s_0 \cdot x]_{x=a}^{b}$
       $= (d - c) \cdot s_0 \cdot (b - a)$.
(3) Since s_0 is a constant, $\int_{y=c}^{d} s_0 = [s_0 y]_{y=c}^{d} = (d - c) \cdot s_0$.
(4) Since $s(x, y) = s_0$ for (x, y) in R.

Similarly, we can show that

$$\int_R s = \int_{y=c}^{d} \left(\int_{x=a}^{b} s(x,y) \right).$$

The theorem is therefore proved for functions which are constant on the entire rectangle of integration.

(ii) Consider now the general case. Let s be a step function on R which is constant on the open subrectangles $R_{jk} = (x_{j-1}, x_j) \times (y_{k-1}, y_k)$ of the partition $\{x_0, x_1, \ldots, x_m\} \times \{y_0, y_1, \ldots, y_n\}$. Using the definition of the integral of a step function and applying the result of part (i), we have

$$\int_R s(x,y) =^1 \sum_{j=1,k=1}^{m,n} s_{jk} A_{jk}, \quad A_{jk} = \text{area } R_{jk}$$

$$=^2 \sum_{k=1}^{n} \left(\sum_{j=1}^{m} s_{jk} A_{jk} \right)$$

$$=^3 \sum_{k=1}^{n} \left(\sum_{j=1}^{m} \int_{R_{jk}} s(x,y) \right)$$

$$=^4 \sum_{k=1}^{n} \left(\sum_{j=1}^{m} \left(\int_{y=y_{k-1}}^{y_k} \left(\int_{x=x_{j-1}}^{x_j} s(x,y) \right) \right) \right)$$

$$=^5 \sum_{k=1}^{n} \left(\int_{y=y_{k-1}}^{y_k} \left(\sum_{j=1}^{m} \int_{x=x_{j-1}}^{x_j} s(x,y) \right) \right)$$

$$=^6 \sum_{k=1}^{n} \left(\int_{y=y_{k-1}}^{y_k} \left(\int_{x=x_0}^{x_m} s(x,y) \right) \right)$$

$$=^7 \int_{y=y_0}^{y_n} \left(\int_{x=x_0}^{x_m} s(x,y) \right)$$

$$=^8 \int_{y=c}^{d} \left(\int_{x=a}^{b} s(x,y) \right). \quad \blacksquare$$

(1) Definition of $\mathcal{R}^2 \to \mathcal{R}$ integral.
(2) Systematizing the preceding summation by first determining subtotals for a given y-column and then the grand total of all y-columns.
(3) Using the fact that $s_{jk} A_{jk} = \int_{R_{jk}} s(x,y)$.
(4) Applying the result of part (i).
(5) Using the additivity property of the $\mathcal{R}^1 \to \mathcal{R}$ integral $\int_{y=y_{k-1}}^{y_k}$.
(6) Using the interval additivity of the $\mathcal{R}^1 \to \mathcal{R}$ integral $\int_{x=x_{j-1}}^{x_j}$.
(7) Using interval additivity of the integral $\int_{y=y_{k-1}}^{y_k}$.
(8) Replacing x_0, x_m, y_0, y_n by a, b, c, and d, respectively.

The preceding theorem for $\mathcal{R}^2 \to \mathcal{R}$ step functions shows that the $\mathcal{R}^2$-integral of a step function is equivalent to a twofold repeated $\mathcal{R}^1$-integral. That is, for a step function s defined on $R = [a, b] \times [c, d]$, each of the integrals

$$\int_R s(x,y), \quad \int_{y=c}^{d} \left(\int_{x=a}^{b} s(x,y) \right), \quad \text{and} \quad \int_{x=a}^{b} \left(\int_{y=c}^{d} s(xy) \right)$$

exists and all are equal. Unfortunately, this result does not carry over to general $\mathcal{R}^2 \to \mathcal{R}$ functions. The following example will illustrate this situation.

Example 4. Let $f:\mathcal{R}^2 \to \mathcal{R}$ be given by

$$f(x,y) = \begin{cases} 0, & \text{if } x \text{ is rational} \\ 1, & \text{if } x \text{ is irrational}, y \geq 0 \\ -1, & \text{if } x \text{ is irrational}, y < 0. \end{cases}$$

Then

$$\int_{y=-1}^{1} f(x,y) = \begin{cases} \displaystyle\int_{-1}^{1} 0 = 0, & \text{if } x \text{ is rational} \\ \displaystyle\int_{-1}^{0} (-1) + \int_{0}^{1} (1) = -1 + 1 = 0, & \text{if } x \text{ is irrational} \end{cases}$$

$$= 0, \quad \text{for all values of } x.$$

Consequently, the repeated $\mathcal{R}^1$-integral given by

$$\int_{x=0}^{1} \left(\int_{y=-1}^{1} f(x,y) \right) = \int_{y=1}^{1} (0) = 0$$

is defined. However, we now show that the $\mathcal{R}^2$-integral $\int_{R} f(x,y)$ is not defined for the corresponding rectangle $R = [0, 1] \times [-1, 1]$. Let $R = R_1 \cup R_2$ as shown in Figure 18.5. Every lower step function for f can have value at most -1 on R_1 and value at most 0 on R_2 (Why?). Thus

$$\int_{R} l = \int_{R_1} l + \int_{R_2} l \leq \int_{R_1} (-1) + \int_{R_2} (0) = -1.$$

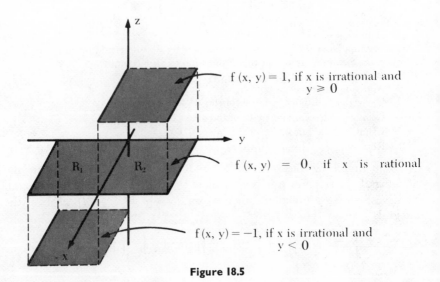

f (x, y) = 1, if x is irrational and y ⩾ 0

f (x, y) = 0, if x is rational

f(x, y) = −1, if x is irrational and y < 0

Figure 18.5

On the other hand, every upper step function u for f must be such that

$$\int_R u = \int_{R_1} u + \int_{R_2} u \geq \int_{R_1} (0) + \int_{R_2} (1) = 1. \quad \text{(Why?)}$$

Thus the upper and lower integrals for f on R are 1 and -1, respectively. Therefore the $\mathcal{R}^2$-integral $\int_R f$ is not defined. For this choice of f we see that the last integral of Theorem 18.11 exists, but the first does not. ∎

Given a function $f : R \to \mathcal{R}$ defined on the rectangle $R = [a, b] \times [c, d]$, the last example shows that the repeated $\mathcal{R}^1$-integral

$$\int_{x=a}^{b} \left(\int_{y=c}^{d} f(x, y) \right)$$

may exist when the $\mathcal{R}^2$-integral $\int_R f$ does not. Hence, we may not treat every $\mathcal{R}^2$-integration problem as if it were just a repeated $\mathcal{R}^1$-integral problem.

Suppose, however, that we know by some means that the $\mathcal{R}^2$-integral $\int_R f$ is defined but we do not know its value. Suppose, further, that we are able to evaluate one of the repeated integrals of Equations (18.3) with s replaced by f. May we then assert that the value of the repeated integral is also the value of $\int_R f$? Happily, this question can be answered affirmatively.

THEOREM 18.12 Let $f : \mathcal{R}^2 \to \mathcal{R}$ be $\mathcal{R}^2$-integrable over the rectangle $R = [a, b] \times [c, d]$. Either of the two formulas

(i) $\int_R f = \int_{y=c}^{d} (\int_{x=a}^{b} f(x, y))$

(ii) $\int_R f = \int_{x=a}^{b} (\int_{y=c}^{d} f(x, y))$

is valid provided that the repeated $\mathcal{R}^1$-integral on the right is defined.

Before giving a proof of this theorem let us see how it can be applied.

Example 5. Theorem 18.13, to be given later, assures us that the function given by $f(x, y) = xy$ is integrable over the rectangle $R = [1, 3] \times [2, 6]$, but does not tell us how to determine the value of $\int_R f$. We therefore consider the repeated $\mathcal{R}^1$-integral

$$\int_{x=1}^{3} \left(\int_{y=2}^{6} f(x, y) \right) = \int_{x=1}^{3} \left(\int_{y=2}^{6} xy \right)$$

$$=^1 \int_{x=1}^{3} \left[x \frac{y^2}{2} \right]_{y=2}^{6}$$

$$=^2 \int_{x=1}^{3} 16x$$

$$=^3 \left[16 \frac{x^2}{2} \right]_{x=1}^{3}.$$

$$=^4 64. \quad ∎$$

(1) Using the $\mathcal{R}^1$-integration to evaluate $\int_{y=2}^{6} xy$. y is the variable of integration as indicated by the subscript on the symbol of integration. x is considered as a constant for this integration.

(2) Evaluating $\left[x\dfrac{y^2}{2} \right]_{y=2}^{6}$.

(3) Using $\mathcal{R}^1$-integration techniques. x is the variable of integration.

(4) Evaluating $\left[16\dfrac{x^2}{2} \right]_{x=1}^{3}$.

This repeated integral is clearly defined. Using Formula (ii) of Theorem 18.12, we conclude that

$$\int_R f(xy) = \int_{x=1}^{3} \left(\int_{y=2}^{6} xy \right) = 64.$$

Let us also look at the alternate repeated integral:

$$\int_{y=2}^{6} \left(\int_{x=1}^{3} xy \right) = \int_{y=2}^{6} \left[y\frac{x^2}{2} \right]_{x=1}^{3} = \int_{y=2}^{6} 4y$$

$$= \left[4\frac{y^2}{2} \right]_{2}^{6} = 64.$$

In this instance we can use Formula (i) to evaluate $\int_R f(x,y) = 64$. ∎

The example just given illustrates what is intended by the symbolism

$$\int_{x=a}^{b} \left(\int_{y=c}^{d} f(x,y) \right).$$

The first integral to be carried out is $\int_{y=c}^{d} f(x,y)$. For that integral, x is to be considered a constant and y is the variable of integration. The result of that integration is a function $F(x) = \int_{y=c}^{d} f(x,y)$ which does not involve y at all but may involve x. The variable y has been "integrated out."

Similar comments apply to $\int_{y=c}^{d} \left(\int_{x=a}^{b} f(x,y) \right)$.

We now give the proof of the theorem.

Proof of Theorem 18.12. Assume that $\int_R f$ and the repeated integral $\int_{y=c}^{d} \left(\int_{x=a}^{b} f(x,y) \right)$ are both defined. For every lower step function l and upper step function u for f we have

$l(x,y) \le f(x,y) \le u(x,y)$ for each (x,y) in R

$$\Rightarrow^1 \int_{x=a}^{b} l(x,y) \le \int_{x=a}^{b} f(x,y) \le \int_{x=a}^{b} u(x,y) \quad \text{for each } y \text{ in } [c,d].$$

$$\Rightarrow^2 \int_{y=c}^{d} \left(\int_{x=a}^{b} l(x,y) \right) \le \int_{y=c}^{d} \left(\int_{x=a}^{b} f(x,y) \right) \le \int_{y=c}^{d} \left(\int_{x=a}^{b} u(x,y) \right)$$

$$\Rightarrow^3 \int_R l(x,y) \le \int_{y=c}^{d} \left(\int_{x=a}^{b} f(x,y) \right) \le \int_R u(x,y).$$

From this last relation we conclude that

$$\underline{\int_R} f(x,y) \le \int_{y=c}^d \left(\int_{x=a}^b f(x,y) \right) \le \overline{\int_R} f(x,y). \quad \text{(Check this!)}$$

But since f is integrable on R, the extreme members of this last chain are both equal to $\int_R f(x,y)$. Consequently, the middle member must also be equal to $\int_R f(x,y)$. This is Formula (i) of the theorem.

In a similar manner we can establish Formula (ii) of the theorem. ∎

(1) Order properties of the $\mathscr{R}^1$-integral with integration variable x. There is one relation for each value of y. Therefore, y is regarded as a constant in the integration.

(2) Order properties of the $\mathscr{R}^1$-integral with integration variable y. ´x has been "integrated out" by the preceding integration.

(3) Applying Theorem 18.11 for the step functions l and u.

NOTE. It is customary to omit the parentheses in the notation for repeated $\mathscr{R}^1$-integrals. Henceforth, we shall use this abbreviated notation:

$$\int_{x=a}^b \int_{y=c}^d f(x,y) \quad \text{shall denote} \quad \int_{x=a}^b \left(\int_{y=c}^d f(x,y) \right)$$

$$\int_{y=c}^d \int_{x=a}^b f(x,y) \quad \text{shall denote} \quad \int_{y=c}^b \left(\int_{x=a}^b f(x,y) \right).$$

Example 6. Given that the function defined by $f(x,y) = xe^y$ is integrable on $R = [0,2] \times [0,1]$, let us determine $\int_R f$ by repeated $\mathscr{R}^1$-integration.

$$\int_R xe^y =^1 \int_{x=0}^2 \int_{y=0}^1 xe^y =^2 \int_{x=0}^2 [xe^y]_0^1$$

$$=^3 \int_{x=0}^{2.} x(e-1) = \left[\frac{x^2}{2} (e-1) \right]_{x=0}^2$$

$$= 2(e-1). \quad \blacksquare$$

(1) Using Theorem 18.12, assuming that the repeated integral exists.

(2) Effecting the integration with respect to y by the usual $\mathscr{R}^1$-integral techniques.

(3) Evaluation at the endpoints of the interval of integration. Note that the y is "integrated out," and a function of only x remains.

EXERCISES

Assume that each of the following $\mathscr{R}^2$-integrals exists and use the method of repeated $\mathscr{R}^1$-integration to evaluate them.

1. $\int_R (x+y), \quad R = [0,3] \times [-2,2]$.

2. $\int_R x \sin y, \quad R = [-1,1] \times [0,\pi]$.

3. $\displaystyle\int_R \frac{y}{x^2}$, $R = [1, 2] \times [0, 2]$.

4. $\displaystyle\int_R (x + x^2)$, $R = [-1, 2] \times [-1, 2]$.

5. $\displaystyle\int_R e^{x+2y}$, $R = [\ln 2, \ln 8] \times [1, 2]$.

6. $\displaystyle\int_R \frac{1}{xy}$, $R = [1, 2] \times [1, 2]$.

7. $\displaystyle\int_R x^2 \cos y$, $R = [0, 2] \times \left[0, \frac{\pi}{2}\right]$.

8. $\displaystyle\int_R \ln xy$, $R = [1, 2] \times [1, 2]$.

ANSWERS

1. $\displaystyle\int_R (x + y) = \int_{x=0}^3 \int_{y=-2}^2 (x + y) = \int_{x=0}^3 \left[xy + \frac{y^2}{2}\right]_{y=-2}^2 = \int_{x=0}^3 4x = 18.$

3. $\displaystyle\int_R \frac{y}{x^2} = \int_{x=1}^2 \int_{y=0}^2 \frac{y}{x^2} = \int_{x=1}^2 \left[\frac{y^2}{2x^2}\right]_{y=0}^2 = \int_{x=1}^2 \frac{2}{x^2} = 1.$

5. $3e^2(e^2 - 1)$.   7. $\frac{8}{3}$.

7. INTEGRABILITY OF CONTINUOUS AND BOUNDED MONOTONIC FUNCTIONS

In the preceding section, we showed how certain $\mathcal{R}^2 \to \mathcal{R}$ functions may be integrated by the technique of repeated $\mathcal{R}^1$-integration. Before the method is applicable we need to be assured that the $\mathcal{R}^2 \to \mathcal{R}$ function under consideration has an $\mathcal{R}^2$-integral defined for it. In this section we will state two theorems which provide a large class of $\mathcal{R}^2$-integrable functions. The first of these is

THEOREM 18.13 (Integrability of Continuous $\mathcal{R}^2 \to \mathcal{R}$ Functions.) Every $\mathcal{R}^2 \to \mathcal{R}$ function which is defined and continuous over a rectangle $R = [a, b] \times [c, d]$ is integrable over R.

We do not give the proof of this theorem but encourage the interested reader to construct his own or to search the literature for one.

There is another class of $\mathcal{R}^2 \to \mathcal{R}$ functions for which the $\mathcal{R}^2$-integral over a rectangle $R = [a, b] \times [c, d]$ is assured. This is the class of piecewise monotonic functions on R. We proceed to define this class.

DEFINITION 18.14 (Piecewise Monotonic $\mathcal{R}^2 \to \mathcal{R}$ Functions.)
Let $f: R \to \mathcal{R}$ be defined on the closed rectangle $R = [a, b] \times [c, d]$.
For each value y_0 of y define the $\mathcal{R} \to \mathcal{R}$ function $S_{y_0}(x) = f(x, y_0)$.
For each value x_0 of x define the $\mathcal{R} \to \mathcal{R}$ function $S_{x_0}(y) = f(x_0, y)$.
  (i) f is said to be *monotonic in x* if and only if S_{y_0} is monotonic for each
      choice of y_0.
  (ii) f is said to be *monotonic in y* if and only if $S_{x_0}(y)$ is monotonic for
       each choice of x_0.
  (iii) f is said to be *monotonic on R* if and only if f is monotonic in both
        x and y on R.
  (iv) f is said to be *piecewise monotonic on R* if and only if f is monotonic
       on each closed rectangle of some partition of R.

Example 7. Let f be the function given by

$$f(x, y) = x^2 \cos y$$

(a) On the rectangle $[0, 1] \times \left[0, \dfrac{\pi}{2}\right]$ we note that:

  (i) $S_{y_0}(x) = x^2 \cos y_0$ is monotonic (increasing) in x;
  (ii) $S_{x_0} = x_0^2 \cos y$ is monotonic (decreasing) in y.
Therefore,

  (iii) f is monotonic (increasing in x, decreasing in y) on $[0, 1] \times \left[0, \dfrac{\pi}{2}\right]$.

(b) On the rectangle $[0, 1] \times \left[\dfrac{\pi}{2}, \pi\right]$ we note that:

  (i) $S_{y_0}(x) = x^2 \cos y_0$ is monotonic (decreasing) in x;
  (ii) $S_{x_0}(y) = x_0^2 \cos y$ is monotonic (decreasing) in y.
Therefore,
  (iii) f is monotonic (decreasing in x, decreasing in y) on

$$[0, 1] \times \left[\dfrac{\pi}{2}, \pi\right].$$

(c) From the above we also know that:
  (iv) f is piecewise monotonic on the rectangle $[0, 1] \times [0, \pi]$.
      (Why?) ■

EXERCISE

Show that the function of Example 7 is piecewise monotonic on the
rectangle $[-1, 1] \times [0, 2\pi]$.

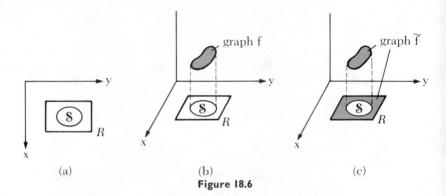

(a)

Figure 18.6

8. $\mathcal{R}^2$-INTEGRALS OVER MORE GENERAL REGIONS

Thus far we have restricted the domain of integration for $\mathcal{R}^2 \to \mathcal{R}$ functions to rectangles; more specifically, to rectangles whose sides are parallel to the coordinate axes. It is an easy matter to extend the definition to include more general domains of integration.

Let S be any subset of $\mathcal{R}^2$ which is contained in some rectangle $R = [a, b] \times [c, d]$. (See Figure 18.6a.) For each function $f: S \to \mathcal{R}$ we introduce the function $\tilde{f}: R \to \mathcal{R}$ such that $\tilde{f} = f$ on S, but $\tilde{f} = 0$ on $R \setminus S$. (See Figure 18.6b and c.) We then define the integrability of f in terms of $\tilde{f}$. Formally this leads to:

DEFINITION 18.15 (Integrals on More General Regions.) Let $f: S \to \mathcal{R}$ be a bounded function on the subset S of $\mathcal{R}^2$. Let S be contained in the rectangle $R = [a, b] \times [c, d]$, and define $\tilde{f}: R \to \mathcal{R}$ such that

$$\tilde{f}(x, y) = \begin{cases} f(x, y), & \text{if } (x, y) \in S \\ 0, & \text{otherwise.} \end{cases}$$

We say that f *is integrable on* S if and only if $\tilde{f}$ is integrable on R. Moreover, if f is integrable on S we denote the *integral of* f *on* S by $\int_S f$ and define it to be the integral of $\tilde{f}$ on R:

$$\int_S f = \int_R \tilde{f}.$$

If S is contained in one rectangle R, it will be contained in infinitely many others as well. Nonetheless, it should be clear that $\int_S f$ will be assigned the same value no matter which rectangle is employed. (See Exercise 11, p. 738.)

The use of repeated $\mathfrak{R}^1$-integration to evaluate an $\mathfrak{R}^2$-integral can also be used for certain more general regions, but not for all. In the following we first describe the regions to which repeated integration can be applied, and then illustrate the technique.

DEFINITION 18.16 (Regions for Repeated Integration.) Let S be a subset of $\mathfrak{R}^2$.

(i) S is called a *region of Type I* if and only if there are continuous functions ψ_1 and ψ_2 such that
$$S = \{(x,y) \mid c \leq y \leq d \quad \text{and} \quad \psi_1(y) \leq x \leq \psi_2(y)\}$$

(ii) S is called a *region of Type II* if and only if there are continuous functions φ_1 and φ_2 such that
$$S = \{(x,y) \mid a \leq x \leq b \quad \text{and} \quad \varphi_1(x) \leq y \leq \varphi_2(x)\}$$

(iii) S is called a region of Type III if and only if it is the union of a finite number of regions of Type I or Type II.

We illustrate below various shapes for each type of region. Of course, there are many other possibilities.

Regions of Type I—subsets of $\mathfrak{R}^2$ which can be regarded as generated by a moving line segment $\overline{x_1 x_2}$ (of perhaps variable length) which is kept parallel to the x-axis.

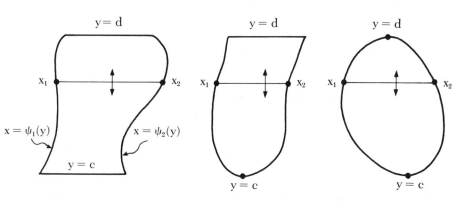

Regions of Type II—subsets of $\mathfrak{R}^2$ which can be regarded as generated by a moving line segment $\overline{y_1 y_2}$ which is kept parallel to the y-axis.

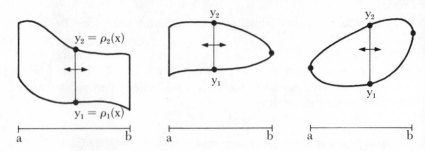

Regions of Type III—subsets of $\mathcal{R}^2$ which are composed of two or more regions of Type I or of Type II.

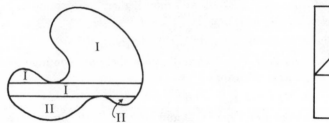

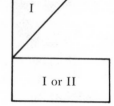

The importance of regions of Type I or Type II is due to the following theorems.

THEOREM 18.17 (Repeated Integration over Regions of Type I or Type II.) Let f be a function defined and bounded on a region $\mathcal{S}$ of $\mathcal{R}^2$.

(i) If $\mathcal{S}$ is a region of Type I given by

$$\mathcal{S} = \left\{ (x,y) \,\middle|\, \begin{array}{l} c \le y \le d \quad \text{and} \\ \psi_1(y) \le x \le \psi_2(y) \end{array} \right\}$$

then

$$\int_{\mathcal{S}} f(x,y) = \int_{y=c}^{d} \int_{x=\psi_1(y)}^{x=\psi_2(y)} f(x,y),$$

assuming that all integrals exist.

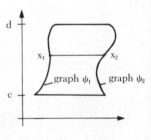

(ii) If $\mathcal{S}$ is a region of Type II given by

then $\mathcal{S} = \left\{ (x,y) \,\middle|\, \begin{array}{l} a \le x \le b \quad \text{and} \\ \varphi_1(x) \le y \le \varphi_2(x) \end{array} \right\}$

$$\int_{\mathcal{S}} f(x,y) = \int_{x=a}^{b} \int_{y=\phi_1(x)}^{\phi_2(y)} f(x,y),$$

assuming that all integrals exist.

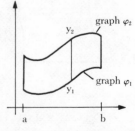

Proof. Let S be of Type I. Then S is a subset of a rectangle $R = [a, b] \times [c, d]$ where $a \leq \min \psi_1(y)$ and $b \geq \max \psi_2(y)$. Introduce the function $\tilde{f}$ defined on R such that

$$\tilde{f}(x, y) = \begin{cases} f(x, y), & \text{when } (x, y) \in S \\ 0, & \text{otherwise.} \end{cases}$$

Now,

$$\int_{x=a}^{b} \tilde{f}(x, y) =^1 \int_{x=a}^{\psi_1(y)} \tilde{f}(x, y) + \int_{x=\psi_1(y)}^{\psi_2(y)} \tilde{f}(x, y) + \int_{x=\psi_2(y)}^{b} \tilde{f}(x, y)$$

$$=^2 0 + \int_{x=\psi_1(y)}^{\psi_2} f(x, y) + 0$$

$$= \int_{x=\psi_1(y)}^{\psi_2(x)} f(x, y).$$

Hence,

$$\int_S f =^3 \int_R \tilde{f}$$

$$=^4 \int_{y=c}^{d} \int_{x=a}^{b} \tilde{f}(x, y)$$

$$=^5 \int_{y=c}^{d} \int_{x=\psi_1(y)}^{\psi_2(y)} f(x, y).$$

Part (i) is thereby established. Part (ii) can be established in a similar manner. ■

(1) Interval additivity for $\mathcal{R}^1$-integrals.
(2) Since $\tilde{f}(x, y) = \begin{cases} f(x, y), & \text{when } \psi_1(y) \leq x \leq \psi_2(y) \text{ and} \\ 0, & \text{otherwise.} \end{cases}$
(3) Definition of $\int_S f$, Definition 18.15, p. 732.
(4) Evaluating $\int_R \tilde{f}$ by repeated $\mathcal{R}^1$-integration. This is permissible since all integrals are assumed to be defined.
(5) Substitution from preceding equality chain.

Example 8. Let us find the integral $\int_S xy$ over the set S pictured in the diagram at the right.
(a) We first shall consider S to be a Type I region so that we may use the method of repeated integration:

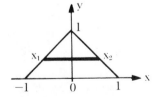

$$\int_S xy =^1 \int_{y=0}^{1} \int_{x=x_1}^{x_2} xy$$

$$=^2 \int_{y=0}^{1} \int_{x=y-1}^{1-y} xy = \int_{y=0}^{1} \left[\frac{x^2 y}{2} \right]_{x=y-1}^{1-y}$$

$$= \int_{y=0}^{1} 0 = 0.$$

(1) For regions of Type I or of Type II the first integration is with respect to the variable which moves you from one end of the generating line to the other. For Type I regions, this is from some x_1 to some x_2 as indicated.

    The second integration involves the remaining variable (y for Type I regions). The limits for this variable are the values between which the generating line must move to sweep out the entire region. In this case the limits are $y = 0$ and $y = 1$.

(2) The generating line segment has its left end point on the line for which $x = y - 1$, and its right end point on the line for which $x = 1 - y$.

(b) Let us now regard $\mathcal{S}$ as the union of two Type II regions as indicated on the right. For $\mathcal{S}_1$ we have

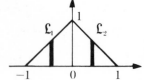

$$\int_{\mathcal{S}_1} xy =^3 \int_{x=-1}^{0} \int_{y=y_1}^{y_2} xy =^4 \int_{x=-1}^{0} \int_{y=0}^{1+x} xy$$

$$= \int_{x=-1}^{0} \left[x\frac{y^2}{2} \right]_0^{1+x} = \int_{x=-1}^{0} \frac{x(1+x)^2}{2}$$

$$= \frac{1}{2} \int_{x=-1}^{0} (x + 2x^2 + x^3) = \frac{1}{2} \left[\frac{x^2}{2} + \frac{2x^3}{3} + \frac{x^4}{4} \right]_{-1}^{0}$$

$$= \frac{-1}{24}$$

For $\mathcal{S}_2$ compute

$$\int_{\mathcal{S}_2} xy =^5 \int_{x=0}^{1} \int_{y=0}^{y=1-x} xy = \int_{x=0}^{1} \left[\frac{xy^2}{2} \right]_0^{1-x}$$

$$= \frac{1}{2} \int_{x=0}^{1} (x - 2x^2 + x^3) = \frac{1}{24} .$$

Thus

$$\int_{\mathcal{S}} xy = \int_{\mathcal{S}_1 \cup \mathcal{S}_2} xy = \int_{\mathcal{S}_1} xy + \int_{\mathcal{S}_2} xy = \frac{-1}{24} + \frac{1}{24} = 0. \quad \blacksquare$$

(3) For a type II region, the first integration is with respect to the variable y which moves along the generating line from some y_1 to some y_2. The limits for the second integration are the extreme values of x between which the generating line must move so as to sweep out the entire region, namely, $x = 0$ to $x = 1$.

(4) On $\mathcal{S}_1$ the end points of the generating line are $y = 0$ and the point on the line $y = 1 + x$.

(5) On $\mathcal{S}_2$ the end points of the generating line are $y = 0$ and the point on the line $y = 1 - x$.

Example 9. We shall evaluate $\int_S xy^2$ where S is the right half of the circle bounded by $x^2 + y^2 = 1$. We shall use repeated integration in which the x integration is effected first.

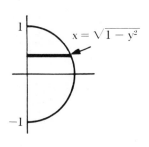

$x = \sqrt{1 - y^2}$

$$\int_S xy^2 = \int_{y=-1}^1 \int_{x=0}^{\sqrt{1-y^2}} xy^2 = \int_{-1}^1 \left[\frac{x^2}{2} y^2\right]_0^{\sqrt{1-y^2}}$$

$$= \frac{1}{2} \int_{-1}^1 (1 - y^2) y^2 = \frac{1}{2}\left[\frac{y^3}{3} - \frac{y^5}{5}\right]_{-1}^1$$

$$= \frac{2}{15}. \quad \blacksquare$$

(1) For this Type I region, the generating line segment has its left end point on $x = 0$, and its right end point on the line $x = \sqrt{1 - y^2}$. The line sweeps from $y = -1$ to $y = 1$.

EXERCISES

In each Exercise sketch the set S and evaluate the $\mathcal{R}^2$-integrals by repeated $\mathcal{R}^1$-integration.

1. $\int_S xy^2,$ $S = \{(x, y) \mid 0 \le x \le 1, x^2 \le y \le x\}.$

2. $\int_S (2x + x^3),$ $S = \{(x, y) \mid y \le x \le \sqrt{y}, 0 \le y \le 1\}.$

3. $\int_S y,$ $S = \{(x, y) \mid \frac{1}{3}y \le x \le 3, 0 \le y \le 9\}.$

4. $\int_S \frac{x}{y},$ $S = \{(x, y) \mid \ln y \le x \le 2, 1 \le y \le e^2\}.$

5. $\int_S x,$ $S = \{(x, y) \mid x^2 + y^2 \le 4, 0 \le y \le 2, x > 0\}.$

6. $\int_S \frac{2y - 1}{x + 1},$ $S = \{(x, y) \mid 0 \le x \le 2, 2x - 4 \le y \le 0\}.$

7. $\int_S (xy + 1),$ $S = $ the set bounded by the lines $x - 2y + 2 = 0,$

$$x + 3y - 3 = 0, \quad y = 0.$$

8. $\int_S (2x + 5y),$ $S = \{(x, y) \mid 3 \le x \le 5; -x \le y \le x^2\}.$

9. $\displaystyle\int_S e^{x+y}$, $S = \{(x,y) \mid 0 \le x \le 3y,\ \ln 2 \le y \le 1\}$.

10. $\displaystyle\int_S y^2 e^{xy}$, $S = \{(x,y) \mid 0 \le x \le 5y,\ 1 \le y \le 2\}$.

11. Let $f:S \to \mathcal{R}$ be a bounded function on the subset S of $\mathcal{R}^2$. Let S be contained in the rectangles R and R_1. Define the functions $\tilde{f}:R \to \mathcal{R}$ and $\tilde{f}_1:R \to \mathcal{R}$ such that

$$\tilde{f}(x,y) = \tilde{f}_1(x,y) = \begin{cases} f(x,y), & \text{if } (x,y) \in S, \\ 0, & \text{otherwise.} \end{cases}$$

Show that $\int_R \tilde{f} = \int_{R_1} \tilde{f}_1$ and thereby prove that $\int_S \tilde{f}$ given in Definition 18.15 does not depend upon the choice of the rectangle containing S.

ANSWERS

1. $\frac{1}{40}$.   **3.** $\frac{81}{2}$.   **5.** $\frac{8}{3}$.   **7.** $\frac{65}{24}$.

9. $2 - \dfrac{e^4}{4} + e$.

9. A VOLUME FUNCTION FOR $\mathcal{R}^3$

Let $f:\mathcal{R}^2 \to \mathcal{R}$ be integrable over the rectangle R in $\mathcal{R}^2$ and be such that $f(x,y) \ge 0$ for each (x,y) in $\mathcal{R}^2$. Consider the solid $\mathcal{M}$ in $\mathcal{R}^3$ which has R as its base and *graph f* as its summit. (See Figure 18.7a.) In Section 5 we noted that the integral $\int_R f$ is intuitively regarded as giving the volume of $\mathcal{M}$: $V(\mathcal{M}) = \int_R f$. We now give the formal definition of a volume function in $\mathcal{R}^3$ for more general solids $\mathcal{M}$. Refer to Figure 18.7b as you read the definition.

DEFINITION 18.18 (Volume in $\mathcal{R}^3$ as an $\mathcal{R}^2$-Integral.) Let $f:\mathcal{R}^2 \to \mathcal{R}$ and $g:\mathcal{R}^2 \to \mathcal{R}$ be integrable over the set S. Let $\mathcal{M}$ denote the set of points (x,y,z) in $\mathcal{R}^3$ lying between *graph f* and *graph g* and above or below the point (x,y) of S. The volume $\mathcal{M}$, denoted by $V(\mathcal{M})$, is defined by

$$V(\mathcal{M}) = \int_S |f - g|.$$

This definition is a straight-forward generalization of the concept of area in $\mathcal{R}^2$ given in Definition 8.26, p. 372. We leave it to the interested reader to show that the volume function thus defined has all the properties listed in Section 1, p. 716.

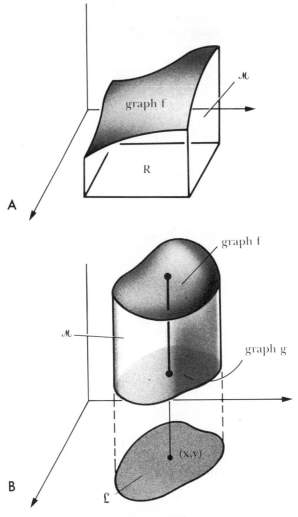

A

B

Figure 18.7

Example 10. Consider the solid $\mathcal{M}$ bounded by the parabolic cylinder $z = 1 - y^2$, the plane $z = 1 - x$ and the plane $x = 1$. The solid is sketched in Figure 18.8a. The upper surface of the solid is the parabolic cylinder given by $z_2 = 1 - y^2$, the lower surface is the plane given by $z_1 = 1 - x$. The volume of $\mathcal{M}$ is given by

$$V(\mathcal{M}) = \int_{S} (z_2 - z_1).$$

where S is the vertical projection of $\mathcal{M}$ on the xy-plane. Inspection of Figure 18.8b shows that S is a parabolic sector bounded by the line for which $x = 1$ and the projection of the curve of intersection of the parabolic cylinder $z = 1 - y^2$ and the plane $z = 1 - x$. This curve is given by $1 - y^2 = 1 - x$, or, equivalently, $x = y^2$. (See Figure 18.8b.) Thus, using repeated integration,

$$V(\mathcal{M}) =^1 \int_{y=-1}^{1} \int_{x=y^2}^{1} (z_2 - z_1)$$

$$=^2 \int_{y=-1}^{1} \int_{x=y^2}^{1} ((1 - y^2) - (1 - x))$$

$$= \int_{y=-1}^{1} \int_{x=y^2}^{1} (x - y^2)$$

$$=^3 \int_{y=-1}^{1} \left[\frac{x^2}{2} - y^2 x \right]_{x=y^2}^{1}$$

$$= \int_{y=-1}^{1} \left(\frac{1}{2} - y^2 + \frac{y^4}{2} \right)$$

$$= \left[\frac{1}{2} y - \frac{y^3}{3} + \frac{y^5}{10} \right]_{-1}^{1}$$

$$= \frac{8}{15} \text{ units}^3. \quad \blacksquare$$

(1) The limits on the integrals are obtained from Figure 18.8b.
(2) Replacing z_2 by $1 - y^2$ and z_1 by $1 - x$.
(3) Carrying out the integration with respect to x, considering y as constant.

The expression $A(y) = \frac{1}{2} - y^2 + \frac{y^4}{2}$ appearing in the preceding example is the result of the integration with respect to x. For each y the number $A(y)$ is the area of a triangular cross-section of $\mathcal{M}$. This cross-section for $y = 0$ is shown in Figure 18.8a.

EXERCISES

In Exercises 1 to 4 sketch the solid $\mathcal{M}$ and determine its volume.

1. $\mathcal{M} = \{(x, y, z) \mid x^2 + y^2 \le 16, x^2 + z^2 \le 16\}$.
2. $\mathcal{M} = \{(x, y, z) \mid x^2 + y^2 \le 25, x \le z \le 5\}$.
3. $\mathcal{M} = \{(x, y, z) \mid y^2 + z^2 \le 2, x^2 - y^2 \le 2\}$.
4. $\mathcal{M} = \{(x, y, z) \mid x^2 + y^2 \le 4z, x^2 + y^2 \le 8y, z \ge 0\}$.

In Exercises 5 to 8 sketch the solid $\mathcal{M}$ which is bounded by the surface(s) described implicitly by the equation(s) given and compute the volume.

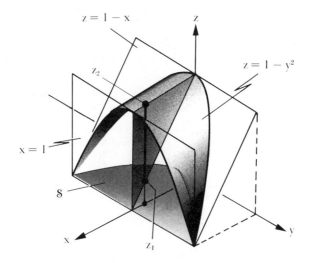

A

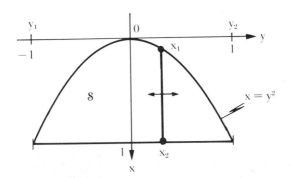

B

Figure 18.8

5. $\mathcal{M}$ = the set bounded by the coordinate planes and the three planes $y = 4$, $x = 3$, $x + y + z = 7$.

6. $\mathcal{M}$ = the set bounded by the coordinate planes, the plane $2x + 3y = 6$ and the surface $z = x + xy$.

7. $\mathcal{M}$ = the set bounded by the ellipsoid $x^2 + y^2 + 4z^2 = 4$.

8. $\mathcal{M}$ = the set bounded by the plane $z - y = 2$ and the paraboloid $4x^2 + y^2 = 4z$.

ANSWERS

1. $V(\mathcal{M}) = 2 \displaystyle\int_{x=-4}^{4} \int_{y=-\sqrt{16-x^2}}^{\sqrt{16-x^2}} \sqrt{16 - x^2} = 341\frac{1}{3}$ units^3.

3. $V(\mathcal{M}) = \displaystyle\int_{y=-\sqrt{2}}^{\sqrt{2}} \int_{z=-\sqrt{2-y^2}}^{\sqrt{2-y^2}} (x_2 - x_1) = 2 \int_{y=-\sqrt{2}}^{\sqrt{2}} \int_{z=-\sqrt{2-y^2}}^{\sqrt{2-y^2}} \sqrt{2 + y^2}$

$\qquad = 4 \displaystyle\int_{y=-\sqrt{2}}^{\sqrt{2}} \sqrt{4 - y^2} = 8 + 4\pi \text{ units}^3.$

(Hint: Use Formula 32, Appendix VI, with $u = y$ and $p = 2$.)

5. $V(\mathcal{M}) = 42 \text{ units}^3.$

7. $V(\mathcal{M}) = \displaystyle\int_{x=-2}^{2} \int_{y=-\sqrt{4-x^2}}^{\sqrt{4-x^2}} \sqrt{4 - x^2 - y^2} = \dfrac{32\pi}{3} \text{ units}^3.$

(Hint: Use Formula 32, Appendix VI, with $u = y$ and $p = \sqrt{4 - x^2}$.)

10. AREA AS AN $\mathcal{R}^2$-INTEGRAL

Let f and g be two $\mathcal{R} \to \mathcal{R}$ functions which are bounded and integrable on $[a, b]$. Further, let $f(x) \geq g(x)$ for each x in $[a, b]$. Let S denote the subset of $\mathcal{R}^2$ which is bounded by the lines $x = a$ and $x = b$ and the graphs of f and g. (See Figure 18.9.) By Definition 8.26, p. 372, the area of S, denoted by $A(S)$, is given by

(18.4) $\qquad\qquad A(S) = \displaystyle\int_{x=a}^{b} (f(x) - g(x)).$

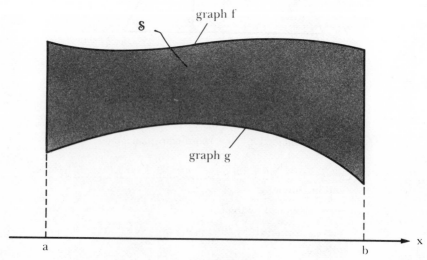

Figure 18.9

Let us introduce the unit function $\mathbb{1}_2$ for $\mathcal{R}^2$. This function is defined by

$$\mathbb{1}_2(x, y) = 1 \text{ for each } (x, y) \text{ in } \mathcal{R}^2.$$

Then we may write

$$\int_{x=a}^{b} (f(x) - g(x)) = \int_{x=a}^{b} \int_{y=g(x)}^{f(x)} 1$$

$$= \int_{x=a}^{b} \int_{y=g(x)}^{f(x)} 1_2$$

$$= \int_{S} 1_2,$$

provided this latter integral exists. Thus, finally, we may write (18.4) in the form

(18.5) $$A(S) = \int_{S} 1_2.$$

Now the integral of Equation 18.5 is an $\mathcal{R}^2$-integral whose existence is assured by the integrability of the functions f and g on $[a, b]$. Consequently, any subset S of $\mathcal{R}^2$ which possesses an area by virtue of Definition 8.26 will have the same area under the following Definition.

DEFINITION 18.19 (Area as an $\mathcal{R}^2$-Integral.) Let S be any subset of $\mathcal{R}^2$ such that the unit function 1_2 (given by $1_2(x, y) = 1$) is integrable over S. S is said to have *area* $A(S)$ where

$$A(S) = \int_{S} 1_2.$$

The area function given by this definition possesses all the properties that

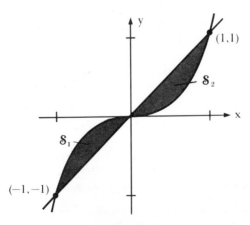

Figure 18.10

were given in Chapter 8, pp. 338–9. Moreover, it is more general than that given by Definition 8.26.

Example 11. Let S be the set in $\mathcal{R}^2$ bounded by the line given by $y - x = 0$ and the cubic curve given by $y = x^3$. (See Figure 18.10.)

By virtue of Definition 18.19 we may write at once:

$$A(S) = \int_S \mathbb{1}_2 = \int_{S_1} \mathbb{1}_2 + \int_{S_2} \mathbb{1}_2.$$

The decomposition of S into the union of the sets S_1 and S_2 enables us to evaluate this integral by means of repeated $\mathcal{R}^1$-integration:

$$A(S) = \int_S \mathbb{1}_2 = \int_{S_1} \mathbb{1}_2 + \int_{S_2} \mathbb{1}_2$$

$$= \int_{x=-1}^0 \int_{y=x}^{x^3} \mathbb{1}_2 + \int_{x=0}^1 \int_{y=x^3}^x \mathbb{1}_2$$

$$= \int_{x=-1}^0 [y]_{y=x}^{x^3} + \int_{x=0}^1 [y]_{y=x^2}^x$$

$$= \int_{x=-1}^0 (x^3 - x) + \int_{x=0}^1 (x - x^3)$$

$$= \left[\frac{x^4}{4} - \frac{x^2}{2}\right]_{-1}^0 + \left[\frac{x^2}{2} - \frac{x^4}{4}\right]_0^1$$

$$= \frac{1}{2} \text{ units}^2. \quad \blacksquare$$

(1) The limits are obtained from Figure 18.10.
(2) The function $\mathbb{1}_2$ is such that $\mathbb{1}_2(x, y) = 1$. Therefore

$$\int_y \mathbb{1}_2 = \int_y 1 = [y].$$

EXERCISES

In each of the following sketch the curve(s) in $\mathcal{R}^2$ described implicitly by the given equation(s) and use repeated integration to determine the area of the set S.

1. S is bounded by $y = x^2$ and $x = y^2$.

2. S is bounded by $x^2 + y^2 = 2y$.

3. S is bounded by $y = \sin \pi x$ and $y = 2x$.

4. S is bounded by $y = e^x$, $x = 0$, and $y = e^2$.

5. S is bounded by $x = \sin y$, $x = 3y$ and $y = \pi$.

6. S is bounded by $xy = 1$, $x = 2$, $x = 4$ and $y = 0$.

ANSWERS

1. $A(\mathcal{S}) = \int_{y=0}^{1} \int_{x=y^2}^{\sqrt{y}} \mathbb{1}_{\,2} = \int_{y=0}^{1} [x]_{x=y^2}^{\sqrt{y}} = \int_{y=0}^{1} (\sqrt{y} - y^2) = \frac{1}{3}$ units².

3. $A(\mathcal{S}) = \int_{x=-\frac{1}{2}}^{0} \int_{y=\sin \pi x}^{2x} \mathbb{1}_{\,2} + \int_{x=0}^{\frac{1}{2}} \int_{y=2x}^{\sin \pi x} \mathbb{1}_{\,2} = \frac{4 - \pi}{4\pi}$ units².

5. $A(\mathcal{S}) = \left(\dfrac{3\pi^2}{2} - 2 \right)$ units².

11. INTEGRATION OF $\mathcal{R}^n \rightarrow \mathcal{R}$ FUNCTIONS

The concept of the integral is readily generalized to $\mathcal{R}^n \rightarrow \mathcal{R}$ functions. One need only proceed in a manner similar to that used for the $\mathcal{R}^2$-integral:

    i) Consider a generalized rectangle, an n-dimensional box

$$\mathcal{B} = [a_1, b_1] \times [a_2, b_2] \times \cdots \times [a_n, b_n].$$

    ii) Define a partition of $\mathcal{B}$.
    iii) Define a step function on $\mathcal{B}$.
    iv) Define the integral of a step function on $\mathcal{B}$.
    v) For bounded functions $f: \mathcal{B} \rightarrow \mathcal{R}$ define the upper and lower integrals of f on $\mathcal{B}$.
    vi) Define the expression "f is integrable on $\mathcal{B}$" and the *integral* $\int_{\mathcal{B}} f$.
    vii) Extend the definition of the integral to more general regions in $\mathcal{R}^n$.
The details of this development parallel those of Sections 1 to 8 for the $\mathcal{R}^2$-integral. In particular, the $\mathcal{R}^n$ integral will have properties similar to those listed in Theorem 18.10, p. 723. One need only replace the rectangle R by the n-dimensional box $\mathcal{B}$ and regard f and g as functions from $\mathcal{B}$ into $\mathcal{R}$.

There is an important property that the $\mathcal{R}^2$- and the $\mathcal{R}^n$-integral share with the integral of $\mathcal{R} \rightarrow \mathcal{R}$ functions. We state it now, without proof, for later reference.

THEOREM 18.20 (Riemann Condition for $\mathcal{R}^n$ Integrability.) Let $f: \mathcal{B} \rightarrow \mathcal{R}$ be bounded on the n-dimensional box $\mathcal{B}$. Then f is integrable on $\mathcal{B}$ if and only if the following condition holds: for each real number r there is a lower step function l_r and an upper step function u_r for f such that

$$\int_{\mathcal{B}} u_r - \int_{\mathcal{B}} l_r < r.$$

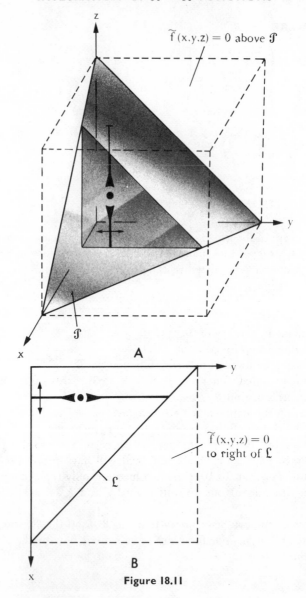

Figure 18.11

12. n-FOLD REPEATED INTEGRALS; VOLUME AS AN $\mathcal{R}^3$-INTEGRAL

Any $\mathcal{R}^n$-integral $\int_{\mathcal{B}} f$ over a box $\mathcal{B}$ may be evaluated by means of an *n-fold repeated integral*

$$\int_{x_1=a_1}^{b_1} \int_{x_2=a_2}^{b_2} \cdots \int_{x_n=a_n}^{b_n} f(x_1, x_2, \ldots x_n)$$

provided that this latter integral and the integral $\int_{\mathcal{B}} f$ *both exist.* If an $\mathcal{R}^n \to \mathcal{R}$ function f is known to be piecewise continuous or piecewise monotonic on $\mathcal{B}$, then f is integrable on $\mathcal{B}$. The proofs of these claims parallel the proofs for the special case of the $\mathcal{R}^2$-integral. We shall use these facts in the next example.

Example 12. Let $\mathcal{S}$ be the subset of $\mathcal{R}^3$ bounded by the three co-ordinate planes and the plane through $(1, 0, 0)$, $(0, 1, 0)$ and $(0, 0, 1)$. Let $f : \mathcal{R}^3 \to \mathcal{R}$ be given by $f(x, y, z) = x$. Since f is obviously continuous on $\mathcal{S}$ we know that $\int_{\mathcal{S}} f$ exists. We shall attempt to evaluate it by repeated integration. We may regard $\mathcal{S}$ as a subset of the 3-dimensional box $\mathcal{B} = [0, 1] \times [0, 1] \times [0, 1]$. (See Figure 18.11a.) Thus we may write

$$\int_{\mathcal{S}} f = \int_{\mathcal{B}} \tilde{f},$$

where

$$\tilde{f}X = \begin{cases} fX & \text{when } X \in \mathcal{S} \\ 0 & \text{otherwise.} \end{cases}$$

But we have

$$\int_{\mathcal{B}} \tilde{f} =^1 \int_{x=0}^1 \int_{y=0}^1 \int_{z=0}^1 \tilde{f}(x, y, z)$$

$$=^2 \int_{x=0}^1 \int_{y=0}^1 \int_{z=0}^{1-x-y} \tilde{f}(x, y, z)$$

$$=^3 \int_{x=0}^1 \int_{y=0}^{1-x} \int_{z=0}^{1-x-y} \tilde{f}(x, y, z)$$

$$=^4 \int_{x=0}^1 \int_{y=0}^{1-x} \int_{z=0}^{1-x-y} x. \quad\blacksquare$$

(1) Assuming the repeated integral exists.
(2) Since $\tilde{f}(x, y, z) = 0$ whenever (x, y, z) is above the plane $\mathcal{F}$ through $(1, 0, 0)$, $(0, 1, 0)$, $(0, 0, 1)$, that is, whenever $z > 1 - x - y$ (since $z = 1 - x - y$ is an explicit representation for $\mathcal{F}$).
(3) Since $\tilde{f}(x, y, z) = 0$ whenever (x, y, z) is such that $(x, y, 0)$ is to the right of the line $\mathcal{L}$ given by $y = 1 - x$. (See Figure 18.11b.)
(4) Since (x, y, z) is restricted to $\mathcal{S}$ by conditions (2) and (3) and thus $\tilde{f}(x, y, z) = f(x, y, z) = x$.

In Section 10 the concept of area in $\mathcal{R}^2$ (originally defined as an $\mathcal{R}^1$-integral) was redefined as an $\mathcal{R}^2$-integral. In Section 9 the concept of volume was defined as an $\mathcal{R}^2$-integral. A discussion similar to that of Section 10 leads to the following definition of volume in $\mathcal{R}^3$ as an $\mathcal{R}^3$-integral.

DEFINITION 18.21 (Volume as an $\mathcal{R}^3$-integral.) Let $\mathcal{S}$ be any subset of $\mathcal{R}^3$ such that the unit function $\mathbb{1}_3$ (given by $\mathbb{1}_3 (x, y, z) = 1$

for every (x, y, z) in $\mathfrak{R}^3$) is integrable over S. S is said to have *volume* $V(S)$ where

$$V(S) = \int_S \mathbb{1}_{\ 3}.$$

EXERCISES

Evaluate the following three-fold repeated integrals.

1. $\displaystyle\int_{x=0}^{1}\int_{y=x^2}^{x}\int_{z=0}^{xy} \mathbb{1}_{\ 3}.$

2. $\displaystyle\int_{x=0}^{\pi/2}\int_{y=0}^{2\cos x}\int_{z=0}^{y^2} y.$

3. $\displaystyle\int_{y=0}^{1}\int_{x=0}^{y}\int_{z=1}^{2} yz.$

4. $\displaystyle\int_{z=0}^{2}\int_{x=1}^{c}\int_{y=1}^{x} \frac{x}{y}.$

5. $\displaystyle\int_{z=-1}^{1}\int_{x=1}^{z^2}\int_{y=0}^{\pi/4} x\cos y.$

6. $\displaystyle\int_{x=1}^{4}\int_{y=1}^{x}\int_{z=x}^{y} (x + z).$

ANSWERS

1. $\frac{1}{24}.$ 3. $\frac{1}{2}.$ 5. $-\dfrac{2\sqrt{2}}{5}.$

19

INTEGRATION IN GENERAL COORDINATE SYSTEMS

1. POLAR COORDINATES IN $\mathscr{R}^2$

In many practical problems it is more convenient to represent points in $\mathscr{R}^2$ by coordinate systems other than the rectangular systems which we have previously adhered to. There are many alternative systems, each having its own range of application. Perhaps the most important of these is the system of polar coordinates which we now describe.

DEFINITION 19.1 (Polar Coordinates in $\mathscr{R}^2$.)

Let $P = (x, y)$ be a point of $\mathscr{R}^2$ other than the origin.

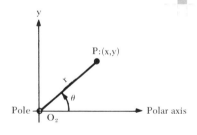

(i) The ordered pair (r, θ) of real numbers is called the *pair of polar coordinates for P* if and only if
(a) r is the distance from the origin to P, and
(b) θ is the radian measure of the angle from the positive x-axis to the line O_2P, where θ is in the interval given by $-\pi < \theta \le \pi$.
(ii) When using polar coordinates in $\mathscr{R}^2$, the origin O_2 is called the *pole* of the coordinate system and the positive x-axis is called the *polar axis* of the coordinate system.

The system of polar coordinates is, therefore, an assignment of a new ordered pair (r, θ) to each point (except O_2) of $\mathcal{R}^2$. We can think of the polar coordinate system as a function T from $\mathcal{R}^2$ into $\mathcal{U}^2 = \{(r, \theta) \mid r, \theta$ are real numbers$\}$. From the preceding definition we find that T is given by

$$(19.1) \qquad \binom{r}{\theta} = T(x, y) = \left(\left(\arccos \dfrac{x}{\sqrt{x^2 + y^2}} \right) \operatorname{sgn} y \right).$$

The symbol "sgn y" is an abbreviation for "sign y" and is defined as follows:

$$\operatorname{sgn} y = \begin{cases} +1 & \text{if } y \geq 0 \\ -1 & \text{if } y < 0. \end{cases}$$

The function T has an inverse which gives the (x, y) coordinates in terms of the polar coordinates. T^{-1} is given by

$$(19.2) \qquad \binom{x}{y} = T^{-1}(r, \theta) = \binom{r \cos \theta}{r \sin \theta}$$

Functions used to relate two different coordinate systems are frequently called coordinate *transformations*. Thus, Equation 19.1 is called the *polar coordinate transformation* for $\mathcal{R}^2$, and Equation 19.2 is the inverse transformation.

Example 1. (a) *Circles with center at the pole* and radius r_1 are given implicitly by

$$r = r_1.$$

(b) *Rays* or half lines from the pole making an angle of θ_1 with the polar axis are given by

$$\theta = \theta_1. \quad \blacksquare$$

Example 2.

(a) Circles passing through the pole and having center on the polar axis are given by

$$r = 2a \cos \theta, \quad a \neq 0.$$

Of course, the pole is not in the set of points satisfying this equation since it has no polar coordinate pair. The radius of the circle is $|a|$. The center is on the polar axis itself or on the axis extending to the left according as a is positive or negative.

That we indeed get a circle with this description is most easily seen by using the polar coordinate transformation to express the above relation in terms of x and y.

$$r = 2a \cos \theta \Leftrightarrow r^2 = 2ar \cos \theta, \quad r \neq 0$$

$$\left. \begin{aligned} &\Leftrightarrow x^2 + y^2 = 2ax \\ &\Leftrightarrow x^2 - 2ax + a^2 + y^2 = a^2 \\ &\Leftrightarrow (x - a)^2 + y^2 = a^2 \end{aligned} \right\} \quad x^2 + y^2 \neq 0.$$

(b) $r = 2a \sin \theta$ represents a circle through O_2 with radius $|a|$ and center on the y-axis. ∎

Example 3. *Conic Sections.* Let $\mathcal{L}$ be a line in $\mathcal{R}^2$ which is perpendicular to the extended polar axis at a point $2p$ units to the left of the pole O_2. Let C be the set of those points $P:(r, \theta)$ such that

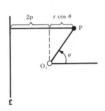

$$(19.3) \qquad d(O_2, P) = e \cdot d(\mathcal{L}, P), \quad e \geq 0.$$

Now $d(O_2, P) = r$ and $d(\mathcal{L}, P) = 2p + r \cos \theta$. Thus (r, θ) is a point of C if and only if

$$r = e(2p + r \cos \theta)$$

or, equivalently,

$$(19.4) \qquad r = \frac{2ep}{1 - e \cos \theta}.$$

When we choose $e = 1$, Equation 19.3 is the defining relation for a parabola given in Chapter 7, p. 310. We ask the reader to show that (i) when $e < 1$ the curve C is an ellipse with one focus at O_2 and (ii) when $e > 1$ the curve C is one branch of a hyperbola with one focus at O_2. In all cases, the line $\mathcal{L}$ is called a directrix for the conic C, and the ratio $e = d(O_2, P)/d(\mathcal{L}, P)$ is called the eccentricity of the conic. ∎

The definition of polar coordinates given in Definition 19.1 ensures that each point (x, y) of $\mathcal{R}^2$ (except O_2) is assigned one and only one polar coordinate pair (r, θ). This is essential to the treatment of integration given in the next section, but is rather limiting in other aspects. For example, it is the restriction of r to positive values only which prevents Equation 19.4 from giving both branches of the hyperbola, when $e > 1$. Accordingly, an extended system of polar coordinates is frequently used.

DEFINITION 19.2 (Extended Polar Coordinates for $\mathfrak{R}^2$.) Let $P = (x, y)$ be any point of $\mathfrak{R}^2$. An ordered pair (r, θ) of real numbers is an *extended polar coordinate pair* for P if and only if

$$x = r \cos \theta \quad \text{and} \quad y = r \sin \theta$$

In extended polar coordinates, no restrictions are placed on r or on θ. Thus the origin has extended polar coordinates $(0, \theta)$ where any value for θ is permissible. Similarly, the point

$$(x, y) = (1, 0)$$

has extended polar coordinates

$$(r, \theta) = \begin{cases} (1, 0) \\ (-1, \pi), \\ (1, 2\pi), \\ \cdot \\ \cdot \\ \cdot \end{cases}$$

The point P of $\mathfrak{R}^2$ having an extended polar coordinate pair involving a negative value for r can be located geometrically as follows:

(i) Determine the half line $\mathcal{L}$ which makes an angle θ with the polar axis

(ii) Extend $\mathcal{L}$ backwards through the pole O_2.

(iii) P is the point on this extension which is $|r|$ units from O_2.

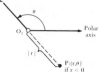

The values of θ may lie outside the $(-\pi, \pi]$ range also, but this should present no difficulty in locating a point P having (r, θ) as its extended polar coordinate pair.

Example 4. Sketch the set of points P whose extended polar coordinates satisfy the equation $r = \sin 2\theta$. We begin by constructing a table of solutions, selecting various values for θ and computing the corresponding value of r.

θ	0	$\pi/6$	$\pi/4$	$\pi/3$	$\pi/2$	$2\pi/3$	$3\pi/4$	$5\pi/6$
2θ	0	$\pi/3$	$\pi/2$	$2\pi/3$	π	$4\pi/3$	$3\pi/2$	$5\pi/3$
$r = \sin 2\theta$	0	0.87	1	0.87	0	$-.87$	-1	$-.87$

In Figure 19.1 we have plotted the points corresponding to the above table and sketched the graph obtained if the intermediate values of θ were used. As θ increases from 0, the point (r, θ) moves along the curve from the pole in the direction of the arrows. As θ increases from $5\pi/6$ to 2π the point

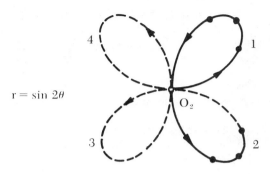

$r = \sin 2\theta$

Figure 19.1

(r, θ) will trace out loop 2 and then loops 3 and 4 as shown in Figure 19.1. Permitting θ to increase still further will cause P to retrace the "four-leafed rose" in the directions of the arrows.

To describe the curve in Figure 19.1 by means of rectangular coordinates, we note that

$$r = \sin 2\theta \Leftrightarrow r = 2 \sin \theta \cos \theta$$
$$\Leftrightarrow r^3 = 2 r \sin \theta\, r \cos \theta$$
$$\Leftrightarrow^1 r^3 = 2xy$$
$$\Leftrightarrow^2 (x^2 + y^2)^3 = 2xy \quad \text{or} \quad -(x^2 + y^2)^3 = 2xy$$
$$\Leftrightarrow (x^2 + y^2)^3 = 2xy \quad \text{or} \quad (x^2 + y^2)^3 = -2xy.$$

(1) Since $y = r \sin \theta$ and $x = r \cos \theta$ (Definition 19.2).
(2) Since r may be negative we have $r = \sqrt{x^2 + y^2}$ or $r = -\sqrt{x^2 + y^2}$.

Since $(x^2 + y^2)^3$ is always non-negative the first equation holds only when $2xy$ is also non-negative, and thus gives only loops 1 and 3 of Figure 19.1. Loops 2 and 4 are given by the latter of the two equations. ■

EXERCISES

In Exercises 1 to 12 sketch the set of points whose extended polar coordinates satisfy the given equation.

1. Cardioid: $r = a(1 + \cos \theta)$.   2. Lemniscate: $r^2 = 2a^2 \cos 2\theta$.

3. Hyperbolic spiral: $r\theta = a$.   4. Parabolic spiral: $r = a\theta^2$.

5. $r = 1 - 2 \cos \theta$.   6. $r = \sin \dfrac{\theta}{2}$.

7. $r = 2 \cos \theta$.   8. $r = 2 \sec \theta$.

9. $r \sin \theta = -1$.   10. $r = \sin 3\theta$.

11. $r = \sin 4\theta$.   12. $r^2 = \sin 3\theta$.

In Exercises 13 to 18, transform the given equation to polar coordinates.

13. $x^2 + y^2 = 4$.   14. $x^2 - 2x + y^2 = 0$.

15. $x^2 - y^2 = 1$.   16. $xy = 3$.

17. $y^2 = 8(2 - x)$.   18. $x^2 - xy - y^2 = 0$.

In Exercises 19 to 22, transform the given equation to rectangular coordinates.

19. $r = a \cot \theta$.   20. $r = \dfrac{a}{1 - \sin \theta}$.

21. $r^2 - 2r \sin \theta - \cos \theta = 0$.   22. $r^2 - 2 \tan \theta - 1 = 0$.

23. Express Equation 19.4 in terms of x and y and show that the equation represents (i) an ellipse with a focus at O_2, if $e < 1$ and (ii) a hyperbola with a focus at O_2, if $e > 1$.

ANSWERS

13. $r = 2$.   **15.** $r^2 \cos 2\theta = 1$.   **17.** $r = r \cos \theta - 4$.

19. $x^2 y^2 + y^4 = a^2 x^2$.   **21.** $(x^2 + y^2 - 2y)^2 (x^2 + y^2) = y^2$.

23. Hint: Use the technique of Example 17, p. 325.

2. INTEGRATION IN POLAR COORDINATES

One of the principal uses of polar coordinates (and other systems of coordinates) is in the simplification of $\mathcal{R}^2$-integration problems. Let $\mathcal{U}^2$ denote polar coordinate space; that is, the set of ordered pairs (r, θ) which are polar coordinates for some point of $\mathcal{R}^2$. Let $T:\mathcal{R}^2 \to \mathcal{U}^2$ denote the polar coordinate transformation given by Equation 19.1. $\mathcal{U}^2$ may be regarded as a subset of $\mathcal{R}^2$, but is best regarded as an entirely separate space, as pictured in Figure 19.2-b. The basic region for integration in $\mathcal{U}^2$ is a coordinate rectangle $R_T = [r_1, r_2] \times [\theta_1, \theta_2]$ (just as the basic region

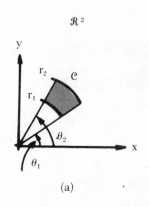

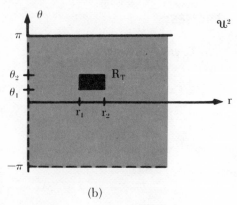

(a)   (b)

Figure 19.2

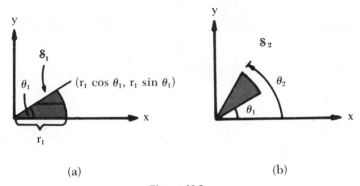

Figure 19.3

for integration in $\mathcal{R}^2$ is a coordinate rectangle $R = [a, b] \times [c, d]$). The set of points in $\mathcal{R}^2$ having polar coordinates in R_T is the set $\mathcal{C} = T^{-1}(R_T)$ pictured in Figure 19.2-a. We refer to each such $\mathcal{C}$ as a T-cell in $\mathcal{R}^2$. We shall show that there is a close relation between integrals over a T-cell $\mathcal{C}$ in $\mathcal{R}^2$ and integrals over the corresponding coordinate rectangle R_T in $\mathcal{U}^2$.

First we show that the area of a T-cell $\mathcal{C}$ (which is, by definition, the integral $\int_{\mathcal{C}} \mathbb{1}_2$) is expressible as an integral over R_T. Consider the area of the set $\mathcal{S}_1$ in $\mathcal{R}^2$ which is bounded by the circle $r = r_1$ and the lines $\theta = 0$ and $\theta = \theta_1$ as shown in Figure 19.3-a.

Now

$$A(\mathcal{S}_1) = \int_{\mathcal{S}_1} \mathbb{1}_2 = \int_{y=0}^{r_1 \sin \theta_1} \int_{x=y \,\mathrm{ctn}\, \theta_1}^{\sqrt{r_1{}^2 - y^2}} \mathbb{1}_2$$

$$= \int_{y=0}^{r_1 \sin \theta_1} \left(\sqrt{r_1^2 - y^2} - y \cot \theta_1 \right)$$

$$= \left[1/2 \left(y\sqrt{r_1^2 - y^2} + r_1^2 \arcsin \frac{y}{r_1} \right) - \frac{y^2}{2} \cot \theta_1 \right]_{y=0}^{r_1 \sin \theta_1}$$

$$= 1/2 r_1^2 \theta_1$$

$$= \int_{\theta=0}^{\theta_1} \int_{r=0}^{r_1} r.$$

The reader should be able to use this result to show that the area of the set $\mathcal{S}_2$ pictured in Figure 19.3-b is given by

$$A(\mathcal{S}_2) = \int_{\theta=0}^{\theta_2} \int_{r=0}^{r_1} r - \int_{\theta=0}^{\theta_1} \int_{r=0}^{r_1} r = \int_{\theta=\theta_1}^{\theta_2} \int_{r=0}^{r_1} r$$

and, consequently, that the area of the T-cell $\mathcal{C}$ in Figure 19.2-a is

$$(19.5) \quad A(\mathcal{C}) = \int_{\mathcal{C}} \mathbb{1}_2 = \int_{\theta=\theta_1}^{\theta_2} \int_{r=0}^{r_2} r - \int_{\theta=\theta_1}^{\theta_2} \int_{r=0}^{r_1} r = \int_{\theta=\theta_1}^{\theta_2} \int_{r=r_1}^{r_2} r$$

$$= \int_{R_T} r \quad \text{where } R_T = T\mathcal{C}$$

$$= \int_{T\mathcal{C}} r.$$

In Section 4 we shall show that this same relation holds for more general domains of integration. That is:

If $\mathcal{S}$ is any domain of integration in $\mathcal{R}^2$ such that $A(\mathcal{S})$ is defined and $T : \mathcal{R}^2 \to \mathcal{U}^2$ is the polar coordinate transformation given by Equation 19.1, then

$$(19.6) \qquad\qquad A(\mathcal{S}) = \int_{\mathcal{S}} \mathbb{1}_2 = \int_{T\mathcal{S}} r.$$

The first integration in Equation 19.6 is in $\mathcal{R}^2$ and is the integral defining $A(\mathcal{S})$, whereas the second integration is in the polar-coordinate space $\mathcal{U}^2$.

Also, in Section 4 we further generalize Equation 19.6 as follows:

Let $f(x, y)$ be integrable over the domain $\mathcal{S}$ of $\mathcal{R}^2$. Let $T : \mathcal{R}^2 \to \mathcal{U}^2$ be the polar coordinate transformation on $\mathcal{R}^2$. Then

$$(19.7) \qquad\qquad \int_{\mathcal{S}} f(x, y) = \int_{T\mathcal{S}} f \circ T^{-1}(r, \theta) r.$$

provided that the second integral exists. This formula corresponds to the change of variable formula for $\mathcal{R}^1$-integrals obtainable from Equation (9.9), p. 395:

$$\int_{x_1}^{x_2} f(x) = \int_{T(x_1)}^{T(x_2)} (f \circ T^{-1}) \cdot (T^{-1})'(u), \quad \text{where } u = T(x).$$

(See Exercise 10.) We shall, therefore refer to Equation (19.7) as the *change of variables formula for polar coordinates* in $\mathcal{R}^2$.

Let us now give examples of the use of the change of variables formula (19.7).

Example 5. We reconsider the determination of the area of sector of a circle of radius r_1 and central angle θ_1. In the preceding development this area was determined by an integration in rectangular coordinates. We now use polar coordinates. Figure 19.4-a shows a sector $\mathcal{S}$ whose area is to be determined. Figure 19.4-b shows the image of $\mathcal{S}$ under the polar coordinate transformation (which we denote by T), that is, the rectangle $T\mathcal{S} = (0, r_1] \times [\theta, \theta_1]$. Thus

$$A(\mathcal{S}) = \int_{\mathcal{S}} \mathbb{1}_2 = \int_{T\mathcal{S}} r.$$

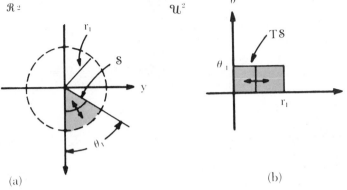

Figure 19.4

We shall evaluate this last integral by iterated integration in $\mathcal{U}^2$. For purposes of integration we note that we may replace TS, which is open at $r = 0$, by the closed rectangle $[0, r_1] \times [0, \theta_1]$. In Figure 19.4-b we have indicated that we shall integrate first with respect to θ and then with respect to r.

(19.8)
$$A(S) = \int_{r=0}^{r_1} \int_{\theta=0}^{\theta_1} r$$

$$= \int_{r=0}^{r_1} [r\theta] \, {}_{\theta=0}^{\theta_1} = \int_{r=0}^{r_1} r\theta_1$$

$$= \left[\frac{r^2}{2} \theta_1 \right]_{r=0}^{r_1}$$

$$= \tfrac{1}{2} r_1^2 \theta_1.$$

This, of course, is the same expression obtained previously. The viewpoint is different and the work simplified. We point out that the integral of Equation 19.8 may be set up directly from Figure 19.4-a without introducing Figure 19.4-b. We regard the area S as swept out by a circular arc. The integration with respect to θ gives the angular extremities of the arc at each r-position. The integration with respect to r gives the extremes between which the arc must move to sweep out the entire region of integration. ■

Example 6. Determine the area of one leaf S of the four leafed rose $r = \sin 2\theta$. The first quadrant leaf is that for which θ is in the interval $[0, \pi/2]$. This leaf is pictured in Figure 19.5-a. We shall set up the iterated polar coordinate integral directly from the $\mathcal{R}^2$ diagram. The leaf S is regarded as "swept out" by a radial line of varying length rotating around the pole. The first integration moves us from one extremity of the radial line, $r = 0$, to the other, $r = \sin 2\theta$. The limits on the second integration are the extreme values of θ between which the radial line must move, $\theta = 0$

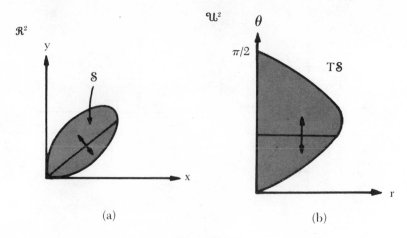

Figure 19.5

to $\theta = \pi/2$. The area is therefore

$$A(S) = \int_{TS} r = \int_{\theta=0}^{\pi/2} \int_{r=0}^{\sin 2\theta} r.$$
$$= \int_{\theta=0}^{\pi/2} \left[\frac{r^2}{2}\right]_{r=0}^{\sin 2\theta}$$
$$= \int_{\theta=0}^{\pi/2} \frac{\sin^2 2\theta}{2} = \left[\frac{\theta}{4} - \frac{\sin 4\theta}{16}\right]_{\theta=0}^{\pi/2}$$
$$= \pi/8.$$

The actual domain of integration in polar coordinate space is shown in Figure 19.5-b. ∎

Example 7. We now determine the volume of the conical wedge $\mathcal{W}$ illustrated in Figure 19.6-a. $\mathcal{W}$ is the set bounded above by the sphere of

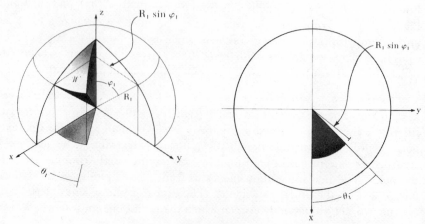

Figure 19.6

radius R_1, center at O_2, bounded below by a cone around the z-axis, vertex at O_2 and semi-vertical angle ϕ_1, and contained between the xz-plane and the plane through the z-axis making an angle of θ_1 with the xz-plane. The projection of this volume on the xy-plane is a circular sector S of radius $R_1 \sin \phi_1$ and central angle θ_1 as shown in Figure 19.6-b. The volume may be computed as the integral

$$V(W) = \int_S z_2(x, y) - z_1(x, y)$$

where $z = z_1(x, y)$ describes the conical surface and $z = z_2(x, y)$ describes the sphere. The sphere is the set of points such that $x^2 + y^2 + z^2 = R_1^2$ or, equivalently, $z^2 = R_1^2 - (x^2 + y^2)$; thus, $z = \sqrt{R_1^2 - (x^2 + y^2)}$ for the upper hemisphere and hence

$$z_2(x, y) = \sqrt{R_1^2 - (x^2 + y^2)}.$$

The point (x, y, z) is on the cone if and only if $z / \sqrt{x^2 + y^2} = \cot \phi_1$. Thus $z_1(x, y) = \sqrt{x^2 + y^2} \cot \phi_1$. Introducing polar coordinates, we may write

$$z_1 = r \cot \phi_1, \quad z_2 = \sqrt{R_1^2 - r^2}.$$

Thus

$$V(W) = \int_S (\sqrt{R_1^2 - (x^2 + y^2)} - \sqrt{x^2 + y^2} \cot \phi_1)$$

$$= \int_{TS} (\sqrt{R_1^2 - r^2} - r \cot \phi_1)$$

$$= \int_{\theta=0}^{\theta_1} \int_{r=0}^{R_1 \sin \phi_1} (\sqrt{R_1^2 - r^2} - r \cot \phi_1)$$

$$= \frac{R_1^3}{3} (1 - \cos \phi_1) \theta_1. \quad \blacksquare$$

EXERCISES

In Exercises 1 to 6 sketch the set S described and use integration in polar coordinates to evaluate the area $A(S)$.

1. S is the set enclosed by the circle $r = 4 \cos \theta$.

2. S is the set enclosed by the cardioid $r = a(1 + \cos \theta)$.

3. S is the set lying inside the circle $r = 4 \cos \theta$ and outside the circle $r = 2$.

4. S is the entire set enclosed by the lemniscate $r^2 = 2a^2 \cos 2\theta$.

5. S is the intersection of the two circles $r = 6 \sin \theta$ and $r = 6 \cos \theta$.

6. S is the set inside the circle $r = 2\sqrt{2} \cos \theta$ and to the right of the *kappa* curve $r = 2 \tan \theta$.

In Exercises 7 to 9 evaluate the integrals by transforming to polar coordinates.

7. $\int_S \dfrac{x}{y}$, $S = \{(x, y) \mid (x, y) \text{ is in first quadrant loop of } r = \sin 2\theta\}$.

8. $\int_S \dfrac{1}{x^2 + y^2}$, $S = \{(x, y) \mid 1 \leq x^2 + y^2 \leq 4\}$.

9. $\int_S xy$, $S = \left\{(x, y) \mid 0 < \sec \theta \leq r \leq 2 \cos \theta, 0 \leq \theta \leq \dfrac{\pi}{4}\right\}$.

10. Use Equation 9.9, p. 395, to obtain the change of variable formula for $\mathcal{R}^1$-integrals:

$$\int_{x=x_1}^{x_2} f(x) = \int_{u=T(x_1)}^{T(x_2)} (f \circ T^{-1}) \cdot (T^{-1})'(u),$$

where $u = T(x)$ denotes the variable transformation.

ANSWERS

1. 4π 3. $A(S) = \displaystyle\int_{\theta=-\frac{\pi}{3}}^{\frac{\pi}{3}} \int_{r=2}^{4\cos\theta} r = \dfrac{4\pi}{3} + 2\sqrt{3}$.

5. $A(S) = \displaystyle\int_{\theta=0}^{\frac{\pi}{4}} \int_{r=0}^{6\sin\theta} r + \int_{\theta=\frac{\pi}{4}}^{\frac{\pi}{2}} \int_{r=0}^{6\cos\theta} r = \dfrac{9(\pi - 2)}{2}$.

7. $\frac{1}{2}$.   9. $\displaystyle\int_S xy = \int_{\theta=0}^{\frac{\pi}{4}} \int_{r=\sec\theta}^{2\cos\theta} r^3 \sin\theta \cos\theta = \dfrac{11}{24}$.

10. We assume that T has an inverse and set $g = T^{-1}$. Since $u = T(x)$ for each x, we then have $x = T^{-1}(u) = g(u)$, for each u. Define $x_1 = T^{-1}(a) = g(a)$ and $x_2 = T^{-1}(b) = g(b)$. Show that $a = T(x_1)$ and $b = T(x_2)$. Now make the appropriate substitutions into Equation 9.9, page 395.

3. GENERAL COORDINATE SYSTEMS IN $\mathcal{R}^2$

Many problems in physics and engineering are most readily dealt with by means of coordinate systems other than rectangular coordinates or polar coordinates. We now introduce the general concept of a coordinate system in $\mathcal{R}^2$. The "change of variables" problem for $\mathcal{R}^2$-integrals will then be treated.

DEFINITION 19.3 (General Coordinate System in $\mathcal{R}^2$). Let $\mathcal{U}^2$ be a copy of $\mathcal{R}^2$:

$$\mathcal{U}^2 = \{(u, v) \mid u \in \mathcal{R} \text{ and } v \in \mathcal{R}\}.$$

A transformation $T:\mathcal{R}^2 \to \mathcal{U}^2$ is called a *coordinate system in $\mathcal{R}^2$* if and only if T is a one-one function, that is:

For each choice of (x, y) in dom T there is exactly one ordered pair (u, v) such that

(19.9) $$\binom{u}{v} = T(x, y).$$

The ordered pair (u, v) is referred to as the *T-coordinate pair for* (x, y).

The preceding definition does not require that every point (x, y) in $\mathcal{R}^2$ be assigned coordinates (u, v) under the transformation T. Nor does it require every pair (u, v) in $\mathcal{U}^2$ to be T-coordinates for some point in $\mathcal{R}^2$. For instance, in the polar coordinate system the origin of $\mathcal{R}^2$ is assigned no polar coordinates and the set of all polar coordinates is restricted to the part of $\mathcal{U}^2$ given by $r > 0$ and $-\pi < \theta \le \pi$.

Figure 19.7 illustrates the ideas involved. The function T assigns to each point (x, y) in *dom T* exactly one point (u, v) of $\mathcal{U}^2$. Consequently, each pair (u, v) in *ran T* is the T-coordinate pair of exactly one point in *dom T*. Thus the inverse transformation $T^{-1}(u, v)$ exists.

Let U and V denote the coordinate functions of T and let X and Y denote the coordinate functions of T^{-1}. We may express each of T and T^{-1} in three ways:

(19.10) $\quad T: \binom{u}{v} = T(x, y), \qquad \binom{u}{v} = \binom{U(x, y)}{V(x, y)}, \qquad \begin{cases} u = U(x, y) \\ v = V(x, y) \end{cases}$

(19.11) $\quad T^{-1}: \binom{x}{y} = T^{-1}(u, v), \qquad \binom{x}{y} = \binom{X(u, v)}{Y(u, v)}, \qquad \begin{cases} x = X(u, v) \\ y = Y(u, v) \end{cases}$

The set $\mathcal{K}$ of points in $\mathcal{R}^2$ having the same v-coordinate, say $v = v_1$, is given

$$\begin{cases} \text{parametrically by} \\ x = X(u, v_1) \\ y = Y(u, v_1) \end{cases} \quad \text{and} \quad \begin{cases} \text{implicitly by} \\ \\ V(x, y) = v_1. \end{cases}$$

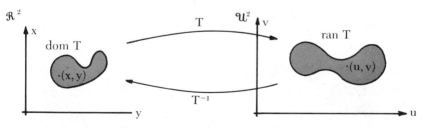

Figure 19.7

When X and Y are continuous this set will be a curve in $\mathcal{R}^2$ such that u varies as we progress along the curve but v is fixed at v_1. We refer to $\mathcal{K}$ as a *u-curve*. A *v-curve* is a set consisting of those points in $\mathcal{R}^2$ for which u has some fixed value, say $u = u_1$. A *v-curve* is described by

$$\begin{cases} x = X(u_1, v) \\ y = Y(u_1, v) \end{cases} \quad \text{or} \quad U(x, y) = u_1.$$

u-curves and *v*-curves play an important role in what follows.

Example 8. Let T be the polar coordinate system:

$$T: \begin{pmatrix} x \\ y \end{pmatrix} \rightarrow \begin{pmatrix} r \\ \theta \end{pmatrix} = \begin{pmatrix} \sqrt{x^2 + y^2} \\ \left(\arccos \dfrac{x}{\sqrt{x^2 + y^2}}\right) \operatorname{sgn} y \end{pmatrix}$$

We then have T^{-1} given by

$$x = r \cos \theta$$
$$y = r \sin \theta.$$

$\mathcal{R}^2$

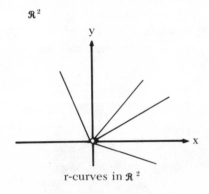

r-curves in $\mathcal{R}^2$

$\mathcal{U}^2$

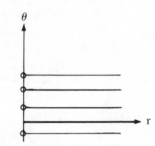

Image of r-curves in $\mathcal{U}^2$

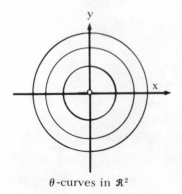

θ-curves in $\mathcal{R}^2$

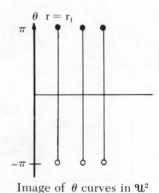

Image of θ curves in $\mathcal{U}^2$

Figure 19.8

r-curves are those curves in $\mathcal{R}^2$ along which θ is fixed; that is, r-curves are the half lines emanating from the origin. θ-curves are those curves in $\mathcal{R}^2$ on which r is a constant; that is, they are circles of positive radius with center at the origin. Reference to Figure 19.8 should clarify the idea of an r-curve as a set in $\mathcal{R}^2$ whose image in polar coordinate space $\mathcal{U}^2$ is a straight line parallel to the r-axis. Similarly, a θ-curve is a set in $\mathcal{R}^2$ whose image in $\mathcal{U}^2$ is a straight line parallel to the θ-axis.

4.   CHANGE OF VARIABLES FOR $\mathcal{R}^2$-INTEGRALS

Let $\mathcal{U}^2$ be a coordinate space for $\mathcal{R}^2$ with respect to the coordinate system T. Consider now the basic domain of integration in $\mathcal{U}^2$—the coordinate rectangle $R_T = [u_1, u_2] \times [v_1, v_2]$ pictured in Figure 19.9. The set $\mathcal{C}$ of points in $\mathcal{R}^2$ having (u, v) coordinates in R_T shall be called a *T-cell* of $\mathcal{R}^2$. Thus, the T-cell corresponding to the R_T given above is that portion of $\mathcal{R}^2$ bounded by the u-curves $v = v_1$, $v = v_2$ and the v-curves $u = u_1$, $u = u_2$. The situation for polar coordinates is pictured in Figure 19.2, p. 754 and for general coordinates in Figure 19.9.

Our present goal is to relate integration over T-cells $\mathcal{C}$ in $\mathcal{R}^2$ to integration over R_T in $\mathcal{U}^2$. We recall that for polar coordinates we have shown that the area of $\mathcal{C}$ may be evaluated as an integral over R_T.

$$(19.12) \qquad\qquad A(\mathcal{C}) = \int_{\mathcal{C}} 1_2 = \int_{R_T} r.$$

To rewrite this relation in a form which we can more readily generalize, we introduce the functions $D_{1_2} A : \mathcal{R}^2 \to \mathcal{R}$ and $D_T A : \mathcal{U}^2 \to \mathcal{R}$ given by

$$D_{1_2} A(x, y) = 1_2(x, y) = 1,$$

and

$$D_T A(r, \theta) = r.$$

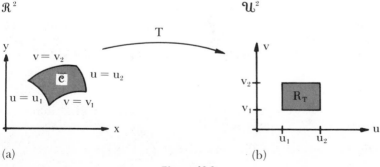

$\mathcal{R}^2$ $\mathcal{U}^2$

(a)   (b)

Figure 19.9

Equation (19.12) then becomes

$$(19.13) \quad A(\mathcal{C}) = \int_{\mathcal{C}} D_{1_2} A = \int_{T\mathcal{C}} D_T A, \quad \text{where} \quad T(\mathcal{C}) = R_T.$$

Thus $D_T A$ is a function which when integrated over R_T, the $\mathcal{U}^2$ image of $\mathcal{C}$, gives the area of $\mathcal{C}$.

We return now to the consideration of a general coordinate system T for $\mathcal{R}^2$. We assume that a function $D_T A$ can be found which is such that Equation 19.13 is valid for every T-cell in $\mathcal{R}^2$. We shall show that this formula is valid for more general domains of integration. In fact, we shall establish a stronger result.

THEOREM 19.4 (Change of Variables in $\mathcal{R}^2$-Integration). Let $T: \mathcal{R}^2 \to \mathcal{U}^2$ be a coordinate system in $\mathcal{R}^2$. If we can find a positive valued function $D_T A: \mathcal{U}^2 \to \mathcal{R}$ such that

$$(19.14) \quad \text{Area of } \mathcal{C} \int_{\mathcal{C}} 1_2 = \int_{T\mathcal{C}} D_T A, \text{ for every } T\text{-cell } \mathcal{C} \text{ in } \mathcal{R}^2,$$

then

$$(19.15) \qquad\qquad \int_S (F \circ T) = \int_{TS} F \cdot D_T A$$

for every S and $F: \mathcal{U}^2 \to \mathcal{R}$, provided that (i) $S \subset dom\ T$, (ii) $F \circ T$ is integrable over S, and (iii) F and $F \cdot D_T A$ are integrable over TS.

NOTE. The power of this theorem should be evident from the following less precise restatement: If an integrating factor $D_T A$ can be found to transform the integral $\int 1_2$ of the unit function on T-cells to an integral in coordinate space, then the same integrating factor can be used to transform $\mathcal{R}^2$-integrals over arbitrary domains to $\mathcal{U}^2$-integrals, subject to the integrability conditions.

Proof. We prove the theorem in three stages of generality.

(i) S *a T-cell; F a step-function on* TS. Let S be a T-cell $\mathcal{C}$ and R_T be the associated coordinate rectangle in $\mathcal{U}^2: R_T = T\mathcal{C}$. Assuming F to be a step function on R_T we consider a set $\{R_{jk} \mid j = 1, \ldots, m; k = 1, \ldots, n\}$ of open subrectangles of a partition of R_T such that F is constant on each R_{jk}:

$$F(u, v) = F_{jk} \quad \text{for } (u, v) \text{ in } R_{jk}.$$

Then the set $\{\mathcal{C}_{jk} \mid \mathcal{C}_{jk} = T^{-1}(R_{jk})\}$ is a partition of $\mathcal{C}$ into disjoint sets and

$$(19.16) \qquad (F \circ T)(x, y) = F(U(x, y), V(x, y)) = F(u, v)$$
$$= F_{jk} \text{ for } (x, y) \text{ in } \mathcal{C}_{jk}.$$

We then have

$$\int_{R_T} FD_TA =^1 \sum_{j=1,k=1}^{m,n} \int_{R_{jk}} F_{jk} \cdot D_TA$$

$$=^2 \sum_{j=1,k=1}^{m,n} F_{jk} \int_{R_{jk}} D_TA$$

$$=^3 \sum_{j=1,k=1}^{m,n} F_{jk} \int_{\mathcal{C}_{jk}} \mathbb{1}_2$$

$$=^4 \sum_{j=1,k=1}^{m,n} \int_{\mathcal{C}_{jk}} F \circ T$$

$$=^1 \int_{\mathcal{C}} F \circ T.$$

Thus, Equation 9.15 is valid when F is a step function and S a T-cell.

(1) Repeated use of the property $\int_{A \cup B} = \int_A + \int_B$ when $\mathring{A}, \mathring{B}$ are disjoint.

(2) Moving the constant F_{jk} outside the integration process.

(3) Using the hypotheses that $\int_{R_T} D_TA = \int_{\mathcal{C}} \mathbb{1}_2$ for each rectangle R_{jk} of the partition.

(4) Moving the constant F_{jk} under the integration process and using Equation 19.16.

(ii) S *a T-cell; general F.* Again, let S be a T-cell $\mathcal{C}$ and let R_T be the associated rectangle in $\mathcal{U}^2$. Let L and U be lower and upper step functions for F on R_T, respectively. For every choice of L and U we have

$L(u, v) \leq F(u, v) \leq U(u, v)$

$$\Rightarrow^1 L \cdot D_TA(u, v) \leq F \cdot D_TA(u, v) \leq U \cdot D_TA(u, v)$$

(19.17) $$\Rightarrow^2 \int_{R_T} L \cdot D_TA \leq \int_{R_T} F \cdot D_TA \leq \int_{R_T} U \cdot D_TA.$$

On the other hand, for every L and U we have

$L(u, v) \leq F(u, v) \leq U(u, v)$

$$\Rightarrow (L \circ T)(x, y) \leq (F \circ T)(x, y) \leq (U \circ T)(x, y)$$

$$\Rightarrow \int_{\mathcal{C}} L \circ T \leq \int_{\mathcal{C}} F \circ T \leq \int_{\mathcal{C}} U \circ T$$

(19.18) $$\Rightarrow \int_{R_T} L \cdot D_TA \leq \int_{\mathcal{C}} F \circ T \leq \int_{R_T} U \cdot D_TA.$$

(1) $D_TA(u, v)$ is non-negative by hypothesis.
(2) Using the order properties of the integral, assuming all integrals are defined.

(3) Using coordinate transformation $\begin{pmatrix} u \\ v \end{pmatrix} = T(x, y)$.

The reader is asked to show (Exercise 4) that the integrability of F on R_T assures us that $\int_{R_T} F \cdot D_T A$ is the only number satisfying the relation 19.17 for *every* choice of L and U. A comparison of Equations 19.17 and 19.18 then enables us to conclude that

$$(19.19) \qquad \int_{\mathcal{C}} F \circ T = \int_{R_T} F \cdot D_T A.$$

Thus Equation 19.15 is valid whenever S is a T-cell in $\mathcal{R}^2$.

(iii) *General S; general F.* The validity of Equation 19.15 for more general domains of integration follows from Part (ii) and the following equality chain:

$$\int_{TS} F \cdot D_T A =^1 \int_{\mathcal{D}} \tilde{F} \cdot D_T A =^2 \sum_{i=1}^{l} \int_{R_{iT}} \tilde{F} \cdot D_T A$$

$$=^3 \sum_{i=1}^{l} \int_{\mathcal{C}_{iT}} \tilde{F} \circ T =^4 \int_{T^{-1}\mathcal{D}} \tilde{F} \circ T$$

$$=^2 \int_{S} \tilde{F} \circ T + \int_{T^{-1}(\mathcal{D}\setminus S)} \tilde{F} \circ T$$

$$=^5 \int_{S} F \circ T. \quad \blacksquare$$

(1) Where $\mathcal{D}$ is a set containing TS and is a finite union of coordinate rectangles: $\mathcal{D} = \{R_{iT} \mid i = 1, \ldots, l\}$. $\tilde{F}$ is defined so that $\tilde{F} = F$ on TS and $\tilde{F} = 0$ on $\mathcal{D}\setminus TS$.

(2) Use of $\int_{A \cup B} = \int_A + \int_B$ when $A \cap B = \varnothing$.

(3) Using Equation 19.19 from Part (ii).

(4) Using property given in (2) and $\{\mathcal{C}_{iT} \mid i = 1, \ldots, l\} = \{T^{-1}R_{iT} \mid i = 1, \ldots, l\} = T^{-1}\{R_{iT} \mid i = 1, \ldots, l\} = T^{-1}\mathcal{D}$.

(5) From definition of F in (1) we know that $\{\tilde{F} \circ T = F \circ T$ on S and $\tilde{F} \circ T = 0\}$ on $T^{-1}\mathcal{D}\setminus S$.

Consider the case in which F is the unit function on TS: $F = \mathbb{1}_2$. Then $F \circ T$ is the unit function $\mathbb{1}_2$ on S, and the change of variable formula gives an expression for the area $A(S)$ of S.

$$A(S) = \int_{S} \mathbb{1}_2 = \int_{TS} \mathbb{1}_2 \cdot D_T A.$$

Now the area of TS in $\mathcal{U}^2$ is given by $A(TS) = \int_{TS} \mathbb{1}_2$. Thus $D_T A$ may be regarded as an *area magnification factor* which converts the area of TS in $\mathcal{U}^2$ into the area of S in $\mathcal{R}^2$:

$$A(S) = \int_{S} \mathbb{1}_2 = \int_{TS} \mathbb{1}_2 \cdot \text{area magnification factor}$$

$$= \int_{TS} D_T A.$$

As such, $D_T A$ represents the rate of change of $A(\mathcal{S})$ with respect to a change in the area of $T\mathcal{S}$. $D_T A$ has, therefore, a rate of change interpretation similar to that given the derivative function f' of an $\mathcal{R} \to \mathcal{R}$ function f. The notation $D_T A$ was chosen with this property in mind.

We now list various forms for the change of variable formula.

CHANGE OF VARIABLES FOR $\mathcal{R}^2$-INTEGRALS.

Let $T : \mathcal{R}^2 \to \mathcal{U}^2$ be a coordinate system in $\mathcal{R}^2$. If we can find an *area magnification factor* $D_T A$ for T-cells:

$$A(\mathcal{C}) = \int_{T\mathcal{C}} D_T A \quad \text{for every } T\text{-cell } \mathcal{C},$$

then for every $\mathcal{S}$ and every integrable function f and F we have

(i) $\displaystyle\int_{\mathcal{S}} F \circ T = \int_{T\mathcal{S}} F \cdot D_T A$

where

$$\begin{pmatrix} u \\ v \end{pmatrix} = T(x, y)$$

$$\int_{\mathcal{S}} F(U(x, y), V(x, y))$$

is given by

$$= \int_{T\mathcal{S}} (F \cdot D_T A)(u, v)$$

$$\begin{aligned} u &= U(x, y) \\ v &= V(x, y) \end{aligned}$$

(ii) $\displaystyle\int_{\mathcal{S}} f = \int_{T\mathcal{S}} (f \circ T^{-1}) D_T A$

where

$$\begin{pmatrix} x \\ y \end{pmatrix} = T^{-1}(u, v)$$

$$\int_{\mathcal{S}} f(x, y)$$

is given by

$$= \int_{T\mathcal{S}} f(X(u, v), Y(u, v)) D_T A$$

$$\begin{aligned} x &= X(u, v) \\ y &= Y(u, v) \end{aligned}$$

provided that $f \circ T^{-1}$ and $F \circ T$ are also integrable on their respective domains.

Example 9. Let T be the coordinate system given by

$$T(x, y) = \begin{pmatrix} u \\ v \end{pmatrix} = \begin{pmatrix} mx + y \\ y \end{pmatrix}.$$

Then

$$T^{-1}(u, v) = \begin{pmatrix} x \\ y \end{pmatrix} = \begin{pmatrix} \dfrac{u - v}{m} \\ v \end{pmatrix}.$$

The T-cells in $\mathfrak{R}^2$ are parallelograms bounded by the lines

$$mx + y = u_1, \quad mx + y = u_2, \quad y = v_1, \quad y = v_2$$

The area of a T-cell $\mathcal{C}$ is the area of a parallelogram

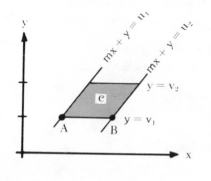 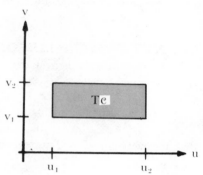

$A(\mathcal{C}) = \text{base} \times \text{height}$

$\qquad = d(A, B) \times (v_2 - v_1)$

$\qquad = \dfrac{(u_2 - u_1)}{m} (v_2 - v_1)$

$\qquad = \displaystyle\int_{u=u_1}^{u_2} \int_{v=v_1}^{v_2} \dfrac{1}{m}$

$\qquad = \displaystyle\int_{T\mathcal{S}} \dfrac{1}{m}$

Therefore $D_T A(u, v) = \dfrac{1}{m}$ is an *area magnification factor* for T-cells. Hence, form (ii) of the change of variables formula is given by

$$\int_{\mathcal{S}} f = \int_{T\mathcal{S}} (f \circ T^{-1}) \frac{1}{m}. \quad \blacksquare$$

For a large class of coordinate systems it is possible to give a general expression for the area magnification factor $D_T A$. We shall state the result and refer the reader interested in a proof to: Apostol, T., *Mathematical Analysis*, Addison-Wesley, 1957, Theorem 10–30.

THEOREM 19.5 (Jacobian Form for Change of Variables in $\mathfrak{R}^2$-Integrals.) Let $f : \mathfrak{R}^2 \to \mathfrak{R}$ be continuous and integrable on $\mathcal{S}$. Let T be a coordinate system in $\mathfrak{R}^2$ such that $\mathcal{S} \subset dom\ T$. Further, suppose the coordinate functions $X(u, v)$ and $Y(u, v)$ for T^{-1} have continuous first partial derivatives and that the Jacobian matrix JT^{-1} is such that

$$\det J_{\binom{u}{v}} T^{-1} = \det \begin{pmatrix} \dfrac{\partial X}{\partial u} & \dfrac{\partial X}{\partial v} \\ \dfrac{\partial Y}{\partial u} & \dfrac{\partial Y}{\partial v} \end{pmatrix} \neq 0 \quad \text{for} \quad (u, v) \in TS.$$

Then

$$\int_S f(x, y) = \int_{TS} f(X(u, v), Y(u, v)) \, |\det J_{\binom{u}{v}} T^{-1}|.$$

That is, we may set

$$D_T A = |\det J_{\binom{u}{v}} T^{-1}|.$$

Example 10. For the polar coordinate system, the coordinate functions of T^{-1} are

$$X(r, \theta) = r \cos \theta \quad \text{and} \quad Y(r, \theta) = r \sin \theta.$$

Thus

$$\frac{\partial X}{\partial r} = \cos \theta, \qquad \frac{\partial X}{\partial \theta} = -r \sin \theta,$$

$$\frac{\partial Y}{\partial r} = \sin \theta, \qquad \frac{\partial Y}{\partial \theta} = r \cos \theta.$$

Consequently,

$$\det J_{\binom{r}{\theta}} T^{-1} = r \cos^2 \theta + r \sin^2 \theta = r,$$

and we may set $D_{pol} A(r, \theta) = |\det J_{\binom{r}{\theta}} T^{-1}| = r$. Observe that this agrees with the result obtained in Section 2. ■

Example 11. For the linear coordinate system of Example 9 we have:

$$X(u, v) = \frac{u - v}{m}, \qquad Y(u, v) = v,$$

$$\frac{\partial X}{\partial u} = \frac{1}{m}, \qquad \frac{\partial X}{\partial v} = -\frac{1}{m},$$

$$\frac{\partial Y}{\partial u} = 0, \qquad \frac{\partial Y}{\partial v} = 1.$$

Thus we may take

$$D_T A = |\det J_{\binom{u}{v}} T^{-1}| = \left| \frac{1}{m} \right| = \frac{1}{m}. \quad ■$$

EXERCISES

In Exercises 1 to 3, T denotes the *oblique plane coordinate system* given by

$$\begin{pmatrix} u \\ v \end{pmatrix} = T\begin{pmatrix} x \\ y \end{pmatrix} = \begin{pmatrix} ax + by \\ cx + dy \end{pmatrix}, \quad \text{where } ad - bc \neq 0.$$

1. Determine the specific form of the inverse transformation T^{-1}:

$$\begin{pmatrix} x \\ y \end{pmatrix} = T^{-1}\begin{pmatrix} u \\ v \end{pmatrix} = \begin{pmatrix} X(u, v) \\ Y(u, v) \end{pmatrix}.$$

2. Show that a T-cell for oblique plane coordinates is a parallelogram $\mathcal{C}$ whose area is given by $A(\mathcal{C}) = \displaystyle\int_{T\mathcal{C}} \frac{1}{|ad - bc|}\, \mathbb{1}_2$ and thereby show that

$D_T A = \dfrac{1}{|ad - bc|}$. Give the change of variables formula for integration in oblique plane coordinates.

3. Verify that $D_T A = \dfrac{1}{|ad - bc|}$ by applying Theorem 19.5 to the Jacobian matrix JT^{-1}.

4. Let $T : \mathcal{R}^2 \to \mathcal{U}^2$ be a coordinate system in $\mathcal{R}^2$. Let R_T be a coordinate rectangle in $\mathcal{U}^2$ and let $F : \mathcal{U}^2 \to \mathcal{R}$ and $D_T A : \mathcal{U}^2 \to \mathcal{R}$ be integrable over R_T. Show that there can be only one real number I such that

$$\int_{R_T} L \cdot D_T A \leq I \leq \int_{R_T} U \cdot D_T A$$

for every choice of upper step function U and lower step function L for F.

ANSWERS

1. $X(u, v) = \dfrac{1}{ad - bc}(du - bv), \quad Y(u, v) = \dfrac{1}{ad - bc}(-cu + av).$

3. $|\text{Det } JT^{-1}| = \left| \dfrac{1}{ad - bc} \right|$

4. Let I and J be two real numbers satisfying the condition given. Let $J \leq I$. Then $I \leq \int_{R_T} U \cdot D_T A$ and $-J \leq -\int_{R_T} L \cdot D_T A$. Hence for every real number ρ we have

$$0 \leq I - J \leq \int_{R_T} (U - L) \cdot D_T A, \quad \text{for every } U, L,$$

$$\leq^1 M \int_{R_T} (U - L), \quad \text{for some real number } M.$$

$$\leq^2 M \int_{R_T} (U_r - L_r) \leq Mr, \quad \text{for some choice of } U_r \text{ and } L_r$$

$$\leq^3 \rho.$$

Thus we conclude that we can only have $I = J$.

(1) $D_T A$ integrable on $R_T \Rightarrow D_T A$ is bounded on R_T
$\Rightarrow (U - L) \cdot D_T A \leq M(U - L)$ for some M.

(2) Applying Riemann Condition of Integrability, Theorem 18.20, p. 745, to F on R_T.

(3) Setting $r = \dfrac{\rho}{M}$.

5. INTEGRATION IN A GENERAL COORDINATE SYSTEM IN $\mathcal{R}^n$

In $\mathcal{R}^3$ and in higher dimensional spaces, coordinate systems other than the rectangular system are often useful. In what follows, we outline the ideas and results that are involved. The development follows that of Section 4 for $\mathcal{R}^2$.

DEFINITION 19.6 (General Coordinate System in $\mathcal{R}^n$). Let $\mathcal{U}^n$ be a copy of $\mathcal{R}^n$: $\mathcal{U}^n = \{(u_1, u_2, \ldots, u_n) \mid u_1 \in \mathcal{R}, i = 1, \ldots, n\}$. A transformation $T : \mathcal{R}^n \to \mathcal{U}^n$ is called a *coordinate system in $\mathcal{R}^n$* if and only if T is a one-one function. The n-tuple $(u_1, \ldots, u_n)$ is called the *T-coordinate n-tuple* for $(x_1, \ldots, x_n)$ if and only if

$$\begin{pmatrix} u_1 \\ \cdot \\ \cdot \\ \cdot \\ u_n \end{pmatrix} = T(x_1, \ldots, x_n)$$

In $\mathcal{R}^2$ the area of a set S is defined to be

$$A(S) = \int_S 1_2,$$

provided that the integral exists. The generalization of the area concept to a subset S of a higher dimensional space is termed the *n-dimensional content*.

DEFINITION 19.7 (*n-Dimensional Content*). Let S be a subset of $\mathcal{R}^n$. We say that S has *n-dimensional content $K(S)$* if and only if

$$K(S) = \int_S 1_n,$$

where 1_n denotes the unit function on $\mathcal{R}^n$

$$1_n(X) = 1 \quad \text{for all } X \text{ in } \mathcal{R}^n.$$

If S is an interval in $\mathcal{R}^1$, the 1-dimensional content of S is the *length* of the interval S. If S is a subset of $\mathcal{R}^2$, the 2-dimensional content of S is called the *area* of S. If S is a subset of $\mathcal{R}^3$, the 3-dimensional content of S is called the *volume* of S.

Let T be a coordinate system for $\mathcal{R}^n$. A subset $\mathcal{C}$ of $\mathcal{R}^n$ is called a T-cell if and only if its image $T\mathcal{C}$ is a coordinate parallelepiped in $\mathcal{R}^n$;

$$T\mathcal{C} = [a_1, b_1] \times [a_2, b_2] \times \cdots \times [a_n, b_n].$$

The change of variables theorem for $\mathcal{R}^n$ may be stated as follows.

THEOREM 19.8 (Change of Variables in $\mathcal{R}^n$-Integration). Let $T:\mathcal{R}^n \to \mathcal{U}^n$ be a coordinate system in $\mathcal{R}^n$. If we can find a positive valued function $D_T A: \mathcal{U}^n \to \mathcal{R}$ such that for every T-cell $\mathcal{C}$ in $\mathcal{R}^n$

(19.20)
$$K(\mathcal{C}) = \int_{\mathcal{C}} \mathbb{1}_n = \int_{T\mathcal{C}} D_T K,$$

then

$$\int_S (F \circ T) = \int_{TS} F \cdot D_T K$$

and

$$\int_S f = \int_{TS} (f \circ T^{-1}) \cdot D_T K$$

for every $S, f:\mathcal{R}^n \to \mathcal{R}$ and $F: \mathcal{U}^n \to \mathcal{R}$ provided that
(i) $S \subset dom\ T$, (ii) the integrals appearing are defined, and (iii) $f \circ T^{-1}$, F and $D_T K$ are integrable over TS.

Application of the preceding theorem to the unit function $\mathbb{1}_n$ yields the following result.

THEOREM 19.9 (n-Dimensional Content in General Coordinates.) Let S be a subset of $\mathcal{R}^n$ which possesses an n-dimensional content $K(S)$. Then

$$K(S) = \int_{TS} D_T K.$$

The change of variables formula in $\mathcal{R}^n$ also admits a Jacobian formulation under certain further restrictions on the functions involved.

THEOREM 19.10 (Jacobian Form for Change of Variables in $\mathcal{R}^n$-Integrals.) Let $f:\mathcal{R}^n \to \mathcal{R}$ be continuous and integrable on S.

Let T be a coordinate system in $\Re^n$ such that $S \subset dom\ T$. Further, suppose the coordinate functions of T^{-1} have continuous first partial derivatives and that the Jacobian matrix JT^{-1} is such that

$$det\ J_U T^{-1} = det \begin{pmatrix} \dfrac{\partial X_1}{\partial u_1} \cdots \dfrac{\partial X_1}{\partial u_n} \\ \cdot \qquad \cdot \\ \cdot \qquad \cdot \\ \cdot \qquad \cdot \\ \dfrac{\partial X_n}{\partial u_1} \cdots \dfrac{\partial X_n}{\partial u_n} \end{pmatrix} \neq 0 \quad \text{for} \quad U \in TS.$$

Then

$$\int_S fX = \int_{TS} (f \circ T)U \cdot |det\ J_U T^{-1}|.$$

That is, we may set

$$D_T A = |det\ J_U T^{-1}|.$$

6. INTEGRATION IN CYLINDRICAL AND IN SPHERICAL COORDINATES

In this section we present two commonly used coordinate systems for $\Re^3$ and the related change of variables formulas for integration.

CYLINDRICAL COORDINATES IN $\Re^3$. The coordinate system given by

$$\begin{pmatrix} r \\ \theta \\ z \end{pmatrix} = T(x, y, z) = \begin{pmatrix} \sqrt{x^2 + y^2} \\ \left(arc\ cos \dfrac{x}{\sqrt{x^2 + y^2}}\right) sgn\ y \\ z \end{pmatrix}, \ x^2 + y^2 \neq 0.$$

is called the *cylindrical coordinate system in* $\Re^3$. The inverse transformation is

$$\begin{pmatrix} x \\ y \\ z \end{pmatrix} = T^{-1}(r, \theta, z) = \begin{pmatrix} r\cos\theta \\ r\sin\theta \\ z \end{pmatrix}, \ \begin{cases} r > 0 \\ -\pi < \theta \leq \pi \end{cases}$$

The geometrical significance of r, θ, and z is pictured in Figure 19.10-a. Since we must have $x^2 + y^2 \neq 0$, no point on the z-axis is assigned cylindrical coordinates. Any interval of length 2π open at one end is admissible for θ. The choice $(-\pi, \pi)$ merely provides a simpler expression for $T(x, y, z)$ than other choices. Simply put, cylindrical coordinates are obtained by replacing the xy–coordinates by the $r\theta$-coordinates and retaining the z-coordinate.

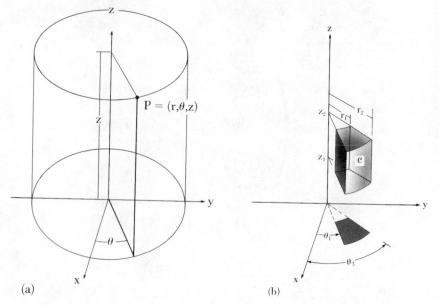

(a)

(b)

Figure 19.10

A T-cell for cylindrical coordinates is pictured in Figure 19.10-b. The volume $V(\mathcal{C})$ of a T-cell $\mathcal{C}$ is readily determined:

$$V(\mathcal{C}) = \text{area of base} \times \text{height}$$

$$=^1 \left(\int_{r=r_1}^{r_2} \int_{\theta=\theta_1}^{\theta_2} r \right) \times (z_2 - z_1)$$

$$= \int_{r=r_1}^{r_2} \int_{\theta=\theta_1}^{\theta_2} \int_{z=z_1}^{z_2} r$$

$$= \int_{T\mathcal{C}} r.$$

(1) Since the base is a T-cell for polar coordinates.

Hence the *volume magnification factor* for cylindrical coordinates is $D_{\text{cyl}}V = r$. The change of variables formula for cylindrical coordinates becomes

$$\int_S f(x, y, z) = \int_{TS} f(r \cos \theta, r \sin \theta, z) \cdot r. \quad \blacksquare$$

Example 10. Evaluate $\int_S z^3$ where S is the subset of $\mathcal{R}^3$ bounded below by the xy-plane and above by the sphere $x^2 + y^2 + z^2 = 4$. (See Figure 19.11-a.) Using cylindrical coordinates (r, θ, z), we may describe the boundaries of S by the equations

$$z = 0 \quad \text{and} \quad r^2 + z^2 = 4, \quad \text{where} \quad -\pi < \theta \leq \pi.$$

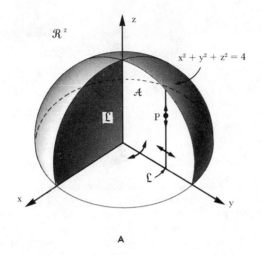

A

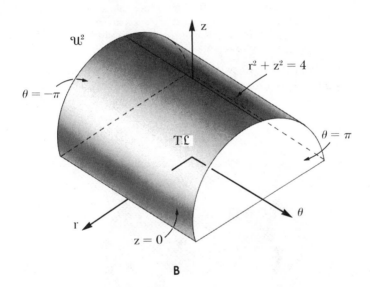

B

Figure 19.11

Thus, TS is the half cylinder pictured in Figure 19.11-b. Since $r^2 = x^2 + y^2$, the integral to be evaluated becomes, in cylindrical coordinates,

$$\int_S z^3 = \int_{TS} z^3 \cdot D_{cyl} V = \int_{TS} z^3 \cdot r.$$

We evaluate the last integral by the method of repeated integration.

Thus

$$\int_S z^3 =^1 \int_{\theta=-\pi}^{\pi} \int_{r=0}^{2} \int_{z=0}^{\sqrt{r^2-4}} z^3 \cdot r$$

$$= \int_{\theta=-\pi}^{\pi} \int_{r=0}^{2} \frac{(r^2 - 4)^2 \cdot r}{4}$$

$$= \int_{\theta=-\pi}^{\pi} \left(\frac{8}{3}\right)$$

$$= \frac{16\pi}{3}. \quad \blacksquare$$

(1) To determine the limits on the integrals directly from Figure 19.11-a we observe:

(a) Fixed values of θ give a planar section $\mathcal{A}$. For $\mathcal{A}$ to sweep out all of S it is necessary for θ to vary over any interval of length 2π.

(b) Fixed values of r give a line $\mathcal{L}$ in $\mathcal{A}$. For $\mathcal{L}$ to sweep out all of $\mathcal{A}$, r must vary from 0 to 2. (Also see (d) below.)

(c) Fixed values of z give a point P on $\mathcal{L}$. For P to sweep out all of $\mathcal{L}$, z must vary from its value in the xy-plane, $z = 0$, to its value on the sphere, $z = \sqrt{4 - (x^2 + y^2)} = \sqrt{4 - r^2}$.

(d) Although the value $r = 0$ is not admissible as a cylindrical coordinate, it is permissible as a limit of integration since the value of the integral depends only upon the values of the integrated function in the interior of the interval of integration.

Example 11. Consider the solid of revolution S shown in Figure 19.12. Using cylindrical coordinates, we may describe S as the portion of $\mathcal{R}^3$

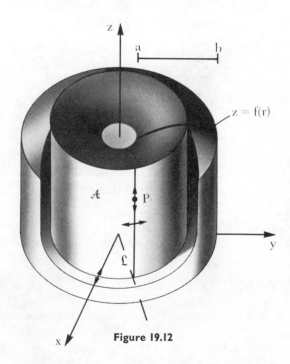

Figure 19.12

between the cylinder $r = a$ and $r = b$, above the plane $z = 0$, and below the surface $z = f(r)$ for some function f. Let T denote the cylindrical coordinate system. The volume of S is given by

$$
V(S) =^1 \int_{TS} D_{\mathrm{cyl}} V = \int_{TS} r
$$
$$
=^2 \int_{r=a}^{b} \int_{\theta=0}^{2\pi} \int_{z=0}^{f(r)} r
$$
$$
= \int_{r=a}^{b} \int_{\theta=0}^{2\pi} rf(r)
$$
$$
= \int_{r=a}^{b} 2\pi rf(r)
$$
$$
= 2\pi \int_{r=a}^{b} rf(r). \quad \blacksquare
$$

(1) Employing Theorem 19.9 for $\mathcal{R}^3$.
(2) (a) Fixed values of r give cylindrical sections $\mathcal{A}$. For $\mathcal{A}$ to sweep out all of S we must allow r to vary from a to b.
   (b) Fixed values of θ give vertical lines $\mathcal{L}$ in the section $\mathcal{A}$. For $\mathcal{L}$ to sweep out $\mathcal{A}$ we must choose an interval of length 2π for θ, say 0 to 2π.
   (c) Fixed values of z give a point P on $\mathcal{L}$. For P to sweep out $\mathcal{L}$, we must choose the limits $z = 0$ and $z = f(r)$.

Note that if we replace r by t the above formula for $V(S)$ becomes

$$
V(S) = 2\pi \int_{t=a}^{b} tf(t).
$$

This is precisely the Formula 12.3, p. 477, derived earlier for volumes of solids of revolution by means of the shell method. This shows the consistency of the treatment by the method of Chapter 12 with the broader definition of volume given in Chapter 18.

SPHERICAL COORDINATES IN $\mathcal{R}^3$. Spherical coordinates in $\mathcal{R}^3$ are given by the transformation

$$
\begin{pmatrix} R \\ \varphi \\ \theta \end{pmatrix} = T(x, y, z) = \begin{pmatrix} \sqrt{x^2 + y^2 + z^2} \\ \arccos \dfrac{z}{\sqrt{x^2 + y^2 + z^2}} \\ \arccos \dfrac{x}{\sqrt{x^2 + y^2}} \operatorname{sgn} y \end{pmatrix} \quad \sqrt{x^2 + y^2} \neq 0.
$$

The inverse transformation is

$$
\begin{pmatrix} x \\ y \\ z \end{pmatrix} = T^{-1}(r, \varphi, \theta) = \begin{pmatrix} R \sin \varphi \cos \theta \\ R \sin \varphi \sin \theta \\ R \cos \varphi \end{pmatrix}, \quad \begin{cases} -\pi < \theta \leq \pi, \\ R > 0, \\ 0 < \varphi \leq \dfrac{\pi}{2}. \end{cases}
$$

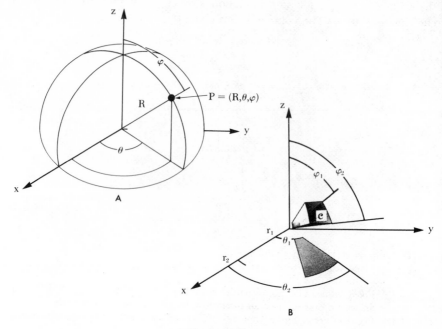

Figure 19.13

Figure 19.13-a shows the geometrical significance of R, φ and θ. Note that points on the z-axis do not have spherical coordinates because of the condition $x^2 + y^2 \neq 0$. A T-cell $\mathcal{C}$ is pictured in Figure 19.13-b. The volume of $\mathcal{C}$ can be obtained by using conical wedges $\mathcal{W}$ such as that pictured in Figure 19.6, p. 758. The volume for such wedges was determined in Example 7 to be

$$V(\mathcal{W}) = \frac{R_1^3}{3} (1 - \cos \varphi_1)\theta_1.$$

This formula is readily expressible as an integral in spherical coordinates:

$$V(\mathcal{W}) = \int_{\theta=0}^{\theta_1} \int_{\varphi=0}^{\varphi_1} \int_{R=0}^{R_1} R^2 \sin \varphi$$

$$= \int_{T\mathcal{W}} R^2 \sin \varphi.$$

For a T-cell $\mathcal{C}$ we would have $V(\mathcal{C}) = \int_{T\mathcal{C}} R^2 \sin \varphi$. Hence, as the *volume magnification factor* we may take

$$D_{\text{sph}} V = R^2 \sin \varphi,$$

and the change of variable formula for spherical coordinates is

$$\int_{\mathcal{S}} f(x, y, z) = \int_{T\mathcal{S}} f(R \cos \varphi \cos \theta, R \cos \varphi \sin \theta, R \sin \varphi) \cdot R^2 \sin \varphi.$$

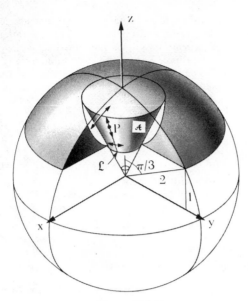

Figure 19.14

Example 12.   Let S be the cap cut from the sphere $x^2 + y^2 + z^2 = 4$ by the plane $y = 1$. (See Figure 19.14.) We compute the volume of S using spherical coordinates.

$$V(S) \overset{1}{=} \int_{TS} D_{\text{sph}} V = \int_{TS} R^2 \sin \varphi$$

$$\overset{2}{=} \int_{\varphi=0}^{\pi/6} \int_{\theta=0}^{2\pi} \int_{R=\sec\varphi}^{2} R^2 \sin \varphi$$

$$= \int_{\varphi=0}^{\pi/6} \int_{\theta=0}^{2\pi} \left(\frac{8}{3} - \frac{\sec^3 \varphi}{3} \right) \sin \varphi$$

$$= \int_{\varphi=0}^{\pi/6} 2\pi \left(\frac{8 \sin \varphi - \tan \varphi \sec^2 \varphi}{3} \right)$$

$$= \left[\frac{2\pi}{3} \left(-8 \cos \varphi - \frac{\tan^2 \varphi}{2} \right) \right]_{0}^{\pi/6}$$

$$= \frac{(47 - 24\sqrt{3})\pi}{9}. \quad \blacksquare$$

(1) Employing Theorem 19.9 for $\mathcal{R}^3$.
(2) (a) Fixed values of φ give conical sections $\mathcal{A}$. For $\mathcal{A}$ to sweep out all of S, φ must vary from $\varphi = 0$ to $\varphi = \pi/6$. (See (d) below.)
    (b) Fixed values of θ give lines $\mathcal{L}$ on $\mathcal{A}$. For $\mathcal{L}$ to sweep out all of $\mathcal{A}$, θ must vary over some interval of length say $\theta = 0$ to $\theta = 2\pi$. (See (d) below.)
    (c) Fixed values of R give a point P on $\mathcal{L}$. For P to sweep out all of $\mathcal{L}$, R must vary from R for the plane to R for the sphere. For the plane we have $z = R \cos \varphi = 1$; thus $R = \sec \varphi$. For the sphere $R = 2$.

(d) Although the values $R = 0$ and $\varphi = 0$ are not admissible as spherical coordinates, they are permissible as limits of integration, since the value of the integral depends only upon functional values in the interior of the interval of integration. For spherical coordinates any 2π interval for θ is admissible. The interval $(-\pi, \pi]$ is chosen in the text simply because the transformation formula is simplest with this choice.

EXERCISES

In Exercises 1 to 4, use integration in cylindrical coordinates to determine the volumes of the solids described.

1. A sphere of radius a.

2. A right circular cone of height h and base radius r.

3. $\mathcal{M} = \{(x, y, z) \mid x^2 + y^2 - 4z \leq 0 \text{ and } z \leq 4\}$.

4. $\mathcal{M} = \{(x, y, z) \mid x^2 + y^2 \leq 5 \text{ and } x^2 + y^2 - z^2 \geq 9\}$.

In Exercises 5 to 8 use integration in spherical coordinates to determine the volumes described.

5. A sphere of radius a.

6. A right circular cone of height h and base radius a.

7. $\mathcal{M} = \left\{(R, \varphi, \theta) \mid 0 \leq R \leq 2a \cos \varphi, 0 \leq \varphi \leq \dfrac{\pi}{2}, 0 \leq \theta \leq 2\pi\right\}$.

8. $\mathcal{M} = \left\{(R, \varphi, \theta) \mid \sec \varphi \leq R \leq 2 \sec \varphi, 0 \leq \varphi \leq \dfrac{\pi}{4}, 0 \leq \theta \leq 2\pi\right\}$.

ANSWERS

1. $\dfrac{4\pi a^3}{3}$ units³.   3. 32π units³.   6. $\dfrac{\pi r^2 h}{3}$ units³.   8. $\dfrac{7\pi}{3}$ units³.

20

L'HOPITAL'S RULE AND IMPROPER INTEGRALS

1. INDETERMINATES OF THE FORM $\frac{0}{0}$; L'HOPITAL'S RULE

Let $f:\mathcal{R} \to \mathcal{R}$ and $g:\mathcal{R} \to \mathcal{R}$ be such that $\lim_{x \to a} f(x)$ and $\lim_{x \to a} g(x)$ exist as finite numbers. In Section 6 of Chapter 3, p. 182, the sum, the product, and the quotient properties of limits were established:

$$(20.1) \qquad \lim_{x \to a} [f(x) + g(x)] = \lim_{x \to a} f(x) + \lim_{x \to a} g(x).$$

$$(20.2) \qquad \lim_{x \to a} f(x) \cdot g(x) = \lim_{x \to a} f(x) \cdot \lim_{x \to a} g(x).$$

$$(20.3) \qquad \lim_{x \to a} \frac{f(x)}{g(x)} = \frac{\lim_{x \to a} f(x)}{\lim_{x \to a} g(x)}, \quad \text{if } \lim_{x \to a} g(x) \neq 0.$$

From the continuity of the logarithm function and the product property (20.2) the reader should be able to establish the exponential property of limits:

$$(20.4) \qquad \lim_{x \to a} [f(x)]^{g(x)} = [\lim_{x \to a} f(x)]^{\lim_{x \to a} g(x)}, \quad \text{if } \lim_{x \to a} f(x) \neq 0.$$

In practice, many limit problems arise which have the sum, product, quotient, or exponential form but are such that an attempt to apply the preceding limit properties leads to meaningless expressions. For example, consider the problems

(20.5)

$$\text{(a)} \quad \lim_{x \to 0} \frac{x}{x}, \qquad\qquad \text{(b)} \quad \lim_{x \to 0} \frac{x^2}{x},$$

$$\text{(c)} \quad \lim_{x \to 0} \frac{x}{x^3}, \qquad\qquad \text{(d)} \quad \lim_{x \to 0} \frac{|x|}{x}.$$

In each case the quotient property of limits is not applicable since the limit of each of the functions in the denominators is zero. If we neglect to check this fact and attempt to apply Property 20.3 blindly we obtain

(20.6)

$$\text{(a)} \quad \lim_{x \to 0} \frac{x}{x} = \frac{0}{0}, \qquad\qquad \text{(b)} \quad \lim_{x \to 0} \frac{x^2}{x} = \frac{0}{0},$$

$$\text{(c)} \quad \lim_{x \to 0} \frac{x}{x^3} = \frac{0}{0}, \qquad\qquad \text{(d)} \quad \lim_{x \to 0} \frac{|x|}{x} = \frac{0}{0}.$$

Thus each problem leads to the meaningless symbol $\frac{0}{0}$. These four problems are examples of what is termed an *indeterminate of the form* $\frac{0}{0}$, a terminology which we now make precise.

INDETERMINATE OF THE FORM $\frac{0}{0}$:

A limit problem of the type

$$\lim_{x \to l} \frac{f(x)}{g(x)} \quad \text{where} \quad \begin{cases} \lim_{x \to l} f(x) = 0 \quad \text{and} \\[2mm] \lim_{x \to l} g(x) = 0 \end{cases}$$

is called an *indeterminate of the form* $\frac{0}{0}$. The limit l may be a real number or may be $+\infty$ or $-\infty$.

Let us reexamine the limit problems given in (20.5). By the use of algebraic simplification it is easy to see that

(a) $\quad \lim_{x \to 0} \dfrac{x}{x} = \lim_{x \to 0} 1 = 1.$

(b) $\quad \lim_{x \to 0} \dfrac{x^2}{x} = \lim_{x \to 0} x = 0.$

(c) $\quad \lim_{x \to 0} \dfrac{x}{x^3} = \lim_{x \to 0} \dfrac{1}{x^2} = \infty.$

(20.7)

(d) $\quad \lim_{x \to 0^-} \dfrac{|x|}{x} = \lim_{x \to 0^-} \dfrac{-x}{x} = \lim_{x \to 0^-} -1 = -1,$

$\qquad \lim_{x \to 0^+} \dfrac{|x|}{x} = \lim_{x \to 0^+} \dfrac{x}{x} = \lim_{x \to 0^+} 1 = 1,$

and thus $\lim_{x \to 0} \dfrac{|x|}{x}$ does not exist.

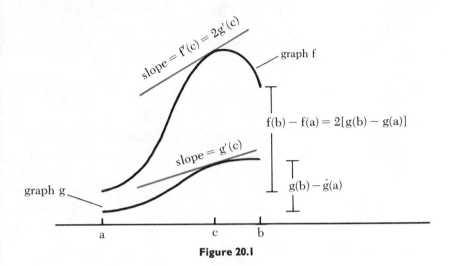

Figure 20.1

Thus, an indeterminate of the form $\frac{0}{0}$ may possess a limit in the re-stricted sense (that is, a real number) such as in (20.7a, b), or a limit in the extended sense $(+\infty$, or $-\infty)$ as in (20.7c), or no limit at all, as in (20.7d). Thus the form $\frac{0}{0}$ gives no clue whatsoever to whether the limit exists or not, nor to the nature of the limit if it does exist. This is why such problems are called *indeterminate* forms. The definition of the derivative of a differentiable function f is an indeterminate of the form $\frac{0}{0}$:

$$f'(t) = \lim_{x \to t} \frac{f(x) - f(t)}{x - t} . \quad \text{(Check this!)}$$

Indeterminates of other types shall be considered in later sections of this chapter.

There are several theorems, collectively referred to as L'Hopital's Rule, which are frequently helpful in evaluating the limits of indeterminates, if they exist. In the first three sections of this chapter, we discuss and apply these theorems.

An extension of the Mean Value Theorem, p. 257, known as Cauchy's Generalized Mean Value Theorem, will be very helpful in the proofs of L'Hopital's Rule. We now develop this extension.

Figure 20.1 represents two functions f and g, both differentiable on (a, b) and such that g is strictly increasing on (a, b) and the increase in f is twice that of g on the interval: $f(b) - f(a) = 2[g(b) - g(a)]$. The next theorem allows us to conclude that in this case there is a point c in (a, b) such that the slope $f'(c)$ of f at c is twice that of $g:f'(c) = 2g'(c)$. In general we have

THEOREM 20.1   (Cauchy's Generalized Mean Value Theorem.) Let $f:[a, b] \to \mathfrak{R}$, $g:[a, b] \to \mathfrak{R}$ be continuous and such that both f and g are differentiable on (a, b) with $g'(x) \neq 0$ for each x in (a, b). Then there exists a point c of (a, b) such that

$$(20.8) \qquad \frac{f(b) - f(a)}{g(b) - g(a)} = \frac{f'(c)}{g'(c)}.$$

NOTE.   If we choose g such that $g(x) = x$, Theorem 20.1 reduces to the ordinary mean value theorem.

Proof.   Since $g'(x) \neq 0$ for each x in (a, b), the mean value theorem (p. 257) requires that $g(b) - g(a) \neq 0$. Hence Equation 20.8 is equivalent to

$$(20.9) \qquad 0 = f'(c)(g(b) - g(a)) - g'(c)(f(b) - f(a)).$$

We define the function $\phi:[a, b] \to$ by

$$\phi(x) = f(x)(g(b) - g(a)) - g(x)(f(b) - f(a)).$$

A point c in (a, b) such that $\phi'(c) = 0$ will satisfy Equation 20.9 and hence also Equation 20.8. Noting that $\phi(a) = \phi(b)$, such a c exists by the ordinary mean value theorem (p. 257) as applied to ϕ. Thus the theorem is proved. ∎

Look now at Figure 20.2 where we have represented two functions f and g such that $\lim\limits_{x \to a+} f(x) = \lim\limits_{x \to a+} g(x) = 0$ and such that $\lim\limits_{x \to a+} f'(x)$ exists and

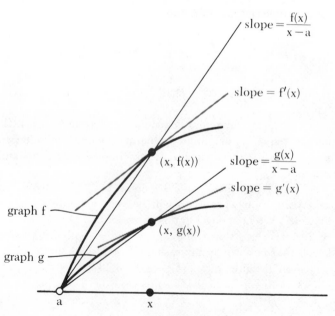

Figure 20.2

is finite (as is also $\lim\limits_{x \to a^+} g'(x)$). From this last criterion one would expect the slope of f at x, namely $f'(x)$, to be approximately the same as the slope of the secant line from $(a, 0)$ to $(x, f(x))$:

$$f'(x) \approx \frac{f(x)}{x - a} \quad \text{(where } \approx \text{ means ``is approximately equal to'').}$$

Similarly, we would expect that

$$g'(x) \approx \frac{g(x)}{x - a},$$

and consequently that

$$\frac{f'(x)}{g'(x)} \approx \frac{f(x)}{g(x)} \quad \text{for } x \text{ close to } a.$$

The following theorem bears out these expectations.

THEOREM 20.2 (L'Hopital's First Rule.) Let $f : (a, b) \to \mathcal{R}$ and $g : (a, b) \to \mathcal{R}$ be differentiable. Let $\lim\limits_{x \to l} \dfrac{f(x)}{g(x)}$ be an indeterminate of the form $\dfrac{0}{0}$, where $l = a^+$, or $l = b^-$, or $a < l < b$. Then

$$\lim_{x \to l} \frac{f(x)}{g(x)} = \lim_{x \to l} \frac{f'(x)}{g'(x)}$$

provided that
  (i) the second limit exists, either as a real number or as an infinite limit, and
  (ii) $g'(x) \neq 0$ $\begin{cases} \text{on } (a, b), \text{ if } l = a^+ \text{ or } l = b^- \\ \text{on } (a, l) \text{ and } (l, b), \text{ if } a < l < b. \end{cases}$

Proof. We assume that $\lim\limits_{x \to l} \dfrac{f'(x)}{g'(x)}$ is finite. We redefine f and g so that $f(a) = g(a) = 0$. For each $x > a$, we use the Generalized Mean Value Theorem to choose c_x such that

$$(20.10) \qquad \frac{f(x)}{g(x)} = \frac{f(x) - 0}{g(x) - 0} = \frac{f(x) - f(a)}{g(x) - g(a)} = \frac{f'(c_x)}{g'(c_x)}$$

where $a < c_x < x$. It follows easily from the definition of limit that

$$\lim_{x \to a^+} \frac{f'(c_x)}{g'(c_x)} = \lim_{x \to a^+} \frac{f'(x)}{g'(x)}.$$

Therefore, from Equation 20.10,

$$\lim_{x \to a+} \frac{f(x)}{g(x)} = \lim_{x \to a+} \frac{f'(x)}{g'(x)}.$$

The proof of the remainder of the theorem is left to the reader. ■

L'Hopital's First Rule enables us to evaluate indeterminates of the form $\frac{0}{0}$.

Example 1. We wish to evaluate $\lim_{x \to 0} \frac{e^x - 1}{\tan 2x}$. We note that this is an indeterminate of the form $\frac{0}{0}$ and that the conditions of L'Hopital's First Rule apply. We obtain thereby

$$\lim_{x \to 0} \frac{e^x - 1}{\tan 2x} = \lim_{x \to 0} \frac{e^x}{2 \sec^2 2x} = \frac{1}{2}. \quad \blacksquare$$

Example 2. We wish to find $\lim_{x \to \frac{\pi}{2}} \frac{\cos 3x}{\sin 2x}$. We note that $\lim_{x \to \frac{\pi}{2}} \cos 3x = \lim_{x \to \frac{\pi}{2}} \sin 2x = 0$ and that $(\sin 2x)' \neq 0$ on $\left(\frac{\pi}{4}, \frac{\pi}{2}\right)$ and $\left(\frac{\pi}{2}, \frac{3\pi}{4}\right)$. Thus, L'Hopital's First Rule applies and we have

$$\lim_{x \to \frac{\pi}{2}} \frac{\cos 3x}{\sin 2x} = \lim_{x \to \frac{\pi}{2}} \frac{-3 \sin 3x}{2 \cos 2x} = \frac{-3}{2}. \quad \blacksquare$$

Notice that although both the functions of Examples 1 and 2 lead to the indeterminate form $\frac{0}{0}$, the limits are different. It may have been tempting in elementary algebra to define $\frac{0}{0} = 1$. We can see now one reason why this is not done.

Sometimes it is necessary to apply L'Hopital's First Rule two or more times.

Example 3. We wish to evaluate $\lim_{x \to 0} \frac{\tan x - \sin x}{x^2}$. This limit is an indeterminate of the form $\frac{0}{0}$. Applying L'Hopital's First Rule we obtain

$$\lim_{x \to 0} \frac{\tan x - \sin x}{x^2} = \lim_{x \to 0} \frac{\sec^2 x - \cos x}{2x}.$$

But this last limit is also an indeterminate of the form $\frac{0}{0}$. We apply

L'Hopital's First Rule a second time.

$$\lim_{x \to 0} \frac{\sec^2 x - \cos x}{2x} = \lim_{x \to 0} \frac{2 \sec^2 x \tan x + \sin x}{2}.$$

This last limit is not indeterminate and is in fact 0. Therefore

$$\lim_{x \to 0} \frac{\tan x - \sin x}{x^2} = 0. \ \blacksquare$$

Each time L'Hopital's rule is applied it is necessary to check that the conditions of the theorem are satisfied.

Example 4. $\displaystyle \lim_{x \to 0^+} \frac{e^x - 1}{x^2} = \lim_{x \to 0^+} \frac{e^x}{2x} = +\infty.$

Note that, had we carelessly applied the rule a second time, we would have obtained

$$\lim_{x \to 0^+} \frac{e^x}{2x} = \lim_{x \to 0^+} \frac{e^x}{2} = \frac{1}{2}, \quad \text{an incorrect result.} \ \blacksquare$$

At each step one should look for algebraic simplifications and for applications of the algebraic properties of limits to isolate terms or factors whose limits are apparent.

Example 5.

$$\lim_{x \to 0^+} \frac{\ln (1 + xe^{2x})}{x^2} = \lim_{x \to 0^+} \frac{2xe^{2x} + e^{2x}}{2x(1 + xe^{2x})} = \lim_{x \to 0^+} \left[\frac{e^{2x}}{1 + xe^{2x}} + \frac{e^{2x}}{2x(1 + xe^{2x})} \right].$$

Now as $x \to 0$, the first term of the above expression goes to one and the second term goes to $+\infty$. Thus the limit of the entire expression is $+\infty$. $\blacksquare$

Example 6. $\displaystyle \lim_{x \to 0} \frac{\sin x \cos x}{\ln(x + 1)} = \lim_{x \to 0} \cos x \cdot \lim_{x \to 0} \frac{\sin x}{\ln(x + 1)}$

$$= 1 \cdot \lim_{x \to 0} \frac{\cos x}{\dfrac{1}{x + 1}} = \lim_{x \to 0} (x + 1) \cos x$$

$$= \lim_{x \to 0} (x + 1) \cdot \lim_{x \to 0} \cos x = 1. \ \blacksquare$$

EXERCISES

Find the following limits.

1. $\displaystyle \lim_{x \to 0^+} \frac{\sin x}{1 - \cos x}.$

2. $\displaystyle \lim_{x \to \frac{\pi}{2}^+} \frac{\cos x}{1 - \sin x}.$

3. $\displaystyle \lim_{x \to 0} \frac{e^x - e^{-x}}{x}.$

4. $\displaystyle \lim_{x \to 0} \frac{\ln 3x}{1 - 3x}.$

5. $\lim\limits_{x \to e} \dfrac{1 - \ln x}{e^2 - x^2}$.

6. $\lim\limits_{x \to 0} \dfrac{e^x - \cos x}{\tan x}$.

7. $\lim\limits_{x \to 0} \dfrac{e^x - \cos x}{\sin 2x}$.

8. $\lim\limits_{x \to 1} \dfrac{\ln x^2}{\sin \pi x}$.

9. $\lim\limits_{x \to 0} \dfrac{\tan x - x}{\sin x}$.

10. $\lim\limits_{x \to 0} \dfrac{\sin 2x + \tan 2x}{x}$.

11. $\lim\limits_{x \to 0} \dfrac{x^2 + 3x}{\sin x}$.

12. $\lim\limits_{x \to \pi} \dfrac{\sin^2 x}{\cos 2x - 1}$.

13. $\lim\limits_{x \to 0} \dfrac{e^{x^2} - e^{-x^2}}{2x^2}$.

14. $\lim\limits_{x \to 0} \dfrac{\sin x - x}{1 - \cos x}$.

15. $\lim\limits_{x \to 0} \dfrac{\tan x - x}{\sin x - x}$.

16. $\lim\limits_{x \to \frac{\pi}{4}^-} \dfrac{\ln (\tan x)}{\cos 2x}$.

17. $\lim\limits_{x \to \frac{\pi}{2}^-} \dfrac{\cos x - x + \frac{\pi}{2}}{x - x \sin x}$.

18. $\lim\limits_{x \to 0} \dfrac{e^{x^2} - \cos x}{x^2}$.

19. $\lim\limits_{x \to 0} \dfrac{\cos\left(\frac{\pi}{2} \cos x\right)}{\sin x}$.

20. $\lim\limits_{x \to \frac{\pi}{2}^-} \dfrac{\cos 5x}{1 - \sin x}$.

21. $\lim\limits_{x \to 3} \dfrac{x^2 - 2x - 3}{x^2 - 9}$.

22. $\lim\limits_{x \to 2} \dfrac{x^2 - 3x + 2}{x^2 - 4}$.

23. Prove Theorem 20.2 for the case where $\lim\limits_{x \to 0+} \dfrac{f'(x)}{g'(x)} = +\infty$.

24. Suppose we wish to evaluate $\lim\limits_{x \to 0} \dfrac{x^4 \cos \frac{1}{x}}{x^3}$. We can of course divide out x^3 and obtain $\lim\limits_{x \to 0} x \cos \dfrac{1}{x}$ which is obviously zero. However, if we attempt to use L'Hopital's Rule we obtain

$$\lim\limits_{x \to 0} \dfrac{x^4 \cos \frac{1}{x}}{x^3} = \lim\limits_{x \to 0} \dfrac{\sin \frac{1}{x}}{3}.$$

But this limit does not exist. What is wrong?

ANSWERS

1. $+\infty$.   **3.** 2.   **5.** $\dfrac{1}{2e^2}$.   **7.** 1/2.   **9.** 0.   **11.** 3.   **13.** 1.

15. -2.   **17.** ∞.   **19.** 0.   **21.** 2/3.

2. OTHER FORMS OF L'HOPITAL'S RULE;

INDETERMINATE FORMS $\dfrac{\infty}{\infty}$, $0 \cdot \infty$, $\infty - \infty$

The next theorem extends L'Hopital's Rule to indeterminates of the form $\dfrac{0}{0}$ when $l = +\infty$ or $-\infty$.

THEOREM 20.3 (L'Hopital's Second Rule.) Let $f : (a, +\infty) \to \mathfrak{R}$ and $g : (a, +\infty) \to \mathfrak{R}$ be differentiable on $(a, +\infty)$. Let $\displaystyle\lim_{x \to +\infty} \frac{f(x)}{g(x)}$ be an indeterminate of the form $\dfrac{0}{0}$. Then

$$\lim_{x \to +\infty} \frac{f(x)}{g(x)} = \lim_{x \to +\infty} \frac{f'(x)}{g'(x)}$$

provided that

(i) the second limit exists, either as a real number or as an infinite limit, and

(ii) $g'(x) \neq 0$ on $(a, +\infty)$.

A similar statement holds if $\displaystyle\lim_{x \to -\infty} \frac{f(x)}{g(x)}$ is an indeterminate of the form $\dfrac{0}{0}$.

Proof.

$$\lim_{x \to +\infty} \frac{f(x)}{g(x)} =^1 \lim_{t \to 0^+} \frac{f\left(\dfrac{1}{t}\right)}{g\left(\dfrac{1}{t}\right)}$$

$$=^2 \lim_{t \to 0^+} \frac{\dfrac{-1}{t^2} f'\left(\dfrac{1}{t}\right)}{\dfrac{-1}{t^2} g'\left(\dfrac{1}{t}\right)}$$

$$= \lim_{t \to 0^+} \frac{f'\left(\dfrac{1}{t}\right)}{g'\left(\dfrac{1}{t}\right)}$$

$$=^1 \lim_{x \to +\infty} \frac{f'(x)}{g'(x)} .$$

(1) Theorem 4.15, p. 212.
(2) Theorem 20.2 and the Chain Rule. ∎

Example 7.

$$\lim_{x \to +\infty} \frac{\ln(1 + e^{-x})}{e^{-2x}} =^1 \lim_{x \to +\infty} \frac{\frac{-e^{-x}}{1 + e^{-x}}}{-2e^{-2x}}$$

$$= \lim_{x \to +\infty} \frac{e^x}{2(1 + e^{-x})} = +\infty. \quad \blacksquare$$

(1) Theorem 20.3.

Other types of indeterminate forms which often arise are of the form $\dfrac{\infty}{\infty}$, $0 \cdot \infty$ and $\infty - \infty$.

INDETERMINATES OF THE FORM

$\dfrac{\infty}{\infty}$: $\lim\limits_{x \to l} \dfrac{f(x)}{g(x)}$, where $\begin{cases} \lim\limits_{x \to l} f(x) = \pm\infty, \text{ and} \\ \lim\limits_{x \to l} g(x) = \pm\infty \end{cases}$

$0 \cdot \infty$: $\lim\limits_{x \to l} f(x) \cdot g(x)$, where $\begin{cases} \lim\limits_{x \to l} f(x) = 0, \text{ and} \\ \lim\limits_{x \to l} g(x) = \pm\infty. \end{cases}$

$\infty - \infty$: $\lim\limits_{x \to l} f(x) - g(x)$, where $\begin{cases} \lim\limits_{x \to l} f(x) = +\infty, \text{ and} \\ \lim\limits_{x \to l} g(x) = +\infty \end{cases}$

The next Theorem extends L'Hopital's procedure to the form $\dfrac{\infty}{\infty}$.

THEOREM 20.4 (L'Hopital's Third Rule.) Let $f:(a, b) \to \mathcal{R}$ and $g:(a, b) \to \mathcal{R}$ be differentiable. Let $\lim\limits_{x \to l} \dfrac{f(x)}{g(x)}$ be an indeterminate of the form $\dfrac{\infty}{\infty}$, where $l = a^+$, or $l = b^-$, or $a < l < b$. Then

$$\lim_{x \to l} \frac{f(x)}{g(x)} = \lim_{x \to l} \frac{f'(x)}{g'(x)}$$

provided (i) the second limit exists, either as a real or infinite limit and
(ii) $g'(x) \neq 0$ on (a, b), or on (a, l) and (l, b), as appropriate.

The preceding statements remain valid if a is replaced by $-\infty$ or b is replaced by $+\infty$.

The proof is rather involved so we omit it. The interested reader is referred to *Real Variables* by John N. H. Olmsted, Appleton, Century, Croft, 1959.

We shall use L'Hopital's Third Rule to evaluate an indeterminate of the form $0 \cdot (-\infty)$.

Example 8.

$$\lim_{x \to 0+} x^2 \ln x = \lim_{x \to 0+} \frac{\ln x}{\dfrac{1}{x^2}}$$

$$= {}^2 \lim_{x \to 0+} \frac{\dfrac{1}{x}}{\dfrac{-2}{x^3}}$$

$$= -\lim_{x \to 0+} \frac{x^3}{2x} = -\lim_{x \to 0+} \frac{x^2}{2} = 0. \quad \blacksquare$$

Example 9.

$$\lim_{x \to +\infty} \frac{e^{2x} - \ln x}{e^x + x} = {}^1 \lim_{x \to +\infty} \frac{e^{2x}}{e^x + x} - \lim_{x \to +\infty} \frac{\ln x}{e^x + x}$$

$$= \lim_{x \to +\infty} \frac{2e^{2x}}{e^x + 1} - \lim_{x \to +\infty} \frac{1}{xe^x + x}$$

$$= {}^2 \lim_{x \to +\infty} \frac{4e^{2x}}{e^x} - 0$$

$$= \lim_{x \to +\infty} 4e^x = +\infty. \quad \blacksquare$$

(1) Elementary properties of limits.
(2) Theorem 20.4.

Indeterminates of the form $\infty - \infty$ can sometimes be evaluated by first changing them to the form $\dfrac{0}{0}$ or $\dfrac{\infty}{\infty}$ and then applying L'Hopital's rules.

Example 10.

$$\lim_{x \to 0} \left(\frac{1}{e^x - 1} - \frac{1}{x} \right) = {}^1 \lim_{x \to 0} \frac{x - e^x + 1}{x(e^x - 1)}$$

$$= {}^2 \lim_{x \to 0} \frac{1 - e^x}{e^x - 1 + xe^x}$$

$$= {}^2 \lim_{x \to 0} \frac{-e^x}{2e^x + xe^x} = \frac{-1}{2}. \quad \blacksquare$$

(1) Algebraic manipulation.
(2) L'Hopital's Rule.

EXERCISES

1. $\lim\limits_{x\to+\infty} \dfrac{e^{2x}}{1 + e^{3x}}$.

2. $\lim\limits_{x\to+\infty} \dfrac{\ln x}{x}$.

3. $\lim\limits_{x\to+\infty} \dfrac{e^x}{x^3}$.

4. $\lim\limits_{x\to\infty} \dfrac{3x^2 + 1}{4x^2 + x - 2}$.

5. $\lim\limits_{x\to+\infty} \dfrac{x^2 \ln x}{e^x}$.

6. $\lim\limits_{x\to+\infty} \dfrac{\ln (1 + e^{-x})}{e^{-x}}$.

7. $\lim\limits_{x\to-\infty} x \ln |x|$.

8. $\lim\limits_{x\to0} \dfrac{\cot x}{\cot 2x}$.

9. $\lim\limits_{x\to0+} \sqrt{x} \ln \dfrac{1}{x}$.

10. $\lim\limits_{x\to0+} x \csc x$.

11. $\lim\limits_{x\to+\infty} \dfrac{x^2}{\ln (1 + x^4)}$.

12. $\lim\limits_{x\to+\infty} \dfrac{\sqrt{1 + x^2}}{\ln x}$.

13. $\lim\limits_{x\to0} \left(\dfrac{1}{x} - \dfrac{1}{\sin x} \right)$.

14. $\lim\limits_{x\to2} \dfrac{4}{(x^2 - 4)} - \dfrac{1}{x - 2}$.

15. $\lim\limits_{x\to1} \left(\dfrac{x}{x - 1} - \dfrac{1}{\ln x} \right)$.

16. $\lim\limits_{x\to1-} \left(\dfrac{x}{x - 1} - \tan \pi x \right)$.

17. $\lim\limits_{x\to+\infty} \dfrac{1}{x \tan \left(\dfrac{\pi}{2} - \dfrac{1}{x} \right)}$.

18. $\lim\limits_{x\to\infty} \dfrac{1 - e^{1/x}}{\sin \dfrac{1}{x}}$.

19. $\lim\limits_{x\to0+} \dfrac{\cos x}{\ln x}$.

20. $\lim\limits_{x\to0+} x \cot x$.

ANSWERS

1. 0.   **3.** $+\infty$.   **5.** 0.   **7.** $-\infty$.   **9.** 0.   **11.** $+\infty$.

13. 0.   **15.** $\frac{1}{2}$.   **17.** 0.   **19.** 0.

3. INDETERMINATES OF THE FORMS 1^∞, 0^0, ∞^0

In this section we introduce the three remaining indeterminate forms.

INDETERMINATE OF THE FORM 1^∞:

$\lim\limits_{x\to l} (f(x))^{g(x)}$,   where $\begin{cases} \lim\limits_{x\to l} f(x) = 1, \quad \text{and} \\ \lim\limits_{x\to l} g(x) = \infty. \end{cases}$

Indeterminates of this type often can be evaluated by the use of the exponential and logarithmic functions and L'Hopital's rule.

Example II. We wish to evaluate $\lim\limits_{x \to 1} x^{\frac{1}{x-1}}$. Using the *exp* and *ln* functions we write

$$\lim_{x \to 1} x^{\frac{1}{x-1}} =^1 \lim_{x \to 1} \exp\left(\frac{1}{x-1} \cdot \ln x\right)$$

$$=^2 \exp\left(\lim_{x \to 1} \frac{\ln x}{x-1}\right)$$

$$=^3 \exp\left(\lim_{x \to 1} \frac{\frac{1}{x}}{1}\right)$$

$$= \exp 1$$

$$=^4 e. \quad \blacksquare$$

(1) Definition of $x^{\frac{1}{x-1}}$.

(2) Using the continuity of the *exp* function.

(3) Using L'Hopital's Rule.

(4) Definition of e.

INDETERMINATE OF THE FORM 0^0:

$$\lim_{x \to l} (f(x))^{g(x)} \quad \text{where} \quad \begin{cases} \lim\limits_{x \to l} f(x) = 0 \\ \lim\limits_{x \to l} g(x) = 0 \end{cases}$$

Example 12. We wish to evaluate $\lim\limits_{x \to 0} x^x$.

$$\lim_{x \to 0} x^x =^1 \lim_{x \to 0} \exp\left(x \cdot \ln x\right)$$

$$=^2 \exp\left(\lim_{x \to 0} x \ln x\right)$$

$$=^3 \exp\left(\lim_{x \to 0} \frac{\ln x}{\frac{1}{x}}\right)$$

$$=^4 \exp\left(\lim_{x \to 0} \frac{\frac{1}{x}}{-\frac{1}{x^2}}\right)$$

$$= \exp\left(\lim_{x \to 0} - x\right)$$

$$= \exp 0 = 1 \quad \blacksquare$$

(1) Usual beginning technique for this type of problem.

(2) Continuity of *exp* function.

(3) Converting $0 \cdot (-\infty)$ form to $\left(\dfrac{\infty}{\infty}\right)$ form.

(4) Using L'Hopital's Rule.

INDETERMINATE OF THE FORM ∞°:

$$\lim_{x \to l} (f(x))^{g(x)} \quad \text{where} \quad \begin{cases} \lim\limits_{x \to l} f(x) = \infty, \quad \text{and} \\ \lim\limits_{x \to l} g(x) = 0. \end{cases}$$

Example 13. We wish to evaluate $\lim\limits_{x \to \frac{\pi}{2}^-} (\tan x)^{\cos x}$. We have

$$\lim_{x \to \frac{\pi}{2}^-} (\tan x)^{\cos x} =^1 \exp \left(\lim_{x \to \frac{\pi}{2}^-} (\cos x) \ln \tan x \right)$$

$$=^2 \exp \left(\lim_{x \to \frac{\pi}{2}^-} \frac{\ln \tan x}{\sec x} \right)$$

$$=^3 \exp \left(\lim_{x \to \frac{\pi}{2}^-} \frac{\dfrac{\sec^2 x}{\tan x}}{\sec x \tan x} \right)$$

$$=^4 \left(\lim_{x \to \frac{\pi}{2}^-} \frac{\sec x}{\tan^2 x} \right)$$

$$=^4 \exp \left(\lim_{x \to \frac{\pi}{2}^-} \frac{\cos x}{\sin^2 x} \right)$$

$$= \exp 0$$

$$= 1 \quad \blacksquare$$

(1) Usual beginning technique for exponential indeterminate forms.

(2) Converting $0 \cdot \infty$ form to $\dfrac{\infty}{\infty}$ form.

(3) Using L'Hopital's Rule.

(4) Trigonometric simplification.

L'Hopital's Rule is not the sole answer to evaluating indeterminate forms. Sometimes it is necessary to rewrite the original problem.

Example 14.   Consider the problem $\lim\limits_{x \to -\infty} \dfrac{e^x}{x^{-1}}$.   This is an indeterminate of the form $\dfrac{0}{0}$.   An attempt to apply L'Hopital's Rule gives

$$\lim_{t \to -\infty} \frac{e^x}{x^{-1}} = \lim_{x \to -\infty} \frac{e^x}{-x^{-2}} = \lim_{x \to -\infty} \frac{e^x}{2x^{-3}} = \cdots$$

It is clear that repeated attempts to apply L'Hopital's Rule in this manner merely trade one indeterminate form for another in an endless chain.

However, rewriting the original problem meets with success:

$$\lim_{x \to -\infty} \frac{e^x}{x^{-1}} =^1 \lim_{x \to -\infty} \frac{x}{e^{-x}}$$

$$=^2 \lim_{x \to -\infty} \frac{1}{-e^{-x}}$$

$$= 0. \quad \blacksquare$$

(1) Converting a $\dfrac{0}{0}$ form to an $\dfrac{\infty}{\infty}$ form.

(2) Applying L'Hopital's Rule.

In the first three sections of this chapter we have introduced limit problems of a type which are called indeterminate forms. These problems have one of the following forms:

(20.11) $$\frac{0}{0}, \frac{\infty}{\infty}, \quad 0 \cdot \infty, \quad \infty - \infty, \quad 1^\infty, \quad 0^0, \quad \infty^0.$$

Each notation representing an indeterminate form involves two symbols. Each form gives the algebraic type of the limit problem—a quotient, a product, a difference, an exponential—as well as the limit of each of the two functions involved—0, ∞, 1. Such forms are called indeterminate because different choices of the component functions yield different results.

The reader should be able to recognize other limit problems which may be represented in a manner similar to (20.11) but which are *not* indeterminate forms:

$$(+\infty) \cdot (+\infty), \quad (+\infty) \cdot (-\infty), \quad \infty + \infty, \quad 1^0, \quad 0^0, \quad \infty^\infty.$$

In fact, one can show that

(20.12) $$\infty \cdot \infty = \infty \quad \infty \cdot (-\infty) = -\infty \quad \infty + \infty = \infty$$
$$1^0 = 1 \quad 0^\infty = 0 \quad \infty^\infty = \infty \quad \infty^{-\infty} = 0.$$

See the next example and Exercise 19.

Example 15.

$$\infty \cdot \infty: \quad \lim_{x \to \infty} x \cdot x^5 = \lim_{x \to \infty} x^6 = \infty.$$

$$\infty(-\infty): \quad \lim_{x \to \infty} x^2 \cdot (-x^2) = \lim_{x \to \infty} (-x^4) = -\infty.$$

$$\infty + \infty: \quad \lim_{x \to \infty} (x^2 + \ln x) = \infty.$$

$$1^0: \quad \lim_{x \to 1} x^{\sin x} =^1 \exp \left(\lim_{x \to 1} (\sin x) \ln x \right)$$

$$= \exp (0) = 1.$$

$$0^\infty: \quad \lim_{x \to 0} x^{\frac{1}{x}} =^1 \lim_{x \to 0} \exp \left(\frac{1}{x} \ln x \right)$$

$$=^2 0.$$

$$\infty^\infty: \quad \lim_{x \to \infty} (e^x)^x = \lim_{x \to \infty} e^{x^2} = \infty.$$

$$\infty^{-\infty}: \quad \lim_{x \to \infty} (e^x)^{-x} = \lim_{x \to \infty} e^{-x^2} = \lim_{x \to \infty} \frac{1}{e^{x^2}} = 0. \quad \blacksquare$$

(1) Usual beginning for exponential problems.

(2) Since, as $x \to 0$, $\dfrac{1}{x} \cdot \ln x \to \infty \cdot (-\infty) = -\infty$ and

$$\lim_{t \to -\infty} \exp t = 0.$$

EXERCISES

Evaluate the following limits.

1. $\lim\limits_{x \to 0} (e^x + x)^{\frac{1}{x}}$.

2. $\lim\limits_{x \to \infty} \left(e^{\frac{1}{x}} + \dfrac{1}{x} \right)^x$.

3. $\lim\limits_{x \to 0} (\cos x)^{\frac{1}{x}}$.

4. $\lim\limits_{x \to \frac{\pi}{2}^-} (\sin x)^{\frac{\pi}{2} - x}$.

5. $\lim\limits_{x \to +\infty} \left(\dfrac{1}{x} \right)^{\frac{1}{x}}$.

6. $\lim\limits_{x \to 0+} (\sin x)^{\sin x}$.

7. $\lim\limits_{x \to 0+} \left(\dfrac{1}{x} \right)^x$.

8. $\lim\limits_{x \to \infty} x^{\frac{1}{x}}$.

9. $\lim\limits_{x \to \infty} \left(\dfrac{1}{x} \right)^x$.

10. $\lim\limits_{x \to 1+} \left(\dfrac{1}{1 - x} \right)^{1 - x}$.

11. $\lim\limits_{x \to \infty} \dfrac{\cos x}{x}$.

12. $\lim\limits_{x \to 0+} (\tan x)^{\frac{1}{x}}$.

13. $\lim\limits_{x \to \infty} \dfrac{e^x}{\sin x}$.

14. $\lim\limits_{x \to 0} e^x \ln\left(\frac{1}{x}\right)$.

15. $\lim\limits_{x \to \frac{\pi}{2}} \dfrac{\cos x}{\tan\left(x - \dfrac{\pi}{2}\right)}$

16. $\lim\limits_{x \to 0} x \sin \dfrac{1}{x}$.

17. $\lim\limits_{x \to 0} x \sin \dfrac{1}{x} \ln \dfrac{1}{x}$. $\left(\text{Hint: } \left|\sin \dfrac{1}{x}\right| \le 1.\right)$

18. $\lim\limits_{x \to 0^+} (e^x - 1)^x$.

19. Prove that $1^0 = 1$, $0^\infty = 0$, $\infty^\infty = \infty$ and $\infty^{-\infty} = 0$. (Hint: See Example 15.)

ANSWERS

1. e^2.   **3.** 1.   **5.** 1.   **7.** 1.   **9.** 0.   **11.** 0.

13. Limit does not exist.   **15.** -1.   **17.** 0.

4. INTEGRALS OF UNBOUNDED FUNCTIONS

Previously we have defined the integral only for functions which are defined and bounded on a closed interval. We now extend the concept of an integral to admit functions which are not bounded on the interval of integration.

Consider the function given by $f(x) = \dfrac{1}{\sqrt{x}}$, $0 < x \le 1$, and sketched in Figure 20.3. This function is undefined at 0 and is unbounded on $(0, 1]$.

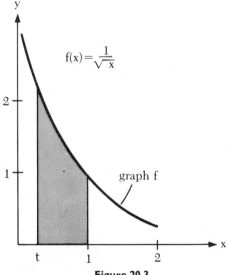

$f(x) = \dfrac{1}{\sqrt{x}}$

graph f

Figure 20.3

Therefore, according to the concept of an integral developed in Chapters 8 and 9 the symbol $\int_0^1 \dfrac{1}{\sqrt{x}}$ has no meaning. However, we do have

$$\int_t^1 \frac{1}{\sqrt{x}} = [2\sqrt{x}]_t^1 = 2 - 2\sqrt{t}, \quad \text{for} \quad 0 < t \le 1.$$

Consequently,

$$\lim_{t \to 0^+} \int_t^1 \frac{1}{\sqrt{x}} = \lim_{t \to 0^+} (2 - 2\sqrt{t}) = 2.$$

From this relation we conclude that the area shaded in Figure 20.3 can be made as close to 2 as we wish by choosing t sufficiently close to 0. It seems reasonable to say that the area under the graph of f from $x = 0$ to $x = 1$ is 2 and to define

$$\int_0^1 \frac{1}{\sqrt{x}} = 2.$$

With the preceding for motivation, we introduce the following definition.

DEFINITION 20.6   Let $f : (a, b] \to \Re$ be bounded and integrable on $[t, b]$ for every t in (a, b). The *integral of f over* $[a, b]$ is defined and denoted by

$$\int_a^b f = \lim_{t \to a^+} \int_t^b f,$$

providing that the limit exists and is finite.

In this case, we say that f is *integrable* on $[a, b]$ and that $\int_a^b f$ *converges*.

If the limit does not exist we say that $\int_a^b f$ *diverges*.

In particular if $\lim\limits_{t \to a^+} \int_t^b f = +\infty(-\infty)$ we say that $\int_a^b f$ *diverges to* $+\infty(-\infty)$.

A similar definition is made if f is defined on $[a, b)$ and is bounded and integrable on $[a, t]$ for every t (approaching b).

Note that if the definition of an integral just given is applied to a function f which is integrable in the sense of Chapter 8, we arrive at the same value for the integral. This happy result is a consequence of the continuity of the integral, Theorem 9.2, p. 380, which assures us that

$$\lim_{t \to a^+} \int_t^b f = \int_a^b f.$$

In order to make distinctions we refer to the integral of a bounded function on an interval $[a, b]$ as a *proper integral*; others are called *improper integrals*.

Example 16.

$$\int_0^2 \frac{1}{\sqrt{4-x^2}} = \lim_{t \to 2^-} \int_0^t \frac{1}{\sqrt{4-x^2}}$$

$$= \lim_{t \to 2^-} \left[\arcsin \frac{x}{2} \right]_0^t$$

$$= \lim_{t \to 2^-} \arcsin \frac{t}{2} = \arcsin 1 = \frac{\pi}{2}. \quad \blacksquare$$

Example 17.

$$\int_0^1 \frac{1}{x} = \lim_{t \to 0^+} \int_t^1 \frac{1}{x}$$

$$= \lim_{t \to 0^+} [\ln x]_t^1$$

$$= \lim_{t \to 0^+} (0 - \ln t)$$

$$= +\infty.$$

The integral diverges. $\quad \blacksquare$

The sketch of the graph of $\frac{1}{x}$ as shown in Figure 20.4 looks quite similar

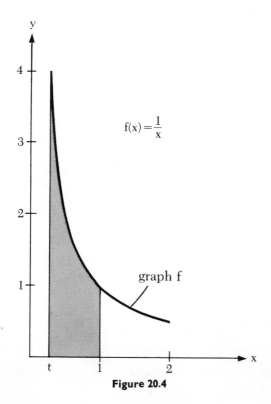

Figure 20.4

to that of $\dfrac{1}{\sqrt{x}}$ in Figure 20.3, yet $\int_0^1 \dfrac{1}{\sqrt{x}} = 2$, while $\int_0^1 \dfrac{1}{x} = +\infty$. Geo-

metrically this means that as $t \to 0$ the graph of $\dfrac{1}{x}$ rises so rapidly that the

shaded area in the figure increases without bound, but the graph of $\dfrac{1}{\sqrt{x}}$

rises slowly enough so that the area is bounded. We shall say that the area

between the graph of $\dfrac{1}{\sqrt{x}}$ and between 0 and 1 is $\int_0^1 \dfrac{1}{\sqrt{x}} = 2$; we do not

assign an area to the region beneath the graph of $\dfrac{1}{x}$ and between 0 and 1.

We can also give a meaning in some cases to the symbol $\int_a^b f$, where f
is undefined at a single point in $[a, b]$,

DEFINITION 20.7 Let $f : [a, c) \cup (c, b] \to \mathcal{R}$ be such that f is
integrable on both $[a, c]$ and $[c, b]$. The integral of f on $[a, b]$ is
defined and denoted by

$$\int_a^b f = \int_a^b f + \int_c^b f.$$

Note that this extends a familiar property of proper integrals to improper
integrals.

Example 18.

Evaluate $\int_{-1}^1 \dfrac{1}{x^2}$. We note that $\dfrac{1}{x^2}$ is undefined at 0. Hence we write

$$\int_{-1}^1 \frac{1}{x^2} = \int_{-1}^0 \frac{1}{x^2} + \int_0^1 \frac{1}{x^2}$$

$$= \lim_{t \to 0^-} \left[\frac{-1}{x} \right]_{-1}^t + \lim_{t \to 0^+} \left[\frac{-1}{x} \right]_t^1$$

$$= \infty + \infty = \infty. \quad \blacksquare$$

Note that, had we carelessly ignored the discontinuity at zero in the
previous example, we might have written

$$\int_{-1}^1 \frac{1}{x^2} = \left[\frac{-1}{x} \right]_{-1}^1 = {}^{\backprime}{-}2,$$

an obviously ridiculous result, since $\dfrac{1}{x^2}$ is positive everywhere it is defined.

We make the natural extensions of our definitions to functions which
have a finite number of discontinuities.

Example 19. Evaluate $\int_3^5 \left(\dfrac{1}{4-x}\right)^{\frac{2}{3}}$. We note that there is a discontinuity at $x = 4$. Hence

$$\int_3^5 \frac{1}{(4-x)^{\frac{2}{3}}} = \lim_{t\to 4^-} [-3(4-x)^{\frac{1}{3}}]_3^t + \lim_{t\to 4^+} [-3(4-x)^{\frac{1}{3}}]_t^5 = 6. \quad\blacksquare$$

In the case of Example 19, we would have gotten the correct answer even if we had ignored the discontinuity at 4. This is due to the fact that the primitive function $-3(4-x)^{\frac{1}{3}}$ is continuous at 4. The next theorem will tell us that we can ignore certain discontinuities in the function being integrated. Note that this is an extension of the Second Fundamental Theorem to possibly unbounded functions.

THEOREM 20.8 Let $f:(a, b) \to \Re$ be continuous. If there exists a function F which is continuous over $[a, b]$ and such that $F' = f$ on (a, b), then $\int_a^b f$ converges and

$$\int_a^b f = F(b) - F(a).$$

Proof. We give the proof for the case in which f is continuous on (a, b), but not necessarily bounded.

Let $a < c < b$. Then

$$\int_a^b f =^1 \int_a^c f + \int_c^b f$$

$$=^2 \lim_{t\to a^+} \int_t^c f + \lim_{t\to b^-} \int_c^t f$$

$$=^3 \lim_{t\to a^+} (F(c) - F(t)) + \lim_{t\to b^-} (F(t) - F(c))$$

$$= \lim_{t\to b^-} F(t) - \lim_{t\to a^+} F(t)$$

$$=^4 F(b) - F(a). \quad\blacksquare$$

(1) Definition 20.7.
(2) Definition 20.5.
(3) Using the Second Fundamental Theorem, p. 384, which requires that f be continuous on the intervals of integration.
(4) Since F is continuous at a and b.

In Example 19, the primitive, $-3(4-x)^{\frac{1}{3}}$, is continuous over $[3, 5]$. Hence the theorem applies and we can ignore the discontinuity at 4. In the case of Example 17, the primitive, $\dfrac{-1}{x}$, is not continuous on $[-1, 1]$; hence the discontinuity at 0 must be taken into account.

EXERCISES

Evaluate the following.

1. $\int_0^3 x^{-\frac{2}{3}}.$

2. $\int_1^3 \frac{1}{\sqrt{x-1}}.$

3. $\int_0^3 \frac{x}{\sqrt{9-x^2}}.$

4. $\int_0^3 \frac{1}{\sqrt{9-x^2}}.$

5. $\int_1^2 \frac{1}{(x-1)^{\frac{1}{3}}}.$

6. $\int_0^1 \frac{1}{\sqrt{1-x^2}}.$

7. $\int_0^1 \frac{1}{\sqrt{x}}.$

8. $\int_{-1}^1 \frac{1}{x^4}.$

9. $\int_1^3 \frac{1}{(2-x)^2}.$

10. $\int_0^\pi \sec^2 x.$

11. $\int_0^2 \frac{1}{\sqrt{2x-x^2}}.$

12. $\int_0^6 \frac{2x}{(x^2-4)^{\frac{2}{3}}}.$

13. $\int_0^1 \frac{\ln x}{x}.$

14. $\int_1^2 \frac{1}{x \ln x}.$

15. $\int_{\frac{1}{2}}^1 \frac{1.}{x (\ln x)^{\frac{1}{5}}}.$

16. $\int_{-\frac{1}{2}}^0 \frac{1}{2x+1}.$

17. $\int_{-2}^2 \frac{1}{\sqrt{4-x^2}}.$

18. $\int_0^2 \frac{x}{(x^2-1)^{\frac{2}{3}}}.$

ANSWERS

1. $3\sqrt[3]{3}.$ **3.** $3.$ **5.** $\frac{3}{2}.$ **7.** $2.$ **9.** $+\infty.$ **11.** $\pi.$ **13.** $-\infty.$

15. $\dfrac{5 (\ln 2)^{\frac{4}{5}}}{4}.$ **17.** $\pi.$

5. INTEGRALS ON UNBOUNDED INTERVALS

We shall now extend our definitions to give meaning to such symbols as $\int_a^{+\infty} f$, $\int_{-\infty}^a f$, and $\int_{-\infty}^\infty f$.

DEFINITION 20.9   Let $f:[a, \infty) \to \mathcal{R}$ be integrable on each interval $[a, t]$, where $t > a$. The *integral of f over* $[a, \infty)$ is defined and denoted by

$$\int_a^\infty f = \lim_{t \to +\infty} \int_a^t f,$$

providing the limit exists and is finite.

In this case we say that f is *integrable over* $[a, \infty]$ and that $\int_a^\infty f$ *converges.*

If the limit does not exist we say that $\int_a^\infty f$ diverges; in particular, if $\lim\limits_{t \to \infty} \int_a^t f = +\infty(-\infty)$ we say that $\int_a^\infty f$ diverges to $+\infty(-\infty)$.

We make the analogous definition for the integral of f over $(-\infty, a]$. Such integrals are also called *improper.*

Example 20.

$$\int_0^\infty e^{-x} = \lim_{t \to +\infty} \int_0^t e^{-x}$$
$$= \lim_{t \to +\infty} [-e^{-x}]_0^t$$
$$= \lim_{t \to +\infty} (-e^{-t} + e^0) = 1. \quad \blacksquare$$

Example 21.

$$\int_{-\infty}^0 x = \lim_{t \to -\infty} \int_t^0 x$$
$$= \lim_{t \to -\infty} \left[\frac{x^2}{2}\right]_t^0$$
$$= -\infty. \text{ The integral diverges to } -\infty. \quad \blacksquare$$

Example 22.

$$\int_0^\infty \cos x = \lim_{t \to +\infty} \int_0^t \cos x$$
$$= \lim_{t \to +\infty} [\sin x]_0^t$$
$$= \lim_{t \to +\infty} \sin t. \quad \blacksquare$$

This limit does not exist. The integral diverges.

The next theorem will prove useful to us in our study of infinite series in Chapter 21.

THEOREM 20.10 Let k be a real number.

$$\int_1^\infty x^{-k} = \frac{1}{k-1}, \quad \text{if} \quad k > 1,$$

diverges to $+\infty$, if $k \le 1$.

Proof. If $k = 1$, $\displaystyle\int_1^\infty x^{-k} = \lim_{t \to \infty} \int_1^t \frac{1}{x} = \lim_{t \to \infty} \ln t = +\infty$, and the given integral diverges to $+\infty$.

If $k \neq 1$

(20.13)
$$\int_1^\infty x^{-k} = \lim_{t \to \infty} \int_1^t x^{-k}$$
$$= \lim_{t \to \infty} \left[\frac{x^{1-k}}{1-k} \right]_1^t$$
$$= \lim_{t \to \infty} \left(\frac{t^{1-k}}{1-k} - \frac{1}{1-k} \right)$$
$$= \frac{1}{k-1} + \frac{1}{1-k} \lim_{t \to \infty} t^{1-k}$$

We set $y = t^{1-k}$. Then, since ln is continuous,
$$\lim_{t \to \infty} \ln y = \lim_{t \to \infty} \ln t^{1-k}$$
$$= \lim_{t \to \infty} (1-k) \ln t = (1-k) \lim_{t \to \infty} \ln t.$$

If $k < 1$, this last limit is $+\infty$; hence $y \to +\infty$ and therefore so does (20.13).

If $k > 1$, the last limit is $-\infty$, so $\lim_{t \to \infty} y = 0$ and (20.13) goes to $\dfrac{1}{k-1}$. ∎

The next definition gives meaning to an integral over $(-\infty, \infty)$.

DEFINITION 20.11 Let $f : \mathcal{R} \to \mathcal{R}$ be integrable (properly or improperly) on $[a, b]$ for each a, b in $\mathcal{R}$ with $a < b$. The *integral of f over* $\mathcal{R} = (-\infty, \infty)$ is defined and denoted by

$$\int_{-\infty}^\infty f = \int_{-\infty}^c f + \int_c^\infty f,$$

where c is any real number, providing both integrals on the right exist and are finite. In this case we say that f is *integrable over* $\mathcal{R}$ and that the integral $\int_{-\infty}^\infty f$ *converges.* Otherwise we say that $\int_{-\infty}^\infty f$ *diverges.*

Notice that the choice of c is arbitrary. It will be left to the reader to show that this is a good definition, that is, $\int_{-\infty}^\infty f$ will not be affected by the choice of c.

Example 23.

$$\int_{-\infty}^\infty \frac{1}{1+x^2} = \lim_{t \to -\infty} \int_t^0 \frac{1}{1+x^2} + \lim_{t \to +\infty} \int_0^t \frac{1}{1+x^2}$$
$$= \lim_{t \to -\infty} [\arctan x]_t^0 + \lim_{t \to +\infty} [\arctan x]_0^t$$
$$= \lim_{t \to -\infty} (-\arctan t) + \lim_{t \to +\infty} (\arctan t)$$
$$= -\left(\frac{-\pi}{2} \right) + \frac{\pi}{2} = \pi. \quad ∎$$

Theorems 20.12 and 20.13 will be needed in Chapter 21. Proofs are left to the reader.

THEOREM 20.12   Let $f:[a, \infty) \to \mathfrak{R}$ be such that $f(x) > 0$ for each $x > 0$. Moreover, let f be integrable on each interval $[a, t]$, where $t > a$. Then

$$\left\{ \int_a^t f \right\} \quad \text{is bounded if and only if} \quad \int_a^\infty f \text{ converges.}$$

THEOREM 20.13   Let $f:[a, \infty) \to \mathfrak{R}$ be integrable on $[a, t]$ for each $t \geq a$. Then, for every $b > a$ we have

$$\int_a^\infty = \int_a^b f + \int_b^\infty f.$$

Thus
(i) $\int_a^\infty f$ converges if and only if $\int_b^\infty f$ converges.
(ii) $\int_a^\infty f$ diverges to $+\infty(-\infty)$ if and only if $\int_b^\infty f$ diverges to $+\infty(-\infty)$.

This last theorem is an extension of the interval additivity property for integrals.

EXERCISES

Evaluate the integrals in Exercises 1 to 14.

1. $\int_2^\infty e^{-x}.$

2. $\int_1^\infty x^{-\frac{4}{3}}.$

3. $\int_1^\infty \frac{1}{\sqrt{x}}.$

4. $\int_1^\infty e^x.$

5. $\int_{-\infty}^0 \frac{1}{1 + x^2}.$

6. $\int_1^\infty \frac{\ln x}{x}.$

7. $\int_{-\infty}^\infty \frac{1}{1 + x^2}.$

8. $\int_0^\infty \frac{1}{x^2}.$

9. $\int_1^\infty \frac{x}{(2 + x^2)^{\frac{5}{2}}}.$

10. $\int_1^\infty e^{-x} \sin x.$

11. $\displaystyle\int_{-\infty}^{\infty} \frac{1}{2x^2 + 2x + 1}$.

12. $\displaystyle\int_{1}^{\infty} \frac{1}{x\sqrt{x-1}}$.

13. $\displaystyle\int_{0}^{\infty} \frac{1}{x\sqrt{1+x^2}}$.

14. $\displaystyle\int_{0}^{1} \frac{1}{x\sqrt{1-x^2}}$.

15. Find the area beneath the graph of $\dfrac{1}{(2x-6)^{1/4}}$ and between 3 and 5.

16. Show that if f is integrable on $(-\infty, \infty)$ and c and d are real numbers,

$$\int_{-\infty}^{c} f + \int_{c}^{\infty} f = \int_{-\infty}^{d} f + \int_{d}^{\infty} f.$$

17. Prove Theorem 20.12.   18. Prove Theorem 20.13.

ANSWERS

1. $\dfrac{1}{e^2}$.   **3.** $+\infty$.   **5.** $\dfrac{\pi}{2}$.   **7.** π.   **9.** $\dfrac{1}{\sqrt{243}}$.   **11.** π.   **13.** $+\infty$.

21

INFINITE SERIES

The topic of infinite series was introduced in Chapter 13. At that point we gave only a brief summary of the subject, omitting many details. In the current chapter we shall cover again much of the same ground, this time in greater depth. Definitions and theorems given earlier will be repeated for ease of reference. Some theorems will be given in a stronger form than given previously; in such cases we shall also give the theorems from Chapter 13 and call attention to the ways in which they have been strengthened.

1. SEQUENCES

DEFINITION 21.1 A sequence of real numbers is a function s, defined for each positive integer n and with range contained in the set of real numbers.

NOTE. We have adopted the notation $\{s(n)\}$, or $\{s(n)\}_{n=1}^{\infty}$, to denote the sequence s. Many other books use the notation s_n to represent $s(n)$. Thus the sequence $\left\{\dfrac{1}{n}\right\}$ might be referred to as the sequence defined by $s_n = \dfrac{1}{n}$. It is sometimes convenient to think of a sequence or series as a function from the *non-negative* integers into the reals. When we have need we shall do this without further comment. The reader will be able to tell by the notation when this is the case.

DEFINITION 21.2 Let $s : \mathcal{R} \rightarrow \mathcal{R}$ be a sequence of real numbers. Let $L \in \mathcal{R}$.

(i) $\{s(n)\}$ *converges to* L if and only if for each $r > 0$ there exists an integer N_r such that $|s(n) - L| < r$ for all $n \geq N_r$. In this case we write

$$\lim_{n \to \infty} s(n) = L.$$

(ii) $\{s(n)\}$ *diverges to* $+\infty$ if and only if for each $M > 0$ there exists an integer N_M such that $s(n) > M$ for all $n \geq N_M$. In this case we write

$$\lim_{n \to \infty} s(n) = +\infty$$

(iii) $\{s(n)\}$ *diverges to* $-\infty$ if and only if for each $M < 0$ there exists an integer N_M such that $s(n) < M$ for all $n \geq N_M$. In this case we write

$$\lim_{n \to \infty} s(n) = -\infty$$

(iv) $\{s(n)\}$ *diverges and oscillates* if and only if none of (i), (ii), or (iii) is true.

THEOREM 21.3 Let s and t be sequences of real numbers. Let $\lim_{n \to \infty} s(n) = L$ and $\lim_{n \to \infty} t(n) = K$, where L and K are real numbers. The following are true.

(i) $\lim_{n \to \infty} (s(n) + t(n)) = L + K$.

(ii) $\lim_{n \to \infty} rs(n) = rL$, for each real r.

(iii) $\lim_{n \to \infty} s(n)t(n) = LK$

(iv) $\lim_{n \to \infty} \dfrac{s(n)}{t(n)} = \dfrac{L}{K}$, if $K \neq 0$.

This is Theorem 13.7.

THEOREM 21.4 Let s be an increasing sequence of real numbers, that is, $s(n) \leq s(n + 1)$ for each positive integer n.

(i) If the set $\{s(n) \mid n \text{ is an integer}\}$ has an upper bound, then

$$\lim_{n \to \infty} s(n) = L$$

where $L = \text{lub } \{s(n) \mid n \text{ is an integer}\}$

(ii) If the set $\{s(n) \mid n$ is an integer$\}$ does not have an upper bound, then

$$\lim_{n \to \infty} s(n) = +\infty.$$

This is Theorem 13.8.

In L'Hopital's Rule, we have available a powerful new tool for the evaluation of sequential limits. In order to make use of it we need the following theorem.

THEOREM 21.5 Let s be a sequence and $f : \mathcal{R} \to \mathcal{R}$ be such that

$$s(n) = f(n)$$

for each positive integer n. Then

$$\lim_{n \to \infty} s(n) = \lim_{x \to \infty} f(x)$$

provided that the second limit exists.

Proof. We consider the case for which $\lim_{x \to \infty} f(x)$ is a finite number L. (The infinite case will be left to the exercises.) We have

$\lim_{x \to \infty} f(x) = L \Leftrightarrow^1$ for each $r > 0$, there exists $M > 0$ such that $|f(x) - L| < r$ whenever $x \in (M, +\infty)$.

$\Rightarrow^2$ for each $r > 0$, there exists an $M > 0$ such that $|s(n) - L| < r$ whenever $n \in (M, +\infty)$

$\Leftrightarrow^3$ for each $r > 0$, there exists an integer N_r such that $|s(n) - L| < r$ for all $n > N_r$.

$\Rightarrow^4 \lim_{n \to \infty} s(n) = L = \lim_{x \to \infty} f(x).$ ∎

(1) Definition 4.14, p. 212.
(2) Since $s(n) = f(n)$ for each n.
(3) Choose N_r to be any positive integer $> M$.
(4) Definition 13.6i, p. 528.

REMARK: The converse of the above theorem does not hold. For example, $\lim_{n \to \infty} \sin n\pi = 0$, but $\lim_{x \to \infty} \sin x\pi$ does not exist.

Example 1.

$$\lim_{n \to \infty} \frac{2n^2}{3n^2 - 2} =^1 \lim_{x \to \infty} \frac{2x^2}{3x^2 - 2}$$

$$=^2 \lim_{x \to \infty} \frac{4x}{6x} = \lim_{x \to \infty} \frac{2}{3} = \frac{2}{3}.$$ ∎

(1) Theorem 21.5.
(2) L'Hopital's Rule.

Example 2. Let us determine $\lim\limits_{n\to\infty} \sqrt[n]{n}$. By the preceding theorem we find that

$$\lim_{n\to\infty} \sqrt[n]{n} = \lim_{x\to\infty} \sqrt[x]{x} = \lim_{x\to\infty} (x)^{\frac{1}{x}}$$

$$=^1 \lim_{x\to\infty} \exp \ln (x^{\frac{1}{x}})$$

$$=^1 \exp \left(\lim_{x\to\infty} \frac{1}{x} \ln x \right)$$

$$= \exp \left(\lim_{x\to\infty} \frac{\ln x}{x} \right)$$

$$=^2 \exp \left(\lim_{x\to\infty} \frac{1}{x} \right)$$

$$= \exp 0 = 1. \quad \blacksquare$$

(1) Usual technique for evaluating an indeterminate of the form ∞^0.

(2) Application of L'Hopital's Rule to the $\dfrac{\infty}{\infty}$ form in preceding step.

This sequence of Example 2 occurs frequently. There are several others which also occur frequently and which we shall list, along with $\lim\limits_{n\to\infty} \sqrt[n]{n}$, in Theorem 21.7. To prove some parts of Theorem 21.7, we shall need Theorem 21.6, which follows.

THEOREM 21.6 Let s and t be sequences of real numbers such that

$$0 \leq t(n) \leq s(n) \quad \text{for each } n.$$

If $s(n)$ converges to zero, then so does $t(n)$:

$$\lim_{n\to\infty} s(n) = 0 \Rightarrow \lim_{n\to\infty} t(n) = 0.$$

The proof is left to the reader.

THEOREM 21.7 Let x and α denote any real numbers, p any positive real number, and n any positive integer.

(i) $\lim\limits_{n\to\infty} x^n = \begin{cases} +\infty & \text{if } x > 1. \\ 1 & \text{if } x = 1. \\ 0 & \text{if } -1 < x < 1. \\ \text{does not exist if } x \leq -1. \end{cases}$

$$\text{(ii)} \quad \lim_{n \to \infty} n^x = \begin{cases} +\infty, & \text{if } x > 0. \\ 1, & \text{if } x = 0. \\ 0, & \text{if } x < 0. \end{cases}$$

$$\text{(iii)} \quad \lim_{n \to \infty} x^{\frac{1}{n}} = \begin{cases} 1, & \text{if } x > 0. \\ 0, & \text{if } x = 0. \\ \text{does not exist, if } x < 0. \end{cases}$$

$$\text{(iv)} \quad \lim_{n \to \infty} \sqrt[n]{n} = 1.$$

$$\text{(v)} \quad \lim_{n \to \infty} \frac{n^{\alpha}}{(1 + p)^n} = 0.$$

Proof. (iv) was proved as Example 2. We shall prove (i) and (v), leaving the proofs of the remaining statements to the reader.

(i) *Let $x > 1$. Then*

$$\lim_{n \to \infty} x^n =^1 \lim_{n \to \infty} \exp{(n \ln x)} =^2 \exp{\infty} =^2 \infty$$

Let $x = 1$. Then $x^n = 1$ and $\lim_{n \to \infty} x^n = \lim_{n \to \infty} 1^n = \lim_{n \to \infty} 1 = 1.$
Let $-1 < x < 1$. Then

$$\lim_{n \to \infty} |x|^n =^1 \lim_{n \to \infty} \exp{(n \ln |x|)} =^3 \exp{(-\infty)} =^3 0$$

Hence $\lim_{n \to \infty} x^n = 0$.

Let $x = -1$. Then $x^n = \begin{cases} +1, \text{ if } n \text{ is even.} \\ -1, \text{ if } n \text{ is odd.} \end{cases}$ The sequence $\{x^n\}$ then

oscillates between $+1$ and -1.
Let $x < -1$. Then

$$\lim_{n \to \infty} |x|^n =^1 \lim_{n \to \infty} \exp{(n \ln |x|)} =^4 \exp{(+\infty)} = +\infty.$$

But since $x < 0$, x^n will oscillate in sign with the oddness or evenness of n as $n \to \infty$. Thus the odd powers x^n diverge to $-\infty$ and the even powers x^n diverge to $+\infty$. Consequently $\lim_{n \to \infty} x^n$ does not exist.

(1) Definition of the exponential x^n or $|x|^n$.
(2) The symbol *exp* ∞ represents a limit of the type $\lim_{x \to l} \exp y$ where $y \to \infty$ as $x \to l$. Since $\exp y \to \infty$ as $y \to \infty$ we may then write $\exp \infty = \infty$. In the above, $y = n \ln x \to \infty$ as $n \to \infty$.
(3) The symbol $\exp{(-\infty)}$ represents a limit of the type $\lim_{x \to l} \exp y$ where $y \to -\infty$ as $x \to l$. Since $\exp y \to 0$ as $y \to -\infty$ we write $\exp{(-\infty)} = 0$. In the above, $y = n \ln |x| \to -\infty$, since $\ln |x| < 0$ whenever $0 < |x| < 1$.
(4) This limit is of the type $\exp \infty$ since $n \ln |x| \to +\infty$ as $n \to \infty$.

(v) Choose k, a positive integer, such that $k > \alpha$. Then since n^x is a strictly increasing function of x for $n > 1$, we have $n^k > n^\alpha$ for each $n > 1$. We shall show that $\lim\limits_{n \to \infty} \dfrac{n^k}{(1 + p)^n} = 0$. The truth of (v) will then follow from Theorem 21.6.

$$\lim_{n \to \infty} \frac{n^k}{(1 + p)^n} =^1 \lim_{x \to \infty} \frac{x^k}{(1 + p)^x}$$

$$=^2 \lim_{x \to \infty} \frac{kx^{k-1}}{(1 + p)^x \ln (1 + p)}$$

$$=^2 \lim_{x \to \infty} \frac{k(k - 1)x^{k-2}}{(1 + p)^x (\ln (1 + p))^2}, \quad \text{if} \quad k - 1 \neq 0.$$

$$= \cdots =^3 \lim_{x \to \infty} \frac{k!\, x^0}{(1 + p)^x (\ln (1 + p))^k}$$

$$=^4 \frac{k!}{(\ln (1 + p))^k} \lim_{x \to \infty} \frac{1}{(1 + p)^x} =^5 0. \quad \blacksquare$$

(1) Theorem 21.5
(2) L'Hopital's Rule
(3) Using L'Hopital's Rule, the process is repeated until the exponent of x is 0.
(4) At this point L'Hopital's Rule no longer applies since the numerator is constant. We use Theorem 21.3 (ii) to move the constant term outside the limit symbol.
(5) By (iv).

The proofs of the following three theorems are left to the reader.

THEOREM 21.8   Alteration of a finite number of terms of a sequence does not affect either its convergence or its limit.

THEOREM 21.9   If a sequence converges to a limit, this limit is unique.

THEOREM 21.10   (Polynomial Test.)   Let

$$p(x) = a_n x^n + a_{n-1} x^{n-1} + \cdots + a_0$$

and $g(x) = b_m x^m + b_{m-1} x^{m-1} + \cdots + b_0$ be two polynomials with non-zero leading coefficients. Then

$$\lim_{x \to \infty} \frac{p(x)}{q(x)} = \lim_{x \to \infty} \frac{a_n x^n + \cdots + a_0}{b_m x^m + \cdots + b_0} = \begin{cases} +\infty, & \text{if } n > m \text{ and } \dfrac{a_n}{b_m} > 0. \\[2mm] -\infty, & \text{if } n > m \text{ and } \dfrac{a_n}{b_m} < 0. \\[2mm] \dfrac{a_n}{b_m}, & \text{if } n = m. \\[2mm] 0, & \text{if } n < m. \end{cases}$$

Example 3. (a) $\displaystyle \lim_{n \to \infty} \frac{n^3 + n + 1}{n} = +\infty$

(b) $\displaystyle \lim_{n \to \infty} \frac{-n^3 + 2n + 3}{n^2} = -\infty$

(c) $\displaystyle \lim_{n \to \infty} \frac{n}{n^2 - 3n} = 0$

(d) $\displaystyle \lim_{n \to \infty} \frac{2n^2 + 1}{3n^2 - n + 1} = \frac{2}{3}$. ∎

EXERCISES

Use L'Hopital's Rule to find the limits of the following sequences.

1. $\displaystyle \lim_{n \to \infty} \frac{n}{\ln n}$.

2. $\displaystyle \lim_{n \to \infty} \frac{(\ln n)^2}{n^2}$.

3. $\displaystyle \lim_{n \to \infty} \frac{\ln n}{\sqrt{n}}$.

4. $\displaystyle \lim_{n \to \infty} \frac{2n + \ln n}{n + 2 \ln n}$.

5. $\displaystyle \lim_{n \to \infty} \frac{n^3 + n^2}{e^n + 1}$.

6. $\displaystyle \lim_{n \to \infty} n e^{-n}$.

7. $\displaystyle \lim_{n \to \infty} \frac{e^n + 2n^3}{2e^n + 2n^2}$.

8. $\displaystyle \lim_{n \to \infty} \frac{2n^2 + 3n}{3n^2 + 1}$.

9. $\displaystyle \lim_{n \to \infty} n \ln \frac{n}{n + 1}$.

Use the polynomial test to evaluate the following limits.

10. $\displaystyle \lim_{n \to \infty} \frac{n^2 + 2n + 1}{n}$.

11. $\displaystyle \lim_{n \to \infty} \frac{2n^2 + 1}{3n^3 + 2n - 2}$.

12. $\displaystyle \lim_{n \to \infty} \frac{2n^2 + 3n + 1}{5n^2 + 1}$.

13. $\displaystyle \lim_{n \to \infty} \frac{n^2 + 4n}{n - 3n^2}$.

Use any means to find the following limits.

14. $\lim\limits_{n\to\infty} \dfrac{\sin n}{n}$. Hint: See Theorem 4.11, p. 204.

15. $\lim\limits_{n\to\infty} \sqrt[n]{\ln n}$.

16. $\lim\limits_{n\to\infty} \dfrac{1+n}{n!}$. Hint: For $n > 3$, $\dfrac{1+n}{n!} < \dfrac{1+n}{(n-1)^2}$.

17. $\lim\limits_{n\to\infty} \left(n + \dfrac{1}{n}\right)$.
18. $\lim\limits_{n\to\infty} \dfrac{\sqrt[n]{2}}{n}$.

19. $\lim\limits_{n\to\infty} \dfrac{\sqrt{1+n^2}}{n}$.
20. $\lim\limits_{n\to\infty} \dfrac{\arctan n}{n}$.

21. $\lim\limits_{n\to\infty} \dfrac{e^n}{\ln n}$.
22. $\lim\limits_{n\to\infty} \dfrac{n!}{n^n}$.

23. $\lim\limits_{n\to\infty} \dfrac{\sqrt[n]{n}}{n}$.
24. $\lim\limits_{\infty\to n} \dfrac{n^2}{\left(\dfrac{3}{2}\right)^n}$.

25. $\lim\limits_{n\to\infty} \dfrac{\sqrt{n}}{2^n}$. (Hint: Use Theorem 21.7v.)

26. $\lim\limits_{n\to\infty} \left(1 + \dfrac{1}{n}\right)^n$.
27. $\lim\limits_{n\to\infty} \dfrac{1}{(n!)^2}$.

28. $\lim\limits_{n\to\infty} \dfrac{2^n}{\sqrt[n]{2}}$.
29. $\lim\limits_{n\to\infty} \dfrac{n^2 \sqrt[n]{n}}{2n^2 + n}$.

30. $\lim\limits_{n\to\infty} \dfrac{\sqrt[n]{4}}{\sqrt[n]{n}}$.
31. $\lim\limits_{n\to\infty} \sqrt[n]{1+n}$.

32. $\lim\limits_{n\to\infty} \left(\dfrac{1}{n^2}\right)^{\frac{1}{n}} = 1$. (This limit is used as an example in the proof of Theorem 13.14.)

33. Prove Theorem 21.5 for the case where L is infinite.

34. Prove Theorem 21.6.

35. Prove Theorem 21.7 ii.

36. Prove Theorem 21.7 iii.

37. Prove Theorem 21.10.

ANSWERS

1. $+\infty$.   **3.** 0.   **5.** 0.   **7.** $\frac{1}{2}$.   **9.** -1.   **11.** 0.   **13.** $\dfrac{-1}{3}$.

15. 1.   **17.** $+\infty$.   **19.** 1.   **21.** $+\infty$.   **23.** 0.   **25.** 0.
27. 0.   **29.** $\frac{1}{2}$.   **31.** 1.

2. THE INTEGRAL TEST FOR SERIES

In this section we shall repeat the definition of a series and some results from Chapter 13. In addition, we shall present several new theorems, chief among them the integral test for the convergence of series.

> **DEFINITION 21.11** Let t be a sequence of real numbers. The sequence S defined by
>
> $$S(n) = \sum_{k=1}^{n} t(k) \quad \text{for each positive integer } n$$
>
> is called the nth *partial sum sequence for t.* A sequence is called a *series* when it is regarded as the nth partial sum sequence of a sequence t. The number $t(n)$ is called the nth *term of the series* S. The sum $S(n)$ is called the nth *partial sum of the series* S.

In Chapter 13, we used the notation $\left\{ \sum_{k=1}^{n} t(k) \right\}$ for the series S of partial sums and we used $\lim_{n \to \infty} \left\{ \sum_{k=1}^{n} t(k) \right\}$ for the limit of this sequence, whenever this limit exists. Henceforth, we shall employ the more conventional notation

$$\sum_{n=1}^{\infty} t(n)$$

to denote both the series S of partial sums and the limit of the sequence of partial sums. Which of these two meanings is intended in a particular situation will be clear from the context.

Theorem 13.10, p. 532, and Theorem 13.11, p. 534, become

> **THEOREM 21.12** (Geometric Series.) Let x be a real number. Then
>
> $$\sum_{n=0}^{\infty} x^n = +\infty, \qquad \text{if } x \geq 1$$
>
> $$= \frac{1}{1-x}, \qquad \text{if } -1 < x < 1$$
>
> $$\text{oscillates and diverges,} \quad \text{if } x \leq -1.$$

> **THEOREM 21.13** If $\sum_{n=1}^{\infty} t(n)$ exists and is finite, then
>
> $$\lim_{n \to \infty} t(n) = 0.$$

The reader will perhaps recall that we used the series $\sum_{n=1}^{\infty} \frac{1}{n}$ to show that the converse of the last theorem does not hold.

Proofs of the following three theorems are left to the exercises.

THEOREM 21.14 The alteration of a finite number of terms of a series does not affect the convergence or divergence, although the limit may be affected.

THEOREM 21.15 Let $\sum_{n=1}^{\infty} k(n)$ and $\sum_{n=1}^{\infty} l(n)$ be two series of real numbers. If $\sum_{n=1}^{\infty} k(n) = K$ and $\sum_{n=1}^{\infty} l(n) = L$, then

(i) $\sum_{n=1}^{\infty} (k(n) + l(n)) = \sum_{n=1}^{\infty} k(n) + \sum_{n=1}^{\infty} l(n) = K + L$, and

(ii) $\sum_{n=1}^{\infty} r \cdot k(n) = r \cdot \sum_{n=1}^{\infty} k(n) = r \cdot K$, for each real number r.

If $\sum_{n=1}^{\infty} k(n)$ and $\sum_{n=1}^{\infty} l(n)$ diverges to $+\infty$ then

(iii) $\sum_{n=1}^{\infty} (k(n) + l(n))$ diverges to $+\infty$.

(iv) $\sum_{n=1}^{\infty} r \cdot k(n)$ diverges to $\begin{cases} +\infty, & \text{if } r > 0 \\ -\infty, & \text{if } r < 0. \end{cases}$

If $\sum_{n=1}^{\infty} k(n)$ diverges then

(v) $\sum_{n=1}^{\infty} r \cdot k(n)$ diverges, if $r \neq 0$.

THEOREM 21.16 A series $\sum_{n=1}^{\infty} t(k)$ of non-negative real terms converges if and only if the nth partial sum sequence $\left\{ \sum_{k=1}^{n} t(k) \right\}$ is bounded. If $\sum_{k=1}^{\infty} t(k)$ diverges, it diverges to $+\infty$.

Our study of improper integrals in Chapter 20 makes possible the following test for convergence of series.

THEOREM 21.17 (Integral Test.) Let $\sum\limits_{n=1}^{\infty} t(n)$ be a series of positive terms and let $f:[a,\ \infty) \to \mathcal{R}$ be such that

f is strictly decreasing and
$f(n) = t(n)$ for each $n \geq a$.

Then

$$\sum_{n=1}^{\infty} t(n) \text{ converges} \Leftrightarrow \int_a^{\infty} f \text{ converges,}$$

or, equivalently,

$$\sum_{n=1}^{\infty} t(n) \text{ diverges (to } +\infty) \Leftrightarrow \int_a^b f \text{ diverges (to } +\infty).$$

Proof. Choose k, an integer, such that $k \geq a$. We shall show that $\int_k^{\infty} f$ and $\sum\limits_{n=k}^{\infty} t(k)$ either both converge or both diverge to infinity. The theorem will then follow since, by Theorem 21.14, we can alter a finite number of terms of a series (in this case those from 1 up to $k - 1$) without affecting the convergence or divergence of the series, and, by Theorem 20.13, p. 805, $\int_a^{\infty} f = \int_a^k f + \int_k^{\infty} f$ converges or diverges according to the convergence or divergence of $\int_k^{\infty} f$.

Assume first that $\sum\limits_{n=k}^{\infty} t(n)$ converges. Define $u(x) = f(n) = t(n)$ for $n \leq x < n + 1$, where $n \geq k$. (See Figure 21.1.) Then

$$\int_k^j u = \sum_{n=k}^{j} t(n)(n + 1 - n) = \sum_{n=k}^{j} t(n)$$

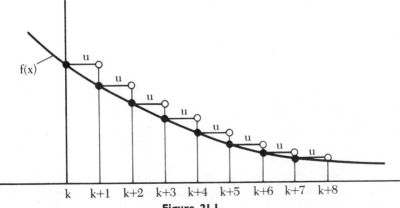

Figure 21.1

for each integer $j > k$. Hence

$$\int_k^\infty u = \lim_{j \to \infty} \int_k^j u = \lim_{j \to \infty} \sum_{n=k}^j t(n) = \sum_{n=k}^\infty t(n).$$

Thus $\int_k^\infty u$ converges. But since $f(x) \le u(x)$ for each x, $\{\int_k^t f \mid t > k\}$ is bounded above by $\int_k^\infty u$. Therefore $\int_k^\infty f$ converges by Theorem 20.12, p. 805.

Assume now that $\int_k^\infty f$ converges. Define the lower step function l for f by $l(x) = f(j + 1) = t(j + 1)$ for $j \le x < j + 1$. Then

$$\sum_{n=k+1}^{j+1} t(n) = \int_k^j l(x) \le \int_k^j f(x) \le \int_k^\infty f(x) \quad \text{for each } j > k.$$

Thus the partial sum sequence $\left\{ \sum_{n=k+1}^{j+1} t(n) \right\}$ is bounded and, by Theorem 21.16, the series $\sum_{n=k+1}^\infty t(n)$ converges. It follows that $\sum_{n=k}^\infty t(n)$ converges.

We have now shown that $\int_k^\infty f$ and $\sum_{n=k}^\infty t(n)$ converge together. It follows that if one diverges so must the other. From Theorem 21.16 $\sum_{n=k}^\infty t(n)$ can only diverge to $+\infty$. It is easily shown that $\int_k^\infty f$ can only diverge to $+\infty$ as well. ■

Example 4. Use the integral test to determine the convergence or divergence of $\sum_{n=1}^\infty \dfrac{1}{(n + 1)^2}$. We test $\int_1^\infty \dfrac{1}{(x + 1)^2} = \lim_{t \to \infty} \left[\dfrac{-1}{x + 1} \right]_1^t = \dfrac{1}{2}$.

Thus by the integral test, $\sum_{n=1}^\infty \dfrac{1}{(n + 1)^2}$ converges. ■

Example 5. Test for convergence $\sum_{n=1}^\infty \dfrac{\ln n}{n}$.

$$\int_1^\infty \frac{\ln x}{x} = \lim_{t \to \infty} \left[\frac{\ln^2 x}{2} \right]_1^t = \lim_{t \to \infty} \frac{\ln^2 t}{2} = +\infty.$$

Thus $\sum_{n=1}^\infty \dfrac{\ln n}{n}$ diverges to infinity. ■

THEOREM 21.18 (*p*-test.) The series $\sum_1^\infty \dfrac{1}{n^p}$ converges if $p > 1$; otherwise it diverges to $+\infty$.

Proof. If $p \ne 1$, we have

$$\int_1^\infty \frac{1}{x^p} = \lim_{t \to \infty} \left[\frac{x^{1-p}}{1 - p} \right]_1^t = \lim_{t \to \infty} \left(\frac{t^{1-p}}{1 - p} - \frac{1}{1 - p} \right).$$

Now if $p > 1$, this limit is $\dfrac{1}{p-1}$, and the integral, and hence (by Theorem 20.17) the series, converges. If $p < 1$, the integral, and hence the series, diverges to $+\infty$.

If $p = 1$, $\int_1^\infty \dfrac{1}{x^p} = \int_1^\infty \dfrac{1}{x} = \lim\limits_{t\to\infty} \ln t = +\infty$, and hence $\sum\limits_{n=1}^\infty \dfrac{1}{n^p} = \sum\limits_{n=1}^\infty \dfrac{1}{n}$ diverges to $+\infty$. ∎

NOTE. The series $\sum\limits_{n=1}^\infty \dfrac{1}{n}$ is called the harmonic series. Hopefully the reader showed its divergence by another method in Exercise 11 of Section 4, Chapter 13, p. 536.

EXERCISES

Use the integral test to determine the convergence of the following series.

1. $\sum\limits_{n=1}^\infty e^{-2n}$.

2. $\sum\limits_{n=1}^\infty \dfrac{1}{1+n^2}$.

3. $\sum\limits_{n=1}^\infty \dfrac{1}{\sqrt[3]{n}}$.

4. $\sum\limits_{n=1}^\infty n e^{-n^2}$.

5. $\sum\limits_{n=1}^\infty \dfrac{1}{n\sqrt{1+n^2}}$.

6. $\sum\limits_{n=1}^\infty \arctan n$.

7. $\sum\limits_{n=2}^\infty \dfrac{1}{n\sqrt{n^2-1}}$.

8. $\sum\limits_{n=1}^\infty \dfrac{e^{\frac{1}{n}}}{n^2}$.

9. $\sum\limits_{n=1}^\infty \dfrac{1}{n \ln n}$.

10. $\sum\limits_{n=1}^\infty \dfrac{1}{\sqrt{n+1}}$.

11. $\sum\limits_{n=1}^\infty \dfrac{1}{2^n}$.

12. $\sum\limits_{n=1}^\infty \dfrac{n}{4+n^4}$.

13. $\sum\limits_{n=1}^\infty \dfrac{n}{(n^2+4)^2}$.

14. $\sum\limits_{n=1}^\infty \dfrac{\ln n}{n}$.

15. $\sum\limits_{n=1}^\infty n e^{-n^2}$.

16. Prove Theorem 21.14.

17. Prove Theorem 21.15 i.

18. Prove Theorem 21.15 ii.

19. Prove Theorem 21.15, iii.

20. Prove Theorem 21.15 iv.

21. Prove Theorem 21.15 v.

22. Prove Theorem 21.16.

ANSWERS

1. Converges.   **3.** Diverges.   **5.** Converges.   **7.** Converges.

9. Diverges.   **11.** Converges.   **13.** Converges.   **15.** Converges.

3. THE COMPARISON TEST

In this section the Comparison Test of Chapter 13, p. 537, will be given in a somewhat stronger form. For the convenience of the reader, we repeat the statement of the earlier theorem, before giving the stronger theorem.

THEOREM 21.19 Let $r(k)$ and $t(k)$ be sequences of non-negative real numbers such that

$$r(k) \leq t(k), \quad \text{for each integer } k > 0.$$

(i) If $\sum_{k=1}^{\infty} t(k)$ converges, then $\sum_{k=1}^{\infty} r(k)$ converges.

(ii) If $\sum_{k=1}^{\infty} r(k)$ diverges (to $+\infty$), then $\sum_{k=1}^{\infty} t(k)$ diverges (to $+\infty$).

The key to the proof of statement (i) of this theorem was our ability to find a real number M $\left(\text{in this case the sum } M = \sum_{n=1}^{\infty} t(n)\right)$ which is an upper bound for $\sum_{n=1}^{\infty} r(n)$. If the convergence of $\sum_{n=1}^{\infty} t(n)$ allowed us to find any other number M_1 which is an upper bound for $\sum_{n=1}^{\infty} r(n)$, the proof would be equally valid. For example, if $r(k) \leq 2t(k)$ for each positive integer k, then $2M$ would serve as the upper bound M_1. Indeed if $r(k) \leq t(k)$ for $k > n_0$, $n_0 \geq 0$, then we could choose $M_1 = 2M + \sum_{K=1}^{n_0} r(k)$ as an upper bound for $\sum_{n=1}^{\infty} r(n)$. This type of relationship between the sequences r and t is formalized in the following definition.

DEFINITION 21.20 Let t be a sequence of non-negative real numbers. The *order dominated set of t* is the set of sequences denoted by

$O(t)$ and defined by

$$O(t) = \left\{ s \, \middle| \, \begin{array}{l} 1. \ s \text{ is a sequence of real numbers} \\ 2. \ |s(n)| \leq ct(n) \begin{array}{l} \text{(for some real number } c \text{ and} \\ \text{for all } n \text{ beyond some } n_0 \end{array} \end{array} \right\}$$

If $r \in O(t)$, r is said to be *order dominated* by t.
If $r \in O(t)$ and $t \in O(r)$, r and t are said to have the *same order*.
If $r \in O(t)$ and $t \notin O(r)$, r is said to be of *lesser order* than t.

Intuitively, $O(t)$ (read "big O of t") is the set of sequences $\{s(n)\}$ whose terms have absolute values which are ultimately dominated by the corresponding terms of some multiple of the sequence $\{t(n)\}$. Since the sequence $\{ct(n)\}$ converges whenever the sequence $\{t(n)\}$ does, we should expect all sequences which are order dominated by t to converge whenever $\{t(n)\}$ converges.

Let us now consider some examples.

Example 6. Consider the two sequences $\{t(n)\} = \{n \ln n\}$ and $\{r(n)\} = \{n\}$. Certainly we expect the first sequence to dominate the second since $\ln n \to \infty$ as $n \to \infty$. To establish this fact we note that when $n \geq 3$ we have $\ln n \geq \ln 3 \geq \ln e = 1$ and thus

$$|r(n)| = n < n \cdot \ln n = t(n).$$

Hence we have $r \in O(t)$.

Is $t \in O(r)$? That is, can we find a number c such that

$$|t(n)| = n \ln n \leq c \cdot n = c \cdot r(n)$$

for all n beyond some n_0? If possible, this would require c to be such that

$$\ln n \leq c$$

for all $n > n_0$. In other words c would be an upper bound for $\ln n$, contrary to the fact that $\ln n \to \infty$ as $n \to \infty$. Therefore $t \notin O(r)$. Consequently r is of lesser order than t. ■

Example 7. Let $\{t(n)\} = \{2n\}$ and $\{r(n)\} = \{n\}$. Clearly $t \in O(r)$ since

$$|t(n)| = 2n \leq 2 \cdot n = 2 \cdot r(n) \quad \text{for all } n.$$

Moreover, $r \in O(t)$ since

$$|r(n)| = n \leq \tfrac{1}{2} \cdot (2n) = \tfrac{1}{2} \cdot t(n) \quad \text{for all } n.$$

The two sequences are therefore of the same order. ■

We are now ready for our stronger version of the comparison test. Note that if we can choose $c = n_0 = 1$ to show that $r \in O(t)$, then Theorem 21.21 reduces to Theorem 21.19.

THEOREM 21.21 (Comparison Test.) Let r and t be sequences of non-negative real numbers such that $r \in O(t)$.

(i) If $\sum_{n=1}^{\infty} t(n)$ converges, then $\sum_{n=1}^{\infty} r(n)$ converges.

(ii) If $\sum_{n=1}^{\infty} r(n)$ diverges to $+\infty$, then $\sum_{n=1}^{\infty} t(n)$ diverges to $+\infty$.

Proof. (i) Assume that $\sum_{n=1}^{\infty} t(n)$ converges. Since $r \in O(t)$, there exists a positive number c and a positive integer n such that

$$r(n) \leq ct(n) \quad \text{for all} \quad n \geq n_0.$$

We may assume that this inequality holds for all positive integers n. (Why?) Consequently, by Theorem 21.19 if $\sum_{n=1}^{\infty} ct(n)$ converges, so does $\sum_{n=1}^{\infty} r(n)$. Theorem 21.3 assures us that $\sum_{n=1}^{\infty} ct(n)$ converges and thus the proof of assertion (i) is complete.

Since series of non-negative terms either converge or diverge to $+\infty$ (Theorem 21.16ii) is simply a contrapositive statement of (i). This completes the proof. ∎

Example 8. Test for convergence $\sum_{n=1}^{\infty} \dfrac{1}{n^{\sin n}}$. We observe that $\sin n \leq 1$ for each n, hence $\dfrac{1}{n^{\sin n}} \geq \dfrac{1}{n}$. By Theorem 21.18 ($p$-test), $\sum_{n=1}^{\infty} \dfrac{1}{n}$ diverges. Therefore, by Theorem 21.19 (Comparison Test), $\sum_{n=1}^{\infty} \dfrac{1}{n^{\sin n}}$ diverges. ∎

Example 9. Test $\sum_{n=1}^{\infty} \dfrac{\ln n}{n^3}$ for convergence. Since $\ln n < n$ for each $n > 0$ we have

$$\frac{\ln n}{n^3} < \frac{n}{n^3} = \frac{1}{n^2}.$$

Since, by the p-test $\sum_{n=1}^{\infty} \dfrac{1}{n^2}$ converges, we have, by Theorem 21.19, that $\sum_{n=1}^{\infty} \dfrac{\ln n}{n^3}$ converges. ∎

COROLLARY 21.22 Two series of non-negative terms which have the same order either both converge or both diverge to $+\infty$.

The following theorem will be helpful in establishing order relations between sequences.

THEOREM 21.23 Let r and t be sequences of positive terms such that $\lim\limits_{n \to \infty} \dfrac{r(n)}{t(n)} = L$ exists.

(i) If $0 < L < +\infty$, then r and t have the same order;
(ii) if $L = 0$, then $r \in O(t)$ and $t \notin O(r)$;
(iii) if $L = +\infty$, then $t \in O(r)$ and $r \notin O(t)$.

Proof. (i) Assume that $0 < L < +\infty$. By definition of limit there exists an n_0 such that

$$\tfrac{1}{2}L < \frac{r(n)}{t(n)} < \tfrac{3}{2}L \quad \text{for all} \quad n \geq n_0$$

or equivalently,

$$t(n) < \frac{2}{L}r(n) \quad \text{and} \quad r(n) < \tfrac{3}{2}Lt(n) \quad \text{for all} \quad n \geq n_0.$$

Thus by Definition 21.20 we know that r and t have the same order.

(ii) Assume that $L = 0$. By definition of limit there exists an n_0 such that

$$\frac{r(n)}{t(n)} < 1 \quad \text{for all} \quad n \geq n_0.$$

Consequently, by Definition 21.20, $r \in O(t)$. For the sake of contradiction assume that $t \in O(r)$. There then exists a positive number c and an n_0 such that $t(n) \leq cr(n)$ for all $n \geq n_0$. Equivalently, $\dfrac{1}{c} \leq \dfrac{r(n)}{t(n)}$ for all $n \geq n_0$. Therefore $\dfrac{1}{c} \leq \lim\limits_{n \to \infty} \dfrac{r(n)}{t(n)} = L$, contradicting our assumption that $L = 0$.

(iii) A proof of (iii) can be derived as a corollary of (ii) and is called for in Exercise 25. ∎

Example 10. Test for convergence $\sum\limits_{n=1}^{\infty} \dfrac{\sqrt{n}}{n+1}$. We note that for large n, the constant 1 in the denominator seems to have little effect on the value of $\dfrac{\sqrt{n}}{n+1}$. This suggests that we compare with $\sum\limits_{n=1}^{\infty} \dfrac{\sqrt{n}}{n} = \sum\limits_{n=1}^{\infty} \dfrac{1}{n^{1/2}}$, which diverges by the p-test. We want to show that $\dfrac{\sqrt{n}}{n}$ and $\dfrac{\sqrt{n}}{n+1}$ have the same order. We find

$$\lim_{n \to \infty} \frac{\sqrt{n}}{n+1} \bigg/ \frac{1}{\sqrt{n}} = \lim_{n \to \infty} \frac{n}{n+1} = 1.$$

Thus by Theorem 21.23(i) the sequences $\dfrac{\sqrt{n}}{n+1}$ and $\dfrac{1}{n^{1/2}}$ have the same order of magnitude and hence both diverge, by Corollary 21.22. ∎

EXERCISES

Test the following series for convergence.

1. $\sum_{n=1}^{\infty} \dfrac{1}{2 + n^2} \cdot$ $\left(\text{Hint: compare with } \sum_{n=1}^{\infty} \dfrac{1}{n^2} \text{ and use Theorem 21.18.}\right)$

2. $\sum_{n=1}^{\infty} \dfrac{\cos n}{n^2} \cdot$

3. $\sum_{n=1}^{\infty} \dfrac{3}{2n + 2} \cdot$ $\left(\text{Hint: Compare with } \sum_{n=1}^{\infty} \dfrac{1}{n} \cdot\right)$

4. $\sum_{n=1}^{\infty} \dfrac{1}{n^n} \cdot$ $\left(\text{Hint: Compare with } \sum_{n=1}^{\infty} \dfrac{1}{n^2} \cdot\right)$

5. $\sum_{n=1}^{\infty} \dfrac{1}{\sqrt{n^2 + 1}} \cdot$ 6. $\sum_{n=1}^{\infty} \dfrac{3}{(1 + n^2)^{4/3}} \cdot$

7. $\sum_{n=1}^{\infty} \dfrac{\ln (n + 1)}{n \ln n} \cdot$ 8. $\sum_{n=1}^{\infty} \dfrac{1}{(n + 1)(n + 2)^{1/2}} \cdot$

9. $\sum_{n=1}^{\infty} \dfrac{3n + 1}{n(n^2 - 2)^{3/2}} \cdot$ 10. $\sum_{n=1}^{\infty} \dfrac{n^2 e^{-2n}}{n^2 + 1} \cdot$

11. $\sum_{n=1}^{\infty} \dfrac{ne^{-n}}{1 + n^2} \cdot$ 12. $\sum_{n=1}^{\infty} \dfrac{1}{(6n^2 + 1)^{2/3}} \cdot$

13. $\sum_{n=1}^{\infty} \dfrac{n + 2}{3n^3 + 2n - 1} \cdot$ 14. $\sum_{n=1}^{\infty} \dfrac{n^3}{n^4 + 1} \cdot$

15. $\sum_{n=1}^{\infty} \dfrac{\sin^2 (n + 1)}{(n + 1)^2} \cdot$ 16. $\sum_{n=1}^{\infty} n \ln \dfrac{n}{n + 1} \cdot$

17. $\sum_{n=1}^{\infty} e^{-n^2}.$ 18. $\sum_{n=1}^{\infty} n^3 e^{-2n}.$

19. $\sum_{n=1}^{\infty} \dfrac{1}{n} \left(\dfrac{\pi}{e}\right)^n .$ 20. $\sum_{n=1}^{\infty} \dfrac{e^n}{(n^2 + 1)^n} \cdot$

21. $\sum_{n=1}^{\infty} e^{n/2}.$ 22. $\sum_{n=1}^{\infty} \dfrac{1}{(\ln n)^2} \cdot$

23. $\sum_{n=1}^{\infty} \dfrac{7 + (-1)^n}{n} \cdot$ 24. $\sum_{n=1}^{\infty} \dfrac{1}{n^2 \ln n} \cdot$

25. Prove Theorem 21.23 iii.

ANSWERS

1. Converges.	**3.** Diverges.	**5.** Diverges.	**7.** Diverges.
9. Converges.	**11.** Converges.	**13.** Converges.	**15.** Converges.
17. Converges.	**19.** Diverges.	**21.** Diverges.	**23.** Diverges.

4. THE ROOT AND RATIO TESTS

The root test was stated and proved as Theorem 13.14, and we shall merely restate it here.

THEOREM 21.24 (Root Test.) Let t be a sequence of non-negative real numbers such that

$$\lim_{k \to \infty} \sqrt[k]{t(k)} = L.$$

Then

$$\sum_{n=1}^{\infty} t(k) \begin{cases} \text{(i) exists and is finite,} & \text{if } L < 1 \\ \text{(ii) may converge or diverge to } +\infty, & \text{if } L = 1 \\ \text{(iii) diverges to } +\infty, & \text{if } L > 1 \text{ or } L = +\infty \end{cases}$$

Example 11. Test $\displaystyle\sum_{n=1}^{\infty} \frac{1}{1 + 2^n}$. By Theorem 21.23, we find the sequence $\left\{\dfrac{1}{1 + 2^n}\right\}$ to be of the same order of magnitude as $\left\{\dfrac{1}{2^n}\right\}$. We apply the root test to $\displaystyle\sum_{n=1}^{\infty} \frac{1}{2^n}$:

$$\lim_{n \to \infty} \sqrt[n]{\frac{1}{2^n}} = \lim_{n \to \infty} \frac{1}{2} = 1/2 < 1$$

Thus both $\displaystyle\sum_{n=1}^{\infty} \frac{1}{2^n}$ and $\displaystyle\sum_{n=1}^{\infty} \frac{1}{1 + 2^n}$ converge. ∎

Example 12. Test $\displaystyle\sum_{n=1}^{\infty} \sqrt{\frac{2}{n}}$. Applying the root test,

$$\lim_{n \to \infty} \sqrt[2n]{\frac{2}{n}} =^1 \sqrt{\lim_{n \to \infty} \sqrt[n]{\frac{2}{n}}}$$

$$= \left(\frac{\lim\limits_{n \to \infty} \sqrt[n]{2}}{\lim\limits_{n \to \infty} \sqrt[n]{n}} \right)^{\frac{1}{2}} =^2 1. \quad \blacksquare$$

(1) Since $\lim\limits_{n \to \infty} \sqrt[n]{n} = 1 \neq 0$, by Theorem 21.7 iv, p. 810.

(2) $\lim\limits_{n \to \infty} \sqrt[n]{2} = 1$ by Theorem 21.7 ii.

Thus the root test fails. However, we may write $\sum_{n=1}^{\infty} \sqrt{\dfrac{2}{n}} = \sqrt{2} \sum_{n=1}^{\infty} \dfrac{1}{n^{1/2}}$. which diverges by the p-test (Theorem 21.18, p. 818).

The ratio test was stated but not proved in Chapter 13.

THEOREM 12.25 (Ratio Test.) Let t be a series of non-negative real numbers such that $\lim_{k \to \infty} \dfrac{t(n+1)}{t(n)} = L$, where $L \in \mathcal{R}$ or $L = +\infty$.

Then

$$\sum_{n=1}^{\infty} t(n) \begin{cases} \text{(i)} & \text{exists and is finite} & \text{if } L < 1 \\ \text{(ii)} & \text{may converge or may diverge to } +\infty, \text{ if } L = 1 \\ \text{(iii)} & \text{diverges to } +\infty & \text{if } L > 1 \end{cases}$$

Proof. (i) Assume that $L < 1$. Let r be such that $L < r < 1$. By definition of limit there exists an integer n_0 such that

$$\frac{t(n+1)}{t(n)} \le r \quad \text{for all} \quad n \ge n_0.$$

An inductive argument shows that

$$t(n_0 + k) \le r^k t(n_0) \quad \text{for each positive integer } k.$$

Since $0 < r < 1$, $\{r^k\}$ is a convergent geometric series. Therefore by the Comparison Test $\sum_{k=1}^{\infty} t(n_0 + k)$ converges, and finally by Theorem 21.8, p. 812, $\sum_{k=1}^{\infty} t(k)$ converges.

(ii) If $L > 1$, we choose an n_0 such that $\dfrac{t(n+1)}{t(n)} > 1$ whenever $n \ge n_0$. For each such n we have $t(n+1) > t(n)$. Hence the sequence t does not approach zero and, by Theorem 21.12, $\sum_{n=1}^{\infty} t(n)$ does not converge. Since $\left\{ \sum_{k=1}^{n} t(k) \right\}$ is an increasing sequence, the series diverges to $+\infty$.

(iii) To prove this statement we must exhibit two sequences r and t satisfying the conditions that $\lim_{k \to \infty} \dfrac{r(k+1)}{r(k)} = \lim_{k \to \infty} \dfrac{t(k+1)}{t(k)} = 1$ and such that one converges and the other diverges. We let $r(k) = \dfrac{1}{k}$, $t(k) = \dfrac{1}{k^2}$.

Then $\displaystyle\lim_{k\to\infty} \frac{r(k+1)}{r(k)} = \lim_{k\to\infty} \frac{k}{k+1} = 1$ and

$$\lim_{k\to\infty} \frac{t(k+1)}{t(k)} = \lim_{k\to\infty} \frac{k^2}{(k+1)^2} = 1.$$

But by the p-test, $\displaystyle\sum_{k=1}^{\infty} r(k)$ diverges, while $\displaystyle\sum_{k=1}^{\infty} t(k)$ converges. ■

Example 13. Test for convergence $\displaystyle\sum_{k=1}^{\infty} t(k)$, where

$$t(k) = \frac{k!}{1\cdot 3\cdot 5\cdots(2k-1)}\,.$$

Now

$$t(k+1) = \frac{(k+1)!}{1\cdot 3\cdot 5\cdots(2k+1)},$$

$$\frac{t(k+1)}{t(k)} = \frac{(k+1)!}{1\cdot 3\cdot 5\cdots(2k+1)}\cdot\frac{1\cdot 3\cdot 5\cdots(2k-1)}{k!}$$

$$= \frac{k+1}{2k+1}$$

and

$$\lim_{k\to\infty} \frac{t(k+1)}{t(k)} = \lim_{k\to\infty} \frac{k+1}{2k+1} = \frac{1}{2}\,.$$ Thus the series converges. ■

Example 14. Test $\displaystyle\sum \frac{2^n n!}{n^n}$ for convergence. Using the ratio test

$$\lim_{n\to\infty} \frac{2^{n+1}(n+1)!}{(n+1)^{n+1}}\frac{n^n}{2^n n!} = \lim_{n\to\infty} 2\left(\frac{n}{n+1}\right)^n$$

$$=^1 2\lim_{n\to\infty} \left(\frac{n}{n+1}\right)^n.$$

At this point, we set $y = \displaystyle\lim_{n\to\infty} \left(\frac{n}{n+1}\right)^n$

$$\ln y = \lim_{n\to\infty} n\ln\frac{n}{n+1}$$

$$=^2 \lim_{x\to\infty} \frac{\ln\left(\dfrac{x}{x+1}\right)}{\dfrac{1}{x}}$$

$$=^3 \lim_{x\to\infty} -x\,\frac{x+1}{(x+1)^2}$$

$$= -\lim_{x\to\infty} \frac{x}{x+1} = -1$$

Hence

$$\lim_{n \to \infty} \left(\frac{n}{n+1} \right)^n = y = e^{\ln y} = e^{-1} = \frac{1}{e}.$$

Thus, the limit of the ratio of successive terms is $\dfrac{2}{e} < 1$. The series converges. ■

(1) Polynomial Test (Theorem 21.10, p. 812).
(2) Theorem 21.5, p. 809.
(3) L'Hopital's Rule.

EXERCISES

Use the root test to determine the convergence or divergence of the series in Exercise 1 to 6.

1. $\displaystyle\sum_{n=1}^{\infty} \frac{1}{2^n}$.

2. $\displaystyle\sum_{n=1}^{\infty} (n^{1/n} - 1)^n$.

3. $\displaystyle\sum_{n=1}^{\infty} 2^{-n^2}$.

4. $\displaystyle\sum_{n=2}^{\infty} \frac{1}{(\ln n)^n}$.

5. $\displaystyle\sum_{n=1}^{\infty} \left(2 - \frac{1}{n} \right)^n$.

6. $\displaystyle\sum_{n=1}^{\infty} \left(\frac{\sin n}{n} \right)^n$.

Use the ratio test to determine the convergence or divergence of the series in Exercises 7 to 12.

7. $\displaystyle\sum_{n=1}^{\infty} \frac{e^n}{n!}$.

8. $\displaystyle\sum_{n=1}^{\infty} n \left(\frac{\pi}{4} \right)^{n+1}$.

9. $\displaystyle\sum_{n=1}^{\infty} n \left(\frac{\pi}{2} \right)^{n+1}$.

10. $\displaystyle\sum_{n=1}^{\infty} \frac{(n+3)(n+4)}{1 \cdot 2 \cdot 4 \cdot 6 \cdots 2n}$.

11. $\displaystyle\sum_{n=1}^{\infty} \frac{n^{100}}{e^n}$.

12. $\displaystyle\sum_{n=0}^{\infty} \frac{(2n+1)!}{n!}$.

Use any available method to test the following series.

13. $\displaystyle\sum_{n=1}^{\infty} n^{\frac{1}{n}}$.

14. $\displaystyle\sum_{n=1}^{\infty} e^{-n^2}$.

15. $\displaystyle\sum_{n=1}^{\infty} \left(\frac{1}{n} - e^{-n^2} \right)$.

16. $\displaystyle\sum_{n=1}^{\infty} \frac{2n+1}{2^n n! (3n+1)}$.

17. $\displaystyle\sum \left(\frac{n}{n^2 + 1} \right)^n$.

18. $\displaystyle\sum_{n=1}^{\infty} \frac{\ln n}{n^3}$.

19. $\displaystyle\sum_{n=1}^{\infty} \frac{2n}{n!}$.

20. $\displaystyle\sum_{n=1}^{\infty} \frac{(2n)!}{(n!)^2}$.

21. $\displaystyle\sum_{n=1}^{\infty} \frac{(n!)^2}{2^{n^2}}$.

22. $\displaystyle\sum_{n=1}^{\infty} \frac{2^{n^2}}{(n!)^2}$.

23. $\displaystyle\sum_{n=2}^{\infty} \frac{1}{(\ln n)^{\frac{1}{n}}}$.

24. $\displaystyle\sum_{n=1}^{\infty} \frac{n}{2^{n-1}}$.

25. $\displaystyle\sum_{n=1}^{\infty} \frac{n!}{2^n}$.

26. $\displaystyle\sum_{n=1}^{\infty} \frac{n!}{n^n}$.

ANSWERS

1, 2, 3, 4, 6, 7, 8, 10, 11, 14, 16, 17, 18, 19, 21, 24, 26 converge.

5. SERIES OF ARBITRARY CONSTANTS

Thus far we have confined our study primarily to series of non-negative terms. In the present section we shall investigate series whose terms may be negative. We shall look first at alternating series. An *alternating series* is one in which the terms are alternately positive and negative.

THEOREM 21.26 (Alternating Series Test.) Let t be an alternating series. Then $\displaystyle\sum_{n=1}^{\infty} t(n)$ converges if $\lim_{n\to\infty} t(n) = 0$ and $\{|t(n)|\}$ is monotone decreasing. Moreover, if $L = \displaystyle\sum_{n=1}^{\infty} t(n)$, then

$$|L - S(n)| \leq |t(n + 1)|,$$

where $S(n) = \displaystyle\sum_{k=1}^{n} t(k)$.

Proof. Assume that $\lim_{n\to\infty} t(n) = 0$ and that $\{|t(n)|\}$ is monotone decreasing. Without loss of generality we may assume that $t(2n) \leq 0$ and $t(2n + 1) \geq 0$ for each n. An elementary computation shows that

(21.1) $\begin{cases} t(2n) + t(2n + 1) \leq 0 & \text{and} \\ t(2n - 1) + t(2n) \geq 0 & \text{for each } n. \end{cases}$

For each integer n let $S(n) = \displaystyle\sum_{k=1}^{n} t(k)$. We then have

(21.2a) $\begin{cases} S(2n) = S(2n - 2) + t(2n - 1) + t(2n) \\ \qquad \geq^1 S(2n - 2) \end{cases}$

and

(21.2b) $\begin{cases} S(2n) = t(1) + \displaystyle\sum_{k=1}^{n-1} (t(2k) + t(2k + 1)) + t(2n) \\ \qquad \leq^1 t(1) + t(2n) \\ \qquad \leq^2 t(1). \end{cases}$

(1) By Inequality (21.1), defining $S(0) = 0$.
(2) Since $t(2n) \leq 0$.

Inequality (21.2) shows that $\{S(2n)\}$ is a monotone increasing sequence bounded below by zero and bounded above by $t(1)$. Therefore, by Theorem 21.4, p. 808, the sequence $\{S(2n)\}$ converges to the limit $L = \text{lub } \{S(2n)\}$. (For later reference we note that we must have $0 \leq L \leq t(1)$).

If we can also show that $\{S(2n + 1)\}$ converges to L, we will know that $\sum\limits_{k=1}^{\infty} t(k) = \lim\limits_{n \to \infty} S(n) = L$ and our proof will be complete. The remainder of the proof is designed to show that $\lim\limits_{n \to \infty} S(2n + 1) = L$.

Let $r > 0$. Since $\lim\limits_{n \to \infty} S(2n) = L$ and $\lim\limits_{n \to \infty} t(n) = 0$, there exists an n_0 such that

$$|S(2n) - L| < r/2 \quad \text{and} \quad |t(2n + 1)| < r/2 \quad \text{for} \quad n \geq n_0.$$

Therefore if $n \geq n_0$,

$$
\begin{aligned}
|S(2n + 1) - L| &=^3 |t(2n + 1) + S(2n) - L| \\
&\leq^4 |t(2n + 1)| + |S(2n) - L| \\
&<^5 r/2 + r/2 \\
&= r.
\end{aligned}
$$

(3) By definition of $S(2n + 1)$ and $S(2n)$.
(4) Triangle Inequality.
(5) Choice of n_0.

and thus by definition of limit we have $\lim\limits S(2n + 1) = L$, completing our proof of the first statement of the theorem.

From the statements following Inequalities (21.2) we know that for the case of an alternating series meeting the conditions of the theorem and the additional restriction that $t(2n) \leq 0$ and $t(2n + 1) \geq 0$ for each n, we have $0 \leq L \leq t(1)$. For the case in which $t(2n) \geq 0$ and $t(2n + 1) \leq 0$ we would obtain $t(1) \leq L \leq 0$. In either case we have

(21.3) $$|L| = \left| \sum_{n=1}^{\infty} t(n) \right| \leq |t(1)|;$$

that is, the absolute value of the sum is at most equal to the absolute value of the leading term of the series.

To prove the last statement of the theorem we observe that $L = S(n) + \sum\limits_{k=n+1}^{\infty} t(k)$. Hence,

$$
\begin{aligned}
|L - S(n)| &= \left| \sum_{k=n+1}^{\infty} t(k) \right| \\
&\leq^6 |t(n + 1)|. \quad \blacksquare
\end{aligned}
$$

(6) Applying the content of Inequality (21.3) to the series in the preceding step. having $t(n + 1)$ as its first term.

Example 15. We shall test $\sum_{n=1}^{\infty} (-1)^n \frac{1}{n}$ for convergence. Since $\lim_{n \to \infty} (-1)^n \frac{1}{n} = 0$ and $\left\{\frac{1}{n}\right\}$ is monotone decreasing, Theorem 21.26 assures us that this series converges. ∎

DEFINITION 21.27 A series $\sum_{n=1}^{\infty} t(n)$ is said to *converge absolutely* if and only if $\sum_{n=1}^{\infty} |t(n)|$ converges.

Example 16. $\sum_{n=1}^{\infty} (-1)^{n+1} \frac{1}{n^2}$ converges absolutely since

$$\sum_{k=1}^{\infty} \left| (-1)^{n+1} \frac{1}{n^2} \right| = \sum_{n=1}^{\infty} \frac{1}{n^2}$$

converges by the *p*-test. On the other hand the series $\sum_{n=1}^{\infty} (-1)^n \frac{1}{n}$ of Example 15 converges but does not converge absolutely, since

$$\sum_{n=1}^{\infty} \left| (-1)^n \frac{1}{n} \right| = \sum_{n=1}^{\infty} \frac{1}{n}$$

is the harmonic series. ∎

The previous example shows that a series may converge without converging absolutely. Such series have the following special name.

DEFINITION 21.28 If a series converges but does not converge absolutely, it *converges conditionally*.

We have already noted that the series $\sum_{n=1}^{\infty} (-1)^n \frac{1}{n}$ converges conditionally.

THEOREM 21.29 A series which converges absolutely must also converge.

(NOTE. This is a restatement of Theorem 13.12, p. 535.)

Proof. Let $\sum_{n=1}^{\infty} t(n)$ be an absolutely convergent sequence. Let $S(n) = \sum_{k=1}^{n} t(k)$. Let $P(n)$ be the sum of the positive terms in $S(n)$ and $Q(n)$

be the sum of the negative terms in $S(n)$. Note that $S(n) = P(n) + Q(n)$ for each n. Thus if we can show that each of $P(n)$ and $Q(n)$ converges we shall know that $S(n)$ converges and thus that $\sum\limits_{k=1}^{\infty} t(k)$ converges.

To this end we define $L = \sum\limits_{k=1}^{\infty} |t(k)|$. Note that $P(n)$ is monotone increasing and $P(n) \leq L$ for each n. Similarly $Q(n)$ is monotone decreasing and $-L \leq Q(n)$ for each n. Therefore each of P and Q converges and our proof is complete. ∎

Example 17. Consider $\sum\limits_{n=1}^{\infty} \dfrac{\cos \dfrac{n\pi}{4}}{n^2}$

$$= \frac{\sqrt{2}}{2} \cdot \frac{1}{1} + 0 \cdot \frac{1}{4} - \frac{\sqrt{2}}{2} \cdot \frac{1}{9} - 1 \cdot \frac{1}{16} - \frac{\sqrt{2}}{2} \cdot \frac{1}{25} + 0 \cdot \frac{1}{36}$$

$$+ \frac{\sqrt{2}}{2} \cdot \frac{1}{49} + 1 \cdot \frac{1}{64} + \cdots$$

The terms are mixed positive and negative but the series is not alternating. Hence we consider the series

$$\sum_{n=1}^{\infty} \left| \frac{\cos \dfrac{n\pi}{4}}{n^2} \right|.$$

Now each term of this series is dominated by the corresponding term of $\sum\limits_{n=1}^{\infty} \dfrac{1}{n^2}$, which converges. Therefore, by the Comparison Test $\sum\limits_{n=1}^{\infty} \left| \dfrac{\cos \dfrac{n\pi}{4}}{n^2} \right|$ converges. Thus our given series converges by Theorem 21.29. ∎

The Ratio Test (Theorem 21.25) can now be extended to series of arbitrary terms.

THEOREM 21.30 (Ratio Test.) Let t be a series of real numbers such that $\lim\limits_{k \to \infty} \left| \dfrac{t(k + 1)}{t(k)} \right| = L$, where $L \in \mathcal{R}$ or $L = +\infty$.

 (i) If $L < 1$, the series converges absolutely.
 (ii) If $L > 1$, the series diverges.
 (iii) If $L = 1$, the series may converge or diverge. (The test fails.)

Proof. (i) is immediate from the definitions and Theorem 21.25. In the proof of 21.25 (ii) we showed that if $L > 1$, then $\lim\limits_{k \to \infty} t(k) \neq 0$. The same proof will work for the present theorem. The same examples used to prove (iii) in Theorem 21.25 apply here also. ∎

Example 18. Examine for convergence $\sum\limits_{n=1}^{\infty} (-1)^{n+1} \dfrac{n^{10}}{\pi^n}$. Applying the ratio test,

$$\lim_{n\to\infty} \left| \frac{t(n+1)}{t(n)} \right| = \lim_{n\to\infty} \frac{(n+1)^{10}}{\pi^{n+1}} \frac{\pi^n}{n^{10}}$$

$$= \lim_{n\to\infty} \left(\frac{n+1}{n} \right)^{10} \cdot \lim_{n\to\infty} \frac{1}{\pi} = \frac{1}{\pi} < 1.$$

The series converges absolutely. ∎

Example 19. Test for convergence

$$\sum_{n=1}^{\infty} (-1)^{n+1} \frac{3 \cdot 6 \cdot 9 \cdots (3n)}{1 \cdot 3 \cdot 5 \cdot 7 \cdots (2n-1)}.$$

$$\lim_{n\to\infty} \left| \frac{t(n+1)}{t(n)} \right| = \lim_{n\to\infty} \frac{3 \cdot 6 \cdot 9 \cdots (3n+3)}{1 \cdot 3 \cdot 5 \cdots (2n+1)} \cdot \frac{1 \cdot 3 \cdot 5 \cdot 7 \cdots (2n-1)}{3 \cdot 6 \cdot 9 \cdots (3n)}$$

$$= \lim_{n\to\infty} \frac{3n+3}{2n+1} = \frac{3}{2} > 1.$$

Thus the given series diverges. ∎

Example 20. Test $\sum\limits_{n=0}^{\infty} \dfrac{(-1)^n}{(2n+2)^2}$. We try the ratio test

$$\lim_{n\to\infty} \left| \frac{t(n+1)}{t(n)} \right| = \lim_{n\to\infty} \frac{(2n+2)^2}{(2(n+1)+2)^2}$$

$$= \lim_{n\to\infty} \frac{4n^2 + 8n + 4}{4n^2 + 16n + 16} = 1,$$

by the polynomial test.

Thus the ratio test fails. However, we can compare with $\sum\limits_{n=1}^{\infty} \dfrac{1}{n^2}$. Applying Theorem 21.23,

$$\lim_{n\to\infty} \frac{1}{(2n+2)^2} \bigg/ \frac{1}{n^2} = \lim_{n\to\infty} \frac{n^2}{4n^2 + 8n + 4} = \frac{1}{4}.$$

Thus $\dfrac{1}{(2n+2)^2}$ is of the same order of magnitude as $\dfrac{1}{n^2}$ and $\sum\limits_{n=0}^{\infty} \dfrac{1}{n^2}$ converges by the p-test. Thus our given series converges absolutely. ∎

EXERCISES

Use the alternating series test to determine whether the series in Exercises 1 to 6 converge.

1. $\sum\limits_{n=1}^{\infty} \dfrac{(-1)^n}{2n+1}$.

2. $\sum\limits_{n=1}^{\infty} \dfrac{(-1)^{n+1}(2n+1)}{3n-1}$.

3. $\displaystyle\sum_{n=1}^{\infty} (-1)^{n-1} \frac{1}{n \cdot 2^n}$.

4. $\displaystyle\sum_{n=1}^{\infty} (-1)^{n-1} \frac{n}{n^2 + 1}$.

5. $\displaystyle\sum_{n=1}^{\infty} (-1)^n \frac{1}{\sqrt[n]{2}}$.

6. $\displaystyle\sum_{n=1}^{\infty} \frac{(-1)^n}{n^n}$.

Test the following series for conditional and absolute convergence.

7. $\displaystyle\sum_{n=0}^{\infty} \frac{(-1)^n}{n^2 + 1}$.

8. $\displaystyle\sum_{n=0}^{\infty} \frac{(-1)^n}{2n - 1}$.

9. $\displaystyle\sum_{n=1}^{\infty} (-1)^{n+1} \frac{(2n + 1)!}{n!}$.

10. $\displaystyle\sum_{n=1}^{\infty} (-1)^{n+1} \frac{1}{\sqrt{n}\sqrt{n + 1}}$.

11. $\displaystyle\sum_{n=1}^{\infty} (-1)^n \frac{n!}{2^n}$.

12. $\displaystyle\sum_{n=1}^{\infty} \frac{(-1)^n}{(2 + \sin n)^n}$.

13. $\displaystyle\sum_{n=1}^{\infty} (-1)^{n+1} \sqrt[n]{n + 1}$.

14. $\displaystyle\sum_{n=1}^{\infty} (-1)^{n+1} \frac{n!}{n^n}$.

15. $\displaystyle\sum_{n=1}^{\infty} (-1)^{n-1} \frac{n^3}{(n + 1)!}$.

16. $\displaystyle\sum_{n=1}^{\infty} (-1)^n \frac{n \ln n}{e^n}$.

17. $\displaystyle\sum_{n=1}^{\infty} \frac{(-1)^n}{\ln (e^n + e^{-n})}$.

18. $\displaystyle\sum_{n=1}^{\infty} \frac{(-1)^{n(n+1)/2}}{2^n}$.

19. $\displaystyle\sum_{n=0}^{\infty} \frac{(-1)^n}{n + 2^n}$.

20. $\displaystyle\sum_{n=1}^{\infty} (-1)^{n+1} \frac{1}{n + \dfrac{1}{2^n}}$.

21. $\displaystyle\sum_{n=0}^{\infty} \frac{\sin \left(\dfrac{n\pi}{2}\right)}{n^2 + 1}$.

22. $\displaystyle\sum_{n=2}^{\infty} (-1)^n \frac{1}{n \ln n}$.

23. $\displaystyle\sum_{n=1}^{\infty} (-1)^n \frac{\ln n}{n}$.

24. $\displaystyle\sum_{n=0}^{\infty} (-1)^n \frac{2^n n!}{2 \cdot 5 \cdot 7 \cdots (3n + 2)}$.

25. $\displaystyle\sum_{n=0}^{\infty} \frac{\cos n}{n^2}$.

26. $\displaystyle\sum_{n=2}^{\infty} \frac{e^n}{\ln n}$.

27. $\displaystyle\sum_{n=1}^{\infty} \frac{\ln n}{e^n}$.

28. $\displaystyle\sum_{n=1}^{\infty} (-1)^n \frac{e^n \ln n}{n!}$.

29. $\displaystyle\sum_{n=1}^{\infty} \frac{(-1)^n n}{e^{n^2}}$.

30. $\displaystyle\sum_{n=1}^{\infty} n \ln \left(\frac{n}{n + 1}\right)$.

ANSWERS

Convergent: 1, 3, 4, 6; Divergent: 2, 5, 9, 11, 13, 26, 30; Conditionally convergent: 8, 10, 17, 20, 22, 23; Absolutely convergent: 7, 12, 14, 15, 16, 18, 19, 21, 24, 25, 27, 28, 29.

6. UNIFORM CONVERGENCE

Thus far in this chapter we have been studying sequences and series of real numbers. In this section we shall begin a study of sequences and series of functions. We shall define both pointwise and uniform convergence of such sequences and series.

DEFINITION 21.31 A sequence of functions is a function s, defined for each positive integer n and with range contained in the set of $\mathcal{R} \to \mathcal{R}$ functions.

The reader should compare the above definition with Definition 21.1, p. 807. Consider now a sequence of functions. If n is a natural number, then $s(n)$ is an $\mathcal{R} \to \mathcal{R}$ function. We shall frequently be interested in the value of this function at some x. If we conform to the usual functional notation, this value is denoted $(s(n))(x)$. Since this notation is cumbersome, and since we seldom need specific reference to the function s we shall adopt an abbreviated notation, which does not refer to s. We shall, in the future, denote $s(n)$ by f_n. Thus $\{f_n\}$ will denote a sequence of functions.

Example 21. For each positive integer n define $f_n : (0, 1] \to \mathcal{R}$ by $f_n(x) = \dfrac{1}{nx}$. Thus $f_1(x) = \dfrac{1}{x}, f_2(x) = \dfrac{1}{2x}, f_3(x) = \dfrac{1}{3x}, \ldots$ Then $\{f_n\}$ is a sequence of functions. Sketches of the graphs of f_1, f_2 and f_3 are found in Figure 21.2. ■

If in Example 21 we evaluate each f_n at $x = 1/2$ we obtain a sequence of real numbers:

$$\{f_n(\tfrac{1}{2})\} = \left\{\frac{2}{n}\right\}.$$

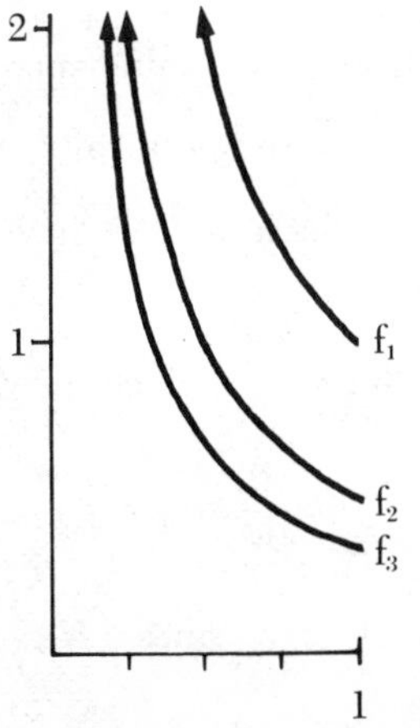

Figure 21.2

We note that $\lim_{n \to \infty} f_n(\tfrac{1}{2}) = 0$. Indeed we have more: $\lim_{n \to \infty} f_n(x) = \lim_{n \to \infty} \dfrac{1}{nx} = 0$ for each x in $(0, 1]$.

Thus we have obtained a new function f by evaluating the limit of the sequence $\{f_n(x)\}$ for each x in $(0, 1]$. In this situation we shall say that the sequence $\{f_n\}$ *converges pointwise* to f.

DEFINITION 21.32 Let $\{f_n\}$ be a sequence of functions and S be a set of real numbers. If for each x in S, $\lim_{n \to \infty} f_n(x)$ exists and is finite, we say that $\{f_n\}$ *converges pointwise* on S. In this case if we define f on S by $f(x) = \lim_{n \to \infty} f_n(x)$, f is called the *limit* of $\{f_n\}$ and we say that $\{f_n\}$ *converges pointwise to f on S.*

According to this definition $\{f_n\}$ converges pointwise to f if for each x in the domain of the f_n and for each $r > 0$, there exists a positive integer N such that $|f(x) - f_n(x)| < r$, whenever $n \geq N$. In the case of the sequence defined by $f_n(x) = \dfrac{1}{nx}$, we showed that $\lim_{n \to \infty} f_n = f$, where $f(x) = 0$ for each x in $(0, 1]$.

Let us examine the sequence $\left\{\dfrac{1}{nx}\right\}$ on $(0, 1]$ more carefully. Let $r > 0$. We ask whether there exists an N such that $\left|\dfrac{1}{nx} - 0\right| < r$ for all $n \geq N$ and all x in $(0, 1]$ simultaneously. If there were such an N, we would have $\dfrac{1}{x} < Nr$ for all x in $(0, 1]$, which is ridiculous since $\lim_{x \to 0+} \dfrac{1}{x} = +\infty$. Therefore the choice of N depends upon both r and x.

On the other hand if we consider the sequence $\left\{\dfrac{1}{nx}\right\}$ on $[\tfrac{1}{2}, 4]$, we discover that if $r > 0$, there exists an N such that $\left|\dfrac{1}{nx} - 0\right| < r$ for all $n \geq N$ independent of x. More specifically, if we choose N to be any integer satisfying $N > \dfrac{2}{r}$ and let $n \geq N$ we have

$$\left|\frac{1}{nx} - 0\right| \leq \frac{1}{Nx} \quad \text{since} \quad \frac{1}{n} \leq \frac{1}{N}$$

$$\leq \frac{2}{N} \quad \text{since} \quad \frac{1}{x} \leq 2 \quad \text{for all } x \text{ in } [\tfrac{1}{2}, 4]$$

$$< r \quad \text{since} \quad N > \frac{2}{r}.$$

This stronger type of convergence is described in the following definition.

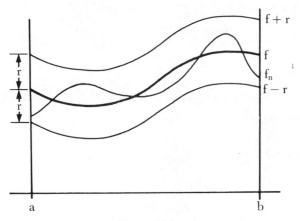

Figure 21.3

DEFINITION 21.33 Let $\{f_n\}$ be a sequence of functions and S be a set of real numbers. If for each $r > 0$ there exists a positive integer N, independent of x, such that $|f_n(x) - f(x)| < r$ whenever $n \geq N$ and $x \in S$, we say that $\{f_n\}$ *converges uniformly to f on S.*

Consider Figure 21.3. To say that $\{f_n\}$ converges uniformly to f on $[a, b]$ means that for each $r > 0$, there is a positive integer N such that whenever $n \geq N$, the graph of f_n lies entirely in the band of width $2r$ between $f - r$ and $f + r$.

The sequence of functions defined by $f_n(x) = \dfrac{1}{nx}$ converges uniformly to zero on $[\frac{1}{2}, 1]$. It converges pointwise, but not uniformly, to zero on $(0, 1]$.

Since a series is but a special type of sequence, it is natural and easy to extend Definitions 21.31, 21.32 and 21.33 to series.

DEFINITION 21.34 Let $\{f_n\}$ be a sequence of functions. The sequence of functions $\{S_n\}$ defined by $S_n = \sum_{k=1}^{n} f_k$ is called the *nth partial sum sequence for $\{f_n\}$.* A sequence of functions $\{S_n\}$ is called a *series of functions* when it is regarded as the nth partial sum sequence of $\{f_n\}$. The function S_n is called the nth *partial sum function* of the series of functions $\{S_n\}$.

The reader should compare this definition with Definition 21.11, p. 815.

A series of functions is frequently denoted by

$$f_1(x) + f_2(x) + f_3(x) + \cdots$$

or

$$\sum_{n=1}^{\infty} f_n(x).$$

If for each x in the domain of the f_n the series $\sum_{n=1}^{\infty} f_n(x)$ converges to $f(x)$ we say that the sum of $\{f_n\}$ is f and write $\sum_{n=1}^{\infty} f_n = f$.

Example 22. Let $f_n(x) = \dfrac{x}{n(n+1)}$ for each positive integer n and for each real x. Then

$$S_n(x) = \sum_{k=1}^{n} f_k(x) = f_1(x) + f_2(x) + \cdots + f_n(x)$$

$$= \frac{x}{1 \cdot 2} + \frac{x}{2 \cdot 3} + \cdots + \frac{x}{n(n+1)}$$

$$\overset{1}{=} \left(x - \frac{x}{2}\right) + \left(\frac{x}{2} - \frac{x}{3}\right) + \cdots + \left(\frac{x}{n} - \frac{x}{n+1}\right)$$

$$\overset{2}{=} x - \frac{x}{n+1}$$

$$= \frac{nx}{n+1}. \quad \blacksquare$$

(1) By the method of partial fractions, we determine that

$$\frac{x}{n(n+1)} = \frac{x}{n} - \frac{x}{n+1}$$

(2) Note that the negative term in each group cancels the positive one in the succeeding group. We say that the series "telescopes."

Thus the sequence $\left\{\dfrac{nx}{n+1}\right\}$ is the nth partial sum sequence for $\{f_n\}$. It is thus the series of functions also denoted by

$$\sum_{n=1}^{\infty} \frac{x}{n(n+1)} = \frac{x}{1 \cdot 2} + \frac{x}{2 \cdot 3} + \frac{x}{3 \cdot 4} + \cdots.$$

Corresponding to Definition 21.32, we have

DEFINITION 21.35 Let $\sum_{n=1}^{\infty} f_n = \{S_n\}$ be a series of functions and S be a set of real numbers. If for each $x \in S$, $\lim_{n \to \infty} S_n(x) = \lim_{n \to \infty} \sum_{k=1}^{n} f_k(x)$ exists and is finite, we say that $\sum_{n=1}^{\infty} f_n$ *converges pointwise.*

In this case if we define f on S by $f(x) = \sum\limits_{n=1}^{\infty} f_n(x)$, f is called the *sum* of $\{f_n\}$ and we say that $\sum\limits_{n=1}^{\infty} f_n(x)$ *converges pointwise to f on S.*

Example 23. Consider the series

$$\frac{x}{1 \cdot 2} + \frac{x}{2 \cdot 3} + \frac{x}{3 \cdot 4} + \cdots$$

of Example 22. We showed that $S_n(x) = \dfrac{nx}{n+1}$. Since $\lim\limits_{n \to \infty} S_n(x) = \lim\limits_{n \to \infty} \dfrac{nx}{n+1} = x$, the given series converges to f, where

$$f(x) = x, \quad x \in \mathcal{R}. \quad \blacksquare$$

Corresponding to Definition 21.33, we have

DEFINITION 21.36 Let $\sum\limits_{n=1}^{\infty} f_n = \{S_n\}$ be a series of functions defined on S. If for each $r > 0$, there exists a positive integer N, independent of x, such that $|S_n(x) - f(x)| < r$ whenever $n \geq N$ and $x \in S$, we say that $\sum\limits_{n=1}^{\infty} f_n$ *converges uniformly* to f.

Example 24. We continue our study of the series

$$\frac{x}{1 \cdot 2} + \frac{x}{2 \cdot 3} + \frac{x}{3 \cdot 4} + \cdots = \left\{ \frac{x}{2}, \frac{2x}{3}, \frac{3x}{4}, \cdots \right\}.$$

Consider Figure 21.4. It is obvious that for any fixed x the sequence converges to $f(x) = x$. However, given $r > 0$ and any $n > 0$, if we go far enough either to the left or to the right, $|S_n(x) - f(x)|$ is greater than r. Thus the given series is convergent but not uniformly convergent on $\mathcal{R}$. But if we restrict our given functions to $[-1, 1]$ we can choose N so that

$$|S_n(1) - f(1)| = \left| \frac{n}{n+1} - 1 \right| = \frac{1}{n+1} < r, \text{ whenever } n > N. \text{ The same}$$

inequality will hold for values of x between -1 and 1. Thus the given series is uniformly convergent on $[-1, 1]$. $\blacksquare$

We give now a useful test for uniform convergence.

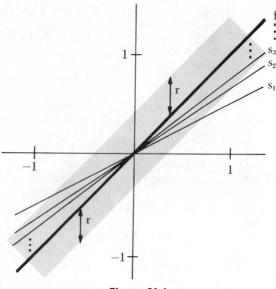

Figure 21.4

THEOREM 21.37 (Weierstrass M-test.) Let $\{f_n\}$ be a sequence of functions: If there exists a convergent sequence $\{M_n\}$ of positive real numbers such that $|f_n(x)| \leq M_n$ for all x in $[a, b]$, then $\sum\limits_{k=1}^{\infty} f_n$ converges absolutely and uniformly on $[a, b]$.

(NOTE. By saying that $\sum\limits_{k=1}^{\infty} f_n$ converges absolutely on $[a, b]$ we mean that $\sum\limits_{k=1}^{\infty} f_n(x)$ converges absolutely for each x in $[a, b]$.)

Proof. Since $|f_n(x)| \leq M_n$ for each n and $\sum\limits_{n=1}^{\infty} M_n$ converges we have by Theorem 21.19, p. 820, that $\sum\limits_{n=1}^{\infty} f_n(x)$ converges absolutely. Hence by Theorem 21.29, p. 831, it converges. Thus $f(x) = \sum\limits_{n=1}^{\infty} f_n(x)$ is defined for each x in $[a, b]$.

Let $r > 0$. To show that $\sum\limits_{n=1}^{\infty} f_n$ is uniformly convergent, we must find a positive integer N, independent of x, such that $|f(x) - S_n(x)| < r$ for all $n \geq N$ and all x in $[a, b]$. Let $M = \sum\limits_{n=1}^{\infty} M_n$ and choose N, a positive integer, such that $\left| M - \sum\limits_{k=1}^{n} M_k \right| < r$, whenever $n \geq N$. Then for $n \geq N$,

$$|f(x) - S_n(x)| =^1 \left| \sum_{k=1}^{\infty} f_k(x) - \sum_{k=1}^{n} f_k(x) \right|$$

$$=^2 \left| \sum_{k=n+1}^{\infty} f_k(x) \right|$$

$$\leq^3 \sum_{k=n+1}^{\infty} |f_k(x)|$$

$$\leq^4 \sum_{k=n+1}^{\infty} M_k$$

$$=^5 \sum_{k=1}^{\infty} M_k - \sum_{k=1}^{n} M_k$$

$$=^6 \left| M - \sum_{k=1}^{n} M_k \right|$$

$$<^7 r. \quad \blacksquare$$

(1) By definition of f and S_n.
(2) Performing the indicated subtraction of the first n terms.
(3) Triangle inequality.
(4) Since each $|f_k(x)| < M_k$.
(5) Adding and subtracting $\sum\limits_{k=1}^{n} M_k$.
(6) Definition of M and the fact that each M_k is nonnegative.
(7) By our choice of n.

Thus, since N was chosen independently of x, the series of functions is uniformly convergent to f. $\quad \blacksquare$

Example 25. We shall show that $\sum\limits_{n=1}^{\infty} \dfrac{1}{n^2 + x^2}$ converges uniformly on $[-1, 1]$. For each x in the given interval, $\dfrac{1}{n^2 + x^2} \leq \dfrac{1}{n^2}$. We choose $M_n = \dfrac{1}{n^2}$ and note that $\sum\limits_{n=1}^{\infty} \dfrac{1}{n^2}$ converges by the p-test (Theorem 21.18, p. 818). Thus by Theorem 21.37 the given series of functions converges uniformly on $[-1, 1]$. $\quad \blacksquare$

Example 26. Show that $\sum\limits_{n=1}^{\infty} (x \ln x)^n$ is uniformly convergent on $(0, 1]$. We note first that $x \ln x \leq 0$ for all x in the given interval. Then a point at which $x \ln x$ is minimum will be a maximum point for $|(x \ln x)^n|$, for each n. By the usual method for finding the minimum value for a function on an interval we find that $x \ln x \geq \dfrac{-1}{e}$ for $x \in (0, 1]$ and for each n. Hence $|(x \ln n)^n| \leq \left(\dfrac{1}{e} \right)^n$, for each x and n. Let $M_n = \left(\dfrac{1}{e} \right)^n$. Now $\sum\limits_{n=1}^{\infty} \left(\dfrac{1}{e} \right)^n$ is a convergent geometric series. Therefore, by the Weierstrass M-test, the given series is uniformly convergent on the given interval. $\quad \blacksquare$

EXERCISES

In exercises 1 to 10, use the Weierstrass M-test to show that the given series are uniformly convergent on the given intervals.

1. $\displaystyle\sum_{n=1}^{\infty} \frac{x^n}{n!}$, $[0, 1]$.

2. $\displaystyle\sum_{n=1}^{\infty} \frac{\sin nx}{n^2}$, $(-\infty, \infty)$.

3. $\displaystyle\sum_{n=1}^{\infty} \frac{n}{e^{nx}}$, $[1, +\infty)$.

4. $\displaystyle\sum_{n=0}^{\infty} \frac{(-1)^n x^{2n}}{(2n)!}$, $[a, b]$.

5. $\displaystyle\sum_{n=0}^{\infty} (-1)^n x^{3n}$, $[-\frac{1}{2}, \frac{1}{2}]$.

6. $\displaystyle\sum_{n=0}^{\infty} n \sin^n (x)$, $\left[-\frac{\pi}{3}, \frac{\pi}{3}\right]$.

7. $\displaystyle\sum_{n=1}^{\infty} n^2 x^n$, $[-\frac{1}{2}, \frac{1}{2}]$.

8. $\displaystyle\sum_{n=1}^{\infty} \frac{x^n}{n^2 + \ln n}$, $(0, 1]$.

9. $\displaystyle\sum_{n=1}^{\infty} \left(\frac{\ln x}{x}\right)^n$, $[1, +\infty)$.

10. $\displaystyle\sum_{n=1}^{\infty} \frac{e^{nx}}{4^n}$, $(-\infty, 1]$.

11. Negate the definition for convergence of a sequence of functions, that is, "$\{f_n\}$ does not converge if . . .".

12. Negate the definition for uniform convergence of a sequence of functions.

7. PROPERTIES OF LIMIT FUNCTIONS

The limit function of a convergent sequence of functions may not be continuous, even if each function of the sequence is continuous.

Example 27. For each n let $f_n(x) = x^n$, $0 \leq x \leq 1$. If $x < 1$ we have $\lim_{n\to\infty} f_n(x) = \lim_{n\to\infty} x^n = 0$, by Theorem 21.7i, p. 810; however, if $x = 1$, $\lim_{n\to\infty} x^n = \lim_{n\to\infty} 1^n = 1$. Thus

$$f(x) = \lim_{n\to\infty} f_n(x) = \begin{cases} 0 & \text{if } 0 \leq x < 1 \\ 1 & \text{if } x = 1. \end{cases}$$

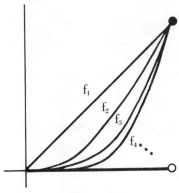

Figure 21.5

We see that f is not continuous, even though each f_n is continuous. The situation is pictured in Figure 21.5.

We note, however, that although the sequence of functions in the preceding example converges to f, it does not converge uniformly. For if we let $r = \frac{1}{3}$, each f_n is outside the band of width $\frac{1}{3}$ about f for points sufficiently close to 1. ∎

The next theorem assures us that if we require a sequence $\{f_n\}$ of continuous functions to converge *uniformly* to f, then the limit function f is also continuous. This gives us a convenient technique for showing that a sequence $\{f_n\}$ does *not* converge uniformly to f. More precisely, if a sequence $\{f_n\}$ of continuous functions converges pointwise to a discontinuous f, then the convergence is not uniform. Take heed: this is *not* equivalent to saying that a sequence of $\{f_n\}$ of continuous functions which converge pointwise to a continuous function f must converge uniformly. The reader is encouraged to find an example of such a sequence of functions which does not converge uniformly.

THEOREM 21.38 Let $\{f_n\}$ be a sequence of functions which converges uniformly to f on a set S of real numbers. If each f_n is continuous on S, then f is also continuous on S.

Proof. To prove that f is continuous on S we have two tools at our disposal; namely, the uniform convergence of the f_n and the continuity of the f_n. Let $x_0 \in S$ and $r > 0$. We wish to find a $\rho > 0$ such that $|f(x) - f(x_0)| < r$ whenever $|x - x_0| < \rho$. Since $\{f_n\}$ converges uniformly to f on S, there exists an integer N such that $|f_n(y) - f(y)| < r/3$ for all y in S and all $n \geq N$. In particular

(21.4) $$|f_N(y) - f(y)| < r/3 \quad \text{for all } y \text{ in } S.$$

Since f_N is continuous at x_0, there exists a $\rho > 0$ such that

(21.5)　　$|f_N(x) - f_N(x_0)| < r/3$　whenever　$|x - x_0| < \rho$.

The following inequality chain will show that for this same ρ it is true that $|f(x) - f(x_0)| < r$ whenever $|x - x_0| < \rho$. Therefore f is continuous at x_0 and our proof is complete.

$$|f(x) - f(x_0)| \leq^1 |f(x) - f_N(x)| + |f_N(x) - f_N(x_0)| + |f_N(x_0) - f(x_0)|$$

$$<^2 \frac{r}{3} + |f_N(x) - f_N(x_0)| + \frac{r}{3}$$

$$<^3 \frac{r}{3} + \frac{r}{3} + \frac{r}{3}　\text{whenever}　|x - x_0| < \rho$$

$$= r. \quad \blacksquare$$

(1) Triangle inequality.
(2) Applying Inequality 21.4 with $y = x$ and $y = x_0$.
(3) By Inequality 21.5.

COROLLARY 21.39　Let $\sum\limits_{n=1}^{\infty} f_n$ converge uniformly to f on a set S. If each f_n is continuous on S, then f is continuous on S.

Proof. Since each f_n is continuous, $S_n = f_1 + f_2 + \cdots + f_n$ is continuous for each n. Thus, since by definition the S_n converge uniformly to f, Theorem 21.38 assures us that f is continuous. $\quad \blacksquare$

We can express this result symbolically by writing

$$\lim_{x \to x_0} \sum_{n=1}^{\infty} f_n = \sum_{n=1}^{\infty} \lim_{x \to x_0} f_n,$$

whenever $\sum\limits_{n=1}^{\infty} f_n$ converges uniformly. In other words, under the hypotheses of the corollary, we can move the symbol "lim" back and forth across the summation symbol. The next theorem will show us that, under certain conditions, the limit symbol can be interchanged with the integration symbol; that is,

$$\lim_{n \to \infty} \int_a^x f_n = \int_a^x \lim_{n \to \infty} f_n.$$

THEOREM 21.40　Let $\{f_n\}$ be a sequence of continuous functions on $[a, b]$ which converges uniformly to f on $[a, b]$. Then

$$\lim_{n \to \infty} \int_a^x f_n = \int_a^x f$$

and the convergence is uniform on $[a, b]$.

Proof. We note first that by Theorem 21.38 f is continuous and hence integrable on $[a, b]$. Define $\{g_n\}$ and g on $[a, b]$ by

$$g_n(x) = \int_a^x f_n \quad \text{and} \quad g(x) = \int_a^x f.$$

Let $r > 0$. We must show that there exists an integer N such that $|g_n(x) - g(x)| < r$ for all $n \geq N$ and all x in $[a, b]$. This task is easily accomplished using the uniform convergence of $\{f_n\}$. Choose an integer N such that

(21.6) $|f_n(t) - f(t)| < r/(b - a)$ for all $n \geq N$ and all t in $[a, b]$.

The following inequality chain will complete our proof by showing that

$$|g_n(x) - g(x)| < r \text{ for all } n \geq N \text{ and all } x \text{ in } [a, b].$$

$$|g_n(x) - g(x)| =^1 \left| \int_a^x f_n(t) - \int_a^x f(t) \right|$$

$$=^2 \left| \int_a^x (f_n(t) - f(t)) \right|$$

$$\leq^2 \int_a^x |f_n(t) - f(t)|$$

$$<^3 \int_a^x \frac{r}{b - a} \quad \text{for all } n \geq N$$

$$=^2 \frac{r}{b - a} (x - a)$$

$$\leq^4 r \quad \text{for all } x \text{ in } [a, b]. \quad \blacksquare$$

(1) Definition of g_n and g.

(2) Basic properties of the integral.

(3) Applying Inequality 21.6 and another basic property of the integral.

(4) Since $\dfrac{x - a}{b - a} \leq 1$ for all x in $[a, b]$.

The following corollary tells us that under the proper conditions, we can interchange the integration symbol with the summation symbol.

COROLLARY 21.41   Let $\displaystyle\sum_{n=1}^{\infty} f_n$ converge uniformly to f on $[a, b]$, where each f_n is continuous on $[a, b]$. Then

$$\int_a^x \sum_{n=1}^{\infty} f_n = \sum_{n=1}^{\infty} \int_a^x f_n.$$

Proof. By definition of uniform convergence of a series, $\{S_n\}$ converges uniformly to f on $[a, b]$ where $S_n = \sum_{k=1}^{\infty} f_k$. Thus, by Theorem 21.40, $\{\int_a^x S_n\}$ converges uniformly to $\int_a^x f$. But $\int_a^x S_n = \int_a^x f_1 + \int_a^x f_2 + \cdots + \int_a^x f_n$, the nth partial sum of the series of functions $\sum_{n=1}^{\infty} \int_a^x f_n$. Hence $\sum_{n=1}^{\infty} \int_a^x f_n$ converges to $\int_a^x f = \int_a^x \sum_{n=1}^{\infty} f_n$. ∎

The reader who solves Exercise 1 at the end of this section will see that the requirement that $\sum_{n=1}^{\infty} f_n$ converge uniformly is necessary.

It is natural to ask whether differentiation can be carried across the limit symbol; that is, is it true that

$$\lim_{n \to \infty} f_n' = \left(\lim_{n \to \infty} f_n \right)' ?$$

The answer, in general, is no (even for uniformly convergent sequences, as is shown by the following example). However, we shall find later that the above statement holds for a very important class of functions; namely, power series.

Example 28. For each n, let $f_n(x) = \dfrac{x}{1 + nx^2}$, $x \in [-1, 1]$. The reader should show that the maximum value of $f_n(x)$ is $\dfrac{1}{2\sqrt{n}}$ and the minimum is $\dfrac{-1}{2\sqrt{n}}$. Since $\lim_{n \to \infty} \dfrac{+1}{2\sqrt{n}} = 0$, $\{f_n\}$ converges uniformly to zero. Thus $\left(\lim_{n \to \infty} f_n \right)' = 0$.

On the other hand, $f_n'(x) = \dfrac{1 - nx^2}{(1 + nx^2)^2}$ and

$$\lim_{n \to \infty} f_n'(x) = \begin{cases} 1 & \text{if } x = 0 \\ 0 & \text{if } x \neq 0. \end{cases}$$

We see that in this case, even though $\{f_n\}$ is uniformly convergent and each f_n is differentiable, $\lim_{n \to \infty} f_n' \neq \left(\lim_{n \to \infty} f_n \right)'$. ∎

EXERCISES

1. Let $\sum_{n=1}^{\infty} f_n$ be the series of functions with sequence of partial sums $\{S_n\}$ where $S_n(x) = nxe^{-nx^2}$, $0 \leq x \leq b$, and let $f(x) = \sum_{n=1}^{\infty} f_n(x)$.

(a) Show that $\displaystyle\int_0^t \lim_{n \to \infty} S_n = 0$, $0 \leq t \leq b$

(b) Show that $\lim\limits_{n \to \infty} \displaystyle\int_0^t S_n = \frac{1}{2}, \ 0 \leq t \leq b$

(c) Show that $\left| f\left(\dfrac{1}{\sqrt{n}}\right) - S_n\left(\dfrac{1}{\sqrt{n}}\right) \right| = \dfrac{\sqrt{n}}{e}$ for each n and conclude that $\displaystyle\sum_{n=1}^{\infty} f_n$ is not uniformly convergent on $0 \leq x \leq b$.

8. POWER SERIES

The reader will recall from Chapter 13 that a series of functions of the form $\displaystyle\sum_{n=1}^{\infty} f_n(a)(x - a)^n$ is called a *power series*. In this section we shall resume our study of this important topic.

We restate Theorem 13.18 for the convenience of the reader.

THEOREM 21.42   (The Root and Ratio Tests.) Let t be a sequence of real numbers and let $\{S_n(x)\}$ be the power series defined by

$$S_n(x) = \sum_{k=0}^{n} t(k)(x - a)^k.$$

If

$$\lim_{k \to \infty} |t(k)|^{1/k} \text{ or } \lim_{k \to \infty} \left| \frac{t(k + 1)}{t(k)} \right| = L \text{ (where } L \in R \text{ or } L = +\infty), \text{ then}$$

(i) $S(x) = \lim\limits_{n \to \infty} S_n(x) = \displaystyle\sum_{n=1}^{\infty} t(n)(x - a)^n$ is defined for all x such that

$$L\,|x - a| < 1;$$

(ii) $S(x)$ is not defined for any x such that

$$L\,|x - a| > 1;$$

(iii) if $L\,|x - a| = 1$, $S(x)$ may or may not be defined.

NOTE.   It is convenient to consider power series as functions from the *nonnegative* integers into the reals. Thus our summation index begins at zero.

The set of all x for which $\displaystyle\sum_{n=1}^{\infty} f_n(x)$ converges will be called the *convergence set* of the series $\displaystyle\sum_{n=1}^{\infty} f_n(x)$.

Example 29.   Find the convergence set for the power series

$$\sum_{n=0}^{\infty} \frac{x^n}{n!} = 1 + x + \frac{x^2}{2} + \cdots + \frac{x^n}{n!} + \cdots.$$

Using the ratio test we find

$$\lim_{k \to \infty} \left| \frac{t(k+1)}{t(k)} \right| = \lim_{k \to \infty} \left| \frac{x^{n+1}}{(n+1)!} \frac{n!}{x^n} \right|$$

$$= \lim_{n \to \infty} \frac{|x|}{n+1} = 0, \quad \text{for all } x.$$

Thus the series converges for all real numbers x. ∎

From the above example and Theorem 21.13, p. 815, we note that

(21.7)
$$\lim_{n \to \infty} \frac{x^n}{n!} = 0, \quad \text{for each } x.$$

This equality is worth remembering.

The next theorem will allow us to apply the results of the previous section to power series.

THEOREM 21.43 If the power series $\sum_{n=0}^{\infty} t(n)(x-a)^n$ converges for $(x-a) = r$ and r_1 is a positive real number such that $r_1 < |r|$, the series converges absolutely and uniformly on $[a - r_1, a + r_1]$.

NOTE. In the proofs of theorems about power series of the form $\sum_{n=0}^{\infty} t(n)(x-a)^n$ we assume for simplicity that $a = 0$. The more general result can always then be obtained by substituting $x - a$ for x.

Proof. Assume $a = 0$. We show first that, for x in $[-r_1, r_1]$, the series converges absolutely. Since $\sum_{n=0}^{\infty} t(n)r^n$ converges, we know, by Theorem 21.13, p. 815, that $\lim_{n \to \infty} t(n)r^n = 0$. Hence there is an integer N such that $|t(n)r^n| < 1$, whenever $n \geq N$. Since $|x| \leq r_1 < |r|$, we have $|t(n)x^n| < 1$ for all $n \geq N$. Hence $|t(n)x^n| = |t(n)r^n| \cdot \left| \frac{x}{r} \right|^n < \left| \frac{x}{r} \right|^n$ for each $n \geq N$.

Now $\sum_{n=0}^{\infty} \left| \frac{x}{r} \right|^n$ is a geometric series with common ratio less than 1; hence it converges. Therefore, by the Weierstrass M-test, $\sum_{n=0}^{\infty} |t(n)x^n|$ converges absolutely and uniformly on $[-r_1, r_1]$. ∎

COROLLARY 21.44 If the power series $\sum_{n=0}^{\infty} t(n)(x-a)^n$ diverges for $(x-a) = r$ and $r_1 > |r|$, then the series diverges for $|x - a| = r_1$.

Proof. Again we consider the case where $a = 0$. If the series converges for $|x| = r_1$, we have, by Theorem 21.43, that it converges for $x = r$. This is a contradiction and the proof is complete. ∎

COROLLARY 21.45 The convergence set of a power series $\sum\limits_{n=0}^{\infty} t(n)(x - a)^n$ is either

(i) the singleton set $\{a\}$,

(ii) an interval (open, closed or half-open) with center at a,

or (iii) the set of all real numbers.

Proof. We assume $a = 0$. Obviously the series converges for $x = 0$. If neither (i) nor (iii) is the convergence set let $r = \text{lub } \{|x|$, the series converges at $x\}$. Then if $|x_1| < r$, Theorem 21.43 guarantees convergence at x_1. If $|x_2| > r$, Corollary 21.44 shows that the series diverges at x_2. Thus the convergence set is either

$$[-r, r], \quad (-r, r], \quad [-r, r) \quad \text{or} \quad (r, r). \quad \blacksquare$$

NOTE. Examples given in Chapter 13 show that all the situations mentioned in Corollary 21.45 can occur.

We say that the radius of convergence of a power series is zero if the convergence set is a singleton, r if it is an interval of radius r, and infinite if it is $\Re$. The following theorem gives us a convenient method for finding the radius of convergence.

THEOREM 21.46 The radius of convergence of the power series $\sum\limits_{n=0}^{\infty} t(n)(x - a)^n$ is given by

$$R = \lim_{n \to \infty} \left| \frac{t(n)}{t(n + 1)} \right|,$$

where $0 \leq R \leq +\infty$, whenever this limit exists.

Proof. We consider the case where $0 < R < +\infty$. Applying the ratio test to the series $\sum\limits_{n=0}^{\infty} t(n)x^n$, we find that this series converges if

$$|x| \lim_{n \to \infty} \left| \frac{t(n + 1)}{t(n)} \right| < 1$$

or

$$|x| < \lim_{n \to \infty} \left| \frac{t(n)}{t(n + 1)} \right| = R.$$

In a like manner we can show that the series diverges if

$$|x| > \lim_{n \to \infty} \left| \frac{t(n)}{t(n + 1)} \right| = R.$$

Thus R is the radius of convergence. ∎

Since we now have more powerful techniques at our disposal than we had in Chapter 13, we shall concern ourselves in exercises and examples not only with finding the radius of convergence of the given power series, but with the entire convergence set; that is, we shall also check end points. We know that the ratio test fails if $L|x| = 1$, where $L = \lim_{n \to \infty} \dfrac{t(n + 1)}{t(n)}$. But for an end point we always have $|x| = R$ and thus the test always fails for an end point, since $R = \dfrac{1}{L}$. It can also be shown that the root test always fails for an end point. Therefore, in order to check end points, we must use some other test such as the comparison test, the integral test, the alternating series test, or others.

Example 30. Find the convergence set for $\displaystyle\sum_{n=0}^{\infty} \frac{(x + 1)^n}{n}$. Applying Theorem 21.46, we find the radius of convergence to be $\displaystyle\lim_{n \to \infty} \frac{n + 1}{n} = 1$. Thus the series converges on $(-2, 0)$. For $x = -2$ we have $\displaystyle\sum_{n=0}^{\infty} \frac{(x + 1)^n}{n} = \sum_{n=0}^{\infty} \frac{(-1)^n}{n}$. This converges by the Alternating Series Test (Theorem 21.26, p. 829). If $x = 0$, we have $\displaystyle\sum_{n=0}^{\infty} \frac{1^n}{n} = \sum_{n=0}^{\infty} \frac{1}{n}$, the harmonic series, which diverges by the p-test (Theorem 21.18, p. 818). Thus the convergence set is $[-2, 0)$. ∎

Example 31. Find the convergence set for

$$\sum_{n=0}^{\infty} (-1)^n \frac{x^{2n+1}}{2n + 1} = x - \frac{x^3}{3} + \frac{x^5}{5} - \frac{x^7}{7} + \cdots.$$

We write this as $x \cdot \left(\displaystyle\sum_{n=0}^{\infty} (-1)^n \frac{(x^2)^n}{2n + 1} \right)$, and treat it as a power series in x^2. Its radius of convergence is $\displaystyle\lim_{n \to \infty} \frac{2n + 1}{2n + 3} = 1$. Hence it converges for x in $(-1, 1)$. At $x = \pm 1$, $x^2 = 1$, and we have $\displaystyle\sum_{n=0}^{\infty} (-1)^n \frac{1^n}{2n + 1}$. This converges by the Alternating Series Test. Thus the interval of convergence is $[-1, 1]$. ∎

EXERCISES

Find the convergence sets for the following power series.

1. $1 - x + \dfrac{x^2}{2} - \dfrac{x^3}{6} + \cdots + (-1)^n \dfrac{x^n}{n!} + \cdots$

2. $-\dfrac{x}{1 \cdot 2} + \dfrac{x^2}{2 \cdot 3} - \dfrac{x^3}{3 \cdot 4} + \dfrac{x^5}{4 \cdot 5} - \cdots$

3. $1 + x + 2x^2 + 6x^3 + 24x^4 + \cdots + n! \, x^n + \cdots$

4. $x - \dfrac{x^2}{4} + \dfrac{x^3}{9} - \dfrac{x^4}{16} + \cdots + (-1)^{n+1} \dfrac{x^n}{n^2} + \cdots$

5. $(x - 1) - \dfrac{(x - 1)^2}{4} + \dfrac{(x - 1)^3}{9} - \dfrac{(x - 1)^4}{16} + \cdots$

6. $(x + 1) - \dfrac{(x + 1)^2}{2} + \dfrac{(x + 1)^3}{3} - \dfrac{(x + 1)^4}{4} + \cdots$

7. $x - \dfrac{x^3}{3!} + \dfrac{x^5}{5!} - \dfrac{x^7}{7!} + \dfrac{x^9}{9!} - \cdots$ (Hint: Factor out x and consider as a series in x^2.)

8. $1 + \dfrac{x^2}{2!} + \dfrac{x^4}{4!} + \dfrac{x^6}{6!} + \cdots$

9. $(x - 2) + 2(x - 2)^2 + 3(x - 2)^3 + 4(x - 2)^4 + \cdots$

10. $x^2 + 2^2 x^3 + 3^2 x^4 + 4^2 x^5 + \cdots$

11. $x - \dfrac{x^3}{3} + \dfrac{x^5}{5} - \dfrac{x^7}{7} + \cdots$

12. $(x + 1) + \dfrac{(x + 1)^3}{3!} + \dfrac{(x + 1)^5}{5!} + \dfrac{(x + 1)^7}{7!} + \cdots$

13. $1 - \dfrac{x}{1 \cdot 3} + \dfrac{x^2}{2 \cdot 4} - \dfrac{x^3}{3 \cdot 5} + \dfrac{x^4}{4 \cdot 6} + \cdots$

14. $\dfrac{x}{3} + \dfrac{x^2}{9} + \dfrac{x^3}{27} + \dfrac{x^4}{81} + \cdots$

15. $(x + 1) + \dfrac{(x + 1)^3}{3} + \dfrac{(x + 1)^5}{5} + \dfrac{(x + 1)^7}{7} + \cdots$

16. $\dfrac{x}{2^2 - 1} + \dfrac{x^2}{3^2 - 1} + \dfrac{x^3}{4^2 - 1} + \dfrac{x^4}{5^2 - 1} + \cdots$

17. $(x - 1) - (x^2 - x) + \dfrac{x^3 - x^2}{2} - \dfrac{x^4 - x^3}{3} + \dfrac{x^5 - x^4}{4} - \cdots$

18. $1 + x + \dfrac{x^2}{\sqrt{2}} + \dfrac{x^3}{\sqrt{3}} + \dfrac{x^4}{\sqrt{4}} + \cdots$

19. $ex + e^4x^2 + e^9x^3 + e^{16}x^4 + \cdots$

20. $1 + 2x + 2^2x^2 + 2^3x^3 + 2^4x^4 + \cdots$

21. $1 + (x + 1) + \dfrac{(x + 1)^2}{2} + \dfrac{(x + 1)^3}{2 \cdot 3} + \dfrac{(x + 1)^4}{2 \cdot 3 \cdot 4} + \cdots$

22. $1 + 2x + \dfrac{2^2x^2}{2!} + \dfrac{2^3x^3}{3!} + \dfrac{2^4x^4}{4!} + \cdots$

23. $1 + \dfrac{(x + 1)}{2} + \dfrac{(x + 1)^2}{2^2} + \dfrac{(x + 1)^3}{2^3} + \cdots$

24. $\dfrac{(x - 2)}{1^2} + \dfrac{(x - 2)^2}{2^2} + \dfrac{(x - 2)^3}{3^2} + \dfrac{(x - 2)^4}{4^2} + \cdots$

25. $\dfrac{x - 1}{1 \cdot 2} + \dfrac{(x - 1)^2}{2 \cdot 3} + \dfrac{(x - 1)^3}{3 \cdot 4} + \dfrac{(x - 1)^4}{4 \cdot 5} + \cdots$

26. $(x + 3) + 2(x + 3)^2 + 3(x + 3)^3 + \cdots$

27. $\displaystyle\sum_{n=2}^{\infty} \dfrac{(-1)^{n+1}1 \cdot 3 \cdots (2n - 3)}{n! \, 2^n} (x - 1)^n$ $\left(\text{Hint: At left end point compare}\right.$ with $\left.\dfrac{1}{2^{n-1}} \cdot \right)$

ANSWERS

1. $\mathcal{R}$; **3.** $\{0\}$; **5.** $[0, 2]$; **7.** $\mathcal{R}$; **9.** $(1, 3)$;
11. $[-1, 1]$; **13.** $[-1, 1]$; **15.** $(-2, 0)$; **17.** $(-1, 1]$;
19. $\{0\}$; **21.** $\mathcal{R}$; **23.** $(-3, 1)$; **25.** $[0, 2]$; **27.** $[0, 2]$.

9. POWER SERIES (CONTINUED)

THEOREM 21.47 The power series $\displaystyle\sum_{n=0}^{\infty} t(n)(x - a)^n$ and the power series $\displaystyle\sum_{n=1}^{\infty} nt(n)(x - a)^{n-1}$, obtained by differentiating the first series, term by term, have the same radius of convergence.

Proof. Again we shall assume $a = 0$. Let the radius of convergence of $\displaystyle\sum_{n=0}^{\infty} t(n)x^n$ be r. If $x \neq 0$ the convergence of $\displaystyle\sum_{n=1}^{\infty} nt(n)x^{n-1}$ is equivalent to that of $\displaystyle\sum_{n=1}^{\infty} nt(n)x^n$ since this last can be obtained by multiplying the first by x. We shall show that $\displaystyle\sum_{n=1}^{\infty} nt(n)x^n$ converges if $|x| < r$ and diverges if $|x| > r$.

Let $0 < x < x_1 < r$. Then by Theorem 21.43, p. 848, $\sum_{n=0}^{\infty} t(n)x_1^n$ converges absolutely. It will be left for the reader to show that $|nt(n)x^n| \in O(|t(n)x_1^n|)$ (Exercise 7, p. 855). Hence, by the Comparison Test (Theorem 21.21, p. 822), $\sum_{n=1}^{\infty} nt(n)x^n$ converges.

On the other hand, if $x > r$ and $\sum_{n=1}^{\infty} nt(n)x^n$ converges, then, since $nt(n)x^n \geq t(n)x^n$ for each n, we have by the comparison test that $\sum_{n=1}^{\infty} t(n)x^n$ converges. This contradicts our choice of x outside the radius of convergence of $\sum_{n=0}^{\infty} t(n)x^n$. This completes the proof. ∎

THEOREM 21.48 Let $\sum_{k=0}^{\infty} t(k)(x-a)^k$ be a power series, and for each x in I, the interval of convergence, let $f(x) = \sum_{k=0}^{\infty} t(k)(x-a)^k$. Then

(i) f is continuous on the interior of I;

(ii) f is differentiable on the interior of I and for all x in the interior of I

$$f'(x) = \sum_{k=1}^{\infty} kt(k)(x-a)^{k-1}$$

(iii) $F(x) = \sum_{k=0}^{\infty} t(k) \frac{(x-a)^{k+1}}{k+1}$ converges in the interior of I and $F'(x) = f(x)$.

NOTE. This theorem includes Theorem 13.20.

Proof of theorem. Let x be in the interior of I. By Corollary 21.45, p. 849, and Theorem 21.43, p. 848, we can choose r such that $a - r_1 < x < a + r_1$ and f is uniformly convergent on $[a - r_1, a + r_1]$. Then, by Corollary 21.39, p. 844, f is continuous at x.

We next establish (iii). We have

$$F(x) =^1 \sum_{k=0}^{\infty} t(k) \frac{(x-a)^{k+1}}{k+1}$$

$$=^2 \sum_{k=0}^{\infty} \int_a^x t(k)(y-a)^k$$

$$=^3 \int_a^x \sum_{k=0}^{\infty} t(k)(y-a)^k.$$

Therefore $F'(x) =^2 \sum_{k=0}^{\infty} t(k)(x-a)^k$

$$=^4 f(x)$$

(1) Definition of F.

(2) First Fundamental Theorem.

(3) Corollary 21.41, p. 845.

(4) By definition of f.

To establish (ii), we have, by Theorem 21.46, that $g(x) = \sum\limits_{k=1}^{\infty} kt(k)x^{k-1}$ has the same radius of convergence as $f(x)$. By (iii)

$$G(x) = \sum_{k=1}^{\infty} (kt(k)) \frac{x^{(k-1)+1}}{(k-1)+1} = \sum_{k=1}^{\infty} t(k)x^k$$

converges on the interior of I and $G'(x) = g(x)$. (To apply the theorem we let $t(0) = 0$.) But f and G differ only by the constant function $t(0)x^0 = t(0)$, so $f'(x) = g(x)$, which completes the proof. ∎

Example 32.   Let $f(x) = \arctan x$. Then

$$f'(x) = \frac{1}{1 + x^2}.$$

Performing the indicated division, we obtain

$$1 - x^2 + x^4 - x^6 + \cdots + (-1)^n x^{2n}.$$

This series can easily be shown by Theorem 21.46 to converge on $(-1, 1)$. By examining the remainder after n terms $(-1)^{n+1}x^{2n+2}$, we can show that the series converges to $f'(x)$ on this interval. By Theorem 21.48 (iii) the series

$$g(x) = x - \frac{x^3}{3} + \frac{x^5}{5} - \frac{x^7}{7} + \cdots$$

is a primitive for f'; hence it differs from $\arctan x$ by a constant c. Thus

$$\arctan x + c = x - \frac{x^3}{3} + \frac{x^5}{5} - \frac{x^7}{7} + \cdots$$

Hence, $0 = \arctan 0 + c = 0$.

Thus $c = 0$ and for $-1 < x < 1$ we have

$$\arctan x = x - \frac{x^3}{3} + \frac{x^5}{5} - \frac{x^7}{7} + \cdots (-1)^n \frac{x^{2n+1}}{2n+1}.$$

In particular,

$$\frac{\pi}{4} = \arctan 1 = 1 - \frac{1}{3} + \frac{1}{5} - \frac{1}{7} + \cdots$$

Using Theorem 21.26, p. 829, to obtain an estimate of the error made in truncating this series, we can compute $\dfrac{\pi}{4}$ to any accuracy we may desire by adding a sufficient number of terms. Thus we can find a decimal approximation for π. ∎

EXERCISES

1. Use the methods of Example 32 to find power series representations and their convergence sets for the following functions:

   (a) $\ln (1 + x)$. (b) $\ln (1 - x)$.

2. Subtract term by term the series found in 1(b) from that found in 1(a) to find a power series expression for $\ln \dfrac{1 + x}{1 - x}$.

3. Find the interval of convergence for the power series found in Exercise 2.

4. Show that any positive number z can be written in the form $\dfrac{1 + x}{1 - x}$, where x is in the interval of convergence found in Exercise 3.

5. Correctly to four decimal places, find x such that $10 = \dfrac{1 + x}{1 - x}$.

6. Use x as found in Exercise 5 and five terms of the series found in Exercise 2 to get an approximation for $\ln 10$. (At this point we have no way of estimating the error committed by neglecting the remaining terms.)

7. Complete the proof of Theorem 21.47 by showing that $|nt(n)x^n| \in O(|t(n)| x^n)$.

ANSWERS

1. (a) $\ln (1 + x) = \displaystyle\sum_{n=1}^{\infty} (-1)^{n+1} \frac{x^n}{n}$, (b) $\ln (1 - x) = -\displaystyle\sum_{n=1}^{\infty} \frac{x^n}{n}$,

3. $(-1, 1)$, **5.** .8182, **6.** 2.24.

10. TAYLOR SERIES

The most useful type of power series is the Taylor series. Indeed, if a power series converges on an interval it is a Taylor series as has been shown in Theorem 13.21.

DEFINITION 21.48 Let $f: \mathcal{R} \to \mathcal{R}$ be infinitely differentiable at a. The power series

$$\sum_{n=0}^{\infty} \frac{f^{(n)}(a)}{n!} (x - a)^n$$

is called the *Taylor series generated by f at a*. We write

$$f(x) \sim \sum_{n=0}^{\infty} \frac{f^{(n)}(a)}{n!} (x - a)^n$$

NOTE. The Taylor series generated by f at 0 is sometimes called the MacLaurin series generated by f.

The reader should compare the above definition with the following one, which is Definition 13.17.

DEFINITION 21.49   Let $f : \mathcal{R} \to \mathcal{R}$ be infinitely differentiable at a and assume that

$$f(x) = \sum_{n=0}^{\infty} \frac{f^{(n)}(a)}{n!} (x - a)^n \text{ converges.}$$

In this case $\displaystyle\sum_{n=0}^{\infty} \frac{f^{(n)}(a)}{n!} (x - a)^n$ is called the *Taylor series representation for* $f(x)$ *about* a.

NOTE: The Taylor series representation for $f(x)$ about 0 is sometimes called the MacLaurin series representation for $f(x)$.

Given a real number x, there are two ways in which the Taylor series generated by f at a as evaluated at x can fail to be the Taylor series representation for $f(x)$ about a; the Taylor series generated by f at a may fail to converge at x, and it may converge to some number other than $f(x)$. We shall give examples in which the first situation occurs; that is, functions for which the Taylor series generated by a function converges only on a finite interval, although the function is defined on a larger interval (see Example 38). Examples can be given to show that the second situation may occur, but we shall not do so here. We can give a necessary and sufficient condition that the Taylor series generated by f at a be a Taylor series representation for f after a restatement of Theorem 13.4.

THEOREM 21.50   (Taylor's Theorem.)   Let $f : \mathcal{R} \to \mathcal{R}$ be such that $f^{(n+1)}$ exists and is continuous on an interval $(a - r, a + r)$. Then for x in $(a - r, a + r)$

$$f(x) = f(a) + f'(a)(x - a) + \frac{f''(a)(x - a)^n}{2!}$$

$$+ \cdots + \frac{f^{(n)}(a)(x - a)^n}{n!} + {}_nR_af(x),$$

where ${}_nR_af(x) = \dfrac{1}{n!} \int_a^x (x - t)^n f^{(n+1)}(t).$

NOTE: ${}_nR_af(x)$ is called the remainder after n terms of the Taylor series generated by f at a.

COROLLARY 21.51   Let $f : \mathcal{R} \to \mathcal{R}$ be such that $f^{(n+1)}$ exists and is continuous on an interval $(a - r, a + r)$ for each positive integer n.

In order that

$$f(x) = \sum_{n=0}^{\infty} \frac{f^{(n)}(a)(x-a)^n}{n!}$$

(that is, the Taylor series generated by f at a is the Taylor series representation for f about a) it is necessary and sufficient that

$$\lim_{n \to \infty} {}_nR_af(x) = 0.$$

Corollary 21.51 is often difficult to apply. A useful sufficient condition for the Taylor series generated by f at a to represent f in an interval is given in the following theorem, which is a restatement of Theorem 13.16.

THEOREM 21.52 Let $f: \mathcal{R} \to \mathcal{R}$ be infinitely differentiable on $(a - r, a + r)$ for some real number a and some $r > 0$. If there exists a real number M such that

$$|f^{(n)}(t)| \le M \text{ for each positive integer } n \text{ and all } t \text{ in } (a - r, a + r),$$

then

$$f(x) = \sum_{n=0}^{\infty} \frac{f^{(n)}(a)}{n!} (x - a)^n$$

for each x in $(a - r, a + r)$.

Example 33. Let $f(x) = \cos x$. We have

$$\begin{array}{ll} f(x) = \cos x & f(0) = 1 \\ f'(x) = -\sin x & f'(0) = 0 \\ f''(x) = -\cos x & f''(0) = -1 \\ f'''(x) = \sin x & f'''(0) = 0 \\ f^{(iv)}(x) = \cos x & f^{(iv)}(0) = 1 \\ f^{(2n)}(x) = (-1)^n \cos x & f^{(2n)}(0) = (-1)^n \\ f^{(2n+1)}(x) = (-1)^{n+1} \sin x & f^{(2n+1)}(0) = 0 \end{array}$$

Thus the Taylor series generated by f at 0 is given by

$$f(x) \sim \sum_{n=0}^{\infty} \frac{f^{(n)}(0)}{n!} x^n$$

$$= 1 + 0 - \frac{x^2}{2!} + 0 + \frac{x^4}{4!} + \cdots + (-1)^n \frac{x^{2n}}{(2n)!} + 0 \cdot x^{2n+1} + \cdots$$

$$= \sum_{n=0}^{\infty} \frac{(-1)^n x^{2n}}{(2n)!}.$$

Now since for each n, $|f^n(x)|$ is either $|\cos x|$ or $|\sin x|$ and is hence ≤ 1, Theorem 21.52 applies and

$$f(x) = \cos x = \sum_{n=0}^{\infty} \frac{(-1)^n x^{2n}}{(2n)!} \quad \text{for all real } x. \quad \blacksquare$$

The direct computation of the Taylor series for a function from the definition is usually cumbersome. Often we can arrive at a power series representation of a function by some other means. The following theorem (a restatement of Theorem 13.21) assures us that if we can find a power series in $(x - a)$ which converges to $f(x)$ for each x in an interval, then this power series is the Taylor series representation of f about a.

THEOREM 21.53 (Uniqueness of a Power Representation.)   Let $r > 0$, let t be a sequence of real numbers and let $f : \mathcal{R} \to \mathcal{R}$ be such that

$$f(x) = \sum_{n=0}^{\infty} t(n)(x - a)^n \quad \text{for all } x \text{ in} \quad (a - r, a + r).$$

Then
(i) f is infinitely differentiable on $(a - r, a + r)$
and
(ii) $t(n) = \dfrac{f^{(n)}(a)}{n!}$.

Consequently $\displaystyle\sum_{n=0}^{\infty} t(n)(x - a)^n$ is the Taylor series representation of $f(x)$ about a.

Example 34.   In Example 32, we showed that, for all x,

$$\arctan x = x - \frac{x^3}{3} + \frac{x^5}{5} - \frac{x^7}{7} + \cdots + (-1)^n \frac{x^{2n+1}}{2n + 1} + \cdots$$

Writing this in the form of a power series in x, we obtain

$$\arctan x = 0 \cdot x^0 + 1 \cdot x^1 + 0 \cdot x^2 + \frac{-1}{3} x^3 + 0 \cdot x^4 + \frac{1}{5} x^5 + 0 \cdot x^6 + \cdots$$

$$+ 0 \cdot x^{2n} + \frac{(-1)^n}{2n + 1} x^{2+1} + \cdots$$

$$= \sum_{n=0}^{\infty} (-1)^n \frac{x^{2n+1}}{2n + 1} .$$

Now by Theorem 21.53, this last is the Taylor series representation of $f(x)$ about zero.   Hence

$$\frac{f^{(2n)}(0)}{(2n)!} = 0, \quad \frac{f^{(2n+1)}(0)}{(2n + 1)!} = \frac{(-1)^n}{2n + 1}$$

or

$$f^{(2n)}(0) = 0, \quad f^{(2n+1)}(0) = (-1)^n(2n)!$$

For example,

$$f^{(4)}(0) = 0, \quad f^{(5)}(0) = -10! \quad \blacksquare$$

The reader should attempt to find these derivatives by direct differentiation in order to see some of the usefulness of Theorem 21.53.

Many Taylor series expansions can be obtained by substituting into or otherwise manipulating a few basic expansions. We collect several useful ones in the following table.

TABLE 21.54

(i) $e^x = \sum_{n=0}^{\infty} \dfrac{x^n}{n!}, \quad x \in \mathcal{R}$

(ii) $\sin x = \sum_{n=0}^{\infty} \dfrac{(-1)^n x^{2n+1}}{(2n+1)!}, \quad x \in \mathcal{R}$

(iii) $\cos x = \sum_{n=0}^{\infty} \dfrac{(-1)^n x^{2n}}{(2n)!}, \quad x \in \mathcal{R}$

(iv) $\ln(1+x) = \sum_{n=1}^{\infty} (-1)^{n+1} \dfrac{x^n}{n}, \quad -1 < x \leq 1$

(v) $\arctan x = \sum_{n=0}^{\infty} (-1)^n \dfrac{x^{2n+1}}{2n+1}, \quad -1 \leq x \leq 1$

(vi) $\dfrac{1}{1-x} = \sum_{n=0}^{\infty} x^n, \quad -1 < x < 1.$

(i) is Example 14, Chapter 13; (ii) is Exercise 1, Section 6 of Chapter 13; (iii) is Example 33 of this chapter; (iv) is Exercise 1 of the previous section; and (v) is Example 34 of this chapter. (vi) may be obtained by performing the indicated division.

Example 35. Find a power series representation for $\sin x^2$. Substituting x^2 for x in Table 21.54(i) we obtain

$$\sin x^2 = \sum_{n=0}^{\infty} \dfrac{(-1)^n (x^2)^{2n+1}}{(2n+1)!} = \sum_{n=0}^{\infty} \dfrac{(-1)^n x^{4n+2}}{(2n+1)!}. \quad \blacksquare$$

Example 36. Find a power series representation for $\cosh x = \dfrac{e^x + e^{-x}}{2}$.

$$e^x = \sum_{n=0}^{\infty} \dfrac{x^n}{n!}$$

$$e^{-x} = \sum_{n=0}^{\infty} \dfrac{(-x)^n}{n!} = \sum_{n=0}^{\infty} (-1)^n \dfrac{x^n}{n!}$$

Adding term by term (Theorem 21.15, p. 816) we obtain

$$e^x + e^{-x} = \sum_{n=0}^{\infty} \left(\frac{x^n}{n!} + (-1)^n \frac{x^n}{n!} \right)$$

$$= \sum_{n=0}^{\infty} \frac{2x^{2n}}{(2n)!} \, .$$

Finally, using Theorem 21.15 (ii), p. 816, we obtain

$$\cosh x = \frac{e^x + e^{-x}}{2} = \frac{1}{2} \sum_{n=0}^{\infty} \frac{2x^{2n}}{(2n)!} = \sum_{n=0}^{\infty} \frac{x^{2n}}{(2n)!} \, . \quad \blacksquare$$

In some of the exercises which follow, we do not ask the reader to show that the series generated by $f(x)$ at a represents $f(x)$ about a. This is because it will be easier to do so after we have developed some additional forms for the remainder ${}_nR_a f(x)$ in Taylor's theorem.

EXERCISES

Apply Definition 21.48 to find the Taylor series generated at a by each of the following functions f. Determine the interval of convergence.

1. $f(x) = e^x$, $a = 1$.

2. $f(x) = \ln x$, $a = 1$.

3. $f(x) = \dfrac{1}{1+x}$, $a = 1$.

4. $f(x) = x^{1/2}$, $a = 1$.

5. $f(x) = e^{-x/2}$, $a = 0$.

6. $f(x) = \dfrac{1}{1+2x}$, $a = 0$.

7. $f(x) = \ln (1 - x)$, $a = 0$.

8. $f(x) = \dfrac{1}{1+x}$, $a = -4$.

9. $f(x) = \cos x$, $a = \dfrac{\pi}{4}$.

10. $f(x) = \sinh x$, $a = 0$.
11. $f(x) = x^2 + 3x + 1$, $a = 1$.
12. $f(x) = x^3 - 3x^2 + 2x - 6$, $a = 2$.

In Exercises 13 to 18, obtain the desired Taylor series by making appropriate substitutions in the formulas of Table 21.54.

13. $\cos 3x$, $a = 0$.
14. $\sin 2x$, $a = 0$.

15. e^{x^2}, $a = 0$.

16. $e^{-x^2/2}$, $a = 0$.

17. $\ln(1 + 2x^2)$, $a = 0$.

18. $(1 + 3x^2)^{-1}$, $a = 0$.

19. Find the first three non-vanishing terms of the Taylor series generated by arctan x at 0 by direct application of Taylor's theorem.

20. Obtain formula (vi) of Table 21.54 by performing the indicated division and showing that the remainder goes to zero on the interval of convergence.

21. Use Exercise 10 and Theorem 21.48 to find the Taylor series generated by cosh x at 0.

ANSWERS

1. $e \cdot \sum_{n=0}^{\infty} \dfrac{(x - 1)^n}{n!}$, $x \in \mathcal{R}$.

3. $\sum_{n=0}^{\infty} \dfrac{(-1)^n}{2^{n+1}} (x - 1)^n$, $(-1, 3)$.

5. $\sum_{n=0}^{\infty} \dfrac{(-1)^n}{2^n n!} x^n$, $x \in \mathcal{R}$.

7. $\sum_{n=0}^{\infty} \dfrac{x^n}{n}$, $-1 \le x < 1$.

9. $\dfrac{\sqrt{2}}{2}\left(1 - \left(x - \dfrac{\pi}{4}\right) - \dfrac{\left(x - \dfrac{\pi}{4}\right)^2}{2} + \dfrac{\left(x - \dfrac{\pi}{4}\right)^3}{6} + \cdots\right.$

$\left. + \dfrac{\left(x - \dfrac{\pi}{4}\right)^{4n}}{(4n)!} - \dfrac{\left(x - \dfrac{\pi}{4}\right)^{4n+1}}{(4n + 1)!} - \dfrac{\left(x - \dfrac{\pi}{4}\right)^{4n+2}}{(4n + 2)!} + \dfrac{\left(x - \dfrac{\pi}{4}\right)^{4n+3}}{(4n + 3)!} + \cdots\right)$,

$x \in \mathcal{R}$.

11. $1 + 3x + x^2$, $x \in \mathcal{R}$.

13. $\sum_{n=0}^{\infty} \dfrac{(-1)^n (3x)^{2n}}{(2n)!}$, $x \in \mathcal{R}$.

15. $\sum_{n=0}^{\infty} \dfrac{x^{2n}}{n!}$, $x \in \mathcal{R}$.

17. $\sum_{n=1}^{\infty} (-1)^{n+1} \dfrac{(2x^2)^{2n+1}}{2n + 1}$.

11. APPLICATIONS OF TAYLOR SERIES

We have seen that a number of elementary functions, indeed all the most familiar ones, can be expressed as Taylor series. Tables of functions such as sin, cos, tan, ln are made by expressing these functions as Taylor series and then taking a sufficient number of terms to insure the desired accuracy. This brings up the problem of estimating the amount of error committed in truncating the Taylor series for a function. It would certainly be undesirable in computing a function value to throw away more than we keep.

To estimate the possible error we may use the formula

$$_n R_a f(x) = \frac{1}{n!} \int_{t=a}^{x} (x - t) f^{(n+1)}(t).$$

However, this is sometimes difficult to apply and we shall develop two other forms of the remainder which are sometimes useful. In order to develop Lagrange's form, we shall need the following theorem.

THEOREM 21.55 (Generalized Mean Value Theorem for Integrals). Let f, $g: \mathcal{R} \to \mathcal{R}$ be continuous on $[a, b]$. If g is non-negative (or non-positive) on $[a, b]$, there is a point c such that $a \le c \le b$ and

$$\int_a^b fg = f(c) \int_a^b g.$$

Proof. We take the case where g is non-negative, the other case being similar. Let m and M be respectively the minimum and maximum values of f on $[a, b]$. Then for each x in $[a, b]$ we have

$$mg(x) \le fg(x) \le Mg(x),$$
$$\int_a^b mg \le \int_a^b fg \le \int_a^b Mg,$$

and

$$m \int_a^b g \le \int_a^b fg \le M \int_a^b g.$$

Now if $\int_a^b g = 0$, this last inequality shows that $\int_b^a fg = 0$ and the theorem holds trivially for any choice of c. Otherwise we can divide through by $\int_a^b g$ to obtain

$$m \le \frac{\int_a^b fg}{\int_a^b g} \le M$$

Thus by the Intermediate Value Theorem (Theorem 4.6, p. 199) there exists $c \in [a, b]$ such that $f(c) = \frac{\int_a^b fg}{\int_a^b g}$ and consequently

$$f(c) \int_a^b g = \int_a^b fg. \quad \blacksquare$$

THEOREM 21.56 (Lagrange's Form for the Remainder in Taylor's Theorem.) Let $f: \mathcal{R} \to \mathcal{R}$ be such that $f^{(n+1)}$ exists and is continuous on an interval $(a - r, a + r)$ and let $x \in (a - r, a + r)$. There exists ζ_n between a and x such that

$$_n R_a f(x) = \frac{(x - a)^{n+1}}{(n + 1)!} f^{(n+1)}(\zeta_n).$$

Proof. We take only the case where $x > a$. We note that $(x - t)^n$ is continuous and nonnegative and $f^{(n+1)}$ is continuous on $[a, x]$. Hence Theorem 21.55 applies and there exists $\zeta_n \in [a, b]$ such that

$$\int_{t=a}^{x} f^{(n+1)}(t)(x - t)^n = f^{(n+1)}(\zeta_n) \int_{t=a}^{x} (x - t)^n$$

Thus

$$_n R_a f(x) = \frac{1}{n!} \int_{t=a}^{x} f^{(n+1)}(t)(x - t)^n$$

$$= \frac{1}{n!} f^{(n+1)}(\zeta_n) \int_{t=a}^{x} (x - t)^n$$

$$= \frac{1}{n!} f^{(n+1)}(\zeta_n) \left[\frac{-(x - t)^{n+1}}{n + 1} \right]_{t=a}^{x}$$

$$= \frac{f^{(n+1)}(\zeta_n)}{n!} \frac{(x - a)^{n+1}}{n + 1}$$

$$= \frac{f^{(n+1)}(\zeta_n)}{(n + 1)!} (x - a)^{n+1}. \quad \blacksquare$$

Example 37. Suppose we wish to compute a table for the cosines of all angles $1°$, $2°$, ..., $45°$ to five decimal places. We use the Taylor series expansion

$$\cos x = \sum_{m=0}^{\infty} \frac{(-1)^m x^{2m}}{2m!}$$

$$= 1 - \frac{x^2}{2!} + \frac{x^4}{4!} - \frac{x^6}{6!} + \frac{x^8}{8!} - \cdots$$

where x is the radian measure of the given angle. We need to determine how many terms of this series are necessary in order to insure the desired accuracy; that is, using Lagrange form of the remainder, we need n (where $n = 2m$) such that

$$|_n R_0 \cos x| = \left| \frac{x^{n+1}}{(n + 1)!} \cos^{(n+1)}(\zeta_n) \right|$$

$$< .000005.$$

Now for each n, $\cos^{(n)}$ is either $\pm\cos$ or $\pm\sin$, hence $|\cos^{(n+1)}(\zeta_n)| \leq 1$. Furthermore, since $45°$ expressed in radians is less than 1, we have $x < 1$ for each choice of x. Hence it suffices to find n such that

$$\left| \frac{1}{(n + 1)!} \right| = \frac{1}{(n + 1)!} < .000005$$

or $(n + 1)! > 200,000$. We find that $8! = 40,320$ and $9! = 362,880$. Hence n must be 8; that is, to insure the desired accuracy we must compute the series for $\cos x$ through the eighth power of x:

$$\cos x = 1 - \frac{x^2}{2!} + \frac{x^4}{4!} - \frac{x^6}{6!} + \frac{x^8}{8!}.$$

As an example we shall compute $\cos 40° = \cos 0.698132$. Let $x = 0.698132$. We find

$$x^2 = .487388 \qquad\qquad \frac{x^2}{2} = .243694$$

$$x^4 = .237547 \qquad\qquad \frac{x^4}{24} = .009898$$

$$x^6 = .115777 \qquad\qquad \frac{x^6}{720} = .000160$$

$$x^8 = .056428 \qquad\qquad \frac{x^8}{40320} = .000001$$

Hence

$$\cos 40° = 1 - .243694 + .009898 - .000160 + .000001$$

$$= .76604, \text{ to five significant figures.} \quad \blacksquare$$

It is in this manner that tables of trigonometric functions were originally computed.

Another form of the remainder in Taylor's theorem which sometimes proves useful is given in the following theorem.

THEOREM 21.57 (Cauchy's Form for the Remainder in Taylor's Theorem.) Let $f : \mathcal{R} \to \mathcal{R}$ such that $f^{(n+1)}$ exists and is continuous on an interval $(a - r, a + r)$ and let $x \in (a - r, a + r)$. Then there exists ζ_n between a and x such that

$$_n R_a f(x) = \frac{(x - a)}{n!} f^{(n+1)}(\zeta_n)(x - \zeta_n)^n.$$

Proof. We take the case where $x > a$. For each n, the function $f^{(n+1)}(t)(x - t)^n$ is continuous on $[a, x]$; hence, by the Mean Value Theorem for Integrals (Exercise 10, p. 868), there is $\zeta_n \in (a, x)$ such that $(x - a)f^{(n+1)}(\zeta_n)(x - \zeta_n)^n = \int_a^x f^{(n+1)}(t)(x - t)^n$. Dividing through by $n!$ we obtain

$$\frac{1}{n!}(x - a)f^{(n+1)}(\zeta_n)(x - \zeta_n)^n = \frac{1}{n!}\int_a^x f^{(n+1)}(t)(x - t)^n.$$

$$= {}_n R_a f(x). \quad \blacksquare$$

Example 38. Determine the interval on which the Taylor series generated by $\ln (1 + x)$ about 0 converges to $\ln x$. (This has already been done by another method, as Exercise 1 of Section 9.)

We have

$$f(x) = \ln (1 + x) \qquad\qquad f(0) = 0$$

$$f'(x) = \frac{1}{1 + x} = (1 + x)^{-1} \quad f'(0) = 1$$

$$f''(x) = -(1 + x)^{-2} \qquad\qquad f''(0) = -1$$

$$f'''(x) = 2(1 + x)^{-3} \qquad\qquad f'''(0) = 2$$

$$f^{iv}(x) = -3 \cdot 2(1 + x)^{-4} \quad f^{(iv)}(0) = -6$$

$$f^{(n)}(x) = (-1)^{n+1}(n - 1)! \, (1 + x)^{-n}$$

$$f^{(n)}(0) = (-1)^{n+1}(n - 1)!$$

$$f^{(n+1)}(x) = (-1)^n n! \, (1 + x)^{-n-1}.$$

Thus

$$\ln (1 + x) \sim \sum_{n=1}^{\infty} (-1)^{n+1} \frac{(n - 1)!}{n!} x^n = \sum_{n=0}^{\infty} (-1)^n \frac{x^n}{n}.$$

The reader should show that this series converges on $(-1, 1]$. Lagrange's form of the remainder is given by

$$_nR_0f(x) = (-1)^n \frac{x^{n+1}}{(n + 1)!} \frac{n!}{(1 + \zeta_n)^{n+1}} = (-1)^n \frac{1}{n + 1}\left(\frac{x}{1 + \zeta_n}\right)^{n+1}.$$

If $1 < x < \zeta_n \le 2$ we have $\dfrac{x}{1 + \zeta_n} < 1$, so

$$|_nR_0f(x)| = \frac{1}{n + 1}\left|\frac{x}{1 + \zeta_n}\right|^{n+1} \le \frac{1}{n + 1}.$$

Hence $\lim_{n \to \infty} |_nR_0f(x)| = 0$. Therefore the series converges to $\ln (1 + x)$ on $(0, 1]$. If $x = 0$ the series converges to $0 = \ln 1$. Finally we shall use the Cauchy form to show that it converges to $\ln x$ on $(-1, 0)$. This form is given by

$$_nR_0f(x) = (-1)^n \frac{x}{n!} \frac{n!}{(1 + \zeta_n)^{n+1}} (x - \zeta_n)^n = \frac{(-1)^n x}{1 + \zeta_n}\left(\frac{x - \zeta_n}{1 + \zeta_n}\right)^n$$

For x in $(-1, 0)$ we have

$$-1 < x < \zeta_n < 0$$

Therefore $0 < 1 + x < 1 + \zeta_n$ and

$$\frac{1}{1 + \zeta_n} < \frac{1}{1 + x}.$$

Thus

$$
\text{(21.8)}\quad
\begin{cases}
|_nR_0f(x)| = \left| \dfrac{x}{1+\zeta_n} \cdot \left(\dfrac{x-\zeta_n}{1+\zeta_n}\right)^n \right| \\[2ex]
\quad = \dfrac{|x|}{1+\zeta_n}\left|\dfrac{\zeta_n - x}{1+\zeta_n}\right|^n \\[2ex]
\quad = \dfrac{|x|}{1+\zeta_n}\left(\dfrac{\zeta_n + x}{1+\zeta_n}\right)^n \\[2ex]
\quad < \dfrac{|x|}{1+x}\left(\dfrac{\zeta_n + |x|}{1+\zeta_n}\right)^n.
\end{cases}
$$

Now since $|x| < 1$ we have

(i) $\zeta_n < \zeta_n |x|$,

(ii) $\zeta_n + |x| < |x| + \zeta_n |x|$

(iii) $\zeta_n + |x| < |x|(1 + \zeta_n)$

and

(iv) $|x| > \dfrac{\zeta_n + |x|}{1+\zeta_n}$.

Substituting this last inequality in 21.8, we obtain

$$|_nR_0f(x)| < \frac{|x|^{n+1}}{1+x}.$$

Thus, since $|x| < 1$,

$$\lim_{n\to\infty} {}_nR_0f(x) = \lim_{n\to\infty}\frac{|x|^{n+1}}{1+x} = 0.$$

Thus $\ln(1+x) = \displaystyle\sum_{n=1}^{\infty}(-1)\frac{x^n}{n},\ -1 < x \le 1.$ ∎

There are many functions whose integrals cannot easily be computed by the Second Fundamental Theorem of Calculus. These are those such as $\sin x^2$ which do not have a primitive that can be expressed in terms of elementary functions. Often we can approximate such integrals to any desired accuracy by means of Taylor series.

Example 39. Compute $\int_0^1 \sin x^2$. From Example 35 we have

$$\sin x^2 = \sum_{n=0}^{\infty}\frac{(-1)^n x^{4n+2}}{(2n+1)!}$$

$$= x^2 - \frac{x^6}{3!} + \frac{x^{10}}{5!} - \frac{x^{14}}{7!} + \cdots$$

Hence,

$$\int_0^1 \sin x^2 = \int_0^1 x^2 - \int_0^1 \frac{x^6}{3!} + \int_0^1 \frac{x^{10}}{5!} - \int_0^1 \frac{x^{14}}{7!} + \cdots$$

$$= \left[\frac{x^3}{3}\right]_0^1 - \left[\frac{x^7}{7 \cdot 3!}\right]_0^1 + \left[\frac{x^{11}}{11 \cdot 5!}\right]_0^1 - \left[\frac{x^{15}}{15 \cdot 7!}\right]_0^1 + \cdots$$

$$= \frac{1^3}{3} - \frac{1^7}{7 \cdot 3!} + \frac{1^{11}}{11 \cdot 5!} - \frac{1^{15}}{15 \cdot 7!} + \cdots$$

We compute

$$\frac{1^3}{3} = 0.66667$$

$$\frac{1^7}{7 \cdot 3!} = 0.23809$$

$$\frac{1^{11}}{11 \cdot 5!} = 0.00076$$

$$\frac{1^{15}}{15 \cdot 7!} = 0.00001$$

Therefore, we conclude that

$$\int_0^1 \sin x^2 = 0.90553. \quad \blacksquare$$

EXERCISES

1. If, as in Example 37, we wish to compute a table for the cosines of all angles $1°, 2°, \ldots, 45°$ to only three decimal places, how many terms of the Taylor series expansion of $\cos x$ will be needed?

2. Compute $\cos 3°$ to three decimal places.

3. If we wish to make a table for the sines of all angles $1°, 2°, \ldots, 45°$, how many terms of the Taylor series expansion for $\sin x$ should be used to assure four place accuracy?

4. Compute $\sin 25°$ to four decimal places.

5. Compute $\ln 1.1$ to four decimal places using the power series for $\ln (1 + x)$.

6. Using the Lagrange form of the remainder, find the number of terms of the power series for $\ln (1 + x)$ needed to evaluate $\ln 1.5$, correct to four decimal places.

The result of Exercise 6 shows that, even though a power series may converge, it may converge so slowly that it is of limited use in practice. The Exercises on page 855 provide a more practical technique, as the following Exercise will illustrate.

7. Using the results of Exercises 2 and 3, p. 855, namely

$$\ln \frac{1 + x}{1 - x} = 2\left(x + \frac{x^3}{3} + \frac{x^5}{5} + \frac{x^7}{7} + \cdots\right), \text{ for } x \text{ in } (-1, 1),$$

determine the number of terms of this series needed to evaluate ln 1.5, correct to four decimal places.

8. Compute ln 1.5 to four decimal places.

9. Compute ln 10, correct to three decimal places.

10. Prove: (Mean Value Theorem for Integrals) Let $f:[a, b] \to \mathcal{R}$ be continuous on $[a, b]$. Then there is a real number c between a and b such that

$$\int_a^b f = f(c) \cdot (b - a).$$

(Hint: Use Theorem 21.55, p. 862.)

ANSWERS

1. Through the sixth power of x.   **3.** Through the seventh power of x.

5. ln 1.1 = 0.0953.   **8.** ln 1.5 = $\dfrac{\ln 1 + 0.2}{1 - 0.2}$ = 0.4055.

APPENDICES

APPENDIX I

SETS AND LOGIC

We assume that the reader is familiar with the language of sets and with logical notation and methods of proof. In this Appendix we present the symbolism and conventions which we follow in the body of the text.

1. SETS AND SET NOTATION

(a) *Symbols for Sets.*

  Ø The empty set; the set with no elements

$$\left.\begin{array}{l} \mathcal{B} \\ \mathcal{R} \\ \mathcal{S} \\ \mathcal{C} \\ \cdot \\ \cdot \\ \cdot \end{array}\right\}$$ Upper case script letters are used (with a few exceptions) to denote the non-empty sets employed in the text.

(b) *Set Descriptions.*

  Verbal —$\mathcal{S}$ is the set of integers between one and ten.
  Roster —$\mathcal{S} = \{2, 3, 4, 5, 6, 7, 8, 9\}$.
  Partial
  Roster —$\mathcal{S} = \{2, 3, 4, \ldots, 9\}$.
  Set
  Builder—$\mathcal{S} = \{x \mid x$ is an integer *and* $1 < x < 10\}$.

(c) *Set Relations.*

  $x \in \mathcal{S}$. x is an *element* (member) of $\mathcal{S}$.
  $x \notin \mathcal{S}$. x is *not an element* of $\mathcal{S}$.
  $\mathcal{S} \subset \mathcal{C}$. Set $\mathcal{S}$ is a *subset* of set $\mathcal{C}$.

      $x \in \mathcal{S} \Rightarrow x \in \mathcal{C}$.

  $\mathcal{S} \not\subset \mathcal{C}$. Set $\mathcal{S}$ is *not a subset* of $\mathcal{C}$.

      For some x we have $x \in \mathcal{S}$ *and* $x \notin \mathcal{C}$.

  $\mathcal{S} = \mathcal{C}$. Set $\mathcal{S}$ is equal to set $\mathcal{C}$.

      $x \in \mathcal{S} \Leftrightarrow x \in \mathcal{C}$.

  $\mathcal{S} \neq \mathcal{C}$. Set $\mathcal{S}$ is not equal to set $\mathcal{C}$:

      $\mathcal{S} \not\subset \mathcal{C}$ *or* $\mathcal{C} \not\subset \mathcal{S}$.

(d) *Set Operations.*

$\mathcal{S} \cap \mathcal{C}$.   The intersection of $\mathcal{S}$ and $\mathcal{C}$:

$$\mathcal{S} \cap \mathcal{C} = \{x \mid x \in \mathcal{S} \text{ and } x \in \mathcal{C}\}.$$

$\mathcal{S} \cup \mathcal{C}$.   The union of $\mathcal{S}$ and $\mathcal{C}$:

$$\mathcal{S} \cup \mathcal{C} = \{x \mid x \in \mathcal{S} \text{ or } x \in \mathcal{C}\}.$$

$\mathcal{S} \setminus \mathcal{C}$.   The set difference:

$$\mathcal{S} \setminus \mathcal{C} = \{x \mid x \in \mathcal{S} \text{ and } x \notin \mathcal{C}\}.$$

2. STATEMENT PATTERNS

A statement is a sentence which is assigned one of two logical values: either T(true) or F(false). Table I.1 lists the statement patterns which are

	Table I.1 Statement Patterns and Their Truth Tables				
Logical Name	**Statement Pattern**	**Truth Table**			
Basic Statements	$q.$ $p.$	T T	T F	F T	F F
Negation	*Not-q.* It is false that q.	F	T		
Conjunction	p *and* q.	T	F	F	F
Disjunction	p *or* q.	T	T	T	F
Implication	$p \Rightarrow q.$ If p then q. p implies q. q is necessary for p.	T	F	T	T
Reverse Implication	$p \Leftarrow q.$ p if q. q is sufficient for p.	T	T	F	T
Bi-implication	$p \Leftrightarrow q.$ p and q are equivalent. q is necessary and sufficient for p.	T	F	F	T
Universal Quantification	For every x in $\mathcal{S}$ we have $p(x)$.	True only if each replacement of x in $p(x)$ by an element s of $\mathcal{S}$ produces a true statement $p(s)$; false otherwise.			
Existential Quantification	There is an x in $\mathcal{S}$ such that $p(x)$.	True only if at least one element s of $\mathcal{S}$ produces a true statement $p(s)$; false otherwise.			

	Table I.2 Statement Patterns and Their Negations	
Statement Pattern	**Negation**
Not-q.	*q.*
p and q.	*Not-p or not-q.*
p or q.	*Not-p and not-q.*
$p \Rightarrow q.$	*p and not-q.*
$p \Leftarrow q.$	*Not-p and q.*
$p \Leftrightarrow q.$	*(p and not-q) or (q and not-p).*
For every x in S we have $p(x)$.	There is an x in S such that *not-p(x)*.
There is an x in S such that $p(x)$.	For every x in S we have *not-p(x)*.

most frequently used in the text and the truth table for each pattern. The symbols p and q represent statements. Each statement pattern contains one or more basic statements, p, q, or $p(x)$. The logical value of each pattern is determined by the logical values of each of the basic statements. For statements involving p and q only, there are four cases to consider, one for each possible pairing of logical values for p and q. The first two rows of the body of the table list the four possible pairings. For each case the corresponding logical value of each statement pattern appears in the column directly below that pairing.

On many occasions it is necessary to consider negations of various statements. Table I.2 lists the basic statement patterns and their negations.

3. IMPLICATION CHAINS

In the body of the text we use a device which hopefully makes proofs easier to follow. This device is the implication chain. A pattern for an implication chain is

$$p \Rightarrow q$$
$$\Rightarrow r$$
$$\Rightarrow s.$$

This chain is an abbreviation of the statement

$$(p \Rightarrow q) \quad and \quad (q \Rightarrow r) \quad and \quad (r \Rightarrow s).$$

The reader should attempt to justify each step in each implication chain. As an aid to the reader we include justifications for some of the steps. The Arabic superscripts on implication arrows refer to justifications written just below the chain. For example,

$$p \Rightarrow^1 q$$
$$\Rightarrow^2 r$$
$$\Rightarrow^3 s.$$

(1) Justification for the statement $p \Rightarrow q$.
(2) Justification for the statement $q \Rightarrow r$.
(3) Justification for the statement $r \Rightarrow s$.

Implication chains with the bi-implication symbol ($\Leftrightarrow$) also appear.

4. PATTERNS OF PROOF

It is not possible to give an exhaustive list of the various patterns of proof. However, there are three basic patterns which are used frequently in the text.

(a) *Direct Argument.* To prove that $p \Rightarrow q$ by the method of direct argument we need merely justify each step of an implication chain beginning with p and ending with q:

$$p \Rightarrow a \Rightarrow b \Rightarrow c \Rightarrow \cdots \Rightarrow r \Rightarrow q.$$

(b) *Contrapositive Argument.* This method uses the fact that $p \Rightarrow q$ and $not\text{-}q \Rightarrow not\text{-}p$ are logically equivalent statements. Thus to prove that $p \Rightarrow q$ is valid we need only show that

$$not\text{-}q \Rightarrow not\text{-}p$$

is valid.

(c) *Indirect Argument.* This method uses the fact that only false statements can lead to a contradiction. To prove $p \Rightarrow q$ by this method we show that $not\text{-}(p \Rightarrow q)$ leads to a contradiction:

$$not\text{-}(p \Rightarrow q) \Rightarrow (m \text{ and } not\text{-}m).$$

We conclude that $not\text{-}(p \Rightarrow q)$ is false and hence $p \Rightarrow q$ is true. From Table I.2 we see that the negation of $p \Rightarrow q$ is $not\text{-}q$ and p. Thus in an indirect argument we usually show

$$(not\text{-}q \text{ and } p) \Rightarrow (m \text{ and } not\text{-}m).$$

PROPERTIES OF THE REAL NUMBERS

This Appendix is a summary of those properties of the real number system which are needed in this text.

We assume that the reader is familiar with the two basic operations of the real number system:

1) the operation of addition, symbolized by "$+$",

2) the operation of multiplication, symbolized by "$\cdot$";

and the two basic relations on the real number system:

3) the relation of equality, symbolized by "$=$",

4) the relation of order, symbolized by "$<$".

In describing the various properties, the set of real numbers is denoted by $\mathcal{R}$. An element of $\mathcal{R}$, that is, a real number, is designated by small italic letters a, b, c and so forth.

A property which is stated without any restriction on the letters appearing therein is understood to be valid for *all* real number replacements.

1. FIELD PROPERTIES OF $\mathcal{R}$

The operations of addition and multiplication possess the following properties.

ADDITION PROPERTIES

 i) Closure: $a + b \in \mathcal{R}$.
 ii) Commutativity: $a + b = b + a$.
 iii) Associativity: $a + (b + c) = (a + b) + c$.
 iv) Identity: There is a single real number 0 such that

$$a + 0 = a$$

 for all a in $\mathcal{R}$.
 v) Inverse: For each a in $\mathcal{R}$ there is a unique number $-a$ in $\mathcal{R}$ such that

$$a + (-a) = 0.$$

MULTIPLICATION PROPERTIES

   vi) Closure: $a \cdot b \in R$.

   vii) Commutativity: $a \cdot b = b \cdot a$.

  viii) Associativity: $a \cdot (b \cdot c) = (a \cdot b) \cdot c$

   ix) Identity: There is a single real number 1 such that

$$a \cdot 1 = a$$

for all a in $\mathcal{R}$.

   x) Inverse: For each a in $\mathcal{R}$ other than 0 there is a single real number $\dfrac{1}{a}$ such that

$$\left(\frac{1}{a}\right) \cdot a = 1.$$

ADDITION-MULTIPLICATION PROPERTIES

   xi) Distinct Identities: The numbers 0 and 1 are distinct, $0 \neq 1$.

  xii) Distributivity: $a(b + c) = ab + ac$.

2. PROPERTIES OF EQUALITY

   i) Reflexivity: $a = a$.

   ii) Symmetry: If $a = b$, then $b = a$.

  iii) Transitivity: If $a = b$ and $b = c$, then $a = c$.

  iv) Additivity: If $a = b$ and $c \in \mathcal{R}$, then $a + c = b + c$.

   v) Multiplicativity: If $a = b$ and $c \in \mathcal{R}$, then $a \cdot c = b \cdot c$.

3. PROPERTIES OF ORDER

The symbol "$>$" is read "is greater than". The symbol "$<$" is read "is less than". We write "$a > b$" if and only if "$b < a$". If $a > 0$, we say that "a is positive". If $a < 0$, we say that "a is negative."

   i) Trichotomy: For each pair a and b of real numbers exactly one of the following is true:

$$a < b,$$
$$a = b,$$
$$a > b.$$

   ii) Transitivity: If $a < b$ and $b < c$, then $a < c$.

  iii) Additivity: If $a < b$ and $c \in \mathcal{R}$, then $a + c < b + c$.

  iv) Multiplicativity: If $a < b$ and $c > 0$, then $a \cdot c < b \cdot c$.

   v) Antimultiplicativity: If $a < b$ and $c < 0$, then $a \cdot c > b \cdot c$.

4. COMPLETENESS PROPERTY

Let S be a set of real numbers.

DEFINITION II.I (Upper Bound.)   A number c is an *upper bound* for S if and only if for each x in S, we have $x \leq c$:

c is an upper bound for $S \Leftrightarrow x \leq c$ for each x in S.

DEFINITION II.2 (Least Upper Bound.)   A number C is a least upper bound (lub) for S if and only if C is an upper bound for S and is less than any other upper bound for S:

$$C \text{ is a lub} \Leftrightarrow \begin{cases} (1) \ \ C \text{ is an upper bound for } S, \text{ and} \\ (2) \ \ \text{for each upper bound } c \text{ of } S \text{ we have } C \leq c. \end{cases}$$

THEOREM II.3 (Uniqueness of Least Upper Bound.)   If a set S of real numbers has a least upper bound C, then C is the only least upper bound for S.

The proof follows from the order properties for the real numbers.

Note that the preceding theorem does not assure that a given set has a least upper bound. It merely states that *if* there is a least upper bound at all, there is only one.

The set of all rational numbers (numbers expressible as the quotient of two integers) possesses all of the above mentioned properties. There are non-empty sets of rational numbers which have an upper bound but which do not possess a least upper bound.

Example I. Let $S = \{x \mid x$ is rational and $x^2 \leq 2\}$. Certainly $S \neq \varnothing$ since $1 \in S$. Moreover, 2 is an upper bound for S, since if $x \in S$ then $x^2 \leq 2 < 4$ and hence $x < 2$. You will search in vain for a *rational number* which is a least upper bound for S.

The property of the real number system which distinguishes it from, say, the rational number system is the following:

COMPLETENESS PROPERTY (Least Upper Bound Property). If S is a non-empty set of real numbers which has an upper bound, then S has a least upper bound.

Example 2. Let S be the same set as that given in Example 1. The least upper bound of S is usually denoted by the symbol $\sqrt{2}$. The reader may be familiar with the proof that $\sqrt{2}$ is not rational.

The following two definitions provide useful terminology.

DEFINITION II.3 (Lower Bound).   A number c is a *lower bound* for a set S if and only if for each x in S we have $c \leq x$:

$$c \text{ is a lower bound for } S \Leftrightarrow c \leq x \text{ for all } x \text{ in } S.$$

DEFINITION II.4 (Greatest Lower Bound).   A number C is a *greatest lower bound* (glb) for a set S if and only if C is a lower bound for S and is greater than any other lower bound for S.

$$C \text{ is a glb for } S \Leftrightarrow \begin{cases} (1) \ \ C \text{ is a lower bound for } S, \text{ and} \\ (2) \ \ \text{for each lower bound } c \text{ for } S \text{ we have } c \leq C. \end{cases}$$

The uniqueness of the greatest lower bound can also be proved from the order properties. The existence of a greatest lower bound follows from the Least Upper Bound Property. Thus we have

THEOREM II.5 (Existence and Uniqueness of the Greatest Lower Bound).   If S is a non-empty set of real numbers which has a lower bound, then S has a unique greatest lower bound.

5. THE PRINCIPLE OF FINITE INDUCTION

PRINCIPLE I.   Let S be a set of real numbers such that

     i) $1 \in S$,
and ii) if $n \in S$, then $(n + 1) \in S$.

Then S contains every positive integer.

PRINCIPLE II.   Let $\mathcal{R}$ be a finite set of real numbers such that

     i) $1 \in S$,
     ii) n_0 is the largest integer in S.
and iii) if $n \in S$ and $n < n_0$ then $(n + 1) \in S$.

Then S contains every positive integer less than or equal to n_0:

$$S = \{1, 2, \ldots, n_0\}.$$

   Example.   Let $S = \left\{ n \mid n \text{ is a positive integer and } 1 + 2 + \cdots + n = \dfrac{n(n + 1)}{2} \right\}$. Using Principle I, we show that every positive integer is in S.

i) $1 = \dfrac{1(1+1)}{2}$. Therefore $1 \in S$.

ii) Assume that $n \in S$. Then $1 + 2 + \cdots + n = \dfrac{n(n+1)}{2}$.

Adding $n + 1$ to both sides of this equation, we have

$$1 + 2 + \cdots + n + (n+1) = \frac{n(n+1)}{2} + (n+1)$$

$$= \frac{(n+1)[(n+1)+1]}{2}.$$

Thus we see that $(n+1) \in S$ whenever $n \in S$. Therefore, by Principle I every positive integer is in S.

Principle II could be used to prove that if

$$\mathcal{C} = \left\{ n \mid n \text{ is a positive integer, } n \leq 72, \text{ and } 1 + 2 + \cdots + n = \frac{n(n+1)}{2} \right\}$$

then $\mathcal{C}$ is the set $\{1, 2, \ldots, 72\}$.

APPENDIX III

TRIGONOMETRY

Consider the circle of radius 1 with center at the origin. (Refer to Figure 1.) Let P be the intersection of this circle with the positive x-axis. If t is any real number, we define the *sine of t* and the *cosine of t* as follows: measure $|t|$ units along the circle from the point P (measuring counterclockwise if $t \geq 0$ and clockwise if $t < 0$); let $Q = (x, y)$ be the point on the circle arrived at by thus measuring t units. We then define $cosine\ (t) = x$ and $sine\ (t) = y$.

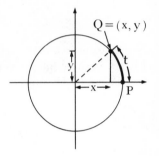

Figure I.

Four other functions are defined in terms of the sine and cosine functions:

$$\text{tangent } (t) = \frac{\text{sine } (t)}{\text{cosine } (t)}, \text{ whenever cosine } (t) \neq 0,$$

$$\text{secant } (t) = \frac{1}{\text{cosine } (t)}, \text{ whenever cosine } (t) \neq 0,$$

$$\text{cosecant } (t) = \frac{1}{\text{sine } (t)}, \text{ whenever sine } (t) \neq 0,$$

$$\text{cotangent } (t) = \frac{\text{cosine } (t)}{\text{sine } (t)}, \text{ whenever sine } (t) \neq 0.$$

The following abbreviations are frequently used for these six functions:

sin = sine	sec = secant
cos = cosine	csc = cosecant
tan = tangent	cot = cotangent.

Furthermore, we usually write "sin t" in place of "sin (t)."

Recall that the circumference of the unit circle is 2π. Thus, for $t = \frac{\pi}{2}$ we have $Q = (0, 1)$, for $t = \pi$ we have $Q = (-1, 0)$, for $t = -\frac{\pi}{2}$ we have $Q = (0, -1)$, for $t = \frac{5\pi}{2}$ we again have $Q = (0, 1)$, and so forth.

Certain values for the sine, cosine and tangent functions are easily computed. The following table lists these values. Since the tangent function is not defined at $\pi/2$ and $3\pi/2$, no value is listed in the table.

	0	$\pi/6$	$\pi/4$	$\pi/3$	$\pi/2$	π	$3\pi/2$	2π
sin	0	$\dfrac{1}{2}$	$\dfrac{1}{\sqrt{2}}$	$\dfrac{\sqrt{3}}{2}$	1	0	-1	0
cos	1	$\dfrac{\sqrt{3}}{2}$	$\dfrac{1}{\sqrt{2}}$	$\dfrac{1}{2}$	0	-1	0	1
tan	0	$\dfrac{1}{\sqrt{3}}$	1	$\sqrt{3}$	—	0	—	0

We divide the values of t in the set $(0, 2\pi)\backslash\left\{\dfrac{\pi}{2}, \pi, \dfrac{3\pi}{2}\right\}$ into four quadrants: the *first quadrant* is $(0, \pi/2)$, the *second quadrant* is $(\pi/2, \pi)$, the *third quadrant* is $\left(\pi, \dfrac{3\pi}{2}\right)$ and the *fourth quadrant* is $\left(\dfrac{3\pi}{2}, 2\pi\right)$. The following table gives the signs of the sine, cosine and tangent functions in these four quadrants.

	I $(0, \pi/2)$	II $(\pi/2, \pi)$	III $(\pi, 3\pi/2)$	IV $(3\pi/2, 2\pi)$
sin	$+$	$+$	$-$	$-$
cos	$+$	$-$	$-$	$+$
tan	$+$	$-$	$+$	$-$

The following five sets of equations are basic trigonometric identities. These equations are valid for all values of s and t for which the functions are defined.

(1) $\begin{cases} \sin^2 t + \cos^2 t = 1 \\ \tan^2 t + 1 = \sec^2 t \\ 1 + \cot^2 t = \csc^2 t \end{cases}$

(2) $\begin{cases} \sin(-t) = -\sin t \\ \cos(-t) = \cos t \end{cases}$

(3)
$$\begin{cases} \sin\left(\dfrac{\pi}{2} - t\right) = \cos t \qquad\qquad \sin(t + 2\pi) = \sin t \\[4mm] \cos\left(\dfrac{\pi}{2} - t\right) = \sin t \qquad\qquad \cos(t + 2\pi) = \cos t \end{cases}$$

(4)
$$\begin{cases} \sin(t + s) = \sin t \cos s + \cos t \sin s \\ \sin(t - s) = \sin t \cos s - \cos t \sin s \\ \cos(t + s) = \cos t \cos s - \sin t \sin s \\ \cos(t - s) = \cos t \cos s + \sin t \sin s \end{cases}$$

(5)
$$\begin{cases} \sin(2t) = 2 \sin t \cos t \\ \cos(2t) = \cos^2 t - \sin^2 t \\ \qquad\quad = 2\cos^2 t - 1 \\ \qquad\quad = 1 - 2\sin^2 t \\ \sin\left(\tfrac{1}{2}t\right) = \pm\sqrt{\dfrac{1 - \cos t}{2}} \\ \cos\left(\tfrac{1}{2}t\right) = \pm\sqrt{\dfrac{1 + \cos t}{2}} \end{cases}$$
$$\left(\begin{array}{l}\text{use } + \text{ or } - \text{ as appropriate for} \\ \text{the quadrant in which } \tfrac{1}{2}t \text{ lies.}\end{array}\right)$$

PROOFS OF SOME IMPORTANT THEOREMS

In this Appendix we give the proofs of some theorems which were stated in the body of the text without proof. The first of these is Theorem 4.6 (Intermediate Value Theorem) on page 199.

THEOREM 4.6 (Intermediate Value Theorem). Let $f: \mathcal{R} \to \mathcal{R}$ and let $[a, b] \subset dom\, f$. Assume that f is continuous at each point of $[a, b]$. If $f(a) < f(b)$ and c is a number such that $f(a) < c < f(b)$, then there exists an x_0 such that

i) $a < x_0 < b$

and

ii) $f(x_0) = c$.

Proof. Consider the set $\mathcal{A}$ consisting of those points x in $[a, b]$ for which $f(x)$ is less than c:

$$\mathcal{A} = \{x \mid x \in [a, b] \quad \text{and} \quad f(x) < c\}.$$

The set $\mathcal{A}$ is clearly bounded above by b and is non-empty since $a \in \mathcal{A}$. Using the Completeness Property given in Appendix II we know that $\mathcal{A}$ has a least upper bound; call it x_0: $x_0 = \text{lub } \mathcal{A}$. We now show that this choice of x_0 meets the conditions of the theorem.

(i) First we show that $a < x_0$. Supppose, to the contrary, that $x_0 = a$. Then for every x in (a, b) we have $f(x) \geq c$. It follows that for every $\rho > 0$, the interval $(a, a + \rho)$ will contain at least one point x_ρ such that $f(x) \geq c$. Hence the interval $(a, a + \rho)$ contains at least one point x_ρ such that $f(x_\rho) - f(a) \geq c - f(a)$. Since $f(a) < c$, the number $r = c - f(a) > 0$. The previous discussion then shows that for the number r the interval $(a, a + \rho)$ contains at least one point x_ρ such that $|f(x_\rho) - f(a)| \geq r$. In other words, for the number $r > 0$ there is no $\rho > 0$ such that $|f(x) - f(a)| < r$ for *every* x in $\mathcal{B}(a; \rho) \cap dom\, f$. This contradicts the continuity of f at a and thus shows that we do not have $a = x_0$.

(ii) An argument similar to that given in (i) will show that $x_0 < b$.

(iii) Finally we show that $f(x_0) = c$ by demonstrating that the assumption that $f(x_0) < c$ or that $f(x_0) > c$ is not possible.

Suppose that $f(x_0) < c$. We then have the following implication chain:

$$f(x_0) < c \Rightarrow c - f(x_0) > 0$$

$\Rightarrow^1$ there is a $\rho > 0$ such that $|f(x) - f(x_0)| < c - f(x_0)$
whenever $|x - x_0| < \rho$

$$\Rightarrow^2 \left| f\left(x_0 + \frac{\rho}{2}\right) - f(x_0) \right| < c - f(x_0)$$

$$\Rightarrow^3 f\left(x_0 + \frac{\rho}{2}\right) - f(x_0) < c - f(x_0)$$

$$\Rightarrow f\left(x_0 + \frac{\rho}{2}\right) < c$$

$$\Rightarrow \left(x_0 + \frac{\rho}{2}\right) \in \mathcal{A}$$

$\Rightarrow x_0$ is not an upper bound for $\mathcal{A}$.

(1) Since f is continuous at x_0.

(2) Since $\left(x_0 + \dfrac{\rho}{2}\right) - x_0 = \dfrac{\rho}{2} < \rho$.

(3) $|A| < B \Rightarrow -B < A < B \Rightarrow A < B$.

Thus the assumption that $f(x_0) < c$ conflicts with the definition of x_0 as the least upper bound of $\mathcal{A}$.

On the other hand, suppose that $f(x_0) > c$. Then we have:

$$f(x_0) > c \Rightarrow f(x_0) - c > 0$$

$\Rightarrow^4$ there is a $\rho > 0$ such that $|f(x) - f(x_0)| < f(x_0) - c$

whenever $|x - x_0| < \rho$

$\Rightarrow^5 c - f(x_0) < f(x) - f(x_0)$ whenever $|x - x_0| < \rho$

$\Rightarrow c < f(x)$ whenever $x_0 - \rho < x < x_0 + \rho$

$\Rightarrow f(x) > c$ whenever $x_0 - \rho < x < x_0$

$\Rightarrow f(x) > c$ whenever $x_0 - \rho < x$

$\Rightarrow x_0 - \rho$ is an upper bound for $\mathcal{A}$

$\Rightarrow x_0$ is not the *least* upper bound for $\mathcal{A}$.

(4) Since f is continuous at x_0.
(5) $|A| < B \Rightarrow -B < A < B \Rightarrow -B < A$.

Thus, the assumption that $f(x_0) > c$ also conflicts with the definition of x_0 as the least upper bound of $\mathcal{A}$.

Since the assumption that $f(x_0) \neq c$ has proved untenable, we must have $f(x_0) = c$. ∎

We shall prove Theorems 4.7 and 4.8 (pp. 200–201) only for the case of $\Re \to \Re$ functions. The general case involves ideas which are too sophisticated for this book. We combine the two theorems into the following single theorem. The proof is by S. J. Bernau.

THEOREM 4.7–8 Let $f: \Re \to \Re$ and let $[a, b] \subset dom f$. Let f be continuous at each point of $[a, b]$. There are real numbers x_1 and x_2 in $[a, b]$ such that

$$f(x_1) \leq f(x) \leq f(x_2)$$

for each x in $[a, b]$.

Proof. We first show that there is a point x_2 in $[a, b]$ at which $f(x)$ takes on a maximum value: $f(x) \leq f(x_2)$ for every x in $[a, b]$. To this end we introduce the set of points x in $[a, b]$ such that the value of f at x is at least as large as the value of f at any point to the left of x:

$$\mathcal{M} = \{x \mid x \in [a, b] \quad \text{and} \quad f(u) \leq f(x) \quad \text{for all } u \text{ in } [a, x]\}.$$

The number b is clearly an upper bound for $\mathcal{M}$. On the other hand, $\mathcal{M}$ is not empty since $a \in \mathcal{M}$. By the completeness property (Appendix II), $\mathcal{M}$ has a least upper bound; call it x_2:

$$x_2 = \text{lub } \mathcal{M}.$$

We show that x_2 meets the conditions of the theorem; that is, $f(x) \leq f(x_2)$ for each x in $[a, b]$.

Let $x \in [a, b]$. Define the set $\mathcal{B}_x$ by

$$\mathcal{B}_x = \{t \mid t \in [a, b] \quad \text{and} \quad f(t) \geq f(x)\}.$$

$\mathcal{B}_x$ is a non-empty set since $x \in \mathcal{B}_x$ and $\mathcal{B}_x$ is bounded below by a. By Theorem II.5 (Existence and Uniqueness of the Greatest Lower Bound), the set $\mathcal{B}_x$ has a greatest lower bound; call it y:

$$y = \text{glb } \mathcal{B}_x.$$

The set $\mathcal{B}_x$ consists of those points of $[a, b]$ at which the value of f is at least as large as $f(x)$. We will show that y shares this property:

$$\text{(i) } f(x) \leq f(y).$$

Moreover, we shall also show that

$$\text{(ii) } y \in \mathcal{M},$$

and that

$$\text{(iii) } x_2 \in \mathcal{M}.$$

These statements enable us to show that $f(x) \leq f(x_2)$. From statement (ii) we proceed as follows:

$$y \in \mathcal{M} \Rightarrow y \leq \text{lub } \mathcal{M}$$
$$\Rightarrow y \leq x_2$$
$$\Rightarrow^1 f(y) \leq f(x_2)$$
$$\Rightarrow^2 f(x) \leq f(y) \leq f(x_2)$$
$$\Rightarrow f(x) \leq f(x_2)$$

(1) Since, according to (iii), $x_2 \in \mathcal{M}$.
(2) Using (i).

Proof of (i). We must show that $f(x) \leq f(y)$. Consider any number $r > 0$. Since f is continuous, there must be a real number $\rho > 0$ such that $|f(y) - f(u)| < r$, whenever $u \in [a, b] \cap (y, y + \rho)$. Since $y = \text{glb } \mathcal{B}_x$ there must be an element z in $[a, b] \cap [y, y + \rho]$ such that $z \in \mathcal{B}_x$; that is, we have $f(z) \geq f(x)$ in addition to $|f(y) - f(z)| < r$. Thus we have

$$f(y) = f(z) + (f(y) - f(z))$$
$$\geq f(z) - |f(y) - f(z)|$$
$$\geq f(z) - r$$
$$\geq f(x) - r.$$

This relation holds for every positive value for r. We conclude that we must have $f(y) \geq f(x)$. Thus (i) is established.

Proof of (ii). We must show that $y \in \mathcal{M}$. We note that

$$y = \text{glb } \mathcal{B}_x \Rightarrow y \text{ is a } lower \text{ bound for } \mathcal{B}_x$$
$$\Rightarrow^1 u \notin \mathcal{B}_x \text{ whenever } a \leq u < y$$
$$\Rightarrow^2 f(u) < f(x) \text{ whenever } a \leq u < y$$
$$\Rightarrow^3 f(u) \leq f(y) \text{ whenever } u \in [a, y]$$
$$\Rightarrow^4 y \in \mathcal{M}.$$

(1) Definition of a lower bound.
(2) Definition of $\mathcal{B}_x$.
(3) Property (i).
(4) Definition of $\mathcal{M}$.

Proof of (iii). We must show that $x_2 \in \mathcal{M}$. Consider any real number $r > 0$. Since f is continuous there exists a number $\rho > 0$ such that $|f(z) - f(x_2)| < r$ whenever $z \in [a, b] \cap (x_2 - \rho, x_2]$. Let u be any element of $[a, x_2)$. Since x_2 is the *least* upper bound of $\mathcal{M}$ there must be some z in $[u, x_2] \cap (x_2 - \rho, x_2]$ such that $z \in \mathcal{M}$. Thus we have $f(u) \leq f(z)$

in addition to $|f(z) - f(x_2)| < r$. We may then write:

$$f(u) \le f(z)$$
$$= f(x_2) + (f(z) - f(x_2))$$
$$\le f(x_2) + |f(z) - f(x_2)|$$
$$\le f(x_2) + r.$$

This relation holds for every positive value of r. We conclude that $f(u) \le f(x_2)$. But the above argument applies for each u in $[a, x_2)$. It follows that

$$f(u) \le f(x_2) \quad \text{whenever} \quad u \in [a, x_2].$$

Thus $x_2 \in \mathcal{M}$.

The existence of a number x_1 in $[a, b]$ satisfying the conditions of the theorem can be demonstrated by a similar argument in which $\mathcal{M}$ is replaced by the set

$$\mathcal{M} = \{x \mid x \in [a, b] \quad \text{and} \quad f(u) \ge f(x) \quad \text{for all } u \in [a, x]\}. \quad \blacksquare$$

Before proving Theorem 6.7, p. 257, we shall prove a preliminary theorem.

THEOREM IV.I (Rolle's Theorem). Let $f: \mathcal{R} \to \mathcal{R}$ such that f is continuous on $[a, b]$ and is differentiable on (a, b), and such that $f(a) = f(b) = 0$. Then there is at least one real number c such that

i) $a < c < b$ and

ii) $f'(c) = 0$.

Proof. If f is constant on $[a, b]$ the theorem is trivially true. Otherwise, using Theorem 4.7, p. 885, let $x_1, x_2 \in [a, b]$ such that $f(x_1) \le f(x) \le f(x_2)$ for any x in $[a, b]$. Then either $f(x_1) \ne 0$ or $f(x_2) \ne 0$. We suppose the latter. (If this is not the case, the argument needed will be quite similar to the one we shall give.) We let $c = x_2$ and observe that $c \notin \{a, b\}$ since $f(c) \ne 0$. Thus f is differentiable at c. If $f'(c) > 0$ then by Theorem 6.3, p. 251, f is strictly increasing at c; hence, by Definition 6.1, p. 249, there exists a number d such that $c < d < b$ and such that $f(x) > f(c)$ for every $x \in (c, d)$. We choose $y \in (c, d) \cap (c, b)$. Such a y exists since $c < b$. Thus $y \in (a, b)$ and $f(y) > f(c)$. But this contradicts our choice of c. Therefore $f'(c) \not> 0$. In a similar way we find that $f'(c) \not< 0$. Thus $f'(c) = 0$ as specified. $\blacksquare$

THEOREM 6.7 (Mean Value Theorem for Derivatives). Let $f:\mathfrak{R} \to \mathfrak{R}$ such that f is continuous on $[a, b]$, and let f' exist on (a, b). Then there is at least one real number c such that

i) $a < c < b$ and

ii) $\dfrac{f(b) - f(a)}{b - a} = f'(c)$.

Proof. We define a new function g by

$$g(x) = \frac{f(b) - f(a)}{(b - a)} (x - a) + f(a) - f(x).$$

Then g is continuous on $[a, b]$ and differentiable on (a, b) with $g(a) = g(b) = 0$. Hence, by Theorem IV.1, there is a real number c such that $a < c < b$ and $g'(c) = 0$. But

$$g'(x) = \frac{f(b) - f(a)}{b - a} - f'(x),$$

so

$$g'(c) = \frac{f(b) - f(a)}{b - a} - f'(c) = 0.$$

Hence

$$f'(c) = \frac{f(b) - f(a)}{b - a}$$

as required. ∎

APPENDIX V

LIST OF SYMBOLS

[A list of logical symbols appears in Appendix I.]

APPENDIX VI

TABLE OF PRIMITIVES

1. $\int cf = c\int f.$

2. $\int (f + g) = \int f + \int g.$

3. $\int f' = [f].$

4. $\int f = [F] \Rightarrow \int f(u)u' = [F \circ u].$

5. $\int (f' \circ g)g' = [f \circ g].$

6. $\int f \cdot g' = [fg] - \int f' \cdot g.$

7. $\int u^r u' = \begin{cases} \left[\dfrac{u^{r+1}}{r+1}\right], & r \text{ any real number other than } -1. \\ [\ln |u|], & r = -1. \end{cases}$

8. $\int (au + b)^r u' = \begin{cases} \left[\dfrac{1}{a(r+1)} (au + b)^{r+1}\right], r \neq -1. \\ \left[\dfrac{1}{a} \ln |au + b|\right], r = -1. \end{cases}$

9. $\int \dfrac{uu'}{au + b} = \left[\dfrac{u}{a} - \dfrac{b}{a^2} \ln |au + b|\right].$

10. $\int \dfrac{u^2 u'}{au + b} = \left[\dfrac{1}{a^3} (\tfrac{1}{2}(au + b)^2 - 2b(au + b) + b^2 \ln |au + b|)\right].$

11. $\int \dfrac{uu'}{(au + b)^2} = \dfrac{1}{a^2}\left[\ln |au + b| + \dfrac{b}{au + b}\right].$

12. $\int \dfrac{u'}{u(au + b)} = \left[\dfrac{1}{b} \ln \left|\dfrac{u}{au + b}\right|\right].$

13. $\int \dfrac{u'}{u^2(au + b)} = \left[-\dfrac{1}{bu} + \dfrac{a}{b^2} \ln \left|\dfrac{au + b}{u}\right|\right].$

14. $\int \dfrac{u'}{u(au+b)^2} = \left[\dfrac{1}{b(au+b)} - \dfrac{1}{b^2} \ln \left| \dfrac{au+b}{u} \right| \right].$

15. $\int \dfrac{u'}{u^2(au+b)^2} = \left[-\dfrac{b+2au}{b^2 u(au+b)} + \dfrac{2a}{b^3} \ln \left| \dfrac{au+b}{u} \right| \right].$

16. $\int \sqrt{au+b}\, u' = \left[\dfrac{2}{3a} \sqrt{(au+b)^3}\, \right].$

17. $\int u\sqrt{au+b}\, u' = \left[\dfrac{2(3au-2b)}{15a^2} \sqrt{(au+b)^3}\, \right].$

18. $\int u^m \sqrt{au+b}\, u' = \dfrac{1}{2m+3} \left([2u^{m+1}\sqrt{au+b}\,] + b \int \dfrac{u^m}{\sqrt{au+b}}\, u' \right).$

19. $\int \dfrac{u'}{\sqrt{au+b}} = \left[\dfrac{2\sqrt{au+b}}{a} \right].$

20. $\int \dfrac{u^m u'}{\sqrt{au+b}} = \left[\dfrac{2u^m \sqrt{au+b}}{(2m+1)a} \right] - \dfrac{2mb}{(2m+1)a} \int \dfrac{u^{m-1}}{\sqrt{au+b}}\, u'.$

21. $\int \dfrac{u'}{u\sqrt{au+b}} = \left[\dfrac{2}{\sqrt{-b}} \arctan \dfrac{\sqrt{au+b}}{\sqrt{-b}} \right],$ for $b < 0.$

$\qquad\qquad = \left[\dfrac{1}{\sqrt{b}} \ln \left| \dfrac{\sqrt{au+b}-\sqrt{b}}{\sqrt{au+b}+\sqrt{b}} \right| \right],$ for $b > 0.$

22. $\int \dfrac{u'}{(au+b)(cu+d)} = \left[\dfrac{1}{bc-ad} \ln \left| \dfrac{cu+d}{au+b} \right| \right], \quad bc - ad \neq 0.$

23. $\int \dfrac{uu'}{(au+b)(cu+d)}$

$\qquad = \dfrac{1}{bc-ad} \left[\dfrac{b}{a} \ln |au+b| - \dfrac{d}{c} \ln |cu+d| \right], \quad bc - ad \neq 0.$

24. $\int \dfrac{u'}{p^2+u^2} = \left[\dfrac{1}{p} \arctan \dfrac{u}{p} \right]$

$\qquad\qquad = \left[-\dfrac{1}{p} \operatorname{arccot} \dfrac{u}{p} \right].$

25. $\int \dfrac{u'}{p^2-u^2} = \left[\dfrac{1}{2p} \ln \left| \dfrac{p+u}{p-u} \right| \right].$

26. $\int \dfrac{uu'}{au^2+c} = \left[\dfrac{1}{2a} \ln |au^2+c| \right].$

27. $\int \dfrac{u'}{u(au^2+c)} = \left[\dfrac{1}{2c} \ln \left| \dfrac{u^2}{au^2+c} \right| \right].$

28. $\int \sqrt{u^2 \pm p^2}\, u' = \frac{1}{2}[u\sqrt{u^2 \pm p^2} \pm p^2 \ln |u + \sqrt{u^2 \pm p^2}|].$

29. $\int \sqrt{p^2 - u^2}\, u' = \frac{1}{2}\left[u\sqrt{p^2 - u^2} + p^2 \arcsin \frac{u}{p}\right].$

30. $\int \sqrt{au^2 + c}\, u'$

$= \left[\frac{u}{2}\sqrt{au^2 + c} + \frac{c}{2\sqrt{a}} \ln |u\sqrt{a} + \sqrt{au^2 + c}|\right],$ if $a > 0.$

$= \left[\frac{u}{2}\sqrt{au^2 + c} + \frac{c}{2\sqrt{-a}} \arcsin\left(u\sqrt{\frac{-a}{c}}\right)\right],$ if $a < 0.$

31. $\int \dfrac{uu'}{\sqrt{au^2 + c}} = \left[\dfrac{1}{a}\sqrt{au^2 + c}\right].$

32. $\int u^2 \sqrt{au^2 + c}\, u'$

$= \left[\frac{u}{4a}\sqrt{(au^2 + c)^3} - \frac{cu}{8a}\sqrt{au^2 + c} - \frac{c^2}{8a\sqrt{a}} \ln |\sqrt{a}\, u + \sqrt{au^2 + c}|\right],$

$\text{if } a > 0.$

33. $\int \dfrac{u'}{\sqrt{u^2 \pm p^2}} = [\ln |u + \sqrt{u^2 \pm p^2}|].$

34. $\int \dfrac{u'}{\sqrt{p^2 - u^2}} = \left[\arcsin \dfrac{u}{p}\right]$

$= \left[- \arccos \dfrac{u}{p}\right].$

35. $\int \dfrac{u'}{u\sqrt{p^2 \pm u^2}} = \left[-\dfrac{1}{p} \ln \left|\dfrac{p + \sqrt{p^2 \pm u^2}}{u}\right|\right].$

36. $\int \dfrac{u'}{u\sqrt{u^2 - p^2}} = \left[\dfrac{1}{p} \arccos \dfrac{p}{u}\right]$

$= \left[-\dfrac{1}{p} \arcsin \dfrac{p}{u}\right].$

37. $\int \dfrac{\sqrt{u^2 + p^2}}{u}\, u' = \left[\sqrt{u^2 + p^2} - p \ln \left|\dfrac{p + \sqrt{u^2 + p^2}}{u}\right|\right].$

38. $\int \dfrac{\sqrt{u^2 - p^2}}{u}\, u' = \left[\sqrt{u^2 - p^2} - p \arccos \dfrac{p}{x}\right].$

39.

$$\int \frac{u'}{au^2 + bu + c} = \begin{cases} \left[\dfrac{1}{\sqrt{b^2 - 4ac}} \ln \left| \dfrac{2au + b - \sqrt{b^2 - 4ac}}{2au + b + \sqrt{b^2 - 4ac}} \right| \right], & b^2 > 4ac. \\[4mm] \left[\dfrac{2}{\sqrt{4ac - b^2}} \arctan \dfrac{2au + b}{\sqrt{4ac - b^2}}\right], & b^2 < 4ac. \\[4mm] \left[\dfrac{-2}{2au + b}\right], & b^2 = 4ac. \end{cases}$$

40. $\displaystyle\int \frac{u'}{\sqrt{au^2 + bu + c}}$

$$= \begin{cases} \left[\dfrac{1}{\sqrt{a}} \ln \left| 2au + b + 2\sqrt{a}\,\sqrt{au^2 + bu + c} \right|\right], & a > 0, \\[4mm] \left[\dfrac{1}{\sqrt{-a}} \arcsin \dfrac{-2au - b}{\sqrt{b^2 - 4ac}}\right], & a < 0. \end{cases}$$

41. $\displaystyle\int (\sin u)u' = [-\cos u].$

42. $\displaystyle\int (\sin^2 u)u' = \left[\dfrac{u}{2} - \dfrac{\sin 2u}{4}\right].$

43. $\displaystyle\int (\sin^3 u)u' = [-\cos u + \tfrac{1}{3}\cos^3 u].$

44. $\displaystyle\int (\sin^4 u)u' = \left[\dfrac{3u}{8} - \dfrac{3\sin 2u}{16} - \dfrac{\sin^3 u \cos u}{4}\right].$

45. $\displaystyle\int \frac{u'}{\sin u} = \left[\ln \left| \tan \dfrac{u}{2} \right| \right]$

$\qquad = [\ln |\csc u - \cot u|].$

46. $\displaystyle\int \frac{u'}{\sin^2 u} = \int (\csc^2 u)u' = [-\cot u].$

47. $\displaystyle\int (\cos u)u' = [\sin u].$

48. $\displaystyle\int (\cos^2 u)u' = \left[\dfrac{u}{2} + \dfrac{\sin 2u}{4}\right].$

49. $\displaystyle\int (\cos^3 u)u' = [\sin u - \tfrac{1}{3}\sin^3 u].$

50. $\displaystyle\int (\cos^4 u)u' = \left[\dfrac{3u}{8} + \dfrac{3\sin 2u}{16} + \dfrac{\cos^3 u \sin u}{4}\right].$

51. $\int \frac{u'}{\cos u} = \left[\ln \left| \tan \left(\frac{u}{2} + \frac{\pi}{4} \right) \right| \right]$

　　　$= [\ln |\tan u + \sec u|].$

52. $\int \frac{u'}{\cos^2 u} = \int (\sec^2 u) = [\tan u].$

53. $\int (\sin au \cos bu) u' = -\frac{1}{2} \left[\frac{\cos (a - b)u}{a - b} + \frac{\cos (a + b)u}{a + b} \right], \quad a^2 \neq b^2.$

54. $\int (\sin^n u \cos u) u' = \left[\frac{1}{n + 1} \sin^{n+1} u \right], \quad n \neq -1.$

55. $\int (\cos^n u \sin u) u' = \left[-\frac{1}{n + 1} \cos^{n+1} u \right], \quad n \neq -1.$

56. $\int \frac{u'}{\sin u \cos u} = [\ln |\tan u|].$

57. $\int \frac{u'}{\sin^2 u \cos^2 u} = [\tan u - \cot u].$

58. $\int \frac{u'}{\sin^m u \cos^n u}$

　　　$= \left[\frac{1}{n - 1} \frac{1}{\sin^{m-1} u \cos^{n-1} u} \right] + \frac{m + n - 2}{n - 1} \int \frac{u'}{\sin^m u \cos^{n-2} u}, \quad n > 1.$

　　　$= \left[\frac{-1}{m - 1} \cdot \frac{1}{\sin^{m-1} u \cos^{n-1} u} \right] + \frac{m + n - 2}{m - 1} \int \frac{u'}{\sin^{m-2} u \cos^n u}, \quad n > 1.$

59. $\int (\tan u) u' = [-\ln |\cos u|].$

60. $\int (\tan^2 u) u' = [\tan u - u].$

61. $\int (\tan^3 u) u' = [\frac{1}{2} \tan^2 u + \ln |\cos u|].$

62. $\int (\tan^n u) u' = \left[\frac{1}{n - 1} \tan^{n-1} u \right] - \int (\tan^{n-2} u) u',$

　　　　　n an integer $> 1.$

63. $\int (\cot u) u' = [\ln |\sin u|]$

　　　$= [- \ln |\csc u|].$

64. $\int (\cot^2 u) u' = \int \frac{u'}{\tan^2 u} = [-\cot u - u].$

65. $\int (\cot^3 u)u' = [-\frac{1}{2}\cot^2 u - \ln |\sin u|]$.

66. $\int (\cot^n u)u' = \int \dfrac{u'}{\tan^n u}$

$$= \left[\frac{-1}{n-1}\cot^{n-1} u\right] - \int (\cot^{n-2} u)u',$$

n an integer >1.

67. $\int (\sec u)u' = [\ln |\sec u + \tan u|]$

$$= \left[\ln \left| \tan \left(\frac{u}{2} + \frac{\pi}{4}\right)\right|\right].$$

68. $\int (\sec^2 u)u' = [\tan u]$.

69. $\int (\csc u)u' = [\ln |\csc u - \cot u|]$

$$= \left[\ln \left| \tan \frac{u}{2}\right|\right].$$

70. $\int (\csc^2 u)u' = [-\cot u]$.

71. $\int (\tan u \sec u)u' = [\sec u]$.

72. $\int (\tan^n u \sec^2 u)u' = \left[\dfrac{1}{n+1}\tan^{n+1} u\right], \quad n \neq -1$.

73. $\int (\tan u \sec^n u)u' = \left[\dfrac{1}{n}\sec^n u\right], \quad n \neq 0$.

74. $\int (\cot u \csc u)u' = [-\csc u]$.

75. $\int (\cot^n u \csc^2 u)u' = \left[\dfrac{-1}{n+1}\cot^{n+1} u\right], \quad n \neq -1$.

76. $\int (\cot u \csc^n u)u' = \left[-\dfrac{1}{n}\csc^n u\right], \quad n \neq 0$.

77. $\int (u \sin au)u' = \left[\dfrac{1}{a^2}\sin au - \dfrac{1}{a}u \cos au\right]$.

78. $\int (u^2 \sin au)u' = \left[\dfrac{2u}{a^2}\sin au + \dfrac{2}{a^3}\cos au - \dfrac{u^2}{a}\cos au\right]$.

79. $\int (u^n \sin au)u' = \left[-\frac{1}{a} u^n \cos au \right] + \frac{n}{a} \int (u^{n-1} \cos au)u'.$

80. $\int (u \sin^2 au)u' = \left[\frac{u^2}{4} - \frac{u \sin 2au}{4a} - \frac{\cos 2au}{8a^2} \right].$

81. $\int (u \cos au)u' = \left[\frac{1}{a^2} \cos au + \frac{1}{a} u \sin au \right].$

82. $\int (u^2 \cos au)u' = \left[\frac{2u}{a^2} \cos au - \frac{2}{a^3} \sin au + \frac{u^2}{a} \sin au \right].$

83. $\int (u^n \cos au)u' = \left[\frac{1}{a} u^n \sin au \right] - \frac{n}{a} \int (u^{n-1} \sin au)u',$

    $n > 0.$

84. $\int (u \cos^2 au)u' = \left[\frac{u^2}{4} + \frac{u \sin 2au}{4a} + \frac{\cos 2au}{8a^2} \right].$

85. $\int (\arcsin u)u' = [u \arcsin u + \sqrt{1 - u^2}].$

86. $\int (\arcsin u)^2 u' = [u(\arcsin u)^2 - 2u + 2\sqrt{1 - u^2} \arcsin u].$

87. $\int (\arccos u)u' = [u \arccos u - \sqrt{1 - u^2}].$

88. $\int (\arccos u)^2 u' = [u(\arccos u)^2 - 2u - 2\sqrt{1 - u^2} \arccos u].$

89. $\int (\arctan u)u' = [u \arctan u - \frac{1}{2} \ln |1 + u^2|].$

90. $\int (\text{arccot } u)u' = [u \text{ arccot } u + \frac{1}{2} \ln |1 + u^2|].$

91. $\int (\text{arcsec } u)u' = [u \text{ arcsec } u - \ln |u + \sqrt{u^2 - 1}|].$

92. $\int (\text{arccsc } u)u' = [u \text{ arccsc } u + \ln |u + \sqrt{u^2 - 1}|].$

93. $\int e^u u' = [e^u].$

94. $\int b^u u' = \left[\frac{b^u}{\ln b} \right], \quad b > 0.$

95. $\int u e^{au} u' = \left[\frac{e^{au}}{a^2} (au - 1) \right].$

96. $\displaystyle\int ub^{au}u' = \left[\dfrac{ub^{au}}{a \ln b} - \dfrac{b^{au}}{a^2 (\ln b)^2}\right], \quad b > 0.$

97. $\displaystyle\int (e^{au} \sin bu)u' = \left[\dfrac{e^{au}}{a^2 + b^2} (a \sin bu - b \cos bu)\right].$

98. $\displaystyle\int (e^{au} \cos bu)u' = \left[\dfrac{e^{au}}{a^2 + b^2} (a \cos bu + b \sin bu)\right].$

99. $\displaystyle\int (\ln au)u' = [u \ln au - u].$

100. $\displaystyle\int \dfrac{u'}{u \ln au} = [\ln |\ln au|].$

101. $\displaystyle\int (u \ln au)u' = \left[\dfrac{u^2}{2} \ln au - \dfrac{u^2}{4}\right].$

102. $\displaystyle\int (u^2 \ln au)u' = \left[\dfrac{u^3}{3} \ln au - \dfrac{u^3}{9}\right].$

103. $\displaystyle\int (\sinh u)u' = [\cosh u].$

104. $\displaystyle\int (\cosh u)u' = [\sinh u].$

105. $\displaystyle\int (\tanh u)u' = [\ln \cosh u].$

106. $\displaystyle\int (u \sinh u)u' = [u \cosh u - \sinh u].$

107. $\displaystyle\int (u \cosh u)u' = [u \sinh u - \cosh u].$

INDEX